新英汉建筑装饰工程辞典

A New English-Chinese Dictionary of Architectural Decoration

李育才　卜纯英　主编

中国建筑工业出版社

图书在版编目(CIP)数据

新英汉建筑装饰工程辞典/李育才,卜纯英主编.—北京:中国建筑工业出版社,2006
ISBN 7-112-08245-5

I.新... II.①李...②卜... III.建筑装饰—词典—英、汉 IV.TU767-61

中国版本图书馆 CIP 数据核字(2006)第 031399

责任编辑:杨 军 费海玲
责任设计:崔兰萍
责任校对:张景秋 关 健

新英汉建筑装饰工程辞典
A New English-Chinese Dictionary
of Architectural Decoration

李育才 卜纯英 主编

*

中国建筑工业出版社出版、发行(北京西郊百万庄)
新 华 书 店 经 销
北京蓝海印刷有限公司印刷

*

开本:850×1168 毫米 1/32 印张:31⅝ 字数:1750 千字
2006 年 10 月第二版　　2006 年 10 月第二次印刷
印数:4501—8500 册　定价:**88.00** 元
ISBN 7-112-08245-5
(14199)

版权所有　翻印必究
如有印装质量问题,可寄本社退换
(邮政编码 100037)

本社网址:http://www.cabp.com.cn
网上书店: http://www.china-building.com.cn

《新英汉建筑装饰工程辞典》
编委会

主　编： 李育才　　卜纯英
副主编： 李振宇
编　委： 李育才　　卜纯英　　李振宇　　吴劭文
　　　　　　乔　敏　　潘建农　　洪孝安　　付子荣
　　　　　　李世振　　胡礼宁　　林　莉　　夏　红
　　　　　　项荣相　　张　明　　江金蓉　　陈　洁
　　　　　　何海燕　　朱如璧　　马　民　　薛小兰
　　　　　　周　红　　宋　云　　赵家庆　　戴文鸿
　　　　　　应之祥　　卜石其　　韩　梅　　张修文
　　　　　　吴　平　　顾淑霞　　殷永红　　蔡　正
　　　　　　胡亦斌　　李则熹

前 言

随着我国国民经济的蓬勃发展,改革开放的不断深化,从东南沿海到内地边陲,高楼大厦拔地而起,道路交通四通八达,建筑装饰装潢业,已呈现出了一个飞速发展时期。现今我国又加入了WTO,建筑装饰业必须走出国门,打入世界先进建筑行列。为此,建筑装饰业的科技人员,必须了解世界各地区的建筑风格和现代各国的文化习俗与风土人情,具有广博先进的专业知识。

为了进一步推动我国建筑装饰装潢业再上一个新台阶,加快中外交流与合作,满足建筑、建材、装饰装修行业科技人员、施工人员、大专院校师生和外语翻译工作者学习和研究的需求,我们编纂了《新英汉建筑装饰工程辞典》一书。其内容主要包括建筑规划与设计、建筑施工与管理、建筑材料与机具、装饰装潢技艺、艺术造型与雕塑、环境艺术与环境心理学、经济管理学等等。英汉专业词语约7万条,还辅以部分简明易懂的插图。内容翔实丰富,图文并茂,是当今一部新型实用的建筑装饰工程工具书,以供广大读者参考查阅。

由于建筑装饰涉猎广博,而编者的水平有限,本辞书编辑时间也紧促,错误与疏漏之处在所难免,希望广大读者提出宝贵意见与批评,待该书再版时重新修订,特此表示衷心感谢!

2006年3月

使 用 说 明

1.《新英汉建筑装饰工程辞典》的词目一律按英文字母顺序排列，复合词、缩略词语也一律顺排。

2. 主干词大多有国际音标，并置于方括号"[ ]"内，有些主干词是组合词或音节划分比较清楚，未注明音标。

3. 方括号之后注明主干词的语法术语缩写字：n.（noun）名词，a.（adjective）形容词，v.（verb）动词，ad.（adverb）副词，vi.（intransitive verb）不及物动词，vt.（transitive verb）及物动词，pl.（plural）复数以及表示主干词词形词尾变化，例如：gulley（＝gully）排水沟和agitate-tated，-tating 搅拌，拌和。

4. 注释词义的同义词，用逗号"，"分开，近义词用分号"；"分开。复合词与前一单词或前一复合词第一部分相同时，相同部分用"～"号代替。

5. 工程术语方括号"[ ]"内的词或字可以省略或与前部分的词、字互换，圆括号"（ ）"内的词或字是注释。

6. 月牙形括号"【 】"表示外来语，并已标明。个别缩略词语已附词语之后。本辞典所附插图均系示意图，紧贴所示词语。

目 录

前言

使用说明

辞典正文 ·· 1~1005

A

abac *n.* 座标网;列线[诺谟]图
　～ scale 列线图标度线
abaca [ˌɑːbəˈkɑː] *n.* 蕉麻,马尼拉麻
abaci [ˈæbəsai] abacus 的复数
abaciscus [ˌæbəˈsaiskəs] *n.* 嵌饰,(用于嵌工的)小块花砖
abaculus = abaciscus
abacus [ˈæbəkəs] *n.*
(古希腊建筑圆柱顶部的)顶(冠)板,柱冠;列(曲)线图;算盘
　～ column 带有冠板的柱
　～ flower 圆柱顶(冠)板花纹

abacus flower

use [work, more counters of] an abacus 打算盘
abandon [əˈbændən] *vt.* 放[抛,废,舍]弃
abandonment [əˈbændəmənt] *n.* 放[废,抛]弃;委托[付];(保险)投保
　～ of a right 弃权
abas [əˈbæs] *n.* 诺谟图,列(曲)线图
abask [əˈbɑːsk] *ad.* (在阳光下)暴晒,(在火炉旁)取暖
abate [əˈbeit] *v.* abated, abating;减少[小,轻,退],降[压]低,抑制;除去,消除;作废,废除,中止,撤销,成为无效;abater, *n.*; abatable, *a.*
abatement [əˈbeitmənt] *n.* 减少[小,轻,退,压,价];消除[却],抑制,中断,失效,作废,撤销
　～ of dust(ing) 除尘,空气净化
　～ of nuisance 消除污染[障碍,公害]
　～ of pollution 污染消除
　～ noise 噪声抑制
　～ smoke 除烟法

abatjour [əˈbeitʒuːr] *n.* 天窗,遮阳板;百叶窗的反光片
A-battery *n.* A 电池组,甲电池
abattoir [ˌæbəˈtwɑːr] *n.* 挡风装置
abatvent *n.* 固定百叶窗;透气帽,障风装置
abatvoix [əˈbeitvɔiks] *n.* 反射板,吸声板
abaxial [æbˈæksiəl] *a.* 离开轴心的,轴外的,远轴的
abbe number 色散系数
abbertite *n.* 黑沥青
abbey [ˈæbi] *n.* 大修道院,僧院,庵堂;曾为僧院或尼庵大教堂
　～ block 教堂[寺院]建筑
　～ building 教堂[寺院]建筑
　～ church 修道院教堂
　the Abbey (= Westminster Abbey) 伦敦威斯敏斯特大教堂(西敏寺)
abbreviate [əˈbriːvieit] *vt.* 将……缩短,省[节,简]略
abbreviated [əˈbriːvieitid] *a.* 简化[写]的;小型的
　～ drawing 简图
　～ expression 缩略词(式)
　～ formula 简写式
　～ indication 缩写字样
　～ ladder 小型爬梯
　～ signal code 传输电码
　～ version 节[简]本
　be ～ to 简写为
abbreviation [əˌbriːviˈeiʃən] *n.* 缩短,缩写,缩略语
　service ～ 业务略语
ABC 初步,入门,基础知识

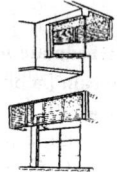
abatjour

The ~ of architectonis 建筑学入门
ABC-process *n.* 污水化学沉淀法,铝、血、碳处理法
abdicate ['æbdikeit] *v.* -cated, -cating; 放弃;让位,辞职
~ the office 辞职
~ M in favor of N 把 M 让给 N
A-bedplate *n.* A 型底座
abeam [ə'bi:m] *ad.* 正横,横向
Abel *n.* 阿贝尔(英国化学家 1827~1902)
~ C. C. testing apparatus 阿贝尔闪点检验仪
~'s close test 阿贝尔密闭实验(测定液体燃料和润滑油闪点的方法之一)
abele [ə'bi:l, 'eibl] *n.* 银白杨
Aberdeen [æbə'di:n] *n.* 亚伯丁(苏格兰东部一城市)
~ granite (英)亚伯丁花岗石(灰色或淡红色的花岗岩)
aberrance [æ'berəns] = **aberrancy** [æ'berənsi] *n.* 离开正道,越轨,脱离常规;逸出
aberrant [æ'berənt] *a.* 离开正道的,脱离常轨的,异常的
aberration [ˌæbə'reiʃən] *n.* 偏差,越轨,失[反]常,畸变,变型[体];象[色]差
aberrationless [ˌæbə'reiʃənlis] *a.* 无象差的,正常的,无畸变的
abet [ə'bet] *vt.* 唆使,怂恿,支持
abetment [ə'betmənt] *n.* 唆使,煽动,怂恿;助长
abeyance [ə'beiəns] *n.* 暂搁[缓],暂时无效,中止,停顿,未定,(化学)潜态
be in ~ 暂停,(暂时)搁置
abfarad [æb'færəd] *n.* CGS 电磁制法拉,绝(对)法(拉),电磁制电容单位=10^9 法拉
abhenry [æb'henri] *n.* 绝对亨(利)(电磁制电感单位=10^{-9}亨利),CGS 电磁制亨利
abidance [ə'baidəns] *n.* 居住;固守,遵守(常与 by 连用)

~ by rules(terms) 遵守规章(条例)
abide [ə'baid] *v.* abided, abiding; 持[继]续,保留;坚持,遵守;忍受;居住,住在; abider, *n.*
~ by 遵守,坚持,服从,容忍,守约
~ by the contract 遵守合同
abiding [ə'baidiŋ] *a.* 永恒的,不变的; abidingly, *a.*; abidingness, *n.*
~ -place 住宅,寓所
abies [ə'bais] *n.* 冷杉(属)
~ firma 枞木,冷杉
~ oil 松节[香]油,冷杉油
~ sachalinensis 库页冷杉,枞木
abietic acid *n.* 松香酸,枞酸($C_{20}H_{30}O_2$)
abietic anhydride *n.* 松香酐
abietic resin *n.* 松(树)脂,枞(树)脂
ability [ə'biliti] *n. pl.* -ties;能力,性能,效率,技能
~ to climb gradients 上坡[爬坡]能力
~ to flow 流动[散]性,(混凝土混合物)和易性
~ to harden 硬化性能
absorbing ~ 吸收能力
adhesive ~ 粘着力
interchange ~ 互换性
load-carrying ~ 起[载]重量,起重[载重,载荷,负载]能力,容量
~ for[in]M M (方面)的能力[本领]
~ to +inf 做……的能力[本领]
of ~ 有才干[本领]的
to the best(utmost)of one's ~ 尽力地,尽最大努力,竭尽全力
abiotic [ˌæbiəu'tik] *a.* 非生物的,无生命的;-abiotically, *ad.* abiosis [ˌæbi'əusis] *n.*
~ environment 非生物环境,无机环境
~ environmental factor 无机环境因素
~ factor 非生物因素[因子]
ablare [ə'blɛə] *a.* 喧闹的; *ad.* 喧哗地,吵闹地

ablate [æb'leit] *vt.* -lated,-lating；使（一部分）除[开]去，烧蚀，消融，腐蚀[融化，蒸发]掉；风化，剥[脱]落

ablation [æb'leiʃən] *n.* 消融(作用)，风化，剥落，磨削；脱离
~ material 烧蚀材料
~ moraine 融蚀水碛
~ shield 防融蚀的屏蔽，烧蚀防护罩
~ velocity 融蚀速度
~ zone 融蚀范围[层,带,区]

ablative ['æblətiv] *a.* 消融的，烧蚀的；*n.* 烧蚀材料
~ mode of protection 烧蚀防护法
~ protection 烧蚀防护，防烧蚀

ablator [æb'leitə] *n.* 烧蚀材料[剂]

ablaze [ə'bleiz] *a.*；*ad.* 着火(的)；闪耀(的)，发光(的)
be ~d with lights 灯火辉煌
set ~ 使燃(烧)

able *a.* 有能力的，有才干的；能……的

able-bodied *a.* 强壮的

able-minded *a.* 能干的

abluent ['æbluənt] *a.* 洗涤的，洗净的；*n.* 洗涤剂

ablution [əb'lu:ʃən] *n.* 清(吹)除，清洗；洗涤(液)；洗身，淋浴；有洗浴设备的房间；厕所，盥洗室
~ board 泄水板，滴水板
~ board of sink 排水口，泄水板
~ fountain 喷泉式洗涤槽(为水磨石环槽,同时可供六人使用)

ablutionary [əb'lu:ʃənəri] *a.* 洗净的，淋身的

abnegate ['æbnigeit] *vt.* 拒绝，否认，抛弃，放弃(权利,主张,要求等)

abnegation [ˌæbni'geiʃən] *n.* 拒绝，否认，抛弃，弃权；克制，自制

Abney level ['ebni;levl] *n.* (测量山坡高度或坡度用的)轻便水准仪，手水准仪，爱尼氏测斜仪

abnormal [æb'nɔ:məl] *a.* 反[异,非]常的，不正常的，不规则的
~ addressing (计算机)异常寻址

~ attribute 异常属性
~ audibility zone 非标准可听度范围
~ bridging 非正常桥(跨)接
~ curve 非正规[异常]曲线
~ density 异常密度
~ drop 不正常降低
~ end 异常终止[结束]
~ fault 异常[逆]断层
~ loading 异常[不规则]荷[负]载
~ noise 异常噪声
~ odor 异常臭气
~ overload 异常过载，事故过载
~ reflections 不规则反射
~ refraction 不正常折射
~ return address 异常返回地址
~ scattering 反常散射
~ setting 异[反]常凝结
~ soil 非正常土壤
~ state 异常状态
~ tide 异常潮，稀遇潮
~ stress 异常应力
~ voltage 反常电压
~ wear 异常[不规则]磨损

abnormity [æb'nɔ:miti] *n.* 异[反]常(现象)，不规则，畸形，异形

abode [ə'bəud] *n.* 住宅[所]，居住，较久的居住；*v.* abide 的过去式和过去分词
make one's ~ 居住
take up one's ~ 定居，住进，与……同住

abohm [ə'bəum] *n.* CGS 电磁制欧姆，绝对欧姆(电磁制电阻单位=10^{-9}欧)

aboideau = aboiteau [ˌaibwa'tou] *n.* 低地防潮(水)堤坝，堰，沼地挡潮闸

aboil [ə'bɔil] *a.*；*ad.* 沸腾的(地)

abolish [ə'bɔliʃ] *vt.* 废止[除]，取消，消除

abolishment [ə'bɔliʃmənt] *n.* 消除，取消

abort [ə'bɔ:t] *v.*；*n.* 故障，失灵，紧急停车；(计算机)异常终止[结束]；中止(实验,任务,工作)

A

~ escape system 紧急逃逸系统
~ handle 应急把手
~ light 紧急故障信号
~ sensing 故障测定
~ situation 故障位置

abortive [ə'bɔ:tiv] *a.* 故障的,应急的,(中途)失败的,无效的,没结果的,未成功的,未发生作用的; abortively, *ad.*; abortiveness, *n.*

abound [ə'baund] *vi.* 富裕,充满,丰富,多
~ in 富于,盛产,繁生
~ with 多,有很多

abounding [ə'baundiŋ] *a.* 丰富的,大量的; aboundingly, *ad.*

about-sledge [ə'baut,sledʒ] *n.* (铁工用的)大铁锤

above [ə'bʌv] *ad.* 在上,上方[面,文,述]; *prep.* 在……之上[以上];超过,超过……范围,高于,胜过,大于; *a.* 上述的,上面的; *n.* 上述者
~ -critical state 超临界状态
~ -curb 路边(侧面)标高以上的
~ earthpotential (对地)电位,(对地)电势
~ facility 地面设备[装置],地上构筑物
~ grade 在地平面上,高于原[订]定级别(指质量或技巧)
~ grade(pipe)line 地面管道
~ ground 正统的,官方的;在地上,地面的,地面上的
~ high water mark 在高水标位以上
~ hydrant 地面式消火栓(给水栓)
~ masonry 地面上的砖砌构筑物[砌体]
~ mentioned 上述的,前述的
~ norm 超标准的,超定额的
~ normal 超常的
~ (pipe)line 地面管道,架空管道
~ reproach 无可指责
~ sea level 海拔
~ stairs (在,向)二楼(的),(在,向)

楼上(的)
~ tank 地面贮[蓄]水池,贮罐
~ thermal 超热的
~ water 水面[上]的,吃水线上的
~ work 地面工程

A-bracket *n.* 人字架

abradability [ə'breidə,biliti] *n.* 磨蚀[损]性,磨损度

abradant [ə'breidənt] *n.* 研磨[磨蚀]剂,金钢砂; *a.* 研磨用的

abrade [ə'breid] *v.* 擦去[掉,破];磨(损,蚀,掉,光);研磨;(用喷砂机)清除;刮擦
~ platform 浪蚀台地
~d quantity 减量,损耗量,磨损量

abrader [ə'breidə] *n.* 磨损实验机,磨石,磨光(砂轮)机,研磨机

abradibility *n.* 可研磨性

Abrams' cone 混凝土强度试验锥

Abram's fineness modulus 粗度模量,粒度模数

Abram's law 混凝土强度试验规范[则]

Abram's method 混凝土强度试验法

Abram's test (对混凝土骨料进行的)氢氧化钠检验[试验]

abrase [ə'breiz] (= abrade) *vt.* 摩擦,擦去[掉],刮掉
~d glass 磨光玻璃

abraser [ə'breizə] *n.* 测定物质抗磨性的器械

abrasion [ə'breiʒən] *n.* 刮掉[去];磨耗[蚀,去光],研磨,冲[海,浪,剥,水]蚀;裸露,露出;去垢
~ drill 回旋钻
~ fatigue 磨损疲劳
~ finishing 磨光
~ hardness 磨耗[耐磨,研磨]硬度
~ loss 磨耗损失
~ machine 耐磨试验机
~ particle 磨蚀微粒
~ resistance 抗磨力[性],磨耗阻力,耐磨度
~ resistant 耐磨的;不掉色

~ resisting steel 耐磨钢
~ strength 耐磨度,抗磨强度
~ test 耐磨[磨损]试验
~ tester 磨损试验机
~ value 磨耗量[值]
~ wear 磨耗量;磨损
abrasion-proof *a.* 耐磨的
abrasion-resistant steel *n.* 耐磨钢
abrasive [əˈbreisiv] *n.* (研)磨(材)料[剂],擦粉;磨擦力; *a.* 磨料的;磨损[蚀]的,研磨的,磨掉的
~ band 研磨带
~ blasting 用磨料进行喷砂处理
~ block 研磨块
~ brick 研磨砖
~ cement 磨料粘结剂
~ cloth 砂布
~ compound 研磨材料
~ disc(=disk) 砂[磨]轮,研磨盘
~ dresser 砂轮修整器
~ dust 磨屑
~ fabric 金刚砂布
~ grain 磨料粒度
~ grinding wheel 磨[砥,油]石
~ hardness 磨蚀[磨耗,研磨]硬度
~ machining 磨削加工,强力磨削
~ material[media] 磨料
~ paper 砂纸
~ particle 磨料粒度
~ powder 研磨[磨料,金刚]粉
~ resistance 抗磨力[性],耐磨(损)能力
~ sand 磨砂,流砂
~ sawing machine 砂轮切割机
~ steel shot 铁砂喷射处理
~ stick 油石,磨条
~ surface 磨蚀[研磨]面,(路面的)磨耗层
~ tools 研磨工具
~ type 研磨式的
~ wear(ing) 磨耗[损,蚀]
~ wheel 磨[砂]轮

~ wheel cutting-off machine 砂轮切割机
abrasiveness [əˈbreisivnis] *n.* 磨耗[损],磨蚀性,研磨[耐磨]性
abrator *n.* 喷砂清理;喷砂清理机
~ head 喷砂头,喷嘴抛丸头
hanger ~ 悬挂式抛丸清理机
wheel ~ 喷丸器,抛丸清理装置
abrazite *n.* 水钙沸石
abreast [əˈbrest] *ad.* 并列[排,联],相并;平行,等速前进
~ of(with) 保持与……并列,跟上,适应
abreuvoir *n.* (砌体)灰缝
a line ~ 并列[横排]成一线,砌墙麻线
abri [ɑːˈbriː] *n.* 岩洞[穴],防空洞
abridge [əˈbridʒ] *vi.* 删节,缩短,省略,简化,摘要,削减
~ trial load method 截面荷载分配法
~d drawing 略图
~d edition 节略版[本],节本
~d general view 示意图
~d indication 简略字样
~d multiplication 简捷乘[除]法
~d spectrophotometer 滤色光度计
be abridged from 是根据……删节[简化]的
abridg(e)ment [əˈbridʒmənt] *n.* 节略,缩短;摘要,概述;节本
abrim [əˈbrim] *ad.* 满满地,满溢地
abroad [əˈbrɔːd] *ad.* 室[户]外,遍布,到处,传开,流行,广泛地;在国[海]外,到国[海]外
at home and abroad (在)国内外
from abroad 国外进口的/go abroad 出口
abrogation [ˌæbrəˈgeiʃən] *n.* 废除,取消
ABRSV = abrasive 磨料
ABRSV RES = abrasive resistant 耐磨蚀的
abrupt [əˈbrʌpt] *a.* 突然[变]的,意外

的；陡的，急剧[转]的，不连续的；
abruptly, *ad.*
~ change of cross-section 断面突变
~ change of voltage 电压突变
~ curve 急变[陡变]曲线
~ discharge 猝然排出量
~ junction 突变[阶跃]点[结]
abruption [ə'brʌpʃən] *n.* 分[断，破]裂，拉[中，隔]断，断路，突然分离[断开]
abruptness [ə'brʌptnis] *n.* 陡（峭）度，陡峭性
abscess ['æbses] *n.* （金属中的）砂眼，气孔[泡]，夹渣内孔
abscind [æb'sind] *vt.* 切断，割去
abscissa [æb'sisə] *n.* 横线，横（坐）标
abscission [æb'siʒən] *n.* 截去，切除，脱离
~ layer 离层
ABSco-polymers *n.* 氢基乙烯，丁二烯和苯乙烯的共聚物（热塑料管的原料之一）
absence ['æbsəns] *n.* 缺少[乏，席]，没有，不（存）在，
~ of draught [draft] 不通风
~ of glare 无光线，不透光
~ of load (ing) 无负[荷]载，卸[释]荷
~ of streaks 水泥和添加剂混合时无[缺乏]痕线
~ of style 风格上不协[谐]调的建筑物，无风格
~ of wear 无消耗[损失]
~ rate 缺勤率
absent ['æbsənt] *a.* 不存在的，缺少[乏]的；[æb'sent] *v.* 不在，离开，缺勤[席]
~ order 缺序，不规则
~ without excuse 擅自缺席，擅离职守
absinthe-green ['æbsinθ'gri:n] *a.* 苦艾绿的，淡绿色的
absinthe green 苦艾绿，淡绿色

absinthe yellow *n.* 灰绿的黄色
absis [əb'sis] *n.* （教堂的）半圆形后室；半圆壁龛
absite [æb'sait] *n.* 钍钛铀矿
absolute ['æbsəlu:t] *a.* 绝对的，纯粹的，无条件的，无限制的；确实的，无疑的；独立的
~ accuracy 绝对精确度
~ address 绝对地址
~ addressing 绝对选址
~ advantage 绝对利益
~ alcohol 无水酒精
~ altimeter 绝对测高计
~ altitude 绝对高度，海拔，标高
~ atmosphere 绝对大气压
~ boiling point 绝对沸点
~ calibration 绝对校准
~ capacity 绝对容量
~ coefficient of expansion 绝对膨胀系数
~ configuration 绝对[独立]构型
~ construction 独立结构
~ data 绝对数据
~ deflection 绝对变位，绝对挠度
~ deformation 绝对失真[变形]
~ demineralization （水的）绝对软化[无盐化]，完全除盐
~ density 绝对密度
~ detection limit 绝对探测极限
~ determination 确定[绝对]的测量[测定法]
~ displacement 绝对位移
~ dry condition 绝干[全干]状态
~ dry specific gravity 绝干[全干]比重
~ dry weight 绝干重量
~ dynamic modulus 绝对动态模量
~ elevation 绝对高程，标高，海拔
~ error 绝对误差
~ ether 无水醚
~ ethyl alcohol 无水乙醇
~ expansion 绝对膨胀
~ extension 绝对延伸

~ filter 绝对过滤器,高性能空气过滤器
~ fine aggregate percentage 绝对细骨料率
~ hardness degree 绝对硬度
~ heating effect 绝对热效率,绝对采暖效果
~ height 绝对高度
~ humidity 绝对湿度
~ instruction 绝对指令
~ language 机器语言,绝对语言
~ level 绝对水平,绝对标高
~ magnitude 绝对量
~ manometer 绝对压力计
~ maximum 绝对最大极限
~ maximum rating 绝对最大额定值
~ measurement 绝对测量
~ methanol 无水甲醇
~ modulus 绝对模量
~ moisture content 绝对湿[潮]度
~ orientation 绝对定向
~ priority 绝对优先权
~ parauax 绝对视差
~ precision 绝对精度
~ pressure 绝对压力
~ profit 绝对利润
~ porosity 绝对孔隙度
~ refraction factor 绝对折射率[系数]
~ retention time 绝对滞留[保留]时间,绝对停留时间
~ scale 绝对尺度
~ sensitivity 绝对灵敏度
~ specific gravity 净重,绝对比重
~ specificity 绝对专一性
~ stability 绝对稳定性
~ stability constant 绝对稳定常数
~ strength 绝对强度
~ symmetrical balance 绝对对称平衡
~ system of measures 绝对度量制,绝对度量单位制
~ temperature 绝对温度
~ temperature scale 绝对温标
~ valency 绝对[最高]价
~ value 绝对值
~ velocity 绝对速度
~ viscosity 绝对黏度,绝对滞度
~ viscosity coefficient 绝对黏滞系数
~ volume (method) 绝对体积[容积](法)
~ volt 绝对伏特
~ watt 绝对瓦特
~ zero 绝对零点
by ~ necessity 按绝对需要
absolutely ['æbsəlu:tli] *ad.* 绝对(地),无条件(地),完全地;确实,当然;实际,真正
~ dry 绝干,全干,干透;无水的
~ drywood 全干木材
~ impossible 绝对不可能的
absolution [ˌæbsə'lu:ʃən] *n.* 免除,赦免,解除责任
absolve [əb'zɔlv] *vt.* -solved,-solving;免(除,于),解除,赦免
absonant ['æbsənənt] *a.* 不合拍[理]的,不和谐的
absorb [əb'sɔ:b] *vt.* 吸收[取,液],减振,缓冲;承担(费用),承受冲击
~ed energy 吸收能
~ed heat 吸收的热
~ed in-fracture energy 冲击韧性[强度],冲击功
~ed layer 吸收[附]层
~ed power 吸收功率
~ed radiation 吸收辐射
~ed water 吸收[附]水
~ed water in wood 木材中的自由水分
absorbability [əbˌsɔ:bə'biliti] *n.* (可,被)吸收性,吸收[取]能力,吸收量
absorbable [əb'sɔ:bəbl] *a.* 可[被]吸收的,容易被吸收的
absorbance [əb'sɔ:bəns] = absorbancy [əb'sɔ:bənsi] *n.* 吸收率[比],吸收系

数(的常用对数)
~ index(=absorptivity) 吸收系数,吸光系数[光谱]
absorbate [əb'sɔ:beit] ***n.*** 被吸收(的)物(质),吸收质
absorbefacient [əb,sɔ:bi'feiʃənt] ***a.*** 吸收性的; ***n.*** 吸收剂
absorbency [əb'sɔ:binzi] ***n.*** 吸收能力,吸墨性
absorbent [əb'sɔ:bənt] ***a.*** 能吸收的,有吸收能力的; ***n.*** 吸收[附]剂,吸收体[质,管]
~ aggregate 吸水集料,吸附(成套)设备
~ bed 吸收[附]层,吸收床
~ blanket 吸声[隔声]垫
~ brick 吸声[隔声]砖
~ carbon 活性碳
~ ceiling sheet 吸声[隔声]顶棚板条
~ cotton 脱脂棉
~ filter 吸附滤池[器]
~ filter medium 吸附(滤池)滤料
~ insulation 吸收绝缘,海绵状绝缘
~ material 吸收[声]材料
~ oil 吸收油
~ paper 吸水纸
~ power 吸收能力
~ refrigerator 吸收式制冷机
~ unit 吸声[隔声]装置[设备]
~ wall 吸声[隔声]墙
~ wall block 吸声[隔声]墙板[块]
~ wall brick 吸声[隔声]墙砖
~ wall paper 吸声[隔声]墙纸
~ wall tile 吸声[隔声]墙空心砖
~ type filter 吸附型滤池[器]
~ acoustical ~ 吸声材料[剂]
absorber [əb'sɔ:bə] ***n.*** 吸收器[剂,装置],过滤器;中和剂;减振[缓冲,阻尼]器
surge ~ 避雷器
suspended ~ 空间吸声体
telescopic shock ~ 筒式减振器
thermal[heat] ~ 吸热器

vibration ~ 减振器,消振器
water jet dust ~ 水射吸尘器
absorbility [əb'sɔ:biliti] ***n.*** 吸附[收]能力;吸收率
absorbing [əb'sɔ:biŋ] ***a.***; ***n.*** 吸收(的),减振(的)
~ ability 吸收能力
~ apparatus 吸收[附]仪(器)
~ backing 吸声[减声]填料
~ board 吸声[隔声]板
~ boom 吸油栅
~ (building)unit 吸声[隔声]组合设备[装置]
~ capacity 吸收能力[容量]
~ cassete 吸声[隔声]镶面[板]
~ ceiling 吸声[隔声]顶棚
~ ceiling board 吸声[隔声]顶棚木板
~ ceiling paint 顶棚吸声涂料[油漆]
~ ceiling system 吸声[隔声]顶棚系统[装置]
~ ceiling tile 吸声[隔声]顶棚贴面板
~ chamber 吸声[隔声]室
~ coffer 吸声[隔声]板
~ construction material 吸声[隔声]结构材料[建筑材料]
~ construction(method) 吸声[隔声]构造[结构]
~ (control)glass 隔声玻璃
~ covering 吸声[隔声]涂[贴,镶]面层
~ facing 吸声[隔声]涂面[贴面层]
~ felted fabric 吸声[隔声]毡
~ felted fabric ceiling 具有吸声[隔声]毡的顶棚
~ fibre board 吸声[隔声]板[纤维板]
~ foil 吸声[隔声]箔
~ gas 吸收气体
~ glass 吸声[隔声]玻璃
~ hung ceiling 吸声[隔声]吊顶
~ lining 吸声[隔声]涂面[贴面]
~ load 吸收负载
~ masonry wall 吸声[隔声]砖砌墙

~ material 吸声[隔声]材料
~ metal ceiling 金属吸声[隔声]顶棚
~ pad 吸声[隔声,减振]垫
~ paint 吸声[隔声]涂料[油漆]
~ pan(el) 吸声[隔声]镶板[方格板]
~ pipettes (气体)吸收球管
~ plaster 吸声[隔声]粉刷层[灰膏]
~ plaster aggregate 吸声[隔声]粉刷灰浆骨料
~ plaster ceiling 具有吸声[隔声]粉刷层的顶棚
~ powder 吸附[收]粉
~ power 吸收能力[效率,容量]
~ sheet 吸声[隔声]板条
~ sprayed on plaster 采用喷射施工的吸声[隔声]粉刷层
~ surface 吸收[附]面
~ (sur)facing 吸声[隔声]贴面
~ suspended ceiling 吸声[隔声]吊顶
~ system 吸声[隔声]系统[装置]
~ tile 吸声[隔声](空心)砖
~ tile ceiling 吸声[隔声]空心砖顶棚
~ tower 吸收塔
~ waffle 吸声[隔声]方形板块
~ wood fibre board 吸声[隔声]木质纤维板

absorbite [əbˈsɔːbait] *n.* 活性碳

absorptance [əbˈsɔːptəns] *n.* 吸收比[性,系数,能力]
   internal ~ 内吸收系数
   radiant ~ 辐射吸收系数
   screen ~ 屏蔽吸收系数,屏幕吸收系数
   spectral ~ 光谱吸收系数

absorptiometric [əbˌsɔːpʃiəˈmetrik] *a.* 吸收(比色)计的,溶[吸]气计的,调液厚器的

absorptiometry [əbˌsɔːpʃiˈɔmitri] *n.* 吸收(能力)测量,液体溶气法,吸收测量法[学]

absorption [əbˈsɔːpʃən] *n.* 吸附,吸收[取,液](作用);吸水(性)

~ band 吸收(光,谱,光谱)带
~ capacity 吸收能力[性能,特性]
~ chamber 吸收室
~ coefficient 吸收系数
~ colouring 吸收着色
~ costing 吸收成本计算法,分摊成本计算法
~ cross-section 吸收截面
~ curve 吸收曲线
~ damper 减振器,消声器
~ enhancement effect 吸收增强效应
~ factor 吸收系数[率]
~ factor of solar radiation 太阳辐射吸收率
~ heat 吸收热
~ heat pump 吸热泵
~ hygrometer 吸收式湿度表
~ law 吸收律
~ level 吸收(能)级
~ limit 吸收极限[范围]
~ meter 调液厚器,(液体)溶[吸]气计,吸收[比色,光度]计
~ of moisture 吸湿[潮]
~ of shocks 缓冲,减振
~ power 吸(声)能力[性能,特性]
~ property 吸声能力[性能,特性]
~ quality 吸声[收]能力[性能,特性]
~ rate 吸收[水]率
~ refrigerating compressor 吸收式制冷机[冷气机]
~ refrigerator 吸收式制冷机
~ strength 吸收强度
~ surface 吸附表面
~ system (空气)吸湿[去湿]系统
apparent ~ 表面[表观]吸收
dust ~ 吸尘
low optical ~ 少量光[微量光]吸收
luminous ~ 光吸收
marginal ~ 临界[边缘]吸收
moisture ~ 水分吸收,吸湿(作用)
selective ~ 选择(性)吸收[吸光]
shock ~ 减振(法)

specific ~ 吸收比[系数,率]
ultraviolet ~ 紫外线吸收
water ~ 吸水率
absorptive [əb'sɔːptiv] *a.* 吸收(性)的,吸水(性)的,有吸收能力的
~ backing 吸声垫[衬]板,石膏吸声板
~ blanket 吸声[隔声]隔板
~ board 吸声隔板
~ brick 吸声[隔声]砖
~ (building) unit 吸声组装部件[构件]
~ capacity 吸收能力[容量]
~ capatity for rain water 雨水滞留能力[容量]
~ cassette 吸声(小块)镶面板
~ ceiling 吸声顶棚
~ chamber 吸声室
~ character 吸收特性
~ coffer 装饰吸声板
~ construction material 吸声[隔声]构造[建筑]材料
~ (control)glass 吸声[隔声]玻璃
~ cover(ing) 吸声护面[涂层,衬砌]
~ facing 吸声护面[涂层,衬砌]
~ felt(ed fabric) 吸声毡
~ fibre board 吸声纤维板
~ foil 吸声箔
~ index 吸收指数[系数,率]
~ lining (混凝土模板用)吸水衬里;吸声护面[涂层,衬砌]
~ masonry wall 吸声[隔声](砖砌)墙
~ material 吸声[隔声]材料
~ pad 吸声垫[隔板]
~ paint 吸声油漆[涂料]
~ pan(el) 吸声[隔声](薄)板
~ plaster 吸声灰膏[粉刷]
~ powder 吸附粉末
~ quality 吸收性
~ sheet 吸声木条
~ (sur)facing 吸声护面[涂层,衬砌]
~ system 吸声装置[设备,系统],吸

收系统
~ tile 吸声砖[瓦]
~ treatment (建筑)吸声处理
~ type filter 吸附型过滤器
~ unit 吸声装置[单元],吸声构造[组装构件]
~ waffle 吸声(薄)板
~ wall 吸声墙
~ wood fibre board 吸声木质纤维板
abstergent [əb'stəːdʒənt] *n.* 去垢剂,洗涤剂;*a.* 去垢的,有洁净作用的
abstersion [æb'stəːʃən] *n.* 洗净,净化
abstersive *a.* 使……洁净的,去垢的
abstinence ['æbstinəns] 或 **abstinency** ['æbstinənsi] *n.* 节制[约]
abstract [æb'strækt] *vt.* 提[萃]取,抽[取]出,除[移]去,散开;摘要,提要,概括;['æbstrækt] *n.* (将热或水)引出,蒸馏;提[萃]取物,摘[提]要,简介;概括,抽象(观点,物),抄录
~ art 抽象艺术
~ code 抽象码,理想代码
~ designs 局部花纹图案,节略花纹图案
Abstract Expressionism 抽象画派风格或画法
~ form 抽象形式
~ heat 散热
~ labor 抽象劳动
~ machine 抽象(计算)机
~ mechanism 抽象模型[原理,机构]
~ model 抽象模型
~ set 电视演播室布景
~ structure 抽象结构
automatic ~ 自动抄录
indicative ~ 要点抄录,简介,内容提要
information ~ 内容提要,报告性文摘
~ one's attention from 从……上转移某人注意
in the ~ 抽象地,理论上
make an ~ of 把……要点摘录下来
abstraction [æb'strækʃən] *n.* 引出,抽

象作用[观念];提取,除去,萃取,分离,退流,(水)袭取(作用)
~ heat ~ 除[散,排,减]热,热的排除[散失,取出]
abt. =about
abuilding [ə'bildiŋ] *a.* 正在建造[建立,发展]的
abundance [ə'bʌndəns] *n.* 丰度[富],充裕,过满,极多,充足,分布量,比较多的数量
abundant [ə'bʌndənt] *a.* 丰富的,充裕[足,分]的;大量的,许多的;abundantly, *ad.* ;abundantness, *n.*
~ number 过剩数
~ proof 充分证据
be ~ in 富有[于],……丰富的
abunits *n.* CGS 电磁制单位
abusage [ə'bju:sidʒ] *n.* 乱[误]用
abuse [əbju:z] *n.* ;*v.* abused, abusing; *n.* ,滥用,违反操作规程,弊病,*vt.* 滥用,误用,糟踏
abut [ə'bʌt] *n.* ;*v.* abutted, abutting; 邻[连]接,毗连[临]近;(紧)靠(……上),靠着,支撑;*n.* 桥台[基],端[尽]头,支点[架,座,承,撑],柱[拱]脚,扶壁,接(合)点,接界;止动,止动点,止动器
~ ted surface 相接[贴合]面
~ ting buildings 毗连房屋[建筑物]
~ ting joint 对接接头,端接接头
~ ting lane 相邻车道
~ ting lot 相邻地段
~ ting property 相邻(地)产,(相邻地皮指与道路邻接的房产,地产)
abutment [ə'bʌtmənt] *n.* (支,拱,墩式台)座,(拱)脚,桥台[墩],支[墩]柱,扶垛,支(撑)面,接合点,接界
~ gallery 岸墩[桥台]廊道
~ hinge 岸墩[支墩,桥台]接合(点)铰[座]
~ masonry 支墩[支座,支柱,桥台]砖台[砌块]

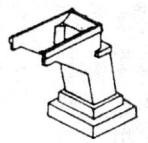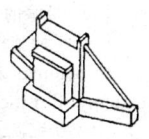

abutment

~ pad 岸墩[支墩,桥台]基座[底座]
~ piece 桥台,支座,(靠)岸(桥)墩
support ~ 支座
abuttal [ə'bʌtl] *n.* 桥台[基],支柱,承重;邻接,接界;界线,(两地相邻的)境界,地界
A. C. =alternating current 交流(电)
~ arc-welder 交流电弧焊机
~ generator 交流发电机
~ motor 交流电动机
acacia [ə'keiʃə] *n.* 刺槐,金合欢,相思树,阿拉伯橡胶树;阿拉伯树胶;黄绿色
~ ecosystem 金合欢属植物生态系统
academia [ˌækə'di:miə] *n.* 学术界;学术环境
academic [ækə'demik] =academical *a.* 高等[专科]院校的,研究院的,学院[会]的,学术的;单纯理论的,*n.* (大)学生,大学教师,学会会员
~ body 学术团体
~ city 学术城
~ degree 学位
~ discussion 学术讨论
~ painting 学院派绘画
~ style 学院派风格
~ style of building 建筑物的学院派风格,学院式的建筑风格
academician [əˌkædə'miʃən] *n.* (学会)会员,院士
academicism [ˌækə'demisizəm] *n.* 学院风气,学院主义,学院式,墨守成规或传统;纯粹推测性的思想与态度
academist [ækə'demist] *n.* 学院派艺术家
academy [ə'kædəmi] *n.* 学院,专科院校;研究院,学会

~ blue 带绿的蓝色
~ board 油画板
~ of science(AS) 科学院
Academy of Fine Arts（绘画，雕塑，建筑）美术研究院[学院]
~ of management 管理研究院
acadialite [əˈkeidiəlait] *n.* 红菱[红斜方]沸石
acajou [ˈækəʒuː] *n.* 槚如树；槚如树脂[胶]；几种桃花心木的任一种
acantha [əˈkænθə] *n.* 棘，棘突
acanthconite [ˌəkænθˈkəunait] *n.* 绿帘石
acanthus [əˈkænθəs] *n. pl.* -thuses；（古希腊科林思柱头上的）叶形（蕃草叶，卷草叶，莨苕叶）装饰；叶板

acanthus

~ frieze（科林思柱与横梁之间）雕画带
~ leaf（科林思柱）叶板，叶瓣
~ scroll（科林思柱）涡卷形装饰
~ spinose（科林思柱）针叶装饰
accede [ækˈsiːd] *vi.*, -ceded,-ceding 同意，允诺，接受；就位；参加，加入
accelerant [ækˈselərənt] *n.* 加速剂，促凝剂，催化剂，触媒
~ coatings 速燃层
accelerate [ækˈseləreit] *v.*, -ated,-ating；加[催，变]速，增加速度，促进
~d ag(e)ing 加速老化[时效]，人工加速老化[时效]
~d binder content determination 胶[粘]合剂含量快速测定
~d cement 快凝水泥
~d circulation 加速环流[循环]，机械循环
~d clarification 加速[机械]澄清
~d consistency test 稠度[黏度]快速检验
~d corrosion test 腐[侵，锈]蚀快速试验
~d curing 加速养护
~d depreciation 加速折旧
~d gum 速成胶质
~d stock 加速胶料
~ing additive 速凝剂,（外加）促进剂
~ing admix(ture)（外加）促进剂,快速掺和剂
~ing agent（混凝土）促凝剂,促进剂,催化剂
~ing convergence 加速收敛
~ing conveyor 加速输送机[带]
~ing electrode 加速电极
~ing field 加速场
~ing flow 加速流
~ing force 加速力
~ing jet 加速喷嘴
~ing load 加速荷载
~ing pump（采暖系统）热加速泵
~ing reducing valve 快速减压阀
~ing region 加速区
~ing relay 加速[多级式]继电器
acceleration [ækˌseləˈreiʃən] *n.* 加速（度，作用），加快，剧升
~ area（车行道的）加速区段
~ of setting 缩短硬化[凝固,凝结]时间,加速硬化[凝固,凝结]
~ response 加速度[过载]反应
absolute ~ 绝对加速度
allowable ~ 容许加速度[值]
angular ~ 角加速度
average ~ 平均加速度
axial ~ 轴向加速度,纵向加速度
brief ~ 瞬时有效加速度
casual ~ 随机加速度
centrifugal ~ 离心加速度
centripetal ~ 向心加速度
circular ~ 圆周加速度
constant ~ 等加速度
convective ~ 对流加速度
Coriolis ~ 科里奥利加速度
cumulative ~ 累积加速度
accelerator [ækˈseləreitə] *n.*（混凝土）的促凝剂,加速[促进,促凝,速滤,催

化]剂;(采暖系统)热水循环泵;加速电机启动器,加速[机械]澄清机
~ -decelerator 速度控制器,加速(器)-减速器,风门控制器
~ dosage 促进剂用量[配量]
~ dynamic test 加速器动态试验
~ pedal 加速踏板
~ ratio 促进剂(用量)比率[配比]
concrete ~ 混凝土促凝[硬]剂
delayed action ~ 缓凝剂
flocculating ~ 絮凝剂
polymerization ~ 聚合加速剂
setting ~ 促凝剂
accelerogram [æk'selərə'græm] **n.** 自计加速图
accelerograph [æk'selərə'grɑːf] **n.** 自动[计]加速仪,加速自(动)记录计[器,仪]
high natural frequency ~ 高固有频率自记加速计
lateral ~ 侧向过荷自记器,侧向加速自记器[仪]
linear ~ 线性加速自记器[仪]
normal ~ 法向过荷[加速]自记器[仪]
accelerometer [æk,selər'ɔmitə] **n.** 加速(度)表[计,仪,器],过荷载传感[指示,自记]器,测震仪
~ chart 加速计[测震仪,过荷传感器]图表
~ type seismometer 加速度地震检波器
angular ~ 角加速度测量仪,角加速度表
counting ~ 带计算器的加速表
diaphragm ~ 膜式加速度,膜合加速计
laser ~ 激光加速仪
lateral ~ 侧向加速表,侧向过荷加速表
linear ~ 线性加速表,过荷传感器
integrating ~ 综合加速测量仪,积分加速计
normal ~ 法向加速表

recording ~ 自动加速表,过荷自记表
self-recording ~ 自记式加速计
shock ~ 冲击负荷加速表
Accelo-filter [æk,selə'filtə] **n.** (高速滤池用)加速滤器
accent ['æksənt] **n.** 重音(符号),音[声]调; [æk'sent] **vt.** 加强,强调;使特别显著
~ colour 重点色彩
~ letter 重音字母
~ light 强光灯,加强光灯
~ed term 重点项目
accentuate [æk'sentjueit] **vt.** -ated, -ating; 强调,加重,增强
~d contrast 加重对比度
accentuation [æk,sentju'eiʃən] **n.** 预增频,预加量,加重,强调
contour ~ 提高(图像中物体)轮廓的明显性
accentuator [æk'sentjueitə] **n.** 选频校正器,加重器[电路];加强电路,振幅加强线路
accept [ək'sept] **vt.** 接收[受],验收(格);承认,允许;承兑
~ed bank 承兑银行
~ed bill[draft] 已承兑票据[汇票]
~ed chips 合格木片
~ed load 使用[承受]荷载[负荷]
~ed product 合格品
~ed stock 合格浆料
~ for carriage 承运
currently accepted 目前[现行]通用的
generally accepted 普遍承认的
acceptability [əkseptə'biliti] **n.** 可接收性,可承兑性,合格,满意
~ criterion 验收规范[标准]
acceptable [ək'septəbl] **a.** 可接受的,容许的,验收的,合格的; acceptability, **n.** acceptably, **ad.**
~ concentration 容许浓度
~ daily intake(=ADI) 污染物浓度的每日容许排放[摄取]量

A

~ dose 容许剂量
~ environment 验收环境
~ environmental limit 容许环境(吸入)极限
~ explosive 合格炸药
~ failure rate 允许故障率
~ indexing 可接受标记
~ level 容许水平
~ life 有效使用寿命
~ limit 容许极限
~ material 合格原料[材料]
~ noise level 容许噪声标准,容许噪声级
~ quality level 验收[正品]质量标准(A.Q.L)
~ principle 公认原理
~ product 合格产品
~ program 可接受程序
~ test 验收[合格]试验
~ velocity 容许速度
~ for 适用于,可为……所接受的

acceptance [ək'septəns] *n.* 验[接]收,收录,容纳,承认,认可;承兑,认付
~ boundary 接收范围[边界]
~ certificate 验收合格证,接受证书
~ check 验收
~ commision 承兑手续费
~ committee 验收委员会
~ condition 验收[合格]条件
~ contract 承兑合同
~ credit 承兑信用证
~ criterion 验收标准[规范]
~ deviation 验收(容许)偏差,验收公差
~ documents 验收文件[资料]
~ domain 接受域
~ drawing 验收图
~ fee 认付费
~ gauge 验收规
~ house 承兑所,票据承兑银行
~ inspection 验收检查[查对],接受验收

~ level 验收标准
~ line (银行)承兑限额
~ margin 验收公差
~ market 证券市场
~ off-size 验收公差
~ permissble variation 验收允许公差
~ range 作用距离[范围]
~ requirements (工程)验收要求,接受规格
~ sample 承认[接收]货样
~ sampling 选样认可
~ specification 验收规范
~ survey [验收]竣工测量
~ test 验收试验
~ tolerance 验收容许公差[偏差]
~ trial 验收试验[检验]
~ value 验收值
find ~ with 得到……允许[承认]

acceptor [ək'septə] = accepter *v.*; *n.* 接受器,受主,票据承兑人;谐振回路,带通电路
~ circuit 接受器电路,带通电路
thermal ~ 热受体

access ['ækses] *n. pl*, -es 入(出)口,引道[桥],通[进]路;调整孔(计算机)存取,取数,(数据,信息)选取;接触[使用]方法;增加[长];捷径
~ activity 存取活动
~ balcony (楼层)入口(处)走廊[过道,通道]
~ board 便桥,入口辅板,跳板,踏板
~ capability (计算机操作系统)存取能力
~ chamber 清扫间[室]
~ circuit 存取电路
~ code control 存取访问码控制
~ connection 进口连接
~ cover 检修盖
~ cycle 存取周期
~ dimension (入口的)进口直径[宽度]
~ door 便门,检修门[孔],入孔盖

~ driveway 进入干道的专用支路
~ duct 进线管道
~ eye 检查孔
~ floor 双层地板,夹层地板,活地板
~ front 入口正面
~ gallery (楼层)入口(处)走廊[过道,通道]
~ galley 雨水口
~ hatch 观察孔,检查孔[口],入孔
~ hole 检查孔,入孔,(计算机多孔磁心用)取数孔
~ hook 铁爬梯
~ ladder (游泳池内用)爬梯
~ method 存取法
~ mode 读出方式
~ monument 入口界石
~ opening 观察孔,检修[查]孔
~ panel 观测台,观察板
~ path 存取通路
~ pit 清扫
~ plate 清扫盖
~ point (道路立体交叉)进入点,入口处
~ port 存取口
~ prevention planting 禁止栽植
~ ramp (道路交叉处)入口坡道,车行道坡道,接坡
~ road 地区街道,次要道路,便道,进路,入境道路
~ rule 存取规则
~ screw 清扫口堵头[丝堵]
~ selector 存取选择器
~ side 进口[入口]端
~ silo 进人井
~ speed 存取速度
~ time 取数时间
~ tower 入口塔(台)
~ tube 检查管
~ tunnel 进出口[交通]隧道
~ way 入口引道,进车道
~ well 交通井,进人井
~ width 存取位数

~ zone 接近段
arbitrary ~ 随意[任意]存取
(be) easy [hard; difficult] of ~ 易[难]接近[到达,理解]的
direct ~ 直接存取
disk ~ 磁盘存取
display ~ 显示存取
immediate ~ 即时[快速]存取
magnetic drum ~ 磁鼓存取
non-random ~ 非任意抽取,有序存取
sequential ~ 顺序存取,按序存取
have [get, obtain] ~ to 可以使用[获得,利用,接触]
serial ~ 串行存取[取数]
simultaneous ~ 同时存取[取数]
zero ~ 立即存取
accessary =accessory
accessibility [æk,sesi'biliti] **n.** 可达[及]性,可接近性,(新仪表使用前的)检查[查看]步骤[方法]
accessible [æk'sesəbl] **a.** 可以[容易]接近[到达,使用]的,易受影响的
~ boundary 可达边界
~ point 可达点
~ state 可接近(状)态
~ stationary points 可达稳定点[逗留点]
accession [æk'seʃən] **n.** 接近,到达;增加(物),加入;**vt.** 把(新书)登记入册
~al **a.** 附加的
accessory [æk'sesəri] **n.** (**pl.**)附[零,配,备]件,附属品,辅助设备,构筑物[装备,部件,仪表];添景物;**a.** 附[从]属的,附带[加]的,次要的,辅[补]助的
~ block 附属建筑物,辅助厂房
~ building 附属建筑物,辅助厂房
~ case 附[零]件箱
~ design specification 附件设计要求[任务书],辅助设计要求
~ equipment 附属设备
~ ingredient 辅助剂,添加剂

accident

~ part 配件,附件
~ pigment 辅助色素
~ risk 附加保险
~ structure 附属结构[设施]
~ supply system （工地）附属供给系统
anchoring accessories 锚定件的加固钢筋,地脚钢筋
engine accessories 发动机辅助装置
roofing accessories 屋面附件
vacuum accessories 真空设备用附件
accident ['æksidənt] *n.* （偶然）事故，（偶发）故障,损伤,破坏,偶然[不测,突发]事件；凹凸不平,褶皱；*a.* 紧[救]急用的；
~ ambulance station 事故救护站
~ and indemnity 意外事故及损失赔偿
~ brake 紧[救]急制动器
~ -cause code 工伤事故原因准则,事故发生原因及规程
~ congestion 事故阻塞
~ control ratio 事故控制比
~ cost 事故费用[支出]
~ crane 救护吊车[起重机]
~ death rate 事故死亡率
~ damage 意外损坏
~ defect 事故损坏
~ error 偶然误差
~ exposure 事故苗子,易出事（地）点
~ hazard 意外（危险）事故
~ hospital 事故救护医院
~ indemnity 意外保险
~ insurance 事故保险
~ jam 事故阻塞
~ locating 出事地点
~ prevention 安全措施,事故预防
~ proof 安全的,保险的
~ -prone （粗枝大叶而）特别易出事故的
~ protection 安全措施,事故防护[预防]
~ psychology 事故心理

~ rate 事故发生率
~ record 事故统计
~ reduction bonus 减少事故奖
~ report 事故报告
~ signal 事故[保障]信号
~ site 事故发生地点
~ survey 事故调查
~ toll 事故发生率
blasting ~ 爆破事故
cause an ~ 造成[引起]事故
climate ~ 气候偶变[突变]
have an ~ 出事故,失事
fatal ~ 死亡事故
human element ~ 责任事故
roof-fall ~ 塌顶事故
without ~ 没有发生事故
accidental [æksi'dentl] *a.* 偶然的,临时的,随机的,附带的；*n.* 附带事物,偶然事件；*ad.* **accidentally**
~ aggregation 偶然聚积
~ air （混凝土中）截留的空气
~ air admission 空气偶然进入
~ contamination 偶然污染
~ current 偶发[事故]电流
~ discharge 事故排放[排出]口
~ ejecta 意外喷出物
~ error 偶然误差[差错],随机误差
~ losses 意外[未预见]损失
~ pollution 意外污染
~ release 事故性排放
~ state 事故状态
accidented ['æksidentid] *a.* 表面起伏不平的,表面不规则的,面不平的
~ topography 凹凸不平的地形
acclimate [ə'klaimit] *v.* （使）服水土，（使）适应新环境[气候],驯化
acclimatize [ə'klaimətaiz] = acclimate *v.* ,-tized,-tizing
acclive [ə'klaiv] *a.* 倾斜的,有坡度的
acclivitous [ə'kllivitəs] *a.* 向上倾斜的
acclivity [ə'kliviti] *n. pl.* -ties,向上的斜坡

acclivous [əˈklaivəs] *a.* (向上)斜的,向上倾斜的,慢坡的
~ column 柱头上呈蜗卷式的柱
accolade [ˌækəˈleid] *n.* 拱门窗上的葱形线饰,S形穹窿

accolade

accolle(e) 缠绕在颈部的,在颈部连接的[接触的]
Accoloy *n.* 镍铬铁耐热合金
accommodate [əˈkɔmədeit] *vt.* , -dated, dating. 调节,(使)适应,供应[给],提供
accommodation [əˌkɔməˈdeiʃən] *n.* (临时)宿舍,住处,寄宿处,营房,宿营地;调节,适应,调视;装备,附件;容纳,接纳;贷款
~ area 住处[寄宿]面积
~ allowance 住房津贴
~ bill 通融期票[汇单]欠单
~ bridge 专用[特设]桥,便桥
~ coefficient 适应[调节]系数
~ crossing 平交道口
~ density 住宿人员密度
~ endorsement 通融(支票)背书[批注]
~ kite 欠单
~ ladder 扒梯,舷梯
~ net (大的有横木)木踏板绳梯[舷梯]
~ reflex 调节反射
~ road 专用公路[道路],房屋(的)后街
~ trailer 建筑工地的活动住房;居住拖车
ocular ~ 目镜[视觉]调节
office ~s 办公用具
stress ~ 接紧装置,张紧夹具
accommodator [əˈkɔməˌdeitə] *n.* 调节器[者,装置],贷款人
accompaniment [əˈkʌmpənimənt] *n.* 伴随[附属]物,装饰物
accompanist [əˈkʌmpənist] *n.* 伴随

[陪衬]物;伴奏者
accompany [əˈkʌmpəni] *vt.* , -nied, -nying.
伴随[生],与……同时进行[发生]
~ing diagram 附图
~ing element 伴同元素
~ing mineral 伴生矿物
~ing sound 伴音
accomplish [əˈkɔmpliʃ] *vt.* 完成,达到,实行,实现
~ work 做功
~ed (已)完成的,竣工的,熟练的
accomplishment [əˈkɔmpliʃmənt] *n.* 完成量,进度;实行;成就,成绩;(*pl.*) 技艺,才能
accord [əˈkɔːd] *v.* ; *n.* 一致,符[迎]合;和谐;协定[议],条约
~ with 与……一致[符合,相协调]
accordance [əˈkɔːdəns] *n.* 协调,匹配,相适应
~ with [to] 按照,根据,与……一致
accordant [əˈkɔːdənt] *a.* 一致的,匹配的,调和的,相合[和]的
~ connection 匹配连接
~ junction 平齐汇流
~ summit level 平齐峰顶线
~ with [to] 与……一致[相合,相调和]
~ to schedule 按预定计划
according [əˈkɔːdiŋ] *a.* 相符的,一致的
~ to modular coordination 与模数协调相符的[一致的]
~ to circumstances 根据情况
~ to schedule 按预定计划
accordion [əˈkɔːdiən] *n.* 可折叠的,折成褶的,褶(状)的; *n.* 手风琴,(印制电路的)"Z"形插孔,折式插孔
~ cam 提花三角
~ coil 折叠式线图
~ contact "Z"形角接触簧片,风琴式触点簧片
~ curtain 折屏障
~ door (金属)褶[折叠]门

account

~ fold 折子式褶页
~ hood 折棚横顶罩
~ -like wrinkles 折叠状褶皱
~ partition 折隔扇[板]
~ plate 折叠式图板
~ plateroof 折叠式屋顶

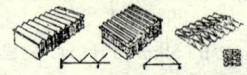

accordion plateroof

account [ə'kaunt] n. 计算,估计,考虑;理由,缘故;账,账户[目,单],计算书,报表;说明,解释 v. 认为,说明(原因、用途),(数量)占;解决;算账

~ current(a /c) 来往账户
~ day 付欠款(结账)日
~ executive 业务经理
~ for 计算(出),计及
~ note 账目单
~ number 账号
~ paid 账项付清
~ payable 应付账款
~ receivable 应收账款
~ sales 清单(材料账)
~ transaction (建筑)造价预算,估计
~ valuation 估价[计]
asset reduction ~ 资产减项
bankchecking ~ 银行支票户头
card of ~ 账目表
giving ~ 支出账目
subscriber's ~ 用户账单

accountability [əˌkauntə'biliti] n. (有)责任

~ analysis 可计量分析

accountable [ə'kauntəbl] a. 负有责任的,可解释[说明]的

~ to M for N N 方对 M 方负责

accountancy [ə'kauntənsi] n. 会计(工作)

accountant [ə'kauntənt] n. 会计(员),出纳(员)

chartered [certified public] ~ 会

计师
mechanized ~ (机械)计算装置

accouplement [ə'kʌplmənt] n. 圆柱成对密立(式),系木[材]配合,配合,匹配,连接

accredit [ə'kredit] vt. 鉴定……为合格;信任[托];认可,特许,归在,把……归咎;委派

~ ed 被认可的,被普遍采纳的

accreditation [əˌkredi'teiʃən] n. 鉴定,任命,认可,备案

accrescence [ə'kresns] n. 成长,增长

accrete [ə'kri:t] v. , -creted,-creting, 增大,生长,堆[累]积,共(同)生(长); a. 合生的,共生的

accretion [ə'kri:ʃən] n. (自然)增大;堆积(物),添加(物),累积;增值,价值附加,结(连,长)合,粘连,(pl.),

Accrington brick (英)阿克林顿红砖(用含氧化铁的黏土制成的高强度砖)

accroides [ə'krɔidis] n, 禾木胶

~ gum 禾木胶
~ resin 禾木树脂

accrual [ə'kru:əl] n. 增加[殖],增加物[额]

accumbent [ə'kʌmbənt] a. 屈身的,斜倚的,(植物)侧倚的,对位的

~ posture 横卧的姿势

accumulate [ə'kju:mjuleit] v. , -lated,-lating. 累积,聚积,[集],存储,蓄能[压],堆积;accumulable, a.

~ d capital 累积资本
~ d deficiency 累积差值
~ d deficit 累计亏损
~ d deflation 累计耗竭
~ d depreciation 累计折旧
~ d (plastic) deformation 累积(塑性)变形

round-off accumulating 舍入误差的累加[积累]

accuracy ['ækjurəsi] n. 正确性,准确[度],精度

accurate ['ækjurit] a. 准确的,已校准

的；**accurately**, *ad.* **accuratenes**, *n.*
~ mass measurement 精确质量测定
~ pointing 准确定向，点测
accustomizatiom [əkʌstəmi'zeiʃən] *n.* 习惯化，适[顺]应
ac/dc 交流/直流
ace [eis] *n.* 一点，少许，能手，专家；微料；*a.* 最高的，第一流的，优秀的，能干的
~ run 头轮
~ up 顶好
not an ~ 毫无
within an ~ of 差一点儿，几乎
Acele *n.* 阿西耳（醋酯纤维）
acendrada *n.* 白泥灰岩
acentric [ei'sentrik] *a.* 无中心的，离开中心的
acerbity [ə'sə:biti] *n. pl.* -ties. 酸，涩度[味]，苦味，尖刻（薄）
acerdol [ə'sə:dəl] *n.* 高锰酸钙
acerose ['æsə,rəus] *a.* 针叶树的，针状[形]的；*n.* 针叶树
acerous ['æsərəs] *a.* 针状[形]的
acervation [əsə'veiʃən] *n.* 堆积
acescency [ə'sesənsi] *n.* 微酸味
acetal ['æsitæl] *n.* 乙缩醛（二乙醇），醛缩醇
~ copolymer 乙缩醛共聚物
~ dehydase 乙醛酶
~ dehyde 乙[醋]醛
~ dehyde resin 乙醛树脂
acetamide [,æsi'tæmaid] *n.* 乙酰胺
acetanilid(e) [,æsi'tænilaid] *n.* 乙酰（替）苯胺，退热冰
phenyl ~ 苯替乙酰胺，苯乙酰替苯胺
acetate ['æsiteit] *n.* 醋酸盐[脂]，乙酸盐[酯，根，苯]，醋酸纤维素
~ base film 醋酸基胶片
~ silk 醋酸(纤维素)丝
~ wire 醋酸纤维绝缘(的)电线
acid potassium ~ 酸式醋酸钾
acid sodium ~ 酸式醋酸钠
ammonium ~ 醋酸铵

cellulose ~ 醋酸纤维素
acetic [ə'si:tik] *a.* 醋酸的，酸的
~ acid 乙酸，醋酸
~ acid rubber 醋酸橡胶
~ aldehyde 乙醛
~ ester 醋酸酯
~ ether 醋酸乙醚
acetimeter [,æsti'mitə] *n.* = acetometer 醋酸（比重）计
acetimetry [,æsi'timitri] *n.* 醋酸定量[用碱滴定法确定醋中的醋酸量]，醋酸测定（法）
acetolysis [æsi'tɔlisis] *n.* 醋解（作用），醋酸水解
acetonitrile [,æsitə'naitril] *n.* 乙腈，氰甲烷
acetonyl *n.* 丙酮基，乙酰甲基
acetose ['æsitəus] = acetous
acetous ['æsitəs] *a.* 醋的，醋酸的
acetoxy *n.* 醋酸基，乙酰氧基
acetum [ə'si:təm] *n.* 醋
acetyl ['æsitil] *n.* 乙酰（基）
acetylene [ə'setili:n] *n.* 乙炔，电石气，炔烃
~ burner 乙炔焊枪，乙炔燃烧器，乙炔灯
~ cutter 乙炔切割器，乙炔割刀
~ flame 乙炔火焰，乙炔灯
~ gas 乙炔气体
~ headlight 乙炔照明[信号，头前]灯
~ lamp 乙炔灯
~ sludge 电石渣
~ welder 乙炔焊机，气焊机
~ welding torch 乙炔焊枪
dissolved ~ 液化乙炔（在铜瓶内溶于丙酮）
achieve [ə'tʃi:v] *vt.* -chieved,-chieving 完成，达到（目的），实现，得到；**achiever**；*a.* **achievable**
~d reliability 实际可靠性，能达到的可靠性
achievement [ə'tʃi:vmənt] *n.* 完成，达

到,成就,功[成]绩
~ of one's object 达到目的
achlusite *n.* 钠滑石
achmatite *n.* 绿帘石
achnakaite *n.* 黑云[倍]长岩
Achnanthes *n.* 曲壳硅藻属,弯杆藻
achroite *n.* 无色电气石,白碧玺
achromatic [ˌækrəu'mætik] *a.* 消色(差)的,无色的,非彩色的
~ colour 中和色
achromaticity [əˌkrəumə'tisiti] = acharomatism [əˈkrəumətizm] *n.* 消色差(性),无色,非彩色
ACI = American Concrete Institute 美国混凝土学会
acicular [əˈsikjulə] *a.* 针状的
~ constituent 针状组成
~ crystals 针状结晶体
~ pigment 针状颜料
~ structure 针状组织
~ type zinc oxide 针状氧化锌
aciculate [əˈsikjulit] *a.* 针状的
~ crystals 针状结晶
acid [ˈæsid] *n.* 酸(类,性物); *a.* 酸性的
~ acceptor 酸性中和剂
~ accumulator 酸性蓄电池
~ -alkali cell 酸碱电池
~ ammonium carbonate 酸式碳酸铵
~ asphalt 酸性地沥青
~ attack 酸侵蚀
~ base balance 酸碱平衡
~ base indicator 酸碱指示剂
~ -base titration 酸碱滴定
~ battery 酸性蓄电池
~ binding agent 酸性粘合剂
~ blue 酸性蓝
~ brick 酸性耐火砖
~ brittleness 酸脆性;酸洗脆裂
~ bronze 耐酸青铜
~ carbonate 酸式碳酸盐
~ clay 酸性黏土

~ color 酸性染[颜]料
~ cured varnish 酸凝漆,酸硬化清漆
~ content 含酸量
~ corrosion 酸蚀,酸性腐蚀
~ cured resin 酸凝树脂
~ embossing 酸性压花[纹](该法采用氟氢酸在玻璃表面形成闪光装饰花纹)
~ embrittlement 酸脆性
~ equivalent 酸当量
~ etching 酸洗,酸侵蚀,酸蚀压花纹
~ -fast 耐酸的,抗酸的
~ firebrick 酸性耐火砖,硅质耐火砖
~ fog 酸雾
~ ground glass (用氢氟酸加工的)毛玻璃
~ lining 酸性衬里,耐酸衬砌
~ proof 耐[防]酸的
~ proof concrete 防酸混凝土
~ -proof material 防酸材料
~ proof mortar 防酸砂(灰)浆
~ protection 防酸,抗酸
~ refractory [高硅]酸性耐火材料
~ -resisting concrete 防酸混凝土
~ -resisting material 防酸材料
~ resisting mortar 防酸砂(灰)浆
~ resisting paint 耐酸漆
~ rock 酸性岩
~ slay 酸性炉渣[矿渣]
~ sludge 酸性污泥,酸渣
~ sludge asphalt 酸性(地)沥青油渣
~ slurry 酸性水泥浆[砂浆,灰浆]
~ streaking (在玻璃表面上进行)酸(性)蚀斑纹
~ strength 酸强度
~ -treated finish (混凝土的)酸性处理整面法
~ value 酸值
~ vapor 酸性蒸汽
~ violet 酸性紫
~ -washed plaster 酸化粉刷处理
~ waste 酸性废水[物],废酸

acetic ~ 醋酸,乙酸
anhydrous ~ 无水酸
azotic ~ 硝酸
carbonic ~ 碳酸
carboxylic ~ 羧酸
chlorazotic ~ 王水(硝酸与盐酸混合物)
chlorhydric ~ 盐酸
chloro-nitric ~ 王水,硝酰氯
chromic ~ 铬酸
concentrated ~ 浓酸
cresylic ~ 甲酚
diazoic ~ 重氮酸
diluted ~ 稀酸
ethanedioic ~ 乙二酸,草酸
ethanoic ~ 醋酸,乙酸
form(yl)ic ~ 甲酸,蚁酸
fumarol(o) ~ 硼酸
glacial acetic ~ 冰醋酸
humic ~ 腐殖酸,胡敏酸
hydrochloric ~ 盐酸
hydrosulfuric ~ 氢硫酸,硫化氢
inorganic ~ 无机酸
metacarbonic ~ 碳酸
metasilicic ~ 硅酸
methanoic ~ 甲酸
mineral ~ 无机酸
nitric ~ 硝酸
nitrous ~ 亚硝酸
nucleic ~ 核酸
oxalic ~ 草酸
phenic ~ 苯酚
phosphoric ~ 磷酸
plumbons ~ 铅酸,(二价铅)
pyrites ~ 亚硫酸
resinic ~ 树脂酸
sulfuric ~ 硫酸
sulfurous ~ 亚硫酸
waste ~ 废酸

acidamide *n.* 酰胺 acidate *v.* 酸[酰]化

acidic [ə'sidik] *a.* 酸性的,酸式的,硅石多的

~ bleaching solution 酸性漂白液
~ resin 酸性树脂
~ rock 硅质岩
~ waste 酸性废物
~ water 酸性水

acidiferous [ˌæsə'difərəs] *a.* 含酸的,生酸的

acidimeter [ˌæsi'dimitə] *n.* 酸度计,pH计,酸(液)比重计

acidimetry [ˌæsi'dimitri] *n.* 酸量滴[测]定法

acidity [ə'siditi] *n.* 酸性[度]
~ coefficient 酸性系数
~ constant 酸度常数
~ index 酸度指数
~ test 酸度测定

acidize ['æsidaiz] *v.* -ized,-izing. 酸处理,酸化

acidoid ['æsidɔid] *a.* 似酸的;有变酸倾向的;*n.* 类酸物质

acidolysis [ˌæsi'dɔlisiz] *n.* 酸解(作用)

acidometer [æ'sidɔmitə] = acidimeter

acidometry [æ'sidɔmitri] = acidimetry

acidosis [ˌæsi'dəusis] *n.* 酸中毒

acidualation [əsidju'leiʃən] *n.* 酸化(作用),酰代

acidum [ə'sidjum] *n.* 酸
~ carbolicum 石碳酸

acidur *n.* 阿西杜尔铁硅合金(16% ~ 17%硅,余量为铁)

acierage ['əsiəreidʒ] *a.* 表面钢化;渗碳

acieral ['æsiərəl] *n.* 铝基合金(3% ~ 6%铜,0.1% ~ 1.4%铁,0 ~ 1.5%锰,0.5% ~ 0.9%镁,0 ~ 0.4%硅)

acieration [æsiə'reiʃən] *n.* 碳化,渗碳;金属镀铁硬化,渗碳法

acies *n.* 缘,边缘,缘端

aciform ['æsifɔ:m] *a.* 针状的,锐利的

aciniform [ə'sinifɔ:m] *a.* 葡萄状的,多核的

aci-nitro *n.* 异硝基,硝酸基

acinose ['æsinəus] = acinous ['æsinəs]

A

    a. 细粒状
    ~ structure 粒状结构
acker ['ækə] *n.* 水的波纹[涟漪]
Ackermann ribbed floor 阿克曼肋形楼
    面(尺寸为 10cm×30cm×25cm)
ackey *n.* 硝(酸)硫(酸)混(合)酸浸渍液
    (电镀用)
acknowledge [ək'nɔlidʒ] *vt.* -edged,
    -edging,确认,肯定,证实,承认[宣布]
    收到;-acknowlegeable,*a.*-acknow-
    edger,*n.*
acknowledgement [ək'nɔlidʒmənt] *n.*
    承认,认可;收货回单,(书刊前)谢启
    ~ lamp 证实指示灯
    ~ of receipt 收妥通知,收妥承认
    in ~ of 感[答]谢
acknowledger [ək'nɔlidʒə] *n.* 瞭望
    装置
aclastic *a.* 无折光的
ACLD=**aircooled** 气冷式
acle *n.* 沉重坚硬耐久的阿克来木材,(一
    种菲律宾乔木)
acline [ə'klain] *n.* (未经褶折的)水平
    地层,无倾线,零等倾线
aclinic [ə'klinik] *a.* 无倾角的,水平的
Acme=**Acme(screw) thread** (英制)梯
    形螺纹(顶角为 29°)
    ~ thread tap 梯形丝锥
acme ['ækmi] *n.* 顶点[上],极点[度],
    弧点,最高点
    ~ harrow 阔刀齿耙
    ~ screw thread 平口螺纹
acmite ['ækmait] *n.* 锥辉石
    ~ augite 霓辉石
    ~ -granite pegmatite 绿辉花岗伟
    晶岩
ACN = **advance change notice**(有关)
    更改的先期通知
acnodal ['æknəudəl] *a.* 弧点的
    ~ cissoid 弧点蔓叶线
acnode ['æknəud] *n.* 弧(立)点,顶点,
    极点
acock [ə'kɔk] *a.* ; *ad.* 向上卷,翘起的,

    反卷
acockbill [ə'kɔkbil] *n.* 悬吊
acorite *n.* 锆石
acorn ['eikɔ:n] *n.* 柱头楦顶装饰;橡实
    管;橡尖端,橡实[子]
    ~ barnacle 藤壶(附着于岩石,码头,
    船底的甲壳动物)
    ~ nut 螺母[帽]
    ~ worm 柱头虫
acousimeter [ə'ku:simitə]= acoumeter
acoustextile [ə'ku:stekstail] *n.* 吸声
    织物
acoustic(al) [ə'ku:stik(əl)] *a.* 声学的,
    传声的,吸声的,音响的
    ~ absorbent 吸声材料
    ~ absorbing medium 吸声介质
    ~ absorption 声吸收
    ~ absorption coefficient 吸声系数
    ~ absorption loss 声吸收损耗
    ~ absorptivity 吸声系数[率]
    ~ alarm 声报警器
    ~ altimeter 声响测高计
    ~ attenuation 声衰减
    ~ backing 隔声[吸声]填料
    ~ baffle 声障板
    ~ barrier 声屏障
    ~ bearing 声定位,声测向
    ~ blanket 吸声[隔声]垫
    ~ board 吸声[隔声]板
    ~ brick 吸声[隔声]砖
    ~ buoy 声浮标
    ~ capacitance 声容
    ~ cassette 吸声贴面板
    ~ ceiling 吸声[隔声]顶棚
    ~ celotex(tile) 甘蔗纤维吸声板
    ~ cladding 吸声涂层
    ~ coffer 吸声[隔声]夹板层
    ~ comfort 声传感,[传声]舒适感
    ~ complicance 声顺,声容抗
    ~ conductivity 声导率,传声性
    ~ conglomeration 声波凝聚
    ~ construction 隔声[传学]构造

[结构]
~ correction 音质校正,声学改正
~ damper 消声[吸声]器
~ damping 声阻尼
~ dazzle 声干扰,异常聚声,狂响聚声
~ delay line 声延迟线
~ depth finder 回声测探仪
~ design（建筑隔声,吸声）声学设计
~ detector 声波探测器
~ direct on-finder 声响探向仪
~ discontinuty 声不连续性
~ dispersion 声频散
~ disturbance 声扰动[干扰]
~ duct 声波导,声道
~ -electric factor 声电系数
~ enclosure 隔声罩
~ energy density 声能密度
~ engineering 声学工程
~ environment 声学环境
~ Faraday rotation 声法拉第旋转
~ fatigue 声疲劳,声致疲劳
~ feed back 声反馈
~ felt 吸声[隔声]毡
~ fiber board 吸声纤维板
~ field 声场
~ filter 消声器,滤声器
~ fog 声雾
~ form board 声衬板,隔声板
~ free field 自由声场
~ frequency 音[烧]频（30 赫 ～ 20 千赫）
~ foil 吸声[隔声]箔
~ fuse 感声引信
~ generator 发声器
~ glass 隔声[声控]玻璃
~ hangovers 声（的）迟滞
~ insulation 隔声,声绝缘
~ material 声学[吸声,隔声]材料
~ noise 噪声
~ optic deflector 声光偏转器
~ oscillograph 声学示波器
~ paint 吸声涂料

~ panel 隔声板,吸声板
~ perforated gypsum board 吸声穿孔石膏板
~ performance 吸声[隔声]特性
~ permeability 透声率,透声
~ phase 声相[位]
~ pick-up 拾声器
~ plaster 吸声灰膏[粉刷,抹灰]
~ plaster ceiling 吸声石膏顶棚
~ pollution 声响[噪声]污染
~ -proof deck 隔声顶棚
~ propagation 声传播
~ quality design 音质设计
~ quartz 传声石英
~ reactance 声抗
~ reflection 声反射
~ refraction 声折射
~ resistance 声阻
~ resonance 声共鸣
~ scattering 声散射
~ septum 隔声板
~ shielding 声屏蔽
~ sounding 声波测探
~ spectrograph 声谱仪
~ spectroscope 声谱仪
~ spectrum 声谱
~ speed 声速
~ (strain) ga(u)ge 声（学）应变仪
~ tex 吸声板
~ tile 吸声（贴）砖,隔声板[砖面]
~ transmission 声传递[发射]
~ transmissivity 声透射性,声透射比
~ treatment 隔声[吸声]措施
~ vault 吸声穹窿
~ velocity 声速
~ wall 吸声[隔声]墙
~ wave 声波
~ wedge 吸声木楔[尖劈]
acoustician [ˌækuːsˈtiʃən] *n.* 声学工程师,声学家
acoustics [əˈkuːstiks] *n.* (用复数)建筑

物的传声性,(用单数)音质,声学
~ engineer 声学工程师
~ expert 声学专家
~ insulation 隔声,噪声防护
~ insulating material 隔声材料
architectural ~ 建筑声学
auditorium ~ 会场声学
musical ~ 音乐声学
room ~ 室内声学

acoustimeter [əkuːsˈtimitə] *n.* 声级[强]计,噪声计

acoustodynamic *a.* 声动力的
~ effect 声动力效应

acousto-elasticity [əˈkuː-stəu‚elæsˈtisiti] *n.* 声弹性

acousto-electric *a.* 声电的
~ effect 声电效应

acoustolith tiles 吸声贴面砖

acoustomagneto-electric *a.* 声磁电的
~ effect 声磁电效应

acoustometer [əˈkuːstəmitə] = acoustimeter

acoustomotive [əˈkuːstəməutiv] *a.* 声波的
~ pressure 声压

acoustooptic(al) *a.* 声光的
~ effect 声光效应
~ material 声光材料
~ deflection 声光偏转
~ modulator 声光调制器

acquisition [ækwiˈziʃən] *n.* 探测,发现;显示,认别;采[收]集,获得;挖截;获得[添加]物
~ cost 获得成本,购置成本,照原价
~ mode 搜索[探测]方式
~ of (building) land 获得建筑用地
~ of development rights 扩界申请,获得扩界权
data ~ 数据采集;数据集合
information ~ 信息获取
infrared ~ 红外探测[捕获]
laser ~ 激光探测[捕获]

optical ~ 光学探测[捕获]
target ~ 目标搜索[探测]

acquittal [əˈkwitəl] *n.* 尽责,履行;偿还,付清

acquittance [əˈkwitəns] *n.* 清欠收据;还清,解除

Acrawax *n.* 合成脂肪酸酯

acre [eikə] *n.* 亩,英亩;(*pl.*)土地,地产
~ -foot 英亩-英尺
~ -inch day 灌溉量英亩-英寸/日

acreage [ˈeikəridʒ] *n.* 英亩数（土地日）,面积

acrolein [əˈkrəuliin] *n.* 丙烯醛
~ resin 丙烯酸树脂

acrolite [æˈkrəulait] *n.* 酚与丙三醇合成的树脂

acrolith [ˈækrəliθ] *n.* 石首石肢木身雕像

Acron *n.* 铝基铜硅合金（铝 95%,铜 4%,硅 1%)

acronym [ˈækrənim] *n.* 字首简称[缩写]词,(字首字母)组合词

acrophobia [ækrəˈfəubiə] *n.* 高空恐怖(症),高处恐怖症

acropolis [əˈkrɔpəlis] *n.* (古希腊城堡的)卫城

across [əˈkrɔs] *prep.*;*ad.* 跨越,横[穿]过,横切,横断,交叉,成十字形;横[宽]直径;从……这头到那头,并联,加分路
~ back 背宽
~ bulkhead 横向隔墙
~ corners 对角
~ cutting 横向切割
~ grain 横(木)纹
~ rule 全域规则
~ the board cut 全面削减
~ the grain 横断面
~ the line 跨接线

acrostolion *n.* 破浪材,船首饰物

acroter(ia) *n.* 山尖(装)饰,山墙(装)饰物

acroteric *a.* 末梢的,周围的

acroter(ion) *n.* 顶花,山墙(装)饰物(古

典建筑人字山头一个角或其顶端的雕塑饰物），雕像柱脚

acroter(ion)

acroterium = acroterion
acrotorque ['ækrətɔːk] *n*. 最大扭力
acryl [ə'kril] *n*. 丙烯(醛基)
~ glass (丙烯酸脂类)有机玻璃
~ lacquer 丙烯酸清漆
Acrylafil *n*. 玻璃纤维增强聚丙乙烯酸系塑料
Acrylaglas *n*. 玻璃纤维增强聚丙烯酸系塑料
acrylaldehyde *n*. 丙烯醛
acrylamide *n*. 丙烯酰胺
~ base(injecting paste) 丙凝(注浆材料)
acrylasar *n*. 玻璃纤维增强聚丙烯酸系塑料
acrylate *n*. 丙烯酸盐
~ -based 丙烯酸盐基
acrylic [ə'krilik] *n*. 丙烯酸(衍生物)的，聚丙烯的，丙醛烯的
~ acid 丙烯酸
~ acid amide 丙烯酸酰胺
~ base 丙烯酸树脂基团
~ (building)mastic 丙烯酸(防潮)玛琋脂
~ (bulk)compound 丙烯酸表面密封材料
~ caulking(compound) 丙烯酸树脂表面密封材料(防潮用)
~ coating 丙烯酸树脂清漆
~ ester 丙烯酸盐[脂]
~ ester-based 丙烯酸盐基
~ fibre 丙烯酸系纤维
~ glass 丙烯酸有机玻璃
~ glue 丙烯酸胶

~ latex 丙烯酸胶乳
~ panel 丙烯(塑料)板
~ plastering 丙烯涂灰[涂面]
~ plastic(s) 丙烯酸塑料
~ polymer 丙烯酸聚合物
~ resin 丙烯酸(类)树脂
~ (resin) concrete 丙烯酸树脂混凝土
~ (resin)emulsion 丙烯酸树脂乳胶[乳化液]
~ rubber 丙烯酸(类)橡胶
~ sealant 丙烯酸树脂表面密封剂
~ sheet 丙烯酸有机玻璃板
~ waste water 丙烯酸废水
Acrylite [ə'krilait] *n*. 聚丙烯酸脂塑料
acryloid *n*. 丙烯酸(树脂溶)剂
act [ækt] *n*. 作用，证明书，报告；法规[令]，条例，决议(书) *v*. 起作用，生效，行动，表现(得)作用(力)，作用(负荷，荷载)
London Building ~s 英国建筑法规
safety responsibility ~ 安全责任条例
sewage utilization ~ 污水利用条例
town planning ~ 城镇规划法规
~ up to …… 遵照办事
~ing head 作用[有效]水头
~ing manager 代经理
Actanium *n*. 镍铬钴低膨胀合金(钴40%，镍15.5%，铬20%，铁15%，钼7%，锰2%，碳0.15%，铍0.03%)
actification [æktifi'keiʃən] *n*. 再生[复活](作用)
actified [æ'ktifaid] *a*. 再生的
~ solution 再生(溶)液
actifier *n*. 再生器[柱，塔]
actinic [æk'tinik] *a*. (有)光化(性)的，光化(学)的；actinically, *ad*.
~ glass 光化[闪光]玻璃
~ rays 光化射线
actinism ['æktinizəm] *n*. 光化性[作用]，射线化学[作用]，感光度
actinium [æk'tiniəm] *n*. 锕[元素名符

actinogram

为 Ac]
actinogram [æk'tinəugra:m] *n.* 光化计记录;X 光像
actinograph [æk'tinəgra:f] *n.* (日光)光化力测定器,(日)光能(量)测定仪,日光强度自动记录器,光化线强度记录器;辐射仪,日射计,自记曝光计
actinouran(ium) *n.* 锕铀 AcU(铀同位素,U^{235})
action ['ækʃən] *n.* 反[效]应,作用,操作,影响;作用[主动]力,作用量,行程
~ center 作用[动作]中心,万能数据机床
~ current 作用电流
~ cycle 作用[动作]周期,(数据处理的全过程),工作周期
~ length 作用长度,有效长度
~ limit 处理[作用]界限
~ line 作用线
~ period 作用期
~ space 作用范围
~ token 作用特征[标记]
~ variable 作用变量
amplified ~ 放大作用
arch(cing) ~ 拱(的)作用
eccentric ~ 偏心动作[作用]
lever ~ 杠杆作用
local ~ 局部作用
mass ~ 质量作用,浓度效应
truss ~ 桁架作用
wobbler ~ 偏心作用,摇动,摆[振]动
activating channel 活化槽
activation [ækti'veiʃən] *n.* 活化(作用),活化(化),激活[化];(活水)曝气处理法,启[驱]开起]动,触发
~ adsorption 活化吸附
~ analysis 放射性[激活,活化]分析
~ sand 活化砂
activator ['æktiveitə] *n.* 活化[激活,催化]剂,助媒剂,提高灵敏度装置;粘附性活性剂
auxilary ~ 辅助激活剂
dominant ~ 主激活剂

reduction ~ 还原活化剂
active ['æktiv] *a.* 活性[泼]的,活化的,放射性的;快速的,有效[功源]的;常用的,运行的,实际的,现行的;actively, *ad.*; actieness. *n.*
~ acidity 活性[有效]酸度
~ alkali 活性碱
~ antenna 有源天线,有效天线
~ assets 活动财产[资产]
~ balance 盈亏结余,动态平衡
~ calcium oxide(of lime) 活性(石灰的)氧化钙
~ filler 活性填料
~ fire defense 自动消防(包括发出火警信号和传感系统的装置),自动灭火系统装置(自动喷洒,二氧化碳灭火系统);应急灭火装置;消防设备辅助装置
~ gases 活性气体,腐蚀性气体
~ jamming 人为干扰
~ maintenance 有效维修
~ pigment 活性颜料
~ pollution 放射性污染
~ porosity 有效孔隙度
~ pressure 主动压力
activise ['æktivaiz] *vt.*,-ized,-izing,激起,使行动起来
activity [æk'tiviti] *n. pl.*-ties,活(动)性,活力[量,度]放射性[强度],主动性,作用;功[效]率;占空系数;文件的)有效性,有效部分
~ address code 单位地址代号
~ address directory 单位地址簿
~ analysis 活性分析,经济活动分析
~ chart 活动示意图
~ number (土的)活性指数(塑性指数与黏土成分之比)
~ room (住宅)活动室
~ space 活动空间[范围],儿童游戏室
contamination ~ 污染放射性
dummy ~ 虚[伪]工作,伪作业
environmental ~ 周围[环境]介质射性

hydraulic ～（胶凝材料的)水硬性
actual ['æktjuəl] *a.* 实际的，有效的，现行的
～ acid 实际酸度
～ address (＝absolute address) 实际地址，有效地址
～ allowance 实际公差[偏差]
～ argument 实际变量
～ balance 实存
～ budget 决算
～ burden rate 实际制造费用[间接费]分配率
～ capital 实际资本
～ carrier 实际承运人
～ cash balance 实际现金余额
～ cash value 实际现金价值
～ clearance 实际净空[间隙]
～ construction time 实际建筑期限[工期]
～ consumption 实际消费[耗水量]
～ cost 实际成本
～ costing 实际成本计算
～ cost method 实际成本法
～ cost system 实际成本制
～ cross-section 实际(横)断面[截面，过流断面]
～ damage 实际损害
～ debts 实际债务
～ delivery 实际交货
～ demand 实际需求
～ depreciation 实际折旧
～ deviation 实际偏差
～ dimension 实际[真正]尺寸[大小]
～ distance 实际距离
～ efficiency 实际效率
～ error 实际误差
～ expected standard cost 实际预算标准成本
～ feed 实际给料[装填]
～ filling depth 实际填充深度
～ finish time 实际完成[完工]时间
～ gain 有效[实际]增益

～ gap 实际缝隙[间隙]
～ garbage 实际废料
～ goods 现货
～ hand-over 实际交付
～ height 实际高度，真高
～ implementation 实际实现
～ indicator card 实际示功图
～ instruction 有效[实际]指令
～ interest rate 实际利率
～ inventory 实地盘存
～ labor rate 实际工资率
～ land form 实在地形
～ lead (打桩机)实际导程
～ life 实际[有效]寿命
～ limit 实际限度
～ living standards 实际生活水平
～ load 有效载荷
～ loss 实际亏损
～ lower bound 实在下界
～ measurement 实(际)测(量)，真实量度
～ moment 实际力矩[弯矩]
～ normal cost 实际正常成本
～ order 实际执行的合同[订货单]
～ performance 实际完成情况
～ personnel 实际人员情况
～ power 有效功率
～ pressure 实际压力
～ price 实际价格
～ profit 实际利润
～ purchase price 实际购入价格
～ quantity 实际[数]量，实际值
～ quotation 实际交易值
～ rate 实际汇率
～ reflux 实际回流
～ reserves 实际贮量
～ runtime 实际运行时间
～ safe load 实际安全载荷
～ size 实际大小[尺寸]
～ staff 实际人员情况
～ standard 实际标准，现行标准
～ state 现状

~ strength 实际强度
~ stress 实际应力
~ stuff 现货
~ temperature 实际温度
~ thickness 实际厚度
~ thrust 实际推力
~ total loss 绝对损失
~ user 实际用户
~ value 实际价值
~ weight 实际重量
~ working machine-days 实际工作台日数
~ working pressure 实际工作压力
~ yield 实际生息率
~ zero point 绝对零点,基点
actuality [ˌæktʃu'æliti] *n.* 实在[际]；*pl.* 现状,事实,实际情况
actualization [ˌæktjuəlai'zeiʃən] *n.* 实现[行],现实化
actualize ['æktʃuəlaiz] *vi.*, -ized, -izing. 实现[行],使现实化
acuity [ək'ju(:)iti] *n.* 锋[锐]利,剧烈；分辨能力
~ auditory ~ 听觉(敏)锐度
~ contour ~ 轮廓锐度
~ stereoscopic ~ 体视锐度
~ visual ~ 视觉敏锐度
acumen [ə'kju:mən] *n.* 尖,敏锐
acuminate [ə'kju:minit] *a.* 锐利的,(有)尖(头)的；['əkju:mineit] *v.* -nated,-nating,弄[变]尖,变锐利
acumination [əˌkju:mi'neiʃən] *n.* 尖锐,锋利,尖头
acutance [ə'kju:təns] *n.* 锐度(曲线)；(胶片的)锐敏[清晰]度
acute [ə'kju:t] *a.* 尖的,(尖)锐的；锐角的；剧烈的,严重[急剧,强烈]的；**acutely,** *ad.* **acuteness,** *n.*
~ angle 锐角
~ arch 锐拱
~ irradiation 强烈照射
~ triangle 锐角三角形
~ly-swept 大后掠角的

acyl ['æsil] *n.* 酰(基)；酰酰(基)
~ -amide 酰胺
aczol ['ækzəl] *n.* (用作木材材料防腐的)苯氧化锌及苯氧化铜的含氨防腐剂
aczoiling ['ækzɔiliŋ] *n.* (电杆)防腐
A/D 模拟数字
~ encoder 模数编码器

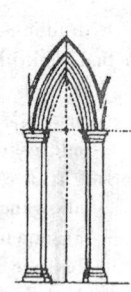

acute arch

Adam ['ædəm] *n.* 亚当(英国建筑师及家具设计师)
~ architecture 亚当建筑(英乔治三世时亚当建筑师采用的罗马古典建筑装饰的建筑形式)
~ style 亚当风格
adamant ['ædəmənt] *n.* 硬石,铁石,金刚石,刚玉；极硬物
~ metal 以锡为主的锡-锑-铜轴承合金
~ steel 铬钼特殊耐磨钢
adamantine [ˌædə'mæntain] *a.* 金刚石制的,花岗石般的
~ clinker 坚硬炼渣
~ luster 金刚光泽
~ spar 刚玉
adamas ['ædəməs] *n.* 金刚石
Adamson joint (锅炉)阿达姆松接头
~'s ring 阿达姆松联接环
adamite ['ædəmait] *n.* 水砷锌矿；人造刚玉(高炭)镍铬耐磨铸铁
adapt [ə'dæpt] *v.* (使)适应[合],配上；修改,改编[作]
~ing piece 配件
~ing pipe 连接[承接]管,套管
(be) ~ed for(to) 适合[宜]于,适应
adaptability [əˌdæptə'biliti] *n.* 适应[适用]性；灵活性
adaptation [ˌædæp'teiʃən] *n.* 适应,顺应；调配作用；改编(本)
~ lighting 适应照明

chromatic (colour) ～ 彩色适应[配合]
luminance ～ 光适应
adapter [ə'dæptə] ***n.*** 接头,特殊铸件;承接管,异径接头;连接装置,转接器;衬套;控制阀
～ bend 异径弯头
～ connector 接头,连接器
～ glass 玻璃接头,配接玻璃
～ junction box 分线盒
～ plate 衬板,垫板
～ skirt 连接裙[套]
～ sleeve 接头套筒,连接套管
casing ～ 套管异径接头,管柱接头
socket ～ 灯座接合器
adaption [ə'dæpʃən] = adaptation
adaptive [ə'dæptiv] ***a.*** 适合[应]的,自适(应)的,应[适]用的
～ control 自适应控制
～ optimal control 自适应最优控制
～ optimization 自适应最优化
adaptor = adapter

casing adapter

add [æd] ***v.***; ***n.*** 加(上),增[附追]加;加(力)

～ed losses 附加损耗
～ed mass 附加质量
～ed value 增值
～ing rate 掺加率,掺加比
addaverter ***n.*** 加法转换器
addend [ə'dend] ***n.*** 加数,附加数
addenda [ə'dendə] (addendum 的复数)
addendum [ə'dendəm] ***n.*** 附录,附加物;齿顶;齿顶高
～ angle 齿顶角
～ circle 齿顶圆
addition [ə'diʃən] ***n.*** 增加,附[追]加;加成,掺加;用量,加入量;附加工作;(结构力学中的)附加应力;(水泥中的)混合材料;扩建,扩建工程;(建筑学中)附加构造
～ agent 掺和[附加,配合]剂
～ compound 加成化合物
～ constant 加常数
～ displacement 附加位移
～ method 加入法,重加法
～ polymer 加聚物
～ polymeriation 加聚作用
～ rate 掺和率,补充率
～ reaction 加成反应
age-inhibiting ～ 防老化添加剂
alloy(ing) ～ 添加合金,合金元素
heat ～ 供热,加热,预热
interground ～ 破碎杂料
limestone ～ 石灰石混合料
repeated ～ 叠加,重复相加
additional [ə'diʃənl] ***a.*** 附加的,添加的,额外的
～ allowance 增加津贴
～ article 增订条款
～ bar 附加钢筋,辅筋
～ budget 追加预算
～ building 附加房屋,辅助建筑
～ charge 附加费
～ coating 附加涂层
～ cost 额外费用,追加成本
～ cover 追加保险额

~ duty 追加关税
~ equipment 辅助设备
~ estimates 追加概算
~ expenses 追加费用
~ investment 追加投资
~item 增列项目,补充项
~ labour costs 附加人工成本
~ living expenses 追加生活费
~ loading 附加载荷
~ loss 附加损耗
~ margin 追加保证金
~ necessary condition 附加必要条件
~ noise 附加噪声
~ outlay 额外开支
~ payment 额外付款
~ period 延长期限
~ premium 附加保险费
~ reinforcement 附加[辅助]钢筋
~ service 附加业务
~ steel 辅助[附加]钢筋
~ storage 辅助存储器
~ stress 附加应力
~ survey 补充测量
~ tax 附加税
~ works 追加工程

additive ['æditiv] ***n.*** 添加物[剂],掺和[加]剂; ***a.*** 附加[增加]的;
~ cement 有掺和料的水泥
~ colour mixing 加色混合,合成颜色
~ soil stabilization 加有稳定剂的地基[土壤],水泥灌浆
antifoam(ing) ~ 防沫[消泡]添加剂
antiacid ~ 抗酸添加剂
antiicing ~ 防结冰添加剂
antiknock ~ 抗爆添加剂
antioxidant ~ 防氧化添加剂
antisludge ~ 抗淤沉[去垢]添加剂
antiwear ~ 抗磨剂
detergent ~ 去垢添加剂
extreme pressure ~ (耐)特压添加剂

addressing ***n.*** 寻址
adduct [ə'dʌkt] ***n.*** 加合物;引证
adduction [ə'dʌkʃən] ***n.*** 氧化(作用),加合(作用);引证[用]
A-derrick ***n.*** 塔式[人字,架式]起重机,A形绞盘车
adept [ə'dept] ***a.*** 熟练[内行]的; ***n.*** ['ædept] 专家,内行,能手
adequate ['ædikwit] ***a.*** 适当的,足够的,满足要求的
~ disclosure 充分反映
~ distribution 均匀分布
~ lubrication 适当的润滑作用
~ management 适当管理
~ sample 充足样本
ader-wax (粗,生)地蜡,含硫地蜡
adfluxion ***n.*** 汇流,集流
adglutinate [əd'glʌtineit] ***v.*** ; ***n.*** 烧结;胶结(产物);凝集
adgru = advisory group 顾问[咨询]组
ADH = adhesive 胶粘的,粘着的
adhere [əd'hiə] ***v.*** -hered,-hering,粘[附]着(于),粘附(于);遵守,依附
adhering moulding mateial 涂模材料
adhering zone 附着区域
adherence [əd'hiərəns] ***n.*** 粘着,附着,附贴,遵守,保持
electrostatic ~ 静电附着[吸附]
adherend [əd'hiərənd] ***n.*** 粘附体,被粘物
adherent [əd'hiərənt] ***a.*** 粘着的,附着的;焊接住 的; ***n.*** (被)胶粘材料,(被)粘附体
adhesion [əd'hi:ʒən] ***n.*** 粘附;粘着力,胶粘度;粘附现象;粘合[胶粘](作用);(电线)接头; ***a.*** 粘合的,胶粘的
~ agent 粘结[合]剂,粘结强度增加剂
~ capacity 胶粘度
~ failure 粘合破坏
~ force 粘着力,(橡胶)密着力
~ heat 粘合[附]热

~ layer 粘附层
~ loss 粘附[胶粘]损失
~ plate 胶合板
~ power 胶粘度,粘附能力
~ -preventing film 粘附保护膜
~ preventives 防粘剂
~ promoting agent 抗剥落添加剂,粘附强度增强剂,粘附[胶粘]活化[激活,催化]剂
~ property 粘附(能)力,粘附性,胶粘度
~ quality 粘附(能)力,粘附性,胶粘度
~ -strength 粘着力,粘着[胶粘]强度
~ tension 粘合张力
adhesional [əd'hi:ʒənl] *a.* 粘附的
~ energy 粘合[结]能
~ strength 附着力,粘着[胶粘]强度
~ wetting 附着润湿
adhesive [əd'hi:siv] *a.* 粘着的,粘附的,分粘的;*n.* 粘着剂,胶粘剂,胶
~ action 粘合作用
~ agent 粘合[粘结,胶粘]剂
~ anchor 胶粘锚固[接],锚栓,地脚螺丝
~ backed tape 胶带[布]
~ band 涂胶封条
~ bitumen primer 冷底子油
~ bond 粘结,粘着力
~ bond strength tester 粘合强度试验机
~ -bonded joint 粘结缝,粘接接头
~ capacity 粘着力,胶粘度,粘附性
~ cement 胶结[质]水泥,粘性灰泥[胶泥,腻子],人工橡胶填充剂,胶浆(子),胶水
~ coat 粘合(涂面)层,(镀)层
~ composition 粘胶[胶合]剂
~ compound 粘附[粘合]剂
~ dispersion 粘性[胶合]散[分散]
~ effect 粘着[附]作用
~ emulsion 胶粘乳浊液
~ failure 脱胶

~ film 粘性膜
~ force 粘结[附,着]力
~ interlayer 胶浆夹层
~ joint 粘合,粘结接合,粘合接头
~ lamnated wood 粘合叠层木,胶合层材,多层胶合木
~ layer 粘结[附]层
~ masking tape (装修技术中)饰[罩]面(胶带[布])
~ material 粘合[胶合]材料
~ measuring tape 胶粘尺[测尺]
~ meter 粘合计
~ mortar 粘结[胶合]沙浆[灰浆,灰泥],胶泥
~ paper tape 胶纸带
~ paste 浆糊,粘合[结]剂
~ phase 粘合剂层
~ powder 粉末胶(大部分为植物胶或纤维素胶)
~ power 粘着(能)力,粘附力
~ property 粘结[胶接]性能,粘结(能)力,粘结度
~ putty 胶结[质]水泥,粘性灰泥[腻子,胶泥],人工橡胶填充剂
~ rubber 粘合[胶合]橡胶
~ separation 粘附脱层
~ slate 粘板岩
~ strength 粘结[胶合]强度[力]
~ suspension 粘性悬浮物
~ system 粘合[胶合,化学键接]结构
~ tap(e) 胶带[布],绝缘带
~ tar composition 焦油沥清粘性成分
~ tension 粘着强度,粘附张力
~ test 粘合试验
~ varnish 粘合清漆
~ water 薄膜[吸附]着水
~ wax 胶粘[粘着]蜡,封蜡
assembly ~ 装配粘胶剂
cold setting ~ 冷凝固胶粘剂
fire retardant ~ 抗热粘合剂
hot settng ~ 热固胶粘剂
pressure-sensitive ~ 压合胶粘剂

adhesivemeter [ədˌhiːsiˈvemitə] *n.* 粘着力计,胶粘计
adhesiveness [ədˈhisivnis] *n.* 粘着学, 粘附性;粘着度;胶粘度
~ test 粘着性试验
adhint *n.* 粘合接头,胶接
adiabat [ˈædiəbæt] *n.* 绝热线,等熵线
adiabatic (**al**) [ˌædiəˈbætik (əl)] *a.* 绝热的,等熵的;**adiabatically** *ad.*
~ apparatus 绝热装置
~ chart 绝热(变化)图
~ coefficient of compression 绝热压缩系数
~ expansion 绝热膨胀
~ saturated change 绝热饱和变化
~ saturated temperature 绝热饱和温度
~ system 绝热系统
~ wall 绝热墙
adiabatics [ˌædiəˈbætiks] *n.* 绝热曲线
irreversible ~ 不可逆绝热线
adiactinic [ˌædiækˈtinik] *n.*;*a.* 不透射线的,绝光化(性)的(物质)
adiathermancy [ˌædiəˈθəːmənsi] *n.* 绝热性,不透红外线性,不透热性
adiathermic [ˌædiəˈθəːmik] *a.* 绝热的
~ body 绝热体
adicity *n.* 化合价,原子价
adinol(**slate**) 或 adinole 或 adinolite *n.* 纳长英板岩
adjacency [əˈdʒeisənsi] *n. pl.* -cies 邻接[近];邻近距离
adjacent [əˈdʒeisənt] *a.* 邻近的,邻接的,毗连的;-**adjacently**, *ad.*
~ accomodation 厢房,附属建筑物
~ angle 邻角
~ area 毗邻区
~ arm 相邻臂
~ contract section 交界[邻接]收缩断面
~ blocks 相邻[毗连,毗连]的建筑房屋[街坊]
~ buildings 相邻[毗连]建筑物
~ channel 相邻通道
~ lane 相邻车道
~ map 邻接图,附图
~ matrices 相邻矩阵
~ opening 邻跨[孔]
~ orthogonal 相邻的正交线
~ peak (相)邻峰
~ pile 邻桩,附桩
~ plank 踢脚板,门头线,门框边延
~ position 相邻位置
~ property 邻近产业[房产,地产]
~ region 相邻区
~ room 邻屋,相邻房间
~ span 邻跨[孔]
~ strip 沥青[混凝土]覆面邻接带
~ structure 毗连[相邻]的建筑物,位于交通线附近的建筑物
~ wall 邻墙
~ water 毗邻水域
~ window 邻窗
adjective [ˈædʒiktiv] *a.* 附属的,辅助的;有关程序的;形容词
~ law 程序法
adjoin [əˈdʒɔin] *v.* 邻接,联接,接界;附加
adjoining [əˈdʒɔiniŋ] *a.* 毗邻[连]的,邻[紧]接的,伴随的
~ blocks 相邻[毗连]房屋[街坊]
~ buildings 邻屋,毗邻街坊
~ concession (交通线)附近居民的许可[同意]
~ course 邻接[结合]层
~ opening 邻跨[孔]
~ owner (交通线)附近居民
~ permises 毗邻地区[地产]
~ property 邻产
~ rock 围岩
~ room 邻屋
~ span 邻跨[孔]
~ wall 邻墙
adjoint [əˈdʒɔint] *a.* 伴随(的),共轭

adjacent plank

的,相结合的,修正的
~ branch 毗连支路
adjudicate [ə'dʒu:dikeit] *v*. 判定,宣判,裁判
adjudicating panel 招标(评议)委员会
adjudication [ə,dʒu:di'keiʃən] *a*. 招标,有奖征求设计方案
adjudicator [ə'dʒu:dikeitə] *n*. 裁判员,审[译]议员
adjunct ['ædʒʌŋkt] *a*. 辅助的,附加的; *n*. 附[配]件,附属物,附加件
adjust [ə'dʒʌst] *v*. 调整,校正,调节;整理
~ accounts 复算账目
adjustability [ə,dʒʌstə'biliti] *n*. 可调(整,节)性
adjustable [ə'dʒʌstəbl] *a*. 可调(整,节)的,可校正的;adjustably *ad*.
~ air-gap 可调气隙
~ angle square 可调角尺
~ bar 可调杆
~ beam saddle 可调梁座

adjustable angle square

~ bench level 可调台水平仪
~ bolt 调节螺栓
~ speed motor 调速电动机
~ square 可调尺
~ stem 校正杆,可调基杆
~ tap wrench 可调丝锥扳手
~ thrust journal 可调止推轴颈
~ triangle 坡度尺,斜度尺
~ valve 调节阀
~ vane 可调节叶片
~ vent flap 可调整通风折翼(扳)
~ voltage divider 可调分压器
~ voltage stabiliser 可调稳压器
~ wrench 活动扳手
adjuster [ə'dʒʌstə] *n*. 调整器[设备],窗撑杆,装配[安装]工
slack ~ 拉紧装置[接头,螺钉],空隙调整器

zero ~ 零位调节器
adjustment [ə'dʒʌstmənt] *n*. 调整[准,节],校正;安装,装配适应;平差
adjutage ['ædʒʌtidʒ] *n*. 管接,管嘴;喷射管,排水筒,放水管;延长臂;横梁;填充物
adjutant *a*. 辅助的, *n*. 副官,助手
adjuvant *a*. 辅助的, *n*. 助手
adlet [ædlit] *n*. 小广告
admeasure [æd'meʒə] *vt*.,-ured, -uring, *n*.;量,且度,测定,分配
admeasuring apparatus 测像仪
admeasurement [æd'meʒəmənt] *n*. 测量,度量,丈量;分配,配合
administer [əd'ministə] *v*. 管理,支配,实施,行使职权;辅助,补给
administration [əd,minis'treiʃən] *n*. 管理,经管;执行;施行;管理机构,行政机关,政府
~ and training 管理与训练
~ account 管理科目
~ block 行政办公楼
~ branch 管理分局
~ budget 行政预算
~ building 行政办公楼
~ center 行政中心
~ complex 行政办公建筑区
~ cost 管理费
~ data processing 行政管理数据处理
~ expense 管理费用
~ section 管理部门,事务部门
~ tower 高层建筑行政办公楼
~ unit 行政办公楼
administrative [əd'ministrətiv] *a*. 管理的,行政的
~ audit 管理[行政]审计
~ authority 管理当局[部门,权限]
~ budget 管理费预算,行政预算
~ cost 行政管理费
~ council 行政理事会
~ data 管理资料,管理数据
~ decision 管理决策

A

~ division 行政区,管理区段
~ expense budget 管理费用预算
~ management 行政管理
~ map 行政区划图
~ measure 行政措施
~ organ 行政机关
~ organization 管理机构
~ overhead 管理间接费
~ services 行政性业务
~ system 先进管理系统
Administrative Procedure Act 管理程序条例
administrator [əd'ministreitə] *n.* 管理人员,行政人员
admiro *n.* 阿德密拉铜锌镍合金(48% Cu,35% Zn,10% Ni,2% Al,3% Mn,2% Fe)
admissible [əd'misəbl] *a.* 容许的,可容纳的,可纳的;-admissibly, *ad.*
~ concentration 容许浓度
~ error 容许误差
~ load 容许负荷[载荷]
~ strain 容许应变
~ stress 容许应力
admission [əd'miʃən] *n.* 允许,许可;进气;供给
~ intake 进气[水]
~ passage 进路,进气道
~ pipe 进入[气,水,油,]管
~ port 进气口
~ space 装填体积
~ stroke 进气冲程
~ valve 进气[水,油]阀
~ velocity 入口[进入]速度
admit [əd'mit] *v.* 进气,供给,容[允许]
~ ting pipe 输入管,进入[气]管
~ ting port 进气口
admittance [əd'mitəns] *n.* 公差;通导,导纳,进[输]入,通道
No admittance (except on business) 闲人免进,非公莫入
admix [əd'miks] *v.* 混合[杂],掺和

admixture [əd'mikstʃə] *n.* 掺和,混合;掺和料[剂],添加剂,(混凝土)掺和剂,杂质
air-entrained ~ (混凝土的)加气剂
antifreeze ~ (混凝土)防冻添加剂
cement-dispersion ~ 水泥磨细[分散]附加剂
early-strength ~ (混凝土)早强剂
water-proofing ~ 防水剂
water-repelling ~ 防水剂
water-resisting ~ 防水添加剂,抗水[耐水]剂
workability ~ 改善和易性的添加剂
adobe [ə'dəubi] *n.* 砖坯,风干土坯(砖,瓦),龟裂土,多孔黏土
~ block 土坯块
~ brick 风干砖坯
~ clay 多孔[灰质]黏土,龟裂土
~ masonry 土砖圬工
~ soil 灰质土,龟裂土,制砖土
adopt [ə'dɔpt] *vt.* 采用,接受,正式通过,仿效,沿用
adoption [ə'dɔpʃən] *n.* 采用,采纳用,选择,正式通过
~ of agenda 通过议程
adoptive [ə'dɔptiv] *a.* 适于采用的,倾向采用的,接受的;adoptively, *ad.*
adorn [ə'dɔ:n] *vt.* 装饰,修饰 adorner, *n.*
adornment [ə'dɔ:nmənt] *n.* 装饰(品,物)
adrate ['ædreit] *n.* 附加税;[美]广告费
adret [æ'drei] *n.* 山能受阳光照射一面,山阳
adretto *n.* 阳坡
adsorb [æd'sɔ:b] *vt.* 吸附[收,引],聚集于某一平面上
~ ed film 吸附膜
~ ed layer 吸附层
~ ed substance (被)吸附[沉积]物(质)
~ ing colloid flotation 吸附胶体浮选
~ ing matter 吸附材料[物质,体]

~ing substance 吸附剂
adsorbability [æd'sɔːbə'biliti] *n.* 吸附性
adsorbate [æd'sɔːbeit] *n.* 被吸附物(质),吸附质
adsorbent [æd'sɔːbənt] *n.* 吸附剂[体、媒]; *a.* 吸附的
~ coated glass strip 涂吸附剂的玻璃条
~ deactivator 吸附减活剂
~ equipment 吸附装置
~ filtering medium 吸附过滤介质
~ resins 吸附树脂
~ temperature 吸附温度
adsorptive [æd'sɔːptiv] *n.* 被吸附物(质); *a.* 吸附的
~ action 吸附作用
~ attraction 吸附力
~ capacity 吸附能力[能量]
~ filtration 吸附过滤
~ force 吸附力
~ hull 吸附[收](薄)膜
~ value 吸收值
adularia [ˌædjuˈlɛəriə] *n.* 冰长石,低温钾长石,月长石
adult ['ædʌlt] *n.*; *a.* 成年(人,的),成熟(的)
~ education 成人教育
~ suite (家庭)套房[客房](成人和儿童卧室分开,有独立的卫生设备)
adulterant [ə'dʌltərənt] *n.* 混杂料,掺杂料; *a.* 掺杂用的
~ process 掺杂过程
adulterate [ə'dʌltəreit] *vt.*, -ated, -ating;掺杂; *a.* 掺杂的,品质低劣的
adulteration [əˌdʌltəˈreiʃən] *n.* 掺杂,改装,劣货,贬列
adumbral [æd'ʌmbrəl] *a.* 遮阳的,阴暗的,蔽荫的
adumbrate [æ'dʌmbreit] *vt.*, -brated, -brating;素描,画轮廓,遮蔽[荫]使暗;预示
adumbration [ædʌm'breiʃən] *n.* 草图,素描,轮廓,阴影,预示,暗示
adustion [ə'dʌsʃən] *n.* 可燃性;烘焦,烘焦部分
advance [əd'vɑːns] *v.*,-vanced,-vancing;推[掘]进;提[抬]高;预付[支];提前; *a.* 前的,先的,预先的
advancing longwall system 突出建筑,建筑物的突出部分(外伸悬臂梁等)
advancement [əd'vɑːnsmənt] *n.* 改[推、促]进,进步,发达;预付
advances *n.* 预付款,垫付款;预付,预支
~ for construction 施工预付款
~ to subcontractors 预付转承(包)者款
advantage [əd'vɑːntidʒ] *n.* 优点,方便,有利条件,利益 *vt.* 有利(益,助)于,能帮助,促进;得益
gain ~ 获益
operating ~ 操作(上的)优点
with ~ 有利地,有效地
advection [æd'vekʃən] *n.* 平[对]流,横向流动,平流热效;平[转]移
~ fog 平流雾,混合[侧流]雾
~ heat 对流热
~ layer 平[对]流层
~ region 平[对]流区
adventure [əd'ventʃə] *n.* 商业的投机(资);冒险,投机(资)
~ undertaking 冒险(投资)事业
joint ~ 联合企业(投资)
advertise ['ædvətaiz] *v.* 登广告,宣传,通知
advertising agency [afirm] 广告公司
advertising expense 广告费
advertising kiosk 广告亭
advertising lighting 广告照明
advertising sign 广告牌
advertising tower 广告塔
advertising wall 广告墙
advertisement [əd'vəːtismənt] *n.* 广告,公告,通知
~ boarding 杂贴广告临时棚栏
~ lighting 广告照明

advertiser ['ædvətaizə] *n.* 信号装置；登广告者

advice [əd'vais] *n.* 通知,劝告,建议；(*pl.*)消息,报道
~ note 通知书[单]
letter of ~ 通知单

advise [əd'vaiz] *v.*, -vised,-vising；通知,报告；商量,劝告
ill- ~d 失策的,愚蠢的
well- ~ 深思熟虑的

adviser [əd'vaizə] *n.* 或 advisor 顾问

advisory [əd'vaizəri] *a.* 忠告的,咨询的；advisorily, *ad.*
~ body 顾问团
~ commission 顾问委员会
~ engineer 咨询工程师
~ group 顾问小组
~ service 咨询业务
~ speed sign 推荐车速标志

advocator ['ædvəkeitə] *n.* 拥[辩]护人,倡导者

adytum ['æditəm] *n.* （古希腊庙宇中）密室;私室;内殿

adz(e) [ædz] *n.* 扁斧,横斧,锛子
~ block 刨身,刨刀座
~ -eye hammer 小铁锤
~ finish 斧削面,用扁斧砍木材表面
~ plane 刮刨
~ work （木材）砍削处理（用扁斧对木材表面进行处理修光夹代替刨面）
~ing gauge （轨枕）削平槽规

A-E =architect-engineer 建筑工程师

aedicula [ˌiːˈdikjulə] = aedicule

aedicule ['iːdikjuːl] *n.* 壁龛,小神龛；小建筑物
~ alter 壁龛隔墙

Aegean [iːˈdʒiːən] *n.* （希腊之东的）爱琴海；*a.* 爱琴海的
~ art 爱琴艺术
~ civilization 爱琴文化

aedicule

aeolation [ˌiːəˈleiʃən] *n.* 风化作用

aeolian [iːˈəuljən] *a.* 风成的,风蚀的
~ abrasion 风蚀作用
~ basin 风成[风蚀]盆地
~ deposit 风成沉积
~ erosion 风蚀（作用）
~ rock 风成岩
~ sand 风积沙
~ soil 风积土
~ vibration 风吹震动

aeolic [iːˈɔlik] *a.* 风成的,风积的

aeolipile [ˌiːəˈlipil] *n.* 蒸汽（喷）射流的反作用；汽转球

aeolotropic [ˌiːələˈtrɔpik] *a.* 各向异性的,偏等性的,非均质的

aeon *n.* 亿万年,永世,万古；[天]十亿年,京年；-ian *a.*

aeral *n.* 埃拉尔铝合金

aerarium [iːˈrɛəriəm] *n.* （古罗马）国库

aerate ['eiəreit] *v.*, -ated,-ating. 曝气,通风,充气,供气；松散,分解
~d block 空心混凝土块
~d concrete 加[引]气混凝土
~d conduit 通气管道
~d filler 疏松填料[如木屑,锯木屑]
~d flame 充气焰,富空气焰
~d gypsum board 加气[轻质]石膏板
~d material 加[充]气材料
~d mortar 加气灰膏[灰浆,砂浆]
~d plastic(s) 充气塑料,泡沫塑料
~d sintered(concrete)aggregate 加气烧结（混凝土）集料
~d tile 空心砖

aerating chemical 充气剂
aerating device 通气装置
aerating powder 发泡剂
aerating system 通气系统

aeration [ˌeiəˈreiʃən] *n.* 曝[充,通,换]气,通气；起气泡；风化；分解；分散

aerator ['eiəreitə] *n.* 去土器,除泥器；曝气[充气]器；松砂机；曝气设备,鼓风机

~ nozzle 曝气喷嘴
~ pipe 曝气管
~ tank 充气槽
aeremia ['eiəremiə] *n.* 气泡栓塞,气栓,气塞
aereous ['eiəriəus] *a.* 青铜色的
aerial ['ɛəriəl] *a.* 气体的,天空的,架空的; *n.* 天线的;(救火用)云梯
~ ladder 架空消防梯,云梯
~ plug-in point （住宅内）公用天线接线点[天线插座]
~ pollution 空气污染
~ socket 天线插座
~ view 鸟瞰图
K- ~ K 型天线
open ~ 室外天线
aeriductus [ˌeiəri'dʌktjus] *n.* 气门,尾状管
aeriferous [ɛə'rifərəs] *a.* 含有[传送]空气的
aerifier *n.* 通气管
aerify ['ɛərifai] *vt.*, -fied, -fying; 充气;掺入;气化,使呈气态
aero ['ɛərəu] *n.* 气体,空气,飞机[船],飞行; *a.* 飞机的,飞行器的,航空术的
~ -colloid 气凝[溶]胶体
~ concrete 加气[多孔]混凝土
~ design 美学[造型]设计
~ -drome 飞机场,航空站
~ -dromometer 气流速度表,风速计
~ -duster 喷粉器
~ -filter 空气过滤器,喷射过滤器
~ -fin 散热片
~ -flocs 气凝剂
~ -view 空[鸟]瞰图
aerobic [ɛə'rɔbik] *a.* 好气[氧]性的,需气[氧]的,有氧的
~ atmosphere 需氧[好氧]环境
aerobond [ɛə'rɔbənd] *n.* 环氧树脂
aeroconcrete [ˌɛərə'kɔnkri:t] *n.* 或 aerocrete 加气混凝土

aerodrome ['ɛərədrəum] *n.* （飞）机场
~ beacon 机场灯塔[灯标]
~ reference point 机场控制点
aerodynamics [ˌɛərədai'næmiks] *n.* 空气动力学
applied ~ 应用空气动力学
design ~ 设计空气动力学,气动力计算数据,空气动力计算
engineering ~ 工程空气动力学
supersonic ~ 超音速空气动力学
trisonic ~ 三种声速范围的空气动力学
aeroelastics [ˌɛərəu'læstiks] *n.* 或 aeroelasticity 气动弹性(力)学
aerofluxus ['ɛərəflʌksəs] *n.* 排[泄]气
aerofoil ['ɛərəfɔil] *n.* 翼,风板,支持面
~ fan 轴流通风机
aerohydrodynamics [ˌɛərə'haidrədinæmiks] *n.* 空气流体动力学
aerohydromechanics ['ɛərəˌhaidrəumi'kæniks] *n.* 空气流体力学
aeromechanics [ˌɛərəmi'kæniks] *n.* 空气[气体]力学
aerophore ['ɛərəfɔ:] *n.* 通风面具
aeroplane ['ɛərəplein] *n.* 飞机
~ shed[hangar] 飞机库[棚]
~ view 空瞰图,鸟瞰图
aeroprojection method 立体投影法,测图法
aerosol ['ɛərəsɔl] *n.* 气溶胶,悬浮微粒,烟雾剂
~ layer 烟雾层
~ particle （空气中）悬浮微粒
~ sampling device 空气采样器[装置]
~ spraying 气溶胶喷涂
aerosolize ['ɛərəsɔlaiz] *vt.*, -ized, -izing; 使成烟雾状散开
aerosphere ['ɛərəsfiə] *n.* 大气层,空气圈

aerostatic [ˌɛərəˈstætik] *a.* 空气[气体]静力学的
~ press 气压机,空压机
aerostatics [ˌɛərəˈstætiks] *n.* 空气[气体]静力学
aeruginous [iəˈruːdʒinəs] *a.* 铜绿色的,氧化铜的
aerugo [iˈruːɡəu] *n.* 铜绿(色);氧化铜,铜锈;金属锈斑
aesthetic [iːsˈθetik] *a.* 审美的,美术的,美学的
~ appeal 美学的作用[魄力,吸引力]
~ area 风景区,街景区
~ aspect 美学[审美]角度[方面]
~ charm 美学的诱惑力[吸引力]
~ concept 美学概念
~ idea 美学设想[概念]
~ judgment 审美能力
~ measure (色彩调配的)美观程度
~ road 风景[美观]道路
~ sense 美感
~ theory 美学理论
aestheticism [iːsˈθetisizəm] *n.* 唯美主义
aesthetics [iːsˈθetiks] *n.* (审)美学
affairs division 事务处
affect [əˈfekt] *vt.* 影响,起作用,感动
~ zone (焊接的)变质区
affection [əˈfekʃən] *n.* 影响,障碍;特[属]性;感情
affiliate [əˈfilieit] *v.*, -ated,-ating; *n.* 联号,分支; *v.* (使)加入,合并;追寻根源
~d company 附属公司,分公司
affinage *n.* 精炼
affinity [əˈfiniti] *n.* 相似,亲和性,化合性,吸引力
~ coefficient 亲和系数
~ law 相似定律
absorption ~ 吸附力
chemical ~ 化学亲合性(能)
cohesive ~ 凝聚力
elective ~ 选择亲和力

affirmative [əˈfəːmətiv] *a.* 肯定的,赞成的,正确的; *n.* 确认,断言
~ propsition 肯定命题
~ sign 正号
~ vote 赞成票
affix [ˈæfiks] *n.* 附件,附加物,添加剂,小装饰物; *v.* [əˈfiks] 使固定,粘上;附加,添加,签署,盖章
~ one's signature 署[签]名
affixture [əˈfikstʃə] *n.* 结合物,联接[合]附加物,添加物;粘[贴]上
afflatus [əˈfleitəs] *n.* 神感,灵感
poetic ~ 诗的灵感
afflux [ˈæflʌks] *n.* 流入[向],汇[集]流,进气;壅水,(水流)提高度
afford [əˈfɔːd] *vt.* 供给,给予;生产;有力量置备
afforest [əˈfɔrist] *vt.* 造林,绿化
A-frame *n.* A型构架,三角支架
~ dam A型构架块,活动A型钢架关闭装置[制动机构]
aft [ɑːft] *n.* 尾[后]部; *a.*; *ad.* 在[从]后部(的),在[近,到,向]船尾的
~ bay 下游建筑部分
after [ˈɑːftə] *ad.* 在后,续后; *prep.* 在……以后,依照;后来的,后面的; *a.* 后期的
~ care 房屋竣工后维修
~ -charge 附加费率
~ drawing 后拉伸
~ drying 再次干燥
~ etching 残余腐蚀
~ flow 塑性变形,(停止加载后)残余塑性变形,蠕变
~ flush 冲洗滞留(大便器冲洗后,水流仍从水箱中缓慢流出,此况下需更换密封圈);连续冲洗
~ frame 后框(架),补架
~ hardening 后期硬化
~ heating 后加热
~ image 残像,余像,视觉残留
~ -sales service 返修,保修,售后服务

~ -shock 余震
agalite ['ægəllait] *n.* 纤(维)滑石
agalmatolite ['ægəlmætəlait] *n.* 蜡石
~ firebrick 蜡石耐火砖
~ fireproofing materials 蜡石耐火材料
agallochum ['ægələtʃʌm] = agalloch = agalwood *n.* 沉香木,加罗木
agate ['ægət] *n.* 玛瑙
agba ['ægbɑ:] *n.* 非洲阿勃木
age [eidʒ] *n.*; *v*, aged, aging or ageing; *v.* 养护;硬化,老[熟,陈]化;(墙等)风蚀; *n.* 龄期,年代
~ -class 龄级
~ coating 老化覆层
~ coating of lamp 灯泡老化层
~ hardening 时效硬化
~ inhibiting 防老化的
~ inhibiting addition 防老化添加剂
~ -proof 抗老化的
~ resistant steel 抗氧化钢
~ -resistor 抗老化剂
~ strength relation 龄期强度关系
~d column 老化柱
~d person's home 养老[敬老]院
~d properties 老化后性能
~ing crack 自然裂纹,时效裂纹
~ing index 老化指数
~ing resistance 抗老化性能
artificial~ing 人工老化
non~ing 未失效的
agency ['eidʒənsi] *n.* 代办[理];机构,公司;作用,介质,工具
~ fee 代理费
~ agreement 代理协定,代理关系协定,代理合同
~ commission 代理手续费
~ contract 代理契约
~ relation 代理关系
sole ~ 独家经理
agenda [ə'dʒendə] *n.* (agendum 的复数)运行程序语言,操作规程;议事日程,备忘录
agent ['eidʒənt] *n.* 作用力;附加[作用]剂,介质;(公司)承包人,代理[代办]人
addition ~ 添加剂
adhesion ~ 胶粘剂
agglomerating ~ 凝结[胶凝]剂
air-entraining ~ 加气剂
anticorrosive ~ 防蚀剂
anti-foam ~ 防沫[消泡]剂
antirusting ~ 防锈剂
antisludging ~ 去垢添加剂,抗淤沉添加剂
antitarnishing ~ 防锈剂
antiwear ~ 抗磨剂
blowing ~ 发泡[起泡]剂
bonding ~ 粘合剂
cementing ~ 粘合剂,胶结料
cooling ~ 冷却剂
curing ~ (混凝土)养护剂
dehydrating ~ 脱水剂
desiccating ~ 干燥剂
discolouring ~ 脱色剂
drying ~ 干燥剂
fire extinguishing ~ 灭火剂
paying ~ 付款银行
rigidity ~ 硬化剂
rust protection ~ 防锈剂,抗腐剂
shipping ~ 运货代理商
softening ~ 软化剂
tracer ~ 指示剂,示踪剂
wedding ~ 乳化剂,结合剂
ager ['eidʒə] *n.* 调色装置,熟化[蒸化]器
agger ['ædʒə] *n.* 土堆(尤指古罗马军营外围所筑的土堤);古罗马的道路
agglomerant [ə'glɔmərənt] *n.* 粘结剂;附聚剂,凝聚剂; *a.* 附聚的,烧结的,凝结的
agglomerate [ə'glɔməreit] *n.*,*v.*,-ated,-ating; *v.* (使)聚集,结块,成团; *n.* 附聚物,烧结矿,块集岩
~ -foam concrete 烧结矿渣泡沫混

凝土

agglomeration [əˌgləməˈreiʃən] *n.* 住宅区,居民点;附聚作用;企业集团

agglutinant *n.* 烧结[凝集,促集]剂
  fine ~ 细聚集体

agglutinate [əˈgluːtineit] *v.*, -nated, -nating;附着,胶粘,烧结; *a.* 附着的,胶着的; *n.* 粘结集块岩

aggradation [ˌægrəˈdeiʃən] *n.* 沉积,沉淀;淤[填]积;(地下土层永冻深度面的)递升;加积作用

aggregate [ˈægrigit] *v.*,-gated, -gating, *n.*; *a.*; *v.* (使)聚集,共计(为); *n.* 骨料[集料,填充料],集合[组合]体;总计; *a.* 集合的
  ~ base course (建筑)骨料基层
  ~ batcher 集料[骨料]配料器
  ~ batch(ing) plant 集料配料装置[设备]
  ~ bin 集料斗[仓]
  ~ bituminizing pant 沥青集料搅拌机
  ~ blending 集料拌合
  ~ breakage 集料压碎量
  ~ chips 石屑
  ~ composition 集料颗粒组成
  ~ concept 统计概念
  ~ cost 总费用
  ~ dimension 总尺寸
  ~ grain 骨料颗粒
  ~ granulosity 骨料粒质
  ~ height 总高
  ~ steel area 钢筋总截面积
  artificial ~ 人工级配集料
  bituminized ~ 敷(铺)沥青集料
  breeze ~ 煤渣[矿粉]集料
  calcareous ~ 石灰石[钙质]集料
  chert ~ 燧石集料
  cinder ~ 矿[炉]渣集料
  clinker ~ 熟料,烧结料,熔渣
  close graded ~ 密级配集料
  coarse ~ (粒径粗于6mm的)粗集料
  concrete ~ 混凝土集料
  crushed rock ~ 碎石集料
  densely graded ~ 密级配骨料
  discontinuous ~ 间断[不连续]级配集料
  discrete ~ 松散集料
  divided coarse ~ 多粒级集料
  expanded perlite ~ 膨胀珍珠岩集料
  fine ~ 细集料
  gap graded ~ 间断级配集料
  gravel ~ 砾石集料
  grouted prepacked ~ 预埋集料(后用水泥浆灌注成混凝土)
  hydrophilic ~ 亲水性集料
  hydrophobic ~ 憎水性集料
  inert ~ 惰性集料
  loose ~ 松散集料
  marble ~ 大理石石屑
  mortar ~ 灰浆集料
  multiple size ~ 多径级集料
  open graded ~ 天然级配集料
  porous ~ 多孔集料[填料]
  prepacked ~ 预填集料
  rounded ~ 砾石
  single-sized ~ 等径集料
  ungraded ~ 无级配集料
  uniformly-graded ~ 均匀级配集料
  unreactive ~ (与水泥)无反应的集料

aggregation [ˌægriˈgeiʃən] *n.* 聚集(作用);聚集体,集合体
  ~ plain 集积平原

aggregative [ˈægrigeitiv] *a.* 聚[集]的,综合的;集合体的,集团的
  ~ company 集团公司
  ~ index number 综合指数
  ~ model 综合模式

aggressive [əˈgresiv] *a.* 侵蚀[腐蚀]性的
  ~ action 腐蚀[侵蚀]性
  ~ agent 侵蚀剂
  ~ atmospheric conditions 侵蚀性大气情况
  ~ carbonic acid 侵蚀性碳酸

~ characteristics 侵蚀性特性
~ matter 侵蚀性物质
~ tack hour (涂层)接触干燥时间
~ water 侵蚀性水,侵入水

agilawood ['ædʒiləwud] *n.* 沉香(奇南香木)

aging ['eidʒiŋ] *n.* =ageing 时效;养护;老[硬]化;分期
~ -proof agent[resistant] 抗老化剂,防老化剂
~ test 老化试验

agiotage ['ædʒətidʒ] *n.* 汇兑行情[业务],兑换;买卖股票

agitate ['ædʒiteit] *v.*,-tated,-tating 搅动[拌],混合,拌合;激发[励],磁[]
~ conveyor 搅拌(混凝土)送料车
agitating-device 搅拌装置
agitating arm 拌和[搅拌]臂[杆]
agitating lorry 搅拌式(混凝土)运输车
agitating tank 搅拌槽
agitating truck 搅拌式(混凝土)运输车
agitating vane 搅拌浆叶

agitation [ædʒi'teiʃən] *n.* 搅拌,混合;激发[励,磁]
~ dredging 搅动挖土法
~ error 骚动误差
~ -froth process 搅动生泡法
~ leach 搅拌沥滤
~ vat 搅拌瓮
~ vessel 带搅拌器的容器
mechanical ~ 机械搅拌
thermal ~ 热搅动[骚动],热激发

agitator ['ædʒiteitə] *n.* 搅拌器[机,装置];搅动[拌合,混合]器

aglatite ['æglətait] *n.* 变锂辉石

aglite ['æglait] *n.* 烧结黏土、膨胀黏土(由黏土和粉状焦碳混合,烧成多孔状烧结块,加以

Agnus Dei

破碎);轻质骨料

Agnus Dei ['ɑːgnus'deil] *n.* 神之羔羊饰

agonic [ei'gɔnik] *a.* 无偏差的,不成角的
~ line 无偏(差)线,零磁偏线

agora ['ægərə] *n.* (*pl.*-rae)(古希腊贸易及集会用)广场,会场

agpaite ['ægpəait] *n.* 钠质火成岩类

Agra ['ægrɑː] *n.* (印度)阿格拉(城市名)
~ work 阿格拉细工(印度一种镶钻石料细工)

agraf(f)e [ə'græf] *n.* 铁箍,钩子,铁夹子;搭扣

agravic [ə'grævik] *n.* 无重力区,(总重力场等于零)无重力状态; *a.* 无重力的,非引力的

agravity (zero-gravity)失重(状态)

agree [ə'griː] *v.* 同意,相同,(意见)一致
~d insured value 约定保险价值
~ with 与……相符,与……一致

agreement [ə'griːmənt] *n.* 协议,契约,合同,协定
~ certificate 官方的证书
~ year 协议[合同]年度
final ~ 最后协议

agribusiness ['ægribiznis] *n.* 农业综合企业

agricultural [ægri'kʌltʃərəl] *a.* 农业[用]的
~ building 农产品加工建筑
~ service building (附属在田庄的)杂用建筑(如仓库、厨房、马厩等)

agrillaceous *a.* 泥质的

agt = agent 剂,媒介物;代理商(人),代表

aguada ['eigjueidə] *n.* 集水洼地

aguano ['eigjuænə] *n.* 大叶桃花心木

agustite *n.* 磷灰石

ahead [ə'hed] *ad.* 在前,向前;有盈余,超过,胜过
~ running 顺车

~ turbine 推涡轮
A-horizon A 层位,表土层
ahu 用作边界、路标和纪念牌的土丘[石头堆]
ahuehuete 尖叶落羽杉
aid [eid] *n.* 助剂,辅动设备,辅助装置;仪器;整修工具;辅[援]助;*v.* 帮[辅]助
　~ design 辅助设计
　~ station 急救站
　antiplastering ~ 抗粘剂
　economic ~ 经济援助
　sedimentation ~ 絮凝剂
AID = American Institute of Decorators 美国室内装饰家学会
aiguille [ˈeigwiːl] *n.* 钻岩用钻头;钻孔机
aile [eil] = aisle 通道,走廊,侧廊
aillette [ˈeilet] *n.* 建筑物的侧翼
aim [eim] *n.* 目的[标];宗旨,方针,计划;*v.* 瞄准,企图
aimantine [ˈeiməntin] *n.* 磁铁矿
aiphyllus [ˈeifiləs] *n.* 常绿林
air [ɛə] *n.* 大气,空气;通风[气];风干,烘干;*v.* 通风;广播;*a.* 通空气的;航空的
　~ bell 气泡,砂眼
　~ blanketing 空气夹层
　~ bleeder 放气管[阀],通气小孔
　~ blister 气泡,砂眼
　~ -borne sound insulation 空气消声[减声],空气噪声防护
　~ -breather 通气孔,通风装置;空气吸潮器
　~ brick 风干砖,多孔砖,空心砖
　~ brush 气刷,喷漆器
　~ bubble 砂眼,气泡
　~ buffer 空气缓冲器,空气隔层
　~ cavity 气眼,气穴
　~ ceiling 通风顶棚
　~ ecll 气泡,砂眼
　~ chamber 空气室,气室,气舱

~ channel 风道,排气道
~ checks 气泡,麻孔
~ chute 通风道
~ collector 集气罐[器]
~ column 空气柱
~ compressing machine 空气压缩机,气压机
~ compressor 空气压缩机
~ condenser 空气冷凝器
~ conditioned building 装有空调的建筑
~ conditioned ceiling 通风顶棚
~ conditioned space 空调室
~ conditioner 空调机[器]
~ conditioning chamber 空调室
~ conditioning conduit 空调管道
~ conditioning duct[shaft] 空调管道井[廊井]
~ conditioning engineering 空调工程[技术]
~ conditioning equipment 空调设备
~ conditioning inlet 空调口[门]
~ conditioning installation 空调装置
~ conditioning plant 空调设备[机房]
~ conditioning space 空调空间[范围]
~ conditioning station 中心空调站
~ conditioning system 空调系统
~ cured concrete 空气养护混凝土
~ curtain 空气幕,风帘,风障
~ demolition pick(hammer) 风镐
~ diffuser 空气扩散器[扩散管,散流器]
~ diffusing 空气扩散
~ diffusing equipment 散流设备,空气扩散装置
~ diffusing outlet 空气扩散流口
~ digger 风铲[锹],风动铲具
~ discharge 排气
~ discharge equipment 排气设备
~ dried material 风干材料

~ dried timber 风干圆木
~ dried brick 风干砖
~ drier 空气干燥器
~ drifted drill 风(动)钻(岩机)
~ drill (hammer) 风(锥)钻
~ dry density (混凝土管)风干密度
~ drying(clear)varish 风干清漆
~ duct installation 风道安装
~ duct riser 送风立管
~ duct system 风道系统
~ ducting 风道
~ ductwork 风道(工程)
~ -ejecting fan 抽气机,排气通风机,排气扇
~ ejection 气动脱模
~ entrainer 起泡剂,加气剂[器]
~ entraining agent (混凝土)加气剂
~ entraining concrete 加[引]气混凝土
~ entrainment mortar 加气灰浆
~ entrapment 滞留空气,内部气泡
~ entrapping structure 多孔结构
~ equipment 压缩空气设备[装置]
~ escape 排气口,漏气
~ escape valve 排气阀
~ exhaust 抽气,排气
~ exit 排风[出气]口
~ extractor 抽气器,抽风机
~ evaporation 空气蒸发
~ -fast 不透风[气]的
~ feed 送气,供(空)气,压气供应
~ -fence 导流栅
~ filter 空气过滤器
~ fixture 通风[排风]装置
~ float 气动抹灰工具[泥刀]
~ floated powder 气溶粉尘[微粒]
~ floor heating 热风地板采暖
~ flow 气流,气量,风量
~ flue 风道,烟道
~ flunnel 风道,风管
~ foam 泡沫
~ foil 翼形
~ force 风力

~ -free concrete 密实混凝土
~ glass block 通风玻璃块
~ governor 送风调节器,压缩空气调节器
~ grate 通风笼子[格栅]
~ grille 通风花窗[格栅]
~ grinder 气动砂轮机
~ grit 通风格栅
~ gun 喷漆枪,风动铆枪,喷雾器

air gun

~ hammer (rock)drill 风锤钻
~ handling 通风,换气
~ handling area 通风面积[范围]
~ (-handling)block 通风空心砌块
~ (-handling)brick 通风空心砖
~ -handling cavity 通风孔[道]
~ handling ceiling 通风顶棚
~ handling connection 通风连接管
~ handling ducting 管道通风
~ handling equipment 通风设备
~ handling glass brick 通风玻璃砖
~ handling installation 通风装置
~ handling(lighting)fixture 通风(照明)装置
~ handling line 通风管道
~ haudling opening 通风孔
~ handling panel 通风(镶)板
~ handling piece 通风(管系)配件
~ handling pipe 通风管
~ -hardenable 气硬的
~ hardening lime 气硬性石灰
~ hardening steel 气冷钢,正火钢
~ header 空气立管
~ heading (隧道)通风坑道
~ heat 空气加热
~ heating 空气加热,热风采暖
~ heating installation 空气加热设备[装置]
~ heating plant 空气加热设备
~ heating radiator 热风采暖器[散热器]

~ heating system 空气加热系统
~ hoist 空气升液机,气动葫芦,气动卷扬机,风动起重机
~ holder 气罐,储气罐
~ hole 风洞,通气孔,通风孔,砂眼
~ hole grate 通气孔格栅
~ hose 风管,送风[输气]软管,橡胶风管
~ humidification 空气加湿
~ humidifier 空气加湿器
~ humidity 空气湿度
~ hunger 缺氧,空气缺乏
~ hydraulic jack 气液千斤顶
~ hydraulic accumulator 空气水压蓄力器
~ impermeability 不透气性,气密性
~ impervious liner 气密层,内衬层
~ induction 空气诱导,进气口
~ infiltration 气渗[作用],空气透入
~ -inflated building 气胎[充气]式建筑
~ inlet 进风口,进气孔
~ inlet pipe 进气管
~ input 进气量,风量
~ installation 通风装置[系统]
~ insulation 空气绝缘
~ intake 进风口,进气孔,通风道[管]
~ layer 空气层,通气层
~ lay-in ceiling 通风[顶棚]天花板
~ leakage 漏气,泄气
~ level 气泡水准仪
~ light 散光,漫射光
~ lock 气塞[阻];气窝[泡],气锁,砂眼;空气封闭;(冷藏)回笼间,空气室外间;(气压沉箱)气闸室
~ louver 空气调节孔,放气孔
~ lubricator 压缩空气润滑器
~ luminaire (fixture) 通风照明装置
~ machine 扇风机,风扇
~ main 主风道
~ manifold 风管,输气管,压气管
~ marking[marks] 小气泡麻孔

~ mat filter 孔垫式滤器
~ meter 混凝土加气量测定仪,气流计,小风速计
~ moistening 空气加湿
~ moisture 空气湿度,空气含湿量
~ moisture comfort curve 空气湿度舒适曲线
~ monitor 大气污染监测器
~ monitoring network 大气污染监测网
~ mortar 加气砂浆
~ motion 空气运动[流动]
~ noise 空气噪声
~ oasis 空气净区
~ opening 通风[进气]孔,风口,排风口,通风道
~ outlet 空气出口,排风口
~ output (风机)送风量,出气量,排气量
~ packing 气密填料
~ painter 喷漆器
~ panel 通风(镶)板
~ permeability 空气渗透率,透气性
~ pervious material 透气材料
~ picker 风镐,(混凝土)振捣器
~ piece 通风(系统)配件
~ pile hammer 气动打桩锤
~ pipe 空气管,通风管,风管
~ pipe tray 通风管支架[托架]
~ piping 风口[道,洞],通风口,空气管道布置
~ piping filter 空气管道过滤器
~ pit 通风井,通气坑
~ placing 气动浇灌(混凝土)
~ plant 通风系统[装置]
~ plug 气孔塞
~ -pocket 气穴[泡],砂眼,残存空气,气囊,垂直气流
~ pollution 空气[大气]污染
~ pollution agent 大气污染因子
~ pollution code 防止空气污染规范
~ pollution control association (APCA) 空气污染控制协会

~ pollution index 空气污染指标
~ pollution prevention 防止空气污染
~ pressure 空气压力,气压,风压
~ pressure water system 气压给水系统
~ -proof 不透气的,气密的;密封
~ propelling 压缩空气驱动
~ purge 空气吹扫
~ purification 空气净化
~ purifier 空气净化器
~ purifying unit 空气净化机组
~ purity 空气纯度[洁度]
~ quality 空气质量
~ quality criteria 空气质量准则
~ quality standard 空气质量标准
~ quantity 空气量
~ -rate 通风量
~ reamer 气压扩孔机
~ rectification 空气吹脱,收制(沥青)
~ reducing agent 除气剂
~ refined asphalt 吹制炼石油沥青
~ regenerating device 换气设备
~ register 调[配]风器
~ regulation 风量调节
~ regulator 风量调节器,空调器
~ release valve 放气阀,(泄)压阀,放空阀
~ relief 放[排]气口[管]
~ relief valve 减压[安全]阀
~ renewal 换气
~ resistance 空气阻力[阻抗],抗气性,不透气度
~ resisting liner 内衬层,气密层
~ retaining substance (混凝土)加气剂
~ return vent 回风口
~ sand blower 喷砂机,吹砂器
~ saw 气锯,气压锯
~ -scape 空中鸟瞰图
~ scoop 进气口,导气罩
~ set 常温凝固的,自然硬化的;气硬,自凝

~ set pipe 空气冷凝管
~ -setting mortar 自凝泥浆
~ shots 气泡,砂眼
~ shutter 空气调节阀(闸门)
~ shut-off valve 空气关闭阀
~ sliding window 通风推拉窗
~ soil hammer 气动夯土机
~ sound attenuator [absorber] 空气消声器
~ spade 风(动)铲
~ spots (模制品)表面不平
~ stack 通风立管,风道
~ stave 通风侧板
~ sterilization 空气消毒
~ storage 空气贮备,露天堆放场,(混凝土)空气储存
~ suction pipe 吸气管
~ suction valve 吸气阀
~ supply duct 送风管,供气管
~ supply hose 送风胶管[软管]
~ supply outlet 送风口,供气口
~ supply panel 送风孔板
~ supply pipe 送风管
~ supply system 送风系统[设备]
~ supply valve 送[供]气阀
~ supported building 气胎式建筑
~ supported fabric dome 充气胶布帐篷
~ system 通风系统[装置]
~ tack cement 封气粘胶水
~ tamper 风动夯[捣棒]
~ tent 气垫[充气]帐篷
~ termination network 屋面避雷网

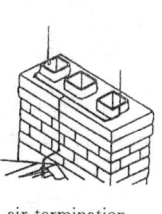
air termination network

~ -tight 不透气的,密封的;气密,密封
~ -tight door 气密门
~ tight material 气密材料
~ -tight seal 密封
~ tool 气动工具

~ traps 气泡,麻孔,气窝
~ trowel 气动抹灰工具[装置]
~ vent 排气口,通气口,通风孔,通气管,空气出口
~ vessel 贮气器[罐],空气室
~ vibrator 气动振动器
~ view 空瞰图,鸟瞰图
~ vitiation 空气污染[恶化]
~ void(s) 孔隙率,含气率,气孔
~ volume 空气体积,风量
~ -water heat pump 气-水热泵
~ wash 空气冲[清]洗
~ way 空气通道,风洞,气眼
~ wetting 空气湿润
~ window 通风窗,气窗
~ zoning 分区送风
compressed ~ 压缩空气
dusty ~ 含尘空气
exhaust ~ 排气,废气
fixed ~ 不流动空气
light ~ 软风
liquefied ~ 液化空气
open ~ 露天的
scavenging ~ 清洗积垢空气
standard ~ 标准大气
steam ~ -heater 蒸汽热风器
tubular ~ -heater 管式热风器
vitiated expired ~ 污蚀排泄气

airy [ˈɛəri] *a.* 通风的,空气的,空中的
Airy cirle 弥散[爱里]圆
Airy point 爱里点(两点支承梁,自重变形为最小的支点)

aisle [ail] *n.* 侧[耳]房,通道,侧廊;(影剧院观众席)走道
~ bay 侧房,耳房
~ gallery 侧廊,外走廊
~ passage 走廊过道,厢房通道
~ pier 侧房扶壁,走廊(支)柱
~ roof 侧[厢]房顶[屋面],教堂半圆后殿屋顶
~ -vault 侧房[厢房]穹窿[拱顶],教堂半圆后殿拱顶
~ wall 过道墙,桥台斜翼墙

~ way 走道
~d church 有厢房[侧厅]教堂
~d hall 教堂中厅(座位所在部分),正厅
common ~ 公共通道
cross ~ 交叉通道
minimum intersecting ~ 车辆直角拐弯时走道最小宽度
passage ~ 小通道

Aitoff projection 艾托夫投影图法,艾托夫等积投影法

ajar [əˈdʒɑː] *a.*; *ad.* 不调和;(门)半开,(门)微开
~ hook 门挺钩,半开门钩

ajutage [ˈædʒutidʒ] *n.* 喷水管,放水管;排水筒;承接管;送风管

akanthus *n.* 阔叶饰,叶板

akroter(ion) *n.* =acroter(ion) 山尖饰,山墙饰物

akroter(ion)

ala [ˈeilə] *n.* (古罗马住宅通向天井的)耳房,边房

Alabama design method 美国亚拉巴马州(柔性路面)设计法

alabaster [ˈæləbɑːstə] *n.* 雪花石膏(一种似大理石的软质白石,用做装饰品),蜡石
~ column 雪花石膏柱
~ glass 乳色[雪花,乳白]玻璃
oriental ~ 条带状大理岩,方解石
specular ~ 透光蜡石

alameda [ˌæləˈmeidə] *n.* 林荫路,散步道路

alamosite *n.* 铅辉石

alanate *n.* 铝氧化物

alar [ˈeilə] *a.* 翼状的,有翼的
~ septum 侧隔壁

alarm [əˈlɑːm] *n.* 警报装置(器),信号装置; *v.* 报警,告急
~ annunciator 警报信号器

~ apparatus 报警(电)器
~ buzzer 报警蜂鸣器
~ circuit 警报线路
~ clock 闹钟
~ display 报警显示
~ facilities 报警装置[设备]
~ gauge 锅炉警报器,警号气压计
~ indicator 警报指示器
~ installation 报警装置
~ lamp 报警灯
~ light 警报灯
~ manometer 示警气压计,气压报警计
~ panel 报警器控制板
~ reaction 紧急反应,警戒反应
~ signal 警报信号
burglar(y) ~ 防窃报警器
fire ~ 火警信号,火灾信号装置
flag ~ 报警信号器
heat [temperature] ~ 过热报警信号器
high water ~ 最高水位指示器
lamp failure ~ 灯光熄灭信号
overflow ~ (水箱水位)超位报警器,满溢信号器
pressure ~ 压力报警器
smoke ~ 烟感报警器
temperature ~ 温感报警器
time ~ 限时报警器
alary [ˈæləri] *a.* 翼状的,翼形的,扇形的
alaskite *n.* 白[花]岗岩
albedo [ælˈbiːdəu] *n.* 扩散反射系数(反射光强与入射光强之比),反照率
Albert cottage 阿伯特馆(英国第一次伦敦国际博览会上展出的工人典型住宅)
albertite [ˈælbətait] *n.* 黑沥青
albery [ˈælbəri] *n.* 壁橱,壁龛
albescent [ælˈbesənt] *a.* 浅白色的,带白色的,发白的
albino [ælˈbainəu] *n.* 缺乏正常色素的动物或植物;白变种
~ asphalt 白沥青

albite [ˈælbait] *n.* 钠长石
albocarbon [ælbəˈkɑːbən] *n.* 萘
albondur *n.* 纯铝包皮,超硬铝板
albronze [ˈælbrɔnz] *n.* 铝青铜,铜铝合金
album *n.* 图(纪念,邮集)册,像片簿,唱片套(集)
alburn(um) [ˈælbəːn(əm)] *n.* 白木质,边材;-ous *a.* 白木质的
alcazar [ˌælkəˈzɑː] *n.* (西班牙或阿拉伯的)要塞,宫堡[殿],公馆
alclad [ˈælklæd] *a.* 铝色的,金属线上(片上)涂有铝[铝合金]的;镀铝的; *n.* 铝衣合金,包铝合金;纯铝包皮超硬板
~ alloy 镀纯铝铝合金
dural ~ 包[硬]铝
alcohol [ˈælkəhɔl] *n.* 酒精,乙醇
~ fastness 耐醇性
~ insoluble matter 乙醇不溶物质,涂料中不溶于乙醇的成分
~ resistance 耐醇性
~ slug process 酒精处理方法(用以增加油的回收率)
~ solution 酒精溶液
~ thermometer 酒精温度计
~ torch lamp 酒精喷灯
~ varnish 乙醇清漆,凡立水
absolute ~ 无水酒精
anhydrous ~ 无水酒精
aqueous ~ 含水酒精
benzyl ~ 苯甲醇
ethyl ~ 乙醇,酒精
methyl ~ 甲醇,木精
alcosol [ˈælkəsəul] *n.* 醇溶胶
alcove [ˈælkəuv] *n.* 壁室;凹室;庭园中的纳凉亭;岸壁
aldehyde [ˈældihaid] *n.* (乙)醛
~ resin 聚醛树脂
amine ~ resin 胺醛树脂
alder [ˈɔːldə] *n.* 桤木,赤杨
Alee-tree 行道树
Aleppo pine 阿拉伯松
alerce [ˈæləs] = alerse *n.* 山达木,(分

枝不规则的巴塔戈尼亚的)木材树

alert [əˈləːt] *a.* 警惕的,机警的; *n.* 警戒[报]; *v.* 发出警报
～ing signal 报警信号

A-level A 式水平仪

alexandrolite [æligˈzɑːndrəlait] *n.* 铬黏土

algam [ˈælgæm] *n.* 锡;铁皮

algeldrate [ˈældʒildreit] *n.* 水合氢氧化铝

algorithmic language 算法语言

Alhambra [ælˈhæmbrə] *n.* (13～14 世纪西班牙)阿尔汗布拉宫

alias [ˈeiliæs] *n.* 换接口,交换点;亦称,别名

alidade [ˈælideid] *n.* 视准仪,测高仪,照准仪(用于平板测量器)
～ stadia method 照准仪视距法
peep-sight ～ 测斜照准仪
plane-table ～ 平板照准仪
sectional ～ 断面照准仪
sightrance ～ 测斜照准仪

alien [ˈeiljən] *vt.* 转让,移交; *a.* 外国人的; *n.* 外国人
～ structure 转让[拆迁]建筑

alifera [əˈlaifərə] *n.* 上侧片

aliform [ˈælifɔːm] *a.* 翼形的

alight [əˈlait] *a.* 点燃的,明亮的; *v.* 降落,落下来

align [əˈlain] *v.* 找平,调直,定线;调整

aligninum [əˈlaininəm] *n.* 木纤维敷料

alignment [əˈlainmənt] *n.* 定线[位];校整[直];顺序,直线,对准
～ design 路线设计,定线设计
～ gauge 定位[校准,水平]仪
～ plate 定位板
～ stake 定位桩,路线桩
horizontal ～ 水平线路,平面定线,水平校准
longitudinal ～ 纵向校准

aligreek [ˈæligriːk] *n.* 希腊式的格子花[回纹形装饰]

aline [əˈlain] *v.* alined,alining; =align

A-line 支承力影响曲线

aliquant [ˈælikwənt] *a.*; *n.* 除不尽的(数)

aliquation [ˌæliˈkwɔʃən] *n.* 层化,起层,熔析

aliso *n.* 阿里索桤木

alit(e) [əˈlit] *n.* A-水泥石,阿里特水泥(水泥熟料中的硅酸三钙),A 盐,铝铁岩
～ cement 阿里特水泥

alitizing [əˈlaitiziŋ] *n.* 铝化(处理),表面渗铝,镀铝

alive [əˈlaiv] *a.* 通电的,电流通过的,作用着的,运行中的;活泼的,存在的,充满着……的
～ circuit 带电线路
keep ～ 点火电极

alizarin(e) [əˈlizərin] *n.* 茜(草色)素
～ dyes 茜草色染料
～ lake 茜素沉淀色料
～ pigment 茜素颜料

alkalescence [ˌælkəˈlesns] *n.* 微碱性,碱化

alkali [ˈælkəlai] *n. pl.* -lis or -lies; 碱,强碱;(*pl.*)碱金属
～ ag(e)ing 碱性时效
～ -aggregate reaction [expansion] (水泥的)碱-集料反应[膨胀]
～ basalt 碱性柱石岩[玄武岩]
～ blue 碱性蓝
～ cellulose 碱性纤维素
～ content 碱含量
～ earth 碱土
～ granite 碱性花岗岩

alkalinity [ˌælkəˈliniti] *n.* 碱度,碱性,含碱量

alkyd *n.* 醇酸(树脂)
～ resin 醇酸[聚酯]树脂

alkyl [ˈælkil] *n.* 烷基,烃基

all [ɔːl] *a.* 完全的,全部的,所有的; *pron.* 全体,总数; *ad.* 全部,同一
～ -aluminium(building) 全铝的(结构,建筑物)

~ aluminiun conductor 全铝导体
~ application digital computer 通用数字计算机
~ brick block 全砖砌建筑
~ clay body 陶器挂釉底料[素坯]
~ colo(u)rs 全色
~ concrete (pavement) 全混凝土(路面)
~ directional (interchange) 全定向的(道路立体枢纽)
~ -dry cement mill 全干法工艺水泥厂
~ -electric (signalling system) 全电气化的(信号标志系统)
~ fillet weld 满角焊
~ glass(facade) 全玻璃的(立面)
~ hydraulic(crane) 全液压式的(起重机)
~ -in-aggregate (未经筛分的)天然[机轧]混合集料
~ -in-bid (工程设计与施工)投标
~ -in contract 全部承包合同
~ -in gravel 混合砾石
~ -in-one paver 联合铺路机
~ -insulated 全绝缘的
~ -level sample 全级试样(取自各个水平位置)
~ -mains 可由任意电源供电的,有通电用源的
~ -metal (building) 全金属的(建筑物)
~ -over design 连续花纹
~ -paper laminte 层压板[制件]
~ -pass network 全通网络
~ -paved (intersection) 全铺式(交叉口)
~ -plastic (structure) 全塑的(支承结构)
~ -purpose (computer) 通用的(计算机)
~ -relay 全继电器式的
~ -risks erection insurance 建筑工程保险

~ -round 共计
~ rubber 通用(型)橡胶,全胶的
~ -service gas mask 防毒面具
~ sliming 金矿泥[泥浆]化
~ -steel(construction) 全钢的(建筑结构)
~ terrain vehicle (ATV) 越野车
~ -timber(door) 木门
~ -time high [low] 最高[最低]纪录
~ up weight 总重量
~ veneer (construction) 全胶合板的(镶板的)(构造)
~ -weather (airfield) 全天候(飞机场)
~ -welded (construction) 全焊接(结构)
~ -wet cement plant 全湿法工艺水泥厂
~ -wheel drive 全轮驱动
~ -wood door 全[实]木门
allautal *n.* 纯铝包皮,铝合金板
allen wrench 六角弯管套筒扳手
al(l)ette [ˈælit] *n.* 古罗马[新古典]式建筑
allevardite [əˈlevədait] *n.* 钠板石
alley [ˈæli] *n.* 林间道;小路[巷],胡同,通廊;行道树
~ -house 小巷内店铺[房屋]
~ planting 并植(等距成行种植的树木)
~ stone 矾石
~ way 小巷,胡同
blind ~ 死胡同
pipe ~ 管道
alligation [æliˈgeiʃən] *n.* 混合法计算;和均性;合金,混合(物)
alligator [ˈæligeitə] *n.* 鳄式[缝式]碎石机,破碎机;自翻式起重机
alligatoring *n.* 涂膜皱皮[鳄纹],(轧制表面)裂痕,龟[皱]裂,鳄嘴裂口
~ lacquer 裂纹漆
allocate [ˈæləkeit] *vt.*, -cated, -cating;定位,配给,分配,配置

allocation [ˌælə'keiʃən] *n.* 分配,分布,配置,地址分配
~ of materials 物质分配
~ of shares 股份分配
~ plan 配置[分配]方案
~ sheet （原料）分配表
capital ~ [调]资金分配
circuit ~ 电路分配
frequency ~ 频率分配
load ~ 负载[荷载]分配
resource ~ 资源分配

allotment [ə'lɔtmənt] *n.* 配额；市郊小菜园[果园]，居民自种小园（市政当局分配给个人作为屋前花园用地）

allotriomorphic [ˌælətri'mɔːfik] *a.* 无本形的,他形的,不整形的
~ granular 人造花岗岩石面,不整形粘状

allotropic(al) [æləˈtrɔpik(əl)] *a.* 同素异形[性]的

allow [ə'lau] *v.* 允许；提供；清算；承认；酌量加[减]
~ed frequency 许用频率
~ed time 预定完工标准时间
~ed value 容许值

allowable [ə'lauəbl] *a.* 容[允]许的,承认的,正当的
~ bearing pressure 容许支承力
~ compression ratio 容许压缩比
~ deviation[error] 容许误差
~ load 容许荷载[负载]
~ pressure difference 容许压（力）差
~ relative deformation 容许相对变形
~ torque 容许扭矩[转矩]

alloy ['ælɔi] *n.* 合金；纯度,成色；混合物,杂质；*v.* [ə'lɔi] 熔合,合铸,掺杂（金属）
~ cast iron 铸铁合金
~ cladding 包合金
~ constructional steel 合金结构钢
~ed nickel steel 镍合金钢
~ housing 合金外壳
~ iron 铁合金
~ layer 合金层
~ pipe 合金管
~ reinforcing steel 合金[特种]钢筋
~ steel 合金钢
~ structural steel 合金结构钢[型钢]
acid resisting ~ 耐酸合金
aluminium ~ 铝合金
anticorrosion ~ 抗蚀合金
antifriction ~ 抗摩合金

alluvium [ə'luːviəm] *n.* 冲积层[土壤]
~ grouting 冲积层压实[密实]

ally [ə'lai] *v.,* -lied,-lying, *n. pl.* -lies；*v.* 联[结]合,联系；*n.* 同盟者[国],伙伴,助手

allyl ['ælil] *n.* 烯丙基
~ plastic(s) 烯丙（基）树脂类[塑料]
~ resin(s) 烯丙（基）树脂类

almery *n.* ['ɔːlməri] 壁橱；教堂放圣器的木橱

almighty [ɔːl'maiti] *a.* 万能的
~ adhesives 万能粘合剂

almost ['ɔːlməust] *a.,ad.,* 几乎是……的,差不多的；几乎,差不多
~ cylindrical shell 有初变形的圆柱面壳
~ hemispherical dome 初变形的半球空壳

alms-house [ɑːmzhaus] *n.* 养老院,救济院

alnico ['ælnikəu] *n.* 铝镍钴磁钢,(作永久性磁铁用)

Alnus *n.* 桤木,赤杨
~ japonica 赤杨
~ maritima 东北赤杨,(海滨)桤木
~ tinctoria 辽东桤木

Alodine process 阿洛载法（一种铝表面处理化学氧化法）

aloxite [ə'lɔksait] *n.* 铝砂,钢玉磨料
~ tube 铝砂管

Alpaka [æl'pækə] *n.* 镍白铜（铜50%，锌25%，镍25%）

Alpax ['ælpæks] *n.* 铝硅合金

Alpert bakable valve 全金属耐烘烤阀

Alperth cable 聚乙烯绝缘铝芯电缆
alpha ['ælfə] *n.* α,希腊文字母中的第一字母
~ active 放射性的
~ brass α-黄铜
~ display 字母数字显示
~ iron α 铁
~ particle α 粒子
~ rays α 射线
alphameric [ælfə'merik] *a.* 字母数字（混合编制）的
~ digit 字母数字数码
~ tube 字母数字显示管
alpine ['ælpain] *a.* 高山的
Alpine fir 阿尔卑斯山冷杉
~ garden 高山花园
~ larch 高山落叶松
~ light 紫外线
Alpine piedmont （山区）山前地带
~ road 高山［山岭］道路,高山公路
~ tunnel 高山［山区］隧道
~ whitebark pine 美国白皮松
Alrak-method （铝及铝合金）表面防蚀化学处理法
alstonite ['ɔːlstənait] *n.* 碳酸钙钡矿
alta-mud *n.* 膨润土
altana ['æltænəl] *n.* 柱承式阳台
altar ['ɔːltə] *n.* 隔［胸］墙；祭坛；圣坛；台阶
~ baldachin 祭坛［圣坛］华盖
~ carpet 圣坛地毡
~ ciborium 祭坛［圣坛］华盖
~ facing 祭坛饰罩
~ fresco 祭坛［圣坛］壁画
~ -front(al) 祭坛饰罩
Altar of Land and Grain 北京社稷坛
~ piece 祭坛上方及后面雕画饰物,祭坛画［雕刻,屏风］

~ rails 祭坛四周栏杆
~ screen 祭坛上方及后面雕画装饰
~ shrine 祭坛神龛
~ stair 祭坛楼梯
~ tomb 纪念坛
Circular Mound ~ 环丘坛
sacrificial ~ 祭坛
altazimuth [ælt'æziməθ] *n.* 地平经纬仪,高度方位仪
alter ['ɔːltə] *a.* 改造［建］
alterability [ˌɔːltərə'biliti] *n.* 可变性,可变更性
alteration [ˌɔːltə'reiʃən] *n.* 改建［造］,交替,变换,改变
~ design 比较设计方案
~ work 改造［建］工程
~ work hour 轮换工作时间
alternate ['ɔːltəneit] *v.*,-nated,-nating, *n.*；交替［错,变］轮流,更迭,比较方案
~ angle 交错角
~ balance line 交替平衡线
~ bay construction [method] （混凝土）建筑路面的间隔［隔仓］施工法（混凝土分块间隔跳仓施工）
~ bays 隔仓（用于建筑混凝土路面）
~ bending strength 交变弯曲［弯曲疲劳］强度
~ block progressive [traffic] system 隔区段进行交通管制系统
~ course 交变层
~ current 交流电
~ depths 交替水深,共轭水深［深度］
~ design (layout) 比较设计（方案）
~ device 交换器,分离器
~ exterior angle(s) 外错角
~ folding 交替折叠
~ form 替换式
~ freezing thawing 交替冰融
~ immersion (corrosion) test 交替［变］浸没（腐蚀）试验
~ interior angle(s) 内错角
~ joint 错缝,［错列式］接缝,错列接
~ lane construction 路面分道［分

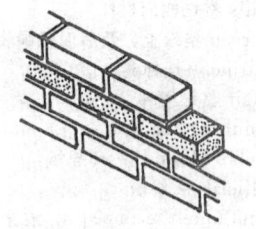
alternate joint

块,交替]浇筑
～ lay（钢丝左右交替捻成钢绳）的交捻
～ load 交替荷载,交变负荷
～ locations(of route selection)（道路选线的)比较位置,比较路线
～ marker 交替路线指示标
～ material 替换[代用]材料
～ one-way traffic control 交替单向通行控制
～ products 代制品
～ progressive system 绿灯交替显示的推进式信号系统
～ route 比较路线
～ stages 交变级,交变水位高度
～ stages of flow 水流的共轭水位
～ strain 交替[反复]应变
～ strength 交变强度
～ stress 交变[交替]应力
～ system(联动信号系统之一的)交替系统
～ tension 交替[交变]拉力

alternating [ˈɔːltəːneitiŋ] *a.* 交替的,交变的,更迭的
～ bending test 反复弯曲实验
～ component of pressure （声学)交变压力
～ coursed ashlar masoury 交错层状琢石砌体
～ curing 交替养护
～ current 交流电
～ current arc welding 交流电弧焊
～ currcent equipment 交流电设备[装置]

～ current generator 交流发电机
～ current method 交变电流勘探法
～ current two-speed elevator 二级速度交流电梯[升降机]
～ deformation 交替形变
～ flexure test 反复弯曲实验
～ load 交变负荷[载]
～ motion 往复运动
～ notch bending test 切口交变挠曲试验
～ notch flexure test 切口交变挠曲试验
～ pressure 交变压力
～ strain 交替应力
～ stress 交变应力
～ torque 交变转矩

alternation [ɔːltəːˈneiʃən] *n.* 房屋改建;更改,更换,循环
～ of wetting and drying 干湿循环[交替]
～ of beds 交互层
～ of cross section 横断面交替
～ of freezing and thawing 冰融循环[交替]
stress ～ 应力循环[交替]

alternative [ɔːlˈtəːnətiv] *a.* 交替的,变更的,比较的; *n.* 比较方案,二者之一
～ bid 比较(投)标价(书)
～ construction method 比较施工方法
～ decision 选择性决策
～ design 比较设计(方案)
～item 选择项目
～ interchange 互通式立(体)交叉
～ location line 比较路线
～ manner 迭代方法
～ material 代用材料
～ method 交替[补充,变异]法
～ optimal 择其最优
～ (parallel) side street 平行主干线的复线,集散道路
～ plan 比较规划方案,备择规划方案
～ price 比较价格

~ project 比较设计(方案)
~ proposal 可供选择的方案
~ route 比较路线
~ scheme 比较方案
alternator [ˈɔːltəneitə] *n.* 交流发电机
single-phase ~ 单相交流发电机
synchronous ~ 同步交流发电机
three-phase ~ 三相交流发电机
altichamber [ˈæltitʃæmbə] *n.* 气压试验室,气压检定箱,高空试验[模拟]室
altigraph [ˈæltigrɑːf] *n.* 高度记录仪,气压计
altimeter [ˈæltimiːtə] *n.* 测高计,高度表,测高仪
altimetric point 高程点
altitude [ˈæltitjuːd] *n.* 海拔高度,高度

altitude level

~ chamber 负压工作箱,高空试验[模拟]室
~ difference 高(度)差
~ gauge 测高计
~ level 测高水准仪,高度计
~ valve 水位控制阀
critical ~ 临界高度
geometric ~ 几何高度
altometer [ælˈtɔmitə] *n.* 经纬仪
altorelievo [ˌæltəuriˈliːvəu] *n. pl.* -vos;(意大利)高(凸)浮雕,圆雕
alucol [əˈljuːkɔl] = aludrox *n.* 氢氧化铝,氢氧化镁混合物
Aludirome [əˈljuːdirəm] *n.* 铁铬铝合金,铁铬铝电炉丝
Aludur [əˈljuːdjuə] *n.* 铝镁合金,阿鲁杜合金
Alufer [əˈljuːfə] *n.* 铝合金,包铝[铝合金]钢板
alugel [əˈljuːdʒəl] *n.* 氢氧化铝
alum(en) [ˈæləm] *n.* 矾,明矾,硫酸铝
~ clay 矾土
~ earth 矾土
~ flour 矾粉

~ -saturated class D (gypsum) plaster 德国大理石水泥经矾土处理石膏
~ schist 明矾片岩
~ shale 明矾页岩
~ shale concrete 明矾页岩混凝土
~ slate 明矾板岩,矾石板
~ solution 明矾溶液
~ stone 明矾石
~ treatment 明矾处理
ammonia ~ 氨明矾
white ~ 明[白]矾
alumag [ˈæljuːmæg] *n.* 铝镁合金,硅酸镁,氢氧化铝混合物
Alumal [ˈæljuːməl] *n.* 铝锰合金
alumatol [ˈæljuːmeitəl] *n.* 硝铵-铝粉炸药
alumian [əˈljuːmiən] *n.* 水钠[无水]矾石
alumigel [əˈljuːmidʒəl] *n.* 氢氧化铝
alumina [əˈljuːminə] *n.* 矾土,铝氧土,氧化铝(Al_2O_3)
~ -bearing material 含铝材料
~ -borosilicate glass 硼硅酸铝玻璃
~ brick 矾土[高铝]砖,[耐火砖]
~ cement [矾土]高铝水泥
~ ceramics 高铝陶瓷
~ (fire)brick 矾土耐火砖
~ fireclay brick 矾土耐火砖
~ gel 氧化铝凝胶
~ mortar 矾土砂浆
~ plant 矾土厂
~ porcelain 矾土陶瓷,高铝陶瓷
~ powder 铝氧粉,氧化铝粉
~ ratio (水泥的)铝率
~ refractory (product) 矾土耐火制品
~ white 矾土白(用天然氧化铝作白色颜料)
aluminate [əˈljuːmineit] *n.* 铝酸盐
~ cement 铝酸盐水泥
Aluminite [əˈljuːminait] *n.* 明矾石,铝矾石

aluminium [ˌælju'minjəm] (= aluminum) *n.* 铝(Al)

aluminize [ə'lju:minaiz] *vt.*, -nized, -nizing; 镀铝,铝化

aluminoferric [ə'lju:minəferik] *n.* 铁矾土; 铝铁剂

aluminosilicate [əˌlju:minə'silikit] *n.* 铝硅酸盐
~ brick 硅酸盐耐火砖
sodium ~ 钠铝硅酸盐

alumino-thermic *a.* 铝热的
~ reaction 铝热反应
~ welding 铝热焊接

aluminous [ə'lju:minəs] *a.* 明矾的,矾土的,含有明矾的
~ cement 矾土水泥
~ clinke(r) 高铝矿渣
~ (fire) brick 矾土(耐火)砖
~ fireclay brick 矾土耐火砖
~ refractory (product) 矾土[高铝]耐火制品[材料]
~ silicate brick 矾土硅酸盐砖

aluminum [ˌælju'minjəm] *n.* 铝
~ absorbent ceiling 铝板贴面吸声顶棚[吊顶]
~ acoustic tiled ceiling 铝板贴面吸声顶棚[吊顶]
~ alloy 铝合金
~ alloy casting 铝合金铸件
~ alloy for temper 锻用铝合金
~ architecture 铝(结构)建筑学
~ austral window 铝制推拉窗
~ base alloy 铝基合金
~ beam 铝(制)梁
~ bearing structure 铝支承结构
~ (blasting) cap 铝(制)雷管
~ body 铝车身
~ bonded roof covering 铝砌合屋顶屋面
~ brass 铝黄铜
~ bridge 铝桥
~ bronze 铝青铜
~ bronze paint 铝青铜涂料

~ (building) entrance door 铝制(建筑)外门
~ (building) member 铝制(建筑)构件
~ building product 铝制建筑制品
~ (building) sheet 铝制(建筑)板材
~ (building) unit 铝制(建筑)构件
~ cable 铝芯电缆
~ cable sheath 铝心电缆甲套
~ casting 铸铝,铸铝件
~ casting alloy 铸铝合金
~ ceiling 铝(板)顶(棚)
~ -cell arrester 铝[管]避雷器
~ cement 矾土水泥,高铝水泥
~ channel 槽型铝材
~ checker(ed) foil 花纹铝箔
~ chloride 氯化铝
~ chrome steel 铝铬钢
~ component 铝制建筑构件
~ compound 铝化合物
~ conductor 铝芯导线[体]
~ -copper-iron 铝铜铁合金
~ corrugated (building) sheet 铝制波纹片[薄板]
~ corrugated profile 铝制波纹形
~ corrugated section 铝制波纹型材,波纹截面铝材
~ corrugated unit 铝制波纹形构件
~ curtain wall 铝制幕墙
~ deck (roof) 铝板屋顶
~ deck unit 铝顶板构件
~ decorative section 铝装饰型材
~ detonator 铝雷管
~ die casting 压铸铝
~ door 铝(制)门
~ doors and windows 铝合金门窗
~ eave(s) gutter 铝制檐沟
~ eave(s) trough 铝制檐沟
~ electrode methed (加固地基用)铝电极法
~ extrusion 挤压铝制品
~ facade 铝制结构立面[正面]

~ face 铝制结构正面[立面]
~ facing 铝饰镶面[贴面]
~ fence 铝制围栏
~ finger plate (门锁的)铝指板(防污板)
~ fillings 铝制小五金
~ flake 高岭土,薄铝片
~ flashing 防雨[挡水,披水]铝板
~ floor covering 楼面铝覆面层
~ flooring (finish) 楼面铝覆面层
~ foil 薄铝片,铝箔
~ foil backing 铝薄背衬
~ foil insert(ion) 铝箔绝热
~ folded plate roof 铝制折板屋顶[屋面]
~ form(work) 铝制模板
~ frame 铝制框架
~ front 铝制结构立面[正面]
~ gasket 铝密封垫圈
~ gold 铝金(含铝黄铜)
~ grease 铝皂润滑脂
~ grid 铝(制)窗格[格子]
~ grid ceiling 铝制网格天花板
~ grille 铝制格栅[格子窗]
~ hand railing 铝制扶手
~ handware 铝制小五金
~ hinge 铝制铰链
~ hydrate [hydroxid(e)] 氢氧化铝
~ iron 铝铁
~ leaf 铝箔,薄铝片
~ leaves 铝箔,薄铝片
~ lining 铝衬裹[底]
~ lintel (门窗)铝制过梁
~ magnesium alloy 铝镁合金
~ -manganese alloy 铝锰合金
~ mesh 铝(制)网
~ mould 铝模
~ nail 铝钉
~ ornamental section 铝装饰型材
~ outrigger base 铝支柱底板
~ oxide 氧化铝
~ paint 铝涂料[漆],银粉漆
~ paint coating 银粉漆涂层

~ panel ceiling 铝板顶棚[顶板]
~ partition 铝板隔墙
~ paste 银灰漆,铝涂料
~ patent glazing bar (用于玻璃采光窗的)铝制玻璃格条
~ perforated ceiling 多孔铝制天花板
~ pigment 铝颜料
~ pipe 铝(制)管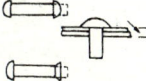
~ plate 厚铝板
aluminum pipe
~ plating 镀铝
~ post 铝柱
~ powder 铝粉,银粉(颜料)
~ primer 铝(粉)打底涂料
~ product 铝材,铝制品
~ profile(d) panel 成型铝板
~ profile(d) sheet 成型薄铝板
aluminum rivet
~ pull (handle) 铝门拉手
~ purlin 铝檩条
~ railing 铝扶手[栏杆]
~ rivet 铝铆钉
~ rod 铝棒[杆]
~ rolling grille 滚动式[卷帘式]铝格栅
~ rolling plant 铝辊式破碎机,铝碾压机
~ rolling shutter 卷帘铝质百叶窗(门)
~ rolling slat 铝制卷板条
~ roofing 铝制屋顶
~ sheath 铝包皮,铝护套
~ sheet(ing) roof 铝板屋面
~ sandwich panel 夹层铝板
~ sash 铝窗框
~ screw 铝螺丝
~ seal(ing) sheeting 密封铝皮
~ section 铝结构型材
~ shape 铝结构型材
~ sheet 铝(薄)板[片,皮]
~ sheet facing 铝皮镶面[包面,贴

面,饰面]
~ sheet lining 铝皮镶面[贴面,饰面]
~ sheet surface 铝皮镶面[贴面,饰面]
~ shielding 铝屏蔽
~ shutter 铝(制)百叶窗
~ shuttering 铝(制)模板[模壳]
~ siding 铝墙板
~ silicate 硅酸铝
~ silo (存放散装水泥)铝仓筒
~ sink (unit) 铝合金洗涤盆[冲洗台]
~ skin 铝包皮,铝涂料层
~ skyscraper 超高铝制烟囱
~ slatted blind 铝板防护板[档板,铝百叶窗(门)]
~ slatted roller blind 铝板卷升式百叶窗
~ sliding door 推拉式铝门
~ smelting plant 铝熔炼厂[设备]
~ soap 铝皂
~ soft wool 软铝棉
~ solder 铝焊料
~ sound control ceiling 铝(制)声控顶棚
~ stabilizer base 铝制支架[支柱]底板
~ stearate 硬脂酸铝
~ steel 铝钢
~ street lighting column 街道照明铝制灯柱
~ strip 铝板条,铝带
~ sulphate (sulfate) 硫酸铝
~ sunblind 铝制遮阳板[百叶窗]
~ sunbreaker 铝制遮阳板
~ surfacing 铝(板)铺面[饰面]
~ swing door 双开式弹簧铝门
~ T(ee)-bar T 形铝杆
~ track 铝轨
~ trim 铝结构型材
~ tube 铝管
~ unit 铝结构型材[构件]
~ vapo(u)r barrier 铝汽封[隔汽层,防潮层]
~ wallpaper 铝墙纸
~ welding rod 铝焊条,铝合金焊条
~ window furniture 铝窗装置[配件]
~ wire 铝(导)线
~ wire cloth 铝丝布
~ wood composite system 铝木组合结构
~ wool 铝纤维
~ wrought alloy 锻制铝合金
~ zinc 锌铝
beaten ~ 铝薄片
cast ~ 铸[生]铝
flake ~ 片状铝粉
steel-cored ~ 铝线绕制钢心电缆

alumiseal [ə'lju:mi'si:l] *n.* 铝密封
alumite ['ælju:mait] *n.* 明矾石,氧化铝膜处理法,(表面有电解氧化膜的)耐酸铝;铝氧化膜;防蚀[防腐,耐热]铝
~ tile [吸声,隔声]氧化铝板
~ wire 防蚀铝线
alumstone *n.* 明矾石
alundum [ə'lʌndəm] *n.* 铝氧粉,刚铝石;人造刚玉(用黏土、木炭和铁屑在电炉中烧成)
~ cement 刚铝石水泥,氧化铝水泥
~ tile 铝石砖
alunite ['æljunait] *n.* 明矾石
~ rock 明矾岩
alupag *n.* 菲律宾大戟木
alure ['æljuə] (= allure) *n.* 院廊,通道,廊道

alure

alusil alloy 铝硅合金
alutile ['æljutai:l] *n.* (吸声,隔声)氧化铝板
alveolate [æl'vi:əlit] *a.* 蜂窝状的,有小

孔的
alveolusity [ˌælviəˈlʌsiti] *n.* 蜂窝
A. M. peak (交通)上午高峰
amalaka *n.* (印度教和耆那教建筑中高塔顶部的)馒头形顶
amalgam [əˈmælgəm] *n.* 汞膏,汞齐[汞合金]混合物
～ fluorescent lamp 汞剂荧光灯
amalgamate [əˈmælgəmeit] *v.* 混合,合并,与汞[水银]混合
～ d dwelling-house (一栋多户集合居住的)集合住宅
Amalgamated Society of Engineers (＝ASE)工程师联合会

amalgamated dwelling-house

amalgamation [əˌmælgəˈmeiʃən] *n.* 混合,合并
～ of contractors 建筑联合企业
～ process 汞剂法
amass [əˈmæs] *vt.* 积蓄,聚积 *vi.* 集合
A-mast A 形轻便井架
amber [ˈæmbə] *n.* 琥珀,琥珀色;*a.* 琥珀制的,琥珀色的
～ brown 琥珀褐色,淡褐色
～ glass 琥珀玻璃
～ light 黄色灯,琥珀色灯
～ mica 金云母,琥珀云母
～ oil 琥珀油
～ period (交通灯的)黄灯时间
～ tar 琥珀柏油
～ varnish 琥珀清漆
carbon ～ glass 琥珀(有色)玻璃
amberina [ˌæmbəˈriːnə] *n.* 艺术玻璃
amberlite [ˈæmbəlait] *n.* (苯酚甲醛)离子交换树脂(商品用名)
～ion exchange resin 琥珀离子交换树脂
～ -IRA 阴离子交换树脂
～ -IRC 阳离子交换树脂
～ -XAD 大网状树脂
amberlyst [ˈæmbəlist] *n.* 大孔树脂(商品用名)
amberoid [ˈæmbərɔid] *n.* 合成琥珀
amberplex [ˌæmbəˈpleks] *n.* 一种离子交换膜(商品用名)
～ A 阴离子交换膜
～ C 阳离子交换膜
amberwood [ˈæmbəwud] *n.* 酚醛树脂胶合板
ambetti [ˈæmbiti] *n.* 装饰窗玻璃,晶粒玻璃
ambient [ˈæmbiənt] *n.* 环境;周围温度,环绕空间; *a.* 包围的,被覆的,周围的
～ conditions 环境[周围]情况[条件]
～ environment 周围环境
～ field 背景场
～ light 环境光线
～ noise 环境噪声
～ pressure 周围(环境)压力
～ temperature 周围[环境]温度,室温
ambiguity [ˌæmbiˈgju(ː)iti] *n. pl.* -ties 不明确性,含糊性,双关性
～ function 不明职权
ambiguous [æmˈbigjuəs] *a.* 含糊的,不明确的,可疑的
～ case 分歧[可疑]情况
～ count 错误计数
ambipolar [ˌæmbiːˈpəulə] *a.* 双极的,二极的
ambit [ˈæmbit] *n.* 范围,周围,区域,轮廓,外形
Amboina wood (印尼)安伯伊那木(作为高级家具用材)
ambo(n) [ˈæmbəu] *n.* (早期基督教教堂)读经台,高座
～ ciborium 读经台华盖[龛室]

ambrain [ˈæmbrein] *n.* 人造琥珀
Ambraloy *n.* 铜合金
ambrite [ˈæmbrɔid] *n.*

灰黄琥珀
ambroid ['æmbrɔid] *n.* 人造［合成］琥珀
ambroin ['æmbrɔin] *n.* 绝缘材料,合成琥珀
~ cement 假琥珀胶
ambrosine ['æmbrəsain] *n.* 褐黄琥珀,化石树脂
ambry ['æmbri] *n.* 壁柜,食品柜;备餐室;教堂壁龛
ambulance ['æmbjuləns] *n.* 救护车,野战医院
~ corps 流动救护队
~ station 救护站
ambulante [æmbju'lɑ:nt] *n.* 18世纪的轻便茶桌
ambulator ['æmbjuleitə] *n.* 测距器［仪］
ambulatory ['æmbjulətəri] *n. pl.* -ries;（有顶棚的,室内的）回廊,步道,步廊;*a.* 行走的,动的
~ aisle 走廊［道］,通道
~ bin 活动料槽,流动谷仓
~ chattels 动产
~ church 十字形圆顶教堂
~ vault 回廊［走道］拱顶
Ambursen-type dam 安布逊式坝,平板支墩坝
Ambursen weir 安布逊溢流堰［量水堰］
AME = angle measuring equipment 测［量］角装置
amelioration [ə,mi:lj(ə)'reiʃən] *n.* 修正,改正,改善
amend [ə'mend] *vt.* 改正,改良;改善,变更
~ed plan 修正图,修正计划
~ing clause 修正条款
amendment [ə'mendmənt] *n.* 更改（图纸等）;变更,修［校］正;修正图,修正方案

ambulatory

~ advice 修改通知书
~ commission 修改手续费
amenity [ə'mi:niti] *n. pl.* -ties;（环境,气候,风暴）舒适性,适宜性;令人愉快的事物
American [ə'merikən] *a.* 美国的,美洲的;*n.* 美国人,美洲人
~ architecture 美国建筑
~ basement（美国）包括房屋主要入口的地面上的底层
~ beech 美洲大叶山毛榉
~ bond 美国式砌墙法
~ brick 美国砖（56mm×203mm）
~ caisson 沉箱
~ cloth 美国油布
~ Concrete Institute（ACI）美国混凝土学会
~ dewberry 美国悬钩子
~ Engineering Standards Commtitee（=AESC）美国工程标准委员会
~ ga(u)ge 美国量规（美国尺寸标准）
~ lumber standard 美国木材标准
~ -Indian architecture 印地安人建筑
~ larch 美洲落叶松
~ National Standard Institute（=ANSI）美国国家标准学会
~ parliament mansion 美国国会大厦

American parliament mansion

~ pitch pine 美国多脂松
~ pitch pine strip floor(ing)(finish) 美国多脂松狭条地板
~ plane 美国悬铃木
~ Society of Civil Engineers（=ASCE）美国土木工程师协会
~ Society of Mechanical Engineers（ASME）美国机械工程师协会
~ standard Association(ASA)

美国标准学会
~ standard channel 美国标准槽钢
~ standard steel section 美国标准型钢材
~ standard fittings 美国标准型配件

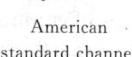
American standard channel

~ Standard Wire gauge (ASW) 美国标准线规
~ system of drilling (美式)钢索冲击钻探法
~ timber 美国木材
~ vermilion 美洲朱砂(鲜红颜料)
~ water turbine 混流式水轮机
~ wheel 美国水轮机(开始为内向流,而后为轴向流的一种水轮机)
~ white pine 美国白松

americium [ˌæməˈriʃəm] *n.* 镅(Am)
~ leather 油布

amerindian architecture 印地安人建筑

ameripol [əˈmeripɔl] *n.* 人造耐油橡胶

amesdial [ˈæmisdail] *n.* 测微仪,千分表

Ames surface gauge 爱姆司平面量规

amethyst [ˈæmiθist] *n.* 紫石英,紫晶

amethystine [ˌæmiˈθistain] *a.* 紫石英色的,紫晶色的;紫水晶制的;含紫水晶
~ quartz 紫晶

Amfion *n.* 一种均质离子交换膜

amianite *n.* 一种用石棉作填料的塑料

amiant [ˈæmiənt] *n.* 石棉,石绒,细丝石棉

amianthus [ˌæmiˈænθəs] *n.* 石麻[绒],石棉

amic acid 酰胺(基)酸

amicron [eiˈmaikrɔn] *n.* 次微[胶]粒;超微子(直径小于 10^{-7} 厘米)

amid-stream entrance 流内进口(与流向正交的进口)

amide [ˈæmaid] *n.* 氨化物,酰胺

amidpulver [ˌæmidˈpʌlvə] *n.* 硝铵,硝酸钾,炭末炸药

Amiens [ˈæmjənz] *n.* 亚眠(法国北部一城市)
~ cathedral (1200 ~ 1269 年法国)亚眠大教堂(为哥特式建筑)

Amiesite [ˈæmiːsait] *n.* 阿米赛特(一种由碎石、稀释剂、膏料沥青及少量硝石灰拌成的冷铺沥青混合料)

amilan [ˈæmilən] *n.* 聚酰胺树脂,聚酰胺纤维

aminate [ˈæməneit] *n.* 氨化产物,氨化

amination [ˌæməˈneiʃən] *n.* 胺化(作用),氨基化(作用)

amine(s) [ˈæmiːn(z)] *n.* 胺
~ cellulose 氨基纤维素
~ formaldehyde resin 胺醛树脂
~ treatment (污泥消化池中的)胺处理
~ resin 氨基树脂

aminex [ˈæmineks] *n.* 多孔性阴离子交换树脂

amino [ˈæminəu] *n.* 氨基
~ acid 氨基酸
~ aldehyde resin 氨(基)醛树脂
~ alkyd resin coating 氨(基)醇酸树脂涂料
~ -plast(ics) 氨基塑料
~ resin(s) 氨基树脂(类)

ammeter [ˈæmitə] *n.* 电流表[计],安培计
tongs-type ~ 钳式安培计

ammoniac [əˈməuniæk] *a.* 氨的; *n.* 氨树胶
sal ~ 氯化铵

ammoniacal [ˌæməuˈnaiəkəl] *a.* 氨的,含氨的
~ copper arsenite (木材防腐用)亚砷酸铜氨液
~ liquor 氨液
~ nitrogen 氨氮,氨态氮

ammonification [əˌməunifiˈkeiʃən] *n.* 成氨[作用],加氨

ammonite [ˈæməait] *n.* 硝石炸药

ammonium [əˈməunjəm] *n.* 铵(NH_4)
~ chloride 氯化铵,卤砂

~ hydroxide 氢氧化铵
~ nitrate 硝酸铵
~ nitrate gelatince 胶质硝酸铵炸药
~ salt 铵盐

amorphous [əˈmɔːfəs] *a.* 非晶质的,无定形的,模糊的;amorphously, *ad.*
~ carbon 无定形碳
~ glass 非结晶形玻璃
~ precipitation 非晶型淀析[沉淀]
~ solid 非晶体,无定形固体
~ wax 软蜡

amortisseur *n.* 消声器,减震器,缓冲器,阻尼器

amortization [əˌmɔːtiˈzeiʃən] *n.* 消声;阻尼,减震;缓冲作用;折旧,偿还,分期偿还
~ cost 偿还资金,偿还费
~ charge 折旧提成
~ fund 清偿费,赔偿费
~ method 摊派法
~ of fixed assets 固定资产摊提
~ payment 摊派提款
~ period 偿还期,偿还期限
~ rent 摊提租金

amortize [əˈmɔːtaiz] *v.* 缓冲,减震;分期偿还;amortizable, *a.*
~d depreciation 分期偿还,折旧

amortizement [əˈmɔːtaizmənt] *n.* 扶壁的斜坡顶,建筑物的冠顶部分

amosite [ˈæməsait] *n.* 铁石棉(填充剂),长纤维石棉

amount [əˈmaunt] *n.* 材料用量;物质(总)量,总值;范围,程度
~ at risk (保险中的)风险额
~ declared 申报金额
~ involved 涉及金额
~ of agitation 搅拌强度,搅拌量
~ of air exhaust 总排气量
~ of bend(ing) 挠曲[弯曲]量
~ of blast 送风量
~ of change scale 等差尺度
~ of compression 压缩度[量]
~ of contraction 收缩量
~ of crown 拱高度,拱高
~ of deflection 变位度,挠度,弯曲度,垂度
~ of deposit 预存费总数
~ of eccentricity (筛分机的)摆动量
~ energy 能量
~ of evaporation 总蒸发量
~ of handling 装卸费,搬运费
~ of heat absorption 总吸热量
~ of inclination 倾角,倾斜角
~ of information 信息量
~ of labour 劳动人数,定额用工
~ of leakage 漏水量
~ of looseness 间隙,松动
~ of monthly rainfall 月降雨量
~ of movement 移动[位移]量
~ of photometry 光度值,测光值
~ of payment 付款额
~ of porosity 孔隙度
~ of precipitation 沉淀量
~ of prestress 预应力值
~ of rainfall 雨水量
~ of reinforcement 配筋率,钢筋百分率
~ of strength 强度值
~ of traffic 交通[运输]量
~ of ventilation 换气[通风]量
~ of wear 磨损率[量]
~ of work done 工程完成量,工程完成额
~ used 消耗(费)量
gross ~ 约计,概数
net ~ 细数,净量

amozonolysis *n.* 氨(解)臭氧化反应(作用)

ampacity [æmˈpæsaiti] *n.* 安载流容量

amparo blue 鲜蓝色

Ampco *n.* 铝铁青铜

ampcoloy *n.* 耐蚀耐热铜合金

amperage [æmˈpiridʒ] *n.* 电流量,电流强度,安培数

ampere [ˈæmpɛə] *n.* 安培

~ balance 安培秤
~ -feet 安培-英尺
~ -hour 安培小时
~ -meter 电流计,安培表
~ -turn 安培匝数
~ -volt 伏安
amphenol connector 电缆接头,接线端子
amphibious [æm'fibiəs] *a.* 水陆两用的,有两种特性的；amphibiously, *ad.*, amphibiousness, *n.*
~ shovel 两栖铲土机
~ site 水陆建筑场地
~ truck 水陆两用载重车
amphibole ['æmfibəul] *n.* 闪石
~ asbestos 闪石石棉
~ magnetite 闪石磁铁矿[石]
amphibolic [ˌæmfi'bɔlik] *a.* 闪石的
~ gneiss 闪石片麻岩
amphibolite [ˌæmfi'bəulait] *n.* 闪岩
amphidetic [ˌæmfi'di:fik] *a.* 两面伸展的
~ ligament 全韧带
amphi-position *n.* 跨位
amphiprostyle [æm'fiprəustail] *n.* 前后有排柱而两边无柱的建筑(两排柱式建筑)

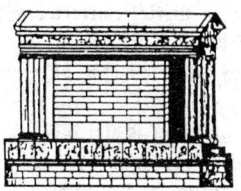
amphiprostyle

amphiprostylos [æm'fiprəustailəs] *n.* 前后有排柱而两旁无柱的建筑
amphistylar ['æmfistailə] *a.* 两排柱式的(建筑)
amphitheater ['æmfiθiətə] *n.* [古罗马]圆形露天剧场,露天剧场中半圆形的观众席；(戏院中)半圆形梯形楼座；四面环有小山的平地

ampholytoid [ˌæm'fəulaitɔid] *n.* 两性胶体
amphora ['æmfərə] amphitheater
n. pl. -ras, -rae；双耳长颈瓶
amphoteric [ˌæmfə'terik] *a.* 两性的,
~ colloid 两性胶体
~ ion exchange resin 两性离子交换树脂
amp-hr＝ampere hour 安培小时
ample ['æmpl] *a.*, -pler, -plest；广大的,丰富的；ampleness, *n.*
~ lobby 大前厅,宽广的走廊
~ flow 丰水
~ power 大功率
~ supply 供应充足
amplification [ˌæmplifi'keifən] *n.* 放大,扩大[张]；放大率
~ coefficient 放大系数
~ factor 增幅率
~ range 放大范围
~ ratio 放大率
~ stage 放大级
amplifier ['æmplifaiə] *n.* 放大器,扩音机
amplitude ['æmplitju:d] *n.* 幅,振幅,距离,范围,作用半径,偏角,方位角；充足
~ attenuation constant 振幅衰减常数
~ of accommodation 调节幅度
~ of noise 噪声幅度
~ of oscillation 摆幅
~ of pressure 压力振幅
~ of stress 应力幅度
~ of swing 摆动幅度
~ of the tide 潮幅
~ of vibration 振幅
~ of wave 波幅
ampoul(e) ['æmpu:l] *n.* 安瓿,小玻璃瓶
Amsterdam [ˌæmstə'dæm] *n.* 阿姆斯

特丹(荷兰首都)
~ Bosh 阿姆斯特丹大公园
~ group(20 世纪初期荷兰)阿姆斯特丹建筑学派

amusement [ə'mju:zmənt] *n.* 娱乐,消遣
~ center 娱乐中心
~ district 娱乐区
~ facilities 文娱[娱乐]设施
~ hall 娱乐厅[场]
~ park 娱乐公园,露天游艺场,公共游乐场

amylaceum [ˌæmi'leiʃəm] *n.* 葡萄糖,淀粉

An = normal atmosphere 标准大气压

anabranch ['ænəbrɑ:ntʃ] *n.* 再流入主流的支流;被沙地所吸收的支流

anacampsis [ˌænə'kæmpsis] *n.* 弯曲,(光,声)折射

anacamptometer *n.* 反射计

anaclinal [ˌænə'klainəl] *a.* 逆四周地层的倾斜方向而下降的

anacline *n.* 正倾型
~ stream 逆斜河

anaconda *n.* 安纳康达(以三氧化砷为主要成分的木材防腐剂)

anaerobic [ˌəˌneiə'rɔbik] *a.* 厌氧性的,嫌氧性的
~ adhesives 厌氧性粘合剂[胶粘剂]
~ sediments 缺气沉积

an(a)esthetic [ˌænis'θetik] *a.* 麻醉的; *n.* 麻醉剂
~ room 麻醉室

anaflow [ˈænəfləu] *n.* 上升气流

anaglyph [ˈænəglif] *n.* 浅型[彩色]浮雕装饰,浮雕
~ map 浮雕图
~ picture 浮雕图,彩色浮雕像

anaglyph map

anaglyphic [ˌænə'glifik] *a.* 浮雕的; *n.* 浮雕装饰,浮雕装饰艺术

~ strip 条形浮雕

anallatic *n.* 光学测远机
~ lens 视距镜
~ point 准距点,标准距离点
~ telescope 视距望远镜

anallobar [ə'næləbɑ:] *n.* 增压区,气压上升区,正变压线

analmatic *n.* 自动检查分析装置

analog ['ænələɡ] = analogue *n.* 模拟系数[装置,设备,计算机];相似物,类似物
~ computer 模拟计算机
~ data 模拟资料[数据]
~ -digital converter 模拟-数字变换器
~ distributor 模拟量分配器
~ result 模拟(试验)结果
~ signal 模拟信号
~ slope detector 模拟斜率检测器
~ technique 模拟技术
~ type 模拟型
electrical ~ 电气模拟装置
structural ~ 结构模拟

analogous [ə'næləɡəs] *a.* 类似的,相似的; analogously, *ad.* analogousness, *n.*
~ column 比拟柱

analog-to-digital converter 模拟-数字转换器

analogy [ə'nælədʒi] *n. pl.* -gies;模拟;相似,类似[比]
~ computer 模拟计算机
~ data 模拟数据
~ equipment 模拟设备
~ method 模拟法
column ~ 柱比法
direct ~ 正相似
stress-current ~ 应力-电流模拟
stress-voltage ~ 应力-电压模拟
stretch ~ 张力[延伸]相似[模拟]
pressure gradient ~ 压力梯度相似
Reynolds ~ 雷诺相似

analyse ['ænəlaiz] = analyze **vt.**, -lysed, -lysing; 分[解]析,(分析)研究
analysis [ə'næləsis] **n. pl.** -ses ;分析, 验定,解析;概要,纲领
~ by finite differences 有限差分法
~ by measure 量测分析
~ by titration 滴定(分析)法
~ life 分析年限(在经济分析中,对各项财产、设施所设计的服务年限)
~ of elasticity 弹性分析
~ of existing circumstance [conditions] 现状分析
~ of prices 造价分析
~ of strain 应变分析
~ of variance 离散[方差]分析
acoustic ~ 声学分析
applied ~ 应用分析
approximate ~ 近似分析
broken-line ~ 折线分析法
circuit ~ 线路分析,网络理论
colorimetric ~ 比色分析
confluence ~ 合流分析
contour ~ 外形[轮廓]分析
diffusion ~ 扩散分析
dimensional ~ 量纲[因次]分析
dynamic ~ 动态[动力]分析
error ~ 误差分析
factor ~ 因子分析
failure ~ 失效分析
fault ~ 故障分析
frequency-response ~ 频度特性法[分析]
grading ~ 粒度分析
grain-size ~ 颗粒组成[粒度]分析
graphic(al) ~ 图解分析
gravimetric(al) ~ 重量定量分析
harmonic ~ 调和分析
image ~ 图像分析
input-output ~ 输入输出分析,经济企业关联服务分析,投资生产分析
linear ~ 线性分析
logic ~ 逻辑分析
matrix ~ 矩阵分析[运算]
mesh ~ 筛分分析,筛分试验
network ~ 网络分析
photometric ~ 光度分析
precipitation ~ 沉淀分析(法),沉淀滴定(法)
proximate ~ 近似分析,组分分析
qualitative ~ 定性分析
quantitative ~ 定量分析
regression ~ 回归分析
screen ~ 筛分分析,筛分实验
sieve ~ 筛分分析[实验]
statistical ~ 统计分析
stress ~ 应力分析
system ~ 系统分析
target ~ 目标选择[分析]
tensor ~ 张量分析
titrimetric ~ 滴定分析(法)
track ~ 径迹分析
transient ~ 瞬态分析
vector ~ 矢量[向量]分析
volumetric ~ 容量[滴定,体积]分析(法)
weight ~ 重量分析

analyst ['ænəlist] **n.** 分析工作者,化验员
~ -programmer 程序分析员

analytic(al) [,ænə'litik(əl)] **n.** 分析的,解析的;analytically, **ad.**
~ balance 分析天平
~ data 分析数据[资料]
~ determination 分析测定
~ function 解析函数
~ geometry 解析几何学
~ precision 分析精确度
~ quantity 分析量,分解量
~ reagent 分析试剂
~ scale 分析天平
~ standard 分析标准
~ test 分析试验

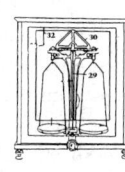

analytic(al) balance

~ unit 分析组件[单元,单位]
analyticity [ænə'li'tisiti] *n.* 解析性,分析性
analytics [ˌænə'litiks] *n.* 分析学,解析学,逻辑分析的方法
analyze ['ænəlaiz] *vt.*, -lyzed,-lyzing;分析,解析,研究
analyzer ['ænəlaizə] *n.* 分析器,测定器,分析员,检偏器,分析程序
digital differential ~ 数值积分器
pulse ~ 脉冲分析仪
set ~ 接收机检验器
sonic ~ 声波探伤仪
anamorphosis [ˌænə'mɔːfəsis] *n.* 变形,失真,畸形

anamorphosis

anastomose [ə'mæstəməuz] *v.* -mosed,-mosing;使吻合,接合,交叉合流
ancestor ['ænsistə] *n.* 祖先,起源,雏形,最初效应[现象]
anchor ['ɔŋkə] *vt.* 加固;系住;锚固,抛锚;*n.* 锚栓,地锚,地脚螺丝;拉杆[桩,柱],锚杆支撑;电枢;街铁;簧片;anchorable,*a.*;anchorlike,*a.*
~ agitator 杆柱[拉柱,固式]搅拌器
~ and collar (水轮机)导向轴承
~ angle 锚固角度,拉杆角度
~ arim 锚臂
~ bar 锚筋[杆]
~ beam 锚梁
~ behavior 固结状态
~ block 管道固定支座,锚墩[块]
~ bolt 地脚螺丝;锚栓
~ brick 有钉钉的木砖,受钉块
~ bulkhead 锚定(的)挡墙,锚定岸臂
~ buoy 标示下锚点的浮筒
~ channel 锚固槽钢
~ capstan 起锚车[铰盘]
~ cell 锚箱
~ chain 锚链
~ chock 锚楔
~ clamp 拉线夹
~ coat 结合层,打底胶浆初层
~ column 锚柱,竖旋桥支柱
~ dart 锚形箭头饰
~ deformation 锚固变形
~ ear 柱环
~ effect 固着效果
~ escapement 锚式擒纵机构
~ force 锚固拉力
~ gate 锚式闸门
~ -ground 锚地,泊碇地
~ grout 锚固灌浆
~ hinge 锚栓[拉杆]铰链
~ hold 抓牢,紧握
~ hole 锚栓孔
~ lift 抓钩提升器
~ line 锚索[链]
~ link roller 锚链转盘
~ log 锚桩,锚定件
~ loop 锚环[孔]
~ mixer 锚式搅拌器
~ nut 地脚螺母
~ opening 锚孔
~ paddle mixer 锚浆式搅拌器
~ picket 锚桩
~ pier 锚墩
~ pile 锚桩
~ pin 锚销,连接固定销
~ plate 锚定板
~ point 锚固[拉紧]点
~ post 锚桩,固定支柱
~ pressure (预应力筋)锚固压力
~ pull 锚固拉力
~ ring 锚定环
~ rod (钢板桩的)锚杆
~ rope 锚索[链],拉绳
~ screw 锚固螺栓,基础螺栓
~ sheeting 锚定式板桩墙
~ slip 锚固滑移
~ slot 锚缝[槽]

~ span 锚跨
~ stake 锚橛[桩]
~ stay 锚索
~ stone 固定石,底砾
~ strap 锚(系)板
~ strut 拉桩[线]支栓
~ support 分段支撑物
~ system 锚定系统
~ tie 锚拉杆,锚[定拉]条
~ tower 锚定[高架]塔
~ trumpet 锚定喇叭管
~ wall 锚墙
~ winch 锚杆收紧器
~ wire 锚缆,兜索,带索,挂钩短索,桩线
~ yoke 锚轭,锚索环
~ed bearing 锚固支承座
~ed cable 锚定缆索
~ed end (预应力混凝土)锚固端
~ed pretensioning (预应力混凝土的)锚定先张法
~ed suspension bridge 锚定式悬索桥

anchored suspension bridge

~ing block 锚定桩,拉杆锚桩
~ing bond (预应力钢丝端部)锚固粘结
~ing by friction 摩阻锚固
~ing element 锚定[固]构件
~ing failure 锚固破坏
~ing force 锚固力
~ing foundation 锚固基础
~ing jack 锚定桩
~ing length 锚固长度
~ing material 结合剂,增粘剂
~ing method 锚固方法
~ing picket 锚柱,路线标桩
~ing pipe 地脚螺栓套管
~ing rail 锚固槽钢
~ing strength 锚接,[锚定结合]强度

~ing system 锚固系统
~ing tube 地脚[基础]螺丝套管
anchorage ['æŋkərɪdʒ] *n.* 锚定[固],锚着[定],固定支座
~ abutment 锚定支座[斜撑,支墩,扶垛]
~ basin 停泊处
~ beam 锚梁
~ bearing 锚座
~ block 锚块
~ by friction 摩阻固定
~ chain 锚链
~ chamber [吊桥]锚定室
~ deformation 锚定变形[滑动]
~ device 预应力混凝土锚具
~ distance 锚固长度
~ element 锚定构件
~ foundation 锚固基础
~ length 锚固长度
~ loss 锚固损失
~ mast 锚(定)柱
~ picket 路线标桩,锚桩
~ rope 锚索[链]
~ shaft (吊桥)锚固竖井
~ shoe 锚座,锚靴
~ slip 锚固变形[滑移]
~ steel 锚钢
~ stone 底砾
~ stress 锚固应力
~ system 锚固系统
~ tower 锚定塔
ancient ['eɪnʃənt] *a.* 古代的,上古的,旧的 *n.* 老人;**anciently**, *ad.*
~ architecture 古代建筑
~ art 古代艺术
~ capital 古都
~ character 古代[老式]特色
~ city 古城
~ decoration 古式装饰
~ furniture 古式家具
~ geological gorge 老[古]河床
~ light 古式窗

ancient furniture

ancient meteorolgical observatory

~ meteorolgical observatory 古代气象台
~ monument 古迹,遗址
~ object 古董
~ painted pottery 古代彩陶
~ picture 古画
~ palace 古宫
~ relics 古迹,古代遗物
~ rent 过去租金
~ Roman 古罗马
~ sediment 古迹

ancientry [ˈeinʃəntri] *n.* 古风,古代风格[艺术品];远古
ancillary [ænˈsiləri] *a.* 辅助的,附属的; *n.* 辅助设备,助手
~ attachment 特殊附件
~ block 辅助部件
~ building 辅助建筑物
~ equipment 辅助装置
~ measurement 辅助量测(温度,气压)
~ shoring 辅助支撑
~ work 辅助工作[工程]

ancon [ˈæŋkɔn] *n. pl.* ancones;肘托,悬臂托梁,卷涡形托梁,肘状支柱

-anconeal, *a.*
andalusite [ˌændəljuˈsait] *n.* 红柱石
~ gneiss 红柱石片麻岩
Andaman padauk 安达曼紫檀木材
andesine [ˈændəsin] *n.* 中长石
andesite [ˈændizait] *n.* 安山石
~ ash 安山石灰
~ (paving)sett 铺路安山石毛石
~ tuff 安山凝灰岩
andiron [ˈændaiən] *n.* 炉壁内柴架
andron [ˈændrɔn] *n.* (古希腊罗马的)男用房间(宴会厅),集会厅,(古罗马)两庭中间的走道
androsphinx [ˈændrəsfiŋks] *n. pl.* -sphinxes, -sphinges;男性人头狮身像
anechoic [æˈnekəuik] *a.* 无回声的,消声的
~ chamber 消声室,吸声室
~ hall 无回声厅
~ paint 吸声漆
~ room 消声室
~ trap 消声槽
anelasticity [ˌænilæsˈtisiti] *n.* 内摩擦力;滞弹性
anelectric [æniˈlektrik] *n.* 非电化体,不能因摩擦起电的物体
anemograph [əˈneməgræf] *n.* 风速计,风力计 anemographic, *a.*
anemometer [ˌæniˈmɔmitə] *n.* 风速表,气流计
anemoscope [əˈneməskəup] *n.* 风向仪,风信仪
anemostat *n.* (通风系统的)稳流管,扩散管
anemovane *n.* (接触式)风向风速器
aneroid [ˈænərɔid] *n.*; *a.* 无液晴雨表;无液晴雨表
~ barometer 膜盒气压表
~ battery 干电池
~ chamber 气压计盒
~ manometer 无液压力计
anfractuosity [ænˌfræktjuˈɔsiti] *n. pl.*

-ties;弯曲(沟渠,路,河);曲折,错综
angel ['eindʒə] *n.* 天使[神]
　evil(or fallen) ～ 魔鬼
　good(or guardian) ～ 吉神;守护神
　potrait of ～ 天使像
Angkor vat [柬]吴哥寺,吴哥窟
angle ['æŋgl] *n.*
角,角度;角[铁];
观点; *a.* 角的;
v. 从角度移动;
钓鱼

AngKor Vat

～ abrader 角式磨耗实验机
～ at center 中心角
～ -balancing method 转角平衡法
～ bar 角铁[钢]
～ bar joint 角铁板连接
～ bead 角条,护角
～ bead tile fitting 半圆形护角水泥板[瓷砖]
～ beam 角钢梁,对角支撑(杆),隔梁
～ beam probe 斜探头
～ beam searching unit 斜探头
～ bend 护角条,角形弯管[接头]
～ bender 钢筋弯折机,弯管机
～ bending machine 钢筋弯折机
～ blade 角形平铲
～ block 基石,屋角石块;角铁,弯板;(家具等)内角加强用三角木条
～ board (木板墙)斜薄板
～ bond 墙勒脚外角砌合
～ brace 角(铁)撑,水平隔撑,角隅斜梁
～ bracing 角铁撑
～ bracket 斜托座,角铁托,角形托座
～ branch 肘管,弯(支)管
～ brick 角砖
～ bulbiron 球头角钢
～ butt strap 斜口平接板
～ buttress 转角扶垛,转角扶壁
～ capital 角柱柱头
～ catch 窗插销

～ centrifuge 斜角离心机
～ change 角变位
～ chapel 角形小教堂
～ check valve 直角单向[止回]阀
～ chimney 转角烟囱
～ clamp 角夹
～ cleat 连接角钢(梁柱式主次钢梁连接用)
～ clip 角铁系
～ closer 镶边砖石,墙角半砖
～ cock 直角旋塞
～ column 角柱
～ crane 斜座起重机,三角架起重机
～ -cross-ties 角钢横系杆
～ -cut (接头处的)斜切
～ cutter 角钢切割机,角铣刀
～ displacement 角位移
～ dozer 斜板推土机,侧铲推土机
～ drill 角钻
～ drive grinder 斜向传动碾磨机
～ dunting 石料粗加工
～ factor 角系数
～ file 三角锉刀
～ fillet (节点)三角形盖板,三角焊缝
～ fire place 角形壁炉
～ fishplate 角铁接合[直角型鱼尾板]
～ float 直角镘刀
～ gauge 角规,量角器;倾角计,角度计
～ gate valve 弯头闸阀,角闸阀
～ grinder 直角研磨机
～ guard 三角保护(器)
～ guide 角形导架
～ guide pin 定角位销
～ head 弯头,角条
～ hinge 角[斜]撑,角铰
～ hip tile 角形坡脊瓦
～ indicator 转角指示器
～ iron 角钢,三角铁

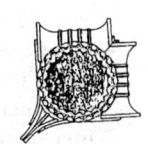
angle capital

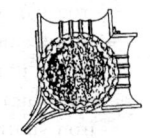
angle float

~ iron bar 角钢条
~ iron frame 角钢构架[框架]
~ (iron) purlinc(e) 角钢檩条
~ iron rotor 角钢转子
~ iron smith 角钢锻工
~ iron stiffening 角钢支撑
~ (iron) with equal legs 等边角钢
~ joint 角接头,隅接
~ lacing 角铁联系
~ lap 磨角
~ -lever 斜杠杆
~ lintel (门窗)角形过梁
~ load 角荷载
~ loads method 角荷载法
~ measurement 测角,角度测量
~ measuring equipment 角测量装置
~ meter 测角器,倾斜计
~ mirror 直角镜
~ modillion 檐托饰
~ moding press 直角形平板机
~ (moulding) press 压角机
~ of advance 提前[超前]角
~ of alternation 偏离角
~ of altitude 仰角
~ of anterior chamber 前房角
~ of ascent 螺旋角,上升角
~ of attack 冲[迎]角,作用角
~ of avertence 偏角
~ of bank 倾斜角
~ of beding 垫层倾角
~ of bite 挟角
~ of boom swing 吊机臂旋转角
~ of break 破坏[断裂,崩裂]角
~ of chamfer 坡口斜角,凹线角
~ of chord 弦角
~ of contact 接触角,包角
~ of contingence 切线角
~ of convergency (水流)收敛角
~ of crater 自然倾角

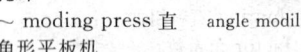
angle modillion

~ of current 水流角
~ of curvature 曲率角,曲度角
~ of cutting 切割[开挖]角
~ of declination 偏转角
~ of deflection 偏转[变位]角
~ of deformation 变形角
~ of delay 滞后角
~ of depression 俯角
~ of deviation 偏转角
~ of diffraction 绕射角,衍射角
~ of dispersion 散射角
~ of draw 牵引角
~ of dump 倾卸角
~ of elevation 仰角
~ of emergence 出射角
~ of extinction 消光角
~ of fill slope 路堤[填上]边坡角度
~ of flange 边缘弯角
~ of flare 承口角(管口扩张角),扩散角
~ of flexure 屈挠角度
~ of friction 摩擦角
~ of gradient 坡度[倾度]角
~ of heel 倾侧角
~ of helix 螺旋斜角
~ of import 攻[冲击]角
~ of incidence 入射(斜)角
~ of inclination 倾斜角,倾角
~ of isocline 等斜角,等倾角
~ of internal friction 内摩擦角
~ of jib swing 吊杆[起重臂]旋转角
~ of lag 滞后角
~ of lay 布线角度
~ of lead 超前角
~ of loosening (松动)开挖角
~ of oscillation 摆动角
~ of pitch 坡度角,螺距角,曲距角,俯仰角
~ of preparation 焊接坡口角度
~ of projection 投影[射]角
~ of pull 牵引角
~ of reflection 反射角

~ of refraction 折射角
~ of repose 静止角,自然倾斜角
~ of resistance 抗滑角
~ of rest 静止角,自然倾斜角
~ of roll 倾斜角
~ of roof (pitch) 屋面倾斜角
~ of rotation 转动[旋转]角
~ of rupture 破裂角
~ of shear 切变角,剪切角
~ of shift 位移角
~ of sides (孔型)侧壁斜角
~ of sight 视角,俯仰角
~ of skew-back 拱脚,斜石块倾斜角
~ of slide 滑动角
~ of sling 吊索与滑轮接触角
~ of spread 扩散角,展开角
~ of strike 岩层走向方位角
~ of sweep back 后掠角
~ of taper 斜削角
~ of tear 撕裂角
~ of throat 入口张(开)角
~ of tilt 倾角
~ of toe-in 内倾角
~ of torsion 扭转角
~ of traction 牵引角
~ of turn 转弯角度
~ of twist 扭转角
~ of vee 坡口角度
~ of view 视场角
~ of visible sky 可见天空角
~ of visual field 视场[视野]角
~ of wall friction (土力学)壁面摩掺角
~ of wedge 楔角
~ parking 倾斜式停车场;斜向停车
~ pavilion 耳房[侧房,厢房],休息室[更衣室]
~ pipe 弯管
~ plane 角刨,斜刨法
~ plate 角型板
~ plug 弯曲插头
~ polisher (polishing machine) 角抛(磨)光机

~ post 角钢支柱
~ press 角压线
~ protractor 量(斜,分)角规
~ pulley 换向滑轮
~ quoin 楔块,楔形石,屋角石块;突[屋]角
~ rafter 角椽,角钢椽条,斜脊椽
~ rib 角肋
~ ridge 角椽,角钢椽条
~ roller 角度矫正机
~ roof truss 三角形屋顶桁架
~ rotation 旋转角
~ seat 角钢支座
~ section 角材;角形断面

angle roof truss

~ separator (钢梁)角形连接构件
~ shear 角钢剪切机
~ sheet iron 角状扁铁
~ site 铅矾,硫酸铜矿
~ splice bar 角形鱼尾板,制作连接板用的异型钢材
~ splice joint 角形拼合接头,角钢拼接
~ splice plate 角形鱼尾板,角钢连接角
~ staff 护角线,角形支柱,角饰
~ staple 角扒钉,角肘钉
~ station (索道)弯道站
~ steel 角钢,角铁
~ steel purlinc(e) 角钢檩条
~ steel (section) 角钢,角材
~ stiffening (对)角支撑[加固]
~ stone 角石
~ stop 角钢挡板[制动装置]
~ strut 角撑,角铁[角材]支柱
~ -table 牛腿托座,角撑架,承托
~ tee 分路,分支
~ thermometer 直角[L型]温度计
~ threshold 角钢门槛
~ tie 角钢拉杆[柱杆],角撑,(加固用)隔板

~ tile 屋角石块[石砖],基石
~ tower 角楼,角塔
~ truss 三角形桁架
~ turret 角楼
~ type joint bar 角形鱼尾板,L形连接板
~ valve 角阀,直角形阀门
~ washer 斜垫圈,楔形垫片
~ web 斜腹板[横挡,隔板]
~ welding 角焊,L形焊接,贴角焊接
~ with sharp corners 锐边角钢[角型材]
acute ~ 锐角
adjacent ~ 邻角
adjusted ~ 平差角
advance ~ 提前[超前]角
alterate ~ (交)错角
aperture ~ 孔径角
apex ~ 顶角
apsidal ~ 横心角
auxiliar ~ 补角
azimuth ~ 方位角
back ~ 后视角
bank ~ 倾斜角,拱倾角
bearing ~ 方位角
bevel ~ 斜面角
bipartition ~ 对分角
blade ~ (水轮机)叶片安装角
brace ~ 撑杆角铁
bracket ~ 托架角铁
bulb ~ 圆头[球缘]角钢
butting ~ 平接角钢
central ~ 圆心[中心]角
chamfer ~ 倒棱角
chord ~ 翼弦安放角
clip ~ 扣角钢
complementary ~ 补角
connecting ~ 连接角钢
connection ~ 连接角钢
contact ~ 接触角

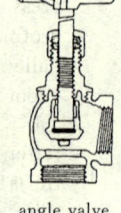

angle valve

conversion ~ 转换角
coordinate ~ 坐标角
correction ~ 修正角
corresponding ~ 对应角,同位角
critical ~ 临界角
cut-off ~ 截止角
defleetion ~ 变位角,偏移角
deflector ~ 偏导角
dip ~ 倾斜角,倾角
discharge ~ 出流角
displacement ~ 位移角,失配角
divergence ~ 发散角
eccentric ~ 离心角,偏心角
elevating ~ 竖角,垂直角
end lip ~ 楔角
entrance ~ 入口角
equal ~ 等边角钢
exterior ~ 外角
external ~ 外角
fastening ~ 联系角铁
flange ~ 翼缘角钢,凸缘角钢
ground ~ 地面角
gusset ~ 扣角钢,结点角钢
half ~ 半角
heeling ~ 横倾角,侧倾角
horizontal ~ 水平角
included ~ 夹角,焊缝角
inflow ~ 流入角
injection ~ 喷射角
interior ~ 内角
internal ~ 内角
intersection ~ 交叉角
joint ~ 连接角钢
lug ~ 辅角钢(节点板上的短角钢)
mitre ~ 45°角
oblique ~ 偏斜角
obtuse ~ 钝角
offset ~ 偏斜角
opposite ~ 对顶角
optical ~ 视角
parallactic ~ 视差角
pipe ~ 管子弯头,管子箍

pitch ~ 俯仰角
plane ~ 平面角
plunge ~ 俯角,降角
rectilinear ~ 直线角
reducer ~ 直角异径弯管
reference ~ 基准方位角
refraction ~ 折射角
right ~ 直角
rolled ~ (轧制)角钢
sash ~ 窗框角铁
scattering ~ 散射角
seat ~ 座角钢
sharp ~ 锐角
shielding ~ 遮光角
skew ~ 斜拱角,相交角
sliding ~ 滑移[滑动]角
slope ~ 坡度角
straight ~ 平角
stiffening ~ 加劲角铁
striking ~ 入射角
subtended ~ 包角
tilt ~ 倾角
unequal ~ 不等边角钢
valve seat ~ 阀座斜角
vertical ~ 顶角
visual ~ 视角
angles back to back 背靠背组合的角钢
Anglo [ˈæŋɡləu] *a.* 英国的
 ~ classic architecture 英国古典建筑
 ~ -classic style 英国古典式
 ~ -Norman architecture 安格鲁-诺尔曼式建筑
 ~ -Palladian architecture 安格鲁-巴拉迪欧式建筑
 ~ -Saxon architecture 安格鲁-撒克逊式建筑
 ~ -Saxon masonry (work) 安格鲁-撒克逊式砖石建筑
 ~ Tudor architecture 英国都铎式建筑

Anglo Tutor architecture

angola copal (制漆用)安哥拉柯巴脂(一种制漆原料)
angstrom [ˈæŋstrəm] *n.* 埃(Å),长度单位(=10^{-19}米)
anguclast [ˈæŋɡjukləst] *n.* 角碎屑
angular [ˈæŋɡjulə] *a.* 角(形,状)的,有棱的;倾斜的,斜角的
 ~ acceleration 角加速度
 ~ aggregate 有棱角混凝土骨料
 ~ aperture 斜角孔,开角
 ~ bearing 向心止推轴承
 ~ bevel gear 斜交伞齿轮
 ~ bitstock 弯把手摇钻
 ~ capital 角(锥)形柱头
 ~ column 角形柱
 ~ contact 斜接,斜角连接
 ~ coordinate 角坐标
 ~ crack (露天开挖)梯段裂缝
 ~ deflection 角挠度,角变形
 ~ deformation 角变形
 ~ degree 角度
 ~ deviation 角偏移[偏转]
 ~ displacement 角位移
 ~ distance 角距离
 ~ distortion 角变形
 ~ field 视场[野,界]
 ~ fish plate 角口接板,转角鱼尾板
 ~ force 角向力,偏向力
 ~ fracture 斜面断口
 ~ fragment 棱形碎屑,角形碎片[石片]
 ~ gap 角形裂隙
 ~ grain 有棱角颗粒,尖角颗粒

angular bitstock

Anglo-Saxon architecture

~ height 高低角
~ instrument 测角仪
~ measure 角规,角尺
~ minute 角分(=1/60度)
~ misalignment 管子接偏
~ momentum 角(转)动量,动量矩
~ motion 角运动
~ movement 角运动
~ oscillation 角摆动
~ particle 有棱角颗[砂,石]粒
~ perspective 斜透视
~ pipe union 弯管接头
~ pitch 角矩,倾斜角
~ point 角顶点尖顶
~ retaining wall 倾斜式挡土墙
~ rotation 角转动[位移]
~ sand 棱角状砂,角粒砂
~ second 角秒(=1/60角分)
~ strain 角应变
~ surface 斜面
~ table (工具机上的)三角桌
~ thread 三角螺纹
~ uncomformity 角度不整合
~ velocity 角速度
~ vertex 角顶点

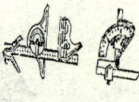
angular measure

angular pipe union

angularity [æŋgju'læriti] *n.* 棱角;曲线[率];弯曲角;倾度,倾斜角;翘曲度
~ correction 角修正
~ index (集料的)棱角指数
anharmonic [,ænhɑ:'mɔnik] *a.* 非(简)谐的,非调和的
~ ratio 非谐比,非调和比
anhedritite [æn'hi:drtait] *n.* 硬石膏
anhydrate [æn'haidreit] *vt.* 去水,脱水
anhydric [æn'haidrik] *a.* 无[脱]水的,无结晶水的

anhydride [æn'haidraid] *n.* 无水物,酐
anhydrite [æn'haidrait] *n.* 无水石膏;硬石膏($CaSO_4$)
~ hand 无水带,硬石膏带
~ binder 硬石膏粘结材料
~ block 硬石膏砌块
~ board 硬石膏板
~ cement 硬石膏水泥
~ jointless floor(ing) 硬石膏无缝地板
~ lime mortar 无水石膏石灰砂浆
~ plaster (用硬石膏与催化剂制成的)硬石膏灰泥(灰浆粉刷)
~ scred 硬石膏灰浆泥抹[刮板,冲筋],硬石膏无缝地板[面层]
~ sheet 无水石膏板
~ tile 无水石膏(砌)块
~ tile partition (wall) 无水石膏砌块隔墙
anhydrous [æn'haidrəs] *a.* 无水的
~ calcium sulphate plaster 烧石膏灰浆
~ gypsum 无水石膏
~ (gypsum)plaster 无水[烧]石膏灰浆,无水石膏粉刷
~ lime 无水石灰,烧石灰
~ material 无水材料
~ sodium carbonate 无水碳酸钠
~ sulphate of calcium 无水硫酸钙
anhydrox *n.* 抗石膏(污染的泥浆处理剂)
anhyetism [æn'haitizm] *n.* 缺雨区,缺雨性
anicut (=annicut) *n.* 堰
aniline ['ænili:n] *n.* 苯胺,苯胺染料,阿尼林
~ black 苯胺[颜料黑]
~ colors 苯胺色素
~ dye 苯染料
~ finish 苯胺涂饰剂
~ nitrate 硝酸苯胺
~ point 苯胺点,苯胺溶液临界温度
~ purple(=mauvein) 苯胺紫

~ red(=fuchsin) 苯胺红
~ resin 苯胺树脂
~ violet 苯胺紫
~ yellow 苯胺黄

anilioplast [æˈniliəplæst] *n.* 苯胺塑料

animal [ˈæniməl] *n.* 动物,兽; *a.* 动物的,兽的
~ adhesive 动物胶
~ charcoal 兽炭,骨炭
~ distemper 动物色胶[刷墙粉]
~ drying oil （油漆用）动物干燥油
~ flower 形似动物之花
~ glue 动物胶
~ hair felt 兽毛毡
~ oil 动物油
~ ornament 动物装饰品
~ parchment 兽皮纸
~ pigment 动物颜料
~ relief 动物浮雕
~ resin 动物树脂
~ sculpture 动物雕塑
~ shelter 厩,圈,棚
~ wax 动物蜡

animate [ˈænimeit] *vt.*, -maed, -amting; 使逼真,使有生气; *a.* 有生命的; animately, *ad.*; animateness; *n.*
~ cartoon 卡通电影

animation [ˌæniˈmeiʃn] *n.* （建筑艺术）生动,活泼

anion [ˈænaiən] *n.* 阴离子
~ exchange 阴离子交换
~ exchange resin 阴离子交换树脂
~ weathering 葱状风化

anionic [ˌænaiˈɔmik] *a.* 阴[负]离子的
~ asphalt emulsion 阴离子（地）沥青乳液
~ dye 阴离子染料
~ emulsifier 阴离子乳化剂
~ emulsion 阴离子乳液
~ permeable membrane 阴离子透膜

anisobaric [ˌænaisəuˈbærik] *a.* 不等压的

anisoelastic [ˌænaisəuiˈlæstik] *a.* 非(等)弹性的

anisoelasticity *n.* 非等弹性

anisometric [ˌænˌaisəˈmetrik] *a.* 不等轴的; 不等角的

anisotonic [ˌænaisəˈtɔnik] *a.* 不等渗的,异渗的

anisotropic [ˌænaisəˈtrɔpik] *a.* 各向异性的,非均质的; anisotropy, *n.*
~ consolidation 各向异性固结
~ elastic body 各向异性弹性体
~ fluid 各向异性流体
~ force system 各向异性力系
~ material 各向异性材料
~ media 各向异性介质
~ membrance 各向异性膜
~ metal 非均质金属
~ plate 各向异性板
~ shell 各向异性层
~ stress 各向异性[不均]应力

anisotropism *n.* 各向异性,非均质性

anisotropy [ˌænaisəˈtrɔpi] *n.* 各向异性现象

ankle [ˈæŋkl] *n.* 踝
~ boot 中筒靴,高帮鞋
~ -deep mud 深及踝部的泥泞

anneal [əˈniːl] *vt.* 退火,焙烧,加热缓冷（热处理）
~ed aluminum 软铝
~ed aluminum wire 软铝线
~ed cast-iron 韧(性)铸铁
~ed casting 退火铸件
~ed copper wire 软铜丝
~ed steel 退火钢,韧钢
~ed tensile strength 退火后的拉力强度

annealing [əˈniːliŋ] *n.* （低温）退火,热处理,加热缓冷; 退火油
~ color 烧色,退火色
~ crack 退火裂
~ oil 退火油

annex(e) [əˈneks] *n.* 附加建筑,附属建筑物[房屋],增建建筑,边[群]房,配

房；*vt.* 合并，附加
~ed table 附表
~ed triangulation net 附连三角网
annexation [æneˈseiʃən] *n.* 连接关系，附加(物)；合并，归并
annihilator [əˈnaihileitəl] *n.* 灭火器，减振器，消灭器
fire ~ 灭火器
annotated [ˈænəuteitid] *a.* 有附注的，有说明的
~ bibliography 附有说明的资料目录
~ list of items 附有说明的项目表
annotation [ˌænəuˈteiʃən] *n.* 注解[释]
announce [əˈnauns] *vt.* -nounced, -nouncing, 正式宣告，发布
announcer [əˈnaunsə] *n.* 宣布，公告；广播员
~'s booth 广播室，播音员室
annual [ˈænjuəl] *a.* 每年的，年年的，一年的；*n.* 年报[刊]，年；年金，年租
~ accounts 年度决算
~ appropriation 年度拨款
~ artificial lake 年调节水库
~ audit 年度审计
~ average daily traffic 年平均交通量
~ average sediment yields 年平均沉积量
~ balance(sheet) 年终决算书(表)
~ basis 年度基准
~ borrowing plan 年度贷款计划
~ budget 年度预算
~ closing 年终结算
~ cost 年费用[成本]，常年费用
~ consumption 年消耗量
~ data 年度资料
~ depreciation 年折旧
~ discharge 年径流(量)
~ efficiency 年效率

annihilator fire

~ financial statement 年度决算书
~ flood 年洪水量，年最大流量
~ flow 年流量，年径流量
~ frost zone 年冻层
~ growth rate 年增长率
~ holding cost 年储存成本
~ improvement factor 年增长系数
~ increment 年增量
~ inequality 年差异[变化]
~ inspection 年度检查，年度检修
~ leave 每年一次的休假
~ load 年负载曲线
~ load factor 年平均负荷因素
~ loss 年度亏损
~ maintenance 年度保养维修[养护]
~ mean daily flow 年平均日(车)流量
~ mean water discharge 年平均(水)流量
~ meeting 年会
~ objectives 年度目标
~ output 年产量
~ overhaul 年度检修
~ pay 年薪
~ precipitation 年降水[雨]量
~ premium 年保险费
~ production 年产量
~ rainfall 年(降)雨量
~ rate method 年率法
~ report 年度报告，决算书
~ ring (of timber) (木材)年轮
~ ring density 年轮密度
~ (storage) reservoir 年调节水库
~ sale volume 年营业额
~ statement 年度决算书
~ target 年度指标
~ traffic 年运输，年交通量
~ upkeep 年度保养维修
~ working program 年度工作计划
~ yield 年产量

annual rings

~ zone（树木）年轮
statistic ~ 统计年鉴
annuity [ə'nju(:)iti] *n. pl.* -ties 年积金，养老金
annul [ə'nʌl] *vt.*，-nulled,-nulling 取消，注销，宣告作废[无效]
annular ['ænjulə] *a.* 环形的，环的，轮状的；-annularly, *ad.*
~ arch 环形拱
~ bearing 环形轴承
~ bit 环形钻头
~ burner 环状喷灯
~ contact 环形接触
~ cross-section 环形断面[截面]
~ dam 挡水墙，胸墙
~ disk 环形盘
~ distange 环孔
~ drainage 环形排水系统
~ fin 环形散热器
~ flow 环形流
~ gasket 环形密封垫，环形衬片
~ girde 环形梁
~ grid 环形格栅
~ groove 环形槽
~ -jet nozzle 环状喷嘴
~ knurl 滚花
~ membrane 环形膜
~ of timber（木材）年轮
~ opening 圆孔
~ passage 环形通道，环形风道
~ piston valve 环形活塞阀
~ plate 环形板
~ producer 圆形发生炉
~ ring 环[圆]圈
~ ring valve 环形阀
~ saw 圆锯
~ section 圆断面
~ slab 环形板
~ space 环(形缝)隙
~ strain 环变形
~ tubes 套管
~ valve 针形[针舌]阀

~ vault 环形[椭圆]穹顶[拱顶]，筒[圆]形穹顶
~ wheel 内齿轮
~ zone 环形区

annular vault

annularity [,ænju'læriti] *n.* 环状[形]
annulate ['ænjuleit] *a.* 有环的，有环纹的；*n.* 环状构造
annulated ['ænjuleitid] *a.* 有环的，有环纹的，有环组成的
~ bit 环形钻头
~ column 环柱
~ shaft（有环纹的）环饰柱
annulation [,ænju'leiʃən] *n.* 环状结构，环状物，环纹
annule [ə'nʌl] *n.* 环带
annulet ['ænjulit] *n.* 柱环饰（柱身上的环状线脚）

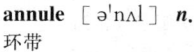
annulets

annulment [ə'nʌlmənt] *n.* 废止，取消，注销
annulus ['ænjuləs] *n. pl.* -li ; 环带，环（形缝）隙
graphite ~ 石墨环
water ~ 环形水道
annunciation *n.* 布告，通知，公布
annunciator [ə'nʌnsieitə] *n.* 信号装置，电铃号码箱，呼唤器
alarm ~ 事故指示装置，警报信号器
drop ~ 掉牌通报器
anodal [ə'nɔdəl] *a.* 阳极的
anode ['ænəud] *n.* 阳[正]极，极板，氧化极
~ casting machine 阳极电镀机[装置]
~ chamber 阳极室
~ compartment 阳极室
~ copper 阳极铜
~ current 阳极电流
~ drop 阳极（电势）降

~ metal 阳极金属
~ plate charging machine 阳极板充电机
anodic [æˈnɔdik] *a.* 阳极的
~ cleaning 阳极清洗
~ coating 阳极氧化镀层
~ current density 阳极电流密度
~ deposition 阳极沉积
~ film 阳极(氧化)镀层
~ finish 阳极(氧化)镀层
~ oxidation 阳极氧化
~ polishing 阳极(电)抛光
~ protection 阳极保护
~ slime 阳极沉渣
~ treatment method 阳极处理法
anodise = anodize [ˈænədaiz] *v.* 阳极氧化[电镀,防腐]
~d aluminium 阳极极化铝
~d finish 阳极化抛光
anolyte [ˈænəˌlait] *n.* 阳极电解液
anorthite [æˈnɔːθait] *n.* 钙长石
~ diorite 钙长石闪绿岩
anorthosite [æˈnɔːθəsait] *n.* 斜长岩
anoxybiotic [əˌnɔkzaiˈbaiətik] *a.* 绝氧[缺氧]的
ANSI = American Natconal standards Institute 美国国家标准协会
antacid [ˈæntˈæsid] *n.* ; *a.* 抗酸剂,中和酸的
anta(e) *n.* (墙砌出部分的)壁角柱,壁柱,(古希腊神殿)墙端柱
~ capital 壁角柱柱头
antaefixae [ænˈtiːfiksiː] *n.* 瓦檐饰
antagonism [ænˈtægənizəm] *n.* 对抗[拮抗,消效]作用
antagonistic(al) [ænˌtægəˈnistik(əl)] *a.* 对抗的,相反的
antalkali [æntˈælkəlai] *n.* 解碱剂,中和碱的
antecedent [ˌæntiˈsiːdənt] *n.*

anta(e)

前项,前事; *a.* 以前的,先成的;先行的,前期的
~ engineering 前期工程
~ moisture 降水[雨]前土壤的湿度,前期水分
~ -precipitation index 前期降水指数
~ rainfall 前期降雨
~ river 先成(的)河
~ stream 先成河
antechamber [ˈæntiˌtʃeimbə] *n.* 前堂,前厅;接待室,预燃室
~ with built-in wardrobe 带衣帽间的前厅[接待室]
antechapel [ˈæntiˌtʃæpəl] *n.* 教堂前厅[门厅]
antecourt [ˈæntikɔːt] *n.* 前庭
antefix(tile) [ˈæntifiks] *n.* (古希腊和古罗马建筑的)瓦当,瓦檐饰
antehall [ˈæntihɔːl] *n.* 前厅
anteklise [ˈæntiklaiz] *n.* 台拱,台背斜
antenna [ænˈtenə] *n.* 天线; *a.* 天线的
~ gallery 天线平台
~ installation 天线装置
~ mast 天线杆[塔]
~ socket 天线插座
all-wave ~ 全波天线
anti-interference ~ 抗干扰天线
antistatic ~ 抗静电干扰天线
balanced ~ 对称[平衡]天线
ball ~ 球形天线
barrel ~ 桶形天线
beamed ~ 定向天线
bilateral ~ 双向天线
block ~ 共用天线
closed ~ 闭路[环形]天线
combined ~ 共用天线
common ~ 共用天线
community ~ 共用天线
coat-top ~ 平顶天线
ground ~ 地面天线
hoop ~ 圆柱形天线
indoor[inside,internal] ~ 室内天线

open ～ 户外天线
pylon ～ 铁塔天线
roof ～ 屋顶天线
room ～ 室内天线
tower(-type) ～ 铁塔[塔式]天线

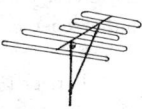
open antenna

antennafier *n.* 天线放大器
antennifer ['æntinifə] *n.* 支角突
antepagments [ˌænti'pægmənt] *n.* 加在房屋上的镶边饰
antependium [ˌænti'pendiəm] *n.* 祭台前的屏饰物[帷幕]
anteporch ['æntipɔːtʃ] *n.* 外门廊
anteport ['æntipɔːt] *n.* 外门[槛]
anteroom ['æntirum] *n.* 接待室,前厅,(通入正室的)较小外室
ante-solarium *n.* 朝阳阳台
antetemple *n.* (寺庙)前殿
antetheca *n.* 前壁
antetype ['æntitaip] *n.* 原型,前型
ante-venna *n.* 遮阳幕[板]
anthem *n.* 颂歌
national ～ 国歌
anthemion [æn'θiːmiən] *n.* (古希腊建筑)棕叶饰,花丛状装饰
～ ornament (古希腊建筑)棕叶装饰

anthemion

anthracite ['ænθrəsait] *n.* 无烟煤,硬[白]煤
～ coal 无烟煤
～ -sand-filter 煤砂双层滤料滤池
～ stove 无烟煤煤炉
artificial ～ 人造无煤烟
low rank ～ 低级无烟煤
pelletized ～ 无烟煤煤砖
anthraconite [æn'θrækənait] *n.* 黑沥青灰岩,黑方解石
anthraxolite *n.* 碳沥青
anthraxylon *n.* 纯木煤
antiager *n.* 抗老化剂
anti-abrasive *a.* 耐磨损的

anti-acid *a.* 抗酸的
antiaircraft [ˌænti'ɛəkrɑːft] *a.* 防空的
～ defence 防空工事
～ tower 对空观测台
anti-air-pollution system 防止空气污染系统
antiattrition *n.* 减(少)磨(损)
antibacklosh spring 消隙弹簧
antibacterial [ˌæntibæk'tiəriəl] *a.* 抗菌的
～ paint 抗菌涂料[漆]
anti-blushing [ˌænti'blʌʃiŋ] *n.;a.* 防变色,抗混浊
～ agent 防变色剂,抗混浊剂
anti-bomb ['ænti:bɔm] *a.* 防弹的
～ glass 防弹玻璃
anti-borer ['ænti'bɔːrə] *n.* 防虫蛀
～ carpet 防蛀地毯
～ plywood 防蛀胶合板
anti-breaker [ˌænti'breikə] *n.* 防碎装置
anti-bubbing agent 消泡剂
anti-chain *n.* 反链
anti-checking *a.* 防龟裂的
～ agent 防龟裂剂
～ iron 防裂钩,扒钉
antichlor *n.* 脱氯剂
anticipant [æn'tisipənt] *a.* 预期的;期望的. *n.* 预期者,先发制人者
anticipate [æn'tisipeit] *vt.* 预期,超前
～d acceptance 先期承兑
～d discount 预期折扣
～d load 预期负荷
～ price 预计价格
～d volume of traffic 预计交通量
anticlastic *a.* 鞍形面的
～ surface 鞍形面,互反曲面
anticlimber *n.* 防攀(登)金属网
anticlinal ['ænti'klainəl] *a.;n.* (地质上)背斜
～ axis 背斜轴
～ fissure 脊缝

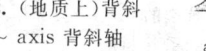
anticlinal axis

~ fold 山脊，鞍脊
~ ridge 背斜脊
~ valley 背斜壳
anticlockwise ['ænti'klɔkwaiz] *a.*；*ad.* 逆时针方向的，左旋的
~ rotation 逆时针旋转
anti-clogging agent 防阻塞剂，防结渣剂
anti-coagulant [ˌæntikəu'ægjuːlənt] *a.* 抗凝剂
~ action 阻凝作用
~ separator 防堵塞分离器
~ step 防堵塞措施
anti-collision ['æntikə'liʒən] *n.* 防撞
~ device 防撞装置
anti-condensation *n.* 防凝结水
~ ceiling 防凝水天花板
~ lining 防凝水内衬
~ paint 防凝水油漆
~ plaster 防凝水熟石膏[灰泥]
~ protective measure 抗凝水保护措施
anticorrosive [ˌæntikə'rəusiv] *a.* 防腐蚀的，防蚀的；*n.* 防腐剂
~ agent 防腐[锈]剂
~ blanket 防腐层[覆面层]
~ coat 防腐涂(料)层
~ coating 防腐涂饰，防蚀层
~ coating with oxides 氧化防腐膜
~ composition 防腐漆
~ foil 防腐箔
~ grout 防腐水泥灰浆
~ insulation 防腐绝缘层
~ mortar 防腐砂浆
~ paint 防腐漆，防锈漆[涂料]
~ pigment 防腐[锈]颜料
~ prime coat 防腐打底涂(料)层
~ primer 防腐打底涂料
~ slurry 防腐水泥灰浆
~ treatment 防腐蚀处理
~ valve 防腐蚀阀门
anticrack *a.* 抗裂的
~ coating 抗裂涂料层
~ reinforcement 抗裂钢筋

anticrease ['ænti'kriːs] *a.* 防皱的
anticreeper *n.* (钢轨)防爬器，防漏电装置
~ barrier 防爬障栅
~ device 轨道防爬设备
~ shield 防爬挡板
~ stake 防爬桩[橛]
anticrustator *n.* 防水锈剂，防垢剂，表面沉垢防止剂
anticum *n.* 壁(角)柱，(寺庙)门廊
anticyclone *n.* 反气旋，反旋风，反低气压
anti-dazzle *v.*；*n.* 防眩
~ glass 防眩玻璃
~ lighting 防眩灯光
~ -screen 防眩档板，遮阳板
antidetonation *n.* 抗爆(震)
antidrag ['ænti'dræg] *a.* 减阻(力)的
antidrip *v.* 防滴漏
antidrum ['ænti'drʌm] *v.* -med，-ming；隔声
antidrumming coat 隔声涂刷[层]
antidrumming treatment 隔声处理
antidrying surface 防干裂面
antidusting *n.*；*a.* 抗尘作用，抗尘性(的)
antifatigue *n.* 耐疲劳，抗(疲)劳剂
antifix *n.* 反曲线(状)挑檐饰
antiflatulent *n.* 除[排]气剂
antiflex cracking 抗折裂，抗弯裂
antiflood *n.* 防洪
~ interceptor 防泛滥截流管[渠]
~ wall 防洪墙
anti-fluctuator ['ænti'flʌktjueitə] *n.* 缓冲器
antiflux *n.* 防熔剂
antifly glass 防碎玻璃
antifoam ['æntifəum] *n.*；*a.* 防沫的
~ additive 防沫添加剂
~ agent 防沫剂
~ aid 防沫添加剂
antifog ['ænti'fɔg] *v.*；*n.* 防雾

antiflood wall

antifoulant *n.* 防污底漆
anti-fouling *n.* 防污
　～ agent 防污剂
　～ composition 防污剂
　～ motor 防污电动机
　～ paint 防污漆,防污涂料
antifreeze ['ænti'fri:z] *n.* ;*a.* 防冻(剂,液,的);抗凝剂
　～ admixture 防冻掺合剂
　～ agent 防冻剂
　～ layer 防冻层
　～ liquid 防冻液
　～ mixture 防冻剂
　～ powder 防冻粉末
　～ solution 防冻溶液
anti-freezer *n.* 防冻剂
antifreezing *n.* ; *a.* 防冻(的)
　～ aid 防冻添加剂
　～ brick 防冻黏土砖
　～ case 防冻箱
　～ (clay)brick 抗冻(黏土)砖
　～ coat 防冻层
　～ dope 阻冻剂
　～ extinguisher 防冻灭火器
　～ hydrant 防冻消火栓
　～ solution 防冻溶液
　～ mixture 防冻剂
antifriction ['ænti'frikʃən] *n.* ; *a.* 润滑剂,减磨,耐磨(的)
　～ alloy 抗磨合金
　～ bearing 滚珠[减磨]轴承
　～ grease 轴承润滑脂
　～ material 耐磨材料
　～ metal 耐磨金属
　～ roller 滑轮,导轮
　～ worm conveyor 抗磨蜗杆运输机
antifrost *a.* 防霜冻的
　～ additive(for concrete) (混凝土)防霜冻添加剂
　～ layer 防霜[冻]层
antifrother ['ænti'frʌðə] *n.* 防起泡剂

anti-fungus *a.* 防霉的
antigas ['ænti'gæs] *a.* 防毒气的
　～ defence 防毒,毒气防御
　～ mask 防毒面具
antiglare *a.* 防眩的,遮光的
　～ device 防眩装置
　～ fence 防眩栅栏
　～ shield 防眩档板
antigradient *n.* 逆梯度,负梯度
anti-ground 防接地的,消除地面影响的
antigravity ['ænti'græviti] *n.* 抗重力
　～ device 抗重力装置
　～ filtration 抗重力过滤
　～ screen 反重力筛分机
anti-G-Valve *n.* 抗重力阀
anti-halation *n.* 消晕作用
antihum *n.* 静噪器; *a.* 消声的
antihunt ['ænti'hʌnt] *n.* ;*v.* ;*a.* 防震(的),缓冲(的),稳定的
anti-icer ['ænti'aisə] *n.* 防冰设备,防冻装置
anti-incrustator *n.* (锅炉)防垢剂
anti-jamming *n.* ;*a.* 反[抗]干扰(的)
antiknock [ˌænti'nɔk] ; *n.* 防爆(的),抗震(的)
　～ additive 抗爆添加剂
　～ agent 抗爆剂
　～ compound 抗爆剂
　～ dope 抗爆添加剂
　～ fuel 防爆燃料
anti-lithic ['ænti'liθik] *n.* 去垢剂
antilogarithm ['ænti'lɔgəriθəm] *n.* 反对数,真数
antimacassar *n.* 沙发[椅子]套子
antimagentic ['æntimæg'netik] *a.* 抗磁(性)的
antimony ['æntiməni] *n.* 锑(Sb)
　～ glance 辉锑矿
　～ oxide1 锑白,氧化锑
anti-noise *n.* 吸声,防噪声
　～ paint 吸声油漆
　～ screen 吸声屏

anti-overloading *a.* 防过载的
antioxidant ['ænti'ɔksidənt] *n.* 抗氧化剂,防老化剂
anti-percolator *n.* 防渗装置
anti-pollution [æntipə'lju:ʃən] *n.* 防污染
　～ measure 防污染措施
antipriming pipe 多孔管,汽水隔离管
antiputrefactive *a.* 防腐的
antiquarium *n.* 古物室
antique [æn'ti:k] *n. a.* 古物[董],黑体字,旧[老]式的,过时的
　～ brick 古砖
　～ bronze colour 古铜色
　～ building 古老建筑
　～ crown 古皇冠
　～ furniture 旧[老]式家具
　～ ornament 古老装饰品
　～ temple 古庙

antique crown

antiradiation *n.* 反辐射
　～ protection 辐射防护
anti-rattler *n.* 减声[防震,隔声,消声]器
antiresonance ['æbti'rezənəns] *n.* 反共振[共鸣],并联共振
anti-roll bar 抗侧倾杆
anti-rot *a.* 防腐的
　～ substance 防腐材料
antirust ['ænti'rʌst] *a.* 防锈的
　～ action 防锈作用
　～ agent 防锈剂
　～ coat 防锈(涂)层
　～ compound 防锈混合剂
　～ enamel 防锈搪瓷[釉瓷]
　～ grease 防锈(润滑)脂
　～ oil 防锈油
　～ paint 防锈漆
　～ paper 防锈纸
　～ pigment 防锈颜料
　～ solution 防锈溶液
　～ varnish 防锈清漆

antirusting 防锈的
antirusting paint 防锈漆
antiscale *n.* 防垢,防垢剂
anti-scour *n.* 抗冲刷
　～ sill 消力槛
anti-seepage *n.* 防渗
anti-seismic *a.* 抗地震的,防震的
　～ structure 抗震建筑[结构]
antiseptic *n.* ; *a.* 防腐剂[的],杀菌剂[的]
　～ varnish 防腐清漆
anti-settling *n.* 防沉
anti-shock *n.* 防冲装置
anti-shrink *a.* 耐[抗]缩的,防缩的
anti-sidetone *a.* 消侧声的
anti-siphonage *n.* 反虹吸作用
　～ pipe 反虹吸作用管
antiskid *n.* ; *a.* 防滑(的)
　～ device 防滑装置
　～ factor 防滑系数
　～ groove 防滑花纹沟
　～ paint 防滑漆[涂料]
　～ property 防滑性能
　～ rib(bed)tile 防滑(肋)板,防滑花纹瓷砖
　～ road surface treatment 路面防滑处理
anti-slip *n.* 防滑性
　～ aggregate 防滑集料
　～ paint 防滑漆
　～ rib(bed)tile 防滑肋板[花纹瓷砖]
antisludge *n.* 抗淤积,去垢,抗垢剂
antisolar *a.* 防太阳光的,遮阳的
antispark *a.* 防[消]火花的
anti-splitting bolt (龙骨垫木)防裂螺栓
antispray *a.* 防溅的
　～ film 隔沫层,隔沫薄膜
　～ guard 防油溃溅护板
antisqueak *n.* 消声器
antistall *n.* 防止失速
antistatic *a.* ; *n.* 抗静电(的)
　～ device 抗静电装置

~ rubber 抗静电橡胶
anti-stripping *n.* 抗剥离(沥青及充填物间的粘结力)
　　~ additive 抗剥落掺和料
　　~ agent 抗剥落剂
antisun *a.* 防止日光照射的
　　~ cantilever roof 抗日光悬臂式屋顶
　　~ glass 抗日光玻璃
　　~ material 抗日光照射材料
antisweat covering 防露覆盖[保温]层
antisymmetric *a.* 反对称
antisyphonage pipe 倒虹吸管
antitrade (**wind**) 反信风,反贸易风
anti-trigonometric function 反三角函数
antium *n.* 门廊,(圆)柱廊,门前阶梯
anti-vacuum *n.* 反压力
anti-vibration *n.* 防振
　　~ clamp 防振夹具
　　~ handle 防振拉手
　　~ mounting 防振台
　　~ pad 防振衬垫[填料]
antivibrator *n.* 防震器
antiwarp wire 反翘线
antiwelding *a.* 抗焊接的
anti-wind *n.* 防缠绕
anti-wrinkling *n.* 耐[防]皱
anvil ['ænvil] *n.* 砧,铁砧
　　~ break 砧嘴
　　~ block 砧座[台]
　　~ chisel 砧凿
　　~ face 砧面
　　~ plate 砧面垫片
　　~ seat 砧座
　　~ swage 砧模
　　~ vise 砧钳
　　~ with an arm 鸟嘴砧
　　bench ~ 台砧
　　black smith's ~ 锻工铁砧
　　file cutting ~ 錾锉砧
　　hammer ~ 锤砧,铁砧
　　rising ~ 活动台砧
　　stock ~ 台砧

straightening ~ 校正砧
swage ~ 锻接砧,冲压砧
anywhere ['eniwɛə] *ad.* 任何地方
　　~ carpet 通用地毯
apart [ə'pɑːt] *ad.* 拆开,离开,个别
　　~ check 取样检验
apartment [ə'pɑːtmənt] *n.* 公寓;单元式宿舍,成套房间;居住地
　　~ area 公寓式住宅区
　　~ block 街坊,公寓建筑[大楼]
　　~ building 公寓建筑,公寓大楼
　　~ dweller 居民,住户
　　~ entrance 楼层住宅入口
　　~ floor 住宅楼层
　　~ hotel 公寓旅馆
　　~ house 公寓房屋,公共住宅
　　~ house of employees 职工住宅
　　~ house of corridor access 廊式公寓
　　~ -house of direct access 服务性空间共用的公寓;门厅、楼梯共用的公寓

apartment house of corridor access

　　~ housing 多户[公共]住宅,多户住宅建筑
　　~ in clogs 底层设有商店等公共建筑的住宅
　　~ kitchen 住宅[公寓]厨房
　　~s combining shop and dwelling units 底层为商店等的集体宿舍[公寓]
　　~ skyscraper 公寓摩天大楼
　　~ storey 公寓楼层
　　~ tower 公寓大楼塔楼
　　~ unit 公寓套房,居住单元
　　~ unit entrance 公寓居住单元外门[入口]
　　~ unit entrance door 公寓居住单元大门[外门]

～ unit floor space 公寓居住单元面积
～ unit kitchen 公寓居住单元厨房
apartotel [əˈpɑːtɔtl] *n.* 公寓酒店
apatite [ˈæpətait] *n.* 磷灰石
aperient water 苦矿水,矿质水
aperiodic(al) [ˈeipiəriˈɔdik(əl)] *a.* 非周期性的
  ～ disturbance 非周期扰动
  ～ function 非周期函数
  ～ instrument 不摆式仪表
  ～ motion 非周期性运动
  ～ osciuation 非周期性振荡
aperiodicity [ˈeipiəriˈɔdisiti] *n.* 非周期性
aperometer [ˈeipiəriəˈmiːtə] *n.* 孔径计
aperture [ˈæpətjuə] *n.* 孔口,口径,隙缝,壁孔,孔眼,宽度
  ～ and stops 孔径与光阑
  ～ area 开孔[孔口]面积
  ～ dimension 孔径
  ～ efficiency 开口[孔径]面积效率
  ～ gap 孔隙
  ～ lens 孔隙透镜
  ～ mask 多孔障板
  ～ of beam 射束孔径
  ～ of bridge 桥孔
  ～ of door 门孔[洞]
  ～ of screen 筛孔
  ～ of sight 观测孔
  ～ of window 窗口
  ～ ratio 相对孔径,孔径比
  ～ size 筛孔径
  ～ type fluorescent lamp 狭缝式(集中照射)荧光灯
  ～ width 筛网孔净宽
  ～ control-grid 控制栅孔径
  effective ～ 有效孔径[口径]
  full ～ 全口[孔]径
  measuring ～ 测量孔径
  real ～ 有效[实际]孔径
  screen ～ 筛孔
  usable ～ 有效孔径

working ～ 有效孔径
apex [ˈeipeks] *n.* 尖,顶点,背斜脊,褶皱线
  ～ angle 顶角
  ～ block 拱顶砖石,镶墙边的砖石[拱冠]
  ～ hinge 顶铰
  ～ hog 拱曲顶
  ～ joint 链接合
  ～ load 顶点荷载
  ～ mould (拱冠剖面)边线;分水岭地形
  ～ of arch 拱顶
  ～ of bend 变头顶点,弯曲顶部
  ～ of deposit 沉积层的顶点
  ～ sag 拱顶下垂
  ～ stone 房脊石,[山墙]顶石
  lower ～ of fold 褶皱底
  upper ～ of fold 褶皱顶

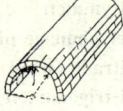
apex of arch

aphanitic [əˈfænitik] *a.* 隐晶质的,非显晶(质)的
aphanophyre 隐晶斑岩
aphlogistic [əˈflɔːdʒistik] *a.* 不能燃烧的;无焰燃烧的
Aphloia theaformis 茶红木
aphodus *n.* 排水沟
apical [ˈæpikəl] *a.* 顶点的,顶上的,尖的
  ～ angle 顶角
  ～ system 顶端的,顶系
apitong [ˈæpitɔŋ] *n.* 大花脑树(木材)(装饰用木材)
aplite [ˈæplait] *n.* 半花岗岩,红钴银矿;红晶岩
apodyterium [ˌæpəudaiˈtiəriəm] *n.* (古希腊、罗马)浴场的更衣室
A-pole A 形柱,简便构架
apomecometer *n.* 测距仪
apophyge [ˈæpəufidʒ] *n.* 柱座[柱头]凹线
apophysis [əˈpɔfisis] *n.* 岩枝[支]

aposandstone 石英岩

apostilb (＝asb) *n.* 阿波西提(亮度单位)

apotheca [ə'pɒikə] *n.* 储存室,藏酒室

apophyge

apothem ['æpəθem] *n.* 边心距

apotype *n.* 补型

APP ＝ approved by 经……批准,批准人

app ＝ ① appendix 附录;② apparatus 仪器

apparatus [ˌæpə'reitəs] *n. pl.* -tus, -tuses;仪器,器械,设备,装置
～ casting 综合仪器箱
～ engineering 仪器设备工程(学)
～ floor 仪器设备层,技术层
approved ～ 安全[防爆]设备
auxiliary ～ 辅助设备
charging ～ 装料设备
clamping ～ 夹具
combustion fuel gas ～ 燃烧排气测定器
commanding ～ 操纵设备
communication ～ 通讯设备
compensation ～ 补偿器
compressed-air painting ～ 压缩空气喷漆器械
condensing ～ 冷凝器
control ～ 调节[控制]器
cooling ～ 冷藏器,冷却器
draft ～ 通讯设备
drying ～ 烘干[干燥]器
dust exhaust ～ 吸尘装置
emergency ～ (紧急)备用设备
facsimile ～ 传真机
feed ～ 给水器,进给装置
filtering ～ 过滤器[装置]
fire ～ 消防器材,灭火器
first-aid ～ 急救设备
float feed ～ 浮筒式给水器
flow measuring ～ 液体流量表,流量测定仪
fool proof ～ 防止误操作的设备
gas measuring ～ 气体流量计
gate ～ 叶片式调节阀门,导向器
gripping ～ 扣紧工具,夹具
guide-vane ～ 叶片式调节阀门,导向器
heat exchanging ～ 热交换器
heating ～ 加热器
hoisting ～ 起重设备
holding ～ 压紧装置,断绳防坠器
immersible ～ 能连续在水下操作的电气装置
inclined shaft mucking ～ 斜井装岩机
induction heating ～ 感应加热器
interlocking ～ 连锁装置
lifting ～ 提升设备
locking ～ 锁紧装置
measuring ～ 测量仪器
porosity ～ 气孔测定仪
pressure jet ～ 机械[压力]喷嘴
projecting ～ 投影仪
rheostatic heating ～ 变阻式加热器
ring and ball ～ 沥青软化点测定器
sanding ～ 喷砂装置,抛砂机
shaking ～ 搅拌装置
smoke burning ～ 锅炉烧烟器
spraying ～ 喷雾器
state ～ 国家机器
steam heating ～ 蒸汽供暖设备
stirring ～ 搅拌器
tension ～ 拉紧装置
water jet suction ～ 射流泵
wetting ～ 给湿装置,加湿装置

apparent [ə'pærənt] *a.* 明显的,表观的,显[形]似的
～ angle 视角
～ area 视面积
～ available area 表观可用面积
～ brightness 表观[视在]亮度

~ bulk modulus 表观体积模量
~ coefficient of compressibility 表观压缩系数
~ coeficient of heat transmission 表现传热系数,标称传热系数
~ colour 表色,视在颜色
~ conversion 表观转化率
~ density 视[表观]密度
~ depression of the horizon 地平俯角
~ dip 视倾角
~ distance 视距
~ easement 外附附属建筑
~ efficiency 视效率
~ easement (~ servitude) 外观附属建筑物
~ elastic modulus 表观弹性模量
~ error 视误差
~ free nappe (水流出口的)自由射流[水舌]
~ gravity 表观[视]比重
~ ground water velocity 地下水表观速度[流速]
~ hardening curve 表观硬化曲线
~ height 视[表观]高度
~ horizon 地平线,视地平,可见水平
~ pitch 视[表观]螺距
~ porosity 视[表观]孔隙率
~ power 表观[视在]功率
~ resistance 表观[视]阻力
~ semidiameter 视半径
~ shear-strength 似抗剪强度
~ shear stress 表观剪应力
~ size 视[表观]尺寸
~ solid volume 固体表观体积
~ specific gravity 视[表观]比重
~ stress 视应力
~ surface 视表面
~ temperature 表观温度
~ velocity 表观速度,视速度
~ viscosity 表观粘性,视在粘性
~ volume 表观体积[容积],松装体积
~ weight 表观[视]重量,毛重

~ wind 视风,相对风
appeal *vi. n.* ①要求,呼吁,控诉
②吸引,引起……兴趣
③有感染力
appear *vi.* ①出现,问世,出版,发表
②看来,好像,显得
appearance [ə'piərəns] *n.* 外貌,外观,形状
~ of fracture 裂面外貌,破碎现象
~ test 外部[外观]检查
architectural ~ 建筑外形[外观]
external ~ 外观[貌,形]
append [ə'pənd] *vt.* 附上,增补,悬挂
notes~ed 附注
appendage [ə'pendidʒ] *n.* 附属部分,附属物,附件
appendix [ə'pendiks] *n. pl.* appendixes 或 appendices；附录[言],附属物,输送管
appentice [ə'pentis] *n.* 厢[耳]房,门[窗]前之雨篷
appliance [ə'plaiəns] *n.* 量具,仪器,设备,装置

appliance

~ outlet 设备[电源]插口[接头]
~ ventilation duct 仪器设备通风管
building ~ 建筑设备
casing ~ 下套管用具
charging ~ 装料[进料]设备
dust-separating ~ 灰尘分离装置
gripping ~ 捕集设备夹具
lifting ~ 升降设备,提升设备
measuring ~ 测量器具[设备]
portable ~ 便携式仪表
pumping ~ 排水设备,提升设备
safety ~ 安全设备
applicable ['æplikəbl] *a.* 适合[宜]的,合用的；applicability, *n.*
~ surfaces 可贴曲面,互展曲面
applicant ['æplikənt] *n.* 申请者
~ proposing to build 建筑申请者
applicate ['æplikeit] *a.* 紧贴于表面的

application [ˌæpliˈkeiʃən] *n.* 施加, 应用; 申请; 申请书
~ by brush(ing) 用刷子刷(油漆涂料)
~ consistency 操作连续性, 使用的一致性
~ drawing 申请[上报]用图纸
~ form 申请书[表格]
~ guide 使用导则, 应用手册
~ method (油漆技术中的)涂刷方法
~ of a surface 曲面的贴合
~ of binder(s) 施加粘合剂[粘结剂]
~ of load 施加荷载
~ of mortar 抹灰, 粉刷, 打底
~ of plaster (采)用粉饰
~ of stress 施加应力
~ of wax resist (采)用蜡保护层
~ program 应用[操作]程序
~ software 应用[算题]软件
~ temperature 使用温度, 泼油温度
~ thickness 使用厚度
~ valve 控制阀
bituminous ~ 敷沥青
cold ~ 冷用
computer ~ 计算机应用
hot-melt ~ 热熔施工
part-time ~ 短时[暂时]使用
point of ~ 作用点, 施力点
real time ~ 实时应用[使用]
service ~ 使用[装机]申请书
stand by ~ 备用机工作, 多机使用
stress ~ 施加应力, 加负荷
surface ~ 敷面
topical ~ 局部施用

applicator *n.*
① 敷贴(料)器, 撒药(粉)机
② 扣环起子
③ 高频发热电极

applied [əˈplaid] *a.* 应用的,
~ column 墙柱, 附[半]柱
~ elasticity 应用弹性学

applied column

~ geology 应用地质学
~ load 外施[施加]荷载
~ mechanics 应用力学
~ moment 外施[施加]力矩
~ moulding (门窗,家具的)木压条
~ overhead rate 装饰用木线脚, 间接费用分配率
~ stress 作用[外加]应力
~ thrust 作用[外加]推力
~ voltage 外加电压
~ work 施加的[外施的]功

applique [æˈpli:kei] *n.* 嵌花, 附饰物
~ figuring 贴花花纹, 刺绣花纹

apply [əˈplai] *v.*, -plied,-plying; 涂, 镀, 贴; 作用, 施加; 应用
~ force 作用[施加]力
~ oil 上油, 加润滑油
~ work 做功

appoint [əˈpɔint] *v.* 供给; 约定; 设备, 装备, 任命

appointed *a.* 决定的, 被任命的, 设备好的

appointee *n.* 被任命者, 被指定人

appointment [əˈpɔintmənt] *n.* 任命; 职位; 约会; 家具, (旅馆, 船上)设备
~ call 定人定时呼叫

apportionment [əˈpɔːʃənmənt] *n.* 分配, 分摊
~ chargers 分摊费用
~ cost 摊派成本
~ method 分摊法
~ tax 摊派税

appose [æˈpəuz] *vt.*, apposesd, apposing; 并列, 置于对面[附近];-apposable, *ad.*

apposition [ˌæpəˈsiʃən] 并列, 相邻; 添附

appraisal [əˈpreizəl] *n.* 检验, 鉴定; 评[估]价
~ method 估价法
~ of plant assets 设备资产[固定资产]估计
~ survey 估价调查

~ detailed 详细估价
appreciation [əˌpriːʃiˈeiʃən] *n.* 鉴别；评价，估价
apprehension [ˌæpriˈhenʃən] *n.* 理解；看法，观点
apprentice [əˈprentis] *n.* 见习生，学徒工，技工助手
~ draughtsman 绘图学员
~'s school 技工学校
~ (work) shop 学徒实用工场
apprenticeship [əˈprentisʃip] *n.* 学徒身份[年限]
~ training 学徒培训
appression [əˈpreʃən] *n.* 被压紧的状态，有重量，重力感
approach [əˈprəutʃ] *n.* 通道，桥梁[隧道]两端的堤，栈桥或其他建筑，引路[道，桥，槽]；*v.* 接[逼，行]近；近似；探讨处理
~ adit 施工（导流）隧洞
~ alignment 桥头引道接线
~ angle 趋近角
~ bank 引道[桥]
~ block 接近闭塞区段
~ channel 引渠，进港航道
~ cutting 引桥[桥头]挖方，引道挖方,（桥梁，隧道）两端开挖
~ embankment 桥头路堤，引道路堤
~ end 引道尽头，接近端
~ fill 引道填筑[填方]
~ flap （移动式桥）引道（折）板
~ flume 引水槽
~ grade 引道坡道
~ gradient 引道坡度

~ of bridge 引桥路,通往桥的道路
~ point 接近点
~ ramp 引道坡，引桥坡道
~ road 引道[路]
~ shoulder 引道路肩
~ sign 引道[接近]标志
~ span 引(桥)跨，岸跨
~ speed （交叉路口）驶近速度,闭合速度
~ surface 机场跑道进入面
~ track 铁路专用线
~ trench 交通壕
~ trestle 引道[排架]栈桥
~ velocity head 行近流速水头
~ viaduct 引道高架桥
~ zone （车辆在接近隧道时光度变化的）接近区段，引道区
bridge ~ 引桥路
elevated ~ 高架桥引道
frequency-respones ~ 频率计算法
highway ~ 公路引道
hot-press ~ 热压法
intersection ~ 交叉路口进口，道口引道
mathematical ~ 数学近似
rapid ~ 快速进给
road ~ 桥梁引道
second ~ 二次近似
system ~ 系统方法[研究]
approachability *n.* 可(易)接近
approachable *a.* 易接近的
approbation [ˌæprəuˈbeiʃən] *n.* 认可，品种鉴定，资格审查，委员会结论
appropriate [əˈprəupriit] *a.* 恰当的；*v.* 拨作……费用，拨用
~ assistance 必要的协助
~ authority 相关当局,有关主管当局
~ chart 专用图
~ price 调拨价格
~d materials 拨定材料
appropriation [əˌprəuprieiʃən] *n.* 拨款，经费，专用

approach ramp

~ limitation 拨款限额
~ period 拨款期限
~ process 拨款程序
~ title 拨款科目
~ warrant 拨款核定通知书
appropriator [ə'prəuprieitə] *n.* 拨给者,专用者
~ of water 水权享用者
approval [ə'pru:vəl] *n.* 许可,批准,核准
~ authority for projects 项目核准权
~ budget 核定预算
~ test 鉴定试验,合格性检验;验收[检查]试验
approved [ə'pru:vd] *a.* 允许的,批准的,许可的,有效的
~ budget 核定[法定]预算
~ (deposited) plans 批准的图纸
~ estimates 核定概算
~ marking (放炮器)安全标志
~ method 批准[验收]方式
~ rule (建筑部门,权威机构)实行的规则
~ program 核定计划
approver *n.* 批准者,赞成者
approximant *n.* 近似值(式,结果)
approximants [ə'prɔksimənts] *n.* 近似式
generalized ~ 广义近似式
approximate [ə'prɔksimit] *a.* 近似的; [ə'prɔksimeit] *v.* -mated, -mating; 近似,接近
~ amount 概算
~ calculation 概算,近似计算(法)
~ cost 估算[近似]成本
~ estimation sheet 概算[粗略]估价单
~ evaluation 近似计值
~ method 近似法,
~ solution 近似解
~ theory 近似理论
~ value 近似[标准,参考]值

approximation [əˌprɔksi'meiʃən] *n.* 近似,略计;近似法,概算,略计
~ by least squares 用最小二乘方的近似法
~ curve 近似曲线
~ hypothesis 近似假设
~ theory 近似理论
best fit ~ 最优近似
diffusion ~ 扩散近似
engineering ~ 工程近似法
exponential ~ 指数近似
appurtenance [ə'pə:tinəns] *n.* 附属建筑[设备];附[零,管]件;附属物
appurtenant [ə'pə:tinənt] *a.* 附属的; *n.* 附属物
~ structure 附属结构
~ works 辅助工程[设备],附属工程
appx = appendix 附录
Apr = April 四月
apr = apprentice 学徒,徒工
apricot *n.* 杏(树),杏黄色
April *n.* 四月
apriorism *n.* 先验论
APrk = Air Park 停机场
apron ['eiprən] *n.* 挡[底,跳,裙,遮挡]板;护坦[桥,墙],散水,窗台板;前台;停机坪
~ area 停机坪
~ arm (轮动式铲运机)裙臂
~ board 裙板
~ cable (轮动式铲运机)裙索
~ control valve (轮动式铲运机)控制阀
~ conveyor 板式输送带

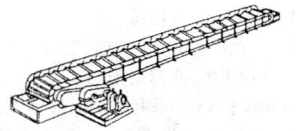

apron conveyor

~ (-conveyor) feeder 链板式加料机
~ facing 窗肚墙[拱肩墙]面装修,小梁侧封板

~ flashing [piece] 遮檐板,拨水板
~ flood lighting 前场[舞台]泛光照明[强力照明]
~ lining 窗肚[拱肩]墙装修,小梁侧封板,楼梯装修中的护墙裙板
~ marking 停机坪标志
~ moulding 门中挡装饰
~ panel (窗户)拱肩镶板
~ piece 遮檐板,拨水板;支承楼梯平台或一系列斜踏步的小梁
~ plate 闸门,裙板,挡板
~ ring 裙圈
~ rolls 运输机皮带滚轴
~ slab 护坦板
~ slope 迎面坡
~ track 轻便轨道
~ wall 前护墙
~ wall facing 前护墙装修
~ wall lining 前护墙封板
~ wall panel 前护墙镶板
~ wheel 履带
deflector ~ 导向板
dumping ~ 卸载溜槽
feed ~ 输送机进料板,锯板送料,板式送煤机
fender ~ 保护板,挡泥板
mud ~ 挡泥板
parking ~ 停车场
rear ~ 海漫,反挡板
solid ~ 重型[强固]铺板
spillway ~ 溢洪道护坦
splash ~ 挡溅板,挡泥板

apropos *a.;ad.* 恰(适)当,中肯(的),及时(的),顺便

aprt =airport 飞机场

aprx =approximate 近似,约计

aps ['æps] *n.* 欧洲山杨

apsacline [,æp'sɑ:klain] *n.* 倾斜型

apse [æps] *n.* 半圆壁龛,教堂后部东面,半圆形室
~ aisle 教堂内由唱诗班通道伸向半圆室或周围的通道[回廊]
~ -buttress 半圆室扶垛

apse

~ -buttressed 用半圆壁龛支撑的
~ window 教堂半圆室窗户

apsidal ['æpsidl] *n.* 半圆形[多角形]建筑;*a.* 半圆室的,轨道拱点的,与轨道拱点有关的
~ angle 拱心角,毗拱角
~ choir (教堂)半圆形歌唱班席位
~ distance 拱(点力心)距
~ entrance hall 半圆形门厅
~ motion 拱线运动
~ surface 长短径曲面

apsides [æp'saidi:z] *n.* 拱点;半圆形屋顶[穹窿,壁龛]

apsidiole [æp'sidiəul] *n.* 教堂小型半圆形后殿

apsilate ['æpsileit] *a.* 有纹饰的

apsis ['æpsis] *n.* 教堂半圆形后殿,拱点
~ arch 教堂半圆形后殿拱
~ arch impost 教堂半圆形后殿拱墩
~ window 教堂半圆形后殿窗户

apteral ['æptərəl] *n.* (古希腊、罗马)无侧柱教堂;*a.* 两侧无柱式的,无侧柱的
~ temple (古希腊)无侧柱的寺庙

aptitudal *a.* 适当[宜]的
~ scale 适当[适宜]规模
~ station 恰当的[适宜]的布置

aptitude ['æptitju:d] *n.* 适当,恰当,适应
~ factor (木材)适应系数

apyre ['æpirə] *n.* 红柱石

apyrite ['æpirait] *n.* 红电气石

apyrous [ei'pairəs] *a.* 防火的,耐火的,不易燃的,*n.* 耐火性

aqua ['ækwə]【拉】*n. pl.* aquas, aquae. 水,液体,溶液 *a.* 水色的
~ ammonia 氨水
~ communis 普通水
~ distillata 蒸馏水
~ fluvialis 河水
~ fortis 浓硝酸
~ frigida 冷水
~ green (同 water green)水绿
~ marina 海水
~ pluvialis 雨水
~ pure 纯水
~ regia 王水
~ sterilisa 灭菌水
~ storage tank 储水池[槽]
~ system 水压贮存系统
~ tepida 温水
~ vitae 烈性酒,烧酒,酒精
strong ~ 浓氨水
aquadag ['ækwədæg] *n.* 胶体石墨;石墨黑胶
aquage [ə'kweidʒ] *n.* 水路
aquagraph *n.* 导电敷层
aqualung *n.* 水中呼吸器,水肺
aquamarine [ˌækwəmə'ri:n] *n.* 蓝绿石,海蓝宝石;
~ glass 海蓝宝石
aquamarsh [ˌækwə'ma:ʃ] *n.* 水沼泽
aquapulper [ˌækwə'pʌlpə] *n.* 水力碎浆机
aquarelle [ˌækwə'rel] *n.*【法】水彩画
aquarium [ə'kwɛəriəm] *n.* 水族馆,养鱼缸

aquarium

aquaseal *n.* (电缆绝缘涂敷用)密封剂 *a.* 密封的

aquastat *n.* 水温自动调节器
aquatic [ə'kwætik] *a.* 水的,含水的,水化的,水生的
~ building(s) 游泳[浴场]构筑物
~ life [organism] 水生生物
~ plants 水生植物
aquatint ['ækwətint] *n.* 蚀镂铜板画,飞尘腐蚀
Aquazur filter 阿奎左尔重力式滤池;下行式滤池
aqueduct ['ækwidʌkt] *n.* 高架渠,渡槽;输水管,桥管;水道,渠道,运河

aqueduct

~ arch 输水道拱
~ bridge 渡槽,渠桥
pipe ~ 渡槽(管)
aqueous ['eikwiəs] 水成的,含水的
~ ammonia 氨液[水]
~ corrosion 水腐蚀,潮湿腐蚀
~ gels 水胶炸药
~ humor 水样
~ lava 泥流岩
~ metamorphism 水变质(作用)
~ phase 水样
~ resin emulsion 水合乳化树脂
~ rock 水成岩,沉积岩
~ soil 含水土壤,沉积土
~ solution 水溶液
~ tint 水彩
~ vapo(u)r pressure 水汽压
~ wood preservative 木材防水剂
aquiclude ['ækwiklu:d] *n.* 半含水层,含水土
aquifer ['ækwifə] *n.* 含[蓄]水层
bed of ~ 含水层

aquifuge *n.* 不透水层
aquosity [əˈkwɒsiti] *n.* 潮湿,含水性
arabesque [ˌærəˈbesk] *n.* 花叶饰,阿拉伯式花纹,蔓藤花纹
~ decoration 阿拉伯式花纹装饰,花叶型装饰
~ ornament 阿拉伯式花饰品

arabesque

Arabian [əˈreibjən] *a.* 阿拉伯的
~ architecture 阿拉伯式建筑(含伊斯兰教建筑和撒拉逊式建筑)
~ capital 阿拉伯式柱顶
~ style 阿拉伯式(建筑)

Arabian capital

Arabic [ˈærəbik] *a.* 阿拉伯的
~ arch 马蹄形拱
~ gum 阿拉伯树胶
araeostyle [əˈriəstail] *n.* 疏柱式建筑(柱距约为柱径的四~五倍)
~ temple 疏柱式建筑寺庙
araeosystyle [əˈriːəˈsistail] *n.* 对柱式建筑(柱距为柱径的二~四倍交替排列成对柱列式)
aragonite [əˈrægənait] *n.* 文石,霰石,碳酸钙($CaCO_3$)
aragotite [əˈrædʒətait] *n.* 霰石沥青,美国加利福尼亚州天然沥青,黄沥青
arake [ˈærik] *a.* 倾斜于垂直线的
Araldite [ˈærəldait] *n.* 以环氧树脂为粘料的一种特制混凝土粘结砂浆,环氧类树脂
arbite *n.* 一种安全炸药
arbiter *n.* 仲裁人,公断人
arbitrary [ˈɑːbitrəri] *a.* 任意的,随机的,不定的;专制的,武断的
~ constant 任意常数
~ decision 任意决定,武断

~ deformation 任意变形
~ moment diagram 任意弯矩图
~ proportions method (混凝土)习用(体积比)配料法,经验配合法
~ scale 任意标度[刻度]
arbitration [ˌɑːbiˈtreiʃən] *n.* 鉴定,检验,仲裁
arbor [ˈɑːbə] *n. pl.* arbores;轴;树,乔木;枝编棚架,圆亭,荫道
~ press 手动[芯轴]平板机
~ walk 蔓藤棚架下步道
arboretum [ˌɑːbəˈriːtəm] *n. pl.* -ta, -tums;树木园,植物园
arborization [ˌɑːbəriˈzeiʃən] *n.* 树枝状
arbour [ˈɑːbə] *n.* 棚架,凉亭
arc [ɑːk] *n.* 弓形,弧拱,电弧,击穿; *a.* 电弧的,圆弧的,拱形的
~ air cutting 电弧切割
~ air gouging 电弧气动钻孔[割槽]
~ bearing plate 弧形支座
~ booster (焊接)起弧稳定器
~ -boutant 飞拱
~ brazing 电弧钎焊,合金电弧焊
~ chute 电弧隔板
~ cutting 电弧切割
~ flame 电弧焰
~ heating 电弧加热
~ horn 角形避雷器
~ lamp 弧光灯
~ point welding 电弧点焊
~ shears 电弧切割
~ spot welding 电铆焊,电弧点焊
~ spraying 电弧喷镀
~ weld steel pipe 电弧焊接钢管,弧焊钢管
~ welding 电弧焊
~ welding electrode 电弧焊条
alternating current ~ 交流电弧
electric ~ 电弧
arcade [ɑːˈkeid] *n.* 有拱顶的走廊,有拱的长形房屋,有拱的[曲线形的]通道[大街];连拱廊
~ apex 连拱顶

~ cornice (连)拱式挑[飞]檐
~ crown (连)拱顶
~ impost (连)拱墩
~ key (连)拱顶石
~ lobby (连)拱形走廊
~ pier (连)拱墩
~ rib (连)拱肋
~ sidewalk 拱廊人行道
~ top (连)拱顶
~ vertex (连)拱顶
blank ~ 盲拱,实心连拱
blind ~ 盲拱廊
false ~ 假拱廊
interlacing ~ 内叉拱廊
intersecting ~ 交叉拱廊
public ~ 公共拱廊步道
wall ~ 实心连拱

arcaded [ɑː'keidid] *a.* 有拱廊的;连拱(式)的,拱廊式的
~ court 拱廊,连环拱廊
~ facade 拱形建筑物正面
~ gallery 拱形看台[楼座,走廊,廊道]
~ ground floor 拱形底层
~ tribune 拱形看台[讲台]
~ window 拱窗

arcading *n.* 柱间拱,(仅作装饰用的)连拱饰

arcading

arcane *a.* 秘密的,神秘的
arc-back *n.* 逆弧
arc-cast *a.* (电)弧熔的
arcature ['ɑːkətʃə] *n.* 小拱廊;(不开洞的装饰用)连拱,小型连拱(如在栏杆中的拱);盲拱,实心连拱
arch [ɑːtʃ] 拱,弓架结构,弓形,半圆形,拱廊[门,洞];穹窿,圆顶;背斜;穹起
~ abdomen 腹拱
~ action 弯拱作用
~ adutment 拱台,圈拱座柱
~ analysis 拱静力学分析
~ and pier system 拱墩系统
~ apex 拱顶
~ axis 拱轴
~ back 拱背
~ band 横向拱
~ bar 拱板
~ barrel 拱形筒壳
~ bay 拱跨
~ beam 拱梁
~ bearing 拱支座
~ block 拱(圈)块,楔形拱面石,木拱楔块
~ bond 拱砌合
~ bound 拱镶边
~ brace 拱支撑
~ braced roof 拱支架屋顶
~ bracing 拱支撑
~ brick 拱砖,楔形砖,过火砖
~ brick for manholes 人孔(楔形)拱砖
~ bridge 拱桥
~ bridge construction 拱桥建筑
~ building 拱形建筑

arch ceiling

~ camber 拱势,拱矢,起拱
~ cantilever bridge 拱式悬臂桥
~ casting 浇筑混凝土拱

~ ceiling 拱形天花板[顶棚]
~ center(ing) 拱[鹰]架
~ center line 拱中心线
~ centre 拱中心
~ chord 拱弦
~ compression 拱压力
~ concreting 浇筑混凝土拱,拱形混凝土预制件
~ construction 拱形结构,拱的施工
~ -core 拱心
~ cornice 拱形飞檐
~ cover 拱形屋盖
~ cover(ing) 拱形屋面,拱板,拱盖
~ crown 拱顶,拱冠
~ culvert 拱涵,拱形涵洞
~ curvature 拱曲率[弯曲度]
~ dam 拱坝
~ -deck buttress dam 拱面支墩坝
~ depth 拱结构高度
~ door 拱门
~ element 拱形结构[支架],拱环[圈,单元]
~ extrados 拱背
~ face 拱面
~ falsework 拱形脚手架
~ fixed at both end's 固定(的)拱
~ flat 平拱
~ foot bridge 人行拱桥
~ force 拱推力
~ form traveler 拱形起重车
~ frame 拱形框架[构架]
~ girder 拱形桁架,拱梁
~ -gravity dam 重力拱坝
~ haunch 拱肩石
~ hinged at ends 双铰拱
~ hinged at the abutments 铰连在支座上的拱形桁架
~ impost 拱座[脚,基,墩,端托],拱底石
~ intrados 拱腹(线)

arch construction

~ in trellis work 格形拱
~ invert 倒拱,仰拱,倒拱形的沟底
~ key (stone) 拱顶[冠]石
~ length 拱长
~ line 拱曲线[弧线]
~ load 拱荷载
~ material 拱材料

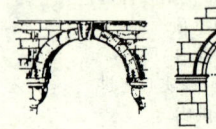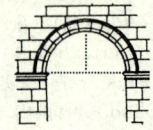
arch

~ moulding 拱饰,拱线脚
~ of Augustus 奥古斯都拱门
~ of Constantine at Rome (在)罗马的康士坦丁拱门
~ of triumph 凯旋门
~ order 拱柱式
~ outline 拱形曲[弧]线
~ parabola 拱抛物线
~ pattern 弓形纹
~ plane 拱平面
~ ponor 拱门落水洞
~ pour(ing) 浇筑混凝土拱,拱形混凝土预制构件
~ pressure 拱压力
~ radius 拱半径,弯曲半径
~ reinforcing 拱的加劲
~ rib 拱肋
~ rib footing 拱肋基础
~ ring 拱圈,拱环
~ rise 拱高[矢,矢高]
~ roof 拱形屋顶
~ scaffolding 拱架
~ section 弧段,弧形部分
~ sets 钢拱形棚子
~ shell 拱形[弧形]薄壳
~ soffit 拱复(线)
~ span 拱跨,桥拱
~ spandrel 拱臂
~ spillway 拱形溢洪道

~ springer 拱脚[座],起拱石
~ stair(case) 拱[弧]形楼梯
~ stay 拱的加劲杆
~ stiffener 拱的加劲杆
~ stiffening 拱的加劲
~ stone 拱石
~ stress 拱应力
~ structural analysis 拱结构分析
~ structure 拱形建筑物[结构]
~ support 拱座
~ system 拱系统
~ thrust 拱推力
~ tie 拱拉杆
~ timbering 拱架
~ top 拱顶
~ truss 拱形桁架
~ vertex 拱顶
~ viaduct 高架[跨线]拱桥,拱形栈桥
~ wall 拱墙
~ way 拱道,拱路

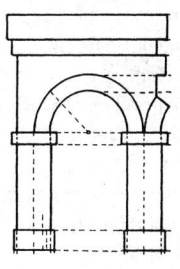

archway

~ window 拱窗
~ -wise 拱廊似的,成弓形的
~ with apex hinge 带顶铰的拱
~ with crown hinge 带顶铰的拱
~ with joggled joints 带啮合接头的拱
~ with key hinge 带顶铰的拱
~ without articulation 无铰拱
~ wtih tie 带拉杆拱
~ with top hinge 带顶铰拱
~ with vertex hinge 带顶铰拱
acute ~ 锐拱,椭圆断面拱

askew ~ 歪曲拱
axed ~ 斧斩拱面
back ~ 拱背
barrel ~ 筒式拱桥
basket handle ~ 三心拱
bell ~ 钟状拱
blunt ~ 垂拱
boomerang ~ 木三铰拱[框架]
bowstring ~ 弓弦杆,具有拉杆的拱
braced ~ 桁拱
braced-rib ~ 肋桁拱
brick ~ 砖拱
brick-on-end soldier ~ 竖砖拱
bridge ~ 桥拱
built-up ~ 组合拱
bull's eye ~ 牛眼(形)拱
camber(ed) ~ 平圆三心拱,平拱
Carnarvon ~ 平半圆混合拱
catenary ~ 反悬链式拱
centre ~ 中心拱
circular ~ 弧拱
circular constant-thickness ~ 等厚圆形拱
cloister ~ 回廊拱
common ~ 粗拱
composite ~ 尖拱,混合式拱
compound ~ 组合拱
concentric ~ 同心拱
contrasted ~ S形[葱形]拱
convex ~ 凸形拱
corbel ~ 撑架拱,突拱
coved ~ 大弧拱
cuneatic ~ 楔形拱
curtain ~ 间壁拱
cusped ~ 尖拱
cycloidal ~ 圆滚线拱
decorated ~ 装饰式拱
depressed ~ 坦圆拱
diminished ~ 平圆拱
discharging ~ 肋拱
double-hinged ~ 双铰拱
drop ~ 垂拱

dumb ~ 假拱
elliptic ~ 椭圆拱
elliptical pointed ~ 椭圆尖头拱
end ~ 端拱
equilateral ~ 等角拱
face ~ 前拱
fairway ~ 通航拱跨
false ~ 假拱
filled spandrel ~ 实肩拱
five-centered ~ 五心拱
fixed ~ 固定拱
fixed end ~ 定端拱
flat ~ 平拱
flattened ~ 平拱,漫尖拱
flood ~ 排洪拱（桥梁）
floor ~ 地板拱

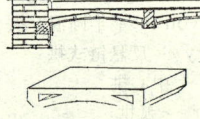

floor arched

flying buttress ~ 扶拱,拱扶垛
foiled ~ 叶形饰,[有叶形饰]拱
four centered ~ 四心拱
framed ~ 构成拱
French ~ 法(国)式拱
full-centred ~ 半圆拱
Georgian ~ 平拱,平圆拱
glued laminated timber ~ 胶合叠板拱
groin(ed) ~ 交叉拱,交叉穹肋
hance ~ 平圆[三心]拱
haunched ~ 加腋拱
high-crowned ~ 高顶拱
hinged ~ 铰接拱
hingeless ~ 无铰拱
horseshoe ~ 马蹄拱
imperfect ~ 斜圆拱
inclined ~ 倾斜拱
inclined barrel ~ 斜筒拱
inflected ~ 弯成拱
interlacing ~ 内叉拱

interrupted ~ 间断拱
inverted ~ 仰拱,倒拱
jack ~ 平拱
keel ~ 葱形拱
laminated ~ 叠层拱
laminated glued timber ~ 胶合叠板拱
lancet ~ 尖拱
lenticular ~ 双叶拱,两面凸的拱
linear ~ 抗力线拱
main ~ 主拱,竖井拱
masonry ~ 圬工拱
memorial ~ （纪念性）拱门
mosque ~ 马蹄拱
multicentered ~ 多心拱
multifoil ~ 多(孔)叶拱
Norman ~ 诺曼式拱
oblique ~ 斜拱
obtuse ~ 钝拱
ogee ~ 葱形拱
ogival ~ 尖拱
one-centered ~ 单心拱
one-hinged ~ 单铰拱
open ~ 明拱
open-spandrel ~ 空腹式[空心侧]拱
parabolic ~ 抛物线拱
pediment ~ 三角拱
perpendicular ~ 垂直式拱
pier ~ （桥）墩拱;柱支拱
plain ~ 粗拱
pointed ~ 哥特式拱,尖拱
polygonal ~ 多角拱
proscenium ~ 舞台大拱
Queen Anne ~ 威尼斯拱,半圆拱与双平拱的组合拱

Queen Anne arch

rail ～ 钢轨拱形支架
rainbow ～ 虹拱
raised ～ 突起拱
raking ～ 跛拱
rear ～ 背拱
recessed ～ 叠内拱
relieving ～ 辅助拱
ribbed ～ 肋拱,扇形拱
right ～ 正拱
rigid ～ 刚性拱
rising ～ 跛拱
rolled steel beam ～ 辊压钢梁拱
Roman ～ 半圆拱
roof ～ 拱顶(框)架,屋顶拱
rough ～ 粗拱
rough brick ～ 粗砖拱
round ～ 半圆拱
round trifoliated ～ 三叶形拱
row-lock ～ 竖砌砖拱
safety ～ 分载拱
scheme ～ 平弧拱
scoinson ～ 三角拱
sconcheon ～ 背拱
segmental ～ 弓形拱
semicircular ～ 半圆拱
semigothic ～ 半尖拱,半哥特拱
shell ～ 薄壳拱
shouldered ～ 肩形拱
side ～ 侧拱,边拱
single-hinged ～ 单铰拱
skew ～ 斜拱
skew-barrel ～ 斜圆筒形拱
smoke ～ 烟拱
soil ～ 分载反拱
soldier ～ 门窗顶上的立砌砖拱,立砌平拱
solid ～ 实体拱
solid spandrel ～ 实肩拱
solid-rib ～ 实肋拱

Roman arch

spandrel ～ 桁构拱
spandrel-braced ～ 肩桁拱,固肩拱
spandrel filled ～ 实肩拱
splayed ～ 八字形拱
sprung ～ 正[弓形]拱,拱式炉顶
squinch ～ 突角拱
steel ～ 钢拱支架,金属拱形支架
stepped ～ 阶形拱
stilted ～ 上心拱
stone faced ～ 石面拱
straight ～ 平拱
straining ～ 扶拱
superimposed ～ 上叠拱
surbased ～ 弓弦三心[平圆]拱
surmounted ～ 超半圆拱
suspended ～ 悬托拱
tented ～ 帐篷式拱
three-center(ed) ～ 三心拱
three-hinged ～ 三铰拱
three-pinned ～ 三铰拱
tied ～ 弦系拱
transverse ～ 横拱
trefoil ～ 三叶形拱
triangular ～ 三角拱
trimmer ～ 壁炉前拱
triple articulation ～ 三节拱
triple hinged ～ 三铰拱
Triumphal ～ 凯旋门

Triumphal Arch

trumcated angular ～ 锥角拱
trumpet ～ 喇叭(形)拱
truss(ed) ～ 桁构拱,拱式桁架
Tudor ～ 都德式拱

tuyere ~ 风口拱墙
twilight ~ 暮光弧
two-centered ~ 双心拱
two cusped ~ 双尖拱
two-hinged ~ 双铰拱
two-hinged braced ~ 双铰弓弦拱
two pinned ~ 双铰拱
umbrella ~ 隧道护拱
voussoir ~ 楔块拱
watertight facing ~ 防渗护面拱
wave ~ 波形拱
wheel ~ 轮罩拱
window niche ~ 窗龛拱
wood ~ 木制大拱
yieldable ~ 可变式拱架,镶压性拱形支架
yieldable steel ~ 镶压性金属地架,可变式钢质拱架
yielding ~ 镶压性拱形支架,缓冲式拱架
yielding steel ~ 镶压性金属支架,缓冲式钢质拱架
zigzag ~ 曲折拱
arch =architecture 建筑(学)
archaic [ɑiˈkeiik] *a.* 古代[式]的,古代风格的
Archaic temple at Ephesus (小亚西亚)以弗所古庙
~ architecture 古建筑学
~ sepulchre 古墓,拱形古墓穴
archecentric [ˈɑːtʃˈsentrik] *a.* 建筑中心的,建筑的;与建筑有关的;指定一种建筑的
arched [ɑːtʃt] *a.* 弓形的,拱形的,弧形的;弓架结构的
~ abutment 拱形桥台[支座]
~ boom 弧形弦杆[吊杆]
~ boom angle iron 弧形角铁弦杆
~ brick roof covering 砖拱屋盖,砖砌窑洞屋盖
~ bridge 拱桥
~ buttress 拱扶垛
~ cantilever bridge 悬臂式拱桥

~ concrete dam 混凝土拱坝
~ concrete roof 混凝土拱形屋顶
~ corbel table (由一排弧形牛腿支承的)挑出面层
~ (culvert)pipe 拱形涵管
~ diagonal 弓形斜(支)撑
~ dome 圆顶穹隆,拱形圆屋顶
~ entrance 拱形[弧形]入口
~ flange 弧形弦杆

arched dome

~ floor 拱形楼板
~ frame 拱形框架[构架]
~ girder chord 拱形弦杆
~ girder hinged at the abutments 与支座铰连的拱梁
~ girder length 拱形梁长度
~ girder load 拱形梁荷载
~ girder of constant cross section 等截面拱形梁
~ girder parabola 拱梁抛物线
~ girder plane 拱梁平面
~ girder with braced spandrels 有支撑共肩的拱形桁架
~ girder with diminished horizontal thrust 有衰减水平推力的拱形桁架
~ girder with intermediate tie 有中间系杆的拱形桁架
~ girder with invariable horizontal thrust 有不变水平推力的拱形桁架
~ girder with parabolic chord 有抛物线型弦杆的拱形桁架
~ girder with polygonal outlines 有多边形状的拱形桁架
~ girder with tie 带拉杆拱梁
~ girder with tieback 带拉杆的拱形桁架
~ girder without horizontal thrust 无水平推力的拱形桁架,通过拉杆来抵消水平推力的拱形桁架
~ of Augusuis 奥古斯都拱门
~ of titus 罗马泰塔斯凯旋门

~ opening 弧形洞
~ pipe 拱形涵洞
~ plate 弧形板
~ portal 拱形桥门[入口,隧道口]
~ principal 拱形大桁梁
~ recess 拱形壁龛,弧形(阀门)门槽
~ roof 拱形筒顶
~ roof truss 拱形屋顶桁架
~ stair(case) 拱[弧]形楼梯
~ stone bridge 曲拱石桥
~ structure 拱形建筑[结构]
~ style (of architecture) 拱形建筑风格
~ system for absorption of thrust 推力缓冲[吸收]拱系统
~ truss(ed girder) 拱形桁架式大梁
~ type piling bar 拱形(钢)板桩
~ vault 拱顶
~ viaduct 拱形高架[跨栈]桥
~ window 拱窗
~ work 拱形结构

arched viaduct

archetype [ˈɑ:kitaip] *n.* 模型,标本,原型;archetypic, *a.*;archetypical, *a.*;archetypically, *ad.*

Archimedean [ˌɑ:kiˈmi:diən] *a.* 阿基米德的
~ drill 螺旋钻
~ screw elevator 垂直螺旋输送机
~ screw water lift 螺旋提水器,螺旋泵

Archimedes' 阿基米德的
~ Law 阿基米德定理
~ principle 阿基米德原理
~ spiral 阿基米德螺旋线[涡线]

arching [ˈɑ:tʃiŋ] *n.* 拱作用,架拱,拱形支护
~ effect 拱作用
~ factor 拱度,弯拱因素
stone ~ 石拱圬工

architect [ˈɑ:kitekt] *n.* 建筑师;*vt.*

设计师
~ and designer 建筑设计师
~ -in-charge (施工现场)总建筑师
~ in private practice 私营建筑师
~ partnership 建筑师合伙[合作]关系
~'s fee 建筑师酬金[费用]
~'s office 建筑师事务所[办公室,画室,工作室]
~'s sketch 建筑师草图
~'s table 绘图桌
chief ~ 总建筑师
chief resident ~ 主任建筑师
civil ~ 民用建筑师
consulting ~ 顾问建筑师
supervising ~ 监理建筑师

architective *a.* 关于建筑的
architectonic(al) [ˌɑ:kitekˈtɔnik(əl)] *a.* 建筑学的,构造的,结构的,属于[关于,依照]建筑学技术原理的;architectonically, *ad.*
~ geology 构造地质学

architectonics [ˌɑ:kitekˈtɔniks] *n.* 建筑原理,建筑学,构造设计

architectural [ˌɑ:kiˈtektfərəl] *a.* 建筑学的,建筑的
~ acoustics 建筑声学
~ aggregate 建筑用混凝土骨料
~ alloy 建筑(用)合金
~ aluminium 建筑(用)铝
~ appearance 建筑外形
~ assistant 建筑绘图员
~ Association 建筑协会
~ award 建筑(获奖)奖赏
~ bid 建筑投标
~ bronze 建筑(用)青铜,铜锌铅合金
~ (builder's) fitting 建筑(施工)五金配件[装配器具]
~ (builder's) fittings 建筑施工小五金[硬件]
~ (builder's) furniture 建筑施工装配器具[五金配件]
~ (builder's) hardware 建筑(施工)

小五金[硬件]
~ building material 装饰(用)建筑材料
~ cast concrete product 装饰(用)预制混凝土制品
~ casting 装饰(用)预制混凝土构件
~ climatic zoning 建筑气候分区
~ competition 建筑方案评比,建筑方案竞赛
~ complex 建筑群[总体]
~ composition 建筑布置[构图,构造方式]
~ concept 建筑构思[初步设计]
~ concrete 装饰用[光面]混凝土
~ concrete aggregate 装饰用混凝土骨料
~ concrete casting 装饰用预制混凝土构件
~ concrete finishing 混凝土一次抹面,混凝土原浆抹面
~ concrete product 装饰用预制混凝土制品
~ conservation 建筑文物保护
~ construction(al) material 装饰(用)建筑材料
~ control 建筑管理规则
~ course 建筑研究室
~ critic 建筑评论家
~ criticism 建筑评论
~ decoration 建筑装饰(学)
~ design 建筑设计
~ designing documents 建筑设计文件
~ detail 反映建筑风格构造,装饰构造部分
~ device 反映建筑风格构造[构件],装饰构件
~ draughtsman 建筑绘图员
~ drawing 建筑图;建筑制图
~ education 建筑(学)教育[训练]
~ element 装饰构件,反映建筑风格构造部分

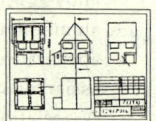
architectural drawing

~ engineering 建筑工程(学)
~ extruded section 装饰(用)挤制叶型
~ fashion 建筑模型
~ feature 建筑风格,建筑表面装饰特点
~ finish(ing) 建筑表面装饰
~ firm 建筑公司,建筑(师)事务所
~ fitting 建筑施工装置[器具]
~ fittings 建筑施工小五金[硬件]
~ flute 柱体凹槽
~ form 建筑形式[模型]
~ furniture 建筑施工装置[器具];设计成与房间建筑特征相配的家具
~ garden 建筑庭园
~ glass article 建筑装修用玻璃制[成]品
~ glass plant 建筑用[装饰用]玻璃生产厂
~ granite 建筑用装饰[装修]用花岗石
~ grill work 建筑格架结构
~ historian 建筑历史学家
~ history 建筑(历)史
~ hygiene 建筑卫生学
~ instruction 建筑(学上的)规程[细则,课程]
~ iron mongery 建筑五金器具
~ journal 建筑杂志
~ journalism 建筑新闻事业(包括出版、刊物、管理、编辑等)
~ laminated glass 建筑叠层[安全]玻璃
~ lamp 装饰灯
~ league(Arch. Lg.) 建筑联合会
~ lighting 建筑照明[采光]
~ masonry(work) 砖石建筑
~ mechanics 建筑力学
~ metal 建筑(用)金属材料
~ modelling (混凝土)预制建筑构件的生产
~ model 建筑模型

~ module 建筑模数
~ motif 建筑特色
~ order 建筑柱式
~ organism 建筑组织
~ ornament 建筑装修［装饰］
~ panel 建筑墙板［板材］
~ perspective 建筑透视图
~ planning 建筑设计［规划］
~ porcelain 建筑陶器
~ (pre)cast concrete product 预制建筑混凝土制品［成品］
~ product 建筑［装饰］用制品
~ profile 建筑分布图
~ programme 建筑规划［计划］
~ projected window 一种较高级的凸窗(用于外观效果重要的建筑)
~ psychology 建筑心理学
~ safety glass 建筑安全玻璃
~ sculpture 建筑雕塑［雕刻］
~ section 建筑型材［构件］
~ seminar 建筑研究班
~ shape 建筑型材［构件］

architectural profile

~ shatterproof glass 建筑安全玻璃
~ sketch 建筑草图
~ slate 建筑(用)石板
~ space 建筑空间
~ structure material 装饰用建筑材料
~ style 建筑风格［式样］
~ style garden 建筑式庭园
~ surveying 建筑测量
~ team 建筑师小组
~ terra-cotta 建筑用陶砖［陶土艺术

制块］
~ theorist 建筑理论家
~ theory 建筑理论
~ tradition 建筑传统
~ training 建筑教育［训练］
~ treatment 建筑(艺术的)处理
~ trend 建筑动向［发展趋势］
~ trim 建筑装修
~ unit 建筑型材［构件］
~ work 建筑工程［施工］
~ working drawing 建筑施工图
architecturally *ad.* 建筑学上，建筑构造上
~ beautiful square 具有建筑特色的地方
~ -enframed window 建筑上配的框架窗户
~ treated 建筑上经过处理的；具有建筑艺术的
architecture [ˈɑ:kitektʃə] *n.* 建筑学；建筑；构造，结构
~ adaptable to conversions 建筑重建［改造］
~ adaptable to extension 建筑物扩建
~ of Asia minor 小亚细亚建筑
~ of Australia 澳大利亚建筑
~ of Austria 奥地利建筑
~ of the Mogul[mughol] Empire (印度)莫卧儿帝国［蒙兀儿王朝］建筑
~ of the United States of America 美国建筑
arcuated ~ 拱式建筑
Byzantine ~ 拜占庭建筑
civic ~ 城市建筑
civil ~ 民用建筑
classic ~ 古典建筑
commemorative ~ 纪念性建筑
decorated ~ 盛饰建筑，尖拱式建筑
domestic ~ 住宅建筑
domical ~ 圆顶式建筑
flamboyant ~ (火)焰式建筑
Gothic ~ 哥特式建筑

 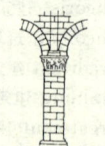

onion architecture   Romanesque architecture

monolithic ~ 整体建筑
neo-classic ~ 新古典建筑
onion ~ 葱头式建筑
palatial ~ 宫殿式建筑
Romanesque ~ 罗马风格建筑 rural ~ 乡村建筑

architrave ['ɑ:kitreiv] *n.* 柱顶过梁;门窗头线条板,贴脸板;框缘,下楣(柱)
~ block 门头线墩子,门窗线条板的座块
~ cornice 门窗框上槛的檐板
~ jambs 门窗线条板边框
~ of a door 门缘饰,门头线条板
~ of proscenium 舞台口边框
banded ~ 带饰门头线

archives ['ɑ:kaivz] *n. pl.* 档案室[馆]
~ building 档案大楼

archivist *n.* 档案保管员

archivolt ['ɑ:kivəult] *n.* 拱缘装饰,穹窿形,拱门饰,拱门侧墙装饰

archivolt

archless ['ɑ:tʃlis] *a.* 无拱的

archlimb ['ɑ:tʃ'laim] *n.* 顶翼,背斜翼,拱翼,穹翼

archtype sheet pile 拱形(钢)板桩

archy ['ɑ:tʃi] *a.* 拱形的,曲线形的

arc-hyperbolic [ɑ:khaipə:'bɔlik] *a.* 反双曲的

arciform ['ɑ:sifɔ:m] *a.* 拱形的,弓状的

arcing ['ɑ:kiŋ] *n.* 焊缝凸起(部分),飞[燃]弧
~ back 回弧,逆弧
~ current 电弧电流

~ ground 电弧接地
~ horn 角形避雷器
~ ring 环形消弧器
~ voltage 发弧电压

arcograph ['ɑ:kəgrɑ:f] 圆弧规

arcola ['ɑ:kəulə] *n.* 小锅炉

arcosic grit *n.* 长石砂岩

arctic ['ɑ:ktik] *a.* 寒冻的,北极的
~ construction 寒冻区建筑

arcual ['ɑ:kjuil] *a.* 拱式的,弓形的
~ construction 拱形构造
~ structure 拱形建筑物
~ style 拱形建筑风格
~ system 拱系统,拱作用
~ system for absorption of thrust 推力缓冲[衰减,吸收]拱系统

arcuate(d) ['ɑ:kjueit(id)] *a.* 拱式的,弓形的; arcuately, *ad.*
~ architecture 拱式建筑
~ construction 拱式建筑[构造]
~ building 拱式建筑[房屋]
~ structure 拱式建筑[结构]
~ style 拱式建筑风格
~ system 拱系统

arcuation [ɑ:kju'eiʃən] *n.* 拱状,弧状;一连串的拱门
~ construction 拱式构造

arcwalling *n.* 弧形掏槽

Ardand type polygonal roof 木拱脚悬臂托梁屋顶

ardometer [ɑ:'dɔmitə] *n.* 高温计,表面温度计

are [ɑ:] *n.* 公亩(=100平方米)

area ['ɛəriə] *n.* 地下室前洼地,场地;基面,面积;地区,区域,领域;范围;区段;庭院;areal, *a.*
~ coefficient 面积系数
~ control 地区管制
~ co-ordinates 面积坐标
~ covering structural elemet 平面承重结构
~ covering structure 平面承重结构建筑物

~ curve 面积曲线
~ -depth curve 面积-深度曲线
~ dispatcher 工地调度员
~ distance method 面积距离法
~ district （土地利用规划的）面积分区
~ drain(age) 地面排水沟,露天排水斗
~ -elevation graph 面积-高程曲线图
~ factor 面积系数
~ flow meter 面积流量计
~ grating 阴井箅盖[格栅]
~ grating cover(ing) （阴井）格栅保护层
~ levelling 面水准测量
~ light 地下室窗采光井
~ load 均布[单位面积]荷载
~ measurement （表）面积测定
~ meter 截面流量计
~ method 面积法
~ moment 面积矩,力矩[弯矩,挠矩]面积
~ moment method 力矩面积法
~ of accumulation 堆积区
~ of activity （在一幢公寓楼内的）起居室[客厅]
~ of adhesion 粘着[咬合,联接]面积
~ of base 地基(底)面积
~ of bearing 支承面积
~ of building 建筑面积
~ of cohesion 粘着[结合,联接]面积
~ of contact 接触面(积)[范围]
~ of cooling 受冷[冷却]面积
~ of counter pressure 反压力面积
~ of cross section 横断面面积
~ of cup 压痕面积（试验硬度时用）
~ of depression 沉降区
~ of dispersion 扩散面积
~ of diversion 分水面积
~ of faulting 断层区
~ of fire grate 炉箅面积
~ of getting 影响面积

~ of glazing 装配玻璃的面积
~ of grate diagam 炉箅面积,火床面积
~ of heating surface 加热面积
~ of indentation 压痕面积
~ of influence 影响面积
~ of influence line 影响线面积
~ of injection orifice 喷口面积
~ of load distribution 荷载分布面积
~ of loading 荷载[负载]面积
~ of maximum rainfall 最大降雨面积
~ of moment 力矩[弯矩]面积
~ of passage 通路面积,有效截面（如阀门）
~ of perpetual shadow 永久阴影面（建筑物常年没有日照部分）
~ of pile head 桩顶面积
~ of pressure 承压[受压]面积
~ of reinforcement 钢筋面积
~ of rivet shaft 铆钉杆(截)面积
~ of safe operation 安全（施工）工作区
~ of solid angle projection 立体角投射面积
~ of steel 钢筋面积
~ of structural steel 型钢截面积
~ of structure 构造[建筑]面积
~ of subsidence 下沉面积
~ of the inlet 入口面积
~ of thrust surface 推力面积
~ of waterway 过流[水流]断面
~ of well influence 井影响面积[范围]
~ of winning （抽水）影响面积
~ of working 影响面积
~ ratio 面积比
~ redistribution 面积的再分配
~ requirement 需要面积[工作面]
~ rug （房间用）小地毯
~ separator 区域分隔带
~ substation 地段变电站
~ to be dredged [挖土]区[范围]

area

~ traffic control 地区交通控制
~ triangulation 面积三角测量
~ volume ratio 面积体积比
~ wall 地下室前空地周围的挡土墙
~ with main services 主干管服务面积
active ~ 有效[工作]面积
aggregate ~ of wires (钢丝绳)钢丝总面积
apparent gap ~ 空隙表面面积
artesian ~ 自流水[井]地区
average end ~ (计算土方量)平均端面积
backfilling ~ 充填区
balanced ~ 平衡面积
base ~ 底部断面面积
bearing ~ 支承[承载]面积
blade ~ 叶片面积
blind ~ (墙外)不露天空地
buffer ~ 缓冲区
building ~ 建筑面积
built-up ~ 组合面积
carrying ~ 承压[支撑]面积,升力面积
catchment ~ 集水[汇水,流域]面积；受雨区
chamber surface ~ 燃烧室表面积
channel ~ 隧道面积
charging ~ 装料台
compression ~ 受压面积,压缩区
contruction ~ 建筑[施工]面积
contact(ing) ~ 接触面积
critical throat ~ 喷嘴临界面面积
cross-connecting ~ 交接区[面]
cross-sectional ~ 横断面面积
developed ~ 展开面积
dewatering ~ 疏干区,排水区
discharge ~ 出口(截面)面积
distribution ~ 配线区
drag ~ 迎面阻力面积
drainage ~ 汇水面积
draining ~ 排水面积
effective ~ 有效面积

elemental ~ 单位面积,面积单位
enclosed ~ 有界[有限]区域
epicentral ~ 震中区
equivalent ~ 当量[等效]面积
equivalent frontal ~ 最大截面的等效面积
excess ~ 剩余面积
exclusion ~ 禁止区
exit ~ 喷管出口截面面积
fenced-off ~ 禁区
filling ~ 填充区
filter ~ 过滤面积
filtering ~ 过滤面积
floor ~ 楼板面积,(设备)占地面积
floor ~ of building 房屋面积
free ~ 有效截面积
friction ~ 摩擦面积
frontal ~ 最大截面；迎面面积
front face ~ 正面面积
gross ~ 全[总]面积
hot ~ 受热面,加热段
impeller inlet ~ 叶轮进口面积
injector hole ~ 喷嘴孔的通过截面面积
inlet-duct ~ 进口(截面)面积
intake ~ 进口(截面)面积
isolated ~ 独立区,隔离带
jet ~ 喷管出口截面面积
loading ~ 负载面积
local ~ 市区,局部地区
matching bond ~ 匹配焊接面积
meizoseismal ~ 强震区
metallic ~ of wire rope 钢丝绳有效金属断面
minimum nozzle ~ 喷管通道最小截面
net ~ 净[工作]面积
nozzle ~ 喷管通路面积
nozzle throat ~ 喷管喉部面积
operating ~ 操作[工作]面积(台)
passage ~ 流道断面,横截面面积
potential ~ 勘探面积[地区]

primary service ～ 基本服务区

residential

reference ～ 基准面,参考面;计算面积;规定面
reinforcing steel ～ 钢筋截面积
residential ～ 居住[住宅]区
rubbing ～ 摩擦面积
scenic ～ 风景区[胜地]
sectional ～ 截[断]面积
service ～ 有效(作用)区,作用区,服务区

scenic area

setting ～ 沉降截面[区]
shaded ～ 阴影面积
shadow ～ 阴影区,静区
shear(ing) ～ 受剪面积
slewing ～ 转动面积
slipping ～ 滑动表面,滑移面
specific surface ～ 比表面面积
spray ～ 喷涂面积
stack ～ 烟囱[炉身]面积
steel ～ 钢筋截面积
substrate bonding ～ 衬底键合面积
suburban ～ 市郊区
surface ～ 表面面积
supporting ～ 支承面
tension ～ 张拉[受拉]面积
throat ～ 喷口面积,喷管临界截面积
throat opening ～ 喷管喉部面积
total blast nozzle ～ 风嘴总面积
total surface ～ 总表面积
total tuyere ～ 风口总面积

transformed ～ 换算面积
unit ～ 单位面积
unshaded ～ 非阴影区
working ～ 工作区,操作地带
areal [ˈɛəriəl] *a.* 面积的,表面的,区[地]域的,地区
～ coordinates 重心坐标,面坐标
～ density 表面密度
～ effect of colour 装饰色彩的面积效应
～ eruption 区域喷溢
～ map 地区图
～ metric 面积度规
areaway [ˈɛəriəwei] *n.* 地下室窗前的采光井;地下室前空地;建筑物之间的过道[通道]
～ wall 采光井墙
arecoline hydrobromide 臭氰酸
arefaction [ˌæriˈfækʃən] *n.* 除湿,干燥
arena [əˈriːnə] *n.* 舞台,(古罗马)圆形竞技场;表演场设在观众席中央的剧场

arena

～ theatre 圆形剧场
～ -type stage 圆形剧场
arenaceous [ˌæriˈneiʃəs] *a.* 多砂质的,散碎的
～ deposit 砂屑岩,粗砂质
～ limestone 石灰砂岩,硅砂砖
～ limestone chip(ping)s 石灰砂岩屑[碎片]
～ quartz 石英砂
～ region 砂质地区[段]
～ rock 砂质岩
～ shale 砂质页岩,片砂岩
～ texture 砂质[松散]结构
arenite [ˈærinait] *n.* 砂质岩,砂粒碎屑岩
areolar tissue 蜂窝状组织

areolat(ed) [æˈriəleit(id)] *a.* 小空隙的,网眼状的

areolation [æˌriəˈleiʃən] *n.* 形成网眼状空隙,网眼状结构

areometer [ˌæriˈɔmitə] *n.* 液体比重计,浮秤,比浮计
~ analysis (液体)比重计分析

areometry [ˌæriˈɔmitri] *a.* 稠液[液体]比重测定法

areo-pycnometer *n.* 稠液[液体]比重计

argent *n.* ; *a.* 银(色,似,制)的

argental [ɑːˈdʒentəl] *a.* 银质的,含银的; *n.* 银汞膏

argentalium [ˌɑːdʒənˈtæliəm] *n.* 银铅

argentiferous lead 银铅

argentite [ˈɑːdʒəntait] *n.* 辉银矿

argil [ˈɑːdʒil] *n.* 陶土,白土,矾[酒]石

argilla [ˈɑːdʒilə] *n.* 泥土,铝氧土,陶[高岭]土

argillaceous [ˌɑːdʒiˈleiʃəs] *a.* 泥质的,含陶土的
~ cement 黏土水泥
~ conglomerate 泥质砾岩[集成物]
~ gravel 泥质砾石
~ gypsum 泥质石膏
~ iron ore 泥质铁矿
~ ironstone 泥质铁矿
~ limestone 泥质石灰岩,黏土质石灰岩
~ mud 泥质石灰岩,黏土浆
~ rock 泥岩,黏土岩
~ sand 泥质砂
~ sandstone 泥质砂岩
~ schist 泥质片岩
~ shale 泥质页岩
~ slate 泥质板岩

argilliferous [ˌɑːdʒiˈlifərəs] *a.* 含黏土的,黏土似的,产生黏土的

argillite [ˈɑːdʒilait] *n.* 泥质板岩,厚层泥岩,粘枝岩

argillo-arenaceous 泥砂质

argillo-calcareous 泥灰质

argillo-calcite 泥砂质石灰岩

argon [ˈɑːgɔn] *n.* 氩
~ arc(welding) 氩弧焊
~ lamp 氩(气)灯
~ welder 氩弧焊机

argument *n.* 幅角,幅度;自变量[数]争论,论证
~ of complex number 复数的幅角
~ of function 函数自变量
~ of vector 矢幅角

arid [ˈærid] *a.* 干旱的,干燥的;-aridness, *n.* ; -aridly, *ad.*
~ period 干季,旱季
~ region 干燥区,干旱地带
~ soil 干燥土

aridextor [ˈæriˈdekstə] *n.* 产生侧[横]向力的操纵机构[装置]

Ariron [ˈæriron] *n.* 耐酸铸铁

arising [əˈraisiŋ] *n.* 上升; *a.* 干旱的,干燥的
~ of joint edges (混凝土路面)缝边上升[胀缝,走拱]

arithmetic(al) [ˌæriθˈmetik(əl)] *a.* 算术的,算术上的
~ average 算术平均数
~ mean 算术平均数[值],等差中项
~ mean temperature difference 算术平均温度差
~ mode 运算方法[模式]
~ operation 算术运算
~ product (算术)乘积
~ progression[series] 算术[等差]级数
~ scale 标准[通用]的比例[尺度]
~ section[unit] 运算器
~ software 运算软件
~ solution 数值解
~ statement 算法语言

arithmometer [ˌæriθˈmɔmitə] *n.* (四则)运算计算机,计算器

ark [ɑːk] *n.* 柜,高处壁橱[龛];大而笨的房子

Arkansas soft pine 阿肯色(州)软松

arkose(sandstone) 长石砂岩,花岗质

砂岩

arm [ɑ:m] *n.* 支托[架],(杠杆)臂;扶手,把201 手柄;侧房;支管[线,路,流];排木,(脚手架上)支承脚手板的水平木桥;水湾,河套,港汊
~ -brace 撑脚
~ chair 扶手椅
~ crane 悬臂式起重机
~ elevator 臂式升降机
~ lie 攀条
~ of angle 角边
~ of balance 平衡臂
~ of couple 力偶臂
~ of crane 起重机悬臂,吊车臂
~ of force 力臂
~ of mixer 混合机浆臂,拌合机浆叶
~ of stability 稳定臂[力臂],回复力臂
~ of wheel 轮辐
~ rest 悬臂支承[支座],靠手,扶手
~ stirrer 浆叶式搅拌器
~ straight paddle mixer 直臂旋浆拌合机
~ saw 摇臂锯
~ -tie 斜撑,交叉撑,横臂拉条,撑杆
actuating ~ 驱动杆,工作臂
agitator ~ 搅动杆[叶轮]
ally ~ 架,支架,托架,悬臂
anchoring ~ 锚臂
balance ~ 平衡臂,均衡梁;称杆
bracket ~ 托架臂
bucket ~ (挖掘机的)斗柄
extension ~ 延伸臂
gib ~ 悬臂,(起重机的)斜支柱
mixing ~ 搅拌机轮叶
over ~ 横杆[臂],悬臂[梁]
overhanging ~ 外伸臂
righting ~ 回复[稳定]力臂
rock(er) ~ 摇臂
rotor ~ 转动臂

gib arm

stirrer ~ 搅拌机叶片,搅拌器搅板
support ~ 支架臂,支撑臂
valve driving [motion] ~ 阀驱动臂
wooden ~ 木担

armament ['ɑ:məmənt] *n.* 装备
~ factory 装备[脚手架]生产厂

armature ['ɑ:mətjuə] *n.* 骨架,框架;钢筋,加固[加强]料附件;衔铁,引铁;电枢

armboard ['ɑ:m'bɔ:d] *n.* 起纹板,搓花板

armchair ['ɑ:m'tʃɛə] *n.* 扶手椅,单人沙发

Armco *n.* 波纹白铁管[金属板]

armoire [ɑ:m'wɑ:] 装饰精美的大橱,大衣橱

armo(u)r ['ɑ:mə] *n.* 甲板,铠板

armo(u)red ['ɑ:məd] *a.* 装甲的,铠装的
~ cable 铠装电缆
~ coat 保护层
~ concrete 钢筋混凝土
~ concrete screed 钢筋混凝土找平层
~ concrete slab 钢筋混凝土板
~ conduit 包皮管道
~ corner 铠装棱角[边帮]
~ course (道路路面)保护层
~ curb 包装缘石,镶铁路缘
~ door 装甲[防火]门
~ (fire proof)door 装甲[防火]门
~ glass 钢化[联弹,铠装]玻璃
~ hose 缠(金属)丝软管,包皮管
~ joint 包铁接缝[接头]
~ material 铠装材料
~ paving slab (工业厂房地坪)装甲铺面板
~ paving tile (工业厂房地坪)装甲铺面砖[板]
~ plate 装甲板,铁板
~ pump 铠装泵
~ rolling contact joint 铠装球形活动接头

~ thermometer 铠装温度计
~ wire 铠装电线
~ wood 包铁(皮的)木材,加金属箍的木材
armo(u)ring [ˈɑːmərɪŋ] *n.* 装甲,铠装;布(钢)筋
  double ~ (钢筋混凝土中)复式钢筋
  iron-wire ~ 铁丝铠装
  metal ~ 金属铠装
  rigid ~ 刚性布筋
  steel tape ~ 钢带铠装
armo(u)rite rubber lining 橡胶衬砌
armo(u)rplate glass 钢化玻璃
armo(u)rply 装甲胶合板,金属面层胶合板
Arms bronze 特殊铝青铜
Arnold sterilizer 阿诺德灭菌器,常压蒸汽灭菌器
arnott *n.* 榧木(地板用木板)
ARODYN = aerodynamics 空气(气体)动力学
aroma *n.* (1)芳香,香味(料) (2)风格
aromatic-free 不含芳烃的,不芳香的
arone *n.* 芳酮
around *ad.; prep.* (1)在周围,经过 (2)在各处 (3)根据,以……为基础
around-the-clock *a.* 昼夜不停的,连续二十四小时的
arousal *n.* 唤起(醒)
arouse *vt.* 唤醒,引起
arr = arrival 到达
arrange [əˈreɪndʒ] *v.* -ranged, -ranging; 布置,装配,安装; arrangeable, *a.*; arranger, *n.*
arrangement [əˈreɪndʒmənt] *n.* 布置,排列,安装,安排;设备,装置
  ~ diagram 布置图
  ~ of bars 钢筋排列

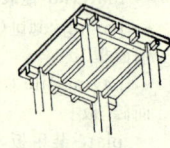
arrangement of beam

~ of beam 桁梁排列
~ of framework 桁架[构架]布置[排列]
~ of lanterns 道路照明[路灯]布置
~ of levers 操纵[控制]杆装置
~ of piles 桩的布置
~ of pole attachment 电杆附件
~ of props 支撑[架]布置
~ of reinforcement 钢筋排列
~ of station line 车站线路布置
~ of wire 布线
~ plan 布置图,配置图
back fire ~ 回火制止器,防回火装置
baffle ~ 挡板[折流板]排列法
clamping ~ 夹紧装置
coupling ~ 连接装置,连接机构
diagramatic ~ 示意[简化]布置图
dimmer ~ 遮光板
end(-to-end) ~ 纵向排列[配置]
fitting ~ 安装系统图
general ~ 总体布置,总平面布置
guiding ~ 引导[导向]装置
noise suppression ~ 噪声吸收器,防噪装置
pipe ~ 管系布置
pump ~ 水泵[抽送]装置
random ~ 无规则排列
reversing ~ 换向[转换]装置
spatial ~ 空间排列[布置]
starting ~ 起动装置
suspension ~ 吊架,锭吊
transposition ~ 交叉配置
tube ~ 管道布置
valve ~ 活门装置
winding ~ 提升设备
arranger [əˈreɪndʒə] *n.* 传动装置
arras [ˈærəs] *n.* 壁毡,花帷,饰墙毛毡
array [əˈreɪ] *n.* 排列,配置;数组,数列;修饰,装扮
arrears [əˈrɪəz] *n.* 未完成的工作;保养;养护;迟滞,拖延
arrest [əˈrest] *v.* 停止,阻止;制动,延迟

[滞];n. 制动装置;arrestable, a.
balance pan ~ 天平托盘
dust ~ 集尘
pan ~ 托盘

arrested [əˈrestid] a. 锁定的,制动的,抑制的
~ anticline 平缓背斜
~ decay 抑制分解阶段
~ failure 制止损坏,限制故障的扩展

arrester [əˈrestə] n. 避雷器,过压保险丝;制动器,锁定装置
airexpansion lightning ~ 气膨胀避雷器
aluminium cell lightning ~ 铝管避雷器
arc ~ 弧形避雷器
auto-valve lightning ~ 自动阀型避雷器
bird cage lightning ~ 鸟笼形避雷器
carbon ~ 碳精避雷器
comb lightning ~ 梳形避雷器
discharge ~ 放电避雷器
disc type lightning ~ 盘形避雷器
dust ~ 集尘[捕尘,挡尘,吸尘]器
earth ~ 一端接地的火花隙避雷器
flame ~ 灭火器,火焰消除器
flashback ~ 回火熄灭器
graded shunt ~ 多级分路避雷器
lightning ~ 避雷针[器]
needle gap lightning ~ 针隙避雷器
outdoor lightning ~ 室外避雷器
pin point ~ 针尖放电避雷器
sound ~ 隔声装置
spike ~ 尖头避雷器
spray ~ 喷水式避雷器
vacuum ~ 真空避雷器
water-column ~ 水柱避雷器
water-jet ~ 水柱避雷器

arrestment [əˈrestmənt] n. 停止;制动[停止]设备

arrests n. (加热,冷却的)临界点

arriere-voussure 墙面内的辅助拱,在厚墙内部或后部支承部分墙体的拱心

arris [ˈæris] n. 尖脊,边棱;棱角;道路接缝的圆角;两面接触线
~ compression 边缘压缩[力]
~ cover angle 边缘保护角
~ cover strip 边缘保护带
~ crack 边缘裂缝
~ edge 斜边
~es of joint (混凝土路面)接缝圆角
~ fillet 棱角线条[嵌块],尖嵌条
~ gutter V形檐槽
~ hip tile 回坡顶垂脊脊瓦
~ length 边缘长度
~ of joint 接缝圆角
~ of slab 板肋,板棱
~ pressure 边缘压力
~ protection 边缘保护
~ rail 三角轨
~ rounding 边缘倒棱,锐棱倒圆,进口锐边磨圆(为改善孔口流量系数)
~ section 边缘断面
~ stress 边缘应力
~ trowel 边角抹子[泥刀]
~ ways 六角形瓦铺屋面,龟甲[六角]形石板,金属板屋面
~ -wise 成对角方向(铺砌,砌砖)排成锯形
~ -sing tool 边缘抹子[泥刀],(新浇混凝土边缘)磨圆器

arrish rail 三角拉杆
arrish tile 棱瓦
arrival [əˈraivəl] n. 到达,到
~ card 更改地址通知单
~ current- 终端(输入)电流
~ level 行近平面
~ lounge (机场)休息厅
~ platform 末站月台,到达站台
~ time 到达时间
~ track (铁路的)到达线
~ wave 来波

cash on ～ 货到付款
new ～ 新到货物
arrondi *a.* 曲线[弧]形的
arrow ['ærəu] *n.* 箭头；指针，测针；标杆；*vt.* 标以箭头 arrowless, *a.* arrowlike, *a.*
～ diagram 网格图
～ head 箭头，楔形符号，箭头[三角形]线迹
～ headed 箭头[楔形]的
～ head twills 人字斜纹
～ height 矢高,臂高
～ loop 炮眼
～ point bracing 测量砚标支撑
～ sign 箭头,指针
～ -type network 箭头式网络
traffic ～ 交通箭头标志
arroyo [əˈneiəu] *n. pl.* -os；干涸河道；旱谷；小河，小溪
ars [ɑːz] *n.* 【拉】艺术
arsenate [ˈɑːsinit] *n.* 砷酸盐
～ dinitrophenol（木材防腐用）砷酸二硝基酚
arsenic [ˈɑːsnik] *a.* 含砷的；*n.* 砷（As）
～ acid 砷酸
～al copper 砷铜合金
～ pigment 含砷颜料
～ sulphide 雄黄,硫化砷
～ trioxide 三氧化二砷
white ～ 砒霜
art [ɑːt] *n.* 艺术；技术，技艺[能]；独创性
～ and craft room 美术工艺室
～ collection 艺术收藏品
～ college 艺术学院
～ connoisseur 艺术鉴赏家[鉴定家]
～ effect 艺术效果
～ form 艺术形式
～ gallery 画廊,美术陈列馆
～ glass 美术[工艺]玻璃

～ glassware 艺术玻璃器皿
～ linen 装饰[用]亚麻平布（作椅垫，台布,刺绣地布等用）
～ lover 艺术鉴赏家
～ museum 艺术博物馆
Art Nouveau(19 世纪末法国和比利时的)新艺术运动[风格派]
Art Nouveau faience 新艺术风格的（上釉）陶瓷
～ of building 建筑[艺]术
～ of fortification 要塞[防御]建筑艺术
～ of sculpture 雕塑艺术
～ of the individual 个性艺术
～ of vaulting 圆顶建筑艺术
～ paper 铜板纸
～s center 艺术厅
～ school 艺术学校
～ square 图案方毯
～ ticking 印花床垫；印花枕头布；印花床单
～ treasure 文物
～ work 布线图,原图,工艺图[品]
ancient ～ 古代艺术
architectural ～ 建筑艺术
Bezantine ～ 拜占庭艺术
fine ～s（雕塑,建筑,绘画等）美术
formative ～ 造形艺术
Gothic ～ 哥特艺术
Greek ～ 希腊艺术
industrial ～ 工艺美术

Gothic art

Minimal ～ 抽象派艺术(用简单几何形体构成的雕塑)
plastic ～ 雕塑[造型]艺术
sculptural ～ 雕刻艺术
useful～s 手艺,工艺
work of ～ 艺术品
artbond [ˈɑːtbɔnd] *n.* 粘氯乙烯薄膜钢板
art-director 艺术指导
artefact [ˈɑːtifækt] = artifact *n.* 人工制品,仿制品

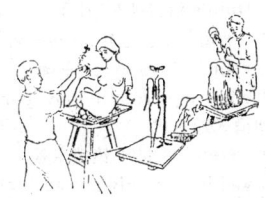
sculptural art

Artemision [ˌɑ:təˈmisiːɔn] *n.* （古希腊）月神庙

arterial [ɑ:ˈtiərəl] *a.* 干线的，主干的，干道的
~ canal 运河干河
~ drainage 排水干管[干渠]；干管[干渠]排水系统，有支渠排水系统
~ grid 干线网
~ highway 干线公路[道路]，干路
~ pair 双线干道
~ pattern 干道类型
~ railway 铁路干线
~ road 主干道，干(线公)路
~ street 城市干道，主要街道
~ system of distribution 配水干管[线]系统
~ traffic 干线交通

arteries of communication 交通干线，交通网

artery [ˈɑːtəri] *n.* 交通要道，干线，大道，动脉
economic ~ 经济命脉
ground water ~ 地下水干道
main traffic ~ 交通干线

artesian [ɑ:ˈtiːzjən] *a.* 自流的，承压的
~ aquifer 承压[自流]含水层
~ basin 承压水[自流泉]盆地
~ bored well 钻出的自流井
~ capacity (井的)自流量，自流产水率
~ condition (地下水)承压条件，承压[自流]情况[状态]
~ discharge 自流井出水量[流量]
~ flow 自流水流
~ flow area 自流区域
~ fountain 自流喷泉
~ ground-water 自流[井]水
~ head 自流[承压]水头
~ loss 自流(水)损失
~ pressure 自流压力[水压]
~ (pressure) head 自流水头
~ pressure surface 自流水压面
~ spring 自流泉
~ spring tank 涌水池
~ surface 自流井压力水位
~ waste 自流排水
~ water 自流井水
~ water power 自流水动力
~ well 自流井
~ well capacity 自流井出水量
~ well pump 深井泵，自流井泵

Arthrospira *n.* 节旋藻

article [ˈɑːtikl] *n.* 条款，项目；制品，产品；论文 *vt.* (1) 把……逐条(登载，罗列)，分条列举；(2) 用条款约束
~ of contraband 违禁物品
all-rubber ~ 全胶制品
definite ~ 定冠词(即 the)
feature ~ 特论
fragile ~ 易碎品，脆弱体
dipped ~ 浸渍制品，无缝制品
hollow ~ 空心制品
rubber and canvas ~ 橡皮帆布制品

articulate [ɑ:ˈtikjuleit] *v.* -lated,-lating；环接，连接，接合；*a.* 铰接的；articulability, *n.*; articulable, *a.*; articulacy, articulateness, *n.*; articulately, *ad.*; articulative, *a.*
~d 铰接的,有活节的
~d arch(ed) girder 铰接拱梁
~d arm 软节臂，蛇节杆
~d arm-type stone grinder 软节臂式石头打磨机
~d bar 铰接杆件
~d beam 联接梁
~d bed-plate 铰接底板
~d boom 铰接悬臂

~d bridge 铰接桥

~d buttress dam 带活接支撑的垛坎[扶壁式坝]

articulated bridge

~d chute （浇灌混凝土用）象鼻管，溜管

~d concrete 铰接式混凝土（板），活节混凝土块（钢筋连通的串块，可随坡度ध坑地自由变形）

~d concrete matting 活节混凝土块罩面（护坡）

~d concrete mattress 活节混凝土褥垫（护坡）

~d connection 活节接合

~d connection rod 活节连杆

~d construction 装配式构造，活节式构造

~d coupling 活节联结器，活头车钩

~d dump truck 有铰接的翻斗车

~d flat-slab buttress type dam 活节平板支墩坝

~d gate shoe 支铰，座铰

articulated dump truck

~d jib 活节臂

~d joint 关节[活节]接合，铰接

~d mat (type) concrete pavement 活节铺块式混凝土铺面

~d mirror 万向转镜

~d pin 活节销

~d plate 活接[连接]板

~d portal frame 铰支门形架

articulated trailer

~d purlin(e) 活节檩条

~d (roadway) arch 铰接拱

~d rocker （桥梁）铰接摇座

~d slab 蛇节连接板

~d steel plate 铰接钢板

~d system 铰接[活接]系统

~d traffic 拖挂车运输

~d trailer 活接拖车[挂车]

~d turntable 铰接转盘[转车台]

~d vehicle 活接[连接]车辆

~d vibrating trough conveyor 铰接振动槽式输送机

~d wood(en) slat 铰接木板

~d yielding (roadway) arch 可变形铰接拱

articulating boom 活接头弦杆

articulation [ɑːtikjuˈleiʃən] *n.* 关[活]节，铰，铰链轴心；（声的）清晰度

~ block 铰块

~ by ball and socket 球窝关节

~ test 清晰度试验，连接检验

artificer [ɑːˈtifisə] *n.* 技师；技术员；工长；技工；设计者

artificial [ˌɑːtiˈfiʃəl] *a.* 人工的，人造的；artificially, *ad.*

~ abradant 人工金刚砂[磨料]

~ abrasive 人工金刚砂[磨料]

~ adhesive 人工合成胶粘剂

~ ageing 人工硬化[老化]

~ aggregate 人工混凝土骨料（如煤渣，矿渣，陶粒）

~ anhydrite 人工无水石膏

~ anthracite 人造无烟煤

~ asbestos 人造石棉

~ asphalt 人造石油沥青[柏油]

~ atmosphere 调节空气

~ bitumen 人造沥青

~ block dyke 人造方块堤

~ (bonding) adhesive 合成树脂胶

~ (bonding) agent 合成树脂胶

~ (bonding) medium 合成树脂胶

~ carbonation 碳化处理

~ cement 人造水泥

~ cementation 人工压浆处理，人工胶接

~ cementing 人工压浆处理

~ cementing agent 人造水泥

~ cementing method 人工压浆处理法，岩石加固法

~ coarse aggregate 人造粗骨料

~ cold 人工冷却
~ (concrete) aggregate 人造混凝土骨料
~ consolidation 人工地基密实
~ contaminant 人工污染
~ cooling 人工冷却
~ corundum 人造金刚砂
~ cotton 人造棉
~ crystal 人造晶体
~ error 人为误差
~ daylight 太阳灯,人工日光
~ draft 人工通风
~ drainage 人工排水
~ draught 人工通风
~ drying out 人工干燥
~ drying oven 人工干燥炉[室]
~ earthing 人工接地
~ environment 人工环境
~ fibres 人造纤维
~ filter media 人工滤料
~ filtrable membrane 人工滤膜
~ fine aggregate 人造细骨料
~ fine grain 人造细粒
~ flushing 人工冲洗
~ freezing of ground 地基冰结(过程),地基冰冻法
~ glue 合成树脂胶
~ grass 人工草皮
~ grindstone 人造磨(刀)石
~ ground 人为接地
~ groundwater 人工地下水
~ harbour 人工港
~ hydraulic mortar 人工水凝灰浆
~ ice rink 人造溜冰场
~ illumination 人工照明
~ indoor illumination 室内人工[光源]照明
~ indoor lighting 室内人工光源照明
~ intelligence (AI) 人工智能
~ island 人工岛
~ island method 人工筑岛法,(深基础施工时下沉井,沉箱用)

~ leather 人造革
~ leather hanging 人造革挂毯
~ levee 防洪堤
~ lighting 人工照明
~ light source 人工光源
~ light-weight aggregate 人造轻质骨料
~ limestone 人造石灰石
~ load 虚负载,模拟负荷
~ marble 人造大理石,仿云石
~ masonry unit 人造圬工用砖石
~ mica 人造云母
~ navigable waterway 人工航道,运河
~ navigation canal 人工航道[通航运河]
~ pond 人工池
~ pozzolan (-icmaterial) 人造火山灰(建材)
~ radiation 人工辐射
~ rainfall 人工降雨
~ recharge 人工重装,人工灌注
~ recharge of ground water 人工地下水补给[回灌]
~ replenishment 人工填充(砂等)
~ resin 合成树脂
~ (resin based) bonding adhesive 合成树脂胶
~ (resin-based) bonding agent 合成树脂胶
~ (resin-based) cement(ing) agent 合成树脂胶
~ (resin-based) glue 合成树脂胶
~ rubber 人造[合成]橡胶
~ sand 人造砂,砂砾
~ sandstone 人造砂石
~ silk 人造丝
~ slate 人造石板
~ skating rink 人工溜冰场
~ slipway 人工浮道
~ sludge drying 人工污泥干燥
~ solidification 人工压浆处理,人工胶接

~ stabilization 人为稳定
~ stone 人造石
~ stone block 人造石块
~ stone floor cover(ing) 人造石[水磨石]楼面或地面
~ stone finish 水磨石饰面

artifical stone floor cover(ing)

~ stone floor(ing) finish 人造石楼面修整
~ stone pavement 人造石铺砌层
~ stone shop 人造石工场
~ stone skin 人造石面层
~ stone stair 人造石楼梯
~ stone tile 人造石砖
~ stone tile floor(ing) (finish) 人造石花砖地面装修
~ stone tile floor cover(ing) 人造石花砖饰面[贴面,楼面]
~ stone waterproofer 人造石防水
~ stone work 人造石构件
~ storage 人工蓄水
~ stream 人工河流
~ subgrade 人工路基[地基]
~ sunlight lamp 人造太阳灯
~ sub-irrigation 人工地下灌溉,人工过滤场
~ travertine 人造凝灰石
~ ventilation 人工通风[换气]
~ venting 人工通风
~ watercourse 人工水道
~ watering 人工浇水,人工雨
~ waterway 人工运河[航道]
~ weathering 人工风化[风蚀]
~ zeolite 人造沸石

artificially [ˌɑːtiˈfiʃəli] **ad.** 人工[造]地
~ aged aluminium 人工老化铝
~ graded aggregate 人工筛分[级配]的骨料
~ refractory construction material 人工耐火建筑材料
~ recharged ground water 人工补充的地下水

artisan [ˈɑːtiˈzæn] **n.** 手工业工人,技工,工匠;-artisanal, **a.**; artisanship, **n.**

artist [ˈɑːtist] **n.** 艺术家,画家
~'s studio 画家[艺术家]工作室

artistic [ɑːˈtistik] **a.** 艺术的;精美的; artisitical, **a.**; artisitically, **ad.**
~ adviser 艺术顾问
~ ceramics 美术陶瓷
~ circle 艺术范围[领域]
~ design 艺术造形[设计]
~ expression 艺术表现
~ form 艺术形式
~ formation 艺术造形
~ (form of) expression 艺术表现形式
~ glaze tile 精美琉璃瓦
~ monument 艺术纪念牌
~ tapestry 工艺美术壁毯

artistic expression

artistry [ˈɑːtistri] **n. pl.** -ries;新技艺;艺术性质;出于艺术家之手

arts and crafts 工艺,造型美术,手工艺品

Artz press sheet 特殊薄钢板

aruhuesiru **n.** (预填集料、灌浆混凝土的)外加剂,掺合剂

as- 词头]在……状态下
~ -built drawing 竣工图,建筑情况图[记录图]
~ -built for settling the accounts 竣工图,结算图
~ -cold 处于冷却状态的
~ -completed drawing 竣工图,建筑(情况)记录图

~ -completed drawing setting the accounts 结算图,竣工图
~ -constructed drawing 竣工图
~ -extruded 挤(压)出来的,压出的
~ -welded condition 焊接状态

asarotum [ˌæsəˈrəutəm] *n. pl.* -ta; 拼花地面,油彩地(古罗马建筑的彩色路面或桥面油彩地)

asbestic [æsˈbestik] *a.* 石棉的
~ half-round tile 石棉(水泥)脊瓦,半圆形石棉瓦

asbestiform [æzˈbestifɔːm] *a.* 石棉状的,石棉构造的
~ half-round tile 半圆形石棉瓦
~ tile 石棉瓦

asbestine [æzˈbestain] *a.* 石棉状的,不燃性的 *n.* 滑石棉

asbeston [æzˈbestən] *n.* 防火布

asbestonite *n.* 石棉制绝热[保温]材料

asbestophalt [æzˈbestəfælt] *n.* 石棉地沥青

asbestos [æzˈbestəs] *n.* 石棉,石绒; -asbestous,-asbestoid,asbestoidal, *a.*
~ apron 石棉底板[档板,盖板]
~ article 石棉制品
~ base(d)(asphalt-) bitumen felt (ed fabric) 石棉沥青油毛毡
~ base laminate 层压石棉板
~ base(d) asphalt paper 石棉沥青油毛毡
~ bitumen 石棉沥青
~ blanket 石棉毡,石棉保温层
~ block 石棉块
~ board 石棉板
~ boarding 石棉铺钉板
~ brake lining 石棉闸衬
~ brick 石棉砖
~ building board (sheet) 建筑石棉板
~ calcium silicate board 石棉硅酸钙板
~ canvas 石棉布
~ cardboard 石棉纸板
~ card liner 石棉衬垫[衬圈]

~ cement 石棉水泥[胶浆]
~ cement article 石棉水泥制品
~ cement board 石棉水泥板
~ cement box (roof) gutter 箱形石棉水泥天沟
~ cement (builiding) component 石棉水泥建筑构件
~ -cement (building) member (unit) 石棉水泥建筑构件
~ -cement cellulose 石棉水泥纤维素
~ -cement cistern 石棉水泥蓄水[雨]池[水槽]
~ -cement closure 石棉水泥封口[封闭]
~ -cement component 石棉水泥建筑构件
~ -cement conductor 石棉水泥雨水落水管
~ -cement corrugated board (panle) 石棉水泥波纹板
~ -cement corrugated roof cladding 石棉水泥波纹屋顶
~ -cement corrugated roof(ing) board 石棉水泥波纹屋面板
~ -cement corrugated sheet (ing) 石棉水泥波纹板
~ -cement discharge pipe 石棉水泥排水管[活水管]
~ -cement distance piece 石棉水泥隔板
~ -cement downcomer 石棉水泥水落管
~ -cement downpipe (downspout) 石棉水泥水落管
~ -cement drain (age) pipe 石棉水泥排水管[活水管]
~ -cement duct 石棉水泥管
~ -cement eave(s) gutter (trough) 石棉水泥檐沟
~ -cement extract ventilation unit (extraction unit)石棉水泥抽气通风装置
~ -cement facade 石棉水泥铺面

[饰面]
~ -cement facade slab 石棉水泥饰面板
~ -cement facing 石棉水泥粉[饰]面
~ -cement facing sheet 石棉水泥饰面板
~ -cement fall pipe 石棉水泥落水管
~ -cement fascia board (for flat roof(s))(用于平屋面的)石棉水泥挑口板
~ -cement fence 石棉水泥栅栏
~ -cement fitting 石棉水泥(管)配件
~ -cement flat board 石棉水泥(平)板
~ -cement flat (run) panel (sheet) 石棉水泥(镶面)板
~ -cement flat siding 石棉水泥平[墙]板(外墙防雨用)
~ -cement flooring 石棉水泥地板材料
~ -cement flower container 石棉水泥花盆
~ -cement flue 石棉水泥烟道

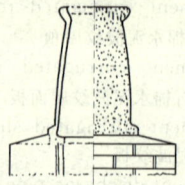

asbestos-cement flue

~ -cement fluted board 石棉水泥槽纹板
~ -cement foul water pipe 石棉水泥污水管
~ -cement fountain basin 石棉水泥喷水[泉]池
~ -cement front 石棉水泥粉面[饰面]
~ -cement gutter 石棉水泥檐沟
~ -cement joint 石棉水泥接头[接缝]
~ -cement leader 石棉水泥雨水落水管
~ -cement lining 石棉水泥衬垫[衬里]

~ -cement (lining) sheet 石棉水泥衬板[封檐板]
~ -cement member 石棉水泥建筑构件
~ -cement mortar 石棉水泥砂浆
~ -cement panel 石棉水泥板
~ -cement partition (wall) 石棉水泥隔墙
~ -cement pipe 石棉水泥管
~ -cement pressure pipe 石棉水泥压力管
~ -cement product 石棉水泥制品
~ -cement profile(d) board (sheet) 石棉水泥异型板
~ -cement refuse water pipe 石棉水泥排水管
~ -cement ridge capping (covering) tile 石棉水泥脊瓦
~ -cement ridging tile 石棉水泥脊瓦
~ -cement roof cladding (covering) 石棉水泥屋面覆盖层

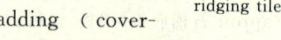

asbestos-cement ridging tile

~ -cement roof(ing) 石棉水泥屋面
~ -cement roof(ing) board (panel, sheet) 石棉水泥屋面板
~ -cement rubber tile 石棉水泥橡胶瓦
~ -cement separator 石棉水泥隔板
~ -cement septic tank 石棉水泥化粪池
~ -cement sewage (sewer) pipe 石棉水泥污水管
~ -cement shake 石棉水泥瓦
~ -cement sheet 石棉水泥衬板[封檐板]
~ -cement sheeting 石棉水泥(镶面)板
~ -cement shingle 石棉水泥瓦
~ -cement siding 石棉水泥防雨板[壁板,墙板]

~ -cement siding shake 石棉水泥墙面板
~ -cement slate 石棉水泥屋面石板瓦
~ -cement slate roofing (铺)石棉水泥平瓦[板]屋面
~ -cement solid board 石棉水泥实心板
~ -cement surfacing sheet 石棉水泥饰面板
~ -cement tile 石棉水泥瓦
~ -cement unit 石棉水泥建筑构件
~ -cement valley boards 石棉水泥斜沟底板
~ -cement valley gutter 石棉水泥斜沟槽
~ -cement vent (ilating) pipe 石棉水泥通风管
~ -cement ventilator 石棉水泥材料通风装置
~ -cement wall board 石棉水泥墙板
~ -cement wall panel 石棉水泥墙板
~ -cement wall sheet (shingle) 石棉水泥墙板
~ -cement ware 石棉水泥制品
~ -cement window sill 石棉水泥窗台板
~ cloth 石棉布,石棉织品
~ clothing 石棉衬里[保温]
~ concrete 石棉混凝土
~ concrete pipe 石棉混凝土管
~ concrete slab 石棉混凝土板
~ cord 石棉绳
~ cord covering 石棉绳缠层
~ core 石棉型心
~ corrugated board 石棉波纹板
~ corrugated panel 石棉波纹板
~ cover 石棉盖[套]
~ covering 石棉覆盖(层)
~ curtain 石棉(防火)幕
~ cushion 石棉垫
~ dressing 石棉加工
~ dust 石棉粉尘
~ fabric 石棉纤维
~ felt 石棉毡
~ fibre 石棉纤维[绒]
~ fibre board 石棉纤维板
~ fibre insulation 石棉纤维保温
~ fibre-reinforced sheet(ing) 石棉纤维加筋(墙)板
~ filler 石棉填料[垫片]
~ filter 石棉滤纸[滤器]
~ filter cloth 石棉滤布
~ firebrick 石棉耐火砖
~ flat (building) board 石棉(建筑)平板
~ flat (building) sheet 石棉平板
~ float 石棉绒
~ flour 石棉粉末
~ foamed concrete 石棉泡沫混凝土
~ gasket 石棉垫片
~ gland packing 石棉压盖填料
~ gutter 石棉雨水天沟
~ hose 石棉软管
~ insert (ion) 石棉垫
~ insulating board (sheet) 石棉保温板
~ joint runner 石棉接口[接头]浇道

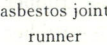

asbestos joint runner

~ layer 石棉层
~ laminate 石棉层压板
~ lining 石棉衬(垫)
~ mill board 石棉板,石棉书皮纸板
~ mineral 石棉矿石
~ mortar 石棉灰浆
~ packing 石棉填料[衬垫,填密件]
~ pad 石棉垫
~ paint 石棉涂料
~ panel 石棉板
~ paper 石棉纸
~ paper gasket 石棉纸垫片
~ pipe 石棉管
~ plaster 石棉粉饰[灰浆]
~ plastic floor(ing) 石棉-塑料地板[楼面板]

~ plate 石棉板
~ plywood 石棉层压板
~ powder 石棉粉末[灰]
~ product 石棉制品
~ protected metal roof(ing) 石棉保护金属屋顶
~ quarry 石棉采矿场
~ rainwater gutter 石棉屋面雨水天沟
~ rock 石棉岩
~ (roof) gutter 石棉(屋面)雨水天沟
~ roof shingle 石棉屋面瓦
~ roofing 石棉屋面材料,石棉毡浸沥青卷材
~ roofing sheet 石棉屋面板
~ rope 石棉绳

asbestes roofing sheet

~ roving 石棉粗纱
~ rubber tile 石棉橡胶板
~ rubber gasket 石棉橡胶密封垫片
~ rubber sheet 石棉胶板,夹胶石棉板
~ rubber tile 石棉橡胶瓦
~ sheet 石棉板[片]
~ sheet packing 石棉密封垫片,石棉片密封
~ shingle 石棉瓦
~ slate 石棉岩板[水泥板]
~ sliver 石棉条
~ tape 石棉(扁)带
~ tester 石棉试验机
~ texolite 石棉胶布板
~ thread 石棉线[绳]
~ tile 石棉瓦,石棉水泥板
~ tile works 石棉瓦厂
~ -veneer plywood 石棉夹心胶合板

~ -vinyl composition 石棉乙烯制品
~ -vinyl floor covering 柔性地板复面层
~ -vinyl floor(ing) finish 柔性楼面[楼板]整修[修面]
~ -vinyl mass (material) 石棉-乙烯(制品)材料
~ -vinyl tile 石棉乙烯饰面板,柔性饰面板
~ wallboard 石棉墙板
~ washer 石棉垫圈
~ wire 石棉线
~ wire gauze 石棉衬(铉)网
~ wool 石棉绒
~ woven fabric 石棉布
~ yarn 石棉绒
blue ~ 青石棉
flaked ~ 石棉粉
platinized ~ 载铂石棉
pulverized ~ 石棉粉
sheet ~ 石棉片
sprayed ~ 喷涂石棉隔声材料
asbolane [æzˈbəlein] *n.* 钴土
asbolite [ˈæzbəlait] *n.* 钴土
ASCE＝American Society of Civil Engineers 美国土木工程师学会
ascend [əˈsend] *vi.* 上升,登;ascendable,ascendible, *a.*
ascending [əˈsendiŋ] *a.* 向上的,上升的
~ air (散热器)上升气流
~ angle 爬高角
~ branch 向上分支,上行支管
~ development 上行展开(法)
~ grade 升[上]坡,上坡度
~ method 上行法
~ node 升交点
~ pipe 直管,(水泵)注水管,压入[增压]管
~ power 升力
~ tube 上行管
~ velocity 上升速度
ascension [əˈsenʃən] *n.* 上升[浮],

往上
~ al ventilation 上升[热压]通风
~ force 升力

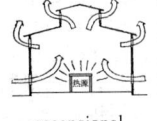
ascensional ventilation

ascensor [ə'sensə] *n.* 竖式缆索铁路

ascent [ə'sent] *n.* 阶梯;坡度,斜度[率];爬高,上升;壅[回]水
capillary ~ 毛细上升
gentle ~ 缓坡
rapid ~ 陡坡

aseismatic [ˌeisais'mætik] *a.* 耐震的,不受震动的;抗地震的
~ design 抗震设计
~ region 无(地)震区
~ structure 抗震结构
~ structure with shear wall 抗震剪力墙结构
~ structure with shear wall core 抗震剪力墙筒体结构
~ structure with shear wall frame interaction system 抗剪力墙框架相互作用体系(的)抗震结构
~ structure with shear wall system 抗剪力墙体系(的)抗震结构

aseismic [ei'saismik] *a.* 无地震的,抗[耐](地)震的;aseismicity, *n.*
~ region 无地震区

ash [æʃ] *n.* 粉尘,尘埃,火山灰,煤渣灰;梣木,槐木
~ bed 火山灰层
~ box 灰箱
~ bunker 灰斗[槽]
~ can 灰桶,废料箱
~ car 灰车,垃圾车
~ cellar 灰坑
~ channel 灰道
~ concrete 灰渣混凝土
~ content 灰分,含灰量
~ conveyer 灰渣输送带
~ crusher 碎渣机
~ door 灰门

~ drop 炉灰道
~ dump 灰堆,由壁炉[炉子]到底部灰坑的通孔
~ ejector 排灰器
~ erosion 炉内结渣
~ formation 灰层,灰渣层
~ free coal 不含灰分煤
~ furnace 灰窑
~ fusibility 灰分熔度
~ fusion temperature 灰溶温度
~ grey 灰白色
~ -handling pump 灰尘处理泵
~ hole 出灰洞,灰洞
~ -hopper 底卸式灰斗
~ lift 提灰机
~ lime block 灰渣石灰砖(砌)块
~ mortar 毛石砂浆
~ mota 粗硬的黄麻纤维
~ pan damper 炉灰挡板
~ pipe 灰管
~ pit 灰坑
~ pit door 除灰门,灰坑门
~ removal 清[排]灰,脱灰
~ -rich fuel 多灰燃烧
~ separator 飞灰分离器
~ slate (火山)灰岩板
~ structure 火山灰结构
~ tile 灰渣石灰(砌)块
~ tray 灰盘
~ -tuff 灰质凝灰岩
~ -valve 排灰阀
~ zone 灰层,火山灰降落范围

asher *n.* 堆灰场,烧灰场

ashing *n.* 抛光,磨光

A-ship *n.* 核动力船

ashlar ['æʃlə] =ashler *n.* 装饰屋内墙面的石板,薄方石,毛石,琢石;屋顶层中在楼板梁和椽子之间的短立筋
~ bond 琢石砌合
~ bonder 琢石砌体砌合丁砖
~ bond stone 琢石束石
~ brick 琢石砖,面砖

~ buttress 琢石扶壁
~ facing 琢石镶[饰]面
~ joint 琢石接缝
~ line 琢石线,外墙面线
~ masonry (work) 琢石砌体建筑,琢石圬工
~ paving 琢石路面,铺砌琢石路面
~ pavement 琢石层
~ piece 砌琢石,琢石镶面
~ pier 琢石墩[柱]
~ slab 琢石板
~ structure 琢石结构[建筑物]
~ vault 琢石穹窿
~ wall 琢石墙
~ window 琢石窗

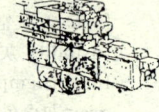
ashlar masonry

ashlaring ['æʃləriŋ] n. 琢石镶面,砌琢石(墙),贴琢石(墙),用方石建筑
ashler ['æʃlə] n. 琢石,方石
ashlering ['æʃləriŋ] n. 阁楼立柱;砌琢石
ashtray n. 烟灰缸,烟灰盘
ashtree n. 水曲柳
ashy ['æʃi] a., ashier, ashiest; 灰(色)的
~ shale 火山灰泥板岩,灰色泥板岩,灰色页岩
Asia ['eiʃə] n. 亚洲
Asiatic sweetleaf 白檀
asisculis [ə'siskjulis] n. 石匠锤
askew [əs'kju:] a. 斜的; ad. 歪
~ arch 斜拱,歪曲拱
~ bridge 斜桥
ASL= above sea level 海拔高度
= applied science laboratory 应用科学实验室
= astrosurveillance science laboratory 天文探测科学实验室
aslant [ə'slɑ:nt] ad.; prep. 成斜角地,(倾)斜地
aslope [ə'sləup] ad.; 斜坡地,倾斜地; a. 有坡度的,倾斜的

aspect ['æspekt] n. 外表,形态;方向[位];方式
~ ratio 形状比,长度比,(叶片)纵横比,(纤维混凝土中纤维)长度与直径之比
aspen ['æspən] n. 白杨,杨木
~ wood 山杨木
asperity [æs'periti] n. 粗糙度[性],不平滑
aspersus ['æspəsəs] a. 有粗点的
asphalt ['æsfælt] n. 石油沥青,柏油 vt. 涂柏油,用沥青铺(面,路)
~ addition 石油沥青掺合剂
~ adhesion-preventing agent (石油)沥青防粘剂
~ adhesive (石油)沥青胶合剂
~ -aggregate mix(ture) (石油)沥青-集料混合料
~ -asbestos composition (石油)沥青-石棉拌合料
~ -asbestos fiber cement (石油)沥青-石棉纤维水泥
~ -asbestos mastic 沥青-石棉玛琋脂
~ -asbestos material 沥青石棉材料
~ base coating 沥青底涂层
~ base (course) 沥青基[打底层]
~ base oil 沥青冷底子油
~ base paint 沥青打油子涂料
~ based mastic joint sealer 沥青基玛琋脂填缝料
~ based paste (石油)沥青基涂料
~ based rust protective paint 沥青防锈涂料
~ batcher 沥青配料计量器
~ binder (course) 沥青结合料
~ bitumen 沥青
~ block 沥青块
~ block pavement 沥青块铺砌路面
~ board strip 沥青填缝条
~ bonding adhesive 沥青胶粘剂
~ bridge carriageway pavement 桥梁车行道沥青铺面层
~ brequette 标准沥青水泥试块

~ building paper 沥青油毡
~ building-up roof(ing) 沥青组合屋面
~ cake 沥青砂胶块
~ carpet coat 沥青油毡层
~ clinker 沥青熔渣
~ coat 沥青涂层
~ coated aggregate 沥青拌制的集料
~ -coated chip(ping)s carpet 涂沥青片石铺面层
~ -coated gravel 涂沥青砾石
~ coated pasteboord 涂沥青硬纸板
~ -coated sand 涂沥青砂
~ -coat(ing) 沥青涂层[面层]
~ coat(ing) material 沥青覆涂胶料
~ coat(ing) roof(ing) 涂沥青的屋面
~ colour coat 有色沥青封层,沥青色的面层
~ composition 沥青拌合料
~ concrete base 沥青混凝土底层[垫层,防水层]
~ concrete carpet 沥青混凝土(铺)面层
~ concrete mixer 沥青混凝土拌合机
~ concrete pavement 沥青混凝土路面
~ concrete surfacing 沥青混凝土铺面
~ cork surface 沥青软木贴面
~ covering 沥青面层
~ damp-proof course 沥青防潮层
~ deck pavement 桥梁行车道沥青铺面层
~ distributor 沥青喷洒机
~ dry penetration surfacing 表面经沥青处理的碎石
~ ductility testing machine 沥青延度试验仪

asphalt-felt roof covering

~ emulsion 乳化沥青
~ facing 沥青涂面[覆面]层
~ felt 油毛毡,(衬垫用)沥青毡

~ -felt roof covering 屋面覆面[防水]油毛毡
~ filler 沥青填料[填充物]
~ -filler mix(true) 沥青填料混凝土
~ film 沥青薄膜
~ finisher 沥青整修机
~ floor cover(ing) 沥青楼面[地面]
~ floor finish 沥青地面
~ fog coat 沥青雾层(不撒石屑极薄表面处理层)
~ grade 沥青等级[种类]
~ gravel 沥青砾石
~ groove joint 沥青槽接
~ grout 沥青砂胶[胶泥]
~ grouted surfacing 沥青碎石路面
~ grouted macadam 沥青灌浆碎石
~ grouting 沥青灌浆
~ gunite process 气动沥青喷浆法
~ gutter 沥青铺面排水沟[天沟]
~ hardness degree 沥青硬度值
~ heater 沥青加热器[锅]
~ highway emulsion 公路用乳化沥青
~ (-impregnated) sand 浸沥青的砂
~ (-impregnated) sealing rope 经沥青处理的嵌缝[封口]绳绨
~ (impregnated) strip 经沥青处理的板条
~ insulating coat 绝热沥青涂层
~ insulating slab 沥青绝缘板
~ (joint) pouring compound 沥青(接头)灌缝混合料
~ jointed pitching 沥青砌石护坡
~ jute pipe 沥青黄麻管[麻布管]
~ -laminated 沥青层的
~ lamination 沥青层压
~ layer 沥青层
~ limestone 沥青石灰石
~ liner 沥青衬里[材料,垫料]
~ lining 沥青衬里[防水层]
~ macadam 沥青碎石[碎石路]
~ macadam surfacing 沥青碎石铺面

asphalt

~ macadam mix 沥青混凝土与碎石灌沥青面层混合铺设法
~ mastic 沥青玛琋脂[砂胶]
~ mastic flooring 沥青砂胶铺面
~ mat 沥青面层
~ material 沥青材料
~ mattress 沥青垫层,沥青毡
~ measuring pump 沥青计量泵
~ membrane 沥青膜
~ metering pump 沥青计量泵
~ -mix design method 沥青混合料(配合比)设计法
~ mixer 沥青拌合机
~ -mixture 沥青混合料
~ mortar 沥青砂浆

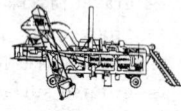
asphalt mixer

~ mortar finish 沥青砂浆面层
~ mulch 沥青盖料
~ overlay 沥青涂面层
~ paint 沥青漆[涂料]
~ paper 沥青纸
~ pavement 沥青路面
~ paver 沥青路面铺料机
~ paving block 沥青铺砌块
~ perforated sheet(ing) 沥青多孔板
~ plank floor 沥青预制板地板
~ plank wearing course 沥青板磨耗层
~ plant 沥青混合料拌合厂
~ pollution 沥青(烟雾)污染
~ pouring compound 沥青嵌缝[填充]料
~ pouring rope 沥青嵌缝[填]料,沥青铅封口用油麻丝
~ powder 沥青粉
~ pre-impregnation 沥青预浸渍[处理]
~ prepared roofing 沥青屋面

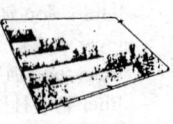
asphalt prepared roofing

料,沥青层面防水油毡
~ pre-saturating 沥青预浸渍处理
~ prime coat 沥青透层,路面头道沥青
~ -primed base 浇过沥青透层的基层
~ primer 沥青透屋,路面打底[头道]沥青
~ product 沥青制口
~ proportioning pump 沥青计量泵
~ protection coat 沥青防护层
~ protective coating(material) 沥青防护涂料
~ putty 沥青油灰
~ ready roofing 沥青屋面料,沥青屋面防水毡
~ road 沥青[柏油]路
~ road burner 熨沥青路面机
~ (road) surfacing 沥青路铺面
~ roof(ing) 沥青[砂胶]屋面
~ roofing felt 沥青屋面防水毡
~ -rubber caulk 沥青橡胶油膏
~ -rubber mass 沥青胶料
~ -rubber material 沥青材料[胶料]
~ -rubber strip 沥青防水胶带
~ sand 沥青浸渍灌浆砂
~ sand mastic 沥青砂胶[玛琋脂]
~ sandstone 沥青砂石
~ -saturated felt 纸胎油毡
~ -saturated felt(ed fabric) mat 沥青油毡垫层[隔层]
~ -saturated glass fiber felt 沥青浸渍玻璃纤维油毡
~ -saturated loose fabric 沥青浸渍网状屋面板
~ -saturated mineral wool felt 沥青浸渍矿棉纸油毡
~ -saturated paper 沥青浸渍纸
~ -saturated rock wood felt 浸沥青页棉毡
~ screen 沥青罩面
~ seal coat 沥青封层
~ seal(ing) 沥青防水[密封]层
~ sealing compound 沥青防水[密

~ seal(ing) rope 沥青封口油麻丝
~ sheet roofing 沥青纸毡屋面
~ shingle 沥青木瓦

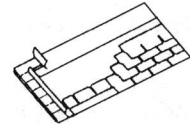

asphalt shingles

~ slab 沥青板
~ slurry 沥青浆
~ slurry seal 沥青浆密封
~ smoother 沥青平[刮]路机
~ -soil stabilization 沥青加固[灌浆加固]地基
~ spray bar 沥青喷洒机,喷油管
~ spreader 沥青摊铺机
~ stabilization 沥青稳定土[稳定法]
~ stone 沥青石
~ street 城市柏油[沥青]路
~ strip 沥青板条
~ subbase 路面沥青层[下垫层]
~ subseal(ing) 沥青基层处理[封底]
~ (surface) coating 沥青涂层
~ surface treatment 沥青表面处理(浇面)
~ surfacing 沥青路面[涂面]
~ tack coat 沥青粘层
~ tamper 沥青捣棒
~ tape 沥青带
~ tar 沥青柏油
~ tar pitch 硬焦油沥青
~ test 沥青试验
~ tile 沥青砖
~ tile base 沥青(预制块)地基
~ top dressing 沥青浇面(料)
~ topping 沥青浇面
~ -type wood fiber board 沥青型木纤维板
~ under seal-work 沥青封底工作,铺筑底层沥青混凝土
~ voids ratio 沥青空隙比
~ wall tile 沥青墙砖[墙贴面砖]
~ wearing course 沥青磨耗层
~ with rock flour 石粉沥青
~ wood fiber-board 沥青木纤维板
~ wool felt 沥青(毛)毡
artificial ~ 人造(地)沥青[柏油]
building ~ 建筑石油沥青
cold-mix ~ 冷沥青混合料
cold process ~ 冷浇沥青
compressed ~ 压制(地)沥青
crude ~ 生[粗]沥青
emulsified ~ 乳化沥青
hard ~ 硬(质)沥青
heavy ~ 黏性沥青,重(质)沥青
hot-mix ~ 热沥青,混合物
natural ~ 天然沥青
oil ~ (石)油沥青
original ~ 天然沥青
oxidized ~ 氧化沥青
paving ~ 铺面沥青
penetration ~ 渗透用沥青
petroleum ~ 石油(地)沥青
plate ~ 沥青板
poured ~ 浇灌性地沥青
rapid-curing ~ 快干铺路沥青
refined ~ 精制地沥青
road ~ 筑路沥青
rock ~ 天然(地)沥青
rolled ~ 滚压沥青
roofing ~ 铺顶沥青
sheet ~ 砂质地沥青,沥青板
steam ~ 蒸汽处理沥青
straight ~ 纯地沥青
straight-run ~ 直馏沥青
wax containing ~ 普通[含蜡]石油沥青

asphaltene [æsˈfælti:n] ***n.*** (石油)沥青质,沥青烯

asphalter [æs'tæltə] *n.* 沥青铺层,沥青铺砌工
asphalf-free *n.* 不含沥青的
asphaltic [æs'fæltik] *a.* 含沥青的,柏油的
~ adhesive (composition) 沥青胶合料
~ binder (course) 沥青结合料
~ -bitument 沥青
~ -bitumen addition 沥青掺合料
~ -bitumen-asbestos composition [compound] 沥青石棉混合料
~ -bitumen-asbestos mastic 沥青石棉玛琋脂
~ -bitumen based building mastic 沥青玛琋脂接缝材料[填料]
~ -bitumen-based plaster 沥青浆
~ -bitumen-bonding compound 沥青胶合料
~ -bitumen briguette 标准沥青水泥试块
~ -bitumen carpet 沥青油毡[面层]
~ -bitumen coated 涂沥青的,沥青灌浆的
~ -bitumen concrete 沥青混凝土
~ -bitumen deck surfacing 桥梁行车道沥青铺(面)层
~ -bitumem facing 沥青铺层[灌浆]
~ -bitumen felt roof 沥青油毡屋面
~ -bitumen-filler mix(ture) 沥青混合填料
~ -bitumen grave 沥青砾石
~ -bitumen joint 沥青(接)缝
~ -bitumen (joint) runner 沥青封口[填缝]油麻丝
~ -bitumen laminated 沥青叠层的,[胶合的]
~ -bitumen liner 沥青衬砌[衬垫]
~ -bitumen macadam 沥青碎石
~ -bitumen mortar 沥青砂浆
~ -bitumen pavement 沥青铺面
~ -bitumen perforated sheet 沥青多孔板

~ -bitumen pre-saturation 沥青预浸渍
~ -bitumen primer 沥青打底涂料
~ -bitumen protection coat 沥青保护涂层
~ -bitumen putty 沥青油灰
~ -bitumen rag felt 建筑用沥青粗制毡
~ bonding composition 沥青胶合材料
~ briguette 沥青标准,水泥试块
~ cementing agent 沥青粘结剂
~ coating 沥青涂层
~ concrete 沥青混凝土
~ concrete base 沥青混凝土基础[底层防水层]
~ concrete carpet 沥青混凝土铺面
~ concrete corewall 沥青混凝土防渗心墙
~ concrete (mineral) skeleton 沥青混凝土(矿质)骨架
~ concrete mixer 沥青混凝土搅拌机
~ concrete pavement 沥青混凝土铺面
~ concrete paver 沥青混凝土铺料机
~ -dispersion 沥青耗散
~ emulsion 乳化沥青
~ facade slab 沥青贴面板
~ facing 沥青铺面
~ felt 沥青油毡
~ felt panel 沥青毡板
~ filler 沥青填料
~ gravel 沥青砾石
~ grouted macadam 沥青灌浆碎石

asphaltie lining

~ insulating slab 沥青绝缘板
~ layer 沥青层
~ limestone 沥青石灰石
~ lining 沥青衬里[衬砌]

~ macadam 沥青碎石
~ macadam pavement 沥青碎石路面
~ mattress 沥青隔板[垫板],沥青隔层[垫层]
~ mortar 沥青砂浆
~ mortar masonry (work) 沥青砂浆砖石建筑[砌体]
~ overlay 沥青涂层
~ pavement 沥青铺面
~ paving-mix (ture) 沥青铺面混合料
~ penetration macadam 沥青灌浆碎石
~ plank wearing course 沥青板混凝土磨耗层
~ pouring rope 沥青浇口[接口]油麻丝
~ product 沥青制品
~ residual oil 沥青油渣
~ resin 沥青树脂
~ roofing 沥青屋面
~ sand 沥青砂
~ sand (stone) 沥青砂(石)
~ sealing 沥青密封[防水]
~ seal (ing) paste 沥青密封浆料
~ slab 沥青板,沥青层
~ surfacing work 沥青覆面[涂面]工作
~ terrazzo tile 沥青水磨石砖
~ tile 沥青砖[瓷砖]
~ tile floor cover(ing) 沥青(瓷)砖地面
~ varnish 沥青漆
~ wall tile 沥青贴墙砖
~ waterproof coat 沥青防水层
~ wearing course 沥青磨耗层
asphaltine [æs'fæltain] *n.* 沥青质
asphalting [æs'fæltiŋ] *n.* 浇灌沥青
asphaltos [æ'fəltəs] *n.* 地沥青
aspherical *a.* 非球面的
asphericty *n.* 非球面性
asphyxia *n.* 窒息(状态),假死,昏厥

asphyxiate *vt.* 使(人)窒息,闷死
asphyxiation *n.* 窒息,闷死
asphxiator [æs'fiksieitə] *n.* 二氧化碳灭火器;排水管道漏水试验器;窒息装置
aspirail [æs'paireil] *n.* 通风孔
aspirating [ˌæspə'reitiŋ]
  a. 吸气,吸出
~ cyclinder 吸气缸
~ engine 吸气[抽吸]发动机
~ hole[mouth] 吸气孔[口]
~ pipe 进[吸]气管
~ pressure 吸气[进气]压力
aspiration [ˌæspə'reiʃən] *n.* 吸气,吸入;吸尘作用;吸尘装置[器]
~ condenser 吸入冷凝器
~ inlet 吸气[进气]口
~ piping 吸气[抽气,吸入]管道
~ pressure control 吸气[进气]压力控制
~ probe 抽气式探针
~ psychrometer 吸气湿度计,通风干湿球湿度计
~ pump 吸扬式泵
~ temperature 通风温度计
~ ventilation 排气[负压]通风法
aspirator ['æspəreitə] *n.* 吸尘器,吸气器,抽气器,水射抽气器
~ combined with dust collector 吸气积尘器
~ pump 吸气[抽气,抽水]泵
chimney ~ 抽风器,烟囱抽风罩,导风板,导流片
as(-)raised gravel (未经)筛洗(的)天然砾石
assay [ə'sei] *v.* 分析,化验,试验,定量分析 *n.* 试样,样品,试验
~ curve 试验[化验]曲线
blank ~ 空白试验
high ~ 高指标样品
wet ~ 湿分析(法)湿验定

assemblage [ə'semblidʒ] *n.* 装配,安装;集合,系综
~ of curve 成组曲线
~ of forces 力系
~ point（桁架拉杆交叉）节点[接头]
~ point connection（桁架拉杆交叉）点连接

assemble [ə'sembl] *v.* -bled,-bling. *v.* 装配[备]安装,集中[合],组合;*n.* 组(装元)件
~ by welding 铆合,铆接

ass(e)mbled [ə'sembld] *a.* 已安装的,装配好的;装配的,装配式的
~ plate-column structure 装配式板柱结构
~ reinforced concrete structure 装配式钢筋混凝土结构

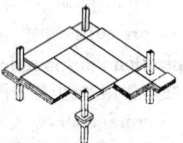
assembled plate-column structure

assembler [ə'semblə] *n.* 汇编程序;收集器;装配工人

assemblies [ə'semblis] *n.* 层积构件,层积木,叠层木

assembling [ə'sembliŋ] *n.* 安装,装配,汇编,编制程序
~ bolt 装配螺栓
~ machine 装配机器
~ mark 装配记号
~ method 安装方法
~ part 组装部件
~ process 装配工艺[过程]
~ rate 装配速度
~ site 安装位置
~ speed 装配速度
~ work 安装工作

assembly [ə'sembli] *n. pl.* -blies. 安装,装配,组合;汇编,配件,系集
~ and repair 装配与维修
~ area 编组场
~ average 汇集平均值
~ belt 总装配带技术
~ charts 装配图
~ cost 装配成本
~ diagram 装配图
~ drawing 总装配图,装配图
~ for installation 安装图
~ jig 拼装点,拼装机架
~ key 装配楔块,紧固楔
~ language 汇编语言
~ layout（室内）平面装配图
~ line 流水作业[装备,生产]线
~ line method 流水作业法
~ mark 装配记号,加工记号,(构件)制作记号
~ parts 组装部[构]件,组合零件
~ place 会场
~ plant 装配工厂

assembly place

~ plate 装配板,装饰板
~ program 汇编程序
~ rate 装配速度
~ room 会议室
~ schedule 装配程序表
~ shop 装配车间[工地]
~ speed 装配速度
~ stage 装配阶段
~ stress 装配应力,起始应力
~ subroutine 汇编子程序
~ time 积压[堆积受压,堆积生效]时间
~ type structure 装配式建筑物,预制结构
~ unit 装配单元
~ (work) shop 装配车间
~ yard 装配工场
cage ~ 升降台

counting ~ 计算装置
detail ~ 细部装配
die ~ 模具,压模装置
final ~ 总装,输出装置
heating-ventilating ~ 采暖通风两用机组
machine ~ 机器组合
pump drive ~ 泵的传动组,涡轮泵组
turbo-pump ~ 涡轮泵组

assessment [ə'sesmənt] ***n.*** 评价,评定,预测,评估
~ district 估值区(域)

assets ['æsets] ***n.*** 资产,财产
~ account 资产账户
~ income 资产收益
~ cover 资产担保
~ coverage 资产担保率
~ revaluation 资产重估
~ stock 资产股份
fixed ~ 固定资产
liquid ~ 流动资产
quick ~ 流动资产

assign [ə'sain] ***v.*** 选[确]定,分配,给予
~ed risk 分担风险
~ed volume 分配量

assignment [ə'sainmənt] ***n.*** 委派,分配;(财产,权利)转让,转让契约
~ lamp 呼叫[联络]灯
~ problem 分派问题
~ statement 赋值语句,计算语句
~ switch 呼叫开关
analog ~ of variables 变量的模拟赋值
facility ~ 设备分配[转让]

assignor ***n.*** 转让人,(专利)转让者

assistant [ə'sistənt] ***a.*** 辅助的,副的,助理的 ***n.*** 辅助物,辅助染色剂;助理[手,教]
~ chief engineer 副总工程师
~ engineer 助理工程师
~ -foreman 工长,领班
~ manager 经理助理,副经理

assisting grade (平地调车场)辅助坡度,推送坡度

assistor [ə'sistə] ***n.*** 辅助装置,加力[速]器,助推器
brake ~ 制动加力器

assize [ə'saiz] ***n.*** 法定标准,法令,条列

Assmann (人名)阿斯曼
~ psychrometer 阿斯曼湿度计
~ 's aspiration psychrometer 阿斯曼通风干湿球湿度计

associated [ə'səuʃieitid] ***a.*** 辅助的,伴生的,毗连的,连带的,组合的
~ contractor 副承包人
~ material 复合材料

association [əˌsəusi'eiʃən] ***n.*** 协会,团体,缔合(作用)
~ headquarters 协会[团体]本部[总部]
~ of architects 建筑师协会
~ of house lessor 房主协会

associative [ə'səuʃieitiv] ***a.*** 联合的,组成的,相连的
~ facitiation 联想易化
~ law 结合律
~ memory 相联存储,联合存储器

assort ***v.*** 分类(配),配合(齐,集)相配(称),协调

assorted [ə'sɔ:tid] ***a.*** 杂集的,各种各样的,混合的;配合的
~ colours 杂色

assortment [ə'sɔ:tmənt] ***n.*** 分类,分配,各色俱备之物

assume ***vt.*** (1)假定,设想
(2)采取,呈现、装作
(3)承担,担任,接受

assumed [ə'sju:md] ***a.*** 假定的
~ design stress-strain 假定设计应力应变
~ load(ing) 假定(计算)荷载[负荷]
~ position 假定方位
~ stress approach 应力估计

assumption [ə'sʌmpʃən] ***n.*** 假定,假设
~ value 假定值

assurance [əˈʃuərəns] *n*. 保证,担保,
保险
　　~ coefficient 安全系数
　　~ factor 安全系数,保证度
assure *vt*. 保证,担保,使确信
assurgent [əˈsəːdʒənt] *a*. 向上升起的,
向上浮起的
Assyrian architecture (公元前 1275 ~
538 年的)亚西利亚建筑
　　~ ornament 亚西利亚装饰
　　~ style 亚西利亚式(建筑)

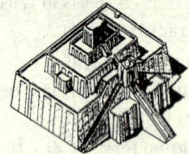

Assyrian architecture

astable [æˈsteibl] *a*. 不稳定的,非稳
态的
astatic [æˈstætik] *a*. 不稳定的,无定向
的,无静差的
　　~ buckling load 非静止[不稳定]压
曲临界荷载
　　~ regulator 无静差调节器
Astatine [ˈæstətiːn] *n*. 砹
astel [ˈæstil] *n*. 挡土墙,屋面顺槽中的
拉条,平巷顶板背[支撑板]
astelic [æˈstelik] *a*. 无中柱的
astely [ˈæsteli] *n*. 无中柱式
asthenosphere *n*. 软流[岩流]圈
astigmatizer [æsˈtigmətaizə] *n*. 夜间
测距(光)仪
astillen [ˈæstilin] *n*. (坑道中的)过梁,
黏土层
astomatal [ˌæstəˈmætəl] *a*. 无气孔的
astoop [əˈstuːp] *n*. 在倾斜的位置上的
astraddle [əsˈtrædl]
　　ad. ; *prep*. 跨,张,放
宽,跨(骑)着,两脚分
开站着(of, on),
把……置于跨下

astragal

astrafoil *n*. 透明箔

astragal [ˈæstrəgəl] *a*. 半圆装饰;(柱
头,柱脚的)圆部面小线脚,半圆线脚双
门盖条;圈带,压缝条,(门窗的)合缝挺
　　~ cornice 圆线型飞檐
　　~ frieze 半圆饰雕带
　　~ plane 圆缘刨
　　~ tile 半圆装饰砖
　　~ tool 半圆刀具
　　~ window 横木窗
astral [ˈæstrəl] *a*. 星的,女儿星的,多
星的
　　~ lamp 无影灯
　　~ crown 星冠
astrasil [əˈstræsil] *n*. 一种夹层材料
astriction [əˈstrikʃən] *n*. 收缩,限制;
约束
astride [əˈstraid] *a*. ; *ad*. 跨越的[地]
astringency [əˈstrindʒənsi] *n*. 黏(滞)
性,收敛性,涩味
astringent [əˈstrindʒənt] *a*. 收敛性的
　　~ clay 涩黏土
　　~ substance 收敛性物质
astroid [ˈæstrɔid] *n*. 星形线
astronomical [ˌæstrəˈnɔmikəl] *a*. 天文
学的
　　~ clock 天文钟
　　~ distance 天文距离
　　~ figures 天文数字
　　~ latitude 天文纬度
　　~ longitude 天文经度
　　~ observatory 天文台
　　~ time 天文时
　　~ transit 子午仪
　　~ unit 天文单位
astronomy [əˈstrɔnəmi] *n*. 天文学
astylar [eiˈstailə] *a*. 无柱式的,无支
柱的
　　~ back 无柱背面
astyllen [æsˈtailən] *n*. 阻水小坝
aswivel [ɑːsˈwaivl] *a*. 转旋的,转动的
asymmetric [ˌæsiˈmetrik] *a*. 不对称的;
asymmetrically, *ad*.
　　~ joint 不对称接合

~ structure 不对称结构

asymmetry [ə'simitri] *n.* （构图的）不对称性

asymptote ['æsimptəut] *n.* 渐近线
  ~ circle 渐近圆
  ~ expansion 渐近展开(式)
  ~ integration 渐近积分
  ~ solution 渐近解

asymptotic [ˌæsimp'tɔtik] *a.* 渐近的,渐近线的
  ~ stability 渐近稳定性
  ~ value 渐近值

asynchronous [ei'siŋkrənəs] *a.* 异步的,非同期的
  ~ motor 异步电机

at [æt] *prep.* 在,从……之处,到达,经由
  ~ -grade intersection 平面交叉
  ~ ground level 底层的,第一层的,同地面[路面]一样高的
  ~ -rest pressure 静压力,静止土压力
  ~ rest pressure test 静压力[静止土压力]试验

ATA (= Atmosphere Absolute) 阿泰,绝对压力单位

atch = attachment 附件,附属装置

atelier ['ætəliei] *n.* [法]画室,工作室

athenaeum [ˌæθi:'ni(:)əm] *n.* 图书馆,文艺(学术)协会

athermancy [ei'θə:mənsi] *n.* 不透辐射热性,不透红外线性质,绝热性

athermanous [ei'θə:mənəs] *n.* 不透红外线的,不导热的,绝热的,不透(辐射)热的
  ~ body 不透(辐射)热物体

athermic ['æθəmik] *a.* 绝热的,不导热的
  ~ effect 绝热效应

athlete *n.* 运动员

athletic [æθ'letik] *a.* 运动的
  ~ field 运动场

athwart [ə'θwɔ:t] *ad.* 横穿过,对穿过

atilt [ə'tilt] *a.*；*ad.* 倾斜的(地),侧倾的(地)

atlantes [ət'læntiz] *n.* 男像柱

atm = atmosphere 大气 = atmospheric pressure (标准)大气压

atmidometer *n.* 蒸发表,汽化[气压]计

atmology [æt'mɔlədʒi] *n.* 水蒸气学,水汽学

atmolysis [æt'mɔlisis] *n.* 微孔分气法,透壁分气法

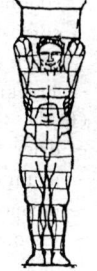

atlantes

atmometer [æt'mɔmitə] *n.* 汽化计,蒸发计

atmos ['ætməs] *n.* 大气压
  ~ valve 大气阀

atmoseal ['ætməsi:l] *n.* 气封(法)

atmosphere ['ætməsfiə] *n.* 大气,大气压;大气圈;环境
  absolute ~ 绝对大气压
  effective ~ 有效大气压
  international standard ~ 国际标准大气压
  metric ~ 公制[米制]大气压
  moist ~ 潮湿空气
  normal ~ 标准大气压
  physical ~ 物理大气压
  plane-parallel ~ 平面平行大气压
  standard ~ 标准大气压
  standard laboratory ~ 标准实验室大气压
  technical ~ 工程大气压
  working ~ 工作环境

atmospheric(al) [ˌætməs'ferik] *a.* 大气的,气压的
  ~ action 大气作用
  ~ agents 大气因素[作用]
  ~ circulation (type) water cooler 空气循环式水冷器
  ~ -compartment drier 常压干燥室
  ~ condensation 降雨[水],雨量
  ~ condenser 常压[空气]冷凝器

~ contamination 大气污染
~ cooling tower (空气)冷却塔
~ corrosion 大气的侵[腐]蚀(作用)
~ crack 风化裂纹
~ cracking 自然龟裂,大气龟裂(作用)
~ drier 空气干燥器
~ drum drier 鼓风式常压干燥器
~ envelope 大气包围层
~ exposure test 空气曝露[耐气候性]试验
~ layer 大气层
~ mosture (capacity) 大气湿度
~ oxygen 大气氧
~ pipe 放空管路,通大气管路
~ pollution 大气[空气]污染
~ precipitation 大气降水(如雨,雪)
~ pressure 大气压力
~ -pressure kiln 常压炉,常压加热炉
~ -pressure equipment 低压仪器
~ -pressure gas system 低压煤气系统
~ -pressure boiler 低压蒸汽锅炉
~ -pressure saturated steam 低压饱和蒸汽
~ -pressure spraying 低压喷洒[喷雾]
~ -pressure steam-cured (混凝土的)低压蒸汽养护
~ -pressure steam heating 低压蒸汽采暖
~ -pressure steam pipe 低压蒸汽管道
~ -pressure wet steam 低压饱和蒸汽
~ radiation 大气辐射
~ refraction correction 大气折射修正
~ relief valve 向空[安全]泄放阀
~ steam 低压[常压]蒸汽
~ steam curing (混凝土)低压蒸汽养护
~ still 常压管式加热炉
~ storage tank 无压[敞口]贮箱[罐]

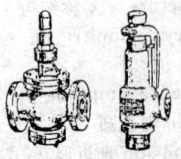

atmospheric relief valve

~ temperature 大气温度
~ transmittance 大气透明度
~ valve 放空[空气]阀
~ vessel 无压容器
~ visibility 大气能见度
~ water cooler 大气水冷器
atoll ['ætɔl] *n.* 环状珊瑚岛,环礁
~ texture 环形结构
atom ['ætəm] *n.* 原子,微小部分
atomic [ə'tɔmik] *a.* 原子的,极微的
~ blast excavation 原子爆破开挖
~ bunker 防原子弹地堡[钢筋混凝土地下掩体]
~ -energy valve 核电站挡板[滑板]
~ fuel 原子燃料
~ -hydrogen welding 氢原子焊接
~ power (station) 原子能(电站)
~ reactor containment structure 原子反应堆建筑物[构筑物]
~ shelter 原子庇护所,(防)原子弹掩蔽室
~ station 原子能发电站
atomics *n.* 原子(工艺术)学,核子学,原子(学)论
atomiser = atomizer
atomization [,ætəmai'zeiʃən] *n.* 喷雾,雾化,扩散
~ glazing 雾化上釉[抛光]
~ pattern 扩散[散流]图像
atomize ['ætəmaiz] *v.*, -ized, -izing 使雾化,粉化
~ by compressed air 压(缩空)气喷雾
~d fuel (spray) 雾化燃料
~d powder 雾化粉
~d suspended oxidation techique 悬浮废液雾化法

~d water 雾化水,水雾
~d water jet 水雾喷射
~d water spray 水雾化[喷洒,喷雾]
atomizer [ˈætəmaizə] *n.* 喷雾器,雾化器,喷(水)嘴
~ aperture 喷雾嘴
~ burner 燃烧喷嘴
~ by compressed air 压气喷雾器
~ chamber 雾化室
~ device 喷管,喷射[喷雾]装置
air blast ~ 空气喷射喷雾器
exhaust heated ~ 排气加热喷雾器
fan ~ 扇式喷雾器
fuel ~ 燃料喷雾器
notched flat headed ~ 缺口平头喷雾器
nozzle ~ 喷嘴喷雾器
oil ~ 喷油器
paint ~ 油漆喷雾器
pintle ~ 针栓喷雾器
sleeve ~ 套式喷雾器
slot ~ 缝隙喷雾器
spray ~ 喷淋喷雾器
steam ~ 蒸汽喷雾器
steam jet ~ 汽流喷雾器
tubular ~ 同心缝隙[管状]喷雾器
atomizing [ˈætəmaiziŋ] *a.* 喷雾的；*n.* 雾化(作用),扩散,喷雾[射]
~ burner 雾化燃烧器
~ cone 喷(雾)锥
~ fineness 雾化细度
~ humidifier 喷雾加湿器
~ jet 喷雾嘴,水枪喷嘴
~ nozzle 雾化喷嘴
~ oil burner 喷雾燃油器
~ pipe 喷管
~ pressure 雾化[喷射]压力
~ pump 雾化式泵
~ spraying 雾化喷洗[涂]
ato-muffler *n.* 减声器,消声器
atommy [ˈætəmi] *n. pl.* -mies. 微粒,尘埃,原子；矮人,侏儒
at-once 立即

~ payment 立即付款
atone [əˈtəun] *v.* , atoned, atoning 补[赔]偿, 偿还, 弥补; atoner, *n.*; atoningly, *ad.*
atrium [ˈeitriəm] *n. pl.* -tri *a.* 天井前厅,正厅;(古罗马住宅中央的)厅,(通往古罗马长方形会堂或住所一个)露天庭院(三面或四面存廊);通向其他房间的一个方厅(现在住宅中用作起居室)
atrophy [ˈætrəfi]
n. ; *v.* , -phieed, -phy-ing. 退化,萎缩,衰退
A-truss A 型桁架
attach [əˈtætʃ] *v.*
附着,附属,接近,相连

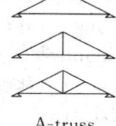
A-truss

attache【法语】*n.* (使馆)馆员,专员,武官,参赞
attached [əˈtætʃid] *a.* 附着式的,悬挂的,连接的
~ building 附联式房屋[建筑]

attached building

~ column 附柱,半柱
~ decoration 附加装饰,(墙上)假窗
~ gable 饰[假]山墙
~ garage 附加[属]车库
~ pent-house 靠近主房的披屋
~ pier 块垛,堵墙,支墩,壁柱
~ pier capital 附墙方柱头[顶]
~ power 备用电源
~ pump 附备泵
~ sign 附设标志
~ing plug (小型)电源插头
~ing rubber to metal 橡胶与金属相结合
attachment [əˈtætʃmənt] *n.* 附[配]件,附属装置[设备];附着,连接
~ clip 卡钉,夹子

~ driving shaft 辅助传动轴
~ for cranes 起重机附加设备
~ for excavators 挖土[掘]机附加装置
~ plug 连接插头[销]
~ screw 定位螺钉,装配螺钉
attack [ə'tæk] *v.*; *n.* (1) 浸蚀,腐蚀,起化学反应
(2) 攻击,侵袭,破坏
(3) 投入,开始(工作)
~ing power 浸蚀性,腐蚀性
chemical ~ 化学侵蚀
gas ~ 气体腐蚀
oxidative ~ 氧化侵蚀
attainment [ə'teinmənt] *n.* 达到,成就,收获
~ of superelevation 超高缓和段[渐变段]
~ of widening (道路)加宽渐变段,加宽缓和段
attemper [ə'tempə] *vt.* 稀释;调节(温度)
attemperation [ətempə'reiʃən] *n.* 温度调节,温度控制
attemperator [ə'tempəreitə] *a.* 温度调节计,降温[恒温]器;保温水管,控制温度用的旋管冷却器
attempt [ə'tempt] *vt.*; *n.* 打算,试图,努力;攻击 attempability, *n.* attempter, *n.* attemptable, *a.*
Atten = attenuation 衰减[耗],减幅
= attenuator 衰减器,消声器
attend *v.* (1)出席,参加;(2)伴随,随从;(3)照顾,护理
attendace [ə'tendəns] *n.* 值班,维护,保养
~ book 签到簿
~ cost 维修费
attendant [ə'tendənt] *a.* 伴随的,随侍的,值班的; *n.* 值班员,服务员, attendantly, *ad.*
~ -controlled lift 值班员控制的电梯,服务员操作的电梯

~ operated control 电梯服务员[值班员]操作控制
~ parking 有服务员的停车场(代客停车,取车)
engine ~ 机工,司机
attention [ə'tenʃən] *n.* 注意,维修,保养; attentional, *a.*
~ device 维护设备
attenuate [ə'tenjueit] *v.*, -ated, -ating. *v.* 打散,散射,衰减,减少,稀释; [ə'tenjuit] *a.* 稀薄的,弱的,衰减的
attenuation [ətenju'eiʃən] *n.* 衰减,减少,稀释,冲淡,扩散
~ band 衰减频带
~ by absorption 吸收衰减
~ characteristic 衰减特性
~ coefficient 衰减[扩散]系数
~ comparator 衰减比测仪
~ compensation 衰减补偿
~ constant 衰减常数
~ cross-section 衰减截面
~ duct 隔声管道
~ equalizer 衰减平衡器
~ ratio 衰减比,阻尼比
acoustical ~ 声的衰减
balance ~ 平衡[对称]衰减
critical ~ 临界衰减
dissipative ~ 损耗性[散射性]衰减
echo ~ 回声衰减
effective ~ 有效衰减
free-space ~ 自由空间衰减
geometrical ~ 几何衰减
noise ~ 噪声衰减
range ~ 距离衰减,衰减范围
regularity ~ 正规衰减,固有衰减
space ~ 空间衰减
vibration ~ 振动衰减
attenuator [ə'tenjueitə] *n.* 衰减器,消声器
piston ~ 活塞式衰减器
atteration [ætə'reiʃən] *n.* 冲积土[层],表土,泥砂
Atterberg (人名)阿特贝尔格

~ limits of soil 阿特贝尔格稠性限度[稠度极限](即土壤的特性湿度指标)
~ limits value 阿氏限度值,土的特性湿度限值
~ test 阿氏试验,土壤含水限度试验
~'s scale 阿氏土粒分组
attestation [ætes'teiʃən] *n.* 证明书,证据;证实 attestative, *a.*
attic ['ætik] *n.* 屋顶室[间],顶楼,阁楼; *a.* (Attic)雅典的
Attic base (古希腊柱的)座盘
~ cladding element 阁楼骨架[外墙],[覆面]构件

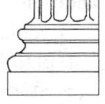
Attic base

~ fan 阁楼(通风)风机
~ floor 顶楼层
~ joist 屋顶层搁栅
~ main system 上行下给式[顶楼干管式]系统
~ order (古典建筑柱型的)雅典的列柱式,顶层角柱式
~ room 顶层房间
~ stair(case) (通向)顶楼(的)楼梯
~ story 阁楼,屋顶层内的空间
~ tank 屋顶[楼顶]水箱

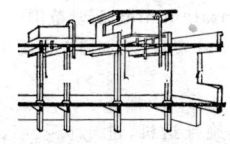
attic tank

~ vent block 屋顶通风空心砖
~ vent opening 屋顶通风孔
~ vent tile 屋顶通风空心砖
~ ventilation 阁楼通风,屋顶通风
~ ventilator 屋顶通风机[通风孔,通风口,气窗]
~ window 屋顶通风窗[气窗]
false ~ 假屋顶层[屋顶,顶室]
Attika [ætikə] *n.* (古典建筑的)顶层
attitude ['ætitju:d] *n.* 状态,姿势;空间方位角;态度;attitudinal, *a.*
attle [ætl] *n.* (采石场经筛后剩余的)石屑,废[矿]屑
attorney [ə'tə:ni] *n.* 代理人,委托者,律师
attract [ə'trækt] *v.* 吸引
~ armature relay 衔铁吸合式继电器
~ed traffic volume 吸引交通量
attraction [ə'trækʃən] *n.* 吸引力,引力 -attractionally *ad.*
~ force 引力,吸引力
~ of gravitation 地心引力
chemical ~ 亲合力
magnetic ~ 磁力
mechanical ~ 机械引力
mutual ~ 互相引力(吸引)
attractive [ə'træktiv] *a.* 具有引力的,引力的,吸引的 attriactively, *ad.*
~ distance 吸引距离,服务距离[范围]
~ effect 重力影响
~ force 吸引力
~ interaction 互相吸引
~ radius (设施服务)吸引半径
~ sphere (设施服务)吸引范围,服务对象范围
attribute ['ætribju:t] *v.*, -uted, -uting. *n.* 标志,属性,特征,性质; [ə'tribju:t] *vt.* 归因,由于,属于
~ of scenery 地形性状
attrite [ə'trait] *vt.* 擦去,消除,磨耗[擦,碎]; *a.* 磨损的
attrition [ə'triʃən] *n.* 磨损[擦,耗],消耗
~ hardness 耐磨硬度
~ loss 磨耗[减]量
~ mill 碾磨机,圆盘磨碎机
~ rate 损耗率,磨损程度
~ resistance 抗磨耗[性]
~ resistant 耐磨的
~ test 磨耗试验
~ tester 磨损试验机

~ testing machine 磨损试验机
~ value 磨损值
attritus [ə'traitə] *n.* 杂质煤
Aubert gate 奥别尔特闸门（翻倒式闸门之一）
auburn ['ɔ:bən] *n.* 赭色，褐色，栗色；*a.* 栗色的，赭色的
auction ['ɔ:kʃən] *n.*；*vt.* 拍卖
~ hall 拍卖大厅
audibility [ɔ:di'biliti] *n.* 可听度，可听见
~ factor 可听度，可听系数
~ range 可听范围
~ value 可听值
audible ['ɔ:dəbl] *a.* 可听的，音响的
~ alarm unit 声音[音响]报警装置
~ call 可听到信号，音响呼叫
~ limit 可听限制（声音强弱和频率高低的上下界限值）
~ noise 声频噪声
~ range 音响[可听]范围
~ signal 音响[可听]信号
~ sound 可听声音
~ tone 可听声音
~ -visual aid 光-声辅助方法
audience ['ɔ:djəns] 接见，会见，观众，听众
~ chamber 接见室，会见室
~ hall 观众厅，接见厅
audifier ['ɔ:difaiə] *n.* 声[音]频放大器
audigage ['ɔ:digeidʒ] *n.* 携带式超声波测厚仪
audio ['ɔ:diəu] *a.* 声[音]的
~ amplifier 声[音]频放大器
~ generator 声频发生器
~ noise meter 噪声计
~ signal 可听声信号
~ tape 录音磁带
audiometer [ɔ:di'ɔmitə] *n.* 听度计，声音测量器
audiometry [ɔ:di'ɔmitri] *n.* 测听技术，听力测定[法]
audio-visual [ɔ:diəu'viʒuəl] *a.* 视听的

~ aids 直观教具
~ classroom 设有视听设备的教室，有放映幻、影片、录音设备的教室
~ center 视听教育中心，有较高级视听设备和器材进行社会教育的机构
~ instruction 直观教学
audit ['ɔ:dit] *vt.*；*n.* 审计，检查
~ -in-depth 分层检查
~ trail [数据]检查跟踪
audition [ɔ:'diʃən] *v.*；*n.* 听力；试听
~ room（用于乐器，唱片的选择及欣赏的）试听室
~ studio 试听播音室
auditor ['ɔ:ditə] *n.* 审计员，查账员
auditorium [ɔ:di'tɔ:riəm] *n. pl.* -riums, -ri *a.* 会堂，音乐厅，礼堂，观众席
~ acoustics 会场声学
~ seating 观众席（位）
~ space 观众席[厅]
auditory ['ɔ:ditəri] *n. pl.* -ries. *n.* 中殿；礼堂；听众席；听众；*a.* 听觉的，耳的
~ acuity 听力，听觉敏锐度
~ area 可听[听觉]范围
~ field 听野
~ impression 听觉
~ localization 声源定位
~ sensation area 听觉范围
Auer metal 奥厄合金（稀土金属65%，铁35%）
auger ['ɔ:gə] *n.* 螺旋钻，螺旋推运机，进料螺旋

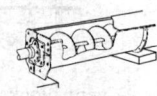
auger

~ backfiller 万能螺旋钻式回填机
~ bit 麻花钻，钻头
~ boring 钻孔，螺钻，钻探
~ brick machine 带式制砖机
~ delivery 螺旋输送
~ drill 螺钻
~ drilling 钻孔，螺钻，钻探
~ drive 螺旋传动[驱动]
~ extension 螺旋接柄

~ feeder 螺旋进[加,给]料器
~ -hole (螺)钻孔
~ -hole charge 钻孔炸药
~ -mining 大直径钻孔法
~ -shaft 钻井
~ shell 螺钻
~ stem 螺旋钻杆[棒]
~ -type extrusion (unit) 螺旋式剂压机
~ twist bit 螺旋钻
~ vane test (土壤)十字板试验
~ with hydraulic feed 水压[液压]推进式钻机
~ with valve 阀式抽泥器
~ed pile 螺旋钻桩
auget [ˈɔːdʒit] *n.* 雷管,爆破管
augite [ˈɔːdʒait] *n.* 斜辉石
~ diorite 辉石闪长石
~ granite 辉石花岗石
~ granophyre 辉石花斑岩
~ melaphyre 辉石暗玢石
~ porphyry 辉石斑岩
~ rock 辉石岩
~ syenite 辉石正长石
augment *v. n.*
(1) 增大[加,长]
(2) 扩大[张],添增
augmentation [ˌɔːgmenˈteiʃən] *n.* 增大,加强,增加率
augmenter = augmentor
augmentor [əˈgmentə] *n.* 增强[压]器
aul [ˈɔːl] *n.* 胶榾木,欧洲榾木
a(u)mbry [ˈæmbri] *n. pl.* -bries. 壁橱,(教堂)小室

a(u)mbry

aural [ˈɔːrəl] *a.* 听力的,听觉的;气味的,气息的;先兆的

~ conditions 伴音[音响]情况
~ impression 听觉
~ null 无声,消息
~ sensation 听觉
~ signal 音响[频]信号
~ type beacon 发声式信标
aurum [ˈɔːrəm] *n.* 金(Au)
~ foliatum 金箔
auspicious [ɔːsˈpiʃəs] *a.* 幸运的,昌盛的
~ beginning 吉利的开始
~ news 好消息
austenitic [ɔːstəˈnitik] *a.* 奥氏体的
~ manganese steel 奥氏体锰钢,高锰钢
~ stainless steel 奥氏体不锈钢
~ steel 奥氏体钢
austere *a.* -ly *ad.* ;-ness *n.*
(1) 严格的,(2) 朴素的,节约的
austerity [ɔːsˈteriti] *n.* 紧缩,节制
austral [ˈɔːstrəl] *a.* 南(方)的,向南的
~ window 南向窗,滑动窗,推拉窗
Austrian [ˈɔːstriən] *a.* 奥地利的
~ cloth 奥地利高级毛织物,奥地利粗支条窗帘布
~ method of timbering 奥地利式隧道支撑法
~ method of tunnel driving (隧道掘进)奥地利施工法,上下导坑先墙后拱施工法
authalic projection 等积投影
authentic [ɔːˈθentik] *a.* 可靠的,可信的;有根据的,权威性的
~ sample 可靠[真实]试样
author *n.* 作(著)者,作家,创始者;*v.* 写作,著出,编辑,创造
authority [ɔːˈθɔriti] *n. pl.* -ties. 权威;当局,管理机构
~ engineer 管理工程师
~ -owned 官方所有的
local ~ authorities 地方当局
the competent ~ authorities 主管机关

authorized [ˈɔ:θəraizd] *a.* 公认的, 经认可的, 经审定的
~ agent 指定代理人
~ controlled material 经审定控制的物资
~ press 极限[规定,允许]压力
~ signature 印鉴

auto [ˈɔ:təu] *a.* 自动的, *n.* 汽车
~-agglutination 自体凝集作用
~-alarm 自动警报器
~-analyzer 自动分析器
~-balancer crane 自动平衡[配重]起重机
~ bond 自动接合[链接,焊接]
~-bridge-factory 自动化桥梁厂
~ coagulation 自动凝聚[结]
~ coarse pitch 自动增大螺距装置
~ collimator 自动照准仪, 自动准直仪
~ compensation 自动补偿
~ condensation 自动凝结[凝聚];自冷凝
~ conduction 自动传感,自感
~ consequent stream 自顺向流
~ convection 自动对流
~ correction 自动校正
~ crane 汽车起重机, 汽车吊

auto crane

~-cut-out 自动断路器
~ cycle 自动循环
~ decomposition 自动分解
~ digestion 自身消化, 内源呼吸
~ door (光电控制的)自动开关门
~-draft 自动制[绘]图
~-elevator device 自动找平装置
~-enlarging apparatus 自动放大器

~ feed 自动进料[给料]
~ fleet 汽车队
~ flocculation 自絮凝作用
~ former 自耦变压器
~ genetic drainage 利用冲蚀面地形排水
~ industry 汽车工业
~-level 自动水准仪
~-lift 自动升降机[电梯]
~-lift-trip scraper 铲斗自动起落刮土机
~-man 自动-手动开关
~-mason 自动砌墙机
~ pack 自动填塞
~ parts 汽车零件

autobus

autobus [ˈɔ:təubʌs] *n.* 公共汽车
autocar [ˈɔ:təukɑ:] *n.* 汽车,机动车
autocartograph [ˈɔ:təukɑ:təgræf] *n.* 自动测图仪
autocatalysis [ˌɔ:təukəˈtælisis] *n.* 自动催化(作用),链式反应扩大
autoclave [ˈɔ:təkleiv] *n.* 高压锅[釜],高压消毒蒸锅;热压处理;用高压锅消毒
~ curing (混凝土)高压蒸汽养护
~ expansion test 蒸汽压力膨胀试验(用于水泥安定性测定)
~d aerated concrete 高压蒸汽养护的加气混凝土
~d asbestos cement caltium silicate board 热压处理的石棉水泥硅酸钙板
~d cellular concrete 高压加气混凝土
~d light-weight concrete 高压轻质混凝土
~d molding 蒸压罐模制法
autogenic [ˈɔ:təuˈdʒenik] = autogenous
autogenor *n.* 自动生氧器

autogenous [ɔːˈtɔdʒinəs] *a.* 自[气]焊的, 自动的
~ cutter (乙炔)气割机
~ fusing (乙炔)气割
~ gouging attachment 自动气刨附加装置
~ grinding 自动研磨
~ hardening 自动[火烧]硬化
~ healing 自动强化, 愈合, 自合
~ rivet cutter 铆钉气割机
~ shrinkage 自然收缩
~ soldering 气焊, 氧铁软焊
~ welded steel pipe 焊接钢管
~ welding 气焊法, 熔接法

autograph [ˈɔːtəɡrɑːf] *n.* 自动绘图, 亲笔, 手稿

autographic [ˈɔːtəɡræfik] *a.* 亲笔的, 自署的
~ record 自动记录
~ recording apparatus 自动记录仪

autographometer [ɔːtəɡæˈfɔmitə] *n.* 自动图示仪, 地形自动记录器

autohesion [ɔːtəˈhiːʒən] *n.* 自粘(作用, 力)

autohoist [ˈɔːtəhɔist] *n.* 汽车起重机

autoignition [ɔːtəiɡˈniʃən] *n.* 自动点火, 自燃
~ conflagration 自燃火灾
~ temperature 自动点火温度

autolay [ˈɔːtəlei] *n.* 自动开关[敷设], 自动扭绞

autoline [ˈɔːtəlain] *n.* 自动线, 汽车路

autoline

auto loader [ˈɔːtələudə] *n.* 自动装卸机[车], 自动装填[送料]机

autolysis [ˈɔːtəlisis] *n.* 自溶作用

automat [ˈɔːtəmæt] *n.* 自动装置, 自动机, 自助食堂[餐厅]

automate [ˈɔːtəmeit] *v.*, -ated, -ating. 使自动化
~d door 自动门
~d guideway transit 自动导向交通系统
~d measurement 自动量算
~d storage 自动贮藏
~d sliding door 自动滑门
~ traffic control 自动化交通控制

automatic(al) [ɔːtəˈmætik] *n.* 自动装置; *a.* 自动的
~ acceleration 自动加速
~ adjustment 自动调整
~ advance element 自动提(前)点火装置
~ advance breaker 自动提前断电器
~ air brake 自动气闸[风闸]
~ air heater 自动空气加热器
~ air valve 自动排气阀
~ alarm 自动报警器
~ -ally operated valve 自动操纵阀
~ arc welding 自动电弧焊
~ ash remover 自动清灰装置
~ ball valve 自动球阀
~ batch mixing 自动配料混合
~ batch plant (混凝土生产厂)大型混凝土生产设备, 自动分批拌合设备
~ batcher 自动配料计量器
~ block system 自动闭塞装置[闭锁系统]
~ boiler control 锅炉自动控制
~ bowl latch 铲斗自动闭锁
~ brake adjuster 自动制动调节器
~ bucket level(ing) device (斗式装载机)铲斗自动调平装置
~ bucket positioner (装载机)铲斗自动定位器[装置]
~ buring appliance 自动燃烧设备[装置]
~ by-pass valve 自动旁通阀, 自动溢流阀
~ cake dsicharge type filter press 自动卸料压滤机
~ central mixing plant 自动中心拌合厂
~ change-over damper 自动变向闸,

自动变向阻尼器
~ check 自动检测
~ checkout and evaluation system 自动检测与鉴定系统
~ checkout equipment 自动检测装置
~ check valve 自动止回阀
~ chill casting-machine 自动硬模铸造机
~ circuit-breaker 自动断电器[电路开关]
~ cistern 自动冲洗水箱
~ clock switch 时控开关
~ closing 自动闭合[接通]
~ closing device 自动开关装置
~ coding 自动编码
~ combustion control 燃烧自动控制
~ compactor 自动内燃夯土机
~ (concrete) block-making machine 混凝土块自动浇制机

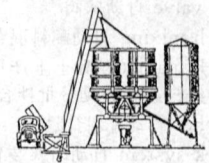

automatic concrete mixing plant

~ concrete column pourer 混凝土柱自动浇铸机
~ concrete mixing plant 混凝土自动拌合设备
~ concrete pipe machine 混凝土管自动浇铸机
~ control 自动控制[调节,管理]
~ control board 自动控制盘[板]
~ control device 自动控制装置
~ control system 自动控制系统
~ control valve 自动控制阀
~ controller 自动调节器
~ conveyor 自动输送机
~ coordinate recording unit 自动座标记录装置
~ counterweight gate 有平衡器的自动阀门
~ counting 自动计数[计算]
~ coupling 自动挂钩[联结器]
~ crab 自动起重小吊车
~ cross-cut chain saw 自动横切割链锯

automatic cross-cut chain saw

~ curing machine (混凝土)自动养护机
~ cut-out 自动断路器
~ cutting machine 自动气割机
~ data processor 自动数据处理机
~ defrosting 自动除霜
~ depot 混凝土自动运送工场
~ design 自动设计
~ discharge 自动排放
~ discharge gauge 自动流量计
~ discharge valve 自动排水阀
~ distributor 自动洒布机
~ dosing 自动投量[投料,加药]
~ door closer 自动关门装置
~ (door) operator 自动开门装置
~ door stay 自动门钩
~ draft tube fill line 自动尾水管[充水管路]

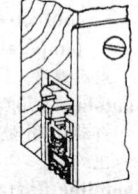

automatic door operator

~ drive 自动传动
~ dump 自动倾卸
~ dump truck 自动卸货车
~ dumping batch seale 自动卸料配料称
~ electric water heater 自动电热水

加热器
~ expansion valve 自动膨胀阀
~ feed 自动进料[送料,进刀,推进]
~ feed unit (钻机)自动推进装置
~ feed water control (住宅)自动供水控制
~ feed water pump 自动给水[供水]泵
~ filter press 自动压滤机
~ furnace 自动燃烧装置
~ fire alarm 火灾自动报警器
~ fire door 自动防火门
~ fire shutter 自动燃烧风门片[板]
~ fire sprinkler 自动喷洒,自动灭火装置
~ fire warning device 火警自动报警装置
~ float 自动浮标
~ floater-controlled bleeder tap 自动控制浮子放水阀(塞)
~ flush back 自动反冲洗
~ flushing cistern 自动冲洗水箱[贮水器]
~ fork lift 自动叉车
~ frequency control 自动频度控制
~ furance 自动燃烧装置
~ gate 自动闸门
~ gauge 自动计量仪
~ geyser 自动热水器;自动喷泉
~ governor 自动调节器[调速器]
~ grab 自动抓斗[夹具]
~ grade control system 级配自动控制系统
~ (gravity) die casting machine 自动硬模铸造机
~ grinder 自动研磨机
~ grouting machine 自动灌浆机
~ heat detector (消防)温感自动报警器
~ highway crossing gate 公路与铁路交叉口自动启闭栅栏
~ hydroelectric station 自动化水力发电厂

~ humidifier 自动加湿机
~ immersion heater 浸没式自动加热器
~ intake device 自动进气[水]装置
~ lamp changer 自动航标灯交换器
~ level 自动调平[找平],自动找平水准仪
~ line plotter 自动线绘图机
~ load maintainer 自动荷载稳定器
~ machine 自动机
~ measuring 自动测量[配料,计量]
~ measuring plant 自动计量[配料]设备
~ mixing valve 自动混合阀
~ monitor 自动监测仪
~ mo(u)ding plant 自动造形设备
~ noise controller 自动噪声控制器
~ noise suppressor 自动噪声抑制器
~ padding machine 自动试浆机
~ pipe (making) machine 自动制管机
~ pipette 自动吸(移)管
~ pontoon 自动浮船式闸,自动浮桥
~ positioning equipment 自动定位装置
~ pressure reducing valve 自动减压阀
~ primer 自灌水器,自动雷管
~ priming 水泵启动,自动充水
~ programming 程序自动化
~ proportioning 自动测量[计量,配料]
~ plotter 自动绘图仪
~ puller 自动拔出器,自动拉出器
~ pump 自动(启动)水泵
~ pump station 自动(启动)泵站
~ purifier 自动净水器
~ pushbutton control (电梯自动操作的)内控制按钮
~ ram pile driver 自动冲锤打桩机
~ recording instrument 自动记录仪
~ reducing valve 自动减压阀
~ regulating apparatus 自动调节

装置
~ regulator 自动调节阀[器]
~ remote control 自动距离控制
~ riveter 自动铆(钉)机
~ return of water 自动回水管路,蒸汽冷凝水管路
~ return trap (蒸汽采暖)汽水分离器,回水盒
~ roadbuilder 自动筑路机
~ run 自动运行
~ safety valve 自动安全阀
~ sampler 自动采样器
~ sand distributor 自动沙料撒布机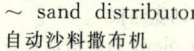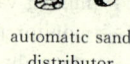
automatic sand distributor
~ sensibility control 自动灵敏度控制
~ sequential operation 自动顺序操作
~ shaft kiln 自动立窑
~ shift 自动连接[接通]
~ shutter door operator 自动门开启装置
~ shuttle valve 自动关闭阀
~ side tipping wagon 自动侧卸铁路货车
~ siphon 自动虹吸
~ slab and tile grinder 自动石板[瓷砖]研磨机
~ smoke detector (消防)自动烟感器
~ softening installation 自动软水装置
~ speed change valve 自动喷射速度调节阀
~ speed control 自动转速控制
~ spillway 自动溢洪道[坝]
~ spraying 自动喷涂
~ spring loaded valve 自动弹簧阀
~ sprinkler 自动喷水器,(消防)自动喷洒装置
~ sprinkler system 自动洒水系统,

自动喷水管路
~ stack 自动搁架[自动控制或驱动控制的非固定式书库或仓库的搁架]
~ starter 自动启动器
~ station 自动(水力)发电站
~ steam generator 自动蒸汽发生器

automatic station

~ stereoplotter 自动立体测图仪
~ stoker 自动加煤机
~ stop valve 自动止水阀
~ stoper (电梯)自动停止器
~ surfacer (木工用)自动刨床,自动进料刨床
~ switch 自动开关
~ syphon 自动虹吸
~ temperature control 自动温度控制
~ thermal regulator 自动温度调节器
~ thermostat 自动恒温器
~ throttle valve 自动节流[调整]阀
~ timer 自动定时器
~ timing device 自动定时装置,操作时间自动调节器
~ tipper 自动倾卸机构[装置,翻斗车]
~ traffic countor 交通量自动计数器
~ transmission 自动变速[转动]装置
~ trimmer 自动修整机
~ unit 自动开关组合
~ unloading machine 自动卸货机
~ valve 自动阀
~ vent 自动放气阀
~ ventilation 自动通风
~ vibration monitor 自动振动探测计
~ volume batching plant 自动体积计量设备
~ water closer 自动冲洗大便器
~ water gauge (混凝土搅拌机)自动水位表[量水表]

~ weigh(ing) plant 自动称重设备
~ welder 自动焊机
~ welding 自动焊接
~ welding procedure 自动焊接方法
~ window 自动开关窗
automation [ˌɔːtəˈmeiʃən] *n.* 自动控制,自动化
~ center 自动化中心
automatization [ɔːtəmətaiˈzeiʃən] *n.* 自动化
automatograph [ˌɔːtəˈmætəɡrɑːf] *n.* 自动记录仪
automobile [ˈɔːtəməbiːl] *n.* (小)汽车, 机动车
~ chassis 汽车底盘
~ diesel engine 汽车柴油发动机
~ engine 汽车发动机
~ noise 汽车噪声
~ race course 赛车跑道
~ road 汽车专用道, 汽车公路
~ traffic 汽车交通量

automobile diesel engine

automonitor [ˌɔːtəˈmɔnitə] *n.* 自动(程序)监控器,自动监视程序
autonomics [ˌɔːtəˈnɔmiks] *n.* 自调系统程序控制研究
autonomous [ɔːˈtɔnəməs] *a.* 自治的,自重的
~ channel 自主通道
autooxidation [ˌɔːtɔksiˈdeiʃən] *n.* 在大气中自动氧化(作用)
~ phase 自动氧化相[期]
auto-pack 自动填塞
auto-panel [ˈɔːtəpænl] *n.* 自动控制指示板
auto-park [ˈɔːtəpɑːk] *n.* 停车场,汽车停放场
auto-partrol(grader) 巡路平机,养路用平路机
autopiler [ˈɔːtəpailə] *n.* 自动编译程序装置

auto pista [西] *n.* 高速公路
auto plotter [ˈɔːtəplɔtə] *n.* 自动绘图仪[编表机]
autoplugger [ˌɔːtəˈplʌdʒ] *n.* 自动充填器,自动锤
auto-polymerization [ˈɔːtəpɔliməraiˈzeiʃən] *n.* 自动聚合(作用)
autopore [ˈɔːtəpəː] *n.* 大管(孔)
auto precipitation [ˌɔːtəprisipiˈteiʃən] *n.* 自沉淀析出(作用)
auto programming [ˌɔːtəˈprəuɡræmiŋ] *n.* 自编程序(控制)
auto-punch *n.* 自动冲压硬度试验机
auto purification [ˌɔːtəˌpjuːrifiˈkeiʃən] *n.* 自净作用,自动净化
autoroute [ˌɔːtəˈruːt] *n.* 多车道高速公路
auto scope [ˈɔːtəuskəup] *n.* 点火检查示波器,自检器
autoset level 自动找平水准仪
auto stability [ˌɔːtəstəˈbiliti] *n.* 自动稳定(性)
auto stable [ˌɔːtəˈsteibl] *a.* 自动调节的,自动稳定的
auto stacker [ˌɔːtəˈstækə] *n.* (大城市中心)多层停车场[库]
auto stairs *n.* 自[活]动梯
auto stopper [ˌɔːtəˈstɔpə] *n.* 自动控制器[停止装置]
auto strada [auːtəuˈstrɑːdɑː] *n.* 高速公路干线,汽车公路

auto strada

auto-terminal [ˌɔːtəuˈtəːminəl] *n.* 汽车终点站[总站]
auto transformer [ˌɔːtəutrænsˈfɔːmə] *n.* 自耦变压器
autotruck [ˈɔːtəuˌtrʌk] *n.* 载重汽车,运

货汽车

autotruck

~ -trol flat bed plotter 自动控制平底绘图仪

auto vac ['ɔ:təvæk] *n.* 真空灌[箱]

auto valve ['ɔ:təu'vælv] *n.* 自动阀,自动活门

autoverify ['ɔ:təu'verifai] *v.*, -ified, -ifying. 自动检验

autovon = automatic voice network 自动电话网

autovulcanization *n.* 自动硫化

autoweighing *n.* 自动计量

autoxidation *n.* 自动氧化

autumn ['ɔ:təm] *n.* 秋; *a.* 秋季的,在秋天
~ construction 秋季施工
~ garden 秋景园
~ timber 晚材,秋(季采伐木)材
~ wood 秋材
~ wood ratio 秋材[晚材]率

autunite ['ɔ:tənait] *n.* 钙油云母

auxiliary [ɔ:g'ziljəri] *n. pl.* -liaries. 辅件; *a.* 辅助的
~ air 辅助[补给]空气
~ air port 辅助风口
~ air receiver 辅助储气器
~ air valve 辅助气阀
~ apparatus 辅助设备
~ bar 辅助钢筋,辅助杆
~ base (line) 辅助基线
~ boiler 辅助锅炉
~ bridge 便桥,临时桥,辅助桥
~ building [construction] equipment 辅助建筑[施工]设施[设备]
~ burner 辅助燃烧器
~ chart 辅助图
~ circuit 辅助电路
~ column 辅助柱
~ compressor 起动[辅助]压缩线
~ condenser 辅助冷凝器
~ construction equipment 辅助建筑[施工]设备
~ contour 辅助等高线
~ cut (平板玻璃裁剪)辅助切割
~ device 辅助装置
~ drilling device 辅助钻探装置
~ drive 辅助传动(装置)
~ elevation 辅助视图
~ engine 辅助发动机
~ equipment 辅助设备
~ facility 附属设备
~ fault 伴生[副]断层
~ force 随动[伺服]传动力,辅助力
~ fracture 伴生断裂
~ fuel 辅助[副]燃料
~ function 辅助功能
~ gallery 辅助坑道
~ heater 辅助加热器
~ illumionation 辅助照明
~ lamp 备用灯,辅助灯
~ landing groud 辅助停机坪[场]
~ lane 辅助车道
~ lighting 辅助[备用]照明
~ line 辅助线
~ marker 路线指示标,辅助标
~ material 辅助原[材]料
~ piping 辅助管路
~ plant 辅助设备[车间,工厂]
~ point 辅助点,补点,补站
~ power plant 辅助[备用]发电站[厂]
~ power supply 备用电源
~ (pre)stressing 辅助预加应力
~ product 辅助材料[物料]
~ projection drawing 辅助投影图
~ properties 次要性能

~ pump 辅助[备用]泵
~ rack 辅助横移
~ rafter 便缘
~ ram 辅助活塞[柱塞]
~ regulator 辅助调节器
~ reinforcement 辅助钢筋
~ scaffold(ing) 辅助钢筋脚手架
~ service (沿路线的)附属服务设备[设施]
~ shaft 副(竖)井
~ sign 辅助标志
~ spring nut 辅簧螺母
~ stair 便梯
~ steel 辅助钢筋
~ store 辅助存储器
~ stressing 辅助预加应力
~ substation 备用变电所
~ supply 辅助[备用]水源
~ tensioning 辅助预加应力
~ truss member 辅助桁架杆件
~ turbine 辅助汽轮机
~ valve 辅助阀
~ variable 辅助变数
~ vault 辅助穹顶
~ view 辅助视图
~ washing unit 辅助冲洗设备
~ work 辅助工作[工程]

auxochrome [ˈɔːksəˌkrəum] *n.* 助色团; *a.* -mic 助色的

AV＝ (1) actual velocity 实际速度
(2) air vent 空气出口

av＝ (1) average 平均值,海损
(2) avoirdupois (英国)常衡(制)(1磅＝16英两)

avail (1) *v.* 有益于,有用(效),帮助
(2) *n.* 效用,利益,帮助

availability [əˌveiləˈbiliti] *n.* 有效性,利用率,可用性
~ factor 有效利用系数
~ ratio 有效度比

available [əˈveiləbl] *a.* 有效的,可用的,现有的,有现货的,有货供应的

~ accuracy 有效精度
~ base 符合要求的基层
~ capacity 有效容量
~ chlorine 有效氯
~ depth 资用[有效]水深
~ dilution 有效稀释
~ draft 可用风量,可用通风压头,有效通风
~ energy 可用能量,有效能
~ factor 可用率
~ for prompt delivery 可立即供水[电,货]的
~ from stock 可从库存中取用的[拨给的]
~ gain 可用增益,有效增益
~ head 有效压头[水头],资用水头
~ heat 可用热量,有效热
~ heating value 有效热值
~ machine time 机器[设备]工作时间
~ moisture 有效水分[湿度]
~ office space 办公使用面积
~ power 有效功率
~ pressure 可用[有效]压力
~ storage capacity 资用[有效]储量
~ surface 有效表面
~ temperature drop 有效温度降
~ thrust 有效推力,可用推力
~ time 有效时间
~ water 有效[用]水,工业[工程]用水
~ water supply 有效供水
~ work 资用功(量)

avalanche [ˈævəlɑːnʃ] *v.*, -lanched, -lanching. *n.*; *v.* 崩落,崩坍
~ baffle 坍方防御建筑,崩落阻挡物
~ brake 排除雪崩用设备
~ debris 崩坍堆积物(包括冰雪,岩石等物)
~ defence 坍落防御设施,防坍方栅栏
~ gallery 坍方[雪崩]防御廊
~ of sand and stone 砂石流
~ preventing forest 雪崩防护林,坍

方防御林
~ protection works 雪崩阻滞工程
~ shed 坍方防御板
avalite *n.* （钾）铬云母
avant-corps [ævɑːŋ'kɔːz] *n.* 从房屋主体凸出的一部分，前亭，建筑物前廊
avant-courier （法语）*n.* 先驱（*pl.*）先锋队
avant-garde (**style of**) **architecture** 建筑学创新[独特]风格
aventurine [ə'ventjəlrin] *n.* 洒金玻璃（含金色细粒的不透明褐色玻璃）；砂金石
~ feldspar 太阳石[琥珀]，猫眼[金缘，日长]石
~ glass 金星玻璃（嵌有黄钢粉的茶色玻璃）
~ glaze 金星釉
avenue ['ævinjuː] *n.* 林荫道，通道，过道，车道，街道
~ of approach 解决途径
~ of sphinxes 斯芬克斯林荫路[大道]
~ planting 道路两边树木
~ to success 成功途径
~ tree 行道树
aver（averred；averring）*vt.* 断言，主张，证明
average ['ævəridʒ] *a.* 平均的，*n.* 平均数
~ acceleration rate 平均加速度率
~ annual daily volume 年平均日交通量
~ annual flood 平均年洪水流量
~ annual precipitation 平均年降水量
~ annual rainfall 年平均雨量
~ annual working capacity 年平均能量[工作量]
~ boiling point 平均沸点
~ composition 平均组成
~ compressive strength 平均抗压强度

~ cost per unit 单位平均成本
~ consumption 平均用水量[消费量]
~ daily consumption 平均日消费量[用水量]
~ daily flow 平均日流量
~ daily traffic(＝A.D.T) 平均日交通量
~ daily volume 平均日交通量
~ depth(＝*a.d.*) 平均深度[厚度]
~ deviation 平均误差[偏差]
~ discharge 平均流量[出水量]
~ efficiency index 平均效率指数
~ end area method （土方计算）平均端面积计算法
~ error 平均误差
~ freight rate assessment 评定平均运费率
~ grade 平均纵坡
~ grading 平均级配
~ grain diameter 平均粒径
~ grain size (of sand) （砂的）平均粒径
~ haul （土方工程）平均运距[输送，运输距离]（挖方至填方的运距）
~ headway 平均车间距或时距
~ highway speed 平均道路车速
~ increment 平均增量
~ illumination 平均照度
~ infiltration 平均渗透率[渗入量]
~ inflow capacity 平均截流量
~ investment 平均投资额
~ least dimension （骨料）平均最小尺寸
~ length 平均长度
~ mileage 平均里程
~ number of inhabitants per building 居住密度，平均居住人数
~ outgoing quality 平均抽查质量
~ output 平均生产率
~ over-all (travel speed) 平均总速率
~ particle diameter 平均粒径
~ per capita consumption 平均每人用（水）量

~ precipitation 平均降雨量[水量], 平均沉淀量
~ precipitation over area 平均面积降雨量
~ pressure 平均压力
~ rainfall 平均降雨量
~ reflectance 平均反射率
~ remaining durable years （建筑物）平均剩余耐用年限
~ sample 平均取样
~ serviceable year 平均使用年限
~ sewage rate per day 平均日污水量
~ sound pressure level 平均声压级
~ speed 平均速度
~ speed difference 平均速度差
~ storey number 平均层数
~ strength 平均强度
~ temperature 平均温度
~ temperature difference 平均温差
~ tensile strength 平均抗拉强度
~ thickness 平均厚度
~ time between maintenance 维修平均间隔时间
~ traffic 平均交通量
~ unit cost 平均单价
~ velocity 平均速度
~ viscosity 平均粘度
~ volume 平均体积
~ volume of grain 平均颗粒体积
~ water consumption perday 平均日用水量[供水量]
~ width 平均宽度
~ wind velocity 平均风速
~ work load 平均工作负载
~ yearly loading capacity 年平均负荷容量
~ yearly rainfall 平均年降雨量
~ yield 平均收益,平均产量（如出水量）
progressive ~ 累加平均数
quadratic ~ 二次平均
time ~ （对）时间（的）平均值,时均值
weighted ~ 加权平均值

averager [ˈævəidʒə] *n*. 中和剂[器]
averaging [ˈævəidʒiŋ] *n*. 求平均值,取平均值,~ out 整平,平衡
avert *vt*. (1)避开,躲避;(2)掉转,防止,避免
avertence [əˈvəːtəns] *n*. 偏转,背转
avidity [əˈviditi] *n*. 亲和力强度,活动性
avirulent [æˈvirjulənt] *a*. 无病毒的,无毒性的;avirulence, *n*.
avoidable [əˈvɔidəbl] *a*. 可避免的,可取消的,avoidably, *ad*.
~ error 可避免的误差
avoidance [əˈvɔidəns] *n*. 避免,取消;无效
~ of cracking 防止裂缝[开裂]
Avogadro's （人名）阿伏加德罗
~ hypothesis 阿伏加德罗假设
~ law 阿伏加德罗定律
~ number 阿伏加德罗数（6.023×10²³）
avoridupols [ˌævədəˈpɔiz] *n*. 常衡（十六英两为一磅）
~ pound 常衡制磅
~ system 常衡制
avometer [əˈvɔmitə] *n*. 万用电表
avulsion [ˈəvʌlʃən] *n*. 冲裂,撕裂
avulsive cutoff 冲裂割断
await [əˈweit] *v*. 等候,准备以待
~ orders 待命
~ parts(AWP) 维修用备件
awaiting-repair time 等待修复时间
award [əˈwɔːd] *v*. 授予,给予,判给; *n*. 决定[断],裁决,奖(品)
~ of contract 签订合同
~ -winning 获奖的
awash [əˈwɔʃ] *a*.; *ad*. 与水面平齐（的）,被波浪冲打（的）,被水打湿（的）
A-waste(s) 放射性废料
away *ad*. 离开,远离,(去掉),(失)去,下去,不断 *a*. 在外的
AWC =absolute worst case 绝对最坏情况

awe *n.*; *v.* (使)敬畏(畏惧)
awhile *a.*; *ad.* 暂时,片刻,少顷
awkward ['ɔːkwəd] *a.* 粗劣的,难辨的,有毛病的
  ~ bend (道路上的)别扭弯道,行车困难的弯道
awl [ɔːl] *n.* 钻子,针锥
  brad ~ 锥钻,打眼钻
  marking ~ 划针
  scratch ~ 划线针,划线台
awning ['ɔːniŋ] *n.* 遮篷,凉篷
  ~ blind 遮阳百叶窗
  ~ curtain 遮阳帷幔
  ~ rafter 天篷脊梁,遮篷椽子
  ~ sash (铰链式)窗扇
  ~ (sash) window 篷式天窗
  ~ (type) window 遮阳式[遮篷式]窗

awning blind

awning window

axe = ax [æks] *n.* 斧
  ~ hammer 剁斧槌
  ~ head 斧头
  ~d arch 斧拱
  ~d work 琢石
  broad ~ 木工(用)斧,宽[阔]斧
  hammer ~ 锤斧
  hand ~ 手斧
  pick ~ 丁字斧
  tool ~ 凿石斧[锤]
axes ['æksiːz] *n.* 轴,心轴
axial ['æksiəl] *a.* 轴向的
  ~ cam 凸轮轴
  ~ blower 轴流
  ~ clearance 轴向空隙
  ~ column 轴向柱,木楼梯梁支柱,螺旋楼梯中心柱
  ~ compression 轴向受压,轴向压力
  ~ compressor 轴流式压缩机
  ~ cylinder swash-plate pump (旋转)斜盘式轴向柱塞泵
  ~ deformation 轴向变化
  ~ diffusion 轴向扩散
  ~ diffusion casing 轴向扩散式机壳[外壳]
  ~ direction 轴向
  ~ dispersion 轴向渗透
  ~ elongation 轴向伸长
  ~ expansion 轴向膨胀
  ~ fan 轴流风机
  ~ fixity 轴向固定[嵌固]
  ~ flow 轴向流
  ~ flow blower 轴流风机
  ~ -flow compressor 轴流式压缩机
  ~ flow fan 轴流式通风机
  ~ flow impulse turbine 轴流冲击式涡轮机
  ~ flow pump 轴流泵,螺浆泵
  ~ flow turbine 轴流涡轮机
  ~ flow turbo compressor 轴流涡轮压缩机
  ~ force 轴向力
  ~ force diagram 轴向力图
  ~ jet velocity 轴向喷流速度
  ~ length 轴长
  ~ load 轴向荷载
  ~ loading 轴向加载
  ~ mixing 轴向混合
  ~ mount 轴向安装
  ~ normal stress 轴向正应力
  ~ piston engine 轴向活塞引擎
  ~ piston machine 轴向活塞机
  ~ plane 轴面
  ~ plunger pump 轴向柱塞泵
  ~ porosity 轴向孔隙率
  ~ pressure 轴向压力
  ~ (pre)stressing 轴向预加应力
  ~ pretensioning 轴向预加拉力
  ~ principal stress 轴向主应力
  ~ ratio 轴比
  ~ restraint 轴向固定[嵌固]
  ~ rib 脊肋
  ~ rigidity 轴向刚度
  ~ road 轴向[辐射]道路

~ seal 轴向密封
~ slip 轴向滑脱
~ stress 轴(向)应力
~ stretching 轴向延伸[伸张,伸展]
~ symmetric(al) flow 轴对称流(动)
~ symmetry 轴向对称
~ tensile strength 轴向抗拉强度
~ tension 轴向拉力,轴心受拉
~ tensioning 轴心预加拉力
~ thrust 轴向推力
~ thrust balancing 轴向推力平衡
~ thrust bearing 轴向推力轴承
~ torque 轴向转(力)矩
~ vector 轴向向量
~ velocity 轴向速度

axially [ˈæksiəli] *ad.* 轴向地,与轴平行地
~ loaded 有轴向负载的,轴向承载的
~ loaded column 轴向受力柱
~ symmetric load 轴向对称荷载
~ symmetric shell 轴向对称的薄壳

axinite [ˈæksinait] *n.* 斧石
axiom [ˈæksiəm] *n.* 定理,原则,公理
axiomatize *v.* (使)公理化
axiradial compressor 轴流式压缩机
axis [ˈæksis] *n. pl.* axes. 轴,轴心线,对称轴
~ bearing 轴承
~ cylinder 轴柱,轴突
~ direction 轴线方向
~ of abscissa 水平轴,x 轴,横坐标
~ of affinity 亲合力轴
~ of arch 背斜轴,穹起轴
~ of coordinates 坐标轴
~ of homology 透视轴
~ of inertia 惯性轴
~ of ordinates 纵坐标轴
~ of reference 参考轴线,坐标轴,基准线
~ of revolution 旋转轴
~ of rotation 旋转轴
~ of sight 视轴,照准轴
~ of strain 应变轴
~ of stress 应力轴
~ of swing 摆动轴
~ of symmetry 对称轴
~ of time (潮汐曲线)时间轴
~ of visual cone 视轴
~ of weld 焊缝中心线,焊接轴
~ rotation 转动轴
anticlinal ~ 背斜轴
arch ~ 拱轴
auxiliary ~ 辅助轴线
beam ~ 梁轴线,射(流)束轴
body ~ 物体(的)轴线,固定轴
cardan ~ 万向节轴,卡登轴
cartesian ~ 直角坐标轴,笛卡尔坐标轴
central ~ 中心轴
chord ~ 固定轴
concurrent ~ 相交轴
conjugate ~ 共轭轴
coordinate ~ 坐标轴
direct ~ 纵轴
earthquake ~ 震轴
Earth's ~ 地球坐标图
eddy ~ 旋涡轴线
elastic ~ 减震[缓冲]轴
flight path ~ 气流坐标轴,速度坐标轴
gravity ~ 重力[重心]轴
guidance ~ 瞄准轴,导向轴
horizontal ~ 水平轴线
inertia ~ 惯性轴
lateral ~ 横轴线,旁轴线
longitudinal ~ 纵轴(线)
major ~ (椭圆的)长轴
mass ~ 重心轴线
mean ~ (椭圆)中轴
minor ~ (椭圆)短轴
moment ~ 挠矩[弯矩]轴
neutral ~ 中性(中和)轴
nofeathering ~ 结构轴(线)
non-polar symmetry ~ 非极性对

axisymmetric

称轴
normal ~ 垂直轴(线),法轴
oblique ~ 斜交轴,斜轴线
offset ~ 位移轴
parallel ~ 平行轴
principal ~ 主轴
principal ~ of ellipse 椭圆主轴
principal ~ of inertia 惯性[惯量]主轴
real ~ 实轴
rectangular ~ 直交轴
reference ~ 坐标轴(线),参考轴
rolling ~ 滚动轴线
rotation ~ 旋转轴
semi-major ~ 长半轴[径]
semi-minor ~ 短半轴[径]
sensitive ~ 量测轴(线)
slewing ~ 旋转轴(线)
solid ~ 实心轴
space ~ 空间座标轴
spin ~ 转轴
spinning ~ 回转轴
spiral ~ 螺旋轴(线)
synclinal ~ 向斜轴(线)
time ~ 时[极]轴
torque ~ 扭矩轴
transverse ~ 水平轴,横轴(线)
trunnion ~ 枢轴
visual ~ 视轴(线)
vortex ~ 旋涡轴线
x ~ 横坐标
y ~ 纵坐标
yawing ~ 偏转轴线
axisymmetric ['æksisimetrik] *a.* 轴(线)对称的
~ bending 轴线对称弯曲
~ shell 轴对称壳体
~al stress 轴向对称应力
axle ['æksl] *n.* 轴
~ arrangement 轴向布置[排列]
~ bar 铁轴棒
~ beam 轴(向)梁

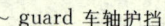
axle

~ bearing 轴承
~ box 轴承,轴套
~ brass 铜衬轴
~ count 轴(向)计数
~ guard 车轴护挡
~ housing 轴向住房建筑
~ housing bracket cap 轴向住房建筑伸臂柱头
~ housing trunnion 轴向住房建筑枢轴
~ king pin 轴向中心销
~ load 轴荷载
~ load scale 轴荷载天平[称]
~ neck 轴颈
~ of drop hammer 穿心锤轴,落锤心棒
~ oil 轴心油
~ pulley 轴开关滑轮
~ spacing 轴矩
~ steel reinforcing bar 轴用钢配筋
~ tree 轴杆,车轴
~ weight limit 轴载重极限
~ yoke 轴轭,轴叉
adjustable ~ 自动调整轴
articulated rear ~ 铰接后轴
beam ~ 梁形横轴
bearing ~ 支承轴
cardan ~ 万向[铰接]轴
couple ~ 连动轴
crank ~ 曲柄轴
datum ~ 底坐轴
end ~ 端轴
hollow ~ 空心轴
idler ~ 中间心轴
intermediate ~ 中间轴
lateral ~ 横轴
live ~ 转动[传力]轴
revolving ~ 回转轴
rigid ~ 刚性轴
sagging ~ 下挠轴
shifting ~ 移动轴

stationary ~ 静止[固定]轴
stub ~ 短心轴
supporting ~ 支承轴
telescopic ~ 套筒轴
trailing ~ 支承轴
turning ~ 转轴
axmen *n.* 斧工
Axminster carpet 阿克斯敏特式地毯
axode *n.* 瞬轴面
axometer *n.* 测(光)轴计
axonometric [ˌæksəˈnɔmitrik] *a.* 三向投影的,不等角投影的
~ drawing 轴测图
~ projection 三向图,不等角投影图,轴测投影
axonomentry [ˌæksəˈnɔmitri] *n.* 轴测量法,三面正投法,均角投影图,三向图
axostyle [ˈæksəˈstail] *n.* 轴杆,轴柱
axostylus [ˌæksəˈstailəs] *n.* 轴杆
azalea [əˈzeiljə] *n.* 杜鹃花
azimuth [ˈæziməθ] *n.* 地平经度,方位(角)
~ alignment 方位校准
~ (bearing) angle 方位角
~ by-pass 左右测管[旁通道]
~ circle 方位圈,地平径圈
~ compass 方位罗盘
~ counter 方位测量器
~ mark 方位标
~al equidistant projection 等距方位投影
~al projection 方位投影
azotea *n.* 在住宅或其他建筑顶上的平屋顶或平台
azotic [əˈzɔtik] *a.* 含氮的
~ acid 硝酸(HNO_3)
azotification [əzɔtifiˈkeiʃən] *n.* 固氮(作用)
Aztec architecture (14 世纪)古墨西哥建筑
azure [ˈæʒə] *n.* 蓝(色),天蓝(色);*a.* 蓝(色)的
~ black 黑天蓝色(的)
~ blue 天青蓝(灰绿蓝色)
~ spar 天蓝石
~ stone 琉璃,石青
azurite [ˈæʒərait] *n.* 蓝铜矿,石青
AZUSA (1) aximuth, speed and altitude (radio course directing set)方位-速度、高度(无线电定向装置)
(2) AZUSA tracking station AZUSA 跟踪站
(3) AZUSA tracking system AZUSA 跟踪系统

B

B [biː]第二已知数(2)B 字形
B=(1)ballistic 弹道(式)的,冲击的;(2)base 碱,基础;(3)board 配电盘,仪表板
b=(1)barn 靶(恩)(核子有效截面单位);(2)(美国)十亿,10^9(英国)万亿,10^{12}
BA=(1)Bachelor of Arts 文学士;
(2)British Academy 英国研究院
(3)British Airways 英国航空公司
Babbit(t) [ˈbæbit] *n.* 巴氏合金,轴承铅;*vt.* 衬以巴氏合金
~ bearing 巴氏合金轴承
~ bushing 浇铅轴衬
~ -lined bearing 巴氏合金衬管轴承
~ metal 巴氏合金,轴承[耐磨]合金
~ing jig 轴承装配工具
Babel [ˈbeibəl] *n.* 古巴比伦之城及该城所建之塔
babel-quartz [ˈbeibəlˈkwɔːts] *n.* 塔状

石英

baboon [bə'bu:n] *n.* 建筑物或装饰品上风格奇特的图案

baby ['beibi] *n. pl.* -bies, *a.*; *v.*, -bied, -bying; *n.* 小型,微型,婴儿; *a.* 小型的,微型的;婴儿的;babylike, *a.*

~ blue 浅蓝色
~ budget 小额预算
~ bulldozer 小型推土机
~ compressor 小型压缩机
~ concrete mixer 小型混凝土搅拌机
~ drill 小钻,钻床
~ farm 托儿所,育婴院
~ knife switch 小型闸刀开关
~ light 小型聚光灯
~ orchid 宝宝兰
~ rail 小钢轨
~ roller 小型压路机
~ room 婴儿室,保育室;儿童病室
~ square 小方木
~ spotlight 小型聚光灯
~ switch 小型安全开关

Babylonian [,bæbi'lənjə] *a.* 巴比伦(古代巴比洛尼亚的首都)的,巨大的,奢华的

~ achitecture 巴比伦建筑(公元前2630 ~ 前1275年)
~ quartz 塔状石英

bacalite *n.* 淡黄琥珀

bacca *n.* (拉丁语)浆果

bacco 或 **baccy** *n.* 烟草

bachelor ['bætʃələ] *n.* 未婚男子,单身汉;学士学位

~ apartment (美)单身公寓
~ flat (英)单身公寓
~ hostel 单人宿舍
~ kitchen 单人用厨房
~ of Architecture(B. Arch.) 建筑学士
~ of civil engineering 土木工程学士
~ quarters 单人宿舍
~'s chest 单人用五斗橱
~'s dwelling unit 单身住宅单元

back [bæk] *n.* 背,后面;基底,衬垫;凸面,梁上表面;岭,脊; *vt.* 支持,拥护; *vi.* 向后退;堵(水);加贴面[里衬];装上(椅)背; *a.* 背后的,后部的;反向的,逆转的,倒的; *ad.* 在后面,倒,后退;backless, *a.*

bachelor's chest

~ access 背径小路
~ angle 后视角
~ arch 拱背
~ axle 后轴
~ bar 壁炉横棍(壁炉内挂容器用)
~ beach 后滨
~ bead 填角焊缝
~ bearing 反方位,反象限角
~ blocks (都市中的)贫民窟
~ boiler 家用热水炉
~ barce 反斜撑
~ brushing 挑顶
~ cargo 归程货物
~ catch 门后扣钩(保护开启状态)
~ chipping 背铲
~ clamping 反向箱位
~ coat 底面涂层
~ connection 内侧连接
~ country 偏僻村镇
~ court 后天井
~ crossing 里层交错板
~ cushion 靠垫
~ deals 井壁临时衬板
~ drop 背景
~ door (entrance) 后门
~ draft 逆流通风;倒行[车,转]
~ drainage 背面排水
~ draught 逆流通风,反向气流
~ ed-off cutter 铲齿铣刀
~ edge 后脊
~ edging 砖瓦清边法
~ elevation 后视图

~ face 后面,背面,反面
~ fall 斜坡
~ fill 回填(土)
~ -fill consolidation 回填夯实,复土压实
~ fillet 平嵌线(墙面与隅石交接处)
~ filling 填充,填实
~ filling soil 回填土
~ -filling tamper 回填压实机,填土夯实机

back fillet

~ fin 后脊,裂纹
~ (-)fire 逆火,回火,反燃
~ flap 里扉,里垂帘,百叶窗
~ -flap hinge 明铰链
~ flow preventer 回流防止器
~ flow valve 止回阀,逆止阀
~ flushed waste disposal 倒灌污水
~ front 房屋的背面
~ garden 后园
~ gauge (房)边距
~ gear 背轮,减速齿轮
~ gouging 背刨
~ -grouting 二次灌浆
~ guide 后导板
~ -guy 拉索;支撑
~ haul 回运(土方),回程运输
~ haul cable 后曳缆
~ hearth 壁炉背
~ hole 倒眼(爆破用的一种炮眼)
~ hook 内挂钩
~ house 后屋,户外厕所
~ hub 后视标杆
~ incline 后斜板
~ injection 反注入
~ inlet 后部进水口
~ joint 留槽待填
~ -kick 逆转,回转
~ land 腹地,后方
~ lane 后巷
~ leg 斜支柱,背面支柱
~ light 背光,逆光,后灯

~ lining 背衬
~ log 存货,储备金
~ marsh 后背湿地,腹地湿地
~ mixing 回混,返混
~ (mixed) plaster 底灰
~ mortaring 抹里皮
~ motion 倒车;倒转
~ moulding 底饰
~ nail 平钉
~ nut 背紧螺母,支承螺母
~ observation 后向观测
~ of beyond (英)内地,僻远地区
~ of levee 堤背
~ of tool 刀(具)背
~ of vault (穹窿)拱背
~ out 逆序操作
~ plastering 背面抹灰
~ plate 背板
~ pressure 回压(力),反压(力)
~ pressure manometer 后压测压计
~ pressure regulator 反压调节器
~ pressure turbine refrigerator 背压式涡轮制冷机
~ pressure valve 止回阀,逆止阀
~ pull 反拉力,反张力
~ putty 打底油灰
~ rake 后倾角
~ range 尾距
~ residence block 偏僻住宅
~ residential block 偏僻住宅
~ ripping 二次挑顶
~ road 便道,村间道路
~ roll 支承辊
~ rolling 重复滚压
~ room 里屋,密室
~ run 背面焊缝;反转,逆行
~ scatter 反向散射
~ seams 地毯背面接缝
~ seat gasket 底座垫圈
~ service (road) 背街服务性道路,内街运输路
~ set 制动器,倒流,逆水,涡流

~ shaft 后轴,副轴
~ shooting 逆向爆炸,回程爆炸
~ shop 修理厂,修理车间
~ shutter 里百叶门或窗
~ site 不临街的建筑基地
~ sloper 量坡规
~ slum 贫民窟
~ stair(s) 内部[后]楼梯
~ street 后街
~ test 弹性复原试验
~ swing 回摆
~ tilt 后倾(角)
~ trough 背槽
~ turn 反回音
~ up block 支撑块
~ up brick 衬里砖,墙里砖
~ upholstering 靠背
~ up lining brick 衬砌砖
~ up material 填充材料
~ up strip 衬板,衬里,背垫条
~ up system 备用系统
~ veneer 衬里层 back veneer 胶板
~ vent 虹吸排气管,背通气管
~ view 背视图
~ view mirror 望后镜
~ wall 背墙,挡土墙
~ ward welding 后退焊,右向焊
~ wash(ing) 反冲洗
~ water 循环水,再用水
~ water trap 回水存水弯
~ welding 背焊(件)
~ woods 偏远的森林地区
~ yard 后院
~ ed-up-weld 后托焊接
~ ed-with blocks 块体衬砌
arc ~ 逆弧
flash ~ 退火,回火;反冲,回击
folding ~ 折叠式椅背
high ~ 高靠背(座位)

hog ~ 底板突出部分,等倾线之顶点
hole ~ 钻眼底,炮眼底
horse ~ 底板突出部分,马脊岭
kettle ~ 穹形顶板
kick ~ 逆转,回收
metal ~ 金属壳[背衬垫]
recycle ~ 反向循环
saddle ~ 顶板圆弯;屋脊状支架;马鞍形山脊
skew ~ 拱座,拱脚,起拱点
window ~ 窗腰
backacter ['bækæktə] *n.* 反铲挖土机,反向铲,开挖机
backacting shovel 反铲(挖土机),反向铲
backboard ['bækbɔ:d] *n.* 背板,靠背,后挡板
backbone ['bækbəun] *n.* 构架,支柱
~ road 主干道,主要干路
backcloth ['bækklɔθ] *n.* 天幕,背景幕
backer ['bækə] *n.* 背衬,石板瓦
~ brick 背衬墙;墙心砖
~ pump 备用[前级]泵
backfiller ['bæk,filə] *n.* 覆土机,回填机
~ blade 回填铲,回沟机平铲
~ tamper 回填机,捣固器
backflow ['bækfləu] *n.* 回流
~ connection 回流管
background ['bækgraund] *n.* 背景;基础(底);环境;本底;底色
~ brightness[luminance] 本底亮度
~ heater 隐闭式供暖器
~ information 背景资料
~ job 后备(后台)作业
~ lighting 背景照明
~ luminance 背景亮度
~ music 背景音乐
~ noise 背景噪声,本底噪声,声底值
~ processing 基本处理
~ projection 背景投影
~ reflectance 背景反差
~ survey 本底调查

backhand ['bækhænd] *a.* 反向的,反手的,间接的,二手的
~ welding 右向焊,反手焊,后退式气焊

backhoe ['bækhəu] *n.* 反向铲,反铲铲土机
~ boom 反铲臂
~ bucket 反铲铲斗
~ excavator 反铲挖掘机
~ front end loader 后挖前卸式挖装机
~ loader 反铲装载机
~ shovel 反铲挖掘机

backing ['bækiŋ] *n.* 背衬,衬里;回填土,支撑;底,垫,后退,逆行
~ belt 倒车皮带
~ block 支撑块
~ blowing (of a well) 压气反洗
~ board 背纸板,底托板
~ brick 墙心砖,背衬砖
~ coat 底面涂层
~ concrete 大体积混凝土,垫层混凝土
~ -filling material 砌体填充材料
~ light (舞台)背景灯光
~ log (码头)档边木
~ material 衬底材料
~ metal 金属垫块
~ -off (应力的)消除;铲齿,铲
~ of veneer 胶合板内层
~ of wall 墙托
~ of window 窗托
~ -out punch 退钉器
~ paper 垫纸
~ pass 底焊焊道
~ plank 衬板
~ plaster 底层灰泥
~ plate 垫模板
~ ring 衬环,垫圈,垫环
~ sheet 衬纸,衬板
~ stone 衬里石,背石
~ strip 垫板
~ tile 衬环,垫圈
~ up 封底焊
~ -up brick 里壁砖
~ up back iron 刨刀护铁
~ weld 背焊

backlash ['bæklæʃ] *n.* 后冲;间隙;拉紧
~ sping 消隙弹簧

backless *a.* 无(靠)背的

backlining *n.* 衬板,衬衬(料)

backlist *n.* ;*vt.* ((把……)列入)多年重版书目

backlog (1)*n.* 积累,储备
(2)*v.* (把……)积压起来

back-mixing *n.* 回(返)混

back(-)out *v.* ;*n.* 退火,反回逆序操作,拧松,取消,放弃

backpitch ['bækpitʃ] *n.* 反螺旋;背节距

backplan ['bækplæn] *n.* 底视图

backplane ['bækpein] *n.* 底板

backplastered [ˌbæk'plɑːstəd] *a.* 背面抹灰的;板条抹灰的

backproject ['bækprədʒekt] *n.* 背面投射

backrest ['bækrest] *n.* 靠背

backrope ['bækˌrəup] 斜桁撑杆后支索

backsaw ['bæksɔː] *n.* 镶边短锯[手锯]

backsawing ['bæksɔːiŋ] *n.* 弦锯

backseat ['bæksiːt] *n.* 后座,次要位置

backsight ['bæksait] *n.* 后视,后视读数

backsiphonage ['bæksaifənidʒ] *n.* 倒虹吸作用,反虹吸(能力)
~ pipe 反虹吸作用管

backslope ['bækslɒup] *n.* 后坡,背坡

backsloper ['bækˌslɒupə] *n.* 量坡规,推土机的推板

backspace ['bækspeis] *v.* ,-spaced,-spacing;返回,后移

backstage ['bæksteidʒ] *n.* (剧场)后台(区); *a.* 在后台的,幕后的; *ad.*

在[往]后台,在剧院的化妆室中
backstay ['bækstei] *n.* 背撑,拉索,后拉缆
~ cable 拉索
backstep ['bækstep] *n.* 分段退焊;后部台阶
~ sequence 分段退焊次序
~ welding 分段退焊,反手焊
backstop ['bækstɔp] *n.*, *v.*, -stopped, -stopping; *n.* 挡板,托架(阻止球滚得太远的)挡球网,(棒球)本垒后方的网或栅;阻挡 *v.* 支持,加强
back(-)up ['bækʌp] *n.* 支撑,后援;溢出,堆积,阻塞;备用品; *a.* 备用的,预备的,支持的
~ man 辅助工人
~ plate 垫片
~ ring 保护垫圈,支承环
~ tile 用空心砖砌混水墙
~ wall 支撑墙
~ washer 支撑垫圈
manual ~ 手动备用调节装置
backward ['bækwəd] *a.*; *ad.* 向后的,倒行的; backwards; *ad.* backwardly, *ad.*
-backwardness, *n.*
~ curved blade 后弯曲叶片
~ drainage 逆向排水
~ erosion 反向冲刷
~ inclination angle 后倾角
~ linking 后向信号联结
~ sight 后视
~ stroke 回冲程
~ type centrifugal fan 叶片反曲离心式通风机
~ wave ascillation 逆波摆动
~ welding 后退焊
backwash ['bækwɔʃ] *n.* 回流,倒流,退浪冲刷; *v.* 冲洗,逆流;backwasher, *n.*
~ cycle 反冲周期
~ rate 反冲速率
~ storage 冲洗水箱
~ water 反洗[冲]水(滤池)

~ water requirement 反冲洗水需要量
backwashing ['bækwɔʃiŋ] *n.* 反冲,洗(滤池);反溅
~ method 反洗法,反冲法
backwater ['bæk,wɔːtə] *n.* 回水,循环水,再用水
~ curve 壅水曲线,回水曲线
~ effect 壅水效应,回水效应
~ function 壅水函数,回水函数
~ gate 回水闸
~ height 壅水高度
~ length 回水长度
~ pressure 回水压力
~ storage 回水库容(量)
~ trap 回水存水湾
~ valve 回水逆止阀
~ zone 回水区

backwater valve

bad [bæd] *n.*; *ad.*; *a.*, worse, worst; 不好的,有害的;不利的
~ contact 接触不良
~ land 崎岖地
~ slip 强烈滑坡
~ top 不稳固顶板
badge [bædʒ] *n.* 标记,符号
~ reader 标记阅读器
badger ['bædʒə] *n.* 管道清理器;榫接边
~ plane 宽槽推刨
badigeon ['bædidʒən] *n.* 油灰;嵌填膏泥
badly ['bædli] *ad.*, worse, worst; 不良地,很坏地;严重地
~ graded sand 级配不良的砂
baffle ['bæfl] *n.*; *v.*, -fled, -fling; *n.* 砥挡板,隔墙;导流片,缓冲板;分流墩,遏声器; *v.* 阻碍,隔声; bafflement, *n.*; baffler, *n.*; baffing, *a.*; baffingly, *ad.*
~ area 阻隔区
~ board 挡板,隔声板,阻流板
~ chamber 隔板集尘区

~ gate 单向门
~ junction 缓冲式交叉
~ paint 伪装掩护色用涂料
~ pier 砥墩,消力墩
~ still 门槛,消力槛
~ structure 消力结构,消力设施
~ wall 隔墙,障壁
~d mixer 搅拌机
~d mixing chamber 隔板式混合池
thermal ~ 隔热板
baffler ['bæflə] *n.* 挡板,阻隔板,阻尼器;消声器;导流板;节流阀
bag [bæg] *n.*; *v.*, bagged, bagging; *n.* 袋,外壳,贮油器; *v.* 装入袋中
~ collector 集料袋
~ concrete 袋装混凝土
~ conveyor 袋输送器
~ elevator 袋货升降机
~ filter 袋(式过)滤器
~ -filter-type collector 袋滤式集尘器
~ goods 袋货
~ machine 袋装机
~ of cement 袋装水泥
~ packer 袋装机
~ plug 袋式堵头
~ pump 风箱泵
~ strapping 家具扩边阔带,家具包装带
~ trap 袋形存水湾
~ wall 砖窑隔火墙
~ ged lime 袋装消石灰
~ ged aggregte 袋装骨料(集料)
~ ging 包装;制袋材料
~ ging machine 装填机
~ ging-off 装袋
~ ging platform 装包平台
tool ~ 工具袋
bagasse [bə'gæs] *n.* 甘蔗渣(软质纤维板的材料)
~ board 甘蔗渣板
~ sheet 甘蔗渣板
bagger ['bægə] *n.* 挖泥(机),掘沟机;泥

bag trap

斗;装袋机
bagman *n.* 行商,推销员
baghouse ['bæghaus] *n.* 集尘室,沉渣室,滤袋装置
~ precipitator 滤袋除尘器
bagnio ['ba:njəu] *n. pl.* bagnios;浴室,浴堂
bagut(te) [bə'get] *n.* 半圆饰,小圆凸线;细小线条
Bahn metal 轴承合金(铅 98.64%,铝 0.2%,钌 0.65%~0.73%,钠 0.58%~0.66%)
bail [beil] *n.* 戽斗,吊桶;绳套;钩环,吊环,耳;栅栏; *v.* 汲出
bailer ['beilə] *n.* 抽泥筒,泥浆泵,水斗,提砂筒
~ grab 捞砂筒的捞钩
~ line 捞砂绳索
~ sheave block 卷扬机滑车
bailey ['beili] *n.* 城廓;外栅;监狱
~ bridge 活动便桥
~ form construction 城廓形建筑
~ wall 水冷外壁
Bailey [beili](人名)贝利
~ bridge 贝利桥(英国战时就地装配小跨钢桁梁桥)
~ span 贝利式桁架桥跨
~ truss 贝利式桁架
bain-marie *n.* 水浴器;水浴
bajada [bə'dʒeidə] *n.* 山麓冲积平原
bake [beik] *v.*, baked, baking; *n.* 烘焙,烘干,烧硬
~ house 面包房,食品铺
~ (-)out 烘烤,退火
~ drying oven 烘干炉,烘箱
~d clay 烧干土,陶土
~d enamel 烧搪瓷,烧珐琅
baking coal 烤焦煤
baking dryness 烘干[烤干]涂料
baking finish 烘漆,拷漆

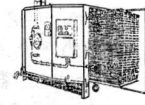

bake drying oven

baking-hot 炎[极]热的
baking paint 烘漆,拷漆
baking soda 碳酸氢钠,小苏打
baking varnish 清烘,[拷]漆
baking zone 烘干层
bakelite ['beiklait] *n.* 醛塑料;(酚醛)电木,胶木
~ cement 酚醛树脂粘结料
~ mouldings 电木模型
~ varnish 酚醛树脂清漆
~d wood 胶木
baker ['beikə] *n.* 烘炉,烤炉
bakery ['beikəri] *n. pl.* -eries；面包房,面包店
~ room 面包房
balance ['bæləns] *n.；v.* -anced,-ancing；平衡；天平,秤；平衡力；结算差额表；差额,余额
~ approach 均衡办法
~ arm 秤杆,天平磅
~ bar 平衡杆,秤杆
~ beam 平衡杆,平天梁
~ box 均衡箱
~ bridge 开启桥,平衡桥
~ brow 踏板
~ cock 摆夹板
~ cone (平衡)锥式液限仪
~ crane 平衡起重机
~ cuts and fills 均衡挖填,挖填土方平衡
~ gate 平衡门
~ indicator 平衡指示器,天平指针
~ level 水准器
~ lever 平衡杠杆
~ piston 平衡活塞
~ reading 天平示数
~ sash 平衡式窗扇
~ screw 摆螺针
~ sheet 平衡表
~ valve 平衡阀
~ weight 平衡重
~ wheel 摆轮

~d arch 平衡拱
~d cable crane 平衡缆式起重机
~d cantilever erection 平衡悬臂架设
~d construction 均衡施工
~d crew 均衡工班组
~d cross-section (厚边混凝土路面)平衡横断面
~ development 均衡发展
~d door 平衡式上下推拉门
~d draft 均衡通风
~d earthwork 平衡土方量
~d engine 平衡式发动机
~d erection 吊挂装配(桥梁)
~d gate 平衡闸门
~d grade (填挖)平衡坡度
~d grading 挖填平衡的坡度设计；土方平衡
~d hoist 平衡起重机
~d load 均衡荷载
~d output 平衡输出；对称输出
~ of precision 精密天平
~d profile 平衡纵断面(填挖相等的纵断面)
~d rigid frame 均衡刚架,平衡构架
~d section 平衡截面
~d steel ratio 平衡配筋率
~d steps 均衡踏步
~d system of ventilation 均衡制通风系统
~d valve 平衡阀
balancing arm 平衡臂
balancing chamber 平衡室
balancing calculation 平衡计算
balancing damper 平衡调节器
balancing lever 平衡杠杆
balancing method 平衡法
balancing moment 平衡弯矩法
balancing of stress 应力补偿
balancing piece 平衡块
balancing reservoir 调节水库,平衡池
balancing tank 平衡池

balancing test 平衡试验
balancing water content 平衡含水量
back ~ 平衡锤;后平衡
beam ~ 杠杆式天平
colour ~ 颜色调谐
compensation ~ 补偿平衡
sash ~ 吊窗衡重
static ~ 静力平衡
thermal ~ 热平衡
unstable ~ 不稳定平衡
weight ~ 砝码

beam balance

balancer ['bælənsə] *n.* 平衡器,配重,均压器,稳定器
balas *n.* 浅红晶石
balata *n.* 巴拉塔树胶
balata belt 巴拉塔胶(皮)带
balatte *n.* 石灰石板
balconet [ˌbælkə'net] *n.* 眺台式窗栏,装饰横档
balcony ['bælkəni] *n. pl.* -nies;阳[凉]台,楼座包厢
~ access 挑廊式入口,眺台入口
~ (access) type 外廊式
~ balustrade 阳台栏杆
~ beam-column 阳台梁柱
~ column 阳台柱
~ door 阳台门
~ facing 阳台饰面
~ front 剧院眺台前沿
~ lining 阳台饰面
~ tile 阳台瓷砖
~ truss 眺台桁架
~ window 阳台窗
baldachin ['bɔːldəkin] *n.* (教堂)龛室,祭坛华盖,墓华盖
bale [beil] *n., v.* baled, baling; *n.* 包,捆,束; *vt.* 将……打包; baleless, *a.*; baler, *n.*
~ breaker 拆包机
~ dumping device 捆包(货)物倾卸设备
~ press 填料压机

~'s catch 自动扣
~ ties 打包铁皮带[钢丝]
~ truck 载包车
~ wire 成捆线材
balection moulding 镜框饰,突出嵌线的装饰线条
balistraria [ˌbælə'streəriə] *n.* 弓箭室;射箭孔(城堡上)
balk [bɔːk] *n.* 横木梁,大梁;障碍; *v.* 停止,中止,阻止,妨碍,错过;balkier, *n.*;balkingly, *ad.*
~ board 防护板
~ roofing 重屋面

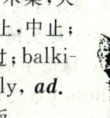
balistraria

ball [bɔːl] *n.* 球,滚珠,团块;海岸沙洲; *v.* 使成球形
~ and ring method 球环测定法(沥青软化点试验用)
~ and roller bearing 滚珠和滚柱轴承
~ and spigot 套接,接插承接合
~ and socket joint 球窝接头
~ bank indicator 球形示倾器,球形倾斜仪
~ bearing 滚珠轴承,球轴承
~ "bearing" action (路面磨损微粒的)滚动作用
~ bearing hut hinge 滚珠轴承合页
~ bushing 球轴套
~ catch 门碰球
~ check 球阀
~ check valve 球止回阀
~ clay 陶土,球土,泥球
~ cock 浮球阀,球旋塞
~ cock assembiles 球阀装置
~ collar thrust bearing (环形)滚珠止推轴承
~ conveyor 球式输送机
~ durometer 球形硬度计
~ float 浮球,球状浮体,浮球阀
~ float level controller 浮球液面控制器
~ float valve 浮球阀

~-flower 圆球饰

ballflower

~ game ground 球场(总称)
~ grider 球磨机
~ guide 滚珠导筒
~ hammer 球头锤
~ handle 球形手柄
~ hardness (布氏)球印硬度
~ hardness number (布氏)球印硬度值
~ hardness testing machine (布氏)球印硬度试验机
~ hooting 上滑道集材
~ indentation test (布氏)球印硬度试验
~ joint 球窝接合,球形接头,环节
~ joint manipulator 球承式机械手
~ jointed screw 万向球铰螺杆
~ knob 球状把手
~ lever 浮球杆
~ method 球印(硬度)试验法
~ mill 球磨机
~ milling 球磨
~ mill refiner 球磨精制机
~ nozzle 球形管嘴
~ nut 滚珠螺母
~ peen (pane, pein, pean) hammer 圆头锤
~ penetration (test) 球体贯入试验
~ piston pump 球塞泵
~ pivot 球枢
~ plug 球塞
~ pump 球形泵
~ race 滚珠座圈
~ relief valve 球形安全阀
~ resolver 球形解算器
~ ring 球环
~ screw 滚珠螺杆

~ seat 球座
~ socket 球窝
~ stone 球石(碱层的细石灰石块体)
~ stud 球头螺栓
~ table 球式工作台
~ tap 球塞,浮球阀
~ track [轴承]滚道
~ transfer table 球式转运台
~ trap 球形存水弯
~ tubemill 球管磨机
~-type interceptor 球形截水器
~ valve (浮)球阀
~ valve sampler 球阀式取土器
~ wire nail 球头圆铁钉
ash ~ 凝灰球
bearing ~ 轴承滚珠
check valve ~ 止回阀球
fulcrum ~ 支承球
relief valve ~ 保险阀球
stop valve ~ 止阀球
ballast ['bæləst] *n.* 道碴;压舱物;镇流器 *v.* 铺道碴;使稳定;ballaster, *n.*; ballastic, *a.*
~ aggregate 道碴,石碴
~ bed 道碴路基,石碴道床
~ box 通碴槽
~ boxing (在土路基中)为摊铺道碴而建筑的路槽
~ bridge floor 铺碴路面
~ cleaner 清碴器
~ cleaning 道碴清筛
~ concrete 石碴混凝土,镇重混凝土
~ crusher 道碴轧碎机
~ density 道碴密度
~ guard (桥梁)道碴围挡
~ hammer 碎石锤
~ mattress 沉碴垫层
~ plow 道碴犁
~ pocket (混凝土中)道碴窝
~ pump 压载水泵
~ rake 道碴耙
~ road 道碴[石碴,碎石]路

~ scarifier 扒碴机
~ shoulder 道[石]碴路肩
~ spreader 撒道碴车,道碴撒铺车
~ surfaced 麻面
~ tamper 砸道机;夯道机;石碴捣固工具,捣碴机
~ tank 压载水箱
~ templates 道碴肩板
~ truck 石碴车
~ tube 镇流管
~ wall 桥台台帽上填土与上部构造间隔墙,子墙
~ weight 镇重,配重
~ed deck 铺碴桥面(铁路桥)
~ed deck bridge 铺碴(上承)桥
~ed floor 铺碴桥面
~ing material 道碴(材料)
~ing up 铺完石碴,施加压重
~ing-up 压载调整

ballerina check 斜交方形花纹,交替斜线格纹

balloon [bəˈluːn] *n.* 氢气球;飞船;球形大玻璃瓶;球饰; *v.* 充气
~ construction 轻捷构造,球形水塔
~ framed construction 轻捷骨架构造
~ framing 轻捷构架
~ payment 分期付款中最后一笔特大的偿还款
~ tire (tyre) 低压轮胎

ballot [ˈbælət] *n.; v.* (无记名)投票
~ against 投票反对
~ for 投票赞成

balsa [ˈbɔːlsə] *n.* 轻质木材
~ wood 轻[筏]木

Baltimore [ˈbɔːltimɔː] *n.* 巴尔的摩
~ truss 平行弦再分式桁架

balun [ˈbælən] *n.* 巴伦仪,平衡-不平衡转换器

baluster [ˈbæləstə] *n.* 栏杆小柱;(*pl.*) 栏杆
~ column 栏杆柱
~ railing 立柱栏杆

~ side 栏杆边

balustrade [ˌbæləsˈtreid] *n.* 栏杆,扶手;-balustraded, *a.*
White marble ~ 汉白玉栏杆

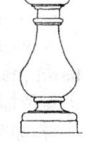
baluster

bamboo [bæmˈbuː] *n. pl.* -boos; *a.; n.* 竹,竹材; *a.* ,竹的,竹制的
~ -blind door 篱笆门,竹帘门
~ bolt 竹销
~ bridge 竹桥
~ carving 竹刻
~ corridor 竹制回廊
~ drain 竹暗渠
~ enclosure 编竹围墙
~ fence 竹篱
~ filament 竹丝
~ framing 竹(材)结构
~ furniture 竹家具
~ lath 竹板条
~ -lath transom 竹条气窗
~ lofty restaurant 竹轩(餐厅)
~ mat 竹凉席
~ mosaic 竹片马赛克
~ nail 竹钉
~ pavilion 竹(制)亭
~ plantation 竹园
~ -pole scaffold 竹脚手架
~ plywood 竹制胶合板,胶合竹板
~ purlin 竹檩条
~ rafter 竹椽,竹楼
~ rail (推拉门用)竹制轨槽
~ rail fence 竹篱,编竹围墙
~ (reinforced) concrete 竹筋混凝土
~ reinforcement 竹筋
~ roof 铺竹屋顶
~ scaffolding 竹脚手架
~ scale (tape) 竹尺
~ spatula 竹刮刀,竹压片,竹片
~ steel 竹节钢
~ wood 竹林

~ -wood inlay 竹木镶嵌
~ work 竹材工作
~ worker 竹木工

band [bænd] *n.* 带，箍；条纹，板带层；扁带饰；带；扁钢；波段，波带；光（谱）带，频带；*v.* 绑扎，联合；bander, *n.*；bandless, *a.*
~ and gudgeon 长铰，大门铰链
~ brake 带闸

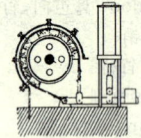

band brake

~ clutch 带式离合器
~ conveyer 皮带运输机，带式输送器
~ course (扁)带层
~ decoration 带饰
~ dryer (污泥)带式干化机
~ elevator 带状升降机
~ feeder 带式给料机
~ hinge 扁铰链
~ iron 扁铁，带钢
~ level 频带级
~ matrix 带形矩阵，带阵
~ merit 带宽指标
~ mill 带锯锯木厂
~ moulding 带饰
~ pass 传递带
~ pass amplifier 带通放大器
~ -pass filter 带通滤波器
~ plate 带板
~ platform (舞台上的)伴奏席
~ rope 扁钢丝绳
~ sandstone 层状砂岩
~ saw 带锯
~ saw blade 带锯条
~ saw(ing) machine 带锯机
~ saw sharpener 带锯磨齿机
~ screen 带条筛，转筛
~ screw 带式螺旋
~ shoe 管箍座
~ speed 带速度
~ spring 板簧
~ steel 条钢，扁钢
~ tape 卷尺
~ tool 带式机床
~ tubing 软韧橡皮管
~ tyre [tire] 实心轮胎
~ wheel 带锯轮
~ width (频)带宽(度)
clay ~ 黏土夹层
conveyor ~ 轮送带
dirt ~ 污层；夹石层
filter pass ~ 滤波器通带
filter stop ~ 滤波器阻带
frequency ~ 频带
head ~ 拱饰门
iron ~ 铁箍
leading ~ 仰斜矿房
metering ~ 计量卷尺，皮尺
safety ~ 安全带
saw ~ 锯带
slab ~ 薄板梁
wheel ~ 轮箍，轮胎

banded ['bændid] *a.* 结合的，箍的；带状的，有条纹的
~ architrave 带饰门头线
~ coal 带状煤
~ column 箍柱，拼合柱
~ granite 带状花岗岩
~ impost 拼合拱墩
~ limestone 带状石灰岩
~ shaft 箍柱
~ steel pipe 加箍钢管
~ structure 带状结构

banded impost

band(e)let ['bændlit] *n.* 细带饰，环柱扁带饰

Banderole ['bændərəul] *n.* 扁带状雕刻装饰，刻扁

banding *n.* 钉夹板，镶边，嵌条；打捆，箍紧；带状物
~ plane 线脚刨

bandstand ['bændstænd] *n.* (室外)音

乐台
bandy ['bændi] *n. pl.* -dies; *v.* -died, -dying; *a.*; *n.* 曲线; *v.* 交换, 传布; *a.* 带状的; bandiness, *n.*
～ clay 带状黏土
～ leg 一种向外翻转的家具脚
banian ['bæniən] *n.* 榕树
banister ['bænistə] *n.* 栏杆小柱, (*pl.*) 栏杆, 扶手
banjo ['bændʒəu] *n.* 箱, 盒, 机匣; 轴 [油, 齿轮]箱; 钻车, 凿岩机; 一种五弦琴
～ axle 整体式桥壳
～ bolt 空心螺栓
～ fixing 对接接头
～ frame (放样)曲线规
～ union 鼓形管接头
bank [bæŋk] *n.* 堤, 岸, 沙洲, 浅滩, 河畔; 倾斜, 边坡, 坡度; 银行, 仓库; 电梯群; 组合; *v.* 堆积, 筑堤; 使倾斜, 超高; 存(款)于银行; 排成序列
～ account 银行往来账
～ angle 倾斜角, 超高角
～ barn 倾斜谷仓
～ bars 井壁衬板
～ book keeping machine 银行记账机
～ building 银行大楼; (积砂)造岸
～ caving 堤岸坍陷
～ clearance 离岩净空(宽度)
～ cubic yard (爆破岩石)实体立方码
～ cutting 开挖岸坡
～ deposit 岸边沉积, 银行存款
～ debits 银行取款
～ draft 银行汇票
～ erosion 堤岸冲刷
～ face 阶段工作面
～ field 沙洲
～ fittings 银行台柜装置
～ full stage 满流水位, 平岸水位
～ gravel 河岸砾石, 河卵石
～ head 坑口, 井口
～ indicator 倾斜指示器

～ investment 银行投资
～ light 排灯, 聚光灯
～ line 岸线, 路面侧坡
～ -line profile 路基边缘线纵断面
～ -line survey 岸线测量
～ measure 填方数量
～ money 银行票据
～ note 钞票, 纸币
～ of china 中国银行
～ of deposit 开户银行
～ of a cut 剖线边沿
～ paper 银行承兑票据
～ pier 桥台, 岸墩
～ protection 护岸(工程), 护坡(工程)
～ protection works 护岸工程
～ rate 倾斜率, 贴现率
～ revetment 护岸
～ run 河岸边的
～ -run aggregate 河岸[岸边]集料
～ run (sand) 原岸[河]砂
～ -rupt 破产
～ seat 碇桩
～ side 岸边
～ slope 岸坡, 倾斜角边坡
～ stabilization 岸坡稳定
～ strengthening 堤岸加固, 护坡
～ suction 岸吸作用(船在窄槽中航行, 水流两旁不匀而使船偏离航线)
～ turn 倾斜转弯
～ -winding (coil) 交叠多层绕组
～ wire 触排导线
banked *a.* 筑有堤的; 倾斜的; 成组的
～ battery 并联电池组
～ bend 筑成从内侧至外侧向上倾斜的弯道
～ boiler 热备用锅炉
～ bowl (高速弯道)超高碗形曲面
～ crown on curves 曲线超高, 弯道上的单坡路拱
～ curve 超高曲线, 横向倾斜曲线
～ turn 超高弯道, 倾斜转弯

~ -up water 回水,壅水
~ -up water level 回水水位

banker [ˈbæŋkə] *n.* 银行业者;造型台;挖土工人;石灰池;堤防土工;人工搅拌台

banket [ˈbæŋkit] *n.* 弃土堆,填土;护坡道,护脚,含金砾岩层

bankette [bæŋˈket] *n.* 弃土堆,填土;护坡道

bankfull *a.* (水位)齐岩的
~ discharge 齐岸流量
~ stage 齐岸水位,平槽水位

banking [ˈbæŋkiŋ] *n.* 筑堤;填土,堆积;超高,倾斜;金融;银行业
~ agreement 银行议定书
~ centre 金融中心
~ curve 超高曲线
~ hall 银行大厅
~ hours 银行营业时间
~ house 银行
~ pin 限位钉
~ power 银行投资能力
~ screw 限位螺钉

banknote *n.* 纸币,钞票
bankroll *n.*；*vt.* 资金;提供资助
bankrupt *n.* 破产者 *a.* 破产的 *vt.* 使破产
bank-winding (coil) 叠绕(线圈),简单绕组
banlieu(e) [bænˈljuː] *n.* 郊区,城郊住宅区
banner [ˈbenə] *n.* 标识[签],旗帜
~ bracket 旗帜托座
~ line 标识线

ban(n)ister [ˈbænistə] *n.* 栏杆(小柱);扶手;椅背上档的直条

Banpo Museum 西安市半坡博物馆

banquet [ˈbæŋkwit] *n.*,*v.*, -queted, -queting;.宴会;*v.* 宴请;banqueter, *n.*
~ hall 宴会厅
~ room 宴会室

banquette [bæŋˈket] *n.* 凸部,台坎;护坡道;弃土堆;填土;踏垛;桥上人行道;窗口凳,窗座
~ of levee 填堤
~ slope 踏垛坡

bantam [ˈbæntəm] *n.* 小型设备
~ car 轻便越野汽车
~ mixer 非倾倒式拌合机

banyan tree *n.* 榕树

Baoguo Temple 四川峨嵋报国寺

baptismal [bæpˈtizməl] *a.* 洗礼的
~ room 洗礼堂

baptist(e)ry [bæpˈtistəri] *n. pl.* -teries ;浸礼教的洗礼堂,浸礼池[所]

bar [bɑː] *n.*，*v.*，barred, barring, *prep.*；*n.* 条,棒,杆,钢筋,铁条;门闩,横木;窗框格;阻碍物,门楼;沙洲[坝];酒吧,餐柜;巴(压强单位 1 巴 = 10^6 达因/厘米^2) *vt.* 拦阻,妨碍; *prep.* 除外,除……之外;barless, *a.*；barrable *a.*
~ and sill method 上导洞法(隧道施工)
~ and soda-fountain sinks 酒吧间及冷饮间(用)洗盆
~ arrangement 配筋,钢筋布置
~ arrangement drawing 配筋图
~ beach 沙坝滩
~ bend table 弯钢筋台
~ bender 钢筋弯曲机,钢筋挠曲器
~ bending 弯钢筋
~ bending machine 钢筋弯曲机
~ chain 杆链
~ chair 钢筋座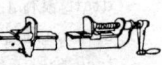
~ chart (统计用)条线表,柱状图

bar clamp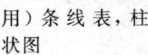

~ clamp 钢筋夹
~ claw 闩爪
~ code 条型码
~ conveyor 杆式输送机
~ cropper 钢筋剪断机

bantam car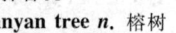

~ cutter 钢筋切断机
~ detector 钢筋探测器
~ door 酒吧间门
~ dowel 短钢筋,合缝接钢,插铁
~ drill 杆钻
~ feeder 杆式给料器
~ folder 弯折机
~ gauge 棒规
~ graph 条线图
~ grate 条杆筛,篦条
~ gravel 河滩砾石
~ grit （棒）条筛
~ grizzly 铁栅筛,篦子筛,棍铁花格
~ iron 铁条
~ joist 轻钢搁架
~ keel 方龙骨,矩形龙骨
~ linkage 连杆机械
~ list 钢筋表
~ loading 杆(构件)荷载
~ lock 插销门锁
~ magnet 条形磁铁
~ mat 钢筋网;钢条网
~ -mat reinforcement 钢筋网
~ mill 小型轧机
~ molding 钢筋(装饰)线条
~ moment 杆弯矩
~ placer 钢筋工
~ plain 沙坝平原
~ past 榫接柱
~ port 候潮港
~ rack 栅门齿条,粗格栅
~ reinforcement 钢筋
~ river mouth 拦门洲
~ room 酒吧间
~ schedule 钢筋表
~ screen 铁栅筛,格栅
~ separator 钢筋定位器(分隔器)
~ -setter 钢筋工
~ shear 钢筋剪切机
~ sheet 钢筋网
~ sieve 铁栅网
~ spacer 钢筋分隔器

~ spacing 钢筋分布(间距)
~ steel 条钢,钢条
~ stock 钢筋储备
~ stress 杆构件应力
~ support 钢筋座
~ theory 沙坝阻隔论
~ timbering 顶梁支护
~tin 锡条[块],白铁皮
~ tracery 铁棱窗(花格)
~ wood 条木

bar tracery

arch ~ 拱板,拱杆
balance ~ 平衡杆
base ~ 杆状基线尺
bearer ~ 托梁
bearing ~ 承重杆,支撑钢筋
bend ~ 挠钢,元宝钢
bent ~ 曲钢筋
bent-up ~ 弯上钢筋
bolt ~ 插销
boring ~ 钻杆
bright-drawn ~ 冷拉光条
building ~ 建筑钢筋
bumper ~ 缓冲梁,减震梁
cantilever ~ 悬臂顶梁
carrying ~ 支架顶梁
check ~ 试件,试样;试验杆
claw ~ 撬棍
compression ~ 压杆
continuous draw ~ 连续拉杆
corrugated steel ~ 竹节钢筋,波形钢
cross ~ 横杆,横木,顶梁
crow ~ 撬棍
crown ~ 顶杆;炉顶支柱
curb ~ 路缘杆
diagonal ~ 斜筋,斜拉杆
diamond ~ 菱形钢筋构件,多角形筛条
distribution ~ 分布钢筋
dolly ~ 支架,底座;把手
door ~ 门闩

dowel ~ 销钉;混凝土路面接缝条
draft ~ 拉杆,牵引杆
drag ~ 拉杆,牵引杆
fire (grate) ~ 炉条
fish-bellied ~ 鱼腹杆,等强度梁
glazing ~ 玻璃格条
grate ~ 炉条
grid ~ 格形炉栅
grizzly ~ 格筛条,栅条
guard ~ 护栏
hooked ~ 钩筋
inner ~ 隔板,挡板
lace ~ 格条,炉栅
level ~ 水平尺
lower arch ~ 下拱杆
marginal ~ 边部钢筋
measuring ~ 丈量杆
mega ~ 兆巴
merchant ~ 多种用途型钢
parallax ~ 视差尺
pedestal tie ~ 支座系杆
peripheral ~ 周边拉杆
plain ~ 光面钢筋;扁钢
plated ~ 熟铁条
plugging ~ 打夯机;捣砂锤
pry ~ 杠杆
shrinkage ~ 收缩钢筋
shutter ~ 百页闩
sill ~ 底杆
sine ~ 正弦规
sinker ~ 冲击式钻杆
steadying ~ 支柱,轴架
steel ~ 钢筋
stirrer ~ 搅拌棒
subtense ~ 横测尺
surface ~ 测平杆
tampering ~ 捣实锤,捣棒;夯棍
transom ~ 门窗横档
truss ~ 桁架杆
twisted ~ 螺旋钢筋
water ~ 止水条
barb [bɑːb] *n*. 倒沟,毛刺;*vt*. 装倒钩[刺]
~ bolt 带刺螺栓,基础螺栓
~ wire fence 有铁蒺藜的围栏
~ed nail 刺钉
~ed wire 刺铁丝
~ed wire reinforcement 刺铁钢筋
barbican [ˈbɑːbikən] *n*. 碉堡,桥头堡,外堡,望楼,更楼
~ entrance 望楼门

barbican

Barcelona [ˌbɑːsiˈləunə] *n*. 巴塞罗那(西班牙东北海港)
~ chair 镀铬钢材皮垫椅,巴塞罗那式椅子
barchan [bɑːˈkɑːn] *n*. 新月形沙丘
bare [bɛə] *v*., bared, baring, *a*. 裸露的,光秃的;空的,无装饰的;仅有的,少的 *vt*. 揭开,剥去,露出,解冻;bareness, *n*.
~ and line chart 划线图表
~ cable 裸线
~ copper wire 裸铜线
~ cut slope 新开挖的边坡
~ electrode 裸焊条,无药焊条
~ -faced tenon 裸面榫头
~ filler rod 裸焊条
~ -faced tongue 裸面雄榫
~ foot 无榫骨架
~ grizzly 支架筛,格子筛
~ ground wire 裸接地线
~ pavement maintenance (使路面外露的)除雪养护
~ pipe 不绝缘管,裸管
~ radiater 裸露散热器
~ weight 皮重,空重
~ wire 裸电线

bare-footed *a.* 赤脚的
bare-handed *a.* 赤手空拳(的),手无寸铁的,不戴手套(的)
bare-headed *a.* 光着头的,不戴帽子(的)
barely *ad.* (1)仅仅,几乎没有,好容易,才 (2)裸,无遮蔽地 (3)公开地,露骨地
bare-turbine *n.* 开式涡轮机
bargain ['bɑ:gin] *n.* 契约,合同,成交;廉价品 *vi.* 谈判,协议,订约;议价;成交;bargainer, *n.* 讨价还价
　~ centre 买卖市场(中心)
　~ cost 议价成本
　~ money 定金,保证金
　~ price 交易价格,廉价
bargainee [bɑ:gi'ni:] *n.* 买主
bargainor ['bɑ:ginə] *n.* 卖主
barge [bɑ:dʒ] *n.* ;*v.* , barged, barging;*n.* 驳船,平底货船;汽艇;夹墙烟道(凸出山墙部分);*v.* 用舶船运载;闯入
　~ board 挡风板,山墙封檐板
　~ couple 檐口人字木
　~ course 山墙檐瓦
　~ crane 浮式起重机,浮吊
　~ derrick 浮式人字起重机
　~ stone 山墙凸石,封檐石
　~ed-in fill 吹填土
barita ['beərait] *n.* 重晶石,氧化钡
baric ['bærik] *a.* 气压(计)的;大气压力的;(含)钡的
　~ flow 气压流
　~ gradient 压力梯度
　~ system 气压系统
barilla *n.* 苏打灰
baring ['beəriŋ] *n.* 揭开,掘开,剥离覆盖层
barite ['beərait] *n.* 重晶石
　~ aggregate 重晶石骨料
　~ cement 重晶石水泥,含钡水泥
　~ concrete 重晶石混凝土
barium ['beəriəm] *n.* 钡(Ba)
　~ carbonate 碳酸钡
　~ cement 含钡水泥
　~ crown glass 钠钙玻璃,铬酸钡玻璃,无铅玻璃
　~ flint glass 燧石玻璃,铅玻璃
　~ glass 钡玻璃
　~ kitchen (医院)钡餐室,调钡室
　~ (mixed) plaster 含钡灰泥(医院特殊抹灰用),防射线抹灰(用含钡砂浆)
　~ oxide 氧化钡
　~ (strontium) cement 钡(锶)水泥
　~ sulphate 硫酸钡
　~ titanate 钛酸钡(用于振动探测仪探头)
bark [bɑ:k] *n.* 树皮;三桅帆船;*v.* 剥树皮;吠,吼叫;barkless *a.*
　~ -bound 皮封的
　~ pocket 夹皮
　~ press 压皮机
　~ seam 夹皮
　~ing drum 筒式剥皮机
　~ing iron 树皮剥刀
　~ing machine 剥皮机
barker *n.* 剥皮器
barley *n.* 大麦
barling *n.* 脚手杆
barn [bɑ:n] *n.* 谷仓,堆房;车库;靶恩(核子有效截面单位 $=10^{-24}$ 厘米^2)
　~ door 仓库大门,挡光板
　~ siding 农仓(披叠)板墙
　~ truss 三铰构架式桁架
　~ waste water 畜圈废水
　~ yard 农仓空场,畜棚场
baroceptor [bærəu'septə] *n.* 气压传感器
barodynamics [bærəudai'næmiks] *n.* 重型建筑动力学
barometer [bə'rɔmitə] *n.* 气压计,晴雨表
　~ altitude 气压计高度
　~ reading 气压读数
barometric(al) [bærə'metrik(əl)] *a.* 气压表的,气压计的

barometrograph

~ damper 气流调节阀
~ determination of altitude 气压测高
~ discharge pipe 大气排泄管
~ efficiency 气压效率
~ gradient 气压梯度
~ height 气压高度
~ leg 气压柱
~ levelling 气压水准[高程]测量
~ low 低气压
~ maximum 高气压
~ pressure 大气压,气压
~ surveying 气压水准[高程]测量

barometrograph [bærə'metrəgra:f] *n.*
= barograph 自录气压计,气压自动记录仪

barometry [bə'rɔmitri] *n.* 气压测定法

baromil ['bærəmil] *n.* (气压)毫巴(测气压的单位)

baroque [bə'rəuk] *a.* 巴洛克,巴洛克式(装饰以曲线特多为特色)艺术或建筑的;形状不规则的,过分装饰的;巴洛克式的作风艺术等;baroquely, *ad.*

~ architect 巴洛克建筑师
~ architecture 巴洛克建筑风格
~ art 巴洛克艺术
~ building 巴洛克建筑
~ cathedral 巴洛克大教堂
~ Church 巴洛克教堂

Baroque Church

~ fountain 巴洛克式喷泉
~ fresco painting 巴洛克壁画
~ garden 巴洛克式庭园
~ master 巴洛克风格艺术大师

~ palace 巴洛克式宫殿
~ period 巴洛克建筑时期(17～18世纪中叶)
~ sanctuary 巴洛克式圣堂
~ sculpture 巴洛克雕塑
~ square 巴洛克式广场
~ statue 巴洛克塑像
~ style 巴洛克式(建筑)

barothermograph ['bærəu'θə:məgra:f] *n.* 自记气压温度计

barothermohygrograph [,bærə'θə:mə'haigrəgra:f] *n.* 自记气压温度湿度计

Barotor(machine) 高温高压卷染机

barotropic *a.* 正压的

barrack ['bærək] *n.* 临时工房,工棚
~ block 临时工房
~ building 营房建筑
~ room 工棚房

barrage ['bæra:ʒ] *n.* 拦河坝,堰坝
~ -fix 固定坝
~ -mobile 活动堰
~ -type spillway 堰式溢洪道

barranca [bə'ræŋkə] *n. pl.* -cas 深峡谷;深切沟,峭壁

barrel [bærel] *n.*;*v.*, -rel(l)ed,-rel(l)ing; *n.* 筒,筒体;拱底面,泵室;桶(容积单位); *vt.* 装桶
~ arch 筒(形)拱
~ boiler 筒形锅炉
~ bolt (band) 圆形插销,管销
~ bulk 一桶(松散物料的体积度量,=松方 0.14 立方米)
~ convertar 卧式吹炉
~ culvert 筒形涵洞
~ drain 筒形排水渠
~ elevator 桶升降机
~ fitting 管筒式接头
~ frame 筒形结构
~ handrail 筒形扶手
~ hook 筒钩
~ hoop 桶箍
~ hoist 桶升降机

~ mast 筒柱
~ mixer 鼓筒式搅拌机
~ newel 旋转楼梯筒柱
~ nipple 筒形螺纹接套
~ pier 筒形桥墩
~ purlin 筒檩条
~ raft 浮筒筏
~ railing 筒形栏杆
~ roll 筒形辊,横滚
~ roof 筒形屋顶
~ roof ribs 筒形屋顶肋
~ saw 筒(形)锯
~ shell 筒形薄壳
~ shell roof 筒形薄壳屋顶

barrel shell roof

~ skeleton 筒形骨架
~ vault 筒形穹窿
~ vault shell 筒(拱)壳(体)
~ wall lamp 筒形壁灯
arch ~ 拱筒
boiler ~ 锅炉圆筒
pump ~ 泵筒[壳]
sinking ~ 凿井吊桶
winch ~ 绞车卷筒

barrel vault

barren ['bærən] *n. pl.* 不毛之地,瘠地; *a.* 荒芜的,贫瘠的;无益的;多孔的(岩石)
~ rock 废(脉)石,多孔岩
~ sand 荒砂

barriada *n.* 城市贫民区

barricade [ˌbæriˈkeid] *n.; v.* ,-caded, -cading; *n.* = barricado [ˌbæriˈkeidəu] *n. pl.* -does, *v.* ,-doed,-doing; *n.* 栅栏,路障;栅栏,隔板,防御墙; *v.* 阻塞,设路障

barrier ['bæriə] *n.* 栅栏,障碍;阻挡层;界线;关卡;潜堰;挡水堤; *pl.* 中世纪骑士比武场的围栅;(赛马的)出发栅; *vt.* 阻碍,用栅围住
~ bar 滨外沙坝
~ basin 堰塞盆地
~ centerline stripe 栏式中央分隔带
~ curb 栏式缘石
~ flat 堰洲坪
~ island 堰洲岛
~ layer 阻挡层,势垒层
~ line 拦阻线,制止线
~ material 防潮材料,隔板
~ post 栅栏柱
~ sheet 挡水板,护墙板
~ shield 屏蔽墙
~ shielding 阻障隐蔽,阻碍物防护
~ spit 沙嘴,堰洲嘴
~ spring 堤泉,堰塞泉
~ stripe 栏式分隔带
artificial ~ 人工障碍物
impermeable ~ 不透水层
sonic ~ 音障,声垒,隔声板
sound ~ 音障,声垒
thermal ~ 热障,热绝缘
vapour ~ 阻凝(蒸汽)层

barrow ['bærəu] *n.* 【英】大山,丘陵;古墓;手推车; *v.* 用手车运输
~ excavation 露天挖掘
~ gang 手(推)车队
~ men 手车工人
~ pit 采石[采料,取土]坑
~ run 手(推)车马道(建筑施工用)
~ runner 手(推)车用跳板
~ truck 手推车
drum ~ 放线车
tip ~ 倾卸手推车

barsowite ['bɑːsəait] *n.* 钙长石

bartizan ['bɑːtizən] *n.* 瞭望台,城墙外之吊搂;bartizaned, *a.*

barway ['bɑːwei] *n.* 有拦路横木的场内小路

barycenter
['bæri'sentə] n.
重心,质(量中)心

baryum ['bɛəriəm] n.
barium 钡

barye ['bɑːri] n. 微巴
(压强单位＝1 达因/厘
米²);巴列(气压单位＝
1 达因/厘米²)

bartizan

baryta [bə'raitə] n. 氧化钡,钡氧
～ crown 钡冕(玻璃)
～ feldspar 钡长石
～ (light) flint 含钡(轻)火石玻璃
～ phosphate 磷钡盐
～ yellow 钡黄,铬酸钡

baryte(s) [bə'rait(iːz)] n. 重晶石
～ concrete 重晶石混凝土
～ mortar 重晶石砂浆
～ mortar finish 重晶石砂浆抹面[饰面]
～ powder 重晶石粉

basad ['beisæd] ad. 朝底地,朝基础地

basal [beisl] a. 基础的,基层;n. 基板
～ conglomerate 底砾岩
～ crack 底面裂缝

basalt ['bæsɔːlt] n. 玄武岩,柱石岩,玄武岩制品
～ -cement tile 玄武岩水泥砖
～ clay 玄武土
～ chip concrete 玄武岩碎屑混凝土
～ concrete slab 玄武岩(碎石)混凝土板
～ floor tile 玄武岩地板
～ lava 玄武岩
～ meal 玄武岩石粉
～ pea gravel 玄武岩细砾[绿豆砂]
～ -porphyry 玄武斑岩
～ powder 玄武岩石粉

base [beis] n., v., based, basing; n.
基础,柱础;地基,基底,底座[板];基点[线,数];基面[准,地]; v. 基于,以……为根据,设有基地; a. 基础的,地基的; basely, ad. beaseness, n.

～ asphalt 沥青底子
～ -bar 基线杆(尺)
～ bend 支座弯头
～ block 柱脚石,门基石

base block

～ burner 自给暖炉
～ coat 底涂层
～ court 里庭,后院
～ elbow 支座弯头
～ exchanging compound 碱交换剂
～ failure 基础破坏,地基塌陷
～ flashing 基础防水板

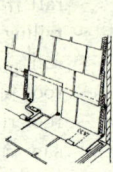
base flashing

～ heater 沿踢脚板铺设的供暖管道设备
～ heating 护壁板供暖
～ knob 门碰
～ lacquer 底漆
～ level 基面,基准面;基本水位,基准高程
～ map (城市规划的)基本地图,工作草图
～ material 基料[材,质]
～ moment 底部弯矩
～ moulding 底座线脚
～ plate 底板,支承板
～ plate heater 底板散热器
～ plate heating 底板供暖
～ plate radiator 底板散热器
～ plywood 底层胶合板
～ sheet 基本图
～ shoe 踢脚板压条
～ slag 碱性矿碴(熔渣)
～ wall 底层墙
～ widening 基层加宽
～ width 基准宽度
auxiliary ～ 辅助基线
axle ～ 轴间距离,轴距
column ～ 柱座
concrete ～ 混凝土基,混凝土基层
ebonite ～ 胶木座
foundation ～ 墙基

jack ~ 插孔板
mixed ~ 混合底子,杂料基层
paint ~ 底漆,打底
pan-shaped ~ 凸缘底板
rhombic ~ 斜方底面
shank ~ 支撑面
Varnish ~ 漆底
baseboard ['beisbɔːd]
n. 踢脚板,护板壁,基础板
~ heater 沿踢脚板铺设的散热器
~ heating 护壁板供暖
~ radiator 沿踢脚板铺设的散热器
~ register 踢脚板内调温装置
~ type unit 沿踢脚板铺设的管道设备

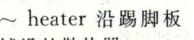
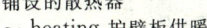
baseboard

basecoat ['beiskəut] *n.* 底涂层
~ (mixed) plaster 底涂灰泥
~ stuff 底涂料
basement ['beismənt] *n.* 地下室,地窖;基础
~ air shaft 地下室通风井
~ area 地下掩蔽部
~ boiler room 地下锅炉房
~ car park 地下停车场
~ complex 基底杂岩
~ concourse 地下中央大厅
~ door 地下室门
~ drainage 地下室排水,地窖排水
~ dwelling 地下住宅
~ entrance 地下室入口
~ floor 地下室层
~ garage 地下车库
~ level 底层,地下室层面
~ light well 地下采光井
~ main system 地下室干管系统
~ shelter 地下避难所
~ slab 地下室地板
~ soil 基土,土路基
~ stair(case) 地下室楼梯
~ storey 地下室层面,底层
~ wall 基础墙,地下室墙
~ window 地下室窗

basic ['beisid] *a.* 基本的,基础的;碱性的; *n.* 基本,要点,要素;基础(训练)
~ design 基本设计
~ dimension 基准尺寸
~ diorite 基性闪长岩
~ dye 碱性染料
~ life-related facilities 生活基础设施(住宅,交通,上下水道,电气,煤气,道路,公园绿化等)
~ refractory 碱性耐火材料
~ refractory brick 碱性耐火砖
basicity ['beisisiti] *n.* 碱度,碱性度,基性度,容碱量
basify ['beisifai] *vt.*, -fied,-fying;使碱化
basilica [bə'zilikə] *n.* (古罗马)长方形会堂或交易所;皇宫;长方形建筑物(尤指教堂)
~ church 长方形教堂
basis ['beisis] *n. pl.* -ses [-siːz]基础;底,本,价,准;根据;原理;主要成分;算法
ash-free ~ 除灰计算
basket ['bɑːskit] *n.* 铲斗,挖泥机;岩心管;篮(筐,笼,篓);花篮状柱头,篮形线圈; *vt.* 装入篮内
~ capital 花篮状柱头
~ chair 柳条椅
~ coil 篮(笼)形线圈
~ core 捞抓土样
~ crib 篮状排架
~ handle 篮状扶柄
~ handle arch 篮状拱
~ hitch 篮式套
~ purchase 整套购买
~ rod 柳条(编篮筐料)

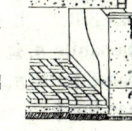
basement wall

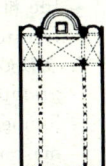
basilica

basket capital

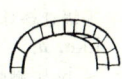
basket-handle arch

~ weave pattern 席纹图案
bas-relief ['bæsrili:f] *n.* 低[浅]浮雕
bassanite *n.* 烧石膏
basse-taille *n.* 浅浮雕
basstonite *n.* 棕黑蛭石
basswood ['bæswud] *n.* 菩提树,椴木材
bast [bæst] *n.* 碳质黏土[页岩],劣质煤
~ -bass 椴木韧皮,椴木绳索
~ -fiber 韧皮纤维
bastard ['bæstəd] *a.* 假的,劣质的,不纯粹的,非标准的,粗的;*n.* 假冒品,劣等货;坚硬巨砾,硬石
~ ALGOL 变形的 ALGOL(算法语言)
~ ashlar 粗饰琢石,毛石墙琢石
~ -cut file 粗齿锉
~ file 粗齿锉
~ freestone 劣质建筑毛石,劣质易碎石,劣质砂石
~ granite 假花岗岩
~ lumber 弦面材
~ masonry 混杂圬工
~ oak 美国栎
~ pointing 粗嵌缝
~ sawed board 粗锯板
~ sawing 顺纺锯木,粗锯木
~ size 等外品
~ stucco 粗粒水泥粉刷
~ trass mortar 粗面凝灰岩灰浆
~ tuck pointing 粗嵌灰缝
bastide *n.* 法国南部的农村小住宅,中世纪法国的防御城镇
bastile ['bæs'ti:l] *n.* 堡塔,监狱;法国巴黎的巴士底狱
baston *n.* (柱)座盘饰凸圆线脚
bat [bæt] *n.*,*v.*, batted, batting, *n.* 半砖,硬(土)块,泥质页岩,铆钉墩头,蝙蝠,[棒球]球棒;*v.* 用棒击球;-bat-like, *a.*
~ bolt 棘螺栓
~ handle 球棒形手把
~ handle swilch 手柄开关

~ -wing antenna 蝙蝠翼天线
batcher ['bætʃə] *n.* 配料计量器,分批箱;混凝土分批搅拌机
~ bin 量斗,投配器,分批箱
~ hopper 装料斗,装载漏斗,定量装料仓
~ plant 分批投配设备
~ scale 自动式(混料)计量器,进料量斗
batchwise ['bætʃwaiz] *a.*;*ad.* 分批的;断续地
~ operation 分批操作
batea *n.* 尖底淘金盘
bateau [bæ'təu] *n. pl.* bateaux 平底船,搭浮桥的船
~ bridge 浮桥
batement light 坡窗
bath [ba:θ] *n. pl*. baths [ba:ðz];浴池[盆,室,场];*v.* 洗澡,沐浴
~ brick 砂砖(用于磨刀或打磨金属面的)
~ chair 有篷轮椅
~ closet 沐浴更衣室
~ cock 浴用龙头
~ faucet 浴池水龙头
~ house 公共浴室,澡堂
~ mat 浴室小地毯
~ room 浴室
~ room fittings 浴室配件
~ rug 浴室地毯
~ stone 奶色石
~ trap 浴盆存水弯
~ tube 浴用管子
~ tub 浴盆
~ tub plug 浴缸排水塞子
~ water 洗涤用水
~ing box 更衣棚
bathe (1) *v.* 浸泡,(冲)洗;(2)*vi.*;*n.* 游泳,洗浴
batholite *n.* (bathylite)岩基,岩盘
batholith *n.* (bathylith)岩基,岩盘
bathometer [bə'θɔmitə] *n.* 测深仪[器]
bathymeter [bæ'θɔmitə] *n.* 深海测深

仪[器]
bathymetric(al) [bæθi'metrik(əl)] *a.* 测[等]深的,深海测深法的
~ survey 水深测量

bathymetry [bə'θimətri] *n.* 海深测量法,海洋测深学

batt [bæt] *n.* 黏土质页岩,沥青质页岩
~ insulation 沥青隔热层

batten ['bætn] *n.* 板条,挂瓦条,狭板;万能曲线尺;警戒孔,相同位穿孔; *vt.* 用板条钉住
~ and button 木板接合(法)
~ board 板条蕊胶合板
~ door 板条门
~ down 封舱
~ ends 短板条(铺地板用)
~ floor 木条地板
~ nails 板条钉
~ of roof truss 屋架水平撑条
~ plate 缀合板,联系板
~ plate column 缀合柱
~ seam 棒状折叠缝
~ seam roofing (金属板屋面的)棒状折叠缝式铺法
~ed grating 板条格栅
~ed panel 有压缝条的木墙板
~ed partition 板条间壁
~ed wall 板条墙
~ing 补隙板,用板条钉住
roof ~ 屋面挂瓦条
slating ~ 挂瓦条

batter ['bætə] *n.* (墙面向上渐薄引起的)垂向倾斜,坡度[面],斜坡,倾斜度;软泥; *v.* 倾斜;敲碎,打坏,磨损
~ board 龙门板,定斜板
~ brace [桥架]斜撑,斜杆
~ -curved wall 倾斜墙
~ drainage 斜坡排水,斜水沟
~ post 斜柱
~ed pilaster 斜面

batter boards

壁柱
~ed pile 斜桩
~ed wall 倾斜墙

battery ['bætəri] *n. pl* -teries;蓄电池,电池组;一套[组,排]
~ cut-out 自动开关
~ plan 同类房间成组布置的平面
~ type (混凝土板)成组立模制作方式

batting ['bætiŋ] *n.* 打击;毛絮,棉胎
~ block 石膏硬块
~ tool 阔凿,打击工具

battlement ['bætlmənt] *n.* (常用复数)城垛,雉堞

Bauhaus [bauhaus] *n.* 包豪斯建筑学院,德国建筑学派之一

bauk=baulk [bɔ:k] *n.* 横梁,大梁;障碍; *v.* 停止,阻止,妨碍

Baumann hardness meter *n.* 鲍曼式硬度计

Baume [bəu'mei] *a.* 玻美比重计的
~ degree 玻美度
~ hydrometer 玻度比重(浮)计,玻美表
~ scale 玻美比重标,玻氏比重计

Bauschinger's expansion tester 鲍辛格式膨胀(收缩)测定器

bawn *n.* 围墙

bay [bei] *n.* 开间,跨度,间隔,桥跨,排架间距;房屋的凸出部分,墙壁以凹进处(如为开窗用);海湾,河湾;挡水堤坝堤防;底板,支柱,机架; *v.* 阻止,以堤防阻水
~ leaf garland 桂冠花饰
~ stall 凸墙处坐位
~ window 凸窗,开窗壁槽

bayonet ['beiənit] *n.* ; *v.* , -neted,-neting; *n.* 接合销钉,卡口; *v.* 插入
~ base 卡口底座
~ cap 台灯插头盖
~ catch 卡口式连接

bay window

~ connection 卡口连接
baywindow *n.* 凸窗(房屋凸出部分所开的窗);开窗壁槽
baywood ['beiwud] *n.* 大叶桃花心木
baza(a)r [bə'zɑː] *n.* 集市,商场,百货[廉价]商店,商品陈列馆
beaching ['biːtʃiŋ] *n.* 海滩堆积;砌石护坡
~ of bank 护岸工程
beachy *a.* 有沙滩的,岸边浅滩的,近岸的
beaconage ['biːkənidʒ] *n.* 信号系统,负责保养信号系统的机构
bead [biːd] *n.* 珠,滴;焊珠,叠珠焊缝;墙角护条,压缝条,凸圆线脚;磁珠[环],垫圈,卷边;串珠饰; *v.* 成珠,起泡;用小珠装饰
~ and quirks 圆凹饰
~ and reel 珠链饰
~ butt and square 平圆方角接
~ chain 珠链
~ flush 串珠饰滚边
~ joint 圆凸勾缝
~ mould (柱头或柱脚凸起的)圆带线条
~ moulding 串珠线条,圆线条
~ pointing 凸圆勾缝
~ sleeker 修边饰器
corner ~ 墙角线条
draught ~ 止水条
glazing ~ 嵌玻璃条
guard ~ 护木条,封窗条
guide ~ (垂拉窗的)内导条
parting ~ (窗框)隔扇条
ploughed ~ 沟槽凸饰
quirk ~ 槽隔小圆
quoin ~ 隔角隔条
rebate ~ 转角圆线
return ~ 曲角线
saddle ~ 马鞍形窗间条
sash ~ 窗扇条
spreading ~ 宽焊道
slaff ~ 隔缘线

stop ~ 窗止条
string ~ 直的圆凸勾缝
sunk ~ 凹入线脚
beaded ['biːdid] *a.* 珠状的,粒状的,带珠的
~ ceiling 球饰平顶
~ esker 连珠蛇形丘
~ joint 圆凸缝
~ molding 墙上圆线条
~ on the edge and centre 边缘及中心珠状饰
~ paint 涂料,油漆
~ pipe 凸缘管
double ~ 双珠饰
beader ['biːdə] *n.* 弯边装置,卷边工具
beading ['biːdiŋ] *n.* 串珠状缘饰;形成珠状;卷边,压出凸缘,直接焊接
~ fillet 凸圆线脚,半圆饰边
~ plane 圆缘刨
~ weld 凸焊
pearl ~ 串珠饰
beak [biːk] *n.* 鸟嘴形,柱的尖头
~ mo(u)lding 鸟嘴饰,鸟嘴线脚
~ing joint 尖口接头,削榫
beakhead (ornament) 鸟嘴[头]饰(诺尔曼门道上的富丽装饰)
~ molding 鸟嘴饰,鸟嘴线脚
~ ornamental 鸟嘴头饰
beak(-)iron *n.* 鸟嘴(丁字,小角)砧
little~ 台砧
mo(u)lding ~ 鸟嘴饰,鸟嘴线脚

beak mo(u)lding

be-all *n.* 全部
Beallon *n.* 铍铜合金
Bealloy *n.* 铍铜合金的母合金,铜铍中间合金(铍 4%,硅 0.12% ~ 0.18%,铁 0.02% ~ 0.05%,铝 0.02% ~ 0.05%,其余铜)
beam [biːm] *n.* =B;BM 梁,桁条,杆;光束[柱];船宽,机身宽度
~ and filler floor 梁和填充式楼板

~ and girder construction 交梁结构
~ and girder floor 交梁楼板
~ and slab structure 梁板结构
~ arm 叉形梁臂,横梁叉
~ balance 杠杆式天平,杠杆式秤
~ bottom 梁底,梁拱腹
~ bottom elevation 梁底标高
~ butt joint 梁接头(口)
~ calliper (大)卡尺
~ channel 槽形梁,槽钢
~ column 梁柱(承受弯矩及压缩的构架件)
~ column construction 梁柱结构
~ compasses 长臂规,长径规
~ control 光亮度控制,射束控制
~ crossing 交叉梁
~ deflexion [光]束偏转
~ depth 梁高
~ distance 梁间间距
~ end face 梁端面
~ fill(ing) 梁间墙
~ flange 梁翼
~ floor 单层楼板,小梁楼板
~ forms 梁模板
~ grid 梁格栅,原木栅栏
~ grill(ag)e 格排梁
~ guard rail 梁式护栏
~ hanger 梁托,油柱挂
~ haunching 梁托臂[加腋]
~ head 梁端
~ hinge 梁铰
~ lamp 光(射)束灯
~ layout 梁的布置
~ lead 梁式引线
~ lock 梁箍
~ luminous flux 光束光通量
~ mounting 梁架
~ knee 梁肘材
~ of non-uniform section 变截面梁
~ of one bay 单跨(简支)梁
~ of one span 单跨(简支)梁
~ of three spans 三跨(连续)梁
~ of two spans 二跨(连续)梁
~ of uniform strength 等强度(变截面)梁
~ of uniform depth 等高梁
~ on alternate frame 间肋梁
~ on elastic foundation 弹性地基梁
~ oscillation 梁的振动
~ pass 梁形孔型
~ pitman 锚筋,拉杆
~ pitman bearing 锚筋支座
~ plan 梁布置图,梁平面图
~ pocket 梁口
~ pourer 梁的混凝土浇注工
~ profile 梁的断面
~ pump 摇臂泵
~ reinforcement 梁的配筋
~ rib 梁肋
~ scale 杆式磅秤
~ seat 梁座,梁垫
~ separator 梁间隔材
~ sheet coat (混凝土)梁护面
~ shuttering 梁模板
~ side 梁侧面,边模板
~ slab 梁板
~ slab bridge 梁板桥
~ soffit 梁底,梁拱腹
~ span 梁跨度
~ splice 梁的接头
~ splitter 光束分离器
~ support 梁支座,梁支承,梁支架
~ test 梁抗弯试验,断裂试验
~ texture 梁式结构
~ theory 梁理论
~ tie 锚梁,梁上系杆
~ timber 方木,梁木
~ -to-column connection 梁柱结合
~ vibration 梁的振动
~ web 梁腹,梁的腹板
~ width 梁宽,射束宽度
~ with both ends builts-in 固端梁(两端固定梁)
~ with central prop 三托梁

~ with compression steel 双筋梁
~ with constant cross-sections 等截面梁
~ with double reinforcement 双筋梁
~ with fixed ends 固端梁
~ with haunches 加腋梁
~ with one overhanging end 悬臂梁
~ with simply supported ends 简支梁
~ with single reinforcement 单筋梁
~ with uniform cross-sectinos 等截面梁
~ wood 方木,梁木
anchor ~ 锚梁
angle ~ 角钢檩条
arched ~ 拱形梁,拱副梁
articulated ~ 联接梁
balance ~ 天平梁;平衡梁
bell ~ 炉钟梁
binding ~ 联梁
bond ~ 结合梁
bow ~ 弓形梁
bowstring ~ 弓弦梁
box ~ 匣形梁,箱形(截面)梁
brake ~ 制动梁[杠杆]
brick ~ (加筋)砖梁
bridge ~ 桥式梁
bridging ~ 横[渡]梁
broad flange ~ 阔翼梁
buffer ~ 缓冲杆
built-in ~ 固端梁
built-up ~ 组合梁
bumper ~ 缓冲梁
camber ~ 曲线梁,上弯梁
cant ~ 转动梁
cantilever ~ 悬臂梁
capping ~ 压檐梁,帽梁
cased ~ 匣形梁
channel ~ 槽形梁

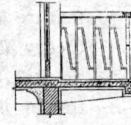
cantilever beam

cased beam

circular ~ 圆梁
Clarke ~ 组合式木梁
claw ~ 钳杆
collapsible ~ 装配可拆式梁
collar ~ 系梁
collimated ~ 准直束,准直柱
column ~ 柱形梁
combination ~ 组合梁
compensating ~ 平衡梁
composite ~ 混成梁,合成梁
compound ~ 合成梁,混成梁
conjugate ~ 共轭梁
continuous ~ 连续梁
continuous ~ on many supports 多孔[跨]连续梁
coped ~ 切口梁(以备联结其他构件)
crane ~ 吊车梁,起重机梁
cross ~ 横梁
dam ~ 迭[叠]梁
deck ~ 上承梁
deep ~ 深梁,高梁
deepened ~ 加深梁
differ flange ~ 宽缘工字梁
directed ~ 定向光束
divergent ~ 发散线束
double reinforcement ~ 双筋梁
double-strut trussed ~ 双柱桁架梁
dragen ~ 枪式小梁
drop ~ 垂梁
electron ~ 电子束[注,流]
ell- ~ L 形梁
end floor ~ 端横梁
equalizer ~ 平衡梁
equalizing ~ 平衡梁
erector ~ 千斤顶,起重臂
false ~ 假梁
fascia ~ 圈梁
fascia ~ connec-

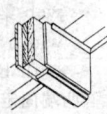
crane beam

fascia beam connection beam

tion ~ 圈梁结合
fender ~ 护舷梁
fish-bellied ~ 鱼腹梁
fished ~ 接合梁,夹接梁
fixed ~ 固定梁
flanged ~ 工字梁
flitched ~ 合板梁,桁木梁
flitch-trussed ~ 桁木桁架梁,虚构梁
floor ~ 地（楼）板梁,板式梁
footing ~ 底脚梁
foundation ~ 基础梁
free ~ 简支梁
freely supported ~ 简支梁
gantry ~ 刚架横梁
grillage ~ 格排梁
H ~ 工字梁,H形梁
half ~ 半横梁
hammer ~ 椽尾[小梁],锤柄
haunched ~ 托臂梁,加腋梁
head ~ 扶手
hollow-trussed ~ 虚构梁
I- ~ 工字钢（梁）
indented ~ 错口式组合梁
jesting ~ 装饰梁
joggle ~ 拼梁
joining ~ 系梁
junior ~ 轻型梁,次梁
keyed ~ 拼梁
laced ~ 缀合[花格,空腹]梁
laminated ~ 叠层梁
lattice ~ 格构梁
lenticular ~ 组合[鱼形]梁
lentiform ~ 组合[鱼形]梁
longitudinal ~ 纵梁
main ~ 主梁,主射束
marginal ~ 边缘梁
masonry ~ 圬工梁
multiple ~ 复光柱[束]
multispan ~ 多跨梁

floor beam

needle ~ 针梁,簪梁
non-uniform ~ 变截面连续梁
notched ~ 开槽梁
oscillating ~ 摇臂
over ~ 悬梁,过梁
overhanging ~ 伸臂梁,悬臂梁,悬挑梁
partition ~ 间壁梁
pier capping ~ 墩盖梁
pile capping ~ 桩承台梁
pilot ~ 缓冲梁
plated ~ 叠板梁
preflex ~ 预弯梁
propped ~ （临时）支顶梁
propped cantilever ~ 有支悬臂梁,有支伸臂梁
quarter ~ 四开梁,1/4梁
queen-trussed ~ 双柱桁架加强梁
rectangular ~ 矩形梁
rolled ~ 辊压梁
rolled steel ~ 辊压钢梁
roof ~ 屋顶梁
sandwich ~ 层结梁,多层组合梁
seat ~ （桥的）座梁
secondary ~ 副梁,次梁
shallow ~ 浅梁
simple ~ 简支梁
single ~ 单跨梁
single-strut trussed ~ 单柱桁架梁
single-working ~ 单臂平衡杆
sleeper ~ 枕梁
socle ~ 悬臂梁
sommer ~ 大梁,檩条
spandrel ~ 外墙托梁
split ~ 裂隙梁
spring ~ 单性梁,弹簧杆
springs and compensating ~ 弹簧平衡杆
straning ~ 跨腰梁
strut ~ 支墩梁
strut-framed ~ 撑托梁
tail ~ 半端梁

tapered ~ 楔形梁
tee- ~ T形梁
test ~ 梁状(混凝土)试件
through ~ 连续梁
tie ~ 系梁,连系梁
top ~ 顶梁
transverse ~ 横向梁
trussed ~ 桁架梁
twin ~ 并置梁
two-way ~ 双向梁
uniform ~ 等截面梁
vacuum lifting ~ 真空提升梁
vibrating ~ (混凝土摊铺机的)振动梁
walking ~ 摇梁,动梁,平衡梁
wall ~ 墙托梁
way ~ 桥面纵梁
whole ~ 圆木,滚子
wide-flangle ~ 宽缘梁
wind ~ 抗风系杆
wing ~ 翼杆
working ~ 工作[平衡]杆
beamhouse [,bi:m'haus] *n.* 准备车间,浸灰间
beaming [bi:miŋ] *n.* 辐[照]射,聚束;*a.* 放光的;beamingly, *ad.*
~ effect 束效应
bean [bi:n] *n.* 管接头,短管;油嘴,阻流器;豆,粒煤
~ ore 豆[褐]铁矿
bear [bɛə] *n.*;*v.*, bore or 【古】bave, borne or born, bearing; *v.*, 支承,负担,忍受;提供,传播;使跌价,卖空,空头;*n.* 打孔器,小型冲孔机;bearlike, *a.*
~ account 空头账户;投机买卖
~ expenses 负担费用
~ interest 负担行息(出息)
~ market 下跌行情;空头市场
~'s breech 叶形装饰(古希腊科林思柱头)
~ trap dam 开合式闸坝[熊阱堰]
bearable ['bɛərəbl] *a.* 支持得住的,可忍受的;
-bearably, *ad.*
~ load 可[支]承荷载
bearer ['bɛərə] *n.* 支承,支座,托架,垫块;运载工具,搬运工人,持有人;无记名
~ bar 托梁,支承梁
~ frame 支承框架
~ ofa gutter 檐槽托
~ plate 支承板,底板
~ supporting bracket 垫块,托座
blade ~ 刀口[棱形]支撑
foundation ~ 地基梁
intermediate ~ 中间托梁;桩间板壁
bearing ['bɛəriŋ] *n.* 支承[座],承载[受];轴承;(pl.)方位[向],象限角;关系,影响;行为,举止
~ bar 承重杆,支撑钢筋
~ beam 支撑梁,受力梁,垫梁
~ block 承重块,支承垫块,轴承座
~ blue 蓝铅油,普鲁士蓝
~ bracket 托座,支架;轴承座
~ brass 轴承巴氏合金,轴砂黄铜
~ brick 承重砖
~ cable construction 悬索结构
~ (capacity) failure 承压破坏
~ carrier 承载构件,支架
~ compass 定位罗盘
~ column 承重柱
~ floor (clay) block 承重楼板面[砖]
~ floor (clay) brick 承重楼板面[砖]
~ force 支承力,承压力
~ hanger 轴承吊架
~ housing 轴承箱
~ journal 支承轴颈,支承暗销
~ layer 持力层
~ line 方位线
~ load 地基压力,轴承荷载
~ lubricant 轴承润滑剂

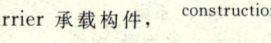
bearing cable construction

~ mark 方位标记
~ masonry (work) 承重[支承]圬工
~ material 轴承材料
~ measurement 方位测量
~ metal 轴承合金,轴瓦
~ module 轴承系数[模数]
~ nut 轴承螺帽
~ pad 乌金轴瓦,承压垫板
~ panel system 大板建筑,预制承重墙板式建筑
~ partition (wall) 承重隔墙
~ pedestal 轴承架
~ picket 标杆,标桩,花杆
~ pile 支承桩,承重桩
~ plane 承载面,支承面
~ plate 承重垫板,承重板
~ plate test 承压板试验
~ point 支承点
~ post 支承柱
~ power 承载能力,承载量
~ preloading 预应力承载
~ pressure 支承[轴承]压力
~ property 承载性能
~ puller 轴承拉出器
~ quality 承载能力,承载量
~ race 滚珠座圈
~ rate 方位变化率
~ ratio 承载[重]比
~ ratio test 承载比试验
~ reaction 支承反力
~ resistance 承载能力,抗压力,轴承阻力
~ rib 承重肋,凸缘
~ ring 支承环,[井筒]支承框架
~ rod 支柱,承重杆
~ saddle 支座锚固,轴承座
~ screw 支承螺栓
~ set 支承装置
~ shaft 支承轴
~ shell 轴承壳
~ shim 轴承垫片
~ shoe (桥梁)支梁靴
~ skeleton 承重框架
~ skeleton construction 承重框架结构
~ skeleton member 承重框架构件
~ skeleton structure 承重框架结构
~ sleeve 轴承套筒
~ socket 轴承承窝
~ spigot 轴承座孔[插口]
~ spring 托簧,承簧
~ stone 基石,底座石
~ stool 承重块支座
~ strain 承压变形,支承应变
~ stratum 承重层,承载地层
~ strength 支承[轴承]强度
~ stress 支承应力
~ structure 支承结构
~ support 支承
~ surface 支承[承压]面
~ surface area 支承[承压]面积
~ system 承载体系
~ test 承载[承压]试验
~ unit 方位测定装置
~ unit capacity 单位承载力
~ unit stress 承压单位应力
~ value 承载力[值]支承量
~ wall 承重墙
~ wall structure 承重墙结构
adjustable ~ 可调整的支承;可调整轴承
allowable ~ 容许支承压力
anchorage ~ 锚座,锚杆
azimuth ~ 方位
back ~ 反象限象,后轴承
center ~ 中心支[轴]承
check ~ 校核方位
coefficient of ~ capacity 承载力系数
course ~ [航]线方位
elevation ~ 仰角方位
end ~ 端支承,端承座
expansion ~ 活动支承,[温度]伸长支承

fixed ~ 固定支承
floor joist ~ 板格支承,地板搁栅梁
forward ~ 前象限角
free ~ 铰座,球形支座
free end ~ 松端支承,伸出支座轴承
freely movable ~ 活动支座
fulcrum ~ 支点承座
grid ~ 格网象限象
guide ~ 导引方位角,导引轴承
hinged ~ 铰承座
knife-edge ~ 刀形支承,刀口承
knuckle ~ 铰座,球形支座,关节轴盔
movable ~ 活动支承
multiple ~ 多次定位[方位]
pendestal ~ 架座
pillar ~ 墩支座
pin rocker ~ 铰接支座
pivot ~ 中心支承
pivotal ~ 铰链[摆动]支座;枢支承;回转轴承
pivoting ~ 中心支承,枢支座
quadrantal ~ 罗盘方位,象限方位角
rail ~ 轨枕;轨承;轨底支承面
relative ~ 相关方位
reverse ~ 后方位,后象限角
rocker bar ~ 铰链支座;平衡轴承
roller ~ (桁架)伸缩支座;滚轴支座,滚柱轴承
shifting ~ 活动支座
spherical ~ 球形支座
true ~ 真象限象
tumbler ~ 球支座;摆动轴承;滚球轴承
wall ~ 墙承重
whole-circle ~ 全圆周方位角

beast column 兽形柱
beat [biːt] *v.*, beat, beaten or beat, beating;敲,打,拍;打碎,锤薄,搏动,偏摆;难倒;风吹,雨打,日晒;打拍子;*n.* 敲打;拍音;差拍;频率;脉冲;跳动;还价;beatable, *a.*
~ cob work 捣土工作
~ frequency 拍频率
~ frequency oscillator 拍频振荡器
~ tone 拍音
~en aluminum 铝板
~en-cob construction 夯土建筑,干打垒
~en gold 金箔
~en path 走出来的小路
~ing crusher 打碎机,破碎机
~ -ing machine 打浆机

beater [biːtə] *n.* 夯具,夯实机,锤,捣棒,搅拌器,拍打器;打浆机,冲击式破碎机
~ drag 打浆计算器
~ pick 夯土镐,捣固道渣镐
~ pulverizer 锤式粉碎机
~ roll 打浆滚

beat-frequency *a.* 拍[差]频的
beating *n.* (1)打,拍,打浆,搅打[拌],打制[扁]
(2) 锻伸[长],锤击展薄
(3) 跳[搏,冲]动
(4) 击败
beat-up *a.* 年久失修的,残破的
beatway 用楔锤破地
beaufet *n.* 简便食堂,小吃店
Beaufort (wind) scale 蒲福氏风级
beaumontage 填孔料(填木孔用的树脂胶泥)
beautificatian [ˌbjuːtifiˈkeiʃən] *n.* 修饰,美化
beautify [ˈbjuːtifai] *v.*, -fied,-fying; 美化,装饰
beauty [ˈbjuːti] *n. pl.* -ties 美,美观,美好
~ art 美容术,化妆术
~ parlor 美容院
~ salon 美容室,化妆室
~ spot 风景,名胜,美景
beaver [ˈbiːvə] *n.* 海狸;狸皮;(中世纪的)头盔脸罩
~ board 轻质木纤维板,人造纤维板
~ type 斜锥式;颊甲式

beck [bek] n. 小河,溪流;大缸[桶]
becket ['bekit] n. 吊绳,环索,金属环
beckern n. =bick-iron 鸟嘴砧,丁字砧
becking n. [轮箍坯]辗轧,扩孔锻造
　～ hitch 环索结
　～ mill 扩孔机
Beckmann thermometer 贝克曼温度计
become [bi'kʌm] v., became, become, becoming;成为,变得;适合;结果是
　～ operative 生效
　～ (or fall) due 期满
bed [bed] n.;v., bedded, bedding; n. 床,底,基,层,垫;河床,路基,试验台;底盘;地基,基础;地层,垫层,填充物
　～ accretion 河床淤高
　～ bolt 底脚螺栓
　～ box 台座,机床脚
　～ bracket 床铺托
　～ building stage 造床水位
　～ -built water stage 造床水位(河道)
　～ center (医院)病床工作中心,寝具消毒整理中心
　～ chamber 卧室
　～ charge 底料
　～ closet 藏床壁橱,卧具室
　～ clothes 床单
　～ course 垫层
　～ cover 床帷
　～ dling concrete 垫层混凝土
　～ course 垫层
　～ current 底流,底层水流,地下(壤中)水流
　～ dowel 砌体中的暗销
　～ dune 河底沙丘,砂纹
　～ erosion 河床侵蚀
　～ factor equation 河床因数方程
　～ filling 被褥填絮
　～ form 河槽[河床]形状
　～ frame 基架,床架
　～ groin 填积堤
　～ hedgehopping 垫高,垫层厚度
　～ hook 底座凹槽,床槽

　～ house 简易旅馆
　～ jacket 女睡衣短外套
　～ joint 层间接缝(圬工),平缝
　～ lamp (light) 床头灯
　～ layer 底砂,推移质

bed joint

　～ load 底砂,(河床上的)推移质(粗砂和石砾)
　～ -load caliber 输砂量
　～ -load discharge 底砂[推移量]输送量
　～ load diversion 底砂分流
　～ load ejector 底砂排射器
　～ -load formula 床砂公式
　～ -load function 床砂函数
　～ -load movement 底砂移动,推移质运动
　～ load rate 河床载率
　～ -load sampler 底砂取样器
　～ load sampling 底砂取样
　～ -load transport 底砂搬运,推移质搬运
　～ making 寝具整理工作(服务工作)
　～ material 河床质
　～ material load 推移质
　～ material sampler 河床质采样器
　～ mortar 坐浆,垫层砂浆
　～ moulding 檐板下的线脚,粉刷线脚的底层

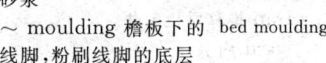
bed moulding

　～ mud 底泥
　～ of passage 过渡层
　～ oil paint 油性底层涂料
　～ pan 便盆
　～ piece 垫[底,座]板,(机)床身
　～ plate [底,床,座,台]板;地脚板;道岔垫板;炉底
　～ -plate foundation [底]板式基础
　～ plug 床边插头

~ production 底砂产生,推移质泥沙形成
~ puddle clay 夯实夯土层
~ recess 底座凹槽,床凹槽
~ ripple 沙坡
~ rock 底[基]岩,岩床
~ room 卧室,卧房
~ sampler 底砂取样器
~ sedimentation 底部沉积(层),河床沉积
~ separation 分层,夹层
~ shear 河床切应力
~ -side cabinet 床头柜
~ sitter 起卧室(卧室兼起居室)
~ sitting flat 卧室及起居两用的单间宿舍
~ sitting room 起卧室
~ -sitting-room dwelling unit 住宅居住单元
~ slope 河床坡度
~ spread 床罩
~ stability factor 河床(河道)稳定系数
~ stead 床架,构架
~ stone 垫石[底石,底梁]
~ stressing 应力工作台
~ system 以床为寝具的居住方式
~ table 诊察台
~ terrace 台地,阶地,地[岩]层,梯田
~ timber 垫[枕]木
~ town 卧城
~ trap 推移质拦截井
~ ward 病房
~ ways 床身导轨
~ width 底宽
ballast ~ 石碴床
box ~ 箱形底座[火架]
card ~ 卡片座[底座]
coarse ~ 粗填料基层;厚层
crib ~ 基础垛盘
die ~ 底模,模[垫]座
dirt ~ 岩[夹]石层

disappearing ~ 壁内折叠床
disrupted ~ 破坏层
double ~ 双人床
engine test ~ 发动机试验台
erecting ~ 安装[现]场,装配平台
filter ~ 滤床[层],过滤层
fixed ~ 固定床[层]
flat ~ 水平层[矿]层
fluidized ~ 流动层,流化床
folding ~ 折叠床
folding seat ~ 两用床
forked ~ 叉形座
foundation ~ 基床
ground ~ 地基
in-a-door ~ 门内折叠床
intercalated ~ 夹层
initial ~ 底料层
optical test ~ 光学试验台
percolation ~ 渗床,生物滤池
pervious ~ 透水层
prestressing ~ (钢筋)预加应力台
railway ~ 铁路路基
road ~ 路基,路床
rock ~ 岩基[床]
rocket test ~ 火箭试验台
sand filter ~ 沙滤层
single ~ 单人床
sofa ~ 沙发床
subjacent ~ 下卧层
superimposed ~ 覆盖层
test ~ 试验台
transfer ~ 机动台架,移送机
wall ~ 壁内折叠床
bedded ['bedid] *a.* 成层的,搁置的
~ bar 埋置钢筋
~ deposit 层状沉积
~ rock 层状岩
bedder ['bedə] *n.* 垫石;底石(油磨的)
bedding ['bediŋ] *n.* 基床[层,座,础];垫层,管基,衬垫;层理,成层;埋藏
~ composition 垫层成分,基层结构
~ concrete 垫层混凝土

～ course 垫层
～ course material 垫层材料
～ course mortar 坐浆,砂浆垫层
～ fault 层面断层,顺层断层
～ fisslity 层面断开性,顺层裂开性
～ gravel 垫层砾石
～ -in 研磨,研配,卧模
～ in soil 土路基,土坝(构造)
～ into 嵌入,砌入
～ joint 层面[顺层]节理
～ lamellae 层面壳层
～ material 垫层材料
～ mortar 砂浆(垫)层
～ of brick 砖铺基床[垫层],砖的大面
～ of land 土地平整
～ of pipes 管道垫层
～ plane 垫层面;层面;顺层面,层理面
～ plant 储料场
～ putty 衬垫层油灰
～ sand 地基砂层,垫层用砂
～ schistosity 层面片理,顺层片理
～ slip 层面滑动,顺层滑动
～ stone 垫石,扁平石
～ surface (岩石)层面
～ value 基床[地基]系数
concrete cradle ～ 混凝土承座基床
ordinary ～ 普通基床
stone ～ 石垫层

bedeck [biˈdek] *vt*. 装[修]饰,点缀
bedment [ˈbædmənt] *n*. (矫正)垫板(压缩试验用)
bedplate [ˈbedpleit] *n*. 底座,基础,底板;地脚板,道岔垫板,炉底
～ foundation 座板基础
bedrock [ˈbedˈrɔk] *n*. 底[基]岩,岩床;基础;*a*. 基本的
bedroom [ˈbedrum] *n*. 卧室;近郊居住区
～ basin 卧室洗涤盆
～ block (building, house, uint) 住宅区

～ closet 卧室壁橱,藏衣室,衣橱间,小间,套间
～ closet bank 卧室壁木厨台架,卧室小间仓库
～ decoration 卧室装饰
～ door 卧室门
～ floor 卧室楼层
～ lamps 卧室灯具
～ story 卧室楼层
～ window 卧室窗
bedspace *n*. (酒店,医院,宿舍等)床位总数
bedspread [bedspred] *n*. 床(罩)单
bedspring [ˈbedspriŋ] *n*. 弹簧床,床弹簧
bedstand [ˈbedstænd] *n*. 床前桌,试验台
bedstead [ˈbedsted] *n*. 床架
bedtock *n*. 【英】床架
bedye *vt*. 着色,施彩色,染[色];漆
bee [bi:] *n*. 蜜蜂;集[聚]会
～ house 养蜂场
～ line 直(距)线,捷径,最短距离,空中距离
～'s honeycomb brick 蜂窝砖
beech [bi:tʃ] *n*. 山毛榉
～ parquet 山毛榉镶木地板
～ strip floor cover(ing) 山毛榉板条楼面
～ strip floor(ing) 山毛榉板条楼面
beechwood [ˈbi:tʃwud] *n*. 山毛榉木材
～ mallet 山毛榉木锤
Beefwood *n*. 硬红木
beehive [ˈbi:haiv] *n*.;*a*. 蜂窝(状的);蜂巢,蜂房;集气架
～ cooler 蜂窝式冷却器
～ house 蜂窝式房屋(用石块叠砌的一种半球形原始构筑物)
～ kiln 蜂窝式窑
～ tomb 地下蜂窝式墓室
～ type radiator 蜂巢式散热器
beer [biə] *n*. 啤酒
～ cellar 啤酒地窖

~ garden 屋外花园酒店
~ hall 啤酒店,小酒店
~ house (英)啤酒店
~ shop 啤酒店
beerbachite *n.* 辉长细晶岩
Beer's law 比尔定律(光的透射与吸收定律)
beeswax ['bi:zwæks] *n.* 蜂[蜜]蜡; *v.* 涂蜜蜡,上蜡
beetle ['bi:tl] *n.*; *v.*, -tled, -tling; *n.* 木夯[槌],夯具,捣棒;搅打;脲醛树脂;甲虫; *v.* 夯实,用大槌捶打; *a.* 突[凸]出的
~ cement 脲醛树脂粘合剂
befall (befell, befallen) *v.* 发生(于),落到,降临,临到……头上
befanamite *n.* 钪石
befit (befitted; befitting) *vt.* 适合(适宜)于,为……所应做[有]的
befitting *a.* 适(应,恰)当的,适宜的
befog (befogged, befogging) *vt.* 把……笼罩在雾中,把……弄模糊
before (1) *prep*, 在……以前,在……前面,向,(宁肯)……而不,(优)先于
(2) *ad.* 以前,从前,在前面,在前头,向前
(3) *conj.* 在……以前,(宁愿)而不,与其……(宁愿)
(4) *n.* 先 before all 首先,before long 不久,before now 从前,before then 在那时以前
befoul *vt.* 弄脏[污],污蔑,诽谤
beg (begged, begging) 请[恳、乞]求
beget (begot, begotten) *vt.* 产生,引起,招致
begin [bi'gin] *v.*, began, begun, beginning;着手,开始,动手,创建
beidellite ['beidəlait] *n.* 贝得石
beige [beiʒ] *n.* 灰褐色,本色毛呢,哔叽; *a.* 灰褐色的
Beijing Capital Airport 北京首都国际机场
Beijing Exhibition Centre 北京展览馆
Beijing Planetarium 北京天文馆
Beijing Tian An Men (Gate of Heavenly Peace) 北京天安门

Beijing Capital Airport

Beijing Tian An Men

Beijing Worker's Gymnasium 北京工人体育馆
bel [bel] *n.* 贝(尔)(声音强度单位)
belabo(u)r [bi'leibə] *v.* 尽力;重打
belated [bi'leitid] *a.* 过期的,遗留的,延误的; belatedly, *ad.*, belatedness, *n.*
belay [bi'lei] *n.*; *v.*, -layed, -laying; *n.* 系绳处,S 形挽桩; *v.* 拴绳,牢栓绳子
~ing cleat 加强板
~ing pin 系索栓,套索桩
Belfast roof 贝式弓弦屋盖
Belfast (roof) truss 贝式弓弦桁架
belfry ['belfri] *n. pl.* -fries;钟楼[塔、亭],瞭望楼
Belgian ['beldʒən] *n.* 比利时人; *a.* 比利时的
~ architecture 比利时(式)建筑
~ black 比利时(士尔纳)大理石
~ method 先拱后墙法
~ method of timbering 比利时式隧道支撑法
~ roof fruss 比利时式屋架
~ sandwich cable method 比利时预应力钢筋混凝土法

~ truss 比利时式屋架(斜腹杆三角形桁架)
belite ['bi:lait] *n.* B盐,丑式盐,二钙硅酸盐(水泥);B岩,斜方硅灰石,β硅钙石
~ cement 二钙硅酸盐水泥,B-水泥
beljankite（creedite） *n.* 铅氟石膏
bell [bel] *n.* 钟,铃;钟口,承口,喇叭(漏斗)口;圆屋顶;钟形柱头;炉盖(高炉); *v.* 装上铃;使成漏斗形

bell

~ and flange piece 承口法兰短管
~ and flange reducer 承盘大小头
~ and plain end joint 平接
~ and socket joint 承插[接]口
~ and spigot 承插式接口
~ and spigot joint 承插式接头,钟口接头

bell and spigot joint

~ and spigot reducer 承插式大小头
~ arch 钟形拱
~ caisson 钟形沉箱
~ canopy 钟顶天棚,钟形帐篷
~ capital 钟state形柱帽
~ chamber 钟楼,钟室
~ frame 钟形构架
~ gable 吊钟尖塔,钟角楼,吊钟山墙
~ hammer 棱结花边杆
~ hanger 吊钟钩
~ hole 钟形穴坑
~ -hole bucket 钟穴形(挖掘)铲斗
~ housing 钟形壳[箱,罩]
~ jar 钟形缸,钟罩
~ joint 插承接合,套管接合
~ metal 铸钟铜,钟青铜
~ mouth 承口,钟形[喇叭,漏斗]口
~ mouth orifice 喇叭孔口
~ -mouthed opening 钟形[漏半,喇叭]口
~ -mouthed pipe 承接管,喇叭口管

~ pipe 承插式接头管
~ push 电铃按钮(门铃按钮)
~ roof 钟形屋顶
~ room 钟楼,钟室
~ rope hand pile driver 拉绳打桩机
~ shaped capital 钟形柱头
~ -shaped dome 钟形穹窿顶

bell roof

~ -stone ware pipe 承插式粗陶瓷接头管
~ stupa (mound) 钟形印度塔墓
~ tower 钟楼
~ (-)turret 钟楼,钟角楼
~ (-type) trap 钟形存水弯;虹吸闸门;钟形下水道
~ valve 钟开阀
alarm ~ 警铃
diving ~ 潜水钟,潜水室
muzzle ~ 承口,套口
belled [beld] *a.* 有承口的[钟形口,套接的]
~ mouth 喇叭口,承口
~ -out pit 扩大竖井
~ -out section 扩大截面
belling ['beliŋ] *n.* 制造管子的喇叭口
~ bucket 钟形铲斗,钟形钻孔锥
bellpull ['belpul] *n.* 铃扣,门铃拉索
Bellrock hollow plaster slab 空心石膏板
belly ['beli] *n. pl.* -lies; *v.*, -lied, -lying; *n.* 隆腹形;钟腰,炉腰;(孔型边)凸部,凹起 *v.* 鼓起
belong *vi.* (1)属于;(2)应归入(处在,位于)
belongings *n.* (*pl.*) 所有物,行李,附属物
below [bi'ləu] *prep.* 在……下面,在……以下,低于,少于; *ad.* 在下面,向下; *a.* 下列的,零下的
~ curb 路缘石标高以下
~ grade masonry (work) 地面标高

以下圬工[砖石砌体]
~ ground masonry (work) 地面以下圬工[砖石砌体]
~ proof 废品,不合格
~ stair 楼下层
~ -the-line expenditure 投资预算支出
belt [belt] *n.* 皮[引,地]带,带状物;区域,层;环行[公路,铁路,线路]; *v.* 绕上带子,用带结上
Beltrami's theory of failure 贝尔塔密强度理论
beltstower *n.* 抛掷式胶带充填机
belvedere ['belvidiə] *n.* 了望台,观景楼,眺望楼
bema ['bi:mə] *n. pl.* bemata;讲坛,教堂中之高座
ben [ben] *n.* (住宅的)内室,内厅;小山,高地; *prep.* 在……内
bench [bentʃ] *n.* 实验桌,工作台;台座,凳子;护道,底层平台;河岸,河岸阶地;组; *v.* 安置凳子;把……挖成台阶形

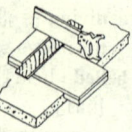
bench hook

~ axe 木工用斧
~ hook (木工)挡木头
~ -table 墙基,台基,柱础
~ wall 承拱墙
~ed excavation 台阶式挖掘
~ed footing 阶梯形底阶,大方脚
bench-table
~ed foundation 阶梯形基础
~ed joint 阶状接缝
~ing 台阶式挖土,管道定边坡
bottom ~ 底层;下部台阶;下梯段
carpenters ~ 木工台
benchmark ['bentʃmɑ:k] *n.* 水准点,基准点;劳动生产率基准
~ data 基本数据
~ statistics 标志性统计数字

bench-scale *a.* 小型的,实验室规模的; *n.* 台称
~ dissolver 小型(电解)溶解器
~ experiment 实验室试验
bend [bend] *n.* ; *v.*, bent, bending; *n.* 弯曲[头,管];转弯,倾向;河曲[湾];可曲波导管; *v.* 弯曲,挠曲,使屈服
~ allowance 钢筋弯曲公差[容许误差]
~ bar 元宝筋,弯起钢筋
~ curvature 曲线[率];反翘,起拱
~ glass 曲面玻璃
~ loss 弯管水头损失
~ metal 易熔合金,弯管合金
~ meter 弯头水表
~ pipe 弯管
~ point 弯曲点
~ test 弯曲[挠曲]试验
~ test piece 受弯试件
~ type expansion joint 弯管式伸缩接头,胀缩弯头
~ up bar 弯起钢筋
~ union 弯管活接头
~ wheel 改向轮
~ wood 弯曲材
close pattern return ~ 闭合式180°弯头
cold ~ 冷弯曲,冷弯
connector ~ 弯头(管子)
double ~ 双弯,双弯头
eighth ~ 45°弯头
elbow ~ 直角弯头
expansion ~ 胀缩弯头
expansion pipe ~ 伸缩管弯
expansion U ~ U形膨胀接头,U形胀缩器
flanged ~ 法蓝弯头
flue ~ 烟道歪斜,烟道弯管
knee ~ 直角弯头
siphon ~ 虹吸弯管
suction ~ 吸水管弯头
tangent ~ 双弯曲
bendability *n.* 可弯[挠,锻]性

bender ['bendə] *n.* 弯曲机,弯管[板]机,折弯机
~ and cutter (钢筋)弯切两用机
bar ~ 弯钢筋机,钢筋弯具
hydraulic ~ 液力压弯机
hydraulic plate ~ 液力弯板机
plate ~ 弯板机
rail ~ 弯轨机
steel ~ 钢筋弯具
tube ~ 弯管机
universal ~ 万能弯管机

bending ['bendiŋ] *n.* 弯曲,挠曲;弯头,弯管;偏移,折射;波束曲折
~ action 弯曲作用力
~ allowance [钢筋]弯曲公差,容许误差
~ apparatus 弯钢筋机,弯曲器
~ brake 板料压弯机
~ capability (capacity) 或弯[挠,段]性
~ coefficient 弯曲[挠曲]系数
~ compression strength 弯曲抗压强度
~ compression zone 弯曲应力区
~ connection 弯曲连接
~ crack 弯曲裂纹
~ creep 弯曲蠕变
~ creep test 弯曲蠕变试验
~ curve 弯曲曲线,挠度曲线
~ cycles 交变弯曲周期[循环]
~ deflection 挠曲变位,挠[弯]度
~ deformation 弯曲[挠曲]形变
~ stress distribution 弯曲应力分布

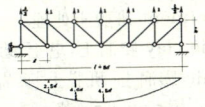

bending stress distribution

~ stress formula 弯曲应力公式
~ table 弯钢筋工作台
~ tension 弯曲拉力
~ test beam 弯曲试验梁
~ through 180° 作 180°弯折
~ tool 弯管工具
~ trestle 弯管工作台
~ twisting coupling 弯曲扭转耦合
~ unit stress 单位弯曲应力
~ up cables (预应力混凝土中的)钢缆弯曲
~ value 抗弯能力
~ vibration 弯曲振动
~ vibration strength 弯曲抗振强度
~ wave 弯曲波
~ width 弯曲宽度
~ yield point 抗弯屈服点
antielastic ~ 反弹性变形弯曲
plastic ~ 塑性弯曲
pure ~ 纯弯
simple ~ 纯挠

bendproof *n.* 抗弯刚度
beneath (1) *prep.* 在……之下,在……(正)下方,低(劣)于 (2)*ad.* 在下(面、方),在底下
Benedict metal 镍黄铜合金
beneficial [ˌbeni'fiʃəl] *a.* 有利的,有益的;有使用权的;beneficially. *ad.*; beneficialness, *n.*
~ cost 受益成本
~ interest 受益利息
~ owner 受益所有者
~ right 受益权
~ spread 利益扩展

beneficiary *n.* 受益(惠)人,收款人
~ of remittance 汇款收款人

beneficiate *vt.* (1)(为改善性能而进行)处理;(2)选矿,选集,精选,富集

benefit ['benifit] *n., v., -fited,-fiting n.* 利[收]益;利润;年金,救济金;保险赔偿费; *v.* 对……有利,有益
~ assessment (公用事业费)受益者分担
~ building society 建筑合作社
~ -cost analysis 效益成本分析,投资效果分析

~ -cost ratio 效益成本比
~ -cost ratio method 效益成本比法
~ factor 受益率
~ flow 利益流量
~ fund reserve 职工福利基金储备
~ principle 受益原则
~ program 福利计划
~ ratio 受益比
~ theory of taxation 赋税受益理论
~ -to-cost relationship 效益-成本关系
~ value 受益值
Benet metal 铝镁合金
Benign *a.* 有益于健康的,温和的,良好的
Benoto (人名)贝诺托
~ grab 贝诺托式抓斗[锤式取土器]
~ machine 贝诺托式(大口径)挖掘机
~ method (process) 贝诺托(大口径现浇混凝土桩)施工法
bent [bent] *n.*; *v.*; *pt*; *pp.* of bend; *n.* 排架,横向构架;V形凿,构柱脚;弯曲,弯头,曲轴,荒地,沼泽;倾向; *a.* 弯曲的;决心的
~ bar 弯曲[挠曲]钢筋
~ bar anchorage 弯筋锚固
~ beam 弯曲梁
~ cap (框架)盖梁
~ centering 排架结构鹰架
~ dent 弯形笱齿
~ down 向下弯(曲)的
~ element 弯[挠曲]钢筋
~ frame [横向]排架
~ glass 曲面[弯面]玻璃
~ lever 直角杠杆,曲杆
~ pier 架式桥墩
~ pipe 弯管,挠板
~ plate 曲板,弯板
~ pyramid 折线形金字塔[棱锥]
~ rod 弯曲钢筋
~ spanner 弯头扳手
~ spar 弯曲翼梁
~ steel 弯[曲钢]筋
~ tap 弯柄螺母丝锥
~ tile 曲瓦,槽瓦
~ timber 弯曲木材
~ tube 弯管,弯头
~ tubular lever 曲管水平仪
~ -up bar 弯起[上弯]钢筋
~ -up end 弯端
~ -up point 弯起点
~ -up portion 弯起部分
~ -up reinforcement 弯起钢筋(混凝土内)
~ wood 弯曲[挠曲]木材
double ~ 双跨排架
expansion ~ 伸缩弯管
square ~ 直角弯头
wind ~ 抗风排架
benthal ['benθəl] *n.* 海底,水底; *a.* 海底,水底的
~ demand 有机沉积需氧量
~ deposits 海底沉积,水底有机沉积
benthos ['benɔs] *n.* 海底生物,底栖生物
bentonite ['bentənait] *n.* 膨润土,膨土岩,班脱岩,皂土,浆土,蒙脱石
~ cement 胶质水泥
~ clay 膨润土,皂土,班脱土
~ concrete 胶质混凝土
~ flocculation 膨润土絮凝(作用)
~ grout 班脱土泥灌浆(填孔用)
~ mud 膨润土泥浆
~ slumy 班脱土泥浆,护壁泥浆
~ stabilizing fluid 膨润土稳定液
~ treatment 班脱土处理
bentonitic clay 膨润土,皂土,班脱土
~ shale 膨润土[班脱土]页岩
benzene ['benzi:n] *n.* 苯
~ hexachloride 六氯化苯
~ insoluble 不溶于苯的
benzidine ['benzidi:n] *n.* 联苯胺
benzine ['benzin] *n.* 汽油,挥发油
~ resisting hose 耐汽油软管

benzoic [ben'zəuik] n. 安息香的
~ acid 安息香酸,苯甲酸
benzoin ['benzəuin] n. 安息香,二苯乙醇酮
benzol(e) [benzəl] n. 粗苯,安息油
~ blends 苯混合物
~ scrubber 洗苯器
benzyl ['benzil] n. 苯(甲)基,苄基
bequest [bi'kwest] n. 遗产;遗赠
Beranck scale 白瑞纳克度标(噪声分类度标)
berengelite n. 脂光沥青
berg [bə:g] n. 冰山;大冰块
bergere [beə'ʒɛə] n. 围手椅
bergfall furrow 石沟
Bergen n. 卑而根(挪威港口)
bergmeal, bergmehl [bə:gmi:l, bə:gmeil] n. 硅藻土,矽藻土
bergschrund n. 大冰隙,冰川后大裂隙
berillia n. 氧化铍
berillite n. 水[白]硅铍石
berillium [bə'riljəm] n. 铍(Be)
Bering sea 白令海
berkelinm ['bə:kliəm] n. 锫(Bk)
berkeyite n. 天蓝石
Berlin [bə:'lin] n. 柏林
~ black 耐热漆,无光墨漆
~ blue 柏林蓝,深蓝色
berlin(e) n. (1)大四轮车(2)细毛线
~ work 毛(线)线刺绣品
berm(e) [bə:m] n. 护坡道,栈道,城墙外狭道;台坎,小平台,后滨阶地,滩肩,崖径
~ construction 护坡道构造
~ crest 滨后脊,滩肩背[外缘]
~ ditch 边坡截水沟,护道排水沟
~ edge 滩肩外缘
~ maintenance tool 护道养护机具
~ spillway 护道排水沟
~ stakes 护道边桩
Bermuda asphalt 百慕大石油沥青
Berner's window 伯尔尼尔窗,大钢窗

Bernolli n. (人名)伯努利
~ eguation 伯努利方程
Berry ['beri] n. pl. -ries (人名)柏利
~ asphalt 柏利地沥青
~ strain gage 柏利式应变仪
berth [bə:θ] n. pl. berths [bə:ðz] 船[泊]位,停泊位,舱位,(车、船)卧铺,床位;住所,职业;安全距离; v. 停泊,入港;住宿
~ structure 锚泊设施[建筑物]
~ed 住宿
berthage ['bə:θidʒ] n. 停泊地;停泊费
bery ['beril] n. 绿柱石[玉]
beryllia [be'riljə] n. 氧化铍
~ refractory (product) 氧化铍耐火材料制品
~ ceramics 氧化铍陶瓷
beryllium [be'riljəm] n. 铍(Be)
~ bronze 铍青铜
~ copper 铜铍合金,铍铜
~ oxide 氧化铍
~ window 铍窗
BESA = British Engineering Standard (s) Association 英国工程(技术)标准协会
beseech v. 恳(哀)求
beseem v. (1)似乎,觉得;(2)适当(宜),合适
besel n. 监视窗[孔],屏
beset(beset, besetting) vt. (1)镶、嵌(珠宝);(2)困扰,围攻;(3)包围(住)
beside prep. (1)在……的旁边;(2)比起……来;(3)除……之外;(4)同……无关
besides prep.;ad. 除[在]……之外,此外,而且,加之,还有,更,又
besiege vt. (1)包围,围困[攻];(2)拥挤在……的周围
besmear vt. (1)抹[涂]遍;(2)弄脏
besmirch vt. 弄脏,染污
bespangle [bi'spæŋgl] v.,-gled, -gling 金银箔饰,使……灿烂发光
Bessel [besəl] n. (人名)贝塞尔

~ equation 贝塞尔方程
~ method 贝塞尔(图上定位)法
~'s ellipsoid (spheroid) 贝塞尔椭圆体

best [best] *a.* [good 和 well 的最高级] 最好的,优质的;最大的,大半的; *n.* 最好的部分;最佳,最大限度,极尽全力; *ad.* 最好,第一
~ approximated 最近似的
~ bid 最高出价(中标价)
~ cokes 最优薄锡层,镀锡薄钢板
~ efforts 最好承担(包工合同)
~ -efforts selling 竭力推销
~ -fit method 最优满足法[至已合法]
~ fitting curve 最符曲线
~ materials 优质材料
~ quality 最好的质量
~ quality brick 优质砖,高级砖
~ results 最好的效果
~ speed 经济速率
~ value 最佳值

bet [bet] *n.*; *v.*, bet or betted, betting; *n.* 漫滩,低泛滥平原; *v.* 打赌
~ ting curve 博弈曲线

beta ['bi:tə] ['betə] *n.*; *a.* (希腊字母)β;第二位的,β位的;晶体管的共发射极电路电流放大系数

bethel ['beθəl] *n.* 圣殿,圣地;【英】礼拜堂

bethelizing ['beθəlaiziŋ] *n.* 木材注油(用杂酚油或煤油灌入木材)

Bethell's process 贝氏防腐法(木材用真空充实防腐油法)

Bethlehem beam 宽缘工字钢梁

beton ['betən] *n.* 【法】混凝土
~ arme'e 钢筋混凝土

betonac *n.* 金属混凝土(混凝土内含有细粒金属骨料)

better ['betə] *a.*, [good, well 的比较级] 较好的,更多的; *ad.* 更好些,更多些; *v.* 改良,改善;超过,胜过; *n.* 较好的条件[事物],较优者

~ input 优质投入
~ -riding 良好行车

betterment ['betəmənt] *n.* 改善[进,良],改建;修善,扩建
~ cost 改善费
~ expense 改良费用
~ survey 改善测量
~ work 改善工程

Betti's law 贝蒂定律,功的互易原理
Betti's reciprocal theorem 贝蒂互易定理

bettle *n.* 夯实,揭棒,木槌

between perpendiculars 垂直间距

bevel ['bevəl] *n.*; *v.*, -el(l)ed, *a.*; *n.* 斜面,斜角,削面;倾斜;斜削;斜角规,歪角曲尺,万能角尺;伞[锥]齿轮; *v.* 斜削,做成斜角;弄倾斜; *a.* 斜的,斜角的,倾斜的
~ angle 斜面角,倾斜角
~ arm piece 斜三通管
~ bearing plate 弧形支座
~ board 斜角(木)板
~ bonder (bondstone) 斜面砌墙石,斜面连接石
~ brick 斜面砖
~ cocking (corking, cogging) 斜面槽形榫
~ edge 斜边,斜缘
~ face 斜面
~ gauge 曲尺
~ gear 伞[锥]形齿轮
~ halving 斜面半搭接
~ header 斜面露头石,丁砖连结石
~ joint 斜面接头
~ moulding 斜面线脚
~ pinion 小锥齿轮
~ protractor 活动量角器,万能角尺

bevel protractor

~ siding 互搭板壁,披叠木板墙
~ square 斜角规,分度规

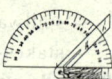
bevel square

~ surface 斜面
~ tie 有斜面的系梁,梯形系杆
~ trimming cutter 斜面修整[剪切]机
~ washer 斜垫圈
~ welding 斜角焊
~ wheel 锥齿轮,斜摩擦轮
~(l)ed brick 斜面砖
~(l)ed closer 斜面砖层收尾砖
~(l)ed cogging 斜鸳鸯榫头
~(l)ed corner 斜角
~(l)ed (corner) halving 斜面接头
~(l)ed door 斜边门
~(l)ed edge 斜缘
~(l)ed gear jack 斜齿轮起重器
~(l)ed glass 斜边玻璃
~(l)ed halving 斜对接
~(l)ed housing 斜槽插接,斜槽眼
~(l)ed hill top 平切丘顶
~(l)ed joint 斜面接头
~(l)ed lip entrance 斜唇形入口
~(l)ed nailing strip 梯形截面受钉木条
~(l)ed rectangular sleeper (tie) 斜角轨枕
~(l)ed siding 互搭板壁
~(l)ed tie 不等厚枕木
~(l)ed washer 斜垫圈,楔形螺栓垫圈
~(l)ing machine 倒角机
~(l)ing plane 榫槽刨
~(l)ing radius 弯曲半径
combination ~ 联合斜面,万能斜面
universal ~ 通用斜角规
bevel(l)er *n.* 磨斜边机,倒角机;磨斜边工人;磨石板机;磨石板工
bevelment *n.* 斜对切,削平
beveloid *n.* 斜面体
beverage ['bevəridʒ] *n.* 饮料
~ bottle 酒[饮料]瓶
~ store 饮料店
bewel *n.* 曲曲,预留曲度
bezant ['bezənt] *n.* (besant, byzant) 圆币[盘]饰,列圆饰,古代金币
bezel ['bezl] *n.* [仪器]玻璃框;企口,斜刃面;凿的刃面
bias ['baiəs] *n.*; *a.*; *ad.*; *v.*, -as(s)ed,-as(s)ing; *n.* 偏,偏离[移,差,向,置,倚];偏压[流,磁];斜线;(英)既定程序,既定的道路; *v.* 使……有偏差,使偏重,加偏压 *a.* 倾斜的;对角的; *ad.* 倾斜地;对角地

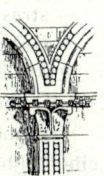
bezant

~ battery 偏压电池组
~ bell 偏动电铃
~ (margina) check 边缘检验,偏压校验
~ current 偏流
~ light 背景[衬托光]
~ lighting 基本[衬底]照明
~ meter 偏流表,偏畸变计
~ -off 偏置截止
~ torque 偏转力矩
~(s)ed error 偏差,偏置误差
~(s)ed estimator 偏估计量
~(s)ed tolerance 偏斜容限(试验行车偏离指定横向地位的容许限度)
~(s)ing 偏置,偏压,位移;磁化
light ~ 光偏移;光线背景;(摄影)轻微漏光
biaxial [bai'æksiəl] *a.* 双[二]轴的,双[二]向的;biaxiality, *n.*
~ bending 双向弯曲
~ compression 双向压缩
~ crystal [双]轴晶体
~ effect (of poisson's ratio) (泊松比的)双轴向影响
~ loading 双轴向负荷
~ strain 二轴应变
~ stress 双轴[二向]应力
~ stress state 双轴向应力状态
~ stress strain relation 双轴应力一应变关系

~ stress system 双轴向应力系统
~ stretch-forming machine 双轴向拉伸成形机
~ tension 双向张力[拉伸]
~ tiltmeter 双轴倾斜仪
bib(b) [bib] *n.* 水龙头,活门;弯管旋塞,活塞;闸板
~ nozzle (美)小水龙头弯嘴,旋塞嘴
~ tap 小水龙头
~ valve 水龙头,弯嘴阀
hose ~ 软管小龙头
sink ~ 小水龙头,洗涤盆龙头
bib(b)cock ['bibkɔk] *n.* (小水)龙头,弯管旋塞,活塞,阀栓
bibbles 软含水层
bibbley-rock *n.* 砾岩
bibelot ['bibləu] *n.* 【法】室内装饰品,床饰,小件古玩,小珍品
bibliolite ['bibiəlait] *n.* 层片岩,书页岩
bibliotheca [ˌbibliəu'θi:kə] *n.* 图书馆,书库
bibulous ['bibjuləs] *a.* 吸水的(高)吸收性的;-bibulously, *ad.*
~ paper 吸水纸,滤纸
bi-cable *n.* 双缆索道
~ aerial tramway 双线架空吊车索道
~ ropeway 双缆索道
~ system 复[双]线式缆道
bicarbonate [bai'ka:bənit] *n.* 碳酸氢盐,重碳酸盐,酸式碳酸盐
~ alkalinity 碳酸氢盐碱度
~ hardness 碳酸氢盐硬度
sodium ~ 碳酸氢钠,小苏打
bice [bais] *n.* 灰蓝色,灰蓝色颜料;绿色,绿色颜料
~ blue 灰绿色蓝,石青色
~ green 孔雀绿
bicircular [bai'səːkjulə] *a.* 二[重]圆的
~ quartic 重[虚]圆只四次线
~ surface 四次圆纹曲面
Bickford's safety fuse 俾氏安全导火线
bick-iron *n.* 丁字砧,铁砧

biconcave ['bai'kɔnkeiv] *a.* 双面凹的
~ lens 双凹镜
biconvex ['bai'kɔnveks] *a.* 两面凸的
~ lens 双凸透镜
bicycle ['baisikl] *n., v.*, -cled, -cling. *n.* 自行车; *v.* 骑自行车; :-bicycle, *n.*
~ and pedestrian path 自行车人行道
~ park 自行车停车场
~ racing track 自行车比赛跑道
~ room 自行车房
~ stand 自行车停车场
~ track 自行车路(径)
~ traffic 自行车交通
~ tube 自行车内胎
~ way 自行车道
bid [bid] *n.; v.* 投标,报价,出价;吩咐,命令;宣布,邀请;寻求,努力,尝试
~ and asked 买进出价和卖出喊价
~ bond 押标金,投标保证金
~ date 投标日期
~ documents 投标文件
~ form 投标形式
~ guarantee 投标担保(人)
~ item 投标项目,条款
~ opening 开标
~ or quotation 报价或估价
~ price 递价,投标价格
~ sheet 投标人名单,投标单
~ ding combination 承包联合,合股
~ ding conditions 投标要求
~ ding contract 投标契约
~ ding period (美)投标期
~ ding procedure 投标诉讼,投标裁决
~ ding sheet 标价单
bidder [bi:dei] *n.* 投标者,投标商
bidet [bi:dei] *n.* 净身盆,坐浴盆
bidonville *n.* 【法】安置区,城郊建筑草率的住宅区;(非非)市郊贫民区
Biedermeierstil 华德迈尔式(19世纪前半期德国奥地利流行的家具风格)
bielzite *n.* 脆块沥青

biennial [bai'eniəl] *a.* 二年一次的,每两年的,双年度的
~ budget cycle 两年预算周期
~ programming 制定两年期计划
biflow filter 双向滤池,双向过滤
bifolding door 双褶门
bifurcate [baifə:keit] *v.*,-cated,-cating；分叉,分为两支；*a.* 分叉的,分为两支的；bifurcately, *ad.*
~d chute 分叉斜槽
~d launder 分叉流槽
~d pipe 分叉管
~d rivet 分叉[开口]铆钉
bifurcation [ˌbaifə:'keiʃən] *n.* 分叉[支,歧,流]；分歧[叉,支]点
~ angle 支管角度,分叉角,支线角
~ gate 分水闸门
~ headgates of laterals 双支渠分水闸
~ ratio 分支比,分叉率
big [big] *a.*, bigger, biggest; *a.* 大的,重要的；宽大的,成功,严重的；跨大的；受人欢迎的；*n.* 大公司；*ad.* [口]非常大量；成功地
~ brick chimney (stack) 扇形砖砌(工业)烟囱
~ business 大企业
~ concrete site 大型混凝土施工现场
~ house 宅第,官邸；(美南、中部)住宅中的生活与社交活动区域
~ log 大原木
~ plaster board 大灰胶纸柏板
~ repair 大修
~ -scale work 大型工程
~ square 大方木
~ tree 巨杉(巨红杉)
~ yield 大[出]水量
biga ['baigɑ:] *n.* 由二马牵引的古代双轮战车(常用于雕塑)
bigalopolis *n.* 【美】大都市
biggin(g) ['bigin] *n.* 【英】任何一种建筑物；【英】无边童帽；一种银制咖啡壶
bight [bait] *n.* 弯曲,角,隅；海湾,新月形海湾；浦,凹；绳圈 *v.* 系牢

biharite ['baihərait] *n.* 黄叶蜡石
bijou ['bigu:] *n.* 珠宝,宝石,小巧精美品
bike [baik] *n.*；*v.*, biked, biking; *n.* [俗]自行车,脚踏车；*v.* 骑自行车
~ lane (way) 自行车专用道
bikeway *n.* 自行车专用道
bilateral [bai'lætərəl] *a.* 两边[面]的；双向[侧]的；对称的
~ lighting 两侧采光
~ negotiations and agreements 双边谈判和协定
bilection (molding) *n.* 【美】凸出嵌线,镜框饰
bilevel *n.* 错落式住宅

bilevel

bilge [bildʒ] *n.*；*v.*, bilged, bilging; *n.* 船腹,底舱,船底,桶鼓出部分；垂度,拱度；*v.* (船底)破漏；搁浅；膨胀
bilinite ['bailinait] *n.* 复铁矾,比林石
bill [bil] *n.* 账单,清单,支票；证书,诉状；广告,招贴；斧,鹤嘴锄,镰刀,锚爪；鸟嘴,岬嘴；*vt.* 记账；招贴通知
~ acceptance 票据承兑
~ account 汇票清算
~ and accounts payable with terms 有条件期付款项
~ of exchange(B/E;BE)汇票,交换券,国外汇票
~ of lading(B/L)提单
~ of materials(B/M)材料清单,材料表
~ of quantities 数量清单,初步估算,建筑工程量表
~ quantity 产品数量表数量
direct ~ 直达提单
straight ~ 记名提单
through ~ 联运提单

billboard [ˈbilbɔːd] *n.*【美】告示板，广告招贴板；锚座

billet [ˈbilit] *n.*；*v.*，-leted，-leting；(金属)坯段，钢坯，短条；锚齿饰；部长宿舍，职位，工作；字条；粗木棒；铁条；*vt.* 分配宿舍；任命；billetee，*n.*，billeter，*n.*

~ moulding 错齿饰

~ timber 劈开的木材

billet moulding

~ -wood 圆材

copper ~ 铜块

billiard [ˈbiljəd] *a.* 弹子戏的；*n.* (撞球中的)得分

~ room 弹子房

~ table 弹子台，台球桌

billing [ˈbiliŋ] *n.* 记账，编制账单；广告，宣传；营业量

Billnear method (混凝土)真空处理[养护]

bi(-)metal [baiˈmetl] *n.* 双金属，复合钢材

~ type dial thermometer 双金属式度盘温度计

bi(-)metallic [baimiˈtælik] *a.* 双金属的，复本位的

~ corrosion 双金属腐蚀

~ element 双金属元件

~ material 双金属材料

~ monetary system 复本位货币制度

~ strip 双金属片[带]

~ thermometer 双金属温度计

~ thermostat 双金属温度调节器，双金属恒温控制器

~ wire 双金属线

bimetallism [baiˈmetəlizm] *n.* 复本位制(金银二本位币制)

bimodal [baiˈməudəl] *a.* 两型的，两种方式的；双峰的

~ curve 双峰曲线

~ distribution 双峰分布

bin [bin] *n.*；*v.*，binned，binning；*n.* 仓库，料仓；料斗，斗；接收器，箱；集料台；活套坑，围栏；*vt.* 贮于箱中或仓中

~ and batcher plant 料仓和分批配料装置

~ and troll(e)y batcher plant 料仓和吊运配料装置

~ bottom 料斗底，仓底

~ card 存料卡，卡片箱

~ compartment 料仓分隔(间)

~ discharge 料仓[斗]卸货，出料

~ door fill valve 料仓装料活门

~ drainage system (材料)自动进出料仓系统

~ drawing channel 料仓倾卸槽

~ drawing device 料仓出料装置

~ feeder 料斗给料机，仓式进料机

~ feeding height 料斗送料高度

~ gate 料斗[料仓]闸门

~ hang-up(s) 料斗阻塞

~ level detector 料位指示器，料位计

~ outlet 料仓出口

~ segregation 料斗内材料分层

~ stopper 料仓闸门，料袋制动器

~ storage 贮料，装仓；仓库储藏，装箱储藏

~ tag 库房标签

~ type retaining wall 仓式挡土墙

~ unloader 料仓卸载装置

~ vibrator 料仓振动器，漏斗式振料器

~ wall 隔仓式挡土墙

~ weigh(ing) batcher scale 料斗定量秤，漏斗秤

ambulatory ~ 移动式料仓

ash ~ 灰坑，灰仓

batch ~ 货架

blending ~ 混合料斗，料仓

feed ~ 装[给]料料

hopper-bottomed ~ 漏斗式料斗

loading ~ 装料料，装载器

measuring ~ 量斗，计量斗

proportioning ~ 配料仓

pudding ~ 石灰消和池

slant(ing)-bottom ~ 斜底仓
stock ~ 料仓,配料仓
tipping ~ 翻斗
binary ['bainəri] *a.*; *n.* 二元的,双的,两重的;二进制[位]的
~ accumulator 二进位累加器
~ adder 二进制加法器
~ addition 二进制加法
~ alloy 二元合金
~ arithmetic 二进制运算
~ bit 二进位
~ coded decimal system 二进制编码的十进制
~ computer 二进位电子计算机
~ digit 二进制数字;二进制数位
~ flig-flog 二进制触发器,二进制双稳态触发器
~ logic gate 二进制逻辑门
~ pair[flip-flop] 二进制触发器
~ rock 二元岩
~ scale 二进制记数法
~ steel 二元合金钢
~ system 二合体系(沥青加矿粉);二进制
~ -vapo(u)r cycle 双汽循环
bind [baind] *v.*, bound, binding;粘[结]合;捆,扎;使凝固,硬结;装钉,镶边; *n.* 粘[结]合;约束,带,索;横撑,系杆,挡条;胶泥,硬黏土,沥青质页岩
binder ['baində] *n.* 粘[胶]合剂;结[粘]合料;沥青结合料;粘结层;胶结土;系梁[杆];丁砖;夹子;绷带,绳索;临时契约
~ -aggregate mix(ture) 粘合剂混合料
~ application 粘合剂洒布,涂敷
~ barrel 粘合剂桶
~ bulk storage and heating installation 沥青集料贮存和加热装置
~ clay 胶黏土
~ coat 粘结层
~ course 粘结层,粘结底层,结合层
~ course mix(ture) 粘结层混合料

~ dispersion 粘合剂分散剂
~ emulsion 粘合剂乳化液,粘合乳剂
~ -filler mix(ture) 粘结剂填充混合料
~ suspension 粘合剂悬浊液
asphalt ~ (地)沥青结合料
bituminous ~ 沥青结合料
calcareous hydraulic ~ 水凝灰浆
open ~ 有隙粘结层
binders *n.* 横向钢筋,箍筋;(环形)钢箍,箍环
binding [baindiŋ] *n.* 粘合,连接,装配,键接;捆,绑;带子;构架; *a.* 粘合的,有束缚力的,捆扎的;有效力的
~ beam 联梁,系梁
~ bolt 紧固螺栓
~ concrete course (layer) 混凝土粘结层,找平层
~ course 粘结层,联系层
~ face 支承面;贴合面
~ gravel 掺粘结料的砾石,含泥砾石
~ head screw 圆顶宽边接头螺钉
~ joist 小托梁,联结搁栅
~ of stones 砌体的结合
~ piece 水平板条,横撑
~ plate 角板,角撑[铁]
~ post 接线柱
~ rafter 承椽木,联结椽
~ rivet 结合[装配]用铆钉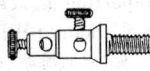

binding post

~ rod 联接杆,系杆
~ screw 夹紧螺钉;接线螺钉
~ strake 加强列板
~ strength 结合强度
~ wire 绑扎用钢丝;绑[扎]线
bing [biŋ] *n.* 材[废]料堆;堆[垛];贮藏箱
~ brick 硬质砂[页]岩,砖瓦
binocular [bi'nɔkjulə] *n.* (常用 *pl.*)双筒望远镜[显微镜];双目镜; *a.* 双眼[目,筒]的

~ hand level 双目手持式水准仪
~ head 双目镜头
binormal [bai'nɔːməl] *n.* 副[次]法线
bio-aeration *n.* 活性曝气(污水氧化处理)法
bioassay [ˌbaiəuæ'sei] *n.* 生物测[鉴]定
biochemistry ['baiəu'kemistri] *n.* 生物化学
bioengineering [ˌbaiəuˌendʒi'niəriŋ] *n.* 生物工程
bioherm ['baiəuhəːm] *n.* 生物岩礁
biolith ['baiəliθ] *n.* 生物岩
biologic(al) [baiə'lɔdʒik(əl)] *a.* 生物(学)的
biomolor *n.* 人工呼吸器
bionics [bai'ɔniks] *n.* 仿生学
bionomy [bai'ɔnəmi] *n.* 生态学;生命学
biont *n.* 生物个性
biosphere ['baiəsfiə] *n.* 生物界[圈],生存范围,生命层
biotite ['baiətait] *n.* 黑云母
~ -gneiss 黑云母片麻岩
~ -granite 黑云母花岗岩
~ -schist 黑云母片岩
biotope ['baiətəup] *n.* 生活小区
biparting *a.* 双扇(对开的)
~ door(BIPD) 双扇(对开)门;(电梯)双向滑动门
bipartite [bai'pɑːtait] *a.* 双向的;两部分构成的;一式两份的;两方之间的
bi-pass *n.* 双行车路,双通
bipod ['baipɔd] *n.* 双脚架,(安装用)人字架,双腿式起重机
Biram's wind meter 伯拉姆式风速计
birch [bəːtʃ] *n.* 桦树,赤杨;桦木 *a.* 桦木的;赤杨的
~ -bark tar 白桦木树皮焦油
~ veneer 桦木夹板
~ wood tar 白桦木焦油
bird [bəːd] *n.* 鸟,禽类;飞机,火箭;导弹

~ head bond 喙形接头
~ 's-beak moulding 鸟嘴式线饰
~ 's-beak ornament 鸟嘴式装饰

bird's-beak moulding

~ 's-eye 鸟瞰的,俯视的;概观的;似鸟眼的;(木材)鸟眼
~ 's-eye gravel 细砾石
~ 's eye perspective 鸟瞰图
~ 's-eye photograph 鸟瞰照片
~ 's-eye view 鸟瞰(图)
~ 's-eye wood 鸟眼纹材
~ 's-nest 了望台
birdbath *n.* 路面凹坑(雨后积水处)
birdsmouth *n.*, *v.* 鸟嘴形;凹角接,承接[角]口

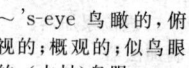
birdsmouth

~ joint 齿接合;插口接;V形缝;角口承接
~ quoin (of masonry wall) 纯角承接屋角石块
~ ed jointing 角口承接(三角形企口)
birefraction [bairi'frækʃən] *n.* 双折射
birefringence [ˌbairi'frindʒəns] *n.* 双折射,重折射
birefringent ['bairi'frindʒənt] *a.* 双折射的
~ plate 双折射板
Birmingham wire gauge 伯明翰线规
biscuit ['biskit] *n., pl.* -cuits or -cuit 陶器素坯,本色陶[瓷]器;淡褐色;片[块,盘]状模制品
~ metal 小块金属,金属块
~ tile 陶瓷砖[瓦]
biscuiting 二次焙烧(上过釉后焙烧)
bisect [bai'sekt] *v.* 二等分,对[平]分;相交,交叉
-bisection, *n.*
~ing compass 分规
~ing line 二等分线
~ing line of an angle 角的平分线,分角线

bisector [bai'sektə] *n.* 平分线，二等分线
  ~ of an angle 角的平分线
bishop ['biʃəp] *n.* 人工夯具，手锤；主教
  ~'s palace 主教宫
Bismarck brown 俾斯麦棕色
bismuth ['bizməθ] *n.* 铋〔Bi〕
  ~ aurife 铋金矿
  ~ blende (eulytite) 闪[硅]铋矿
  ~ glance 辉铋矿
  ~ ocher 【美】铋华，碳酸铋
bismuthinite [biz'mʌθinait] *n.* 辉铋矿
Bison floor 毕生楼板（一种专利的防火楼板）
bisque [bisk] *n.* 陶瓷素坯，本色陶器
  ~ tile 陶瓷砖
bistagite ['bistədʒait] *n.* 透辉岩
bisulfite *n.* 亚硫酸氢盐，酸式亚硫酸盐
  ~ of lime 石灰的亚硫酸氢盐
bisymmetric(al) [baisi'metrik(əl)] *a.* 双对称的
  ~ girder 双对称的大梁
bit [bit] *n.*；*v.*，bitted, bitting；*n.* 钻，锥，凿；钻[刀]，钎头，(截煤机的)截齿，刨刀，刀片；毕特（二进位数的信息单位）；二进制数[数位]；一点点，一小块，片刻，(*pl.*)上下钳口
  ~ blade 钻头刃
  ~ -brace 曲柄钻，摇钻
  ~ clearance 钻头与孔壁间隙，钻头出刃
  ~ cutting angle 钻冠的尖度，钻头磨角，钎子头刃角
  ~ density 二进码密度，存储密度
  ~ -gatherer 挑料工
  ~ grinding 堵钻处理
  ~ keyed lock 钥匙锁
  ~ of axe 斧刃
  ~ of information 信息单位
  ~ of the vice (老)虎钳口
  ~ shank 钎尾，钎柄
  ~ stock 摇柄钻，手摇钻，曲柄钻

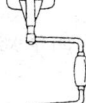
bit stock

  ~ time 一位[二进]时间
  auger ~ 木螺锥；麻花钻
  auger twist ~ 麻花柱钻孔钻，螺旋钻
  brace ~ 钻孔器，手摇钻
  jack ~ 凿岩机钎子，手持式风钻钎子
  plane ~ 刨刀
  plough ~ 凿子
  sash ~ 深孔木钻
  screw driver ~ 螺丝刀
bite [bait] *n.*；*v.*，bit [bit], bitten [bitən] or bit, biting；咬，啮；侵[腐]蚀；(锯、锉的)齿；切削刀，车[刨]刀，刀头[刃]；辊缝 *v.* 夹住[紧]，卡住[紧]；腐[侵]蚀；刺激；穿透；弄穿；bitable, biteable, *a.*
bitt [bit] *n.* 系缆柱[桩]
  ~ bracket 系缆柱支承[垫]板
  ~ pin 系缆柱杆
bittiness *n.* (粉刷)起块
bitty ['biti] *a.* 麻点的(油漆不均匀而产生的表面麻点)
  ~ cream 凝结乳脂
bitulithic ['bitju:'liθik] *n.* 沥青碎石路，沥青混凝土路；*a.* 沥青碎石的，沥青混凝土的
bitum ['bitjum] = bitumen
bitumastic [bitju:'mæstik] *a.* 沥青的；沥青砂胶的，沥青玛琋脂的；沥青厚浆涂料
  ~ enamel 沥青瓷漆
  ~ paint 沥青涂料，沥青漆
bitumen ['bitjumin] *n.* 沥青，天然地沥青
  ~ addition 沥青添加剂
  ~ adhesive 沥青胶粘剂
  ~ -aggregate mix-(ture) 沥青集料[骨料]混合料
  ~ and tar (melting) kettle 沥青和柏油(溶解)烧锅
  ~ base 沥青基，沥青底子
  ~ -based building mastic 含沥青建筑砂胶

bitumen

~ based paste 含沥青胶[膏]
~ -based rust protection (protective) paint(英)含沥青防锈漆
~ batcher 沥青配料计量器
~ batching pump 沥青计量泵
~ binder (course) 沥青粘结层
~ bonding adhesive (agent, medium) 沥青(加固)胶粘剂
~ bound base 沥青结合料基层
~ briquette 沥青模制[标准]试块
~ building paper 建筑沥青毡防水卷材
~ calcium silicate brick (英)沥青硅酸盐砖,沥青浸渍灰砂砖
~ carpet 沥青毡层
~ cement(ing agent) 沥青胶粘剂
~ coat(ing) 沥青涂层
~ -coated chip (ping)s 涂沥青片石
~ -coated chip (ping)s carpet 涂沥青片石毡层
~ -coated gravel 涂沥青砾石
~ -coated material 涂沥青材料
~ coating compound 沥青涂层混合料
~ coating (material) 沥青涂层材料
~ concrete base 沥青混凝土基层
~ concrete carpet 沥青混凝土覆盖层
~ concrete skeleton 【英】沥青混凝土骨架
~ deck surfacing 沥青车行道路面
~ -dipped 浸沥青的
~ distributor 沥青喷洒机
~ emulsion 沥青乳液,乳化沥青
~ emulsion injection 乳化沥青灌注(法)
~ felt roof 沥青油毡屋面
~ -filler mix(ture) 沥青填充混和[拌和]料
~ film 沥青薄膜
~ for dampproof coatings 防潮层沥青
~ globule 沥青球

~ grade 沥青等级[级配]
~ gravel roofing 沥青砾石铺屋面料
~ (-impregnated) calcium silicate brick 沥青(浸渍)灰砂砖
~ (-impregnated) felt 沥青(浸渍)油毛毡
~ (-impregnated) paper 沥青(浸渍)纸
~ (-impregnated) sand 沥青(浸渍)砂
~ (-impregnated) lime-sand (sand-lime) brick 沥青(浸渍)灰砂砖
~ (-impregnated) seal(ing) rope 沥青(浸渍)封口绳
~ (-impregnated) strip 沥青(浸渍)板条
~ impregnating 沥青浸渍的[饱和的,浸透]
~ impregnation 沥青浸渍[饱和]
~ index 纯沥青指数
~ injection control valve 沥青喷射控制阀
~ insulating (insulation) coat 沥青绝缘层
~ insulating (insulation) paste 沥青绝缘膏
~ insulating (insulation) slab 沥青绝缘板
~ joint 沥青接缝
~ (joint) pouring compound 沥青(接缝)浇注混合料
~ joint runner 沥青接缝麻刀[绳]
~ judaism 犹太沥青[死海沥青]
~ -laminated 沥青叠层
~ lasbestos composition (compound, mass, material) 沥青石棉合成物
~ lasbestos fibre cement 沥青石棉纤维水泥
~ -latex emulsion 沥青橡胶乳剂
~ layer 沥青层
~ liner 沥青衬垫
~ lining 沥青衬里[防水层]

~ macadam 沥青碎石
~ measuring (metering, proportioning) pump 沥青计量泵
~ melting boiler 沥青加热炉
~ membrane 沥青防水膜
~ mortar 沥青砂浆
~ of Judea 地沥青
~ pavement mix(ture) 沥青路面混合料
~ polymer binder (BPB) 沥青聚合物结合料
~ pouring rope 沥青麻刀,沥青绳
~ pre-impregnating (pre-saturating) 预先浸渍的
~ pre-impregnation (pre-saturation) 预先浸渍
~ prepared roofing 沥青组成屋面料
~ protective (protection) coat 沥青保护层
~ protective coating (materia) 沥青保护涂层(材料)
~ protective felt 沥青保护油毡
~ pump 沥青泵,碳氢化合物粘合剂泵
~ putty 沥青油灰[腻子]
~ rag-felt (ready roofing) (屋面用)沥青粗制油毡
~ (road) emulsion 沥青道路乳液,冷沥青乳液
~ rock 沥青岩
~ roof sheathing 沥青屋面涂层
~ roof covering 沥青油毡屋面覆盖层
~ roof sheathint 沥青屋面望板
~ roof(ing) 沥青铺屋面
~ roof(ing) cement 沥青屋面冷底子油膏
~ roof(ing) felt 沥青屋面油毡
~ roof(ing) sheet(ing) 沥青屋面护墙板,屋面挡板
~ roof(ing) shingle 沥青屋面板
~ rich carpet 沥青过多的毡层
~ -rubber composition (compound, mass, material) 沥青橡胶合成物
~ -rubber strip 沥青橡胶条
~ -saturated felt (edfabric) mat (pad) 沥青浸油毡衬垫[压缝条]
~ -saturated felt (ed fabric) mat (pad) with cork 沥青浸油毡软木衬垫[压缝条]
~ -saturated paper 沥青浸渍纸
~ sealing compound 沥青塞缝[封缝]混合料
~ seal(ing) 沥青封口膏
~ seal(ing) sheet(ing) 沥青封口[缝]板
~ sheet roofing 沥青毡屋面
~ sheet(ing) 沥青护墙板
~ shingle 沥青屋[墙]面板
~ slurry seal 沥青浆封缝[塞缝]
~ solution 沥青溶液
~ spray bar 沥青喷油管[喷杆]
~ sprayer 沥青洒布机
~ spraying machine 沥青洒布机
~ stabilization 沥青稳定[土壤]
~ strip 沥青条
~ subseal(ing) 沥青基层处理,沥青封底
~ tape 沥青绳
~ -tar blend 石油沥青—焦油沥青混合料(一般指含煤沥青较多的),混合沥青
~ tar mixture 石油沥青—焦油沥青混合物
~ -treated 浸透沥青的,沥青处治的
~ waterproofer (英)沥青防水层
~ wood fibreboard 沥青木纤维板
~ wool felt 沥青羊毛毡
asbestos ~ 石棉沥青
asphaltic ~ (地)沥青胶
blown asphaltic ~ 吹制纯(地)沥青
cut-back ~ 稀释[轻制]沥青
heavy ~ 重质[稠]沥青
viscous ~ 粘沥青,半固体沥青
bituminiferous [bitjuːmiˈnifərəs] *a.* 含油沥青的,油沥青质的

bituminization [biˌtjuːminaiˈzeiʃən] *n.*
沥青处治

bituminize [biˈtjuːminaiz] *vt.*,-nized,-nizing;使含沥青,使与沥青混合

bituminized [biˈtjuːminaizd] *a.* 含[加]沥青的
~ aggregate 涂沥青的骨料
~ cement 沥青膏,加沥青水泥[胶合剂]
~ chip(ping)s (英)涂沥青片石
~ concrete 加沥青混凝土
~ cord 涂沥青绳
~ -fiber pipe 沥青纤维管
~ floor covering material 涂沥青楼面覆盖材料
~ jute hessian cloth 预制沥青粗麻布面
~ mortar 加沥青砂浆
~ paper 含沥青油纸
~ stone 含沥青石

bituminous [biˈtjuːminəs] *a.* 沥青质的,含沥青的
~ adhesive composition (compound) 沥青胶结化合物
~ (agent) binder 沥青结合料
~ base (course) 沥青基[底]层
~ binder (course) 沥青胶结层
~ board 沥青板
~ body coat 沥青主层
~ -bound material 沥青结合料
~ broken-stone pavement 沥青碎石路面
~ brush(able) compound 沥青涂刷混合料
~ (building) mastic 沥青(建筑)玛琋脂
~ built-up roof(ing) 沥青组合屋面
~ carpet coat 沥青毡层
~ cementing agent 沥青粘结剂
~ -coated chips 黑色石屑(拌过沥青的石屑)
~ coating composition (compound mass, material) 沥青涂层混和料 [材料]
~ composition (compound mass, mixture) 沥青混和料
~ -concrete 沥青混凝土
~ concrete paver 沥青混凝土摊铺机
~ cooker (kettle) 沥青烧锅
~ dampproofing and water proofing 沥青防潮防水剂
~ decking 铺薄沥青表层,薄面铺盖
~ distributor 沥青喷布机
~ expansion joint 沥青伸缩缝
~ felt 沥青油毡
~ fiber filler 沥青纤维填(缝)料
~ finisher machine 沥青整面机
~ floor cover(ing) 〔floor(ing) (finish)〕沥青楼层覆面
~ grouting 沥青灌浆
~ insulating paint 沥青防锈绝缘漆
~ insulation
(-grade) board 沥青绝缘板
~ joint filler 沥青填缝料
~ lacquer 沥青溶媒漆
~ level(l)ing course 沥青整平层
~ limestone 沥青(质)石灰石,沥青灰岩
~ limestone pavement 沥青石灰石路面
~ macadam and tarmacadam mixing plant 沥青碎石和柏油碎石拌和装置
~ macadam pavement 沥青碎石路面
~ marl 沥青灰泥
~ mastic concrete 沥青砂胶混凝土
~ mat 沥青垫层,沥青毡层
~ material 沥青材料
~ membrane 沥青防水膜
~ mixing plant 沥青拌和装置
~ (plant) mix(ture) 厂拌沥青混合料
~ prime coat(ing) 沥青底涂层
~ protective coat(ing) 沥青保护层
~ protective membrane 沥青防水

薄膜
~ roof(ing) felt 沥青屋面油毡
~ roof(ing) sheet(ing) 沥青屋面板
~ sandstone pavement 沥青砂石路面
~ seal coat 沥青封层
~ sheet(ing) 沥青护墙板
~ slurry 沥青稀浆
~ -soil aggregate 沥青土集料
~ spray(ing) bar 沥青喷油管[喷杆]
~ spray(ing) machine 沥青喷洒机
~ storage installation 沥青贮存装置
~ storage tank 沥青贮罐
~ structural material 沥青筑路材料
~ surface course 沥青面层
~ surface covering 沥青罩面[罩面料]
~ underseal 沥青底封层
~ varnish 沥青清漆
~ waterproofer coat(ing) 沥青防水层
~ wearing coat 沥青磨耗层,沥青面层

bitusol *n.* 天然沥青,固体分散胶溶沥青
black [blæk] *n.* 黑色,黑颜料[染料],炭黑,黑漆;黑斑,污点;软质黑色页岩;黑人;*a.* 黑色的,吸收全部辐射能的;无镀层的;黑人的 *vt.* 弄黑;*vi.* 变黑
~ and white 黑白画,印刷品,黑白板,黑白的,未着色的
~ and white work 木石结构(木构架中填石和灰泥)
~ annealed wire 黑退火钢丝
~ bamboo 紫竹
~ bar(s) 小型轧材[型钢]
~ base (course) 沥青基层,黑色基层,黑底(砾青混凝土或碎石混凝土基层)
~ blind 无光(不透光)百叶窗
~ bolt 粗制螺栓
~ brick 青砖
~ coating 黑色[黑漆]涂层
~ cinder (水淬不完全的)黑渣硬渣
~ cotton soil 黑棉土(一种膨胀土,分布于印度、非洲等地)
~ culvert 黑铁管涵洞
~ diamonds 黑色金刚石;煤
~ -figure style 黑色图案
~ -figure vase 黑色柱饰(科林恩或混合柱式柱头)
~ friction tap (黑)绝缘胶(布)带
~ gang (土木工程承包的)设备维修小组
~ glass 黑色玻璃
~ granite 黑花岗石,黑岗石
~ hill spruce 云杉
~ iron 黑钢皮,黑铁板,黑铁矿
~ iron plate 黑铁板
~ japan 沥青漆,深黑漆
~ light crack detector 超紫外线或红外线探伤仪
~ line print 黑色图
~ locust 刺槐
~ mahogany 赤红木
~ market 黑市
~ market exchanger 黑市外汇
~ market price 黑市价
~ market operation 黑市交易
~ market sale 黑市销售
~ mortar 黑色水泥石灰砂浆
~ nut 粗制螺母
~ out cloth 遮光布
~ -out paint 遮光漆
~ -out plant 无窗厂房
~ oxide of cobalt (英)锰钴土
~ oxide of iron (pigment) 黑色氧化铁颜料
~ paint 黑色漆
~ paste 沥青膏
~ patches 黑斑点,(钢材上)未[欠]酸洗部分
~ peat 黑色泥炭
~ persimmon juice 黑柿汁,黑柿汁漆
~ persimmon wood 黑柿木
~ pigment 黑色颜料,炭黑

~ pine 黑松(澳大利亚松)
~ pipe 黑铁管,非镀锌管
~ plate 黑钢板[铁皮]
~ pitch 柏油脂
~ prophyry 暗玢岩
~ roofing adhesive 屋面油毡胶粘剂
~ roofing felt 屋面油毡
~ sand 黑砂
~ sheet 黑钢板[铁皮]
~ sheeting felt 沥青油毡(中)间层
~ slag 黑渣
~ speck 黑色斑点
~ tea 红茶
~ tin 锡石
~ -top finishing roller 沥青道路表面修整(压路机)路碾
~ -top spreader without hopper 无料箱沥青路面铺料机
~ varnish 沥青漆
~ wash 造型涂料
Berlin ~ 无光黑漆
blackbase n. 沥青基层,黑色基层
blackbody ['blækbɔdi] n. 黑体,全部吸收辐射能的物体
blackboard ['blækbɔːd] n. 黑板
blacken ['blækən] v., -ened,-ening; 变[涂]黑,黑化,变暗
blackening ['blækəniŋ] n. 涂[烧,变]黑;发黑处理,黑化,炭粉,黑度
blacking ['blækiŋ] n. 黑色涂料,造型涂料,粉磨石墨
~ brush 涂料用的毛刷
~ up 泛黑(沥青路面铺石屑后黑色表面外露)
blackish ['blækiʃ] a. 带黑色的,稍黑的;blackishly, ad.; blackishness, n.
blackjack [blækdʒæk] n. 粗黑焦油,闪锌矿
blackness ['blæknis] n. 毛面,黑色,黑度;凶险
blackout ['blækaut] n. 无光,变黑,遮蔽,封锁[闭],停止,中断
~ building 无窗建筑

~ door 不透光门,暗门
~ installation [system] 不透光设备,暗设备
~ jalousie (louvres, slatted blind) 不透光气窗,不透光百页窗
~ window 无光窗,暗窗
~ 's chisel 锻工凿
~ 's hammer 锻锤
~ 's hardy 锻工用方柄凿
~ welded joint 锻接接头
~ welding 锻接,锻焊
blackspot ['blækspɔt] n. 黑点[斑],斑痕
blacktop ['blæktɔp] a. 以柏油铺装的; vt. 铺以柏油
~ ping 建筑黑色面层
blackwash ['blækwɔʃ] n. 黑色[造型,炭素]涂料
~ glue 镖胶
blade [bleid] n. 刀片[刃];平铲,主椽,桁条;叶[浆]片,锯条,刮刀;刀形开关,闸刀盒;(轨机)卫板,(无心磨板)托板; vt. 装刀片[刮刀],叶片
~ control wheel 平路机刀片操纵轮
~ ed 叶片状,刃状;有叶片的,有刃口的
~ ed bit 多刃钻头
~ ed structure 刃状构造,叶片状组织
~ ing 叶片装置,叶栅
~ ing compaction 刮平压实
~ into a single window (用平路机)将材料堆成单条条堆
~ lift arm 平铲提升臂
~ lift control housing 平铲提升控制箱
~ lift control pinion 平铲提升控制齿轮
~ lift 平铲提升连杆
~ lift mechanism 平铲提升机械装置
~ lock 平铲闭锁装置
~ marker 叶片标号
~ mixer 叶片式混砂机
~ mixing 用刀片拌和

~ of shovel 铲刀
~ of T-square 丁字尺身
~ over the roadway 满堂摊料(用平地机在整个车行道摊铺材料)
~ paddle mixer 叶片式拌和机
~ paddle stirrer 叶片式搅拌机
~ pitch 叶片间距[节距]
~ point (toe) 刀锋[刃]
~ reinforcement 叶片加固
~ reverse control (平地机)刀片水平回转操纵
~ saw 片锯
~ scraper 刮铲式平地机
~ setting 装定叶片角,叶片位置调整
~ shaft 叶片式混砂机传动轴
~ spreader 刮板式(混合料)分布机[摊铺机]
axe ~ 斧刃
back-saw ~ 锯条
circular saw ~ 圆锯片
fan ~ 通风机(风扇,风轮机)叶片
hack-saw ~ 弓锯片
hand hack saw ~ 手弓锯片
planing ~ 刨刀(片)
scraper ~ 刮刀(片)
shear ~ 剪刀片
switch ~ 开关闸刀;闸刀开关铜片;辙[岔]尖,尖轨
thermostat ~ 恒温器开关叶
wiper ~ 刮水片
wiping ~ 密封的框架
work rest ~ 刀形支承,导向尺,垫板

blademan ['bleidmæn] *n*. 平地机手,铲刮工

blader ['bleidə] *n*. 平路机,叶片安装工
~ -shaped diaphragm 叶片状隔板

blae [blei] *n*. 灰青碳质页岩,劣质黏土页岩

blanch [blɑːntʃ] *v*. 漂[煮,发]白;粉饰;镀锡

blank [blæŋk] *n*. 空白,间隔,缝隙;空地[间];无用,无价值;毛坯,坯料,粗制材;半成品; *a*. 空白[着]的;无门[窗]

的;单调的; *v*. 取消,作废,使无效;下料;断开,熄灭
~ account 空白账簿
~ arcade 封闭拱廊,假拱廊,实心连拱
~ arch 假拱,轻拱,拱形装饰
~ bill 空白支票
~ bolt 无螺纹栓
~ bond 保单,总括保证
~ check 空白支票
~ door 假[暗,死]门
~ fenestration 假门窗(布局,组合),暗门窗
~ flange 无孔法兰盘
~ key 钥匙型片
~ masonry wall 无开口圬工墙
~ mounting 毛坯安装
~ -out sign [雕空]漏光式标记
~ panel 空面板,备用面板
~ rosette 实心圆花窗
~ run 空转
~ tracery 实心窗格
~ triforium 教堂拱门上面的实心拱廊
~ wall 无门窗墙,难以通过的障碍物或屏障
~ window 假窗
~ wire 裸线

blanket ['blæŋkit] *n*. (铺面,覆盖,保护,平覆,再生,表面,敷,热,垫)层;膜,套,外壳;毡,毯;垫板,坯板;熄灭装置; *a*. 一般的,总括的,无大差别的; *vt*. 铺毡毯,盖毯子,包裹;覆盖,掩蔽;消除,扑灭
~ area 敷层面积,掩蔽区
~ chest 一种箱式家具
~ coat 粘层
~ license 总许可证
~ mortgage 联合抵押
~ order 总订货单
~ policy 总保单
~ price 共同价格
~ing line 落料生产线
~ing material 铺盖材料

asbestos ~ 石棉毡(隔音用)
baffle ~ 吸音毡
drainage ~ 排水铺面
filter ~ 渗透垫层,滤料垫层
levelling ~ 平整层
pervious ~ 排水层,透水铺盖
protective ~ 防护铺盖
sand[gravel] ~ 砂[砾石]盖层,过滤层

blare [blɛə] *v.*, blared, blaring; *n.*, 亮光,(颜色)光泽;粗大响声,嘟嘟声; *v.* 发亮光;高声发出;发嘟嘟声

blast ['blɑːst] *n.*; *v.* 爆炸[破],炸掉,摧毁;冲击波,强射流;鼓[送]风;喷气[射][喷[吹]砂;鼓风机,喷砂器,压缩器;变晶;blaster, *n.*; blasty, *ad.*
~ air 鼓风,送入空气
~ area 爆破[放炮]区
~ blower 鼓风机
~ box 风箱
~ chamber 爆破室,鼓风室,燃烧室
~ cleaning 喷砂清理
~ cold 吹冷
~ design 爆破设计
~ down 爆破,崩落
~ draft 压力气流[通风]
~ effect 喷气作用
~ furnace brick 高炉砖[鼓风炉]砖
~ furnace brickwork 鼓风炉砌砖,高炉砌砖
~ furnace cement 高炉矿渣水泥
~ furnace cinder 高炉矿渣
~ furnace coal-tar pitch 高炉煤焦油沥青
~ furnace drainage 高炉[鼓风炉]排水
~ furnace heater 鼓风加热器,热风设备
~ furnace lump slag 高炉块渣
~ furnace slag 高炉矿渣
~ furnace slag aggregate 高炉矿渣集料
~ furnace slag cement 高炉矿渣水泥,硅酸盐水泥
~ furnace slag coarse aggregate 高炉矿渣粗骨料
~ furnace slag concrete 高炉矿渣混凝土
~ furnace slag dust〔filler, fill(ing)〕高炉矿渣填充料
~ furnace slag fibre 高炉矿渣纤维
~ furnace slag sand 高炉矿渣砂
~ furnace slag sand concrete 高炉矿渣砂混凝土
~ furnace trass cement 高炉火山灰水泥
~ gate 风阀,风闸
~ heater 热风炉
~ -hole 炮眼,爆破孔;钻孔[眼];风洞;进水口,入水孔
~ hole bit 无岩心钻头,爆破孔钻头
~ hole drill 爆破孔钻机
~ hole drilling 爆破孔钻进,打炮眼
~ main 总风管
~ meter 测风计(测风压,风量)
~ nozzle 喷砂嘴,风嘴
~ of air (空)气流
~ of wind 阵风,风的冲击
~ pit 排气井
~ plate 防烟板
~ pressure ga(u)ge 风压计
~ proof 防爆的
~ -resistant civil defence【美】[defense] structures 防爆土木防护结构
~ resistant construction 防爆结构
~ rig 钻炮眼设备,爆炸井钻具
~ tube 送风管
~ volume 风量
~ wall 防爆墙
cold ~ 冷风;冷吹[鼓]风
dry ~ 鼓[送]风
dry air ~ 乾风
sand ~ 喷砂器

blaster ['blɑːstə] *n.* 喷砂机
sand ~ 喷砂器

blasting ['blɑ:stiŋ] *n.* 爆炸[破],放炮；鼓风,吹洗；喷砂,喷抛处理；震声,炸裂声；在气流中运动,风洞试验；过载失真
- ~ accessory 爆破用具
- ~ action 鼓风作用
- ~ gear 爆破设备,放炮工具
- ~ gelatine 甘油[明胶,胶质]炸药
- ~ layout 爆破(点)布置
- ~ machine 发爆机,电爆器；喷砂机
- ~ mat 防爆帘
- ~ method 爆破方法
- ~ operation 爆破作业,鼓风操作
- ~ shot 喷的铁丸[砂砾]
- ~ test 爆破试验
- air ~ 鼓风；喷[洒]水(用压力喷水器)
- free sand ~ 无遮喷砂法
- grit ~ 喷砂消除法,吹砂,喷砂处理
- minor ~ 二次爆碎；扩孔底,套眼
- radio ~ 无线电(操纵)爆破
- sand ~ 喷砂

blaze [bleiz] *n.;v.*, blazed, blazing; *n.* 火焰,火灾；强烈的光；激[爆]发；*v.* 燃发,放光；激发；刻记号
- ~ the line 标出设计路线
- ~ the trail for 为……铺平道路

bldg = building 建筑物,大楼

bleach [bli:tʃ] *n.* 漂白剂；*v.* 漂白,弄白,脱色；bleachability, *n*; bleachable, *ad.*
- ~ liquor (liquid) 漂白液
- ~ oil 无色滑油,漂白油
- ~ out process 漂白法
- ~ -spot 白斑
- ~ed earth 漂白[漂洗]土
- ~ed glue 漂白胶
- ~ed sand 漂白砂
- ~ed (shel) lac (漂)白虫胶[紫胶]
- ~ed zone 褪色带
- ~ing agent 漂白剂
- ~ clay 漂白土
- ~ing effect 褪色效应
- ~ing powder 漂白粉

bleacher ['bli:tʃə] *n.* 漂白剂,漂白器；漂白工人；(*pl.*)露天简易看台

bleb [bleb] *n.* 气泡；水泡；砂眼,空洞；blebby, *a.* 有水[气]泡的,有气孔的；blebbing, *n.* 起泡,气蚀,穴蚀
- ~ steam 废汽,撒汽

bleed [bli:d] *v.*, bled [bled], bleeding; *v.* 排放(液体或气体)；泄漏,渗[流]出,泛[冒]油,渗色,褪色,油漆流淌；*n.* 泄放孔,放出的液[气]体
- ~ air[gas] 放气
- ~ air pipe 放气管
- ~ hole 排气孔,放出孔
- ~ -off pressure 排出压力
- ~ opening 放出孔
- ~ valve 排水[放气]阀

bleeder ['bli:də] *n.* 放出阀[管],泄放器,放油开关,放水装置；分压器；泄漏电阻
- ~ circuit 分压电路,泄放电路
- ~ cock 放水龙头[旋塞]
- ~ hole 通风孔
- ~ line of tile 排水瓦管
- ~ pipe 放水[气]管,排出管
- ~ screw 螺旋塞
- ~ steam 抽气
- ~ tile 泄水瓦管
- ~ type condenser 溢流式大气冷凝器
- ~ type condensing plant 溢流式大气冷凝装置
- ~ valve 排[抽,放]气阀
- air ~ 放气阀[管]；通气小孔；气泡管

bleeding ['bli:diŋ] *n.* 凝胶收缩；(涂料)色料扩散；放(泛,析)出,放气,渗漏,分级加热法
- ~ cement (混凝土表面)水泥浆沫[浮浆,泛浆]
- ~ cock 排气阀,放水龙头
- ~ joint 石油沥青接缝(油毡之间)
- ~ plug 放水[排气]旋塞
- ~ recovery (制冷机冷媒)排气回收装置

blemish

~ shutter（放气管）节气门,放油[水]管调节门
~ turbine 放[抽]气式透平

blemish [blemiʃ] *n.* 表面缺陷,瑕疵,污点; *vt.* 损伤,毁坏

blend [blend] *v.*, blended of blent, blending;混[掺,融,拌]合,配料; *n.* 混[掺]合物,混合料;合金
~ price 综合价格
~ed aggregate 混合集料
~ed asphalt 掺合[掺配,调配]地沥青
~ed asphalt joint filler 填缝石油沥青掺合料
~ed cement 混合水泥
~ed protland cement 混合硅酸盐水泥
~ed sand 混合砂
~ing bin 拌合斗[仓],混合箱[斗]
~ing bunker 混合料斗[仓]
~ing container 混合容器
~ing conveyor 混合输送机
~ing machine 搅拌机,拌和机
~ing mixture 搅拌机,拌合机,混和阀
~ing plant 搅拌[混合]设备,搅拌器
~ing pump 混合泵

blender ['blendə] *n.* 搅拌机,拌和机,搅拌[混合,掺和]器;混合颜料的工具
~ grinder 磨碎拌和机

blenometer *n.* 弹簧弹力[测量]仪
~ed area 郊外简陋街区;荒废地区;阴影面积

blind [blaind] *n.* 遮帘,幕,百叶窗,屏风,挡板,防护[遮阳]板,障碍物,隐蔽处,尽端,塞子,螺旋帽; *a.* 封闭[不通,堵死]的;瞎[盲,暗]的;未磨光的;隐蔽的,不显露的; *ad.* 盲目地; *vt.* 堵塞(孔,隙),铺填砂石;隐蔽,遮住; blindness, *n.*
~ abutment 隐蔽[暗]支座[桥台,扶壁]
~ arcade 盲[假]拱廊
~ arch 假[实心,

blind arcade

装饰]拱
~ area 房屋外墙以外全部或部分有遮盖以免墙面受潮的空地;封闭地块
~ attic 紧接房屋顶下不装修而封闭的空间
~ axle 静轴,游轴,侧轴
~ balustrade 实心栏杆
~ bond 砖砌体的一种砌合法
~ box 窗帘内箱
~ catch 关闭把手,闭门扣,百叶窗扣
~ column 倚柱,附壁柱
~ door 暗[百叶]门
~ dovetail 暗楔榫,暗鸠尾榫接头
~ drilling 未取任何资料的钻进;(钻液)有进无出钻进
~ edge 暗缝边
~ end 封闭端
~ fast 百叶窗固定器
~ flange 盖板,管口盖凸缘,堵塞[闷头]法蓝
~ header 暗丁砖,半块砖
~ hinge 暗铰链
~ hole 盲孔,漏头钻孔,闷眼
~ joint 无缝[暗]接头
~ lattice 假坚棂条
~ lift 升降百页窗的执手
~ main 尽端干管
~ mortise 凹榫
~ nail(ing) 暗钉
~ nut 螺帽
~ operator 百页窗中控制器
~ panel 不开窗口的墙板
~ patch 堵孔板
~ pier 附墙柱
~ pier capital 附墙柱头
~ pit 盲[暗]井
~ plate 盲板
~ plug 绝缘插头,空插头
~ power 电抗功率,无功功率
~ riser 暗冒口
~ rivet 埋头铆钉
~ rosette 实心圆花窗
~ sewer 污水暗管

~ shaft 窗帘卷轴,盲[暗]井
~ shield 闭口[封闭式,挤压式]盾构
~ side 死角,弱点
~ siding 尽头线
~ slat 百叶窗帘板
~ stop 暗闩
~ stop bar 窗帘止铁
~ stopper 窗帘止卡
~ mortise and tenon (joint) 凸凹榫(接)
~ tracery 实心窗格
~ triforium (gallery) 实心拱廊

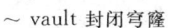
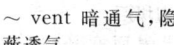
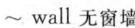

blind mortise and tenon(joint)

~ vault 封闭穹窿
~ vent 暗通气,隐蔽透气
~ wall 无窗墙
~ well 沙底水井
~ window 百叶窗
Persian ~ 百叶窗
sun ~ 遮帘

blinding ['blaindiŋ] *n*. 填充道路面层孔的细石屑,铺撒填缝石屑;贫混凝土垫层[封层];堵塞,挡住,填没; *a*. 眩目的;blindingly, *ad*.
~ concrete course (layer) 混凝土找平层
~ layer (基础)垫层
~ (layer of) concrete 贫混凝土垫层,混凝土找平层
~ material (填充表面孔隙的)细石屑
~ of screen 筛子的阻塞
~ sand 填砂
~ sand stone 嵌缝砂石
~ tile drain 盖土排水瓦管

blindstory ['blaind,stɔ:ri] *n. pl*. -ries, 无窗之楼;暗层

blister ['blistə] *n*. 水[气]泡,气孔,砂眼,小丘,折叠,局部隆起; *v*. 产生气泡,起泡,爆皮,起皮
~ copper 粗铜
~ figure 疱状突起

~ sand 砂疤
~ steel 泡钢(由熟铁渗碳而成的钢),渗碳钢

bloat [bləʊt] *v*. ; *n*. ; *a*. ,膨胀,膨胀的
~ed clay 膨胀(黏)土(一种拌制混凝土用的轻集料)
~ed clay aggregate 膨胀黏土集料
~ed clay concrete 膨胀黏土混凝土
~ed clay concrete solid block 膨胀黏土混凝土砌块
~ed clay concrete solidtile 膨胀黏土混凝土砖
~ed clay (concrete) wall slab 膨胀黏土混凝土墙板
~ed concrete aggregate 膨胀混凝土集料
~ed shale (clay) concrete 膨胀页岩混凝土
~ed slate 膨胀板岩
~ed slate concrete 膨胀板岩混凝土
~ed slate factory 膨胀板岩工厂
~ing agent (chemical) 膨胀剂
~ing clay 膨胀(黏)土
~ing clay aggregate 膨胀黏土集料
~ing clay concrete 膨胀黏土混凝土
~ing clay concrete solid block 胀膨黏土混凝土砌块
~ing clay concrete solid tile 膨胀黏土混凝土砖
~ing clay (concrete) wall slab 膨胀黏土混凝土墙板
~ing concrete aggregate 膨胀混凝土集料
~ing shale (clay) concrete 膨胀页岩混凝土
~ing shale coarse aggregate 膨胀页岩粗骨料
~ing slate 膨胀板岩
~ing slate concrete 膨胀板岩混凝土
~ing slate factory 膨胀板岩工厂
~ing test 膨胀试验

bloater ['bləʊtə] *n*. 膨胀器,膨胀剂
blob [blɔb] *n*. ; *v*. , blobbed, blobbing;

n. 一块[滴];斑点,色斑;(*pl.*)光[气]泡;*vt.* 弄污,弄错

blocage *n.* 毛石砌体

block [blɔk] *n.* 块,方块,砌块,块段,块体,巨块,地块;毛石,石[岩]块;粗料,毛坯,铸造板坯;街坊[区],区(段,组,间);大楼[厦],并列的房子,邻近的一组建筑物,建筑物的一部分;部件,装置,单元;基座工作台,支架,盘木,枕座;滑车[轮],方框图;*vt.* 闭锁[塞],阻塞,阻止,封锁,妨碍;中断,停用,冻结;使成块状

~ allocation 整笔拨款
~ and cross bond 丁顺砖交叉砌合
~ arch 砌块拱
~ architecture 大型砌块住宅
~ beam (预应力混凝土)分块拼装梁
~ bearing 支承轴承,止推轴承
~ bill 宽刃斧
~ bogie 小车,窄轨斗车
~ bond 丁砖与顺砖交叉或隔层砌合
~ brake 闸块式制动器,瓦闸
~ bridging 横撑
~ capital 方块式柱头(罗马风)上下方圆柱头
~ cast 整铸
~ casting 块铸
~ chain 块环链,车链
~ check 棋盘格纹饰
~ chimney 砌块烟囱
~ chopper 凿石工,石块切割机
~ clutch [伸缩]闸瓦离合器
~ column 建筑支柱,支撑
~ complex 建筑组合
~ compression strength 砌块[石块]抗压强度
~ construction 大型砌块建筑
~ construction met-hod 砌块构造方法
~ -cork insulation 软木块绝缘[绝热]
~ curve 连续曲线,实线
~ cutter (cutting machine, splitter) 块料切割机
~ diagram 方框[块状图,方块]图,立体图,草图
~ diagram symbol 方块图符号
~ distribution 集体分配法(结构力学的一种解法)
~ entrance 建筑入口
~ entrance door 建筑入口大门
~ equipment 建筑设备
~ extension 扩建
~ factory 建筑砌块厂
~ finish 混凝土表面磨光(用金刚砂块)
~ flooring 块料地板
~ footing 块形底部[大方脚]
~ for drawing 拉丝卷筒
~ format 砌块规格[大小,尺寸]
~ frame 建筑框架,房屋构架
~ foundation 块形基础,大型块体筑基
~ glass 玻璃块
~ hammer 落锤
~ heel riser 叉尾垫块
~ hole 浅炮眼,二次爆破炮眼
~ holing 二次爆破
~ holing method 分块钻孔法
~ house 木屋,工棚,碉堡,掩体
~ in 画草图,拟大纲,堵塞,封锁
~ -in-course bond 砖楔块砌合
~ -in-course (masonry) 嵌入楔块层,细琢方石砌体
~ joint 预留砖孔或混凝土块孔(作接头用)
~ -layer 砌块层
~ laying 砌块砌筑
~ leaf 砌块门扉
~ level 气泡[平放]水准仪
~ -lifting machine 砌块提升[装

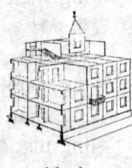

block construction

载]车
~ line 房屋界线,建筑线,街区线
~ lintel 砌块过梁
~ load 基本荷载;巨额负载
~ machine 砌块制造机
~ making 砌块制造
~ making machine 压块机
~ map 立体图,方块图[块状]图,框图
~ masonry (work) 砌块工程,凿石圬工
~ masonry wall 砌块圬工墙
~ mass 大块,岩块,地块,图块
~ moraine 块砾碛
~ nut 防松螺帽
~ of flats 公寓,住宅群
~ of grouped shops 商业中心
~ of large blocks 大砌块住宅
~ of large slabs 大板住宅
~ of large tiles 大空心砖建筑
~ of offices 办公室大楼,写字楼
~ of staggered design 错列式设计住宅
~ of uniform height 等厚块或料
~ of valve 阀锁
~ off 阻塞,堵截,挡住
~ openings 填塞空洞
~ orientation 建筑定位
~ oscillation 建筑振动
~ out 打草图,勾轮廓规划,画略图,拟大纲
~ outs 预留孔堵块(浇混凝土后移去)
~ parquetry floor covering 镶木地板条块复盖
~ partition (wall) 砌块隔墙
~ pavement 块料路面
~ paving (法)块料铺砌
~ plan 区划[分段]图,分区规划,略图
~ plane 短刨(横木纹的刨)
~ planing 分区规划,基地规划,总平面布置
~ press 砌块模压制
~ pulley 滑轮组
~ quay wall 码头挡土墙块,码头岸壁块
~ silo 混凝土砌块筒仓[谷仓,竖井]
~ size 街廓大小
~ spar 块长石
~ spirit level 平放水平仪
~ splitter 砌块切削机
~ splitting machine (石块再加工)切削机
~ square 矩形角尺
~ stack 砌块烟囱
~ station 闭塞站,线路所
~ stone 块石
~ -stone lining 块石衬砌
~ stone pavement 块石路面,块石铺砌路面
~ -stone road [铺]方块石路
~ strength 砌块强度
~ structure 块状结构,断块构造
~ terminal 分线盒,配电盒[箱]
~ type 建筑类型
~ up 堵塞,封闭,隔断
~ -walling 砌块圬工
~ wood pavement 木块(铺砌)路面,木块铺面
~ed course 块状砌层,封闭层
~ed doorway 墙上壁龛,有雕塑的壁龛门道
~ed joint 分段连接[接头]
~ed stone [分]块石
anchor ~ 地锚,地下横木,锚枕,锚墩,镇墩
angle ~ 角铁,弯板
asphalt ~ [地]沥青块
back ~ 衬[垫]板
bearing ~ 轴承座;承重块;矿柱
brake ~ 制动器闸瓦
building ~ 空心砌块;预制组件,标准元件,结构单元,积木式部件
building ~ principle 组[拼]装原理
bull ~ 拉丝机;拉丝模
bumper ~ 弹性垫座
camber ~ 反挠垫块

cement [预制]水泥[空心]型块
cinder ~ 煤渣砖
concrete ~ 混凝土块
cupola ~ 化铁炉异型耐火砖
deep hole ~ 空心砖,深孔石块
deflecting ~ 三角柱体石块[砖块]
die ~ 滑块;板牙;模板
drawing ~ 拉丝模板,拉丝板
fill ~ 填[垫]块
filler ~ 衬块,止水塞
finishing ~ 拉细丝机
fixing ~ (受)钉[块]条
frazzi ~ 空心陶瓷砌块
fuse ~ 保险丝盒
gap ~ 轴瓦
glass building ~ 建筑用玻璃砖
gypsum ~ 石膏砌块
hearth ~ 炉床砖
joint ~ 连接砌块;接块
junction ~ 接头管(户内下水道与街道排水管的接头)
key ~ 拱顶石
packing ~ 垫[填]块
pecker ~ 支承,支架
plaster ~ 石膏块
screw ~ 千斤顶
slag concrete ~ 矿渣混凝土砌块
stepped (-packing) ~ 阶梯形支柱
stock ~ 垫墩
stone ~ 大块石头[岩石],石块
subintradosal ~ 内拱承重块
supporting ~ 支柱[杆,座],垫板
swage ~ 型砧,型模块
tappet ~ 支柱,支架
terminal ~ 接线盒,用户接线盒
timber foot ~ 棚腿垫板,木垫板
title ~ 明细表(制图)
vitrified ~ 硅石砌块
wedge ~ 楔座,楔块
wedging ~ 楔合块
blockage [ˈblɔkidʒ] *n.* 堵塞,封锁,障碍;闭塞,闭锁,截止;拳石块,小方石

~ stone 制造拳石或小方石的石料【美】
blockboard *n.* 蕊块胶合板
block-diagram 方块图,简(草)图
blockhouse [ˈblɔkhaus] *n.* 木舍,圆木小屋,盒,箱,框架;碉堡,掩体;钢筋混凝土的半球形房屋
blocking [ˈblɔkiŋ] *n.* (以块料或类似物件)阻塞,支承,成形,压制;封锁,闭塞;压檐[女儿]墙;大块料,锤碎石块;方块木胶合[作业]法;字组化,模块化;划分采区;冻结

~ agent 封堵剂,堵水剂
~ course 檐头墙,压顶
~ device 闭锁装置
~ flow 阻塞流
~ nut 锁紧螺母
~ of drainage 排水阻碍
~ property 粘结[附]性能
wood ~ 木枕
blockman [ˈblɔkmən] *n. pl.* -men; 做板块的工匠,砌石块的工人,公路铺砌工
blockwood *n.* 木块
~ pavement 木块铺砌路面
blockwork *n.* 砌块工程,凿石圬工
blocky [ˈblɔki] *a.*, blockier, blockiest;块状的,短而结实的; blockily, *ad.*; blockiness, *n.*
~ structure 块状结构
blockyard *n.* 预制混凝土构件场
bloodwood *n.* 红[苏]木
bloom [blu:m] *n.* 钢坯[锭],大方坯;砖墙风化;盛期;华,霜,光圈,晕,图象模糊[浮散]
~ and slab piler 钢坯和板材卸垛机
~ -base 支柱座
~ base plate 支柱底板
~ film glass 加膜玻璃
blossom [ˈblɔsəm] 华(地质)
~ rock 落华石
blot [blɔt] *n.; v.*, blotted, blotting;

blocking course

n. 墨迹,污斑[渍],缺陷; *v.* 吸油,吸去,弄污,抹掉
~ing material 吸油材料

blotch *n.* 疱,疙瘩,污迹,斑(点) *vt.* 弄脏,涂污

blotchy ['blɔtʃi] *a.* 斑污[渍]的

blotter ['blɔtə] *n.* 吸油(集料),吸墨纸[具];
~ aggregate 吸油集料

blout *n.* 块状石英

blow [bləu] *v.* blew, blown, blowing; 吹炼,鼓风,爆炸,炸裂,放炮,冲[打,攻]击,自喷,熔解[化],(保险丝)烧断; 跑掉,浪费,传播; *n.* 鼓风,吹气,冲击;(地下)涌水,沉箱放气;(*pl.*)气孔[穴];(保险丝)烧断;疾[强]风
~ asphalt 吹制沥青,氧化沥青
~ bending test 冲击弯曲试验
~ by (gases) 漏气,渗透,瓦斯喷出
~ -by of piston 活塞漏气
~ -cock 排放旋栓,排气栓
~ engine 鼓风机
~ form 衬砌模架
~ gun 喷枪
~ -out switch 放气开关
~ sand 喷砂,飞砂,吹砂
~ stack 放空烟道
~ stress 冲击应力
~ tank 泄料桶疏水箱
~ -up (混凝土路面板因膨胀而形成的)胀裂,膨胀,拱起,炸毁,爆炸,允气,吹胀
~ -up failures (on the pavement) (路面的)冻胀破坏,破裂
~ -up patches (路面的)冻裂修补处
~ -up performance (of pavement) (路面的)冻胀性状
~ -ups (由冻胀引起的)隆起
~ vent 通气口,排气口
~n-film 多孔膜,吹塑薄膜
~n-out concrete 充气混凝土(多孔混凝土)
~n sponge 海绵胶

~n-up mosaic 放大镶嵌图
~ing agent 发[起]泡剂
~ing current 熔断电流
~ing down 停风,排污
~ing engine 鼓风机
~ing fan 鼓[送]风
~ing opening 送[进]风口
~ing out preventer 防喷阀
~ing vent 送[进]风口
~ing ventilation 压入式通风
air ~ 送[鼓]风
air sand ~ 空气喷砂
glass ~ 吹玻璃
hammer ~ 锤击

blower ['bləuə] *n.* 鼓[吹]风机,风扇; 吹风管;低压空气压缩机;增压器;自喷井;螺旋桨;打眼装药工,放炮工,吹制工,转炉工
~ drain valve 鼓风机放泄阀
~ fan (离心式)鼓[通]风机,风机[扇]
~ line 放气管
~ pipe 吹送管,压入式风管
~ pump 增压泵
air sand ~ 空气喷砂器[机]
auxiliay ~ 辅助风扇,局部扇风机
cleansing ~ 吹净器
ensilage dump ~ 强力鼓风机
exhaust ~ 吸风机,抽气机
fan ~ 扇风机
flue ~ 吹灰机
forced draught ~ 加压通风机,送风机
helical ~ 叶轮式鼓风机;螺旋[轴向]鼓风机
high-pressure ~ 高压鼓风机
paint ~ 压缩空气喷涂器,喷漆枪
portable ~ 局扇,轻便式扇风机
positive ~ 旋转[正压]鼓风机
rotary ~ 回转(活塞)式鼓风机
sand ~ 喷砂机
soot ~ 吹灰机
spiral ~ 螺旋机[叶轮]鼓风机
steam ~ 蒸汽鼓风机

steam-jet ~ 喷汽送风机;喷射[注水]器[抽]
suction ~ 吸[抽]风机
blowhole ['bləuhəul] *n.* 喷[通]风孔;铸孔、砂眼,气泡[孔]
deep-seated ~ 深位汽泡
slag ~ 砂眼,渣孔
surface ~ 表面气孔
blowlamp ['bləulæmp] *n.* 吹管,喷[焊]灯
blowoff ['bləuɔːf] *n.* 吹[溢]出;吹卸器,排泄装置
~ line 排污[放气]管道
~ pipe 排污管,通风管
~ pressure 停吹气压
~ tank 吹泄箱
~ tee 排污三通
~ valve 排出[放泄]阀,排污阀
~ ventilation 排气[抽气]通风
~ water 排污水
blowout ['bləu'aut] *n.* 吹[喷]出,爆裂,漏气;风蚀洼地,残沙丘;废炮;吹熄,停炉
~ current 熔断电流
~ preventer 防喷器,封井器,防喷装置
~ prevention 防喷
~ process 吹动法
~ switch 放气开关
~ valve 排气管,排出阀
~ window 爆裂窗(起安全阀作用)
blowpipe ['bləupaip] *n.* 风[吹]管,通风管;喷焊[割]器;
~ analysis 吹管分析
~ assay 吹管分析,吹管鉴定
~ test 吹管试验
cutting ~ 喷割器
electric ~ 电弧喷焊器
injector ~ 低压喷焊器
multi-jet ~ 多焰喷焊器
oxyacetylene ~ 氧乙炔吹管
oxyhydrogen ~ 氢氧吹管
welding ~ 喷焊器

blowtorch ['bləutɔːtʃ] *n.* 吹管,喷灯,焊枪
blue [bluː] *n.; a.,* bluer, bluest, *v.*, blued, bluing or blueing; *n.* 蓝色,青色,普林士蓝;蓝色染料;蓝天,海洋,蓝(铅)油;*a.* 青的,阴郁的;*vt.* 染成蓝[青]色
~ annealing 蓝退火;软化退火(钢板)
~ annealing wire 发蓝钢丝
~ asbestos 青石棉[钠闪石],蓝石棉
~ beam 蓝色射线
~ bind 硬黏土
~ black 深蓝色的,蓝黑
~ brick 青砖
~ brick paving 青砖铺面
~ carbonate of copper 蓝铜矿(石青)
~ clay 蓝泥,蓝黏土
~ copper 铜蓝;蓝铜矿(石青)
~ copperas (vitriol) 硫酸铜,胆矾
~ flame burner 蓝焰喷灯
~ gas 蓝[水]煤气
~ gas tar 蓝[水]煤气柏油
~ grind stone 青色磨石
~ iron earth 蓝铁矿
~ jack 胆矾,五水[合]硫酸铜
~ john 萤石(氟石)
~ lead 方铅矿,蓝铅,金属铅
~ lias lime 蓝色水硬石灰
~ -light source 蓝光源
~ marl 青泥灰岩
~ metal 蓝铜锍(含铜约62%)蓝(锌)粉
~ mud 青泥
~ organic pigment 蓝色有机颜料
~ pigment 蓝色颜料
~ pigment dyestuff 蓝色染料
~ planished steel 发蓝薄钢板
~ powder 蓝(锌)粉
~ print 蓝图,方案,计划
~ print apparatus 晒图机[设备]
~ print drawing 蓝图
~ print lamp 晒图灯

~ print paper 晒图纸
~ printing apparatus (machine) 晒图机[设备]
~ printing paper 晒图纸
~ rot （木材）青腐
~ sheet 蓝钢皮，发蓝退火的薄钢板
~ stain 蓝变，蓝斑(木材)
~ steel 蓝[皮]钢
~ stone 胆矾，青石，硬黏土
~ -top grade 旧路改建前的纵断面(美)
~ ultramarine 蓝群青
~ vitriol 胆矾
~ water gas 蓝水煤气
~ing 蓝化，烧蓝，涂蓝，着色(检验)
cobalt ~ 氧化钴，钴蓝
dark ~ 深蓝色，暗青色
light ~ 淡青色
powder ~ 氧化钴
Prussian ~ 普鲁士蓝，蓝色颜料

blueness ['blu:nis] *n.* 蓝色，青蓝

blueprint ['blu:'print] *n.*; *vt.* 蓝图，设计图；方案，计划，详细制订；蓝色板[照相] blueprinter, *v.* 晒蓝图工人

blue-ribbon *a.* 第一流的
~ connector 矩型插头座
~ program [计]无错[一次通过]程序

bluestone ['blu:stəun] *n.* 胆矾，蓝矾；(五水)硫酸铜；青石，硬黏土，蓝灰砂岩

bluff [blʌf] *n.* 陡岸，悬崖，天然陡坡，断崖；非流线形体，不良流线体；*a.* 陡削的，绝壁的，扁平的；bluffly, *ad.* bluffness, *n.*
~ failure 陡壁坍毁
~ -racking 空转
~ work 边坡整平工作

bluish ['blu:iʃ] *a.* 带蓝色的，浅蓝色的（亦作 blueish）bluishness, *n.*
~ white 青白色

bluish-grey *a.* 蓝灰色的

blunder [blʌndə] *n.*; *v.* 故障，误差；失策，做错，疏忽；盲目行动；blunderer, *n.*

blunge [blʌndʒ] *vt.*, blunged, blunging；用水搅拌，揉黏土

blunger ['blʌndʒə] *n.* 圆筒掺和机，黏土混合机，搅拌机

blunt [blʌnt] *a.* 钝(圆)头的，无锋的；粗率的；*n.* 钝器，短粗的针；*v.* 弄钝，削成钝角，挫折，减弱；bluntly, *ad.* bluntness, *n.*
~ angle 钝角
~ arch 垂拱
~ bit 钝钻头
~ edge 钝缘
~ end 钝端

blunt arch

~ end of pile 桩的粗端
~ file 直边锉，齐头平锉
~ -hook 钝钩
~ mill file 平行细锉
~ nail point 钉的钝端
~ -nosed cone 钝头锥体
~ nosed pier 平头敦
~ pile 钝头桩
~ed spur 切平山嘴，削钝山嘴

bluntness *n.* 钝(度)，粗率

blur [blə:] *n.*; *v.*, blurred, blurring; *n.* 污点[斑]；模糊；*v.* 弄脏，渗开，使模糊；blurred, *a.* blurredness, *n.*; blurry, *a.*
~ed image 模糊图像
~ed picture 模糊图像[像片]
~ing 模糊，不清晰，污损

blush [blʌʃ] *n.*; *v.*, blushed, blushing；红色，使呈红色；玫瑰色；变红，泛红色；混浊膜（油漆）；blusher, *n.*; blushfulness, *n.*; blushful, *a.*; blushfully, *ad.*

blymetal *n.* 胶合金属板

board [bɔ:d] *n.* 建筑用板，(薄)木板，厚纸板；船舷，甲板；配电盘，操纵台；仪表[插件]板；会议桌，餐桌，伙食；董[理]事会，委员会，政府部、厅、局；(复)舞台

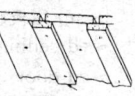

board and batten

~ and batten 板和板条

~ -and brace work 镶接木板
~ body type 板式体形
~ coal 纤维质煤
~ cutting machine 锯板机
~ drop hammer 夹板落锤,木柄摩擦落锤
~ drop stamp 夹板落锤捣碎机
~ eaves 木板挑檐
~ facing 衬板
~ fence 板篱,木栅栏
~ foot 板尺,木料(1尺方1吋厚)
~ forms 木模板,木壁板
~ form (work) 木模板,木壁板
~ hole 承板槽口
~ joint 木板接合
~ insulation 绝[隔]热板
~ lath 石膏板
~ lining 衬板
~ machine 纸板机
~ -marked 形成木板纹表面
~ measure 按板尺计算
~ mill 锯板厂
~ of asbestos cement 石棉水泥板
~ of felt(ed fabric) 油毡板
~ of paper 纸板
~ of work (建筑)工程委员会
~ partition 板壁[墙]
~ -rule 量木尺,板尺
~ scale 按板尺计算
~ scraper 板式刮土机
~ sheathing 盖板,(屋顶瓦下的)夹衬板
~ shuttering 木模板,木壁板
~ tree 板材树
~ type insulant 板状绝热[绝缘]材料
~ up 用板遮住,[围]住
~ed ceiling 板平顶,板吊顶,板顶棚
~ed door 木板门
~ed floor 木楼板[铺木板]地面
~ed leather 磨光皮革
~(ed) parquetry 木板嵌镶工作

access ~ 跳[搭]板;梯子
acoustic celotex ~ 隔音纸板
acoustical fiber ~ 吸音纤维板
apron ~ 裙板
asbestos ~ 石棉板[纸板]
auger ~ 钻架
automatic plotting ~ 自动标定图板
backing ~ 垫板
baffle ~ 隔音板;烟道风帽;折流板
barge ~ 挡风板,(山)头封檐
base ~ 基线板,踢脚板
bastard sawed ~ 粗锯板
binder ~ 刨花板
blasting ~ 放炮时的支架保护板,爆破保护板
boning ~ 测平板
breast ~ 栏板
bridging ~ 搁栅跨板
bucket ~ 戽(斗)板
building ~ 建筑板材,人造板
bulletin ~ 布告栏
cant ~ 带[镶]棱板
calculating ~ 计算台
cap ~ 顶梁(棚子的),柱帽
ceiling ~ 顶棚
celotex ~ 隔音板
dash ~ 遮雨板
deals ~ 厚(松)板
deckle ~ 框板
distributing ~ 配电板
distribution ~ 分配盘,配电[线]盘
drain(ing) ~ 滴水板
drawing ~ (制)图板
eaves ~ 风檐板
elbow ~ 窗台(板)
fascia ~ 挑口板
feather edge ~ 楔边板
fender ~ 挡泥板
fiber ~ 纤维[木丝]板,纤维填缝板
flash ~ 插板,决泻板,闸板(坝顶调节水位用)
floor ~ 楼面板,地板,底板

follow ~ 模[样]板(量具)
foot ~ 踏板
frame ~ 腰线板,檐壁板
gauffered ~ 皱纹纸板
gauge ~ 样[模]板;仪表盘,规准尺
glass-fiber ~ 玻璃纤维板
glued ~ 胶合板
graining ~ 压纹板
guard ~ 护[挡]板,挡边板
guide ~ 导板
gutter ~ 排口板(檐槽)
gypsum plaster ~ 石膏壁板
gypsum wall ~ 石膏壁板
heated ~ 保温板
level ~ 水平板
lifter ~ 提升台
limbers ~ 污水道盖板
loam 刮板
louvered ~ s 百页板,散热片
louvre ~ 条板,散热片
lower edge ~ 下边板
luffer ~ 条板
match ~ 假型板,模板
mould(ed) ~ 样[型,模]板
moulding ~ 样[型,模]板
panel ~ 镶板,配电盘,仪表板
particle ~ 碎料板
partition ~ 隔板
paste ~ 纸板,胶纸板
plank ~ 衬板
plaster ~ 糊墙纸板
plinth ~ 门基板
poling ~ 堰板,撑板
profile ~ 剖面样[模]板
ribbon ~ 条板
roof(ing) ~ 屋面板,顶棚
rough ~ 毛板
sarking ~ 屋面衬板
shear ~ 护坡[砂土]坚板
sighting ~ 觇板
skirt ~ [平皮带运输机]侧壁;踢脚板
slag-gypsum ~ 矿渣石膏板

sound ~ 响板,坚实板,共鸣[振]板
sounding ~ 托梁,响板,测深板
splash ~ 挡板,挡泥板,遮水条
stamping ~ 垫模板
straw ~ 马粪纸,草纸板
string ~ 楼基盖板
switch ~ 开关板,配电盘
thistle ~ 石膏板(排挡间的)
three-ply ~ 三层[夹板]
toe ~ 趾板,搁脚板
tread ~ 踏板
turning over ~ 底板
veneer ~ 胶合板
upper edge ~ 上边板
verge ~ 挡风板,[山头]封檐板
wall ~ 墙板,建筑纸板
wash ~ 壁脚板
window ~ 窗台板
wiring ~ 布线底板,装配底板,配线盘
wood fibre ~ 木纤维板

boarding ['bɔːdiŋ] *n.* 木板,板条;上船(车,飞机),铺[盖,镶,钉]板;起纹,寄宿;*a.* 供膳(宿)的
~ card 旅客乘(飞)机证,搭载客货单
~ house 招待所,宿舍,食堂
~ joist 裸搁栅
~ school 寄宿学校
ceiling ~ 顶棚覆板
coloured porcelain ~ 彩瓷板
cornice ~ 花檐板
cup ~ 橱柜
drawing ~ 绘图板,画板
feather-edged ~ 楔边板安装
glass-fronted bill ~ 橱窗
floor ~ 地板
match ~ 舌槽企口板
roof ~ 屋面板,木板屋顶
side ~ 餐具柜
weather ~ 风雨(护壁,墙面)板
wedge ~ 楔镶板

boards *n.* 舞台面
boardwalk *n.* 步行板,木板人行道,散步

场(木板铺成)

boart [bɔ:t] *n.* 金刚石屑[砂],圆粒金刚石,下等金刚石
　~ -set bit 金刚石钻头

boast [bəust] *n.* 铁棒,钢铁板,厚板; *vt.* 粗琢[凿]石
　ed ashlar 粗凿石圬工,乱纹凿石饰
　~ed finish of stone 粗凿石面
　~ed joint surface 刻平行槽石面,宽凿接缝槽面
　~ed surface 粗凿工作
　~ed work 刻石[槽]工作
　~ing 粗凿石块
　~ing chisel 片石阔凿

boaster ['bəustə] *n.* 阔凿,榫槽刨

boat [bəut] *n.* 船;船形器皿;吊灯槽; *v.* 船运,乘船
　"~ bottom" type of deformation "船底式"变形
　~ house 游艇俱乐部,艇库[房]
　~ level 船形水平尺
　~ nail 方钉
　~ spike 船钉,大方钉
　~ truck 脚轮推车
　~ varnish 船用清漆

boatfall *n.* 吊艇滑车[组]索,钩头篙

boathook *n.* 挽钩,钩头篙

bob [bɔb] *n.*; *v.*, bobbed, bobbing; *n.* 摇摆,浮动,截[剪]短,秤[测]锤,撬;(排水暗沟)扩孔器; *vt.* 上下跳动[疾动];轻敲
　~ (-)sled 二撬拖材车(运木材用)
　~ sleigh 二撬拖材车
　~ -weight 砝码,平衡重[锤],反重
　~ bing 振[摆]动;截短;抛光
　~ bing machine 振动机

bobber ['bɔbə] *n.* 浮沉材,浮材

bobbin ['bɔbin] *n.* 线轴;细绳,(门闩上的)吊带把手,线圈架
　~ bit 槽形扁钻头(加工木材用)
　~ disk 卷盘
　~ tape 扁带,圆带
　ribbon ~ 色带盘

bobbinite *n.* 硫铵炸药(矿山用);筒管炸药

bobby prop 防爆木档墙;顶梁短支柱

bobsled ['bɔbsled] *n.*; *v.*, bobsledded, bobsledding; bobsledder, *n.* 雪撬,长撬;串联雪撬运输

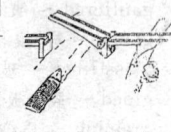
bobtail drawbridge

bobtail ['bɔbtei] *a.*; *vt.* 截[剪]短的;截尾;剪短
　~ drawbridge 截尾仰开桥
　~ ing 沿地面集材
　~ target 隐显目标

bobweight *n.* 配重,平衡重;平衡锤,铅锤

bobwire *n.* 刺铁丝
　~ fence 刺铁丝围栏

bocage [bəu'ka:ʒ] *n.* 【美术】挂毯或花瓶上之树木,支叶等的花纹,【法】陶瓷人像的装饰背景

bocca *n.* 喷火口,小锥体;玻璃熔炉的炉口

boccaro *n.* 宜兴陶瓷

bod [bɔd] *n.* 泥塞[封],砂塞,塞子

bodega [bəu'di:gə] *n.*【西】酒窖,酒店,杂货店;仓库,柜台,酒吧间

bodger *n.* 撬杆;木雕刻工,木车工,椅子车工

bodied linseed oil 聚合亚麻子油

bodied oil 聚合油,乾性油

bodiness 加重[厚],增稠;稠度(胶浆);物体,体积,重量

bodkin ['bɔdkin] *n.* 锥子,粗针

body ['bɔdi] *n. pl.* bodies, *a.*; *v.*, bodied, bodying; *n.* 支柱[架],基础,底盘,基座;底质,粘[稠,浓]度,流动性,身[物,机]体;车[船]身;水域,水体;团体,机构; *vt.* 稠化
　~ axis 物体的轴线,固定轴,联系轴
　~ bolster 车身承梁
　~ bracket 车身托架
　~ brick 优质烧透砖,炉体砖

~ case 壳体,外壳
~ capacity 车身容量
~ centered lattice 体心点阵
~ coat 质地涂层,底子漆
~ colo(u)r 不透明体,主体色;保护[覆盖]色,车身颜色
~ construction 车身构造
~ design 车身设计
~ end furring 车端木梁[填木]
~ end plate 上端梁,顶棚端梁
~ed rail 端墙内镶条
~ flex 人体屈曲度
~ fluid 流体
~ frame 主[机,车]架
~ hardware 车身金属构件
~ hold down bracket 车身托架
~ lining 车身衬料
~ of ballast 道碴层
~ of concrete 混凝土块[体]
~ of fresh water 淡水体
~ of oil 油基,润滑油的底质
~ of paint 油漆的稠[粘]度
~ of pump 泵体
~ of regulator 调节器体
~ of salt water 盐水体
~ of saw 锯架
~ of screw 螺钉体,螺杆
~ of valve 阀体
~ of wall 墙身,砌筑体
~ of water 水体,贮水池,积水,水面,水塘
~ of wheel 轮盘
~ paint 底彩[色]
~ paper (stock) 厚纸
~ pigment 打底[底质]颜料
~ plan 正面图,横断面,横剖型线图,体平面
~ rail 机身梁
~ size hole 穿透孔,通孔
~ solid 固体
~ stress 内[体]应力
~ system 体系

~ transom 底架横梁
~ type 体型
~ vanish 磨光漆,外用清漆,长油性清漆
~ waste 物质损耗
~ing agent 稠化[增稠]剂,基础剂
~ing of oil 油引[发]聚合,油的聚合
~ing speed 粘度增长速度
~ing-up 涂火洒漆,涂罩光漆
bodywood ['bɔdiwud] *n.* 无节新材,主干材
bodywork ['bɔdiwəːk] *n.* 车[机]身制造;车[机]身修理
boehmite *n.* 勃姆石,软水铝石,薄铝矿
Boeton asphalt 菩伊吞地沥青(一种天然地沥青)
bog [bɔg] *n., v.,* bogged, bogging; *n.* 沼泽,沼泽地,泥炭沼泽,泥泞地,泥塘; *v.* 陷入,下陷; bogginess, *n.*; boggish, *a.*; boggy, *a.*
~ blasting 泥沼[炭]爆炸
~ borer 泥沼钻
~ coal 沼煤
~ down 埋陷,陷入泥[沼]中
~ drainage 沼泽地排水(设施)
~ ground 沼泽[泥沼]地
~ -head coal 藻煤
~ hole 垃圾孔
~ iron ore 沼铁矿
~ lime 沼灰土,泥灰岩,湖白垩
~ manganese 沼锰矿,锰土
~ marl 沼泽灰泥,泥灰岩
~ mine ore 沼铁矿
~ moss 泽边苔藓;泥炭藓
~ muck 沼泽腐泥,泥炭,泥煤
~ ore 沼铁矿,沼锰矿,沼锌矿
~ peat 沼泽泥炭,泥泽泥炭土
~ soil 沼泽土
~ type 沼泽式(成土作用)
~ water 沼泽水,泥沼水
~ ging 沼泽土化;陷[沉]入;装载
~ ging down 下陷,陷入泥中

bogaz *n.* 深岩沟

bogen structure 弧形组织

bogginess *n.* 沼泽性,泥沼状态

boggy ['bɔgi] *a.*, -gier, -giest;(多)沼泽的,(地面)湿软的,卑湿的;泥炭的
- ~ country 泥泞[沼泽]地区
- ~ ground 卑湿地
- ~ soil 沼泽土

boghead *n.* 藻[烟]煤
- ~ cannel (shale) 沼油页岩
- ~ coal 藻[烟]煤
- ~ shale 藻油页岩

boghole *n.* 沼穴

bogie ['bəugi] *n. pl.* -gies;小车,手推车;转向车[架,盘],挖土机车架,矿车;悬挂[平衡]装置,(吊车的)行走机构
- ~ bolster 转向架承梁
- ~ car 转向车,转架车
- ~ hearth furnace 活车[底]炉
- ~ skirt 转向架脚板
- ~ truck 转向架,转架车,转车台
- ~ vehicle 转向车
- ~ wagon 有转向架的铁路车辆
- ~ wheel 负重轮
- motor ~ 自动转向架

boglet *n.* 小泥塘[沼泽]

boglime *n.* 泥灰土,泥灰岩

Bohemian [bəu'hi:mjən] *a.* 波希米亚的
- ~ crystal (glass) 波希米亚(晶质)玻璃
- ~ glass 波希米亚玻璃,刻花玻璃,钾玻璃
- ~ vault 波希米亚拱形屋顶

boil [bɔil] *v.*; *n.* 沸腾,煮沸,起泡,蒸发;冒水,冒水翻沙现象,地基管涌现象;沸点;(塑体的)隐匿气孔 boilable, *a.*
- ~ away 煮干,完全蒸发
- ~ dry 煮干
- ~ mud 土移,涌泥流
- ~ of sand 涌砂,管涌
- ~ off 蒸发,沸腾,汽化,浓缩
- ~ out 熬[煎]煮
- ~ over 沸[腾]溢(出)
- ~ -proof 抗煮沸的
- ~ -proof bond 耐高湿砌合
- ~ up 煮开,沸腾,消毒,涌起

boiled [bɔild] *a.* 煮沸的,煮熟的;熟练的
- ~ linseed oil 熟亚麻仁油,干性油
- ~ oil 熟炼[脱水]油,清油
- ~ oil paint 熟炼[干性]油漆
- ~ tar 脱水[熟]焦油
- ~ wood oil 熟桐油

boiler ['bɔilə] *n.* 锅炉,蒸煮[蒸发]器,热水箱;boilerless, *a.*
- ~ accessories 锅炉附件
- ~ (anti-scaling) compasition 锅炉清洗[防垢,防锈]剂
- ~ ash 锅炉灰渣
- ~ auxiliaries 锅炉附件
- ~ bearing 锅炉座
- ~ bedding 锅炉座
- ~ blasting 锅炉爆炸
- ~ blow-down water 锅炉冷凝水,锅炉排污水
- ~ body 锅炉体
- ~ bottom 锅炉底
- ~ bracket 锅炉托架
- ~ brickwork 锅炉的砖衬(用于绝热),炉墙工作
- ~ bridge 火[炉]坝
- ~ capacity 锅炉容量
- ~ case 锅炉套箱
- ~ casing 锅炉外壳(围壁)
- ~ clothing 锅炉砖砌面层,炉套,外火箱
- ~ coaling installation 锅炉加煤装置
- ~ cock 锅炉用旋塞
- ~ composition 锅炉防锈剂
- ~ compound 锅炉防垢剂
- ~ construction 锅炉构造
- ~ control 锅炉调节[控制]

~ corrosion 锅炉腐蚀
~ deposit 锅垢
~ drum 锅炉壳[筒],锅炉[汽]包,煮[沸]鼓
~ dust 锅炉粉尘
~ efficiency 锅炉效率
~ equipment 锅炉设备[装置]
~ feeder 锅炉给水器
~ feed water 锅炉给水
~ feed (water) pump 锅炉给水泵
~ feeding 锅炉供水
~ fittings 锅炉装配附件
~ fixtures 锅炉附属设备
~ flue 锅炉烟道
~ foot 锅炉底座
~ for domestic use 生活用锅炉
~ foundation 锅炉基础
~ fuel 锅炉燃料
~ gauge 锅炉水位计
~ grate 锅炉箅[排]
~ heating surface 锅炉受热面
~ holder 锅炉支架
~ house 锅炉房
~ lagging 锅炉套箱,炉衣
~ making 锅炉制造
~ masonry (work) 锅炉圬工,圬工砌体
~ oil 燃料油,残渣油
~ output 锅炉输出功率,锅炉蒸发量
~ plant 锅炉设备,锅炉间,锅炉厂
~ pressure 锅炉压强
~ proving pump 锅炉检验泵
~ room 锅炉间[房]
~ scaling hammer 锅垢锤
~ scaling tool 去水锈工具
~ seat 锅炉座
~ sequence control 锅炉程序控制
~ setting 锅炉安装,锅炉基础
~ shell 锅炉壳体,锅炉包
~ shop 锅炉车间
~ slag [锅]炉渣
~ stay 锅炉牵[撑]条

~ steam 锅炉蒸汽
~ -steam dome [锅炉]汽包,聚汽室
~ steel 锅炉钢板
~ (supply) water 锅炉用水
~ support 锅炉支架
~ supporting structure 锅炉支架
~ tank 锅炉水柜
~ tower 锅炉水塔
~ trim 锅炉附件联接管
~ tube 锅炉管
~ water treatment 锅炉水处理
auxiliary ~ 备用锅炉
exhaust-heat ~ 余热[废汽]锅炉
high-duty ~ 高能率锅炉
horizontal ~ 卧式锅炉
packaged ~ 整装[块装]锅炉

boiling ['bɔiliŋ] *a.* 沸腾的,煮沸的,极热的,汹涌的; *n.* 砂沸,翻浆,[砂土等]管涌,河水泡水[翻花,翻浪];沸腾,煮沸,喷出; *ad.* 达到沸腾的程度
~ bulb 蒸硫釜[锅]
~ down 煮浓,提尽
~ ebullition 沸腾
~ fastness 耐煮性
~ heat 蒸发[汽化]热
~ hot 酷[滚]热的
~ -house 沸腾室
~ of sand 涌砂,管涌
~ of soil 土沸现象
~ -on-grain 炒砂法
~ point 沸点
~ point thermometer 沸点温度计
~ proof 抗煮沸的
~ range 沸腾(温度)范围,沸点
~ seasoning 煮沸干燥
~ soil 土涌
~ spring 沸泉,蒸汽泉
~ steel 沸腾钢
~ temperature 沸点
~ test 沸水试验,蒸煮[煮沸]试验
~ treatment 煮沸[蒸煮]处理
~ -up soil 冻胀土

~ vessel 煮沸器

boilproof *a.* 耐煮的

bolar *a.* 黏土的

bold [bəuld] *a.* 陡的,险峻的;大胆的,清楚的;boldness, *n.*

~ cliff 绝壁

~ coast 陡岸

~ face letter 粗[黑]体字,粗线体

~ line 粗线

~ shore 陡岸

bole [bəul] *n.* 树干;胶块土,红玄武土,细黏土;墙上的采风通风孔,墙上的小凹处

bolection [bəu'lekʃən] *n.* 凸出嵌线

~ moulding 凸式线脚,镜框式线脚

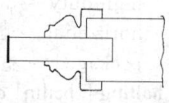
bolection moulding

Boliden salts 卜立顿盐剂(一种木材防腐剂,含加铬砷酸锌等)

bollard ['bɔləd] *n.* 系船柱,防止机动车进入某地区的短桩柱

~ cleat 系柱

Bollman truss 博尔曼式桁架(多弦式三角形桁架)

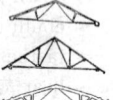
Bollman truss

bologram *n.* 热辐射测量图

bolograph ['bəuləgra:f] *n.* 测辐射热图;辐射热测定器 bolographic, *a.*; bolographically, *ad.*; bolography, *n.*

bolometer [bəu'lɔmitə] *n.* 辐射热测定器 bolometric, *a.*; bolometrically, *ad.*

bolson *n.* 干湖地,封闭洼地,沙漠盆地

~ plain 山间平原

bolster ['bəulstə] *n.* 枕垫,垫块[板],托[肋]木,立柱支承;横撑,支持物;台面,活动台,穿孔台;套管[板],夹圈,*vt.* 支持,支撑,加固,装填

~ plate 承梁板

~ work 隆腹状砌筑,支柱工,承垫工

check ~ 防松承梁

track ~ 转向架承梁

bolt [bəult] *n.* 螺栓[杆];插销,门闩;筛,筛子,锚杆,锚固支护; *v.* 螺栓固定,栓接;闩上,上插销;筛选

~ and nut 带螺帽螺栓

~ and nail connection 螺栓及钉结合

~ bar 螺栓杆,插销

~ cam 活动[进出]三角,螺栓闸门

~ chisel 扁尖凿[錾]

~ clasp 螺栓夹扣

~ clipper 断线钳

~ -connected joint 螺栓接合

~ cutter 断线钳

~ drill 螺栓孔钻,木钻

~ driver 改锥,螺丝刀

~ end 螺栓端

~ eye 螺栓眼[圈]

~ fastening 螺栓连接

~ former 螺栓锻机

~ -gauge 螺栓线

~ handle 机[闩]柄

~ head 螺栓头

~ head with feather 带销螺栓头

~ header 螺栓头镦锻机

~ hole 螺栓孔

~ hook 有螺杆和螺母可用作螺栓的钩子

~ in double shear (受)双剪螺栓

~ in single shear (受)单剪螺栓

~ iron 螺栓铁

~ joint 螺栓结合[联结]

~ -lock 螺栓保险,栓锁

~ neck 螺栓颈

~ nut 螺母

~ -on 螺栓紧固

~ pin 螺栓销,销子

~ pitch 螺栓(间)距

~ sleeve 螺栓套

~ splice bar 四[栓]孔鱼尾板

~ stock 螺栓钢料
~ stress 螺栓应力
~ supports 螺栓支撑,顶板锚杆支护
~ threading machine 螺栓车纹机
~ tightener 螺全纹紧器
~ timber 大块木材,短圆材,锯材
~ together 螺栓连接
~ torque 螺栓起动力矩
~ -up 螺栓紧固
~ -upright 直竖的
~ washer 螺栓垫圈
~ with stop 制止螺栓
~ with winged nut 翼形帽螺栓
anchor ~ 锚杆,杆柱;地脚螺栓;锚栓,镇钉
assembling ~ 装配螺栓
attachment ~ 联接螺栓
cap ~ 倒角螺栓,盖螺栓
carrying ~ 支撑螺栓
check ~ 防松螺栓
coach ~ 方头螺栓
cotter ~ 地脚螺栓,带销螺栓,锚栓
countersunk ~ 埋头螺栓
dormant ~ 沉头螺栓
door ~ 门销
fastening ~ 桩栓
foundation ~ 地脚[基础]螺栓
lifting ~ 起重螺杆
long ~ 地脚螺栓
rag ~ 棘螺栓,地脚螺栓
raised head ~ 凸头螺栓
sash ~ 窗插销
screw ~ 螺栓[杆];全螺纹螺栓
truss ~ 桁架螺栓,支撑螺栓;拉杆,系条
window ~ 窗插销
bolted *a*. 用螺栓固定[连接]的,栓紧[接]的
~ connection 螺栓结合
~ (fish-plate) splice 螺栓(鱼尾板)接合
~ flanges 螺栓结合凸缘,栓结法蓝

~ joint 螺栓接合[联结]
~ -on attachment 螺栓连接
~ -on connection 螺栓连接法
~ pile 螺栓接桩
~ pump 螺栓连接泵
~ splice 螺栓铰接
bolting *n*. 螺栓连接,锚杆支护;筛分[选]
~ cloth 筛子[绢],绢钢
~ down (机台的)螺栓固定
~ machine 筛分[选]机,机械筛;拉杆[锚杆]安装机
~ mill 筛分[选]机
~ steel 螺栓钢
boltel (boutel, boutell, bowtel, bowtell) ['bəultl] *n*. 凸圆饰;集柱中的一个柱身
bolter ['bəultə] *n*. 筛,筛选[分]机,筛石机;选木工;纵切圆锯机
~ -up 外板装配工
rock ~ 锚杆冲凿机
bolt-head 螺栓头,(枪)机头
bolthole ['bəulthəul] *n*. 锚杆孔;螺栓孔;石门,(采矿)联络小巷
~ circle 螺栓孔分布图
~ drilling 锚杆孔凿岩
bolts *n*. 短圆材
bolus ['bəuləs] *n. pl*. boluses;填塞物;胶块土,红玄武土,陶土;团块
~ alba 高岭土
bomb [bɔm] *n*. 炸弹;高压容器,氧气瓶,小型人工喷雾器; *vt*. 轰炸,投弹
-bombable, *a*.
~ breaking layer 爆破层
~ calorimeter 弹式量热器
~ furnace 封管炉
~ gas 瓶装气体,钢瓶气体
~ -proof 防炸的
~ -proof shelter 防空洞
~ resistant 防炸弹的,经得住轰炸的
~ shelter 防空洞,避弹掩壕,防空掩蔽所
~ shelter-garage 防空掩蔽车库

~ shelter door 防空洞门
bombproof ['bɔmpruːf] *n.* 防空洞[壕],避弹室; *a.* 防轰炸的,避弹的
~ glass 防弹玻璃
bond [bɔnd] *n.* 结[砌,胶,化]合;连[焊,胶]接;粘结,握裹,附着;粘结料,粘合剂,结合物,接头; *v.* 粘结,焊接,砌[结]合 bondable, *a.*; bonder, *n.* bondless, *a.*
~ anchorage 粘结锚固
~ area 粘结[结合,附着,粘附]面
~ beam 结合梁
~ beam block (tile) 预制混凝土结合梁块
~ behavior 粘合性能
~ between binder and aggregate 粘结[胶合]剂粘结,结合
~ between concrete and steel 钢筋与混凝土结合力(或握裹力)
~ breaker course 粘连间断
~ -breaking 断裂[开]
~ clay 砌合[胶结]黏土
~ coat 粘结涂层
~ concrete 胶结混凝土
~ course (layer) 粘结[砌合]层,丁砖层
~ creep 粘着徐变
~ external plaster 外部胶结粉刷
~ (external) rendering 外部胶[粘]结粉刷
~ failure 粘着力破坏,结合损坏
~ finish 胶结粉刷
~ fixing (预应力混凝土先张法的)粘着固定
~ flux (粘结)焊剂
~ header 砌合头(石)咬合砌(石)法
~ length (钢筋)锚着[握裹]长度
~ line 粘合[剂]层,胶层;粘结[接合]缝
~ log 水泥胶结测井图,固井声波测井图
~ master 环氧树脂类粘合剂
~ meter 胶接检验仪

~ mortar 胶接砂浆
~ open 焊接断开,焊缝裂开
~ performance 粘附[粘结,附着]性能
~ plaster 粘结灰泥
~ property 粘性,粘结性能
~ rendering 胶结外部粉刷
~ resistance 粘结强度
~ slip 粘结滑动
~ stone 束[系]石
~ strength 粘着强度
~ stress 粘结[握裹]力
~ stucco 胶结外部粉刷
~ study 粘结研究[分析,考察]
~ test 结合力[粘结]试验
~ tester 胶接[粘合力]检验仪
~ testing machine 粘合力试验机
~ tile 搭盖瓦
~ timber 束[系]木,枕距护木
~ type 圬工砌合类型
~ value 粘结度[力]
~ unit stress 单位握裹力,单位粘结应力
~ wheel 结合剂砂轮
angle ~ [墙]外角砌分
arch ~ 拱砌合
block-in-course ~ 券[圈]砌合
boundary wall ~ 界墙砌法(三顺一丁)
brick ~ 砌砖法
cellular ~ 花墙砌法
chair ~ 链式[横直]砌合
chimney ~ 烟囱砌合法
chisel ~ 凿形焊接
common ~ 普通砌砖式
conduct ~ 对[头]接
course ~ 丁砖层
cross ~ 交叉砌合
decorative ~ 装潢砌合
diaper ~ 交错砌合
dog tooth ~ 犬牙式砌合
Dutch ~ 荷兰式砌合

English ~ 英(国)式砌合
facing ~ 砌面
fire ~ 耐火材料的粘结剂
Flemish ~ 荷兰式砌合,佛式砌合

English bond

flying ~ 跳顶砖砌合
garden wall ~ 圆墙砌合
glue ~ 胶合
header ~ 丁砖砌合(法)
header and stretcher ~ 横直砌合
heart ~ 墙心顶块接砌
herringbone ~ 人字(形)砌合
longitudinal ~ 纵列砌合法(美国用)
monk ~ 两顺一丁砌合
mortar ~ 灰浆砌合
out and in ~ 凹凸砌合
rail ~ 轨道夹紧器
Polish ~ 波兰式砌合
raking ~ 对角砌合,斜纹砌合
rat-trap ~ 花墙砌法,多孔砌法
resinoid ~ 树胶粘合剂
rivet ~ 铆接
rubber ~ 橡胶粘结剂
runing ~ 侧砖砌合
shellac ~ 虫胶粘结剂
silver-block ~ 花墙砌法
sloping ~ 斜层砌合
stone ~ 砌石法
stretcher ~ 顺砖砌合
yorkshire ~ 跳顶砖砌合
bondage [ˈbɔndidʒ] *n.* 约束,束缚
bonded [ˈbɔndid] *a.* 粘结的,砌合的,化合的,束缚的;保税的

bonded arch

~ arch 砌合拱
~ block 砌合方块,结合部件
~ brick work 咬合砌砖[体]
~ (clay) brick arch 规准黏土砖拱
~ carpet 粘结地毯,粘线地毯
~ coating 胶粘涂料
~ concrete overlay 结合式混凝土加厚层
~ joints 粘接结合面[接头],胶接
~ resistance 粘着抗力
~ roof 砌合屋顶
~ steel wire 粘着的钢丝
~ stone 束[系]石
~ store 保税仓库
~ structure 粘结强度
~ test 结合力[粘结]试验
~ timber 墙结合木,束木
~ tendon 灌浆的(预应力混凝土中的)钢筋束或钢丝束
~ with asphaltic material 沥青胶结剂
~ with synthetic (resinous material) 合成树脂胶合
~ wood 胶合木板
~ wood construction 胶合木结构
bonder [ˈbɔndə] *n.* 砌墙石,丁砖,锚石,连接石;粘结物,结合物;焊接工
~ brick 丁砖
~ course 丁砖层,系石层
~ wire (钢筋)绑接钢丝
bonderite *n.* 磷酸盐(薄膜防锈)处理层
bonderize [ˈbɔndəraiz] *vt.*,-ized, -izing;磷酸盐处理,深磷处理
bonderizing *n.* (钢丝或钢管拉拨前的)磷化[磷酸盐]处理
bonding [ˈbɔndiŋ] *n.* (砖石砌体)砌[结]合,加固;焊[连,粘]接,胶结;粘结料[剂]
~ adhesive (agent) 结合料,胶粘剂
~ adhesive (agent) based on coal tar 煤沥青粘结剂
~ adhesive (agent) for concrete 混凝土胶结剂
~ adhesive (agent) for laying 敷设[砌筑]粘结剂
~ additive 胶粘剂,活化剂防脱落作用剂[掺合料]
~ agent 胶粘剂
~ anchor 胶结锚固,连接拉杆
~ area 粘结[结合,附着,粘附]面

~ behavior 粘附[附着,粘结]性能
~ brick 束砖,空斗墙连接砖,接合榫砖
~ capacity 粘结[胶结,胶合]性能
~ cement 粘结胶泥,胶质水泥
~ coat 粘结涂层
~ composition 胶粘剂成分[组成]
~ compound 胶结剂成分[组成化合物]
~ concrete 胶结混凝土
~ course 结合[粘结]层,丁砖层
~ dispersion 胶结[粘结]剂分散(作用),弥散(现象)
~ emulsion 粘结乳剂
~ energy 粘合[结]能
~ exerior (external) plaster (rendering) 外部胶结粉刷
~ failure 粘着力破坏,结合损坏
~ finish 胶结粉刷[抹灰]
~ force 结合[联结]力
~ header 砌墙石,丁砖
~ header course 丁砖层,系石层
~ jumper 金属条,跨接线,搭接片
~ layer 粘结层
~ machine 焊接机
~ material 粘接材料
~ medium 胶粘剂
~ medium for concrete 混凝土胶结剂
~ medium for laying 砌筑胶结剂
~ method [砖]砌合法
~ metal 粘接金属
~ mortal 砌筑[胶结]砂浆
~ of brickwork 砖的粘结
~ of metal(s) 金属连[焊,粘]接
~ pad 焊接点[区],焊(结合)片,结合区
~ paper 胶粘纸
~ paste 水泥浆
~ performance 粘附[附着,粘结]性能
~ plane 砌结面,粘着面

~ plaster 粘结灰泥,胶结石膏泥
~ pocket 砖层凸出
~ point 结合点,焊点
~ process 粘结法
~ property (power, quality) 粘结[胶合,粘结]性能
~ putty 粘结胶泥,胶质水泥
~ rendering 外部粘结粉刷
~ resin 胶结树脂
~ rubber 胶结橡胶
~ strength 粘合强度,粘合力
~ strength of glue join 粘合强度
~ stress (混凝土与钢筋的)握裹力,粘结力,粘着应力
~ stucco 胶结外部粉刷
~ suspension 胶结悬浊[浮]液
~ tape 胶带
~ technique 焊接[粘合]技术
~ technology 焊接工艺
~ temperature 粘合[粘结]温度
~ tie 连接拉杆,锚固
~ test 结合力[粘合]试验
~ tool 焊头[具]
~ wire 焊线接合线
face down[up] ~ 侧[正]焊
hand ~ 钉头式焊
stitch ~ 自动点焊,跳焊
bondless *a.* 无粘性的,松散的,不结合的
bondline thickness 粘合[剂]层厚度,胶层厚度
bondstone ['bɔndstəun] *n.* 束[系]石,砌合石
~ course 系石层
bond tester *n.* 胶接检验仪
bondu *n.* 一种耐蚀的铝合金
bone [bəun] *n.* ;*v.* , boned, boning; *n.* 骨,骨状物,骨制品;炭质页岩,骨煤; bonelike, *a.*
~ bed 骨层,骨屑层
~ -black 骨灰(漂白剂),骨炭粉
~ black (涂)骨黑
~ charcoal 骨炭
~ china 轻质瓷器(掺骨灰的)骨瓷

~ coal 骨碳[煤]
~ dry 全干的
~ fat 滑脂
~ glass 乳白玻璃
~ glue 骨胶,牛骨胶
~ grease 骨润滑脂
~ tar 骨焦油沥青,骨柏油
~ tar pitch 骨沥青
boning *n*. 测平法,检验墙身垂度
~ board 板,测平杆,(T形)测平板
~ out 定直线
~ pegs 测平桩(修缘石材用)
~ rod 测平杆,水平尺,整坡杆
bonnet ['bɔnit] *n*. 机罩,阀盖,烟囱罩,保护罩,柱帽,阀帽,帽状物,闸门槽箱,(汽车)引擎盖;(房)遮风;*vt*. 戴帽,加罩;bonnetless, *a*.; bonnetlike, *a*.
~ body seal 阀帽体密封
~ gasket 阀帽全密封垫圈
~ headed door or window 外抱斜削门或窗
~ hip tile 屋脊弯[盖]瓦

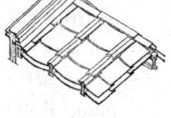
bonnet tile

~ tile (房)罩瓦,脊瓦
bononian stone 重晶石
bonstay *n*. 暗井,盲井
bont *n*. 提升装置(提升用的钢丝索及其附件)
bony ['bəuni] *a*., -ier, -iest;页岩质的(煤炭),骨煤的,劣质的(煤炭);boniness, *n*.
~ coal 骨炭
booby ['bu:bi] *n. pl*. -bies;笨蛋,呆子;boobyish, *a*.
~ hatch 精神病院;高出来的结构,有活节盖的小舱口;马拉篷车
book [buk] *n*. 书,书本,手册,说明书,支票;*v*. 注册,预定;-booklike, *a*.
~ case 书柜

~ clay 书页黏土
~ holder [固定书刊用]夹书架
~ mica 书页云母
~ of estimates 概算书
~ of reference 参考书
~ rack 搁书架
~ repository 书库
~ rest 阅书架
~ shelf 书架[柜]
~ shop 书店;正[主]房
~ slide 活动书架
~ stack 分层书架
~ stack room 固定书架室,书库
~ stand 书架
~ store 书店,书屋
~ table 书架桌
~ tile 带凹凸边的书形空心砖,(屋)脊瓦,平瓦
~ tower 多层图书馆
~ vault 书库,藏书室
hand ~ 手册,参考书
instruction ~ 说明书
bookable ['bukəbl] *a*. 可预购[订]的
bookcase ['bukkeis] *n*. 书架
~ drier 书柜式干燥器
booking ['bukiŋ] *n*. 预约
~ list 预定单
booklet ['buklit] *n*. 小册子,目录单
bookrack ['bukræk] *n*. 书架,阅览架
bookrest [buk'rest] *n*. 阅书架
bookshelf ['bukʃelf] *n. pl*. -shelves;书架,书橱
bookshop ['bukʃɔp] *n*. 书店
bookstand ['bukstænd] *n*. 书架,阅览架
bookstore ['bukstɔ:] *n*. 书店,书屋
book-structure *n*. 页状构造(岩)
boom [bu:m] *n*. 桁梁,构[托]架;弦[横,吊]杆;悬臂,起重臂;(喷混凝土的)机械手;栏木,浮栅,水栅,筏堰,拦河埂,水上航标;*v*. 麦鸣,隆隆声;繁荣,兴旺;-boomless, *a*.
~ and bucket delivery (混凝土)吊

杆与屑斗递送法,吊斗运送法
~ angle 翼缘角钢,箍用条,起重机吊杆的斜角度
~ brace 起重臂桁架式结构,吊臂斜杆
~ bracing 弦杆加固,加劲杆
~ brake 吊臂斜杆,起重臂衍架结构
~ bucket 悬臂式铲斗
~ cable 流材索,缆绳
~ crane 悬臂式起重机
~ crutch 吊杆托架
~ cylinder (装载机的)转臂油缸
~ -dragline 索斗(导索)式挖土机
~ element 弦杆
~ excavator 吊臂挖土机
~ extension 喷杆延长杆
~ head 吊臂上端
~ hoist 臂式吊车[绞车]
~ length 起重机悬臂长度
~ lift cable 起重机悬臂提升索
~ light 吊杆灯
~ line 悬臂绷绳
~ net 栅栏网
~ of arch 拱环,拱形弦杆
~ plate 翼缘板,弦板
~ rope 流材索,缆绳
~ scraper 臂式铲运机[刮料机,刮集器]
~ section (喷雾器的)喷杆组
~ stick 档木缆木
~ stiffening 弦杆加固,加劲杆
~ support 弦杆支柱,梁支架
~ support guy [挖土机]机臂,吊索
~ swing 吊臂回转角,吊机臂转动
~ swing angle 吊臂回转角
~ table 吊杆平台
~ tackle 吊杆滑轮组[索具]
~ type shovel 单斗式挖土机,机械铲
~ type trenching machine 梯形挖沟机,链斗式挖沟机
~ awning ~ 篷式桁架
~ bottom ~ 下弦,底桁
~ bracing ~ 撑梁,联弦杆

~ compression ~ 受压弦杆
~ crane 起重机吊架,吊臂
~ discharge ~ 卸料吊杆[悬臂]
~ gabion ~ 石笼挡栅
~ heavy [jumbo] ~ 重吊杆
~ loader ~ 横梁式装料机
~ log ~ 木浮栅
~ lower ~ 下弦
~ sectional ~ 分节起重臂,分段吊杆
~ spar ~ 主梁
~ tail ~ 尾桁
~ top ~ 上弦杆
~ wind-braceed ~ 抗风撑杆

boombrace ['bu:mbreis] *n.* 伸梁支架
boomer ['bu:mə] *n.* 自动木栅,自动闸门,放木闸门
boort *n.* 圆粒[黑]金刚石,金刚砂
boost [bu:st] *vt.* 升压,加强,加速;发展,提高,增加;推起,推上; *n.* 推动,后援,帮助
~ control 增压调节
~ fan 鼓风机
~ gauge 增压压力表
~ pump 推进泵
~ ed water supply 给[供]水,上水道
~ ing 电压升高,局部通风
~ ing transformer 增压变压器
booster ['bu:stə] *n.* 升[增]压器;局部通风机,鼓风机;升压泵,辅助泵;辅助装置;引爆剂,爆管;转插台
~ amplifier 升压放大器
~ fan 鼓风机,升压风机
~ heater 辅助[中间]加热器
~ installation 增[升]压站
~ pump 增[升]压泵
~ -type diffusion pump 增压式扩散泵
~ transformer 增压变压器
boot [bu:t] *n.* 进[给]料室,给料斗;水落管槽,屋面管(凸缘)套;梁托,浮砖;橡皮套,保护罩
~ water-line paint 水线涂料[漆]
booth [bu:θ, bu:ð] *n.* 摊子,小舍[室],棚,临时构筑物;工作间[室];舱,厢,邮

亭,公用电话亭
control ~ 控制室
paint spray ~ 喷漆房

bootless *a.* 无益的,无用的, ~ly, *ad.*
~ness *n.*

boracic [bəu'ræsik] *a.* 硼的,硼砂的,含硼的
~ acid 硼酸
~ acid flint glass 硼酸火石玻璃

borane *n.* (甲)硼烷,硼化氢

borascu *n.* 硼砂,焦硼酸钠

borate ['bɔːreit] *n.*; *v.*, -rated,-rating; *n.* 硼酸盐,任何含硼酸的盐; *vt.* 用硼酸[砂]处理,混合
~ flint 含硼火石玻璃
~ glass 硼酸盐玻璃
~ of lime 硼酸盐石灰

borated ['bəureitid] *a.* 含硼酸[砂]的;以硼砂或硼酸处理过的
~ concrete 含硼混凝土

borax ['bɔːræks] *n.* 硼砂,硼酸钠
~ glass 硼砂玻璃,四硼酸钠
~ concrete 硼砂混凝土

Borda loss 水管突扩大处,流速水头损失

border ['bɔːdə] *n.* 边缘[界],边界线;路缘,缘饰; *v.* 毗邻,接壤,交界;镶边;接
~ growth 路边绿篱(或植物栽培)
~ lace 装饰花边,衣着饰边
~ pen 绘图笔(画轮廓用)
~ plane 边缘槽刨
~ strip 边缘防护带
~ wire 绑钢丝

bordering *n.* 边界[缘],饰边,饰带;炮泥

borderlight *n.* 顶光,舞台上部横挂的照明灯

borderline ['bɔːdəlain] *n.* 边[界]线,国界;轮廓线; *a.* 边界上的,不明确的
~ curve 轮廓线,边界曲线
~ hole 边界孔

bordroom ['bɔːdərum] *n.* 煤房

bore [bɔː] *n.*; *v.*, bored, boring; *n.* (中心)孔,内径腔,炮眼,钻[扩]孔器;隧道,(施工的)隧洞,石门,涌潮 *v.* 打眼,钻探,开凿;打[穿,钻,镗,扩]孔;boreable,borable, *a.*
~ check 精密小孔测定器
~ diameter 内[净]孔径
~ diaphragm 螺旋桩墙
~ hole 钻孔,炮眼
~ log 钻孔柱状图,测井纪录[曲线]
~ pipe 钻管
~ size 内径
basic ~ 基孔
wind ~ (泵)进气管

bored *a.* 空心的,镗孔的,镗过的

borehole *n.* 钻孔,镗孔,炮眼;钻井
~ axial strain indicator 钻孔轴向应变计
~ cast-in-place concrete pile (混凝土)钻孔,灌注桩
~ deformation meter 钻孔变形计
~ diametral strain indicator 钻孔径向[横向]应变计
~ drilling machine 钻机
~ pressure 钻孔(或炸药)压力
~ sample 钻孔样品
~ sinking 钻孔
~ spacing 钻孔间距
funned-shaped ~ 斗形钻孔
precementation ~ 预灌浆钻孔

borer ['bɔːrə] *n.* 钻孔机,穿孔器,凿岩机,风钻,钻孔,镗工,打眼工,镗床,镗孔工具
breast ~ 胸压手摇钻,手摇钻
increment ~ 探[取]木钻
Percussion ~ 风钻,凿岩机;冲击式钎子[钻机]

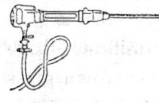
percussion borer

bore-sight *n.* 瞄准线,视轴枪(炝)筒瞄准

boric ['bɔːrik] *a.* 含硼的,含硼素的,硼衍生的
~ acid 硼酸

boring ['bɔːriŋ] *n.* 钻探,钻孔,打眼,钻凿;地质钻孔试验;(*pl.*)钻粉,切屑;

a. 钻[镗]孔的
~ and mortising machine 钻孔机
~ bit 活钻头,活钎头
~ by rotation 旋转式钻孔,旋转钻进
~ crown 钻头
~ depth 钻(镗)孔深度
~ direction 钻孔方向
~ machine 钻探机,打孔机,木钻床,镗床
~ mill 镗床,钻孔机
~ pattern 钻孔排列,钻孔图
~ system 钻孔[凿岩,打眼]方法
~ table 镗床工作台
~ time 钻孔[历时]
~ tool 钻具,镗孔车刀
~ work 钻孔,镗孔工作
water-flush ~ 水[冲]洗钻进
wet ~ 湿式凿岩,湿法打眼
borocalcite *n.* 硼酸方解石,硼钙石
boron ['bɔːrɔn] *n.* 硼(B)-boronic, *a.*
~ -coated 涂硼的
~ doping 掺硼
~ filament 硼丝,硼纤维(增强用)
~ glass 硼玻璃
~ hydride 氢化硼
~ -loaded 涂[含]硼的
~ -loaded concrete 含硼混凝土
~ oxide 二氧化硼
~ steel 硼钢
borosilicate ['bɔːrə'silikeit] *n.* 硼硅酸盐
~ crown glass 硅酸硼冕牌玻璃,硼硅酸盐冕玻璃
~ glass 硼硅(酸盐)玻璃,光学玻璃
borrow ['bɔrəu] *v.;n.* 借,借用,采用,借土,取土;借土坑,取料坑,挖出料,开采料;借位,借款,模仿;borrowable, *a.*; borrower, *n.*
~ light 借光窗,间接采光窗
~ed light 室内窗;间接光窗,采用光
~ed scenery 借景
~ing space 借景
boryslowit 硬地蜡
bosket ['bɔskit] = bosk *n.* 丛[矮]林,小树林
bosom ['buzəm] *n.* 对缝联接角钢,角撑,角板,连接板;内部,中间;胸
~ bar (piece) 衬角钢
~ knee 木肘材
~ -piece 对缝联接角钢,角撑
bosquet ['bɔskit] *n.* 丛[矮]林,小树林
boss [bɔs] *n.* 浮雕,凸饰,突出物,泥灰桶,进气道中心体; *vt.* 浮雕,指挥,控制; *a.* 主要的
~ flange 凸法蓝
~ hammer 大锤,锻工锤,碎石锤

boss

bossage *n.* 浮雕[凸面]装饰,毛面浮雕
~ without bevel 斜削浮雕装饰
bossed [bɔst] *a.* 有浮雕装饰的,凸起的(浮雕)
~ (on) both sides 两侧凸起的(有浮起装饰的)
~ (on) one side (only) 一侧凸起的(有浮起装饰的)
bosselated *a.* 有圆凸的
bosselation *n.* 小圆凸,圆凸
bossi work 大理石镶嵌技术
bossing *n.* 凸起部分,饰以浮雕;锤碾金属,金属片加工;轴包套
~ stick 加工木边槌
window ~ 窗腰
bosslike ['bɔslaik] *n.* 穹状,穹隆状
bossy ['bɔsi] *a.* bossier, bossiest; 凸起的,有浮雕装饰的,时髦的,穹状的

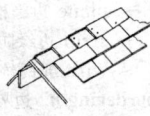

Boston hip

Boston ['bɔstən] *n.* 波士顿
~ hip 波士顿屋脊
~ rocker 讲究的摇椅
botanical [bə'tænikəl] *a.* 植物的,植物学的; botanically, *ad.*
~ garden 植物园
botel [bəutl] *n.* 凸型圆饰

both [bəuθ] *a.*; *ad.*; *pron.* 两,双,二者,双方
~ end threaded 两端带螺纹的
~ ends supported beam 两端支承梁
~ sides 两侧
bottle ['bɔtl] *n.*; *v.*, -tled,-tling; *n.* 瓶,小瓶,罐;容器;瓶子的容量;罩,外壳; *v.* 灌注,装入瓶中
~ air 瓶装气体
~ cement 瓶[罐]装水泥
~ green 深绿色
~ -nose drip 屋檐（瓶鼻式）滴水槽
~ -nose step 圆边踏步级
~ shaped pillar 瓶形柱墩
~ tight 束紧如瓶,密封的
~ trap 瓶形存水弯,瓶式曲管
bottled *a.* 瓶装的
bottom ['bɔtəm] *n.* 基础,地基,底[下,深]部,水底,尽头,末端;(*pl.*)河谷,沿河谷地;(*pl.*)底部沉积物,残留物,残渣,船,艇船体,坐垫,心,中心,原因,真相,船; *a.* 底下的,根本的,最低的; *v.* 做底脚,装底,到底,测量,深浅,查明真相;根据,基于
~ ash 底灰,底积粉煤灰
~ bar （梁板的）底[下]部钢筋
~ bed 底层,底板岩层
~ block 带钩滑轮,底座,窑（炉）底砖
~ board 底部,蕊板,下部搁板,船底活动垫板
~ bracing 底撑
~ bracket 底托架,下机架
~ break 底部断裂
~ brick 底砖
~ case 底座,下箱
~ channel 底沟
~ clearance 径向间隙
~ coat 底[基]层
~ contour 底部等高线,底部地形
~ course 底[基]层
~ (course) concrete 混凝土垫层,底层混凝土
~ covering 底面涂层

~ cut 底部掏槽,底槽,底部冲刷
~ diameter （螺丝）底径
~ die 底膜
~ -discharge pipe 泄水底管,底部泄水管
~ door 底排水孔,清扫孔,底[下]门
~ door rail 下横档
~ drain valve 底部排水阀
~ edge 下[底]缘
~ enamel 底漆
~ end 原木木头
~ exhaust duct 地下排气管道
~ facing machine 光面机(底部转动)
~ flange （梁）下翼缘,底凸缘
~ flange plate 底翼缘板,凸缘底板
~ floor 底板
~ force 下(阴)模
~ form (work)(shuttering) 底模
~ grid 热绝缘底槽板
~ guide 底部导向装置
~ guide track 底部导轨,导向角钢(闸门,大门)
~ lateral bracing 底部横向水平支撑
~ laterals 底部横向水平支撑杆
~ layer 底层
~ liner 底衬
~ of foundation 地基底板,基础底面
~ panel 底面镶板,踢脚板
~ plate 底[垫]板,底盖,基础底板
~ plug 泄水孔塞,钻孔[注水泥]下木塞,下塞
~ radiator bracket 散热器下部托座
~ rail 下横挡,下冒头,楼梯[阳台]扶手,栏杆

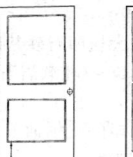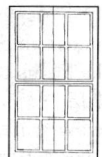

bottom rail(of a door)
bottom rail(of a window)

~ rail of door or of window 下冒头、门(窗)底边
~ rung 楼梯的最下段
~ seal (坞门等的)底部止水条封[封缄][闸门]
~ -set 底层
~ shore 底部支柱
~ side 底部
~ stand 机座,台柱,柱脚
~ step 起步级
~ stone 基石,底岩,耐火黏土
~ stratum 底层
~ surface 底面,底面积
~ surface dummy joint 底面收缩缝[假缝]
~ timber 底木
~ view 底视图
concrete ~ 混凝土底面

bocharde ['bautʃəd] *n.* 凿石锤,(石面)凿毛锤

bouche ['bautʃ] *n.* 钻孔,(枪炮)口; *v.* 钻孔

boucherie process (木材的)硫酸铜防腐法

boucherize *v.* 用蓝矾[硫酸铜]浸渍

boucherizing 落式差(硫酸铜)注入法,包肖尼树液置代法(木材防腐)

bought [bɔːt] *n.* buy 的过去式和过去分词;棚子;弯曲部,曲线,买来的,现成的

boulder ['bəuldə] *n.* 巨[蛮]石,大块石,漂砾;圆[拳]石,大卵石
~ buster 岩石破碎器
~ concerete 卵石混凝土,粗砾石混凝土
~ crusher 粗碎机
~ flint 卵石
~ foundation 大块卵石砂基础
~ -like inclusions (松软岩石中的)石砾状杂层
~ paving 蛮石[卵石]铺面
~ setter 蛮石铺砌层
~ wall 蛮石墙,漂砾墙,冰砾壁,大卵石墙

boulderhead ['bəuldə,hed] *n.* 堤坝前的一排桩

bouldering ['bəuldəriŋ] *n.* 大块石铺面(亦作 bowldering)

boule ['buːli] *n.* 通风木堆;刚玉,人造刚玉

boulevard ['buːlvɑːd]
n. 林荫大道,大街[道]

boulle [buːl] *n.* 镶嵌装饰,镶嵌的家具

boultin(e) *n.* 馒头形饰,凸圆饰四分之三椅柱[壁柱]

boulle

boulton process (木料)去湿法,热槽处理

bounce [bauns] *n.*; *v.*, bounced, bouncing;跳[振,摆]动,反跳,弹起,反射;
~ ing putty 弹性油灰,弹性油泥

bouncing ['baunsiŋ] *n.*; *a.*, 跳[振]动;跳钻,振裂;跳跃的,活泼的;大[重]的;反射式照明法;bouncingly, *a.*
~ putty 弹性油灰[泥]

bouncy *a.* 有弹性的,活跃的

bound [baund] *n.*; *a.*; *v.*, 界线,边界,范围;限制,束缚;受约束的;联[接,粘]合;跳跃,弹回;一定的,有责任的
~ edge 包边,缝边
~ energy 束缚能,结合能
~ medium 粘合介质
~ pile 装有钢筋的钢筋混凝土桩
~ water 束缚水,结合水,胶层面吸附水
~ ed edge 包[镶]好的边
~ ing bar 界边[缘周]角钢
~ ing capacity 粘结力
~ ing layer 边界层
bituminous ~ 沥青结合
cement ~ 水泥结合

boundary ['baundəri, 'baundri] *n. pl.* -ries;界线,边界,界面[标];边线[缘],范围,极限;限制,约束;(轧槽孔型的)轮廓,外形

~ beam 边梁
~ dimension 轮廓尺寸,外形尺寸,边界尺寸
~ face 边界面
~ fence 边界棚栏,边篱
~ gable 边山墙
~ layer 边界层,界面层
~ line of street 道路红线[边界线]
~ member 边缘构件(建筑结构力学)
~ surface 界面,边界面
~ wall 边界墙
~ wall bond 三顺一丁砌合
thermal ~ 热障
bounden *a.* 有责任的,必须担负的
boundless ['baudlis] *a.* 无限的,无边无际的; boundlessly, *ad.*; boundlessness, *n.*
bourg [buəg] *n.* (中世纪早期的)城镇,(法国)市集
bourn(e) [buən] *n.* 边界,界限,范围;目的地
bourock *n.* (苏格兰)茅屋,石堆
bouteillenstein *n.* 暗绿玻璃
boutel(l) *n.* 四分之三倚柱[壁柱]
boutique [bu:'ti:k] *n.* 镶嵌品,装饰品
bovey coal 褐煤
bow [bau] *n.* 拱,弓,虹,弧;弓曲物,锯弓,弯曲;船首,头部; *v.* 弯曲,屈服,压弯; *a.* 弓形的,螺形的
~ and string girder 弓弦式桁梁
~ beam 弓形式梁,弧形梁
~ crook (木料的)翘曲、弓形弯曲
~ girder 弓形大梁
~ member 弓形构件,拱架
~ room 弓形室
~ saw 弓形锯
~ -string girder 弓弦式桁架
~ up 上挠
~ warping 板材翘曲
~ window 凸窗,弓形窗
~ed height 矢[拱]高
~ing 弧状弯曲

bower [bauə] *n.* 凉亭,村舍,树荫处;-bowerlike, *a.*
bowl [bəul] *n.* 碗形结构建筑;(挖土机的)斗;离心机转筒,转子; *v.* 旋转,滚动,滑动;bowllike, *a.*
~ arrangement 碗[盆]状构造
~ capital 碗形柱头装饰,方块式柱头
bowlder ['bəuldə] *n.* 巨[蛮]石,漂砾;圆[拳]石,大卵石
~ shore 大卵石岸
bowline ['bəulin] *n.* 帆脚索,张帆索,结缆结,单结套,弓形线
bowsaw ['bəusɔ:] *n.* (锯曲线用的)弓锯,弧锯
bowstave ['bəusteiv] *n.* 弓材
bowstring ['bəustriŋ] *n.*; *v.*, -stringed of -strung,-stringing; *n.* 弓弦,铰索; *a.* 弓[弧]形的
~ arch(ed girder) 弓弦拱
~ beam 弓弦式大梁
~ girder 弓弦式大梁
~ (roof) truss 弓弦桁式屋架,组合桁架
bowtel(l) *n.* 凸圆饰,四分之三倚柱(壁柱)
box [bɔks] *n.* 盒,匣,箱;外壳,罩,套;窗框,接线盒,轴承箱,(剧场)包厢,公共电话亭,岗亭,邮亭,车厢,牲畜棚舍,【英】小屋或农舍;窗框上窗板可以关进的凹入部分,插销孔;混凝土箱涵;单元组件,部分;黄杨木; *vt.* 装箱,分隔
~ beam 箱[方]形梁
~ beam floor 箱形截面梁楼板[空心梁楼板]
~ bed 有栏杆的床,箱形底床
~ bolt 门锁插销
~ casing 平衡重滑道框
~ chamber 长方室
~ chisel 起钉凿
~ closure 接合螺栓,套环螺栓,系墙螺栓
~ column 箱形柱

~ connection 承插式接合
~ construction type 箱形结构形式
~ cornice 箱形飞檐
~ couch 内装储柜的床
~ design 箱形设计(结构,构造,建筑)
~ driver 套管螺钉起子
~ for mirrors 镜框
~ -form (built) steel column 箱形钢柱
~ forms (formwork, shuttering) 箱形模板
~ frame 箱形构架,隔成小间
~ frame construction 箱形框架结构
~ frame of window 匣形窗框
~ frame type reinforced concrete construction 箱形框架钢筋混凝土结构
~ frame type shear wall 箱形框架式抗震墙
~ girder 箱形(截面)梁
~ green 黄杨绿
~ heart 罩壳心材
~ -in 加边框式
~ key 套筒扳手
~ kiln 箱式干燥窑
~ lattice(d) girder 箱形格构大梁
~ level 圆水准器
~ lock 包壳锁
~ lumen meter 箱式流明[光通]计
~ nut 盖螺母,外套螺母
~ out (模板)留孔[洞]
~ pew 箱形教堂凳
~ pile 箱形钢桩,箱形桩
~ plate girder 箱形板梁
~ -ply (wood) portal (frame) 箱形层压板门式钢架(框架)
~ prestressed concrete structure 箱形预应力混凝土结构
~ profile 箱形断面
~ rib profile 箱形肋断面
~ scarf 直面搭接(木材的)
~ (seat) 剧院包厢席,运动场正面看台

~ section column 箱形截面柱
~ section foundation (structure) 箱形断面基础[基脚、底脚]
~ section frame 箱形框架
~ (section) girder 箱形(截面)梁
~ (section) string(er) 箱形(截面)楼梯斜梁
~ section (sheet) pile 有(外)壳(的)桩
~ settle 木制有扶手的高背长靠椅
~ -shaped column 箱形柱
~ sheeting 平板支撑,箱式支撑
~ shutter 箱形百叶窗板
~ sill 在木板框架中的砖或混凝土基础
~ slab 箱形板(桥梁)
~ slip 黄杨木护面条
~ spanner 套筒扳手
~ spring bed 箱式弹簧床、单垫式床
~ stair 箱形楼梯
~ staple (门挺上的)门锁槽
~ strike 门锁舌片
~ strut 箱门(截面)支撑杆
~ type 箱形的,箱形断面的
~ -type boom 箱形吊杆
~ -type construction 箱形结构
~ wrench 套筒扳手
~ed buttress wall 箱形墙垛
~ed dimension 总[全]尺寸,外形尺寸
~ed frame 箱形框架
~ed heart 罩壳心材
~ed mullion (门窗的)匣形竖框
~ed-off 隔成小间,钉板的,隔开的
~ed steel column 箱形钢柱
alarm ~ 警报装置
branch ~ 分线箱
conductor support ~ (垂直敷设)导线支撑盒
connecting ~ 联结器[箱]
control ~ 操纵台;控制部件
curtain ~ 窗帘匣

distrbution ～ 配电箱,电缆交接箱,分线盒
draft distributing ～ 分风箱
drain ～ 排水箱
four-way (tube) ～ 十字[四通]接头
panel ～ 配电箱[盘]
service ～ 厕所水槽;引入线箱;配电箱
sketch ～ 画箱
spool ～ 箱,盒;轴瓦盆;接头
switch ～ 开关箱,转换开关箱
terminal ～ 接线盒
T-junction ～ T 形套筒,T 形接头
boxboard ['bɔksbɔːd] *n.* 纸[箱]板,大型字体
boxiness ['bɔksinis] *n.* 呈四方状,无装饰或变化的方棱方角的性质
boxing ['bɔksiŋ] *n.* 装箱,做箱的材料;吊窗匣子,模[挡]板
～ arena (field) 拳击场
～ -in 安装,装配[置],砌入,浇注[筑]
～ shutter 箱形百叶窗板
～ tenon 框榫
boxroom ['bɔksrum] *n.*【英】(皮箱,旅行箱等的)收[贮]藏室
boxwood ['bɔkswud] *n.* 黄杨木材,黄杨木
boxwork ['bɔkswəːk] *n.* 箱状构造,蜂窝状网络,网格构造
boys gym(nasium) 体育馆,健身房
boziga *n.* 住[寓]所
bozzetto *n.*【意】初步设计的模型,泥稿,素材
brace [breis] *n.;v.*,braced,bracing; *n.* 支柱[撑,架],支持物;拉杆,系杆,手摇曲柄钻,曲柄;水湾,小港;张力;一双[对];(*pl.*)砌台(砖)体,结合,联结;吊带;*vt.*支持,连接,固定,拉紧
～ (and) bit 手摇曲柄钻
～ angle 撑杆角铁
～ bar 撑杆
～ drill 曲柄钻
～ head (key) 扳头

～ nut 拉条螺母
～ piece 联杆,加劲杆,联板
～ rod 联结杆,撑托
～ summer 支撑梁,双重机梁
～ table 支撑台(桌)
～ wrench 曲柄扳手
angle ～ 角铁撑,角撑
back head ～ 外门板拉条
batter ～ 斜拉条,斜撑
bit stock ～ [木工]摆钻
boom ～ 吊臂斜杆
buttress ～s 加劲梁,垛间支撑
corner ～ 角撑,齿轮传动的手摇钻
cross-arm ～ 角撑,悬臂撑
corss member ～ 横挡撑板
diagonal ～ 对角拉条;支叉撑条
expansion ～ 伸缩拉条
form ～ 装配支柱,模板支撑
hammer ～ 橡尾架斜撑
knee ～ 隅撑,膝形拉条
sash ～ 排架横撑
sill and bolster ～ 承梁拉条
steady ～ 固定销,定位销
wind ～ 抗风联杆,牵[拉]条
braced [breist] *a.* 加支撑的,联结的;拉紧的
～ arch 桁架形拱,拱形桁架
～ core building 格架式筒体建筑
～ door 联结门
～ drain 排水支管
～ frame 撑系[斜撑]框架,刚性构架
～ frame construction 撑系框架结构
～ framing 支撑系结构
～ girder 有刚性腹杆的梁式桁架,联结大梁
～ panel 斜撑节段
～ strut 联结支撑
～ structure 支撑结构
bracer ['breisə] *n.* 拉条,撑块;带,索;支持物
bracing ['breisiŋ] *n.* 支撑,系杆;联结,加劲[肋],交搭,拉紧; *a.* 使拉紧的;

-bracingly. *ad.*; -bracingness, *n.*
~ angle 加强角钢
~ beam 联结条
~ boom 加劲杆
~ cable 拉索
~ cage 支撑铁架
~ column 支撑柱
~ diagonal 连接[加强]斜杆
~ diaphragm 加劲隔板
~ frame 支撑架,加劲框架加固,构架
~ girder 加劲梁
~ member 加劲杆
~ piece 加劲[支撑]杆;斜撑;横[斜]梁
~ plane 支撑面,联结面
~ plate 撑板
~ ring 支撑圈
~ system 杆件系统,斜撑式
~ tied-arch 桁架式连杆拱
~ wire 拉铁丝
~ with verticals 竖杆联结式
external ~ 外拉;外部拉条
ladder ~ 桁架格条
radial ~ 径向支撑
push ~ 推杆
string ~ 纵梁联结系
brack *a.* 不合格的,有缺陷的(木材等);固体中的裂缝;土壤中的碱
bracket ['brækit] *n.* (墙上伸出的)托架,牛腿,隔撑;夹子,卡钉;(铸件)加强筋,筋条,*vt.* 装托架,分类
~ baluster 踏步栏杆柱
~ board construction 悬挑板结构
~ bracing 托座支撑,托架轴承
~ capital 伸臂柱头
~ column connection 梁柱托座结合
~ floor 组合[空心,构架]肋板
~ frame 肘板框架肋骨
~ joint 角接板,带角肘的结合
~ kness 梁肘

bracket

~ light 悬臂式壁灯
~ -like column 托臂式支柱,(具有柱)帽(的)柱
~ metal 支承用铁件,支架铁件,托架轴承(合金)
~ mount 托架,角撑架
~ post 托臂支柱
~ scaffold(ing) 挑出式脚手架
~ -step 挑出踏步
~ stringer 挑出的楼梯纵梁
~ support 托臂支座
~ed stairs 悬臂式楼梯
~ed step 悬臂式楼梯
~ing 托架[座],撑托
bearer supporting ~ 垫块,托座
cantilever ~ 悬臂加厚部;悬臂支托,牛腿
sight ~ 视准尺架
support(ing) ~ 托架
wall ~ 墙上托架;墙支架(装绝缘子用)
bracketless *a.* 无托架的,无肘板的
brad [bræd] *n.* 角钉,平[曲,无]头钉
~ nail 角钉,无[曲]头钉
~ punch 压钉器
~ setter 线脚钉夹钳
bradawl ['brædɔːl] *n.* 锥钻,打眼钻,小锥
brae [brei, briː] *n.* (苏格兰)小山;丘陵;倾斜处,斜[山]坡
Brahman style 婆罗门式(建筑)
braid [breid] *n.* 发辫交织河道;织物;*vt.* 纺织
~ed cable 编缆
~ed door 编竹门,栅栏
~ed rug 编织地毯
~ing 编织装饰
brake [breik] *n.*; *v.*, braked, braking; 制动器,闸,刹车,碎土[重型]耙;金属压弯成形机,丛林; *v.* 制动,刹车,减速
~ mechanism 制动机构
safety ~ 保险[安全]闸
branch [braːntʃ] *n.*; *v.* 枝,分枝[岔];

支路[流],旁路,扇形拱支肋;部门
~ box 分线盒[箱]
~ canal 支渠,支通路
~ centre 小区中心
~ drain 排水支管
~ duct 岔(支)道
~ ell 支弯管
~ factory 分厂
~ joint 三通接头,分支接头[连接]
~ knot 树节(木材缺陷)
~ pipe 支[套]管,三通管
~ road 支路
~ shaped timbering 隧道枝叉型柱支撑
~ to a building 进户支管[支线],用户引入线
~ type switch board 分立型配电盘
~ vent 通气支管
angle ~ 弯管,肘管
blow-off ~ 放水支管
canal ~ 迂回支道
connecting ~ 连接支管
delivery pipe ~ 输水[输气]支管
exhaust connecting ~ 排水[排气]接户支管
exit ~ 支管,出水支管,排出管
flanged ~ 法兰接头支管,凸缘支管
house ~ 入户管线
right angled ~ 直角支管
~ed work 树叶状装饰
branchy [ˈbrɑːnʃti] *a.* -ier,-iest;多枝的,分枝的
~ wood 多节木材,密枝木
brand [brænd] *n.* 商标,牌号,品种;部分被烧过的木材;*vt.* 打烙印,铭刻
~ mark 商标
~ name 商标名称
~ of cement 水泥牌号
famous ~ 名牌
brandering *n.* 板条木椽
brandreth [ˈbrændreθ] *n.* 三脚架,铁架,井栏
brandy [ˈbrændi] *n.* 白兰地酒

branner *n.* 磨光机,磨光工
brash [bræʃ] *n.* 碎石,瓦砾,崩解石块,脆性;*a.* 脆的,易破碎的
~ glass 碎玻璃
~ wood 脆木
brashy [ˈbræʃi] *a.*, brashier, brashiest;脆的,易碎的;brashiness, *n.*
brasq(ue) [bræsk] *n.* 填[衬]料,耐火封口材料
brass [brɑːs] *n.* 黄铜,黄铜制品明亮的金属器材(装置,金属用具,装饰品);*a.* 黄铜制的,黄铜[色]的
~ and bronze 有色金属
~ brazing (黄)铜(钎)焊
~ brazing alloy 铜焊焊料,[黄]铜焊合金
~ casting 黄铜铸件[造]
~ cock angle 铜弯嘴旋塞,黄铜角旋塞
~ dowel 黄铜销钉
~ fittings[hardware] 黄铜配件,小五金
~ jacket 黄铜套
~ plummet 铜线锤
~ sheet 黄铜板[皮]
~ smith 铜活工(人),铜匠
~ solder 黄铜焊,钎焊
~ -ware 黄铜器皿
~ work 铜制品
~ yellow 铜黄色的
brassing *n.* 黄铜铸件,镀黄铜
brassy [brɑːsi] *a.*, brassier, brassiest;黄铜的,黄铜色的,似黄铜的;brassiness, *n.*
brattice [ˈbrætis] *n.*; *v.*, -ticed, -ticing;临时木建筑,(保护机械的)围板;(矿坑通气的)隔间板壁,隔墙,风障
~ cloth (黄麻或粗亚麻)风障,[帘]布
~ sheet 风障[幕]
~ wall 隔[风]墙
air ~ 风帘
bratticing *n.* 花格装饰,墙顶饰,围板,风障,隔风墙

brattishing 透空花格
bray [brei] vt. 捣碎,捣成粉
~ stone 碎石,多孔砂岩
braze [breiz] vt., brazed, brazing; 用黄铜制造[镶饰],用铜(锌合金)焊接
~ over 镀黄铜
~ welding 钎接焊
~d copper tube 黄铜管
~d joint 黄铜接头,硬钎焊接
~d nipple 黄铜喷嘴[管]
sandwich ~ 夹心[层]焊接
brazen ['breizn] a. 黄铜的,黄铜制的; brazenly, ad.; brazenness, n.
brazier ['breizjə] = brasier n. 黄铜匠;火盆,焊炉
~ head rivet 扁头铆钉
~ head screw 扁头螺钉
brazil ['brəzil] n. 巴西木,巴西红色(红色植物染料)
brazilette n. 巴西木的心材
Brazilian [brə'zilzn] a. 巴西(人)的; n. 巴西人
~ pebble 石英,水晶
brazilwood [brə'zilwud] n. 巴西木,苏木
brazing ['breiziŋ] n. 硬(铜)焊
~ alloy 钎焊合金
~ apparatus 铜焊工具
~ burner 铜焊焊枪[喷嘴]
~ filler metal 钎料
~ solder 铜焊料
brea n. 沥青,沥青砂,矿物焦油;树脂
~ bed 沥青砂土层
breach [bri:tʃ] n. 裂缝,堤坝口;小溪,小海峡;违背,毁约,破坏; vt. 攻[冲]破;违反
~ of contract [promise] 违约,违背合同
breadth [bredθ] n. 宽[广,跨]度;幅,横幅;外延
~ of road 路幅
~ of weld metal 焊缝宽度
break [breik] n.; v., broke, broken, breaking;裂缝,裂口,断层,破坏,断[开,路,电]
~ a contract 违背合同
~ angle 错接角(焊)
~ down point 断裂点,破损点,破强度
~ iron (刨刀的)护铁
~ joints 断[错]缝,断裂节理
~ of fore castle 首楼后端
~ of joint 错缝
~ of poop 尾楼前端
~ -off point 折裂点
~ scraper 裂缝探测器
~ shovel 松土铲
~ -wind 防风林[墙];幕
cross-~s 横折,折纹
smooth ~ 平滑断裂
breakable ['breikəbl] a. 易碎[断]的,脆的; n. 易碎品,脆性物
breakage [breikidʒ] n. 破损(处);(材料)破裂碎片;损耗量,破损赔偿费;断路[线]
breakaway ['breikə,wei] n. 脱[分]离,断开;破裂
breaker ['breikə] n. 破[轧]碎机,碎石机;断路器,开关器;(横跨路面的)排水小沟;汽车护胎带
~ bolt 安全螺栓
~ box 开关箱
line ~ 线路开关
breakfront n. 凸肚型橱柜(书柜)

breakfront

breaking ['breikiŋ] n. 破[粉,轧]碎,破坏,断路,切断;分解[层]
~ down 断裂,分解,开木料
~ in 夯实
~ joint(s) 隔层接合,参差接缝,间砌法
~ of joint 错缝
~ staggering to joints (混凝土板的)错列接缝

~ through 穿透

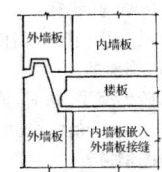

breaking staggering to joints

breakjamb *n.* 门窗框边挺

breakline *n.* 折断[断裂]线；接缝，对接线

breakout ['breikaut] *n.* 中断，爆发，崩落；烧穿，提升钻杆，钻杆起卸法
~ material 易磨损材料
~ tongs 大管[吊]钳

breakover *n.* 圆脊；穿通，转折[换]

breakstone *n.* 碎石

breakthrough ['brei'θru:] *n.* 突破，穿透；临界[转折]点；技术革新，重要发明

breakup ['breikʌp] *n.* 破[分崩，断]裂，中断，缺口；分解，溶化，停止，完结

breakwater ['breikwɔ:tə] *n.* 防波堤，挡水板
~ -glacis 破浪石砌体（桥墩式护岸），防波堤斜坡
~ with sloping faces 斜坡式防波堤
~ with vertical faces 直墙式防波堤
mound ~ 堆石防波堤

breakwind *n.* 防风墙，挡风（罩）

breast [brest] *n.* 胸墙；样底；器物侧面，（扶栏、梁的）下侧，窗下墙，椽木底，窗腰；*v.* 逆……而进
~ beam 腰梁
~ bench 锯架
~ height 胸墙高度，女儿墙高度
~ lining[(sur)facing] 胸高护墙（面）饰窗台线
~ moulding 窗盘线条，窗下墙线脚
~ panel 窗下墙，窗下护墙板
~ rail 腰栏
~ summer 过梁，托墙梁，横楣
~ wall 胸墙，挡土墙
~ work 胸墙（矮的防卸墙）；壁炉胸

breastplate ['brestpleit] *n.* 胸板，挡风板

breastrail ['brestreil] *n.* 腰栏，（窗前的）栏杆

breastrope *n.* 安全索

breathability *n.* 透气性

breather ['bri:ðə] *n.* 通气装置[设备]，通气孔[筒]
~ pipe 通气管
~ place 休息处
~ valve 通气阀，呼吸阀
~ vent 通气口

breathing ['bri:ðiŋ] *n.* 进气，排气；呼吸；片刻，*a.* 活的
~ ability 通气性能
~ apparatus 通气[供气]设备
~ hole 通气孔
~ pipe 通气管
~ property 透气性
~ space 休息场所，休息时间
~ valve 通气阀

breccia ['bretʃiə] *n.* 角砾岩
~ marble 角砾大理石

breech [bri:tʃ] *n.* (水平)烟道
~es fitting 叉（形）管
~es pipe 叉（形）管

breeze [bri:z] *n.* 煤渣（建筑材料）；矿粉；垃圾；微(和)风
~ block 焦渣石，煤渣砌块
~ brick 焦[煤]渣砖
~ concrete 焦渣混凝土

breezeway ['bri:zwei] *n.* 连接两座房屋之间的走廊[阳台]

bressummer *n.* (=breast) 托墙梁，大木，过梁，横楣

briar ['braiə] *n.* 横截锯
~ dress 拨料
~ tooth （锯的）钩齿，偏泽齿

brick [brik] *n.* 砖，块料，方木材；*a.* 砖造的，用砖(铺)砌的；*vt.* 砌砖；bricklike，*a.*

~ aggregate 黏土砖骨料
~ and-a-half wall 一砖半厚墙（一丁加一顺砖的厚度）
~ and-stud 砖木墙壁（木架砖壁）
~ arch 砖拱
~ arch floor 耐拱楼盖
~ architecture 砖砌建筑物
~ axe 劈砖斧,瓦工锤
~ backing 砖衬背
~ baffle 砖隔屏
~ barricade 砖屏
~(-)bat 砖片[块],半砖
~-bat drain 碎砖盲沟
~ beam （配筋的）砖过梁
~ bed (course) joint 砖砌层水平接缝,砖接缝
~ block 砖隔墙
~ bond 砌砖体,砖圬工
~ bonder (bonding header, bondstone) 丁砖,束砖[石],接合砖
~ boundary wall 边界砖墙
~ building 砖结构建筑物
~-built 砖砌体
~ burning 砖的烧制,黏土煅烧
~ cap parapet 砖顶压檐墙,砌压顶女儿墙
~ casing 砖工刷灰
~ cathedral 砖圆层顶,砖穹顶
~ cavity [hollow] wall 砖砌空心墙
~ chisel 修砖錾
~ clay （制砖用）黏土
~ clad building 砖镶房屋
~ coffering 衬砖井壁
~ construction 砖石结构[工程建筑物]
~ core 砖砌拱模
~ course 砖层
~ cube 方砖块
~ cube pavement 方砖块路面
~ culvert 砖（砌）涵（洞）
~ curb 砖缘石,砖道牙
~ divelling 砖建住宅

~ engineering 砖石工程
~ fabric 砖结构[建筑]
~ facade 砖立面[正面,门面]
~ faced 砖饰面的
~ facing 砖砌面层[饰面,护面]
~ floor (cover(ing)) 砖地,铺砌地面
~ flue 砖砌烟道
~ footing 砖基础[底脚]
~ for wedge use 楔形砖
~ gable 砖砌山墙
~ grille 砖砌花格[花格窗]
~ hammer （砌砖用）尖尾手锤
~-in 砖衬的
~ inclusion 夹（入）砖
~ joint 砖层接缝
~ layer 砌砖工,瓦工；砖层
~ layer's trowel 砖工镘
~ laying 砌砖
~ lining 砖（内）衬
~ lintel 砖砌过梁（窗上）
~ mason 砌砖工
~ masonry (work) 砖圬工,砖体
~ masonry structure 砖石结构
~ mortar 砌砖砂[灰]
~ mo(u)ld 砖砌线脚
~ moulded sill 砖砌窗台
~ nogged timber wall 砖填木构架墙
~-nogging 砖填木架隔墙,木架砖墙
~-nogging building 立贴式房屋,木架砖壁房屋
~ on-edge 侧砌砖
~-on-edge coping 侧砖压顶
~-on-edge course 侧砌砖层
~-on-end 竖砌砖
~-on-end soldier arch 竖砖拱
~ panel 预制砖板,护墙砖块
~ panel building unit 护墙砖[预制砖板]建筑单元
~ pier 砖墩
~ plinth 砖底座[勒脚]
~ red 红砖色,桔红色

~ reinforcement fabric 砖砌体钢筋(垫)网
~ relieving arch 砖砌辅助拱
~ rib 砖肋
~ rubble 碎砖骨料,粗面砖块
~ sculpture 砖雕
~ set on edge 侧砌砖
~ setting 砖砌体,砖工,砌砖
~ shaft 砖砌柱身
~ shaped 砖形的,长方形的
~ -shapes 异形砖
~ spread foundation 砖砌大方脚
~ step 砖踏步
~ stopping 砖砌隔墙
~ structure(s) 砖石结构[工程,建筑物]
~ (sur)facing 砖砌面层
~ system building 预制装配式砖石建筑
~ tracery 砖砌窗花格
~ trimmer 砖托梁
~ towel 灰镘(砌石砖镘)砌砖镘刀,灰铲,泥抹子
~ tunnel vault 砖砌筒形穹顶
~ veneer 砖砌镶面层
~ vennered 砖镶面的
~ wall 砖墙
~ wall arch 砖砌实心拱
~ wall fence 砖围墙
~ walling 砖砌井壁
~ ware 陶砖[块]
~ with groove 带槽砖
~ work joint 砖缝
abrasive ~ 耐磨砖;光锉,磨块
acid-proof ~ 耐酸砖
air ~ 空心砖,干砖坯
alumina ~ 高铝砖
backing ~ 充填砖
basic ~ 碱性砖
bauxite ~ 矾土砖
black ~ 青砖

blue ~ 青砖
bonding ~ 束砖;(空心墙两壁间的)系砖
bottom ~ 炉底砖
breeze ~ 焦渣砖
capping ~ 檐口砖
cavity ~ 空心砖
cell insulate ~ 绝热砖
cellular ~ 蜂窝[空心]砖
cementitious ~ 粘合砖;轻[浮]石水泥砖
chamber ~ 火泥砖
checker ~ 格子砖,格砖
checker fire ~ 火口砖
clay ~ 黏土砖
coping ~ 履面砖
cored ~ 含孔砖
cork ~ 软木砖,多孔砖
cove header ~ 切头砖
crown ~ 拱顶砖,楔形砖
curved ~ 曲面砖
diatomite ~ 硅藻土砖
dinas ~ 硅石砖,硅砖
dolomite ~ 白云石砖
double standard ~ (宽度等于标准砖宽的二倍)倍宽砖
Dutch ~ 高温烧结砖,缸砖
edge ~ 角砌砖
emery ~ 金刚砂石,磨[油]石
enamelled ~ 釉瓷砖
external octagon ~ 外八角砖
feather-edge ~ 薄边砖
Figure-drawing brick of Han hynesty 汉书象砖
fire ~ 耐火砖
fireclay ~ 火泥砖(高碱性耐热的)
fixing ~ 软性砖
Flemish ~ 缸砖
fletton ~ 橙黄色页岩砖
foamed slag ~ 泡渣砖
ganister ~ 硅石砖
gault ~ 白垩砖

glaze-covered ~ 釉面砖
glazed ~ 釉面砖
grout-lock ~ 灌浆槽孔砖
hard stock ~ 耐火砖
high refractory ~ 高级耐火砖
insulating ~ 绝缘[隔热]砖
insulating fire ~ 隔热火砖,消防砖
internal bull-nose ~ 内凹砖
internal octagon ~ 内八角砖
inwall ~ 内皮(耐火)砖,内壁砖
jumbo ~ 大砖(普通尺寸为10.2厘米×10.2厘米×30.5厘米)
kiln-marked ~ 窑斑砖
lath ~ 条形砖,板条砖
lime ~ 石灰砖,硅酸盐砖,石灰硅砖
lime-sand ~ 白灰砂砖
limstone ~ 石灰岩砖
lining ~ 衬砌砖
lug ~ 有耳砖,带耳砖
main grizzle ~ 白垩黏土软性砖
metal-cased ~ 金属夹圈砖(耐火材料)
mullite ~ 高铝红柱石砖,莫来石砖
multicolour ~ 五彩砖
notched ~ 开槽砖
perforated cellular ~ 多孔空心砖
plinth ~ 基座砖
quartzit ~ 石英岩砖
radial ~ 辐向[中心]砖
refractory ~ 耐火砖
repressed ~ 加压砖
Rhenish ~ 轻质硅酸盐砖
salt-glazed ~ 瓷砖
sand finished ~ 沙面砖
sanding ~ 刚砂石
sand-lime ~ 灰砂砖
side-cut ~ 削边砖
siliceous ~ 含硅砖
skew ~ 斜削砖
skintled ~ 墙面斜交砖
slag ~ 煤渣砖,(炉)渣砖
soft ~ 软性砖

soft burnt ~ 软烧砖
standard ~ (尺寸为6.4厘米×13.3厘米×20.3厘米)标准砖
super-duty ~ 超级耐火砖
trapezoidal ~ 梯形砖
triple ~ (尺寸为10.2厘米×16.9厘米×30.5厘米)三倍砖
tubing ~ 管砖
twin ~ (尺寸为7.6厘米×26.7厘米×26.7厘米)双倍砖
veneered ~ 釉面砖
vertilating ~ 空心通风砖
vitrified ~ 陶砖
voussoir ~ 拱用砖,楔形砖
water-struck ~ 水湿模砖
brickbat ['brikbæt] *n.* 碎砖,砖片,[砖块]
 ~ drain (填)碎砖排水[盲]沟
bricking *n.* 砌砖,砖工,砖衬
 ~ ring 砖壁座
 ~ scaffold 砌砖吊盘
 ~ -up 用砖塞孔,砖衬
bricklayer ['brikleiə] *n.* 泥瓦工,砌砖工,砖瓦匠
 ~'s charge hand 泥瓦工工头
 ~'s hammer 瓦工锤
 ~'s ladder 木工梯
 ~'s line 泥瓦线,砌砖工线
 ~'s scaffold(ing) 砖工脚手架
 ~'s tools 泥瓦工工具
 ~'s trowel 砖工镘刀
 ~'s work 砖工工作,圬工
bricklaying ['brikleiiŋ] *n.* 砌砖
brickmaker ['brikmeikə] *n.* 制砖工
brickmason *n.* 砌砖工
bricknog (= bricknogging) *n.* 砖[填]木[架]墙壁
birckwork ['brikwə:k] *n.* 砖墙,砖圬工,砖砌工程,砖工;(*pl.*)砖瓦厂,砖瓦工程
 ~ arch 砖拱
 ~ base 砖座[基]
 ~ bond 砖砌体,砖圬工

~ casing 砖块镶面,砖砌面层
~ castle 砖砌城堡
~ corbel 挑出砖牙
~ fire place 砖砌壁炉
~ gable 砖砌山墙
~ joint 砖(接)缝
~ mortar 砖砌筑砂浆
~ pointed arch 砖砌二内心挑尖拱
~ system building 预制装配式砖石建筑
foundation ~ 基础砌砖
gauged ~ 清水砖工
intermediate ~ 中间砖墙
reinforced ~ 钢筋砖圬工
bricky ['briki] *a.* 砖砌的,似砖的(尤指颜色上)
bridestone *n.* 砂岩大圆石
bridge [bridʒ] *n.*; *v.*, bridged bridging; *n.* 桥[梁],电桥; *vt.* 架桥,跨越[接],接通
~ board 楼梯梁,斜梁
~ gallery 过街楼
~ -on-edge 侧砖
~ on-edge-paving 侧砖铺砌
~ -on-end 竖砖
~ -on-end soldier arch 竖砖拱
~ piece 过梁,连接弯管,连接件,(车床的)过桥
~ plate 装卸跳板,架接板
~ ramp 跳板
~ rubble 碎砖
~ screw 夹板螺钉
~ setting 砖砌体,砖工
~ spot welding 带接合板的点焊
~ welding 带接合板的焊接,桥接焊
bridgeboard ['bridʒbɔːd] *n.* 楼梯梁,楼梯帮,斜梁
bridgeway *n.* 桥上道路,楼间架空通道
bridging *n.* 架桥,跨接,桥接
~ effect (对裂缝)遮蔽作用
~ piece 挑(梁)板
~ run 便桥式脚手架
bridle ['braidl] *n.* 托梁,承接梁,拉紧器,

板簧夹;系船索;*v.* 抑制,控制,拘束
~ bar 系杆
~ cable 短索缆
~ chain 悬挂罐笼的保险链
~ joint 啮接(木材)
~ rope 吊绳

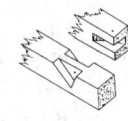
bridle joint

Brig 布里格
~'s standard pipe thread 布氏标准管螺纹
bright [brait] *a.* 光[明]亮的,光滑的,光[泽]的;鲜[透]明的;灿烂的;辉煌的;生气勃勃的,欢快的;聪明的,伶俐的
~ bolt 精制螺栓,光(制)螺栓
~ cherry-red 樱桃红
~ finish 抛光
~ glaze 光泽釉
~ red 鲜红色
~ sap 净面树材,无皮边材
~ -work 五金器具
brighten ['braitn] *v.* 磨[擦,抛]光;使发亮[光]
brightener *n.* 抛光剂,光亮剂,增白剂,增艳剂
brightening *n.* 照明;发光,擦亮,磨光,增亮
brightness ['braitnis] *n.* 光泽,辉煌;明亮,鲜明;亮度,照度
~ contrast 亮度比
~ degree 亮度等级
~ luminance 亮度
~ of daylight 日光的亮度
~ of window surface 窗口亮度,采光口亮度
~ ratio 亮度比
~ scale 亮度标准
brightwork ['braitwəːk] *n.* 五金器具,抛光或镀光的金属制品;光亮的木制品;精加工
brilliance ['briljəns] *n.* 光泽,[亮采],鲜艳,漂亮;光辉,灿烂,辉煌;卓越;辉[亮]度

point ~ 光点亮[辉]度
brilliancy ['briljənsi] *n.* 亮[辉]度;光泽[彩],鲜艳;辉煌,灿烂,卓越;多面型宝石光
brilliant ['briljənt] *a.* 辉[亮]度的;明亮的,辉煌的,卓越的,(重发高音)逼真的; *n.* 宝[钻]石,多面型
~ -cutting 多面型修饰
~ polish 抛光剂
~ varnish 发亮清漆
~ white 炽[亮]白光
brim [brim] *n.* 井栏,缘,边; *v.* 装满,满到边
brimstone ['brimstən] *n.* 硫磺,硫磺石
brindle ['brindl] *n.* 变[杂]色的;(*pl.*)劣砖,二级砖(颜色差的)
brindled ['brindld] *n.* 有斑纹的
~ brick 起条纹的次砖
brine [brain] *n.* 盐[卤,海]水; *v.* 用盐水处理
~ -proof paint 防盐水漆
bring [briŋ] *vt.* brought, bringing;引起,产生;招致,促命名;拿[带]来
~ to grade 定坡,填筑到设计线
~ up 水准测量,(砌墙,砌砖)找平
brise-bise [bri:zbi:z] 半截式花边窗帘
brisk [brisk] 活泼的,轻快的;清新的
bristle [brisl] 硬毛,猪鬃
~ brush 硬毛刷
Britannia [bri'tænjə] 不列颠,英国
~ joint 不列颠式焊接(锡铜锑合金焊接),英式焊接
~ (metal) (不列颠)锡铜锑合金
British ['britiʃ] 英国的,英国人的,大不列颠的
~ gum 糊精,淀粉胶
~ sandard(BS) 英国标准
~ system 英制
~ system of units 英国单位制
brittany (=blue)布列塔尼蓝色
brittle ['britl] *a.* 脆(性)的,易碎的
~ alloy 脆性合金
~ behavior 脆性

~ fractural face 脆裂面
~ heart 木材节疤,脆心材
~ lacquer 脆性漆[涂料]
~ material 脆性材料
~ metal 脆性金属
~ rock 脆性[易碎]岩石
~ varnish 脆性漆
brittleness ['britlnis] *n.* 脆性[度]
blue ~ 蓝脆[性]
red ~ 热脆
broach [brəutʃ] *n.* (教堂的)尖塔,(石工)宽凿;三角锥,铰刀,剥刀,钻孔器,卷筒,铁叉 *v.* (石工)粗刻,砍平;开口,打眼,扩[绞]孔,拉削;提出,讨论
~ post 桁架中竖杆
~ roof 尖塔屋顶
~ spire 尖塔顶,锥形尖塔
~ taper 钻,锥,铰刀
~ed work 在石面刻对角线槽
~ing 拉削,扩[铰]孔
~ing bit 铰刀,扩孔钻头
broad [brɔ:d] *a.* 宽的,宽阔[广]的,广泛的,主要[概括,充足]的; *n.* 宽处,河流的宽阔部分,湖沼受淹的低注地;扩孔刀具 *ad.* 宽阔地
~ axe 阔[宽]斧,木工[用]斧
~ beam 宽缘梁
~ chisel 平凿
~ daylight 全日(光)照,白昼
~ flange beam 阔翼梁
~ flang(ed)(I)beam 宽缘工字钢
~ -footed rail 宽底轨
~ hatchet 短柄宽斧
~ knife 铲刀
~ leaved tree (wood) 阔叶树
~ planning 大区规划,初步规划
~ stone 琢石,铺面,大石板
~ tooler 阔凿
broadax(e) ['brɔ:dœks] *n.* 宽[劈,战]斧
broadcast ['brɔ:dka:st] *v.;n.* 广[传,撒]播 *a.* 广[撒]播的 *ad.* 以广播的方式,撒播

~ room 广播室
~ sodding 泛铺[满堂]草皮
~ station 广播电台
~ studio 播音室
broadcasting ['brɔːdkɑːstiŋ] *n.* 广播, 撒布,撒播
~ center 广播中心
~ house 播音馆[室]
~ room 广播室
~ station 广播电台
~ studio 广[演]播室
broaden ['brɔːdn] *v.* 加[放,变]宽;扩展,使扩大
broadside ['brɔːdsaid] *n.* 侧视图,船舷;侧边[面],横向 *ad.* 侧向地,无计划地
broadsiding *n.* 侧移
broadstep *n.* 楼梯踏板,楼梯(休息)平台
broadway ['brɔːdwei] *n.* 宽阔的道路,公(大)路;(*pl.*)横向
brob *n.* 钩头钉,对接木端固定钉;支柱,斜撑
brocatel(le) *n.* 彩色大理石(黄色带深红色条纹)
broch *n.* (苏格兰)月圜=(奥克尼群岛,设得兰群岛以及苏格兰本土的一种史前的)圆形石塔
brog *n.* 曲柄[手摇]钻,钻子
broggite *n.* 褐地沥青
broken ['brəukən] *a.* 断[破,碎]的,不完整的;凹凸不平的,崎岖的;破产的,倒闭的
~ aggregate 碎砖骨料
~ arch 缺口拱,装饰拱
~ ashlar 不等形琢石块
~ ashlar masonry 不等形琢石圬工
~ bond 不规则砌缝
~ brick 碎砖
~ bricks 碎砖骨料
~ -brick base 碎砖基层
~ brick concrete 碎砖混凝土
~ brick concrete filler brick 碎砖混凝土填充块体

~ circuit 断路
~ color 复色,配合色
~ concrete wall slab 碎砖混凝土墙板
~ concrete 混凝土碎块
~ corner 角裂
~ course 十字砌合层(砌体)断裂(序);断层,隔断砌层
~ granite 碎花岗岩
~ gravel 碎砾石
~ joint(s) 错缝接合,错列接头,间砌法,断缝
~ line crown 折线型路拱
~ marble 碎大理石
~ -out section 破裂断面,切面
~ -out section view 局部断面视图,局部剖视图
~ pediment 缺口三角楣饰,断裂人形山头
~ range masonry 断层石砌
~ rock 碎石
~ -rock pavement 碎石铺砌层,碎石面层

broken pediment

~ seam 间断缝
~ stone 碎石,石碴
~ stone chip 碎石碴
~ -stone pavement 碎石面层,[铺砌层]
~ stone sand 混凝土细砂
bronze [brɔnz] *n.* 青铜,青铜制品,青铜色,青铜色涂[颜]料 *v.* 镀青铜,上青铜色,晒黑
~ -age building 青铜时代建筑
~ -age Palace 青铜时代宫殿
~ alloys 青铜合金
~ anchor 青铜锚固
~ annulet 青铜圆箍线(枯环饰)
~ blue 铜蓝色
~ door 青铜门
~ grille 青铜花[格子窗格栅]
~ hinge 青铜铰链

~ mirror 铜镜
~ mo(u)ld 青铜造型,青铜线脚
~ relief 青铜浮雕
~ (roof)tile 青铜瓦
~ shaff-ring 青铜柱环饰,环形线脚
~ sheet panel 青铜金属护墙板(镶板)
~ statue 青铜像
~ surround 青铜镶嵌[包裹]
~ swing door 青铜双式弹簧门
~ tripod 铜鼎
~ unit [profile section, shape, trim] 青铜外形
~ welding 青铜焊(以青铜为焊接剂)
~ wire 青铜线
~ yellow 古铜黄色

Bronze Age 青铜器时代

bronzy ['brɔnzi] *a.* 青铜(褐)色的,青铜一样的

brooder [bru:də] *n.* 暖房[暖棚],育雏房

brooklyn suspension bridge (纽约)勃洛克林悬索桥(美国)

broom [bru:m] *n.* 路刷,扫帚 *v.* 扫除;用锤击木桩使顶部散裂,松碎
~ closet 清洁工具柜(橱)
~ finish 刷清路面;扫处理,扫面;扫帚痕纹饰面
~ finishing 扫毛
~ head of pile 桩的蓬裂顶部,桩头开花

broomstick ['bru:mstik] *n.* 帚柄

brother ['brʌðə] *n.* 兄弟,同胞
~ chain 吊链
~ of the brush 画工,油漆工

brow [brau] *n.* 边线[缘];悬崖,山顶,陡坡;集材场,装车场;泄水槽,眉(檐)板,眉棱(木材缺陷);井口;眉,眉毛
~ piece 大块木材

browing *n.* 抹灰的垫(中)层

brown [braun] *a.*;*n.* 褐色,棕色,棕色颜(染)料;褐色,棕色,晒黑
~ coat 二道抹灰

~ george 棕色大陶罐
~ glass 棕色玻璃
~ -glazed brick 褐釉砖
~ (iron)oxid(e)(pigment) 褐色铁氧化物颜料
~ lime 次[褐]石灰
~ madder 褐红
~ ochre 赭石,褐铁矿
~ ore 褐铁矿,类矿石
~ oxide of iron (pigment) 褐色铁氧化物颜料
~ paint 褐色厚漆
~ pigment 褐色颜料,褐色素
~ pine 高罗汉松
~ powder 有烟火药,黑火药
~ rof 褐[棕]色腐朽
~ salt glaze on pottery 陶器用褐色釉
~ sienna 赭色
~ soil 棕钙土
~ stain(ing) 棕褐色变(砌体灰缝)
~ stone 褐(色砂)岩
~ stone district 高级住宅区(19世纪)
~ ware (褐色)陶制品(装饰用)
~ wood board 棕纸板
red ~ 紫色
tabacco ~ 棕色

brownness ['braunnis] *n.* 棕色

browning ['brauniŋ] *n.* 二道抹灰,变成褐色,变暗,致黑,褐化
~ plaster 抹浆

brownstone ['braunstəun] *n.* 褐石,褐砂岩

bruite *n.* 水滑石,氢氧化镁石
~ -marble 水镁大理岩

bruise [bru:z] *n.*;*v.* bruised, bruising; *v.* 撞,擦 *n.* 压[捣]碎;挫[撞]伤;路边伤;机械磨损
~ check 碰撞裂纹(玻璃制品缺陷)
~ mark 碰痕

bruising 压[破,捣]碎,压扁;撞[硬]伤
~ glazing 刷釉

Brunswick 布仑司维克
~ black 布仑司维克黑（一种黑色清漆）
~ blue 布仑司维克蓝（铁蓝与硫酸钡的复合颜料）
~ green 布仑司维克绿,永久绿

brush [brʌʃ] *n*. 灌木材,树枝,梢料;刷子,毛[电]刷;毛[画]笔 *v*. 刷,扫,擦;擦光,涂刷[料],刷过
~ application 涂抹[刷],涂刷法
~ applied 用刷涂敷的
~ applied coating 刷敷涂层,刷涂
~ clearing 用刷子清除
~ coat 刷敷涂层,刷涂
~ coating 毛刷涂布,刷涂层,刷涂
~ graining 刷饰木纹状漆面
~ mark 刷痕,刷纹
~ over 用刷上色
~ painting 刷漆,上色
~ polish 上光剂
~ polishing 刷磨光
~ treatment （木材防腐）涂刷处理（法）,涂刷防腐剂
~ work 毛笔画,涂刷（颜色）画法,画风
~ed finish 刷饰
ari ~ 喷雾刷色器,喷射笔
painting ~ 画笔
scrubbing ~ 板刷
steel wire ~ 钢丝刷
wire gauze ~ 钢丝刷
writing ~ 毛笔

brushability *n*. 刷涂性,耐刷能力
brushable *a*. 耐刷能力
~ consistence[consistency] 可涂刷稠度

brushing [brʌʃin] *n*. 刷光,擦亮 *a*. 擦拭的,掠过的
~ compound 刷光涂料
~ glazing 刷釉
~ lacquer 刷漆
~ quality 涂刷质量
~ off 擦洗[掉],刷去[净]

brushmark *n*. 刷痕
brushy [brʌʃi] *a*. 多灌木的,用灌木复盖的,毛刷一样的
brutalism [bruːtəlizm] *n*. （现代建筑形式的）粗野(犷)主义
brutalist *a*. 粗野派[野兽派]建筑的;粗野派[野兽派]建筑家
~ architecture 粗野派建筑

bubble [bʌbl] *n*. 泡,气泡,水准器气泡;泡形物,球形顶;磁泡 *vi*. 冒泡,沸腾,涌出
~ concrete 泡沫混凝土,加气混凝土
~ entinguisher 泡沫灭火机[器]
~ level 气泡水准器
box ~ 球形水准器
circular ~ 圆水准器

bubblement *n*. 起泡状态
bubbler [bʌblə] *n*. 护散器;起泡器,水浴瓶,喷水式饮水口
~ fountain 喷水式饮水器

bubblet *n*. 小气泡
bubbly [bʌbli] *a*. 起（多）泡的,充满泡沫的

buck [bʌk] *n*. 锯架[台],门边立木,大装配架 *v*. 轧碎,锯开,冲,推,猛然开动;消除,补偿
~ frame 装有凹槽的门框
~ saw 架[框,弓]锯
~ -up 拧紧管接头,用铆钉撑锤顶住铆钉头

bucker [bʌkə] *n*. 破（粉,压）碎机;宽头碎矿锤;桶匠
~ -up 铆钉工

bucket [bʌkit] *n*. 吊[水]桶,料罐[斗];戽[挖、铲、吊、抓]斗;（坝）挑流鼻坎,消力戽
~ -and-belt elevator 斗带式提升机
~ elevator 斗式提升机
~ guide 吊桶吊绳,吊桶导向装置
~ lanyard 吊桶绳
~ type privy 桶厕（采用便桶的厕所）
hoist ~ 提升吊桶
windlass ~ 小吊桶,绞盘吊桶

bucketful [ˈbʌkitful] *n.* 一斗[桶]，满桶

bucking [ˈbʌkiŋ] *n.* 二次爆破；人工磨矿；顶撞，反作用
~ bar 打钉杆，铆钉顶棒

buckle [ˈbʌkl] *n.* 扣，带扣，金属扣，皮带扣；紧系，拉杆，拉紧，套筒；凹凸，翘曲，皱纹；中间浪（板材缺陷）*v.* 扣住；使弯曲或变形（如用热或压力等）膨胀，变形，弯曲，扣住，胀砂
~ clamp （皮带等的）扣夹，卡夹
~ fold 弯曲褶曲
~ latch （安全带）搭扣
~ outward 向外翘曲
~ pattern 翘曲图式[形]
~ plate 凹凸板，皱曲钢板
~ -up 安全扣带
~d frame member 弯曲的构件

buckram [ˈbʌkrəm] *n.* （装帧用）厚麻布，（涂胶的）硬麻布

bucksaw [ˈbʌksɔː] *n.* 弓锯，架子锯，大木锯

buckskin *n.* 被除去树皮的木头

buckstay *n.* （拱边）支柱[撑]

bucrane [bjuːˈkrein] *n.* 牛头骨状（雕）饰（在古罗马爱奥尼和柯林斯柱式上的中楣内）

bucranium [bruːˈkreiniəm] *n.* (*pl.* -nia) = bucrane 牛头骨状（雕）饰
~ frieze 牛头骨状雕带

bud [bʌd] *n.* 芽，花蕾
~ capital 柯林斯柱头

Buddha [ˈbudə] *n.* 佛，佛陀，浮屠
~ pine 罗汉松
family hall for worshipping ~ 佛堂
figure(image) of ~ 佛像

bud capital

Buddhist [ˈbudist] *n.* 佛教徒 *a.* 佛的，佛教的
~ architecture 佛教建筑
~ statue 佛像

Buddhistic [buˈdistik] *a.* 佛的；佛教（徒）的

budget [ˈbʌdʒit] *n.* 预算[表]；堆集[积]，存积；根据预算所做的计划；一束[捆]；*v.* 做预算，编入预算；预定，安排
~ chart 预算图表
~ cost 预算成本
~ deficit 预算赤字
~ document 预算书
~ estimate 概算
~item 预算项目
~ making 预算编制
~ program 预算方案
~ restriction 预算限制
~ sheet 预算表
~ statements 预算书
~ed performance 预计完成情况
~ing 预计编制；编制预算
~ing technique 预算编制方法
actual (balanced) ~ 决算

budgetary [ˈbʌdʒitəri] *a.* 预算的
~ control 预算控制
~ cost 预算成本
~ overhead 预算费用
~ performance 预算执行情况
~ planning 预算计划
~ price 预算价格
~ resources 预算经费

buff [bʌf] *n.* 浅黄皮，米色；软皮；抛[磨]光轮 *vt.* 磨[抛]光；缓冲，减振
~ away 擦光，磨去，消除
~ brick 浅黄砖

buffer [ˈbʌfə] *n.* 缓冲器，缓冲减振器，保险杆；缓冲剂[液]；消声器，挡风板；磨光工具 *vt.* 缓冲，减振，阻尼
~ action 缓冲[减振]作用
~ layer 缓冲层
~ rail 护墙栏杆，缓冲窗

buffet [bˈfei, ˈbʌfit] *n.* 食堂，餐室，小卖部；餐具柜，碗橱 *vt.* 打击，抖动，振颤
~ bar 快餐酒吧

~ car 餐车

buffeting *n.* (涡流,空气流促使结构物的)颤振,振动,抖振

buffing [ˈbʌfiŋ] *n.* 抛[打,磨]光
- ~ lathe 磨[抛]光机
- ~ paper 磨皮砂纸
- ~ wheel 抛光轮,砂轮
- colour ~ (镜面)抛光

bug [bʌg] *n.*;*v.*, bugged, bugging; *n.* 缺点[陷];故障;小型汽车; *vt.* 干扰,破坏
- ~ dust 粉尘,钻屑,煤粉
- ~ duster 除粉[尘]器
- ~ holes (混凝土表面的)蜂窝,面

bugduster *n.* 除尘器,清除煤粉工

buhl [bu:l] *n.* 镶嵌工艺品;布尔细工(镶嵌金、银等于木材上)
- ~ and counter 装饰细工
- ~ saw 框架锯

build [bild] *n.*;*v.*, built, building; *v.*;建造[筑,立];设立,创建,营造,组合; *n.* 建筑,构造,造型;工程,建造的式样;砌体的竖缝;体格
- ~ in 固定,砌[嵌]入
- ~ joint 构造缝
- ~ labourer 建筑工人
- ~ -up 建造[立],改建;安装,装配;增[生]长,聚集,积垒;组合[成];熔焊[堆]焊;计算,制图
- ~ -up member 拼装[装配]构件
- ~ -up of pressure 建立[造成]压力

builder [ˈbildə] *n.* 建筑[立、造]者、建筑工人,施工人员,营造业者;次品砖;充填墙用的石料
- ~'s diary 施工日记
- ~'s 'fitting [furniture] 建筑五金
- ~'s fittings [hardware] 建筑小五金,建筑器具配件
- ~'s glass fitting 建筑玻璃小五金配件
- ~'s hand cart 建筑用手推车
- ~'s iron supplies 建筑五金器具
- ~'s ladder 木工梯
- ~'s level 施工水准仪
- ~'s licence 施工许可证
- ~'s lift 施工(用)升降机
- ~'s merchant 建筑材料商,建材公司
- ~'s rough planks 建筑用毛板
- ~'s rubbish 建筑垃圾[废料]
- ~'s square 施工(用)角尺
- ~'s staging 施工台架,笨重脚手架
- ~'s yard 建筑场地,工地,施工现场,建筑者

building [ˈbildiŋ] *n.* 建筑物,房屋,大楼;建[制]造;组装[合];建筑术[业]
- ~ acoustic measurement 建筑声学测量
- ~ activities and losses 建筑功能
- ~ act(s) 建筑法(规)
- ~ adhesive 建筑用胶粘剂
- ~ alteration 建筑物重建[改进]
- ~ and contents 建筑物及其内部设施
- ~ and repair 建造[新建、扩建、修建]
- ~ appliance 建筑用具
- ~ area 建筑面积
- ~ article [goods, products, ware] 建筑产品[制品]
- ~ asphalt 建筑沥青
- ~ barracks (临时)工棚施工
- ~ block 砌块,空心砌块;构件;标准组件,构造单元;建筑区段[街坊]
- ~ block lintel 砌块过梁
- ~ block masonry wall 砌块圬工墙
- ~ block masonry (work) 砌块工程,凿石圬工
- ~ block module 砌块模数
- ~ block system 积木式结构方式,插入式程序系统
- ~ blocks 建筑砌块
- ~ board 建筑板材
- ~ board facing [lining] 建筑衬板
- ~ bonding adhesive [agent] 建筑用胶粘剂
- ~ brick 墙心[衬里]砖
- ~ by laws[code] 建筑规则[法规,规范,条例,标准]

~ carcass [fabric, shell] 建筑骨架
~ cement 建筑水泥
~ centre 建筑中心
~ clinker 建筑用烧结砖
~ code 房屋建筑规范
~ column 建筑物支柱[支承,撑木]
~ complex 建筑物组合件
~ component (预制的)建筑单元
~ construction 房屋构造,房屋建筑学
~ contract 建筑合同[协议书]
~ contractor 工程承包人[商],建筑承包单位
~ control 建筑管理
~ core 混凝土心墙
~ cost(s) 建筑总费用
~ demolition 建筑物拆除
~ density 房屋[建筑]密度
~ department 房屋管理处,建筑施工部门
~ dimension 建筑结构尺寸
~ drain 房屋排水管
~ drain system 房屋排水管系统,排水设备
~ drainage 房屋排水
~ drawing 建筑施工图,建筑制图
~ economy 建筑经济
~ element 建筑物件,建筑功能分区的要求,建筑构成要素
~ engineer 建筑工程师
~ equipment 建筑设备
~ erection system 建筑安装方法
~ evacuation 建筑物拆除
~ expense 房屋费用
~ extension 建筑扩建[展]
~ field 建筑工地、施工现场
~ fire 建筑物火灾
~ firm 建筑公司,建筑企业
~ frame 房屋构架,建筑框架
~ gable 建筑山墙
~ glass 建筑玻璃
~ glazier 建筑玻璃工

~ goods 建筑制品
~ ground 建筑工地[场地]
~ heating 建筑物供暖[加热]
~ heating installation 建筑物供暖[加热]设备
~ in 堵塞,插入,嵌固
~ industry 建筑工业
~ insulant [insulating material, insulation (-grade) material] 建筑绝缘材料
~ insulating article [product] 绝缘产品
~ insulating board [sheet, slab] 建筑绝缘[隔声、绝热]板
~ insulating brick 绝缘砖
~ insulating [insulation (-grade)] 建筑绝缘[热],隔声[热],保温
~ insulating paper 绝缘纸
~ insulation 建筑绝缘[绝热保温],建筑绝缘体,绝缘材料
~ insulator 建筑绝缘体
~ interior 建筑物内部
~ interior decoration 建筑物内部装饰
~ iron 建筑钢
~ labourer 建筑工人
~ land 建筑工地
~ law 建筑法(规)
~ lime 建筑用石灰
~ line 房屋界线,房基线,设计线,房屋边线,建筑线,红线
~ line platform 建筑平台线,房屋建筑线
~ load 建筑物荷载
~ maintenance 建筑物维修
~ masonry wall 建筑物墙体,圬工墙
~ mastic 玛琋脂,填缝料,填缝油灰
~ material 建筑材料
~ member (预制)房屋单元
~ method 建筑(施工)方法
~ mortar 建筑砂浆(灰浆)
~ of historic interest 历史性建筑
~ of large blocks [tiles] 大型砌块

建筑
~ of large slabs 大板建筑
~ of reinforced concrete construction 钢筋混凝土结构建筑物
~ of skeleton construction 构架房屋
~ of wooden construction 木结构建筑
~ office 建筑事务所
~ operation 建筑施工(操作,工房,过程)
~-out network 附加网络
~ owner 营造业主,房工,建筑工程发包单位
~ paint 建筑物油漆[涂料]
~ panel 建筑物护墙板
~ paper 防潮纸,油毛毡
~ part 建筑物细部
~ particulars and plans 建筑施工文件
~ plan 建筑设计,平面图
~ planning 建筑施工计划
~ plaster 建筑石膏粉饰
~ plastic 建筑用塑料
~ plastic film[sheeting] 建筑塑料薄膜
~ plot 房屋地区图,建筑用[基]地
~ preservative 建筑防腐剂
~ principles 建筑原则[法规]
~ product(s) 建筑产[制]品
~ production 建筑生产
~ project 建筑工程方案[计划]
~ program(me) 建筑施工工程序
~ proposal 建筑投标
~ quicklime 建筑石灰,施工用生石灰
~ quilt 隔板,垫板
~ reconstruction 房屋重建
~ regulation(s) 建筑规则[规范,规定,规章,条例]
~ reinforcing material (建筑用的)钢筋
~ repair 建筑物修理[维修,修补,修缮]

~ research institute [station] 建筑研究所
~ research academy (BRA) 建筑科学研究院
~ restriction line 建筑范围[面积]
~ roof 屋面,屋顶
~ rubber compund 橡胶组合,配橡胶料
~ sand 建筑用砂
~ season 建筑施工季节
~ sewer 房屋污水管(由房屋外墙外一米处至公共下水道的一段管道)
~ sequence 建筑程序,施工程序
~ sheet 建筑板材
~ sheet facing[lining, surfacing] 建筑衬板
~ site 建筑工[场]地
~ site installation 工地设备,建筑场地设备
~ skeleton 建筑框[构,管]
~ slab facing [lining, surfacing] 建筑衬板
~ site 建筑工地[现场]
~ speed (建筑)施工速度
~ stair(case) (建筑物)楼梯
~ standard law 建筑标准法
~ standard (specification) 建筑标准(规范)
~ stone 建筑石材,结构基石
~ storm drain 房屋雨水管
~ storm sewer 房屋雨水管
~ (structural) system 建筑结构体系
~ style 建筑风格[形式,式样,格式]
~ subdrain 房屋地下排水管(由房屋地下排水设备至截水井的一段重力排水管)
~ substructure 基础(基准线以下结构)
~ supervision 建筑管理,施工检查[监督]
~ team 建筑施工队
~ technician 建筑技术,施工技术

人员
~ tile 空心砖,砌[石]块
~ tile lintel 砌块过梁
~ tile masonry (work) 砌块工程,凿石圬工
~ tile (masonry) wall 砌块圬工墙
~ timber 建筑(用)木材
~ trap 房屋存水弯
~ traffic 工地交通,现场运输
~ type 建筑形式[类型]
~ unit (预制装配的)建筑单元[构件]
~ -up 装配,安装,建成
~ with dwellings 公寓住宅
~ work 建筑[土木]工程
~ worker 建筑工人,施工人员
~ yard 建筑场地
~ zone 建筑用地[范围]
annexed ~ 群体建筑
attached ~ 附联式房屋,附联建筑物
brick nogging ~ 立贴房屋,木架砖壁建筑
concrete-steel ~ 钢筋混凝土建筑
court ~ 庭院式建筑
courtyard with ~ on four sides 四合院(中国)
detached ~ 独立式房屋,独立建筑,不接邻房屋
dry-brick ~ 干砌砖工房屋,无砂浆砌砖
fabricated ~ 装配式房屋(建筑)
fire-proof ~ 耐火房屋
framed ~ 框架式建筑
frame-panel ~ 间架式建筑
functional ~ 专用房屋
line ~ 行列式建筑
multistoried ~ 多层房屋,多层建筑
science ~ 科学馆
sectional ~ 预制部件构成房屋,装配式房屋
single-aisle ~ 单楹房屋
sporadic ~ 散布式建筑
steel skeleton ~ 钢骨房屋
temporary ~ 临时性建筑[房屋]

tenement ~ 公寓
tier ~ 多层房屋
tower ~ 塔式建筑
two-aisle 两楹间[走廊]房屋
Building Code Requirements for Reinforced Concrete 钢筋混凝土建筑规范
built [bilt] (build 的过去式和过去分词) *a.* 组合的,建造[成]的,堆积的
~ arch 组合拱
~ beam 组合梁
~ -for-purpose tools 专用工具
~ -in arch 嵌入式拱,固定拱
~ -in bathtub 镶入式浴盆
~ -in comfort (in building) (建筑物的)砌入式生活设备
~ -in cupboard 壁橱
~ -in edges 嵌入[固定]边缘
~ -in end 嵌入[固定]端
~ -in fitting 预埋件,埋件镶入器具
~ -in furniture 镶壁家具
~ -in girder 固定端大梁(两端固定)
~ -in gas fire 嵌墙式煤气炉
~ -in gutter 暗水管,内落水
~ -in kitchen (嵌入式)内部厨房
~ -in lamp 墙内灯
~ -in light 固定[内装]照明
~ -in lighting 嵌入照明设施
~ -in lock 暗锁
~ -in nailing block 木落砖
~ -in radiator 墙内散热器
~ -in sections 由部件装配的,分段建造的
~ -in shower stall 内部(嵌入)小淋浴间
~ -in unit(s) 固定设备
~ -in wardrobe 嵌入式衣橱
~ kithchen cabinet 内部厨房,橱柜
~ spar 组合杆件,拼制木柱
~ -up 组合的,建成的,装配的,焊接的,合成,增长,积累
~ -up arch 组合拱
~ -up architrave 组合的(门窗)线

条板
~ -up beam 组合梁
~ -up column 组合柱
~ -up frame 组合构架[框架]
~ -up girder 组合大梁[架]
~ -up glued beam 胶合梁
~ -up joined 组合接头[节点]
~ -up member 装配部件,组合构件
~ -up pillar 组合柱
~ -up plate 组合[双面]模板
~ -up purlin 组合檩条
~ -up roof(ing) 组合屋面
~ -up segmental arch 组合圆弧拱(用木材榫接)
~ -up type 组合式
~ -up type scaffolding 装配式脚手架
bulb [bʌlb] *n*. [抹灰]凸包,隆起,气[水、灯]泡,玻璃泡;球形物,球管,球形零件;测温计[仪表] *v*. 成球形
~ angle section 圆头角材
~ beam 球头工字钢梁,圆缘梁
~ edge (平板玻璃的)厚圆边
~ edge severing device (平板玻璃)切边装置
~ holder 灯座
~ -shaped base 扩大底部的建筑物
~ socket with Edison cap 螺口灯泡[座]
~ socket with swan cap 卡口灯座[泡]
~ tee 圆头丁字铁,轨形截面
~ with bayonet cap 卡口灯泡
~ with Edison(edison) cap 螺口灯泡
~ with swan cap 卡口灯泡
air ~ 气泡
incandescent ~ 白炽灯
neon ~ 霓虹灯
bulbous [ˈbʌlbəs] *a*. 隆起的,球形的
~ dome 球形屋顶,葱头形屋顶
bulge [bʌldʒ] *v*. bulged, buling; *n*. 凸[隆]起,隆丘,凸出,隆胀,(墙面)

不平
~ clearance 膨胀间隙
~ in 凸进
~ out 凸出,凸[隆]起
~d-in 压入,插入的
bulging [ˈbʌldʒiŋ] *n*. 膨胀,凸出,突度;打气,渗水
~ force 膨胀[隆起]力
~ in 使压(挤)入
~ wall 凸肚墙
bulk [bʌlk] *n*. 松散材料,容[体]积,容量,大小,尺寸;大块[量,批];松密度,胀量 *a*. 散装的,块状的,大块的,体积(内)的
~ article 标准产品,大量生产的制品
~ cement 散装水泥
~ concrete 大体积混凝土
~ concrete structure 大体积混凝土结构[建筑物]
~ excavation 大面积(规模)开挖
~ filling 大规模回填
~ gypsum 散装石膏
~ (hydrated)lime 熟[消]石灰
bulkage *n*. 大体积物质,涨方
bulkcargo *n*. 散装货
bulkhead [ˈbʌlkhed] *n*. (沉箱沉井或空心方块中的)隔壁,隔墙,舱壁,隔框,防火门;(平屋顶上的)楼梯间,升降机房;围堰,护[堤,驳]岸;挡土[水]墙,海塘;板桩,堵头板(灌混凝土用) *vt*. 用墙(壁)分隔;bulkheaded, *a*.
~ air grille 挡风格窗,散热器的百叶窗
~ tee 长臂丁字尺
~ed 嵌入的,固定的
fire ~ 挡[防]火墙
flexible ~ 软性隔壁
masonry ~ 砖砌密闭墙
partial ~ 半堵壁
pile ~ 桩构堵壁
bulking *n*. 砂的湿胀性,体积膨胀
~ agent 填充剂,填料;湿[膨]胀剂
~ effect 膨胀性

~ of sand 砂的湿胀[体积膨胀]
bulkload *n.* 毛载,散装,散装货物,粒状物,堆放物
bulky ['bʌlki] *a.* 松散的,体积大的,庞大的,笨重的
~ character 丰满[感]特性,膨松特性
~ color 体色彩
~ joint 粗[凸]缝,凹凸缝
bull [bul] *n.* 庞大物体,庞大物件;公牛 *a.* 大型的;雄的 *vi.* 抬高价格,涨价
~ bay 广玉兰
~ bit 凿形钻头,一字形钻头
~ brick 单圆角砖
~ crack 裂纹
~ float 大慢
~ head 双头式
~ header 露顶侧砖,圆端丁砖
~ head rivet 圆头铆钉
~ headed rail 小圆头栏杆,圆头钢轨,双头钢轨
~ header 圆端丁砖
~ nose 外圆角,外圆角创
~ nose brick 单圆角砖
~ -nose coping 圆角帽梁
~ -nose header 圆端丁砖
~ -nose plane 牛头刨
~ -nosed step 圆头踏步
~ pine 泡松
~ rope 钢丝绳
~ set 小石壁
~ 's-eye 牛眼灯,小圆窗,采光天窗,圆形凸透镜
~ 's-eye arch 牛眼形拱,圆拱
~ s'-eye glass 牛眼玻璃
~ 's-eye window 小圆窗
~ 's-nose 外圆角,外钝角
~ stretcher 露边侧砖,圆端顺砖
~ switch 照明控制开关
~ trowel 圆角铰刀[泥刀、抹子]

bull's-eye

bulldong *n.* 补炉底材料
~ plate 爪板
~ wrench 锷形扳手,管子扳手
~ connector （用于木结构的）齿形板连接器
bulldust *n.* 粗[粉]尘
bullen 圆头钉
~ nail 阔[圆]头钉
bullet ['bulit] *n.* 弹,子弹,弹头,核,核心,铅锤,锚塞,锥形体;针,撞针尖
~ bolt 伸缩插销
~ catch[latch] 自动扣,弹子门闩
~ -defying 防弹的
~ -headed 圆头的,似子弹头的
~ -nose median 弹头式分隔带
~ -nose rod 圆头杆
~ -nosed 圆头的,化弹头的
~ perforator 子弹式穿孔器（机）
~ -proof 防弹的
~ -proof glass 防弹玻璃
~ -resistant 防弹的
~ -resistant glass 防弹玻璃
~ -resistanting 防弹的
~ -resistanting glass 防弹玻璃
~ valve 球阀,球形阀
bull float 大抹子
bullhead ['bulhed] *n.* 小圆头,大头鱼,双头式
~ tee 大头三通
bullion ['buljən] *n.* 纯金[银],金[银]条;条形金属
bullule *n.* 小泡
bulwark ['bulwək] *n.* 堡垒,寨墙,防卸工事;防波堤[板],(*pl.*) 舷墙(甲板上的船舷),栅; *vt.* 防卸[护];用堡垒保护
bumicky *n.* 水泥石粉浆
bump [bʌmp] *v.* 冲撞,碰撞;冲[撞]击力;表面的凸起,隆起,小山,丘陵
~ contact 块形连接
~ joint 法蓝[凸缘]接头
bumper ['bʌmpə] *n.* 撞[冲]击物;防撞[缓冲,减振]器,保险杆,挡板;制砖工,

铆钉工
~ block 缓冲块
~ post 缓冲柱,挡柱
window glass ~ 窗玻璃垫条

buna ['bjuːnə] *n.* 布纳橡胶,丁纳橡胶,合成橡胶
~ -N 丁腈橡胶
~ rubber 丁纳橡胶
~ -S 布纳 S,丁苯橡胶

bun foot 圆球脚

bunch [bʌntʃ] *n.* 束,捆,串,卷,包;线束,股绳;粘合剂,隆起物; *v.* 捆成一束,粘合;隆起
~ light 聚束灯光
~ wire 绞合线,多绞线
~ed wire 绞合线

bunchy ['bʌntʃi] *a.* 隆起的,成束的,穗状的

bundle ['bʌndl] *n.* 束,扎,捆,包;线卷[盘,垛] *v.* 包,捆,扎;粘合
~ finishing 束状纹
~ iron 铁丝卷
~ of columns 柱群,簇桩
~ of piers 桩群,桩组
~ of pillar(s) 群柱,集柱
~ piers 桩群[桩组]
~ pillar 群柱,集柱
~d tube 成束管

bung [bʌŋ] *n.* 木塞,塞子;桶口[盖]; *v.* 塞住
~ hole 桶(侧)口

bungaloid ['bʌŋɡəlɔid] *a.* 平房式的

bungalow ['bʌŋɡəhəu] *n.* (有凉台的)平房,小房;别墅

bungle ['bʌŋɡl] *n.* 粗制滥造,笨拙的工作

bungum *n.* (一种冲积的)粉砂或软泥,新淤泥

bunk [bʌŋk] *n.* 框架床铺
~ apron 上铺挡板
~ bed 双层床

bunkhouse ['bʌŋkhaus] *n.* (建筑工地)简易工棚,简陋小屋

bunkhouse

bunkload 一批圆木

bunning 木台轲,横撑支柱工作平台

Burberry ['bəːbəri] *n. pl.* -ries ,防水布,柏伯丽雨衣

bureau [bjuə'rəu] *n. pl.* -reause, -reaux ;局,科,处,司,社,所;写字台,办公桌,大衣柜,梳妆台

burglar ['bəːɡlə] *n.* 窃贼,夜盗
~ alarm(system) 防窃报警器,防窃报警装置[设备]

burglary ['bəːɡləri] *n. pl.* -ries,盗窃行为
~ protection 防盗窃保护
~ resistance 抗盗窃设备

Burgundian [bəː'ɡʌndiən] *n.* 勃艮地人; *a.* 勃艮地[人]的
~ church 勃艮地教堂
~ Gothic(style) 勃艮地哥特建筑形式[风格]
~ portal 勃艮地大门
~ style of sculpture 勃艮地雕塑形式
~ vault 勃艮地穹窿

Burgundy pitch 白树脂

burgy *n.* 细粉,煤屑

burial ['beriəl] *n.* 埋入[藏],掩埋;(在缓冷坑中)冷却
~ chamber 墓室
~ vault 穹窿形墓穴

buried ['berid] *a.* 埋入[置]的;浸[沉]入的
~ control center 地下指挥所[控制中心]
~ depth 埋置深度,埋深
~ fire hydrant 埋置消火栓
~ pipe 埋设管,

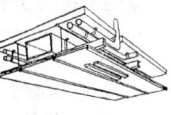

buried pipe

暗管
~ shelter 地下防空洞
~ structure 掩体建筑物,掩没构造
~ tubular conduits 埋设的[隐蔽的、暗装的]设备
~ wiring 暗线
burin ['bjuərin] *n.* 雕刻刀,錾刀,雕刻风格
burl [bə:l] *n.* 木节,树节[疤],树瘤,瘤状花纹
~ veneer （有）涡形饰纹（的）镶面层板
burlap ['bə:læp] *n.* 粗麻布,麻袋;（作室内装饰用的）麻织物
~ bag 麻袋
~ finish 布面装饰
Burmese [bə:mi:z] *n.; a.* 缅甸的,缅甸人（的）
~ architecture 缅甸建筑
~ style 缅甸式（建筑）
burn [bə:n] *n.; v.*, burned or burnt, burning;烧,燃烧 烧毁[焦,着];烧毁区域;烙,烫,气割,使氧化; *n.* 烧,烧制;（苏格兰）小溪[河]
~ -cut 空眼掏槽
~ -on 黏砂
~ -out pipe 通气孔
burnable ['bə:nəbl] *a.; n.* 易[可]燃的,[可]燃物
burned [bə:nd] *a.* 燃烧的,烧毁的,焙[窑]烧的
~ brick 烧黏土砖
~ clay 烧黏土
~ clay brick 焙烧黏土砖
~ clay curved roof(ing)tile 烧制黏土曲瓦
~ clay curved tile roof 烧黏土曲瓦屋面
~ clay hip tile 烧黏土戗脊盖瓦
~ clay light (weight) aggregate 煅烧黏土轻(重)骨料
~ clay product 烧黏土制品
~ clay ridge tile 烧黏土盖瓦
~ dust 烧黏土泥料粉（熟料）
~ filler 烧黏土泥料粉（熟料）

~ lead joint 铅熔[焊]接
~ lime 烧[煅]石灰,氧化钙
~ ocher[ochre] 烧赭石
~ product 烧制[成]品
~ sand 焦砂
~ shale 烧页岩
~ shale product 烧页岩制品
~ sienna 暗橙色,煅黄色,煅富铁黄土
~ umber 烧褐土,煅棕土,中等棕色
~ ware 烧制品
~ wood 木炭
burner ['bə:nə] *n.* 燃烧器[室];燃烧口,火口;喷枪,吹管,香炉;气焊[割]工,烧窑工
~ block 小炉砖,喷嘴砖
~ ring 燃烧口砖
~ tile 燃烧器耐火瓦,炉底,燃烧口砖
bunsen ~ 本生灯
burnetizing *n.* 氯化锌防腐法
burning ['bə:iŋ] *n.; a.*, 燃烧,烧光[毁],氧化,煅烧;燃[焙]烧的,炽热的;紧急的
~ of lime sludge 白泥煅烧
~ of limestone 煅烧石灰石
~ off paint 油漆燃除
~ stone 硫磺（石）
~ to(cement)clinker 烧结水泥熟料
cigarette ~ 端部起燃,端面燃烧
earth ~ 烧土
lead ~ 铅焊
burnish ['bə:niʃ] *v.* 磨[擦]光,打磨,抛光,擦亮,精加工,涂光; *n.* 光泽[亮,滑];burnishable, *a.*;burnishment, *n.*
~ed bolt 精制螺栓
~ed gold 亮金黄色
~ed metal 磨光金（装饰）
~ed staw 亮麦秆色
burnisher *n.* 磨光器,抛光机,磨光[擦亮]工
burnishing *n.* 摩擦[加压]抛光,磨[抛挤]光,擦亮,光泽
~ action 磨光[打磨作用]
~ gold 磨光金

~ machine 抛光机
~ oil 抛光[擦亮]油
~ powder 抛[磨]光粉
~ silver 磨光银
~ surface 磨[抛]光面；平整面
~ wheel [摩擦]抛光轮

burnout ['bə:naut] n. 大火灾；烧毁[尽]，燃耗，燃尽时刻
~ pipe 通气孔
~ proof 耐高温的，防烧蚀的

burnover n. 欠火砖

burnt [bə:nt] v. burn 的过去式和过去分词；a. 烧过的，烧成的，窑烧成的；烧毁[坏]的，过烧的
~ aggregate 烧结骨料
~ brick 烧透砖
~ clay 烧结黏土
~ clay article 烧结黏土制品
~ clay brick 烧结黏土砖
~ clay curved roof(ing) tile 烧黏土屋面曲瓦
~ clay curved tile roof 烧黏土曲瓦屋面
~ clay hiptile 烧黏土戗脊盖瓦
~ clay light (weight) aggregate 煅烧黏土轻(重)骨料
~ clay masonry (work) 烧黏土圬工 (砖建筑)
~ clay product 烧结黏土制品
~ clay ridge tile 烧黏土脊瓦
~ coal 天然焦炭
~ gypsum 烧石膏
~ lime 烧[生]石灰，氧化钙，锻石灰
~ ocher 烧赭石
~ plaster 煅[烧]石膏
~ product 烧制[成]品
~ refractory 烧成耐火材料，熔烧耐火物
~ sand 焦砂
~ shale 烧页岩，页岩残渣
~ shale product 烧页岩制品
~ sienna 岱赭，烧过的富铁黄土，暗红橙色(颜料)，煅黄土

~ umber 烧赭土(颜料)，煅棕土
~ ware 烧成品，烧成器皿
~ weld 氧焊
~ wood 木炭

burr [bə:] n. 过火砖，磨石，三角凿刀，小圆锯；毛头[口]，凿纹，垫圈，轴套，小箍；坚硬石灰岩，岩基；v. 除毛刺[边]，用磨石磨；在……上造成粗糙边缘
~ (buhr) mill 石磨机，盘磨机
~ -drill 圆头锉，钻锥
~ removing machine 去毛刺机

burring ['bə:riŋ] n. 去毛刺，毛口磨光，去毛石

burrow ['bʌrəu] n. 洞穴，避难所，土堆，废[矸]石堆

bur(r)stone ['bə:stəun] n. 磨石，磨盘石，亦作 buhrstone

burst [bə:st] v. burst, bursting, n.；爆炸[发，裂]；破裂，裂口[缝]，决口；冲击，震动

burton ['bə:tn] n. 辘轳，滑车组，复滑车

bury ['beri] vt., buried, burying；埋入，填覆，掩蔽，埋葬；burier, n. 地窖[穴]；软黏土，黏土页岩
~ length 埋置深度

bus [bʌs] n. pl. buses or busses；v., bussed or bused, bussing or busing；n. 公共汽车，汽车；母线，汇流条；手推四轮车；v. 乘公共汽车

bus

~ chamber 母线盒
~ duct 母线管道，汇线管
~ -loading bay 公共汽车停车站

bush [buʃ] n. 丛林，灌木，矮树丛；荒地[野]；衬套，套筒[管]，轴瓦，绝缘管；v. 用金属杯里，加衬套
~ hammered dressing 石锤修琢，剁斧琢面
~ hammered face 石

bush hammer

锤琢面
~ hammered finish 凿[花]锤饰面
~ hammer(ed) plaster 凿锤石膏,灰泥涂层

bushing ['buʃiŋ] *n.* 衬套,套筒,轴衬;螺丝缩节,连接套管;引[导]线
~ block 衬砖
shock absorbing ~ 减震衬套
wall ~ 穿墙套筒

business ['biznis] *n.* 商[营]业;商店[行];业务,职责,事务,交易,买卖;企[实]业,营业所
~ area 商业区
~ block 商业建筑
~ buliding 办公楼
~ centre[center] 商业中心,商[营]业区
~ district 商业区
~ (local) street 商业区街道
~ parade 商业街道
~ premises 商业房屋,房产
~ property 商业地产
~ quarters 事务所,商业区
~ school 商学院,商业学校
~ street 商业街道

bust [bʌst] *n.* 半身塑像或画像;破产,失败;*v.* 破产;*a.* 破产的

buster ['bʌstə] *n.* 钉头切断机,风镐,翻土机,双壁开沟,铆钉铲,切除机;庞然大物

bustle ['bʌsl] *v.*, -tled,-tling, *n.* 忙乱,奔忙,喧闹,活跃
~ pipe 环[圈]管,促动管

busy ['bizi] *a.*, busier, busiest; *v.*, busied, busying; *a.* 繁忙的,热闹的,无空闲的,忙碌的;占线的
~ crossing 繁忙的交叉口,负荷很大的交叉口
~ period (交通)高峰[繁忙]时间
~ season 旺季

but [bʌt] *n.* (苏格兰)房子的外部房间(尤指茅舍之厨房)

butadiene [bju:tə'daii:n] *n.* 丁二烯

~ rubber 聚丁橡胶,丁二烯橡胶

butane ['bju:tein] *n.* 丁烷,罐装煤气,天然瓦斯
~ blowlamp 丁烷喷灯

butanol ['bju:tənəul] *n.* 丁醇(=butyl alcohol)

butler ['bʌltlə] *n.* 司膳的人
~ finish (板材表面)无光(毛面)精整
~'s pantry 膳室
~'s tray 腰圆形木盘,装有折叠腿的托盘
~'s window 送菜窗口

butlery ['bʌtləri] *n.* 餐具室,膳室

Buton(resin) *n.* 布通树脂

butt [bʌt] *n.* 铰链,平(封,连)接,接缝;底部,端面,粗[大]端,桩头;(*pl.*)胸端,靶垛;*v.* 对接,紧靠;冲,撞;干涉
~ and break 对头错开
~ -and butt 两头对接
~ and lap joint 互搭接头
~ and strap hinge 丁字铰链
~ and strapped joint 搭板对接
~ bolt 铰链筒
~ block 对接贴板,对接缝衬垫
~ chain 对接链
~ chisel 平头凿
~ connection 端[对]接
~ contact 对接触点
~ cracks 端头裂纹
~ cut 根端材,树基材
~ end 粗端,木头(木桩)
~ (end lap) 横接缝
~ -end treatment 木尾防腐
~ fusion 熔接
~ hinge 铰链,平接铰链
~ joint 对接头,丁字对接焊缝,平[对]接
~ joint with double straps 双搭板对接
~ joint with single strap 用单搭板对接
~ jointed 对接的
~ jointed seam with strap 搭板对

接缝
~ junction 对[平]接头
~ laying 平接壁板
~ line 接缝
~ log 树基材,根段材
~ of pile 桩的根端,桩头
~ pin 铰链销
~ plate 搭[衬,镶]板,连接板
~ plate splice 钢板对头拼接
~ -prop 立柱的撑杆,对接杆叉,对接支柱
~ rammer 平头锤,平头砂春
~ rivet joint 夹板平铆接
~ riveted joint 铆钉对接
~ riveting 对头[对接]铆接
~ saw 截木锯
~ -scarf joint 对嵌接,铰链嵌接
~ seam 对接[头]接合
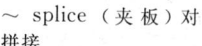
~ splice (夹板)对拼接

butt splice

~ strap 平接板,鱼尾板,连接板
~ strap[ped]joint 鱼尾板接合,搭接连接
~ strip 拼接板,鱼尾板
~ -swelling 干脚,脚材
~ temoin 外露层
~ -type heat insulating material 厚板状隔热材料
~ veneer 薄板树根
~ weld 对接焊,对头[接]焊缝
~ welded joint 对焊接头
~ -welded seam 对焊缝
~ -welded tube[pipe] 对接焊管
~ welded with chamfered ends 斜头平焊接
~ welded with square ends 方头平焊接
~ welding 对(接)焊(金属板边对放,沿接缝焊接)
~ welding process 对头焊接法
cross garnet ~ 十字铰链

double acting spring ~ 双动弹簧铰链
lap(ped)[overlap] ~ 搭接
window ~ 窗铰链
butte [bju:t] *n.* 孤山[峰],小尖山
butted *v.* butt 的过去式和过去分词; *a.* 对接的,粗凿的,铆牢的
~ bridging joint 顶栅搁栅
~ joint 对[平]接
~ tight 撞紧的
~ tube 粗端管,端部加粗管,异壁厚管
butter ['bʌtə] *n.* 焊膏,缓冲器,黄油; *vt.* 涂抹灰浆
~ yellow 奶油黄
~ with mastic 打腻子,用油灰填平(粘牢)
~ zone 缓冲绿地,防护地带
~ed joint 薄浆缝
~ed masonry 挤浆砌筑
buttercover plate 对接[鱼尾]板
buttercup yellow *n.* 锌黄
butterfly ['bʌtəflai] *n.* 蝶形阀门[节流门,节气门];蝶形板曲撑杆,蝶式[形]

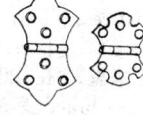

butterfly hinges

~ bolt 蝶形[反叶]螺栓
~ hinge 蝶形铰链,长翼铰链
~ nut 蝶形螺帽
~ screw 蝶形螺钉
~ seam 蝶形接缝
~ shaped roof 蝶形屋顶,V形屋顶
~ self-closing clamper 蝶形自闭门
~ table 蝶形折叠桌
~ valve 蝶形阀,风门,混合气门
buttering *n.* (用镘刀)涂灰浆,隔离层,预堆边焊
~ trowel (砌墙)涂灰镘,(涂)砂浆镘刀
buttery ['bʌtəri] *a.*; *n. pl.* -ies;酒窖,酒类贮藏室;膳务室,伙食房,食品小卖部
~ hatch 配膳室与食堂之间的窗口
button ['bʌtn] *n.* 按钮,钮扣,钮(状)

饰,球形捏手,路钮(路面标线用); a. 钮扣形的; v. 扣住[扣紧]
~ bottom pile 牛形底桩
~ dies 可调圆扳手
~ -head 圆形端头,球头,钮头
~ head bolt 圆头螺栓
~ -head cap screw 圆头螺钉
~ (head) rivet 圆头铆钉
~ head screw 半圆头螺钉
~ head spike 圆头道[长]钉
~ -headed bolt 半圆头螺栓
~ -headed screw 半圆头螺钉
~ plate 凸头钢板
~ socket 按钮灯口
~ stem 推拉用小把手
~ swith 按钮开关
~ telephone 按钮电话,琴键式电话
push ~ 接通按钮
reset ~ 复位按钮
stop ~ 停止按钮
buttress ['bʌtris] *n.* 扶壁,扶垛,撑墙,支墩,支持物,支柱; *vt.* 支持[撑],(用扶壁)支住,加固
~ brace 支撑刚性梁,加强梁垛间支撑梁
~ bracing struts 刚性梁,横撑
~ head 垛头
~ niche 扶壁壁龛
~ pier 扶壁墩
~ shaft 细柱
~ thread 梯形丝扣,锯齿螺纹
~ tower 拱门两旁的塔楼
~ wall 扶垛墙
~ wing 扶壁翼
~ed arch 扶壁式拱
~ed wall 前扶墙,扶撑墙
~ed wing 扶壁侧翼
~ing pier 支墩,拱座
arch(ed) ~ 拱扶垛
flying ~ 飞扶垛
hanging ~ 悬扶垛
hollow ~ 空心支墩

pier(ed) ~ 墩式扶壁,支墩
butyl ['bjuːtil] *a.* 丁基
~ acetate 醋酸异丁酯
~ alcohol 丁醇
~ ruber 异丁(丁基)橡胶
~ rubber base 异丁橡胶基层底板
~ rubber beading 异丁(烯)橡胶墙角护条
~ rubber (building) mastic 异丁橡皮胶(填缝料)
~ rubber foam 异丁橡胶泡沫
butylene ['bjuːtiliːn] *n.* 丁烯,亦作 butene
~ glycol 丁二醇
butyric [bjuːˈtrik] *a.* 酪酸的
~ acid 丁酸
buyer *n.* 买主[方],购买单位
buzz [bʌz] *n.* 蜂音,嗡嗡叫;蜂鸣器,汽笛,电话; *v.* 嗡嗡叫,蜂音
~ planer 刨床
~ saw 圆盘锯
buzzer ['bʌze] *n.* 蜂鸣器,汽笛,工地电话,磨砂轮,轻型穿孔机
buzzy *n.* 伸缩式风钻[凿岩机] *a.* 嗡嗡响的
BXcable (安装用)软电缆
by [bai] *prep.* 在旁边,沿,经,由,向;依据,按照,用,凭,靠; *ad.* 在旁,过去,放开,经过
~ -alley 侧巷
~ contract 承包
~ -heads 间歇自喷,间歇喷的
~ lane 小巷
~ -laws [building code] 建筑规则[规,规范,条例]
~ -pass 夹管,旁通管;支流,侧道,旁路
~ -pass 支路,绕行管,旁道管,回流管路
~ -pass opening 旁通孔
~ -pass pipe 旁通管
~ pass-product 副产品
~ -pass valve 旁通阀,分流阀
~ -passage 便道

～ -pass(ing)door 旁门
～ -pipe 旁通管
～ -room 侧[私]室
～ -sight 目测
～ -street 旁街,支巷
～ -trail 支路
～ usage 照例,按照常规
～ workman 临时工(人),短工
bye [bai] *a.* 次要的,其次的; *n.* 不重要的或次要的东西
～ hole 侧孔
～ -pass 支管路,旁通路
byerlite *n.* 氧化石油沥青
byerlyte *n.* 石油沥青,炼焦烟煤
bylane [ˈbailein] *n.* 小道,僻巷
bylaw [ˈbailɔː] *n.* 法规,附[细]则
bypass [ˈbai-paːs] *n.* 旁路,间道,旁通管; *vt.* 设旁路,绕道

～ valve 旁通阀
bypath [bai-paːθ] *n.* 小路,僻径,支路
byproduct [ˈbaiprɔdəkt] *n.* 副产品
～ material 副产物
bysmalith *n.* 岩柱,岩栓
byway [ˈbaiwei] *n.* 间道,僻路,小径; 次要的方面
byzant [ˈbizənt] *n.* 列圆节,圆盘饰带
Byzantine [biˈzœntain] *a.* 拜占庭建筑的,
～ architecture 拜占庭建筑
～ building 拜占庭建筑
～ capital 梯形柱头
～ church 拜占庭教堂
～ column 拜占庭柱
～ house 拜占庭住宅
～ style 拜占庭式(建筑)
Bz ＝bronze 青铜

Byzantine capital

C

Cab [kæb] *n.* 小室,司机室;出租车
cabane [kəˈbæn] *n.* 翼[支]柱;顶梁[架],锥体形支柱泵
～ radiator 屋[拱]顶散热器
cabin [ˈkæbin] *n.* 小室,小房间,舱; *v.* 分隔(房间),住在小室
～ hook 门窗钩
control ～ 控制室
log ～ 木屋
operator's ～ 操作室
signal ～ 信号房[室]
switch ～ 电缆配线房
cabinet [ˈkæbinit] *n.* (小)室,箱,盒; *a.* 细木工的,小巧的
～ finish (室内)壁板装饰
～ for dry washing (住宅中的)洗衣干燥柜
～ heater 柜式采暖器
～ lavatory 间隔盥洗室

～ maker 细木工,家具工
～ panel 配电盘,接线板
～ projection 斜角立体投影
～ type air conditioner 箱式空调器
～ -type governor 箱式调速器
～ type shower 分间式淋浴室
～ urinal 间隔小便池
～ window 陈列橱窗
～ work 细木[工]作
distributing ～ 分线[配电]盒
drying ～ 烘箱,干燥箱
environmental ～ 人造环境室,模拟环境条件的柜[箱]
humidity ～ 湿度室
silence ～ 隔声[音]室
cable [ˈkeibl] *n.* 钢[缆]索,钢绞线,钢丝绳,电缆; *vt.* 固定,栓住,锚定
～ box 电缆箱,分线盒
～ for lighting 照明电缆

~ laying (placing) 电缆敷设
~ mat heating system 席式[地板]采暖系统
~ moulding (建筑装饰用的)卷缆花饰
~ net structure 钢索网结构
~ network 钢索网格结构
~ (cantilevering) roof 悬吊屋顶
~ roof with saddle shape 悬吊鞍形屋顶
safety ~ 安全[保险]索
suspension ~ 悬缆,吊索

cabling [ˈkeibliŋ] n. 架设电缆;(柱头)卷绳状装饰
~ diagram 电缆线路图

cabtire [ˈkæbtaiə] n. = cabtyre 汽车轮胎; a. 有厚橡皮套(绝缘)的
~ cable 厚橡胶(绝缘)软电缆
~ sheath(ing) 硬橡胶套管[护套]
~ wire 厚橡胶线
vinyl ~ cable 聚乙烯绝缘软质电缆

cacaerometer n. 空气污染检查器

cache [kæʃ] n. (计算机)超高速缓冲存贮器,贮藏处;vt. 贮存;vi. 躲藏

cadaster [kəˈdæstə] = cadastre n. 地藉图,河流志,水册

cadastral [kəˈdæstrəl] a. 地籍的, n. 地籍
~ map 地籍图

cadastration [ˌkædəsˈtreiʃən] n. 地籍测量

caddy [ˈkædi] n. pl. -dies;盒,箱,罐

cadmiferous a. 含镉的

cadmium [ˈkædmiəm] n. 镉(Cd)-cadmic, a.
~ coat (finish) 渡镉层
~ colour (pigment) 镉颜料
~ plated 渡镉的
~ red 镉红
~ titan alloy plated 镀镉钛合金的
~ yellow 镉黄

caesious [ˈsiːziəs] a. 青灰色的

cafeteria [ˌkæfiˈtiəriə] n. 自助食堂,顾客自取饭菜食堂

cage [keidʒ] n.;v., caged, caging; n. 笼,箱;升降机箱[室],运输笼,提升罐笼;钢筋骨架;v. 制动,锁定
~ antenna 笼形天线
~ construction 骨架构造
~ hoist (elevator) 笼式升降机
~ lifter 笼式升降机
~ of reinforcement 钢筋骨架
~ screen 笼筛
~ structure 骨架结构
hoisting ~ 提升[起重]笼
self-dumping ~ 自卸罐笼
supply ~ 供料罐笼
suspension ~ 吊笼
tilting-deck ~ 可翻卸罐笼
winding ~ 提升罐笼

cairngorm [ˈkɛənˈgɔːm] n. 烟水晶

caisson [ˈkeisən] n. 沉箱,船坞闸门,充气浮筒;(天花板的)凹格,藻井;caissoned, a.
~ ceiling 沉箱盖板;古建筑中的藻井
~ soffit 空腹拱
~ work 沉箱施工

caisson soffit

cake [keik] n.;v. (污泥)滤饼,团[结]块,粘[烧]结,胶凝
graphite ~ 石墨片[块]
tough ~ 精铜(含铜约99%)

caking [ˈkeikiŋ] n. 粘结,烧结
~ power 烧结能力

cak(e)y [ˈkeiki] a., cakier, cakiest; 成了块的,凝固了的;

calacata n. 灰纹理大理石

calandria [kəˈlændriə] n. 排管式加热器,加热管组

calc [kælk] n. 石灰(质),钙(质)
~ granite 钙质花岗岩
~ sinter 石灰华,多孔石灰岩
~ spar 方解石
~ tufa 石灰华

calcareous [kæl'kɛəriəs] *a.* 石灰质的，钙质的；calcareously，*ad.* calcareousness，*n.*
~ aggregate 石灰质集料
~ alga 石灰(质)藻
~ alabaster 钙质石膏
~ cement 水硬性石灰
~ clay 白垩黏土，石灰质黏土
~ marl 石灰质泥灰岩
~ mud 石灰粘泥
~ ooze 钙质软泥
~ rock 含钙[石灰]岩
~ silex 硅质灰岩
~ spar 方解石
~ tufa 石灰华

calcia ['kælsiə] *n.* 氧化钙
calcic ['kælsik] *a.* 石灰质的，钙质的
calciferous [kæl'sifərəs] *a.* 含碳酸钙的，含钙的
~ petrosilex 石灰角页岩

calcify ['kælsifai] *v.* 钙化，石灰化
calcilutyte [kælsi'lju:tit] *n.* 灰泥岩，灰质碎屑岩
calcimine ['kælsimain] *n.* 刷墙水粉，可赛银粉，墙粉；*vt.* 刷墙粉于……
calcination [kælsi'neiʃən] *n.* 煅烧，烧成之物
calcine ['kælsain] *n.*；*v.* 煅[焙]烧
~d alumina 煅烧氧化铝[铝氧土]
~d bauxite 煅烧铝矾土
~d dolomite 煅烧白云石
~d flint chips 煅烧燧石屑
~d gypsum 熟石膏
~d lime 生石灰
~d magnesite (magnesia) 煅烧苦土[氧化镁]

calcite ['kælsait] *n.* 方解石
~ aegiapite 方解霓磷灰岩
~ syenite 方解正长岩

calcitic ['kælsitik] *a.* 石灰质的
~ limestone 方解石
~ marble 石灰质大理石

calcitrant ['kælsitrənt] *a.* 耐火的，不易熔化的

calcium ['kælsiəm] *n.* 钙(Ca)
~ acetate 乙酸钙，醋酸钙
~ aluminate 钙矾土，铝酸钙
~ aluminate cement 高铝水泥，矾土水泥
~ aluminate hydrate 水化铝酸钙
~ aluminoferrite 铁铝酸钙
~ bicarbonate 重碳酸钙
~ bleach 漂白粉
~ carbide 电石，碳化钙
~ carbonate 碳酸钙
~ chloride 氯化钙
~ fluoride 莹石，氟化钙
~ hydroxide 氢氧化钙，消[熟]石灰
~ lime 生[钙]石灰，未消石灰
~ oxide 生石灰，氧化钙
~ (quick)lime 生石灰，商品钙
~ silicate 硅酸钙
~ silicate facing brick 石灰砂砌面砖
~ silicate hydrate 水化硅酸钙
~ sulfoaluminate 硫化铝酸钙
~ sulphate 硫酸钙，石膏
~ sulphate plaster 石膏(抹面)灰浆
~ sulphate incrustation 石膏装防火隔声板

calcium sulphate incrustation

calculable ['kælkjuləbl] *a.* 可计算的，可测出的
calculagraph ['kælkjuləgra:f] *n.* 计时器[仪]
calculate [kælkjuleit] *v.* 计算；计划；评估，议价；核算，推测
~d height 计算高度
~d screen cut 计算断面
~d span 计算跨度

calculation [,kælkju'leiʃən] *n.* 计算；估计，预测
~ error 计算误差
~ of cutting and filling 挖方和填方

计算
~ of loading 荷载计算
cost ~ 成本计算
design ~ 设计计算
determinant ~ 行列式计算
earth mass ~ 土方计算
fixed-point ~ 定点计算
graphical ~ 图解计算
matrix ~ 矩阵计算
numerical ~ 数值计算
preliminary ~ 初算,概算
short-cut ~ 简化计算

calculative [ˈkælkjulətiv] *a.* 计算的
~ cost 计算价格,造价

calculator [ˈkælkjuleitə] *n.* 计算器;计算员
analogue ~ 模拟计算器
hand ~ 手摇计算机
wind ~ 风速计算器

calculus [ˈkælkjuəs] *n.* 计算(法),演算;微积分学

caldarium [kælˈdɛəriəm] *n. pl.* -ria [-riə];(古罗马的)高温浴室

calefacient [ˌkæliˈfeiʃənt] *a.*,发热的,供暖的,生热的

calefactory [ˌkæliˈfæktəri] *n. pl.* -ries 暖房,取暖;*a.* 生热的,温暖的

calendar [ˈkælində] *n.* 日历
~ progress chart 工作计划进度表

calf [ka:f] *n.* 冰川上崩落漂流的冰块,小牛(皮)
~'s tongue moulding (建筑)牛舌饰

caliber [ˈkælibə] *n.* = calibre 管内径,直[口]径;卡钳;能力
~ size 管径大小,口径尺寸
heavy ~ 大口径

calibrate [ˈkælibreit] *vt.* -brated, -brating;校准,检查,测定,刻度

calibrating [ˌkæliˈbreitiŋ] *n.* 校准

caliche [kəˈli:tʃi] *n.* 钙积层,泥灰石;生硝,智利硝

caliduct [ˈkælidʌkt] *n.* 暖[热]气管道

calk [kɔ:k] *vt.* 嵌塞,填缝(隙),填实
~ed ends 开尾螺栓[索杆]
~ing groove 嵌槽

call [kɔ:l] *n.*;*vt.* 调用,引入(子程序);要求,称呼,叫做
~ -back pay 加班费
~ bell 信号[呼叫]铃,电铃
~ box 公用电话亭,邮政信箱
~ button 呼人或报警铃的按钮
~ indicator 呼叫指示器
~ lamp 信号灯
~ing for tenders 招标
~ ing indicator 引入指令

calliper [ˈkælipə] *n.* = caliper 圆规,量径器;*vt.* 测量
~ gauge 测径规
~ rule 卡尺
inside ~s 内径卡规[钳]
micrometer ~s 千分卡尺[规]
outside ~s 外径卡规
slide ~s 游标卡尺
telescope ~s 光学测微仪
vernier ~s 游标卡尺,游标测径器

callus-wood *n.* 愈合材

callys 板岩,片岩

calm [ka:m] *a.* 平静的,[镇]静的;*n.* 无风,零级风
~ belt [zone] 无风带
~ing section (柱的)减震部分

caloricity [ˌkæləˈrisiti] *n.* 热容量,热值,发热量

calorie [ˈkæləri] *n. pl.* -ries = calory 卡路里(热的单位)

calorific [ˌkæləˈrifik] *a.* 热的,发热的
~ capacity 热容量
~ conduction 热传导
~ effect 热效应
~ receptivity 感热性
~ value (power) 热值,卡值

calorifier [kæˈlɔrifaiə] *n.* 热风机[炉]，加[顶]热器
calorised steel *n.* 渗铝钢
calorize [ˈkælərɑiz] *vt.* 表面渗铝
 ～ zing process（表面）渗铝，铝化处理法
 ～ zing steel 铝化钢
 pack（powder）～ zing 固体铝化（处理）
calorstat [ˈkælɔːstæt] *n.* 恒温箱[器]
calotte [kəˈlɔt] *n.* 半球面，帽状物，平扁弯顶，拱顶，屋顶塔尖之小圆盖
 oval ～ 椭圆拱顶[曲面]
 paraboloid ～ 抛物曲面拱顶
 spherical ～ 球面拱顶
calpis [ˈkælpiz] *n.* 乳浊液
calsomine *n.* 刷墙粉
calx [kælks] *n., pl.* calxes, calces [ˈkælsiːz]；金属灰，矿灰，石灰
cam [kæm] *n.* 凸轮，偏心轮，样板，仿形板
 ～ bearing 凸梁，弯梁
 ～ bearing clearance 弯梁的净空
camber [ˈkæmbə] *n.* 上挠度，曲率，弧线，拱高；*vt.* 扭曲，弯曲
 ～ arch 弯拱
 ～ barge board（gable）弓形山墙封檐板，下卷式山墙封檐板
 ～ beam 弓背梁，拱形梁，上凸梁
 ～ line 弧线，倾斜线
 ～ piece 拱材
 ～ roof 弓形屋顶
 ～ slip 砌拱垫块
 ～ed ceiling 弓形顶[顶棚]
 ～ed plate 弓形板
 double ～ 双弧线
cambium [ˈkæmbiəm] *n.* 形成层，新生层
 ～ layer 木材青面纤维层
Cambodian style *n.* 柬埔寨式(建筑)
camboge *n.* 穿孔混凝土块
cambogia [kæmˈbəudʒiə] *n.* 藤黄
cambric [ˈkeimbrik] *n.* 细麻布，黄蜡布
 ～ cable 细麻索缆
 ～ insulation 细麻布绝缘
 ～ paper 布纹纸
came [keim] *n.* (嵌窗玻璃用的)有槽铅条
camel [ˈkæməl] *n.* 起重浮箱[筒]；骆背式
 ～ back top chord 桁架拱形上弦
 ～ back type crusher 驼峰形锤式破碎机
cameo [ˈkæmiəu] *n.* 建筑装饰浮雕，浮雕玉石
 ～ glass 浮雕玻璃
 ～ ware 有浮雕陶制小装饰品

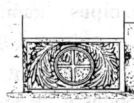
cameo

camera [ˈkæmərə] *n.* 摄影机，镜[暗]箱
 ～ angle 摄影角度
 closed-circuit TV ～ 闭路电视摄像机
 kite ～ 俯瞰图照相机
 laser ～ 激光摄影机
 pickup ～ 摄像机
 projecting ～ 投影机，幻灯
 sound ～ 录音室
 streak ～ 扫描照相机
 (TV) ～ chain(s) (电视)摄像系统
cam-lock [ˈkæmlɔk] *n.* 偏心夹，凸轮锁
camloy *n.* 镍-铬-铁耐热合金
camomatic grinder 全自动凸轮磨床
camouflage [ˈkæmuflɑːʒ] *n., v.*, -flaged, flaging；伪装，掩饰
 ～ coat 伪装涂层
 ～ paint 保护色油漆[涂料]
 fixed ～ 永久性的伪装
camp [kæmp] *n.* 帐篷，营宿；*v.* 营地
 ～ buildings 施工营地
 ～ car 宿车
 ～ chair 轻便折椅
 ～ equipment 外业设备
 ～ sheeting 板桩
 ～ stool 轻便板凳

campanile [kæmpə'ni:li] *n. pl.* -niles, -nili；(教堂的)钟塔[楼]

campbell-stokes' heliometer 康贝尔—斯托克斯式日照仪

campceiling 帐篷式顶棚

camphor ['kæmfə] *n.* 樟脑
~ tree 樟[树]
~ wood 樟木

campo ['kæmpəu] *n. pl.* -pi 市镇中心的广场；大草原

campus ['kæmpəs] *n.* 校园，校内场地
~ plan (大学)校园平面布置图
~ planning 校园规划

camus *n.* 装配式建筑施工法；预制结构施工法
~ system 卡莫斯式体系建筑(法)

canal [kə'næl] *n.* 运河，渠道，沟渠，水道
~ construction 渠道建筑
~ garden 运河庭园(文艺复兴时期的一种几何式花园)

cancel ['kænsəl] *n.；v.*, -cel(l)ed, -cel(l)ing；取消，删去，作废；组成网格状
~ arch 古罗马凯旋门
~ order 注销定货单
~ led structure 格(组)构(架)

cancellation [kænsə'leiʃən] *n.* 网格结构；抵消，作废，删去，化为零
~ clause 撤消条款
~ of building licence 建筑封锁[禁止]期，收回或取消建筑许可证

candela [kæn'di:lə] *n.* 发光强度单位，(新)烛光，堪德拉

candelabrum [kændi'lɑ:brəm] *n.* 大分支烛台[灯台]；华柱
~ base 小型烛台形灯座

candle ['kændl] *n.* 蜡烛，烛状物，烛光(发光强度单位)
~ pitch 硬沥青，柏油脂
~ power distribution 配光
~ tar 柏油，焦油沥青
concentric ~ 同心电极弧光灯
Debrun ~ 直角弧光灯
foot ~ 烛光(照度单位)
international ~ 国际烛光

cane [kein] *n.* 竹料，藤料，棍，棒
~ bolt 竹栓，棒栓
~ chair 藤椅
~ work 藤制工艺品

canephora [kə'nefərə] *n.* 古希腊少女雕像，头顶盛有圣物之篮的少女柱形头雕像(亦作 canephoros, canephor, canephore 或 canephorus)

canephora

cannelure ['kænəljuə] *n.* 细槽，环形槽，纵槽[沟]

cannibalize ['bænibəlaiz] *v t.* 拆下零件，装配[修]；同型装配，轨上装配

cannon ['kænən] *n. pl.* -nons, -non；空心轴，加农高速钢；*v.* 冲击，撞击
~ connector 加农插头与插座
~ plug 圆柱形插头，加农插头
~ tube shield 筒形屏蔽罩
~ proof 防弹的

canon ['kænə] *n.* 标准，规范，规则，定律
~s of art 艺术标准

canonical [kə'nɔnikəl] *a.* 典型的，标准的，公认的，权威的
~ form 范式，规则形式

canopy [kænəpi] *n.* 盖，罩，冠；防雨篷，防雨盖板
~ ceiling 防雨天盖
~ door 上悬挑门，滑动挑门
~ lip 坡屋[雨篷]边缘
~ sidewalk 挑棚[走廊]式人行道
bell ~ 钟顶天幔
dust ~ 防尘罩
grass ~ 草皮[盖]

cant [kænt] *n.* 屋隅，外角；斜面[边]；*v.* 斜切，成斜角；倾斜，倾覆
~ bay 斜砌壁洞，三边形突肚窗
~ board 天沟侧板，侧立板
~ bonder 斜削[向外张开的]砌缝

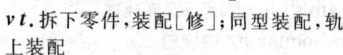

~ brick 斜面砖
~ column 多角柱,菱形柱
~ moulding 斜饰,多角线饰
~ stretcher 斜面砌砖
~ strip 镶边板条
~ tie plate 斜面垫板
~ed 切去棱角的,倾斜的
~ed folding 倾斜折叠
~ed wall 互支角墙,曲角墙

cantilever [ˈkæntiˌliːvə] *n.* 悬[伸]臂,悬臂梁,悬桁
~ arch 悬臂拱
~ arch truss 拱形悬臂桁架
~ arched girder 悬臂拱梁
~ arm 悬臂距
~ bar screen 悬臂铁栅筛
~ beam 悬臂梁
~ block 小牛腿,横支柱,托架,支柱石
~ boom 水平悬臂梁
~ bracket 悬臂托座,悬臂牛腿
~ bridge 悬臂桥
~ construction 悬臂施工法
~ cupola 悬臂式圆[半球形]屋顶
~ diaphragm 悬臂式膜窗
~ dome 悬臂式穹顶
~ element (拱坝的)悬臂梁
~ erection 悬臂架设施工法
~ extension 悬臂引伸
~ folded plate(roof) 悬臂折板结构(屋面)
~ footing 悬臂基础[脚]
~ frame 悬臂框架
~ girder 悬臂大梁
~ hipped-plate roof 悬臂斜脊屋顶
~ masonry(work) 悬臂砌筑
~ prismatic shell roof 悬臂折板薄壳屋顶
~ roof 悬臂式屋顶,翅棚
~ segment (桥梁结构上的)悬臂扇形体
~ span 悬臂(外伸)跨度
~ truss 悬臂桁架

~ vault 小牛腿,托架,横支柱
~ed landing 悬臂楼梯平台
~ed wall 悬臂墙

canton [ˈkæntən] *n.* 外角,凸出的隅角石; *vt.* (划)分
~ed building 隅角建筑物
~ed column 隅角柱饰

canton

canvas [ˈkænvəs] *n.*; *a.* 帆布,油画(画)布
~ awning 帆布凉篷
~ belt 帆布带
~ decking 防水帆布屋面
~ duct 帆布软管[水龙带]
~ gate 帆布挡板
~ hood 帆布罩
~ hose 帆布水龙带

caoutchouc [ˈkautʃuk] *n.* 天然橡胶,生橡胶

cap [kæp] *n.* 帽,盖(板),罩,塞;桩头,桩帽;管盖,管帽; *vt.* 覆盖,胜[超]过
~ bolt 有头[帽]螺栓,盖螺栓
~ collar gasket 螺帽垫圈
~ copper 带状黄铜
~ flange 螺帽垫圈
~ flashing (防漏用)金属盖片
~ form (制造桩帽用的)桩帽模型
~ grouting 顶盖灌浆法
~ lamp 帽灯,头灯
~ nut 螺母,(锁紧)螺帽
~ of pile 桩帽
~ped pile 带桩帽的桩
~ped quartz 冠状石英
~ piece 帽子,帽罩,喷头
~ plug 盖塞
~ rock 冠岩,盖保岩
~ screw 有头[帽]螺钉
~ sheet 最上层屋面板
~ -sill 盖木
~ strip 帽材

~ vault 拱形顶
blast ~ 风帽
breather pipe ~ 通风管盖
dust ~ 防尘罩
end ~ 管端盖帽
funnel ~ 烟囱帽
glass ~ 玻璃罩[帽]
lamp ~ 灯头,管帽
radiator ~ 散热器盖
sand ~ 防砂[尘]罩
screw ~ 螺钉头,螺帽
set ~ 顶梁,棚梁
square-set ~ 方框支架的横梁
turn ~ 烟道风帽
watch ~ 烟囱罩,导烟罩[帽]
capability [keipə'biliti] *n. pl.*-ties 能力,可能性;容量;效力,功率
absorption ~ 吸收能力
crack-arrest ~ 止裂能力
damage ~ 破坏能力
fatigue life ~ 耐疲劳能力,疲劳寿命
heat-transfer ~ 传热能力
load-carrying ~ 承载能力
self-test ~ 自检能力
capacious [kə'peiʃəs] *a.* 容积[量]大的,宽敞的,广阔的-ness *n.*
capacity [kə'pæsiti] *n. pl.* ties 容积[量];生产(能)力,能量,负荷量;权力,资格;电容
~ tonnage 载重量
absorption ~ 吸收能力
bearing ~ 承载能力
carrying ~ 载重量
carrying ~ of pipe 管道过流能力
discharge ~ 通行能力
flow ~ 流量,泄[排]水能力
wearing ~ 耐磨性
working ~ 工作能力[量]
cape [keip] *n.* 海角,岬;披肩,肩衣
~ asbestos 兰石棉
~ chisel 扁尖凿,岬錾
~ diamond 黄金刚石
~ ruby 红榴石

~ top 篷车顶,软车顶
capillator *n.* 毛细管比色计
capital [kæpitl] *n.* 柱头[顶,冠];资本,资方;首都[府]; *a.* 最重[主]要的,基[根]本的;第一流的;资本的

capital

~ assets 固定资产
~ budget 基本建设预算,资本预算
~ carving 柱头雕塑
~ construction 基本建设
~ cost 成本,资本(值),核定投资额
~ destruction 资金耗损
~ employed 使用资本额
~ expenditure 基本建设费用
~ investment 基本建设投资
~ of Byzantine 拜占庭式柱头
~ of composite column 混合式柱头
~ of Corinthion column 科林斯柱头
~ of Doric column 多立克柱头
~ of Gothic column 哥特式柱头
~ of Ionic column 爱奥尼亚柱头
~ of Moorish column 摩洛哥式柱头
~ of Romanesque column 罗曼式柱头
~ of Tuscan column 德斯金柱头
~ ornament 柱头装饰
~ outlay 基建投资,资本支出
~ project funds 基建项目基金
~ repair 大修
~ stock 基金,(资本)股票
~ works 基本建设工程
ancient ~ 古都,旧都
angle ~ 转角柱帽[头]
bracket ~ 伸臂柱头
carved ~ 雕花柱头
column ~ 柱头
cubical block ~ 斗状柱头
double ~ 双柱帽头

capital carving

impost ~ 拱墩帽
monopoly ~ 垄断资本
pilaster ~ 壁柱帽
vaulting ~ 拱顶扁倚柱头
Capitol ['kæpitl] *n.* 古罗马 Jupiter 神殿,美国国会议场
~ Hill 美国国会
caplastometer [ˌkæpləs'tɔmitə] *n.* 粘度计
capping ['kæpiŋ] *n.* 表土,剥离物,盖层岩;(防波堤的)盖顶石;压顶,盖面;桩帽;楼梯扶手
~ beam 压檐梁
~ brick 檐砖,压顶砖
~ piece 压檐木[梁]
pilaster ~ 壁柱帽
ridge ~ 层脊泛水
capron(e) ['kæproun] *n.* 卡普纶,聚己内酰胺纤维
capstan ['kæpstən] *n.* 绞盘,卷扬机,拉丝卷筒
~ bar (转动的)绞盘臂[杆]
~ bolt 绞盘螺栓,带销螺栓
~ nut 有孔螺母
~ winch 绞盘
~ windlass 绞盘式卷扬机
capstone ['kæpstəun] *n.* 拱[压]顶石,顶(层)石
capsulation [ˌkæpsju'lei-ʃən] *n.* 密封,封装
capsule ['kæpsju:l] *vt.* 压缩,节略; *n.* 气压输送机[筒],容器,封壳,摘要
~ metal 铅镁合金(92%铅,8%锡)
plastic ~ 塑料封壳
caption ['kæpʃən] *n.* 标题,说明,目录,插图说明
captive ['kæptiv] *n.* 被困者; *a.* 被吸[困]住的,被捕获的
~ key 弹性销[键]
~ pin 安全销钉,合缝钉
~ screw 系紧螺钉
car [ka:] *n.* 汽车,电车,货车;车厢,电梯,吊篮

~ park 停车场
~ park basement 地下停车场,地下车库
~ parking roof 停车棚
~ port 简易车库,停车场
~ ramp 汽车用的坡道
~ repair shed 修车棚
~ storage park 存车场
caracole ['kærəkəul] *n.* 旋梯; *vi.* 半旋转
caramel ['kærəmel] *n.* 垫块[片],灰浆块隔片;酱色(着色)
caravansary [ˌkærə'vænsəri] *n.* 大旅舍,大客店
carballoy *n.* 卡波硬质合金,碳化钨硬质合金
carbamate ['ka:bəmeit] *n.* 氨基甲酸酯,甲氨酸酯[盐]
carbamide [ka:'bəmaid] *n.* 尿素,脲
~ resin 聚脲树脂,碳酰胺树脂
carbene *n.* 碳烯,碳质沥青
carbide ['ka:baid] *n.* 碳化钙,电石;碳化物,硬质合金
~ alloy 硬质合金
~ brick 碳硅砖
~ cermets 碳化物金属陶瓷
~ lime 电石石灰
~ of calcium 电石
~ of silicon 碳化硅,金刚砂
~ -tipped 硬质合金的
cemented ~ 烧结硬质合金
double ~ 复合碳化物
hard ~ 硬质合金
silicon ~ 碳化硅,金刚砂
tri-ferrous ~ 渗碳体
carbinol ['ka:binɔl] *n.* 甲醇
carbofrax *n.* 碳化硅耐火材料,金刚硅砖料
carbolic acid 石炭酸,苯酚
carbolic oil 酚油,石炭酸油
carbolic soap 酚皂,石炭酸皂
carbolon ['ka:bəuloun] *n*, 碳化硅
carboloy *n.* 碳化钨硬质合金

carbon ['kɑ:bən] *n.* 炭,石墨,碳精电极,碳膜电阻;复写本,副本; *a.* 碳的,以碳处理的
~ adsorption method 活性炭吸附法
~ amber glass 有色玻璃,琥珀玻璃
~ arc 碳弧
~ arc cutting 碳弧切割
~ arc welding 碳弧焊接
~ arc lamp 弧光灯
~ bearing 含[带]碳的
~ black 碳黑
~ brick 碳砖,碳质耐火砖
~ carburizing steel 渗碳钢
~ construction(al) steel 碳素结构钢
~ construction(al) quality steel 优质碳素结构钢
~ dioxide gas extinguisher 二氧化碳灭火器
~ packing 碳素[质]衬垫
~ pick-up 渗碳,碳化
~ refractory 碳素耐火材料
~ remover 碳尘消除剂
~ steel 碳钢
absorbent ~ 活性碳
active ~ 活性碳
decolourizing ~ 脱色碳
graphite ~ 石墨碳
hard ~ 硬碳,固体碳粒
medium ~ steel 中碳钢
vitreous ~ 玻璃态石墨
carbonado [,kɑ:bə'neidəu] *n.* 墨金刚石; *vt.* 在……上面刻出深痕
carbonate ['kɑ:bənit] *n.* 碳酸盐[酯],黑金刚石; *vt.* 碳化,焦化
~ of lime 碳酸钙,石灰石
carbonatite *n.* 碳酸岩
carbonic [kɑ:'bɔnik] *a.* 含碳的
~ limestone 石炭纪石灰岩
~ rock 含碳岩石
~ sandstone 碳质砂岩
carbonizing steel 渗碳钢
carbonitride [,kɑ:bə'naitraid] *n.* 碳氮化合物,氰化物
~d steel 氰化钢
carborundum [,kɑ:bə'rʌndəm] *n.* 人造金刚砂,碳硅砂,碳化硅
~ brick 碳化硅砖
~ disc 金刚砂盘
~ grinding wheel (金刚)砂(磨)轮
~ paper (金刚)砂纸
~ refractory 碳化硅耐火材料
~ saw (金刚)砂锯
~ tile 金刚砂砖
carburet ['kɑ:bjuret] *vt.* -ret(t)ed, -ret(t)ing;使与碳化合,增碳
~ed iron 碳化铁
carburize ['kɑ:bjuraiz] *vt.* =carburise 渗碳
carburizing steed 渗碳刚
pack carburizing 固体渗碳
carcase ['kɑ:kəs] =carcass *n.* 支[骨,框]架;壳体;钢筋;轮胎外胎;定子;尸体;遗骸
~ work 预埋工程(管子或电线)
~ flooring 毛地板
~ roofing 毛屋顶[面]
carcassing ['kɑ:kəsiŋ] *n.* 骨架制作
~ timber 木构架构件
Carcel unit 卡索(灯)光度单位(=6.9英国烛光单位)
carcinotron [kɑ:'sinəutrən] *n.* 回波管
carclazyte *n.* 高岭土,白[陶]土
card [kɑ:d] *n.* 穿孔卡,卡片,程序单,图表;印刷电路板,纹板,花板,标度板;(纸)牌;办法,计划,措施;钢丝刷 *v.* 制成卡片,列入表中
~ compass 平板罗盘
~ middle 衬纸
~ of patterns 装有几个模型的型板
~ of work order 施工说明卡片,操作程序卡片
~ wire 针布钢丝
aperture ~ 穿孔卡
height ~ 高度绘图仪

card compass

indicator ~ 指示图
load ~ 凿孔卡
test ~ ~ 测试图[表]
cardboard [ˈkɑːdbɔːd] *n.* 卡片,(硬质)纸板
~ space former 填充纸板
cardinal [ˈkɑːdinəl] *a.* 主要[基本]的,最重要的;鲜红色的 *n.* 基数(常用复数)
~ point (方位)基点
~ principle 基本原理
care [kɛə] *n.;v.* 管[护]理,维护
~ and maintenance 保养与维修
career [kəˈriə] *n.* 生涯,经历,职[事]业;专业;*a.* 职业性的 *vi.* 飞驰[奔]
~ plan 企业管理制度,专职计划
~ training 专职训练
careful [ˈkɛəful] *a.* 注意的,小心的,仔细的
~ treatment 精心处理
carina [kəˈrainə] *n. pl.* -nae [-niː];龙骨突;脊;隆线
carmine [ˈkɑːmain] *n.;a.* 洋红色(的),胭脂红色(的)
indigo ~ 靛蓝胭脂红,酸性靛蓝
carnation [kɑːˈneiʃən] *n.* 石竹;淡红色,肉色,绘画上的肉色部分
~ red 紫红色
carnauba(wax) [kɑːˈnaubəwæks] *n.* 棕榈蜡
caro bronze 磷青铜(锡 7.5% ~ 9%,磷 0.11% ~ 0.4%)
carpenter [ˈkɑːpintə] *n.* 木工;*v.* 做木工活
~'s bench 木工台
~'s apprentice 木工学徒
~'s art 木工工艺
~'s band saw 木工带锯
~'s level 木工水平仪
~'s shop (yard) 木工场
~'s square 木工尺
~'s tool 木工工具
~'s trade 木工职业,木工手艺
~'s undertaking 木作企业
~'s work 木工工程
~'s and joiners' tools 粗细木工具
carpentry [ˈkɑːpintri] *n.* 木匠业,木作,木工,木器;木结构
~ tongue 木工凿,木雄榫
carpet [ˈkɑːpit] *n.* 毡层,地毯;磨损层;桌毯;绿茵;罩,包围;*vt.* 铺,盖
~ adhesive 地毯粘结剂
~ bed 地毯式花坛
~ (bonding) agent 地毯粘结剂
~ cut 地毯卡槽
~ lining 地毯衬垫
~ squart (同 ~ tile)组合地毯,小方块地毯
~ stretcher 地毯绷紧工具
~ strip 门槛,条形地板
~ treatment 铺筑毡层,表面处理
~ veneer 毡层,表面处理
~ work 铺筑毡层
asphalt ~ 沥青覆盖层
chipping ~ 石屑毡层[铺盖]
sand ~ 铺砂
carport [ˈkɑːpɔːt] *n.* 停车场,简陋的汽车棚
carr =carrier *n.* 承重构件,托架,运载工具
~ bit 冲击式(一字形)钻头,单刀钻冠
carrara marble (意大利产)雕塑用大理石
carrefour [kærəˈfuə] *n.* 十字路口,位于十字路口的广场
carrel [ˈkærel] *n.* (带书架的)阅览桌,图书馆书库中为读者使用的卡座
carriage [ˈkæridʒ] *n.* 车辆,支架,底座,平台,承重装置;运输;运费;楼梯斜梁[搁栅];排水管
~ bolts 方颈螺栓
~ piece 楼梯斜梁
~ porch 停车门廊,车廊
~ return 回车
~ shed 车棚

carrick bend 麻花大绳接头[结]
carrick bitts 支承座立柱
carried ['kærid] *a.* 悬挂式的,被运送的
carrier ['kæriə] *n.* 承运人,承运公司；运载工具,装卸机,运输机；载波[体],吸收剂；梁,承重构件,承重层
~ assy 托架,支座,牛腿
~ bar 承载梁
~ frame 托架,(汽车)底盘框架
~ plate 顶板
carry ['kæri] *v.*, -ried, -rying；携带；含有；传播,搬运；安装,负荷,推行,贯彻,通过；*n.* 进位；指令；传送
carryable [kærieibl] *a.* 可移动的,可携带的,轻便的
carrying ['kæriiŋ] *a.* 装[承]载的,运输的
~ capacity of the environment 环境负荷量,环境的负担能力
~ channel U形承载槽
~ cupola 承载的钟形屋顶
~ rail 承载轨道
~ up 砌墙
~ wall 承重[压]墙
cart [kɑ:t] *n.* 大车,手推车；*v.* 运输[载],用车装运
~ (and) barrow hoist 混凝土手推车提升机
battery ~ 电瓶车
hand ~ 手推车
grout ~ 灰浆(小)车
cartographer [kɑ:'tɔgrəfə] *n.* 制图员,地图绘制员
cartography [kɑ:'tɔgrəfi] *n.* 绘制图表,绘[制]图法,制图学
carton [kɑ:tən] *n.* 纸板(箱),厚纸
cartoon [kɑ:'tu:n] *n.* 草[底]图；漫画；连环画；动画片
cartoscope ['kɑ:təskəup] *n.* 图像显示屏
cartouche [kɑ:'tu:ʃ] *n.* 涡形[卷轴]装饰,(法国)卷边椭圆形牌匾

cartridge ['kɑ:tridʒ] *n.* 夹头；灯座,管壳,支架；编码键筒；盒式存贮器；微型磁带；胶卷筒
~ brass 黄铜,铜锌合金
~ (type) fuse 熔丝管,保险丝管,安全保险丝管
~ paper 厚纸,图画纸
~ powered tool 锚枪
film ~ (放胶片的)暗盒
carve [kɑ:v] *v.* 将……切成碎片；雕刻,切开；开创,创造(常与out连用)
~d altar 雕刻祭[神]台
~d capital 雕饰柱帽
~d decoration (pattern, enrichment) 雕刻装饰
~d decorative feature (finish) 雕刻装饰(终饰)
~d foliage 雕刻的树叶或花形的饰物
~d motif 雕饰图式[题材]
~d ornamental finish (feature) 雕塑装饰
~d pulpit 雕塑的(教堂)讲台
~d relief 浮雕
~d style 雕刻艺术[风格]
~d waterspout (奇形怪状雕像)滴水咀装饰
~d wood(en) wall panel(l)ing 木材雕刻的墙面外装饰
~d work 雕刻工作
carver [kɑ:və] *n.* 雕刻者,雕工
case [keis] *n.* 情况,事例,理由；外壳,箱,盒,室,容器；主体,机身；表面,把……装入盒[箱]套管]内,加固钻孔
~ bay 梁间距,桁间
~ bay part 椽梁间距部分
~ bolt 圆头螺栓
~ handle 装入凹槽的门拉手
~ hardened 表面硬化的
~ hardened glass 表面硬化玻璃,钢化玻璃
~ hardered steel 表面硬化钢,渗碳钢
~ hardness 表面硬度

~ study house 实验性住宅
~ way wiring system 箱式布线法
door ~ 门框
cased [keisd] *a.* 箱形的,有外套(管)的;加框的
~ beam 箱形梁
~ column 箱形[空]心柱
~ frame 箱式窗架
~ glass 套色玻璃
~ hole completion 带孔的预制件
~ post 箱形柱
~ type fan 有保护网的风扇
~ window frame 匣形窗架
casein ['keisi:in] *n.* 酪蛋白,酪素
~ glue 酪蛋白胶
~ paint 酪素涂料
~ plastic 酪素塑料
~ water paint 酪素水溶性涂料
casemate ['keismeit] *n.* 空心造型,穹窿;暗炮塔[台]
casement ['keismənt] *n.* 玻璃窗扉,窗扉;箱,盒,套,框
~ adjuster 窗风撑,门档[钩]
~ bolt 玻璃窗插销
~ cloth 薄窗帘布
~ door 玻璃门
~ fastener 窗插销
~ stay 窗风钩
~ sections 窗框钢
~ screen 平开隔帘
~ window 竖铰链窗,双опе窗
~ with sliding upper-sash 上扇滑动式窗扉
folding ~ 折窗
French ~ 玻璃落地窗
metal ~ 合页金属窗
out-swinging ~ 外开窗扉
pivoted ~ 旋转窗(弧形转动窗)
side hung ~ 推开窗扉
ventilation ~ 通风窗扉
cash [kæʃ] *n.* 软片岩,矸;现金,钱; *vt.* 兑现

~ advance 预付现金
~ deposit as collateral 保证金
~ with order 定货即付款
~ payment 付现金
cashew [kæ'ʃu:] *n.* 漆树,腰果油树
~ -nut aldehyde plastic 漆酚醛塑料
~ resin 漆酚树脂
~ water coat 腰果油水性防锈底漆
casing ['keisiŋ] *n.* 框架,围墙,挡板;套管;箱,壳,罩
~ cap 螺旋管塞
~ elevator 管式电梯[升降机]
~ flange 管子法兰盘
~ glass 镶色玻璃
~ gun 弹筒打孔机
~ nail 圆铁钉
~ of column 柱子护面[围板]
~ ply 骨架层
~ tongs 套筒扳手
~ with bell-mouthed collars 承插式套管
air ~ 气隔层(平屋顶)
back ~ 临时木棚子
brickwork ~ 砖工罩面
cemented ~ 已注水泥的套管柱
concrete ~ 混凝土面饰
protective ~ 防护罩
casino [kə'si:nəu] *n. pl.* -nos;游乐场,俱乐部;意大利别墅
~ folie 娱乐场
cask [kɑ:sk] *n.* 容器,桶;一桶
casket ['kæskit] *n.* 容器,桶,吊斗;(精美的)小盒子
cassava [kə'sɑ:və] *n.* 卡萨瓦淀粉胶结剂;木薯属植物的根[淀粉]
cassel yellow 氯化铅黄
casseroles brown coal 焦褐色颜料
casserole ['kæsərəul] *n.* 瓷勺,柄皿
cassette [kæ'set] *n.* 盒式磁带;箱;X光底片盒
~ ceiling 空心楼板
cast [kɑ:st] *n.*; *v.* 铸造[件];模型;浇捣[注];计算,预测,安排

~ aerated concrete （预制）多孔混凝土
~ aerated concrete wall panel 多孔混凝土墙板
~ alloy 铸造合金
~ alumium 铸铝
~ beam 预制混凝土梁
~ brass 生[铸]黄铜
~ bronze 铸青铜
~ concrete 浇注混凝土
~ concrete umbrella 预制混凝土隔水墙
~ glass 压铸玻璃
~ iron pipe 铸铁管
~ joint 铸焊，浇铸连接
~ lintel 预制混凝土过梁
~ member 预制混凝土构件
~ plastics 铸塑塑料
~ resin 铸塑树脂
~ steel 铸钢
~ stone 铸石
~ ware 注浆陶瓷
inblock ~ 整体铸造
individual ~ 分块铸造
castable refractory concrete 耐火混凝土
castellated ['kæstəleitid] *a.* 城堡形的，建有城堡的；齿形的，有许多缺口的
~ beam 垛形梁
~ nut 槽顶螺母
castellation [ˌkæste-'leiʃən] *n.* 城堡状建筑，穹形齿，垛形墙顶，雉堞墙
casting ['kɑ:stiŋ] *n.* 铸造[件]；投掷；预制混凝土建筑；计算
~ and blading of material 材料的堆筑与整平
~ area 浇注面
~ box 型箱，浇混凝土用的模板
~ cycle 混凝土浇筑周期
~ nut 槽顶[形]螺母
~ plaster 熟石膏
plaster ~ 石膏模型
cast-in-place *a.* 现场浇筑的

~ concrete 现浇混凝土
~ floor 现浇混凝土楼板
~ lining 整体式衬砌
~ pile 钻孔灌注桩
cast-iron ['kɑ:staiən] *n.*；*a.* 铸铁（的），生铁（的）；硬的，无伸缩性的
~ fish-bellied girder 鱼腹式铸铁大梁
~ fitting 铸铁管件
~ pipe 铸铁管
~ radiator 铸铁散热器
~ socket 铸铁套管
~ soil pipe 铸铁污水管
~ trap 铸铁存水弯[管]
castle ['kɑ:sl] *n.* 城堡，大建筑物
~ architecture 宫殿建筑艺术
~ building 宫殿建筑
~ circular nut 六角圆顶螺母
~ nut 蝶形螺母，槽顶螺母
casual ['kæʒjuəl] *a.* 偶然的，随机的，临时的，不规则的
~ expense 临时费用
~ labour 临时工
~ sands 不规则砂层
casualty ['kæʒjuəlti] *n. pl.* -ties；故障，损坏；事故，伤亡；伤员
~ dressing and examination room 急救室
cat [kæt] *n.* 吊锚；履带拖拉机；耐火土；*vt.* 吊锚
~ bar 木插销
~ ladder 爬梯
~'s eye 猫眼
bear ~ 帆布安全带
catachosis [ˌkætə'kɔsiz] *n.* 破碎变质
cataclasis ['kætəklæsis] *n.* 破碎（作用），压碎
cataclasite [ˌkætə'kleisit] *n.* 破裂岩
cataclasm *n.* 碎断
catacomb ['kætəkəum] *n.* 地下墓穴，陵寝
catalin ['kætəlin] *n.* 铸塑酚醛塑料
Catalonian architecture （11世纪西班

牙)加泰隆尼亚建筑
catalysis [kə'tælisis] *n. pl.* -ses [-siz];催化(作用,现象,反应);触媒(作用)
catalyst ['kætəlist] *n.* 触媒,催化剂
catastrophe [kə'kæstrəfi] *n.* 事故,灾难,毁坏,灾变
catastrophic [kætə'strɔfik] *a.* 灾难性的,大变动的,不幸的
catch [kætʃ] *n.*; *v.*, caught [kɔːt], catching;捕捉,制动,按钮,指针;保险器;门扣,门窗拉手
~ bolt 弹簧门锁
~ drain 截[集]水沟,泄水沟,盲沟
~ fire temperature 着火温度
~ frame 格框,栅栏
~ gear 锁紧装置
~ hoom 截木栅
~ net 保护网
~ water ditch 集水排水沟
catcher ['kætʃə] *n.* 收集器,捕捉器
dust ~ 集尘器
whirler-type dust ~ 离心式集尘器
catching ['kætʃiæ] *a.* 有传染性的. *n.* 捕捉,收集
dust ~ 收[集]尘
spray ~ 喷雾附着
catenarian [ˌkæti'nɛəriən] *a.* 垂曲线的,链的
catenary [kə'tiːnəri] *n.*; *a.* 垂曲线(的),链(的)
~ arch 悬链曲线拱
~ curve 垂曲线
~ suspension 悬链
cater-cornered ['kætəˌkɔːnəd] *a.* 对角线的,成对角线的
cathedra [kə'θiːdrəl] *n.* 主教座;主教职位;讲座
cathedral [kə'θiːdrəl] *n.* 大教堂
~ glass 教堂内拼花玻璃,半透明的有花

cathedra

纹的玻璃
~ dome (cupola) 教堂半球形屋顶
Catherine wheel window 玫瑰形窗,车轮形窗
cathodic(al) [kə'θɔdik(əl)] *a.* 阴[负]极的
~ protection 阴极保护法
~ protection parasites 防碍阴极保护的物质
cation ['kætaiən] *n*,阳[正]离子
~ adsorption 阳离子吸附
~ exchanger 阳离子交换剂
cationic [kætai'ɔnik] *a.* 阳[正]离子的
~ additive 阳离子添加剂
~ binder 阳离子粘合剂
~ emulsion 酸性乳剂,阳离子乳液
~ slurry 阳离子沥青乳浆
~ surface active agent 阳离子表面活性剂
cationite ['kætaiənait] *n.* 阳离子交换剂
catoptric [kə'tɔptrik] *a.* 反射(光,镜)的
catoptrics [kə'tɔptriks] *n.* 反射光学
cattle hair felt 牛毛毡
catwalk ['kætɔːk] *n.* 轻便梯;脚手架,跳板,人行栈桥;小通路
Caucasian carpet 高加索(式)地毯,几何图案(式)地毯
caul [kɔːl] *n.* 填块,档[盖]板,抛光板
caulk [kɔːk] *vt.* 堵[嵌]缝,堵塞
~ weld 填缝焊
~ed joint 嵌密缝
caulker ['kɔːkə] *n.* 填缝工具,捻缝工
caulking ['kɔːkiŋ] *n.* 堵缝,铆接[合];凿密法,凿紧
~ hammer 填隙[堵缝凿密]锤
~ing chisel 填隙[捻缝]凿
~ing groove 嵌槽
~ing strip 嵌条
hand ~ 手工凿密
caustic ['kɔːstik] *a.* 腐蚀性的,苛性的;*n.* 腐蚀剂

~ alkalinity 苛性碱度
~ embrittlement 腐蚀性脆化
~ mud 石灰泥浆
~ potash 苛性钾,氢氧化钾
~ soda 苛性钠,烧碱
ground ~ 碱粉,粉状腐蚀剂
Causul metal 镍铬铜合金铸铁
cauterize ['kɔ:təraiz] *vt.* 腐蚀;灼烧
caution ['kɔ:ʃən] *n.* 小心,谨慎;警告
~ money 保证金,抵押金
~ security 保证金
~ signal 警告信号
cave [keiv] *n.* 凹槽[痕];内腔,岩洞,洞穴;*v.* 凹进去,倒坍
~ dwelling 洞[穴]居
~ -in 坍方[落]
~ temple 石窟(寺)
~ d material 碎片,残碴,瓦砾
cavea ['keiviə] *n. pl.* -veae 罗马半圆形露天剧场的阶坡座位
cavern ['kævən] *n.* 大山[岩]洞,洞穴,石窟
~ limestone 多孔石灰岩
~ rock 多孔岩石
cavernous ['kævənəs] *a.* 多孔的,海绵状的;凹的,洞穴状的
~ body 海绵体
~ structure 孔状结构
cavetto [kə'vetəu] *n.* 凹雕,凹弧饰;打圆,载面为 90° 弧的凹线脚

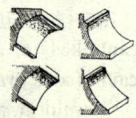
cavetto

~ vault 正方形拱顶,中部为平顶的反水槽式拱顶
cavitate ['kæviteit] *vi.* 抽空,出现涡凹
cavitation [kævi'teiʃən] *n.* 气蚀,空穴,空蚀(作用);混凝土中的大孔洞
~ damage 气蚀破坏[损伤]
~ erosion 气蚀,空隙腐蚀
~ limit 涡凹限度
cavity ['kæviti] *n. pl.* -ties;空腔,中空,空心,岩石中的裂缝

~ block 阴模
~ brick 空心砖
~ concrete wall 混凝土空心墙,混凝土两侧模板墙
~ masonry (work) 空心墙
~ panel 空心板
~ party wall 防火空心墙
~ pocket 空腔[洞],气泡
~ wall 空心墙,双层壁
dies (mould) ~ 型腔
gas ~ 气孔
shrinkage ~ 缩孔(铸造缺陷)
space ~ 空隙
cavo-relievo ['kɑ:vəuri'li:vəu] *n. pl.* -vos, -vi [-vi];凹浮雕(亦称 sunk-relief)
cawk ['kɔ:k] *n.* 氧化钡;重晶石,硫酸钡矿石
cay [kei] *n.* 暗礁,沙洲
~ sandstone 礁砂岩
cedar ['si:dz] *n.* 杉松[木],红松
~ panelled door 杉木镶板门
cedarwood *n.* 杉木,雪松属木材
cedrus ['sedrʌs] *n.* 雪松
ceil [si:l] *vt.* 装顶棚
ceiling ['si:liŋ] *n.* 顶棚,平顶,顶板,棚顶;限额,最高限额
~ board 顶棚,顶棚面板
~ coil 顶棚供暖盘管
~ cornice 顶棚周围檐口式线脚
~ dome (light) 圆屋顶(采光)
~ duct 顶棚通风管
~ enrichment 楼板装饰
~ fan 吊扇
~ filling 附顶灯具
~ girder 顶梁,排梁
~ height 室内净高
~ joist 吊顶木龙骨,吊顶搁栅,顶棚托梁
~ lamp 吊灯,平顶灯
~ light 平顶照明
~ load 顶棚载荷

~ panel 顶棚镶板
~ panel heating 顶棚板面采暖
~ panel strip 古建筑的顶棚枝条,顶棚木压条
~ plan 平顶[天花板]布置图,平顶仰视图
~ plaster 楼板[平顶]粉饰
~ rafter 顶棚搁栅
~ rail 顶棚周边的压条
~ sheet metal 薄钢板制天花板
~ shower 平顶淋浴器
~ strut 顶棚支撑
~ suspension system 顶棚悬吊索
~ without trussing 无梁平顶
beam ~ 露梁平顶
boarded ~ 板平顶
camp ~ 闷顶[阁楼]楼板
coffered ~ 格子平顶
coom ~ 隔音[声]平顶
coved ~ 穹窿天花板
coved and flat ~ 圆脚平顶
flat slab ~ 无梁天花板
groin ~ 格形天花板
hanged ~ 吊顶
celadon ['selədɔn] *n.* 中国青瓷;灰绿色
celite ['selait] *n.* C 盐,寅式,C-水泥石(水泥熟料中的矿物成分);次乙酰塑料
cell [sel] *n.* 单元,元件;晶粒[胞];原[蓄]电池,光电管元件,传感器,压力盒;容器,管,槽,小室
~ box girder 多孔箱形梁
~ call 单元引入,(编程序时的)子程序编码
~ concrete 多孔混凝土
~ in hollow tile 空心砖孔
~ quartz 多孔石英
~ type 程控(运用电脑可任意规定所需的顺序和时间的作业方式)
~ type sound absorber 小格型[蜂窝式]消声器
~ vault 格型拱顶
cella ['sel] *n. pl.* -lae [-li:];古希腊、古罗马庙宇里的内殿

cellar [selə] *n.* 地下室,地窖
~ air shaft 地下室通风采光竖井
~ drain(age) 地下室排水
~ for branches 楼体地下连结通道
~ masonry wall 地下室墙
~ masonry (work) 地下室墙体工程
~ window 地下室窗
cellular ['seljulə] *a.* 蜂窝[多孔,分格,泡沫]状的
~ beam 格型梁
~ block 空心混凝土块
~ brick 多孔砖
~ building 箱式建筑
~ core wall 格体式心墙
~ concrete 泡沫[加气,多孔]混凝土
~ concrete block 泡沫混凝土砌块
~ construction 格型构造[建筑(物)]
~ floor 格型楼板
~ girder 格型梁,空腹[心]梁
~ glass 泡沫玻璃
~ gypsum plaster board 多孔石膏建筑板材
~ pier 格型墩
~ plastics 泡沫塑料
~ resin 泡沫塑料
~ radiator 蜂窝式散热器
~ retaining wall 格间式挡土墙
~ structure (texture) 网格[蜂房]结构
~ thermal insulation 泡沫绝热材料
~ -type building 蜂窝状建筑
cellulate ['seljuleit] *ad.;v.* (使成)蜂窝状
~d ceramics 多孔陶瓷制品
celluloid [seljulɔid] *n.* 赛璐珞,明胶
cellulose ['seljuləus] *n.* 纤维素
~ acetate 醋酸纤维素
~ adhesive 硝酸纤维素胶粘剂
~ coating 纤维素涂料
~ enamel 快干瓷釉
~ filler 纤维素填料
~ lacquer 纤维素漆

~ stopper 纤维素填充料
~ thinner 纤维素稀释剂
~ wool 化学纤维织物

cellulosic [ˌseljuˈləusik] *a.* 纤维素的
~ plastics 纤维素塑料
~ sizing agents 纤维素填料
~ varnish 纤维素涂料

celotex(board) [ˈseləˌteks(bɔːd)] *n.* 纤维(隔音,绝缘)板,隔音材料,色罗提隔音板

Celtic architecture (英国)凯尔特式建筑

Celtic ornament (英国)凯尔特式纹样

cembra pine 瑞士松

cemedin(e) [ˈsemidin] *n.* 胶粘剂,粘合剂

cement [siˈment] *n.* 水泥,胶凝材料,粘结物; *v.* 加水泥于,胶,凝固,巩固
~ asbestos 石棉水泥
~ asbestos board 石棉水泥板
~ asbestos mortar 石棉水泥砂浆
~ asbestos pipe 石棉水泥管
~ bacillus 水泥杆菌,水泥硫铝酸钙
~ base 水泥基础
~ block 预制水泥空心砌块
~ blower 水泥喷枪
~ brand 水泥牌号
~ concrete 水泥混凝土
~ covering 水泥罩面[面层]
~ extender 填充性水泥混合材料
~ facing 水泥盖面
~ fibrolite plate 水泥纤维板
~ fillet 水泥圆线脚,水泥压线条
~ filler grout 水泥稀浆
~ finish 水泥抹面
~ flag pavement 水泥板[铺砌]路面
~ floor 水泥地面
~ gel 水泥凝胶
~ grouting 水泥灌浆
~ gun shooting 水泥浆喷射法
~ lime mortar 水泥石灰浆
~ mark 水泥标号
~ matrix 水硬性胶凝材料,水泥
~ mortar 水泥砂浆
~ mortar rendering (plastering) 水泥砂浆抹面
~ plaster 水泥砂浆
~ product 水泥制品
~ rendering 水泥抹面
~ roof tile 水泥瓦
~ rubble masonry 水泥毛石圬工
~ sand bed 水泥砂浆层
~ screeding 水泥浆找平
~ slurry 水泥稀浆
~ stucco 外墙拉毛粉饰
~ stuff 水泥抹面砂浆
~ tile 水泥地面砖,水泥瓦
~ treated subgrade 水泥加固的基础
~ wash 水泥浆刷面
~ -water ratio 灰水比
~ waterproofer 水泥止[防]水材料
air-entrained ~ 加气水泥
alumina ~ 高铝[矾土]水泥
alumina self-stressing ~ 铝酸盐自应力水泥
aluminate ~ 铝酸盐水泥
aluminate refractory ~ 铝酸盐耐火水泥
aluminous ~ 高铝矾土水泥
alunite expansive ~ 明矾石膨胀水泥
ambrein ~ 人造硬珀胶
asphalt ~ 地沥青胶泥
bag of ~ 袋装水泥
bakelite ~ 树脂胶凝材料
barium silicate ~ 钡水泥
bauxite ~ 矾土水泥
blast(-furnace) ~ 高炉矿渣水泥
blended ~ 微集料水泥
boron containing ~ 含硼水泥
brick ~ 烧黏土水泥
bulk ~ 散装水泥
calcareous ~ 水硬性石灰
coloured portland ~ 彩色硅酸盐

水泥
cupola slag sulphated ~ 石膏化铁炉渣水泥
high-early (strength) ~ 早强[快硬]水泥
high magnesia ~ 高镁硅酸盐水泥
high-quality ~ 高标号水泥
jet ~ 喷射水泥
La Farge ~ 石灰矿渣水泥
lime pozzolanic ~ 石灰火山灰质水泥
lime slag ~ 石灰矿渣水泥
masonry ~ 砌筑水泥
modified ~ 中热水泥
montan ~ 无熟料水泥
natural ~ 天然水泥
neat ~ 净水泥
non-staining ~ 白水泥
ordinary portland ~ 普通硅酸盐水泥
polyvinyl acetate ~ 聚乙酸乙烯水泥
portland ~ 硅酸盐水泥
portland blastfurnace-slag ~ 矿渣硅酸盐水泥
portland fly-ash ~ 粉煤灰硅酸盐水泥
portland-pozzolana ~ 火山灰质硅酸盐水泥
quick-hardening ~ 快硬水泥
rapid hardening portland ~ 快硬硅酸盐水泥
red-mud sulphated ~ 赤泥硫酸盐水泥
regulated set ~ 调凝水泥
Roman ~ 天然水泥,罗马水泥
rubber ~ 橡胶水泥
sacked ~ 袋装水泥
sandy ~ 砂质水泥
self-stressing ~ 自应力水泥
slag-sulphate ~ 石膏矿碴水泥
sodium-silicate ~ 水玻璃胶凝材料
steel furnace slag ~ 钢碴水泥
straight ~ 纯水泥
strontium silicate ~ 锶水泥
sulfate resisting portland ~ 抗硫酸盐硅酸盐水泥
Sulph-aluminate expansive ~ 硫铝酸盐膨胀水泥
super sulphated ~ 石膏矿碴水泥
white portland ~ 白色硅酸盐水泥
zeolite ~ 沸石水泥
cementation [si:emn'teiʃən] *n.* 水泥胶结(作用);渗碳法;水泥灌浆;胶凝
~ index 水泥硬化指数
~ of fissures 裂缝灌浆
cement-copper 沉积[渗碳]铜,泥铜
cemented [simentid] *a.* 胶结的;渗碳的;烧结的
~ chip board 水泥刨花板
~ excelsior board 水泥木丝板
cementing [si'mentiŋ] *n.* 胶结,表面硬化;水泥灌浆
~ agent 粘结剂
~ hole 水泥灌浆孔
~ hose 灌浆软管
~ material 粘[胶]结材料
~ value 粘[胶]结值
cenotaph ['senəta:f] *n.* 纪念塔[碑],衣冠冢
Cenozoic [ˌsi:nə'zəuik] *n.*; *a.* 新生代(的),新世界(的)
~ era 新生代
centage ['sentidʒ] *n.* 百分率
centare ['senteə] *n.* =centiare 1 平方米,1 平方公尺
centaur [sen'tɔ:r] *n.* 半人半马像,人首马身像
centenary [sen'ti:nəri] *n.*; *a.* 一百年(的),一世纪(的);百年纪念(的)
center ['sentə] *n.*; *v.* 中心[央];拱架;对中心;聚集,集中
~ angle 圆心角
~ arch 中心拱
~ bay 中心跨度
~ bearing type 中心支承式
~ bit 中心钻

~ bulb 建筑上半圆花边止水条
~ by-pass 阀的中立旁通
~ distance 中心距,轴间距
~ heading 隧道的中央导洞
~ hinge 门窗枢轴
~ hinge pivot 中支枢轴
~ hung sash 中旋窗框
~ hung swivel window 中旋窗
~ -line 中心线
~ -line grade 中线坡度
~ (-line) stake 中线桩
~ mark 中心标志
~ of top (vertex) 拱顶中央
~ split pipe 天沟,檐槽,半圆形断面的雨水槽
~ spot room (设在剧场后墙中心的)后墙投光室
~ surveying 中线测量[设置]
centering ['sentəriŋ] *n.*; *ad.* 置于中心,向中心,集中;建筑中的拱架,鹰架

centering

~ of arch 拱顶鹰架
~ Rogla 罗格拉鹰架,移动钢桁架式鹰架
travelling ~ 移动式脚手架
centesimal [sen'tesiməl] *a.* 百分(之一)的,百进位的
~ system 百分制
centibar [sentibɑ:] *n.* 厘巴(压力单位 =1/100bar)
centi-degree [sentidigri:] *n.* 摄氏度,℃
centigrade ['sentigreid] *a.* 摄氏温度的,百分刻度的
~ thermometer 摄氏温度计
centimeter ['setimi:tə] *n.* = centimetre 厘米
centimetre-gram-second (system) 厘米-克-秒制
centi-octave [senti'ɔktiv] *n.* 1/100 倍频程,1/100 八度音程
centipois(e) ['semtipɔiz] *n.* 厘泊,10^{-2} 泊(粘度单位)

centistoke [sentis'təuk] *n.* 厘泡(重力粘度单位)
centner ['sentnə] *n.* 生奈尔(一种欧洲商用重量名称,德国、丹麦等国=50公斤;英国=100磅)
centrad ['sentræd] *n.* 厘弧度(百分之一弧度); *ad.* 向中心
central ['sentrəl] *a.* 中心[中央]的;主[重]要的
~ air conditioning equipment 集中式空调装置
~ apparatus room 中央设备室
~ angle 中[圆]心角
~ area (district) 中心区,市中心
~ archway 城门的中央拱廊
~ avenue 中央林荫道
~ axis 中心轴(线)
~ batching plant system 集中投配方式
~ commercial district 中心商业区
~ calm 中心无风[稳静]区
~ control room 中央控制室
~ core wall 中心筒墙
~ -corridor 中间过道,内走廊
~ -corridor residential building 内走廊房屋
~ drain 中央排水渠[管]
~ dust collecting (precipitating) system 集中集尘方式[系统]
~ ducting system 集中式通风管道系统
~ facilities 中心设施
~ feature 立体结构(如拱形结构)
~ heating 集中采暖[供热]
~ porosity 中心缩孔
~ post (木结构佛塔)中心柱
~ puriln(e) 中檩条
~ projection 中心投影
~ ventilating station 供风中心
centralab ['sentrəlæb] *n.* 中心实验室
centralize ['sentrəlaiz] *v.* 使聚于中心;集中管理
~d building 中心枢纽[建筑物]

centrally ['sentrəli] *ad.* 在中心
centre-line-average 平均高度,算术平均值
centrosymmetry [ˌsentrəusi'metri] *n.* 中心对称
centum ['sentəm] *n.*【拉】一百
 per ~ 百分比[率]
centuple ['sentjupl] *n.; vt.; a.* 百倍(的),用百乘
century ['sentʃuri] *n.* 百年,一世纪;一百个
ceraceous [sə'reiʃəs] *a.* 蜡状的,似蜡的
ceralumin [serə'ljumin] *n.* 铝铸造合金
ceram ['seira:m] *n.* 陶瓷(器)
ceramal [si'ræməl] = ceramel, cermet *n.* 金属陶瓷,陶瓷合金;粉末冶金学,烧结金属学
ceramic [si'ræmik] *a.* 陶瓷材料的,陶器[质]的; *n.* 陶瓷制器
 ~ bond 陶瓷粘结剂
 ~ cleat 陶瓷夹板
 ~ coat(ing) 陶瓷涂层
 ~ flooring tile 陶瓷地砖
 ~ lavatory 陶瓷洗手器
 ~ metal 金属陶瓷合金
 ~ mosaic 陶瓷锦砖,陶瓷马赛克
 ~ pickup 钛酸钡陶瓷传感器
 ~ shower tray 陶瓷淋浴盆
 ~ tile 瓷砖,陶瓷面砖
 ~ veneer 陶瓷板镶面
 ~ ware 陶瓷制品
cerargyrite [siraː'dʒirait] *n.* 氯化银(AgCl);角银矿
cerated ['siəreitid] *a.* 涂蜡的,有蜡膜的
cerdip ['səːdip] *n.* 陶瓷浸渍
ceresin(e) ['serisin] *n.* (纯)地蜡
cermet ['səːmet] *n.* 金属陶瓷,陶瓷合金
ceroplastic [ˌsiərəu'plæstik] *a.* 蜡塑的;- ~ s 蜡塑术
certificate [sə'tifikit] *n.; v.* , -cated, -cating; *n.* 证书[明];检验证,合格证,执照 *vt.* 鉴定,证明,认为合格
 ~ of acceptance 验收证书
 ~ of compliance 合格证(书)
 ~ of delivery 交货证明书
 ~ of insurance 保险单
 ~ of manufacturer 出厂证明
 material ~ 部件[材料](检验)合格证
certification [ˌsəːtifi'keiʃən] *n.* 证[明],鉴定,确认
 ~ of fitness 合格证书
 ~ of proof 检验证书
certified ['səːtifaid] *a.* (有书面)证明的,(经过)检定的,(检定)合格的
cerulean [si'ruːliən] *a.; n.* 天蓝色(的)
ceruse ['siəruːs] *n.* (碳酸)铅白,铅[白]粉
cesspipe ['sespaip] *n.* 污水管
chad [tʃæd] *n.* 砂砾,碎片,纸屑
chadded [tʃædid] *a.* 穿孔的
chafe [tʃeif] *n.; v.* 磨损,冲洗
chain [tʃein] *n.* 链,线路,通道,波道 *vt.* 用链连结
 ~ -and-tackle 链滑轮
 ~ belt 链条
 ~ block 拉链起重器,神仙葫芦
 ~ bolt 带链插销
 ~ bond 链式砌合
 ~ contracts 连锁合同
 ~ door fastener 链式门扣件
 ~ hook 链钩
 ~ of hinges 铰接链,活节链
 ~ pendant 灯具吊链
 ~ pendant lamp 吊链灯,悬链吊灯
 ~ pipe wrench 链条管子板手[钳]
 ~ polymer 链形聚合物
 ~ riveting 平行[并列,链式]铆接
 ~ saw 链锯
 ~ timber 木圈梁

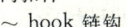
chain door fastener

chair [tʃɛə] *n.* 椅子,托架,垫板, *vt.* 主持人座
 ~ rail 护墙板,靠椅栏

chalcography [kæl'kɔgrəfi] *n.* 雕铜术,铜板雕刻

chalcomorphite [kælkə'mɔːfait] *n.* 硅铝钙石

chalet ['ʃælei] *n.* 瑞士山中倾斜屋顶的木屋[农舍]

chalk [tʃɔːk] *n.* 白垩,粉笔,石灰石 *vt.* 白垩处理,粉[灰]化
~ line 白粉笔画的线
French ~ 滑石

chalking [tʃɔːkiŋ] *n.* 油漆打粉底;起霜,垩化,粉化

chalkstone ['tʃɔːkstəun] *n.* 白垩石,石灰石

chalky ['tʃɔːki] *a.* chalkier, chalkiest 含白垩的,质地颜色像白垩的

chamaecyparis pisifera 花柏

chamber [tʃeimbə] *n.* 小室,腔,容器 *a.* 室内的,小规模的 *v.* 装入室[腔]内
chilling ~ 冷冻室
climatic ~ 人工气候室
environmental ~ 环境控制室,人工气候室

chamet bronze 锡黄铜

chamfer [tʃæmfə] = chamfre *n.* 槽,凹线;斜面;削角,倒棱 *vt.* 在……上面刻槽,斜切,去角
~ strip 施工缝小木条
~ed edge 削角边,斜削边,倒棱
~ed joint 斜削接头
~ed rustic work 削边粗石工
~ed step 削边踏布
~ing plane 削角刨

chamot(te) [ʃə'tɔm] *n.* 熟耐火土,火泥,陶渣,黏土砖
~ brick 黏土砖,耐火砖
~ concrete 耐火泥混凝土
~ mortar 耐火砂浆

chance [tʃɑːns] *n.;v.*, chanced, chancing 概率,机会,几率,或然率
~ cause 机遇因素

chancel ['tʃɑːnsəl] *n.* (教堂的)高[圣]坛

~ arch 高坛拱

chandelier [ˌʃændi'liə] *n.* 枝形吊灯[灯架],花灯

chandi ['ʃændi] *n.* 爪哇祠堂,印尼墓穴
~ Prambanan 9 世纪爪哇岛的普朗巴南寺

change ['tʃeindʒ] *n.;v.*, changed, changing 变化,改变;[兑]交换;不归零法
~ house (工人)更衣室
~ing cabin (room) 更衣室

changeable ['tʃeindʒəbl] *a.* 易[可]变的

channel [tʃænl] *n.* 管[孔,水,渠]道;河床[槽],航道;槽;系统,部分 *v.* 成渠,引导
~ beam 槽形[钢]梁
~ iron U 形铁
~ of approach 引槽
~ scraper and elevator 下水道清理设备
~ slab 槽(形)板
~ steel 槽钢
~ stone 排[雨]水槽
~ tile 槽瓦
~ed plated 波纹板,网纹钢板
air ~ (通)风道
spillway ~ 泄水槽

channeller ['tʃænələ] *n.* 凿沟机

chanoine wicket dam 查诺尼式门堤

chantlate [tʃɑːntleit] *n.* 檐口滴水条

chap [tʃæp] *n.* 裂缝[隙],龟裂 *v.* 使裂开,变为粗糙

chapel ['tʃæpəl] *n.* 小教堂,大建筑物内部的小礼拜堂

chapiter [tʃæpitə] *n.* 柱头,柱上部的大斗

chaplet ['tʃæplit] *n.* 花冠;项圈;串珠饰

chapterhall ['tʃæptəhɔːl] *n.* 大[主]厅,会场

char [tʃɑː] *n.* 木炭,炭化 *v.* 成炭,烧焦

activated ~ 活性炭
character ['kæriktə] *n.* 性质,特性[霉];数字,符号,号码;工作制度,状态
~ display 数字显示器
~ of service 工作制度[状态]
characteristic [,kæriktə'ristik] *a.* 特有的,表示特性的. *n.* 特征,性能;规格;对数的首数
~ dimension (相似理论)特征[基准]尺寸
shop ~ (材料的)工艺性能
charcoal ['tʃɑːkəul] *n.* 木炭,活性炭;
~ filter 活性炭过滤器
~ tinplate 厚锡层镀锡薄钢板
~ wire 超低碳钢丝
charge [tʃɑːdʒ] *n.* 电荷,负荷;装载量;管理,委托;费用 v. 加载,充气,加[装]料,填充;指令,记账,要价
~ account 记账,赊欠
~ against revenue 收益支出
~ construction 充气结构
~ for use 使用费
~ of rupture 破坏载荷
~ of surety 安全[允许]载荷
~ relief valve 充气[水]安全阀
~ sheet 记料单
~ valve 充气[水]阀
~ well 垃圾箱
charging crane 装料吊车(起重机)
charging device 装料设备
charging hopper (装)料(漏斗)
carrying ~ 维修费
inventory ~ 材料的耗费
overhead ~ 总管理机构开支,杂项开支
chargehand ['tʃɑːdʒhænd] *n.* 工长,领班;[凿井]装岩工
charnockite ['tʃɑːnəkait] *n.* 紫苏花岗岩
charring [tʃɑːriŋ] *n.* 炭化防腐,(木材表面)烧成炭层防腐法
superficial ~ 皮焦法
charry ['tʃɑːri] *a.* 似炭的,炭(化)的

chart [tʃɑːt] *n.* 曲线(图),计划图 *vt.* 用图说明[表示];制订计划
~ constant 制图常数
break-even ~ 盈亏[收支](平衡)图
calendar progress ~ 日程表,工作计划进度表
control ~ 管理图,检查图
flow ~ 方框图,程序操作顺序图表
flow process ~ (工艺)流程图
statistical ~ 统计图表
two-way control ~ 复式管理图
charter ['tʃɑːtə] *n.* 许可证,营业执照 *vt.* 特许;包租,包货
chartreuse [ʃɑː'trəːz] *a.;n.*,浅黄绿色[的]
chase [tʃeis] *n* 暗线槽,管子槽,竖沟 *v.* 用梳刀刻螺纹,刻槽
~ mortise 槽榫
chaser [tʃeisə] *n.* 螺纹梳刀,碾压机,揉泥碾
chasm ['kæzəm] *n.* 裂隙[缝,口],断层,峡谷;空隙;空白,中断处
chassis ['ʃæsi] *n.* 底架[座,盘],框架
chateau ['ʃɑːtəu] *n.* 【法】宫殿,城堡,府邸,乡村别墅
chatelet ['ʃætəlei] *n.* 【法】小城堡,小宫殿
check [tʃek] *n.* 支票,收据;核对,监督;制动装置,节制闸,挡水闸,阀门;开裂,细裂纹[缝] *v.* 校对,验算,控制
~ and shake 裂纹[缝],开裂
~ board 挡板
~ book 支票簿
~ crack 细裂缝,收缩裂纹
~ indicator 检查指示灯
~ list 检验单,检查表
~ loading 验算荷载
~ nut 锁紧[防松]螺母
~ of drawing 审图
~ -point 抽点检验
~ programme 检验程序
~ ring 挡圈,锁紧环
~ room 衣帽(存放)间

～ stubs 支票存根
～ surface 表面龟裂
～ table 检查表
blank ～ 空白支票
fire ～ 热裂纹
honeycomb ～ 蜂窝裂缝
line ～ 小检修
observation(al) ～ 外部检查
regular ～ 定时[期]检查
spot ～ 抽查
built-in ～ing 固定检查
programmed ～ing 按程序检查
checker ['tʃekə] *n.* 方砖[格],格子花;检验设备,检验员 *vt.* 使成格子[棋盘格]图样,使交错
～ brick 格子[方格]砖
～ board (masonry) bond 格子形砌合墙体
～ed plate 网纹钢板
checkout ['tʃekaut] *n.;v.* 检验,测试,调整,验算,结账
checkwork ['tʃekwə:k] *n.* 方格式铺砌工作,棋盘形砌木工
cheek (damp) course 避湿层
cheese [tʃi:z] *vt.* 停[阻止]
～ bolt 圆头螺栓
～ head rivet 高平[圆]头铆钉
～ head screw 圆头螺钉
chemboard ['tʃembɔ:d] *n.* 硬质纤维板
chemical ['kemikəl] *a.* 化学的; *n.* 化学试剂,化合物
～ attack 化学侵[腐]蚀
～ building material 化学建筑材料
～ contamination 化学污染
～ durability 化学稳定性,抗化学侵蚀性
～ milling 化学蚀刻[抛光]
～ pulp 化学纸浆
～ purification 化学净化
corrosive ～ 腐蚀(性化学药)剂
chemigum ['kemigʌm] *n.* 丁腈橡胶
chemism ['kemizəm] *n.* 化学作用(化学亲和力)

chemolysis [ki'mɔləsis] *n.* 化学分析[解],化学溶蚀
chemosorbent [‚keməu'sɔ:bənt] *n.* 化学吸附剂
chemosmosis *n.* 化学渗透(作用)
chemotec ['kemətea] *n.* 环氧树脂类粘结剂
Chengde Mountain Resort 承德避暑山庄
cheque [tʃek] *n.* = check 支票,格子花,方格图
chequer ['tʃekə] *n.,vt.* = checker
cherry ['tʃeri] *n. pl.* -ries ;樱桃,樱桃色;cherrylike,a
～ -hard brick 过火砖
chert [tʃə:t] *n.* 燧石,黑硅石,角岩
cherty ['tʃə:ti] *a.* 燧石的,黑硅石的
～ flint 燧石
chest [tʃest] *n.* 资金,公款;箱,柜,容器
chestnut ['tʃesnʌt] *n.* 栗树木;栗色,红棕色,栗褐色
cheval-glass *n.* 穿衣镜
chevet [ʃə've] *n.* 【法】教堂的内室,半圆形或圆形的休息室
chevron ['ʃevrən] *n.* 山形符号,波浪[曲回]纹饰,锯齿形花饰

chevron
～ baffle 百叶障板
～ bar tread 人字条纹
～ drain 侧向排水,人字形排水
chian ['kəiən] *n.* 沥青,柏油
～ varnish 柏油清漆
chicken ['tʃikin] *n.* 鸡,雏鸡
～ architect 总建筑师
～ grit 大理石渣
～ -wire 铁丝(织)网
chief [tʃi:f] *a.* 重要的,主要的,;-chiefless *a.*
～ architect 总建筑师
～ contractor 主[总]承包人
～ designer 设计总负责人,总设计师
～ engineer 总工程师

~ inspector 总检查员，总监察员
~ of section 工段长
~ resident architect（常住工地的）主任建筑师
~ wall 主墙

chill [tʃil] *n.* 寒冷,冷冻 *v.* 使冷冻,使冷凝,使冷却
~ point 凝固点
~ room 冷冻室
~ed glass 钢化玻璃
~ing unit 致冷设备

chimney ['tʃimni] *n. pl.* -neys 烟囱（状物）；冰川井[竖坑]；柱状矿物体
~ bond 烟囱砌法
~ cap 烟囱帽
~ hood 烟囱帽
~ jack 旋转式烟囱帽
~ neck 烟道
~ ventilation（draft）烟囱[自然]通风
brick(work) ~ 砖砌烟道

China ['tʃainə] *n.* 中国
~ wood oil 桐油
~ ink 墨(汁)

China Society of Civil Engineering 中国土木工程学会

china ['tʃainə] *n.* 陶瓷[陶器],瓷器;瓷(质黏)土
~ brick 瓷砖
~ -clay 高岭土,瓷土
~ orange 橙色
~ sanitary ware 卫生陶瓷制品
~ stone 瓷土石

Chinaberry ['tʃainəberi] *n. pl.* -ries（同 Chinatree）紫花树,楝树

chinacyupress ['tʃainə'saipris] *n.* 水松属

chinafir ['tʃainəfə:] *n.* 杉属

Chinagreen ['tʃainəgri:n] *n.* 亮丝草,广东万年青

chinampa ['tʃaiəmpə] *n.* 墨西哥式人造地坪

chinar [tʃi'nɑː] *n.* 法国梧桐,藤悬木

chinaware ['tʃainəwɛə] *n.* 瓷器

chinese [tʃai'ni:z] *a.* 中国的 *n.* 中国人
~ architecture decoration 中国建筑装饰
~ black pine 罗汉松
~ character printer 汉字打印机
~ drill stock 弓形钻,辘铲钻
~ Export Commodities Fair(CECF) 中国出口商品交易会
~ juniper 桧
~ sapium 乌桕
~ screen 中国屏风
~ wax 白[中国]蜡
~ white 锌白,(ZnO)
~ wood oil 桐油

Chinese industrial Standards 中国工业标准

chink [tʃiŋk] *n.* 裂缝[隙]；*v.* 堵缝塞孔,龟裂

chinking ['tʃiŋkiŋ] *n.* 灰泥,油灰,腻子

chinky ['tʃiŋki] *a.* 多孔隙的,有裂缝的

chinoiserie [ʃi:nwɑ:z(ə)'ri:] *n.*【法】(17~18世纪在欧洲出现的)中国风格的建筑或工艺品

chip [tʃip] *n.*；*v.* chipped chippin;碎片,石[木]屑 *v.* 劈碎,碎裂；用铲清理
~ axe 琢石斧
~ breaker [石片]压碎机
~ cap 木刨压板
~ped glass 冰花玻璃
~ped stone 琢石
~ped stone surface 石屑铺面,琢石面
~ping carpet 石屑毡层
~ping chisel（碎）石凿
~ping hammer 手锤,琢石锤

chipboard ['tʃipbɔ:d] *n.* 刨花板,粗纸板,碎纸胶合板

chipper ['tʃipə] *n.* 削片机,凿

chippings ['tʃipiŋs] *n.*（细粒）碎石填缝骨科,撒观集料
~ concrete 碎石集料混凝土

chisel ['tʃizl] *n.*；*v.* ,凿,雕,錾,铲

~ bit 冲击式钻头
~ed slate 凿面石板
butt ~ 平头铲
cross ~ 十字凿
caulking ~ 堵缝凿
cross mouthed ~ 十字冲击钻

chiselly [ˈtʃizli] *a.* = chisley 含砂砾的,粗颗粒的

chloride [ˈklɔːraid] *n.* 氯化物
~ of lime 氯化石灰,漂白粉
~ stabilization (of soil) 氯盐土壤稳定法

chlorimet [ˈklɔːrimət] *n.* 耐蚀合金

chlorinate [ˈklɔːrineit] *n.* 氯化物 *vt.* 使氯化,用氯消毒[处理]
~d lime 漂白粉
~d polyethylene 氯化聚乙烯
~d polyvinyl chloride 氯化聚乙烯
~d polyvinyl chloride 氯化聚氯乙烯,聚过氯乙烯
~d rubber 氯化橡胶

chlorination [ˌklɔːriˈneiʃən] *n.* 氯化作用[处理]

chlorine [ˈklɔːriːn] *n.* 氯(Cl)
~ and ~ derivative disinfectants 氯系消毒剂

chlorite [ˈklɔːrait] *n.* 绿泥石,亚氯酸盐
~ spar 硬绿泥石

chloroform [ˈklɔːrəfɔːm] *n.* 三氯甲烷,氯仿;*vt.* 用氯仿麻醉[杀死]

chloronorgutta [ˌklɔːrənˈɔːdʒətə] *n.* 氯丁橡胶

chloroprene [ˈklɔːrəpriːn] *n.* 氯丁二烯,氯丁橡胶

chock [tʃɔk] *n.* 塞子,垫木 *vt.* 堵[阻]塞,填满,垫住
~ing section 堵塞截面

choir screen (教堂中)围隔唱诗班的屏障或围栏

choke [tʃəuk] *n.*;*v.* 抑制,阻塞,节流,闸门,调节阀[闸],阻气门
~ aggregate 拱顶石,嵌缝石
~ damp 窒息气,二氧化碳气
~ plug 塞头
~ stone 堵缝石,拱顶石

choker [ˈtʃəukə] *n.* 堵缝材料,阻气门

chopping [ˈtʃɔpiŋ] *a.* 风向常变的,破浪凶涌的,急变的 *n.* 斩波,中断
~ bit 冲击钻头

choppy [ˈtʃɔpi] *a.* 多裂缝的,有皱纹的;方向常变的,波浪起伏的
~ grade 锯齿形的纵断面

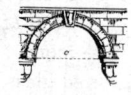
chord

chord [kɔːd] *n.* 弦,桁弦,桁架弦杆
~ member 弦杆
~ of arch 拱弦
~ plate 翼缘板,弦板
~ splice 桁弦接合板
broken top ~ 折线上弦梁,多边形屋架

chorometry [kɔːˈrɔmətri] *n.* 土地测量

Christian architecture 基督教建筑

Christian basilica 长方形基督教堂

chroma [ˈkrəumə] *n.* 色彩浓度,色度

chromansil [krəˈumænsil] *n.* 铬锰硅钢,结构用低合金钢

chromatic [krəuˈmætiks] *a.* 颜色的,色彩的
~ adaptation 色适应
~ contrast 色(彩)度对比
~ difference 色差
~ dispersion 色散(现象)
~ sensitivity 感色灵敏度
~ sheet 镀铬薄板

chromatics [krəuˈmætiks] *n.* 色彩学;chromatist,

chromatographic [ˈkrəu-mətəgrɑːfik] *a.* 色层(分离)的,层析的
~ analysis 色层分析

chrome [krəum] *n.* (Cr) = chromium 铬;铬黄,铬钢 *v.* 镀铬
~ alum 铬(明)矾
~ manganese steel 铬锰钢
~ molybdenum steel 铬钼钢
~ nickel steel 铬镍钢
~ plating 镀铬

~ red 铬红
~ silicon steel 铬硅钢
~ yellow 铬黄
high ~ 高铬钢
chromel ['krəuməl] *n.* 铬镍合金
chromium ['krəumiəm] *n.* 铬(Cr)
~ steel 铬钢
~ -nickel electrode 铬镍焊条
~ -nickel steel 铬镍钢
~ -plated 镀铬的
~ plating 镀铬
chromizing ['krəumaiziŋ] *n.* 铬化[渗铬]处理
chronograph ['krɔnəgrɑːf] *n.* 记时器[仪],时间记录器
chrysotile ['krisətail] *n.* 温石棉,加拿大石棉
chrysotil(ite) ['krisətil (ait)] *n.* (纤维)蛇纹石,水含硅酸镁石棉
chuck [tʃʌk] *n.* 夹具[头],卡盘,吸盘
chuff [tʃʌf] *n.* 废品砖,纹裂砖
chunk [tʃʌŋk] *n.* 块,棒;大的数量
~ glass 碎玻璃
church [tʃəːtʃ] *n.* 教堂,礼拜堂
~ architecture 教堂建筑
churn [tʃəːn] *n.* 搅拌器,冒口,钢丝绳冲击式钻机 *v.* 强烈搅动,翻腾
~ drill 冲击钻,旋冲钻
Churriguer esque style (17 世纪后期～18 世纪初期西班牙)邱利格拉式建筑
chute [ʃuːt] *n.* 溜[斜]槽,斜坡,急流槽 *v.* 用斜槽送输
~ board 滑[斜]槽板
air ~ 通风道
angle loading ~ V 型装载槽
cable ~ 电缆槽[沟]
charge ~ 装料槽
ciboriun [si'bɔːrəm] *n.* 祭台上之天盖
cilery ['siləri] *n.* 柱头卷叶饰
cill [sil] *n.* = sill 槛,窗台;基石,底木;海底山脊,岩床
cimatium ['simətiəm] *n.* 反

cilery

曲线盖板
cimbia ['simbiə] *n.* 束柱带,柱箍条
cimolite ['siməlait] *n.* 漂砾黏土,泥砾土,水磨土
cincfoil ['siŋkfɔil] *n.* = cinquefoil 五瓣[梅花饰]
cinch [sintʃ] *n.* 有把握的事 *v.* 紧握,扭住,捆[绑]紧
~ anchors 加固锚栓
cincture ['siŋktʃə] *n.*;*v.*,-tured,-turing; *n.* 柱带,柱头轮形装饰,柱身上下端环形线脚 *vt.* 用带子缠卷

cincture

Cindal ['sindəl] *n.* 铝基合金
cinder ['sində] 炉[煤,矿,火山]渣;煤屑,炉灰,灰尘,氧化皮 *vt.* 撒炉[煤,矿,火山]渣
~ block 煤渣砖[砌块]
~ brick 矿[炉,煤]渣砖
~ catcher 集尘器,集[渣]器
~ concrete 矿渣混凝土
~ dust 矿渣填充材料
~ floor 炉渣地面
~ sand concrete 炉渣砂混凝土
~ road 煤渣路
cindery ['sindəri] *a.* 含[似]煤渣的,灰烬的
cinefaction [sini'fækʃən] *n.* 灰化,煅灰法
cineholomicroscopy ['sinihɔləmaikrɔskəpi] *n.* 显微全息照相术
cinema ['sinimə] *n.* 电影院;电影,影片
cinepanoramic (house) ['sinpænər'ræmik] *n.* 全景宽银幕电影院
cinerary ['sinərəri] *a.* 灰的
cinereous [sini'əriəs] *a.* 灰白色的,似灰的;已成灰的
cinnabar ['sinəbɑː] *n.* 朱砂,朱红, *a.* 朱红色的
cinnamomum camphora 樟树

cinquecento [ˌtʃiŋkwi'tʃentəu] *n.* 16 世纪意大利艺术
cinquefoil ['siŋkfɔil] *n.* 五叶形的装饰,梅花饰
~ arch 五瓣拱
cipher ['saifə] *n.* 暗[记]号, *v.* 计算,算出译成密码
~ tunnel 装饰立面用的假烟囱,暗烟囱
cipolin(o) ['sipəlin] = cipolline *n.* 意大利产的白绿花纹的大理石
cir = **circa** 大约,前后[用于年代前];= circular 圆形的
circa ['səːkə] *ad.*; *prep.* 大约,前后[用于年代前]
circiter ['səːsitə] = circa *ad.*; *prep.* 大约
circ = circulate 循环,环行 = circumference 四圆周,周边,四周
circle ['səːkl] *n.* 圆[环](形物),圆周运动,圆形广场,轨道,期,循环,范围,领域;刻度盘,道路交叉口 *v.* 旋转,循环
~ brick on edge 环形侧砌砖
~ diagram 圆图
~ distribution 环形配水
~ distribution system 环形配水系统
~ draft frame ball socket 活环接头,环窝关节
~ fluorescent lamp 环形荧光灯
~ of wall plates 承梁板凸缘
~ reverse 回转角
~ trowel 圆泥刀
addendum ~ 外圆(周),水平刻度盘
circlewise ['səːklwaiz] *ad.* 成环状(地),环形地
circline ['səːklain] *n.* 环形
circuit ['səːkit] *n.* 线[网,电]路;流程工序;节围,循环 *v.* 环行,接成网路
circular ['səːkjulə] *a.* 圆[环]形的; *n.* 传单,函件
~ arch 圆弧拱
~ barrel vault 圆筒形穹窿
~ bead 圆角,台肩,弧面秋叶
~ beam 圈梁
~ building 圆形建筑
~ campanile 圆形钟塔[钟楼]
~ conchoid 圆蚌线,螺旋线
~ conduit-type sewer 圆形断面排水管
~ cone 圆锥
~ corridor 回廊,游廊
~ cylinder 圆(筒)柱
~ cylindrical shell 圆柱筒壳
~ dial-type scale 数字显示秤[天平]
~ dome 圆屋顶,圆[穹]顶
~ domical vault 环状穹顶
~ face 圆凸面
~ foundation 圆形基础
~ geometrical stair 圆盘旋梯
~ hut 圆形建筑物
~ joint 环接
~ Mound Altar 天坛圜丘坛
~ newel stairs 环柱旋梯
~ orifice 圆孔,圆洞口
~ path 环形路线
~ pediment 弧形山墙
~ peristyle 圆列柱廊
~ plane 圆刨
~ plate 圆板
~ recess 圆形槽
~ ring girder 环形梁
~ rip saw 纵切圆锯
~ saw 圆锯
~ shake 木材环裂
~ spider type joint 十字叉连接
~ spike 大圆钉
~ sunk face 凹圆面
~ temple 圆形神庙
~ track 循环
~ vault 圆顶穹窿
~ wagon vault 圆筒形穹顶
~ washer 圆垫圈
circulating ['səːkjuleitiŋ] *n.* 循环,环流
~ asset 流动资产

~ capital 流动资金
~ door 旋转门
~ fan 通风扇
~ (real) capital 流动资产,周转资金
circumdenudation [səkʌmdinju:'deiʃən] *n.* 环状侵蚀
circumference [sə'kʌmfərəns] *n.* 周围[界]
circumferential [sə'kʌmfə'renʃəl] *a.* 周围的,环形的,圆周的
~ drainage system 环形排水系统
~ scaffolding 环形悬臂脚手架
circumflexion [sə:kəm'flekʃən] *n.* 弯曲,曲率,弯(曲)度 *v.* 弯成圆形
circumscription [sə:kəms'kripʃən] *n.* 花边;区域,界限,范围
circumvallation [səkʌmvæ'leiʃən] *n.* 城墙[堡],防御墙,壕沟
circumvolution [sakʌmvəlju:ʃən] *n.* 旋绕饰,卷曲,旋转
circus ['sə:kəs] *n. pl.* -cuses; 圆形广场,环形交叉口;圆马戏[竞技]场
cirque [sə:k] *n.* 圆形山谷[冰沆],圆形场地
cissing ['sisiŋ] *n.* 收缩
cistern ['sistən] *n.* 水槽,贮水池;收集地表水和雨水的存水器
~ head 檐槽,屋面排水管上承口
citadel ['sitədl] *n.* 城堡,要塞
citric ['sitrik] *a.* 柠檬性的
~ acid 柠檬酸
citrine ['sitrin] *n.; a.* 黄水晶;柠檬色(的)
citron ['sitrən] *n.; a.* 香橼;香橼色(的)
city ['siti] *n. pl.* cities ;城市
~ air blanket 城市空气覆盖层,城市热空气层
~ area 市区
~ center 市中心
~ council 市政府[厅],市参议会
~ culture 城市文化[物]
~ gate 城门
~ hall 市政厅
~ lay-out 城市规划
~ panorama 城市全景
~ parking lot 市内停车场
~ planning 城市规划
~ wall 城墙
Forbidden ~ 紫禁城,故宫
civic ['sivik] *a.* 城市的;市民的
~ architecture 城市[民用]建筑
~ auditorium 礼堂,宴会厅
~ building 城市房屋
~ crown 花帽箍,叶环饰
~ design 城市建筑艺术
civics ['siviks] *n.* 市政学
civil ['sivl] *a.* 民用[间]的
~ architect 民用建筑师
~ architecture 民用建筑
~ construction 土木建筑
~ defense against air raid 防空
~ defense construction 人防建筑
~ defense (defence) 人防工程
~ defense shelter 人防建筑物
~ engineer(civ. Endr.) 土木工程师
~ engineering(Civ. Eng) 土木工程(学)
~ engineering and building construction 土木建筑工程
~ engineering material 建筑(工程)材料
~ engineering site 建筑工地
~ engineering stresses 建筑结构应力
~ engineering structure 工程建筑
~ engineering works 土木工程
~ minimum 城市生活设施最低水平
clad [klæd] *n.* 覆盖层 *a.* 金属包覆的,镀有金属的 *vt.* 包覆,镀
~ laminate 敷箔[叠压]板
~ metal sheet 复合金属板
~ pipe 复合管
~ plate 复合板,装饰板
~ steel sheet 复合钢板

cladding ['klædiŋ] *n.* 路面;覆盖层,外罩,表面处理;维护结构,房屋骨架填充墙
~ sheet 骨架填充板(材),覆盖板(材)
~ material 镀层,覆盖材料

claim [kleim] *n.;v.*,要求,申请,索赔

clam [klæm] *n.* 夹钳[板]

clamp [klæmp] *n.* 夹紧装置,夹具,抓斗 *vt.* 夹紧,定位
~ burned brick 土窑砖
~ device 紧固[定位]装置
~ drill 夹钻
~ holder 夹持器
~ -off 冲砂
~ rail 固端板,板端镶木
~ screw 制动螺钉
~ timber 夹撑木
door ~ 门扣

clamping ['dlæmpiŋ] *n.* 夹紧,固定;计算机的固定
~ apparatus 夹具
~ bolt 紧固螺栓
~ jaws 夹紧钳
~ lever 夹紧手柄
~ moment 固端变矩
~ nut 锁紧螺母
~ yokes 夹具

clamshell ['klæmʃəl] *n.* (蚌壳式)抓斗,抓斗式挖土机
~ type gate 扇形门

clapboard ['klæpbɔ:d] *n.* 护墙板,楔形板,隔板

claret ['klærət] *n.;a.* 紫红色(的)

clarificant ['klæ'rifikənt] *n.* 澄清剂

clarificate [klæ'rifikeit] *v.* 净化,澄清

clarification [ˌklærifi'keiʃən] *n.* 净[纯]化;说明,解释

clarifier ['klærifaiə] *n.* 净化器,澄清器[剂,池]

clarify ['klærifai] *v.* 净化,澄清

clarity ['klæriti] *n.;a.*,透明(的),清彻(的)
see-through ~ 透明度

clark beam 木组合梁

clasolite ['klɑ:səlait] *n.* 碎屑岩

clasp [klɑ:p] *n.;v.*,铆固,钩住,接合器
~ hook 弯脚钩
~ joint 搭扣接合
~ lock 自动弹簧锁
~ nail 抓钉,扁钉

class [klɑ:s] *n.* 类别,等级,粒度 *vt.* 分类[级,等]
~ of precision 精度等级
A (first) ~ 头等

classic ['klæsik] *a.* 古典[传统,典型]的
~ architecture 古典建筑
~ orders 古典柱型
~ revival 古典复兴式(15世纪的雕刻和19世纪的绘画)
Nanjing fuzimiao ~ style building 南京夫子庙古典式建筑

classification [ˌklæsifi'keiʃən] *n.* 分级,筛分;级[类]别;分类(法)

classify ['klæsifai] *vt.* 分类[级]
classified population 职业人口

classroom ['klɑ:srum] *n.* 教室

clastate *n.* 破碎[岩石]

clastic ['klæstik] *a.* 裂成碎片状的,碎片性的

clastomorphic [klæstə'mɔ:fik] *n.* 碎屑侵蚀变形

clathrate ['klæθreit] *a.* 窗格形的,笼形的,网状的

clause [klɔ:z] *n.* 条款,款项
memorandum (additional) ~ 附加条款
saving ~ 对例外情况的附加条款
special ~ 特别条款

clauster ['klɔ:stə] *n.* 峡谷,堰,禁室

claustra ['klɔ:strə] *n.* 漏空石墙

claustrum ['klɔ:strəm] *n.* 带回廊的修道院

claw [klɔ:] *n.* 爪,钳,钩,销,把手,爪凿,爪形钩钳
~ bar 撬棍,爪杆

~ bolt 爪栓
~ hammer 羊角锒头
~ hatchet 拔钉斧
~ plate 爪板
~ tool 石工齿凿
clay [klei] *n.* 黏土,白土,泥土
~ and straw plaster 柴泥抹面
~ article 烧土制品
~ -bearing 含泥的,含黏土的
~ bed 黏土层[地基]
~ binder 黏土粘合料
~ blanket 黏土密实的防渗墙
~ body 烧制的陶器
~ bond 黏土质粘结剂
~ book tile 书形空心舌槽瓦[砖],黏土屋脊瓦
~ brick 黏土砖
~ -cement grout 黏土水泥浆
~ -cement mortar 黏土水泥砂浆
~ coated finish 滑泻涂料
~ core 黏土心墙
~ course 黏土层
~ (drainage) tile 瓦管
~ facing tile 饰面瓷砖
~ gypsum 土石膏
~ model 泥塑模型
~ mortar (mud) 泥浆
~ packing 黏土压实[铺盖]
~ parting 黏土夹层
~ pipe 瓦管,陶(土)管
~ products 黏土制品
~ puddle 夯[捣]实黏土
~ roof tile 黏土瓦
~ sculpture 泥塑
~ seam 黏土夹层
~ shingle 黏土瓦
~ shower tray 陶瓷浴盆
~ slurry 黏土浆
~ stone 黏土岩
~ substane 标准黏土
~ wall 土墙
~ wall tile 烧土墙砖

acid ~ 酸性黏土
activated ~ 活性黏土
activated montmorillonite ~ 活性微晶高岭土
bonding ~ 造型黏土,接合黏土
burnt ~ 烧黏土
china ~ 白瓷土
fire (refractory) ~ 耐火黏土,耐火泥
glase-pot ~ 陶土
mild (sandy) ~ 亚黏土,瘦黏土
plastic ~ 可塑黏土,陶土
raw ~ 生黏土
clayey [ˈkleiiː] *a.* 黏土质的,含黏土的
~ sand 黏质砂土
~ silt 黏质粉土
~ soil 黏质土,亚黏土
~ strata 黏土层
clay-filler [kleifilə] *n.* 黏土填料
claying [ˈkleiiŋ] *n.* 抹泥
clayite [ˈkleiit] *n.* 高岭石
clayware [ˈkleiwɛə] *n.* 黏土制品
cleading [ˈkliːdiŋ] *n.* 护墙板,衬板
clean [kliːn] *a.* 新鲜的,表面光洁的
ad. 完全,彻底; *v.* 清理,净化,归零
~ bench 清洗台
~ bole 无枝干材
~ -cut timber 光洁木材,无疵木材
~ deal 光洁的木板
~ gap graded aggregate 精选集料
~ oil 轻质[透明]油
~ -out 清洁[扫]口,管道检修口
~ -out cap 清洁孔盖板
~ -out auger 抽汲筒
~ -out clearance 清洁孔的净距
~ -out door 出灰门
~ -out plug 清洁口堵头
~ sand 净砂
~ steel 纯钢
~ timber 无节疤木材
~ -up and move out 现场清理
~ -up scraper 刮土板

cleaner ['kli:nə] *n.* 清洁剂;吸尘器;除垢器;修型工具
　upright vacuum ~ 立式吸尘器
　cylinder vacuum ~ 筒式吸尘器
　all-purpose vacuum ~ 万能真空吸尘器
cleaning ['kli:niŋ] *n.* 清理,平整;(*pl.*)垃圾
　~ apparatus 除尘装置
　~ blower 吹风净化器
　~ cock 排污龙头
　~ effect 冲洗[洗涤]效果
　~ eye 清理孔
　~ hinge 间隙窗铰链
　~ solution 洗液
　~ strainer 滤净器
cleanliness [kli:nlinis] *n.* 清洁度
cleanser ['klenzə] *n.* 清洁工[工人]
cleansing [klenziŋ] *n.* 净化;(*pl.*)垃圾 *a.* 清洁用的
clear ['kliə] *a.* 透明的,通畅的;已卸完货的;已检验过的,完好的 *ad.* 显然,一直,离着 *v.* 清理[除];清算;卸货;归零 *n.* 空隙[间];中空体的内尺寸
　~ buttress spacing 拱跨
　~ -cole 油灰,油漆打底
　~ colour 清色
　~ cover 保护[覆盖]层
　~ -cut 轮廓清晰
　~ dimension 净距[值]
　~ distance 净距[离]
　~ glass 透明玻璃
　~ headroom 净空[高度],头上空间
　~ height 净高
　~ lacquer 透明漆,清喷漆
　~ lamp 透明灯泡
　~ log 锯材原木
　~ mesh 高空[宽,距],内径,直视距离
　~ space 净空间
　~ spacing 净间距
　~ span of grider 大梁净跨
　~ timber (stuff, wood) 无节疤的木料
　~ varnish 透明清漆
　~ width 净宽,内径
　~ wire glass 透明嵌金属钢玻璃
　a ~ space 空地
clearance ['kliərəns] *n.* 净空,距离,间隙,缺口,出清[空];排除障碍
　~ angle 间[余]隙角
　~ height 净空高度
　~ limit 建筑限界,规划建筑线
　~ loss 净空损失
　~ of span 跨度净空
　~ under valley tiles 天[斜]沟两侧瓦下的间隙
　track ~ 规划建筑线
clearcole ['kliəkəul] *n.* (打底子的)油灰
clearing ['kliəriŋ] *n.* 清扫,清除草木
　~ and grubbing 清除地面障碍物
　~ ground 清理地面
　~ of site 场地清理
　~ sheet 交换清单
clearness [kliənis] *n.* 清晰度
clearstory ['kliəstəri] = **clerestory** *n. pl.* -rise;纵向天窗或气楼,开窗假楼,阁楼天窗,高侧窗
　~ lighting 高侧窗采光
cleat ['kli:t] *n.* 踏板上的防滑条,桁条上的加强角片;夹具[板],条缆墩[栓],羊角;高花纹轮胎 *vt.* 栓住,用楔子固定
　~ plane 楔子劈入面
　~ wiring 用瓷夹头固定电线
　belaying ~ 加强板
cleavability ['kli:vəbiliti] *n.* 可裂性,裂开,龟裂
cleavage ['kli:vidʒ] *n.* 劈裂[开],裂缝[纹],矿物的节[解]理
　~ crack 劈裂
　~ fracture 劈碎
　~ plane 劈裂面
　~ structure 劈裂构造

cleave [kli:v] v. 劈[锯,分,破]开;穿过
cleaving hammer 劈锤
cleaving saw 大锯
cleaving timber 锯材[料]
cleet [kli:t] n. 梁柱间的楔子
cleft [kleft] n. 裂缝 a. 裂[劈]开的
~ timber 顺纹劈开的木材
~ welding 裂口焊接,V 形焊接
cleftiness ['kleftinis] n. 裂隙,节理
clench [klentʃ] =clinch v. ;n. 钉牢,钉
[铆]紧,钉头敲弯
~ nailing 弯钉钉合
clepsydra ['klepsidrə] n. pl. -dras,
-drae;(古代计时用的)漏壶,水时钟
clerestory ['kliəstəri] n. 天窗;高侧窗;
长廊;楼座
clerk [klɑ:k] n. 会计,管理员
~ of the works 工程[建筑]管理员,
现场监工
clevis ['klevis] n. U 型联接装置;马蹄
[U]形钩
click [klik] n. 棘轮,掣子;噪声;插销
~ bore 曲柄钻,摆把钻
client ['klaiənt] n. 委托人;顾客,买主
climate ['klaimit] n. 气候,地带,天气
climatic [klai'mætik] a. 气候上的
~ chart 气候图
~ variation 气候变迁
climatology [,klaimə'tɔdʒi] n. 气象
学,风土学,气候学
climatron ['klaimətrɔn] n. (大型的不
分隔的)人工气候室
climbing ['klaimiŋ] a. 上升的,爬坡
[高]的,n. 攀登,爬坡
~ form(work) 滑动上升模板
~ scaffold(ing) 滑升脚手架
~ shuttering 滑升模板
clinch [klintʃ] n. ;v. 钉住,抓紧,敲弯
的钉;铆钉连接;解决
~ bolt 夹紧螺栓
~ joint 互搭接头
~ nail 抱钉
clincher ['klintʃə] n. 夹紧钳,铆钉

~ boarding 鱼鳞板
~ built ceiling boarding 搭接钉的天花板
~ work 搭接工作
cline strata 倾斜层
clinker ['kliŋkə] n. 水泥熟料,熔[煤]渣,烧结块[砖],火山熔岩
~ -bearing slay cement 矿渣水泥
~ brick 缸砖,熔渣砖
~ cement 熟料水泥
~ concrete 矿渣骨料混凝土
~ -free cement 无熟料水泥
~ floor 缸砖地面,熔渣地面
~ masonry 缸砖砌体
~ tile 烧结釉面砖
portland ~ 硅酸盐水泥熟料
clinker-built ['kliŋkəbit] a. 鳞状搭接的 n. 互迭[鱼磷]式外壳
clinkering ['kliŋkəriŋ] n. 烧结
~ contraction 烧结收缩
~ crack 烧结裂缝
~ expansion 烧结膨胀
~ strain 烧结变形
clinkery a. 烧结的
clinking ['kliŋkiŋ] n. 裂缝
clinographic [,klainə'græfik] a. 倾斜的,斜射的
clip [klip] n. ,v. 夹,剪,箍,钳
~ angle(钢结构接合部分的)加固[拼接]角钢
~ bolt 夹紧螺栓
~ bond 箍制砌合,顺砖内角斜切的砌合体

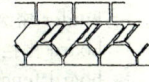
clip bond

~ joint 砖砌体箍制接头
~ joints of rubble masonry 毛石圬工箍制接头
~ plate 夹[压]板
~ tingle 压板铁片
cable ~s 电缆夹,电缆挂钩
pipe ~ 管夹
reataining ~ 固定夹
clipper ['klipə] n. 钳子,剪线钳

clipsham stone 英国产的浅黄色的石灰石

cloaca [kləu'eikə] *n. pl.* cloacae；阴沟，下水道；厕所

cloak [kləuk] *n.* 外衣；*vt.* 隐匿，掩饰；-cloakedly, *ad.*
　～ -room 衣帽间；车站行李寄存处；盥洗室，便所
　～ screen 衣帽间围屏
　～ stand 衣帽架

clock [klɔk] *n.* 仪表，时钟
　～ chamber 教堂钟塔上的钟室
　～ tower 钟楼[塔]

clockwise ['klɔkwaiz] *a.*；*ad.* 顺时针移动方向的[地]，右旋的[地]

clog [klɔg] *v.*, clogged, clogging, *n.* 障碍(物)，阻塞(物)；填塞，障碍，粘住，制动

cloisonne [ˌklɔizə'nei] *n.* 景泰蓝(制品)

cloister ['klɔistə] *n.* (修道院的)回[走]廊
　～ arch 回廊拱
　～ garth 回廊庭园
　～ vault 回廊穹窿，方形平面拱顶

close [kləus] *a.* 密实[闭，集]的，紧[周]密的，详细的，限定的 *v.* 闭合，接通，堵塞，结束 *n.*；建筑物周围的场地
　～ bed 密实层
　～ binder 密级配结合层，紧密结合层
　～ board 围栏木板，封板
　～ board fencing 鱼鳞板围篱
　～ -board roof 整片铺设的木板屋顶
　～ -contact adhesive 表面粘结剂
　～ couple 人字屋架上弦的紧密联结
　～ coupled integral water closet 带水箱的坐式大便器
　～ -coupled roof 有拉杆的人字形屋面
　～ -grained wood 密纹木
　～ -grained wheel 细砂轮
　～ inspection 严格检查
　～ -knit surface 密实面层
　～ meshed 密网眼的
　～ nipple 无隙内接头
　～ rafter 带索杆的人字木，封闭椽
　～ sand 密实砂
　～ supervision 严格管理
　～ texture 密实结构
　～ timbering 木板墙，木护墙板
　～ -up view 近视图，全貌图
　～d butt joint 紧连连接
　～d construction 实体式结构
　～d drain 排水暗管
　～d end spanner 闭口扳手
　～d fault 闭合断层
　～d fold 闭合褶皱
　～d frame 闭合构架
　～d front seat 前圆形便器座
　～d hot water heating system 封闭式热水采暖系统
　～d joint 无间隙的接头
　～d newel 砖砌旋梯的中柱
　～d polygonal rigid frame 闭合多边形刚架
　～d position 封闭状态，关闭位置
　～d ring dowel 闭口环榫，无缝环形暗销
　～d specification 详细规范
　～d stack system 闭架式
　～d stringer stair 遮盖的楼梯梁
　～d system of heating 闭式供暖系统
　～d system of ventilation 封闭式通风系统
　～d-tee joint 紧密T形接合
　～d tie 箍筋

closing board 封板

closing coil 闭合螺管

closer ['kləuzə] *n.* 镶墙边的砖，拱心石，封闭砖石，塞头，端盖
closer
　～ brick 接合砖，(砖墙砌体中的)封口砖
　king (queen) ～ 去角(纵剖)砖

closet ['klɔzit] *n.* 小房[套]间，壁橱；盥

洗室,厕所,抽水马桶大便器
~ basin 盥洗室水盆
~ bowl 大便器
~ cubicle 分隔的厕所间
~ floor flange 大便器地面连接法兰
~ flush pipe 大便器冲洗管
~ horn 大便器排出口
~ tank 大便器水箱
~ valve 大便器冲洗阀
closure [ˈkləuʒə] *n.* 截流,合拢,闭合;围墙,隔板;填塞砖 *vt.* 使……结束
clot [klɔt] *n.* 凝[团]块 *v.* 使凝结[结团];使捅塞
cloth [klɔθ] *n.* 织物
~ dust collector 袋式集尘器
~ filter 布滤器
~ hanging 墙面油漆前糊布
~ tape 布卷尺
abrasive ~ 砂布
crocus ~ 细砂布
laminated ~ 层压布
mica ~ 云母箔
tracing ~ 描图纸
wire ~ 金属丝网
clothes [kləuðz] *n.* 衣服,被褥
~ hook 挂衣钩
~ horse 晒衣架
cloud [klaud] *n.* 烟尘,浮云状物,云 *v.* 使混浊[模糊]
~ed glass 毛玻璃
~ing 云状花纹
Cloud pillar 华表(天安门)
cloustonite *n.* 发气沥青
clout [klaut] *n.* 流量孔板,大头钉垫片,布片,靶心 *vt.* 用布擦,敲一下
~ nail 大帽钉
cloven [kləuvn] *a.* 劈[裂]开的
~ board 削边板
clover [kləuvə] *n.* 三叶草,苜蓿属植物
~ -leaf type plane (教室)拉丁十字形平面
club [klʌb] *n.*;*v.*, clubbed,clubbing; *n.* 棒,俱乐部 *v.* 用棒打;组[结]合

~ hammer 砌砖用的手锤
~ house 俱乐部会所,运动员更衣室
~ room 俱乐部集会室
clump [klʌmp] *n.* 建筑群;树丛,团[木,土]块
clunch [klʌntʃ] *n.* 耐火黏土,硬质黏土,细粒泥质岩
cluster [ˈklʌstə] *n.* 集桩,梅花桩;建筑组群;组件;凝块 *v.* 群聚
~ hont 群集节,木材缺陷
~ house 住宅群
~ plan 群组式平面布置,成组串联式平面布置
~ structure 团粒结构,葡萄状结构
~ed column 集柱,束柱
~ed pier 集墩
clustered column
clutch [klʌtʃ] *n.* 夹紧装置,离合器,套管,扳手,起重机钩爪 *v.* 抓,连接咬合
~ collar 接合套,联接器
clyburn spanner 活扳手
coacervate [kəuˈæsəveit] *vt.*;*n.*;*a.* 凝聚(的)
coach [kəutʃ] *n.* 车辆
~ bolt 方头螺栓
~ screw 方头木螺钉
~ yard 车场
coagel [ˈkəuədʒel] *n.* 凝[聚]胶
coagulability [kəuˌægjuləˈbiliti] *n.* 凝结[聚]能力,凝结性(能)
coagulant [kəuˈægjulənt] *n.* 促凝剂,混凝剂
~ aid 助凝剂
coagulate [kəuˈægjuleit] *v.* 使凝结 *n.* 凝结物 *a.* 凝结的
~d matter 凝聚物
coagulating reagent 促凝剂
coal [kəul] *n.* (烟)煤,炭 *v.* 供煤,烧成炭,加煤
~ briquet(te) 煤砖[块]
~ clay 耐火黏土
~ tar creosote 煤沥青木材防腐剂

burnt ~ 天然焦炭
coaly ['kəuli] *a*. 含[似]煤的,墨黑色的
coarse [kɔːs] *a*. 粗的,未加工的,不精确的,厚的,钝的
~ grained soils 粗粒土
~ plate 厚板
~ sand 粗砂
~ sand wares 粗砂陶
coarseness *n*. 粗度
coat [kaut] *n*. 面[涂,镀,覆盖]层 *vt*. 加覆盖层,上涂料[油漆]
~ closet 存衣壁橱
~ glass 覆层玻璃
~ room 衣帽间
~ed cast-iron pipe 涂层铸铁管
~ed chips (grit) 拌沥青的石屑,预拌石屑
~ed electrode 涂药电焊条
~ed pipe 涂层管
~ed plastic 涂膜塑料
~ed type aggregate 轻质面层骨料
anti-acid ~ 耐酸涂层
anti-freezing ~ 防冻面层
back (base, bottom, first, ground) ~ 底层
brush applied ~ 刷涂
doped ~ 加固涂料
finishing ~ 修饰层
fire ~ 氧化膜
float ~ 抹面层
gel ~ 凝胶漆,表面涂漆
ground ~ 底漆,底涂(层)
protection ~ 防护层
setting ~ 罩面层
tack ~ 结合层
wearing ~ 磨耗[损]层
coating ['kəutiŋ] *n*. 涂料[层]
~ material 涂料
~ steel pipe 镀锌钢管
~ varnish 罩光清漆
damp-proofing ~ 防湿涂层
lime ~ (钢丝的)石灰处理

under ~ 上底漆,底涂(层)
zinc ~ 镀锌
coaxal [kəu'æksl] = coaxial *a*. 同轴的;同心的
~ cable 同轴电缆
~ circuit 同轴电路
~ line 同轴线
~ line termination 同轴线终端负载
cob [kɔb] *n*. 夯土建筑,泥砖,草筋泥 *vt*. 弄[破]碎
~ brick 土砖
~ construction 黏土夯实(施工)
~ wall 土墙
cobalt [kəubɔːlt] *n*. 钴(Co),钴类颜料
~ blue 钴蓝
~ cemented titanium carbide 钴钛硬质合金,钴钛金属陶瓷
~ cemented tungsten carbide 钴钨硬质合金,钴钨金属陶瓷
~ steel 钴钢
cobbing *n*. 人工破碎
cobble ['kɔbl] *n*. 卵[圆]石,铺路圆卵石 *vt*. 用圆石铺砌
~ boulder 鹅卵石
~ gravel 中砾石
~ mix 中块石混凝土(拌合料)
~ pavement 大卵石路面
~ stone 大鹅卵石,圆石块
cobweb ['kɔbweb] *n*. 蜘蛛网状物
~ rubble masonry 蜘网形缝毛石圬工
cochlear *a*. 蜗形的,蜗壳状的
cochleoid *n*. 蜗牛线
cock [kɔk] *n*. 阀[气]门,水龙头;开关,旋塞;指针,风向标;吊车,起重机 *v*. 堆[竖]起
~ hole cover 旋塞孔洞盖
~ tap 龙头
~ wrench 龙头扳手
angle ~ 转角管塞
ball ~ 球阀
T ~ 三通旋塞
three-way ~ 三通旋塞

valve ~ 阀栓
cocker [ˈkɔkə] *n.* 斜撑,采掘面上的支柱
cocking *n.* 斜面,斜度
cockle [ˈkɔkl] *n.* (薄板边缘的)皲裂纹
~ -stair 螺旋梯
cockloft [ˈkɔklɔft] *n.* 顶[阁]楼
cocoon [kəˈkuːn] *n.* 茧状物,防护层, *vt.* 做防护层
~ing 防护措施,防护喷层
COD= ① carrier on-board delivery 货到付款;②cash on-delivery 现款运货
codazzite *n.* 铈铁白云石
code [kəud] *n.* 规范[程],标准;代号,标记 *v.* 制订法规
~s of construction 建筑条例
~ of practice 业务法规,操作规程
~d program 编码程序
~d stop 程序停机
coding [ˈkəudiŋ] *n.* 编码,编制程序
~ bit 代码(信息)单位
~ legend 图例,符号解释[说明]
automatic ~ 自动设计程序
coefficient [ˌkəuiˈfiʃənt] *n.* 系数,因素,程度,率
~ of condensation 冷凝系数
~ of contraction 收缩率[系数]
~ of durability 耐久性系数,疲劳系数
~ of elasticity 弹性系数
~ of expansion 膨胀系数
~ of friction 摩擦系数
~ of heat conduction 导热系数
~ of opacity 不透明度,不透明系数
~ of pitch of building 建筑物间距系数
~ of plasticity 塑性系数
~ of roughness 粗糙系数
~ of safety 安全系数
occupation ~ 使用率
quality ~ 精度系数
coes wrench 活动扳手
coessential [ˌkəuiˈsenʃəl] *a.* 同素[质]的
coffee [ˈkɔfi] *n.* 咖啡,咖啡色
~ shop 咖啡馆
coffer [ˈkɔfə] *n.* 沉箱,围堰,浮船坞;吸声板,平顶[天花板]的镶板,藻井 *v.* 贮藏

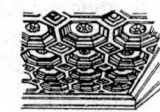
coffering

~ed ceiling 井格天花板
~ed floor 格式楼面
~ing 方格天花板[平顶]衬壁,衬砌
cog [kɔg] *n.* 嵌齿,榫头;岩脉[墙];轮齿 *v.* 装齿轮,打榫
~ging joint 榫齿接合
cogwheel [ˈkɔgwiːl] *n.* (钝,嵌)齿轮
cogging joint
~ gearing 齿轮传动装置
co-hade *n.* 断层倾角
coherence [kəuˈhiərəns] = coherency *n.* 粘聚力[性],内聚力;相干[关]性,连贯
coherent [kəuˈhiərənt] *a.* 一致的,连贯的,合乎逻迹的;附[粘]着的,连贯的
cohesion [kəuˈhiːʒən] *n.* 粘结(力),附着(力),内聚力
~ height (粘土)粘着(极限)高度
cohesionless *a.* 无粘性的,松散的
~ material 无粘性材料
~ soil 砂性土
cohesive [kəuˈhiːsiv] *a.* 有粘结[内聚]力的
~ soil 黏性土
~ strength 粘结强度
coign [kɔin] = coigne *n.* 隅(石),外角;楔
coil [kɔil] *n.* 盘条[管],卷材,线圈 *v.* 卷,盘绕
~ coating 滚涂
~ pipe 盘管
~ spring 螺旋弹簧,盘簧

~ steel 盘圆钢
~ strip 成卷带材
~ed material 卷材
heating ~ 采暖[供热]盘管
rod ~ 线材卷,盘条
coincidence [kəu'insidəns] *n.* 相符,一致;叠[吻]合
coke [kəuk] *n.* 焦炭[煤] *vt.* 焦化,炼焦
~ breeze concrete 焦炭屑混凝土,煤渣混凝土
~ oven tar 焦炉柏油
colalloy *n.* 考拉洛铝镁合金
colas ['kəuləs] *n.* 沥青浮化液
colcrete [kɔlkri:t] *n.* 预填骨料灌浆混凝土
cold [kəuld] *n.*; *a.* 低温(的),冷(的),不加热的
~ air 冷气
~ air circulating system 冷风循环系统
~ air return 冷空气回流
~ application 冷用,冷铺,冷浇筑
~ bending 冷弯
~ bending test 冷弯试验
~ bitumen emulsion 沥青乳液,乳化沥青
~ brittleness 冷脆性
~ check resistance 抗冷裂性
~ chisel 冷錾,白錾
~ colour 淡色
~ crack 冷缝,低温裂缝
~ draw 冷拉[拔]
~ drawn bar 冷拉钢筋
~ drawn low carbon wire 冷拔低碳钢丝
~ drawn pipe 冷拉管
~ drawn (steel) wire 冷拉钢丝
~ driven rivet 冷压铆钉
~ endurance 耐寒性
~ extruded 冷拔的
~ flattened steel bar 冷轧钢筋
~ frame 不供暖的种植玻璃房
~ joint (新旧混凝土之间的)建筑[冷缩]缝
~ mix 冷拌(沥青)混合料
~ proof construction 防寒构造
~ proof dwelling house 防寒住宅
~ proof rebate (门窗构造上的)防寒错口
~ resisting prpoperty 抗冻性,耐寒性
~ room 冷藏室
~ setting 冷塑化,冷凝固
~ shortness 冷脆性
~ stretched steel bar 冷拉钢筋
~ weather construction 冬季施工
colemanite ['kəulmənait] *n.* 硬硼酸钙石
collage [kə'lɑ:ʒ] *n.* (装饰工艺之一的)拼贴技术,美术拼贴;抽象派美术
collapse [kə'læps] *n.*; *v.* 破坏[损],坍陷;事故,故障
~ design (method) 破坏阶段[极限强度]设计法
~ load 破坏[极限]荷载
collapsible [kə'læpsəbl] *a.* 活动的,可折叠的
~ die 组合模,可折模
~ form 活动模板
~ gate 折门
~ steel shuttering 拼装式活动钢模板
collar ['kɔlə] *n.* 系梁,底梁;接头,套管,法兰盘,凸缘;柱环 *v.* 扭住,缠住,刻痕,打眼
~ and tie roof 系梁三角屋架
~ beam 系梁
~ beam roof truss 有系梁的人字木屋架
~ joint 套管接头,人字木屋架系杆节点
~ nut 圆缘螺帽
~ rafter roof 有系梁的人字木屋顶
~ roof 三角屋架
~ tie beam 圈梁
collaring *n.* 灰浆砌檐瓦,作凸缘,标定

炮眼位置

collation [kəˈleiʃən] *n.* 检验,整理,校对

collect [kəˈlekt] *v.* 收集,征收 *a.;ad.* 由收方付款(的),送到即付现款(的)
freight forward ~ 运费由提货人照付

collecting [kəˈlektiŋ] *n.;v.* 收集,聚集,堆积,集中

collection [kəˈlekʃən] *n.* 集水,收集,集中;(征)收,(委)托收(款);选择
~ plan 装配平面图

collective [kəˈlektiv] *a.* 聚集的,集体的,集中的
~ control elevator 集中控制电梯
~ drawings 装配图,图集
~ shelter 公共防空洞[隐蔽处]

collector [kəˈlektə] *n.* 集尘器,捕集器;集气管,总管;收款员
~ pipe 集水管
drain ~ 排水干管
exhaust ~ 排气总管

collegiate [kəˈliːdʒiit] *a.* 大学的,学院的,大学生用的
~ architecture 学院风格的建筑

collet [ˈkɔlit] *n.* (弹性)夹头,(开口)夹套

colligate [ˈkɔligeit] *vt.* 绑,扎,束,缚

colligation [ˌkɔliˈgeiʃən] *n.* 绑扎,束缚

collimation [ˌkɔliˈmeiʃən] *n.* 视准,准直,测试,观测

collimator [ˈkɔlimeitə] *n.* 准直仪[管],平行光管

collision [kəˈliʒən] *n.* 碰撞;振动;打击
~ mat 堵漏毡
~ post 防撞柱

collocation [ˌkɔləˈkeiʃən] *n.* 排列,安排,安(配)置

collochemistry *n.* 胶体化学

colloform [ˈkɔləfɔːm] *n.* 胶体,胶团结构

colloid [ˈkɔlɔid] *n.;a.*,胶体(状)的 *vt.* 使成胶态

emulsion ~ 乳胶体
suspension ~ 悬胶

colloidal [kəˈlɔidəl] *a.* 胶体(状)的
~ clay 胶质黏土
~ matter 胶体物质
~ sol 溶胶
~ solution 胶态溶液
~ state 胶体状态
~ substance 胶体物质

Cologne [kəˈləun] *n.* 科伦(德国西部一城市)
Cologne cathedral church 科伦主教堂

colonial style 17 ~ 18 世纪美国独立前的建筑式样

colonnade [ˌkɔləˈneid] *n.* 柱廊[列],间距相同的一系列柱子
~d 有柱廊的
~d avenue 连拱廊的通道
~d street (罗马建筑艺术中)室内[有遮顶的]街道

colonnade

colonnette *n.* 装饰性小圆柱,小柱

colophonium [ˌkɔləˈfəuniəm] *n.* 树脂,松脂[香]

colorama [ˌkʌləˈrɑːmə] *n.* 彩色光
~ lighting (自由)调色照明

colorimeter [ˌkʌləˈrimətə] *n.* 色度计,比色计

colorimetric *a.* 比色分析的
~ analysis 比色分析法
~ phrometer 比色高温计
~ indicator for pH pH 比色指示剂

colorimetry [ˌkʌləˈrimətri] *n.* 比色法,比色试验

colossal [kəˈlɔsl] *a.* 庞大的,异常的
~ order 巨柱式,两层通高的柱式

Colosseum 古罗马的圆形大剧场(创建于 80 A.D.,此剧场大部分至今尚存)

colossus [kəˈlɔsəs] *n. pl.* -lossi, -lossuses ; 巨型雕塑

colour [ˈkʌlə] = color *n.* 色彩,颜色

[料] *v.* 上[染]色
~ atlas 色卡
~ chart 色彩图表
~ coating glass 涂色玻璃
~ comparison tube 比色管
~ conditioning [dynamics] 色彩调节
~ contrast 色彩对比度
~ coordination 色彩协调[配合]
~ difference 色差
~ evaluation of weight 色彩的重量感
~ filter 滤色片[板]
~ floc 有色绒体
~ harmony 色彩协调
~ harmonic theory 色彩调色理论
~ matching 配色
~ notation system of CIE 国际照明委员会采用的标色系统
~ perception 色知觉
~ perspective 彩色透视画法
~ planning 色彩设计
~ removal 脱色
~ rendering (rendition) 显色性
~ index 显色指数
~ scheme (室内装饰)色彩设计
~ sensation 色感觉
~ sequence (色灯信号)显示顺序
~ slide 彩色幻灯片
~ solid (用三度空间表示色彩之间关系)色立体
~ stimulus 色刺激
~ television 彩色电视
~ transparency 彩色幻灯片
~ tutanage 有色镀锌铁板
~ed bulb 彩色灯泡
~ed cement 彩色水泥
~ed cement concrete 彩色水泥混凝土
~ed mortar finish 彩色砂浆罩面
~ed sculpture 彩塑
~ed sheet glass 彩色窗玻璃
~ed speck (陶瓷制品上的)色斑
~ed wood-cut 套色木刻
colo(u)rwash [ˈkʌləwɔʃ] *n.* 彩色涂料 *v.* 上彩色涂料
colter [ˈkəultə] *n.* 开沟器,铲
column [ˈkɔləm] *n.* 圆[立,支]柱,墩,塔

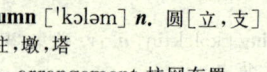

~ arrangement 柱网布置
~ base 柱座[础]
~ body type 柱式体形
~ box 柱形模板

column

~ cap 柱头,立柱罩壳
~ capital 柱头
~ casing 柱筒
~ clamp 柱模板,箍铁
~ of angles 角钢柱
~ of built channels laced 槽钢缀合柱
~ of concrete filled tube 混凝土填塞管柱
~ of trays 层板
~ pier 柱墩
~ pile 柱桩,端承桩
~ radiator 柱式散热器
~ shaft 柱身
~ shuttering 柱体模板
~ socle 柱基座

column socle

~ spacing 柱间距
~ spiral (钢筋混凝土柱的)螺旋箍筋
~ structure 柱式结构
~ strip 柱顶条板
~ tie (钢筋混凝土柱的)箍筋
~ with hinged ends 端部铰支柱
~ with lateral reinforcement 纵向配筋柱
~ with variable cross-sections 变截面柱
beam ~ 梁柱
clustered ~ 簇柱
Corinthian ~ 科林斯柱式
Doric ~ 多立克柱式

eccentrically loaded ~ 偏心受压柱
Flowery ~s 华表
gray ~ 四槽钢柱
Ionic ~ 爱奥尼柱式
iarimer ~ 十字形的工字钢组合柱
spinal ~ 脊柱
twisted ~ 涡卷圆柱
vermilion ~ 朱漆柱
columnar [kə'lʌmnə] *a.* （圆）柱状的，钉状的
~ aggregate 柱状团聚体
~ architecture 柱式建筑
~ deflection 柱纵向挠曲
~ joint 柱状节理
~ order 柱型
~ pier 柱式墩
~ stroked dressing （石面）槽纹修琢
~ structure 柱状结构
columniation [kə,lʌmni'eiʃən] *n.* 列柱【法】
comb [kəum] *n.* 蜂窝，梳齿状，梳状物，刻螺纹的器具
~ed glass 条纹装饰玻璃
~ed plywood 蜂窝胶合板
~ed stucco 蜂窝状拉毛粉饰
~ -grained wood 心木，密纹木
~ plaster 拉毛粉刷
~ plate 篦形防滑板
~ structure 蜂窝状结构
combination [,kɔmbi'neiʃən] *n.* 混[联，结，化]合
~ beam 组合梁
~ board 合成纸板
~ column （型钢，混凝土）混成柱，组合柱
~ construction 混合结构
~ door 组合门，双层门
~ faucet 冷热水混合龙头
~ fixture 整套卫生设备
~ frame 组合框架
~ grate 供热口

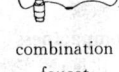
combination faucet

~ lock 暗码锁
~ pliers 剪[钢丝]钳
~ square 组合角尺，什锦角尺
~ waste-and-vent system 排气及通气合用系统
~ window 混合窗
combine [kəm'bain] *v.* 混[结，联，复]合 *n.* 联合收割机，综合工厂
~d building sewer 房屋的合流污水管
~d capping slab 联合盖板
~d dwelling house 商店(兼用)住宅
~d efficiency 综合效率
~d frame 带窗的门框
~d joint 铆焊混合接头
~d lime 混合石灰
~d use district （工业、商业、居住等各种用途的建筑）混合作用区
~d ventilation system 联合通风系统
combing [kəumiŋ] *n.* 木纹状漆面，梳刷
~ of brick face 砖面修刮
combustibility [kəmbʌstə'biliti] *n.* 可[易]燃性
combustible [kəm'bʌstəbl] *a.* 可燃的，易燃的
~ material 易燃材料[物品]
come [kʌm] *v.* 来，到；形成
~ into effect 生效，实施
~ into operation 开工
comfort [kʌmfət] *n.* 室内水暖设施；舒适 *vt.* 使舒适
~ humidity 舒适湿度
~ index 室内空调舒适系数
~ zone chart 快感区图，舒适带图
~ station 公共厕所
~ temperature 舒适温度
comfortability *n.* 舒适性
command [kə'mɑ:nd] *v.*；*n.* 控制，指令
~ definition language 指令语言
commander [kə'mɑ:ndə] *n.* 手夯，木槌
commemorative [kə'memərətiv] *a.* 纪念的，庆祝的

~ architecture 纪念性建筑
~ column 纪念柱
commencement [kəˈmensmənt] *n.* 开工[始]
~ of initial setting 初凝
commercial [kəˈməːʃəl] *a.* 商品[业]化的 *v.* 商业广告节目
~ aluminum 民用铝,商用铝(纯度大于99%)
~ area 商业区
~ building 商业建筑
~ port 商埠
commingler [kəˈmiŋglə] *n.* 搅拌[混合器]
comminute [ˈkɔminjuːt] *vt.* 粉[磨]碎
~d solid 粉碎固体
comminutor *n.* 粉碎机
commissioning *n.* 开工,投入运行
commissure [ˈkɔmisjuə] *n.* 合缝处,石砌体的层缝,缝口
commitment [kəˈmitmənt] *n.* 委托,约定,许诺,保证
commode [kəˈməud] *n.* 洗脸台;小书橱,五斗橱;厕所,盥洗室
~ -type toilet 坐式便溺器
common [ˈkɔmən] *a.* 普通[一般,公共]的
~ aisle 公共通道
~ arch 粗拱
~ area 公用地
~ beam 普通梁,共用的梁
~ bald cypress 落羽松
~ bond 普通砌砖式
~ brick 普通砖
~ concrete 普通混凝土
~ dovetail 普通鸠尾榫
~ facilities 公共设施
~ iron 普通钢材
~ labour 杂工
~ labourer 一般工人

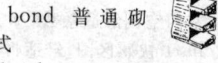
common dovetail

~ lodging house 简易旅社
~ main 总水管
~ mode noise 共态噪声
~ ram 手夯
~ rafter 普通椽木,共用的椽木
~ sewer 共用下水道
~ space 普通[粉饰]灰泥
~ steel 普通钢
~ tap 公用水龙头
~ thread 普通螺纹
~ tile 普通瓦
~ vent 共用通风管
~ wall 公共墙
communal [ˈkɔmjunl] *a.* 共有的,集体的,社会的
~ building 公用建筑
~ garden (公寓式住宅的)公用花园
communicate [kəˈmjuːnikeit] *v.* 连通,互通;联系,交通
~ pipe 连通管
~ ting rooms 有门互通房间
community [kəˈmjuːniti] *n.* 地区,团体
~ planning 地区规划
compact [kəmˈpækt] *a.* 密实的,致密的 *v.* 压实,密[夯]实;成型 *n.* 合同,协定;烧结物,成型件
~ city 小城市
~ grained 密级配的,按最小空隙选择的
~ gypsum 纯白石膏
~ed earth lining 压实底土衬砌
~ed layer 压实层
~ed soil 压实土
~ed thickness 压(夯)实厚度
compaction [kəmˈpækʃən] *n.* 压实,致密
compactness [kəmˈpæktnis] *n.* 密度,比重;密实性,紧密度
compactor [kəmˈpæktə] *n.* 压土机,夯土机
compages [kəmˈpeidʒiːz] *n.* 骨架,结构,组织
compalox *n.* 氧化铝

companile *n.* (车站)钟楼
comparative [kəm'pærətiv] *a.* 比较的
　～ cost 比较造价
　～ cost planning 比较成本分析法
　～ designs 比较设计[方案]
compartment [kəm'pɑːt-mənt] *n.* 隔间;隔板
　～ ceiling 格子平顶
　～ roofing 分隔屋面[顶]
　air ～ 通风[空气]室
　top ～ 顶室[间]
compass ['kʌmpəs] *a.* 圆弧形的, *v.* 指南针,罗盘仪;范[周]围,(*pl.*)圆规,两脚规 *vt.* 计划,围绕
　～ brick 拱砖
　～ caliper 弯脚卡钳
　～ plane 凹刨,曲面刨
　～ rafter 轮椽(山墙装饰)
　～ roof 半圆形屋顶
　～ saw 开孔锯,截圆锯
　～ timber 弯木料
　～ window 圆肚窗,半圆形凸窗
compatible [kəm'pætəbl] *a.* 相容的,能共存的
compatibility [kəmˌpætə'biliti] *n.* 相容性,协调,一致
compensate ['kɔmpenseit] *v* 补[赔]偿
compensation [kəmpen'seiʃən] *n.* 补强,校正;补偿平衡
　～ joint 补强接缝,调整缝
　～ of errors 误差调整,平差
compensatory [kəm'pensətəri] *a.* 赔[补]偿的,报酬的,补充的
competent ['kɔmpitənt] *a.* 耐久的;主管的,胜任的
　～ authorities 主管部门
　～ party 法定一方
competitive [kəm'petitiv] *a.* 竞争的,比赛性的
　～ bid 公开招标的
　～ bidding system 招标制
　～ power 竞争力
　～ price (投)标价,竞争价格

　～ tender 公开投标
compile [kəm'pail] *vt.* 编码,编程序;编辑,汇编
complementary [ˌkɔmpli'mentəri] *a.* 补足[充]的 *n.* 余[补]码
　～ colour 互补色,余色
　～ error 互补误差
　～ space (多)余空间,补空间
complete [kəm'pliːt] *vt.* 实施;完成,结束 *a.* 成套的,完成了的,全部的
　～ bath (设有浴盘、洗脸器、便器等设备的)整套浴室
　～ certificate 竣工说明书
　～ inspection 全面检查
　～ly hydrated cement 完全水化水泥
　～ly mixed basin 完全混合池
　～ly knocked down 完全拆开
　～ overhaul 大修
　～ project management 统筹规划管理法
　～ rationing system 包干制
　～d amount 完成工程量
completion [kəm'pliːʃən] *n.* 完[竣]工
　～ date 完工日期
　～ report 完工报告
complex ['kɔmpleks] *n.* 复合物,合成物,全套设备,综合企业 *a.* 综合的,复杂的
　～ alloy steel 多合金钢
　～ building (房屋的底层与上层不同用途的)复合建筑
　～ chart 综合图
　～ surface 复合地面
　～ tone 杂音
　～ truss 复式桁架
compliance [kəm'plaiəns] = compliancy *n.* 可塑[柔顺]性;一致,相符
compluvium *n.* 天花板上的方形通风孔
compo ['kɔmpəu] *n.* 水泥砂浆,灰浆,混合涂料;耐火混合物;组成;工伤赔偿费
　～ pipe (铅锡)合金管
component [kəm'pəunənt] *n.*; *a.* 构

[元]件(的);成份(的),组成(的);分量[力](的)
~ assembly 零件装配
~ part 构件
~ specification 部件规格
~ system 构件组合系统,堆砌式结构

composite ['kɔmpəzit] *n.* 混合料;复合材料;混合物 *a.* 混[复]合的
~ arch 尖拱,复合拱
~ beam 组合梁
~ capital 罗马复合式柱头
~ column (型钢、混凝土)组合柱
~ construction 组合结构
~ cost 包工包料,工料合算单价
~ die 拼合[块]
~ door 组合门
~ electrode 组合焊条
~ filter 混合式过滤层
~ floor board 复合地板
~ girder 组合大梁
~ joint 混合接头
~ landscape 复合景观
~ material 复合材料
~ order 组合柱式
~ panel 复[组]合板
~ plane wall (多层材料的)复合墙
~ refractory 混合耐火材料
~ rigid frame 组合刚架
~ roof turss 组合屋[桁]架
~ structure 复[混]合结构
~ truss 组合桁架
~ wall 组合墙,混合墙
~ wood 合成木材

composition [ˌkɔmpə'ziʃən] *n.* 结构,构造;组成,成份;布局;结合,合成
~ floor 组合地板
~ in architecture 建筑构图
~ joint 铆焊并用接合
~ metal 合金
~ plane 接合面,复合面

composite capital

composite order

architectural ~ 建筑布局

compost ['kɔmpɔst] *n.* 灰泥,混合涂料 *vt.* 涂灰泥

compound ['kɔmpaund] *n.* 化合物,剂料 *vt.* 合成,配合, *a.* 混合的,合成的
~ arch 复式拱
~ beam 组合梁
~ column 组合柱,集柱
~ girder 组合大梁
~ glass 多层玻璃
~ pier 集墩
~ pipe 复合管线
~ section 组合截面
~ steel 合成钢
~ switch 组合开关
~ type apartment house 组合式住宅
~ wall 多层壁,组合墙
~ web plate 组合腹板
caulking ~ 填缝料
coating ~ 涂料
filling ~ 填(充)料
sealing ~ 腻子,密封胶
stripping ~ 脱模剂

compounding [kɔm'paundiŋ] *n.* 配方[料];混合,配合

compreg *n.* 胶压木,胶压板
~ nated wood 渗(胶)压(缩)木材

comprehensive [ˌkɔmpri'hensiv] *a.* 综合的,广泛的
~ design 综合设计
~ development 综合开发
~ planning 综合规划
~ utilization 综合利用

compress [kəm'pres] *v.* 压[浓]缩;摘要 *n.* 收缩器,打包机;绷带
~ed air painting apparatus 压气喷漆工具[器]
~ed asbestos sheets 石棉纸板
~ fiberboard 压制纤维板
~ed (laminated) wood 强化木材

compressibility [kəmˌpresi'biliti] *n.*

压缩性
compression [kəm'preʃən] *n.* 加压,压力,压缩
~ faucet 压力龙头
~ joint 承压缝,压力接合
~ nut 压紧螺帽
compressive [kəm'presiv] *a.* 有压(缩)力的
compressor [kəm'presə] *n.* 压缩机[物]
~ wire 预应力钢丝
painting ~ 压气喷漆枪
compromise [ˈkɔmprəmaiz] *n.;v.*, -mised,-mising;和解,妥协
~ (form of) architecture 和[协调]的建筑艺术
~ faces 协和面
engineering ~ 工程折衷方案
compulsory [kəmˈpʌsəri] *a.* 强制[迫]的
computation [ˌkɔmpju'teiʃən] *n.* 计算,测定
~ center 计算中心
~ room 计算机室
compute [kəm'pju:t] *n.;v.* 计算,求解
computer [kəm'pju:tə] *n.* 电子计算机,电脑;计算者
~ aided design 计算机辅助设计
~ aided mapping 计算机辅助绘图
~ graphic 计算机制图
~ programming 计算机程序设计
~ software 计算机软件
general purpose ~ 通用计算机
program(me) controlled ~ 程序控制计算机
computerization [kəmˌpju:təraiˈʒeiʃən] *n.* 用(电子)计算机处理[计算],计算机化
concave [ˈkɔnˈkeiv] *a.* 凹的,向外弯曲的; *v.* 凹面(物),拱形,穹拱
~ bit 凹形凿
~ brick (砌拱用的)凹形砖
~ ceiling 拱形层顶[盖板]
~ chamfer 凹圆面
~ - ~ 双凹形的,两面凹的,
~ -convex 凹凸形的
~ crown 凹顶
~ fillet weld 凹形角焊缝
~ joint 凹缝
~ mirror [glass] 凹镜
~ plane 凹底刨
~ tile 牝瓦,凹瓦
~ tooled joint 圆截面缝,勾圆缝
~ upward 上凹线,反弧形
~ weld 凹面焊缝
concavity [kɔnˈkæviti] *n. pl.* -ties;凹曲度,凹面[线]
conceal [kənˈsi:l] *vt.* 隐蔽[藏]
~ed arch 暗拱
~ed cleat 暗固着楔
~ed convector 隐蔽式对流器
~ed dovetail 暗榫
~ed fencing 隐蔽式拦沙障
~ed flashing 暗挡条
~ed girder 隐蔽梁,非露明梁
~ gutter 隐蔽式天沟
~ed illumination 无影[反射,间接]照明
~ed joint 暗缝,隐藏接缝
~ed light 有灯槽的反光灯,暗灯
~ed nailing 隐藏钉,暗钉
~ed piping 暗装管道
~ed radiator 藏置式散热器
~ed rafter 暗椽
~ed shower 暗装淋浴器
~ed wire 暗线
~ed work 隐蔽工程
concentrated [kɔnsentreitid] *a.* 集中的,浓缩的,强烈的
concentration [ˌkɔnsenˈtreiʃən] *n.* 浓度,集中,浓缩
concentric(al) [iənˈsentrik(əl)] *a.* 同(中)心的,集中的
~ arch 同心拱
~ circles 同心圆

~ reducer 同心变径管,同心锥形管
~ structure 同心构造
~ type 环式,同心式
~ wire rope 同心式钢丝绳
concentricity [kɔnsentrisiti] *n.* 同(中)心,同心度,集中;同心环纹
concept [ˈkɔnsept] *n.* 概念,原理,思想
~ phase 初步设计阶段
conceptual [kənˈseptjuəl] *a.* 概念(上)的
~ design 方案设计
~ model 概念性模式
~ phase 初步[草图]设计阶段
concert [ˈkɔnsət] *n.* 音乐会 *v.* 商议,安排
~ hall 音乐厅
concession [kənˈseʃən] *n.* 使用权,租借地;许可,核准;让步,妥协
conch [kɔŋk] *n. pl.* conchs, conches; 半圆形穹窿;具壳海神
concha [kɔŋkə] *n. pl.* -chae; 半圆形屋顶,贝壳状结构
conchoid [ˈkɔŋkɔid] *n.* 螺旋[蚌]线,柱凸肚线,贝壳状断面
concordant [kənˈkɔːdənt] *n.* (预加应力用的)吻合钢索束; *a.* 一致的,协调的
~ profile 吻合线,吻合截面
concourse [ˈkɔŋkɔːs] *n.* 群集场所,中央广场,汇集,集合
concrete [ˈkɔnkriːt] *n.* 混凝土,三合土, *v.* 浇筑混凝土;使凝结 *a.* 混凝土的,凝固的,具体[实在]的
~ accelerator 混凝土速凝剂
~ additive 混凝土添加剂
~ admixture 混凝土外加剂
~ aggregate 混凝土骨料
~ antifreezer 混凝土防冻剂
~ apron 混凝土护墙[护坦,底板]
~ article 混凝土制品
~ back filling 混凝土回填
~ baffle pier 混凝土消力墩
~ bagging 袋装混凝土
~ bar 混凝土配筋
~ bay 混凝土底板
~ beam 混凝土梁
~ bent construction 混凝土构架结构
~ block 混凝土砌块
~ block for shielding 防射线混凝土砌块
~ block gravity wall 混凝土方块重力式墙
~ block works 混凝土砌块工程
~ building unit 预制混凝土房屋构件
~ cap 混凝土柱帽
~ casing 混凝土护[饰]面
~ -casting 混凝土预制件,装配式混凝土构件
~ cast in situ (place) 现(场)浇(注)混凝土
~ collar 混凝土圈梁
~ coloring 混凝土着色
~ compacted by jolting 振实混凝土构件
~ component 混凝土构件
~ construction 混凝土施工,混凝土建筑[结构]
~ core 混凝土筒体
~ core wall 混凝土心墙
~ cover 混凝土保护层
~ crib 混凝土格笼
~ deformation 混凝土变形
~ densifying agent 混凝土密实剂
~ deposite 混凝土浇筑体[混合物]
~ design 混凝土配合比设计
~ drain pipe 混凝土排水管
~ encasement 混凝土外壳
~ facing 混凝土面层
~ finish 混凝土抹面[终饰]
~ floor 混凝土楼板
~ form 混凝土模板
~ foundation 混凝土基础
~ frame 混凝土构架
~ grade 混凝土标号
~ grout 混凝土浆

~ gun 水泥喷枪
~ hardening agent 混凝土增强剂
~ hollow block 混凝土空心砌块[砖]
~ -in-mass 大体积混凝土
~ in situ 现浇混凝土
~ jacket 混凝土外层
~ layer 混凝土层
~ lift 混凝土(薄)层
~ -lined 混凝土衬砌的
~ lining 混凝土衬砌
~ lintel 混凝土过梁
~ mark 混凝土标号
~ masonry 混凝土圬工[砌体]
~ masonry unit 混凝土砌块
~ membrane 混凝土薄层
~ mix 混凝土拌和物
~ nail 混凝土钉
~ of low porosity 少孔隙混凝土
~ of one consistency 等稠度混凝土
~ of stiff consistency 干硬性混凝土
~ oil (混凝土)模板油
~ pillar 混凝土支柱[墩]
~ pipe 混凝土管
~ product 混凝土制品
~ reinforcement steel 混凝土钢筋
~ screed 混凝土找平层
~ sheet piling 混凝土板桩
~ shell 混凝土壳体
~ skeleton 混凝土骨架
~ shield 混凝土衬砌层[护壁]
~ specifications 混凝土规范
~ stair 混凝土楼梯
~ vault 混凝土穹顶
~ wall 混凝土墙
~ with artificial resin admix(ture) 合成树脂混凝土
~ workability 混凝土和易性
concreting in freezing weather 混凝土冬季施工
concreting in lifts 分层浇筑混凝土
air entrained ~ 掺气混凝土
air-placed ~ 喷射混凝土
alkali-fast ~ 耐碱混凝土
armoured ~ 钢筋混凝土
bamboo ~ 竹筋混凝土
bituminous ~ 沥青混凝土
blown-out ~ 充气[多孔]混凝土
calcium silicate ~ 硅酸盐制品
cellular ~ 泡沫混凝土,加气混凝土
ceramisite ~ 陶粒混凝土
cinder ~ 炉渣混凝土
close graded asphaltic ~ 密级配沥青混凝土
coarse graded asphaltic ~ 粗级配沥青混凝土
cold-laid ~ 冷混凝土
compressed ~ 压实混凝土
continuous ~ 连续浇筑混凝土
cyclopean ~ 毛石混凝土
dry ~ 干硬性混凝土
early strength ~ 早强混凝土
exposed aggregate ~ 露石混凝土
fast setting ~ 快凝混凝土
fiber reinforced ~ 纤维增强混凝土
fine sand ~ 细砂混凝土
fire-resisting ~ 耐火混凝土
flowing ~ 流态混凝土
foamed ~ 泡沫混凝土
fresh ~ 新鲜混凝土
gap-graded ~ 间断级配混凝土
gas ~ 加气混凝土
green ~ 新浇[未完全凝固的]混凝土
grouted aggregate ~ 灌浆混凝土
heaped ~ 未摊铺混凝土
heavy ~ 重混凝土
high strength ~ 高强混凝土
hollow ~ 大孔混凝土
hydrated ~ 水化混凝土
job-placed ~ 现浇混凝土
light weight ~ 轻混凝土
liquid ~ 流态混凝土
long-line prestressed ~ 先张法预应力混凝土
magnesite ~ 菱苦土混凝土

mass ~ 大体积混凝土
matured ~ 成熟混凝土
non-fines ~ 无砂混凝土
non-reinforced ~ 无筋混凝土,素混凝土
oil-proof ~ 耐油混凝土
open graded asphaltic ~ 无级配沥青混凝土
pea gravel ~ 细石混凝土
plain ~ 普通无筋混凝土,素混凝土
plastic ~ 塑性混凝土
polymer-cement ~ 聚合物水泥混凝土
polymer impregnated ~ 聚合物浸渍混凝土
polystyrene ~ 聚苯乙烯膨珠混凝土
post-stressed ~ 后张法预应力混凝土
precast ~ 预制混凝土
premixed ~ 预拌混凝土
prestressed ~ 预应力混凝土
prestressed wire ~ 钢弦混凝土
pretensioned ~ 先张法预应力混凝土
pumice ~ 浮石混凝土
pump ~ 泵送混凝土
pumped ~ 泵送混凝土
quality ~ 优质[高级]混凝土
quaking ~ 塑性混凝土
radiation-shielding ~ 防辐射混凝土
rapid-hardening ~ 快硬混凝土
ready mixed ~ 预拌混凝土
refractory ~ 耐火混凝土
reinforced ~ 钢筋混凝土
rubble ~ 毛石混凝土
self-stressing ~ 自应力混凝土
slag slurry ~ 湿碾矿渣混凝土
steamed ~ 蒸汽养护混凝土
steel chips ~ 钢屑混凝土
spray ~ 喷射混凝土
spun ~ 离心成型混凝土
subaqueous ~ 水下混凝土
tamped ~ 捣固混凝土
transit-mix(ed) ~ 运送拌和混凝土

under water ~ 水下混凝土
vacuum-treated ~ 真空处理混凝土
vibrocast ~ 振实混凝土
water-bearing ~ 含水混凝土
water-cured ~ 湿养护混凝土
water glass acid proof ~ 水玻璃耐酸混凝土
water-tight ~ 防水混凝土
concretion [kən'kri:ʃən] *n.* 凝结(物)
concurrent [kən'kʌrənt] *a.* 并存的,合作的,一致的,交于一点的 *n.* 同时发生的事件;共同
concussion [kən'kʌʃən] *n.* 冲击,振动
condemned [kən'demd] *a.* 报废[破损]了的
~ stores 废品
condensability [kən‚densə'biliti] *n.* 可压缩[冷凝,凝结]性
condensate [kən'denseit] *n.* 浓缩(物),冷凝(物) *v.* 凝结,冷凝,浓缩 *a.* 浓缩了的
condensation [kɔnden'seiʃən] *n.* 冷凝[凝结,浓缩,雾化](作用)
~ of moisture 结露,冷凝
~ point 露点,凝固点
~ polymer 缩聚物
~ product 冷凝物,浓缩物
~ rate 浓缩度
~ sinking 气窗上的冷凝水集水槽
~ temperature 凝结温度,凝结点
~ water 冷凝水
~ within structure (围护结构)内部结露
condensator ['kɔndenseitə] = condenser *n.* 冷凝器,制冷装置
condense [kən'dens] *v* 使冷凝[凝结];压缩;精简
~ water 冷凝水
~d specifications 简明技术规范
condenser [kən'densə] *n.* 冷凝[凝结]器;聚光器;电容器;制冷装置
condition [kən'diʃən] *n.* 条件,状态,状况 *vt.* 调节,制约,以……为条件

~s for site planning 总平面设计(地形、地质、水文)条件
~ precedent 先决条件
~ed air 空气调节
~ed ceiling 空气调节的天花板
air ~ (装)空气调节(设备)
ambient ~ 环境条件,大气状态

conditioning [kən'diʃəniŋ] *n.* 调整[节],整修,改善
~ chamber 调节室,加湿[干燥]室
~ of air 空气调节
end ~ 管材端头预加工

conditioner [kən'diʃənə] *n.* 调节器[剂]

conduct ['kɔndəkt] *v. n.* 传导;指导,经营,管理
~ing layer 传导层,导电层
~ing tissue of wood 木材
~ing wire 导线

conductance [kən'dʌktəns] *n.* 传[电]导,导电性
~ factor 传导系数

conduction [kən'dʌkʃən] *n.* 输送,传导

conductor [kən'dʌktə] *n.* 导体[线];管理[指挥]员;避雷针
~ pipe 导管

conduit ['kɔndit] *n.* 管道,阴沟,引水道
~ joint 管道接头
~ pipe 导管,管道
~ system 地下管道系统
~ tube 管道,导管
~ valve 管道阀门
aerial ~ 架空管道
covered ~ 地下管道,暗管

cone [kəun] *n.* 圆锥(体) *v.* 使成锥体,使带斜角
~ headed bolt 圆锥头螺栓
~ head rivet 锥头铆钉
~ hip tile 戗脊锥形筒瓦
~ joint 双锥管接合,圆锥接头
~ nut tie 锥形螺帽拉杆
~ valve 锥形阀
truncated ~ 截锥体,截角锥
~d dowel 锥形暗销

conference ['kɔnfərəns] *n.* (投标)协商,会议,公会
~ block 用于集会,会议的一组建筑物[群]
~ hall 会议厅

confidence ['kɔnfidəns] *n.* 置信度,可靠程度

configuration [kən,figju'reiʃən] *n.* 外貌,形状,地形;化合物的构[组]成
~ of earth 地形

confine [kən'fain] *v.* 限[控]制;*n.* 界限,范围
~d stratum 封闭层,隔水层
confining bed 隔水[封闭]层
confining stratum 隔水[封闭]层

confinement [kən'fainmənt] *n.* 限制

confirm [kən'fə:m] *vt.* 认可,确定,批准,使……坚定

conflagration [kɔnflə'greiʃən] *n.* 大火,火灾,暴燃

conform [kən'fɔ:m] *v.* 依照,遵守,使一致 *a.* 一致的,相似的
~(al) map(ping) 等轴测正投影图,正形投影制图,等角制图法
~ing article 合格品
~ing plate 整形板

conformable [kən'fɔ:məbl] *a.* 一致的,相似的

conformity [kən'fɔ:miti] = conformance *n.* 相似[符],遵守;地层整合
~ certificate 合格证明[书]

conge ['kɔ:ʒei] *n.* 圆弧凹形嵌边饰

congeal [kən'dʒi:l] *v.* 冻[凝]结,变冷
~ed moisture 凝结水
~ed solution 冻结溶液
~ing point 凝固点

congelation [kɔndʒi'leiʃən] *n.* 凝固[结],凝冻,冻坏;凝结物,凝块

congest [kən'dʒest] *v.* 阻塞,拥挤[塞],充满;

conglomerate [kən'glɔmərit] *n.* 天然

聚集物,砾岩 a. 密集的,聚成剂团状的 v. 使积聚成团
conic ['kɔnik] n. 圆锥[二次]曲线; a. 圆锥形的
conical ['kɔnikəl] a. 圆锥(形)的
～ dome (圆)锥形穹顶
～ head bolt 圆锥头螺栓
～ head rivet (截锥)平头铆钉
～ hip 锥形屋脊
～ roof 锥形屋顶
～ vault 锥形穹窿

conical vault

conjugate ['kɔndʒəgit] n.; a. 共轭值(的); v. 共轭,配对
～ angles 共轭角
～ beam 共轭梁
conjunction [kən'dʒʌŋkʃən] n. 逻辑乘法[积];连结[结合]件[点]
conjunctive [kən'dʒʌŋktiv] a. 连接的,联合的
connate ['kɔneit] a. 原生的,同源的
connect [kə'nekt] v. 连接,结合,接通
～ed yoke 系梁,连接横木
～ing angle 结合角钢
～ing bolt 连接螺栓
～ing box 接线盒
～ing chain 连接链
～ing flange 连接法兰
～ing link 连杆
～ing pin 连接销
～ing place 连接板
～ing rod bolt 连杆螺栓
～ing tag 连接销
connection [kə'nekʃən] n. 连接(机构),接合面,接头,连轴节,离合器
～ angle 接合角钢
～ band 接头箍
～ clip 结合扣
～ dimension 连结尺寸
back to back ～ 交叉连接
cross ～ 十字接头,交叉连接
delta [△] ～ 三角连接
drain ～ 排水管接头
earth ～ 接地线
field ～ 现场装配
water ～ 水管结合配件
connector [kə'nektə] n. 连结器[线,管道];插头[座]
～ bend 弯头[管]
～ pin 插销,塞子
～ plug 插头,塞子
conoid ['kəunɔid] n;a. 圆锥形(的),劈锥曲面,圆锥体
～ vault 锥形穹顶
conoscope ['kəunəskəup] n. 锥光偏振[干涉]仪,锥光镜
consecutive [kən'sekjutiv] a. 连续的,顺序的,依次相连的
consent [kən'sent] n. 同意,许可;插座,塞孔
～ to build 施工执照,建筑许可证
consequent ['kɔnsikwənt] a. (地质上)顺向的; n. (数学上的)后项结果,结论
consequential [ˌkɔnsi'kwenʃəl] a. 间接的,相应而生的,逻辑上的
～ loss 间接[从属]损失
conservancy [kən'sə:vənsi] n. pl. -cies. 管理,(河流,航道,港口)管理局,水利委员会;(森林,资源)保护区,水土保持
conservatory [kən'sə:vətri] n. pl. -ries. 温室,暖房,库房;音乐学院 a. 保存的,保管人的
conserve [kən'sə:v] v. 防腐,保存 n. 防腐剂
conserving agent 防腐剂
consideration [kənsidə'reiʃən] n. 考虑研究,条件,理由
design ～ 设计依据[根据]
general ～ 一般原则
consignment [kən'sainm-ənt] n. 委托,交付,托运(的货物)
consistence [kən'sistəns] = consistency n. 稠[浓,密实]度;一致[连续,相

容]性
consistent [kən'sistənt]
 a. 相符[一致]的,坚固的,相容的,始终如一的
console ['kɔnsəul] *n.*
 (螺形)支柱,落地式支架;操纵[控制]台
 ~ bracket 螺形支托
 ~ cabinet 控制室
 ~ receiver (set) 落地式收音机
 ~ screen 落地式屏风
 ~ switch 操作开关

console bracket

console screen

consolidate [kən'sɔlideit] *v.*, 加固,强化;固结,压实,统一
consolidation [kən,sɔli'deiʃən] *n.* 强化,固结(作用);渗压,巩固;捣实
conspectus [kən'spektəs] *n.* 提要,简介,示意图,一缆表
constance ['kɔnstəns] constancy *n.* 恒定,不变,持久性
 ~ of temperature 温度恒定
 ~ of volume 容积不变性
constant ['kɔnstənt] *n.* 常数,恒量,恒定值;*a.* 不变的,经常的,恒久的
 ~ cross-section 固定截面
 ~ dimension column 等截面柱
 ~ duty 不变工况
 ~ section 等截面
 ~ temperature 恒温
 design ~ 设计常数
constantan *n.* 康铜,铜镍合金
 ~ wire 康铜丝
constituent [kən'stitjuənt] *n.* (组成)成分,构成;分量[力] *a.* 组成的
constitution [kɔnsti'tjuːʃən] *n.* 组成,结构;状态,条件,法规,章程
constitutive ['kɔnstitjuːtiv] *a.* 构成的,组织的,基本的
constrain [kən'strein] *vt.* 约束,强[抑]制
 ~ed beam 固端梁
 ~ed procedure 约束方法,限定程序
constraint [kən'streint] *n.* 强制,约束
constrict [kən'strikt] *vt.* 压缩,使收缩[变小]
 ~ed end joint 缩口接头,异径(管子)接头
 ~ed paint sprayer 缩口喷漆头
constriction [kən'strikʃən] *n.* 收缩,缩头;阻塞物,收缩障碍;收敛管道
constringency [kən'strindʒənsi] *n.* 收缩(性)
construct [kən'strʌkt] *vt.* 建筑,施工;制图,设计;*n.* 构造,构造物
 ~ing contractor 建筑承包商
construction [kən'strʌkʃən] *n.* 建筑,建筑工程,工地;结构,构造;建筑方法,设计,施工,制图
 ~ and repair 建造与修理
 ~ assistance vehicle 建筑辅助车辆
 ~ authority 建筑工程局,施工管理局
 ~ budget 施工预算
 ~ business 建筑业
 ~ calendar 施工日程表
 ~ camp 工棚
 ~ company 工程承包单位
 ~ contract 施工合同
 ~ contract award 建筑合同裁定书
 ~ contractor 工程承包商
 ~ control 施工管理
 ~ cost 工程费用,建筑费
 ~ department 建筑[施工]部门
 ~ details 施工详图
 ~ diary 施工记录[日记]
 ~ diversion 施工导流
 ~ drawing 结构图
 ~ engineering 建筑工程学
 ~ engineering corporation 建筑工程公司
 ~ equipment 施工设备
 ~ estimate 施工估价
 ~ fund 建设资金

~ gap 施工缝
~ gauge 规划建筑线
~ industry 建筑业
~ investment 建筑投资
~ joint 施工[工作]缝,构件接缝
~ lift 浇筑层,水平施工缝
~ lime 建筑石灰
~ line 作图线
~ load 施工荷载
~ machinery 施工[建筑]机械
~ management 建筑施工(企业)管理
~ map 建筑[施工]图
~ marker 施工标志
~ material 建筑材料
~ member 构件
~ method 施工方法
~ order 基建任务书
~ overhead 施工费用
~ paint 结构防护漆
~ period 工期
~ phase 施工阶段
~ plan 施工布置[平面](图)
~ plant 施工设备
~ plastics 建筑塑料
~ progress report 施工进度报告
~ shaft 施工竖井井字架
~ site (work) 施工现场
~ specification 施工规范,建筑施工标准
~ sequence 施工程序修筑程序
~ survey 施工测量
~ unit 结构单元,建筑构件
~ work 施工[建筑]工程
~ work in process 在建工程
all metallic ~ 全金属结构
block ~ 大型砌块建筑
brick ~ 砖石结构[工程]
bridge ~ 桥梁工程
building ~ 房屋建筑学,房屋构造
cage ~ 笼式结构
cellular ~ 格形构造[建筑物]
civil ~ 土木建筑

civilian ~ 民用建筑
combination ~ 混合构造
composite ~ 组合结构
concrete ~ 混凝土建筑[施工]
en-block ~ 整体结构
engineering ~ 工程建设
external ~ 外部构造
housing ~ 住宅建筑
masonry ~ 砖石结构,圬工建筑
mill ~ 半防火木结构
monolithic ~ 整体式结构
pan ~ 盘形结构
precast ~ 预制法施工,预制构造物
prefabricated ~ 预制结构
reinforced concrete ~ 钢筋混凝土结构
section(al) ~ 预制构件拼装结构
stage ~ 分期建筑,多层面构造,剧院建筑
thin-shell ~ 薄壳结构
veneer ~ 镶板结构
winter building ~ 冬季(房屋)施工

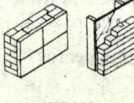
veneer construction

constructional [kən'strʌk-ʃənl] *a.* 建筑(物)的,构造(上)的
~ drawing 施工图
~ element 构件
~ material 建筑材料
~ reinforcement 构造钢筋
~ steel 建筑[结构]钢
constructivism [kən'strʌktivizəm] *n.* 构成主义,构成派,构造论(其特点系以抽象及几何图案构图)
constructor [kən'strʌktə] =constructer *n.* 设计师,建造者,施工人员
consult [kən'sʌlt] *v.* 咨询,参考,请……鉴定
~ing architect 顾问建筑师
~ing engineer 顾问工程师
~ing room 咨询处
consumer [kən'sju:mə] *n.* 用户,消费者

~'s goods 消费[生活]资料,消费品
consumption [kən'sʌmpʃən] *n.* 消耗(量),流量,费用
  energy ~ 能量[电能]消耗
  heat ~ 热量消耗
  power ~ 动力[电能]消耗
consumptive [kən'sʌmptiv] *a.* 消耗[费]的
  ~ use of water 消耗用水
  ~ water use 耗水量
contact ['kɔntækt] *n.;v.* 接触,联系 *a.* 有联系的,保持接触的
  ~ area 接触面积
  ~ cement 接触型胶结剂
  ~ corrosion 接触腐蚀
  ~ joint 对接接头
  ~ plane 接触面
  ~ spring 接触弹簧
  ~ surface 接触表面
  sliding ~ 滑动接触
  travelling ~ 活动接触
contactless *a.* 无接触的,无接点的
contain [kən'tein] *v.* 含有,包含;抑[控]制
container [kən'teinə] *n.* 集装箱,容器;外壳,包裹物
  ~ board 盒纸板
  ~ horizon 含油层
containment [kən'teinmənt] *n.* 容积,负荷额,电容;密封外壳;抑制
  natural ~ 天然保[防]护层(如水、土)
contamination [kəntæmi'neiʃən] *n.* 污染,污染物(质),杂质
  atmospheric ~ 大气[空气]污染
  color ~ 彩色混杂
  environmental ~ 环境污染,周围介质污染
contaminator *n.* 污染物
contemporaneous [kən,tempə'reinjəs] *a.* 同时代的,同生的
content ['kɔntent] *n.* 容量[积],含量;(*pl.*)目录,内容
  ~ indicator 内容[信息]显示器

absolute moisture ~ 绝对湿度
labor ~ 加工工作量,劳动量
moisture ~ 含水量,湿度
contexture [kən'tekstʃə] *n.* 组织,构造
continent ['kɔntinənt] *n.* 大陆,陆地洲
continental [kɔnti'nentl] *a.* 大陆的,洲的
contingence [kən'tindʒəns] = contingency *n.* 偶然[意外]事故;意外[临时]费用
continuity [kɔnti'njuːiti] *n. pl.* -ties. 连续(性)
continuous [kən'tinjuəs] *a.* 连续的,顺次的
  ~ arch bridge 连续拱桥
  ~ beam 连续梁
  ~ chart 带形记录纸
  ~ concreting 连续浇灌混凝土
  ~ construction 连续施工法
  ~ cooling 连续冷却
  ~ core wall 连续心墙
  ~ cylindrical shell 连续筒壳
  ~ duty 连续运行
  ~ footing 连续基脚[底座]
  ~ foundation 连续基础
  ~ girder 连续大梁
  ~ heating 连续供暖
  ~ hinge 连续铰接
  ~ house 联排式多户住宅
  ~ interstice 连续间隙
  ~ lights 连续窗
  ~ production operation 流水作业,连续生产作业
  ~ rigid frame 连续刚架
  ~ running 连续运行
  ~ sash 连续窗,联窗
  ~ slab 连续楼板
  ~ sliding sash 连续拉窗
  ~ stone footing 条石连续底脚
  ~ stringer 连续楼梯梁
  ~ truss 连续桁架
  ~ wall 连续墙

continuousness *n.* 连续性
contour ['kɔntuə] *n.* 等高线,轮廓,形状,断面 *vt.* 画出轮廓,画出等高线 *a.* 仿形的;异形的;沿等高线的
~ chart 等高线图,地形图
~ drawing 白描
~ -etching 外形腐蚀(加工)
~ line 等高[轮廓]线
~ map 等高线图
~ outline 轮廓,略图
~ painting 白描
~ plan 地形图,等高线平面图
contouring *n.* 地形测量
contract [kən'trækt] *v.* 订合同,承包;收缩,限制;['kɔntrækt] *n.* 合同,契约,包工
~ acquired property 合同所规定的性能
~ agreement 承包工程协议书[合同]
~ amount 承包(工程价)额
~ authorization 订约[合同]授权
~ award 合同裁决(书)
~ award date 合同签订日期
~ cancellation 取消合同
~ change motification 合同更改通知书
~ change request 合同更改请求
~ construction 发包工程,包工建筑,承包施工
~ contents 合同内容
~ control 建筑施工管理
~ date 合同[订立]日期
~ document 承包契约,合同证书,合同文件
~ drawing 发包图样
~ expiry 合同期满
~ for projects 承包工程
~ grade 合同品级
~ item 合同项目
~ item material list 合同项目材料清单
~ labour 合同工
~ letting 签订合同,合同奖金
~ manager 承包人
~ market 合同市场
~ number 合同号
~ period 合同期限
~ renewal 变更合同
Contract Rules for Construction Services 建筑合同法(规)
~ specification(s) 合同说明书,发包细则
~ surveyor 工地监工员
~ system 包工制
~ term 合同条款
~ technical requirements 合同技术要求
~ work 发包工程
~ed drawing 缩图
~ed section 收缩断面
~ing agency 承包代理人[经理处]
~ing combine 建筑联合企业
~ing crack 收缩裂纹
~ing firm 建筑合同
advance received on ~ 按合同预收款项
contraction [kən'trækʃən] *n.* 收缩,缩小,缩减量[作用];填料,密封
~ crack 收缩裂缝
~ fissure 收缩裂缝
~ joint 收缩缝[节理]
~ joint grouting 收缩缝灌浆
lateral ~ 横向收缩
contractor [kən'træktə] *n.* 承包商[者],包工(头),订约人;压缩[力]机
~'s agent 建筑[施工]承包人
~'s plant 施工设备
general ~ 总承包人
contraflexure ['kɔntrə'flekʃə] *n.* 反(向)弯曲,反挠,回折
contrarotation ['kɔntrərəu'feiʃən] *n.* 反(向旋)转,反向转动,反旋
contrast ['kɔntrɑːst] *n.* 对比度,反差;对照物;差别;*v.* 对比,比较
~ colours 反衬色,对比色彩
~ of houe 色相对比

～ of tone 色调对比
picture ～ 色调(深浅)对比
temperature ～ 温差
contravariance [ˌkɔtrəˈvɛəriəns] *n.* 反[抗]变性
contravariant [ˌkɔntrəˈvɛəriənt] *n.*; *a.* 反变(的)
contribution [ˌkɔntriˈbjuːʃən] *n.* 组成,成分,所起的作用,影响,贡献;文献
contributory [kənˈtribjutəri] *n.* 起作用的因素; *a.* 起作用的,有助于……的
～ factor 影响因素,辅助系数
contrivance [kənˈtraivəns] *n.* 设计,创造,发明;设备,工具
contriver [kənˈtraivə] *n.* 设计者,创造[发明]人
control [kənˈtrəul] *n.*; *v.* 管理,操作,控制,调节,检查
～ area 控制面积
～ block 消力墩
～ board 配电盘[板],控制盘
～ box 控制箱
～ building 配电室
～ cabinet 控制盘,操纵板
～ card 控制卡片
～ cost 管理费
～ data 控制数据
～ desk 操作台
～ device 控制装置
～ house 管理所,控制室
～ interlock 操纵连锁
～ joint 伸缩缝
～ panel 操作台,配电盘
～ portion 控制部件
～ reception 验收
～ rod 操纵[控制]杆
～ room 控制室
～ system 控制系统
～ swith 控制开关
～led concentration 控制浓度

control joint

～led concrete 均匀控制混凝土
～led crest 设有调节水位装置的坝顶
～led foreign trade 统制对外贸易
～led humidity 控制湿度
～ling dimension 控制尺寸
～ling elevation 控制标高
～ling manhole 控制人孔,检修孔
～ling room 操纵[控制]室
centralized ～ 集中操纵[控制]
distance ～ 遥控,远距离操纵
hand ～ 手工操纵
noise ～ 噪声控制
purity ～ 色纯度调节
routine ～ 常规控制
safety ～s 防护装置,安全措施
controller [kənˈtrəulə] *n.* 管理[检验]员;控制器,传感器
convection [kənˈvekʃən] *n.* (热,电)对流,传递,迁移
convector [kənˈvektə] *n.* 对[换、环]流器
convenience [kənˈviːnjəns] *n.* 方便,便利;(*pl.*) 生活设备

convention hall

convention [kənˈvenʃən] *n.* 会议,集会,习惯,公约
～ hall 会场,宴会厅
conventional [kənˈvenʃənl] *a.* 惯用[常规,平常]的,协定的
～ aggregate 普通[常规]骨料
～ diagram 示意图
～ door 普通门,常用门
～ galvanizing 普通热镀锌
～ method 惯用法
～ mud 普通泥浆
～ representation 习用表示法,传统表示法
conventionalization *n.* (雕刻题材的)程式化,图案化,简化
converge [kənˈvəːdʒ] *v.* 集中,收敛
conversion [kənˈvəːʃən] *n.* 转化[变],

换算;改装;渐变段
~ of timber 锯材
~ table 换算表
convert [kən'və:t] *v.* 转[变]换,改造
~ed amount 折合数量
~ed steel 渗炭钢,硬质钢
~ed timber 加工木材
~ing costs 加工成本
converter [kən'və:tə] *n.* 变流器,转换器,转炉
convertible [kən'və:təbl] *a.* 可逆的,可变(换)的;活动的 *n.* 可改变的事物
convex ['kɔn'veks] *n.* 凸圆体,钢卷尺; *a.* 凸形的
~ concave 凸凹的,一面凸一面凹的
~ fillet weld 凸形角焊缝
~ glass 凸镜
~ joint 凸(圆接)缝
~ lens 凸透镜
~ plane 平凸形
~ shell 凸面壳体
~ tile 牡瓦,凸筒瓦

convex shell

~ tooled joint 圆截面缝
~ weld 凸面焊(缝)
convey [kən'vei] *vt.* 输送,传播
conveyance [kən'veiəns] *n.* 转让(证书);运输(工具),车辆;传播
conveyer [kən'veiə] *n.* 传送带,输送机[器];转让人,交付人
~ system 传送带流水作业法
convolution [kɔnvə'lu:ʃən] *n.* 盘旋,褶积;涡流
~ method 褶合法
cook [kuk] *v.* 蒸煮(过程) *n.* 厨师,炊事员
~ room 厨房
~ shop 饭馆
~ stove 烹调用炉
~ing (木材)蒸煮处理
~ing apparatus 烹调用具
~ing range 炉灶

cooker ['kukə] *n.* 蒸煮器,蒸煮锅;炊具
~ hood 抽油烟机
gas ~ 煤气,炉灶,炊事间
cookhouse ['kukhaus] *n.* 厨房
cool [ku:l] *v.* 冷却,降温; *a.* 冷的,冷藏的,寒色的
~ chamber 冷藏室
coolant ['ku:lənt] *n.* 冷却[散热]剂,载热体
cooler ['ku:lə] *n.* 冷却[凝]器,冷冻机,致冷设备
spray ~ 喷水池
coolie ['ku:li] = cooly *n.* 苦力,小工
cooling ['ku:liŋ] *a.* 冷的,冷却的 *n.* 冷却,致冷
~ agent 冷却剂
~ brittleness 冷脆性
~ fins 散热(翼)片
~ fissure 冷缩裂缝
~ flange 散热凸缘
~ hardness cast iron 冷硬铸铁
~ joint 冷缩节理,收缩缝
~ plate 冷却板,散热片
~ space 收缩缝
~ surface 冷却[散热]面
~ system 冷却[散热]系统
coom [ku:m] *n.* 炭黑,煤烟[灰],锯屑
cooperation [kəuɔpə'reiʃən] *n.* 合作,协同
cooper-Hewitt lamp 玻璃管汞弧灯
coordinate [kəu'ɔ:dinit] *n.* ;*a.* 坐标系,坐标(的),位(的),协调(的) *v.* 使同等,使协调
coordination [kəuɔ:di'neiʃən] *n.* 协调,同等[位]
copal ['kəupəl] *n.* 苯乙烯树脂,制造清漆用的树脂
~ gum 透明树胶
~ varnish 柯巴树脂清漆
copartner [kəu'pɑ:tnə] *n.* 合股者,合伙人
cope [kəup] *n.* 顶盖[层],墙压顶,墙帽;小室 *v.* 覆盖;解决;对抗

~ chisel 凿槽刀
~ cutter 钻头,切削雄榫器
~ line 盖顶边线
~d joint 暗缝
~ing 盖梁,墙帽,桥墩顶帽遮檐,挡板
~ing brick 压顶砖,檐砖
~ing saw （锯曲线用的）手弓锯
~ing stone 盖顶石
~ing wall 坝顶拦墙
gable ~ 山墙压顶,封檐压顶
saddle back ~ 鞍形顶盖

coping

coplanar [kəu'pleinə] a. 共面的,同一平面的
copped [kɔpt] a. 圆锥形的,尖头的
copper ['kɔpə] n. 铜 a. 铜色的 vt. 镀铜,包铜
~ bar 铜条,铜棒
~ bearing 含铜的
~ bond 黄铜焊接
~ coating 镀铜
~ fitting 铜接头
~ facing [plating] 镀铜
~ flashing 铜皮泛水
~ jacket 铜套
~ rainwater articles 铜制屋顶配件
~ sheathing 铜套
~ sulphate 硫酸铜
~ plated steel 镀钢钢板
~ product 铜材,铜制品
~ tube 铜管
~ weld 铜焊
~ welding rod 铜焊条
~ wire 铜丝
pig ~ 生铜
red ~ 紫铜
sheet ~ 铜板
copperas ['kɔpərəs] n. 绿矾,硫酸亚铁 ($FeSO_4 \cdot 7H_2O$)
coppice ['kɔpis] = copse, copsewood n. 矮树丛,小灌木林

copy ['kɔpi] n. pl. copies ,复制品,副[抄]本 v. 复制,抄录
~ machine 复印机
~ pattern 复制图
~ rule 仿形尺
~ing paper 复写纸
~ing press 复印机
foul (rough) ~ 底[原]稿
sample ~ 样本
coquillage n. 贝[介]壳花饰
coquina [kəu'ki:nə] n. （建筑用的）介壳石,贝壳岩
coquinoid limestone 贝壳灰岩
coral ['kɔrəl] n. ;a. 珊瑚(的)
~ reefs 珊瑚礁
coralgal n. 珊瑚沉积
corallite ['kɔrəlait] n. 珊瑚[色大理]石
coralloid(al) [kɔrəlɔid(əl)] a. 珊瑚状的
corbeil ['kɔ:bel] n. 花篮或果篮饰
corbel ['kɔ:bəl] n. 牛腿,托臂,悬臂桁架,突出部分建筑 vt. 以承材支撑
~ arch 突拱,叠涩拱
~ arm 拱,悬臂挑梁
~ back slab 悬挑板,翅板
~ bracket arms cluster 一垛斗拱
~ brick 悬臂砖
~ course 突腰线,肬层
~ gable 踏步式山墙
~ joint 挑头接合
~ling 梁托,结构,撑架结构
~ mould 挑出线脚
~ out 挑头
~ piece 挑出块
~ plate 梁托垫板
~ steps 山墙上踏步式压顶
~ table 挑檐

corbel

corbie ['kɔ:bi] n. 乌鸦式
~ -gable 阶梯状山墙
~ -step 山形墙侧边梯形突出物

corcass *n.* 河岸沼泽地
cord [kɔ:d] *n.* (绳)索 *v.* 用绳(索)系住
~ grommet 索环
~ pendant 电灯吊线,软线吊灯
~ switch 拉线开关
~ weight 绳锤
~ing diagram 接线图
cordage ['kɔ:didʒ] *n.* 索具,绳索,缆索
cordate ['kɔ:deit] *a.* 心(脏)形的
cordon ['kɔ:dən] *n.* 飞檐层;封锁[警戒]线
~ counts 小区交通调查,区界交通量统计
corduroy ['kɔ:dərɔi] *n.* 垛架,木排路 *vt.* 筑木排路 *a.* 用横木铺成的
~ mat 木排[圆木]铺面
~ road 木排路
cordwood ['kɔ:dwud] *n.;a.* 积木式,材堆式
~ construction 积木式建筑
core [kɔ:] *n.* 核心,岩心,坝心,心墙;填充料 *v.* 钻取岩心[土样]
~ bit 空心钻
~ board 夹芯胶合板
~ city 中心城市
~ cover 型心涂料
~ in ~ structure 筒中筒结构
~ pin 中心销
~ plan 高层建筑中心式平面布置
~ structure 筒式结构
~ wall 心墙
~d brick 空砖
~d ceiling 空心楼板
~d foamed slag concrete block 多孔泡沫矿碴混凝土砌块
~d hole 型芯孔
~d panel 心板
~d slab 空心板
~d solder wire 空心焊丝,纤焊丝
~d tile 筒形瓦
clay ~ 黏土心墙
coreless *a.* 空心的
corhart *n.* 耐火材料

Corinthian (古典建筑上的)科林斯式
~ column 科林斯式柱
~ entablature 科林斯式柱顶盘
~ order 科林斯柱型
~ style 科林斯式(样)

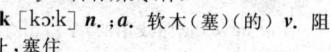
Corinthian column

cork [kɔ:k] *n.;a.* 软木(塞)(的) *v.* 阻止,塞住
~ board 软木板
~ brick 多孔砖,软木块
~ carpet 软木地毡
~ filler 软木填料
~ flooring 软木地板
~ gasket 软木衬垫
~ rubber (密封用)软木橡胶
~ gauge 塞(径)规
~ sheet 软木(薄)板
~ stoper (软)木塞
~ -wood 软木,轻木
mountain ~ 石棉
corkscrew ['kɔ:kskru:] *a.* 螺旋状的 *v.* 曲折而进 *n.* 起软木塞的起子
~ column 螺旋形柱
~ rule 旋转法则
~ stair 盘梯
corn [kɔ:n] *n.;v.* 使成粒状,制成细粒
~ing glass 麻粒玻璃,透射紫外线玻璃
corner ['kɔ:nə] *n.* (墙)角,街道转角;弯头[管]; *v.* 转弯 *a.* 转弯处的,角上的
~ angle 顶[棱]角
~ bar 角隅钢筋
~ bath 墙角浴盆
~ bead 墙角护条,角焊缝
~ block 内角加强用三角条
~ bracing 角斜撑,转角联条
~ cabinet 墙角柜橱[饰架]
~ capital 转角柱头
~ chisel 角凿
~ cutting tile 八角形瓷砖

~ dimension 弯头尺寸,夹角大小
~ drill 角(轮手摇)钻
~ fillet 圆角嵌条,贴条焊缝
~ guard 护角铁
~ halving 转角嵌接
~ iron 角铁
~ joint 弯管[头]连接,角接
~ lot 转角地段,两侧临街地段
~ metal 隅铁
~ mold (梁柱)边角填块
~ pilaster 转角半露方柱,转角壁柱
~ pile 角桩
~ pillar 转角柱墩
~ plate 角板,角撑铁板
~ post 角柱
~ radiator 墙角散热器
~ stay 角牵条
~ stiffener 角部加劲件
~ stone 墙角石,柱石,基础
~ stone laying ceremony 奠基仪式
~ strap 角部连接条
~ weld 90°角接焊缝
cornice ['kɔːnis] *n*. 飞檐,檐板,上眉 *v*. (装)挑檐
~ boarding 檐板
~ bracket 挑檐托座
~ lighting 挑檐照明
~ soffit 檐口平顶
crown ~ 大檐饰
dentil ~ 齿饰檐
open ~ 敞檐
rake ~ 斜檐
string ~ 楼基楣饰,腰线
stucco ~ 灰泥粉饰檐口
wall ~ 墙壁挑檐
corolliform [kəˈrɔləfɔːm] *a*. 花冠状的
corollitic column 叶饰柱身的柱子
coromat *n*. 包在管子外面防止腐蚀的玻璃丝层
corona [kəˈrəunə] *n*. 冠状物,飞檐,挑檐滴水板;日冕;电晕

corona

coronet [kɔrənit] *n*.;*a*. 冠(状的),门头线饰
corporation [kɔːpəˈreiʃən] *n*. 公司,团体
~ cock 煤气管或水管的总开关,公司头目
corral [kɔːˈrɑːl] *n*. 栅栏, *v*. 围成栅栏
corrasion [kəˈreiʒən] *n*. 磨[刻]蚀
correct [kəˈrekt] *a*. 正确的,校正[标准]的,恰当的 *vt*. 校正,补偿,排除,制止
~ level 标准[校正]水准
~ed value 校正值
correction [kəˈrekʃən] *n*. 修正(值),改正,调整;补偿,中和,制止
~ coefficient 修正系数
~ factor 修正系数
~ for grade 倾斜校正
correlation [kɔriˈleiʃən] *n*. 相关性,相关关系[作用],对比
correlativity [kəˌreləˈtiviti] *n*. 相关性,相关程度
correspondence [ˌkɔrisˈpɔndəns] = correspondency *n*. 相符,一致,对比;信函
correspondent [ˌkɔrisˈpɔndənt] *n*. 对应[相对,一致]的; *n*. 顾客,对应物;代理银行
corridor ['kɔridɔː] *n*. 走[回]廊,通道,狭长地带
~ access type 通廊式平面布置
~ type basilica 单廊式长方形会堂
long ~ 长廊
painted ~ 彩绘回廊
corrode [kəˈrəud] *v*. (使)腐[侵,锈]蚀
~ing agent 腐蚀剂
~ away 腐[侵]蚀(掉)
corrodent [kəˈrəudənt] *n*. 腐蚀剂[介质] *a*. 有腐蚀作用的
corrosion [kəˈrəuʃən] *n*. 腐蚀(物),铁锈
~ by gases 气体腐蚀

~ control 防腐蚀方法
~ crack 腐蚀断裂
~ inhibitor 阻蚀剂
~ of metal 金属锈蚀
~ prevention 防蚀[腐]
~ preventive 防腐[蚀]剂
~ proof 抗腐蚀的,不锈的
~ remover 防腐剂
~ resistance 耐蚀性
~ resistant 抗腐蚀性
~ resisting agent 防锈剂
~ resisting steel 不锈钢
~ strength 耐蚀性
bi-metallic ~ 双金属腐蚀,电化学腐蚀
chemical ~ 化学侵蚀
crevice ~ 隙间腐蚀
corrosiron n. 耐腐蚀硅钢
corrosive [kəˈrəusiv] a. (有)侵蚀(性)的
~ action 腐蚀作用
~ strength 耐蚀性
~ water 有侵蚀性的水质
coorosiveness n. 腐蚀性,侵蚀作用
corrosivity [kɔrəuˈsiviti] n. 腐蚀性
corrugate [ˈkɔrugeit] v. 使成波纹状; a. 波纹状的,起伏不平的,竹节状的
~ fastener 波纹扣件
~ steel 竹节钢筋
~d acoustical panel 波形隔[吸]声板
~d asbestos cement sheet 波形石棉水泥板
~ bar 竹节钢筋
~d barrel roof 波形筒状屋顶
~d galvanized sheet iron 镀锌波纹铁皮
~d glass 瓦楞玻璃
~d metal 瓦楞铁(皮),波纹薄铁板
~d metal pipe 金属波纹管
~d metal pipe arch 波纹钢管拱
~d plate 波纹板

corrugated fastener

~d paper 瓦楞纸(板)
~d roofing 波纹屋顶
~d sheet 波纹薄[钢]板
~d siding 波纹披迭板
~d steel 波纹钢板
~d tile 波形瓦
~d vault roof 波形穹屋盖
~d wired glass 波形夹丝玻璃
corrugation [kɔruˈgeiʃən] n. 波纹(状),起伏
corrupt [kəˈrʌpt] a. 不纯的,搀杂的; v. 使混浊,使不纯,腐蚀
corticin(e) n. (用树胶胶成的)软木地毯
cortile [kɔːˈtiːlei] n. 内院,中庭
corundum [kəˈrʌndəm] n. 刚玉、刚石,金刚砂
~ bit 金刚砂钻头
synthetic ~ 人造[合成]刚玉,人[合成]金刚砂
corvic n. 聚氯乙烯树脂
Coslett process n. 钢铁防蚀的磷化处理法
cost [kɔst] n. 成本,费用,价格 v. 值,花费
~ analysis 建筑成本分析
~ -benefit ratio 投资收益比
~ breakdown 成本细目[分析]
~ calculation 成本计算
~ control 成本管理
~ estimate 造价估算,成本估算
~ factor 造价影响因素
~ index 成本[价格]指数
~ keeping 成本核算
~ of construction 施工费用,建筑费
~ of maintenance 维修[保养]费,养护费
~ of operation 管理费
~ of upkeep 维修费
~ of overhaul 检修费
~ planning 成本计划
~ plus a fixed fee 成本加附加费
~ plus fee contract 实费承包工程

～ price 成本价格
～ record 成本账
～ reduction 成本降低,费用
～ sheet 成本账单
～ unit price 成本单价
administrative ～ 行政管理费
annual ～ 常年费用
capital ～ 基建投资费,资本值
comparative ～ 比(较造)价
construction ～ 建筑[工程]费用
extra ～ 额外费用
current ～ 市价
depreciation ～ 折旧费
engineering ～ 工程费(用)
installation ～ 设备投资
operating ～ 生产费用
out-of-pocket ～ 实际费用,实际成本
overhead ～ 杂项开支
staff ～s 工资费用
working ～ 工作费用
removal ～ 拆除费

costalia *n.* 肋板
costing ['kɔstiŋ] *n.* 成本会计,(*pl.*)概[预]算,估价
cot [kɔt] *n.* 小[茅]屋,帆布床,吊床
coteau [kəu'təu] *n. pl.* -teaux 高地,高原,丘陵地区,山谷之边缘
cotidal [kəu'taidəl] *a.* 等[同]潮的
cotonier *n.* 法国梧桐
cottage ['kɔtidʒ] *n.* 村舍,小型别墅,单幢住所
～ hospital 乡村医院
～ roof 无桁架的小跨度屋盖
cottar ＝cottier *n.* 乡村居民
cotter ['kɔtə] *n.* 键销,开尾销;乡村居民 *vt.* 用销固定
～ bolt 带销螺栓
～ hole 销钉孔
～ joint 销[键]接合
～ pin 开尾销,扁销
cotton ['kɔtn] *n.* 棉花,棉织品; *a.* 棉的

～ -covered wire 纱包绝缘线
～ duck 棉织帆布
～ -enamel covered wire 纱包漆包线
～ insulation cable 纱包电缆
～ wood 杨木白杨
～ -tape 纱带
couch [kautʃ] *n.* 长沙发椅,休息处;层
～ board 多层纸板
coulability *n.* 铸造性
coulee ['ku:li] *n.* 斜壁谷,深河谷
council ['kaunsl] *n.* 委员会,协会
  Engineering Council 工程协会
count [kaunt] *n.* ; *v.* 计算,统计,期待
～ing-house 会计室
counter ['kauntə] *n.* 计数器,柜台,相反物 *a.* ; *ad.* 相反(的),副(的) *v.* 抵销,反对
～ approach 防卸工事
～ arch 扶垛拱
～ balanced sash 平衡式上下推拉窗
～ balanced window 平衡窗
～ battens 木地板木键,交叉压条,顺水条
～ brace 交叉撑
～ -camber 预留弯度
～ ceiling (起隔声绝热作用的)悬吊平顶
～ -clock-wise 反时针方向
～ die 底模
～ edging 边接边
～ flashing 泛水盖帽
～ floor 粗地板,垫板
～ light 逆光
～ murement 副壁,复墙内壁
～ nut 埋头螺帽
～ sinking bit 埋头钻
～ tank 计算机储存器
～ top lavatory 嵌入式的洗面器
～ vault 贴附穹顶
～ wall 贴邻边墙
counterbalance [kauntə'bæləns] *n.* 平衡(力),配重;托盘天平 *vt.* 使平衡,抵

销,补偿

counterbore [ˌkauntəˈbɔː] *n.*; *vt.*; 平头钻,扩孔镗孔

counterbrace [ˈkauntəbreis] *n.* 交叉对角撑,副斜杆,副撑臂

counterflange [ˈkauntəflændʒ] *n.* 对接法兰,过渡法兰(盘)

counterfoil [ˈkauntəfɔil] *n.* 存根,票根

counterfort [ˈkauntəfɔːt] *n.* 护墙,后扶垛,拱柱

countermand [ˌkauntəˈmɑːnd] *n.*; *v.* 废除,撤回,消去

counterplot [ˈkauntəplɔt] *n.* 对抗策略

countersink [ˈkauntəsiŋk] *vt.* 钻埋头孔(过过式 countersunk) *n.* 埋头孔,埋头钻

~ drill 埋头钻

countersunk and chipped rivet 埋头铆钉

countersunk bolt 埋头螺钉

countersunk not chipped rivet 半埋头铆钉

counterstrut [ˈkauntəstrʌt] *n.* 桁架抗压杆,受压杆

country [ˈkʌntri] *n. pl.* -tried *a.*; *n.* 国家,祖国;地方,区域 *a.* 地方的,乡间的

~ seat 别墅

county [kaunti] *n. pl.* -ties. 县,乡镇

couplant *n.* 耦合剂[介质]

couple [ˈkʌpl] *n.* 一对,一双,(力)偶,力矩;热电偶,连接器; *v.* 连在一起

~ roof construction 椽架,联[排]椽屋架

~d beam 对(拼)梁,成对组合梁

~d columns 对柱

~d knee braces 夹板斜撑

~d roof 双坡屋顶

~d windows 对窗

coupler [kʌplə] *n.* 连接器;连接管;车钩

coupling [ˈkʌpliŋ] *n.* coupled windows 管接头,联轴节,车钩,连接器; *a.* 耦合的,连接的

~ bar 连接杆,拉杆
~ bolt 连接螺栓
~ collar 连接环
~ flange 连接法兰盘,接合凸缘
~ nut 连接螺帽
~ unit 系[拉]杆,螺丝扣
rigid ~ 固接,刚性联接

coupon [ˈkuːpɔn] *n.* 【法】试样[件];金融[商务]证卷

course [kɔːs] *n.* 过程,方向,线路;层,行列;科目,课程; *v.* 砌(砖),运行

~ bond 丁砖层砌合
~ depth 墙体的层高
~ joint (建筑物的)层间接缝
~ of bricks 砖层
~ of exchange (外汇)兑换率
~ of headers 丁砖层
~ of stretchers 顺砖层
~d ashlar 成层琢石
~d blocks 层砌方块
~d brickwork 成层砖砌体
~d masonry 成砖砌体,层砌圬工
~d pavement 成层铺面
~ rockfill 成层砌石
~d rubble 成层毛石圬工,铺砌毛石
base ~ 基层,底层勒脚层
bearing ~ 承压层,承重层
brick soldier ~ 竖砖层
coursing joint 成行缝
damp(-proof) ~ 防潮层
header ~ 露[丁]头层
subbase ~ 基底层
upright ~ 竖[立]砌层

court [kɔːt] *n.* 庭院,宫廷,球场

~ art 宫庭艺术
~ of honour 纪念馆[堂]
~ painter 宫廷画家
~ yard with building on the four sides 四合院
front ~ 前庭
royal ~ 宫廷

courtesy [ˈkəːtisi] *n.* 允许,优惠,礼貌
～ desk 接待处
courthouse *n.* 法院
courtroom [ˈkɔːtruːm] *n.* 法庭
courtyard [ˈkɔːtjɑːd] *n.* 院子,天井
～ house 四合院住宅
coussinet *n.* 拱基石,基础垫层,爱奥尼帽状盖块
cove [kəuv] *n.* 河湾;凹圆形,穹窿,拱 *v.* 使成穹形,使内凹
～ ceiling 拱形天花板
～ skirting 凹圆踢脚板
～d and flat ceiling 凹圆线脚平顶
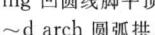
～d arch 圆弧拱
～d ceiling 圆弧天花板

cove ceiling
～d checker ceiling 四周凹圆的格子平顶
～d lattice ceiling 四周凹圆的细格平顶
～d vault 凹圆穹顶
coving 壁炉顶斜面,穹窿
cover [ˈkʌvə] *vt.* 遮盖,包含[括],代替 *n.* 覆盖物,保护层,盖子;保证金
～ block 护面块体
～ fillet 盖条
～ iron 护铁
～ layer 保护层
～ piece 罩壳
～ plate 盖板
～ power 覆盖能力
～ slab 镶面板,盖板
～ stone 盖面石(料)
～ strip 压缝板,盖缝条
～ed gutter 暗沟
～ed pipe 暗管,保温管
～ed radiator 隐藏式散热器
～ed spectator's stand 屋顶了望台
～ed street-way 穿廊式街道
～ed walk 带顶棚的公共通道,骑楼通道

～ed way 廊[暗]道
～ed wire 绝缘线
～ed with gravel 铺道碴
～ing 覆盖层[物],涂[护]层,护壁板
～ing for fire protection 防火保护层
～ing in scale tiles 瓦片屋面
～ing material 罩面材料
～ of joint 接头覆盖,压缝板
roof ～ing 屋面
protective ～ 防护外壳,保护层
wall ～ing 墙面涂料
coverage [ˈkʌvəridʒ] *n.* 面层,铺砌层;保险总额;范围,占地面积
covermeter [ˈkʌvəmiːtə] *n.* 面层测厚仪
coverture [ˈkʌvətjuə] *n.* 包覆,隐藏,掩饰
cowcatcher [kauˈkætʃə] *n.* 排障装置
cowhouse [ˈkauhaus] *n.* 牛舍[棚]
cowl [kaul] *n.* 外壳,罩,烟囱顶上的风雨帽
～ ventilator 带罩通风器
cows *n.* 保护棱角用的排桩
crab [kræb] *n.* 起重小车,猫头吊;绞车,卷扬机;*v.* 偏斜,斜行
～ bolt 地脚螺栓,锚栓
crack [kræk] *n.* 裂缝[纹],龟裂;*v.* 破[开]裂
～ and fissure 裂隙
～ control reinforcement 防止裂缝钢筋[配筋]
～ edge 裂边
～ filler 填缝材料
～ in tension 拉裂
～ opening 裂缝宽度,裂口
～ pouring 灌缝
～ propagation 裂纹扩展
～ sealer 封缝料
～ed 有裂缝的,碎的,裂化的
～ed condition 开裂状态
～ing 开裂,裂隙[纹]
capillary ～ 毛细裂缝
check (contraction, shrinkage) ～

收缩裂纹
hair ~ 发丝裂纹
heat [hot, thermal] ~ 热裂
incipient ~ 起始裂纹
restriction ~ 阻碍型裂缝
cracker ['krækə] *n.* 破碎机,裂化室
crack'free *a.* 无裂纹的
crackle [krækl] *vi.* 产生裂纹, *n.* 小裂纹,碎裂花纹
~d 炸裂花纹
~ lacquer 裂纹漆
~ paint 裂纹漆
~ varnish 裂纹清漆
cracky ['kræki] *a.* 易碎的,多裂缝的
cradle [kreidl] *n.* 支[吊]架,管道的鞍座,支墩;*vt.* 置于支架上
~ scaffold 悬挂式脚手架
~ vault 筒形穹窿
cradling 弧顶架
cradling piece 支架杆件
cradling roof 有支撑架的屋顶
craft [krɑːft] *n.* 技术[巧],手艺
~ room 手工艺室
~ paper 牛皮纸,不透水纸
~ union 同业工会
art and ~s 工艺美术
craftman [krɑːftmən] *n. pl.* -men 技工,工匠
craftmanship [krɑːftmənʃip] *n.* 技能,手艺
crag [kræg] *n.* 峭壁,岩石碎片
crage *n.* (陶瓷)釉面裂纹
craggan *n.* 陶制缸
cragged ['krægid] *a.* 崎岖的,多岩的,碎片的
cram [kræm] *n.*;*v.* 填塞[入];压碎
crammer ['kræmə] *n.* 填塞物
cramp [kræmp] *n.* 铁搭,铁箍,夹钳,钢筋;*vt.* 用钳子平紧,固定
~ folding machine 折边机
~ frame 弓形夹
~ iron 铁钩,两爪钉
~ ring 扣环

crampo(o)n [kræmpən] *n.* 起重吊钩,金属钩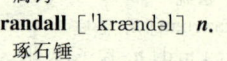
crandall ['krændəl] *n.* 琢石锤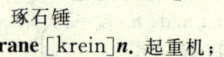

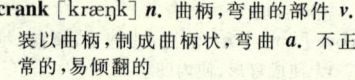
crandall

crane [krein] *n.* 起重机;龙头 *v.* 用起重机搬运
crank [kræŋk] *n.* 曲柄,弯曲的部件 *v.* 装以曲柄,制成曲柄状,弯曲 *a.* 不正常的,易倾翻的
~ brace 钻孔器,手摇钻
~ pin 曲柄销
crannog ['krænəg] *n.* (古代爱尔兰在湖沼地带建的)湖中央古屋或人工岛
cranny ['kræni] *n. pl.* nies, 裂缝[隙] *vi.* 起裂缝
crash [kræʃ] *a.* 应急的;*v.*;*v.* 碎裂,坠毁,失败
~ helmet 防护[安全]帽,防撞头盔
~ pad 防震垫
~ program (project) 紧急措施
~ roll 防震垫
crate [kreit] *n.* 条板箱,输送框,筛状容器;破旧汽车;*vt.* 装箱运输
wooden ~ 木箱
crater ['kreitə] *n.* 焊口[槽],火山口
~ crack 弧坑裂纹
crawl [krɔːl] *n.*;*v.* 蠕动,慢(爬)行
crawler [krɔːlə] *n.* 履带(牵引装置),履带式拖拉机
crawling *n.* 蠕变,爬行
crayon ['kreiən] *n.* 粉笔画,蜡笔画;*vt.* 蜡笔作画,勾轮廓
craze [kreiz] *n.* 裂痕,发丝裂缝,细裂纹;*v.* 开裂
crazy ['kreizi] *a.*, -yier, -yiest 混乱的,弯弯曲曲的,不安全的
~ pavement 散乱碎石铺砌(的路面),碎石路
cream [kriːm] *a.*;*n.* 米黄色(的),胶状(的);*vt.* 搅拌成乳状
~ of latex 胶乳
~ of lime 石灰浆,石灰乳浊液

~ solder 乳酪焊剂
creaming *n.* 乳状液,涂敷脂膏 *v.* 形成乳状液
crease [kri:s] *n.* 褶痕,皱褶；*v.* 使有褶痕
~ pipe bend 褶皱的弯管
creasing machine 褶边机
creasing course 逐层挑出的砖墙,烟囱压顶
create [kri'eit] *vt.* 形成,建立,创造
creation [kri(:)'eiʃən] *n.* 创造[新],创造物,创作品
art ~ 艺术作品
artistic ~ 艺术创作
creative [kri'eitiv] *a.* 有创造力的,创造的
~ design material 造型材料
~ engineering 创造工程学
~ imagination 创造想像
~ playground (培养创造力的)儿童游乐场
~ thinking 创造思维
credit ['kredit] *n.* 贷方,存款,赊购,信贷,债权
creek [kri:k] *n.* 小河(溪),小港弯
creep [kri:p] *n.*；*vi.*；*prep.* creeping 蠕[徐]变,屈服,坍方,滑坡,渗水；蠕动,爬行
creeper ['kri:pə] *n.* 卷叶式浮雕,藤蔓花样浮雕；履带；倾斜式链条输送机
creeping [kri:piŋ] *n.*；*a.* 土的滑塌,坍方；蠕变(的)
creepwash *n.* 滑坡,土流,土崩
crenelate ['kreniιleit] *vt.* 成锯齿状
~d moulding 锯形线脚,齿饰
~d pattern 钝锯齿形

crenelated moulding

creosol ['kri:əsɔl] *n.* (作防腐剂用的)木焦油醇
creosote ['kri:əsəut] *n.* 杂酚油,木材防腐油；*v.* 以杂酚油浸渍处理
~d timber 油浸木材,用杂酚油防腐处理过的木材
crepitation ['krepiteiʃən] *n.* (混凝土)气蚀碎裂现象
crescent ['kresnt] *n.* 月牙饰；*a.* 新月形的,月牙形的
~ (adjustable) wrench 可调扳手
~ arch 月牙拱
~ truss 月牙形桁架
cresol ['kri:sɔl] *n.* 甲酚
~ resin adhesive 甲酚树脂粘合剂
cresset ['kresit] *n.* 标灯,号灯
crest [krest] *n.* (峰)顶,堰顶,山脊,屋脊饰物

crest

~ board 层脊饰板
~ curve 凸形曲线
~ of screw thread 螺纹牙顶
~ slab 顶板
~ table 墙帽
~ tile 屋脊(饰)瓦
crevasse [kri'væs] *n.* (冰河,堤坝的)裂缝[隙],决口；*vt.* 使有裂缝[口]
~ crack 裂缝,龟裂
crevice ['krevis] *n.* 裂隙破口
~ corrosion 裂隙腐蚀
~ formatiom 裂隙[岩]层
crib [krib] *n.* 木(格)笼,叠木框,木架[排]；箱,盒,小室
filled ~ 填石木垛
open ~ 空心木垛
cribbing [kribiŋ] *n.* 棚架,叠木,木骨板壁,垛式[井框]支架
cribble ['kribl] *n.* 粗筛；*vt.* 过筛
cribriform [kribrifɔ:m] *a.* 筛状的
cribweir *n.* 木笼堰
cribwork ['kribwə:k] *n.* 叠木框,木笼,筏状基础

cricket

cricket [krikit] *n.* 防热屋顶,泻水假屋顶,

木制矮脚凳
cricoid ['kraikɔːd] *a.* 轮[环]形的
crimp [krimp] *vt.* 使成波形,使折皱; *n.* 曲贴(角钢); *a.* 脆的,薄弱的
  ～ mesh 波形钢丝网
  ～ seal 锯齿形焊缝
  ～ wire netting 波形钢丝网
  ～ed finish 皱纹状修饰
  ～ed wire 脆性钢丝
crimper ['krimpə] *n.* 钢筋弯曲[折]机,折波钳,卷边机
crimping ['krimpiŋ] *n.* (大直径直缝焊管加工时)卷边,锁缝,(钢丝织网时)压出波浪弯
crimson ['krimzn] *a.* 深红色(的) *v.* 使成深红色
crinkle ['kriŋkl] *v.* 使皱,使起波纹; *n.* 波纹
  ～ finish 皱纹(壁面)漆,波纹面饰
criosphinx ['kraiəˌsfiŋks] *n.* (古埃及)狮身羊头像
cripple ['kripl] *n.*;*v.* 脚凳,踏脚,纵向(绕曲),变形;使受损伤
  ～ scaffold 扶臂脚手架
  ～-timber 变形木材
  ～ window 跛窗
crippling ['kripliŋ] *n.* 断裂,局部失稳破坏,局部压屈
  ～ loading 破坏[临界]荷载
  ～ stress 断裂[临界]应力
crisscross ['kriskrɔs] *a.*;*ad.*;*v.* 成十字形(的),互相交叉(的); *n.* 十字形,方格
  ～ pattern 十字形图案
  ～ structure 方格构造
cristobalite ['kristəubəlait] *n.* 白硅石,方晶石,方石英
criterion [krai'tiəriən] *pl.* criteria, *n.* 标准,规范,准则;指标,尺度,规模;条件,依据
  control ～ 质量控制准则
  design ～ 设计标准
  specified ～ 给定技术条件,明细规范

critical ['kritil] *a.* 临界的,极限的,决定性的; *n.* 临界(值)
crizzling *n.* 表面缺陷
crocidolite [krəu'sidəlait] *n.* 青石棉,钠闪石
crock [krɔk] *n.* 破碎物,碎瓦片,油烟,污物;瓶,缸,罐
crockery ['krɔkəri] *n.* 陶[瓦]器
crocket ['krɔkit] *n.* 卷叶(形花)饰,卷叶形浮雕
crocodile *vi.* 龟裂,形成交叉裂缝; *n.* 轧体前端的分层

crocket

crocus ['krəukəs] *n.* 枯黄色,番红色,磨光粉
crolite *n.* 陶瓷绝缘材料
cromlech ['krɔmlek] *n.* (史前时代的)环列石柱
crook [kruk] *n.*;*v.* 弯[挠]曲,(成)钩(状);使弯曲,弄变
  ～ warping 板材纵向挠曲
  ～ed hole 斜孔
  ～ed nail 钩头道钉
  ～ed timber 弯曲木材
crookedness *n.* 弯曲,曲率
crop [krɔp] *v.* 修剪,剪切,露头; *n.* 叶尖饰;一组,一群,大量;产量
  ～ping out of ground-water 地下水露头
cross [krɔs] *n.* 十字接头,四通,十字架,交叉路, *v.* 成十字交叉,相互交叉,横过; *a.* 横的,逆的,相反的,交叉的,十字的
  ～ aisle 交叉通道,交叉走廊
  ～ arched vaulting 交叉穹窿
  ～ arm 横撑
  ～ bar 横木[杆]闩柱
  ～ battens 交叉条板
  ～ beam 横交叉梁(肋)梁,大梁
  ～ bearer 横梁[撑]
  ～ bending 横向弯曲
  ～ binding 横向支撑[系杆,连接]

~ -bolt 横销
~ bond 交叉砌合,交联键,交叉捆扎
~ brace 交叉支撑,横拉条
~ branch 十字管,四通管
~ breaking 横切,横向断裂
~ bridging 搁栅斜撑,交叉撑

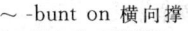
cross bridging

~ -bunt on 横向撑杆,横向剪刀撑楔入
~ cock 四通阀[旋塞]
~ -connection 四通[十字]连接
~ -cut chisel 十字形凿刀,横切凿
~ cut-off wall 横向隔墙
~ -folding 交错褶皱
~ fracture 横向断裂
~ -garnet 丁字形蝶铰
~ girder 横梁
~ grain 木料上的斜纹
~ grained rock 交错层砂岩
~ groove 十字形凹槽,横槽
~ hairs (测镜内的)十字丝[线]
~ hatching 断面线,十字形线
~ head 十字结联轴节,十字头,丁字头,滑块,横梁
~ joint 十字接头,交叉连接,横缝,横节理
~ -like 十字形的
~ link (聚合物)交联,横向连接
~ notching 对开槽
~ piece 横木,过梁,十字管头,联接板
~ pin 插销
~ pin type joint 万向接头
~ pipe 十字形管
~ riveting 十字形铆接
~ sectional view 横断面图
~ shake (板材的)辐裂,横裂
~ slot 十字槽,交叉槽
~ staff 十字杆
~ stay 横[斜]撑
~ strut 横撑[挡]
~ tail butt 十字铰键
~ tongue 横舌榫

~ transom 横撑,连杆
~ type dowel 十字形暗销
~ -under 穿接
~ vault 交叉穹窿
~ wall 隔板,横墙

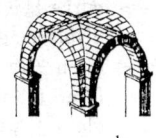
cross vault

crossbill ['krɔsbil] n. 交喙鸟
~ joint 叉口[交喙]接合
crossette [krɔ'set] n. 螺旋状支柱
crossing ['krɔsiŋ] n. 十字路口,渡口,道口,铁路道岔;跨接,横切[断];交叉着建筑物
crossover ['krɔsˌəuvə] n. 观众厅的横向过道,(铁路)转线轨,(横越公路的)陆桥
crosswalk [krɔswɔ:k] n. 人行道
~ line 人行横道线
crosswise ['krɔswaiz] a. 交叉的,十字状的,对角线的;横的,斜的
crotch [krɔtʃ] n. (河道的)岔口,分岔处;叉柱;弯脚钉,弯钩
~ weld 楔接锻接
crotchet ['krɔtʃit] n. 叉柱,小钩,方括弧
crow [krəu] n. 起货钩,铁撬,撬杠
~ bar 撬[铁]棍
~ foot crack 爪形裂缝
crowd [kraud] n. 一群[堆]; v. 积聚,堆积,拥挤
~ed downtown area 繁华商业区,闹市区
crowfoot ['krəufut] n. pl. crowfeet 联结扒钉,吊索,防滑三脚架;图中标尺寸的箭头
~ cracks 爪形皱裂
crown [kraun] n. 冠状物,凸面,路拱,(拱)顶;塔式尖叶饰; vt. 给……装顶,隆起
~ bar 顶杆
~ block 拱顶石
~ brick 拱顶砖
~ cantilever 拱冠悬臂梁

～ cornice 大屋檐
～ curve 冕状曲线
～ displacement 拱顶位移
～ ditch 天沟,坡顶的截水沟
～ filler 上等填料
～ glass 上等玻璃
～ imperial 冠顶
～ joint 拱顶接缝
～ lagging (隧道拱顶部)支拱木
～ line of vault 穹顶线
～ moulding 冠顶饰
～ nut 槽顶螺母
～ of arch 拱顶
～ piece 墙上的垫木,支撑短木
～ plate 梁垫板
～ post 桁架中柱
～ saw 筒形锯
～ section 拱顶截面
～ steeple 冠状尖塔
～ -steps 阶式山墙顶
～ stone 拱顶石,山墙顶石
～ tile 顶[冠]瓦
～ tray 梁垫[枕]
～ing 凸[拱]起;板材中心部分增厚

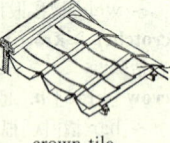
crown plate
crown tile

crozzle *n.* 过烧砖
cruciform [ˈkruːsifɔːm] 或 crucishaped; *a.* 十字形的
～ joint 十字接头,十字形连接
crucks *n.* 曲木屋架,帐幕式棚架木杆
crude [kruːd] *a.* (原)生的,粗的,未提炼的 *n.* 天然物质
～ aggregate 天然骨料
～ asbestos 天然石棉
～ asphalt 原生石油沥青
～ materials 原料
～ rubber 生橡胶
crumb [krʌm] *n.* 屑粒,碎皮 *vt.* 弄碎
～ structure 团粒[屑粒状]结构
crumble [ˈkrʌmbl] *v.* 坍落,弄碎; *n.*
碎土,破碎物
～ structure 团粒结构,屑粒状结构
crumbling away 剥[脱]落,碎裂
crumbly [ˈkrʌmbli] *a.* 易碎的,脆的
crumple [ˈkrʌmpl] *v.* 弄皱,使崩溃; *n.* 皱纹
crush [krʌʃ] *n.*;*v.* 压[碾、粉]碎,磨细;拥挤
～ room 剧院休息室
～ed brick 碎砖
～ed gravel 碎卵石
～ed sand 碎石砂,人工砂
～ed slag 碎矿渣
～ed stone 碎石
～ed stone concrete 碎石混凝土
～ed stone course 碎石层
～ed stone pavement 碎石铺砌层
crusher [ˈkrʌʃə] *n.* (颚式)破碎机,碎石机
～ -run stone 未筛碎石
crust [krʌst] *n.* 面层,表皮;地壳;水垢 *v.* 结皮,用外皮覆盖
crustal [ˈkrʌstəl] *a.* 地壳的,外壳的
crutch [krʌtʃ] *n.* 支撑(物),支柱,道钉,丁字形终端接续套管; *vt.* 支持[撑]
crutcher [ˈkrʌtʃə] *n.* 螺旋式拌和机
cryogen [ˈkraiədʒən] *n.* 冷冻[致冷,冷却]剂
cryogenic [ˌkraiəˈdʒenik] *a.*;低温的,深冷的,冷冻的
cryolite [ˈkraiəlait] *n.* 冰晶石
～ glass 乳白[冰晶]玻璃
crypt [kript] *n.* 地窖[穴],教堂地下的墓室
cryptomeria [ˌkriptəˈmiriə] *n.* 柳杉,日比杉
～ bark roofing 铺杉木皮的屋面
cryptoporticus *n.* 古罗马地道
crystal [ˈkristl] *n.* 晶体,水晶,精制玻璃 *a.* 水晶(制)的,透明的 *vt.* 使结晶
～ finish 水[结晶]纹饰面
～ glass 水晶玻璃

~ lattice 晶体格构
~ structure 晶体结构
Crystal palace 水晶宫，1851年在英国伦敦举办第一届国际博览会的展览馆

crystal finish

crystalline [ˈkristəlain] *a.* 结晶(状)的，透明的；*n.* 晶态，结晶体
~ glaze 结晶釉
~ rock 结晶岩
~ structure 结晶结构
crystallinity [ˌkristəˈliniti] *n.* 结晶度[性]
crystallization [ˌkristəlaiˈzeiʃən] = **crystallisation** *n.* 结晶(作用,过程)，结晶体
crystalloid [ˈkristəlɔid] *n.* 晶体，似晶质；*a.* 晶体的,透明的
crystolon *n.* (研磨用)人造碳化硅
cube [kjuːb] *n.* 立方体 *vt.* 求立方[体积]，三乘，铺立方石
~ globe 立体球形灯罩
~ spar 硬石膏
cubic [ˈkjuːbik] *a.* 立方(体)的,三次的,体积的；*n.* 三次曲线[函数]；体[容]积,立方量
~ centimeter 立方厘米
~ deformation 体积变形
~ effect 立体感
~ expansion 体积膨胀
~ foot 立方英尺
~ inch 立方英寸
~ meter 立方米
cubical [ˈkjuːbikəl] = cubic
~ block capital 方块柱头
~ dilatation 体积膨胀
cubicle [ˈkjuːbikl] = cubi-cule, cubicu-lum *n.* 隔间，小卧室
~ switch 组合室内开关,室内用配电箱
~ switch board 组合开关板
~ switch gear 组合开关装置
~ system 箱形结构系统；(医院内)小间分隔式
cubiform [ˈkjuːbifɔːm] *a.* 立方形的
cubing *n.* 体积测量，收方，以体积计算[量]
cubism [ˈkjuːbizəm] *n.* (美术上)立体派图画作风
cubist [ˈkjuːbist] *n.* 立体派艺术家；*a.* 立体艺术的
~ architecture 立体派建筑艺术
cuboid [ˈkjuːbɔid] *n.* 矩形[长方]体；*a.* 立方形的
cul-de-four【法】*n.* 半穹窿，炉灶后墙
cul-de-lampe【法】*n.* 灯垂饰
Culiculus *n.* 科林斯式柱头的蜗卷叶梗
cull [kʌl] *v.* 选择；*n.* 次品，废品等外材
~ of brick 光端砖,劣质砖
cullet [ˈkʌlit] *n.* 碎玻璃片
culminate [ˈkʌlmineit] *v.* 达到顶点，到中天，结束，完成
~ing point 坡折点
culmination [ˌkʌlmiˈneiʃən] *n.* 顶点，极度
cultural [ˈkʌltʃərəl] *a.* 文化的,栽培的
~ building 文化馆
~ center 文化中心
~ facilities 文化设施
~ landscape 文化景观
~ relics 文物
Cultural Palace of Nationalities 民族文化宫
culture [ˈkʌltʃə] *n.*；*v.* 培养,训练；文化[明]；(测量上的)地物
culver [ˈkʌlvə] *n.* 斑鸠
~ tail 鸠尾榫
culvert [ˈkʌlvət] *n.* 涵洞，暗渠，下水道，阴沟；地下电缆
cum = cubicmeter 立方米
cumulative [ˈkjuːmjulətiv] *a.* 累积[加]的,附加的
cumulus [ˈkjuːmjuləs] *n. pl.* -li. 堆[累]积,积云

cuneate ['kju:niit] *a.* 楔形的
cuneiform ['kju:niifə:m] *a.* 楔形的
~ arch 楔形砌拱
cup [kʌp] *n.* 帽,盖,罩,杯;坩埚;凹圆形,盆地; *v.* 成杯型,成凹形,密封
~ bearing 木插口
~ head rivet 半球头铆钉
~ joint 套接
~ shake 木材环裂
~ping 木材干燥翘曲,翘成杯形
cupboard ['kʌbəd] *n.* 碗[食]橱,柜
bedroom ~ 卧室内壁橱
wall ~ 壁橱
cupola ['kju:pələ] *n.* 圆屋顶,圆顶阁,小型熔铁炉,圆顶鼓风炉;岩钟
~ brick (砌圆顶用的)楔[扇]形砖
cuppy ['kʌpi] *a.* 杯形的,中空的,有杯状孔隙的,(地面上)窟窿多的
cupreous ['kju:priəs] *a.* 铜的,含[似]铜的,铜色的
cupric ['kju:prid] *a.* (正)铜的,含铜的
~ chloride 氯化铜
~ oxide 氧化铜
~ sulfate 硫酸铜
cuprum ['kju:prəm] *n.* 铜
curb [kə:b] *n.* 路(边)缘(石),缘饰;井栏 *v.* 设置路缘石,控[抑]制
~ fender 路栏,路缘护木,壁炉周边装饰
~ plate 侧板
~ roof 复折[斜]屋顶
curbside *n.* 街道边,路边
curd [kə:d] *n.* ; *v.* (使)凝结,凝乳
curdle ['kə:dl] *v.* 凝结,变稠,变质
curdy ['kə:di] *a.* 凝结了的,凝乳状的
cure [kjuə] *n.* ; *v.* 硬化,凝固,处理,(混凝土)养护
~d resin 硬[凝固]树脂
curie ['kjuəri] *n.* 居里(放射性强度单位=3.7×10¹⁰次衰变/秒)
curing ['kjuəriŋ] *n.* ; *v.* 混凝土养护,熟[固]化
~ agent 混凝土养护剂,固[熟]化剂

~ composition 混凝土养护剂
~ cycle 养护[周]期
~ membrane 养护薄膜
~ period 养护周期
~ rack 混凝土养护格栅,养护架
~ strain 硬化变形
~ temperature 养护[固化]温度
~ tent 养[防]护帐篷
dry ~ 干养护
field ~ 现场[工地]养护
moist ~ 湿养护
steam ~ 蒸汽养护
curio ['kjuəriəu] *n. pl.* -rios . 艺术珍品,古董[玩]
curiosity [kjuəri'ɔsiti] *n. pl.* -ties. 珍品,奇物;好奇心、奇特性
curl [kə:l] *v.* ; *n.* 围绕,盘旋,弯曲,(起)波纹;螺旋状物,卷曲物
curlicue ['kə:likju:] = curlycue *n.* 卷曲装饰,花体字
curly ['kə:li] *a.* 有波纹的,卷曲的
~ figure 波状纹理
~ grain 木节,旋涡纹
~ veneer 涡形饰纹镶面板
current ['kʌrənt] *n.* 水[电,气,潮]流,趋势; *a.* 流通[行]的,现行的,本年月的
~ assets 流动资产
~ cost 市价,现时成本
~ lamination 波状纹理
~ practice 现行惯例
~ price 时价,行情
~ regulation 现行条例,现行规章
~ ripple mark 波痕
~ standard 现行标准
curriculum vitae 履历[表]
currycomb ['kʌrikəum] *n.* 耙式整坡器
cursor ['kə:sə] *n.* (卡尺上的)游标,指针,指示器;转动臂,滑块
curtail [kə:'teil] *vt.* 缩短,减缩
~ step (楼梯)起步级
curtailment [kə:teilmənt] *n.* 收缩,削减,省略

import ～ 缩减进口
curtain ['kə:tn] *n.* 帐,幕,窗帘,掩蔽物,隔板 *v.* 掩蔽,挂帘[幕]
～ box 窗帘匣
～ coater 帘式淋涂器
～ fabric 窗帘织物
～ hardware 窗帘用小五金
～ lace 花边窗帘,花边纱窗
～ line 大幕线,舞台地面与幕的交线
～ net 网眼窗帘
～ rail 帘轨
～ ring 窗帘圈
～ rod 窗帘辊
～ wall 幕墙,护墙
～ zone 大幕拉开后的停幕处,拉幕区
safety ～ 防火窗帘
window ～ 窗帘
curtaining *n.* 下垂,垂落(涂后漆膜形成大面积下垂)
curtilage ['kə:tilidʒ] *n.* 庭园,院子,宅地
curvature ['kə:vətʃə] *n.* 曲率[度],弯曲,屈曲
～ distortion 屈曲变形
curve [kə:v] *n.* 曲线,弯曲; *v.* 使弯曲
～ board 曲线板
～ cut-off 裁弯取直
～ diagram 曲线图
～d timber 弯曲木材
～ family 曲线簇
～ fitting 曲线配合
～ gauge 曲线规
～ pen 曲线笔
～ plotter 绘图仪[器]
～ radius 曲线半径
～ ruler 曲线尺
～d beam 曲梁
～d brick 曲面砖
～d chord truss 折弦桁架
～d corrugated sheet 瓦垄薄板
～d earthware pipe 曲瓦管
～d intersection 曲线交叉
～d laminated wood 曲型多层胶合板材
～d line 曲线
～d pediment 弧形山墙
～d pipe 弯管
～d plate 曲面板
～d plywood 曲面胶合板
～d radiator 弧形散热器
～d rib truss 曲弦桁架
～d roof [ing] 曲面屋顶
～d surface 曲面
French ～ 曲线板
curvemeter *n.* 曲线计
curvic *a.* 弯曲的,曲线的
curvity *n.* 曲率
curvilineal [ˌkə:vi'liniəl] = curvilinear *a.* 曲线的,弯叶状的
～ style 曲线式,盛饰式
～ tracery 曲线窗花格
curvometer *n.* 曲线仪[计]
curving *n.* 弯曲
curvity *n.* 曲率
cusee ['kju:sek] = cubic feet per second *n.* 立方英尺/秒
cushion ['kuʃən] *n.* 垫层,缓冲器,拱座石,填料 *v.* 作垫层用
～ block 垫块
～ capital 垫块状柱头
～ coat 垫层
～ material 补强剂,附加剂,缓冲材料
cushioning ['kuʃniŋ] *n.* 弹性垫层,缓冲,减震
～ effect 减震[缓冲]作用

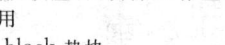

cusps

cusp [kʌsp] *n.* (曲线相交的)尖角饰,(叶形饰的)尖头,尖端(饰),尖头,回复点
cusped [kʌspt] = cuspate *a.* 有尖端的,似尖头的
～ arch 尖拱
～ edge 尖棱
～ point 尖点
custodian [kʌs'təudjən] *n.* 保管[管

理]员
~ fee 管理费

custom [kʌstəm] *n.* 海关,关税;顾客;习惯,风俗
~s duty [due] 关税
~s house 海关(CH)
~s inspection 海关检查

customer ['kʌstəmə] *n.* 用户,顾客

custos ['kʌstɔs] *n. pl.* custodes 监督人,管理者

cut [kʌt] *n.*;*v.* 开挖,挖土(方);路堑,沟坑,渠[河,隧]道;相交[切],切削[断,开],切削加工,雕刻,掏槽,割纹;采伐;分馏;插图 *a.* 断开的,分割的,切削加工过的,有锯齿边的
~ away view 平面图,剖视图,内部接线图
~ and mitred bead 偶角圆线条
~ and mitred string 偶角侧木
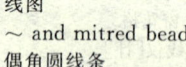
cut away view
~ glass 刻花的厚玻璃(器皿)
~ joint 接缝[头]
~ lock 插销
~ nail 方钉
~ open view 剖视图
~ spike 大方钉
~ stone 琢石
~ stone work 琢石工程
~ stone veneer 琢石镶面
~ string 木楼梯露明小梁
~ to measure 按尺寸下料
~ veneer 切制的镶面层板
paper ~ 剪纸
wood ~ 木刻

cutability *n.* 可切(削)性

cutback ['kʌtbæk] *n.* 稀释,轻制;减少,取消
~ bitumen 轻制沥青
~ group 轻制地沥青混合料

cutfit *n.* 备用工具

cuticle ['kju:tikl] *n.* 表皮,薄膜

cut-in [kʌtin] *n.*;*a.* 切[插]入(的),接通(的),超车
~ bottom 梁底板
~ method 插接法

cutlery ['kʌtləri] *n.* 刀具,餐具

cutlet ['kʌtlit] *n.* (切)片

cutline ['kʌtlain] *n.* 图例说明

cut-off ['kʌtɔ:f] *n.* 齿墙,截水墙;河流的裁弯段,裁弯取直,桩的截断处;切[断]开,断流器,保险装置;停车 *a.* 分界的
~ plane 截断面
~ switch 断路开关
~ wall 截水墙,防渗墙,隔墙,围堰,横堤
~ water 渗透[流,漏]水
mercury ~ 水银开关
safety ~ 安全开关

cutout ['kʌtaut] *n.* 切断,中断,断流器,保险开关[装置];排气门;齿墙,截水墙
~ case 保险盒
~ device 安全保险装置
~ switch 断路开关

cutover ['kʌtəuvə] *n.*;*v.* 接入(电源,电路)转换

cutstuff *n.* 碎木片,碎片

cutter ['kʌtə] *n.* (切削)刀具,切割机;切面砖;(地质上的)倾斜节理,横向裂纹

cutting [kʌtiŋ] *n.* 挖方;切削[割],掏槽,截断;开凿[采],加工,侵蚀 *a.* 供切(削)用的,尖锐的
~ blade 切削刀片,平地机刮板
~ dimension 标[划]线尺寸
~ disc 圆盘刀
~ edge 刃口
~ effect 侵蚀作用
~ float 刮刀
~ gaug 线勒子,木工起线用的划线工具
~ list 钢筋(选择),配筋(表)
~ nippers 老虎钳
~ off 切割,切槽
~ operation 切削操作

~ pliers 剪钳
~ through 刻纹过度
~ tool 切削刀具
cross ~ 横切
length ~ 纵切
cuttings *n.* 岩粉[屑]刨花,钻屑
cutwork [kʌtwəːk] *n.* 切削工作
c/w = **cement-water ratio**（混凝土的）灰水比
C-Washer *n.* C形垫圈
cyanaloc *n.* 氰基树脂防水剂
cyanide ['saiənaid] = cyanid *n.* 氰化物 *vt.* 用氰化物处理
~ copper 氰化物电镀铜
cyanotype [sai'ænətaip] *n.* 晒蓝图,蓝晒法,氰版照相
cybernation [saibəːˈneiʃən] *n.* (用电子计算机)控制
full automation and ~ 完全自动控制
cybernetic [saibəˈnetik] *a.* 控制论的
cycle ['saikl] *n.;v.* 循环,周期,自行车
~ of operation 运行周期
~ per second 赫(兹),周/秒
storage ~ 存储周期
cyclelog ['saikllɔg] *n.* （自动化）程序控制[调整]器
cycleweld ['saiklweld] *n.* 合成树脂结合剂
cyclic ['saiklik] = **cyclical** *a.* 循环的,周期的,环状的
~ surface 圆纹曲面
cycling ['saikliŋ] *n.* 周期性变化,循环变化,定期工作; *a.* 交替的,周期性工作的
cycloid ['saiklɔid] *a.* 旋轮线的,圆形的,摆线
cycloidal [sai'klɔidl] *a.* 旋轮线的,圆形的,摆线的
~ arch 圆滚线拱
~ cylinder 圆形筒柱
cyclometry [sai'klɔmətri] *n.* 测圆法
cyclonic [sai'klɔnik] *a.* 低压的,气旋(似)的,旋风[涡]的

~ collector 旋风[涡]吸尘器
cyclop(a)edia [saiklə'piːdjə] *n.* 百科全书[辞典];丛书
cyclopean [sai'klɔupjən] *n.* 乱[蛮]石堆; *a.* 巨大的,蛮石的,巨大堆积的,镶嵌状的
~ block 巨型毛石方块
~ concrete 蛮[毛]石混凝土
~ masonry 毛石圬工
~ riprap 大块石的毛石堆
~ wall 蛮[乱]石墙
cyclopite [saikləpait] *n.* 钙长石
cyclorama [saiklə'ræmə] *n.* 半圆形透视背景照明,圆景画景
cylinder ['silində] *n.* 圆柱体,圆筒
~ (head) lock 弹簧锁
~ lock 弹子锁
~ number 圈数
~ pile 大直径桩
~ type pipe 筒形管
~ wrench 圆筒扳手
cylindric(al) [si'lindrik(əl)] *n.* 圆柱[筒]形的,柱面的
~ arch 圆筒形拱
~ building 圆形建筑物
~ ceiling 筒形顶棚
~ -illuminance 柱面照度
~ intersecting vault 筒形交叉穹顶
~ lock 圆柱锁
~ pile 筒桩
~ shaft 管柱
~ shell 筒形薄壳
~ surface 圆柱面
~ vault 圆柱穹窿
cylindrocoical *a.* 圆锥形的
~ ball mill 圆锥形球磨机
cylindroid ['silindrɔid] *n.* 椭圆柱[筒],圆柱形面 *a.* 椭圆柱的
cyma ['saimə] *n.* (*pl.* -mas, -mae.) 反曲线,波役线脚,波状花边
~ recta 表反曲线,上凹下凸波纹花边
~ reversa 里反曲线,上凸下凹波纹

花边

cyma recta

cyma reversa

cymatium [si'meiʃiəm] *n.* (*pl.* -tia.) 反曲线状,拱顶花边,波状花边

cymbiform ['simbəfɔ:m] *a.* 船形的
cymometer [sai'mɔmitə] *n.* 频率计,波长计
cymoscope ['saiməskəup] *n.* 检波器
cyplex *n.* 聚酯树脂
cypress ['saipris] *n.* (扁)柏,柏树枝
cyrtostyle ['sə:təstail] *n.* 圆形凸出的门廊,柱子排列成向外突出的弧形的回廊

D

D = desity;depth;密度;深度
D = diameter;dyne 直径;达因(力的单位)
dab [dæb] *n.* 团,块 *v.* 轻拍[敲],锤琢;涂,敷,湿润
  ~ bed finish 细凿琢面
  ~ bed mortar 砂浆饼块
  mortar ~ 砂浆涂抹
  plaster ~ 灰泥涂抹
dabble ['dæbl] *v.* 湿润,溅湿,浸湿,灌注,喷洒
dachiardite *n.* 环晶石
dacholeum *n.* 沥青
dacite ['deisait] *n.* 英安岩
  ~ tuff 英安凝灰岩
D&D = double-acting door 双开[弹簧]门
daddock *n.* 烂木头
dado ['deidəu] *n.* 柱基座,墩身;护壁板,墙裙;插木板之槽
  ~ base 墙脚
  ~ capping 护壁板压顶条,墙裙压顶条
  ~ frame 护壁板架
  ~ head 开槽工具
  ~ head machine 开槽机
  ~ moulding 护壁木条(饰)
  ~ plane 开平底槽的刨子
  ~ rail 装在墙裙上的木[轨]条
  ~ tile 护壁贴砖
daedal ['di:dəl] *a.* 巧妙的,迷宫式的,千变万化的
daedalian [di'deiliən] = daedalean *a.* 迷宫式的,巧妙的,错综复杂的
Daedalus ['di:dələs] *n.* 希腊神话中的建筑师和雕刻家底得勒斯
daffodil ['dæfədil] *n.* 水仙花;*a.* 水仙花色的
dag [dæg] *n.* 悬端,悬片;齿状饰边;石墨粉
dagoba ['dɑ:gəbə] *n.* 印度的佛骨堂;佛教中的舍利塔
White Dagoba 白塔

dagoba

daily ['deili] *ad.*;*a.* 每日的 *n.* 日报
  ~ advance 日进展
  ~ air temperature cycle 气温日循环
  ~ allowance 每日津贴
  ~ balance 每日余额
  ~ capacity 日生产量
  ~ cash balance bock 现金余额日记账
  ~ cash ratio 每日现金需要比率
  ~ compensation 日补偿
  ~ consumption 日消耗量
  ~ demand 水、电的日需要量
  ~ expense 每日费用

~ inspection 日常检查
~ load curve 日负荷曲线
~ load fluctuation 日负荷变动
~ mean 逐日平均
~ mean temperrature 日平均温度
~ necessaries 日常需要
~ output 日产量
~ pay 日工资
~ routine 日常业务
~ service report 每日业务报告
~ sheet 日报表,每日工作记录
~ statement 每日账表
~ statue 每日状况
~ wage 计日工资
~ variation 每日变动
~ work report 每日工作报告单

dais ['deiis] *n.* 讲台,演坛,高台

daisy ['deizi] *n.* (*pl.* -sies.)雏菊
~ Mae 澳大利亚测高计

dale [deil] *n.* 山谷;槽,排水孔

dallage *n.* 大理石面层[铺地],石头面层[铺地]

dalle [dɑːl] *n.* 装饰板;铺路石板

damage ['dæmidʒ] *n.;vt.* 损害,损伤;毁坏,破坏;损失
~ claim 损坏索赔
~ condition 残损情况
~ fastness 耐损(度,性)
~ free 无损坏
~ index 损坏指数
~ line 破损线
~ portion 残损部分
~ repair 损害补偿
~ resistance 耐损(度,性)
~ risk criterion 损害风险准则
~ suit 损害赔偿诉讼
~ survey 事故检定
~ tolerant design 破损设计

damaging ['dæmidʒiŋ] *a.* 破坏性的,严重的
~ impact 破坏性冲击

damascene ['dæməsiːn] *a.* 金属镶嵌的,波纹装饰的;*n.* 金属镶嵌法,波纹装饰,金属镶嵌制品 *v.* 用波纹装饰(金属);用贵重金属镶嵌(钢铁用品)

damask ['dæməsk] *n.* 大马士革钢;红玫瑰色 *vt.* 以金属镶饰;饰以色彩
~ steel 大马士革钢,表面带水纹的刀箭钢

dammar ['dæmə] *n.* 达马(树)脂
~ resin 达马树脂(用于制油漆)
~ varnish 达马清漆

damp [dæmp] *a.* 潮湿的 *n.* 潮湿,湿度;阻尼,衰减 *vt.* 使潮湿
~ air 湿空气
~ course 防潮层
~ proof course 防潮层
~ sheet 风帘,风幛
~ storage 湿度养护

dampen ['dæmpən] *v.* 使潮湿;阻尼,减振,缓冲
~ out 减弱,缓冲,吸收
~ing-effect 制动(减振)作用

damper ['dæmpə] *n.* 阻尼器,制动器,减速器;减振器;湿润器;气流调节器;挡板
~ brake 制动闸
~ circuit 阻尼电路
~ cylinder 减振筒
~ guard 制动器防护装置
~ leg 缓冲支柱,减振支柱
~ regulator 风门调节器,调节阀
~ valve 调节阀
air ~ 空气阻尼器,气压制动器
dynamic ~ 动力减振器
liquid ~ 液压减振器,液体阻尼器
mechanical ~ 机械阻尼器
pole ~ 阻尼[减振,静噪器]
slide ~ 滑动挡板

damping ['dæmpiŋ] *n.;a.* 阻尼的,减幅的,衰减的,减振的;湿润的,湿润加工
~ action 阻尼[制动]作用
~ capacity 减振能力;吸湿能力
~ course 减振层

~ degree 阻尼程度
~ effect 阻尼效果[作用]
~ factor 阻尼因素
~ force 阻尼力
~ layer 阻尼层
~ material 阻尼材料
~ machine 调湿机
~ power(= ~ property= ~ quality) 减振能力,吸湿能力
~ ratio 阻尼比
~ screen 稳水栅

dampness ['dæmpnis] *n.* 潮湿,湿度,含水量

dampproof ['dæmppru:f] *a.* 防潮的,抗湿的,不透水的
~ coating 防潮涂层
~ course 防潮层
~ foundation 防潮基础
~ insulation 隔潮
~ machine 防潮电机
~ material 防潮材料
~ membrane 楼层或平屋顶的防潮层
~ slab 防潮板
~ing admixture 防潮剂
~ing ground floor 防潮底层地坪
~ing wall 防潮墙

dan [dæn] *n.* 空中吊运车,小车;浮标;枸,瓢,斗,桶,担,挑水箱

dance ['dɑ:ns] *n.* 跳舞,舞蹈; *v.* 跳舞 *a.* 跳舞用的
~ floor 舞池
~ hall 舞厅

dancery ['dɑ:nsəri] *n.* 跳舞厅

dancette [dɑ:n'set] *n.* 曲折饰,锯齿形花饰,曲折线脚

dandy ['dændi] *n.*(*pl.*-dies)小型沥青喷洒机;双轮小车;双桅帆船 *a.* 漂亮的;上等的

danger ['deindʒə] *n.* 危险,威协;危险物〔品〕
~ board 危险警牌
~ level 危险程度
~ light 告警信号灯光
~ point 危险点
~ sign 危险标志
~ space 防火间隔
~ zone 危险地带[区]

dangerous ['deindʒrəs] *a.* 危险的
~ building 危险房
~ hill 险坡
~ section 危险截面;危险地段
~ signal 危险信号,停止信号
~ structure 危险结构

dangler ['dæŋglə] *n.* 悬摆物

dank [dæŋk] *n.;a.* 阴湿（的）,潮湿（的）

dap [dæp] *n.* 凹口,切口; *v.* 刻痕,挖槽
~ped joint 互嵌结合

dapper ['dæpə] *n.*（锯木材用）圆锯;支架制备机 *a.* 灵活的,整洁的

dapping ['dæpiŋ] *n.* 刻槽

dappled ['dæpld] 斑驳的,杂色的,有圆形斑点的

dapt [dæpt] *n.* 榫眼

darby ['dɑ:bi] *n. pl.* -bies. 刮尺;镘刀;混凝土镘板

darg [dɑ:g] *n.* 工作,任务;产量定额

dark [dɑ:k] *a.* 黑色的,暗淡的,无光的 *n.* 黑暗,暗处,无光
~ adaptation（视觉）对黑暗的适应性
~ -and-light 明暗,浓淡,深浅
~ background 黑暗背景
~ blind 防光的,不透光的
~ blue 深蓝色
~ -coloured pavement 黑色路面
~ current 暗电流
~ door 遮光门
~ -green 深绿
~ -heat radiation 红外辐射
~ installation(= ~ system) 灯火管制装置〔系统〕
~ jalousie 遮光百叶窗
~ light 不可见光
~ pine 樟木

~ repair 暗修复
~ room 暗室
~ slide 遮光板
~ surrounds 黑暗环境
~ tint face 暗淡面
~ window 遮光窗
dart [dɑ:t] ***n.*** 箭头装饰；箭
~ and egg 蛋与簇形装饰
~ union 活络管子节
dash [dæʃ] ***n.*** 灰浆；溅泼；注入；槌柄；阴影线，破折号 ***v.*** 浇泼，洒浆；划线
~ adjustment 缓冲调节
~ -and-dot line (= ~ -dotted line) 点划线
~ area 阴影部分
~ board 遮雨板，遮水板；防波板；仪器板
~ coat 泼涂层
~ control 按钮[缓冲]控制
~ current 冲击电流
~ finish 浇泼饰面
~ line 虚线，短划线
~ plate 缓冲板
~ pot 缓冲筒，减振器
~ unit 仪表板
dasher ['dæʃə] ***n.*** 挡泥板；遮水板；浆式搅拌器；冲击物
dashpot ['dæʃpɔt] ***n.*** 减振器，缓冲器，阻尼器
air ~ 空气缓冲器
data ['deitə] ***n.*** 数据；资料
~ acquisition system 数据收集系统
~ analysis center 资料分析中心
~ book 数据手册[表]，清单
~ collection system 数据收集系统
~ drawing list 资料图纸清单
~ encoding system 数据编码系统
~ exchange system 数据交换系统
~ form 资料记录表
~ gathering system 数据收集系统
~ handling system 数据处理系统
~ interchange 数据互换
~ management 数据处理[控制，管理]
~ plate 铭牌
~ transmission 数据传输
analog ~ 模拟数据
auxiliary ~ 辅助数据
primary ~ 第一手数据[资料]，原始数据[资料]
qualitative ~ 定性资料
quantitative ~ 定量资料
raw ~ 原始资料
dataller ['deitələ] ***n.*** 计日工
datatron ['deitətrɔn] ***n.*** （十进制计算机中的）数据处理
date [deit] ***n.*** 日期，年代 ***v.*** 记日期；断定……的年代；属于某时期
~ closing 结算日
~ draft 定期汇票
~ due 到期日
~ of completion 完成[竣工]日期
~ of contraction 签约日期
~ of delivery 交货日期
~ of expiration 有效日期
~ of grace 宽限日
~ of maturity [支票，汇票]兑现日期
~ palm 枣椰树
due ~ 到期日
target ~ 预定日期
datebook ['deibuk] ***n.*** 记事册
dated ['deitid] ***a.*** 注明日期的，陈旧的，过时的；***n.*** 保险日期
dateless ['deitlis] ***a.*** 没有日期的，年代不明的
dating ['deitiŋ] ***n.*** 注明日期，断定年代
age ~ 年代测定
datum ['deitəm] ***n.*** (***pl.*** data) 论据；资料，材料；基准面
~ benck mark 水准基点
~ drift 基准偏差
~ horizon 基准地平
~ level 基准面
~ line 基准线
~ mark 基准点，标高
~ plane 基准平面

~ point 基准点
~ surface 基准面
~ water level 基准水平面
fixed ~ 固定基准线
geodetic ~ 大地基准点
daub [dɔ:b] ***n.*** 胶泥,粗灰泥;涂料;***v.*** 涂抹;打泥[色]底,抹胶
mud ~ 泥补裂缝
dauber ['dɔ:bə] ***n.*** 涂沫者,泥木工;涂抹工具;涂料
daubing ['dɔ:biŋ] ***n.*** 粗抹面,石面凿毛;涂料,粗抹灰泥
dauby ['dɔ:bi] ***a.*** 黏性的,胶粘的
daugh ***n.*** 耐火黏土
daughter ['dɔ:tə] ***n.*** 女儿 ***a.*** 如女儿的;第一代的
~ board 子插件
dauk [dɔ:k] ***n.*** 砂质黏土;黏质砂岩
davenport ['dævənpɔ:t] ***n.*** 小型写字台;坐卧两用沙发
dawk [dɔ:k] ***n.*** 黏质砂岩
dawn [dɔ:n] ***n.*** 黎明,破晓
~ effect 曙光效应
~ redwood 水杉
day [dei] ***n.*** 日,白天;紧接地表的岩层
~ beacon 昼标
~ bed 躺椅,两用沙发
~ bill 定期票据
~ blindness 昼盲,夜视症
~ capacity 日生产量
~ flow 日流量
~ gate 安装在保险库内部的格栅
~ labour 计日工作;散工
~ labour system 按日计工制
~ labourer 按日计酬的散工
~ light 自然光,日光
~ -light saving time 夏季[令]时间,日光节约时间
~ man 计日工
~ mark 昼标
~ mark-buoy 不发光浮标
~ of grace 宽限日期
~ of maturity 到期日
~ of reckoning 结账日
~ of supply 日供应量
~ off 休息日
~ parking 日间停车场
~ room 休息室,文娱室
~ school 日校
~ shift 日班
~ signal 书标
~ taler 计日工,临时工
~ -to-day production 日常生产
~ visibility 白昼能见度
~ wage 计日工资
~ work 计日工作
~ work joint 日工作缝
~ worker 临时工,日班工人
~ work rate 计日工资单价,点工单价
daylight ['deilait] ***n.*** 日光,白昼,自然光
~ control 日光控制
~ driving 白昼行车
~ effect 日光作用,白昼效应
~ factor 日照因素
~ filter 日光滤光器
~ hour 日照时数
~ illumination 日光照明
~ lamp 日光灯
~ opening (压机)压板间距
~ photometry 日光光度学
~ saving time 夏令时间,经济时
~ signal 白昼信号,色灯信号
~ source 自然光源
~ing curve 日光曲线
daytime ['deitaim] ***n.*** 日间,白昼
~ shift 日班
dazzle ['dæzl] ***n.*** 眩光
~ -free glass 防眩光玻璃
~ light 眩目灯光,强光
~ paint 伪装色彩
dazzling ['dæzliŋ] ***a.*** 眩目的;dazzlingly, ***a.***
~ white 眩目白色,炽白色
dB =decibel 分贝(声强单位,电平)
D-cracking ***n.*** 混凝土的D开裂纹

de luxe [di'luks] *a.*【法】高级的,豪华的,精装的;*ad.* 豪华地
~ type hotel 豪华级饭店
deaccentuator [di:æk'sentjueitə] *n.* 校平器,频率校正电路
deacidification [ˌdi:əsidifi'keiʃən] *n.* 除酸作用,去酸化
dead [ded] *a.* 死的,静止的,固定的;不通行的,停顿的;不灵的,失效的,废弃的;完全的;绝对的;*ad.* 绝对地,完全地
~ air 闭塞空气,静空气
~ air compartment (空心墙内的)闭塞空间
~ air insulation 闭塞空气绝缘

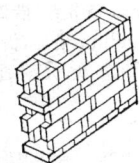

dead air compartment

~ air space 闭塞空间
~ air void 闭塞空隙
~ alloy 低碳合金
~ and dry face 干枯面
~ angle 死角
~ arcade 封闭拱廊
~ area 死角区,截面不受力部分
~ axle 静轴,从动轴
~ beat 非周期的,无振荡的,不摆的
~ bolt 无弹簧锁闩
~ -bright 磨光的,抛光的
~ burned 烧僵的,透彻煅烧的
~ burned dolomite 烧僵白云石
~ burned gypsum 烧僵的石膏(抹灰用的无水石膏)
~ door 假门;船上用于挡风暴的门外遮板
~ earth 固定接点,直通地
~ end 末端,尽头;终端,终点;闭塞的一头,死头;死胡同
~ end corridor 尽端式走廊

~ -end main 尽端干管
~ -end piping 尽端管线
~ -end weld 死端焊
~ finish 普通饰面
~ floor 粗地板,楼板中的垫板
~ fold 摺皱
~ ground 完全接地;无矿岩层
~ handle 常闭式安全把手
~ -hard steel 高强钢,极硬钢
~ hole 死洞,爆炸后的炮眼,残眼,盲孔
~ knot 木料的腐节
~ -level roof 平屋顶
~ -light 固定窗扇,固定天窗
~ line 闭置线路,空线,最后期限
~ load 恒载,静载;静重,自重
~ -load deflection 静载挠度
~ load moment 静载弯矩
~ lock 单闩锁;停顿,僵局
~ loss 纯损失
~ masonry wall 砖石砌筑的实墙
~ melted steel 镇静钢
~ mild steel 极软碳钢
~ plate 障热板,固定板
~ room 静室,消音室
~ setting steel 去氧钢
~ shore 固定支撑
~ smooth file 极细锉,油光锉
~ soft steel 极软钢,低碳钢
~ sounding 隔音层
~ space 无用空间,静区,死水域,死角
~ steel 全脱氧钢,全镇静钢,软钢,低碳钢
~ storage 长期库存,垫底[死]库容
~ wall 暗墙,无窗墙
~ weight 自重,静重
~ white 无光泽的涂料
~ window 假窗
~ wire 已不通电流的电线,死线;固定索
~ wood 枯木
~ work 准备工作

deadening ['dedniŋ] *n.* 吸音,消音;减弱

~ agent 消光剂,黯淡剂
~ dress (吸音的)粗面修琢
~ fabric 隔音布
~ felt 隔音纸
deaerate [di:'ɛəreit] *vt*. 从……中除去空气或气体,通风
~d concrete 去气混凝土
deaerating layer 除气层
deaeration [di:ɛə'reiʃən] *n*. 除气,除氧
deaerator ['di:'ɛəreitə] *n*. 除气器,空气分离器,脱氧器
deafen ['defn] *vt*. 隔音,消音
deafener ['defnə] *n*. 消音器,减音器
deafness ['defnis] *n*. 聋度;听力损失
~ percent 听力损失率
deafening ['defniŋ] *n*. 止响物[装置];隔音;隔声材料 *a*. 令人耳聋的
deair [di:'ɛə] *vt*. 除气,排气
~ed brick 去气砖
~ed clay 去气黏土
~ed concrete 去气混凝土
~ing machine 除气器
deal [di:l] *v*. dealt,dealing, *n*. 铺板;厚松板;针叶树材
~ board 松(杉)木板
~ ends 短(松)板
~ flooring 松木地板
~ wood 松材,针叶木材
back(ing) ~ 顶背板
cut ~ 中厚板
slit ~ 薄板
yellow ~ 黄松板
deambulatory *n*. 教堂中的回廊通道

deambulatory

deambulatory vault 回廊拱顶

deambulatory vault

deamless pipe 无缝钢管
dean [di:n] *n*. 教务主任[长],学院院长
deanery ['di:nəri] *n*. (*pl*. -eries.)院长[教务长]的办公室或宅邸
deash *vt*. 除灰,去灰分
deasphalt *vt*. 脱沥青
deathtrap ['deθtræp] *n*. 危险场所,不安全的建筑物
debark [di'bɑ:k] *v*. 去(树)皮;卸载,起(货,岸)
debit ['debit] *n*. 借方;负债; *vt*. 使负债
~ advice 欠款报单;借项通知单(D、A)
~ and credit memos 借贷凭单
~ balance 借方差额
~ customers 债务者,借款户
~ memorandum 借据,借项清单
~ note 借款清单;欠款通知单,欠货单,欠单
~ side 借方
debiteuse *n*. 土制浮标;(玻璃窑)槽子砖
deblooming [di:'blu:miŋ] *n*. 去荧光
deboost [di:'bu:st] *vt*. 减速,制动
debooster *n*. 限制器,制动器,减压器
debris ['debri:] *n*. 岩屑,瓦砾,碎片;矸石;河床推移砂石;垃圾,残砬;有机物残余
~ cone 砂砾锥
~ from demolition 建筑碎料
~ -storage basin 沉砂区;贮砂库
debt [det] *n*. 债;债务;欠款
~ at call 即期债务
~ capital 借入资本
~ ceiling 债务限额
~ collector 收债人
~ discount 债务折扣
~ financing 借款筹资
~ instrument 债务证券
~ interest 欠息
~ management 债务管理
~ memo 借项凭单
~ of honour 信用借款
~ paying ability 偿债能力

~ relief 债务免除
~ remain 剩余债款
~ service funds 债务支付基金
debtee [de'ti:] *n.* 债权人,债主
debtor ['detə] *n.* 债户,借方
~ -creditor relationship 借贷关系
debug [di:'bʌg] *vt.* 排除错误[故障];调试,调整
debugging *n.* 故障的排除
~ expense 折卸搬迁费用
deburring [di'bə:riŋ] *n.* 修边,除去毛刺
~ machine 修边机
~ tool 修边工具
decade ['dekeid] *n.* 由十所构成的一组;十进制,十年
decadent ['dekədənt] *a.* 衰落的
decagon ['dekəgən] *n.* 十边形,十面体
decahedron [ˌdekə'hi:drən] *n.* 十面体
decametre ['dekəmi:tə] *n.* 十米
decanal [ki'deinl] *a.* 教堂内坛南侧的
~ side 教堂内坛南面
decantation [ˌdi:kæn'teiʃən] *n.* 倾注;沉淀分取法
~ method 沉淀法
decarbonize [ki:-'dɑ:bənaiz] *vt.* 除去碳素
~d steel 低碳钢
decarburizer [di:-'kɑ:bjuraizə] *n.* 脱[除]碳剂

decastyle

decastyle ['dekəstail] *n.* 十柱柱列式,十柱式建筑;十柱式柱廊; *a.* 十柱式的
~ temple 十柱式庙宇
decauville [di:'kɔ:vil] *n.* 轻便铁路,窄轨铁路
~ railway 轻便铁路
~ truck 轻便轨道车
~ tub 斗车,手推车
~ wagon 斗车,手推车
decay [di'kei] *v.*; *n.* 衰落,衰减;衰变;腐烂

~ area 朽坏区
~ factor 衰减因素
~ time 衰减时间
~ed knot (木材的)朽节
deceleration [di:ˌselə'reiʃən] *n.* 减速
decenter [di:'sentə] *v.* 拆卸拱架,拆卸模架;(使)离中心
decentralization [di:ˌsentrəlai'zeiʃən] *n.* 分权化;分散经营
dechlorination *n.* 脱[去]氯
~ agent 脱氯剂
dechromisation *n.* 去[除]铬
decibel ['desibel] *n.* =dB 分贝(音强单位)
~ above reference noise 超过基准噪声的分贝数
~ -meter 分贝计[表]
~ -loss 分贝衰减
deciduous [di:'sidjuəs] *a.* 每年落叶的
~ tree 落叶树
~ wood 落叶树木材
decimeter ['desiˌmi:tə] *n.* 分米
decision [di'siʒən] *n.* 决定;决策
~ analysis 决策分析
~ center 决策中心
~ criteria 决策准则
~ design 优选设计
~ element 判定元素
~ evaluation 决策估计
~ maker 决策者
~ making body 决策机构
~ making process 决策程序
~ model 决策模型
~ tree 决策图表
deck [dek] *n.* 房屋楼层;图书馆多层书库的地板;坝面;桥面;上承;甲板
~ beam 上承梁;顶棚梁;上承式架;甲板梁
~ cant 撑木;平屋顶泛水
~ cantilever 上承式悬臂桥
~ chair 折叠式躺椅
~ curb 屋面缘栏

~ dormer 平屋顶无窗
~ floor 平楼盖；平台甲板
~ -on-hip 四坡屋顶上的平屋顶
~ plate girder 上承板梁
~ post 顶棚柱
~ rail 顶棚纵向梁
~ roof 平台式屋顶；晒台
~ sash 顶棚窗
~ slab 上承板
~ soffit 上承拱腹
~ structure 上承结构
~ truss 上承式桁架
~ type 上承式
roof ~ 屋顶层面
decken structure *n.* 叠瓦构造
decker ['dekə] *n.* 出租房屋；层次结构
decking ['dekiŋ] *n.* 铺面；盖板；支撑板；起重机台面
~ system 屋顶下部结构系统
declaration [deklə'reiʃən] *n.* （海关的）申报，公告，声明（书）
~ form 申报单
customs ~ 报关单
joint ~ 联合声明
declared [di'klɛəd] *a.* 呈[申]报的价格表记的
~ value 申报价格
declination [dekli'neiʃən] *n.* 倾斜；偏差；偏角
declinator ['deklineitə] *n.* 偏差器
declinometer [,dekli'nɔmitə] *n.* 磁偏计；方位计，测斜仪
declivity [di'kliviti] *n.* (*pl.* -ties.) 倾斜；倾斜面；下斜坡，坡度，梯度
decoat *v.* 除去涂层
decoherence [,di:kəu'hiərəns] *n.* 脱散，散屑
decoherer [di:kəu'hiərə] *n.* 散屑器
decolorimeter *n.* 脱色计
decolorization [di:,kʌlərai'zeiʃən] *n.* 脱[褪]色(作用)，漂白[作用]
decolourizer = decoluriser *n.* 脱色剂，漂白剂

decomposed [,di:kəm'pəuzd] *a.* 已分解的；已风化的
~ granite 风化花岗岩
~ rock 风化岩
decompression [,di:kəm'preʃən] *n.* 减压，降压
~ chamber 减压室
decompressor [di:kəm'presə] *n.* 减压器，失压装置
decontaminant *n.* 纯化剂
decontamination ['di:kən,tæmi'neiʃən] *n.* 净化；清污；消毒
~ facility 净污设备
~ factor 净化系数
~ shower 洁净吹淋(装置)
Deco-Polymer flooring 以环氧树脂为底的叠合楼面
decor ['dei'kɔ:] *n.* 室内装饰布置；舞台美术；布景装置
Decorated ['dekəreitid] *n.* （在爱德华一世至三世期间流行的）英国中世纪建筑
decorated ['dekəreitid] *a.* 装饰的，增色的，装潢的
~ arch 装饰拱
~ architecture 盛饰建筑，尖拱式建筑
~ archivolt 拱门饰
~ area 装饰范围[面积]
~ ceiling 装饰顶棚
~ style 盛饰建筑风格；哥特式建筑形式
~ surface 装饰表面
~ tile 装饰瓦
~ with battlements 装饰雉堞墙
decorating *n.* 装饰，装潢
~ art 装饰，[潢，修]艺术
~ mortar 装饰砂浆
decoration [,dekə'reiʃən] *n.* 装饰，装潢
architectural ~ 建筑装饰
interior ~ 内部装饰，内部装潢
stencilled ~ 印花装饰

decorative ['dekərətiv] *a.* 装饰的，装潢的
~ acoustical gypsum pan(el) slab 吸声石膏装饰板
~ appearance 外观装饰
~ arch 装饰拱
~ architecture 装饰建筑
~ area 装饰范围[面积]
~ art 装饰艺术
~ artificial stone 人造装饰石
~ band 装饰带条
~ barrel vault 装饰筒拱
~ board 装饰板
~ bond 装饰砌合
~ bracket 装饰托座
~ brick 装饰砖
~ ceiling 装饰顶棚

decorative ceiling

~ ceiling board (=ceiling sheet) 装饰顶棚
~ coating 表面装饰，装饰涂层
~ column 装饰柱
~ composition floor(ing) 无接缝装饰地板
~ concrete 装饰用混凝土
~ door 装饰门
~ element 装饰元素
~ embossment 装饰浮雕
~ feature 装饰特征
~ finish 面层装饰
~ fittings 装饰小五金
~ fixture 装饰[修]
~ floor cover(ing) 地面装饰
~ foil 叶形装饰
~ form 装饰形式
~ gable 装饰山墙
~ glass 有装饰图案的玻璃
~ glass block 装饰玻璃砖
~ glass worker 装饰玻璃工
~ glued laminated wood 装饰胶合板
~ grille 装饰格栅
~ hardware 装饰小五金
~ in-siu floor(ing) 无接缝装饰地板
~ iron 装饰铁件
~ ironwork 装饰铁制品
~ joint 装饰节点
~ jointless floor(ing) 无接缝装饰地板
~ light fitting 装饰照明设施
~ link 装饰链环
~ metal 装饰金属
~ motif 装饰主题[特色]
~ nail 装饰钉
~ niche 装饰壁龛
~ paint 装饰用油漆
~ panel 装饰镶板
~ period 盛饰时代
~ plastic board 装饰塑料板
~ quality 装饰能力
~ structure 装饰结构
~ style 装饰风格
~ touch 装饰格调
~ trim 装饰性修剪
~ turret 装饰用小塔
~ window 装饰窗
~ window frame 装饰窗框
~ window curtain 装饰窗帘
~ window screening 装饰窗纱
~ work 装饰[修，潢]工作

decorator *n.* 装饰家，制景人员
decorticator *n.* 剥皮[脱壳]机
decreasing [di:kri:siŋ] *a.* 减少的，渐减的
~ amplitude 降[减]幅
~ cost 成本递减
~ vibration 衰减振动
decree [di'kri:] *n.* 命令；法令[规]；布告 *v.* 颁令；判定

emergency ~ 安全技术规程
decrement ['dekrimənt] *n.* 减少;消耗;衰减量
~ curve 减幅曲线
~ field 变址字段,减量字段
decrustation [ˌdi:krʌs'teiʃən] *n.* 脱皮[壳]
decrusting ['di:krʌstiŋ] *n.* 刮洗
decurved [di'kə:vd] *a.* (弧形)向下弯的
dedans【法】*n.* 网球看台
dedust [di:'dʌst] *vt.* 除尘[灰,末],脱尘
deduster *n.* 除尘器
deed [di:d] *n.* 议定书,契据
~ box 文件保险箱,契约箱
~ of purchase 买契
~ of sale 卖契
~ of security 保证书
~ of trust 信托契据
deem [di:m] *vt.* 认为,视为
~ed paid credit 信得过信贷
deenergize [di:'enədʒaiz] = deenergise *vt.* 去能,去能源;切断电路
deep [di:p] *a.* 深的;深色的
~ beam 厚梁,主梁
~ beam slab 厚梁板
~ bead 深[厚]压条
~ borehole 深孔
~ cleaning 深度净化
~ coloured gold 赤金
~ compaction 深度压实
~ consolidation 深度加固
~ -cutting 深刻,深切削
~ (drill) hole 深(钻)孔
~ engraving 深雕刻
~ etch test 深蚀试验
~ fillet welding 深角焊
~ floor 加强肋板
~ foundation 深基础
~ frame 深肋骨,加强肋骨
~ girder 深梁
~ grouting 深孔灌(水泥)浆
~ hole 深孔
~ -hole blasting 深孔爆破
~ hole drill 深孔钻头
~ level 深层
~ lift 深层
~ pier 深基础柱[墩]
~ -ribbed slab 深肋 T 形梁板
~ seam 凹缝(焊管缺陷)
~ -seated 深嵌的,深埋的,深成的
~ seepage 深层渗漏
~ shadow 投影
~ sheet 深冲薄板
~ slab 厚板
~ -slot 深槽的
~ socket wrench 长套管型套筒扳手
deepen ['di:pən] *v.* 加深,变深;强烈
~ed beam 加深[厚]梁
deepening ['di:pəniŋ] *n.* 加深,延深,(港口,河道等)挖深,疏浚
~ stage 加深阶段
deepfreeze ['di:pfri:z] *vt.* 使冷藏 *n.* 深冻冷藏室[库,箱],电冰箱
deface [di'feis] *vt.* (表面)磨损,损伤,销毁,毁坏
defeasance [di'fi:zəns] *n.* 作废,解除(契约)
defeature [di'fi:tʃə] *vt.* 损坏外形;使变形,使不能辨认
defecate ['defəeit] *v.* 澄清;去污,净化
defecator ['defikeitə] *n.* 澄清器[槽],过滤装置
defect [di'fekt] *n.* 缺点;缺陷;亏损;故障
~ detector 探伤仪
~ detecting test 缺陷检查
~ in timber 木材缺陷
defective [di'fektiv] *a.* 有缺点的;有缺陷的 *n.* 次品
~ goods 次品[货]
~ plumbing 有缺陷的卫生设备系统
~ rightness 不紧密
~ water 不纯的水
~ wood 缺陷(木)材
defence [di'fens] = defense【美】*n.*

防卸[护];防卸工程,堡垒
~ parapet 胸墙
~ wall 防护墙
anti-aircraft ~ 防空设施

defensible [di'fensəbl] *a.* 可防范的
~ space（用建筑处理来加强邻里联系、防止偷盗和对付犯罪的）可防范空间

defensive [di'fensiv] *n.*;*a.* 防卸物;防卸（用的）;自卫（的）,守势（的）,辩护（的）
~ gateway 雕堡式城门
~ hedge 防护围墙,篱笆
~ (masonry) wall 雉堞墙

deferment [di'fə:mənt] *n.* 延期,延迟
~ charge 逾期费
~ delivery 推迟交货

deferent ['defərənt] *n.*;*a.* 传送物（的）,导管;圆心轨迹

deferrization *n.* 除[脱]铁

defibrator *n.* 碎木机,木料碾碎机

deficit ['defisit] *n.* 不足（额）,赤字,亏损
~ financing 赤字财政
~ statement 亏损表

defilade [ˌdefi'leid] *vt.*;*n.* 掩蔽;障碍物

defile [di'fail] *v.* 玷污;分行列 *n.* 小路,狭道

definition [ˌdefi'niʃən] *n.* 确定;定义
~ of term 条款解说
~ phase 技术设计阶段,初步设计阶段,技术经济条件确定阶段

deflagrate ['defləgreit] *v.* 爆燃;燃烧

deflate [di'fleit] *vt.* 排气;抽气,减压
~ing valve 放气阀[嘴]

deflation [di'fleiʃən] *n.* 抽去空气;跑气;（通货）紧缩,压缩,紧缩
~ opening 排气孔
~ valve 放[排]气阀

deflect [di'flekt] *v.* （使）偏斜;（使）偏离;（使）转向;折射
~ angle 偏移角,变位角

~ing plate 遮护板;隔板
~ing wedge 偏心楔

deflected [di'flektid] *a.* 偏离的;转向的
~ pile 偏位桩

deflection [di'flekʃən] = deflexion *n.* 挠度;变位;偏转;折射
~ angle 偏转角;变位角
~ measurement 挠度测定
~ of beam 梁的挠度
~ point 变位点
~ test 弯曲试验
column ~ 柱的纵向弯曲

deflectometer [diflek'tɔmitə] *n.* 弯度计,挠度计

deflector [di'flektə] *n.* 折射器;转向器,偏转板,折流板
~ plate 偏转板
draft ~ 挡风器[板]

deflexion [di'flekʃən] *n.* = deflection 挠度;变位;偏转

defloccular [di'flɔkjulənt] *n.* 反絮凝剂,散凝剂;黏土悬浮剂;胶体稳定剂

deflocculating agent 反絮凝剂,散凝剂;黏土悬浮剂;胶体稳定剂

defoam [di'fəum] *v.* 消泡,除沫
~ing agent 消泡剂,抗泡剂

defoamant [di'fəumənt] *n.* 消[去]泡剂

deforest [diː'fɔrist] *vt.* 砍伐的森林

deform [diː'fɔːm] *v.* 使变形;使不成形

deformability [diˌfɔːə'biliti] *n.* 变形性;形变能力

deformable [di'fɔːməbl] *a.* 可变形的,应变的
~ bar 异形钢筋,螺纹钢筋,竹节钢筋
~ body 柔[变形]体
~ damper 变形阻尼器

deformation [ˌdiːfɔː'meiʃən] *n.* 变形;毁坏
~ action 变形作用
~ amplitude 变形幅度
~ angle 变形角

deformed

~ effect 变形效果
~ limit state 变形极限状态
~ method 形变法
~ process 变形过程
~ state 变形状态
~ structure 变形结构
~ zone 变形区
absolute ~ 绝对变形
affine ~ 均匀变形
areal ~ 表面变形
compressive ~ 受压变形
creep ~ 徐变
elastic ~ 弹性变形
lateral ~ 横向变形
load ~ 荷载变形
permanent ~ 永久变形
plastic ~ 塑性变形
shear ~ 剪切变形
tangential ~ 切向变形,(受)剪应变
tectonic ~ 构造变形
thermal ~ 热变形
time ~ 随时间而发生的变形
vertical ~ 竖向变形,沉陷

deformed [di'fɔ:md] *a.* 变形的,畸变的;
~ area 变形区
~ bars 各式[不同形式]钢筋
~ metal plate 变形金属板
~ plate 变形板,凹凸板
~ prestressing steel wire 异形预应力钢丝
~ reinforcing bar 异形[竹节]钢筋
~ reinforcing bar with spiral ribs 螺纹钢筋
~ steel bar 螺纹钢筋

deformeter [di'fɔ:mitə] *n.* 变形计,应变仪,变形测定器

defroster [di:'frɔstə] *n.* 除霜器
hot air heater ~ 暖气除霜器

defrosting [di'frɔstiŋ] *n.* 解冻,除霜

degasify [di:'gæsifai] *vt.* = degas. 除气
degasified steel 镇静钢
degasifying agent 除气剂

degasser [di'gæsə] *n.* 除气器,除气剂,脱气装置

degauss [di:'gæus] *vt.* 去[消]磁
~ing cable 消磁电缆

degausser *n.* 去磁器;去磁电路

degeneration [di,dʒenə'reiʃən] *n.* 退化;降级;衰变
noise ~ 噪声衰减

degradation [,degrə'deiʃən] *n.* 破裂;逆降分解;退降;剥蚀
~ level 破裂平面;剥蚀层

degraded [di'greidid] *a.* 退化的;剥蚀的

degree [di'gri:] *n.* 程度;等级;度数;学位
~ -day 度-日(供暖计算用单位)
~ of accuracy 精确度
~ of cementation 胶结度
~ of centigrade 摄氏度数
~ of clarification 净化程度
~ of compaction 密实度,压实度
~ of consolidation 固结度
~ of convexity 凸度
~ of curvature 弯曲度
~ of density 密实度
~ of distortion 变形程度
~ of durability 耐久度
~ of expansion 膨胀度
~ of Fahrenheit 华氏度数
~ of fastness 坚牢程度
~ of fineness 细度
~ of finish 光洁度
~ of flexibility 挠曲程度
~ of freedom 自由度
~ of hardness 硬度
~ of inclination 倾斜度
~ of irregularity 不平整度
~ of latitude 纬度
~ of longitude 经度
~ of mattness 粗糙程度
~ of mechanisation 机械化程度
~ of moisture 湿度,水分

~ of porosity 孔隙度
~ of prestress 预应力度
~ of pulverization 粉碎程度
~ of purification 净化度
~ of reliability 可靠程度
~ of resistance 耐力程度
~ of reverberation 混响度
~ of roughness 粗糙程度
~ of roundness 圆度
~ of rusting 生锈程度
~ of safety 安全(程)度
~ of saturation 饱和度
~ of security 安全程度
~ of sensitivity 灵敏度
~ of shrinkage 收缩度
~ of size reduction 破碎度
~ of slope 坡度
~ of stability 稳定程度
~ of tightness 密封度
~ of torsion 扭转度
~ of uniformity 均匀度
~ of vacuum 真空度
~ of water-resistance 防水程度
~ of weathering 风化度
~ scale 刻度
~ visual 视度

degum [di:'gʌm] *vt.* 使脱胶,使去胶
dehumidification ['di:hju:ˌmidifi'keiʃən] *n.* 减湿(作用),除湿(作用);干燥,脱水
~ system 干燥[除湿]系统
dehumidifier [ˌdi:hju(:)'midifaiə] *n.* 减湿[干燥]器
dehumidify [di:hju'midifai] *vt.* 减湿,干燥,脱水
~ing capacity 减湿能力
dehydrate [di:haidreit] *v.*;脱[去]水,干燥
~d lime 生石灰,氧化钙
dehydrating agent 脱水剂
dehydration [ˌdi:hai'dreiʃən] *n.* 脱水,去湿,干燥
~ test 脱水试验

dehydrator [di:hai'dreitə] *n.* 脱水剂,脱水器,烘干机
dehydrolysis *n.* 脱水(作用)
deice ['di:'ais] *vt.* 防冻,除冰
deicing ['di:'aisiŋ] *n.* 防冻;除冰
~ agent 防冻[除冰]剂
~ liquid 除[去]冰液
~ salt (道路)防冻盐类
~ salt solution 防冻盐溶液
~ sealant 防冻填封料
~ work 防冻工作
delamination [ˌdi:ˌlæmə'neiʃən] *n.* 分层,剥[层]离,起鳞,裂为薄层
delanium graphite 人造石墨,高纯度压缩石墨
delay [di'lei] *v.*;*n.* 耽搁;延缓;延期
~ action 延迟作用,滞后作用
~ deformation 延迟形变
~ distortion 延迟偏差
~ drop 平稳下降,缓慢下降
~ mixing 延迟拌和
~ time 延时,滞后时间
~ed elastic deformation 滞弹性变形
~ed elasticity 滞弹性
~ed finish 延迟整修处理;延迟完成
deleterious [ˌdeli'tiəriəs] *a.* 有害的;有毒的
~ material 有害材料
~ matter 有害物质
~ substance 有害物质
delf [delf] *n.* 排水器;出水沟;管道采石场;彩色陶器;薄矿层
delicate ['delikeit] *a.* 精美的;细致的,微妙的;敏感的
~ adjustment 精密调整
~ colours 淡色,柔和的颜色
~ comprehension of art 对艺术有敏感的鉴赏力
~ operation 谨慎操作
delignification *n.* 去木质作用,水致侵蚀作用
delime [ki'laim] *vt.* 脱灰
delimit [di:'limit] *vt.* 定界;划界;确

[限]定
delineascope [diːˈliniˌeskəup] *n.* 幻灯;映画器
delineation [diˌlinˈeiʃən] *n.* 描绘;记述;轮廓,草图,示意图
~ marking 划线标志
delineator [diˈlinieitə] *n.* 绘[描]图者;描画器
deliquesce [ˌdiliˈkwes] *v.* 溶化,融解;潮解;冲淡,稀释
deliquescent [ˌdeliˈkwesnt] *a.* 吸湿的,易潮解的
deliver [diˈlivə] *vt.* 递送;交付
~ed heat 供热,放热
delivery [diˈlivəri] *n.* (*pl.* -eries) 递送;交货;释放;排水量,流量
~ and shipment 交货与装运
~ book 交[送]货簿
~ box 放水箱
~ capacity 排水量
~ cock 泄放旋塞,出水龙头
~ conduit 输送(管,渠)道
~ date 交货期
~ end 出料端
~ failure 交货误期
~ gate 斗门;放水门;供料门,出水口
~ head (供水)水头,扬程
~ hose 输水软管
~ lift 压升高度,扬程
~ manifold 供水管线
~ mechanism 输送装置
~ on arrival 货到即提
~ on field 就地交货
~ on term 定期交货
~ order 提货单,出库凭单
~ orifice 送气孔;出水孔;输出孔口
~ pipe 供[输]水管
~ station 发货站
~ term 交货期限
~ time 交付时间
~ track 供料支线;输送轨道
~ tunnel 输送隧道
~ value 排气阀;放水阀

dell [del] *n.* 出水沟;幽谷,谷地
delta [ˈdeltə] *n.* 三角洲;三角形
~ channel 水道
~ connection 三角形接法
~ matching 三角形搭接
deltaic [delˈteiik] *a.* 三角洲的;三角形的
~ region 三角形地带
delubrum [dəˈluːbrəm] *n.* 古代罗马的寺庙
deluge [ˈdeljuːdʒ] *n.;v.* 大洪水,泛滥,(倾盆)大雨,暴雨
~ collection pond 蓄水池
~ system 集水系统
delusterant [diːˈlʌstərənt] *n.* 褪[消]光剂
delustre [diːˈlʌstə] *vt.* 去光泽,褪光
deluxe [dəˈluks] 【法】*a.* 华美的,华丽的;质量高的; *ad.* 豪华地
demand [diˈmɑːnd] *n.;vt.* 需要;要求量
~ analysis 需求分析
~ and supply 需求与供应,供求
~ competition 需求竞争
~ curve 需求曲线
~ element 需求项目
~ expansibility 需求扩张
~ factor 需求因素
~ forecast 需求预测
~ goods 需求货物
~ hydrograph 需用(水,电)量过程线
~ inflation 需求膨胀
~ price 需求价格
~ schedule 需求一览表
firm ~ 固定负荷
maximum ~ 最大负荷
demarcation [ˌdiːmɑːˈkeiʃən] *n.* 边[分]界
demask [diˈmɑːsk] *v.* 暴露,解掩蔽
demeane [diˈmein] *n.* (土地的)所有,领地;范围
demi-bath(tub) 坐浴浴盆
demi-column 嵌墙柱,半柱,壁柱

demi-relievo 浅浮雕
demi-section 半剖面(图),半节[段]
demineralization *n.* 脱矿质(作用);软化;除盐
demineralizer [ˌdiːˈmɪnərəlaɪzə] *n.* (水)软化器
demise [diˈmaɪz] *n.*; *vt.* 转让,租让
　～d premises 转让房地产
demist [diːˈmɪst] *vt.* 为……除霜[雾]
demister [diːˈmɪstə] *n.* 除[去]霜[雾]器
demitint [ˈdemitint] *n.* 中间色调
demolish [diˈmɔliʃ] *vt.* 毁坏,破坏;拆毁[除]
demolishable *a.* 可拆毁的;适宜拆毁的
demolisher *n.* 承包拆屋者
demolition [ˌdeməˈliʃən] *n.* = demolishment 破坏,毁坏;拆毁[除]
　～ contract 拆毁合同
　～ hammer 冲锤
　～ permission 拆毁许可
　～ project 拆毁方案
　～ scheme 拆毁方案
　～ site 拆毁地点
　～ tool 捣碎器
　～ work 拆毁工作
demonstration [ˌdemənsˈtreiʃən] *n.* 示范;论证;图解;举例说明;实验,表演
　～ building 样板建筑
　～ project 示范项目
demoulding [diːˈmould] *v.* 拆[脱]模
　～ing agent 脱模剂
demountable [diːˈmauntəbl] *a.* 可卸下的,可拆开的
　～ building 可拆卸建筑物
　～ connection 可拆卸接点
　～ partition 灵活隔断
　～ division wall 活动隔墙
demulsibility [diːˌmʌlsiˈbiliti] *n.* 反[抗]乳化度[性,率]

demi-columns

den [den] *n.* 休息室;小储藏室;书房;洞窟
denacol *n.* 一种环氧树脂
denary [ˈdiːnəri] *a.* 十的;十(倍,进)的
dendritic [denˈdritik] *a.* 树枝状的
　～ drainage 树枝形排水系统
　～ structure 树状结构
denote [diˈnəut] *vt.* 指[表]示;代表,意味着;概述
dense [dens] *a.* 密实的,密级配的;稠密的;浓厚的
　～ aggregate 密实集料[骨料]
　～ concrete 密实混凝土
　～ -graded 密级配的
　～ grain 密纹
　～ layer 密实层
　～ sand 密实砂
　～ structure 密实结构
　～ wood 密纹木材
densely [ˈdensli] *ad.* 密集地,稠密
　～ inhabited district 人口密集区
densification [ˌdensifiˈkeiʃən] *n.* 压实;人工加固;稠化(作用)
densifier *n.* 增密器,密化器,浓缩器;脱水机;增浓剂
densify [ˈdensifai] *vt.* (使)增加密度;(使)硬化;稠化
densified laminated wood 硬化层压木板
densified wood 高密度木材
densifing agent 防潮剂
densimeter [denˈsimitə] *n.* 密度[比重]计
density [ˈdensiti] *n.* (*pl.* -ties)密度;浓度;稠度;密实
　～ controller 密度控制器
　～ index 相对密度,密度指数
　～ of building 建筑密度
　～ of load 荷载密集度
　～ of occupancy 居住密度
　～ of population 人口密度
　～ test 密度试验
dent [dent] *n.* 凹口[痕] *v.* 弄成缺口;

使成凹痕

dental [dentl] *a.* 牙齿的 *n.* 齿形挡板

dentate ['denteit] *a.* (锯)齿(状)的,有齿的
~ sill 齿形槛

denticle [dentik] *n.* 齿形装饰;小齿状突起
~ frieze 齿形雕带

denticular [den'tikjulə] *a.* 齿形装饰的,小齿形的
~ cornice 齿形装饰的挑檐

denticulation [den,tikju'leiʃən] *n.* 齿形装饰,小牙饰
~ corona 齿饰挑檐滴水板

dentil ['dentil] *n.* 齿形装饰
~ band 齿饰带
~ cornice 齿形挑檐
~ course 齿形突腰线;齿形挑砖砌层
~ frieze 齿饰中楣

dentist ['dentist] *n.* 牙科医生

denudation [,di:nju(:)'deiʃən] *n.* 剥光,裸露,剥蚀(作用),去垢,滥伐
~ level 剥蚀表面

deodar ['diəuda:] *n.* (产于喜马拉雅山的)雪松;雪松木材
~ cedar 雪松木材

deodorant [di:'əudərənt] *a.* 除[防]臭的 *n.* 除臭剂

deodorization [di:,əudərai'zeiʃən] *n.* 除臭(作用,过程)

deodorise (= deodorize) [di:'əudəraiz] *vt.* 脱[除]去臭气,去臭
deodorizing material 除臭材料

deodorizer [di:'əudəraizə] *n.* 除臭剂[器]

deodorizing *n.* 除[防]臭
~ material 除臭材料

deoscillator [di:'ɔsileitə] *n.* 减振[阻尼]器

deoxidation [di:,ɔksi'deiʃən] *n.* 脱氧,还原,除[脱]酸

deoxidizer [di:'ɔksidaizə] *n.* 还原剂,去氧剂

department [di'pɑ:tmənt] *n.* 部分[门];系;科;所
~ in charge 主管部门
~ of architecture 建筑系[部门]
~ of public works 市政工程局
drafting ~ 设计科
engineering ~ 工程部门,技术科

departure [di'pɑ:tʃə] *n.* 离开[去];横距;偏转[差]
~ building 候机楼
~ hall 候机厅
~ platform 出发站台
~ point 出发点,开端

dependability [di,pendə'biliti] *n.* 可靠性,强度,坚固度

dependable [di'pendəbl] *a.* 可靠的
~ capacity 可靠容量

dependency [di'pendənsi] *n.* = dependency 建筑物的一翼;附属物

dependent [di'pendənt] *a.* 依赖[靠]的;关连的
~ building 附属建筑物
~ error 相关误差
~ event 相关事件
~ type 连接式

depeter ['depitə] *n.* (墙壁的)碎石面饰(灰泥未干时,压碎石人内的一种墙壁面饰法)

dephlegmate [di:'flegmeit] *v.* 分馏,局部冷凝

dephlogistication ['di:flə'dʒistikeiʃən] *n.* 脱燃素(作用)

depicture [di'piktʃə] *vt.* 描(绘)述;想像

depingment [di:'pigmənt] *v.* 去[除]色

depiler *n.* 装[进]料台,分送[垛]机

depinker *n.* 抗爆剂

depletion [di'pli:ʃən] *n.* 耗损,亏损;枯竭;降低
~ cost 耗竭成本
~ rate 折旧率;亏耗率

deposit [di'pɔzit] *n.* 沉淀物;存款;矿床;保管处; *v.* 涂,覆,浇筑;放置

~ account 存款账户
~ concrete 浇筑混凝土
~ fill material 放置填充材料
~ for security 保证金
~ on contracts 合同保证金

deposition [ˌdepəˈziʃən] ***n.*** 沉(淀,积); 沉淀物,证据
~ process 沉积过程

depositional ***a.*** 沉积的
~ interface 沉积界面

depository [diˈpɔzitəri] ***n.*** (*pl.* -ries) 仓库; 储藏室

depot [ˈdepəu] ***n.*** 机车库; 仓库; 车站; 航空站
loop ~ 环道车站

depressed [diˈprest] ***a.*** 压下的,降低的,凹陷的
~ area 低[洼]地; 萧条区
~ arch 平坦拱
~ panel form 凸边格形模板

depression [diˈpreʃən] ***n.*** 降低; 凹地; 沉降地; 沉降
~ head 降落水头
~ of ground 地面沉降
~ of order 降阶法
~ of support 支座沉陷

depreter [ˈdepritə] ***n.*** = depeter 粉饰凿面

depth [depθ] ***n.*** 深[高,厚]度; 深处; 深奥; 浓厚
~ contours 等深线
~ -integrating sediment sample 随深度取的砂样
~ of a room (hall) 房间(厅)的进深
~ of arch 拱的厚度
~ of beam 梁的高度
~ of burying 埋地深度
~ of camber 上拱高
~ of colour saturation 色彩饱和度
~ of compaction 压实深度
~ of cover 覆盖深度
~ of erosion 侵蚀深度
~ of field 景深,视场深度
~ of fill 填土高度
~ of foundation 基础理置深度
~ of groove 槽深
~ of impression 印痕深度
~ of passage way 走廊进深
~ of pavement 铺面厚度
~ of vault 穹顶厚度
~ -width ratio 高宽比
wearing ~ 磨损深度

depuration [ˌdepjuˈreiʃən] ***n.*** 净化; 提纯

depurator [ˈdepjuretə] ***n.*** 净化器[剂,装置]

derby [ˈdɑːbi] ***n.*** 刮尺; 金属块,块状金属
~ float (粉刷用)整平刮尺

derivometer ***n.*** 测偏仪

dermateen ***n.*** 漆布,布质假皮

dermatine ***n.*** 人造皮革

derrick [ˈderik] ***n.*** 人字起重机; 起货桅杆
~ stone 粗[大]石块
stiff leg ~ 刚性柱架

derust [diˈrʌst] ***v.*** 除锈
~ agent 除锈剂

desalting [ˈdiːˈsɔːltiŋ] ***n.*** 除盐

desander ***n.*** 去砂器
mud ~ 泥浆去砂器

desaturation [diːˌsætʃəˈreiʃən] ***n.*** 稀释,冲淡; 减低色彩的饱和度

desaturator [diːˈsætʃəˈreeitə] ***n.*** 干燥器[剂]; 稀释剂

descale [diːˈskeil] ***v.*** 去壳,去氧化皮,除锈皮[垢],除鳞
~ rust 脱鳞锈

descending [diˈsendiŋ] ***a.*** 下降的; 下行的; 递降的
~ grade 下降坡度

descent [diˈsent] ***n.*** 除低; 沉陷; 下坡
spiral ~ 螺旋下降
steepest ~ 最陡下降

describe [disˈkraib] ***vt.*** 描绘; 绘图; 记叙; 作……运动

description [dis'kripʃən] *n.* 叙述；描写；图形；说明书
descriptive [dis'kriptiv] *a.* 描写的；叙述的；说明的
~ geometry 画法几何
~ process 描述性工艺过程
deseaming [di'si:miŋ] *n.* 气炬烧剥
desensitization [diːˌsensitai'zeiʃən] *n.* 脱敏（现象）；减感（作用）
desert ['dezət] *n.* 沙漠；荒地[漠] *a.* 沙漠的；荒芜的
~ concrete（用少量水泥作为胶结材料，大块地方粗石作为骨料的）混凝土
desiccant ['desikənt] *a.* 干燥的，去湿的 *n.* 干燥剂
desiccated ['desikeitid] *a.* 干燥的；脱水的
~ wood 干木材
desiccation [ˌdesi'keiʃən] *n.* 干燥[化]
~ crack 干裂缝
~ fissure 干缩裂缝
desiccator ['desikeitə] *n.* 干燥器[剂]，防潮砂
desicchlora *n.* 无水粒状高氯酸钡（干燥剂）
design [di'zain] *n.* 设计；计划；图案 *v.* 设计；预定
~ aids 设计工具；设计参考资料
~ altitude 设计高度
~ analysis 设计分析
~ and construct firm 设计与施工公司
~ approval drawing 批准的设计图
~ assumption 设计假定
~ bedding 图案花坛
~ capacity 设计能力，设计功率
~ change notice 设计更改通知
~ change summary 设计更改一览表
~ chart 设计图表
~ code 设计规范，计算机程序
~ concept change 设计概念的改变
~ consideration 设计依据
~ constant 设计常数
~ criteria 设计准则[规范]
~ data 设计数据，设计资料
~ data sheet 设计资料图表
~ detail 设计细节
~ department 设计部门
~ drawing 设计图，计划图
~ elevation 设计标高
~ engineer 设计工程师
~ example 样板设计
~ factor 设计因素
~ feature 设计特征
~ flow 设计流量
~ handbook 设计手册
~ head 设计水头
~ hypothesis 设计假定
~ layout 设计布置
~ life 设计使用周期，设计寿命
~ limit load 设计极限荷载
~ load 设计荷载
~ mannual 设计手册
~ memorandum 设计记录
~ method 设计方法
~ mix(ture) 配合比设计
~ moment 设计弯矩
~ of mixture 配料设计
~ of pipe system 管网设计
~ office 设计部门（室）
~ paper 绘图纸
~ parameter 设计参数
~ period 设计期限
~ philosophy 设计原理[特点]
~ principles 设计原则[原理]
~ procedures 设计程序[步骤]
~ proposal（~ scheme) 设计方案
~ requirement drawing 设计要求的图纸
~ review 设计审查
~ right 设计权
~ rough plan 设计草图
~ rule 设计规则
~ seismic force 设计地震力
~ scheme 设计方案

~ section 设计组
~ sight distance 设计视距
~ specifications 设计规范[说明书]
~ stage 设计阶段
~ standard manual 设计标准手册
~ strength 设计强度
~ stress 设计应力
~ system 设计系统
~ supplement 设计补充文件
~ table 设计图表
~ task 设计任务
~ test 鉴定试验
~ value 设计值
~ verification 设计验证
acoustical ~ 音质设计
alternate ~ 比较设计
applied ~ 贴花
architectural ~ 建筑设计
art of ~ 装饰艺术,美术
artistic ~ 美术设计
carved ~ 刻花
ceiling ~ 顶棚图案
decorated ~ 装饰设计
empirical ~ 经验设计
modular ~ 标准设计
original ~ 原始设计
preliminary ~ 初步设计
process ~ 工艺流程设计
sample ~ 样品设计
structural ~ 结构设计
technical ~ 技术设计
urban ~ 城市设计
designed [di'zaind] *a.* 计划的;设计的
~ load 设计荷载
~ period 设计期限
~ reliability 设计可靠[耐久]性
designer [di'zainə] *n.* 设计者;制图者
~ of formal gardens and parks 园林建筑师
desilter [di:'siltə] *n.* 沉淀[滤水]池
desire [di'zaiə] *n.;v.* 要求,需要;愿[希,期]望

~d length 要求长度
~d speed 预定速度,理想速度
~d thickness 要求厚度
desk [desk] *n.* 台,桌;面板;实验台;说教坛
~ calculator 台式计算器
~ computer 台式计算机
drawing ~ 绘图桌
writing ~ 写字台
deslicking [di:'slikiŋ] *n.* 防滑
~ material 防滑材料
~ (ceramic) tile 防滑瓷砖
~ treatment 防滑处理
desludger *n.* 除泥[去垢]装置
desorption [di:'sɔ:pʃən] *n.* 解吸附[吸收]作用
despatch [dis'pætʃ] =dispatch *n.;v.* 派遣,发送
~ room 分发[调度]室
desterilize [di:'sterilaiz] *vt.* 恢复使用,解封
destination [ˌdesti'neiʃən] *n.* 目的地,终点;目的,目标
~ board 指路牌
~ sign 目的地指示标志
destruction [dis'trʌkʃən] *n.* 毁坏[灭];拆毁;破坏
~ of bond 粘结破坏
~ test 破坏性试验
destructive [dis'trʌktiv] *a.* 破坏的,毁灭的;危害的
~ effect 破坏作用
~ power 破坏力
detachable [di'tætʃəbl] *a.* 可分(开,离)的
~ device 可拆装置
~ shoe 活动垫座
detached [di'tætʃt] *a.* 分(开,离)的;独立的;已拆卸的
~ building 独立式房屋
~ column 独立柱
~ house 独立式住宅
~ palace 离宫

~ statuary 独立雕像

detachment [di'tætʃmənt] *n.* 分(开,离)

detail ['di:teil] *n.* 细节[部];详图;条款;复杂而精细的装饰 *vt.* 画明细图,细部设计
- ~ account 明细账
- ~ assembly template (DAT) 细部装配样板
- ~ card 细目卡片
- ~ construction 结构详图,细部构造
- ~ drawing 大样[细部]图,详图,零件图
- ~ file 说明资料
- ~ ledger 明细分类账
- ~ of construction 施工[构造]详图
- ~ of design 设计图样
- ~ of facade 立面细节
- ~ paper 大样[底]图纸
- ~ point 碎部点
- ~ specification 详细规格[说明书]
- ~ survey 细[碎]部测量

detailed ['di:teild] *a.* 详[明]细的
- ~ analysis 详细分析
- ~ audit 详细审计
- ~ budget 详细预算
- ~ design 细部设计
- ~ drawing 详图
- ~ investigation 详细勘察,详细调查
- ~ section 细部截面

detailer *n.* 大样[细部]设计员

detection [di'tekʃən] *n.* 检波;探测;检[侦]查

detector [di'tektə] *n.* 探测器;检测器;检波器
- ~ lamp 检漏灯
- earth ~ 接地指示器
- flaw ~ 探伤仪

detent [di'tent] *n.* 插锁,门扣,扳手;停止;稳定装置
- ~ catch 锁闩

detention [di'tenʃən] *n.* 阻止;延迟;滞留,停滞
- ~ period 滞留时间

detergent [di'fə:dʒənt] *n.* 清洁[洗涤,去垢]剂

determinant [di'tə:minənt] *n.* 判定因素;决定要素;行列式 *a.* 决定性的,有决定力的

determinate [di'tə:minit] *a.* 一定的,确定的;有限的
- ~ error 预计[定]误差
- ~ system 确定体系[系统]

determination [di,tə:mi'neiʃən] *n.* 测定;确定,判定

deterministic [di,tə:mi'nistik] *a.* 可定的;明确的;决定性的
- ~ analysis method 确定性的分析方法
- ~ model 确定型模型

detour ['deituə] *n.* 便道;绕行路,迂回路
- ~ arrow sign 迂回线指向标志
- ~ bridge 便道桥
- ~ marker 迂回线指示标
- ~ plan 绕行路计划
- ~ road 迂回路;便路
- ~ sign 迂回路标志

detrimental [,detri'mentl] *a.* 有害的,不利的

develop [di'veləp] *v.* 发展,开发;展开
- ~ resources 开发资源
- ~ed area 已开发土地
- ~ed dye 显色染料
- ~ed length 展开长度
- ~ed surface 展开面
- ~ed width 展开宽度
- ~ing agent 显影剂

development [di'veləpmənt] *n.* 发展,开拓,开发;生[成]长
- ~ area 发展区,建筑区
- ~ capital 开发资金
- ~ cost 开发费用,发展成本
- ~ district 开发区
- ~ expense 发展费用

~ framework 发展体系
~ of land 土地开发
~ plan(ning) 开发规划[设计]
~ program 开发方案
~ project 发展项目
structural ~ 结构改进
deviation [ˌdiːviˈeiʃən] *n.* 离题,偏差
~ angle 偏角
~ track 便道
deviator *n.* 偏差器;变向装置
device [diˈvais] *n.* 设备,装置;方法,手段
~ availability 设备利用率
~ for closing 闭锁装置
automatic call ~ 自动呼叫装置,自动报警设备
copying ~ 仿型装置
devil [ˈdevl] *n.* (铺沥青路面用的)加热器;切碎机;火炉
~ float 常钉抹子,木蟹
devise [diˈvaiz] *v.* 设计,发明,计划
deviser [diˈvaizə] *n.* 设计[发明]者;发生器
devitrification [diːˌvitrifiˈkeiʃən] *n.* 脱玻作用,失去光泽,透明消失,反玻璃化
~ of glass 玻璃闷光,反玻璃化
devitrify [diːˈvitrifai] *vt.* 使不透明;使无光泽
devetrified slag 失去光泽的碴
~ing solder 失透性焊剂
devitroceram *n.* 玻璃陶瓷
devoid [diˈvɔid] *a.* 缺乏的,空的
devolution [ˌdiːvəˈluːʃən] *n.* 转移[让],移交,授与;崩塌;滚落[下]下
dew [djuː] *n.* 露(水),湿润
~ point 露点
~ point depression 露点温差
~ point hygrometer 露点湿度计
~ point temperature 露点温度
~ pond 露池,(人工挖成的)蓄水池
dewater [diːˈwɔːtə] *v.* 脱水,浓缩,增稠
dewaterer *n.* 脱[除]水器
dewatering *a.* 脱[排]水的 *n.* 脱[排]水

~ agent 脱水剂
~ unit 脱水剂[器]
dewax [diːˈwæks] *vt.* 除蜡
~ing plant 除蜡设备
~ing with urea 尿素除蜡
dezincification [diːzɪŋsifiˈkeiʃən] *n.* 除锌作用,脱锌现象
D-handle shovel D形手柄铲
D-horizon *n.* D层(土),丁层(土),母岩层
diabase [ˈdaiəbeis] *n.* 辉绿岩
~ traprock 辉绿暗色岩
diabatic [ˌdaiəˈbætik] *a.* 非绝热的,透热的
diac *n.* 两端交流开关(元件)
diaclase *n.* (*pl.*)构造裂缝,压力裂缝
diaconicum *n.* 古代教堂中的圣器室;收藏室
diacrete *n.* 硅藻土混凝土
diacritical [ˌdaiəˈkritikəl] *a.* 区分的
~ sign 区别记号
diagenesis [ˌdaiəˈdʒenisis] *n.* 成岩作用,原状固结
diaglyph [ˈdaiəglif] *n.* 凹雕
diaglyphic *a.* 凹雕的
~ ornament 凹雕装饰
~ work 凹雕作品
diagnosis [ˌdaiəgˈnəusis] *n.* 诊[判]断,发现,识别
differential ~ 鉴别诊断的
diagnostic [ˌdaiəgˈnɔstik] *a.* 诊断的
~ program 诊断程序
~ routine 诊断程序
diagonal [daiˈægənl] *a.* 对角线的;斜的,斜纹的 *n.* 对角线;斜线,斜行[列]
~ band 对角带
~ bar 斜撑;斜置钢筋
~ bars tread 斜交条纹
~ beam floor 斜梁楼盖结构
~ bond 对角砌合
~ brace 对角支撑,剪刀撑
~ bridging 对角支撑,剪刀撑

~ˈbuttress 斜扶壁墩
~ cable bridge 斜缆桥
~ coffer (slab) floor 斜镶板楼盖
~ cracking 斜向开裂
~ drain 斜渠,斜(水)沟
~ draw bar 斜拉杆
~ flooring 斜铺地板
~ frame 斜构架
~ joint 斜接
~ laying 斜置
~ line 双角线
~ matrix 对角矩阵
~ member 斜撑
~ parking lane 斜列式停车道
~ pattern 斜砌图案
~ paving 斜向铺砌
~ pitch 斜凿纹
~ ramp 斜向坡道
~ reinforcement 斜钢筋,弯起钢筋
~ˈrib 交叉肋,斜肋
~ rod 斜杆
~ screed 斜刮板
~ shear member 受剪对角斜杆
~ sheathing 斜角覆盖层
~ slating 对角铺砌
~ slip fault 斜滑断层
~ square grid 斜交方格网
~ stacking 斜堆
~ stay 斜撑
~ strut 对角支撑,斜撑,斜支柱
~ system 斜列系统
~ tie 斜拉杆,斜系杆
~ tube 斜管
~ waffle (slab) floor 斜镶板楼盖
~ warping 斜弯翘
~ web member 斜腹杆,桁架斜杆
compression ~ 受压斜杆
principal ~ 主对角线
secondary ~ 次对角线
tension ~ 受拉斜杆
diagram [ˈdaiəɡræm] *n.*; *v.* 图(样,表,解)

~ of connection 接线图
arrangement ~ 布置图
assembly ~ 装配图
flow ~ 流程图
functional ~ 作用[方框,功能]图
moment ~ 弯矩图
skeleton ~ 方框[块]图;总图
string ~ 线图
diagrammatic [ˌdaiəɡrəˈmætik] *a.* 图(表,解)的;概略的,大体的
~ arrangement 简[草]图
~ curve 图解曲线
~ decomposition 图表分解法
~ drawing 示意图
~ representation 图示
~ sectional drawing 断面草图
~ sketch 草图,简图
diagraph [ˈdaiəɡrɑːf] *n.* 作图器,分度尺;放大绘图器
dial [ˈdaiəl] *n.*; *v.* 分度盘;日晷;自动电话拨号盘
~ gauge 千分表,量表,测微仪
~ indicator 刻度盘指示器
~ lock 转字锁,对号锁
diallage [ˈdaiəlidʒ] *n.* 异剥石,剥辉石
dialysis [daiˈælisis] *n.* (*pl.* -ses.) 分离[解];渗析
diamagnetism [ˌdaiəˈmæɡnitizəm] *n.* 抗磁性,反磁现象
diameter [daiˈæmitə] *n.* 直径
~ length 直径长度
inside ~ 内径
outside ~ 外径
pore ~ 孔径
diametral [daiˈæmitrəl] *a.* 直径的
~ compression test 劈裂试验
~ curve 沿径曲线
~ gap 直径径向公差
~ pitch 径节[距]
~ plane 切径平面
~ prism [pyramid] 第二正方柱[锥]
diamond [ˈdaiəmənd] *n.* 金刚钻,钻石;菱形,斜方形;菱形饰

~ abrasive 钻石磨蚀剂
~ bar 菱形凸纹钢筋
~ check 菱形格子
~ circular saw 金刚石圆锯
~ cloth 斜纹布
~ crossing 菱形交叉
~ cutter 玻璃切割刀
~ disc 金刚石圆锯
~ dust 金刚粉
~ expanded metal 菱形网眼钢板
~ fabric 斜纹织物
~ fret 斜 X 字回纹饰;斜方形浮雕花饰
~ frieze 菱形雕带

diamond fret

~ glass cutter 钻石玻璃刀
~ grain 金刚砂
~ hammer 菱形锤
~ head buttress dam 方头支墩坝,大头坝
~ interchange 菱形立体交叉
~ mesh 菱形筛眼
~ motif 菱形装饰母题
~ moulding 菱形饰
~ ornament 菱形饰
~ panel 菱形镶板
~ pattern 菱形图案
~ pavement 菱形铺砌面
~ plate 菱形板
~ point 嵌金刚石针;斜方形轨道交叉点
~ point engraving 用金刚石雕刻的玻璃表面图饰
~ point tool 菱形尖端切削工具
~ saw 圆盘式金刚石锯
~ shape slab 菱形(混凝土)板
~ -shaped 菱形的
~ spar 刚玉
~ tool 金刚石工具[刀]

~ type 菱形式
~ vault 带有凹槽的肋穹顶
~ work 菱形砌块
~ woven fabric 斜纹布
diaper [ˈdaiəpə] n. 菱形花纹装饰;菱形织物
~ bond 菱形砌合

diaper

~ ornament 平面图案装饰
~ work of brick-laying 砖砌的菱形图案
diaphaneity [ˌdaiəfəˈniːiti] n. 透明度[性]
diaphragm [ˈdaiəfræm] n. 横隔墙;地下连续墙;隔膜,膜片 vt. 装以隔板
~ orifice 板孔
~ plate 横隔板
~ wall 隔墙;地下连续墙
watertight ~ (刚性)防渗心墙
diaspore [ˈdaiəspɔː] n. 水矾[铝]石
~ clay 水矾土
diastyle [ˈdaiəstail] a. 柱列间隔为三米的 n. 间隔为三米的柱列
diathermal [ˌdaiəˈθəːməl] a. 透热(辐射)的,导热的
diathermanous [ˌdaiəˈθəːmənəs] a. 透热(辐射)的
~ body 导热体
diathermic [daiəˈθəːmik] a. 透热的
~ heating 高频加热
~ membrane 绝热膜
diatom [ˈdaiətəm] n. 硅藻
~ brick 硅藻土砖
~ earth 硅藻土
~ ooze 硅藻(软)泥
diatomaceous [ˌdaiətəˈmeiʃəs] a. 硅藻类的;硅藻土的

~ brick 硅藻土砖
~ -earth filter 硅藻土过滤器
diatomite [daiˈætəmait] ***n.*** 硅藻土
~ brick 硅藻土砖
~ filter 硅藻土滤池
diatoni *n.* 古希腊建筑中的贯穿砌合石，突隅石
diatonous *n.* 突出墙面的砖或石块
diazoma *n.* 环绕古罗马圆形露天剧场的休息平台
dichroic [daiˈkrəuik] ***a.*** 两色的；分色[光]的
~ glass 两色玻璃
~ mirror 分色[光]镜
dicky [ˈdiki] ***n.*** 汽车后部备用的折叠小椅；马车[汽车]的尾座
die [dai] ***vi.***；***n.*** 冲垫；铸模；方形柱脚；墩身
~ casting material 模铸金属
~ head rivet 冲垫铆钉
~ holder 模座
~ metal 模铸金属
~ plate 印模
~ stock 螺丝铰板
diestock [ˈdaistɔk] ***n.*** 螺丝铰板；板牙架
difference [ˈdifərəns] ***n.*** 不同，相异；差额
~ in height 高度差
~ in temperature 温度差
~ of level surface 水准面差
mean temperature ~ 平均温差
minimum perceptible brightness ~ 最小视觉亮度差
differential [ˌdifəˈrenʃəl] ***a.*** 有差别的；差动的；微分的 ***n.*** 差别
~ coasting 双面差厚涂镀
~ cost 差异成本
~ deformation 差异变形
~ shrinkage 不均匀收缩
~ staining 对比着色
~ temperature 差异温度
~ thermometer 微差温度表

differflange I-beam 不等翼工字梁
difficult [ˈdifikəlt] ***a.*** 困[艰]难的，有障碍的
diffluent [ˈdifluənt] ***a.*** 分流性；易溶解的
difform [diˈfɔ:m] ***a.*** 形状不同的；形状不规则的
diffract [diˈfrækt] ***vt.*** 分解[散]；使绕射
~ed ray 衍射线
~ing power 衍[绕]射本领
diffraction [diˈfrækʃən] ***n.*** 衍[绕]射
~ coefficient 衍射系数
~ cross-section (有效)绕射面积
~ grating 衍射光栅
~ pattern 绕射图像
diffuse [diˈfju:s] ***a.*** 扩散的，漫射的；***v.*** 散布，扩散
~ illumination 漫射[散光]照明
~ reflection 漫反射
~ reflection coefficient 漫反射系数
~ sound 漫射[扩散]声
~ sound control 扩散声控制
~ sound reduction 扩散声衰减
~ transmission 扩散透射(漫透射)
diffused [diˈfju:zd] ***a.*** 扩散的，漫射的，广布的
~ air plate 空气扩散板
~ coating 扩散涂料
~ illumination 散光[漫射]照明
~ light 漫射光线
~ reflection 漫反射
~ sound 散音，漫射音

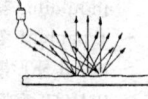
diffused reflection

diffuser ***n.*** 扩散；扩散器；漫射体；洗料池；传播者
~ plate 扩散板，配流板
diffusibility [diˌfju:zəˈbiliti] ***n.*** 扩散率[性]，弥漫性，扩散率
diffusing ***n.*** 扩散，漫射
~ agent 扩散剂

~ block 扩散板
~ ceiling 漫射光顶棚
~ glass 散光玻璃
~ panel 扩散镶板
~ phenomena 扩散现象
~ wall 散光墙体
diffusion [di'fju:ʒən] *n.* 扩散,散布,漫射;普及
~ humidity 扩散湿度
~ hygrometer 扩散湿度表
~ pattern 扩散方式
~ process 扩散过程
load ~ 负荷分布
diffusivity [ˌdifju:'siviti] *n.* 扩散性,扩散能力;扩散系数,扩散率
difunctional *a.* 双作用的,有两种功能的
dig [dig] *v.* 挖[采]掘,开凿;插入
~ through 挖通[穿]
digest [di'dʒest] *v.* 消化,处理;浸渍;摘要 ['daidʒest] *n.* 摘要,文摘,纲领
digestibility [diˌdʒestə'biliti] *n.* 消化性[率]
digestion [di'dʒestʃən] *n.* 消化(作用);蒸煮(作用);侵提
~ period 消化期,消化时间
~ process 消化处理;消化过程
Digigraf *n.* 自动制图系统
digital ['didʒitəl] *a.* 数字(式)的
~ data 数字数据(资料)
~ date acquisition and processing system 数据资料收集和处理系统
~ data handling 数据处理
digitizer ['didʒitaizə] *n.* 数字转换器
diglyph *n.* 双槽面,双槽板
dilapidated [di'læpideitid] *a.* 毁坏的,倒塌的;破旧的
~ building 破烂建筑物,失修房屋
dilapidation [diˌlæpi'deiʃən] *n.* 破烂,倒塌,崩塌
dilatable [dai'leitəbl] *a.* 可膨胀的,可扩大的
dilatancy [dai'leitənsi] *n.* 膨胀性,扩张性

dilatation [ˌdailei'teiʃən] *n.* 扩张,膨胀;舒张
~ joint 伸缩缝,膨胀缝
dilative [dai'leitiv] *a.* 膨胀的
dilator [dəi'leitə] *n.* 膨胀箱;扩张器
dilly ['dili] *n.* (*pl.* -lies;)平板车,手推车;小型车辆
diluent ['diljuənt] *a.* 稀薄的;冲淡的;*n.* 稀释剂,冲淡剂
dilute [dai'lju:t] *a.* 稀薄的,淡的;*v.* 稀释,冲淡
~ concentration 稀释浓度
diluted [dəi'lju:tid] *a.* 稀释的,冲淡的
~ lubricant 稀释润滑剂
dilution [dai'lu:ʃən] *n.* 冲淡,稀释;产权削弱;减损
~ ratio 稀释率
~ water 稀释水
dim [dim] *a.* 暗淡的,模糊的;无光泽的;*v.* 使[变]暗淡,使[变]模糊
~ light 小光灯
~ light bulb 小光灯泡
~ med illumination 小光灯
dimension [di'menʃən] *n.* 大小,尺寸[度];量纲,因次;维,度
~ book 工程尺寸记录册
~ chart 轮廓图
~ control 尺寸控制
~ figure 尺寸图
~ line 尺寸线
~ lumber 分级木材,标准尺寸木材
~ paper 尺度纸
~ saw 浅切锯
~ standard 尺寸标准
~ stock 规格木料
~ stone 规格石料
~ timber 建筑木材
boundary ~ 外形[轮廓]尺寸
characteristic ~ 特性尺寸
coupling ~ 连接尺寸
critical ~ 临界尺寸
fixing ~ 装配[规定]尺寸

general ~ 主要尺寸
dimensional [di'menʃənəl] *a.* 尺寸的,量纲的
~ accuracy 尺寸的精确性
~ change 尺寸变化[更改]
~ coordination 尺寸调整
~ data 尺寸数据
~ deviation 尺寸偏差
~ discrepancy 尺寸误差
~ drawing 有尺度的图;轮廓图
~ limit 尺寸限制
~ range 尺寸范围
~ reference system 尺寸参考系统
~ sketch 轮廓草图
~ sound 立体音响
~ stability 外形稳定性
~ standard 尺寸标准
~ tolerance 尺寸容差
~ unit 尺寸单位
~ variation 尺寸变化
three ~ 三维空间的,立体的
two ~ 平面[二维]的
dimensioning *n.* 量尺寸;定尺度,选定断面
~ of joint(s) 节点尺寸的确定
base line ~ 基线尺寸
dimethyl [dai'meθil] *n.* 乙烷,二甲基
~ aniline 二甲苯胺
dimetric [dai'metrik] *a.* 正方的,四边形的
diminish [di'miniʃ] *v.* 使成尖顶;减少,缩小
~ed angle 缩角
~ed arch 平圆拱
~ed scale 蜗线尺
~ed stile 不等宽门窗边梃
~ing piece 联结不同直径管子的管子配件
~ing rule 仿形尺,柱径渐减的柱样板
~ing scale 缩尺
diminution [ˌdimi'njuːʃən] *n.* 减少;缩小
diminutive [di'minjutiv] *a.* 小的;缩

的,变小的
~ tower 装饰性角楼
dimmer ['dimə] *n.* 减光器,制光装置,小光灯,(灯)罩
~ switch 减光器开关
dim-out ['dimaut] *n.* 灯光管制;遮光设备
dimple ['dimpl] *n.* 凹痕;波纹;建筑材料的嵌线;*v.* 使起波纹
~ed plate 波纹板
dinas *n.* 砂[硅]石
~ bricks 砂砖,灰砂砖
~ rock 硅石
lime ~ 石灰硅石
dinette [dai'net] *n.* 小餐厅
ding [diŋ] *n.* 勾缝;板材的弯折
dinge [dindʒ] *n.* 表面凹陷
dingle ['diŋgl] *n.* 雨篷;防风暴门;贮藏间;幽谷;小排水沟
Dingling Underground Palace of Ming Dynasty (北京)明朝定陵地下宫殿
dingus ['diŋgəs] *n.* 小装置,小机件
dining ['dainiŋ] *n.* 吃饭,进餐
~ alcove 小餐厅
~ hall 餐厅
~ kitchen 厨(房),餐(厅)间
~ nook(reless) 餐座
~ recess 餐座
~ room 餐室,食堂
~ space 用餐空间
~ table 餐桌
dinner ['dinə] *n.* 正餐;宴会
~ cloth 餐桌台布
~ clothes 餐服
~ jacket 小礼服,无尾礼服
~ pail 饭盒
~ plate 餐盘
~-set 成套餐具
~ table 餐桌
~ wagon 带轮的食品输送架
dint [dint] *n.* 压痕,凹处;*vt.* 压痕,打出凹痕
diopside [dai'ɔpsaid] *n.* 透辉石
diopter [dai'ɔptə] *n.* 屈光度;照准仪,

照准器

dioptric [daiˈɔptrik] *a.* 光线折射的;屈光学的
~ apparatus 屈光器
~ glass 光线屈折镜
~ lens 屈光透镜
~ imaging 折射成像
~ strength 焦度
~ system 屈光系统,折射体系

diorama [ˌdaiəˈrɑːmə] *n.* 西洋景;透视画;实景模型

diorite [ˈdaiərait] *n.* 闪绿岩,闪长岩
~ porphyrite 闪绿斑岩,闪长斑岩

dip [dip] *n.* 倾斜;倾角;垂度;弛度;浸渍;液体;树脂;*v.* 浸;汲取;挖掘;下沉
~ angle 俯角,倾角
~ application 浸涂施工
~ -braze 铜浸焊
~ can 选样器
~ circle 测斜仪,磁倾仪
~ coating 浸涂
~ compass 倾斜仪
~ face 浸渍面
~ fault 倾斜断层
~ galvanizing 浸渍镀锌
~ gauge 垂度规
~ method 浸渍法
~ rod 水位指示器
~ soldering 浸入焊接
erection ~ 垂度
horizon ~ 倾斜,(地平)俯角

dipping *n.* 倾斜,下倾;浸渍
~ compass 测斜仪,倾度仪
~ consistence 浸渍稠度
~ engobe 浸没上釉底料
~ glazing 浸没上釉
~ lacquer 浸没上漆
~ method 浸渍方法
~ process 浸渍过程
~ varnish 浸渍清漆

dipteral [ˈdiptərəl] *a.* 双列柱廊的
~ building 双列柱廊式建筑
~ temple 双列柱廊式神庙

dipteros

dipteros [ˈdiptərɔs] *n.* 双列柱廊式(建筑,神庙)

diptych [ˈdiptik] *n.* 双幅联画雕刻;双幅折叠画板,折叠写字板

direct [diˈrekt] *a.* 直接的,捷径的;*v.* 指挥,管理;指导[引]
~ access 直接存取;直接作业,直接出入口
~ access programming system 直接存取程序设计系统
~ acting load 直接作用荷载
~ axis 纵向轴线
~ bearing 直立支承
~ benefit 直接效益
~ charges 直接费
~ comparison 直接比值
~ compression 直接压力
~ connection 干道与进路间的直接联系
~ connection design 立体交叉的直接联系设计
~ -connection interchange 直连式立体交叉
~ consignment 直接交货
~ construction cost 直接施工费
~ corrosion 直接腐蚀
~ cost 直接成本
~ current 直流电
~ current arc welding 直流电弧焊接
~ damage 直接损失
~ drawing change 图纸直接更改
~ department 直接部门
~ diffused light 直接漫射光
~ distribution 直接分配
~ drive 直接传动
~ exchange 直接交换

~ expense cost 直接费用成本
~ field 直接声场
~ financing 直接筹资
~ financing leases 直接融资租赁
~ fire 直接供暖设备
~ fire pressure 直接消防压力
~ fired dryer 直烧干燥器
~ flexure 纯弯曲
~ glare 直接眩光
~ heating 直接供暖
~ illumination 直接照明
~ input 直接输入
~ investment abroad 国外直接投资
~ investment interest 直接投资利息
~ labour 直接劳动[人工]
~ labour cost 直接工值[资]
~ lighting 直接照明
~ labor time 直接工时
~ load 直接荷载
~ loans 直接贷款
~ materials 直接材料
~ menmory access 直接存贮器存取
~ negotiation 直接洽谈
~ observation 直接观测
~ operating cost 直接操作费用
~ oxidation 直接氧化
~ price 直接价格
~ pricing 直接计价
~ quotation 直接估价
~ radiant heater 直接辐射式供暖
~ radiation 直接辐射
~ -return system 直接回水制[系统]
~ run off 直接迳流
~ sound 直达声
~ steam 直接蒸汽;新气,活气
~ stereoscopic vision 直接立体观察
~ wage 直接工资
~ warmer 直接供暖设备
~ yield 直接收益
direction [di'rekʃən] *n.* 方向,方位,趋向;指示;管理,监督
~ angle 方向角
~ arm 指路牌
~ distribution 定向分布
~ indicator 方向指标器
~ light 方向灯
~ of deformation 形变方向
~ of friction 摩擦(力)方向
~ of load application 加载方向
~ of rotation 旋转方向
~ of the applied load 加载方向
~ sign 方向标志,路标
base ~ 基线方向
directional [di'rekʃənəl] *a.* 方向的;定向的;指示方向的
~ antenna 定向天线
~ characteristic 方位特性
~ detector 定向检测器
~ post 导向柱
~ properties 方向性
~ sign 指路[方向]标志
~ split 方向分路
directionality [di₁rekʃə'næliti] *n.* 方向;方向性
directive [di'rektiv] *n.* 指向,射向;指示,指令
directivity [dired'tiviti] *n.* 方向性,指向性
directly [di'rektli] *ad.* 直接地;即刻地
~ -heated 直热式的
director [di'rektə] *n.* 理事,董事;指挥者,管理者
~ board 董事会
~ system 指挥系统
directorate [di'rektərit] *n.* 指导者,董事;董事会,管理局
directorship [di'rektəʃip] *n.* 指挥职能
directrix [di'rektriks] *n.* 准线
dirt [də:t] *n.* 泥土,灰土;污物,垃圾
~ band 泥[砂]屋
~ catcher 除尘器
~ collector 吸尘器
~ -proof 防尘的
~ repellent 防尘剂

~ screed 匀泥尺
~ screening 除尘筛
~ wall 土[泥]墙
gumming ~ 胶泥
dirty ['də:ti] *a*. 脏的,污秽的
~ money 付给在困难或特殊条件下工作的建筑工人的额外报酬
~ cargo 油污货物(如油、漆等)
disalignment *n*. 偏离中心线,轴线不重合,定线不准
disappear [ˌdisə'piə] *vi*. 失踪;消失,消散;
~ing bed 暗式壁柜床
~ing stair 折梯
disassemble [ˌdisə'sembl] *v*. 拆卸,分散,投散
disassembly [ˌdisə'sembi] *n*. 拆卸[开],分解[散]
disaster [di'za:stə] *n*. 灾难[害];事故,故障
disburse [dis'bə:s] *v*. 支[付]出;分配
disbursing *n*. 支出,分配
~ office 收支机构
~ officer 出纳主管
disbursement [dis'bə:smənt] *n*. 支出,付出;付出款
~ voucher 支付凭单
disc [disk] =disk *n*. 圆盘,圆板;唱片;铁饼
~ auger 盘式螺旋钻
~ bit 圆盘钻
~ frieze 圆盘花饰
~ pile 盘头桩
~ screen 圆盘筛
~ water meter 盘式水表
discard [dis'ka:d] *v*. 废除;解雇;舍弃,废弃
discharge [dis'tʃa:dʒ] *v*. 卸载,卸货;排出;发射. *n*. 卸货;流出;流量;放电
~ agent 脱色剂,漂白剂
~ air shaft 通风井,通风竖管
~ area 出口[过水]面积,流束面积
~ bay 泄水池

~ branch 排水支管
~ capacity 泄流能力;排放量
~ cock 放水龙头
~ duct 排气管道,排水管道;排料槽
~ fan 排气风扇
~ flue 排气道,排烟道
~ gutter 排水槽
~ hose 送水软管
~ lamp 放电管,放电灯
~ line 排水线
~ nozzle 喷嘴,出料管
~ of opening 排水口,卸料口
~ pipe 排水管
~ port 放出口
~ spout 喷口,漏嘴
~ system 排水系统
~ trough 排水槽
~ valve 排水阀
discharger [dis'tʃa:dʒə] *n*. 排水装置;放电器,避雷器;卸货人
discharging *n*. 放[排]出;排水;放电;卸货
~ arch 辅助拱
~ tube 出水管,泄水管;放电管
~ valve 排水阀,放水阀
disciform *a*. (椭)圆形的
discoid ['diskɔid] *a*. 圆盘形的 *n*. 圆盘状物体
discolour [dis'kʌlə] *v*. 使变色,使褪色
~ing agent 脱色剂
discolouration [disˌkʌlə'reiʃən] *n*. 变色;褪色;污染,斑渍
discompressor [ˌdiskəm'presə] *n*. 减[松]压器
disconformity [diskən'fɔ:miti] *n*. 不协调,不相称,不一致
disconnect ['diskə'nekt] *vt*. 拆开,分离,断开
~ing chamber 隔离室;检查井
~ing switch 隔离开关
disconnected ['diskə'nektid] *a*. 分离的,脱离的;不连贯的,无条理的
disconnection [ˌdiskə'nekʃən] *n*. 分

离,分开;断开
disconnector n. 断路器;绝缘体;拆开者
discontinuity ['dis͵kɔnti'njuːiti] n. 不连续,间断;不连续性,间断性
～ of marterial 材料的不均匀性
～ switch 切断开关
discontinuous ['diskən'tinjuəs] a. 中断的,不连续的
～ heating 定时供暖,间歇式供暖
～ impost 间断拱墩
discontinuously ad. 间断地,不连续地
discount ['diskaunt] n. 折扣,贴现 vt. 打折扣,贴现
～ cost 折扣费用
～ earned 折扣收入
～ed price 折扣价格
discrete [dis'kriːt] a. 分离[立]的;不连续的;无关联的
～ aggregate 松散集料
～ material 松散材料
discriminability [dis͵kriminə'biliti] n. 可分别性,可辨别性;辨别力,识别力
discriminant [dis'kriminənt] n. 判别式
discriminating [dis'krimineitiŋ] a. 形成区别的,识别性的;有差别的,区别对待的;有辨[识]别力的
～ order 判别,指令
discrimination [dis͵krimi'neiʃən] n. 区别,辨别(力)
～ data processing system 数据鉴别处理系统
discriminator [dis'krimineitə] n. 鉴别器
discriminatory [dis'krimənətəri] a. 能鉴别的,能选择的
disc-seal n.;a. 盘形封口(的)
discussion [dis'kʌʃən] n. 讨论,商议
diseconomy ['disi(ː)'kɔnəmi] n. 不经济的,使成本增加

discontinuous impost

disengage [͵disin'geidʒ] vt. 解约;解除,脱离;放开,释放
disengaging n. 分离;解除
～ device 分离装置
～ gear 分离装置
disengagement [͵disin'geidʒmənt] n. 脱离,解除;解约
disfunctional [dis'fʌŋkʃnl] a. 失去功用[能]的
dish [diʃ] n. 盘,碟;盘形,盘形物
～ garden 盆景
～ gas holder 湿式气柜
～ washer 洗碟机
dished [diʃt] a. 凹状的,盘形的,穹隆形的
～ -out 圆屋顶构架的
～ roof 穹窿形屋顶
dishing ['kiʃiŋ] a. 碟形的,凹陷的 n. 凹图槽;坑塘
dishono(u)r [dis'ɔnə] n. 拒付[收],不兑现 vt. 使作废,拒付[收]
～ed cheque 空头支票
disinfectant [͵disin'fektənt] a. 消毒的 n. 消毒剂
～ paint 杀菌涂料
disinfection [͵disin'fekʃən] n. 消毒,杀菌(作用)
～ plant 消毒设备;消毒厂
disinfector [disin'fektə] n. 消毒器(具),消毒剂
disintegration [dis͵inti'greiʃən] n. 瓦解,分裂;剥蚀;衰变
disintegrator [dis'intigreitə] n. 粉碎机;分离器;轧石机,松砂机
disjoin [dis'dʒɔin] v. 分离,拆散,散开
disjunction [dis'dʒʌŋkʃən] n. 分离;分裂;折断
disk [disk] n. 圆盘[板],盘状物
～ bit 圆盘钻
～ colo(u)rimeter 盘式色度计
～ dowel 圆木榫,圆木暗销
～ fan 圆盘风扇
～ file 磁盘存储器,磁盘文件

~ grinder 盘式研磨机
~ harrow 圆盘耙
~ rivet 圆盘钉
~ sander 地板打磨器,盘式打磨机
~ saw 盘锯
~ spacer 圆隔板
~ water meter 圆盘形水表
dislocation [ˌdislə'keiʃən] *n.* 脱节；扰乱；断层，错位
dismantle [dis'mæntl] *vt.* 拆除(房屋,设备)；拆卸；粉碎，摧毁
dismount ['dis'maunt] *vt.* 拆卸
dismountable [dis'mauntəbl] *a.* 可拆卸的
~ connection 活络节点,灵活节头
~ division wall 灵活隔断
disorder [dis'ɔːdə] *n.* 混乱,杂乱无章；机能失调；无规则，不协调
disparity [dis'pæriti] *n.* (*pl.* -ties) 悬殊；差别；不一致，不均衡
dispatch [dis'pætʃ] *n.*；*vt.* 迅速派遣[发送,办理]
~ clerk 调度员
~ room 调度[发送]室；急件室
dispatching *n.* 控制；调度；派工
dispensary [dis'pensəri] *n.* 药房；门诊部
dispensation [ˌdispen'seiʃən] *n.* 管理方法；体制；分配
dispense [dis'pens] *vt.* 分配；执行，实行
dispenser [dis'pensə] *n.* 分配器，计量器
dispersal [dis'pəːsəl] *n.* 疏散,分布；解散,驱散
dispersant [dis'pəːsənt] *n.* 分散剂
disperse [dis'pəːs] *a.* 分散的 *v.* 分散，散布
~d town 布局分散的城镇
~ing agent 分散剂
~ing characteristic 分散特性
~ing medium 分散介质
dispersion [dis'pəːʃən] *n.* 分散，散开；驱散；消散；色散；离差；漂移；弥散(现象)；悬浮体
~ degree 分散度
~ ratio 分散率
~ resin 离散树脂
dispersity *n.* 分散度,色散度
dispersivity *n.* 分散性,色散性
displace [dis'pleis] *vt.* 移动,变位；取代，置换
displacement [dis'pleismənt] *n.* 位移，移程；取代，置换；排[水,气,油,液]量
~ diagram 位移图,变位图
~ load 位移荷载
~ meter 变位计；位移式水表
~ observation 位移观测
~ of abutment 支座位移
~ of water 排水量
~ velocity 排水速度
displacer [dis'pleisə] *n.* 混凝土用毛石料块；排除物；代用品；排代剂,置换剂
display [dis'plei] *n.*；*vt.* 展览，陈列；显示，表现
~ cabinet(sase; stand) 陈列柜
~ case 陈列柜
~ drawing 展览图
~ in digital form 数字显示
~ lighting 陈列品照明
~ rack 陈列架
~ window 展览橱窗
disposal [dis'pəuzəl] *n.* 排除；处理[置]；清理；排列，布置
~ area 清理场地
~ by dilution 稀释处理
~ costs 清理成本
~ of refuse 垃圾处理，废料处理
~ of sewage 污水处理
~ of water 水的处理
~ point 清理场地
~ site 垃圾倾倒场,废料堆放处
~ system 处理系统
disposition [ˌdispə'ziʃən] *n.* 安排，配置；陈列
~ of equipment 设备配置

~ plan (设备)布置平面图

disproportion [ˌdisprəˈpɔːʃən] *n.* 不平衡;不相称;不成比例

disproportionately *ad.* 不成比例地;不相称地;不平衡地

~ graded 级配不良的

disqualification [disˌkwɔlifiˈkeiʃən] *n.* 不合格;无资格;无能力

disrepair [ˌdisriˈpɛə] *n.* 失修;破烂

disrupt [disˈrʌpt] *vt.* 破裂,断裂,碎裂,破坏

disruption [disˈrʌpʃən] *n.* 断裂,破坏

disruptive [disˈrʌptiv] *a.* 分裂的;破坏(性)的;

~ effects 破坏性作用

dissect [diˈsekt] *vt.* 切开,解剖

dissimilar [disˈsimilə] *a.* 不同的,异样的;不相似的

~ welding rod 与被焊金属成分不同的焊条

dissipated [ˈdisiˌpeitid] *a.* 浪费的;消耗的

~ energy 消耗能量

dissipation [ˌdisiˈpeiʃən] *n.* 耗[消]散;浪费

dissipative [ˈdisipeitiv] *a.* 消[耗]散的,驱散的

dissociation [diˌsəusiˈeiʃən] *n.* 分解[离],离解

~ degree 离解程度

~ temperature 离解温度

dissolubility [disəljuˈbiliti] *n.* 溶解度,可溶性

dissolution [ˌdisəˈluːʃən] *n.* 分离[解]裂开;溶解;融化

dissolve [diˈzɔlv] *n.;v.* 溶解;解散[除];溶化

~d matter 溶解物质

dissolvent [diˈzɔlvənt] *a.* 有溶解力的; *n.* 溶剂

dissolving [diˈzɔlviŋ] *a.* 消溶的,消失的

distance [ˈdistəns] *n.* 距离,间隔;遥远,远处;路程

~ across 横穿距离
~ bar 横撑
~ between buildings 建筑物间距
~ between centers 中心间距
~ between girders 梁的间隔
~ block 定距隔块
~ control 遥控
~ factor between air voids 加气混凝土的气孔间距系数
~ measurement 距离测量
~ observation 远距观测
~ of hydraulic jump 水跃距离
~ piece 间隔物
~ plate 定距隔板;隔片,隔垫
~ pole 标柱
~ range 通达距离
~ thermometer 遥测温度计
unsupported ~ 自由长度,净跨
viewing ~ 观察距离

distant [ˈdistənt] *a.* 遥远的,远隔的

~ control 遥控
~ indication 遥示
~ operation 遥控
~ point 远点
~ sign 前置交通标志

distemper [disˈtempə] *n.* 水粉画颜料,壁画颜料;水浆涂料,刷墙粉;胶画法;胶画 *vt.* 用胶画颜料涂

~ brush 画笔;胶质涂料刷
~ coat 胶质涂料面层

distemperature [disˈtempəritʃə] *n.* 水浆涂料;色粉涂饰

distensible [disˈtensəbl] *a.* 有弹性的,可伸展的,会膨胀的

distension [disˈtenʃən] *n.* 膨胀,伸展,胀大

distiller [disˈtilə] *n.* 蒸馏器;(蒸馏装置的)凝结器

distinction [disˈtiŋkʃən] *n.* 个[特]性;区[差]别;

distorted [disˈtɔːtid] *v.;a.* 扭曲(的);变态(的);曲解(的)

~ model 变态模型

~ scale 变态比例尺
distortion [dis'tɔ:ʃən] *n.* 畸变,扭曲;失真
~ of sound 声音失真
~ resistance 变形抗力
distortional [dis'tɔ:ʃənl] *a.* 歪曲的,变形的
~ deformation 扭曲变形
distress [dis'tres] *n.* 事故,损坏
~ in concrete 混凝土的龟裂
distribute [dis'tribju:t] *vt.* 分[散]布;分配,配线;排列分类;扩充;区分
~d load 分布荷载
~d steel 分布钢筋
~d system 分布系统
distributer [dis'tribjutə] = distributor *n.* 分配器;分布者
~ block 接线板
distributing [dis'tribju:tiŋ] *n.* 分布;分配;分类
~ bars 分布钢筋
~ boom 布料吊杆
~ box 配电[线]箱
~ frame 配线架
~ insulator 配线绝缘子
~ line 分布线
~ net 配水管网;配电网
~ pipe 分配管道
~ plate 承重板
~ water pipe 配水管道
distribution [ˌdistri'bju:ʃən] *n.* 分布;分配;分类
~ and fuse board 配电与保险丝盘
~ bar 配力[分配]钢筋
~ board 配电盘
~ channel 配水渠
~ coefficient 分配系数
~ grid 配水管网;配电网
~ line 分布线
~ main 配水总管
~ map (面砖、板材等)装配详图
~ network 分布网络

~ of bearing pressure 承压力分布
~ of bending 弯曲分布
~ of conncrete 混凝土(从搅拌机到模板的)运输
~ of cracks 裂缝分布
~ of daylight factor 采光系数分布,天然照度系数分布
~ of deformation 形变分配法
~ of demand 用量分配
~ of heat 热量分布
~ of light 光线分布
~ of load 荷载分布
~ of rafter 椽的布置,排椽法
~ panel 配电盘
~ ratio 分配率
~ reinforcement 分布钢筋
~ rod (steel;bar) 配力[分布]钢筋
~ steel 配力[分布]钢筋
~ system 配水系统;配电网
distributive [dis'tribjutiv] *a.* 分配的、普及的;分布的
~ effect 分配作用
distributor [dis'tribjutə] *n.* 配水[电]器;喷洒器;分电盘;分配者
~ condenser 配电器电容器
~ disk 配电盘
~ road 分支道路
district ['distrikt] *n.* 地方,区域
~ board 区域界限
~ cooling 分区冷却
~ heating 分区供暖
~ heating duct 分区供暖管道
~ heating line 分区供暖线
~ heating plant 分区供暖设备
~ school 地区学校
~ surveyor 地区工程检查员
central business ~ 商业中心区
residential ~ 住宅区
rural ~ 乡村,农村
shopping ~ 商业区
urban ~ 市区
disturb [dis'tə:b] *vt.* 扰动[乱],使紊

乱,打扰
disturbance [dis'tə:bəns] *n.* 扰乱;扰动
distyle ['distail] *a.* 双柱式的; *n.* 双柱式
~ in antis 两端壁角柱间有两根柱子的
ditch [ditʃ] *n.* 明沟,沟渠 *v.* 开沟,挖渠;(使)出轨,溅落
~ conduit 明排管道
~ drainage 明沟排水
~ lining 沟衬砌
~ sett paving 沟内毛石铺砌
open road ~ 明路沟
paved ~ 铺砌沟
ditetragon *n.* 双四边形
ditetrahedron *n.* 双四面体
ditriglyph *n.* (古希腊多立克建筑上)三陇板间距
diurnal ['dai'ə:nl] *a.* 每日的;周日的
~ cycle 日循环,昼夜循环
divarication [dai,væri'keiʃən] *n.* 分叉,交叉点
divergence [dai'və:dʒəns] *n.* 分歧;差异,发[扩]散
diverging [dai'və:dʒiŋ] *n.* 分流,发散,分开
~ nozzle 扩张型喷管,喇叭形管嘴
~ tube 扩散管
diversification [dai,və:sifi'keiʃən] *n.* 变化;多样化
diversiform [dai'və:sifɔ:m] *a.* 各式各样的
diversion [dai'və:ʃən] *n.* 转向[移,变]
~ area 分水区
~ box 分水箱
~ curve 转移曲线
diversity [dai'və:siti] *n.* (*pl.* -ties) 差异;多样性;变化
~ factor 分散系数
divide [di'vaid] *n.* 分水岭,分界线; *v.* 分开,分裂
~d ventilation 分道通风
divider [di'vaidə] *n.* 间隔物,分隔墙;分切器

dividing [di'vaidiŋ] *a.* 区分的,划分的;分开的
~ pier 隔墩,分水墩
~ strip 分车[隔]带;划分(线),分隔(线)
~ wall 分水墙,隔水墙
division [di'viʒən] *n.* 分开;划分;分配;部门
~ bar 隔条
~ box 分水[流]箱
~ center 中心结构
~ line 分隔线
~ masonry wall 防火圬工[砖石]墙
~ of construction 建筑部门,建筑科,工程科
~ of labour 分工制
~ of responsibility 分工负责制
~ plate 分隔板
~ surface 分隔[界]面
~ wall 隔墙,防火墙
divisional [di'viʒənəl] *a.* 分开[割]的;分区[部]的
~ organization 分区组织
~ system 分区[段]制
doab ['dəub] *n.* 汇流点;多砂黏土
doak=donk *n.* 脉壁黏土
dobie ['dəubi] *n.* 黏土砖
dock [dɔk] *n.* 船坞,码头 *vt.* 把……引入船坞 *vi.* 入船坞
docking ['dɔkiŋ] *n.* 入船坞; *a.* 入船坞的
~ block 龙骨垫
doctor ['dɔktə] *n.* 输送校正器,调节机;刮刀
~ knife 刮刀
~ blade 刮浆刀,刮片
document ['dɔkjumənt] *n.* 文件,公文,证据
~ glass 吸收紫外线的玻璃
documentation [,dɔkjumen'teiʃən] *n.* 证明文件的提供
dod [dɔd] *n.* 沟管模板
dodecagon [dəu'dekəgən] *n.* 十二边,十二角形

dodecahedron ['dəudikə'hedrən] *n.* 十二面体

dodecastyle [dəu'dekəstail] *n.* 十二柱式 *a.* 十二柱式的
～ temple 十二柱式神庙

dog [dɔg] *n.* 搭扣;扒钉;制动爪
～ anchor 扒钉,两爪钉
～ clutch 爪形离合器
～ hook 爪形钩
～ -ear fold 折边收口
～ -head spike 狗头道钉
～ iron 两爪铁扣
～ -legged stair 无梯井的层间双折楼梯,层间双折楼梯
～ -nail 道钉
～ shores 水平撑柱,支船木
～ spike 道钉
～ tooth bond 犬牙式砌合

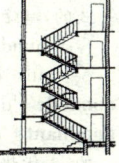
doglegged staircase

dogbolt ['dɔgbəult] *n.* 直角固定螺栓

doglegged ['dɔglegd] (亦作 dogleg) *a.* 弯去如犬后足的;"〈"形的
～ staircase "〈"形楼梯

dogtooth ['dɔgtu:θ] *n.* 齿饰;犬牙
～ moulding 齿状线脚

dogtrot ['dɔgtrɔt] *n.* 有顶通道

dogwood ['dɔgwud] *n.* 草皮;山茱萸;韧硬木

dolerite ['dɔlərait] *n.* 粗粒玄武岩;辉绿岩;深色火成岩

dollar ['dɔlə] *n.* 美元,$ 或 $

dolly ['dɔli] *n. pl.* dollies. 铆顶;垫桩木;小机车;捣碎棒
～ bar 夹铆钉棍

dolmen ['dɔlmen] *n.* 史前时期用石块架成的纪念物;石牌坊;石窟;石台

dolomite ['dɔləmait] *n.* 白云石[岩];石灰岩,大理石
～ brick 白云石砖
～ cement 石灰岩水泥
～ glass 白云石玻璃,镁玻璃
～ marble 白云质大理石
～ plaster finish 白云质石膏饰面

dolomitic [,dɔlə'mitid] *a.* 含白云石的;含镁的
～ lime 含镁石灰
～ marble 含镁大理石
～ sand 白云质砂

domain [də'mein] *n.* 领域;产业;产业所有权

dome [dəum] *n.* 穹顶,圆屋顶;圆盖;油库 *vt.* 覆以圆顶;使成圆顶 *vi.* 呈圆顶状
～ apex 圆顶顶点
～ brick 土窑砖;楔形砖
～ cornice 圆顶挑檐
～ crown 圆顶顶点
～ -drum 穹顶鼓座
～ edge 圆顶边缘
～ form 圆顶形式
～ impost 圆顶拱基
～ light 穹顶天窗孔;顶灯
～ of multiangular plan 多角穹顶
～ of polygonal plan 多边穹顶
～ on tambour 鼓座上的圆顶
～ ring 圆顶环
～ roof 圆盖形屋顶
～ shell 半球形薄壳
～ slab 圆形板
～ surface 圆顶表面
～ top 圆顶顶端
arched ～ 圆屋顶式拱,拱圆顶
bulbous ～ 球形圆顶
hemispherical ～ 半球形圆顶
lobed ～ 扁圆屋顶
onion ～ 葱形圆顶
pendentive ～ 三角穹顶
pointed ～ 尖头圆屋顶
projection ～ 突出圆顶
saucer-shaped ～ 碟形屋顶
spherical ～ 球形屋顶

whispering ~ 低音廊
domed [dəumd] *a.* 有圆顶的;圆盖形的;半球形的
~ basilica church 带有圆顶的巴西利卡式教堂
~ building 带有圆顶的建筑物
~ central-plan church 带有穹顶的集中式教堂
~ diminutive tower 带有圆顶的小塔楼
~ hall 有圆顶的大厅
~ mosque 带有圆顶的清真寺
~ pavilion 圆顶亭
~ shell 半球形薄壳
~ structure 圆顶结构
~ turret 有圆顶的小塔楼
domelike *a.* 穹顶状的
~ structure 穹顶状结构
domes *n.* 月面拱형结构
domestic [də'mesitik] *a.* 家(里,庭)的;本国的,国内的;国产的
~ airport 国内机场
~ appliance 家庭用具
~ architecture 居住建筑
~ baseboard heating 国产隔热板
~ block 居住建筑
~ building 居住房屋
~ chapel 礼拜堂,小教堂
~ construction 住宅建设
~ consumption 生活用水量
~ demand 生产用水量;国内需要
~ electric(al) appliance 家用电器设备
~ filter 家用过滤器
~ gas appliance 家用煤气设备
~ heating plant 家用采暖设备
~ hot water 家用热水
~ instruments 家庭用具,室内陈设
~ kitchen 家用厨房
~ manufacture 国内制造工业
~ marble 国产大理石
~ quarter 住宅区
~ refuse 生活垃圾

~ sauna bath 家庭蒸汽浴室
~ sewage 生活污水
~ sewer system 生活污水管网
~ softener 家用水软化剂
~ stair(case) 家用楼梯
~ stove 家用火炉
~ supply system 家庭供水系统
~ terminal building 国内乘客候机楼
~ utility corridor 家用过道
~ utility room 家庭杂用室
~ water 生活用水
~ water meter 用户水表
~ wood 国产木材
domical ['dəumikəl] *a.* 圆屋顶的;圆顶式的,穹隆式的
~ mound 舍利子塔
~ vault 半球形穹顶
domicile ['dɔmisail] *n.* 家,固定住所
dominant ['dɔminənt] *a.* 支配的,统治的,占优势的;高耸的,居高临下的 *n.* 主因,要素
~ discharge 控制流量
~ shape 主形
~ wind 盛行风
domus ['dəuməs] *n.* 古罗马或中世纪时期的住宅
donjon ['dɔndʒən] *n.* 中世纪城堡的主塔
donk *n.* 脉壁黏土,亚黏土
door [dɔ:] *n.* 门;通道;家户;路
~ abutment piece 门槛
~ accessories 门上附件
~ aperture (opening; stead) 门口
~ arch 门拱
~ axis 门中轴线
~ bar 门闩
~ bell 门铃
~ bolt 门销
~ bottom rail 门扇下横木
~ buck 门边立木
~ bumper 门制止器
~ butt 门铰链

~ casing 门框；门侧装修
~ center line 门轴线
~ chain 门链
~ check 门制止器
~ cheek 门框侧板
~ closer 门碰头，门制止器

door casing

~ curtain (hood) 门帘
~ engine 自动关门机，闭门器
~ finish 门边装修，门的整修
~ fixture 门用零件

door check

~ frame 门框
~ furniture 门用五金
~ gasket 门垫板
~ gear 门装置
~ glass channel 门玻璃槽
~ grille 门下通风笼子，门格栅

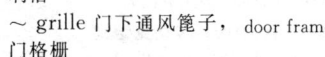
door frame

~ handle (knob) 门把手
~ hanger 门钩
~ hardware 门用五金
~ head 门框上槛
~ hinge 门铰链
~ holder 门扣
~ jamb 门侧柱[边框]
~ knocker 门环
~ latch 门闩锁
~ leaf 门扇
~ ledge 拼板门的横档
~ lintel 门过梁
~ lock 门锁
~ mat 门前擦鞋棕垫
~ mechanism 门装置
~ middle rail 门扇中冒头
~ mirror 门镜
~ muntin 门扇中梃
~ nail 护门帽钉；大头钉

~ niche arch 门龛拱
~ opener 开门装置
~ operator 门的自动开闭装置
~ panel 门心板
~ pier 门墩
~ pillar 门柱
~ pivot 门轴
~ plate 门牌
~ pocket 推拉门门架
~ post 门柱
~ profile 门贴脸
~ pull 门拉手
~ rabbet 止门槽口
~ rail 门横档
~ reveal 门侧
~ roller 拉门滚轮
~ seal(ing) fillet 门密封嵌条
~ spring 门用弹簧
~ step 门阶
~ stile 门扇边挺
~ stone 门前铺石，门前阶石
~ stop 门碰头，止门装置，门制止器
~ stud 门边立木
~ switch 门开关
~ system 门安装
~ top rail 门扇上冒头
~ trim 门贴脸，门框的细木工
~ viewer 门窥视孔
~ window 门窗
~ yard 门前庭院，天井
access ~ 便门，检修门
accordion ~ 摺门
air ~ 气门，通气门；风门
air lock ~ 双层风门
arch ~ 拱[卷]门
batten ~ 板条门
bevelled ~ 斜边门
bi-part ~ 双扉门
blind ~ 百叶门
boarded ~ 直板门
double action 变向门
double folding ~ 双层折门

double sliding ~ 双扇拉门
double swing ~ 双摆门
dwarf ~ 两截门
external ~ 外门
flush ~ 平面门
frame ~ 框架门
French ~ 玻璃落地门
hinged ~ 铰链门
hollow metal ~ 空心金属门
inspection ~ 检查[门]孔
internal ~ 内门
lamp ~ 灯框
lever gear ~ 提升门
manhole ~ 人孔门
one-panel ~ 单扇门
panel ~ 格板门
plain batten ~ 平板门
plate glass ~ 平板玻璃门
revolution ~ 转门
rollershutter ~ 卷帘门
rolling ~ 滑[滚]动门
round-the-corner ~ 转角门
secret ~ 隐蔽门,密门
single action ~ 单向门
slide ~ 拉门,滑槽门
swing ~ 双动自止门,摇荡门,摆动门
trap ~ 水平翻门
underhung ~ 扯门
up and over ~ 上翻门
upward-acting ~ 上开门,吊门
vertical sliding 竖拉门
vestibule ~ 避风门
watertight ~ 防水门
doorcase ['dɔ:keis] *n.* 门框
doorkeeper ['dɔ:ki:pə] *n.* 门卫,守门人
~'s lodge 传达室
~'s room 传达室
doorless ['dɔ:lis] *a.* 无门的
doorman ['dɔ:mæn] *n.* 门房,传达
doorway ['dɔ:wei] *n.* 门口,门道
dope [dəup] *n.* 涂料,蒙布漆;浓液;防爆剂;活化剂 *vt.* 给……上涂料
~ bucket 装封口油灰的斗(管子接头配件)
fireproof ~ 耐火涂料
doping ['dəupiŋ] *n.* 涂漆;浸渍
dopplerite ['dɔplərait] *n.* 弹性[橡皮]沥青;灰色沥青
Doric ['dɔrik] *a.* (古希腊建筑中)多立克式的
~ architrave 多立克柱顶过梁
~ base 多立克柱础
~ capital 多立克柱头
~ colonnade 多立克柱廊
~ column 多立克柱
~ cornice 多立克檐口
~ cyma (tium) 多立克波纹线脚
~ echinus 多立克柱头顶板下 1/4 圆饰
~ entablature 多立克檐座
~ frieze 多立克雕带
~ order 多立克柱式
~ portico 多立克门廊
~ structure 多立克结构
~ style 多立克风格
~ temple 多立克神庙

Doric order

dormant ['dɔ:mənt] *a.* 休眠的,潜伏的 *n.* 横梁,固定的梁;枕木
~ bolt 暗闩,暗门销
~ lock 暗锁,埋头锁
~ screw 埋头螺钉
~ tree 横梁,楣
~ window 老虎窗
dormer ['dɔ:mə] *n.* 屋顶[老虎,采光]窗
~ cheek 老虎窗两侧
~ roof 有老虎窗的屋顶
~ window 老虎窗

dormer window

deck ~ 平屋顶老虎窗
flat roof ~ 平屋顶气窗
gable ~ 山墙顶窗
hip ~ 屋脊窗

dormered ['dɔːməd] *a.* 有老虎窗的
dormette ['dɔːmet] *n.* 躺椅
dormitory ['dɔːmitri] *n.* (*pl.* -ries) 宿舍; 郊外居住区
~ area 宿舍区
~ block 住宿区
~ building 宿舍楼
~ town 居住城镇
dornick ['dɔːnik] *n.* 小石子, 小鹅卵石
dorter (**dortour**) *n.* 宿舍
dosing *n.* 剂[用]量; 配制
~ chamber 配制室
~ ratio 投配比
~ room 投配间
doss [dɔs] *n.* 廉价寄宿房屋
~ house 下等客栈
dossal ['dɔsəl] *n.* 墙帷, 幔布, 帐帘, (祭坛等后面的) 挂布
dosseret ['dɔːsərit] *n.* 拱基垫块; 拜占庭建筑中的副柱头
dot [dɔt] *n.* 点, 小点 *vt.* 打点于 *vi.* 打点
~ and dash line 点划线(即—·—·—)
~ mark 刻印标记
~ ted line 点线, 虚线
~ type 打点式
~ weld 点焊
dote [dəut] *n.* 腐朽物; 腐朽木材
dotter ['dɔtə] *n.* 划点器
double ['dʌbl] *a.* 两[加]倍的; 双的, 双重的; 双关的 *ad.* 双倍地; 双重地; 双双地 *n.* 两倍; 折叠; 双打
~ acting door 双开弹簧门
~ acting hinge 双动[向]铰链
~ -acting pump 双向两用水泵
~ -acting spring butt 双向弹簧铰链
~ -acting switch 双动电开关, 双联开关
~ agitator 双铰刀
~ alternate 双重交替
~ ambulatory 双回廊
~ arched corbel-table 双拱挑出面层
~ archway 双拱门
~ armouring 加双重钢筋, 加复筋
~ arrow sign 双指向标志
~ axe 双刃斧
~ beaded 双珠饰
~ bedroom 双人房间, 双人卧房
~ bell bend 双承弯头
~ bend S 形的双弯头
~ bevel 双斜面
~ bitt 双缆柱
~ (bituminous) surface treatment 保护层, 多层处理
~ block 双轮滑车
~ bowstring truss 鱼形桁架, 双弓弦桁架
~ box 双箱式
~ bracing 双支撑
~ branch pipe 双支管
~ break switch 双断开关
~ bridging 两对交叉撑
~ bull-nose brick 双圆角砖
~ cabin 双间圆木屋
~ cable plane 双索面
~ cable suspension bridge 双索悬索桥
~ cableway 双缆道
~ cant 双斜面
~ capital 双层柱头
~ -casement window 双层开关窗
~ clay pot 空心黏土砌块
~ -closed tube 双闭管
~ coagulation 双重混凝[凝聚]
~ coating 双道涂料, 两层粉饰
~ concave 双凹
~ cone moulding 诺曼底式建筑中拱的线脚
~ cone pile 双锥体料堆

~ convex 双凸
~ course 双层板
~ curvature 双曲率
~ -curvature shell 双曲薄壳

double-curvature shell

~ deal 二英寸厚板
~ decker 双层小公寓;双层公共汽车
~ diamond interchange 双菱形互通式立体交叉
~ door 双道门
~ door bolt 双门闩
~ draft 双联单
~ drainage 双向排水
~ -end tenoner 双端制榫机
~ -end trimmed 双端修饰的
~ false ceiling 双层假平顶
~ -flight stair (case) 双跑楼梯
~ floor 双层地(楼)板
~ floor covering 双层地(楼)板饰面
~ folding door 双层摺门
~ frame 双层框架
~ framed floor 双层框架楼板
~ -frogged brick 双凹槽砖
~ gateway 双重门道
~ girder 双梁
~ glass 双层玻璃
~ -glazed window 双层玻璃窗
~ glazing 双层玻璃
~ groove 双槽
~ hanging (roof) truss 双柱上撑式桁架
~ headed rail 护墙板
~ -hinged arch 双铰拱
~ -hinged flat arch(ed grider) 双铰平拱(形梁)
~ -hinged frame 双铰框架
~ -hinged parabolic arch(ed girder) 双铰抛物线拱(形梁)

~ -hinged rectangular frame 双铰矩形框架
~ house 拼连的两所房屋
~ hung 双吊钩
~ -hung (sash) window 双悬窗,上下扯窗,双铰窗
~ intersection truss 复式交叉桁架
~ intertie 复式交叉系杆
~ iron 工字铁
~ jack rafter 复式小椽
~ joint 双接缝
~ lap tile 双层叠瓦
~ lath 双面灰板条
~ -layer 双层(数,次)
~ -layer construction 双层构造
~ layer of reinforcement 双重钢筋
~ -leaf door 双摺门
~ lining 联合式衬砌
~ lock seam 双锁卷边接合
~ -main system 双干管系统
~ mo(u)ld 复合式线脚
~ -naved church 双中殿教堂
~ nose brick 双圆角砖
~ offset 双支距;双登插管,双偏置管
~ -pane sash 两块玻璃的窗扇
~ parabolic girder 双抛物线梁
~ partition (wall) 空心隔墙
~ pendulum bearing 双摇摆支座
~ -pipe dropping system 上分式双管热水供暖系统
~ pipe heating system 双管供暖系统
~ -pitch roof 复折形屋顶
~ pitch(ed) skylight 双坡天窗
~ plank (plate) wall 双层木板墙
~ plate girder 双层板梁
~ layer 双层的
~ pole scaffold 双杆支脚手架
~ pressed brick 两次压制砖
~ projected window 双凸窗
~ quarter-turn 带有斜踏步的双直角转弯楼梯
~ reinforced beam 复[双]筋梁

~ reinforcement 双面钢筋,双配筋
~ -return siphon 乙字形存水弯管,双弯虹吸管
~ return stair 双分式楼梯
~ rib 双肋
~ riveting 双行铆接
~ rocker bearing 双摇动支座
~ rodded method 双转点法
~ roof 双重屋顶
~ sash window 双扇窗户
~ screed finisher 双找平修整器
~ shear rivet(ed) joint 双剪铆接
~ shell 双层薄壳
~ side coating 两面粉饰
~ -side shuttering 双面模板
~ -side partition (wall) 两面空心隔墙
~ skew notch 双斜槽口
~ skin (construction) 双壳层结构
~ skin floor 双壳层地板
~ sliding door 双扇拉门
~ sound-boarded floor 双层消音地板
~ strength 强力,高强度
~ strength window galss 高强窗玻璃
~ -strut trussed beam 双支撑桁架梁
~ strutted frame 双支撑架
~ swing door 双摆门
~ T(ee) (floor slab) 双 T 形楼板
~ T(ee) form (work) 双 T 形模板
~ T(ee) frame 双 T 形结构
~ T(ee) plate (slab) 双 T 板
~ T(ee) roof slab 双 T 形屋面板
~ T(ee) slab 双 T 板

double tenons
~ tenon 双雄榫
~ threaded screw 双头螺丝
~ -tower facade 双塔正面
~ towered 双塔的,双塔楼的

~ towered gatehouse 两侧有双塔的城楼
~ transept 十字形教堂的双耳堂[袖廊]
~ triangulated system 双支撑
~ tumbler bearing 双摇座
~ vault 双拱圆顶
~ wall 空心墙,双墙
~ wall with concrete fill 混凝土填充墙
~ Warren truss 复式华伦桁架
~ web 双肋式
~ -webbed 双肋的,双腹板的
~ -webbed girder 双腹板梁
~ -webbed plate arch 双腹板拱
~ -webbed plate arch(ed) girder 双腹板拱梁
~ -webbed plate girder 双腹板梁
~ -webbed T-beam 双腹板 T 形梁
~ welded joint 两面焊缝
~ window 双层玻璃窗
~ wing door 双侧门
~ -wing(ed) building 有两翼的建筑物

doubling ['dʌbliŋ] *n.* 加倍,双重;褶叠
~ back 双层油漆
~ on itself 作 180°弯摺
~ piece 风檐板

doubly ['dʌbli] *ad.* 加倍地
~ curved 双曲的
~ curved shell 双曲薄壳
~ ruled surface 双直纹面

doucine *n.* 葱形曲线饰,反曲线形装饰

dough [dəu] *n.* 揉好的陶土,捏塑体
~ mill 调胶[浆]机

Douglas fir pine 美国松,枞木

douse [daus] *v.* 把……浸入水里;浇[洒,泼]

dove [dʌv] *n.* 鸽,鸠
~ color 淡灰色,浅灰而略带紫红的颜色
~ gray 一种紫灰色
~ hinge 鸠尾铰

~ marble 淡灰色大理石
dovetail ['dʌvteil] *n.* 鸠尾[楔形]榫 *vi.* 用鸠尾榫结合;严密嵌合
~ groove 燕尾结合
~ halving 鸠尾对半结合
~ joint 鸠尾结合
~ lot 鸠尾槽
~ machine 制榫机
~ merlon 鸠尾雉蝶墙
~ rib 鸠尾肋
~ saw 鸠尾锯
~ tenon 鸠尾榫
~ed fillets 鸠尾楔形木条地板
~ed growing and tonguing 鸠尾鸳鸯接头
~ing joint 鸠尾榫接头
common ~ 普通鸠尾榫
lap ~ 互搭鸠尾榫
miter ~ 斜楔榫

dovetail halving

dovetailing joint

dowel ['dauəl] *n.* 木钉,合缝钉;暗销[榫];连接筋,合缝钢条 *vt.* 用暗销结合
~ action 暗销作用
~ bar 暗销杆;接头插筋;(水泥混凝土路面的)传力杆
~ bar chair 传力杆支座
~ bit 勺形钻,半圆形木钻
~ brick 木砖
~ hole 插筋孔
~ joint 暗销结合
~ lubricant 暗销润滑剂
~pin 销钉,合缝销
~pin holes 销钉孔眼
~pin joint 销钉结合
~ sleeve 销钉套筒
~ spacer 间隔暗销
~ steel 传力杆,合缝钢条
~ling jig 孔钻
screwed ~ 螺纹木塞;定缝螺钉
spike ~ 钉栓

spiral ~ 定缝螺钉
steady ~ 固定[定位,销紧]销
wooden ~ 木销钉

dowelled *a.* 暗销的,暗销结合的
~ beam 键合梁
~ connection 暗销连接
~ expansion joint 设置传力杆的伸缝
~ joint 传力杆接缝
~ tongue and groove joint 有传力杆的企口接缝
~ transverse expansion or contraction joints 设置传力杆的横向伸缩缝

down [daun] *ad.* 向下;降下;往市区 *a.* 向下的;下行的
~ apron 后挡板
~ buckling 向下弯曲
~ feed riser 下给立管
~ feed system 下给系统
~ -light 向下照光
~ ramp 下坡道
~ spout 水落管;溢流管
~ take 下降管,下导气管

downcast ['daunkɑ:st] *a.* 向下的;垂视的

downcomer ['daunˌkʌmə] *n.* 泄水[下水,下导,排气]管,下气[烟]道

downgrade ['daungreid] *n.* 下坡

downhand welding 俯焊

downhill ['daunhil] *a.;ad.* 向下的(地);下坡的(地)
~ conveying 下向输送
~ pipe-line 下倾的输送管线
~ welding 俯焊

downpipe *n.* 水落管,下流管
circular ~ 圆形水落管
rectangular ~ 矩形水落管

downside ['daunsaid] *a.* 下侧的 *n.* 下侧,底面

downspout ['daunspaut] *n.* 水落管
~ conductor 溢[下]流管

downstage ['daun'steidʒ] *a.* 前台的 *ad.* 在舞台前

downstairs ['daun'stɛəz] *a.* 楼下的 *n.* 楼下 *ad.* 楼下地
downtank *n.* 下流槽；收集器
downtime ['dauntaim] *n.* 停工期，停机时间
downtown ['daun'taun] *a.* 城市商业区的
～ area 市中心区，闹市区
downward ['daunwəd] *a.* 向下的；下降的 *ad.* 向下地
～ system of ventilation 下向通风系统
downwarp *v.* 下翘
dozy ['dəuzi] *a.* （木料，水果等）腐烂的
Dr. Sun Yat-sen's Mausoleum（Tomb）南京中山陵
draft [drɑ:ft] *n.* =draught 草案［图］；设计；气流，穿堂风；通风装置；琢边，凿槽 *vt.* 起草；设计
～ apparatus 通风设备
～ articles 条文草案
～ hole 通风孔
～ power 通风能力
～ tube 通风［吸入，引流］管
～ed general budget 总预算草案
～ed masonry (work) 粗琢方石
～ed stone 琢边块石
air ～ 气流
artificial ～ 机力［人工］通风
down ～ 下向通风
mechanical ～ 机械通风
natural ～ 自然通风
rough ～ 草图
static ～ 静力通风
up ～ 上向通风
draftiness ['drɑ:ftinis] *n.* 处于穿堂风状态；通风
drafting ['drɑ:ftiŋ] *n.* 起草；制图，绘图
～ board 绘图板
～ instrument 制［绘］图仪器
～ machine 制图机
～ office (room) 绘图室
～ pen 绘图笔
～ scale 绘图比例尺
～ set 全套绘图仪器
～ table 绘图桌
draftproof ['drɑ:ft'pru:f] *a.* 隔气的
draftsman ['drɑ:ftsmən] *n.* (*pl.* -men) 起草者；制图者
draftsmanship ['drɑ:ftsmənʃip] *n.* 制图术；起草术
drafty ['drɑ:fti] *a.* 通风的，通风良好的
～ room 通风良好的房间
drag [dræg] *n.* 路刮；钢齿；粗耙；阻力；刹车；拖曳 *v.* 拖，拉，拖曳
～ bar 拉杆，牵引杆
～ conveyor 刮板式输送机
～ flights 刮板式梯格
～ leveling course 整［刮］平层
～ rod 牵引杆
～ truss 阻力桁架
dragging ['drægiŋ] *a.* 拖曳用的；*n.* 刮平；拖曳
dragon ['drægən] *n.* 龙
～ beam 支承脊橡梁
～ tie 角铁联系，隅撑
dragsaw ['drægsɔ:] *n.* 拉锯，曳锯
drain [drein] *n.* 排水；排水管［沟］；消耗 *vt.* 排去……的水 *vi.* 流掉；排水
～ board 滴［泄］水板
～ cock 放水龙头［旋塞］
～ for rain water 雨水进水口；雨水沟渠
～ gutter 天［檐］沟
～ tile 排水陶管
～ trunk 排水干管；溢流道
～ing board 滴水板
air ～ 通风道，防湿沟
land ～ 地面排水沟
sub-soil ～ 地下排水沟
surface water ～ 地［路］面排水沟
drainable *a.* 可排水的
drainage ['dreinidʒ] *n.* 排［泄］水；排水设备［系统］；排出的水，污水

~ board 滴[泄]水板
~ collector 排水干管,污水集水管
~ conduit (duct pipe) 排水管
~ duct 排水管
~ engineering 排水工程
~ hole 泄水孔
~ of foundation 基础排水
~ pipe 排水管
~ tile 排水瓦管
~ water 排水
~ works 排水工程[设施]
artificial ~ 人工排水
balcony ~ 阳台排水
base-course ~ 基层排水
double ~ 双向排水
single ~ 单向排水
trellis ~ 格形排水系统

drainboard *n.* 滴水板

drained [dreind] *a.* 已排水的

drainer ['dreinə] *n.* 排水装置,排水[滤干]器;排水工,下水道修建工

drain layer *n.* 排水管铺设机

drainpipe ['dreipaip] *n.* 排水管

drake [dreik] *n.* 浮石片

dramshop ['dræmʃɔp] *n.* 酒店[吧]

drape [dreip] 窗[布] *n.* 帘;褶皱 *vt.* 披盖;悬挂,装帧 *vi.* 成褶皱状

drapery ['dreipəri] *n.* (*pl.* -peries) 帷幕,帐帘;衣饰;帷幔

draught [drɑ:ft] *n.* 气流;通风,穿堂风;通风装置,草案[图] *vt.* 起草;设计
~ bar 拉杆
~ bead 窗挡风条
~ -board molding 墙上的棋盘格装饰
~ excluder 门的气流隔断
~ hood 通[抽]风罩
~ lobby 气流隔断
forced ~ 强力通风
suction ~ 抽引风力

draughting ['drɑ:ftiŋ] *n.* 起草文件;绘制

draughtsman ['drɑ:ftsmən] *n.* (*pl.* -men.) 起草人;制图员

draw [drɔ:] *n.* 拖,拉,吸;仰开桥的开合部分 *vt.* 绘制,描写;草拟,制定;拉延[制];拔 *vi.* 制图,画画;拖拉
~ back plate 推拉板
~ bar 拉杆,牵引杆
~ bolt 接合螺栓
~ -bore 钻销孔
~ for fixation 安装图
~ -in air 引入空气
~ -in bolt 拉紧螺栓
~ -in ventilation 引入式通风
~ -off cock 泄水[排气]龙头
~ out 拔出,拔桩;拉长,抽取
~ pin 榫销
~ plate 起模[划眼]板
~ runner 滑条
~ tongs 紧线钳
~ to scale 按尺度制图[翻样,放样]
~ up an agreement 订合同
~ vice 拉钳

drawback ['drɔ:bæk] *n.* 障碍;后退
~ lock 内开锁
~ rod 后拉杆

drawbar ['drɔ:bɑ:] *n.* 拉杆,牵引杆
~ coupling 牵引装置

drawbolt ['drɔ:bəult] *n.* 套环螺栓
~ lock 内开锁

drawdown ['drɔ:daun] *n.* 下降,沉陷;水面降落
~ the curtain 开幕,放下窗帘

drawer ['drɔ:ə] *n.* 画家,制图人 [drɔ:] *n.* 抽屉
~ dovetail 互搭鸠尾榫抽屉
~ lock 抽屉锁
~ runner 抽屉滑条

drawhook ['drɔ:huk] *n.* 拉钩

drawing ['drɔ:iŋ] *n.* 制图;素描;牵引,拉拔;回火
~ board 绘图板
~ change 更改图纸

~ change request 更改图纸的要求
~ compass 绘图圆规
~ die 冲模,拔丝模
~ list 图纸清单
~ machine 拔丝机
~ of site 基地平面图
~ pin 图钉
~ room 绘图室;会客[休息]室
~ scale 绘图比例尺
~ strickle 刮板
ancient coloured ~ 古彩画
architectural ~ 建筑制图,建筑画
as-built ~ 竣工图
assembly ~ 总图,装配图
caption of ~ 图签
cold ~ 冷拉
contracted ~ 缩绘
cross sectional ~ 断面图
design ~ 设计图
detail ~ 详图,细部图
enlarged ~ 放大图
erection (installation) ~ 安装图,施工安装图
explanatory ~ 说明图
isometric ~ 等角[轴测]图
key ~ 例图
perspective ~ 透视图
plain view ~ 平面图
rough ~ 草图
sectional ~ 剖[断]面图
structural ~ 结构[构造]图
working ~ 工作[生产]图,施工详图
drawknife ['drɔːnaif] *n.* (*pl.* -knives.) (两端有柄的)刮刀
drawn [drɔːn] *a.* 牵引的;拔出的;拉伸
~ glass 拉制玻璃
~ grain 皱面
~ -out 延长,伸出
~ wire 冷拉[拔]钢丝
~ work 二道墁抹罩在湿底灰上
dream [driːm] *n.* 梦;梦幻
~ hole (仓库等的)风窗,气窗

dreikanter ['draiˌkɑːntə] *n.* 三棱石
drench [drentʃ] *n.* 雨淋;弄湿 *vt.* 浸润[透]
dress [dres] *n.* 服饰 *v.* 修整,装饰,修琢,刨光
~ circle 戏院内的前排包厢
~ ed and headed 刨光并在端部作榫的
~ ed and matched boards 刨光的镶(拼)板
~ ed and matched flooring 刨光的镶木地板
~ ed brick 磨光砖
~ ed fair face 天然石装修
~ ed masonry 敷面圬工
~ ed one side 单面修整
~ ed stone 修琢石
~ ed stuff 修饰过的材料
~ ed timber 刨光木材
~ ed two sides 双面修整
dresser ['dresə] *n.* 雕琢工;整形机;打磨机;选矿机;修整工具;橱柜[桌];化妆台;伙食台
contact point ~ 白金打磨机
crush-form ~ 磨轮压刮整形机
diamond wheel ~ 金刚石磨轮整形机
grindstone ~ 磨石刻槽器
dressing ['dresiŋ] *n.* 修整[琢];表面处理;选矿
~ cab (in) 化妆室
~ compound 水泥
~ cubicle 梳妆室
~ glass 梳妆镜
~ hammer 整面[修整]锤
~ locker 衣帽柜
~ -room 更衣[化妆]室
~ stone 修琢石
~ table 梳妆台
bituminous ~ 沥青敷面
cement ~ 水泥敷面
fine pointed ~ 细琢
hammer ~ 锤整[琢]
pitched ~ 琢边石工

plain ~ 光面修整
pointed ~ 尖凿修整
polished ~ 磨光,抛光,研磨
rock faced ~ 粗琢石,粗琢石工
rock work ~ 粗面石工
surface ~ 表面修琢[加工]
window ~ 橱窗装饰

dressings *n.* (文艺复兴及其派生建筑正面上的)线脚和装饰

drib [drib] *n.* 点,滴;细粒,碎片
 ~ bing 修整道路坑凹;零星修补

dribble ['dribl] *n.* 出水碎石沟;滴水;少量 *vt.* 使点滴流下 *vi.* 流滴

dried [draid] *a.* 干燥的
 ~ wood 干木
 oven ~ 烘干
 smoke ~ 烟熏干燥

drier ['draiə] *n.* 干燥工;干燥机,干燥器,干燥剂
 adiabatic ~ 绝热干燥器
 agitator ~ 搅动干燥器
 atmospheric ~ 空气干燥器
 band ~ 带式干燥器
 cabinet ~ 干燥室[箱,橱]
 liquid ~ 液体干燥剂
 slilca-gel ~ 硅胶干燥剂

drierite *n.* 烧石膏,无水硫酸钙

drift [drift] *n.* 漂流;漂流物;堆积物;拱圈水平推力;横坑道;穿孔器;偏流 *vi.* 漂流;吹积 *vt.* 使漂流;把……吹积
 ~ barrier 漂木屏障
 ~ bolt 穿钉;系栓
 ~ indicator 倾角计
 ~ pin 冲钉(伸张铆钉孔用)
 ~ -ping 锥[楔]塞
 ~ plug 铅管矫直木棒
 ~ punch 锥形冲头

drifter ['driftə] *n.* 冲头;架式凿岩机;穿孔器;漂流物
 ~ drill 取心[穿孔]钻

drifting ['driftiŋ] *a.* 漂流[动]的;无目标的 *n.* 漂流[积];砂[雪]堆;打洞

drill [dril] *n.* 钻机[头];钻,锥 *v.* 钻孔
 ~ bit 钻头
 ~ file 小细锉,圆边锉
 ~ hole 钻孔
 ~ lathe 钻床
 ~ machine 钻机,钻床
 ~ test 钻孔试验

drillable ['drɪəlbl] *a.* 可钻孔的

drilling ['driliŋ] *n.* 钻孔[井],钻探
 ~ bit 钻头
 ~ diamond 黑金刚石
 ~ hammer 打眼锤
 ~ machine 钻床,钻孔机

drimeter ['drimitə] *n.* 干度计,含水量测定计

drinking ['driŋkiŋ] *a.* 饮用的 *n.* 饮,饮用
 ~ fountain 自动饮水器,喷嘴式饮水龙头
 ~ water 饮用水,清水
 ~ water quality standards 饮用水水质标准
 ~ water standard 饮用水标准
 ~ water supply 饮用水供应

drip [drip] *n.* (屋)檐;滴水槽;水滴;水滴声 *vi.* 滴落[下] *vt.* 使滴落
 ~ comdenser 水淋[回流]冷凝器
 ~ cap 滴水挑檐;外窗台
 ~ edge 滴水槽檐
 ~ groove 滴水凹槽
 ~ hole 滴水孔
 ~ impression 滴痕
 ~ joint 屋面金属片互搭接头
 ~ mold 滴水槽
 ~ moulding 滴水线脚
 ~ nozzle 滴水喷嘴
 ~ -proof 防滴[雨]的,不透水的
 ~ tube 滴管

dripping ['dripiŋ] *a.* 滴水的 *n.* 滴落;水滴声
 ~ eaves 滴水檐
 ~ moisture 冷凝水

dirpstone ['drɪpstəun] *n.* 滴水石
drive [draɪv] *n.* 传动；驾驶；粗琢 *v.* 驱，赶；驾驶；铺设；把(钉、桩)打入
  ~ cap 导[引]盖；锤帽
  ~ circuit 驱动电路
  ~ fit 密配合
  ~ head 套管帽
  ~ home 把……打到底；把钉子敲进去
  ~ pin 带动销
  ~ screw 钉状螺钉；传动螺杆
drivehead *n.* 桩帽，承锤头
driveline *n.* 传动系统
driveway ['draɪvweɪ] *n.* 车道，汽车道，公路，马路
driving ['draɪvɪŋ] *a.* 推[传]动的 *n.* 打入；掘进
  ~ axle 驱动轴
  ~ belt 传动带，传动皮带
  ~ fit 密合，装配
  ~ hammer 桩锤，击锤
droke *n.* 灌木丛
drome [drəum] *n.* 飞机场 *vi.* 起飞；飞行
dromos ['drɔməs] *n.* (*pl.* dromoi) 通往地下古墓或古庙的通道；古希腊的一种赛马场
drop [drɔp] *n.* 落下；点滴落差；吊饰；落锤 *v.* 使滴[落]下；降低
  ~ arch 平圆拱，垂拱
  ~ bolt 埋头螺栓
  ~ ceiling 吊顶
  ~ chute 跌落急水槽
  ~ compass 小圈圆规
  ~ door 吊门
  ~ elbow 起柄弯头
  ~ front 活动面板
  ~ handle 悬垂把手
  ~ -in girder 悬[吊]梁
  ~ -leaf table 有活动翻板的桌子
  ~ ornament 悬吊装饰
  ~ panel 无梁楼盖的柱帽
  ~ point 下降点，滴点
  ~ riser 采暖的下供立管
  ~ scene 舞台吊幕，吊下布景
  ~ siding 互搭披叠板
  ~ strake 合并列板
  ~ system 上行[下给]式供暖系统，水循环供暖系统
  ~ table 活动桌
  ~ window 上下滑动吊窗
  ~ped ceiling 吊平顶
  ~ped panel 下垂拖板
  ~ped girder 下垂桁架
  ~ped seat 凹椅座
droplight ['drɔplaɪt] *n.* 吊灯[窗]
dropper ['drɔpə] *n.* 吊索[架]；滴管
dropping ['drɔpɪŋ] *n.* 滴落[下]；降下
drove [drəuv] *n.* 阔凿子
  ~ chisel 阔凿，石匠平凿
  ~ finish of stone 石料的短槽纹修琢
  ~ work 粗琢工作；凿修石面
drown [draun] *v.* 淹没，浸湿，把……淹死
  ~ed lime 熟石灰
  ~ed pipe 沉浸管
drugget ['drʌgɪt] *n.* 粗毛地毯；棉毛混纺织物
drugstore ['drʌgˌstɔː] *n.* 美国的杂货店[药房]
druid stone 平纹石
druids altar 牌坊
drum [drʌm] *n.* 圆筒，鼓状物；穹顶的鼓座；鼓
  ~ -built column 鼓形柱
  ~ dryer 干燥筒，鼓式烘干器
  ~ of column 鼓状柱身
  ~ of dome 穹顶的鼓座
  ~ preheater 鼓式预热器
  ~ roller 滚压机
  ~ saw 筒形锯
  ~ switch 鼓形开关
  ~ -water meter 鼓式水表
drunkery *n.* 酒馆
drw = drawing *n.* 图纸，草图

dry [drai] *a.* 干的,干燥的;干旱的;干枯的,枯竭的 *n.* 干,干燥;干物 *vt.* 使干燥 *vi.* 变干
~ after-treatment 干养护
~ air cooler 干空气冷却器
~ arch 防潮基拱;不加装饰的拱
~ -batched aggregate 干拌合材料
~ bonding strength 干砌合强度
~ bottoming 干铺底基层
~ brickwork 干砌砖
~ brushing 干刷光
~ bulb thermomtter 干球温度表
~ castings 干模铸件
~ ceiling 石膏板顶棚
~ cleaning shop 干洗店
~ cold bending 干冷弯
~ concrete 干硬性混凝土
~ connection 干接管
~ construction 干施工
~ curing 干养护,空气养护
~ fan 风干器
~ felt 干油毡
~ -fine 干抛光
~ finish 干压光
~ galvanizing 干镀锌
~ gas 液化气
~ glazing 无灰装玻璃
~ granulation 干法成粒
~ grinding 干磨
~ iron 低硅生铁
~ joint 非砌合接缝
~ lining 干衬砌
~ masonry 无砂浆砌筑
~ masonry wall 无砂浆砌筑墙体
~ mix 干拌
~ mix concrete 干拌混凝土
~ -mix shotcrete 干拌喷射混凝土
~ mix(ture) 干拌混合物
~ mortar 干砂浆
~ -moulded tile 干塑瓦管
~ offset 干胶印
~ paying 干砌,无灰浆铺砌
~ polishing 干磨光
~ press method (process) 干压法
~ -pressed brick 干压砖
~ pressing 半干压制
~ process-type rotary kiln 干燥作业回转窑
~ -rolled 干碾压的
~ -rolled surface 干压[碾]面层
~ roofing felt 铺屋面用干油毡
~ rubble 干砌毛石
~ rubble construction (masonry) 干砌毛石圬工
~ sand 干型砂
~ steel 低级钢
~ -stone base 干石基层
~ stone wall 干砌石墙
~ wall partition 清水隔墙
~ wash 干洗;干冲积物
~ wedging 干楔固
~ wood 烘干木材

dryer ['draiə] *n.* 干燥器[剂,箱]
~ drum 干燥滚筒

dryhouse *n.* 干燥室

drying ['draiiŋ] *n.* 干燥
~ agent 干燥剂
~ apparatus 干燥器[设备]
~ behaviour 干燥状态
~ by evaporation 挥发干燥
~ capacity 干燥能力
~ crack 干裂缝
~ mechanism 干燥装置
~ process 干燥处理
~ room 干燥室
~ stage 晒台
~ varnish 干清漆
furnace [oven] ~ 烘[烤]干

dryness ['drainis] *n.* 干燥
~ fraction 干燥度

drywall *n.* 干饰面内墙
~ construction 干饰面内墙构造
~ finish 内墙干饰面
~ material 内墙干饰面材料

dual ['dju(:)əl] *a.* 二的,双的;二重的; 二元的;复式的
~ cross-section 双断面
~ -flight stair (case) 双跑楼梯
~ girder 双梁
~ head girder 双顶梁
~ mould 复式线脚
~ portal 双门
~ -purpost mixing plant 两用搅拌设备
~ vent 两用通气管
~ yoke vent 双轭通气管

duality [dju(:)'æliti] *n.* 二重[元]性

dubbing ['dʌbiŋ] *n.* 粉刷找平;塑化剂,油液;防水油
~ out 刮平,打底抹面找平

duck [dʌk] *n.* 轻帆布;水陆两用军车;鸭
~ board 修屋面用步级板
~ 's-egg green 鸭蛋青

duckbill ['dʌkbil] *n.* 鸭嘴兽
~ bit 鸭嘴形钻头

duckboard ['dʌkbɔːd] *n.* 修屋面用临时步级板;铺在堑壕底的板道;挡雪板

duct [dʌkt] *n.* 管[通]道;槽,沟,渠道;预应力筋孔道
~ channel 地沟,沟槽
~ construction 管道施工
~ cover 通道盖
~ design 风道设计
~ entrance 地下管道入口
~ for electric wiring 安装电线管道
~ grouting 导管灌浆
~ riser 竖向管道
air supply ~ 供气道
ceiling ~ 平顶通道
connecting ~ 连接导管
discharge ~ 排料槽;排水管道,排气管道
distributor ~ 配电干线管道
exhaust ~ 排气风筒
fresh air ~ 新鲜空气通道
gas ~ 烟[气]道
intake ~ 进水[气]道

ductile ['dʌktail] *a.* 韧性的,可延展的;柔软的
~ cast iron 球墨铸铁
~ material 韧性材料
~ metal 韧性金属
~ rupture 韧性断裂
~ steel 韧性钢

ductility [dʌk'tiliti] *n.* 延[韧]性

ductilometer *n.* 延度计

ductule ['dʌktuːl] *n.* 小导管排泄管

ductway ['dʌktˌwei] *n.* 管式通道

ductwork ['dʌktˌwəːk] *n.* 通风系统的通风管

dull [dʌl] *a.* 钝的;阴暗的;呆滞的 *vt.* 使迟钝;使阴暗 *vi.* 变得迟钝
~ bit 钝钻头[杆头]
~ cherry-red 淡樱红
~ deposit 毛面镀层
~ edge 钝刀口
~ glass 暗淡玻璃
~ gold 暗金黄色
~ surface 无光面,暗面

dullness ['dʌlnis] *n.* 迟钝;暗淡

dumb [dʌm] *a.* 哑的,无音的
~ arch 假拱
~ iron 填缝铁条

dumboard [ˌdʌm'bɔːd] *n.* 隔声板

dumb-waiter ['dʌm'weitə] *n.* 小件升降机;自动旋转式送货机[食品架]

dummy ['dʌmi] *a.* 假的,做样子的 *n.* (*pl.* -mies) 虚设[伪装,模仿]物
~ building 假建筑
~ door 假门
~ joint 假缝;假结合
~ lock 插锁
~ plate 隔板
~ rivet 假铆钉
~ type contraction joint 假缝式缩缝
~ window 假窗

dump [dʌmp] *n.* 垃圾堆;堆存处;倾卸;

清除;切断电源 *vt*. 倾倒;倾销 *vi*. 倒垃圾;倾销商品
dumpage ['dʌmpidʒ] *n*. 卸货;卸货权[费];垃圾
dumpcart ['dʌmpkɑ:t] *n*. 倾卸车
dumper ['dʌməp] *n*. 垃圾车;自动倾卸车;清洁工人
dumping ['dʌmpiŋ] *n*. 倾倒[卸];堆积;垃圾倾卸处置
~ board 倾卸板
~ device 倾卸装置
dumpy ['dʌmpi] *a*. 粗短的,矮胖的
~ level 定镜水准仪
dunite ['du:nait] *n*. 纯橄榄岩
dunnage ['dʌnidʒ] *n*. 衬板,垫料,填木,枕木
dunstone ['dʌstəun] *n*. 镁灰岩,杏仁状辉绿岩
duodecagon [ˌdju:əu'dekəgɔn] *n*. 十二边形
duodecastyle *n*. 十二柱式
duodecimal [ˌdju(:)əu'desiməl] *a*. 十二的;十二进位制的 *n*. 十二分之一
duomo ['dwɔ:mɔ:] *n*. 教区中的主要教堂,大教堂
duotone ['dju:ətəun] *a*. 双色的 *n*. 双色照片[插图];双色套印法;艺术品复制件
duplex ['dju:pleks] *a*. 双的,二倍[重]的;双联的,复式的 *n*. 跨两层楼的公寓套房;供两家居住的房屋
~ apartment 二联式公寓,占两层楼的一套公寓房间
~ fitting 三通接头
~ -headed nail 双头钉
~ house 联式房屋
~ lock 双重钥匙锁
~ paper 叠层纸
~ planning 成双的房屋平面布置
~ receptacle 双插座
~ slide rule 两面计算尺
~ -type house 并联式住宅
duplicate ['dju:plikit] *a*. 复制的,完全一样的;成对的,副的;二重[倍]的 *n*. 复制品;副本,抄件
duplicating *n*. 复制;描图
~ machine 复写[制]机
~ room 复制室,描图室
duplicator ['dju:plikeitə] *n*. 复印机;复制者
durability [ˌdjuərə'biliti] *n*. 耐久性;坚固
~ factor 耐久性系数
~ index 耐用指数
~ test 耐久试验
duralumin [djuə'ræljumin] *n*. 硬铝
duramen [djuə'reimen] *n*. 中心木质,心材
duration [djuə'reiʃən] *n*. 持续[久];历时,持续时间
~ of service 设备使用年限
durax-cube pavement 嵌花式小石路面
durax paving 嵌花式小方石铺砌[路面]
durax stone block 嵌花式小方石块
durn [də:n] *n*. 门的木框架
durometer [djuə'rɔmitə] *n*. 硬度计
dust [dʌst] *n*. 灰尘,尘土;粉末;纷挠 *vt*. 去掉……上的灰尘 *vi*. 去掉灰尘
~ allayer [extractor; remover] 除尘器
~ alleviation 减尘
~ arrester 防[吸]尘器
~ arrester plant 防[吸]尘装置
~ bag 集尘袋
~ bin 垃圾箱
~ board 挡尘板
~ boot 防尘套
~ cap 防尘盖
~ catcher [collector] 吸尘器
~ chamber 除尘室
~ cloud 尘雾
~ coat 风衣,防尘便外衣
~ -collecting installation 吸尘装置
~ collection 吸尘
~ color 暗褐色

~ contamination 粉尘污染
~ control 防尘
~ control barrier 防尘隔离物
~ cover 防尘罩
~ diluent 粉剂填料
~ exclusion [extraction] 除尘
~ exhausting fan 除尘风扇
~ gold 金粉
~ guard 防尘板[罩]
~ gun 手提喷尘器,粉末喷枪
~ impurity 粉末杂质
~ layer 防尘器
~ laying 防尘的
~ laying agent (preventer); (protection agent) 防尘剂
~ laying material 防[止]尘材料
~ monitoring system 粉尘监控系统
~ mop (擦地板)长柄拖把
~ palliative 减尘剂,灭尘剂
~ -pan dormer 簸箕形老虎窗
~ powder 撒布剂,敷粉
~ precipitator 收尘器
~ proof 防尘
~ protection wall 防尘墙
~ shield 防尘板
~ -tight construction 防尘建筑

duster ['dʌstə] *n.* 除尘器,扫尘;撒粉器
dustfree ['dʌstfri:] *a.* 无尘的
dusting ['dʌstiŋ] *n.* 撒粉;掸扫
dustless ['dʌstlis] *a.* 无尘的
~ screenings 无灰石屑
dustman ['dʌstmən] *n.* 清洁工
dustpan ['dʌstpæn] *n.* 畚箕,粪斗
dustproof ['dʌstpru:f] *a.* 防尘的
~ ceiling 防尘顶棚
dustproofer ['dʌstpru:fə] *n.* 防尘剂
dusttight *a.* 防尘的
dusty ['dʌsti] *a.* 多尘的;粉状的
~ blue 灰兰色
~ green 灰绿色
~ line 粉线
~ material 粉末状材料

~ pink 土灰粉红色
Dutch [dʌtʃ] *a.* 荷兰的;荷兰人的;荷兰式的;荷兰文化的
~ bond 荷兰式砌合
~ brick 荷兰式砖
~ clinker 荷兰缸砖,铺路硬砖
~ door 上下可以分别开关的两截门
~ lap 荷兰式叠搭
~ light 活动玻璃框
~ oven (利用火墙余热的)荷兰式炉
~ Renaissance 荷兰文艺复兴
~ tile 饰瓦,荷兰蓝色釉砖
dutchman ['dʌtʃmən] *n.* (*pl.* -men)塞孔补缺的木料,补缺块
duty ['dju:ti] *n.* (*pl.* -ties)生产率;负荷;功率;税
 operating ~ 操作规程[制度]
dwang *n.* 板条墙立筋间斜撑;大螺帽扳手
dwarf [dwɔ:f] *n.* 矮子 *a.* 矮小的
~ door 矮门,两截门
~ partition 短隔屏
~ partition wall 短隔墙
~ wainscoting 矮护壁
~ wall 地龙墙,桥台台帽前缘的矮墙
dweller ['dwelə] *n.* 居民,居住者
dwelling ['dweliŋ] *n.* 住宅[处],寓所
~ area noise 住宅区噪声
~ construction 住宅建设
~ district 居住区
~ house 住宅,房子
~ kitchen 家用厨房,厨餐间
~ on honeycomb 蜂蜗式住宅
~ place 住处[所,址]
~ unit 居住单元
dye [dai] *n.* 染色;色,色彩;颜[染]料 *vt.* 染,染色于 *vi.* 染色
~ cell 染料盒
dyeability [,daiə'biləti] *n.* 可染性,着色性[度]
dyehouse ['daihaus] *n.* 染厂[坊],染色间
dyeing ['daiiŋ] *n.* 染色,;染色法;染色

工业
dyestuff ['daistʌf] *n.* 染料,着色剂
～ solution 染液
dyewood ['daiwud] *n.* 染料木
dying ['daiiŋ] *a.* 行将完结的,近于结束的
～ away 衰减,消失
dyke [daik] *n.* =dike 堤,堰,坝;防护栏,障碍物 *vt.* 筑堤防护
dynamic [dai'næmic] *a.* 动力[态]的;动力学的 *n.* 动力[态]

～ deflection 冲击挠度[弯沉]
～ elastic behaviour 动弹性状态
～ hardness 冲击硬度,马尔特氏硬度
dynamical [dai'næmikəl] *a.* 动力[态]的;动力学的
dysmorphosis *n.* （malformation）畸形,变形
dysphotic [dis'fəutik] *a.* 照明微弱的
～ zone 弱光带
dystone spar 硅硼钙石
dytory [dai'təri] *n.* 胶体泥浆

E

EAA=Engineers and Architects Associatioon 工程师与建筑师协会
each [iːtʃ] *a.*; *ad.*; *pron.* 各自,各个,每个(地)
～ layer 每层
eagle ['iːgl] *n.* 希腊古建筑正面上方的三角墙;鹰(徽)
eakleite *n.* 硬硅钙石
ear [iə] *n.* 建筑物上为装饰或结构目的而设置的突出部;楔形木[石,金属]块上的凹凸部分;耳(状物);吊耳[钩],(支撑)环,把手;(针)孔;外轮胎;(*pl.*) 耳子(板材或带材的端部缺陷)
～ nut 翼形螺帽
adjusting ～ 可调整的夹持器;(*pl.*)拉线用滑轮
clamp(ing) ～ 夹具,卡具
straining ～ 紧定夹持器,拉紧吊耳
tighting ～ 拉紧吊耳
earing ['iəriŋ] *n.* 耳索;(板材深冲时形成的)花边
earlandite ['iələndait] *n.* 水碳氢钙石;水柠檬钙石
earlumin *n.* 伊尔铝合金
early ['əːli] *a.* 早,初期的,初始的,旧的 *ad.* 早,初,先
～ -age cracking 早期裂纹

～ baroque 早期变态式(装饰过分的)建筑
～ bond 早期结[粘]合
～ Christian church architecture 早期基督教堂式建筑
～ crack(ing) 早期裂缝
～ curing period 早期养护阶段
～ development 早期开发
～ English architecture 早期英国式建筑

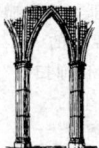 

early English style   early English style base

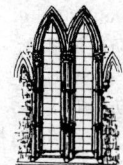

early English style window

～ English cathedral style 早期英国天主教堂式建筑

~ English style 早期英国式
~ English style base 早期英国式建筑柱基
~ English style window 早期英国式建筑窗
~ Gothic style church 早期哥特式教堂
~ Plantagenet style 早期金雀王朝式建筑
~ Pointed 早期(英国)尖拱式建筑
~ Renaissance style 早期文艺复兴式建筑
~ Romanesque style 早期罗马式建筑
~ setting (水泥或混凝土等的)早凝,速凝
~ setting cement 快凝水泥
~ shrinkage 早期收缩
~ stability 早期稳定性
~ stage 早[初]期,初始阶段
~ state 初始状态
~ stiffening 过早变稠;(水泥的)假凝
~ Stone Age 旧石器时代
~ strength 早期强度
~ -strength admixture 早强掺合剂
~ strength cement 早强水泥,快硬水泥
~ strength concrete 早强混凝土
~ wood 春材,早材

earmark [ˈiəmɑːk] *n.* 标记,特征 *vt.* 特别指定

earn [əːn] *v.* 赚[挣,博,赢,应]得;使获得,给带来

earnest [ˈəːnist] *n.* 保证金,定金;预示[兆];真实,实在 *a.* 认真的,紧要的,迫切的,坚决的
~ money (投标)保证金,定金

earning [ˈəːniŋ] *n.* 赚,挣;(*pl.*)工资,收益,报酬,利润
~ assets 收益资产,赚钱资产
~ capacity 收益能力,生产能力,利润率
~ rate 利润[收益]率
~ statement 业务收益表,损益计算书

gross ~ 总[毛]收入

earth [əːθ] *n.* 地球;陆地;泥土、土壤;接地 *v.* 埋入土中;接地
~ arrester 接地火花隙避雷器
~ blanket 粘土防渗面层
~ bus 接地母线
~ colour 矿物颜料
~ column [pillar] 土柱
~ conduit 瓦管,陶管
~ connection 接地线,埋设地线工程
~ contact 接地
~ -continuity conductor 接地导线
~ -damp concrete 干硬性混凝土
~ -dry concrete 干硬性混凝土
~ed pole 接地极
~ electrode (金属板或水管等)接地导体
~ fault 接地故障
~ -fault protection 接地保护[装置]
~ flax 石棉
~ free 不接地的
~ house 土坯房,土屋
~ lodge 木架土屋
~ membrane 黏土防渗层[墙]
~ metal sheet 接地金属板
~ moist concrete 干硬性混凝土
~ mortar 土灰浆
~ pigment 天然矿物质颜料
~ pitch 地沥青,软沥青
~ plate 接地板
~ proof 抗震的
~ silicon 硅石;二氧化硅
~ structure 土工建筑物
~ -type 陶制的
~ wall (construction) method 地下连续墙施工法
~ wax 地蜡
activated ~ 活性土
adamic ~ 红黏土
blue ~ 青[硬]黏土
brick ~ 制砖土

colo(u)ring ~ 矿物颜料
diatomaceous ~ 硅藻土
fuller's ~ 漂白土
gypsum ~ 土(状)石膏
lemnian ~ 水磨土
potter's ~ 陶土
silicon ~ 硅土
earthen [ˈəːθəm] *a.* 土制的,陶制的;土地的
earthenware [ˈəːθənwɛəʳ] *n.* 陶器,瓦器;黏土 *a.* 陶瓷的;黏土的
~ body 陶坯体
~ clay 陶土
~ glazed finish 陶面釉
~ mosaic 陶瓷锦砖
~ pipe 陶管,瓦管
~ small-sized mosaic 小块陶瓷锦砖
~ tile 陶制面砖;瓦片
~ tile pavement 陶面砖铺地
~ wall tile 陶制墙面砖
glazed ~ 釉面陶器
earthing [ˈəːθiŋ] *n.* 接地(线),埋设地线工程
earthquake [ˈəːθkweik] *n.* 地震
~ bracing 抗震斜撑
~ construction 抗震结构(物)
~ design 抗震设计
~ -proof construction 抗震结构
~ -proof foundation 抗震基础
~ -proof joint 防震缝
~ resistant design 抗震设计
~ resisting wall 抗震墙
earthwork [ˈəːθwəːk] *n.* 土方量,土工程
~ engineering 土方工程
earthy [ˈəːθi] *a.* 土(状)的,土质的;接地的;地电位的
ease [iːz] *n.* 流畅;舒服,容易 *v.* 缓和,减轻;使舒适
~ the grade 缓和坡度
~d edge 光圆边缘
easing *n.* 围墙[建筑物墙]的下部;松型

(防热裂);精修块体材料;缓和曲线
~ wedge 易脱[对垫]楔块
easel [ˈiːzl] *n.* 绘图桌,画架,展示架,张贴架,黑板架
~ painting 架上绘画
~ picture 架上画
folding ~ 折叠画架
studio ~ 画架
easement [ˈiːzmənt] *n.* 缓和曲线,自然弯线;采光权,地役权;附属建筑物
~ curve 缓和[过渡]曲线
east [iːst] *n.* 东(方),东部 *a.* 东(方)的 *ad.* 向东(方)
~ end 东端
East Indian laurel 东印度月桂树(内部装饰用木材)
~ -ward position 坐西向东,东向配置
~ window 教堂东端(主祭坛处)之窗
eastern [ˈiːstən] *n.* 东方人 *a.* 东(方)的,朝东的
~ apsis 教堂东端的半圆形殿
~ choir (= ~ church) 东方教堂式建筑
~ choir tower 东方教堂尖塔
~ crossing tower 东方教堂的中央(平面十字交叉处)尖塔
~ exedra 教堂东端的前廊
~ hemlock 加拿大铁杉
~ indian walnut 山合欢木
~ larch 美洲落叶松
~ pediment (古希腊建筑)正面入口上方三角墙;(其他建筑)门廊上方的三角顶
~ quire 东方教堂式建筑
~ red cedar 东方红杉
~ spruce 东云杉(木)
Eastern Tombs (清朝)东陵
~ W.C. pan 东方蹲式便盆
easy [ˈiːzi] *a.* 容易的,简易的;平顺的,安适的
~ chair 安乐椅
~ -cleaning hinge 长脚铰链

~ curve 平缓曲线
~ -machining steel 易加工的钢材
~ starter 简易起动装置
eat [i:t] v. 吃；腐[侵，蛀]蚀；消耗，销磨
~ing room 餐室
eaves [i:vz] n. (pl.)屋檐，山墙斜面的底部
~ board 橡上封檐板条，连檐垫板
~ catch 檐口挂瓦条
~ ceiling 廊檐顶棚
~ channel 檐槽
~ course 檐口瓦层
~ detail 屋檐详图
~ drop 屋檐滴水
~ fa(s)cia 封檐板，檐口饰带
~ flashing 檐口泛水
~ fillet 檐瓦条木
~ gallery 檐廊
~ gutter 檐沟[槽]，檐头雨水管
~ height 檐高（地面至檐口的高度）
~ lath 檐口板条
~ lead 铅制屋檐水槽
~ mo(u)lding 檐饰线条（泥灰或木工嵌线）
~ plate 檐口垫木，封檐板
~ pole 檐檩
~ rafter 檐椽
~ soffit 挑檐平顶板，檐口托板
~ strut 檐口支撑
~ tile 勾滴瓦
~ trough (trowel) 檐沟[槽]
dripping ~ 滴水檐
everhanging ~ 飞檐

eaves fa(s)cia

ebb [eb] n. 落潮，退潮；衰退 vi. （潮）退落；衰退
ebonite ['ebənait] n. 硬质胶；硬质橡胶；胶木
~ board 胶木板，硬质橡胶板
~ driver 胶柄螺钉旋具
cellular ~ 蜂蜗硬质胶，微孔硬胶
microporous ~ 微孔橡胶
ebony ['ebəni] n. (pl. -onies) 黑檀，乌木 a. 黑檀色的；黑檀制的
~ asbestos 黑石棉
~ wood 乌木，黑檀木
eccentric [ik'sentrik] n. 偏心圆[轮，器] a. 偏心的，异常的
~ angle 偏心角
~ brace 偏心支撑（加劲）
~ connection 偏心连接
~ joint 偏心接合
~ loaded column 偏心承载柱
~ riveted joint 偏心铆接
~ table joint 偏企口接合，偏心嵌接
~ tee 偏心异径三通
eccentricity [ˌeksen'trisiti] n. 偏心率[度，距]；反常，壁厚不均度
~ correction 偏心校正
ecclesia [ikli:zjə] n. (pl. -ae)教堂
ecclesiastical [iˌkli:zi'æstikəl] a. 教堂的，教会的
~ architecture 教堂建筑
~ basilica 教堂的长方形会堂
~ building style 教堂式建筑
~ monument 教堂纪念碑[古迹]
ecclesiology [iˌkli:zi'ɔlədʒi] n. 教堂建筑[装饰]学
echauguette n. 古堡角楼（带观察孔和枪眼，常设在转角或大门附近）
echelon ['eʃəlɔn] n. 梯队，梯阵 v. 排成梯形
~ form 梯形
~ matrix 梯阵
~ operation （铺筑沥青混合料的）阶梯操作法
~ strapping 梯形绕带
echinus [e'kainəs] n. (pl. -ni) 凸圆线脚（古希腊多立克式柱头上的四分之一圆线脚）

echinus

echinus and astragal

~ and astragal 柱头凸圆线脚
echo ['ekəu] *n.* (*pl.* echoes) 回波,回声,反射波 *v.* 发出回声,产生共鸣;重复,模仿;反射
~ chamber 回声[反响]室
~ -location 回声定位法
~ ranging 回声测距法
~ talker 回送干扰
Echo Wall 天坛回音壁
eclectic [ek'lektik] *a.* 折衷的;选编的
~ structure 折衷式建筑
~ (al) -garden 折衷式庭园
eclecticism [ek'lektisizəm] *n.* (建筑装饰风格上的)折衷主义
eclipse [i'klips] *n.* 日[月]蚀;晦暗;掩蔽 *vt.* 蚀,遮掩
eclipsing effects 重叠效应
eclogite ['eklədʒait] *n.* 榴辉岩
Ecole Cistercienne【法】(11世纪法国早期哥特式建筑的)西斯丁学派
Ecole d'Anjou 英国安茹王朝式建筑
ecological [ˌekə'lɔdʒikəl] *a.* 生态(学)的
~ architecture 符合生态学法则的建筑
~ art 生态艺术
~ balance 生态平衡
~ distribution 生态分布
~ environment 生态环境
~ factor 生态因素
ecology [i'kɔlədʒi] *n.* 生物[社会]生态学
economic [ˌi:kə'nɔmik] *a.* 经济(学)的;实用的;节俭的
~ accounting 经济核算
~ advisability 经济合理性
~ area 经济区
~ benefit 经济效[利]益
~ contract 经济合同
~ cooperation 经济合作
~ development zone 经济开发区
~ effects 经济效益[果]
~ order quantity (EOQ) 经济(最佳)订货量
~ worth 经济价值
economical [ˌi:kə'nɔmikəl] *a.* 经济(学)的,节约的
~ city 经济城市
~ durable years 经济的耐用年限[使用期]
~ of space 经济利用空间
economy [i(:)'kɔnəmi] *n.* (*pl.* -mies) 经济,节约;经济制度
ecru ['eikru:] *n.* 淡褐色 *a.* 淡褐色的,亚麻色的;未漂白的
ectotheca *n.* 外壁
ecumenical [ˌi:kju(:)'menikəl] *a.* =ecumenic 普遍的,全球的,世界范围的
edge [edʒ] *n.* 边(缘);刀口;界限,边界;侧面 *v.* 镶边,修边;沿边,渐近,侧近,装刃
~ action 边缘作用
~ angle 边缘角,棱角
~ arch 边拱
~ bar 缘杆,边铁
~ beam 边梁
~ -bend (=crook【美】)弯钩(状)
~ boring 边孔
~ brick 角砖
~ buckle 边扣环;边缘卷曲
~ butt joint 边抵边接合;抵头接合
~ cam 端面凸轮
~ column 边柱
~ condition 边界条件
~ course (砖)侧砌
~ crack 边缘裂缝
~ cross member 边缘横杆
~ damage 破边,边伤
~ distance 边到边距离
~ dowel 边缘插筋
~ finish(ing) 边缘加工
~ -fixed 边缘固定的
~ form (浇筑混凝土用的)边模
~ fracture (木材的)边裂
~ glass 棱镜

~ gluing 侧面[边端]胶结
~ grain （木材的）直行[径面]纹理
~ grain lumber 径[四]开木料
~ grain shingle 径切木板
~ grinding machine （石料）边缘研磨机
~ hinge 明铰链
~ iron 角铁;铁制边缘;草地修器
~ joint 边缘[端边]连接;成角边接合

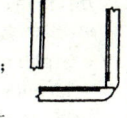
edge joint

~ joint of flooring board 纵接地板
~ knot （木材的）侧面节疤
~ knurling machine (tool) 滚（压）花机（工具）
~ lot 边缘建筑基地
~ member 边缘杆件
~ mo(u)lding 边缘线饰
~ moment 边缘[末端]弯矩
~ nailing 隐钉（钉由板边钉入并为邻板所掩盖）
~ piping 滚边,镶边
~ planer 刨边机;倒角
~ planing 刨边
~ plate 护边板
~ preparation 修边准备;（焊接前的）边缘整修
~ protection 边缘防护
~ protection bead 护角压条
~ protection strip 护角条
~ protector 护边物
~ pull 嵌入平拉门边框的把手
~ punched card 边缘穿孔卡片
~ purlin(e) 边檩
~ rail 边缘横杆
~ reinforcement 边缘钢筋;边缘加固料
~ roll 边卷装饰线脚
~ rounding 弄圆边角
~ runner 磨轮,碾子
~ scraper 边缘刮刀
~ shafts （诺曼第式建设中）支撑体的柱身

~ shape 边缘装饰线轮廓
~ shot (board) 修边的木板
~ slab 边板
~ stiffening 边缘加劲[固]
~ stress 边界[缘]应力;棱边应力
~ strip （夹板门的）围条
~ -to-edge distance 边到边距离
~ -to-face flocculation 边对面絮凝
~ to enailing 隐钉;斜嵌钉法
~ tool 削边刀;有刃的工具;修饰边缘的工具
~ trimming plane 修缘刨
~ venting 屋顶透气孔
~ view 边视图
~ warping 边缘翘曲
~ weld 边缘焊;端面焊
beaded ~ 加厚边;折边;紧固凸缘
bearing ~ （圬工中的）承梁侧
buff ~ （板的）光边
caulking ~ 嵌缝边
chamfered ~ 削边,倒棱
clamped ~ 夹固边
cutting ~ 切削刃,刃口;边缘,刃脚
delivery ~ 倾斜角
eased ~ 缓弧边
exhaust ~ 排气缘
feather ~ 薄边;（刀的）薄缘;刮尺,直规
flanged ~ 凸缘;折缘;法兰边
following ~ 后沿,后缘
front [leading, rising entering, advancing] ~ 前缘
full ~d （木材）整边
inner ~ 内缘
jagged ~ 锯齿状边缘
lagging ~ 后沿
luminous ~ 发光边
nozzle ~ 喷管切口
outer ~ 外边界
plain ~ 光面边缘
planed ~ 刨（成）边
punch ~ 冲头镶块

reference ~ 基准缘;参考端
rolled ~ 辊(成)边
root ~ (焊缝)底缘
rough ~ (木材)毛边
rounded ~ 圆缘
safe ~ 安全边
scraped straight ~ 刮准直边
simply supported ~ 简支边
straight ~ 直尺;直缘
strengthened ~ 加厚边

edger *n.* 修边器
~ grain 径切纹理

edgewise ['edʒwaiz] *a.* 沿边的; *ad.* 把边缘朝外或朝前
~ brick paving 沿边砖铺砌
~ placing 沿边铺设
~ view 边视图

edging ['edʒiŋ] *n.* 边缘线饰;金属或木质镶边带;板边修圆
~ board 镶边板
~ mo(u)ld 边缘装饰线脚
~ strip (起保护作用)镶边条
~ trowel (用于新鲜混凝土的)修边馒刀
~ mill 轧边机

edifice ['edifis] *n.* 大厦,大型建筑物;体系

education [ˌedju(:)'keiʃən] *n.* 教育,训练;教养,修养,教育学;教育程度
aesthetic ~ 美学教育
art ~ 艺术教育

educational [ˌedju(:)'keiʃənl] *a.* 教育(上)的;有教育意义的
~ block 文教建筑群
~ building 文教建筑物
~ centre 文教中心
~ facilities 教育设施

eduction [i(:)'dʌkʃən] *n.* 引出;推导[断];析出物;废气
~ pipe 排泄[气]管

eductor [i'dʌktə] *n.* 排放装置;喷射井点
water ~ 喷水器

edulcorate [i'dʌlkəreit] *vt.* 纯化,除去杂质,洗净;(信息的)精选,清除

Edwardian style *n.* (英国)爱德华式建筑

effect [i'fekt] *n.* 结果,效果[效应,效力],作用;感触,印象; (*pl.*)财物,动产;影响,效验;大意 *vt.* 实现,产生
~ a policy 投保,取得保证单
~ lacquer 美饰漆,真空涂漆
art ~ 艺术效果
back ground ~ 背景效果
converse ~ 逆效应
cubic ~ 立体效果
curdle ~ (瓷釉的)凝结效果
luminous ~ 照明效果
ornamental ~ 装饰效果
scenic ~ 舞台效果
size ~ 尺寸效应
space ~ 空间效果
surface ~ 表面效果

effective [i'fektiv] *a.* 有效的;生效的;实际的
~ age 有效使用期
~ angular field 有效视场[域]
~ aperture 有效孔径
~ area 有效[使用]面积
~ cohesion intercept 有效粘聚力
~ coil (弹簧的)有效圈
~ column length 有效柱长
~ cross-sectional area 有效截面面积
~ date 有[生]效日期
~ drainage porosity 有效孔隙率
~ efficiency 有效功能[效率]
~ flange width 有效翼缘宽度
~ height 有效高度
~ humidity 有效湿度
~ length 有效长度
~ life 有效寿命
~ management 有效管理
~ measures 有效措施
~ opening 有效孔径
~ perceived noise level (EPNL) 实

际感觉噪声级
~ permeability 有效渗透性
~ pitch 有效节距,有效螺矩
~ pore space 有效孔隙
~ pore volume 有效孔隙体积
~ size 有效尺寸,有效粒径
~ slenderness ratio 有效细长比
~ span 有效[计算]跨度
~ superficial porosity 有效表面孔隙率
~ temperature 有效[实感]温度
~ thickness 有效厚度
~ throat (depth) 有效焊缝厚度,焊缝计算厚度
~ value 有效值
~ width 有效宽度
~ working area 有效工作面积
effervesce [ˌefəˈves] *vi.* 沸腾;起泡(沫);冒泡
efficacy [ˈefikəsi] *n.* (*pl.* -cies)功[能];效能[力]
efficiency [iˈfiʃənsi] *n.* 功[效]能;效率;有效性;经济性;供给能力
~ apartment 小型公寓套房(一至二居室)
~ type apartment house 集中式公寓住宅(电梯间、楼梯间布置在住宅的中央)
efficient [iˈfiʃənt] *n.* 因素,因子;作用力 *a.* 有效的,有用的
~ protocol 有效协议
~ range 有效[工作]范围
effigy [ˈefidʒi] *n.* (*pl.* -gies)肖像;雕像;画像;俑
efflorescence [ˌefloːˈresns] *n.* 风化,粉化;晶化,凝霜;泛碱,渗斑;开花,花簇
~ resistance 抗风化(能力)
efflorescent [ˌefloːˈresnt] *a.* 风化的;晶化的;开花的
~ -proof 抗风化的

effluent [ˈefluənt] *n.* 流出物;河湖的支流;流出的废水,净化的污水 *a.* 流出的,放出的,渗漏的
~ conduit 出水管道
~ discharge conduit 排泄水管
~ pipe 出水管
~ trough 出水槽
~ water quality standard 排水水质规定标准
efflux [ˈeflʌks] *n.* 流出,流出物;射流;时光流逝;期满
effort [ˈefət] *n.* 努力;成果,作品;作用力,力量;工作(项目),(研究)计划
design ~ 设计工作
effuse [eˈfjuːz] *v.* 流出;泻出;喷出;散布 *a.* 舒展的
effusive [iˈfjuːsiv] *a.* 流出的,喷出的
egg [eg] *n.* 蛋,卵;卵形物
~ -albumin 蛋白胶合剂,白朊胶合剂
~ and anchor mo(u)-lding (西方古建筑中)卵形与锚形交替的线饰
~ and arrow mo(u)lding (西方古建筑中)卵形与箭形交替的线饰
~ and dart mo(u)lding (西方古建筑中)卵形与飞标形交替的线饰
~ and tongue mo(u)lding (西方古建筑中)卵形与舌形交替的线饰
~ china 薄瓷器
~ crate 蛋形格栅;顶棚嵌灯下的方格架[花格灯罩]
~ -crate canopy 花格式出檐;编格式遮篷;板条格式挑棚
~ -crate louver 花格式百叶窗
~ -shaped clothoid 卵形缓和曲线
~ -shaped cross section 卵形截面
~ -shaped gallery 卵形长廊
~ -shaped pipe 卵形管道
~ sleeker 蛋形墁刀
~ -shell 蛋壳(工艺品);易碎的
~ -shell china 薄胎瓷
~ -shell finish 蛋壳状装饰
~ -shell gloss 蛋壳色光泽,暗光

~ -shell lustre 蛋壳色光泽,暗光
~ -shell paint 蛋壳彩画
~ tempera 蛋青画
egress ['i:gres] *n.* 外出,出口;出路 *vi.* 出现,出去
Egyptian [i'dʒipʃən] *a.* 埃及的;埃及人的;埃及语的 *n.* 埃及人;古埃及语
~ alabaster 埃及条纹大理石
~ architecture (古代)埃及建筑
~ art 埃及艺术(对古代埃及建筑、雕塑、绘画的总称)
~ capital (古代)埃及式柱头
~ minaret 伊斯兰教寺院的尖塔
~ ornament 埃及装饰(多以象形文字式的符虎,有翅的圆球,刻有圣甲虫的宝石,象征性的动物、树叶,特别是落叶和棕榈叶为主要内容)
~ prayer-tower 埃及的祈祷塔
~ style (古代)埃及式柱头
eiderdown ['aidədaun] *n.* 一种厚绒布
eidograph ['aidəugrɑ:f] *n.* 绘图用缩放仪,缩放绘图仪
Eiffel Tower *n.* (法国巴黎)埃菲尔铁塔
eight [eit] *n.* 八 *a.* 八个
~ columned 八柱式(建筑)
~ -hole brick 八孔砖
~ shape 8 字形
~ -sided aisle 八边侧廊
~ -sided base 八边形基座
~ -sided building 八边形建筑
~ -sided cupola 八边形穹顶
~ -sided 八边形的(中世纪)城堡主楼
~ -sided lantern 八角灯
~ -sided mosaic tile 八边形陶瓷锦砖
Eight treasuries 八宝(佛教图案)
ejection [i(:)'dʒekʃən] *n.* 喷[投,抛,排]出;喷出[排泄]物
~ nozzle 喷嘴
~ opening 喷出口
ejector [i'dʒektə] *n.* 喷[发]射器;喷射泵;喷射井点
~ plate 顶板

~ rod 顶杆
~ type ventilator 射流式通风机
ekistical [i'kistikəl] *a.* 城市与区域计划的
ekistics [i:'kistiks] *n.* 人类环境生态学
elaboration [i,læbə'reiʃən] *n.* 精心制作[苦心经营,精心装饰]的成果;精密[致]
elaeolite [i'li:əlait] *n.* 脂光石
elastic [i'læstik] *n.* 橡皮筋,松紧带 *a.* 有弹性的,灵活的
~ abutment 弹性拱座
~ arch 弹性拱
~ beam 弹性梁
~ bending 弹性弯曲
~ breakdown 弹性失效
~ -brittle material 弹-脆性材料
~ characteristics 弹性特性[征]
~ condition 弹性状态[条件]
~ construction 弹性结构(体系)
~ design (method) 弹性设计(方法)
~ equilibrium 弹性平衡
~ factor 弹性因素
~ fastening device 弹性接合装置
~ fibre 弹性纤维
~ frame 弹性框架
~ joint 弹性接合[接头,节点]
~ limit 弹性极限[界限,限度]
~ mass 弹性体
~ material 弹性材料
~ packing 弹性填料,弹性垫
~ -plastic interface 弹塑性界面
~ -plastic material 弹塑性材料
~ rubber 弹性橡胶
~ sheet 弹性薄板
~ space 弹性隔块[垫条]
~ support 弹性支承[支座]
~ surface 弹性(曲)面
~ washer 弹性垫圈
~ wedge 弹性槽楔,空心槽楔
~ zone 弹性区域
elastica [i'læstikə] *n.* 弹力;弹性;弹性

(线)体;梁弹性弯曲的形状
elastically *ad.* 弹性地
～ built-in 弹性固定的
～ fixed 弹性固定的
～ supported 弹性支承的
elasticity [ˌelæsˈtisiti] *n.* 弹性(学);弹性原则;灵活性
elasticizer [iˈlæstisaizə] *n.* 增韧[塑]剂
elastomer [iˈlæstəmə] *n.* 任何具有橡胶特性之物;合成橡胶;高弹性体;弹胶物;弹性灌浆料
～ adhesives 弹胶[合成橡胶]粘合剂
～ -based contact solution 高弹性触压胶浆
～ dispersion 弹胶[合成橡胶]分散体
～ paste 弹胶[合成橡胶]膏体
～ powder 弹胶[合成橡胶]粉
～ seal(ing) 弹胶[合成橡胶]密封剂
elastomeric [iˈlætəˈmerik] *a.* 有橡胶特性的
～ bearing 弹性[合成橡胶]支承
～ butyl caulk 丁基橡胶嵌缝
～ joint sealant 弹胶[合成橡胶]接缝剂
～ pad bearing 弹胶[合成橡胶]衬垫支承
～ sheet 弹胶[合成橡胶]薄片
～ (tank base) joint 弹胶(箱式)接头
elastoplastic [iːˈlæstəplæstik] *n.* 弹性塑料 *a.* 弹塑性的
～ body 弹塑性体
～ deformation 弹塑性变形
～ material 弹塑性材料
elaterite [iˈlætərait] *n.* 弹性沥青
elbow [ˈelbəu] *n.* 弯头[管],肘管 *v.* 变成肘状
～ bend 弯头[管],肘管
～ board 窗台板
～ catch 肘形搭钩
～ cock 肘形龙头
～ equivalent 弯管当量长度
～ joint 弯管接头,肘接
～ jointed lever 肘节杆

～ lamp bracket 肘形灯托架
～ piece 肘形配件,弯头
～ pipe 弯管[头]
～ rail 窗台板,弧形栏
～ tube 肘管(直角弯管)
～ed leader 水落管弯头,鹅颈管
elconite [ˈelkənait] *n.* 钨铜合金(焊条用)
electric [iˈlektrik] *a.* 电力[动、气]的,生电的
～ air cleaner 电动空气净化装置
～ air heater 空气电热器
～ alarm system 电动警报系统
～ apparatus 电气设备
～ appliance 电气设备[仪表]
～ arc lighting [lamp] 弧光灯
～ arc welding 电弧焊
～ automatic control system 电力自动控制系统
～ bath 电热浴池
～ bell 电铃
～ blanket 电热毯
～ bulb 电灯泡
～ cabinet 配电箱[柜]
～ calorific installation 电热装置
～ commutator 换向器,转换开关
～ conductor 导(电)体,电线[缆]
～ connector receptacle 电插座
～ control room 配电室
～ cord 电线[缆]
～ curing (混凝土)电热养护
～ (disappearing) stair-way 电动扶梯
～ domestic installation 家用电器设备
～ door opener 电动开门器
～ drying oven 电烘箱
～ dumbwaiter 电动货梯
～ elevator 电梯
～ fan 电扇,鼓[通]风机
～ -filament lamp 白炽电灯
～ fittings 电器

~ fixture 电力(照明)设备
~ float 电馒刀
~ -floor panel 电热式地板,(采暖)辐射板
~ heat 电热
~ heater 电热器,[家用]电炉
~ heating 电热采暖
~ (hot) water heater 电热水器
~ illumination 电力照明
~ infrared heater 电热红外线采暖器
~ installation 电力[气]设备[装置]
~ insulation 电绝缘
~ insulation varnish 电绝缘涂料[漆]
~ interlocker 电动连锁器
~ iron 电烫斗[烙铁]
~ lamp 电灯
~ lift 电梯
~ light 电灯[光]
~ lighting 电力照明
~ luminaire 电光源
~ main 输电干线
~ meter 电度表,电量计
~ paint spray gun 电动喷漆枪
~ point welding 电点焊
~ polisher 电动抛光机
~ porcelain 电工瓷,绝缘瓷
~ precipitation 电集[除]尘
~ protection 电器保护
~ radiant heat 电辐射热
~ range 电炉灶
~ receptacle 电插座
~ refrigerator 电冰箱;电冷藏库
~ riveting 电铆
~ safety lamp 安全灯
~ saw 电锯
~ screwdriver 电动螺钉旋具
~ shovel 电动铲
~ (shuffer door) operator 门的电动启闭器
~ sign 电光标志,照明广告
~ siren 电[气]警报器
~ soldering 电焊

~ space heating 电热采暖器
~ spot welding 电点焊
~ stairway 电动扶梯
~ stirrer 电动搅拌器
~ switch 电开关
~ treatment bath 电疗浴缸
~ trowel 电动馒刀
~ varnish 高度绝缘清漆,电漆
~ wall panel 电热式墙板
~ water heater 电热水器
~ welding 电焊(接)
~ wire 电线
~ wiring 安装电线,电气线路
~ wiring regulation 装线规则
~ work 电气工程
~ wrench 电动扳手

electrical [iˈlektikəl] *a.* 电(动)的;与电有关的
~ connector 插塞,接线盒
~ discharge machining 电蚀加工
~ dust collector 电集尘器(吸尘器)
~ heating installation 电热设备
~ immersion heater 浸没式电热器
~ insulator 电绝缘器
~ porcelain 绝缘瓷瓶
~ radiator 电热辐射器
~ resistance weld 电阻焊
~ shock prevention 触电防护

electricity [ilekˈtrisiti] *n.* 电,电力,电流,电子流;电学
~ grid 供电系统,配电网
~ meter 电度表
~ substation 变电所

electrification [iˌlektrifiˈkeiʃəb] *n.* 充电,带电;电气化;使用电力
~ work 电气化工程

electrify [iˈlektrifai] *vt.* 充电,使带电;使电气化

electrobath [iˈlektrəbɑːθ] *n.* 电镀[解]浴,电解液

electrocast *n.;v.* 电熔,电铸
~ block 电熔铸块

~ brick 电熔铸砖
~ refractories 电熔耐火材料
electrocement *n.* 电制水泥
electrochemical [i'lektrəu'kemikəl] *a.* 电化学的
~ finish 电化学涂层
electrocoating *n.* 电涂；电泳涂漆
electroconductibility *n.* 导电性
electrocopper *n.* 电解铜
electrocorrosion *n.* 电腐蚀
electrode [i'lektrəud] *n.* 电极；电焊条
~ arm 电极
~ boiler 电极热水器，电热锅炉
~ tip （点焊）电极尖端
~ water heater 电极式热水器
eletrofluorescence [i‚lektrəu‚fluə'resns] *n.* 电致发光
electroform [i‚lektrəu'fɔ:m] *n.* 电铸，电解成形
electrogalvanizing *n.* 电镀锌（法）
electrogilding [i‚lektrəu'gildiŋ] *n.* 电镀金，电镀术
electro gramophone *n.* 电唱机
elecctrograph [i'lektrəugra:f] *n.* 传真电报机，电铸版机，示波器，电记录器，电雕刻器
electro heating *n.* 电热（装置）
eletro-insulation *n.* 电绝缘；电绝缘体[材料]
electrola *n.* 电唱机
electrolier [i‚lektrəu'liə] *n.* 枝形大吊灯（架），装潢灯，集合灯，电烛台；（悬挂数灯的）电气信号器
~ switch 装潢灯闪烁器
electroluminescence [i‚lektrəu‚lumi'nesns] *n.* 电致发光，电荧光
electrolysis [ilek'trɔlisis] *n.* 电解（作用），电蚀，电分析
electrolytic [i‚lektrəu'litik] *a.* 电解（质）的，可电解的
~ bath 电解槽
~ cell 电解池

~ coagulation 电解凝聚[结]法
~ colo(u)ring 电解着色法
~ etching 电解蚀刻（法）
~ galvanizing 电解镀锌法
~ polishing 电解抛光
~ refining 电解精炼
~ tin plate 电镀锡薄板，电镀马口铁
electrolyze [i'lektrəulaiz] *vt.* 电解[离]
electromagnet [i'lektrəu'mægnit] *n.* 电磁体[铁]
electromagnetic [i'lektrəumæg'netik] *a.* 电磁（铁）的；电磁学的
~ blow-out 电磁火花熄灭器，电磁灭弧器
~ type relay 电磁式继电器
electromatic *a.* 电力自动的
~ drive 电动式自动换排挡
electron [ilektrɔn] *n.* 电子
~ gun 电子枪
~ metal 镁铝合金
electronic [ilek'trɔnik] *a.* 电子的，由电子作用的
~ computer 电子计算机
~ controller 电子控制机
~ crowbar 电子保安器
~ lighting 电子照明
~ punch 电子穿孔机
electronics [i‚lek'trɔniks] *n.* 电子学；电子设备；电子艺术
electropaint *n.* 电涂层 *v.* 电涂
electropercussive [i‚lektrəupə:'kʌsiv] *a.* 电冲击的
~ welding 冲击电焊
electroplate [i'lektrəupleit] *n.* 电镀器，电铸版 *vt.* 电镀
~d coating 电镀层
~d finishes 电镀抛光
~d nails 电镀钉
electropneumatic [i‚lektrəunju:'mætik] *a.* 电动气压式的
~ interlocker 电动气压连锁装置
~ switch 电动气压开关

electropolish [i'lektrəupɔliʃ] *vt.* 用电解法磨光(金属),电抛光
electropsychrometer *n.* 电测湿度计
electrosilvering *n.* 电镀银
electrostatic [i'lektrəu'stætik] *a.* 静电的
~ cleaner 静电清洁器
~ coating 静电喷涂
~ dust collector 静电集尘器
~ filter 静电过滤器
~ paint-spraying 静电喷漆(法)
~ precipitator 静电集尘[聚灰]器
electrotinning *n.* 电镀锡
electrowelding *n.* 电焊
electrum [i'lektrəm] *n.* 琥珀色之金银合金
element ['elimənt] *n.* 元素,要素,成份;基础,原理,初步,纲要;单元,单体,构件,零部件;电池
~ connection 构件的连接
~ design 构[元、零]件设计
binding ~ 粘结剂
elementary [ˌeli'mentəri] *a.* 基础的,基本的;初步的,未成熟的;元素的;单元的
~ area 单元面积
~ line 原线
~ school building 小学校舍
elemi ['elimi] *n.* (*pl.* -mis)榄香
~ balsam 榄香油[胶]
~ resin 榄香树脂
elephant ['elifənt] *n.* 刻[开]槽机;波纹铁,瓦垄铁;象;累赘
~ trunk 混凝土输送管,象鼻形溜管;巨大柱身
elevate ['eliveit] *vt.* 举起,提高,架高
~d duct 高架管道
~d footway 高架人行道
~d pedestrain crossing 人行天桥
elevating boom 提升吊杆
elevation [ˌeli'veiʃən] *n.* 建筑物立面图;高程,海拔标高;上升,提高
~ angle 仰角,高度角;垂直角;竖向角

~ drawing 立面图
~ planning of a building 建筑立面设计
~ view 立[正]视图,立面图
elevator ['eliveitə] *n.* 升降机、电梯(美)
~ apartment 高层公寓
~ automatic dispatching device 电梯自动调度装置
~ bank 电梯组
~ cage 升降机箱,电梯笼
~ hall 电梯厅[间]
~ hoistway 升降机[电梯]井
~ lobby 电梯厅[间]
~ machine room 升降机机房
~ residence buiding 高层住宅建筑
~ stage 升降式舞台
eleven [ilevn] *n.* 十一个 *a.* 十一
eligible ['elidʒəbl] *n.* 合格者 *a.* 合格的,适当的;可以采纳的
~ source 合格货源
elimination [iˌlimi'neiʃən] *n.* 除去,消去;排除,排泄;淘汰(数学)消元法
~ method 消元法
eliminator [i'limineitə] *n.* 分离[排出,消除]器;除[挡]水板;空气净化器;阻塞滤波器,干扰消除器,等效天线
~ plate 除水板,(洗涤室)阻水板
moisture ~ 脱湿[干燥]器
elinvar ['elinva:] *n.* 镍、铬恒弹性钢
Elizabathan architcture *n.* (16世纪英国的)伊丽莎白式建筑

Elizabethan architcture

ell [el] *n.* L 形之物;与主楼垂直的侧楼;弯管,弯头
~ -beam L 形梁
~ girder L 形衍架梁
ellipse [i'lips] *n.* 椭圆(形)
~ of elasticity 弹性椭圆
~ speaker 椭圆形扬声器
ellipsograph [i'lipsəgra:f] *n.* 椭圆规

ellipsoid [i'lipsɔid] *n.* 椭圆,椭球 *a.* 椭圆(形)的,椭球(形)的
～-dome 椭圆形穹顶
～ vault 椭圆形拱顶

elliptical [i'liptikl] *a.* 椭圆(形)的
～ arc 椭圆弧
～ arch 椭圆拱,扁拱
～ cylinder 椭圆柱
～ dome 椭圆形穹顶
～ file 椭圆形锉,半圆锉
～ pointed arch 椭圆尖头拱
～ section 椭圆形截面
～ staircase 椭圆梯
～ vault 椭圆拱顶

ellipiticity [ˌelip'tisiti] *n.* 椭圆率[状],椭圆度

elm [elm] *n.* 榆树,榆木

elongate ['i:lɔŋgeit] *v.* 延[伸,拉]长 *a.* 延长的,细长的
～d particle 针状颗粒
～d trough 延长凹槽

elongation [ˌi:lɔŋ'geiʃən] *n.* 延长,伸长;延长的部分;伸长率

elucidate [i'lu:sideit] *vt.* 阐明,说明

elusion [i'lu:ʒən] *n.* 逃避,避免

elutriation [i'lu:trieiʃən] *n.* 淘洗[分,选,析],洗提[涤,净]

elutriator [i'lu:trieitə] *n.* 淘析器(测定土粒级配);冲洗器,提洗器,砂子洗净器

Emanual style *n.* (16世纪初葡萄牙的)埃曼努尔式建筑

emarginate [i'mɑ:dʒəneit] *a.* 边缘有凹痕的
～d tail 凹尾

embattle [im'bætl] *vt.* 筑以城垛,饰以锯齿形
～d moulding 锯齿形线饰
～d parapet wall 锯齿形胸墙

embed [im'bed] *vt.* 埋入[置,藏],嵌入,插入
～ded column 暗柱(部分埋入墙面)
～ded footing 埋[嵌]入式基脚,嵌固基础
～ded in concrete 埋入混凝土内
～ded mesh 预埋网
～ded panel 埋置[嵌装]板
～ded panel system 壁板式(供暖)系统
～ded parts 埋设[预埋]件
～ded penstock 埋藏式压力水管
～ded pipe 暗装管道,地下暗管
～ded steel 埋设[预埋]钢筋

embellish [im'beliʃ] *v.* 装饰,修饰,美化

embellishment [im'beliʃmənt] *n.* 装饰,修饰,润色;装饰品,艺术加工
～ work 装饰工程

emblazon [im'bleizən] *vt.* 饰以纹章,使炫耀,盛装

emblem ['embləm] *n.* 象征,标志,纹章,徽章,图案表示

emblema [em'bli:mə] *n.* (*pl.* -mata) 浮雕装饰

emboss [im'bɔs] *vt.* 做浮雕[凸饰];使隆起,(纸上)压花纹
～ed aluminium foil 浮雕铝箔
～ed design 浮雕设计
～ed glass 浮雕玻璃
～ed (hard) board 浮雕(硬)木板
～ed linoleum 压花漆布
～ed panel 浮雕面板
～ed pattern 浮雕花式
～ed plywood 浮雕胶合板
～ed sheet 压花金属薄板
～ed texture 压花织物
～ed tile 浮雕面砖
～ed wall paper 压花墙纸
～ed work 浮雕细工

embossing *n.* 浮雕,压[滚,印]花(法)
～ die 压花模

embossment *n.* 浮雕(花样,细工),浮花装饰

embow [em'bəu] *v.* 弯成弧形,成弓状

embrasure [im'breiʒə] *n.* 门窗的内宽外窄的开口;碉堡的炮眼;漏斗状斜面墙

embrittlement [em'britlmənt] *n.* 脆裂[变,化],脆性[度]
embroidery [im'brɔidəri] *n.* (*pl.* -deries)刺绣(艺品)
emerald ['emərəld] *n.* 翡翠,绿宝石;翠绿色
 ~ feather 文竹
 ~ green 翡翠绿,翠绿色;祖母绿
 oriental ~ 绿刚玉
emerge [i'mə:dʒ] *vi.* 出[呈]现,显[曝]露;排[射]出
emergency [i'mə:dʒnsi] *n.* (*pl.* -cies)紧急[突然,意外]事件,急需,备用 *a.* 紧[应]急的,备用的,安全的,临时的
 ~ accommodations 应急住所
 ~ apparatus 应急备用设备
 ~ basin 备用池
 ~ construction 防险建筑物
 ~ corridor (防火灾用)应急走廊
 ~ decree 安全技术规程
 ~ dwelling 避难住所
 ~ exit 太平门,疏散口,安全出口
 ~ facilities 应急措施设备
 ~ feature 应急用装置
 ~ house 应急临时住宅
 ~ illumination 应急照明
 ~ key 应急按钮
 ~ kits 应急工具
 ~ ladder 应急梯
 ~ lamp 事故备用灯
 ~ lighting 事故照明
 ~ management 应急措施
 ~ measures 应急措施
 ~ outlet 事故[安全]出(水)口
 ~ safety switch 应急安全开关
 ~ stair 太平梯,防火灾用疏散梯
 ~ temporary construction 应急临时建筑物
emery ['eməri] *n.* 金刚砂
 ~ belt 金刚砂带
 ~ cloth 金刚砂布
 ~ grinder 金刚砂磨石
 ~ grit 金刚砂粒
 ~ powder 金刚砂粉
 ~ (sand) cloth (金刚)砂布
 ~ (sand) paper (金刚)砂纸
 ~ stone 金刚砂石,刚玉石
 ~ wheel (金刚)砂轮
eminent ['eminənt] *a.* 高的,显著的,突出的
emissarium *n.* (古罗马的)地下溢洪渠,地下水道
emissary ['emisəri] *n.* 分[排]水道,支流
emission [i'miʃən] *n.* 发出[射,散];发布,发行;发射物,流出物
 ~ angle 光投射角
 heat ~ 热辐射
emissivity [,emi'siviti] *n.* 发射率,辐射系数
emit [i'mit] *vt.* 发射[出];发布,发行(纸币)
emollescence *n.* 软化(作用)
emolument [i'mɔljumənt] *n.* 薪金,报酬,津贴
empire ['empaiə] *n.* 帝国;绝缘(漆)
 ~ cloth 绝缘油布,漆[胶]布
 Empire style (拿破仑称帝后)法国建筑,家具的设计式样
 ~ tube 绝缘套管
empirical [em'pirikl] *a.* 经验(主义)的,凭实验的
 ~ data 经验数据
 ~ method 经验方法
emplecton (=emplectum) *n.* 空斗石墙
employee [,emplɔi'i:] *n.* 雇员,职工,受雇者
 ~ benefit fund 职工福利基金
 ~ bonus 职工酬劳金
 ~ insurance fund 职工保险基金
employer [im'plɔiə] *n.* 雇主,老板,使用者
employment [im'plɔimənt] *n.* 职业,工作;使用,雇用;服务,就业
 ~ centre 就业中心

~ contract 雇用合同
~ department 就业部门,劳资科
emporium [em'pɔːriəm] *n.*(*pl.* -riums,-ria)商业中心区,商场,大百货商店
empress ['empris] *n.* 女皇,皇后
~ slate 大石板瓦
empty ['empti] *n.* 空车[箱,罐] *v.* 变空,流[注]入 *a.* 空(虚)的
~ -cell (pressure) process 空细胞法(木材防腐的压力处理法)
emulational [ˌemjuˈleiʃeiʃənl] *a.* 模仿[拟]的,仿真[效]的
~ model 仿真模型
emulsify [iˈmʌlsifai] *vt.* 使乳化,使成乳剂
~ fied asphalt 乳化石油沥青
~ fied bitumen 乳化沥青
~ fied oil 乳化油
~ fied reclaimed rubber coating 回收橡胶的乳化涂层
~ fied silicone oil 乳化硅油
~ fied substance 乳化剂
~ ing agent 乳化剂
emulsion [iˈmʌlʃən] *n.* 乳浊液,乳剂[胶]
~ adhesive 乳胶粘结剂
~ base 乳胶基底
~ binder 乳胶粘结剂
~ binding medium 乳胶粘结剂
~ bonding agent 乳胶粘结剂
~ breaker 乳胶分解剂
~ carrier 载胶体
~ cement 乳胶粘结剂
~ cleaning agent 乳液清洁剂
~ coating 乳胶涂层(混凝土养护用)
~ glue 乳胶体
~ membrane 乳胶薄膜
~ of tar 乳化焦油沥青
~ paint 乳胶漆,乳剂涂[颜]料
~ polish 乳液抛光剂
~ separation 乳胶分离

~ slurry 乳浆,乳液稀砂浆
~ treatment (沥青)乳液处理
~ tube 乳胶管
~ varnish 乳胶清漆
~ water 水乳液
~ wax 乳状蜡
emulsoid [iˈmʌlsɔid] *n.* 乳胶体[液],乳浊体
emulsor [iˈmʌlsə] *n.* 乳化器[剂]
enamel [iˈnæməl] *n.* 瓷釉 *n.* 珐琅,搪瓷;搪瓷制品;亮[光]漆 *vt.* 涂瓷釉[搪瓷,亮漆];使发光泽
~ blue 大青色,藤青
~ clay 釉瓷黏土
~ colour 搪瓷颜色,釉瓷色
~ covering 漆(包)皮
~ glazed coating 瓷釉涂层
~ (insulated) wire 漆包(绝缘)线
~ lacquer (纤维素)瓷漆
~ leather 漆皮
~ paint 瓷[亮]漆
~ paraffin wire 蜡浸漆包线
~ stove 烧釉炉
~ ware 搪瓷器
~ led brick 釉瓷砖,琉璃砖
~ led cable 漆包线,绝缘线
~ led cast iron 搪瓷铸铁
~ led glass 釉面玻璃
~ led ironware 搪瓷铁器
~ led leather 漆皮
~ led pressed steel 搪瓷钢板
~ led product 搪瓷制品
~ led reflector 搪瓷反射罩
~ led tile 釉瓷瓦,琉璃瓦
acid-resisting ~ 耐酸搪瓷
air-drying ~ 风干瓷漆
baking ~ 烘干瓷漆
drawing ~ with copper body 铜胎画珐琅
filigree ~ with copper body 铜胎掐丝珐琅
translucent ~ 半透明珐琅

high -gloss ~ 高光瓷漆
porcelain ~ 搪瓷
semi-gloss ~ 半光瓷漆
vitreous ~ 透明釉,釉瓷
vitrified ~ 搪瓷
en-block [en'blɔk] *n.* 单块,整体
~ construction 整体[单块]结构
encapsulate [en'kæpsəleit] *vt.* = encapsule 密封,封装,装入胶囊,压缩
encarpa [en'ka:pəs] *n.* = encarpus 垂花装饰,花果形装饰
encasement [in'keismənt] *n.* 装箱,包装(物),外壳,镶板,饰面
encase [in'keis] *vt.* 包裹[围],埋[镶,嵌]入,装箱
~d knot 枯木节疤,暗节
encasing [in'keisiŋ] *n.* 砌[饰]面;护壁,模板;外壳,罩子;装[埋,嵌]入,包裹
encashment [en'kæʃmənt] *n.* 兑现,付现
encaustic [enj'kɔ:stik] *a.* 蜡画的,烧入色彩的 *n.* 蜡画(法),烧彩画(法)
~ brick 彩砖,釉面砖,琉璃砖
~ painting 蜡画,烧瓷画
~ tile 彩瓦,釉面瓦,琉璃瓦
enceinte [en'seint]【法】*n.* 围廓[地]
enchain [in'tʃein] *vt.* 锁住,束缚,抓牢
enchase [enj'tʃeis] *vt.* 镶嵌,镂刻,浮雕
ench(e)iridion [enkaiə'ridiən] *n.* (*pl.* -ridions, -ridia) 手册,便览
encircle [in'sə:kl] *vt.* 环绕,包围
enclasp [in'kla:sp] *vt.* 紧握,环抱
enclose [in'kləuz] *vt.* 围绕;封入,附寄;封闭
~d arcade 环围式拱廊[骑楼]
~d building 封闭式建筑
~d court 四周有墙或建筑物围绕的庭院,大杂院
~d ground 隐蔽地区
~d knot 木材暗节
~d porch 封闭式门廊
~d slot 封口槽
~d space 建筑空间

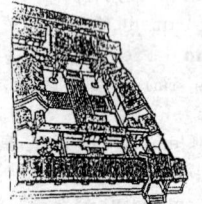
enclosed court

~d stair 隔绝楼梯
enclosing construction 围护结构
enclosing masonry wall 外围砖[石]砌墙
enclosure [in'kləuʒə] *n.* 围绕[场,墙,圈];外壳,盒,套,罩;界限
~ plan 周边式布置
~ planting 围植
~ wall 大围墙
flameproof ~ 防火外壳
radiator ~ 散热器罩
shielded ~ 屏蔽室
shower bath ~ 淋浴小间
stairway ~ 楼梯间
encounter [in'kauntə] *n.*; *v.* 遭遇,碰撞,冲突,对抗
encroachment [in'krəutʃmənt] *n.* 侵占[犯,蚀];侵蚀地
encrust [in'krʌst] *v.* 包以外壳,固结成皮,结垢,镶饰
encyclop(a)edia [en'saikləu'pi:diə] *n.* 百科全书
end [end] *n.* 端点,末尾;目的[标],结局;限度,极限,边界 *v.* 结束,终止
~ acroterion (希腊建筑)正门三角墙端点上的饰物
~ -anchored tendon 两端锚固的预应力钢筋
~ area method 端面积法
~ arch 端拱
~ band 端带[箍]
~ bar 端部钢筋
~ bay 边跨
~ beam 端梁

~ bearing 端支座[承]
~ board 端[侧]板
~ bracing 端部横[斜]撑
~ bracket 端部牛腿,尾轴承架
~ built-in 端末砌[嵌]固[插入]
~ bulkhead 端隔墙
~ butt joint 端头对接
~ cam 端面凸轮
~ cap 端帽
~ cap orifice 带帽管口
~ check (木材)端[辐]裂
~ column 端[边]柱
~ -cut brick 端切[砍头]砖
~ diaphragm 端部隔板
~ elevation 端立面,侧视图
~ filler (檐沟的)端盖
~ fixity 端部固定
~ girder 端部大梁
~ grain 端面纹理,木材横切面
~ joint 端[对,平]接
~ lap joint 搭接
~ liner 底衬
~ -matched lumber 端头拼接板
~ -milling 立铣
~ mould 端模
~ mode 末端节点
~ nozzle (檐沟的)端部下水口
~ of brick 砖端
~ of contract 终止合同
~ ornament 墙端[尾顶]上的装饰
~ pin 尾销
~ plate 心[镶、端]板,侧板
~ post 端柱,(桁架)端压杆
~ ridge tile 端脊瓦
~ section 端截面
~ shake 木材端裂[辐裂]
~ skew brick 斜端砖
~ slip of bar 钢筋端部滑移
~ span 端[边]跨
~ split 木材端裂[辐裂]
~ stiffener 端部加劲杆
~ tenoner 端部开榫机

~ tie bar 末端拉杆
~ tile 端盖瓦
~ to joint 对[平]接
~ view 端视,侧视图
~ wall 端[侧]墙
~ -ways 竖的,以端向前(或朝上)的
~ wrench 平扳手

endemic [en'demik] *a.* 地方性的,风土的

endless ['endlis] *a.* 无穷尽的,环状的,循环不止的
~ belt 环带
~ chain 循环链
~ joint 环状接合
~ saw 环锯,带锯

endorsement [in'dɔːsmənt] *n.* 票据等后面的签名,背书;认可,批注
~ comfirmed 已确认的背书
~ for collections 托收背书
~ guaranteed 已担保的背书

endothermic [ˌendəu'θəːmik] *a.* 吸热的

end-over-end mixer 立式圆筒混合机

endowment [in'daumənt] *n.* 捐赠(基金或财产)

endurance [in'djuərəns] *n.* 持久力,耐久性
~ behaviour 耐久性能
~ crack 疲劳断裂[裂缝]
~ degree 耐久程度
~ fracture 疲劳断裂
~ hours 持续(工作)时间
~ period 持续时间
~ phenomenon 耐久现象
~ strength 耐久强度

enforce [in'fɔːs] *vt.* 强制,实施,执行
~ a rule 实施规则

engaged columns

engage [in'geidʒ] *v.* (机械)衔接,啮合;从事;预约[定],毕(身显)露,附联
~d column 嵌墙柱,壁柱,半露柱

~d pier 附墙方柱

engagement [in'geidʒmənt] *n.* 保证,约会,契约,雇用,(机械)衔接
~ screw 衔接丝杆

engineer [ˌendʒi'niə] *n.* 工程师,机械师,技师 *vt.* 设计;监督;建造;指挥
~ in chief 总工程师
~'s scale 比例尺
architecture ~ 建筑工程师
civil ~ 土木工程师
decoration ~ 装饰[璜]工程师
general ~ 总工程师

engineering [ˌendʒi'niəriŋ] *n.* 工程(学),工程技术,管理,操纵
~ administration manual 工程管理手册
~ brick 高强抗蚀砖
~ construction 工程建设
~ contract 工程承包合同
~ design 工程设计
~ design plan 工程设计方案
~ design review 工程设计的审查
~ drawing 工程制图
~ instructions 技术说明书[细则]
~ legislation 工程法规
~ management 工程管理
~ manual 工程手册
~ materials 工程材料
~ material specification 工程材料规范
~ plastics 工程塑料[材]
~ property 工程性能[特性]
~ report 工程技术报告
~ scale 比例尺
~ specification 工程规范,技术指标
~ standard 工艺[技术]标准
~ structure 工程结构[建筑物]
~ supervision 工程视察[检查],技术监督
~ technical design specification 工程技术设计规范
acoustical ~ 声学工程
architectural ~ 建筑工程
civil ~ 土木工程,民用建筑
construction ~ 建筑工程
electrical ~ 电气[机]工程
environmental ~ 环境工程
illuminating ~ 照明工程
maintenance ~ 保养工程,维修技术
management ~ 管理技术
municipal ~ 市政工程
production ~ 工艺设计,生产过程组织技术
safety (-first) ~ 安全技术
sanitary ~ 卫生工程
service ~ 维修[运行]工程

English ['iŋgliʃ] *a.* 英国(人)的 *n.* 英文 *vt.* 译成英文
~ architecture 英国式建筑
~ basement 英国式房屋底层
~ bond 英国式砌砖法(顶砖层与顺砖层交错)
~ -Chinese garden (18 世纪英国流行的)中英混合式庭园
~ cross bond 英国十字缝砌砖法(即荷兰式砌法)
~ garden wall bond 英国式庭园墙砌砖法(三顺一顶砌法)
~ Gothic architecture 英国哥特式建筑
~ landscape style garden 英国风景式庭园
~ Romanesque 英国罗马式(建筑)
~ roof tile 英国式屋面瓦
~ shingle 英国式屋面瓦
~ truss 英国式桁架(三角桁架)
~ spanner 英式[活动]扳手
~ units 英制单位

engobe *n.* 釉底料

engorgement [en'gɔːdʒmənt] *n.* 舱口,装料孔

engrail [in'greil] *vt.* 镶双锯齿形花边,将边刻成锯齿[波纹]状

engrain [in'grein] *vt.* 深染,使染成木纹色
~ lining paper 木纹色衬纸

~ wallpaper 木纹色墙纸

engrave [in'greiv] *vt.* 雕刻,铭记
~d letters 刻字
~d sign 雕刻标志

engraving [in'greiviŋ] *n.* 雕刻[镌版]术;木刻[铜]版;木刻[版]画

engross [in'grəus] *vt.* 大字写,正式誊写;以垄断方式收买,独占

enhance [in'ha:ns] *vt.* 增加,提高,加强
~d type 增强型

enhydrous [en'haidrəs] *a.* 内部含有水分的

enlarge [in'la:dʒ] *v.* 扩[增,放]大
~d culvert inlet 流线型洞口
~d detail 大样图
~d drawing 放大图

enlargement [in'la:dʒmənt] *n.* 扩大[充、建],放大;增补(物),扩建部分,放大的照片;详述
~ to a building 建筑的扩建部分

enneagon ['eniəgɔn] *n.* 九角[边]形

enneahedron [,eniə'hedrɔn] *n.* 九面体

enneastylos [,eniə'stailəus] *n.* 九柱式的建筑

enoscope ['enəskəup] *n.* 折光镜,L形视车镜

enrichment [in'ritʃmənt] *n.* 浓缩,富集(化);增添装饰
degree of ~ 浓度

enrockment [in'rɔkmənt] *n.* 基底填[抛]石;堆[抛]石体

enrol(l)ment [in'rəulmənt] *n.* 登记[注册],登记人数

ensemble [a:n'sa:mbl] 【法】*n.* 全[整,总]体,综[集]合,建筑群
~ average 总体平均值

ensue [in'sju:] *vt.* 随起,因而发生,结果

entablature [en'tædlətʃə] *n.* (西方古典柱式的)檐部(包括檐口、檐壁、额枋),柱顶(线)盘,上横梁
chamfered ~ 雕槽柱顶盘
composite ~ 混成型柱顶盘
Corinthian ~ 科林斯柱顶盘
Doric ~ 多立克柱顶盘
Ionian ~ 爱奥尼柱顶盘

entablement [en'teiblmənt] *n.* 柱顶线盘,承雕像的平台,柱上楣构

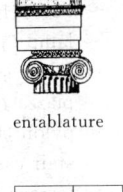
entablature

entailment [in'teilmənt] *n.* 雕刻装饰,细雕

entanglement [in'tæŋglmənt] *n.* 纠纷,缠结,陷入困境;(*pl.*)有刺铁丝网,障碍物

entasis ['entəsis] *n.* (西方古典式柱身的)收分线,微凸线

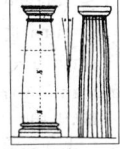
entasis

enter ['entə] *v.* 进入,参加;注册,登记;申报海关
~ a bid 投标

enterclose ['entəkləuz] *n.* 隔墙,通道,穿堂

enterprise ['entəpraiz] *n.* 企业,公司
~ cost 企业成本

entertainment [,entə'teinmənt] *n.* 游艺,娱乐;款待,招待
~ centre 娱乐中心

entitle [in'taitl] *vt.* 定名称,称呼

entourage [,ɔntu'ra:ʒ] *n.* 【法】周围,环境,配景

entrain [in'trein] *v.* 使坐火车,携[夹]带,输送,吸入
air ~ing agent (混凝土中的)加气剂
air ~ing concrete 加气混凝土

entrance ['entrəns] *n.* 入[进,河]口,大门
~ and exit 进出口
~ arch 进口门拱
~ corridor 进口走廊
~ court 前院
~ curve 入口[驶入]曲线
~ door 进口大门,外门
~ driveway 进口车道

~ facade 进口门面
~ foyer 进口门厅[休息室]
~ gellery 进口廊道
~ hall 门厅
~ lamp 门灯
~ lobby 前厅
~ piazza 进口长廊,外廊
~ pylon 进口塔门
~ stair-case 进口楼梯
~ terrace 进口平台
~ turn 进口回车道[场]
service ~ 用户入线
entrap [in'træp] *vt.* 诱陷,夹[裹]住,截留,滞阻
~ped moisture 吸着水分
entrepot ['ɔntrəpuə] *n.* (*pl.* -pots)【法】仓库,货物集散地,商业[贸易]中心
entrepreneur [ˌɔntrəprə'nə:] *n.*【法】(*pl.* -neurs)企业家,创业者,承包人,主办人
entresol ['ɔntresɔl] *n.*【法】(*pl.* -sols)假[半,阁]楼(在两层楼房之间),夹层
entry ['entri] *n.* (*pl.* -tries)进[输,记]入;进[入,河]口,门;(书,表中的)条目,记载
envelope ['envilǝup] *n.* 包络[迹,线],外[机]壳,封套[袋],气囊,信封
environ [in'vaiərən] *n.* (*pl.*) 附近,近郊,郊区 *vt.* 环绕,包(围)
environment [in'vaiərənmənt] *n.* 环境,围[环]绕
~ control system 环境控制[管理]系统
~ model 环境模型
~ system 环境体系
architecture ~ 建筑环境
induced ~ 外界感应环境
social ~ 社会环境
environmental [inˌvaiərən'mentl] *a.* 环境[绕]的,周围的
~ adaptation 环境适应性
~ amenity 环境的舒适

~ architerture 环境[周围]建筑
~ assessment 环境评价
~ assimilating capacity 环境容量
~ atmosphere 环境气氛
~ climate 环境气候,室内小气候
~ colours 环境色
~ condition 环境条件
~ control chamber 人工气候室,环境控制室
~ design 环境设计
~ evaluation 环境评价
~ factor 环境因素
~ green space 环境绿地,防护绿带
~ health 环境卫生
~ noise 环境噪声
~ planning 环境规划
~ protection 环境保护
~ psycholgy 环境心理学
~ research 环境研究
~ sanitation 环境卫生
~ space 环境空间
~ standard 环境标准
~ temperature 环境温度
Eocene ['i:ɔsi:n] *n.* ;*a.* 始新世(的)
eolation [ˌi:əu'leiʃən] *n.* 风蚀(作用)
eolotropic *a.* 各向异性的
ephebeion [ˌefə'baiɔn] *n.* (*pl.* -beia)(古希腊罗马建筑中的)运动场所
ephemeral [i'femərəl] *a.* 短暂的,瞬息的
epicycloid [ˌepi'saiklɔid] *n.* 外摆线,圆外旋轮线
Epidauros theater 埃比道洛斯剧场
epidote ['epidəut] *n.* 绿帘石
epigenesis [ˌepi'dʒenəsis] *n.* 外力变质(作用)
epigraphy [i'pigrəfi] *n.* 碑铭学,金石学
epikote *n.* 环氧(类)树脂(商品名)
epinaos [ˌepi'neiɔs] *n.* (*pl.* -naoi)(古希腊神庙中的)屋后走廊,后部小室
episcope ['episkəup] *n.* 反射映画器,不透明物的投影器

epistyle ['epistail] *n.* = architrave 轩缘, 楣梁; 盘各门框及窗等的嵌线
epitaph ['epita:f] *n.* 墓志铭, 碑铭[文]
epitaxy [epi'tæksi] *n.* 外延, 取向附生[接长]
epitome [i'pitəmi] *n.* 摘要, 概括, 缩影
epoch ['i:pɔk] *n.* 纪元, 时代[期]
　～-making 划时代的
epoxide ['epɔksaid] *n.* 环氧化合物, (*pl.*)环氧衍生物
　～ alloy 环氧树脂合金
　～ cement 环氧胶结剂
　～ resins 环氧树脂
epoxy ['epɔksi] *n.* (*pl.* -oxies)环氧基树脂; *a.* 环氧基的
　～ alloy 环氧树脂合金
　～ adhesive 环氧粘结剂
　～ -asphalt 环氧沥青
　～ binder 环氧胶结剂
　～ cement 环氧胶结剂
　～ bonded fiber-glass board 环氧树脂纤维玻璃板
　～ coating 环氧涂层
　～ concrete 环氧树脂混凝土
　～ crack filler 环氧填缝料
　～ foam 环氧泡沫
　～ glass 环氧玻璃
　～ grout 环氧薄浆
　～ insulation 环氧树脂绝缘
　～ joint 环氧连接
　～ membrane 环氧薄膜涂层
　～ mortar 环氧砂浆
　～ paint 环氧树脂涂料
　～ paste 环氧膏
　～ -phenolic binder 环氧酚醛胶结剂
　～ plastics 环氧塑料
　～ plasticizers 环氧增塑剂
　～ resin 环氧树脂
　～ resin adhesive 环氧树脂粘合剂
　～ varnish 环氧漆
　～ wearing surface 环氧磨耗层
epoxy-bonded *a.* 环氧树脂粘合的
epoxyn *n.* 环氧树脂类粘合剂

epsilon [ep'sailən] *n.* 希腊字母(大写 E, 小写 ε)
　～ symbols ε 符号
Epsom salt *n.* 泻盐(硫酸镁结晶)
Epur'e [,epju:'rei] (asphalt)*n.* 精制湖(地)沥青
equal ['i:kwəl] *n.* 对手, 相等物 *v.* 等于, 使相等 *a.* 相等的, 同样的, 公平的, 均匀的
　～ altitude method 等高法
　～ angle (iron) 等边角钢
　～ division 平均分配
　～ leg angle 等边角钢
　～ sides 等边
equalization [,i:kwəlai'zeiʃən] *n.* 平均, 均等[衡], 调平[均]
equalize ['i:kwəlaiz] *vt.* 使平等[均], 使均衡
equalizer *n.* 平衡器, 平衡杆[梁], 均压线
equator [i'kweitə] *n.* 赤道
equatorial [,ekwə'tɔ:riəl] *a.* 赤道的 *n.* 赤道仪
equiangular [,i:kwi'æŋgjulə] *a.* 等角的
　～ figure 等角形
　～ polygon 等角多边形
　～ spiral 等角螺线
equiaxial [,i:kwi'æksiəl] *a.* 等轴的
equidimension [,i:kwidi'menʃən] *n.* 等尺寸, 同大小, 等因次
equidistance [,i:kwi'distəns] *n.* 等距离
equiform ['i:kwifɔ:m] *a.* 同形的, 相同功用的, 相似的
　～ group 相似(变换)群
equilateral [,i:kwi'lætərəl] *n.*; *a.* 等边
　～ arch (哥特式)等边拱, 等边二心拱
　～ polygon 等边多边形
　～ roof 等角屋顶
　～ triangle 等边三角形
equilibrium

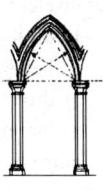

equilateral arch

[ˌiːkwiˈlibriəm] n. (pl. -ums -ria) 平[均]衡
~ brightness 平衡亮度
~ condition 平衡条件
~ configuration 平衡形状
~ diagram 平衡图
~ humidity 平衡湿度
~ index 平衡指数
~ method 平衡方法
~ polygon 平衡多边形,索多边形
~ temperature 平衡温度

equiluminous [ˌiːkwiˈluːminəs] a. 等照度的
~ surface 等照度面

equimultiple [ˈiːkwiˈmʌltipl] n. 等倍数[量] a. 乘以同数的,等倍数的

equipage [ˈekwipidʒ] n. 设[装]备,成套器皿

equipment [iˈkwipmənt] n. 设[装]备,工[机]具,部[附]件,运输配备
~ drawing 设备图
~ installation 设备安装

equipoise [ˈekwipɔiz] n. 平衡(物,力),均势,配重;称锤;砝码 vt. 使平衡

equiponderant [ˌiːkwiˈpɔndərənt] a. 平[均]衡的,等量的
~ state 均衡状态

equipressure [ˌiːkwiˈpreʃə] n. 等压
~ surface 等压面

equity [ˈekwiti] n. (pl. -ties) 公平,公正,资产净额,股权
~ financing 集资
~ ownership 资产所有权

equivalent [iˈkwivələnt] n. 相等物,当量 a. 相等[当]的,等效的
~ beam 等效梁
~ direct radiation 等效散热(面积)
~ material 等代[替代]材料
noise ~ 噪声等值

era [ˈiərə] n. 纪元,时代

eraser [iːˈreizə] n. 橡皮,挖字刀,黑板擦,消磁头

erect [iˈrekt] v. 竖立,架设,安装 a. 直立的
~ing bill 安装材料单
~ing frame 脚手架
~ing jib 装配夹具
~ing welding 安装[现场]焊接

erection [iːˈrekʃən] n. 直立,竖起;建筑,架设,安装,装配;建筑物
~ allowance 装配容许偏差
~ bolt 安装螺栓
~ bracing 安装支撑
~ by overhang 吊挂装配
~ clearance 安装净空
~ column 安装[临时]柱
~ diagram 装配图
~ drawing 装配图,施工图
~ equipment 安装设备
~ hinge 装置铰链
~ insurance 安装保险
~ joint 装配接头
~ loop 安装环,吊环
~ method 安装方法
~ of framing 构架安装
~ on site 现场安装
~ procedure 安装程序
~ reference plane 安装[装配]基准面
~ scaffold(ing) 安装用脚手架
~ schedule 安装进程[计划]
~ sequence 安装顺序
~ tolerance 安装公差,允许误差
~ truss 安装桁架
~ weight 安装重量
~ work 安装工程,装配工作

erector [iˈrektə] n. 安装[装配]工,安装升降架

ergogram n. 示功图,测力图

ergograph n. 示功器,测力器

Erichsen test n. 埃里克森试验,材料的拉伸性能试验

erode [i'rəud] *v.* 侵[腐,冲]蚀,冲刷
erodible *a.* 易蚀的
 ~ material 易受蚀材料
erosion [i:'rəuʒən] *n.* 腐[冲,侵,剥]蚀,风化
 ~ protection 侵蚀防护,防冲(刷的铺砌)
 ~ resistance 抗侵蚀能力
 ~ resistant lacquer 防腐漆
 ~ resisting insulation 耐腐蚀绝缘
 ~ scar 侵蚀痕
erratic [i'rætik] *a.* 不稳定[规则]的;移动的 *n.* 漂砾石
error ['erə] *n.* 误差,错误
 ~ allowance 误差容许限
 ~ control 误差控制
 ~ excepted 允许误差
 ~ of calculation 计算误差
 ~ of measurement 测量误差
 ~ of position 位置误差
eruptive [i'rʌptiv] *a.* 爆发的,喷出的 *n.* 火成岩
escalade [eskə'leid] *n.* 爬云梯,活动人行道 *vt.* 用梯子攀登[爬墙]
escalator ['eskəleitə] *n.* 自动楼梯
 ~ arrange 自动楼梯的布置
escape [is'keip] *n.* 漏水[气],排气[水]管,出口 *v.* 逸漏,排泄
 ~ cock 放气[水]龙头[旋塞]
 ~ corridor 安全[防火]通道
 ~ door 安全出口,太平门
 ~ gate 放水[安全]闸门
 ~ exit 太平门,疏散口
 ~ hatch 安全闸门,太平门
 ~ hole 排出口,放水孔
 ~ ladder 安全梯
 ~ opening 安全门
 ~ roof (walk) way 安全(防火)屋顶走道
 ~ route 安全撤离路线
 ~ stair (case) 太平梯
 ~ window 太平窗

 ~ -way 安全通道
 ~ works 泄水建筑物
 fire ~ 防火安全出口
 gas ~ing 漏气
escar ['eskə] *n.* 蛇形丘
escarp [is'ka:p] *n.* 壕沟内壁 *v.* 筑陡坡
escarpment [is'ka:pmənt] *n.* 陡坡,悬崖,急斜面,断层,马头丘
esconson *n.* 窗边框内侧
escort ['eskɔ:t] *n.* 护卫(者,队),仪仗队 [is'kɔ:t] *vt.* 护送[航],押运
escribe [es'kraib] *vt.* 旁切
escrow ['eskrəu] *n.* 待完成的担保证书
 ~ agreement 有条件的转让契约
escutcheon [is'kʌtʃən] *n.* 饰有纹章的盾;钥匙孔盖,门锁的覆板;遮护板
 ~ pin 盾形针
 ~ plate 门把手垫板
 key ~ 键纹板
 lamp switch ~ 灯开关片
esker ['eskə] *n.* =eschar, eskar 冰河沙堆,蛇形丘
espagnolette *n.* 长窗钩,(法国式窗子用的)插销
 ~ bolt 长插销
espalier [is'pæljə] *n.* 树篱[墙],花木攀架
 ~ lath 花棚板条
esplanade [,esplə'neid] *n.* 广场,(海边)游憩场,(外岸的)斜堤
esquisse [es'kwi:s] *n.* 草拟图稿,底稿,梗概
essential [i'senʃəl] *a.* 基本[重要,必要]的 *n.* 本质,要素
 ~ colour 主色
 ~ parameter 基本参数
 ~ part 基本部分
 ~ tone 基本色调
esserbetol [e'sə:bitəl] *n.* 聚醚树脂
establish [is'tæbliʃ] *vt.* 建(设,确)立,证实,开业,定居
 ~ed customs 常规,成例
 ~ed policy 一贯政策

establishment [isˈtæbliʃmənt] *n.* 建[确]立,创办;机构,企业
branch ~ 分局,附加设备
estate [isˈteit] *n.* 地[财]产,庄园,领地;旅行车
~ agent 地产经纪人
~ planning 住宅群公用地综合规划
housing ~ 住宅区
personal ~ 动产
real ~ 不动[房地]产
estavel *n.* 地下河,涌泉
ester [ˈestə] *n.* 酯;(*pl.*)酯类
~ gum 松(香酸)酯胶,甘油三松香酸酯
~ gum varnish 酯树胶清漆
esthetic [i:sˈθetik] *a.* 美(学)的,富有美感的
~ area 观光[风景]区
~ aspect 美学观点
~ effect 美学效果
estimate [ˈestimit] *n.* 估[预,概]算,估计,判断 [estiˈmeit] *v.* 评价[定],估计
~ accuracy 估算[概算]精度
~ of cost 估价
~ price 估价
~d amount 预算[计划]工程量
~d completion data 估计完工日期
~d cost 预算价值[费用],估计成本
~d price method 预定价格法
terms of ~ 估算项目
written ~ 施工预算书
estimating *n.* 编制预算
estrade [esˈtra:d] *n.* 讲台[坛]
estuarine [ˈestjuəˈrain] *a.* = estuarial 入海河口的,海湾的
~ sand 河口砂
etalon [ˈeitəlɔn] *n.* 标[校]准器,标准量具[样件],基准,规格
etch [etʃ] *v.* 蚀刻,侵[浸,腐,刻,镂]蚀
~ cut method 腐蚀切割法
~ pattern 腐蚀图形
~ pit 腐蚀坑[痕,陷斑]
~ primer 磷化底漆,反应性底漆
~ -proof 抗腐蚀
~ reagent 酸洗[腐蚀]剂
~ed glass 磨砂[无光]玻璃
eternit *n.* 石棉水泥
~ corrugated sheet 石棉水泥波纹板
~ pipe 石棉水泥管
~ roofing 石棉水泥屋面
ethane [ˈeθein] *n.* 乙烷
~ acid 醋酸,乙酸
ethanol [ˈeθənɔul] *n.* 乙醇,酒精
ethenoid resin *n.* 乙烯树脂
ethoxyline resin *n.* 环氧树脂
ethyl [ˈeθil] *n.* 乙(烷)基,防爆剂
~ acrylate 丙烯酸乙酯
~ cellulose plastics 乙基纤维类塑料
~ ester 乙酯
~ gasoline 乙基汽油(有防爆作用)
ethylene [ˈeθili:n] *n.* 乙烯,乙撑,次乙基
~ chloride 氯乙烯
~ copolymer 乙烯共聚物
~ oxide 环氧乙烷
~ -propylene rubber 乙丙橡胶
~ tube 乙烯管
ethyne [ˈeθain] *n.* 乙炔
Etruscan architecture *n.*（古代意大利的)伊特拉斯坎建筑
~ art 伊特拉斯坎艺术
ettringite *n.* 钙矾石,三硫型水化硫铝酸钙结晶(俗名水泥杆菌)
European [ˌjuərəˈpi:ən] *n.* 欧洲人;*a.* 欧洲(人)的
eustylos *n.*（古希腊,古罗马神庙的)柱式(其柱间距为柱径的2¼倍)
eutectic [juːˈtektik] *n.* 低共熔混合物,共晶体 *a.* 易熔的,共晶的
~ alloy 共晶[易熔]合金
eutectoid [juːˈtektɔid] *n.* 共析混合物,共析体 *a.* 共析的
~ alloy 共析合金

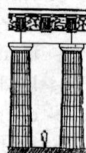
eustylos

~ steel 共析钢
evacuate [i'vækjueit] *v.* 排泄,清除,抽空,撤离
evaluation [i,vælju'eiʃən] *n.* 评价[估],鉴定;计算数值
~ chart 评估图
~ criteria 评价准则
evaporation [i,væpə'reiʃən] *n.* 蒸发,脱水
even ['i:vən] *a.* 平坦的,均匀的,对等的,偶数的 *v.* 使平整[衡,坦],使对等
~ dye 均匀染料
~ grain 均匀木纹
~ illumination 均匀照明
~ joint 平接
~ surface 平整的表面
event [i'vent] *n.* 事件,结果,现象
~ schedule 事项进度表
ever ['evə] *ad.* 曾经,无论何时,永远
~ green 常青(树)
~ increasing 不断增进[增强]的
everyday ['evri'dei] *a.* 每日的,日常的
~ routine 日常工作,每日的例行公事
evidence ['evidəns] *n.* 证[根,数]据,迹象,证明书
evolute ['i:vəlu:t] *n.* 渐屈线,展开线,法包线,波形装饰
exact [ig'zækt] *a.* 正[准,精]确的;严格的;精确[密]的
~ analysis 精确分析
~ fit 精确配合
~ height 准确高度
~ value 精确值
exaggerate [ig'zædʒəreit] *v.* 夸大[张];扩大,增加
examination [ig'zæmi'neiʃən] *n.* 检验[测],审查,考验
~ of materials 材料检验
final ~ 最后检验
visual ~ 目视检验,直观检查
examine [ig'zæmin] *vt.* 检验[查],审[调]查,研究,测验
~ and approve 核准

~ and receive 验收
example [ig'za:mpl] *n.* 例证[题,如],样品[本]
~ of application 应用实例
excavate ['ekskəveit] *vt.* 挖掘[方],开挖
exceed [ik'si:d] *v.* 超[越]过,过剩[度]
~ capacity 超过范围[可能]
~ing length 超长[余量]尺寸;毛长
excelsior [ek'selsiɔ:] *n.* 木丝,锯屑,细刨花
~ board 细刨花板
excentral [ek'sentrəl] *a.* = excentric(al) 偏心的
excentricity [iksen'trisiti] *n.* 偏心距[率度]
exception [ik'sepʃən] *n.* 例[除]外;反对,异议
~ clause 例外条款
~ list 例外清单
exceptional [ik'sepʃənl] *a.* 例外的,特别的,异常的
~ advantages 特别利益
~ use 特殊用法
excess [ik'ses] *n.* 过多[度,分];超额,过剩 ['ikses] *a.* 过分[剩]的,额外的,附加的
~ budget 超出预算
~ demand 超额需求
~ earnings 超额收益
~ glaze 釉缕,过厚釉
~ noise 过量噪声
~ of export 出超
~ of import 入超
~ reserve 超额准备[储备金]
excessive [ik'sesiv] *a.* 过度[量,分,甚,大]的,极度的,非常的
~ camber 过大路[梁]拱,过大拱度
~ cost 超额成本
~ expenditure 超额开支
exchange [iks'tʃeindʒ] *v.* 交[互,调,转,兑]换;交易 *n.* 交[转,兑]换;交换机[台],汇兑(率),交易所,电话局

~ resin （离子）交换树脂
value in ~ 交换价值
value of ~ 交换价格
exchanger [iks'tʃeindʒe] *n.* 交换器[机,剂],换[散]热器
~ -type subcooler 热交换式低温冷却器
mixed-bed ~ 混合床树脂交换器
exchequer [iks'tʃekə] *n.* 国库,资财,财源
excircle [ek'sə:kl] *n.* 外圆
excise [ek'saiz] *v.* 删去；收税 *n.* 消费[执照]税,货物税
excite [ik'sait] *vt.* 激发[励,磁,起],刺激,使感光
exclude [iks'klu:d] *vt.* 排除[去],隔绝[断],拒绝
~d space 占住空间
excluder [iks'klu:də] *n.* 排除器[剂]
~ pigment 防锈颜料
dust ~ 防尘器
exclusion [iks'klu:ʒən] *n.* 拒绝,排[斥],隔绝,禁区,限外之物
~ area 禁区,无人居住区
exclusive [iks'klu:siv] *a.* 排他[外,除,斥]的,专用[有]的；全部的；*n.* 专利权
~ distribution 总经销（商）
~ district 专用区
~ residential district 住宅专用区
~ right 专利权
~ use 专用
excogitate [eks'kɔdʒiteit] *vt.* 想出,设计,创造,发明
excogitation [ekskɔdʒi'teiʃən] *n.* 设计,发明；计划,方案
excubitorium [eks,kju:bi'tɔriəm] *n.* (*pl.* -toria) (古罗马）守夜人的职位,站岗人的岗位；（僧院中）守夜人的房间
execute ['eksikju:t] *v.* 实[执]行,完成,实现；签字盖章,批[签]发
~ a contract 在合同上签字
~ a plan 实现计划

~ an order 接受定货
execution [,eksi'kju:ʃən] *n.* 执[实]行,完成,签名盖章
~ cycle 执行[完成]周期
~ programme of works 工序,工程进度表
~ scheme drawing 施工计划图
executive [ig'zekjutiv] *a.* 执行的,行政的 *n.* 行政部[机关],执行者,行政官员
~ branch 行政部门
~ contract 执行合同
~ cost 施工费用
~ program 管理指令,执行程序
exedra ['eksədrə] *n.* (*pl.* -drae) 户外的半圆形长凳；（古希腊,古罗马的）龛座,有凳门廊
exemption [ig'zempʃən] *n.* 免[解]除,免税
~ certificate 免税单
~ from taxation 免税
~ period 免税期
exercise ['eksəsaiz] *n.*；*v.* 练[实]习,训练,运动；履行
exfoliation [eks,fəuli'eiʃən] *n.* 剥落[离],分层,层离
~ phenomenon 剥离现象
exhaust [ig'zɔ:st] *v.* 取[用,耗]尽,排除[空,气,干] *n.* 排气装置,废气
~ air 排净空气
~ air duct 排风管道
~ blower 抽气[吸风]机
~ duct 排气通道
~ fan 排气风扇[机]
~ grill(e) 排气格栅
~ hole 排气孔
~ hood 排气罩
~ muffler 排气消声器
~ nozzle 排气(喷)嘴
~ opening 排气孔[口]
~ pipe 排气管
~ port 排气口[门]
~ steam heating 废气采暖

~ temperature 排气温度
~ vent 排气孔
~ ventilation 排气通风
~ water pipe 排水管

exhibition [ˌeksibiʃən] *n.* 展览,陈列,显示;展览会,陈列品,博览会
~ building 展览馆
~ hall 展览馆[厅]
~ room 展室

exist [igˈzist] *vi.* 存在,生存,有
~ing building 已建[现有]的建筑
~ing circumstance 现状[况]
~ing construction map 建筑结构分类现状图
~ing utility 现有公用事业设施

exit [ˌeksit] *n.* 出口[路],太平[安全]门 *v.* 退场,离去
~ branch 出口管
~ corridor 出口廊道
~ device 出口设施
~ direction sign 出口方向标志
~ door 出口[太平]门
~ facilities 出口设施
~ indicators 出口指示器
~ lighting 出口指示灯
~ opening 太平门,屋顶窗
~ passageway 出口通道
~ signs 出口标志
~ stairway 出口楼梯
emergency ~ 紧急出口,太平[安全]门

exogenous [ekˈsɔdʒinəs] *a.* = exogenic 外生[成,来,因]的;起于外部的,生于外因的

exorbitant [igˈzɔːbitənt] *a.* 过[度,高,大]的,非法的
~ prices 非法[过高]价格
~ profit 超额利润

exothermic [ˌeksəuˈθəːmik] *a.* = exothermal 放[发、散]热的

exotic [egˈzɔtik] *a.* 外来的,外国产的 *n.* 舶来品,外来语
~ fashions 异国风尚

expand [ikˈpænd] *v.* 扩张[大,展],膨胀,开展,延伸,发展
~ed aggregate concrete 膨胀性骨料混凝土
~ed cement 膨胀水泥
~ed clay 膨胀黏土
~ed cork sheet 膨胀软木片
~ed glass 多孔玻璃
~ed joint 伸缩缝[接头]
~ed joist 伸展搁栅
~ed mesh screen (反眩光)金属网眼屏
~ed metal 多孔(拉制)金属网,网眼钢板
~ed metal lath 灰幔钢网
~ed pearlite 膨胀珍珠岩
~ed plastics 多孔塑料
~ed polystyrene 膨胀性聚苯乙烯
~ed rubber 多孔橡胶
~ed shale 膨胀性页岩
~ed -shale aggregate 膨胀页岩骨料
~ed sheet metal 网眼薄钢板
~ed slag beads 矿渣膨珠
~ed slate 烧胀页岩
~ed vermiculite 膨胀蛭石
~ed view 展开图,透视图
~ing and noncontracting cement 膨胀性无收缩水泥
~ing agent 膨胀剂
~ing drill 扩孔钻
~ing mortar 膨胀性砂浆
~ing pliers 扩边钳

expansion [iksˈpænʃən] *n.* 扩张[大,展],膨胀,延伸,延伸率;扩大部分,扩张物;展开式
~ agent 膨胀剂
~ anchor 膨胀锚固
~ and contraction 伸缩
~ apparatus 膨胀仪
~ bend 伸缩弯管
~ bit 伸缩钻头,扩孔器
~ bolt 伸缩螺栓
~ clearance 膨胀间隙

~ concrete 膨胀混凝土
~ couplings 伸缩接头[联接器]
~ crack 膨胀裂缝
~ gallery 廊道扩大段
~ hinge 伸缩铰链
~ joint 伸缩接头,伸缩缝
~ joint sealing 伸缩缝盖缝条[止水]
~ loop 伸缩弯管,膨胀圈[环]
~ plate 伸缩板
~ ratio 膨胀率[比]
~ shim 填隙片
~ slot 伸胀孔
~ spacing 胀缝隙
~ strip 伸缩缝嵌条
~ U pipe U 形伸缩管
~ washer 伸缩垫片

expansive [iks'pænsiv] *a.* 膨胀的,宽阔的
~ soil 膨胀土

expect [iks'pekt] *v.* 预期,期待,希望
~ed quality level 预期品质水准
~ed value 期望值

expend [iks'pend] *v.* 花[销]费,耗尽

expenditure [iks'penditʃə] *n.* 经费,开支,费用,支出额
~ of capital 投资费用
~ on construction 建造[施工]费用
capital ~ 基本建设费用

expense [iks'pens] *n.* 费用,代价,消费,经济负担
~ of idleness 窝工费用
~ of production 生产费用
~ of superision 监督管理费用
~ standard 费用标准[定额]

experience [iks'piəriəns] *vt.* 经历,感受 *n.* 阅历,经验

experimental [eks͵peri'mentl] *a.* 实验的,根据经验的,试验性的
~ design 试验(性)设计,经验设计
~ error 实验误差
~ facility 实验装置
~ formula 经验公式

~ method 实验方法
~ model 实验模型
~ plot 试验小区

expert ['ekspə:t] *n.* 专家,检验[鉴定]人 [ek'spə:t] *a.* 熟练的,有专长的
~ evidence 鉴定
~ knowledge 专门知识
~ skill 专门技能

expiration [͵ekspaiə'reiʃən] *n.* 期满,终止
~ notice 期满通知书

explanatory [iks'plænətəri] *a.* 解释的,说明的
~ drawing 说明图
~ legend 说明图例

explicit [iks'plisit] *a.* 明确[显]的
~ cost 外现[显性]成本,货币支付成本

exploration [͵eksplɔ:'reiʃən] *n.* 勘探[查],调查,研究,确定

exploratory [iks'plɔ:rətəri] *a.* 勘探的,考察的,调查的,试探性的

explore [iks'plɔ:] *v.* 勘探,查勘,探测,调查

explosion [iks'pləuʒən] *n.* 爆发[炸,破];剧增
~ protection door 防爆门

exponent [eks'pəunənt] *n.* 指数,幂,阶;样品[本]

export [eks'pɔ:t] *v.* 输出,出口,外销 ['ekspɔ:t] *n.* 输出物,出口货

expose [iks'pəuz] *vt.* 暴[揭]露,曝光,陈列
~d aggregate concrete 露石混凝土
~d aggregate finish 水刷石饰面
~d body (陶瓷)脱釉,剥落釉
~d conduit 明管
~d face 外露面
~d finish tile 外饰面砖
~d joint 明缝[接头],清水砖缝
~d joist 明搁栅
~d nailing 外露钉
~d pipe 明管

~d post 明露（承重）柱
~d wall 外露墙
exposure [iks'pəuʒə] *n.* 暴[揭]露，曝光，照相底片，陈列品，方位[向]
~ control 曝光控制
~ intensity 照射[曝光]强度
~ interval 曝光间隔
~ of lining 露砖
express [iks'pres] *vt.* 表示[达] *a.* 运送的，特别快的 *n.* 快车[递]
~ liner 豪华快轮
~ pile 大头桩
expropriation [eks'prəuprieiʃən] *n.* 征用，没收，侵用
~ of land 土地征用
expropriate [eks'prəuprieit] *vt.* 没收，征用（土地，财产），剥夺所有权，侵占；
expulsion [iks'pʌlʃən] *n.* 驱逐，逐出，排放出，喷溅
~ of water 脱水，去水
exsiccator ['eksikeitə] *n.* 干燥器[剂]，除湿器
extend [iks'tend] *v.* 伸长，延展，扩大[建]，传播，填充
~ed foundation 扩展式基础
~ed range 扩充域[范围]
~ ed shank bib [cock] 长把旋塞[龙头]
extender [eks'tendə] *n.* 扩张[延伸]器；填充料[剂]；油漆调和料
~ pigment 体质颜料，填充性颜料
extensibility [ikstensə'biliti] *n.* 伸长率，延伸度；延展度
extension [iks'tenʃən] *n.* 伸长，扩张[建]，推广，增加的部分
~ bolt 门插销，附加门栓
~ casement hinge 伸张架式铰链
~ flush bolt 伸缩式埋头螺栓
~ hinge 阔隙窗铰
~ pipe 延伸管
~ rod 伸缩标尺，伸长杆
~ rule 伸缩尺
~ spring 拉簧

~ table 可伸长式桌
extensive [iks'tensiv] *a.* 广阔[泛]的，大量的
~ order 大批定货
~ town planning 大区域城镇规划
extent [iks'tent] *n.* 程度，限量，广度，范围
exterior [eks'tiəriə] *n.* 外部[面，表] *a.* 外部[在，来]的
~ apartment area (EAA) 公寓外面的面积（阳台等）
~ balcony 外阳台
~ coating 外壳[套，罩]，被覆
~ column 外柱
~ common area (ECA) 户外公用面积（场地）
~ concrete form 外部混凝土模板
~ door 外门
~ drainage 室外排水
~ durability 外露耐久性
~ extent 外广延（度）
~ finishing 外部装修[面饰]
~ installation 室外安装
~ lighting 室外照明
~ marble 外用大理石
~ noise 外部噪声
~ orientation 外定向[方位]
~ paint 外用[表层]油漆
~ panel 端格间板
~ perspective centre 外透视中心
~ pipe system 外部管道系统
~ plywood 外用胶合板
~ -protected construction 围护结构
~ resident area (ERA) 外居住面积（等于公寓外面的面积 EAA＋户外公用面积 ECA）
~ sewer system 外部排水系统
~ space 外部空间
~ stairway 室外楼梯
~ structure 外部结构
~ stucco 外墙拉毛粉刷
~ surface 外表面

~ trim 外部门头装饰
~ -type plywood 外装修用胶合板
~ varnish 外用清漆
~ view 外视图
~ wall 外墙
~ yard 室外庭院

external [eks'tə:nl] *a.* 外部[面,界]的,客观的,外来的
~ angle 外[凸]角
~ cable 室外电缆
~ cooling 外部冷却
~ corrosion 外部腐蚀[锈蚀]
~ crack 表面裂缝
~ device 外部装置
~ diameter 外径
~ dimension 外形尺寸
~ distance 外距
~ door 外门
~ drencher 外部防火淋水器
~ economics 外部的有利条件
~ environment 外部环境
~ equipment 外部设备
~ force 外力
~ form 外观[形]
~ friction 外摩擦
~ gauge 外径规
~ heater 外加热器
~ mitre 外隅角,外斜角接合
~ observation 外部观测
~ plumber 屋外管子工
~ skin 外层墙
~ temperature 室外温度
~ thread 外螺纹
~ validity 对外有效度
~ wall 外墙
~ window sill 外窗台

extinct [iks'tiŋkt] *a.* (已)熄灭的,灭绝的,已废的,已失效的

extinction [iks'tiŋkʃən] *n.* 熄灭;吸光;吸声
~ coefficient 吸光系数,吸声系数

extinguish [iks'tiŋgwiʃ] *v.* 熄灭,消除,使衰减,使无效
portable ~er 轻便灭火器

extinguishment [iks'tiŋgwiʃmənt] *n.* 消灭,废除,失效
~ by smothering 窒息[断氧]灭火
~ equipment 消防设备

extra ['ekstrə] *a.* 额外的,特别的,多余的,备用的 *n.* 附加物[费],增刊
~ -fine steel 优质钢
~ fine-thread 超细牙螺纹
~ hard steel 特硬钢
~ heavy pipe 特厚管
~ -light 特轻的
~ -light drive fit 特轻配合
~ loss 额外损失
~ low carbon steel 极低碳素钢
~ mild steel 特软钢
~ space 特殊空间
~ ply 附加层
~ quality 特优质量
~ -rapid-hardening portland cement 特快硬波特兰[硅酸盐]水泥
~ reinforcement 额外增强
~ slack running fit 松动配合
~ soft steel 特软钢
~ -strong pipe 特厚壁钢管,特强管

extract [iks'trækt] *vt.* 抽[拨,排]出,提炼[分离]出,开(平)方,求(平方)根 ['ekstrækt] *n.* 抽出[提取]物,摘录,选集
~ system 抽风系统
~ ventilation 抽气通风
~ ventilator 抽风机

extraction [iks'trækʃən] *n.* 抽[拨,排]出(物),提炼,摘录,开方[求根]法
~ fan 排气风扇
~ pipe 抽气管
~ ventilation system 抽气式通风系统

extractor [iks'træktə] *n.* 提取[萃取,抽取]器,排气辅助器;拔出[脱膜,退壳]器
air ~ 抽气[抽风,排气]机
dust ~ 收[捕]尘器

screw ～ 起螺丝器
split pin ～ 起开尾销器
stud ～ 双头螺栓拧出[入]器
tool ～ 拔除工具器
water ～ 干燥[水分离]器, 脱水机

extrados [eks'treidɔs] *n.* (*pl.* -dos,-doses) 拱背(线), 外拱圈, 外弧面

～ springing line 拱外弧起拱线

extrados

extraordinary [iks'trɔ:dnri] *a.* 非常的, 异常的, 特别的; 特派的

extrapolation [ˌekstrəpəu'leiʃən] *n.* 外推[插,延]法, 推延[断], 归纳

～ method 外推[插]法

extratropical [ˌekstrə'trɔpikəl] *a.* 热带以外的

extremum [iks'tri:məm] *n.* 极值, 最大[小]值

extrudate *n.* 压出物, 压出型材

extrude [eks'tru:d] *v.* 挤压[出]

～d electrode 压涂的焊条
～d section 挤压型钢
～d shape 挤压成型
～d structural pipe 挤压结构钢管

extruder [eks'tru:də] *n.* 挤压机, 螺旋挤出机, (土样)顶样器

extrusion [eks'tru:ʒən] *n.* 挤[压]出(物)

～ jack 顶样器
～ moulding 挤压成型, 挤压模塑法
～ of metal 金属的挤压成型
～ of plastics 塑料挤压成型
～ process 挤压成型法
～ product 挤压制品
～ under vacuum 真空挤压成型

extrusive [eks'tru:siv] *a.* 挤[喷,压]出的; *n.* 喷出物

～ window sash (plastics) 挤压出的窗框(塑料)

exudation [ˌeksju:'deiʃən] *n.* 渗[流]出(物), 泌水作用

exurb ['egzə:b] *n.* 城市远郊富裕阶层居住地区

eye [ai] *n.* 眼(状物), 耳[针]孔, 耳环, 索眼; 信号灯, 光电管; *v.* 观看, 打孔眼

～ bar 眼杆, 带环拉杆
～ bar hook 眼杆钩
～ bar packing 眼杆填料[垫板]
～ base 眼距, 眼基线
～ diameter 入口直径
～ diaphragm of caisson 沉箱的眼窗隔墙
～ height 视线高度
～ hole 检[窥]视孔, 猫眼
～ joint 眼榫接头, 眼圈[铰链]接合
～ lens 目镜
～ line (透视图上的)视平线
～ -measurement 目测
～ nut 吊环螺母
～ observation 目测
～ of axe 斧柄眼
～ of cyclone 气旋[旋风]眼
～ of dome 穹顶孔眼
～ of hatchet 斧柄眼
～ of storm 风暴眼
～ pit 视孔
～ point 视点
～ reach 视野, 眼界
～ ring 套环
～ screw 眼圈螺钉
～ sketch 目测草图
～ -sore 碍眼物
～ splice 索端结扣眼圈, 环接索眼
～ strain 视觉疲劳
～ visible crack 肉眼可见裂缝
～ window 眼形窗

eyebolt ['aibəult] *n.* 环首螺栓, 眼头钉

～ and key 插销[环首]螺栓
～ for bilge block 侧垫木穿钉

eyebrow ['aibrau] *n.* 眉毛, 眉窗, 滴水

～ dormer 矮老虎窗

eyelet ['ailit] *n.* 眼孔, 小金属圈; *vt.* 开眼孔

～ bolt 活节螺栓
～ wire 带环电线
～ work 冲孔,打眼
eyepiece ['aipi:s] *n.* (望远镜,显微镜的)目镜
eykometer *n.* 泥浆凝胶强度和剪切力测定仪
eyot [eit] *n.* 河[湖]心小岛

F

fabric ['fæbrik] *n.* 结构,构造,组构;建筑物,构造物;织造品,纤维织物,布;建筑材料;建筑方法
～ belt 纤维带
～ dust 纤维性尘埃
～ filter 纤维过滤器
～ reinforcement 钢筋网
～ wall covering 贴墙织物
asbestos ～ 石棉纤维
coated ～ 漆布(一种代用革)
cotton ～ 棉织物,棉布
duck ～ 帆布织物,工业用织物
gauze ～ 金属织网;纱布
phosphated ～ 磷酸盐化织物,磷酸盐处理织物
proof ～ 胶布
reinforcing ～ 钢筋网
steel ～ 钢筋网
synthetie ～ 合成纤维织物
water-proof ～ 防水布,防雨布
welded steel ～ 焊接钢筋网
wire ～ 钢丝围栅
fabricant ['fæbrikənt] *n.* 制造者,制作者
fabricate ['fæbrikeit] *vt.* 建造,制造;装配,安装;织造
fabricated *a.* 装配的,构造的
～ bar 网格钢筋,钢筋网
～ building 装配式房屋
～ construction 装配式施工
～ structure 装配式结构
～ section 拼合截面,安装件
～ shapes 加工型材
fabricating *n.* 建造,建造[筑]
～ cost 建筑造价;安装费用
～ yard 工场;施工现场,建筑工地
fabrication [ˌfæbri'keiʃən] *n.* 制作,建[构]造,装[制]配;捏造;建造[制作,捏造]
fabricator [fæbri'keitə] *n.* 制作者,装配者;制造用的工具;修整工;金属加工厂
Fabrikoid ['fæbrikɔid] *n.* [美]法布里科德(一种防水织物);仿造布革
fabroil *n.* 纤维胶木
facade [fə'saːd] *n.* 正面,建筑物立面
～ wall 正面墙
face [feis] *n.* 面,正[前]面;表面,工作面;采掘面;票[面]面 *v.* 对,面向;砌面,加贴面;刨光,削平
～ airing 工作面通风
～ arch 前拱
～ ashlar 琢石面,饰面琢石
～ -bonding 正(面)焊(接),叩焊
～ brick 贴面砖
～ contact 按钮开关(接点),按压接触
～ cutter 端(平面)铣刀
～ cutting 车平面
～ -down bonding 面朝下焊接(法)
～ formwork 饰面模板
～ grinding 磨端面,平面磨削
～ guard 护面罩
～ hammer 琢面锤
～ hardening 表面硬化
～ joint 表面接缝[接合];出面灰缝
～ -lift 改建,翻新;上油漆;表面修饰
～ measure (建筑物)立面宽度

~ mix 饰面混合材料
~ moment （框架）端部节点弯距
~ mo(u)ld 面模,样板
~ of brick 砖面
~ of drawing 图纸的正面
~ of the screen 屏蔽面
~ off 倒角
~ plate 面板;卡盘
~ putty （露）面（油）灰（嵌门窗玻璃的）
~ side 正面;外露面;(木工)木面
~ slab 镶面板
~ spanner 端面扳手
~ -sperigging 插型钉
~ stone 饰面石
~ stringer 出面的楼梯小梁
~ surface 消光面
~ tile 饰面砖
~ timbering 工作面支撑（架）
~ up 面朝上;涂色（美化用）
~ -up bonding 面朝上焊接（法）
~ veneer 表面装饰薄板,镶面板
~ waling 工作表面横撑,表面围檩
~ wall （镶）面墙
~ width 表面宽度
~ work （墙等的）抹［饰,护］面;抹面工作
~d plywood 贴面胶合板
~d wall 饰面墙

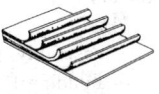
faced plywood

air ~ 空气表面;背水面,下游面
anvil ~ 砧面
arcwall ~ 弧形面
dent ~ 弯曲面
bevel ~ 斜角面
binding ~ 支撑［承］面;贴合面
building ~ 房屋外貌
depth ~ 铺块厚度;面层厚度
diagonal ~ 倾斜（工作）面,对角（工作）面
front ~ 前面
gripping ~ 支承面（螺栓连接）

inner ~ 里面;正面
joint ~ 接合面
lateral ~ 侧面
outer ~ 外面,反面
prismatic ~ 柱面
pyramidal ~ 锥面
wall ~ 外墙面

faceplate ['feispleit] *n.* 划线平板;面板,花［卡］盘

facet ['fæsit] *n.* 柱槽筋（槽间凸出的平面竖线脚）,凸线;刻面;倒角 *v.* 在……上刻面

facett *n.* 柱槽筋;刻面

facia ['fæʃə] *n.* (*pl.* faciae)商店招牌
~ board 招牌,牌匾,挑口板

facial ['feiʃəl] *a.* 正面上的,表面的
~ angle 面角

facilitate [fə'siliteit] *vt.* 使容易;使便利

facility [fə'siliti] *n.* (*pl.* -ties)设备,装备［置］,工具;机关［构］;可能性,方便,灵活
~ control console 设备控制台
air-cleaning ~ 空气净化装置
charging ~ 装料设备
communication ~ 通信设备
lighting ~ 照明设备
maintenance ~ 辅助设备;维护设备
storage ~ 器材库,贮藏室

facing ['feisiŋ] *n.* 饰面,保护面;铺面;镶面;贴边,［镶面］材料;车端面,切割面
~ aggregate 饰面骨料
~ bond 顺砌（砖）
~ brick 饰面砖,贴面砖
~ concrete 铺面混凝土
~ expansion joint 面层伸缩缝
~ hammer 琢面锤
~ material 饰面材料
~ machine 刨床
~ slab 镶面板
~ stone 饰面石,贴面砖
~ tile 饰面砖

~ wall 护面墙
ashlar ~ 琢石砌面
brick ~ 砖(露)面
cement ~ 水泥盖面
finish ~ 精加工端面
friction ~ 摩擦面
granulated stone ~ 颗粒石铺面
hard ~ 镀以硬质合金;硬盖面(层);表面(淬火)硬化
random ashlar ~ 乱(砌)琢石面

facsimile [fæk'simili] *n.* 传真;摹写,影印本;电传真;复制 *vt.* 精摹;复制; *a.* 如复制的;制造复制本的
~ radio 无线电传真
~ telegraph 传真电报

fact [fækt] *n.* 事实,真情;真实,真相
~ finding 实况调查,实地调查

factice *n.* 油胶[膏]
black ~ 墨油胶;硫化油胶
dark ~ 暗黑胶,深胶
white ~ 白胶,浅胶

factor ['fæktə] *n.* 因素;因数[子];系数;率;指数,乘数;分解因子
~ of expansion (热)膨胀系数
~ of safety 安全度[因数]
acoustic(al) absorption ~ 吸声系数
acoustical reduction ~ 隔音[声降]系数
acoustical transmission ~ 透声[传声]系数
assurance ~ 安全[保险]系数
cement-water ~ 灰水比
common ~ 公因子[数];一般[公共]因素
daylight ~ 天然采光系数
decontamination ~ 净化因数
design load ~ 设计荷载系数
energy ~ 能量因数[因子];质量[品质]因素
engineering ~s 工程因素,技术条件
enironmental ~ 环境因素,周围条件
expansion ~ 膨胀系数

human ~ 人为因素
illumination ~ 照明系数
imperviousness ~ 不透水系数
leakage ~ 漏逸[渗漏,漏水]系数
luminance ~ 亮度因数
luminosity ~ 发光率
merit ~ 质量优良程度,优质率
noise ~ 噪声系数
personal ~ 人为因素
quality ~ 品质[质量]因数
regulation ~ 调节系数;稳定因数
safety ~ 安全系数
severity ~ (测定工艺过程操作强度的)强度系数[因数]
shadowing ~ 阴影系数,影因数
shape ~ 形状系数
stability ~ 稳定(度)系数;稳定安全系数;刚性系数
strength ~ 强度因数[度]系数
structure ~ 结构因数
tear ~ 抗裂系数,抗磨因数
utilization ~ 利用系数[率]
vignetting ~ 遮光系数
visibility ~ 可[能]见度
void ~ 空隙比

factory ['fæktəri] *n.* (*pl.* -ries) 工厂;制造厂
~ assembly 工厂装配
~ building 厂房
~ building constructions 厂房建筑[结构]
~ lumber 加工木材
~ made house 工厂预制房屋
~ overhead 生产(间接)费用
~ planks 加工木板,制门窗用板材

factotum [fæk'təutəm] *n.* 杂工

factual ['fæktjuəl] *a.* 事实的,确实的
~ proposal 实际的建议

factualism ['fæktjuəlizm] *n.* 尊重事实(的作风)

factum ['fæktəm] *n.* (*pl.* facta) 事实,实情;论据

facture ['fæktʃə] *n.* 制造,建筑;制[成]

品,制作法;表现技法;建筑表面处理手法

fade [feid] *v.* 褪色;渐沉消失;减弱,衰落;失去光泽
~ -away 逐渐消失
~ -in 淡入(图像逐渐出现),淡显,渐显
~ out 淡出(图像逐渐消失),渐隐,衰落
~ over 淡入淡出

fadeproof ['feidpru:f] *a.* 不褪(色)的

fader ['feidə] *n.* 音量渐减器;照明渐减器;音量(或光量)调节器

fagging *n.* 撑板,挡土板;绝热[保温]套;防护套

fag(g)ot ['fægət] *n.* 柴捆[把];成束的熟铁板条;捆成一捆

Fahrenheit ['færənhait] *n.*;*a.* 华氏温度计(的),华氏刻度(的)
~ degree 华氏温度

faience [fai'a:ns] *n.* 彩(色)瓷[陶](砖或器)
~ masaic 嵌花地砖,釉陶锦砖
~ tile 釉陶锦[面]砖

fail [feil] *n.*;*v.* 失败,失效,破坏,毁坏;故障;缺少,不足;衰弱,减弱
~ in bending 弯(曲)损(坏)
~ in bond 粘结损坏[破坏]
~ open 出故障时自动打开
~ soft 有限可靠性

fail-safe *a.* 可靠的,不出故障的,具有失效保护的
~ control 失效保险控制
~ system 失效保险系统

failing ['feiliŋ] *n.* 失败;过失;损坏
~ load 破坏荷载
~ stress 破坏应力

failure ['feiljə] *n.* 故障,事故;失效[事,败];破[损]坏;破产,倒闭;折损
~ analysis 故障分析
~ load 破坏荷载
~ plane 破坏[裂]面
~ surface 破坏面

~ zone 破坏区[范围]
bending ~ 弯曲破损,挠曲破坏
catastrophic ~ 灾难性的毁坏;重大事故
chance ~ 偶然[意外]故障
elastic ~ 弹性破坏
equipment ~ 设备故障
fatigue ~ 疲劳破坏
induced ~ (环境)诱导故障
repeated stress ~ 疲劳断裂(反覆加载破坏)
structural ~ 结构损坏
transverse ~ 横向断裂

faint [feint] *a.* (微,衰)弱的,(暗)淡的,细[轻]微的
~ difference 细微的差别
~ red 淡红色

fair [fɛə] *a.* 公平的,正直的;美好的;明晰的 *n.* 光顺;校平;定期集市,交易会;展览会
~ average quality 中等质量,大路货
~ raking cutting 修整切削
~ ends 琢石(砖)露头
~ faced concrete 光面混凝土
~ in place 现场安装
~ valuation 合理的估价
~ wear and tear 合理磨损

Faircrete concrete (= Fiber-air-entrained concrete) *n.* 纤维加气温凝土

fairfaced brickwork 清水(勾缝)砖砌体

fairlead ['fɛəli:d] *n.* 引线孔,引线管;转动击船柱;机绳滑轮
aerial ~ 天线引线管

fairleader *n.* 卷扬机械,辗子

fairlight *n.* 门顶窗,气窗

fairness ['fɛənis] *n.* 公正,公平;晴(朗),洁白

fairway ['fɛəwei] *n.* 无障碍的成开阔通道;水路,航道(油气)通路

fake [feik] *n.* 冒牌货,伪造;盘索,线圈 *a.* 伪造的 *vt.* 伪造;盘卷

fakes *n.* 云母板状岩,云母质砂岩,砂质页岩

fall [fɔːl] *n.* 坡降,斜度;降雨量;起重机绳索;落差;秋天;瀑布 *v.* 落下,下降,跌落,减退
~ back 倾斜面;泄水;落下;后退,退路
~ bar 轴转关门闩
~ line 吊绳
~ pipe 水落管
~ rope 吊绳
~ tube 排水管
cat ~ 吊锚索
cross ~ 纵向坡度,横斜度
laid to ~s (抹面)做出坡度
natural ~ 天然落差[水头]
fallacy ['fæləsi] *n.* (*pl.* -cies) 错误,谬误
fallaway *n.* (各级)分开,散开,排出
~ section 分离段
falling ['fɔːliŋ] *a.* 降落的;倾斜的,带有斜度的 *n.* 降落,下落;落体运动
~ debris 倒塌房屋的瓦砾
~ door 降落式门,吊门
~ gradient (道路)下降坡度
~ main (下水道)下行主管
~ mo(u)ld 曲型线条,弯曲形式
~ weather 雨季,雪季
~ weight (打桩用)落锤
fallow ['fæləu] *a.* 淡黄色的;未开垦的,休闲的(土地)
fallway ['fɔːlwei] *n.* 楼面井,运货竖道[竖井]
false [fɔːls] *a.* 假的,不可靠的;不成立的;辅助的;临时的
~ arcade 假拱廊
~ arch 假拱
~ attic 假(屋)顶层
~ balustrade 假扶手,假拦杆
~ base 活底座
~ beam 假梁(不承重的梁)
~ bearing 间接支承
~ bedding 假层理,交错层
~ bill 假票据
~ bottom bucket 活底料斗
~ ceiling 假平顶,假顶棚
~ door 假门
~ ellipse 圆弧椭圆,三心装饰拱
~ ellipse arch 装饰成的三心拱,近似椭圆拱
~ floor 假楼板
~ form 假模,临时性模板
~ header 假丁砖(半砖头)
~ joint 假接缝
~ member (桁架中的)伪杆
~ rafter 假椽
~ roof 吊(悬屋)顶,假屋顶
~ set (水泥)假凝(结);轻型临时支架
~ stull 临时模撑
~ window 盲窗,配景窗
falsework ['fɔːlswəːk] *n.* 脚手架,工作架,模板,临时支撑
arched ~ 拱(庸)架,拱里脚手架
erect ~ 架立脚手架
flying ~ 悬空式工作架
falsification [ˌfɔːlsifi'keiʃən] *n.* 伪造;证明为假,证明为无根据;(测量结果的)畸变,失真
falsify ['fɔːlsifai] *v.* 伪造;曲解;反证
falter ['fɔːltə] *v.* 摇晃,颤抖
family ['fæmili] *n.* 家庭;族;类;系
~ cirele 剧场小包厢
fan [fæn] *n.* 风扇[箱],通风机;叶片,翼;扇形地 *v.* 扇,通风吹向,展开
~ atomizer 扇式喷雾器
~ blade 风扇叶(片)
~ blower 鼓风机
~ casing 通风机罩
~ connector 扇形连接器
~ convector heater 扇式对流加热器
~ drift 通风道
~ Fink truss 扇形芬克式桁架
~ groining 扇形穹顶
~ guard 脚手架上防止杂物落下的挡板

fan Fink truss

~ house 扇风(通风)机房
~ in let 通风机进气口
~ key cock 阔把龙头
~ ring 风扇外环
~ roof 扇形(花格交叉)拱顶
~ room 通风机室
~ sewer system 扇形断面下水道系统
~ shutter 通风调节板
~ tracery 扇形(花)格架
~ tracery vault 扇形格式穹顶
~ truss 扇形桁架
~ vault(ing) 扇形穹顶
~ window 通风窗,扇形窗
~ work 扇形花格架
blast ~ 鼓风机
discharge ~ 抽风机;排气风扇
donkey ~ 辅助扇风机
draft ~ 通风风扇,吸风机
dust-collecting ~ 吸尘器
exhaust ~ 排气扇,抽风扇
fresh air ~ 净气风扇
gas purifier ~ 净气风扇
induced ~ 抽风机
oscillating-base ~ 摆动吹风机
pivoting ~ 摆动吹风机
positive ~ 强制送风机
suction ~ 吸风机

fancier [ˈfænsiə] *n.* 设计师;空想家

fang [fæŋ] *n.* 风井,通风平洞,通风巷道;钩、凿;铁篱笆
~ bolt 锚栓,地脚螺栓

fanlight [ˈfænlait] *n.* 门上(的)扇形窗;气窗;腰头窗,楣窗

fanlight

~ opening gear 腰窗(开闭的)传动装置
~ quadvant 腰窗挡

fanner [ˈfænə] *n.* 风扇,通风机

fanning [ˈfæniŋ] *n.* 用通风器吸尘;通[扇]风;(呈扇形)展开

~ strip 扇形(端子)板,扇形片

fan-shaped *a.* 扇形[状]的
~ arc 扇形弧
~ gate 扇形门
~ tenon 扇形榫头

fantail [ˈfænteil] *n.* 扇形尾,鸠尾榫,扇形饰
~ arch 扇形拱
~ joint 鸠尾连接
~ roof 扇形屋顶

fantascope *n.* 幻视器

fantasound *n.* 立体声

far [fɑː] *a.* 远(处)的,遥远的
~ end (线路)远端
~ out 极端的,先锋派的
~ point 远点
~ seeing 能够看得很远的;眼光远大的
~ seeing plan 远景规划
~ sight 远见[望]

farad [ˈfærəd] *n.* 法拉(电容量单位)
international ~ 国际法拉

fare [fɛə] *n.* 运费
~ register 计费器

Fareham reds *n.* 法列哈姆式红石砖

farm [fɑːm] *n.* 农场,农田 *v.* 耕种
~ building 农村建筑
~ house 农舍

farm-out *n.* 分包出任务;转交;处理

farmstead *a.* 农场建筑物

farmyard *n.* 农家庭院,农业场地

fascia [ˈfeiʃə] *n.* 托板;封檐底板;(古典柱式建筑)柱顶横梁的横带;窗过梁饰带;招牌
~ beam 装饰梁
~ board 檐头板,挑口板

fascine [fæˈsiːn] *n.* 束柴;柴笼;柴排;梢(捆);柳辊;粗杂材;束枝栅工
~ cradle 束木支架
~ layer 梢料填层

fashion [ˈfæʃən] *n.* 形状;形[样,方]式;风格;时新式样 *vt.* 制作,做成……的形状,加工修饰

fashioner

~ colour 流行色
~ed iron 型铁
~ part 异形配件
latest ~ 最时新花样
random ~ 乱七八糟的

fashioner ['fæʃənə] *n.* 设计者;制造者;造型者

fast [fɑ:st] *n.* 紧洁物(如锁、闩、钩等) *a.; ad.* 坚牢的;固定的;不褪色的;快的(地),迅速的(地)
~ colour 不褪的颜色
~ -drying material 快干材料
~ -drying paint 快干漆
~ flange 固定法兰
~ hardening concrete 快硬混凝土
~ -pin hinge 紧固销铰链
~ setting 快凝
~ -setting cement 快凝水泥
door ~ 门闩

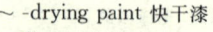
fast-pin hinge

fasten ['fɑ:sn] *v.* 扣牢,栓住,使固定;(门等)关紧

fastener ['fɑ:snə] *n.* 紧固零件(如螺钉、销钉等);闭锁;接线柱,线夹,扣钉;闸阀
casement ~ 窗插销
sash ~ 窗风钩
screw belt ~ 螺钉带扣
snap ~ 按扣
window ~ 窗闩销;插销
zip ~ 拉链

fastening ['fɑ:snɪŋ] *n.* 紧结物(如锁、闩、钩等);紧固(件);扣件;连接;连接法
~ angle 连接角铁
~ bolt 连接螺栓
~ piece 连接件
~ pile 系定桩
~ screw 紧固螺钉
~ wire 紧固[绑扎]用钢丝

fastigium [fæ'tidʒiəm] *n.* (*pl.* -iums, -ia)(房屋)顶尖,屋脊,山墙

fastness ['fɑ:stnis] *n.* 坚牢,不褪色性
acid ~ 耐酸性
alkali- ~ 耐碱度
colour ~ 颜色坚牢度
dust ~ 耐尘性
light ~ 耐光性,不褪色性
rubbing ~ 耐摩擦程度
wet ~ 耐湿性

fat [fæt] *n.* 富灰浆;脂,脂肪;油类 *a.* 肥的,多脂肪的;多沥清的; *v.* 变肥胖
~ asphalt mixture 多(地)沥青混合料
~ board 灰浆板
~ concrete 富混凝土
~ -extracted 脱脂的
~ lime 富石灰
~ mortar 富砂浆
~ paint 厚漆,浓漆
~ sand 粘土型砂,肥砂
~ soil 沃土
~ surface 多油面层(即沥青多的面层)
~ wood (多脂)松木

fatal ['feitl] *a.* 致命的,致死的
~ accident 死亡事故

fatigue [fə'ti:g] *n.* 疲劳,疲乏;*v.* 老化;使疲劳; *a.* 疲劳的
~ allowance 疲劳限度
~ auditory 听觉疲劳
~ crack (材料的)疲劳裂缝
~ endurance limit 疲劳耐久极限
~ failure 疲劳破坏
~ fracture 疲劳断裂
~ limit 疲劳限度[强度]
~ strength 疲劳强度

faucet ['fɔ:sit] *n.* 旋塞;水龙头;出水嘴
~ ear 管子吊环
~ end 龙头管端
~ joint 套筒接头;水龙头接头
~ pipe 套接管;水龙头管
air ~ 气嘴,气旋塞
measuring ~ 量水龙头

self closing ～ 自闭龙头
fault [fɔ:lt] *n.* 错误,缺点;故障,疵病,缺陷;漏电;断层
～ bench 断层阶地
～ control 事故控制
～ current 故障电流
～ detection 探伤
～ finder 探伤器
～ free 无故障的
～ indicator 探伤器
disconnection ～ 断电故障
image ～ 图像失真
numerical ～ 数值误差[错误]
faulty ['fɔ:lti] *a.* 有缺点的,不合格的
～ concrete 劣质混凝土
～ operation 错误操作
Fawcettis floor *n.* 陶土槽板楼盖
fax [fæks] = facsimile *n.* 传真,电视画面;摹本;传真通讯;无线电传真机
～ chart 传真图
fay [fei] *v.* 紧密连接,紧配合
faying surface 接触[接合,搭接]面
feasibility [ˌfi:zə'biliti] *n.* (*pl.* -ties) 可行性,可能性,现实性
engineering ～ 技术(上的)可行性
feasible ['fi:zəbl] *a.* 可(实)行的;可能(有)的;合理的
～ direction 可行[容许]方向
～ domain 可能领域,可行范围
～ program 可行方案[程序]
feather ['feðə] *n.* 滑键;羽毛
featherlike, *a.*
～ board 薄边板
～ check 发丝裂缝,羽毛状裂缝
～ cutter 企口凿
～ -edge(d) board 薄边板
～ -edge brick 削边砖
～ grain 羽状纹理
～ joint 插楔[铰链]接合
～ key 滑键
～ piece 榫舌,暗销
～ tongue 斜削销
～ way 滑键槽

～ -weight paper 轻磅纸
feathering ['feðəriŋ] *n.* 羽状物;叶瓣饰;羽翼式铺开;三角形窗的尖顶(装饰用)

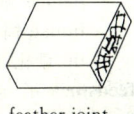
feather joint

feature ['fi:tʃə] *n.* 特征[色];性能[质];地形[势], *vt.* 以……为特色, *vi.* 扮演角色
checking ～ 检验特性
distinctive ～ 特征,特点
high-wearing ～ 高耐[抗]磨性
structural ～ 构造细部[特点]
unique ～ 特色,特点
fee [fi:] *n.* 费,税,薪金, *vt.* 交费给;雇用
～ method 收费法
～ of permit 牌照税,执照税
feebly ['fi:bli] *ad.* 无力地,无效地;微弱地
～ cohesive soil 弱粘(聚)性土
～ hydraulic lime 弱性水硬石灰
feed [fi:d] *v.* 供给;输送;供[加]料;供水,馈电;进刀 *n.* 饲料
～ apparatus 给水器;进料装置
～ cock 给水龙头[开关]
～ head 进料口[头]
～ line 进给线,给水管
～ pipe 送料[供水]管
～ -sponge 海绵金属料
～ water 供水,给水;饮用水
～ -water filter 给水过滤器
～ way 供给装置
feedback = feed-back ['fi:dbæk] *n.* 反馈,反复,反应;重整
～ compensation techniques 反馈校正技术
～ information 信息反馈
feeder ['fi:də] *n.* 进[加]料器;给水器;支线[流],馈电线
feeding ['fi:diŋ] *n.* 增大,滋长;供水,供电;饲育,供养
～ center 供食中心

feedstock

~ device 加料[进给]装置
~ house 饲(畜)舍
~ up 硬化;稠化

feedstock *n.* 原料

feeler ['fi:lə] *n.* 测隙规;厚薄规,塞尺;触探器,探针
~ gauge 厚薄规,塞尺

feigned [feind] *a.* 假的,虚伪的
~ column 假柱

feeler gauge

feldspar ['feidpa:] *n.* 长石

feller ['felə] *n.* 伐木机,伐木工;整平者[平缝工,接缝工]

felloe ['feləu] *n.* 轮辋,车轮外缘(扇形)轮缘
~ band 钢带,载重带

fellow ['feləu] *n.* 伙伴,同行;会员
~ -trader 同行,同业

felt [felt] *n.* 油(毛)毡,毡垫圈;隔热板;隔音板
~ and gravel roof 油毡撒绿豆砂屋顶,油毡铺砾屋面
~ carpet 油毡覆(盖)面层
~ -cloth 毡布;毛巾;薄毡料
~ deadener 吸声毡
~ packing 填密毡
~ paper 绝缘[油毡]纸
~ retainer 毡护圈
~ strip 油毡条
~ tarpaulin 防雨毡
~ washer 毡垫圈,毡衬垫
acoustic ~ 吸声毡
asbestos (=asbestus) ~ 石棉毡
asphalt saturated ~ 沥青浸润毡
bituminous ~ 石油沥青毡
drier ~ 干燥毡,干燥毛布
grease retainer ~ 护脂毡圈
hair ~ 油毛毡
impregnated ~ 浸渍毡
king pin ~ 中心立轴毡
pitched ~ 柏油毡

press ~ 压毡
roofing ~ 屋面毡
treated ~ 油毡

feltside *n.* (纸)正面,毛布面

female ['fi:meil] *n.* 凹形物,凹槽 *a.* 女性的,雌性的
~ cap 凹形盖
~ fitting 内螺纹配件;凹接头,凹形配件
~ joint 承插式接头
~ plug 插座
~ screw 阴螺纹;内螺纹螺栓
~ spanner 套筒扳手
~ thread 阴螺纹,内螺纹

femerell *n.* 屋顶(排气)笼,天窗

fence [fens] *n.* 围墙,篱笆,栅栏;排柱,密集支柱 *v.* 防御
~ gate 栅门
~ line 栅栏线
~ nail 修篱笆用的钉子
~ off 用栅隔开
~ post 栅栏柱,围(栏)柱
grill(e) ~ 格式栅栏
guard ~ 护栅
lath ~ 板条围栏
lattice ~ 格构围栏,格栅
post-and-block ~ 柱夹块式围墙,柱[板]围墙
protection ~ 护栏
site ~ 工地围栏
timber ~ 木围栏
wire ~ 铁丝网

fender ['fendə] *n.* 围护物;火炉围栏;防冲桩;护(舷)木;挡土板
~ apron[board] 保护板,挡泥板
~ beam 护木
~ bracket 保护板架
~ pier 护墩
~ post 护柱
~ structure 围护结构
~ wall 防护墙
~ timber 木护栏

fenestella [ˌfenisˈtelə] *n.* (小型)窗扇,

小窗洞;窗形壁龛
fenestra [fi'nestrə] *n.* (*pl.* -tral) (采光)小洞,小窗
~ rotunda 正圆窗
fenestrate [fi'nestreit] *a.* 安窗的;有窗的

fenestella

fenestration [ˌfenis'treiʃən] *n.* 门窗布局设计,外墙窗洞组合,面上(如墙上成薄膜上)的洞口或缝隙;开窗法;穿孔,穿通
fenestrato *n.* 组合窗
fenestrule *n.* 小窗
feretory ['feritəri] *n.* (*pl.* -ries) 神龛,圣骨龛
ferro alloy 铁合金
ferro cement 铁矿渣[高铁]水泥
ferro-chromium 铬铁合金
ferro-concrete 钢筋混凝土,钢骨水泥
ferro-crete *n.* 含铁硅酸盐水泥(快凝水泥),快硬水泥
ferrodo *n.* 磨擦材料;摩擦插片
ferro-glass *n.* 钢化[装饰,钢丝]玻璃
ferro-manganese 锰铁(合金)
ferro-portland cement 含铁波特兰水泥
ferro-silicon 硅铁(合金),硅钢
ferro-vanadium steel 钒钢
ferrule ['feruːl] *n.* 箍,套圈,金属箍,环圈;密套环; *vt.* 装以金属箍
pipe ~ 管箍,水管口密套
plate ~ 管板密套
wire ~ 线箍
festoon [fes'tuːn] *n.* 垂花雕饰,花彩,花彩装饰垂链
~ lighting (电)灯彩
~ -shape curve 花环曲线;垂花曲线
fettle ['fetl] *n.* 状态;修补(炉衬);清理(铸件);铲除(炉壁渣子)
fettling ['fetliŋ] *n.* 补炉材料
~ bench[table] 清理台
~ knife 修补刀
fiber = **fibre** ['faibə] *n.* 纤维;硬纸板;刚纸
~ box 纸板箱
~ board 纤维[硬化]纸板
~ cam 纤维板箱
~ cement 纤维水泥
~ conduit 硬(纸)导管
~ glass 纤维玻璃
~ glass tile 纤维玻璃钢瓦
~ pipe 纤维管
~ reinforced composite 纤维增强复合材料
~ reinforced materials 纤维加强材料
~ saturation point (木材)纤维饱和点
~ sheet 纤维纸板
~ stress 纤维应力
~ tube 纤维管,硬纸板管
artificial ~ 人造纤维
asbestos ~ 石棉纤维[纸板]
cellulose base ~ 纤维素纤维
ceramic ~ 陶瓷纤维
cross ~ 十字[交叉]纤维
gelatinous ~ 凝胶纤维
glass ~ 玻璃丝(纤维)
isotropic ~ 均质纤维
laser ~ 激光玻璃[光学]纤维
optical ~ 光学纤维
parchmentized ~ 纤维;硬化纸板
polyester ~ 聚脂纤维
polyethylene ~ 聚乙烯纤维
quartz ~ 石英纤维
staple ~ 常产纤维,人造棉(毛),人造(短)纤维
synthetic ~ 合成[人造]纤维
twisted ~ 合股纤维
vulcanized ~ 纤维板;硬化纸板;钢纸板
fiberboard ['faibəbɔːd] *n.* 纤维板
fibrage *n.* 纤维编织
fibration *n.* 纤维性;纤维组织;[构造],纤维化
fibrator 纤维(素)

fibred *a.* 纤维状的;纤维质的
fibri(la) ['faibri(ə)] *n.* (*pl.* -lae) 微丝,原纤维
fibrofelt *n.* (绝缘用)纤维毡
fibrotile *n.* 石棉水泥波形瓦
fibrous ['faibrəs] *a.* 纤维状(的),由纤维构成的;含纤维的
~ concrete 纤维混凝土
~ fracture 纤维裂缝
~ glass 纤维玻璃
~ insulation 纤维隔热层,纤维隔离层
~ plaster 纤维灰(泥)膏
~ plastering 纤维灰泥抹面
~ reinforcement 纤维增强
~ slab 纤维板
fictile ['fiktail] *a.* 黏土制的;陶器的;可塑造的,可塑性的;人工成形的,塑造的
fictitious [fik'tiʃəs] *a.* 想像所得的,想像上的;虚假的,虚构的,伪装的
fid [fid] *n.* 钉子,硬木钉,销子,螺钉;双端[柱]螺栓;测针;支撑材,固定材
fiddle ['fidl] *n.* 台座,台架,支柱;(纱)罩
~ drill 弓钻
fidelity [fi'deliti] *n.* (*pl.* -ties) (细节的)准确度,逼[保]真度
acoustic ~ 声音逼真度
chromatic ~ 色彩逼真度
colour ~ 色彩逼真度
fiducial [fi'dju:fjəl] *a.* 基准的,信托的
~ axis (基)准轴(线)
~ line 基准线
~ point 准点
fiduciary [fi'dju:ʃjəri] *n.* (*pl.* -aries) 受托人,信托者, *a.* 受托人的,信托的
~ contract 信托契约,信用合同
~ contribution 信用投资
~ loan 信用贷款
~ estate 信托资产
~ work 信托业务
field [fi:ld] *n.* 场,电[力]场;工地,现场,田野,范围,界,领域;视界[域];字段,信息[符号]组
~ automation 现场作业自动化
~ bolt 现场安装螺栓
~ bolted 现场栓接
~ cable 被覆线
~ condition 工地[现场]条件
~ connection 工地装配,现场联结
~ construction 现场施工,施工工地
~ curing 工地[现场]养护
~ data 应用[工作]数据
~ engineering 安装技术[工程]
~ joint 现场接合,安装接头
~ location work 实地定线
~ maintenance 现场安装
~ manual (FM) 施工手册
~ map 实测原图
~ mix 工地[现场]拌和;工地拌和混合料;工地配合比
~ of view 视场[野,域,圈]
~ reveting 工地铆接
~ situation 现场情况
~ sketching 现场绘制草图,草测,目测
~ survey 现场调查
~ tile 砌底砖
~ work standards 现场操作规范
apparent ~ 外视场,外表场,有效视场
color ~ 色视场
common ~ 公用区,公用单元
link frame ~ 圆柱形线弧
~ of vision [view, observation] 可见区
viewing ~ 视界,视野
visual ~ 视野
figuline *n.* 陶[瓷]器 *a.* 陶制的,塑造的
figurate stone 琢纹石
figurative design 象征性设计
figuratrix 特征表面
figure ['figə] *n.* 图形,图,附图;数字[值,码],位数;图解;形状;影像 *v.* 描绘;用图表示;用数字表示;计算
carved ~ 雕刻图样

circumscribed ～ 外切[接]形
congruent ～s 全等[叠合]图形
curvilinear ～ 曲线图形,曲线形
geometrical ～ 几何图形
irregular ～ 不规则图形
linear ～ 线性图形
perspective ～ 透视图形
plane ～ 平面图形
projecting ～ 投影图形
rectangular ～ 矩形
rectilinear ～ 直线图形
similar ～s 相似图形
solid ～ 立体形
symmetric(al) ～ 对称图形
figured ['figəd] *a.* 图示的;图解的;有图案的,有花纹的
～ bar iron 型钢,异形钢
～ glass 压花玻璃
～ plate glass 图案玻璃,(压)花玻璃
～ rolled glass 图案玻璃,(压)花玻璃
～ sheet glass 图案玻璃,(压)花玻璃
～ twills 花纹,斜纹
～ veneer (管)图案(的)镶板
～ wood 富纹木
figurer *n.* 陶器的图案描绘者
filament ['filəmənt] *n.* 线,丝,灯丝
～ lamp 白热丝灯
filamentar(y) [ˌfiləˈmentəri] *a.* 丝的,纤细的
filamentous [ˌfiləˈmentəs] *a.* 丝的,纤细的
filar ['failə] *a.* 丝状的;丝的,浅的
file ['fail] *n.* 档案,卷宗;文件夹;行列;外存储器,外件;资料数据集,磁带集;文件,锉(刀) *v.* 登记,记录;锉
～ card 档案卡片
～ carrier 锉柄
～ chisel 锉錾,锉凿
bastard [rəughcut] ～ 粗锉
round ～ 圆锉
saw ～ (修)锯锉
secontcut ～ 中锉

smooth ～ 细锉
square ～ 方锉
triangular ～ 三角锉
filigree glass 银丝玻璃,嵌丝玻璃
filing ['failiŋ] *n.* (文件的)整理汇集;档案;锉,琢磨
～ cabinet 档案柜,卡片箱
～ room 档案室
fill [fil] *n.* 填[土,方,料](路)堤;装填物 *v.* 充填,填塞,注满,充满;填筑
～ block 填(塞)块
～ construction 填土施工[工程]
～ -in 临时填补物;镶嵌
～ insulation 填塞绝缘
～ light (=fill-in light) 辅助[补充]光;柔和光
～ opening 加料口,注入孔
～ plate 填板
～ed board 夹心(层)板
～ed bitumen 加填料沥青
～ed composite 充填材料
～ed insulation 绝热填料
～ed pipe column (混凝土)填实管柱
～ed spandrel arch 实肩拱
back ～ 回填,充填
coarse ～ 大块充填料
deep ～ 深填方
fine ～ 细充填料
loose ～ 疏松充填
loose rock ～ 松堆石,疏松填石
rock ～ 堆石,填石
stone ～ 填石
filler ['filə] *n.* 填(缝)料;填板,镶入板;填充数;漏斗,注入孔,(加)油口;充填[装载]工,装填者;补充[候补]人员
～ aggregate 填冲骨料
～ cap 漏斗盖
～ joist 加固搁栅
～ metal 填充金属,填隙合金
～ piece 填隙片
～ plate 填板
～ ring 垫圈
～ rod 焊[嵌]条

~ wall 柱间[填充]墙
active ~ 活性填料
asphalt ~ 沥青填料
bituminous fibre ~ 沥青纤维填缝料
cement grout ~ 水泥灌浆
cork ~ 软木衬垫
cotton ~ 棉织衬垫
fabric ~ 纤维填料
fender ~ (保)护板嵌料
hard ~ 硬填缝料
inactive ~ 钝性填料,非活性填充剂
joint ~ 填缝料
paint ~ 油漆底层
pitch ~ 脂填缝料;沥青填料
reinforcement ~ 增固填料
ribbed-slab ~ 肋板嵌块
windshield glass ~ 挡风玻璃嵌料
wood block ~ 木嵌块
fillet ['filit] *n.* 嵌条,嵌线;平缘,突出横饰线;楞条;圆抹角,倒角,填角焊缝
~ and groove joint 企口接口
~ gauge 圆角规
~ in normal shear 正面焊缝(垂直于切力的焊缝)
~ in parallel shear 侧面焊缝(平行于切力的焊缝)
~ moulding 平条(线脚)贴角条
~ of screw 螺纹圈
~ plane 圆缘刨
~ weld 填角焊
~ welding 贴角焊,条焊
corner ~ 圆角嵌条;隅缘线,窗外缘饰
cover ~ 盖条饰
dividing ~ 分离嵌条,阻挡层
half-round ~ 馒形线饰
inside ~ 内角焊缝
light ~ 浅角焊缝
skew ~ 墙帽斜浅脚
tension ~ 拉缘条
tilting ~ 瓦座;倾斜缘条

filleting *n.* 角隅填密法;嵌缝法
filling ['filiŋ] *n.* 充填,填塞;充填料,填土
~ compound 填料
~ element 填(塞)料
~ knife 填缝刀
~ material 填(塞)料
~ piece 填隙片
back ~ 回填
beam ~ 梁间墙,梁端间隙填料
clumn ~ 填柱物;柱的填料
dielectric ~ 绝缘填料
pitch ~ 煤沥青[柏油渣]填料
rock ~ 堆石,填石
stone ~ 填石,装石
fillister ['filistə] *n.* 凹刨,(嵌镶玻璃等之)凹槽;凹形饰缘;(门窗上的)线脚 *v.* 刨槽,开槽
~ head screw 有槽圆头螺钉
~ (ed) joint 凹槽接合,凹槽缝
film [film] *n.* 薄层[膜];遮蔽物;浆沫胶[软,底]片,胶卷;影片; *v.* 生薄膜,覆薄膜
~ adhesive 薄膜胶粘剂
~ glass 薄膜玻璃
~ glue 薄膜胶
~ of paint 漆皮,漆膜,漆层
~ strength 薄膜强度
filter ['filtə] *n.* 滤器[池,纸] *v.* 过滤,滤清,渗入
~ course 过滤层
~ drain 滤层排水;滤水暗管
~ layer 过滤(渗透,透水)层
~ paper 滤纸
~ sand 滤砂
~ screen 滤网
filum [failəm] *n.* (*pl.* -la)丝,纤维;线状组织,线状结构
final ['fainl] *n.* 末了,结局; *a.* 最后的,最终的;决定的
~ acceptance 竣工验收
~ assembly 最后组装[装配]

~ budget 核定预算
~ cost （工程）决算，终值，最后成本
~ cleanup 最后清理工作
~ design 最终设计
~ drawing 最终图样
~ estimate 结算
~ finishing 最终修整(饰)
~ process 最后程序
~ report 决算报告
~ return 最后回报，最终利润
~ set （水泥成混凝土）终凝
~ setting time 终凝时间
~ settlement 最终沉降
~ shaping 最后定型
~ statement 决算表
~ user 最终用户

finance [fai'næns] *n.* 财政，财务，金融 *v.* 投资于，理财

financial [fai'nænʃəl] *a.* 财政的，财务的，金融的；会计的
~ analysis 财务分析
~ duty 财政关税

financing *n.* 资金供应[筹措]，供资

find [faind] *v.* 发现[觉]，看[遇，碰]见；寻获[找]，觅[获]得，搜索，探测，选择，定位；断[决]定 *n.* 新产地，发现(物)

finder ['faində] *n.* 探测器，寻视器，测距器，选景器；方位仪，定向器；瞄准器，寻线机

fine [fain] *a.* 细的，稀薄的，精细[密]；上等的，优良的；干净的，纯的 *vt.* 精炼 *vi.* 变精细 *n.* 罚款
~ aggregate 细骨料
~ arts 艺术（绘画，雕刻，建筑等）
~ concrete 细骨料混凝土
~ crack 细裂缝
~ crushing 细轧，细碎，精碎
~ earth 细粒土
~ -graded 细级配的
~ -graded asphaltic concrete 细级配沥青混凝土
~ -graded aggregate 细级配骨料
~ grain 细粒

~ grained sand 细砂
~ -grained wood 细纹木
~ grinding 细磨，磨细
~ gravel 细砾石
~ machining 精加工
~ material 细（颗）粒材料
~ metal 纯金属
~ pointed dressing 细琢（石面），细凿修整
~ pointed finish 细凿修整
~ pointed stone 细凿石
~ purification 精制
~ rack 细格栅
~ sand 细（粒）砂
~ sheet asphalt 细粒片地沥青（混合料）
~ soil 细粒土
~ -sorted material 细粒材料
~ stone 细碎石 (AASHO 规定粒径 3/8—10 号筛)
~ stuff 细料
~ thread 细（牙螺）纹
~ thread screw 细纹螺钉
~ wood 细纹木

fined *v.* 使精细[稀薄]；精制；变好[纯，精致，稀薄]

finedraw ['fain'drɔː] *vt.* 拉细丝；细(密)缝

finefied ['fainifaid] *a.* 装饰了的

finery ['fainəri] *n.* 装饰，盛装，(pl.) 装饰品

fines *n.* 筛屑，碎屑，粉末

finger ['fiŋgə] *n.* 指（头），手指，抓手；指状物；指(示)针，箭头，阀，活门，闸门；销；测厚规
~ board 指向牌
~ brush 指形刷
~ chute 指形滑槽
~ clamp 指形压板
~ cutter 指形铣刀
~ joint 指形接合，梳形接合

finger joint

~ lever 指形手柄
~ plate （门的）防污板,指板,推手板
~ post 指路牌,路标
~ raise 指形格条天井
~ stone 小石块
finial ['fainiəl] *n.* 顶尖,最高点;叶尖[尖顶]饰;收尾或顶盖的装饰成细部

finials

fining ['fainiŋ] *n.* 净化,澄清
~ agent 澄清[净化]剂
finish ['finiʃ] *n.*; *v.* 修整,饰,精加工,完工,最后加工,最后一道工序;完成,结束,竣工,成品;表面光洁度;抛光;(表面)涂层,保护层,漆面;装饰材料,面漆
~ casing 框饰,台口线
~ carpentry 木工装修作业
~ coat 罩面层（油漆成抹灰）
~ coat paint 罩面漆
~ grading 最后整平,（跨面）最后整型
~ rolling 终压,最后碾压;打光压实
~ to gauge 按样板加工
~ to size （按尺寸）加工
~ed cement 成品水泥
~ed goods 成品
~ed line 竣工线,终止线
~ed sheet 精整薄板
bright ~ 抛光;光亮精整;镜面抛光
brushed ~ 刷面装饰
cabinet ~ 细木壁饰,壁板终饰
cement ~ 水泥饰面
concrete ~ 混凝土饰面
crimp-proof ~ 防皱整理
dash ~ 浇泼粉面

dead ~ 无光泽抛光
dry ~ 干浆料
dull ~ 无光泽修饰
fine ~ 高级精加工,细加工,高光洁度
fine-pointed ~ 细凿修整
frosted ~ 霜花面饰;毛花整理;磨砂
gloss ~ 光泽整理
ground ~ 磨光
hard ~ 硬质罩面,石灰膏抹灰
hardwood ~ 硬木表饰
high ~ 光制,粘磨,研磨
lapping ~ 研磨
looking-glass ~ 镜面抛光
medium ~ 中级精加工;中等光洁度
mill ~ 轧制光度
minute ~ 抛光至镜面光泽
mirror ~(ing) 镜面精加工[光洁度]
monolithic ~ 整体修饰
non-creasing ~ 防皱整理
pean-hammered ~ 斧锤琢面
pigmented ~ 着色地面
pointed ~ 点凿面
polished ~ of stone 石面打光
pressed ~ 压光面饰
protective ~ 表面[防腐]处理
rough ~ 粗饰,初级修整;低光洁度
rubbed ~ 磨出面
sand ~ 砂饰面
sand blast ~ 喷砂(表面)处理
sawed ~ of stone 锯成石面
scraped ~ 刮磨
scrubbed ~ 刷石子面
spaded ~ 捣拍（混凝土）修整;泥铲压抹修整
surface ~ 表面光洁度,表面精整
tooth chisel ~ 齿凿面
vermiculite plaster ~ 虫蚀状粉饰面
washed ~ 粉刷面;刷石子面
finisher ['finiʃə] *n.* 修整器,平整机;精加工工具;修整工
cement ~ 水泥平整器
concrete ~ 混凝土修面机

paving ~ 整面机,铺面修整机
finishing [ˈfiniʃiŋ] ***n.*** 修整,终[粉]饰;竣工;结尾[束];饰面
~ chip 精加工
~ cloth 擦布
~ coat 罩面层
~ hardware 精制小五金
~ knife 修整刀具
~ layer (路面)罩面,盖层;终饰层
~ machine 修整机,整面机
~ material 装饰材料
~ method 抛光方法
~ nail 装修钉
~ operation 修整[终饰]工作
~ paint 饰面漆
~ polish 装修抛光
~ procedure 修整手续[程序]
~ process 修整[终饰]工作
~ screen 细[终]筛
~ touch 最后修整
~ varnish 饰面清漆
~ work 修饰工作
finite [ˈfainait] ***a.*** 有限[尽]的
~ deformation 有限变形
~ space 有限空间
~ stripe method 有限条分法
Fink truss 芬克式桁架
fire [ˈfaiə] ***n.*** 火,发光[炽热]体,燃烧,着火 ***v.*** 着火,发射;
~ alarm 火警
~ apparatus 灭[救]火器,消防设备
~ area 防火带中用防火墙分隔开的区段
~ back 壁炉背墙
~ barrier(s) 防火间隔,挡火墙
~ blanket 防火毯
~ brick (耐)火砖
~ brick arch 耐火砖拱
~ bridge 火砖拱
~ bulkhead 防火墙
~ cement 耐火水泥
~ clay (耐)火泥,(耐)黏土
~ clay sleeve 火泥釉砖
~ company 消防队,火灾保险公司
~ damper 火灾阻止器[遮断器]
~ demand 消防用水
~ division wall 隔火墙
~ door 防火门,保险门,炉门,火门
~ escape 安全梯,安全出口,太平梯
~ exit 太平门,安全出口
~ -exit bolt 安全门闩
~ extinguisher 灭火器[机]
~ foam 泡沫灭火剂
~ hose 灭火水龙带
~ -hose station 消防站
~ house 消防车库,消防站
~ hydrant 消防栓,消防龙头
~ insurance 火灾保险
~ office 火灾保险公司
~ partition 防火隔墙
~ plug 消火栓
~ precaution 防火措施
~ prevention 防火
~ proof 防火的,耐火的
~ protection 防火
~ pump 消防泵
~ resistance 耐火性
~ resisting concrete 耐[抗]火混凝土
~ retarding glazing 耐火玻璃
~ sand 耐火砂
~ shutter 防火百叶窗
~ standpipe 消防竖管
~ stone 耐火石,(打)火石
~ stop 挡火物
~ trap 无太平门的建筑物
~ truck 消防车,救火车
~ valve 灭火阀,消火阀
~ wall 隔火墙
~ window 防火窗
firedog [ˈfaiədɔg] ***n.*** 炉壁柴架
fireplace ***n.*** 壁炉

fireplate *n.* 防火板
fireproof ['faiəpru:f] *a.* 耐[防]火的,不燃的,耐热的
～ aggregates 耐火集料
～ bllilding 耐火房屋
～ bulkhead 防火隔墙
～ construction 防火结构[建筑]
～ door 耐火门,防火门
～ floor 耐火楼板
～ material 耐火材料
～ structure 耐火结构
～ wall 防火墙,耐火墙
fireproofing 耐火材料;耐火处理[工序]
～ tile 耐火砖(瓦)
firm [fə:m] *a.* 坚固[硬]的,坚定的,牢固的 *n.* 实业,公司,厂商
～ clay 硬粘土
～ing agent 固化剂
firmware ['fə:mwεə] *n.* 固件,稳固件[设备]
firring *n.* 灰板条,长楔形木板[条],板条面壁
first [fə:st] *n.* 第一,最初,首位[先] *a.* 第一的,最初[早]的,首要的
～ class 一级,最佳级
～ class certificate 一级品证明书
～ coat 底涂,头道抹灰[涂层],打底
～ cost 初期投资,初次[基建]费用,建造[购置]费,生成[最初]成本,原价
～ crack 初始裂缝
～ floor【美】第一层(楼面),【英】第二层(楼面)
～ grading 初次整平[整型]
～ polish 底层磨光
～ stone 基石
～ story 第一层(楼)
fish [fiʃ] *n.* (*pl.* fishes or fish)鱼;鱼尾板,[片];接合板;夹片;钓锚器,悬鱼饰[板];*v.* 用夹板[鱼尾板]连接[加固,补强]
～ bar 夹杆
～ joint 夹[鱼尾]板接合
～ -plate splice 鱼尾板拼接
～ piece 鱼尾[接合]板
～ plate 鱼尾[接合]板
fishbolt *n.* 夹紧螺栓,接合用压紧螺栓,鱼尾(板)螺栓
fission ['fiʃən] *n.*;*v.* 分裂,裂开,剥离
fissure ['fiʃə] *n.* 缝隙,裂缝[口,隙,纹] *v.* (使)裂开,(使)破裂
～d structure 裂缝结构
～ water 裂隙水
compression ～ 挤压裂缝
contraction ～ 收缩裂缝
desiccation ～ 干缩裂缝
fault ～ 断层裂纹
roof ～ 顶板裂缝
tectonic ～ 构造裂缝
fit [fit] *v.* 装配,配合;调整;镶嵌;符合;磨合;吻合;密接 *a.* 合适的,适合
～ joint 套筒接合
～ strip 镶条
fitch [fitʃ] *n.* 长柄漆刷,小毛刷
fitment ['fitmənt] *n.* 家具,设备;(*pl.*)装修;附件,配件
fittage *n.* 装配任务;间接费用;杂费
fitter ['fitə] *n.* 装配工,钳工,修理工,设备安装工
assembling ～ 装配工
assistant ～ 安装助手
engine ～ 机器装配工
glass ～ 玻璃工
machine ～ 机器装配工
pipe ～ 管子工
saw ～ 锯工
sheet metal ～ 板金工
fitting ['fitiŋ] *n.* 拟[符]合;配[另,附,连接]件,装配部件,接头,套筒;装配,安装,结合;选配
～ and fitments 房屋设备安装及装修
～ -up 结构部件成型,结构部件安装
～ -up-gang 安装队
angle ～ 弯头
duplex ～ 三通接头
exhaust ～ 排气管接头

female ~ 阴螺纹管接头
male ~ 阳螺纹管接头
plumbing ~ 卫生管道配件
sanitary ~s 卫生设备
soldered ~ 焊接头
Tee ~ T形三通管
watertight ~ 防水配件
window ~s 窗配件

five [faiv] *n.* 五 *a.* 五个
~ centered arch 五心拱
Five Dragon pavilion 北京北海五龙亭
Five phoenix tower 北京故宫五凤楼

fix [fiks] *v.* 安装[设];定位;固定;定影[象];方位点,标定点;坐标;交会;设备安装后的调整,修理 *n.* 定方位
~ position 定位
~ screw 固定螺丝
~ stopper 定位销
absolute ~ 绝对定位

fixation [fik'seiʃən] *n.* 安置;固定;定影[形]凝固

fixative ['fiksətiv] *n.* 固定剂 *a.* 固定的,防褪色的

fixed [fikst] *a.* 固定的,不变的
~ beam 固定梁
~ bearing 固定支座
~ capital 固定资本
~ charge 固定费[开支]
~ cost 固定成本,固定费(用)
~ -end arch 定端拱,无角拱
~ end beam 固端梁
~ -end colunm 固端柱
~ fence 固定栅栏
~ form 固定模板
~ light 固定窗
~ sash 固定窗框[窗扇]
~ shutters 固定百页窗
~ window 固定窗

fixing ['fiksiŋ] *a.* 固接[定];定位[向,影];接头;整[修]理;装配[修]
~ agent 固定剂

~ block 受钉块,木落砖(可钉钉)
~ brick 受钉块,木落砖
~ collar 加固圈
~ device 固定装置
~ fillet 受钉嵌条
~ moment 固定力矩
~ mortar 固定砂浆
~ ring 固定环
detachable ~ 可拆卸式装配
elastic ~ 弹性固接
rigid ~ 刚性固接

fixings *n.* 嵌固件

fixity ['fiksiti] *n.* (*pl.* -ties)稳[固]定性,不变性,永恒性;硬性,(水的)硬度;不挥发性;刚性

fixture ['fikstʃə] *n.* 装置物;设备;支架;工作夹具;固定值
~ block 卡[夹]块
~ branch 固定分支
~ joint 固定连接
bathroom ~(s) 浴室装置
light ~ 灯具
master ~ 基准型架
sanitary ~ 卫生器皿
welding ~ 焊接夹具

flag [flæʃ] *n.* 标志[识],特征,记号;石板,板层;旗(帜) *v.* 铺(砌)石板
~ alarm 报警信号器
~ stone 石板
~ stone paving 石板铺砌,用石板铺路;石板路面
concrete ~ 混凝土板
covering ~ 盖板

flagging ['flægiŋ] *n.* 铺砌石板,石板路,板[薄]层砂岩
~ stone (铺路)石板,扁石

flake [fle:k] *n.* 薄(石、雪、鳞)片,(纤维素塑料)乾坯料;(可移动的一段)栅栏,篱笆;一卷绳索 *v.* 起壳;成片降落[剥落];成薄片
~d asbestos 薄片石棉
~ glass coating 薄玻璃涂层
~ graphite 片状石墨

~ texture 片状结构
~ tool 刮削器
~ white (碳酸)铅白
flakeboard(s) 木屑刨花板,碎料板
flakelet ['fleiklit] *n.* 小薄片
flaker ['fleikə] *n.* 刨片机
flakiness ['fleikinis] *n.* 薄片(性质或状态),片层分裂
~ ratio 宽厚比
flaking ['fleikiŋ] *n.* 制片,耐火材料剥落,片落(油漆或粉刷成薄片剥落),成片薄落,起鳞
~ mill 制片机
flaky ['fleiki] *a.* 片状的,成片的,成层的;易成片剥落的,易破碎成薄片的
~ fir 鳞皮冷杉
~ material 片状材料
~ resin 片状树脂
flambeau ['flæmbəu] *n.* (*pl.* -beaux, -beaus)(装饰用)烛台,火焰台,火炬形饰;火炬,火把;燃烧废气的烟囱
flamboyant [flæm'bɔiənt] *a.* 灿烂的,辉耀的,华丽的;火焰似的;火焰式曲线窗格的
~ architecture 火焰式建筑
~ period 火焰式曲线时期(建筑)(1450～1530)

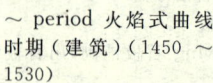
flamboyant tracery

~ structure 辉耀构造
~ style 火焰式
~ tracery (火)焰式窗格
flamboyanttree 凤凰木
flame [fleim] *n.* 焰,火焰[舌];燃烧;光芒[辉];形状(颜色)似火焰的物器 *v.* 燃烧
~ ablation 熔化烧蚀
~ arc lamp 弧光灯
~ colored (光辉)桔黄色
~ damper 灭火器
~ resistance 抗燃性,耐火性
~ -resisting 耐[防]火的
~ retardant paint 耐燃漆
~ retardants 阻燃剂
~ -retarded resin 阻燃树脂
~ spraying 热喷喷镀法,火焰喷(涂)
~ tree 凤凰木
flameproof ['fleimpru:f] 防火[爆]的,耐火,不易燃的
~ paint 耐火漆
flamethrower *n.* 火焰喷射器
flameware *n.* 耐热玻璃器皿
flammable ['flæməbəl] *a.* 可燃的,易燃的
flanch [flæntʃ] =flange *n.* 凸[突]缘;法兰
Flandipak 费兰狄帕克数字(显示)板
flange [flændʒ] *n.* 凸[突]缘,(梁)翼缘,板边;法兰(盘);镶边,凸饰边,凸褶 *v.* 折边,作凸缘,装凸边[法兰(盘)]

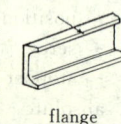
flange

~ and spigot bend 法兰套管弯头
~ angle 凸缘角铁,翼角钢
~ bend 法兰弯头
~ bolt 凸缘螺栓
~d beem 工字木梁
~d connection 法兰[连接]
~d cross 法兰四通
~d joint 法兰接头
~ nut 法兰螺帽,凸缘螺母
~ pipe 凸缘[法兰]管
~ plate 翼[凸]缘板
~ slab 翼缘板
~ splice 翼缘镶板
~ steel 凸缘钢
~ union 法兰接头,折缘管
follower ~ 填料压盖法兰
flangeway ['flændʒwei] *n.* 轮缘槽
flanging ['flændʒiŋ] *n.* 折边,[缘];外(缘)翻边
~ machine (卷)折边机
~ press 压边机

flank [flæŋk] *n.* 翼；侧面，[翼，墙]边；山侧斜地；厢房；外侧车道；齿侧面 *v.* 位于侧面
~ angle（螺纹）侧面角
~ wall 侧[山边]墙
~ing window 边窗

flannel ['flænl] *n.* 法兰绒，绒布

flannelet(te) [,flæn'let] *n.* 棉法兰绒，绒布

flanning ['flæniŋ] *n.* 窗框[壁炉]两侧斜边，*a.* 外张的，八字形的

flap [flæp] *n.* 折翼，可偏转的翼片，襟翼，阀瓣；片状物；铰链板；挡水板，阻力板；盖边 *v.* 摆动，拍打
~ door 吊门，活板门
~ gasket 平垫圈
~ tile 折瓦，照瓦
bottom ~ 折叠底（板）
chimney ~ 烟道门
port ~ 排水孔，带铰链的盖
rubber ~ 橡胶刮板
safety ~ 安全瓣，安全节气门

flare [flɛə] *n.* 喇叭管[口]；端部斜展，锥度，锥形孔；闪烁亮光；火苗[舌，焰] *v.* 闪耀；船倒外倾；张开
~ angle 斜张角（建筑物中线和墙之间夹角），扩张角
~ header 半头黑砖
~ opening 喇叭口
~ tube fitting 喇叭管接头
~ wingwall 斜翼墙，喇叭形翼墙
~d access 喇叭口形入口[路口]
~d outlet 喇叭形出水口
~d tube 扩口管

flash [flæʃ] *n.* 闪光[现，烁]；光泽，亮度，（水库）泄水，冲砂，堰，水闸；极薄的镀层；毛刺，飞边；溢料；手电筒；暴风雨 *a.* 迅速的，快速的，瞬时的，闪光的
~ arc 闪光电弧；火花弧
~ barrier 隔弧板
~ card 闪光卡，闪片
~ coat 闪光焊覆层，表面处理盖层，喷浆盖层（混凝土）

~ dry 急骤干燥
~ glass 有色玻璃
~ hider 消焰器，灭火帽
~ lamp 闪光灯
~ light 闪光信号灯
~ mold 溢出式塑模，平线脚
~ paper 闪光纸
~ plating 薄镀（层）
~ point 闪点，着火[闪燃]点
~ set （混凝土）急凝，瞬时凝结，快速凝固
~ signal 闪光信号
~ weld 闪光焊，闪速对焊
~ed glass 贴色[套色]玻璃

flashback ['flæʃbæk] *n.* 回火，逆火[燃弧]

flashbulb *n.* 闪光灯，镁光灯

flasher ['flæʃə] *n.* 闪烁光源[开关]；（玻璃）镶色工；闪光装置，自动闪光器[设备]
~ unit 闪光标记
signal ~ 信号闪光灯

flashguard *n.* 防弧器

flashgun *n.* 闪光灯；闪光枪

flashing ['flæʃiŋ] *n.* 挡水板；防水片，（防漏用）盖片，防雨板，灌水，（水力）充填；泛水，暴涨，快速[急骤]蒸发，闪蒸，镶色烧砖法，玻璃镶边；闪光
~ arrow 闪光指示箭头
~ beacon 闪光标灯
~ block 陶瓦块
~ board 防雨板
~ compound 防水填缝料
~ -off （耐火材料的）软熔，烧熔边缘（焊接），熔化
~ point 闪光点，引火点，燃点
~ ring 管道套圈
~ tile 瓦管
apron ~ 遮檐[披水]板
cover ~ 披水板

flashlight ['flæʃlait] 手电筒，闪光信号灯，脉冲灯

flashy ['flæʃi] *a.* 闪光的，一瞬间的，骤

发的
~ load 瞬间荷载
flask [flɑ:sk] n. 造型砂箱,长颈烧瓶
~ board 托模板,底板
~ clamp 砂箱夹
~ pin 砂箱的定位销
~ rammer 平头捣锤
flat [flæt] n. 平面、平坦形、平地;扁钢;一层楼公寓,[建筑中的一层、一套或一个]阁楼房间,分成若干单元的一座建筑;玻璃压板,压板;单一定额;统售价格 a. 平的,平淡的;扁的,薄的;(市场)萧条的 v. (使)变平
~ arch 平拱,扁拱
~ band 素石
~ bar 扁钢,条钢
~ bar keel 平板龙骨
~ bearing 平导板,双脚支柱,扁柱
~ beater 平板打夯机
~ bed truck 平板车,平板式运货汽车
~ bit tongs 扁嘴钳
~ block 装配平台
~ bone 扁平骨
~ -bottomed 平底的
~ brick 扁砖
~ car 平(板)车,敞车
~ chisel 扁凿
~ cold-rolled sheets 冷轧薄板
~ cost (预算)直接费;成本费
~ crank 平曲柄
~ crown 平坦路拱
~ fillet weld 平角焊
~ gauge 样板,填板规
~ grain 顺缘木纹,弦面纹理
~ head 平头,扁头
~ interlocking tile 连锁平瓦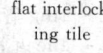
flat interlocking tile
~ invert 平缓仰拱
~ jack 扁千斤顶
~ joint 平缝,平节理
~ -joint pointing 勾平缝
~ jumper 扁凿
~ key 平键
~ link chain 扁环节链,板链
~ moulding 平线脚装饰
~ paint 平光漆
~ pattern 平型
~ pointing 勾填平缝
~ relief 平浮雕
~ roof 平屋顶
~ roof construction 平屋顶构造
~ rope 扁绳
~ -sawn 弦截,平纹锯法
~ sawed lumber 平锯[顺锯]木材
~ scraper 平面刮刀
~ seam 平合缝
~ sheet 平面图
~ silver 成套银餐具
~ slab 平板;无梁板
~ slab capital (column) construction 无梁板柱构造
~ -slab construction 无梁板结构
~ slab floor 无梁楼盖
~ slab structure 无梁板结构
~ spring 板弹簧
~ steel 扁钢
~ tile 平瓦
~ tile roof 平瓦房顶
~ type piling bar 平板桩
~ varnish 无光漆
~ vaulted ceiling 平弧顶棚
~ wall brush 扁平墙壁刷
~ washer 平垫圈
~ weld(ing) 平焊
~ ting agent 消[平,减,区]光剂
lead ~ 铅皮平屋顶
round-edged ~ 圆边扁钢
flatbed ['flætbed] n. 平板;平板卡车[挂车];平河床 a. 平(板)的
flatcar ['flætkɑ:] n. 平板车,敞车
flatiron ['flæt,aiən] n. 条铁,扁铁;熨斗
flatland n. 平地;假设的二维世界
flatlayer a. 平层的
flatlet ['flætlit] n. 单间紧凑的小公寓

flatness ['flætnis] *n.* 平直度;均匀性;平面[平滑]度

flatten ['flætn] *v.* 压平,修平,平直[整]

flattening ['flætniŋ] *n.* 压[修]平,弄直,平整,整[变,铲]平,矫平[直];补偿;扁率[度]

~ -out 辗压,平整,压平

flattop *n.* 平顶(建筑物)

flatware ['flætwɛə] 浅皿[容器];(美)盘碟类,扁平餐具

flatwise ['flætwaiz] *n.* 平放,放平的样子;平面向下 *ad.* 扁平地,平放地

flaunching ['flɔ:ntʃiŋ] *n.* 突缘

flaw [flɔ:] *n.* 裂缝;缺陷,瑕疵;疵点;掠断层,横推断层 *v.* 破裂,有裂缝;产生缺陷,有瑕疵

flawless ['flɔ:lis] *a.* 无裂缝[纹],无瑕疵的;完善[美]的

flawy ['flɔ:i] *a.* 有裂缝[瑕疵]的;遭受突然阵风的

flax [flæks] *n.* 亚麻,胡麻

earth — 石棉,石绒

fiber — 亚麻

fossil — 石棉

fleam [fli:m] *n.* 锯齿夹角

~ -tooth 等腰三角形锯齿,尖角锯齿

fleche [fleiʃ] *n.* [法]尖顶塔;尖顶点;离岸沙坝

fleck [flek] *n.* 斑点,雀斑

Flemish ['flemiʃ] *n.*;*a.* 佛兰芒人[的]

~ bond 荷兰式砌合法(同层丁顺砖交错)

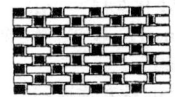

Flemish bond

~ brick (铺面用)黄色硬砖

~ foot 带有 C 型或 S 型旋纹的酒桶状家具脚

~ garden bond 每层三顺一丁砌合(一砖厚)

~ garden wall bond 荷兰园墙式砌合法

Flemish garden wall bond

fletton *n.* 美国彼得郡(斑红色)砖,费莱顿砖

fleuron ['fluərɔn] *n.* 百合花饰;花形图案装饰

flex [fleks] *n.*;*v.* 挠[弯]曲;折曲电[皮,花]线,塞绳

~ hook 电源线挂钩

flexible ['fleksəbl] *a.* 可弯曲的,柔顺的,有弹性的,有挠性的;可塑造的;软脊

~ bend 柔性弯头

~ cable 柔(性)索缆[钢索,电缆],软电缆

~ cladding 柔性面层

~ collodion 挠性胶棉

~ column 柔性柱

~ conduit 软(导)管,蛇皮管

~ cord 电线,皮[软]线

~ coupling 柔性联轴节

~ -fender system 柔性护[舷]木系统

~ glue 软胶

~ layer 柔性层

~ overlay 柔性盖层

~ pavement 柔性铺面[面层,路面]

~ pipe 软管

~ spline 柔性塞缝片;柔性夹板

~ tubing 柔(性)连接管

~ wall 柔性墙;悬臂挡墙

~ wire 花[皮]线

flexibility [,fleksə'biliti] *n.* 弯曲[柔软,伸缩,挠]性,柔度;(光的)折射性,(棒的)细长[度];适应[机动,通融,灵活]性

~ agent 助柔[增韧]剂

fleximeter *n.* 挠度计

flexion ['flekʃən] *n.* = flection 弯曲;弯曲部

flexlock *n.* 柔性止水缝

flexuosity [flekʃu'ɔsiti] *n.* (*pl.* -ties) 弯曲,屈曲,弯曲性

flexural ['flekʃjurəl] *a.* 弯[挠]曲的,挠性的
~ centre 弯曲中心
~ deflection 挠曲变形
~ failure 弯曲破坏

flexure ['flekʃə] *n.* 挠[弯]曲;平面弯曲,单斜挠曲,(地质)单斜挠褶

flicker ['flikə] *n.*;*v.* 闪烁[光],火花;突然而短暂的动作;一时激动
~ing lamp 闪光灯
brightness ~ 亮度闪烁
chromaticity ~ (= colour ~) 彩色闪烁
image ~ 图像闪烁
luminance ~ 亮度闪烁
signal lamp ~ 信号灯闪烁器

flier ['fiaiə] = flyer *n.* 梯级,踏步板;补充目录,书内的附加页

flight [flait] *n.* 阶梯[楼梯]梯级;螺丝槽行程,螺纹;行[射]程;刮板
~ plan 梯级规划

flint [flint] *n.* 火石,燧石;坚硬之物
~ aggregate 坚硬的骨料
~ brick 坚硬[燧石]砖
~ clay 硬质[燧石]粘土
~ glass 铅玻璃,火石[无色,燧石]玻璃
~ glass paper 粗砂纸
~ gray 燧石灰色
~ paper 研光纸
~ sand 火石砂
~ wall 火石墙,坚硬的墙

Flintkote 费林特玛琣脂面层

flinty *n.* 燧石的;坚硬的;透明的
~ ground 坚硬地面,石质硬底
~ slate 坚石板,燧石板岩
~ soil 含硅土,坚硬土
~ steel 硬钢,硅钢

flip [flip] *v.* 交换,倒转,翻动;轻碰撞; *n.* 轻弹,跳动

flit [flit] *n.*;*v.* 掠过,离开,飞来飞去
~ -plug 可拆卸插头

flitch [flitʃ] *n.* (组梁)贴板,料板,桁板,厚条板,背板;(有树皮的)厚木材,对开圆木,粗加工大木料,大方木
~ beam 组合板梁,(钢木)夹合梁,贴[桁]板梁
~ chunk (组合)厚板
~ girder 组合板大梁
~ plate 组合板,脊板;组合板夹铁,(钢木夹合梁的)夹合钢板

flitter ['flitə] *n.* 闪光颜料;金属小片或金属碎片,很细的金属碎屑
~ gold 黄铜箔

Flixor【瑞士】弗利克索尔(聚酰胺纤维)

float [fləut] *n.* 浮体[筒,船坞];漂浮物;抹子,镘刀,镘;磷灰石粉;路面整平器;浮动[余裕]时间; *v.* 滑翔,下滑;发行(公债,债券等);创设(公司、商号等);筹款
~ board 承板,轮翼
~ coat 抹面层
~ -cut file 单纹锉刀
~ finish 抹光,用镘修整
~ material (浮选中的)轻料,浮料
~ stone 浮石,泡石,(铁)磨砖石
~ tank 浮式箱(带浮子调节器的水箱)
~ work 抹灰工作,抹面
~ed coat 抹灰层
~ed concrete 抹平[面]的混凝土,镘整混凝土
~ed finish 抹灰饰面
~ed surface 抹平面

floater ['fləutə] *n.* 浮子,漂浮物,浮体;抹灰工具;抹灰工;流动(临时)工;筹资开办人

floating ['fləutiŋ] *a.* 抹[镘]平;变[流,移]动的;漂浮,浮式[运]的; *n.* 浮雕

~ bay 抹灰面积[格间]
~ blanket 悬浮层
~ brick 轻质砖(白云石质的石灰砖)
~ coat 二道油漆[抹层]
~ construction 浮动结构[建筑]
~ cover 浮式顶盖
~ floor 浮式[夹层]地板
~ gang 流动工队
~ garden 漂浮庭园
~ gauge 浮表[规,标尺]
~ holder 活动夹具
~ log 漂浮圆木
~ -post type 浮动柱型
~ rail 活动护栏
~ -rolled steel 辊轧扁钢
~ rule 抹灰面用嵌线条条;(抹灰面)刮尺[杆],镘板
~ screed 冲筋(标志抹灰厚度的窄条)
~ scum-board (抹灰用)托灰板

floc [flɔk] *n.* 絮片;絮状沉淀,絮凝物

flocculate [ˈflɔkjuleit] *v.* 絮凝,绒聚;(土等)结成小团块

flocculated structure 絮凝结构

flocculation [ˌflɔkjuˈleiʃən] *n.* 絮凝作用
~ agent 絮凝剂

flocculent [ˈflɔkjulənt] *a.* 絮凝[结]的,绒聚的
~ structure 絮凝[密族]结构

flocculus *n.* 絮状[凝聚]体,絮团(絮凝粒;棉絮;片状物)

flock [flɔk] *n.* 毛[棉]屑,绒,短纤维;凝聚[絮状]体
~ bed 毛(棉)屑垫床
~ paper 糊墙(花)纸
cotton ~ 棉绒

flong [flɔŋ] *n.* 作纸型用的纸

flood [flʌd] *n.*;*v.* 洪水,涨潮,泛滥;浸溃
~ -light 泛光照明,强力照明

floor [flɔː] *n.* 楼面[板,层],底[垫,地]板;桥面;台面;覆盖层;最低额,底价;*v.* 铺地板
~ arch 平背拱,平拱
~ area 建筑各层面积
~ beam 桥面[楼面]横梁;楼面[地板]梁
~ board 地板;楼面板
~ brad (扁头的)地板钉
~ brick 铺地砖
~ breaks 底板裂隙
~ burst 底板隆起[突出]
~ chart 布置图
~ chisel 铺地板錾紧凿,地板填隙铁
~ closer 地板闭锁器
~ clamp 紧板马铁(夹钳)
~ cloth 铺地面织物;门前擦鞋墩毯;擦地板布
~ construction 楼盖构造,楼或桥面构造
~ depth 楼面高度
~ drain 楼面排水;地面排水
~ dresser 地板装修器
~ filling (plugging, pug) 楼板填料(隔音材料,泥料)
~ finish 楼面修整
~ framing 楼板骨架
~ -framing plan 楼板结构平面图
~ girder 楼盖主梁;桥面板主梁
~ grinder 地板磨光机
~ guide 楼板[滑门]门槽(滑门的地板导槽)
~ hanger 支承地板托梁的镫铁
~ hardener 水泥地面坚硬材料(如铁屑)
~ head 肋板头
~ heating 楼板暖气
~ heave 底板隆起
~ hinge 装在门底与地板之间的铰链
~ hopper 底(部)卸(料)
~ joist 楼盖格栅,楼地板搁栅
~ knob 门碰头,门挡
~ lamp 落地灯,座灯
~ level 底板,底层,最低水平
~ light 落地灯,地板窗

floor knob

~ line 楼地板线；楼面[层]线
~ load 楼面荷载
~ machine 地板机，木地板刨平机
~ pack 底板分块
~ panel （由墙，柱或梁支承的）预制装配式楼板单元
~ plan 平面[楼面]面布置图
~ plank 桥面板；地板
~ plate 垫板；楼面板；地[底]板
~ polish 地板蜡
~ polisher 楼板擦光器；擦地板工
~ push 脚踏[闸刀]开关
~ receptacle 地板插座

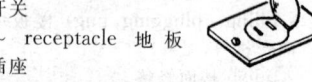

floor receptacle

~ sill 底座（梁）栅子的底梁；消力坎
~ slab 地板，楼板，桥[路]面板
~ space 楼面[房屋]面积；底面积
~ squeeze 底板隆起
~ stop （地板上的）门挡
~ strutting 楼盖梁加劲条[支撑]
~ suspender 楼板[桥面]吊杆
~ switch 楼层开关
~ system （桥）楼面系统
~ -through 公寓房间
~ tile （铺）地面砖
~ topping 楼板面层
~ tramming 楼板骨架
~ type 落地（固定）式
~ varnish 地板清漆
~ ventilation 地板通风
~ wax 地板蜡
access ~ 活地板

arched ~ 拱形楼盖；砌反拱的卷道底板
beam and girder ~ 交梁楼板
brick arch ~ 砖拱楼面
cirque ~ 环形楼层
clinker ~ 缸砖地面
continuous ~ 无缝[连续]地板
counter ~ 粗地板（为表层地板的基底）
deep ~ 加强筋板层
double-framed ~ 梁格栅面
false ~ 辅助底板；格栅板，假楼板
first ~ 第二层楼（英）；第一层[楼，层]（美国）
flat slab ~ 无梁楼板，平板式楼板
floating ~ 浮式地板
garret ~ 阁楼地板
girderless ~ 无梁楼板
granitoid ~ 仿花岗石地板
grooved and tongued ~ 槽舌[企口]地板
ground floor 底层面，一楼，底楼
gypsum ~ 无缝石膏地面
herring bone parquet ~ 人字拼木地板，席纹地面
hogbacked ~ 波浪形底板
inlaid ~ 镶嵌地板
inlaid strip ~ 镶条（嵌）地板
insulated ~ 保温地板
insulcrete ~ 绝缘（混凝土）地板
intermediate ~ 中间楼面
jointless ~ 无缝地板
laminated ~ 叠层地板
lower ~ 下层楼
main ~ 主要水平，主层
marble ~ 大理石地板
naked ~ 光秃地板，未装修的地面
open ~ 露格栅楼板，露梁楼板
parquet ~ 镶木地板
plank ~ 木板地面
plaster ~ 抹灰地板
plastic ~ 塑料地板，塑胶铺面

principal ~ 主层楼面
raised ~ 活地板
rib ~ 肋形楼板
ribbed reinforced concrete ~ 带肋钢筋混凝土楼板
rough ~ 毛[粗]地面
solid ~ 实心楼板
straight joint ~ 一顺地板
strip ~ 狭条地板
stull ~ 横撑支柱假顶,横撑上的铺板
terrazzo ~ 水磨石地面[铺面]
tile lintel ~ 空心砖加筋密肋楼板
tiled ~ 花砖地面
timber ~ 木地板;木材铺面
timber framed ~ 木架楼板
tongue ~ 企口板地板
trough ~ 槽形铺面
upper ~ 上层楼面
wearing ~ 耐磨地板
wood ~ 木地板
floorage ['flɔːridʒ] **n.** 地板面积,建筑面积,做地板的材料
flooring ['flɔːriŋ] **n.** 铺地面,铺面;地板(材料);桥面安装
~ block 铺地(板)块
~ board 铺面板
~ nail 铺面螺钉
block ~ 分块地板[铺面]
cork ~ 软木地板
dressed and matched ~ 刨光企口地板
linoleum ~ 油毡[漆布]铺地
mastic ~ 胶脂地面
square-set ~ 方框支架楼板
flop [flɔp] **n.** 拍击的动作或声音;睡眠的地方;无效 **v.** 突然转变;突然下降
~ house 廉价住所,低级旅馆
flopper ['flɔpə] **n.** 薄板上皱纹
floppy ['flɔpi] **a.** 松懈的;下垂的
flora ['flɔːrə] **n.** (pl. floras, florae) 窗外花台;地区性植物,植物志
floral ['flɔːrəl] **a.** 花卉的
~ designs 花纹(花卉)图案,花样,花纹
~ form 花式
~ motifs 花卉花纹
Florence ['flɔrəns] **n.** (意大利) 佛罗伦萨 (市)
~ leaf 装饰用的黄色合金(金属)箔
florentine ['flɔrəntain] **n.** 佛罗伦萨人 **a.** 佛罗伦萨城的
~ arch 佛罗伦萨式拱(两心拱)
~ blind 佛罗伦萨式(窗)遮帘
~ Renaissance 佛罗伦萨文艺复兴式
~ mosaic 佛罗伦萨式马赛克
floriated ['flɔːrieitid] **a.** 花形的,具有花卉装饰的,华丽的
floriation
[ˌflɔːri'eiʃən] 花饰
florid ['flɔrid] **a.** 华丽的,鲜红色的
~ Gothic 华丽哥特式

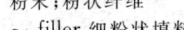
florid Gothic

flour ['flauə] **n.** 粉,粉末;粉状纤维
~ filler 细粉状填料
~ paste 浆糊
fossil ~ 硅藻土
mountain ~ 石粉
rock ~ 岩粉
slate ~ 石粉,页岩粉末
sillica ~ 石英粉
wood ~ 木屑,木粉
flourish ['flʌriʃ] **n.** 花饰;华丽的装饰或修饰
ceiling ~ 顶棚花饰
flow ['flau] **n.** 水流;气流;流量;涨潮;溢出;流程;流变,(土的)塑变;流挂性 **v.** 溢出;泛溢
~ capacity 泄水能力,排水能力
~ cascade 梯级,跌水
~ cast 流型
~ chart 流程图;生产过程图;程序框图表;流量图,操作程序图
~ coat 浇[流]涂
~ diagram 流程图,生产过程图;程序

框图,工艺流程图,流动曲线图
~ mark(s) 波流痕,波纹
~ of funds 资[基]金流转
flowage [fləuidʒ] *n.* 流动;流出;溢出;流动状态,流动特性
flowchart *n.* 程序方框图,操作程序图;[工艺]流程图;流量图
operations ~ 操作流程图,操作程序图
flower ['flauə] *n.* 花饰物;花卉,精华
~ bed 花坛
~ garden 花园
~ nursery 花圃
~ vase 花瓶
~ window 用花装饰的窗子
flowing ['fləuiŋ] *a.* 流动的,流畅的,自喷
~ line 出水管线,排水管线
~ tracey 气流窗花格

flowing tracey

fluate *n.* (防止建筑石料表面风化用的一种)氟化物
fluctuate ['flʌktjueit] *v.* 涨落,起伏,波动,脉动
fluctuation [ˌflʌktjueiʃən] *n.* 涨落,起伏;波动,不稳定变动;脉动
brightness ~ 亮度波动
temperature ~ 温度变化
flue [flu:] *n.* 烟[风、管]道,暖[通、导]气管;毛屑,乱丝
~ bridge 通道拱顶
air ~ 气[风、烟]道
discharge ~ 排气管
exit ~ 排气管;出口烟道
gas ~ 烟道
smoke ~ 烟囱,烟道
stack ~ 烟道
suction ~ 吸气管
upper ~ 上烟道;排气道
ventilating ~ 通风管
fluff [flʌf] *n.* 绒毛 *v.* 起毛,使疏松[松散]
~ point 疏松点

fluffy ['flʌfi] *a.* 木面起毛的;松软的;绒毛(状)的
fluid ['flu(:)id] *n.* 液体;流质 *a.* 流体的,流动的
~ -tight 不透[漏]水的
antistatic ~ 防静电液
soldering ~ 焊(接)液
welding ~ [熔]焊剂
fluing *n.* 厚墙的窗洞斜边
~ arch 斜面拱
fluke [flu:k] *n.* 锚爪,锚钩
flume [flu:m] *n.* 水槽,渡槽;斜槽;水道,山夹沟,涧 *v.* 滑运,槽运
reinforced-concrete ~ 钢筋混凝土渡槽
fluorescence [fluəˈresns] *n.* 荧光;荧光性
laser ~ 激光发光[荧光]
lasing ~ 激光发光[荧光]
fluorescent [ˌfluəˈresnt] *a.* 发荧光的,有荧光性的
~ dye 荧光染料
~ fittings 荧光灯装置
~ lamp 荧光灯,日光灯
~ lighting 荧光灯
~ lighting fixture 荧光灯;荧光照明设备
~ paint 荧光漆,发光漆
~ pigment 荧光颜料
~ screen 荧光屏
~ tube 荧光管(灯),日光灯管
fluorin(e) ['fluəri:n] *n.* 氟(F)
fluorite ['fluərait] *n.* 氟石,荧石
fluoroscope ['fluərəskəup] *n.* 荧光镜,荧光屏
flush [flʌʃ] *n.* 齐平 *v.* 弄平;冲水,冲砂 *a.* 同高的,齐平的
~ bead 平的串珠线脚,平焊缝,凸圆线脚
~ boarding 横钉平接护墙板
~ bolt 平头螺栓,埋入插销
~ coat (沥青)面层,整面
~ coater 喷洒机

~ color 底色
~ -cut joint 平头接合
~ door 平[光]面门
~ -filled joint 平嵌灰[接]缝
~ joint 平(灰)缝
~ panel 平镶板
~ plate 平槽板
~ plug socket 嵌入式插座
~ receptacle 墙插座
~ rivet 平头铆钉
~ -switch 嵌入[齐面]式开关,平装开关
~ toilet 抽水马桶
~ valve 冲洗阀
~ weld 平焊缝
bead ~ 串珠饰滚边
flusher *n.* 冲洗装置,喷面装置
flushing [ˈflʌʃiŋ] *n.* 冲洗,洒水
~ chamber 冲洗室
~ cistern 冲洗水箱
~ lever 冲洗扳手
~ trough (公共厕所用)自动冲洗槽
~ valve 冲洗阀
flushometer *n.* 冲水阀,冲洗阀
flute [fluːt] *n.* 凹槽,槽沟 *v.* 制作凹槽,开槽
~ cast 槽模
~ profile 槽[沟]形
fluted *a.* 有凹槽的
~ column 有凹槽的柱
~ formwork 有凹槽的模板
~ glass 槽纹玻璃
~ moulding 凹槽线脚
~ nut 槽顶螺母
~ sheet 槽纹板;波纹片
~ twist drill 麻花钻
fluter *n.* 做凹槽的工人;做凹槽的工具
fluting [ˈfluːtiŋ] *n.* 柱槽(建筑);开槽,表面发裂,带槽的材料
~ plane 凹刨,槽刨
flutter [ˈflʌtə] *v.* 震[扰],颤)动,干扰
~ valve 翼形阀,波动阀
flux [flʌks] *n.* 通量;(沥青)稀释剂;助熔剂;焊剂 *v.* 流化,熔化;加稀释剂,加焊药;
~ material 焊剂;[助]熔剂
~ oil (沥青)稀释油,稀释剂
~ed asphalt 软制(地)沥青
~ed bituminous material 软制沥青材料
~ed native asphalt 软制天然地沥青
~ing asphalt 软制(地)沥青
~ing oil 稀释[软制]油
light ~ 光通量
soldering ~ 软[助]焊剂;焊液
fly [flai] *v.* 飞,飞扬 *n.* 飞轮
~ ash 飞[浮]尘,飞灰;粉煤灰,烟灰
~ -cutting 快速[飞刀]切削
~ drill 飞轮手钻,手拉钻
~ gate 两开门,溜槽口活门
~ nut 蝶形螺帽[细]
~ over crossing 立体交叉
~ -past 跨度
~ rafter 飞椽

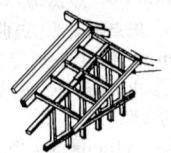

fly rafter

~ rail 折叶板撑;舞台天桥上的栏杆
~ screen 防蝇纱窗
~ way 舞台道(自舞台向观众席伸出)
~ wire 板缝盖网
flying [ˈflaiiŋ] *a.* 飞的,悬空的,浮动的
~ bond 跳丁砖砌合(在一定距离插进一块丁砖)
~ buttress 拱式扶垛,飞(扶)拱

flying buttress

flywheel

~ cut-off 剪铁条机
~ disk 有翼(的)园形装饰
~ facade 女儿墙(房屋正面超出屋顶的墙)
~ flemish bond 每层两顺一丁砌砖法
~ scaffolds 悬空脚手架
~ shore 横撑

flywheel ['flaiwi:l] *n.* 飞轮,手轮
foam [fəum] *n.* 泡沫 *v.* 起泡沫,喷泡沫
~ asphalt 泡沫地沥青
~ concrete 泡沫混凝土
~ extinguisher 泡沫灭火器
~ glass 泡沫玻璃
~ed plastics 泡沫塑料
~ed -slag concrete 泡沫溶渣混凝土,泡沫湿碾矿硅混凝土
~ing agent 发泡剂,泡沫剂
foamite *n.* 泡沫灭火剂,灭火药沫
Foamseal *n.* 富姆泡沫止水剂
focal ['fəukəl] *a.* 焦点的,(集中)在焦点上的 *n.* 焦点面;[聚]焦曲线
focalization *n.* 焦距调整;定焦点
focalize ['fəukəlaiz] *v.* 使集中在焦点上;对焦点;聚焦
foci ['fəusai] (focus 的复数)焦点
focus ['fəukəs] *n.* (*pl.* -cuses or -ci) 焦点;焦距;聚焦;震源;对光 *v.* 定焦点
~ lamp 聚焦灯
focuser *n.* 聚焦放大镜;聚焦装置
focusing ['fəukəsiŋ] *n.* 对光调焦,调整焦距
~ device 聚焦器
~ screw 对光[聚焦]螺旋
image ~ 图像聚焦
fog [fɔg] *n.* 雾;图像模糊,模糊度
~ coat 喷雾涂层,雾化沥青封面
~ nozzle 喷雾管嘴
~ room [喷]雾室,湿治室(混凝土养护作用)
fogger *n.* 润湿器
foil [fɔil] *n.* (金尾)薄片;衬托物;叶形

饰 *v.* 衬托;加叶形饰
~ -polystyrene laminate
聚笨乙烯[牛皮纸]
金属箔胶合板 foils
fold [fəuld] *v.* 卷折,折叠 *n.* 倍数;褶皱,折合
~ -plate roof 折板屋顶
~ structure 折板结构
~ed over joint 折接
~ed plate 折板
~ed plate structure 折板式结构
folding ['fəuldiŋ] *n.* 褶皱(作用);弯曲,折页
~ bed 折叠床
~ casement 折(玻璃)窗
~ chair 折椅
~ concrete form 折叠式混凝土模板
~ door 折门,叠门,复窗门;(两个套间之间的)推拉门
~ gate 折叠大门
~ panel door 折叠式镶板门
~ partition 折叠隔屏;折壁,折叠式板壁,折叠隔断
~ pocket measure 折尺
~ rule 折尺,折尺
~ sash 折窗框
~ screen 折屏风
~ stair 折梯,活动楼梯
~ wedges 松紧(架)楔
foliage ['fəulidʒ] *n.* 叶饰;叶
acanthus ~ 叶形板
foliated *a.* 叶状的;层状的
~ arches 叶状拱
~ capital 叶饰柱头
~ copper 薄铜片
foliation [,fəuli'eiʃən] *n.* 叶瓣饰
~ plain 层状沉积平原;花叶形装饰
folium ['fəuliəm] *n.* (*pl.* -lia, -liums) 叶形线,(岩石的)薄层
follower ['fɔləuə] *n.* 跟踪机构;随动件,随动装置;复示器,输出器
graph ~ 图形复制器

font [fɔnt] n. 喷水池,泉源[水];字型,字盘
fontanel [ˌfɔntəˈnel] n. 囟,囟门
Foochow pole 福州筒木
foot [fut] n. (pl. feet) 英尺;渣滓;脚爪,支点,底座,底部,交点脚步 v. 总结;结算
~ base 勒脚线条
~ bath 脚盆
~ bench 搁脚凳
~ block 线脚托座,墙裙座,木柱座;踢脚板固定块
~ board (plate) 踏板
~ bolt 脚踏门锁,门底脚螺栓
~ brake 脚踏制动器
~ guard 护脚档
~ hold 立足点[处]
~ iron 铁爬梯
~ -lambert 英尺-朗伯(亮度单位)
~ lights 舞台脚光
~ mat 基础钢筋网;底垫层
~ note 脚注;附记
~ path(s) 人行道
~ path paving 人行道铺面
~ pedal 脚踏开关
~ piece 底木,支架底梁
~ plank 脚手板,跳板,桥面步行板
~ plate 脚踏[踢脚]板支柱垫板
~ point (垂线的)垂足
~ rule 英尺(美制量尺)
~ scaper 脚刮板
~ screw 地脚螺钉
~ stall 基墩;柱墩;柱塞,柱墩的勒脚
~ step 踏步,[楼梯]台阶,脚步,步距
~ stone 基石
~ -switch 脚踏开关
~ waling 垫板
~ wall 基础墙
bare ~ 无榫骨架
board ~ (英制)木料尺;板尺
column ~ 柱底脚
linear ~ 纵尺

foot bolt

link ~ (月牙板的)滑环脚
table ~ 台脚
footag n. 总尺码,总长,长度(尺),面积(平方尺)
footboard n. 脚踏板;踏步;台阶梯级
footcandle n. 英尺、烛光
footcloth n. 地毯
footer [ˈfutə] n. 步行者;高度
foothold n. 支柱,支架,支点
footing [ˈfutiŋ] n. 底脚[层],基座[脚],总额
~ beam 基础梁
~ dressing 墙脚处理,基础整修
~ forms 基脚模板
~ foundation 底座[底脚],基础;柱基
~ load 基脚[基础]荷载,底脚荷重
~ of wall 墙底脚,墙基;墙脚
~ piece 底脚块,底板
column ~ 柱底脚
combine column ~ 联合柱座;双柱底脚
concrete ~ 混凝土基底
connected ~ 结合底脚
continuous ~ 连续底脚
grillage ~ 格排底座
isolated ~ 独立底脚
offset ~ 单面地脚,不对称地脚
pole ~ 杆基
spread 扩展底座[地脚]
stepped ~ 阶形底座[基础]
stone ~ 石底座[基础]
footline n. 栏杆下横档;下栏索;末行
footplate n. 踏板
footrill n. 通道,平巷;水平坑道,大卷
footring n. (壳)脚卷梁,穹顶脚圈梁
footstep [ˈfutstep] n. 阶梯,台阶,梯级;轴承架,垫轴台
footstock [ˈfutstɔk] n. 顶座,尾架,尾座
~ lever 踏杆
footway [ˈfutwei] n. 人行道;梯子间(井筒)
force [fɔːs] n. 力;强度;强化,加强 v.

加力
~ -account construction 计工制工程,计工建筑
~ account work 包工工程,计工制工作
~d draft 强力通风
~ drying 强制干燥,促干
~ plane 粗刨
~ plate 阳模托板
~ plug 阳模
top ~ 上[阳]模

forced [fɔːst] *a.* 强迫的,不得已的
~ air heating 压力热风供暖
~ air supply 压力供气,强制式通风[供气]
~ draft 强制通风,鼓风,压力通风
~ fan 压力通风风扇,压力送风机
~ ventilation 压力通风

forceps ['fɔːseps] *n.* (*pl.* -ceps,-cipes) 镊子,钳子,焊钳

forcing [fɔːsiŋ] *n.* 强迫,施加压力, *a.* 强迫的,施加压力的
~ house 温室,花房,玻璃房
~ pipe 压力(水)管
~ screw 紧固螺钉

ford [fɔːd] *n.* 渡口;过水路面,浅滩,河滩
paved ~ 铺砌过水路面

fore [fɔː] *a.* 前面的,在前的,先前的 *ad.* 在前面
~ apron 前台(舞台前部);房前院地
~ court 前院
~ gate 总入口,前大门
~ ground 前景
~ mast 前柱
~ plane 粗刨
~ sight 前视

forearm ['fɔːrɑːm] *n.* 前臂 *v.* 预作准备

forecast ['fɔː-kɑːst] *n.* 预报,估计,预防;预测值

forecasting ['fɔːkɑːstiŋ] *n.* 预测(法)
~ of cost 成本预测

foredoor *n.* 住房的前门

foregift ['fɔː-gift] *n.* 定金,承租者预付的租金

foreground ['fɔːgraund] *n.* 前景;显著的地位;前台

forehammer *n.* 大[手]锤

forehand ['fɔːhænd] *n.* 领先的地位;前方的,居前的
~ welding 正手焊法

foreign ['fɔrin] *a.* 外国的,外来的;无关的,异样的,不适合的
~ body 掺合物
~ capital 外资
~ firm 外国企业
~ investment 国外[外商]投资
~ market 国外市场
~ material 外来材料,异物
~ matter 杂质;外来物

forelock ['fɔːlɔk] *n.* 开口销,扁销,栓,楔,开口键
~ key 销,开口键

foreman ['fɔːmən] *n.* (*pl.* -men)工长,技术员,领工员,领班
~ ship 基层管理能力
gang ~ 班[组,队]长
general ~ 总领工员[工务员]

foremost ['fɔːməust] *a.* 最初的,第一流的,主要的 *ad.* 最先

forepart ['fɔːpɑːt] *n.* 前面,前部;(时间的)前段

forepole ['fɔːpəul] *n.* [防陷]护桩 *v.* (开挖前)先打挡土木桩(成挡板)

forepump *n.* 预抽(真)空泵,前级泵

foreroom ['fɔːrum] *n.* 前堂,客厅;房屋的最前房间

foreset 工作面支架,临时超前支架

foreshorten [fɔː'ʃɔːtn] *vt.* (绘画中)按照透视法缩短[小]

foreshortening [fɔː'ʃɔːtniŋ] *n.* 缩短(投影),视[像]收缩;透视的夸大,透视画法

foreside ['fɔːsaid] *n.* 前沿,前部

foresight ['fɔːsait] *n.* 前视;预见,向前看

forest [ˈfɔrist] *n.* 森林,森林地带 *v.* (植树)造林
　~ product(s) 木材,木料
forestick *n.* 挡木
forwarmer *n.* 预热器
foreword [ˈfɔːwəːd] *n.* 序,前言,绪言,献词
forge [fɔːdʒ] *n.* 锻铁炉;锻工场 *v.* 锻造;伪造
　~ hammer 锻锤
　~ press 锻压机
　~ welding 锻焊
forgeable [ˈfɔːdʒəbl] *a.* 可锻造的
　~iron 可锻铁
forging [ˈfɔːdʒiŋ] *v.* 锻 *n.* 锻件
fork [fɔːk] *n.* 叉,叉形物,叉路;做成叉形;耙
　~ catch (门的)叉形拉手
　~ chuck 叉形卡盘
　~ connection 叉状[分岔,插头]连接
　~ gauge 叉规,分叉标准尺
　~ joint 叉形接头,叉形连接
　~ -lift 叉架起货机,叉车(堆垛用)
　~ link 叉形杆
　~ spanner 叉形扳手
　~ staff plane 突圆线刨
　~ tongs 叉式钳
　~ wrench 叉形扳手
　~ed joint 叉形接[头]
　~ed lever 叉杆
　~ed loop 叉形环,双环
　~ed loop bar 叉形眼杆
　~ed pipe 叉形管
　~ed spanner 叉形扳手
　~ed tenon 叉形榫
　suspension ~ 悬架

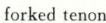
forked tenon

form [fɔːm] *n.* 模板[壳,型](灌混凝土用);形[方,样,型]式,结构 *v.* 形成,产生,构成;建立
　~ anchor (活动模板的)锚定板
　~ board(ing) 模(壳)板
　~ brace 装配支柱,模板支撑
　~ cage 模板骨架
　~ clamp 模板夹具
　~ coating 模板涂油[涂料]
　~ control template 仿形靠模板
　~ diameter 外形直径
　~ hanger 模板支撑
　~ line 拟构等高线,地形线
　~ liner 模板垫条
　~ lining 模板内衬[衬垫,涂层]
　~ panel 模(壳)板;钢制模板块
　~ sheathing 模板[壳]衬板
　arch ~ 拱型
　canonical ~ 范式,基本形式
　collapsible ~ 活动模板
　concrete ~ 混凝土模板
　land ~ 地形
　master ~ 原模
　order ~ (空白)定单
　porcelain ~ 陶瓷管
　prefab ~ 预制模板
　sheet steel ~ 钢板模板
　steel ~ 钢模板
　wooden ~ 木材模板

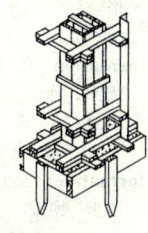
form

formal [ˈfɔːməl] *a.* 形式(上)的,外形(上)的;正式[规]的;整齐的
　~ agreement 正式协议
　~ approval 正式批准
　~ claim 正常索赔
　~ contract 正式合同[契约]
　~ effective 正式生效
　~ garden 规则式庭园
　~ notice 正式通知
　~ resemblance 形式上的相似
formale [ˈfɔːmeil] *n.* 聚乙烯
　~ copper wire 聚乙烯(绝缘)铜线
formalin [ˈfɔːməlin] *n.* 福尔马林,甲醛溶液
formall process 落锤深冲法
format [ˈfɔːmæt] *n.* 形[格,版]式,规

格,结构;大小,尺寸
packed ~ 合并[组合]形式
formation [fɔː'meiʃən] *n.* 道路基面,路床面;(水文上的)构造水;岩层,层系,组[构,生]成
 ~ rule 形成规则
formative ['fɔːmətiv] *a.* 形[构]成的,造型[型]的
 ~ arts 造型艺术
 ~ technology 造型工艺
formbuilding *n.* 构型法
former ['fɔːmə] *a.* 从前的;在前的 *n.* 构成者,创造者;模型;量规
 ~ plate 仿形[靠模]样板
 Former Imperial Palace 北京故宫
 wooden ~ 木模
formula ['fɔːmjulə] *n.* (*pl.* -las, -lae) 公式,式子;配方;计算方案
formuate alternative 制定可行(性)方案
formulation [ˌfɔːmjuˈleiʃən] *n.* 组成,成分;列方程式;系统阐述
formvar *n.* 聚醋酸甲基乙烯酯(绝缘材料牌号)
formword *n.* 灌注水泥的模架
formwork ['fɔːmwəːk] *n.* 模板[槽],支[立]板,模板工程;桥梁施工临时支架,量[定]规
 ~ drawing 模板图样
 inner ~ 内模(壳)
 sliding ~ 滑动模板
 travelling ~ 移动式[活动]模板
fornicate ['fɔːnikeit] *a.* 拱形的,穹窿形的
fornication [ˌfɔːniˈkeiʃən] *n.* 拱顶或拱的构造
fornicolumn *n.* 穹窿柱

fornicolumn

fornicommissure *n.* 穹窿连合[体]
fornix *n.* (*pl.* fornices) 穹窿,穹
fortalice ['fɔːtəlis] *n.* 小堡垒,外堡
fortification [ˌfɔːtifiˈkeiʃən] *n.* 筑城;防御工事;堡垒,要塞
fortify ['fɔːtifai] *v.* 设防;加强
fortified *a.* 加强的,加固的
 ~ paint 加固涂料
forum ['fɔːrəm] *n.* (*pl.* forums)(古罗马)会场,讨论会研究会,座谈会
forward ['fɔːwəd] *a.*;*ad.* 向前(的),前进(的)前方(的);预约的 *v.* 促进,运送;转送 *n.* 机头部;远期外汇
 ~ delivery 远[定]期交货
 ~ feed of material 材料的预约供应
 ~ heat shield 前置防热板
 ~ nodal point 前节点
 ~ order (货物)托运单
 ~ overlap 前后重叠
 ~ planning 预先计划
 ~ sight 前视
 ~ vision 前视
fosse [fɔs] =fossa *n.* 沟,渠,护城河;冰川和冰碛之间的洼地;坑
fossil ['fɔsl] *n.* 化石; *a.* 化石的,陈旧的
 ~ flax 石棉
 ~ flour 化石粉
 ~ meal 硅藻土,化石粉
 ~ resin 化石树脂
 ~ wax 矿物蜡,地蜡
fossula ['fɔsjulə] *n.* (*pl.* -lae.) 内[狭]沟
fotoceram *n.* 感光(微晶)玻璃,光敏玻璃陶瓷
foul [faul] *n.* 污物 *a.* 泥泞[阻塞,填塞住]的
 ~ air flue 浊气道
 ~ anchor 绞缠锚,锚饰
 ~ clay 不合格(的)砖泥
found [faund] *v.* 铸[熔]造;打基础;建立,创造
foundation [faunˈdeiʃən] *n.* 基础,地

基;底[基]座;根本[据];财团,基金会;基色
~ analysis 基础分析
~ base 基底,墙基
~ beam 基础梁
~ block 坡脚砌块,扩脚块体
~ bolt 地脚[基础]螺栓
~ brickwork 基础砖工,基础砖砌体
~ filling 木框石心基(础),木笼装石基(础)
~ course 勒脚[基础]层
~ cylinder 基础圈柱
~ design 基础设计

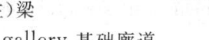

~ girder 基础(主)梁

concrete strip foundation

~ gallery 基础廊道
~ joint 基础缝
~ mat 基础底板,基褥,基垫层
~ material 基础材料
~ plan 基础平面图
~ plate 基础板,底板
~ ring 底圈
~ screw 地脚螺钉
~ stone 基石(房屋基础中的石块);根基;基础
~ wall 基墙
concrete strip ~ 混凝土条形基础
frame ~ 架座
isolated ~ 独立基础[地脚]
macadam ~ 碎石基层
founding *n.* 做基础;熔制,熔铸
foundry ['faundri] *n.* 铸造车间[厂],铸工;铸件;铸造,翻砂
~ facing (石墨)涂料
~ iron 铸铁
~ pig iron 生铁,铸件
~ sand 铸模砂
fountain ['fauntin] *n.* 喷水[泉],喷水池,人造喷泉;喷水器;中心注管;*v.* 喷洒(水)

~ aerator 喷泉式曝气池,喷水通气器
~ basin 喷水[泉]池
~ brush 在把柄上装有墨槽或油漆槽的划线刷或漆刷
~ flow 喷泉流
~ head 喷水头
~ jet 喷泉射流
artesian ~ 自流喷泉
bubbler ~ 喷水式饮水口
circular wash ~ 圆形喷水式洗手器
drinking ~ 喷水饮水器
eruptive ~ 喷泉
four [fɔ:] *n.* 四,四个
~ -center spiral 四心螺线
~ centered arch 四心桃尖拱
~ corners 十字路口,交叉路口
~ -jaw chuck 四爪盘
~ panelled door 四镶板门
~ -part vault 四分穹窿

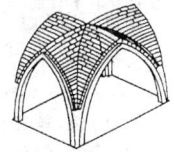

fourpart vault

~ paws 四爪链钩
~ poster 四柱大床
~ seater 四座式
~ -square lock 正方形锁
~ -unit sliding door 四扇一组的拉门
~ way 四通的
~ -way connection 四通接头

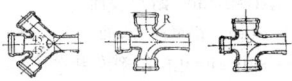

fourway connection

~ way piece 四通管
~ way reinforcing 四向配筋
~ -way switch 四路开关
~ -way valve 四通阀

fowl [faul] *n.* (*pl.* fowls or fowl) 家禽
fox [fɔks] *n.* 狐狸,狡猾的人 *v.* (使)变[褪]色,(使)变
　～ bolt 开尾地脚螺栓,端缝螺栓
　～ wedge 扩裂[紧榫]楔
foxiness ['fɔksinis] *n.* 褐斑;变色;木材腐朽
foxtail ['fɔksteil] *n.* 钉楔,薄键;炉渣,沉渣;短柄刷;狐尾草
　～ wedging 狐尾楔栓,开尾榫
foyer ['fɔiei] *n.* 剧场休息室;聚会场所;炉灶;(直接通向楼梯或私人起居用房内部的)门厅
fractable *n.* (山墙端)盖顶(石)

fractables

fraction ['frækʃən] *n.* 分数,小数,百分率;部分;粒(度)级,粒径组合;(曲线的)折点,拐点;碎片,细粒 *v.* 使成为部分;分馏硫
　～ void 疏松度
fractional ['frækʃənəl] *a.* 分数的,小数的,部分的
　～ distance point 分距点
　～ error 部分误差,相对(比例)误差
　～ load 部分荷载
fracture ['fræktʃə] *n.* 裂缝[痕],断裂图[面];断层 *v.* 破裂;断裂
　～ed surface 断裂表面
　～ plane 破裂面;断面
　～ safe design 断裂安全设计
fragile ['frædʒail] *a.* 脆的,脆性的,易碎的
　～ cargo 易碎[损]货物
　～ material 脆性材料
fragility [frə'dʒiliti] *n.* 易碎[裂]性,脆弱性

fragment ['frægmənt] *n.* 断[碎]片,碎块[屑];段落,片[分]段 *v.* 成为碎片,打碎
fragmentation [frægmen'teiʃən] *n.* 破碎,碎裂;破碎作用
frail [freil] *a.* 脆弱的,易碎的
　～ construction 单薄的结构,不牢固的结构
frame [freim] *n.* 框[骨,构,机]架,结构,构造;木层;画幅[框];肋骨;体制;系统 *vt.* 构成[造];组织;设计;装配;装框架 *vi.* 发展,进行
　～ and braced door 框构门
　～ and brick veneer construction 框架嵌砖结构
　～ and panel construction 框架墙板结构
　～ and shear wall structure 框架剪力墙结构
　～ board 腰线板
　～ construction 框架结构,构架建筑
　～ diagram 框架图
　～ door 直拼斜撑框构门
　～ for fittings 门窗框,门窗档
　～ girder 构架(大)梁,框架横梁
　～ grounds 门窗框木砖
　～ house 框架房屋
　～ -panel building 骨架板材式房屋
　～ saw 框锯,架锯
　～ -space 构架间距
　～ stand 框架座
　～ timber (yoke) 模板框架横木
　～ work 构架(工程),框架工程
angle iron ～ 角铁框架
antenna ～ 天线架
assembly ～ 装配架
auxilirary ～ 辅助架
bar ～ 杆架
bear ～ 支架
binder ～ 吊架,挂脚
bow saw ～ 弓锯框
box ～ 匣形构架
box-type ～ 箱形架

bracing ~ 加固构架
built-up ~ 组合构架
cant ~ 斜架,斜肋骨
cased ~ 箱形框;箱形窗框
cased window ~ 空心窗框,箱形窗
composite truss ~ 混成桁架
cross braced ~ 横支架
deal ~ 木锯架
drop hanger ~ 吊架
girder ~ 横梁(构架)
half-lattice ~ 半格构架
log ~ 垂直锯(木)架;多锯机
longitudinal ~ 纵向构架
main ~ 主框架,承力构架
portal ~ 门框;龙门架
reinforced ~ 加固架,钢筋架
reinforced-concrete ~ 钢筋混凝土构架
rigid ~ 刚架
rolled section ~ 型钢框架
sash ~ 窗架
skeleton ~ 钢骨构架
solid window ~ 实心窗架
space 空间[间隔]构架
stiff ~ 刚(性构)架
stiffening ~ 加劲框架
supporting ~ 支架
timber ~ 木构架
transverse ~ 横向构架
trussed ~ 桁构架

framed *a.* 榫构合,构架的,构成的,构架式
~ arch 构架拱,桁拱;弓形桁架
~ building 构架建筑
~ connection 梁柱框架式联结
~ floor 框架楼板[桥面]
~ girder 桁架梁,构桁
~ grounds 门窗框木砖
~ partition 构架面板隔墙
~ structure 构架[框架]结构,骨架
~ wall 构架墙,立筋墙

framer ['freimə] *n.* 组织者,制造[编制]者,预制支架工

frameset *n.* 框式支架

framework ['freimwə:k] *n.* 骨[框]架;构架工程;结构;体制;组织
structural ~ 结构骨架
supporting ~ 支承构架
tectonic ~ 构造格式[轮廓]

framing *n.* 构筑框架;构思;计划;图像定位
~ anchor 骨架锚固件
~ chisel 斜面凿
~ component 框架构件
~ scaffold 脚手架
~ square 木工角尺
~ stair well beam 楼梯井框架梁
~ steel 结构钢
~ table 构架台
balloon ~ 轻捷骨架
braced ~ 联结构架
deep ~ (横架式)宽肋骨架
timber ~ 木构架
triangular ~ 三角架

frangibility [ˌfrændʒi'biliti] *n.* 易碎[折]性;脆性[度]

frangible ['frændʒibl] *a.* 破碎的,折断的,松散的,脆的

Frankfort black 德国黑颜料;法兰克福黑

franking *n.* 斜角接头

fraueneis *n.* 透明石膏

fray [frei] *v.* 磨损,擦破[断];擦磨 *n.* 磨损处

free [fri:] *a.* 自由的;游离的;免费的;任意的,空着的,无阻碍的 *ad.* 免费地,自由地 *vt.* 使自由
~ access floor 活动地板
~ beam 筒支梁
~ box wrench 活套筒扳手
~ charge 免费
~ edge 自由边缘,无支承边
~ end (梁的)自由端;活动支座
~ end bearing 自由端支承;筒支(承)
~ floor 活动铺板

~ hand drawing 徒手画,草图
~ -hand sketch 徒手画,草图
~ height 净空高
~ joint 万向节
~ lime 游离石灰
~ module 自由模
~ -open-textured 松散结构的
~ retaining wall 自由式挡土墙
~ schema 自由模式
~ space 自由空间,(可自由分隔的)灵活空间
~ -standing heater 独立式供暖器
~ standing bell tower 独立钟楼
~ stuff 软性木材,无缺陷木材
~ style 自由式
~ surface 自由(表)面
~ tar 游离柏油,游离柏油沥青
~ top 筒支顶盖

freely ['fri:li] *ad.* 自由地;无拘无束地,随意地;直率地;免费地
~ supported 简支的
~ supported beam 简支梁

freestone ['fri:stəun] *n.* 石灰石,毛[砂,乱]石
~ masonry 毛石砌体

freeze ['fri:z] *v.* 冻结,凝固;冷却;冷藏;烧焊
~ drying 冷冻干燥
~ proof 防冻的,抗冻的
~ -thaw stable 耐冻,耐熔

freezer ['fri:zə] *n.* 冷却[制冷]器,冷藏箱[库];制冷者
chest ~ 冷冻柜
home ~ 家用冰箱
reach-in ~ 冻结柜

freezing ['fri:ziŋ] *n.* 冻结法(土壤加固结之一) *a.* 冰冻的,制冷的
~ locker 冷藏库
~ resistance 抗冻性,耐寒性

freight [freit] *n.* 运费;货物[运]船运(货) *vt.* 运货
~ handling 货物运输,货物装卸[搬运]

~ house 货栈,仓库
~ shed 货棚,仓库

freightage ['freitidʒ] *n.* 货物[运],运费,装货

French [frentʃ] *n.;a.* 法国(的)
~ arch 法(国)式拱
~ architecture 法(国)式建筑
~ blue 法(国)蓝,群青
~ bolt 法兰西螺栓
~ casement 玻璃落地窗
~ chalk 滑石粉
~ door 法式两用门(嵌装方格玻璃的双扇门)
~ drain (毛石)排水沟,(用碎石或砾石填满的)盲沟,暗沟
~ fliers 法国式梯级
~ grey 浅灰色
~ leg 一种构造轻巧的双曲形家具腿
~ method roofing (板材屋面的)菱形(斜向)铺法
~ order 法国柱式
~ polish 上(法国)罩光漆,法国抛光漆[剂]
~ red 红彩
~ Renaissance 法国文艺复兴式
~ roof 折线形屋顶;法国式屋顶

French roof

~ rose 法国蔷薇
~ sash 落地[铰链]窗
~ scarf joint 斜嵌接
~ scaffold 移动脚手架
~ scroll 法国涡饰
~ stucco 法国粉浆(拉毛)饰面
~ tile 法国槽瓦,马赛瓦
~ truss 法国式桁架
~ varnish 法国漆,虫胶漆
~ white 净白铅

~ window 落地长窗
frenchman ['frentʃmən] *n.* 接头修整工具,[砖缝]嵌填工具
frequency ['fri:kwənsi] *n.* (*pl.* -cies) 频率,周率;(统计中的)次数
Frequentit *n.* 费利宽帝(一种绝缘材料)
fresco ['freskəu] *n.* (*pl.* -coes 或 -cos) 湿壁画;壁画;湿墙加色漆
fresh [freʃ] *n.* (淡水)小河;淡水泉[池];泛滥;新鲜 *a.* 新鲜的,淡的
~ air duct 净气管
~ air inlet 净气进口,(下水道或暗沟的)进风孔
~ air intake 净气进口
~ air louvers 净气百叶窗
~ concrete 新浇[拌]混凝土,未结硬混凝土
~ material 新(材)料
~ mulch 初成残草覆盖,新草覆盖
~ paint 新油漆
~ water 淡水
freshen ['freʃn] *vt.* 使新鲜 *vi.* 变为淡水,变新鲜
freshly *ad.* 新(近,鲜地),刚才
~ mixed 新拌合的
~ set mortar 初凝砂浆
fret [fret] *n.* 回纹(装)饰,字花纹 *vt.* (用)回纹装饰;磨耗,倾蚀,使粗糙

fret

~ lead 嵌彩色玻璃的窗铅条骨架
~ saw 钢丝锯,雕花锯
~ -work 回纹(透雕)细工,格字通气孔
diamond ~ 斜字回纹饰
dove ~ 鸠尾回纹饰
Greek ~ 希腊回纹饰
running ~ 连续回纹饰
symmetrical ~ 对称回纹饰

fretted ['fretid] *a.* 有格子纹的,格子细工的
~ ceiling 有格子纹的顶棚
~ rope 磨损了的绳子
friable ['fraiəbl] *a.* 脆性的,易碎的
~ gypsum 脆性石膏
~ material 脆性[易碎]材料
friction ['frɪkʃən] *n.* 摩擦;摩擦力,摩擦离合器
~ angle 摩擦角
~ catch 摩擦挡,(门)碰轧
~ clamp 直钳
~ coat 防摩涂层
~ -free 无摩擦的
~ surface 摩擦面
~ tape 摩擦带,绝缘胶布
~ welding 摩擦焊
wall ~ 壁面摩擦
frictionless ['frɪkʃənlis] *n.* 无摩擦的,光滑的
~ hinge 无阻铰链
frieze [fri:z] *n.* (古典建筑柱石横梁与桃檐之间的)雕带,横饰带;墙顶与顶棚之间作为装饰用之横条,或雕有图案,花纹等)腰线;腰带
~ panel 束腰板,雕刻的条板
~ rail 雕刻条板的下横档;中帽头
friezette *n.* 棱纹家具布
frig(e) [fridʒ] *n.* (英俗)冰箱,冷冻机,冷藏库
frill [fril] *n.* 张开的或打褶的装饰边,类似饰边的东西
frigid ['fridʒid] *a.* 严寒的,极冷的;冷淡的
~ climate 严寒气候
~ zone 寒带
frigidaire *n.* 电冰箱
frigorimeter *n.* 低温计,冷冻仪
fringe [frindʒ] *n.* 条纹,缘饰;边缘,干扰带;镶边,边纹,散乱边纹;引起电视画面损坏的不规则边纹;小额福利 *vt.* 镶边 *a.* 边缘的,次要的
~ contrast 条纹衬度

~ cost 附加成本
achromatic interference ~ 消色差干涉条纹
diffraction ~ 衍[绕]射条纹
frit [frit] *n.* 釉料,玻璃料 *v.* 熔合,凝结,烧结
　~ seal 熔接密封
　~ ted glass 熔结[多孔]的玻璃
　~ ted glaze 熟釉
　~ ted porcelain 烘炙陶瓷
froe(frow) *n.* 劈板斧
frog [frɔg] *n.* 蛙;岔心,辙叉(铁道)
　~ brick 凹槽砖
front [frʌnt] *n.* 正[前]面;前方[线];端部;波前[锋];装饰正面;正面座位 *a.* 前[正]面的 *ad.* 在前面,向前 *v.* 面向
　~ angle 前角
　~ brick 正面砖
　~ court 前院空地
　~ door 前[正]门
　~ edge 石板瓦前缘
　~ elevation 正面[前视]图
　~ elevation view 正视[主视,正面]图
　~ entrance 正面入口
　~ fillet weld 正面角焊缝
　~ garden 宅前花园
　~ glass 遮光[挡风]玻璃
　~ office 公司商号的办公室,旅馆前厅办公室
　~ outline 前视轮廓图
　~ -panel 面板
　~ perspective 正面透视
　~ room 房子前部的房间(如客厅)
　~ screed (抹灰的)前样板,匀泥尺
　~ shoe 前瓦形支块,前托块
　~ side spot (观众厅两侧的)耳光室
　~ spot 前光(观众厅的灯光总称)
　~ stair 前楼梯
　~ stead 屋前基地,庭院
　~ steps 前门台阶
　~ string 露明楼梯斜梁
　~ view 前视[正面,主视]图
　~ wall (桥台)胸墙,前墙
　~ yard 前院
　shop ~ 铺面
　store ~ 铺面
　street ~ 沿街屋前空地,沿街建筑立面
frontage ['frʌntidʒ] *n.* 屋前空地;(房屋的)正面;正面的宽度;(建筑物)前面,前方,临街面;滩岸
　~ line 临街建筑线
frontal ['frʌntl] *a.* 正[前]面的 *n.* (房屋的)正面,门窗上部的三角形或弧形顶饰
　~ datum plane 正基准面
　~ line 正平线
　~ plane 正平面
　~ -profile line 铅垂线
　~ projection 正面投影
　~ slope 房屋正面的坡屋面
frontier ['frʌntjə] *n.* 国境,边界;边缘
frontispiece ['frʌntispi:s] *n.* 装饰性正门三角形或弧形楣饰,弧形门楣;卷首插图
frontlet ['frʌntllit] *n.* 额饰,正面物
fronton ['frʌntən] *n.* 三角形或弧形门楣,山形墙;回力球场
F-roof plate F形屋面盖板
frost [frɔst] *n.* 霜,冰冻,严寒 *v.* 霜冻,降霜,冻结;冻伤;使(玻璃等)失去光泽
　~ -heart 冻心材
　~ prevention 防冻
　~ protection 防冻
　~ resistance 抗冻性
　~ -resistant brickwork 防冻砖砌体
　~ work 冻裂;(水汽凝结成的)霜花,霜花纹装饰
　~ed finish 雪花面饰
　~ed glass 毛玻璃,雪花玻璃(磨砂玻璃)
　~ed lamp 磨砂灯泡

~ed rustic work 霜花粗面装饰
~ed work 仿雪花装饰
frosting ['frɔstiŋ] *n.* 无光泽面；糖霜
~ glass 毛玻璃
~ varnish 无光漆
Frostis cement 弗劳斯特水泥(用白垩土和黏土制成的)
froth [frɔθ] *n.* 泡沫,砂沫,浮渣；道路翻浆 *vt.* 起泡
~ agent 起泡剂
~ fire extinguisher 泡沫灭火机
~ pit 气泡斑
~ rubber 泡沫橡胶
frothing ['frɔθiŋ] *n.* 起泡,发泡
frottage *n.* 【法】磨损,擦伤；擦净[亮]，擦地板
fruit [fru:t] *n.* 水果,果实；成果 *v.* 结果实
~ parlour 水果店兼冷饮店
frustum ['frʌstəm] *n.* (*pl.* frusta, frustums) 棱锥台,(平)截头锥体；柱身
~ (of) a pyramid 平截头棱锥体
frutex ['fru:teks] *n.* 灌木
fuchsin(e) ['fuksin] *n.* 品红,洋红
~ test 电瓷浸品红甲醇溶液试验,(陶瓷、电瓷)吸红试验
fugacity [fju(:)ɡæsiti] *n.* (易)逸性,(易)逸度；有效压力
fugitive ['fju:dʒitiv] *a.* 易消失的,游逸的
~ binder 短效粘结剂
~ colour 易褪的颜色
~ dye 短效染料
fulcrum ['fʌlkrəm] *n.* 支撑,支轴,支杆
~ bar 支杆
full [ful] *a.* 完全的,充分[满,足]的；留有加工余量的 *n.* 完全；全部；充分 *ad.* 完全,充分,极,很
~ bond 顶砖砌合
~ brick 整(块)砖
~ -cell process [木材]浸渍防腐法
~ -cell treatment [木材]浸渍防腐法

~ centre arch 半圆拱
~ centre calotte 半圆穹顶
~ colo(u)r 彩色,五彩
~ disk 整(圆)盘
~ duration 全部持续时间
~ -finish 双面整理
~ framing 满堂支架
~ section 全剖视
~ sectioned view 全剖视图
~ size design 足尺设计图
~ size drawing 足尺图
~ surface hinge 大面铰链
~ thread 全螺纹
~ time 全部工作时间
~ view 全视图,全景
~ width 总宽度,全[平]幅
fuller ['fulə] *n.* 填料工；(半圆形的)套柄铁锤,敛缝,小沟[槽] *v.* 锤击,堵缝
~ board 压(制)板,填隙压板
~ earth 漂白土
~ faucet 富勒式水龙头；弯管龙头
fully ['fuli] *ad.* 完全地,全部地
~ -enclosed 全封闭的
~ -graded aggregate 全级配骨料
~ locked 密封的
~ refined 精制的
~ restrained beam 固端梁,砌梁
fumatory ['fju:mətəri] *n.* 熏蒸所,熏蒸消毒室 *a.* 熏蒸的
fume [fju:m] *n.* 烟,烟雾；汽,气；微粒污染物
~ hood 通风[去烟]橱
fumigant *n.* 熏蒸消毒剂
fumigator *n.* 烟熏器,熏蒸消毒器
function ['fʌŋkʃən] *n.* 功能[用],作用,职能；函数；函项(数量) *vi.* 起作用,有效
~ building 专用房屋
~ unit 功能部件
functional ['fʌŋkʃənl] *a.*; *n.* 泛函数,函数(的)；功能(的),有作用(的)
~ absorber 空间吸声体

~ architecture 实用建筑
~ building elements 建筑功能构成要素
~ colouring （建筑）功能配色
~ design 功[性]能设计
~ diagram 功能图,方块图
~ distribution 功能性分配,职能分配
~ joint 构造缝,工作缝
~ organization 职能组织,职能机构
~ principle 实用原则
~ requirement 功能要求(规格)

functionalism *n.* （建筑）功能主义
~ architecture 功能主义建筑

fund [fʌnd] *n.* 现款,资金,基金,款项,费用 *vt.* 储备
~s flow 资金流转
~ procurement 资金调度
~ shortage 资金不足
loan ~ 贷款资金

fundament ['fʌndəmənt] *n.* 屋基;基础;基本原理
~ standard 基本标准

fundamental [ˌfʌndə'mentl] *a.* 基[楚]的,主要的,原则性的,根本的 *n.* 基[根]本,基本要素,原理[则];基波[频]—一次谐波
~ construction 基本建设
~ factor 基本因素
~ norms 基本规范
~ principle 基本原理

fundamentum *n.* 基本法则

fungicidal [ˌfʌndʒi'saidl] *a.* 杀菌的
~ paint 杀菌涂料

fungicide ['fʌndʒisaid] *n.* 杀菌剂

fungistat ['fʌndʒistæt] *n.* 抑菌剂

funicular [fju:'nikjulə] *a.* 纤维的,细绳子的;用绳索或铁索绷紧的 *n.* 缆索道
~ curve 索状曲线,悬链线
~ polygon 索状多边形

funnel ['fʌnl] *n.* 漏斗,烟囱,烟筒;通风管;仓斗,风帽 *vi.* 从漏斗中漏过,向……集中

~ -form 漏斗状的
~ -shaped 漏斗状的
~ stand 漏斗架
~ -like 漏斗形的
air ~ 通风筒

fur [fə:] *v.* 在……钉板条 *n.* 软毛,皮毛,(锅炉中)水垢,水锈

furbish ['fə:biʃ] *vt.* 擦亮,磨光;刷新,翻新;烧蓝,(钢铁)发蓝处理

furlong ['fə:lɔŋ] *n.* 浪(英制长度单位＝1/8英里)

furnace ['fə:nis] *n.* 炉,燃烧室;蒸汽采暖锅炉 *v.* 熔炼
~ brick 耐火砖(砌高炉用)
~ heating 火炉供暖
~ stack 炉身
holding ~ 混合[保温]炉
wind ~ 通风炉

furnish ['fə:niʃ] *vt.* 供给,提供;布置;陈设,配备,家具
~ed room 备有家具的(出租)房间

furniture ['fə:nitʃə] *n.* 家具;器具;装置,设备,贮藏物;填充材料;缆具
~ connection 家具接榫
~ layout drawing 家具布置图

furred [fə:d] *a.* 钉板条的
~ ceiling 吊顶,贴条吊顶
~ wall (有粉刷面的或嵌钉薄板的)混水墙

furring ['fə:riŋ] *n.* 钉板条,抹灰板条,衬条;衬里;水垢
~ brick 贴面砖,护面砖
~ nails 抹灰柱条(钉)
~ piece (平屋顶梁上形成坡度的)垫木
~ strips 钉罩面板的木条,横筋
~ tile 贴面瓷砖,衬里陶砖
wall ~ 墙板条
wood ~ 钉木板条

fuse [fju:z] ＝fuze *n.* 保险丝,熔丝,熔断器;导火线,引信 *v.* 熔化,装引信
~ alloy 易熔合金
~ block 保险丝盒
~ box 保险丝盒

～ carrier 保险丝座
～ -element 保险丝
～ head 保险插销
～ -holder 保险丝盒[座]
～ metal 易熔金属,保险丝合金
～ -resistor 保险丝电阻器
～ switch 熔线开关
～ wire 保险丝,熔断线
safety ～ 保险丝
fused ['fju:zd] *a.* 熔化[合]的
～ alumina 熔融氧化铝
～ cement 熔凝水泥
～ quartz 熔凝石英
～ silica 熔凝氧化硅
～ sprinkler head 熔断式喷水龙头
fusing ['fju:ziŋ] *n.* 熔化,熔断
～ soldering 熔焊
fusiform ['fju:zə∫ɔ:m] *a.* 流线形的,双端尖的,纺锤状的
fusion ['fju:ʒən] *n.* 熔化
～ -cast refractory 熔铸耐火材料
～ cutting 熔化切割
～ thermit welding 热剂熔焊
fusorole *n.* 盘珠饰
fust [fʌst] *n.* 柱身,壁柱
fusuma ['fu:su:ma:] *n.* 日本式隔扇,日本房屋中的拉门
futility [fju:'tiliti] *n.* 没有效益,无价值
futtock ['fʌtək] *n.* (复)肋材
future ['fju:t∫ə] *n.;a.* 未来(的),将来(的)
～ enlargement 远景扩建
～ expansion area 远景发展地段
Futurism *n.* (20世纪建筑的)未来派
fuzz [fʌz] *n.* 外来的微噪音
～ stick 干木片,枯枝
fuzzines *n.* 模湖,不清晰(图像)

G

gab [gæb] *n.* (偏心盘杆的)凹节,凹口[槽]
gabarit ['gæbərit] *n.* 外轮廓,外形尺寸,曲线板,样板
gabbro ['gæbrəu] *n.* (*pl.* -bros) 辉长岩,基性岩,飞白岩
gabion ['geibiən] *n.* (装土、石用的)罗筐,笼
～ boom 石笼挡栅
gable ['geibl] *n.* 山墙,三角墙
～ arch 山形[人字]拱
～ board (山墙)封檐板
～ coping 山墙[斜面]压顶石
～ cross 山头十字架
～ end 山墙端,建筑物侧面
～ frame 人字[山墙式]构架
～ moulding 山墙饰线
～ pole 山头桁条
～ roof 人字[三角]屋顶
～ roof truss 人字屋架
～ shoulder 山墙托肩
～ side 山[侧]墙面
～ springer 山墙托臂
～ tile 山墙顶盖瓦
～ wall 山墙
～ window 山墙窗
～ with corbel steps 阶式山墙
gablet ['geiblit] *n.* 花山头,小山墙
gad [gæd] *n.* 测杆,小钢凿,销,键,量规 *vt.* 钻孔,劈裂
～ picker 鹤嘴锄,凿子
～ tongs 平口钳
gadder ['gædə] *n.* 凿孔机
gadget ['gædʒit] *n.* 精巧的小机械,小配件
gadgeteering [ˌgædʒi'tiəriŋ] *n.* 小器具设计,零件设计
gadroon *n.* 刻纹圆线条装饰
gaff [gæf] *n.* 吊钩,钩杆,斜桁
gaffer ['gæfə] *n.* 工头,班组长

gag [gæg] *n.* 压板,塞盖,整轨锤 *vt.* 封闭,堵塞,关阀,矫正
~ press 压直机,矫正压力机
hold down ~ (剪切机的)压紧装置

gage [geidʒ] *n.* 抵押品;挑战 *vt.* 以(某物)做担保

gagger ['gægə] *n.* 造型[铸模]工具,(型材)辊式矫正机

gain [gein] *vt.* 获得,增益,到达;开槽,榫接 *vi.* 增进,得利 *n.* 榫槽;增益[放大]系数;(*pl.*)收益,盈余
~ control 增益控制
~ experience 获得经验
~ in mass 质量增加
~ joint 榫接口

gain joint

gaine [gein] *n.* 套,罩,箱,壳

gaize [geiz] *n.* 生物蛋白岩,海绿云母细砂岩

galilee ['gælili:] *n.* (教堂西端的)门厅,前廊

gall [gɔ:l] *n.* 磨损处,擦伤处,瑕疵 *v.* 磨损[耗],擦伤,塑变

gallery ['gæləri] *n.* (*pl.* -leries) 走[画]廊,美术陈列馆,楼座,(横)坑道,风道,集水道
~ apartment house 有外廊、可独立进入的公寓
~ corridor type 外廊式(建筑)
~ system 行人廊道体系

galleting ['gælitiŋ] *n.* 碎石片嵌灰缝

gallonage ['gælənidʒ] *n.* 加仑量,汽油消耗量

gallows ['gæləuz] *n.* (*pl.* -lowses) 架状物,门形吊架,构架
~ bit 双柱吊架
~ bracket 构架斜撑
~ frame 门式吊架,龙门起重架
~ frame derrick 门式起重机
~ timber (矿坑的)撑木,木构架

galvanize ['gælvənaiz] *vt.* 通电流于,电镀[镀锌]于
~d copper wire 镀锌铜线
~d corrugated sheet 镀锌波纹铁皮
~d iron sheet 镀锌铁皮,白铁皮,马口铁
~d iron wire 镀锌铁丝
~d nail 镀锌钉
~d steel pipe 镀锌钢管,煤气用白钢管
~d steel wire rope 镀锌钢丝绳
~d wrought-iron pipe 镀锌熟铁管
galvanizing by dipping 热浸镀锌

galvanometer [ˌgælvə'nɔmitə] *n.* 电流计,电表

galvano-voltameter *n.* 伏安计

gamble ['gæmbl] *v.* 赌博,投机 *n.* 冒险(的事业)

gambling house 游艺场

game [geim] *n.*;*v.* 游戏,比赛,对策,博奕,(*pl.*)运动会 *a.* 勇敢的
~ court 运动[游戏]场
~ house 游艺场
~ room 文娱室
~ with perfect information 全信息对策

gang [gæg] *n.* 群,队,帮;机组,同轴 *v.* 联结,成套排列[运转],组合 *a.* 联动[组合]的
~ adjustment 同轴[联动]调整
~-board 脚手架板,(上下船的)跳板,木条,跳板
~ boss 工[班]长
~ control 同轴[联动]控制
~ cutter 组合铣刀
~ dies 多头冲模,复式模
~ foreman 领班,工长
~ form 成套[组合]模板
~ master 工长
~-plank 跳板,脚手架板
~ saw 排[框]锯
~ socket 连接插座
~ switch 联动开关
~ tool 组合刀具

~ -way 过道,跳板,渡桥,座间通道,主巷道

ganister ['gænistə] *n.* 黏土质硅岩,致密硅岩,火泥
~ brick 硅砖
~ sand 硅粉,石英砂

Gantt chart *n.* 施工进度表

gap [gæp] *n.* 间[空]隙,裂缝,缺口;区间,差额 *vt.* 造成缝隙
~ allowed for expansion 伸缩间隙
~ arrester 火花隙避雷器
~ at joint 接头缝隙
~ -bar 接缝杆[栓]
~ -filling glue 填缝粘合胶
~ gauge 厚薄规,塞尺
~ sheaf 沟层
~ shears 凹口剪切机
~ test 间隙检验
~ weld 特殊点焊,断续焊接
~ width 间隙宽度

garage ['gæra:ʒ] *n.* 汽车库,修车厂,飞机库,掩体 *vt.* 置于汽车间,送入汽车厂
~ lamp (带金属护网的)安全灯

garbage ['ga:bidʒ] *n.* 垃圾,废料,污物;无价值的东西
~ bin 垃圾[清洁]箱
~ can 垃圾桶
~ chute 垃圾溜槽[井筒,管道]
~ shaft 垃圾筒

garden ['ga:dn] *n.* 花[果,庭]园,园林,(*pl.*)公园 *v.* 从事园艺,造园
~ apartment 花园公寓
~ architecture 庭园[园林]建筑(学)
~ city 花园城市
~ craft 庭[造]园术
~ design 庭园[园林]设计
~ -engine 庭园用小型抽水机
~ furniture 庭园[室外]家具
~ hose 花园浇水带
~ house 园舍[亭]

Garden of Harmonious Interest 北京颐和园谐趣园

Garden of Harmonious Virtue 北京颐和园德和园
~ ornament 庭园小品点缀
~ party 游园会
~ plot 庭园地
~ sprinkler 庭园洒水器
~ suburb 城郊田园住宅区
~ terrace 庭园露台

Garden to Linger In Suzhou 苏州留园
~ village 田园村落
~ wall 花园围墙
~ (wall) bond 园墙砌法(每皮顺砖间断地插入丁砖)
~ ing 园艺
~ wicket 庭园便门
~ ing plan 园林平面布置(图)

garderobe ['ga:d͵rəub] *n.* 衣柜,私室,(中世纪建筑物)厕所

gargoyle ['ga:gɔil] *n.* (怪兽状的)滴水嘴,出水口,怪形雕刻像
~ lining 出水口内衬

gargoyle

garland ['ga:lənd] *n.* 花圈[冠],类似花圈的装饰品,花叶果形装饰性凸雕 *vt.* 饰以花圈

garment ['ga:mənt] *n.* 外衣[观,表];包皮,饰面,外涂层 *vt.* 使穿衣服
~ press 压力矫正机
~ tag 外表特征

garner ['ga:nə] *n.* 谷仓,仓库 *vt.* 收[储]藏

garnet ['ga:nit] *n.* 柘榴石,石榴子石 *a.* 深红色的
~ hinge 丁字铰链
~ laser 柘榴石激光器
~ paper 石榴石粉砂纸

garnish ['ga:niʃ] *n.* 装饰品,扣押财产[工资] *vt.* 加装饰

garniture ['ga:nitʃə] *n.* 装饰品,陈设品

garret ['gærət] *n.* 顶[阁]楼,屋顶层,了望台 *v.* 填塞石缝
~ floor 顶楼层
~ window 天窗,老虎窗
garrison ['gærisn] *n.* 卫戍部队,警备区,要塞 *vt.* 驻防,守卫
~ house 堡垒式房屋
garth [ga:θ] *n.* 庭园,内院,场地
gas [gæs] *n.* 气体,煤[毒,天然]气,瓦斯 *v.* 充[供,排]气
~ absorption refrigerator 煤气吸收式制冷机
~ apparatus 煤气装置[器具]
~ appliance 煤气用具
~ arc lamp 煤气[弧光]灯
~ -ash concrete 加气灰渣混凝土
~ barrier 排气装置
~ bracket 煤气灯管
~ -burner 煤气喷灯[嘴],煤气灶
~ central heating 煤气集中供热
~ checking (油漆饰面的)气致皱纹
~ cleaning device 煤气净化装置
~ concrete 加气[泡沫]轻混凝土
~ -filled lamp 充气灯泡
~ -fired boiler 煤气锅炉
~ fitter 煤气装修工
~ fitting work 煤气安装工程
~ fittings 煤气设备[配件]
~ flame welding 气焰焊接
~ -forming styrene concrete 泡沫苯乙烯混凝土
~ -free 不含气的
~ furnace 煤气(发生)炉
~ gouging 气割开槽,气刨
~ hole 排气孔
~ hose 煤气软管,乙炔气软管
~ -jet 煤气喷嘴[火焰],气焊枪
~ lamp 煤气灯
~ light 煤气灯(光)
~ log (煤气暖炉的)圆材状燃管
~ -main 煤气总管
~ nozzle 气嘴,焊炬[枪]喷嘴

~ -oven 煤气炉
~ permeability 透气性
~ pipe 煤气管
~ pipe hose 煤气软管
~ pliers 气管钳
~ pore 气孔
~ radiator 煤气辐射供暖器
~ shaft 气窗
~ slot 气缝
~ stochs and dies 煤气管螺丝扳牙和扳手
~ tap 煤气嘴,管螺纹丝锥
~ -tight 气密的,不透[漏]气的,防毒气的
~ tongs (煤气)管钳
~ torch 气焊焊枪,煤气喷灯
~ vent 排[通]气道
gash [gæʃ] *n.* 深痕,裂口,齿隙 *vt.* 深砍,砍伤
~ angle 齿缝角
~ fracture 张开破裂
~ -vein 裂缝脉
gasket ['gæskit] *n.* 衬垫,垫圈[片],填隙料,密封带[套管]
~ cement 衬片粘胶
~ material 填料,垫衬材料
~ packing 板式填料,垫片
~ joint 填实接缝
~ paper 纸衬[垫]
~ ring 垫圈,环形垫片
gaskin ['gæskin] *n.* =gasket
gatage *n.* 门叶[闸门]开度
~ indicator 门叶开度指示器
gate [geit] *n.* 大[闸,阀]门,铸口,流道,整流栅,门脉冲,锯架 *vt.* 装门,开启
~ action 开闭作用
~ arch 门拱
~ arm 栏路木,栅门臂
~ bar 门闩
~ beam 横肋梁
~ circuit (计算机的)门电路
~ flap 门叶,舌瓣

~ head 门道
~ hinges 大门铰链
~ hoist 启门机
~ hook 门钩,耳轴
~ leaf 闸板,门叶[扇]
~ leakage 闸门漏水量
~ (legged) table 折叠式桌子
~ light 门灯
~ of honour (庆典用的)正门
~ pier 门墩柱
~ post 门柱
~ rail 门轨
~ saw 排锯
~ seat 门座[槛]
~ sill 门槛
~ slot 门槽
~ spring (自关)门弹簧
~ stop 门钩[档]
~ strut 门柱
~ table 折叠式桌子
~ tie 大门斜撑
~ tower 门楼[塔]
~ trunnion 门轴支座
~ -way 门道,入口
~ weir 门堰
carry ~ 进位门
chute ~ 入孔盖
folding ~ 折叠门
time ~ 时间选通门

gathering ['gæðəriŋ] *n.* 集合[聚,会],积累
~ track 汇集轨道

gauffer ['gɔːfə] =goffer *n.*; *vt.* 起皱,做出波纹[皱纹,浮花]
~ machine 压纹机
~ed paper 皱纹纸

ga(u)ge [geidʒ] *n.* 标准度量,计量器,轨[行]距,水尺 *vt.* 划分,精确计量,校准,标准化
~ auger 匙形钻,规准螺旋钻
~ board 规准板[尺],样板,仪表操纵板
~ brick 标准砖

~ brick work 规准[清水]砖工
~ distance 规准距离
~ hole 定位[工艺]孔
~ invariance 规范不变性
~ lath 挂瓦条
~ mortar 掺熟石膏砂浆,速凝砂浆
~ pile 定位桩
~ pin 测量头,定尺寸销
~ ring 环规
~ rod 规准杆,标杆
~ rule 轨距规
~ setting 卡规校准
~ stuff 装饰石膏
~d arch 规准[清水砖砌]拱
~d brick 规准[清水]砖
~d data 测定的数据
~d mortar 速凝砂浆
~d plaster work 标准抹灰工作

gauging board (上水泥、灰浆粉刷的)附着板

gauging plaster 罩面层用的石膏灰浆
altitude ~ 测高计
caliber ~ 测径规,厚薄规
clearance ~ 塞尺,测厚规
distance ~ 测距规
float ~ 浮规
heavy ~ 大型量规
height ~ 高度规,高度游标尺
internal ~ 塞规,内径规
plug ~ 圆柱塞规
slide ~ 游标卡尺
wire ~ 线规

gault [gɔːlt] *n.* 重黏土
~ brick 白垩砖
~ clay 重黏土,泥灰质黏土

gauze [gɔːz] *n.* 纱[滤,金属丝]网
~ filter 网式滤器,滤网
~ strainer 滤网
~ wire 细目丝网
wire ~ (金属,铁)丝网

gavelock [ˈgævəlɔk] *n.* 铁橇[度,钎]
gazebo [gəˈziːbəu] *n.* (*pl.* -bos,-boes) 露[阳,眺,信号]台,屋顶塔楼,凉亭
gear [giə] *n.* 齿轮[链],传动装置,排挡数,起落架;财货,动产 *v.* 开[传]动,啮合,装备
gel [dʒel] *n.* 凝胶(体),冻胶,胶滞体 *vi.* 胶凝[化]
～ coat 凝胶漆,涂模用胶膜
～ (ling) point 胶凝[化]点
～ -space ratio 凝胶空隙比
～ state 凝胶状态
～ water 凝胶水
silica ～ (氧化)硅胶,硅酸盐冻胶
gelatin(e) [ˈdʒeləˈtiːn] *n.* 明胶,动物[骨]胶,凝胶体,半透明滤光板
～ filter 胶质滤光片
～ glue 动物胶
gelation [dʒeˈleiʃən] *n.* 凝[冻]结,胶凝体,凝胶化(作用)
～ temperature 胶凝[糊化]温度
～ time 胶凝时间
gemel [ˈdʒeməl] *n.* 铰链 *a.* 成对的,双生的
～ arch 对拱,铰链拱
～ hinge 环钩铰链
～ window 二开窗,对[孪]窗
～ed columns 对[孪]柱
general [ˈdʒenərbəl] *a.* 普遍[概括,一般]的,总的,首席的 *n.* 一般,全体,将军,总则 *vt.* 指挥
～ acceptability criterion 一般可接受的准则
～ act 总议定书
～ arrangement 总体布置,总协定
～ assembly drawing 总装图
～ budget 总预算
～ characteristic 通性,一般特性
～ conditions 一般条件,合同的总则
～ conditions of construction 营造总则,施工概况
～ contract 总承包[契约]
～ contractor 总承包人[商]
～ description of construction 施工[构造]说明书
～ diffused lighting 一般[全面]漫射照明
～ dimension 总[概要]尺寸
～ drawing 总[全,概]图,通用设计
～ estimate 总概算
～ extension 全面扩建,均匀[总]伸长
～ features of construction 施工概要
～ final accounts 总决算
～ flowchart 综合流程图
～ foreman 总工长
～ illumination 全面照明
～ layout 总体布置,总平面图
～ location sheet 地盘[位置]图
～ plan 总体规划,计划概要,总图
～ planning 总体规划
～ principle 原[总,通]则,普通原理
～ requirement 一般规格,一般要求
～ rules(GR) 总则,总规则
～ scale 通用比例尺
～ sign and symbol 通用标志和符号
～ symbols 通用符号
～ ventilation 全面通风
～ view 全视[景]图,大纲,概要
generalize [ˈdʒenərəlaiz] *v.* 总结,归纳,概括,推广,普及
～d coordinates 广义坐标
～d displacement 一般位移,广义变位
generate [ˈdʒenəreit] *vt.* 产[发]生,导致,形成
～ pressure 产生压力
generating *a.* 生成的,发电的
generation [ˌdʒenəˈreiʃən] *n.* 产[发]生,制造,形成;世代
second ～ 第二代(产品)
genetic [dʒiˈnetik] *a.* 原生的,遗传的,发生[展]的
～ analysis 成因分析
gentle [ˈdʒentl] *a.* 温和[柔]的,易控制的,轻微的 *vt.* 驯驭,抚慰

~ bend 平缓的拐弯
~ curve 弯度不大的曲线
~ slope 平缓[缓和]坡度
geodesic [ˌdʒiːəuˈdesik] (= geodetic) *a.* 测地学的,大地测量学的 *n.* (最)短程线,测(大)地线
~ construction 外壳承载式结构
~ structures 大量重复构件的结构
geographic [ˌdʒiəˈgræfik] *a.* 地理(学)的
~ features 地势
~ landscape 地理景观
~ map 地(形)图
~ position 地理位置,地理坐标
geologic [ˌdʒiəˈlɔdʒik] *a.* 地质(学)的
~ condition 地质条件
~ development 地质展示图
~ drawing 地质图
geometric [ˌdʒiəˈmetrik] *a.* 几何(学)的,图形的
~ body 几何体
~ decoration 几何构形装饰
~ design 几何形状设计,线形设计
~ figure 几何图形
~ form 几何形
~ garden 规则式[几何形,图案式]庭园
~ ornament 几何形装饰
~ pattern 几何形花纹[图案],几何模型
~ proportion 几何[等比]比例
~ stairs 螺旋形楼梯
~ style 几何图形式
~ symmetry 几何对称
~ tracery 几何图形窗饰[花格]

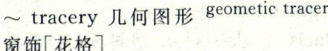
geometic tracery

Georgian architecture *n.* (18世纪英国及北美)乔治王朝式建筑
geostatic [ˌdʒiːəˈstætik] *a.* (耐)地压的,土压的
~ arch 耐地压的拱

germ [dʒəːm] *n.* 细[病]菌,胚芽,根源
~ repellent paint 杀菌涂料
German [ˈdʒəːmən] *n.* 德国人,德语 *a.* 德国(人,语)的
~ architecture 德国式建筑
~ Baroque 德国巴罗克式(建筑)
~ feather boarding 德式横钉木板墙
~ Gothic 德国哥特式(建筑)
~ kieselguhr 德国硅藻土
~ Renaissance 德国文艺复兴式(建筑)
~ Romanesque 德国文艺复兴式(建筑)
~ silver 德国(镍,锌,铜的)合金,银,锌白铜
gesso [ˈdʒesəu] *n.* 石膏底粉饰,(雕塑用)石膏粉
getter [ˈgetə] *n.* 吸气剂[器] *v.* 吸收[消除]气
~ material 吸气剂
~ pump 吸[抽]气泵
geyser [ˈgaizə] *n.* 喷泉,间歇泉;[ˈgiːzə](厨房、浴室的)热水器
~ basin 间歇喷泉带
~ cone 间歇喷泉锥
~ jet 喷泉水柱
push-through ~ 通过式热水器
geyderite *n.* [ˈgaizərait] 硅华
Ghati gum 或 **Ghatti gum** 印度树胶
ghost [gəust] *n.* 幻[重]象,叠影
~ image 重[幻]象,叠影
giant [ˈdʒaiənt] *n.* 巨人[物],水枪,大喷嘴,冲矿机 *a.* 巨大[型]的
~ order 巨柱式(两层通高的柱式)
gib [gib] *n.* 起重杆,吊杆,吊机臂;扁栓,夹条,凹字楔,榫 *vt.* 用扁栓[夹条]固定
~ and cotter 合楔
~ head 螺栓头,钩[销子]头
adjustable ~s 调整镶条
cross head — 十字头扁栓
gibbet [ˈdʒibit] *n.* 起重杆,吊杆,起重机臂;撑架,托座,绞架 *vt.* 吊于绞架上
gig [gig] *n.* 提升机,绞车,吊桶;旋转

物;快艇
~ saw 直立往复式带锯
gild [gild] v. 镀金,装饰; n. 行[协]会
~ -hall 会馆,公所
~ed frame 镀金镜框
~ing metal 仿金合金(铜、锌合金)
gill [gil] n. 鱼鳞[波形]板,百叶窗;肋条,加固筋;散热片
Gillmore needle n. 吉尔摩仪(一种水泥凝结时间测定仪)
gilsonite ['gilsənait] n. 硬[黑]沥青(地沥青石的一种)
gimbal ['dʒimbəl] n. 万向接头,(pl.)平衡环[架],框架
gimlet ['gimlit] n. 木工手钻,螺丝锥 vt. 用手钻钻,用锥子锥
~ bit 螺旋钻
~ for nail 钉孔钻
~ point 手钻钻头
gin [dʒin] n. 三角起重架[机],打桩机
Gingerbread Style n. (19世纪美国)一种装饰华丽的建筑风格
gingko ['giŋkəu] n. (pl. -koes) 银杏,白果树
ginkgo ['giŋkgəu] n. = gingko (pl. -goes) 银杏,白果树
girandole ['dʒirəndəul] n. 多枝烛架,壁灯架,旋转烟火[喷水]
gird [gə:d] v. 束紧[缚],围绕 n. 横梁,保安带,木材防腐包带
girder ['gə:də] n. 桁架,主梁,撑杆,槽钢
~ (and beam) connection 大小梁连接
~ bent 带实腹横梁的框架,梁式排架
~ brace 桁[梁]撑
~ bridge 板梁桥
~ casing 大梁护面,梁模
~ construction system 梁式结构体系
~ dog 吊梁钳,起梁钩
~ frame 桁架梁
~ grillage 梁式承台[格排]

~ mould 梁的覆面层
~ pole 桁架杆柱
~ post 大梁支承柱[杆]
~ space 桁[横梁]间隔
~ span 梁跨
~ steel 钢梁,工字梁
~ structure (大)梁式结构
~ system 桁梁系统
~ truss 桁架梁,梁构桁架
~ wed 大梁腹板
braced ~ 有刚性腹杆的梁式桁架
hollow ~ 空心梁
lattice ~ 格构梁
open web ~ 空腹梁
plate ~ 板梁
truss ~ 桁架梁
girderage ['gə:dəridʒ] n. 大梁搭接体系
girdle ['gə:dl] n. 抱柱带,环带[圈],赤道 v. 环绕,包围
girt [gə:t] v. 用带尺量周围,环绕 n. 转[墙]梁,柱间连系梁,带尺,周长,大小尺寸
~ strip 棚板
chicken ~ 石米,瓜子石
girth [gə:θ] n. 围长,曲线周长,带尺,围梁,翼缘,大小尺寸 v. 围绕,量围长,展开
~ joint 环向接头,周接
~ rail 圈栏
~ seam 圈缝
~ sheets 圈板
~ welding 环缝焊接
give [giv] v. 给[赠]予,付出,委托,产生,发出
~ -and-take lines 计算不规则形面积的取舍线,协调性
given size 规定尺寸
glacis [ˌglæsis] n. 缓斜坡,斜堤[岸],缓冲地区
glade [gleid] n. 林间空地[通道],沼泽地
glance [gla:ns] n. 一瞥,闪光,辉矿类 v. 看一眼,闪耀,掠过

~ pitch 辉沥青,光泽地沥青石
gland [glænd] ***n.*** 密封压盖,填料盖,塞栓,电缆封头
~ bolt 压盖螺栓
~ cover 密封套[压盖],填料盖
~ nut 压[锁]紧螺母
~ packing 压盖填料,密封垫
water sealed ~ 水封套
glare [glɛə] ***n.*** 强[眩,闪]光 ***v.*** 发眩光,闪耀
~ index 眩光指数
~ing light 眩耀光
~ -proof glass 防眩光玻璃
~ -reducing glas 减眩玻璃
~ shield 遮光罩,闪光屏挡
~ zone 眩光带[分布区]
glass [glɑːs] ***n.*** 玻璃(制品,体),透镜,(***pl.***)眼镜 ***vt.*** 镶以玻璃,磨光 ***vi.*** 呈玻璃状 ***a.*** 玻璃制的
~ architecture 玻璃建筑
~ blanket 玻璃棉毡
~ block 玻璃砖[块],镜片
~ brick 玻璃砖
~ case 玻璃橱
~ cement 玻璃胶
~ cloth 玻璃丝布
~ cock 玻璃塞
~ -concrete construction 玻璃混凝土结构
~ cotton 玻璃棉
~ dead seal 玻璃封口
~ decoration 玻璃装饰
~ door 玻璃门
~ed surface 磨光面
~ embossing 玻璃浮雕
~ enclosure 玻璃罩
~ epoxy (环氧)玻璃钢板
~ felt 玻璃棉毡
~ fibre 玻璃纤维[丝]
~ fibre board 玻璃纤维板
~ fibre cloth 玻璃纤维布
~ fibre mat 玻璃纤维垫[板]

~ fibre material 玻璃纤维材料
~ fibre (reinforced) cement (GFRC) 玻璃纤维增强水泥
~ fibre reinforced polyester corrugated sheet 玻璃纤维增强聚酯波纹板
~ floor 玻璃地板
~ flume 玻璃水槽
~ foam 泡沫玻璃
~ glazed 浓釉的
~ grinder 玻璃磨光机
~ -hard-steel 特硬钢
~ hollow block 玻璃空心砖
~ house (玻璃)温室,暖[花]房
~ insulation material 玻璃绝缘材料
~ jalousies 玻璃百页窗
~ laminate 安全玻璃
~ lined 搪玻璃的,搪瓷的
~ mosaic 玻璃马赛克[锦砖,镶嵌砖]
~ pane 窗玻璃
~ panel 建筑上的嵌玻璃部分
~ paper 玻璃砂纸
~ plate 玻璃厚板[底片]
~ (plate) diapositive 玻璃正片
~ (plate) negative 玻璃底片
~ pliers 玻璃钳
~ -pot clay 陶土
~ prism 棱镜
~ putty 镶玻璃用油灰[腻子]
~ rabbet 镶玻璃槽口
~ reinforced plasties 玻璃钢
~ roof 玻璃屋顶
~ roving 玻璃纤维无捻纱
~ roof 玻璃滑槽
~ sand 玻璃砂
~ silk 玻璃丝[棉]
~ slate 玻璃板瓦
~ specification 玻璃规格
~ stop 玻璃嵌条
~ strand 玻璃纤维丝
~ substitute 玻璃代用品
~ thread 玻璃丝
~ tile 玻璃瓦板

~ rube 玻璃管
~ wall facing 玻璃墙饰面
~ -ware 玻璃器皿,料器
~ window 玻璃窗
~ wire 玻璃线
~ wool 玻璃棉[绒]
~ woven fabric 玻璃布
~ work 玻璃制造[器皿,制品,工作]
anti-reflection ~ 无反射玻璃
armoured ~ 钢丝玻璃
bullet-proof ~ 防弹玻璃
cast ~ 铸造玻璃
cathedral ~ 拼花玻璃
channel 槽形玻璃
cladding ~ 幕墙玻璃
clear ~ 净片玻璃
corrugated wire ~ 波形夹丝玻璃
crown ~ 硬性光学玻璃,冕玻璃
cut ~ 雕花玻璃
diffusing ~ 漫射玻璃
double glazing ~ 双层中空玻璃
electric heat ~ 电热玻璃
enamelled ~ 釉面玻璃
fibre ~ 玻璃纤维
figured rolled ~ 压花玻璃
flat ~ 平板玻璃
flint ~ 燧石玻璃,软性光学玻璃
float ~ 浮法玻璃
foam ~ 泡沫玻璃
frosted ~ 毛[磨砂,霜化]玻璃
ground ~ 毛[磨口]玻璃
hard ~ 硬[耐火]玻璃
heat-absorbing ~ 吸热玻璃
heat reflecting ~ 热反射玻璃
horizon ~ 水平镜
ice ~ 冰花玻璃
laminated ~ 夹层玻璃
lead ~ 铅玻璃
magnifying ~ 放大镜
mosaic ~ 锦玻璃
obscure ~ 毛面玻璃
opal ~ 乳白玻璃

optical ~ 光学玻璃
organic ~ 有机玻璃
photochromic ~ 光致变色玻璃
plate ~ 板玻璃
polished ~ 磨光玻璃
polished silvered plate ~ 镜子(磨光涂银玻璃板)
reeded ~ 槽纹玻璃
reflector ~ 反光镜
rolled ~ 轧制玻璃
safety ~ 安全玻璃
sheet ~ 平板玻璃
sight ~ 观察[窥视]孔
spun ~ 玻璃丝[纤维]
structural ~ 建筑玻璃
tinted ~ 有色玻璃
toughened ~ 钢化[淬火]玻璃
translucent ~ 半透明玻璃
triplex ~ 三层璃板
window 窗玻璃
wired ~ 嵌丝玻璃

glassine ['glaːsiːn] *n.* 半透明玻璃纸,耐油纸

glassy ['glaːsi] *a.* 玻璃质的,透明(如玻璃)的,平稳如镜的
~ material 玻璃质[透明]材料
~ millboard 光面纸板
~ state 玻璃态
~ surface 光泽[如镜]面

Glauber's salt *n.* 芒硝,结晶硫酸钠,元明粉

glauconite ['glɔːkənait] *n.* 海绿石

glaze [gleiz] *n.* 釉(料,面),瓷釉,珐琅(质),光辉,光滑面,冰暴,雨松 *v.* 装配玻璃,上釉,抛光,研磨上光
~ colours 釉色
~d brick 釉面[琉璃]砖,瓷砖
~d door 镶玻璃的门
~d earthenware 釉面陶器
~d facing tiles 釉面[玻璃]瓦
~d paper 釉[蜡光]纸
~d partition 玻璃隔断[间壁]
~d pig 脆性生铁

~d pottery 釉面陶
~d roof 釉面[嵌玻璃]屋顶
~d tile 玻璃瓦,釉面砖
~d tile pipe 釉面陶管
~d wallboard 釉面墙板
~d wall tile 釉面墙砖
~d work 釉面砖坭工
glazier ['gleizjə] *n.* 装玻璃工,釉工;轧光机
~ 's chisel 玻璃工用的凿子
~ 's diamond 割玻璃用的金刚钻石刀
~ 's pliers 玻璃钳
~ 's point（镶玻璃用的）金属刮刀,扁头针
~ 's putty 镶玻璃用的油灰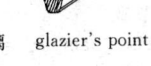
glazier's point
~ 's sprig 镶玻璃用的扁头针
glazing ['gleiziŋ] *n.* 装配玻璃,上光[釉]
~ bar 玻璃格条
~ bead 镶玻璃压条
~ brad 镶玻璃(金属)压条
~ by dipping 沉浸式上釉(药)
~ by dusting 涂粉式上釉(药)
~ by spraying 喷雾法上釉(药)
~ clip 固定玻璃用的弹簧夹
~ compound 镶玻璃用的填隙料
~ fillet 玻璃嵌条
~ furnace 釉面炉
~ glass 窗用玻璃
~ machine 抛[磨]光机
~ molding 玻璃压条,玻璃线脚
~ paint 上光涂料
~ pin 固定玻璃用钉
~ plate glass（橱窗用）厚玻璃
~ putty 镶玻璃用油灰
~ size 玻璃尺寸
~ temperature 上釉温度
~ work 玻璃安装工程
glide [glaid] *n.* 滑移[翔],滑道,滑裂带

v. 使滑动,渐变[消]
~ direction 滑动方向
~ mirror 滑移面
~ slipe 下滑道
gliding fracture 韧性断裂
gliding window 滑移式窗
glimmerite *n.* 云母岩
global ['gloubəl] *a.* 球形的,地[全]球的,整体[全局]的
~ coordinate system 球[总]坐标系统
~ error 全面误差
~ property 整体[全局]性质
~ quota 总括配额
globe [gloub] *n.* 球体[形物],球形灯罩[泡],地球(仪) *v.* 使成球状
~ chair 球形椅子
~ holder 球形灯座
~ joint 球关节,球窝接头
~ lighting 球罩灯照明
~ photometer 球形光度计
~ T 球形三通
~ tap 球形[浴缸]水龙头
~ thermometer 球形温度计(测辐射热用)
~ tube 球形管
~ -type luminescence 环型发光
~ valve 球形阀
globular ['glɔbjulə] *a.* 球形[面]的,小球的,世界范围的
~ chart 球面投影地图
~ discharge 球形放电,球状电闪
~ projection 球状投影
glomerate ['glɔmərit] *n.* 砾岩,团块 *v.* 聚合,粘结 *a.* 团聚的,密集的
gloss [glɔs] *n.* 光泽(面),珐琅(质),抛光膏;注释,词汇 *v.* 上釉[光],评注
~ electroplating 光亮电镀层
~ enamel 光亮釉
~ galvanization 光亮镀锌
~ index 光泽度
~ -meter 光泽计

~ oil 光泽油,松香清漆
~ paint 光泽漆[涂料]
~ retention 保光性
~ type finish 光泽饰面
~ varnish 光泽清漆
glost [gləst] *n.* 釉
~ firing 烧釉
glove [glʌv] *n.* 手套 *vt.* 套手套
~ box 手套箱,干燥箱,放射物操作箱
~ compartment 工具袋,小型工具箱
protective ~ 防护手套
rubber ~ 橡皮手套
glow [gləu] *n.* 白[灼]热光 *vi.* (无焰)燃烧,发热[光]
~ bulb 辉光灯泡
~ corona (辉光)电晕
~ (discharge) tube 辉光(放电)管
~ lamp 辉光灯,白炽灯
~ starter 辉光(日光灯)启动器
~ switch 辉光[引燃]开关
glue [glu:] *n.* (骨,动物,牛皮)胶,胶粘剂 *vt.* 上胶,粘结[贴]
~ bond 胶粘剂
~ colour 胶质颜料
~ gun 喷胶枪
~ joint 粘合接头,胶粘点[层]
~ laminated timber 多层胶合木料
~ -line 胶缝
~ lumber 胶合木材
~ mixer 胶合木材
~ off 胶粘
~ putty 胶质油灰
~ spread 涂胶(量)
~ water 胶水
~d adhesion 胶着接合
~d board 胶合板
~d connection 胶连接
~d construction 胶合结构
~d joint 胶接,胶合节点[接头]
~d laminated board 胶合层板,层积材
~d laminated tember arch 胶合叠层板拱
~d slab flooring 胶板地面[楼板]
~d timber construction 胶合木结构
~d wood 胶合木[板]
glycerin(e) ['glisərin] *n.* = glycerol 甘油[醇],丙三醇
glycol ['glaikɔl] *n.* 乙二醇,甘醇
glyph [glif] *n.* 竖面浅槽饰,来腰竖沟;雕像
glyptal ['gliptəl] *n.* 甘酞[丙苯]
~ resin 甘酞树脂
glyptic ['gliptk] *a.* (玉石)雕刻的. *n.* 雕刻,镂镂
glyptograph ['gliptəgra:f] *n.* 宝石上雕刻的花饰,刻有花饰图案之物
gnarl [na:l] *n.* 木节[瘤] *v.* 扭曲,形成节瘤
gneiss [nais] *n.* = gneissose = gneissy 片麻岩
gnomonic [nəu'mɔnik] *a.* 日晷仪的,心射的
~ projection 心射图法,球心投影
goal [gəul] *n.* 目标[的],终点;球门
~ programming 目标规划
~ setting 目标制定
gob [gɔb] *n.* (填筑用的)杂石,砖块;空岩洞;(*pl.*)大量
gobbet ['gɔbit] *n.* 一块[堆,部分];引文,片断
gobo ['gəubəu] *n.* 遮光片[罩,布],扩音筒除杂音遮布
~ flag 镜头遮光罩
go-cart *n.* 手推车
go-devil ['gəu,devl] *n.* 管子清洁器,刮管器;手推车,撬棍
godown ['gəudaun] *n.* 仓库,栈房
gof(f)er ['gəufə] *n.* = gauffer 皱褶 *vt.* 打褶皱
~ machine 压纹机
~ed iron 皱纹铁
~ed paper 皱纹纸
~ed roller 压纹滚
goffering *n.* 形成皱纹[浮花]

goggle ['gɔgl] *n.* (*pl.*)护目[风,墨]镜 *v.* 睁视
～ tenon 短榫
welding ～s 焊工护目镜

going ['gəuiŋ] *n.* 离去,出发;进行状况;工作方法[条件];阶梯;通路 *a.* 进行[运转]中的
～ project 进行中的项目
～ rod 定距标杆

gold [gəuld] *n.* (黄)金,金箔[粉,线,币],(化学元素)金(Au) *a.* 金色[制]的
～ beryl 金色绿柱石
～ -blocking 烫金
～ dust 砂金,金粉
～ foil 金箔
～ leaf 金箔[叶]
～ -plated 饰金,包金
～ -plating 镀金
～ size 贴金漆,涂金胶
～ solder 金焊料
placer ～ 砂金

golden ['gəuldən] *a.* 金制的,含金的,黄色的;可贵的
～ age 黄金时代
Golden Gate Bridge (美国)金门大桥

golf [gɔlf] *n.*;*vi.* (打)高尔夫球
～ club 高尔夫球场
～ links 高尔夫球场

gong [gɔŋ] *n.* 铜锣,铃盅

goniometer [ˌgəuni'ɔmitə] *n.* 测[量]角仪
～ eyepiece 测角目镜
～ system 测向装置,测角系统

good [gud] *a.* 优良的,坚固的,安全的 *n.* 好处;用途(*pl.*)货物,商品
～ colour 颜色均匀
～ gradient 平缓坡度
～ stream shape 良好的流线型
～ sudsing 易起泡沫的
～ two sides 两面光的(胶合板)
～ visibility 良好能见度

～s in bulk 散装货

gooseneck ['gu:snek] *n.* 鹅颈管,S形弯管,水落管弯头,弹簧式弯头车刀
～ connection 鹅颈接头
～ faucet 鹅颈式龙头
～ type 鹅颈式的

gore [gɔ:] *n.* 三角地带,穹顶的楔形构件 *vt.* 缝以长三角布
～ lot 三角地带
～ sign (三角形)分道点标志

gorge [gɔ:dʒ] *n.* 峡(谷);障碍物;小凹圆线脚;凹槽 *v.* 塞饱
～ cut 小凹圆线脚,凹槽

gorgerin *n.* 柱颈

gorgonion (=gorgoneum) *n.* (古希腊)妖女饰

gospel ['gɔspəl] *n.* 福音
～ hall 福音堂

Gothic ['gɔθik] *n.* 哥特式[尖拱式]建筑 *a.* 哥特式的
～ arch 哥特式尖拱
～ architecture 哥特式建筑,高直建筑
～ art 哥特式艺术
～ chippendale 哥特式家具
～ church 哥特式教堂
～ column 哥特式柱座
～ equilaterial pointed arch 哥特式等边尖顶拱
～ pitch 哥特式屋面坡度(60°)
～ Revial(18～19世纪欧美)哥特式复兴时代(建筑),新哥特式(建筑)
～ roof 哥特式屋顶
～ style 哥特[高直]式建筑

gouache [gu'a:ʃ] 【法】*n.* 树胶水彩
～ paint 树胶水彩颜料

goudron *n.* 焦油,沥青

gouge [gaudʒ] *n.* 半圆[弧口]凿,凿槽;断层泥 *vt.* 凿孔;打眼;刨槽
～ bit 弧口钻
～ carving 凿槽
～ hole 半圆凿穴
～ slipe 弧口凿磨石

turning ～ 弧口旋凿
govern ['gʌvən] v. 管理,控制,调节;限定,支配
～ing torch 气刨枪
positive ～ing 强迫[直接]调整
government ['gʌvənmənt] n. 管辖,政府[权];行政管理
～ house 政府办公楼
～ license 政府许可证
～ property 国有土地,公有财产
～ standards manual 政府(美国)标准手册
grab [græb] n. (挖土机)抓斗,开挖机;夹钳[具];爬杆脚扣 v. 抓取;攫夺
～ bar 拉手
～ handle 抓柄
～ iron 铁撬棍
～ rail 扶手杆
～ sample 取集的样品
～ skipper 拨钩器
grace [greis] n. (票据到期后的)宽限,缓期 vt. 增光[色,荣]
～ period 宽限期
gradation [grə'deiʃən] n. 分类[等,级],级配,(颜色)层次,渐变
～ of aggregate 骨料级配
～ of colour 色彩层次
grade [greid] n. 等[年]级,坡[梯,程]度;百分度;成绩 v. 分级[等,度];平土方
～ A 一级,最佳级
～ beam 斜坡梁,地基梁
～ crossing 平面交叉口
～ estimation 等级评定
～ intersection 平交口
～ labelling 商品质量的标签说明
～ level 地面线
～ of material 材料等级
～ of steel 钢(级)号
～ of tolerance 公差等级
～ separation 等级分类,立体交叉
～ washer 倾斜垫圈
～d aggregate 级配骨料

～d coating 分层涂层
～d joint 递级接头
～d material 级配材料
～d seal 递级封接
～d sizes (骨料)分级规格尺寸
～d standard sand 标准级配砂
～d width 修整宽度
grading specification 按质分等的标准
grader ['greidə] n. 平地机,分类器,筛选机
gradient ['greidjənt] n. 坡[梯,倾斜]度,斜率,比降;斜面,坡道 a. 倾斜的
～ of slope 倾斜率
～ -projection method 梯度投影法
～ tints 高层分层设色
gradin(e) ['greidin] n. 阶梯的一级,阶梯座位的一排;祭坛后方之供台;(雕刻用)齿凿
gradual ['grædjuəl] a. 逐渐的,(坡度)平缓的
～ contraction 渐变收缩[缩小]
～ enlargement 逐变扩大
～ failure 渐变失效
～ load 渐加荷载
～ losses 渐变损失
graduate ['grædjuit] n. ['grædjueit] vi. 毕业,得学位 vt. 授与学位,准予毕业;刻度于,校正,分等级 n. 毕业生;量杯,分度器 a. 毕了业的,已得学士学位的,研究院的;刻度的,分等的
～ scale 分度[比例]尺
～d cable 测绳
～d circle 刻度盘,分度圈
～d L-square 刻度角尺
～d measuring rod 水准标尺
～d plate 刻度盘
～d ring 刻度盘,分度圈
～d rod 分度标杆,测(深)杆
～d staff gauge 水准标尺
graduation [ˌgrædju'eiʃən] n. 分[刻]度;校正[准],分级;毕业,得学位;(pl.)表示经纬度或数量等的线

~ of curve 曲线修匀
~ tower 梯塔
graffito [grəˈfiːtəu] ***n.*** (***pl.*** -ti)(古罗马古迹上的)粗刻[画]或文字
graft [graːft] ***n.***; ***v.*** 接枝, 接[融]合; 贪污, 受贿; 敞口铁铲
~ copolymer 接枝共聚物
~ polymer 接枝聚合物, 融聚物
~ rubber (高分子)接枝橡胶
~ing material 接[融]合材料
~ing tool 土锹, 平锹
grafter [ˈgraːftə] ***n.*** 平铲[锹]; 接枝者
grahamite [ˈgreiəmait] ***n.*** 脆沥青
grail [greil] (= gravel) ***n.*** 细砾石, 砂砾, 鹅卵石; 杯, 盘
grain [grein] ***n.*** 颗[晶]粒; 纹理, 粒面, 组织, 构造; 谷物 ***v.*** 使成粒状, 使表面粗糙[漆成纹理]
~ alcohol 乙醇, 酒精
~ effect 压纹效应
~ fineness 颗[晶]粒细度
~ fineness number 砂子细度, 平均粒度
~ of rice (瓷器中的)透明花纹
~ of wood 木材纹理
~ orientation 晶粒取向
~ed stone facing 粗粒石面
~ed tinplate 糙面镀锡薄钢板
~ing board 压纹板
~ing machine 压纹机
~ing roller 压纹滚
~ing sand 细砂
gram [græm] ***n.*** 克(质量单位)
grand [grænd] ***a.*** 主[重]要的, 宏伟的, 巨大的, 豪华的, 漂亮的, 全部的
~ arcade 大拱廊, 中堂拱廊(哥特教堂中堂与侧廊之间的拱廊)
~ average 总平均
~ entrance 大[正]门
~ -father chair 高背椅子
~ hotel 高级旅馆, 大饭店
~ piano 大[三角]钢琴
~ relief 隆浮雕

~ sight 壮观
~ -stand (运动场等的)正面大看台
~ sweet 游览广场, 娱乐场所
~ trier 花坛顶层
~ total 总和, 共计
granite [ˈgrænit] ***n.*** 花岗岩[石]
~ block 花岗石块
~ blockwork 花岗石块加工
~ chips 花岗石碎片[石屑]
~ cube floor 花岗石块地面
~ facing 花岗石饰面
~ enamel 花岗岩纹搪瓷
~ -gneiss 花岗片麻岩
~ pavement 花岗石铺面
~ plank 花岗石板材
~ -prophyry 花岗斑岩
~ screenings 花岗石筛屑
~ sett 花岗石小方石块
~ ware 有花岗岩纹的器皿
granitic [græˈnitik] ***a.*** (似)花岗石的, 坚硬的
~ plank 花岗石板材
~ plaster 人造花岗石面, 汰石子粉刷
granitite ***n.*** 黑云花岗岩
granitoid [ˈgrænətɔid] ***n.*** 人造花岗石面, 洗石子粉刷 ***a.*** 似花岗石的
granolith [ˈgrænəliθ] ***n.*** 花岗岩碎石混凝土, 人造铺地石, 人造石铺面
~ic finish floor 人造石地面
~ic layer 人造石铺面层
granular [ˈgrænjulə] ***a.*** 粒状的, 结晶的
~ activated carbon 粒状活性碳
~ base 粒料基层
~ crystalline 粗晶体
~ -fill insulation (松散颗粒填充的)隔热层
~ limestone 粒状石灰岩, 云石
~ material 颗粒材料
~ stone 散粒石材
granularity [grænjuˈlæriti] ***n.*** 粒度; 成粒性

granulate ['grænjuleit] v. 使成粒状；使表面粗糙；轧碎
~d cork 粒状软木
~d mica 粒状云母
~d stone facing 粒状石面
granulation [ˌgrænju'leiʃən] n. 造粒，粒化(作用)；粉碎
granule ['grænju:l] n. 颗粒，粒状物
~ roundstone 小粒圆石
granulometric a. 颗粒的
graph [græf] n. 图表[解，像]，曲线图；网络 vt. 用图表[曲线]表示；以胶版印刷
~ of errors 误差曲线图
construct a ~ 绘制曲线图
coordinate ~ 坐标制图机
graphic ['græfik] a. 图解[示]的；自动记录的；印刷的
~ alphanumeric display 图形显示器
~ analysis 图解
~ arts 图表[印刷]艺术，书画刻印艺术
~ chart 图表[解]，曲线图
~ construction 作图
~ design 图表设计法
~ display 图表显示
~ expression 图示[解]
~ granite 纹理花岗岩
~ log 柱状剖面图
~ method 图解[示]法
~ presentation 图示
~ processing 图表处理法
~ progress chart 图示工程进度表
~ symbol 图例，图示符号
graphics ['græfiks] n. 图形学；图解计算法，制图学
graphite ['græfait] n. 石墨，碳精
~ brick 石墨砖
~ bronze 石墨青铜
~ fibre 石墨[碳]纤维
~ flake 片状石墨粉粒
~ lubricant 石墨润滑剂

~ paint 石墨涂料[油漆]
~ rosette 菊花形石墨
~ schist 石墨片岩
grapnel ['græpnəl] n. 四爪锚，(锚形)铁钩
grappier n. 石灰渣
grapple ['græpl] v. 抓[钩锚]住 n. 抓斗[钩]，抓扬机；钩竿
~ hook 抓升钩
grappler ['græplə] n. 钩[抓]东西的人[物]；(pl.) 爬杆脚扣
grasper n. 抓紧器
grass [gra:s] n. 草(地，坪)；茅草；地表面
~ cloth (覆盖墙面用的)植物纤维疏松织物，草簾
~ cover 草皮
~ cutter 割草机
~ed area 铺草(皮)地区
~ green 草绿色
~ -hopper 输送装置
~ -land 草地[原]
~ plot 小草坪
~ protection 植草保护，草皮护坡
~ -roots 草根，表土层；基层(群众)
~ rope 纤维绳
~ rubber 草(木橡)胶
~ strip 植草地带
~ surface 植草皮地面
~ table 泥土台
~ tree gum 禾木胶
~ tex (运动场用)沥青与植物纤维混合铺地面
~ thatch 茅草屋面
~ verge 路旁草坪
~ water way 草地泄水道
grate [greit] n. 炉栅[条，篦]；花格；格[光]栅；晶格 v. 磨擦[损]；轧碎；装格栅
~ bar structure 条格结构
~ inlet 格栅式进水口
~ of sink 污水池格栅

~ type inlet 帘格式进水口
grating beam 槛木,排架座木
grater ['greitə] *n.* 磨碎[光]机;粗齿锉刀,擦子;装铁栅栏的工人
graticule ['grætikju:l] *n.* 十字线,方格图,交叉丝;标[分度]线
gratuitous [grə'tju(:)itəs] *a.* 无偿[免费]的,无故的
~ contract 单方受益的合同
grave [greiv] *n.* 墓(碑) *a.* 严肃的,沉重的 *v.* 雕刻,铭记
~ trap 舞台升降平台
gravel ['grævəl] *n.* 卵[砾]石 *vt.* 铺砾石 *a.* 粗糙的,刺耳的
~ aggregate 砾石骨料
~ (-aggregate) concrete 砾石(骨料)混凝土
~ base 砾石基层
~ box 砾石笼
~ coating 砾石面层
~ concrete 砾石混凝土
~ covering 砾石盖面
~ drive 砾石路
~ fill 砾石垫层
~ mulch 砾石遮护料
~ packing 砾石回填[衬垫]
~ pebbles 小卵石
~ plant 砾石筛选厂
~ road 砾石路
~ -sand cushion 砂石垫层
~ screen 砾石筛
~ sidewalk 砾石人行道
~ stop 屋顶周围凸缘(防止砾石从屋面滚下)
~ stratum 砾石层
~ walk 砾石小路,砂砾路
~ washer 洗砾机
~ washing screen 洗砾筛
gravimeter [grə'vimitə] *n.* 重力仪,比重计,重差计
graving ['greiviŋ] *n.* 船底的清理及涂油;雕刻品,版画
~ tool 雕刻刀

gravitation [,grævi'teiʃən] *n.* (万有)引力,重力,地心吸力;倾向,趋势
~ filter 重力滤器,过滤澄清器
specific ~ 比重
gravitational [,grævi'teiʃənl] *a.* (万有)引力的,重力的,地心吸力的
~ differentiation 重力分异
~ separation 重力分离
~ system of unit 重力单位制,工程制
~ ventilation 重力[自然]通风
gravity [,græviti] *n.* (*pl.* ties)万有引力,地心吸力,重力;比重
~ arc welding 重力式电弧焊
~ bulkhead 重力式挡土墙
~ dock wall 重力式坞墙
~ fender 重力式护木
~ heater 重力循环加热器
~ heating 重力式循环供暖
~ hinge 重力铰链
~ mixer 重力[阶梯]式拌和机
~ retaining wall 重力式挡土墙
~ tank trunk 重力式自动洒水车
~ weight 铅锤
~ wall 重力(挡土)墙
~ weight 铅锤
~ weld 倚焊
~ welding 重力式电弧焊
centre of ~ 重心
grease [gri:s; *v.* gri:z] *n.* 黄油,油脂,润滑脂;硝化甘油 *v.* 涂油
~ coating 油脂涂层
~ cock 润滑脂旋塞
~ groove 润滑脂槽
~ gun 滑脂[注油]枪
~ hole 油脂[注油]孔
~ intercepter 油脂分离器,隔[除]油器
~ lubricator 滑脂润滑
~ nipple 滑脂嘴
~ oil 润滑油
~ proofness 防油性[度]

~ removal 去[除]脂
~ retainer 护脂(毡)圈
~ seal 油封
~ tap 润滑孔[嘴]
albany ~ 润滑脂,黄油
aluminium ~ 铅皂润滑脂
graphite ~ 石墨(滑)膜
silicon ~ 硅润滑脂

great [greit] *a.* 巨[重]大的,很多的,全部
~ calorie 千卡
~ gross 十二罗(计数单位,＝1728个)
~ hundred 十打(计数单位,＝120个)
Great Hall of the People 人民大会堂(中国)
Great Temple of Parthenon 帕提农神庙
Great Wall (中国)万里长城

greatest *a.* (great 的最高级)最大的
~ limit 最大极限

Grecian ['gri:ʃən] *a.* 希腊的 *n.* 希腊人
~ architecture 希腊式建筑
~ pitch 希腊式屋面坡度(15°)

Greco-Roman [gri:kəu'rəumən] *a.* 希腊罗马的,受希腊罗马影响的
~ art 希腊罗马艺术
~ style 希腊罗马式(建筑)

gredag *n.* 石墨油膏[脂],胶体石墨

Greek [gri:k] *a.* 希腊(人,语)的 *n.* 希腊人,希腊文[语]
~ architecture (公元前9～1世纪)希腊建筑
~ cross (各边等长的)希腊十字,希腊正教教堂的十字形平面
~ fret 希腊回纹饰
~ order 希腊柱型
~ ornament 希腊装饰(多在柱顶盘上用对称的树叶和玫瑰花形装饰)
~ Revival 希腊复古式(建筑)
~ roof tile 希腊式屋面瓦
~ theatre 古希腊露天剧院

green [gri:n] *n.* 绿色(颜料,染料);草坪[原];(*pl.*)蔬菜 *a.* 绿色的;新鲜的,未成熟[加工,烧结]的,潮湿的
~ alder 绿桤木
~ area plan 绿化区总平面图
~ belt 绿化地带
~ (bloom) oil 绿油
~ brick 砖坯
~ concrete 新拌[浇]混凝土
~ copper ore 孔雀石
~ cut 带水斩毛
~ enamel 绿瓷漆
~ facilities 绿地设施
~ fence 植株篱笆
~ glass 瓶料玻璃,绿色劣质玻璃
~ glue stock 生胶料
~ heart wood (造船用)绿心硬木
~ house 温室,暖[花]房
~ light 绿灯,放行,准许
~ log 新伐原木
~ lumber 新伐[生]木材
~ masonry 未硬化的砌体,新筑圬工
~ mortar 未硬化的砂浆
~ ocher 绿颜料
~ paint 绿颜料
~ parlour 陈设花卉植物的大厅
~ permeability 透湿气性
~ public space 公共绿地
~ resin 未固化的树脂
~ roof 新露顶板
~ room 剧院后台休息室,工厂内未加工品的贮存室
~ salt glaze 绿色釉
~ sand 新取[湿]砂,海绿石砂
~ sheet 预算明细比较表
~ stone 闪绿岩,软玉
~ surface 新铺面层
~ tack 初步粘合
~ timber 新伐[湿]木材
~ unit 新制品[构件,块体]
~ vitriol 绿矾,七水硫酸铁
~ ware 半成品,生坯,陶坯

~ water 跃波
~ way 绿荫路
~ wedge 楔形绿地
~ wood 新伐[湿]木材
Greenwich ['grinidʒ] *n.* 格林尼治(英国国立天文台所在地,为世界各地经度的起算点)
~ civil time(G. C. T.) 格林尼治民用时间
~ (mean) time(G. M. T.) 格林尼治平均时间,世界标准时
~ meridian 格林尼治子午线
~ standard time 格林尼治标准时间
grey [grei] *a.* 灰色[半透明]的 *n.* 灰色颜料[衣服];黄昏,黎明
~ body 灰体
~ brick 青砖
~ cast iron 灰口铸铁
~ cement 灰色水泥
~ desert soil 灰色沙漠土,灰钙土
~ pig iron 灰口生铁
~ plaster 青灰色灰浆
~ scale 灰色标度
~ soil 灰土
~ tile 青瓦
~ -wacke 灰色杂砂岩
grid [grid] *n.* 栅格[极],网格[架],帘格;地图的坐标方格
~ azimuth 平面[坐标]方位角
~ chart 方格图
~ connection 栅极接线
~ coordinates 平面[网格]坐标
~ floor 格形地面
~ formation 网格式结构布置
~ iron 铁格框
~ line 网格线
~ mesh 栅网
~ plan 网格平面布置
~ roller 方格压印滚筒,网格辗
~ type coil 格式盘管
~ valve 栅形阀
~ window 网格窗口

griddle ['gridl] *n.* 大孔[选矿]筛
gridiron ['gridˌaiən] *n.* (铁)框格,梁格结构,格子船台,修船架,管道网 *a.* 方格形[棋盘式]的
~ pattern 方格[棋盘]型
griffe [grif] *n.* 圆柱基部的虎爪形装饰,虎爪式柱座
griffin ['grifin] *n.* = gryphon 狮身鹰头翼兽形建筑装饰
grill(e) [gril] *n.* (门窗的)铁栅栏,网格,铁丝格子 *vt.* 装铁栅
~ fence 格式栅栏
~ louver 格形百叶窗
~ room (西式)小食堂,烤炙室
~ -work 格架
radiator ~ 散热器护栅
wire ~ 金属丝网格
grillage ['grilidʒ] *n.* (基础)格床[排]
~ beam 格排梁
~ girder 格排梁
checker work ~ 砖格
reinforced ~ 钢筋网格
grime [graim] *n.* 尘垢,烟灰,灰尘,浮土,污秽物 *vt.* 用灰尘弄脏,使积灰
~d with dust 被灰尘弄脏,覆上尘垢
grind [graind] *v.* 研磨,磨[抛]光;旋转 *n.* 研磨(声,机)
~ belt 研磨带
~ lap 磨石盘
~ machine 研磨[砂轮]机,磨床
~ material 磨料
~ media 研磨剂
~ mill 研磨[磨碎]机,磨坊
~ stone 磨石,砂轮
~ing aids 助磨剂
~ing compound 磨剂,金刚砂
~ing crack 磨痕,研磨裂纹
~ing disk 砂轮
~ing fluid 润滑液,研磨冷却液
~ing installation 研磨设备
~ing rod 研磨用钢棒
~ing slip 磨刀凿用的油石

grindability

～ing stone （天然）磨石砂石（轮）
～ing surface 磨面
～ing wheel 磨[砂]轮
grindability *n.* 可[易]磨性
grinder ['graində] *n.* 磨床,研磨机,砂轮;磨工
  disc ～ 盘磨机
  emery ～ 金刚砂磨石[床]
  rotary ～ 圆台平面磨床
grip [grip] *n.* 紧握[夹];握裹[粘着]力;夹钳,抓[把]手 *v.* 紧握[夹];扣[箍,粘]住;啮合
  ～ bolt 夹持[握固]螺栓
  ～ coat （搪瓷）底层
  ～ nut 夹紧螺帽,防松螺帽
  ～ ring （固定试件用的）夹圈
  ～ vice 夹紧虎钳
  ～ ping device 夹具,固定[抓取]器
  ～ ping head 握固头,夹头
  ～ ping pattern （轮胎）防滑花纹
  ～ ping tongs 平口钳
gripper ['gripə] *n.* 夹钳[具],抓爪[手]装置,牙扳夹头
  ～ die 夹紧模
grit [grit] *n.* 粗砂(岩),砂砾;金属屑,硬渣,磨料 *v.* 打磨;铺砂
  ～ blasting 喷砂处理
  ～ carborundum 金刚砂砾
  ～ content 含砂量
  ～ finish 磨砂处理,磨光
  ～ gravel 砂砾
  ～ material 砂砾材料
  ～ -stone 砂砾石,粗砂岩,天然磨石
  steel ～ 硬砂砾,钢砂
gritcrete *n.* 砾石混凝土
gritstone *n.* （粗）砂岩[砾],天然磨石
gritter *n.* 铺砂机
gritty ['griti] *a.* 含[有,像]砂的,砂砾的
  ～ consistence 含砂度
  ～ finish 石屑铺面
  ～ soil 粗砂土,砂质土
  ～ surface dressing 石屑铺面

grizzle ['grizl] *n.* 未烧透的砖,灰色次砖;灰色
  ～ bricks 欠火砖
grizzly [grizli] *n.* 铁栅筛,铁笼子筛 *a.* 带灰色的
  ～ bar 铁栅筛栅条
  ～ screen 格筛,铁栅筛
grocery ['grəusəri] *n.* 杂货店
  ～ store 杂货店
groceteria [ˌgrəusə'tiəriə] *n.* （自助）食品杂货店
grog [grɔg] *n.* 耐火材料,陶渣;熟料;烈酒（*pl.*）泥块
  ～ brick 耐火砖
  ～ refractory 耐火材料
  ～ shop 酒店
groin [grɔin] *n.* 穹棱;丁坝,防波堤;交叉拱 *vt.* 做成穹棱[交叉拱];造防波堤[丁坝]
  ～ arch 交叉拱,穹[弧]棱

groins

  ～ ceiling 穹棱顶棚
  ～ centering 交叉拱架
  ～ rib 穹棱肋
  ～ vault 交叉筒拱,十字拱顶
  ～ point 穹棱交点
  ～ roof 穹窿屋顶
  ～ slab 井字形梁板

groined vault

  ～ed point 穹棱交点
  ～ed roof 穹棱屋顶
  ～ed rib 交叉肋
  ～ed vault 交叉拱
grommet ['grɔmit] *n.* = grummet 金属孔眼,索环;绝缘孔圈;麻丝垫环
  cord ～ 索环
  drainage ～ 漏水垫圈
groove [gru:v] *n.* 沟,槽;凹线,企口;焊接坡口 *vt.* 开槽,挖沟,做企口

~ and tongue 槽舌榫,企口
~ connection 企口[槽式]连接
~ for sash 吊窗槽
~ joint 槽式接合,凹槽缝
~ of pulley 滑轮槽
~ of sheave 滑轮槽
~ of thread 螺纹槽
~ planer 槽刨,企口刨
~ shape 槽形
~ shed insulator 复檐绝缘器
~ -type contraction joint 凹槽形伸缩缝
~ weld 开槽焊,坡口焊
~d and tongued flooring 企口地板,槽舌地板
~d and tongued joint 企口[槽舌]接合
~d-back brick 沟纹砖
~d glass 槽纹玻璃
~d panel 槽纹镶板
~d pile 企口板桩
~d plywood 企口胶合板
~d rail 有槽导轨
~d seam (金属板的)折缝[咬口]接合
~d slab 企口板,槽纹板
~d tile 有槽瓦板
grooving plane 开槽刨
grooving saw 开槽锯

groover [gru:və] n. 开槽机[刨],挖槽工具

gross [grəus] a. 总[毛]的 n. 全体,总数,毛重;罗(=12打) vt. 总共赚得
~ apartment unit area 公寓单元总面积
~ area 总面积
~ benefits 总收益
~ building area 建筑占地总面积
~ building density 总建筑密度
~ capacity 总容量
~ charge 总支出
~ commercal area 总商业面积,总交易面积,总生产面积
~ coverage 总建筑占地面积系数
~ floor area 建筑总面积
~ floor space index 总占地指标
~ income 总收益
~ investment 投资总额
~ margin 毛利
~ profit 总[毛]利润
~ residential area 总居住面积
~ section 总[毛]截面

grotesque [grəu'tesk] a. 丑怪的 n. 怪诞的装饰图形,(哥特式教堂屋顶上的)怪兽像饰

grotto ['grɔtəu] n. (pl. -toes,-tos)岩穴;(人工开挖的)洞室,石窟

ground [graund] n. 地面[基],土壤;接地;广场,底层,背景 v. 放在地上;打基础,建立在基础上;搁浅 a. 地面的;磨光[细,碎]的
~ base 地[土]基
~ -batching plant 泥浆拌和机
~ beam 地基梁
~ bolt 地脚螺栓,锚栓
~ brick 碎砖,砖粉
~ brush 大漆刷
~ clearance 离地净高
~ coat 底层涂料,底漆
~ colour 底色
~ connection 接地线,磨口接头
~ elevation 地面高程,标高(图)
~ equipment 地面设备
~ facilities 地面设施
~ finish 磨光
~ floor 底层,一楼,地面层
~ floor area 底层面积,地面面积
~ floor construction 底层地面构造
~ floor plan 底层平面图
~ floor price 最低价格
~ glass 磨砂玻璃,毛玻璃[屏]
~ graphite 石墨粉
~ height 地基高度
~ hydrant 地面消火栓
~ joint 接地连接,磨口接头
~ joist 地格栅

~ level 地平面,地面标高
~ line 地平面[线],基准线
~ parallel 地面纬圈,地平线
~ pipe 地下管道,接地导管
~ plan 水平投影平面图;地面图,初步方案
~ plane (透视图)地平面,屏蔽面
~ plate 接地板,柱脚垫木,埋板
~ plot 地形平面图,地基图
~ point 地面[形]点,接地点
~ pumice 浮石粉
~ quartz 石英粉
~ quick lime 磨细生石灰
~ rock 石屑,磨细岩石
~ sill 地[潜、底]槛,卧木
~ stone 基础石,磨细石料
~ story 一层,底层
~ switch 接地开关
~ table 地面标高
~ terminal 接地端
~ wall 地下[基础]墙
~ wire 接地电线
~ing plug 接地插座

group [gru:p] *n.* 群,组,类,集团 *v.* 分类,组合
~ assembly parts 组装零件
~ comparison 分类比较
~ dwelling 住宅群
~ house 住宅组群,院落
~ of piles 桩群
~ system 组合制[系统]
~ technology 组合工艺(学),成组技术
~ed columns 群柱
~ ed commercial district 商业密集区
~ed site 住宅组群用地

grouser ['grausə] *n.* 锚定桩;轮爪,履带齿片

grout [graut] *n.* 薄[灰,水泥]浆,胶结泥;(*pl.*)渣滓 *vt.* 灌[涂]灰浆,粉饰
~ blanket 灌浆层
~ consumption 耗浆量
~ filler 灌浆填缝料,填缝薄砂浆
~ hole 灌[喷]浆孔
~ recess 灌浆凹面设施
~ repairing 灌浆修补
~ed aggregate concrete 骨料灌浆混凝土
~ed brick 浆砌砖
~ed concrete 压浆混凝土
~ed joint 灌浆接缝
~ed riprap 浆砌乱石
~ed rubble 浆砌块石
~ing agent 灌浆材料,胶结剂
~ing sand 灌浆用砂

growth [grəuθ] *n.* 生[增]长,发展;产物
~ constant 增长常数
~ factor 增长系数,放大因子
~ of concrete 混凝土膨胀
~ shake 木心环裂

groyne [grɔin] *n.* 防砂堤 *vt.* 筑防砂堤

grub [grʌb] *v.* 掘除,除根;*n.* 残根
~ saw 锯石用手锯
~ screw 平头螺钉,木螺钉

guarantee [ˌgærən'ti:] *n.* 保证书[人],担保品 *vt.* 保证,承认,许诺
~ against defects 质量保证,保修

guaranty ['gærənti] *n.* 保证,担保;抵押品,保证书 *vt.* 担保

guard [gɑ:d] *v.* 保卫[护],防止 *n.* 防护[隔离]装置,挡泥板,保险器;警卫员,哨兵
~ against damp 防潮
~ bead 护条,滑窗内轨
~ board 护[挡]板
~ boom 防护栅
~ cable 防护[安全]索
~ cradle 保护网
~ fence 防护栅,护篱
~ lock 防护[挡潮]闸,保险锁
~ net 防护网,控制栅极
~ pin 叉头钉
~ rail 护轨[栏]
~ signal 告警[安全]信号

~ stake 防护桩
~ stone 护[侧]石
~ time 防护时间
~ timber 护木
~ wall 防护墙
dust ~ 防尘板[罩]
eye ~ 护目板
heat ~ 绝热体
life ~ 救生员,(火车)排障器
lightening ~ 避雷器
manhole ~ 人孔拦栅
safety ~ 保险板,安全设备
gudgeon ['gʌdʒən] *n.* 轮[耳]轴,舵枢;螺栓;旋转架,托架
reversing screw ~ 回动螺旋,十字螺帽
guest [gest] *n.* 客人,旅客
~ hall 大客厅
~ house 宾馆,招待所
~ room 客房
~ rope 辅助缆索,扶手绳
guglia *n.* 方尖柱[碑]
guhr *n.* 硅藻土
guichet [giːˈʃei]【法】*n.* 小门;售票口
guidance [ˈgaidəns] *n.* 指[引,制]导,导航,操纵;导槽[板,轨]
guide [gaid] *n.* 指[引,制]导,导航,瞄准;操纵;导向装置;指南,路标,手册;指导者 *vt.* 指引;操纵
~ angle 导向角钢[装置]
~ arm 导杆
~ bar 导杆
~ bead (活叶窗)导条,(窗)滑轨
~ block 导块[瓦]
~ body 导向架
~ book 指南,入门,参考手册
~ chase 导框(排字或砌块用)
~ coupling 导管连接器
~ duct 导风管
~ elbow 定向弯头
~ frame 导架
~ groove 导槽
~ line 导[标]线,准则,指导方针

~ pile 定位[导]桩
~ pin 定位[导]销
~ principle 指导原则
~ rod 导向杆
~ rule 准则,导则,指南
~ sign 导向[指示,定位]标志
~ specifications 指导性规范
~ tube 导[套]管
~ wall 导水墙,导流岸壁[堤],护井
~ way 导向槽[道],导轨
~ wire 导向钢丝绳,导线
~ work 导航建筑
guiding lamp 导灯
guiding rule 样板,规准,(抹面用)靠尺
wire ~ 钢绳导道,电焊丝导向轨
guild [gild] *n.* 行会,同业工会,协会,互助会
~ -hall 行会会馆,同业公所,市政厅
~ -ship 同业公会

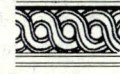
guilloche

guilloche [giˈləuʃ] *n.* 扭索纹建筑装饰
guillotine [ˈgilətiːn] *n.* 截断机,轧刀 *vt.* 截断
~ window 吊升窗
gulf [gʌlf] *n.* 海湾,深渊,鸿沟,旋涡
~ red 铁红
gullet [ˈgʌlit] *n.* 水落管,凹槽[沟];水道,海峡 *v.* 开槽,修整锯齿
~ plate 喉板
~ saw 钩齿[齿槽]锯
gully [ˈgʌli] *n.* 排水[集水,冲刷]沟,檐槽,雨水进口,溪谷 *vt.* 开沟
~ cover 沟[窨井]盖
~ drain 下水道,排水渠
~ grating 进水井盖[帘格]
~ -head tile 沟头瓦
~ -stabilization structure 沟壑稳定建筑物
gum [gʌm] *n.* 树胶[脂],橡皮;橡胶树;木焦油 *v.* 上胶,胶合
~ arabic 阿拉伯胶

~ cement 橡胶粘合剂
~ compound 纯胶料
~ elastic 弹性树胶
~ mastic 玛琋脂[树胶]
~ pole 圆木
~ resin 树胶脂,橡胶树脂
~ spirit 松节油
~ spirit cement 松节油灰泥(防水用)
~ tape 树胶粘条
~ tree 桉[橡胶]树
~ turpentine 松节油
Arabic ~ 阿拉伯树胶
artificial ~ 人造树脂
black ~ (美国)黑胶
gumbo ['gʌmbəu] *n.* (*pl.* -bos) 细[肥,强]黏土
~ clay 重[肥,碱性]黏土
gummy ['gʌmi] *a.* 胶黏的,树胶状的,含[涂]有树胶的
~ appearance 树胶状的,黏稠的
~ sand 含黏土多的砂
gun [gʌn] *n.* 手[喷]枪,炮;喷雾器;铆钉[注油]枪 *v.* 打枪
~ adapter 油枪嘴
~ -driven rivet 枪铆铆钉
~ finish 表面喷枪修整
~ hose 喷枪软管
~ jet 喷枪,喷水器
~ metal 炮铜,锡锌青铜
~ spraying 喷枪喷洒[喷涂,喷雾]
~ -stock stile 枪托状门窗边梃
cement ~ 水泥喷枪
spray ~ 喷[漆,涂,浆,雾]枪
gunite ['gʌnait] *n.* 喷枪;喷射法,压力喷浆 *v.* 喷射[涂]
~ coat 喷浆面层
~ layer 喷浆层
~ lining 喷浆衬砌
~ material 喷浆材料
~ work 喷浆工作
~d concrete 喷浆[射]混凝土
guniting machine 喷浆机

gunny ['gʌni] *n.* (*pl.* gunnies) 粗麻布,麻袋
~ sack 粗麻布袋
Gunter's chain *n.* 测链(长 66 英尺)
gurgoyle ['gəːgɔil] = gargoyle
gusset ['gʌsit] *n.* 节[接]点板,角撑板,加力片 *vt.* 装角撑板
~ angle 节点[隅撑]角钢
~ connection 节点板联接
~ felt 箱衬纸板
~ piece 连接板
~ plate 节点[联接]板,缀板
~ stay 节点角撑,角板撑条

gusset

gut [gʌt] *n.* 狭窄水道,海峡[岬],山间隘道;水蒸汽小管 *vt.* 破坏……内部
gutta ['gʌtə] *n.* (*pl.* guttae)(希腊柱式檐壁上的)圆锥[雨珠]形建筑装饰;杜仲胶
~ -percha 杜仲[马来]树胶
~ -percha ring 树胶圈

gutta

gutter ['gʌtə] *n.* 排水沟,檐[出料]槽,通风口,雨水口,漏斗 *v.* 开沟,装檐槽,天沟
~ bearer 檐槽托
~ bed 天沟挡水板
~ board 檐槽挑口板
~ bolt 天沟螺栓
~ bracket 檐沟托
~ hanger 檐槽吊[托]钩
~ hook 檐槽吊[托]钩
~ member 水落饰件
~ offtake 路沟泄水口
~ plank 檐槽板,槽盖
~ spout funnel 水落斗
~ tile 沟瓦
box ~ 方形檐沟

half round ~ 半圆檐沟
PVC ~ 聚氯乙烯檐沟
guttings *n.* (路面用)细石屑
guy [gai] *n.* 拉线[杆,索],风缆,牵索 *vt.* 用支索撑住,使稳定
~ cable 拉索,风缆
~ clamp 拉线夹板
~ line 拉线,牵索,风缆
~ rod 拉杆
~ rope 拉线,牵索,风缆
~ tightener 紧索轮
~ wire 拉线,牵索,钢缆
~ed structure 拉索结构
anti-rolling ~ 防滚索
back ~ 后拉线
gymnasium [dʒim'neiziəm] *n.* (*pl.* -siums 或 sia)体育馆,健身房
gynaeceum [ˌdʒaini'siəm] *n.* (*pl.* -cea) (古希腊、罗马建筑中)妇女专用的房间
gypsite ['dʒipsait] *n.* 土(状)石膏
gypsum ['dʒipsəm] *n.* 石膏;灰泥板 *v.* 用石膏处理
~ aluminate expansive cement 石膏矾土膨胀水泥
~ anhydrite 无水石膏,硬石膏
~ base board 石膏灰泥纤维板
~ block 石膏块材
~ board 石膏板
~ board sheathing 石膏衬板
~ calcination 石膏煅烧
~ cement 石膏水泥
~ concrete 石膏混凝土
~ earth 石膏土
~ element 石膏制品
~ fireproofing 石膏防火盖面
~ lath 石膏板条
~ lath nail 石膏板条钉
~ model 石膏模型
~ mortar 石膏灰浆
~ mould 石膏外模
~ neat plaster 石膏净粉饰
~ panel 石膏心墙板
~ plaster 石膏灰泥
~ plastering 石膏粉饰
~ product 石膏制品
~ -retarded cement 石膏缓凝水泥
~ roof plank 石膏屋面板
~ sand 碎屑石膏
~ sheathing 石膏盖板,外墙防水膏板
~ slag cement 石膏矿渣水泥
~ slurry 石膏泥浆
~ tile 石膏瓦[面砖]
~ wallboard 石膏墙板
~ wood-fibered plaster 石膏木丝灰泥
gypsy ['dʒipsi] *n.* (绞车上的)铰绳筒
gyre [dʒaiə] *n.* 旋转运动[体],大漩涡,回流,旋风 *v.* 旋[回]转
gyro ['dʒaiərəu] *n.* (*pl.* -ros)回转仪[罗盘],陀螺仪[罗盘]
~ -compass 回转[陀螺]罗盘
~ horizon 陀螺地平仪
~ -mixer 回转拌和机
~ -rotor 回转体,陀螺转子
~ sextant 陀螺[回转式]六分仪
gyrograph ['dʒaiərəgræf] *n.* 转数记录器,旋转测度器
gyroscope ['gaiərəskəup] *n.* 回转[环动,陀螺]仪,回旋器,旋转机
directional ~ 定向陀螺仪
free ~ 三自由度陀螺仪
rate ~ 二自由度陀螺仪,速率[阻尼,微分]陀螺
vertical ~ 垂直陀螺仪
gyroscopic [dʒaiərə'skɔpik] *a.* 陀螺[回转]的
~ compass 方向陀螺仪,陀螺[回转]罗盘
~ level 回转式水平仪

H

habit [ˈhèbit] *n.* 服饰,衣着,服装的式样;习惯,习性 *vt.* 装扮,穿着

habitable [ˈhæbitəbl] *a.* 可居住的,适于居住的
~ attic 屋顶[顶楼]住室
~ house 可居住房屋
~ room 起居室

habitacle [ˈhæbitəkəl] *n.* 壁龛

habitant [ˈhæbitənt] *n.* 居民,居住者

habitat [ˈhæbitæt] *n.* 动物的栖息地,植物的产地,场[住]所

habitation [ˌhæbiˈteiʃən] *n.* 居住,住宅,住处,聚居地,住房

habitual [həˈbitjuəl] *a.* 日[平,通]常的,惯例[常]的
~ work 日常工作

hachure [hæˈfjuə]【法】*n.* 影线(表示地形、断面图等的)(*pl.*)蓑状线;痕迹 *vt.* 用影线表示,用蓑状线表示
~ lines 影线

hacienda [ˌhæsiˈendə]【西】*n.* 农场,牧场,种植园,大庄园,工场

hack [hæk] *n.* 砍,劈,削;刻痕,切口;鹤嘴锄,十字镐;晒架;晒砖场;出租车;雇工 *v.* 劈,(乱)砍,削平;碎[破]土;排列(砖坯)使干燥 *a.* 受雇的,出租的;用旧的
~ -file 手锯,刀锉
~ hammer 劈石斧,斧形锤
~ iron 十字镐,矿用丁字镐,鹤嘴锄,錾
~ lever 架杆
~ saw 弓形锯,钢锯
~ -saw blade 弓锯片
~ed bolt 凹痕螺钉
~ing off (墙面)打毛

hackbarrow [ˈhækˌbærəu] *n.* 运砖手推车,砖坯车,运砖架

hackberry [ˈhækberi] *n.* (*pl.* ries) 朴树,朴树材

hacked *a.* 劈[砍,削]的
~ bolt 锚栓,棘[底脚]螺栓;凹痕螺钉
~ -type hammer 斧式锤片,(粉碎机的)楔形锤片

hacker [ˈhækə] *n.* 砍伐工;掘根用的钩耙或手操工具

hacket *n.* 木工用斧

hackia 砍斧木

hacking [ˈhækiŋ] *a.* 劈,砍伐;拉毛;刻记[痕];堆高砌坯;干砖;开槽,开片;碎土,锄地[草]
~ knife 砍刀
~ off (墙面)打毛
~ -out knife 油灰刮刀

hackle [ˈhækl] *n.* 手工梳麻台,梳麻[棉]机,栉梳,锯齿形 *vt.* 乱砍,砍光;栉梳,梳棉[麻]

hackly [ˈhækli] *a.* 锯齿形的,参差不齐的,粗糙不齐的

hackmatack [ˈhækmətæk] *n.* 西方落叶松的木材,欧洲刺柏,美洲落叶松,杜松

hacksaw [ˈhæksɔː] *n.* 弓形钢锯,钢锯
~ blade 弓锯片
~ frame 钢锯架,弓锯架

hade [heid] *n.* 断层余角,断层伸向[角] *vi.* 垂直倾斜

Hadfield steel 一种高锰钢

Hadfield manganese steel 哈氏高锰钢,奥氏体高锰钢

haecceity (= hecceity) *n.* 个性,特征

haematinum *n.* 血色玻璃

haematite [ˈhemətait] *n.* 赤铁矿

haft [hɑːft] *n.* (锄、斧、刀等的)柄,把手;旋钮 *vt.* 给……装柄

hag [hæg] *n.* 槽,缝,切口,凹口,刻痕,

缺口;采伐线;沼地;河流的悬垂边沿 v. 砍,伐木,刨煤

haggle ['hægl] n. 争论;讨价还价;乱砍,粗切 vi. (在价格,条件等方面)争论不休 vt. 乱砍,乱劈,粗切

haggy a. 潮湿和崎岖的

hagioscope ['hægiəskəup] n. 斜孔小窗,窥视窗,(中世纪教堂墙壁上的)倾斜小窗孔

ha-ha ['ha:ha:] n. (修筑在花园界沟中不遮挡视线的)矮篱[墙],(界沟中的)隐蔽围墙

hair [hɛə] n. 头[毛]发,毛(状物);麻刀;游丝,微动弹簧;些微,一点儿;(pl.)十字线,叉线,瞄准线

～ brush 鬃帚,毛刷
～ checking 发(状幅)裂,细裂纹
～ compass 微调圆规
～ crack 发(状裂)缝,细裂缝
～ cross (目镜中的)十字线,瞄准线
～ felt 油毛毡
～ fibered 含有纤维作为一种粘合料的
～ -fibered cement mortar【美】麻刀灰泥
～ fibered plaster 麻刀灰泥
～ -fibered lime mortar【英】麻刀灰泥
～ grease 毛填料润滑脂
～ gypsum【美】麻刀石膏
～ gypsum-lime mortar【英】麻刀石膏灰泥
～ like 像毛的,毛(似)的,细的
～ lime-mortar【美】麻刀灰泥
～ line 发丝,细缝;十字[瞄准]线
～ -line crack 发样裂纹
～ line seams 发纹
～ mortar 麻刀灰浆
～ (-)pin 细销,发卡
～ salt 发盐,铁[羽]明矾
～ seam 毛发[裂]缝
～ sieve 细孔[马尾]筛
～ spring 游丝,细弹簧

～ spring divider 细弹簧分规
～ -thin 细如发发的
cotton ～ 棉线

hairen a. 毛制的

hairy ['hɛəri] a. (多)毛的,毛状的

hake [heik] n. (pl. hakes, hake)砖(瓦)坯干燥槽架,格架;牵引调节板

Halcomb n. 哈尔库姆合金钢

halcyon ['hælsiən] a. 平稳[静]的,安静的;富足的

hale [heil] vt. 拖曳,猛拉 a. 强壮的
～s 把手

half [ha:f] n. (pl. halves)半,一半,二分之一;一部分;相当地;不完全,不充分;与主节面成45°的工作面 a. 半个的,一半的 ad. 一半地,部分地

～ anchor pole 半锚杆
～ anchor towers 半拉杆塔,半锚塔,半锚式杆架
～ -and- half joint 对拼接头
～ -baked 半焙烧[干]的
～ baked brick 半焙烧砖
～ -balance 不完全平衡
～ balk 破成两半的坑木,对开圆木;顶板支梁
～ baluster 半栏杆小柱
～ basin 半水槽[池]
～ bat 半砖,封口[堵头,封缝]砖
～ bat thick 半砖厚
～ bay 半跨,半柱距[间格]
～ -beam (船舱上的)半梁,半横梁
～ -black 半加工[处理]
～ -blank 半加工[处理]
～ block 半瓦,半瓷砖,半空心砖,半瓦管
～ boiling 半脱胶
～ breadth 半宽度
～ -brick wall 半砖墙
～ -burnt (clay) brick 半烧透砖
～ chimney block 对分烟囱砖
～ column 半径,半身(附墙)柱
～ -countersunk 半埋头(的)

~ -crystal (glass) 半结晶玻璃
~ -cupola 半圆屋顶,半穹顶
~ -cylinder 半圆柱体
~ -diamond indention 倒三角形
~ -dislocation 半位错
~ -dome 半圆屋顶,半穹顶
~ -door 半截门
~ -dressed stone 粗琢石
~ -elliptic 半椭圆形的
~ engaged column 半露柱,半柱
~ -ever green 半常绿的
~ -faced 显出一个侧面的,三面封闭仅保留前部敞开的
~ -finished 半精加工的,半完成的
~ -finished product 半成品
~ -fired (clay) brick 半烧透砖
~ -flat 半层,半套房间
~ flight (楼梯的)半跑段
~ floor 半(楼)层
~ -front view 半前视图
~ frame 半框架,一梁一柱式支架
~ -gate 半开(闸门),补偿栅
~ -glass door 上半截装配玻璃的门
~ -groove 半企[切]口,半凹槽
~ -hard 中硬(金属)
~ -hard rubber 半硬质胶
~ -hard steel 半硬钢,中硬钢
~ header 收尾的半砖,半圆木大顶梁
~ height block 4 英寸厚的混凝土块体
~ hip 半斜(屋)脊
~ -joist T 形搁栅,T 型钢梁
~ -landing 楼梯平台,全宽梯台
~ lap joint 半叠接
~ lap scarf joint 半叠嵌接,半叠楔面接合
~ -lattice frame 半格构架,半柜格式构架
~ lattice girder 半格组梁,半格构梁
~ length 半身椽(的)
~ -light 暗淡的光线
~ -log 对开圆木

~ main rafter (伸不到屋脊的)半橼条,半人字木
~ -metal 半[类]金属
~ -mitre joint 半斜接
~ monitor 锯齿形顶
~ moon 半月,半月形
~ -nut 对开螺母
~ -open-jointed 半开式接缝的
~ open interval 半开区间
~ -pace 梯台,上楼梯转 180°的平台;凸窗高台
~ -pier 半窗间壁墙,半(支)柱
~ -pitch roof 半斜屋顶
~ plane 半平面
~ -plain work (石面)半琢平工作
~ polished plate glass 半磨光平板玻璃
~ -prestressed beam 半预应力梁
~ -principal (伸不到屋脊的),半椽条,半人字木
~ -rear view 半后视图
~ relief 半凸浮雕
~ restrained 半约束[固定]的
~ ripper 细木锯
~ ripsaw 细木锯
~ (roof) tile 半(屋面)瓦
~ round 半圆形,半月形的
~ -round arch 半圆拱
~ -round apse[apsis] 半圆形室,半圆形前廊
~ -round barrel vault 半圆筒形拱顶
~ -round bastion 半圆(形)堡垒
~ -round ceiliny 半圆形顶棚
~ -round cross-vault 半圆形交叉筒拱
~ -round cylindrical roof 半圆柱形屋顶
~ -round exedra 半圆形室,半圆形前廊
~ -round file 半圆锉
~ -round gutter 半圆形檐槽,半檐沟
~ -round iron 半圆钢

~ -round niche 半圆形壁龛
~ -round nosing 半圆突边,半圆形突缘饰
~ round pointing 半圆缝
~ -round profile 半圆形轮廓
~ -round rivet 半圆铆钉
~ -round screw 半圆头螺栓
~ -round spade 半圆铲,半圆形铁锹
~ -round stretcher 半圆形露侧砖[石]
~ -round tower 半圆形塔[架]
~ -round tunnel vault 半圆形隧洞拱顶
~ -round window 半圆形窗
~ -round wood screw 半圆头木螺钉[旋]
~ rounds 各种直径的半圆钢
~ sectional side elevation 半剖视侧图
~ sectional view 半剖视图
~ set T 字形支架
~ shell 半壳
~ -sip saw 细齿锯
~ -size 半粒径(的),半大颗粒(的),缩小一半的
~ -size scale 半尺寸,比例尺 1∶2,二分之一缩尺
~ solid floor (= lightening plate frame) 空心肋板
~ space 半无限体;半度空间;梯台,休息平台
~ span roof 单坡屋顶
~ split pipe 半圆管道
~ split pipe duct 半圆槽
~ split pipe (gutter) 半圆檐槽
~ -story 屋顶层,夹层
~ sunk 半埋[插]入的
~ sunk rivet 半埋头铆钉
~ tenon 粗短榫,半榫
~ thread 半螺纹
~ -through arch 半穿式拱
~ -through girder 下承板梁
~ -through truss 半穿式桁架,下承矮桁架

~ -tile 半瓦,半瓷[空心]砖,半瓦管
~ tile wall 半截贴面[砖]墙
~ -timber construction 砖木混合结构
~ -timber house 半露木房屋
~ -timber wall 半露木墙
~ -timbered building 半露明木(骨)架建筑
~ -timbered wall 半露明木(骨)架墙

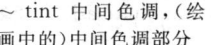
half-timbered building

~ tint 中间色调,(绘画中的)中间色调部分
~ (-)tone 半色调[度],中间色调
~ top view 半剖俯视图
~ tunnel vault 半隧洞[圆]顶
~ -turn stair 两次直角拐弯楼梯
~ -wrought timber 半成品木材

halfer *n.* 单脚

hall [hɔːl] *n.* 大厅,会[礼,讲]堂;门厅,通道,走廊;机[厂]房,车间,办公室;大学的宿舍
~ access type 厅式平面(各室入口开向大厅的)布置方式
~ bedroom 门厅卧室,廊底小卧室
~ -choir (教会)有唱诗班席位的大厅;排练厅
~ -church 哥特式教堂
~ clock 大挂钟
~ closet 大楼内的套间[壁橱,盥洗室,厕所]
~ constuction 大楼式建筑
~ cupola 半球形顶棚的大楼
~ dome 半圆形顶棚的大楼
~ -dungeon 城堡的大厅;监狱
~ form 大楼结构[类型,模壳]
~ -keep 城堡的大厅;监狱
~ -nave (车站教的)中央广场;早期教堂的中殿(座位所在部分)

hall of ambassador 大使馆
Hall of Ancestral Worship 北京故宫奉先殿

Hall of Benevolence and Longevity 颐和园仁寿殿
Hall of Central Harmony 北京故宫中和殿
hall of columns 圆柱大厅
Hall of Happiness and Longevity 颐和园乐寿堂
Hall of Imperial Model 北京故宫皇极殿
Hall of Jade Ripples 颐和园玉澜堂
Hall of Judg(e)ment 法庭[院]
~ of justice 法院
Hall of Mental Cultivation 北京故宫养性殿
Hall of Prayer for Good Harvests (北京)天坛祈年殿
Hall of Preserving Harmony 北京故宫保和殿
Hall of Supreme Harmony 北京故宫太和殿
Hall of the Dragon King 颐和园龙王堂
Hall of Union 北京故宫交泰殿
~ -quire (教会)有唱诗班席位的大厅;排练厅
~ -stand 衣帽架
~ system apartment (各户出入口向大厅的)厅式公寓
~ -transept (教堂的)交叉甬道
~ -tree 衣帽架
~ -type building 厅式大楼(建筑)
~ way 门厅,穿堂,过道,回廊
assembly ~ 会堂
audience ~ 听衆堂,接见厅
entrance ~ 门厅
lecture ~ 讲堂,演讲厅
music ~ 音乐厅
palatial ~ 大殿
stair ~ 楼梯间,有楼梯的前厅
station ~ 车站候车厅,车站大厅
the Great Hall of the People (in Beijing) (北京)人民大会堂
hallcist *n.* 古代作坟墓用的以石板筑成的地下长方形巨廊
halloysite *n.* 埃洛石,多水高岭土[石]
hallway ['hɔːlwei] *n.* 门厅,过道,回廊
halo ['heiləu] *n.* (*pl.* -los or -loes) 晕,晕圈,光环 *v.* 使成晕圈,以光圈围绕
~ effect 光[晕]圈效应
halogen ['hælədʒen] *n.* 卤素
~ process 卤化(电镀锡)法
halometer [hæ'lɔmitə] *n.* 盐量计
halomorphic soil [ˌhælə'mɔːfiksɔil] *n.*
= halogenic soil 盐渍土
halt [hɔːlt] *n.* 临界点;停止,防止;小站,招呼站 *v.* 站住,停止
~ing place 野餐场所,【英】候车站,【美】路边休息处,休息区
halter ['hɔːltə] *n.* 平衡棒[器];(马的)笼头,缰绳;绞索 *vt.* 束缚,抑制
halve [hɑːv] *vt.* 对半分,平分,二等分,平均分担;把……开半对搭,重接
bevel halving 斜面半塔接
corner halving 角嵌[镶]接
dovetail halving 鸠尾榫嵌接
square-corner halving 角平嵌接
halved *a.* 二等分的;在接合处将(两木)各剖去厚度之半而接合的;二开片
~ door 半截门
~ joint 对搭接合[头],对塔接,嵌接,相嵌结合,重接,合嵌接
hamlet ['hæmlit] *n.* 小村(庄)
hammam ['hæmæm] *n.* 土耳其(式)浴室或澡堂
hammer ['hæmə] *n.* 锤,头;汽锤,铆枪;撞针;冲击式凿岩机 *v.* 锤击[打,炼],敲打;吊销执照
~ beam 托臂梁,椽尾(小)梁
~ beam roof 托臂梁屋顶

hammer beam

hammer-beam roof

~ beam truss 托臂梁桁架
~ brace 椽尾梁斜撑
~ -dressed quarry stone 锤琢毛[原]石
~ dressing 用锤整修,锤琢[整];锤整的
~ faced stone 锤琢石
~ finish 锤琢修整(石面)
~ forging 锤锻,锻造锻件
~ glass 锻玻璃
~ hack (琢石的)斧形锤,劈石斧
~ handle 锤柄[把]
~ pick 尖锤,鹤咀镐
~ post 悬臂托柱
~ riveting 锤打铆接
~ shaft 锤柄
~ smith 锻工,铁匠
~ stem 锤杆
~ stone 石锤
~ support 锤杆座
~ tail 锤杆
~ tongs 锻锤夹钳
~ valve 锤形(减压)阀
~ welding 锻接[焊]
brick ~ 瓦工锤
ca(u)lking ~ 填隙[堵缝]锤
carpenter's ~ 木工锤
chipping ~ 錾平锤
dead flat ~ 矫平面锤
dressing ~ 敷面锤
drop ~ 落锤
joiner's ~ 细木工锤
jump ~ 夹板锤,摩擦锤
mason ~ 瓦工锤
moulder's ~ 造型锤
pavior's ~ 铺砌锤
pick ~ 鹤咀锤[锄],风镐,十字稿
pneumatic ~ 风动錾[铲,镐]
polishing ~ 抛光锤
ripple ~ 摆动锤
scabbling ~ 粗琢锤
set ~ 击平锤

slate ~ 石板瓦工锤
sledge ~ 大铁锤
sleeking ~ 抛光锤
soldering ~ 钎焊;焊铁(焊接器),烙铁,焊烙铁
spiking ~ 道钉锤
treadle ~ 踏锤
wiper ~ 杵锤
hammered ['hæməd] *a.* 用锤打的,锻造的
~ finish 锤琢(石面)
hammerer *n.* = hammerman 锻工,铁匠
hammerhead ['hæməhed] *n.* 锤[榔]头;弦槌
~ chisel 锤头凿
~ key 锤头形键槽
hammerless [hæməlis] *a.* 无冲击的,无击锤的
hammersmith ['hæməsmiθ] *n.* 锻工,铁匠
hance [hæns] *n.* 拱腰[腹](椭圆拱脚处的最小半径弧),梁腋;窗户上遮阳光的水平挡板;把楣和侧柱连接起来的小拱
~ arch 平圆拱,三心拱
hand [hænd] *n.* 手,手工,人工;指针;手柄;雇员,工人,船员;手艺,才能,技巧;管理,掌握,支配;一手宽(约4吋);侧,方面 *vt.* 传递,交给;扶持 *a.* 手的,用手的;手制
~ adjustment 人工调整
~ applied (mixed) plaster 人力拌制的灰浆
~ banisters 扶手栏杆,(楼梯)扶手
~ barrow (双轮)手推车,独轮手车,塌车;担架
~ basin 洗手盆
~ bench 手工台
~ bender 手动弯曲机
~ bending machine 人工弯筋机
~ blocked (wall) paper 手工印花墙纸
~ brick 手工砖

~ -broken metal (英)小石子
~ -broken stone [英]小石子
~ broom 扫帚
~ buggy 手推[摇]车,手拉车
~ chipping 人工修整[錾平,凿平]
~ cleaning 人工清除[理]
~ cleft (wood) shingle 手工劈制的(木质)屋顶板
~ coat 手工刷浆
~ concrete mixing 人工混凝土拌和
~ control system 人工控制系统
~ control unit 人工控制机构
~ driven rivet 手铆铆钉
~ -erected scaffold 人工架设的脚手架
~ finish 手工修整[精修,最后]加工
~ finisher 混凝土手工修整器
~ fitting 手工装配[组装,安装]
~ float (手)镘板
~ float trowel 镘刀,抹子
~ float light 手提泛光[强力]照明灯
~ floating 手镘,人工镘平[镘涂]
~ forging 手工锻造
~ -formed brick 手制砖
~ grab 扶手
~ gravure 手工凹印
~ grinder 手工磨机[砂轮(机)]
~ grip 手柄
~ guard 扶手
~ guard bar 扶手栏杆
~ guided 人工控制
~ hammer 手锤
~ haulage 人力运输[推车]
~ -held welder 手动焊接机
~ hole (地下电缆)手孔,检修[装配,注入]孔;手工凿岩[掏槽]
~ hole cover 检查[手]孔盖
~ -holing 人工打眼,手工掏槽
~ laying 人工敷设[衬垫]
~ level 手持水准仪,手水准;水平尺
~ made 手工制造的;手工制品
~ made brick 手工砖

~ made paper 手工纸
~ -mixed concrete 人工拌制的混凝土
~ -mixing 手拌(法),人工拌和(法),人工搅拌
~ molding 手工线条[脚]
~ money 定金
~ motion 手动,手开动
~ -mo(u)ld 手工铸模,手工压模
~ mo(u)ld tile 手工[制]型瓦
~ -operated 人工驱动[操作]的
~ -operted auger 手摇螺钻机
~ operated screed 手工整平板
~ operations 手动装置
~ -packed bottoming 人工毛石铺底
~ -packed hardcore;[-packed stone] 人工堆石
~ -packed rock 人工砌石
~ -packed rockfill (= hand-packed rubble) 人工抛石[堆石]
~ -packed rubble 人工铺砌块石
~ -painted picture tile 手工绘制瓷砖
~ -pitched base 手铺小石块底层
~ placed riprap 人工堆筑的抛石护坡
~ placement 手工摊铺
~ plane 手刨
~ plate (门上)推手扳
~ plate shears 手工平板切割机
~ portable riveter 人力轻便铆钉器
~ power 手拉[摇]
~ press 手(动)压机,手动印刷机[滚]
~ print 手纹[印]
~ printed (wall) paper 手工印刷的(糊)墙纸
~ -rail 栏杆,扶手
~ rail baluster 扶手栏杆(小柱)
~ rail banister 扶手栏杆(小柱)
~ rail plane 扶手刨
~ rail profile 扶手外形
~ rail scroll 栏杆末端的蜗形

~ rail section 栏杆分段[零件]
~ rail shape 栏杆钢材
~ rail stanchion 栏杆柱
~ rail standard 栏杆支柱
~ rail support 扶手托座
~ raked 手耙的
~ ram 手夯,人工夯
~ ram(mer) 手夯,人力夯,手夯锤(槌)
~ rendering 手工粉刷[抹灰]
~ reset 人工重调;手工工具支托;靠手;手动复位
~ rest 手工工具架
~ riveting machine 手铆机
~ rock facing 人工块石护面
~ rubble 手工砌筑毛[块]石
~ sample 小样,手工[人工]取样
~ sampling 手工取样[取样,采样]
~ saver 保护手套,手护
~ -saw 手锯,板锯,手板锯
~ saw blade 手锯条
~ screen 手工筛;焊工面罩
~ screw 手动千斤顶[起重机];手旋螺钉;木框螺栓夹板
~ sieve 手工筛
~ split (wood) shingle 手工劈制的木片瓦[屋顶板]
~ split (wood) siding shingle 手工劈制的木质板壁
~ stirrer 手工搅拌器
~ swabbing 手工刷浆[刷色]
~ sweeping 人工清扫
~ tap 手用螺钉攻
~ -taut 用手劲拉紧
~ tight 用手劲拉紧
~ tool 手工工具
~ trammel 木工规,长臂圆规
~ -vice 手虎钳,手钳
~ welder 人工焊机
~ (-)work 手工(加工),精细工艺
~ -wound 手绕的
~ -wrought 手工制成的
handbook ['hændbuk] *n.* 手册,指南

handicraft ['hændikrɑ:ft] *n.* 手工业;手艺;手工艺品;技工
~ -type metal grille 工艺型金属栅[花格窗]
handing 手感;交出,传递
~ equipment 装卸设备
handiwork ['hændiwə:k] *n.* 手工(业),手工艺制品
handle ['hændl] *n.* 手柄,把手;驾驶盘,手轮;(焊)钳;一种啤酒量具(大约一品脱);经销,买卖 *vt.* 运用,操纵;处理,管理
~ bar 手柄杆,操纵柄
~ blank 宜于做把手或手柄的一段木材
~ lead 测深铅锤
handler ['hændlə] *n.* 管理者;(信息)处理机(器);处理程序;装卸装置;近距离操纵机械手
handling ['hændliŋ] *n.* 处理,管理,操纵;装卸;吊运;修改,中间加工
~ and loading 搬装
~ container 集装箱
~ frame 操作架
ash ~ 除灰
bulk ~ 散装运输(指松散货物)
cable ~ 敷设电缆
crop ~ 清理切边,排除切头
safe ~ 安全运输[转]
handmade ['hænd'meid] *a.* 手动的;手工制的(物品);人造的
~ brick 手工砖
handpick [hænd'pik] *vt.* 用手拣[摘]
handplaced *a.* 手放[堆,铺]的,人工铺砌的
~ rock 干砌块石
handset ['hænd'set] 手机,送受话机 *vt.* 用手排(铅字) *a.* (铅字)手排的,用手排铅字印刷的
handstone *n.* 小石子,鹅卵石,手工操作的碾磨石
~ subbase 小石子底座
handy ['hændi] *a.* 便于使用的,方便

的;手边的,合手的,轻便[巧]的,可携带的 **ad.** 在附近,就近
~ man 巧匠,多才多艺的人
hangarage *n.* [英]飞机库[飞机棚]
hangarette *n.* 小型机库
hanger ['hæŋə] *n.* 挂钩;吊架;吊轴承;支架,托梁;垂饰;顶板;悬崖,绝壁;上盘;(固定屋檐落水槽的)金属吊钩)
~ bolt 吊架螺栓
~ frame 吊架
~ iron 挂铁
~ rope 吊缆[索]
~ rot 吊杆
shot ~ blast 连续喷丸清砂
spring ~ 弹簧支柱
step ~ 踏阶吊铁
swing ~ 摆动吊梁

hanger rot

hang [hæŋ] *n.* 悬挂[吊装]方式;下垂物[状态];大意,要点,决窍,用法;斜坡,倾斜;间断[隙],暂;断层余角 *v.* 悬,挂;装饰,贴(糊墙纸等);安装(活动的东西);拖延
~ five 五趾吊
~ ten 十趾吊
~ wall paper 糊墙纸
hanging ['hæŋiŋ] *n.* 悬吊[挂];悬式[空,垂];斜坡;顶[上]盘,顶板;(*pl.*)(指帘帷等)悬挂在墙上的东西 *a.* 悬式[空]的,垂下的
~ arch 悬拱
~ beam 吊梁
~ bolt 挂钩,吊架,悬杆[架]
~ bracket 悬臂支架
~ buttress 悬式扶垛,悬挂支墩
~ cable 吊索[缆]
~ chain 悬链
~ floor 悬式楼面板
~ garden 空中花园
~ guard 上护板
~ guide 悬挂导架[杆]
~ gutter 悬吊槽
~ leader 悬式导架
~ paper 墙纸

~ paste 悬胶
~ pillar 悬柱
~ post truss 吊柱桁架
~ rail 铰链横挡,钉铰链的横木
~ rainwater gutter 悬式截集雨水的檐槽
~ roof 悬挂屋盖
~ sash 吊窗,吊(窗)格框
~ scaffolding 悬式[空]脚手架
~ shelf 悬式格架
~ socket 悬式灯座
~ stairs 悬空(式)楼梯,墙上突出外端无支撑的楼梯
~ step(s) 悬空式梯台[踏步],半悬梯级
~ stile 铰链挺,窗挡的竖边
~ story 悬式楼面(板)
~ tile 挂瓦
~ truss 悬吊式桁架
~ wall 上盘,顶板
~ window 吊窗
hangrod *n.* 水平挂衣杆
hangtag *n.* 使用保养说明标签
hank [hæŋk] *n.* 卷线轴;(缠线用的)工字形框;优势,控制 *vt.* 使成为一绞一绞
~ of cable 电缆盘,一盘电缆
hanse (=hance,hanch)舵,弦墙等急转折部分;梁腋,拱腋[腰]
Hanshan Temple 苏州寒山寺
haphazard ['hæp'hæzəd] *n.* 偶然[意]性;无规则绕线 *a.* 偶然[任意]的,无计划的,不测的 *ad.* 偶然地,[任意]地,乱七八糟地
~ building 不规则建筑
haplite *n.* 简单花岗岩,细晶岩
haplobasalt *n.* 人造玄武岩
harbo(u)r ['ha:bə] *n.* (小)海港,港湾[港口]码头;避难所 *v.* 隐匿,窝藏;包庇;包含;停泊,暂住;聚集
~ building 海港建筑物
~ city 海港[港口]城市
~ construction 港湾建筑

~ construction site 港湾建筑工地
~ engineering 港湾工程[学],港口工程[学]
~ extension 港口扩建(工程)
~ structure 港湾建筑
~ work(s) 海港工程,港口建筑物,港口工程,港口设施
artificial ~ 人工港

hard [ha:d] *n.* (英)硬海滩,登陆处 *a.* 硬[结实,坚固]的;困难[艰苦,费力]的,刻苦的;严格[苛刻]的;确实的,不容怀疑的 *ad.* 硬,牢,坚硬[牢固]地;紧紧[接近]地;立即地;困难[艰苦]地;猛烈地
~ alloy 硬(质)合金
~ asphalt 硬沥青(贯入度 10 以下)
~ board 硬质纤维板,高压板(一种由废木屑压成的夹板),硬纸板
~ brick 硬砖,过火砖
~ brittle material 脆性材料
~ -burned 过火的,烧硬的,炼制的,(石灰)煅烧过度的
~ -burned brick 炼[硬,炼制]砖,烧透砖,过火砖
~ -burned free lime 生石灰
~ -burned gypsum 高温焙烧石膏
~ -burned refractory ware 硬烧耐火制品
~ -burnt plaster 无水石膏
~ calcareous slate 硬石灰板
~ concrete 干硬性混凝土
~ copper 冷加工铜,硬铜
~ core 作路面或基础底层的,天然岩石碎块,碎砖块,石填料;硬核(心);核心硬件
~ (-)drawn 硬[冷]拉的
~ (-)drawn aluminium wire 硬(拉)铝线
~ -drawn wire 冷拉钢丝
~ dry 干硬(状态)(涂下一层油漆前的油漆薄膜)
~ facade 坚硬表面
~ faced 硬表面的

~ facing 表面硬化[淬火],硬质焊敷层
~ fibre 硬纤维板[材料,结构]
~ fibre board 硬质纤维板
~ finish 干硬[细灰泥]抹面,硬质墙[面]层
~ finish plaster 缓硬石膏
~ finished plaster 需加催硬剂的无水煅烧石膏
~ floor paint 地板漆
~ flux (使沥青提高稠度的)硬掺合料
~ glass 硬玻璃
~ glazed coating 硬珐郎(质)涂[镀]层
~ gloss 硬珐郎(质)
~ -grade billet 硬钢杆,硬钢坯段料
~ -grained 粗粒状的,粗木纹的
~ ground 岩[硬]底,硬防蚀剂
~ hat 安全[矿工;建筑工人]帽
~ head 硬头(锡铁合金,锡精矿还原熔炼的一种副产品),硬质巴比合金;难熔的矿石
~ head sponge 硬海棉
~ image 对比度强的[黑白分明的]图像
~ -laid 硬[紧]搓的
~ lead sheet 硬铅皮[薄板]
~ mass 人造宝石,冒充宝石用的硬玻璃
~ mastic asphalt 硬质砂胶沥青
~ metal 硬质合金,硬金属
~ (metal) alloy 硬质[高强度]合金
~ mortar 干硬砂浆,硬性灰浆
~ paste 硬质粘土
~ pavement 硬(质)路面
~ pitch 硬焦油脂,硬沥青
~ refractory ware 硬烧耐火器材
~ rime 霜淞
~ rock 硬岩
~ rock chip(ping)s 硬碎石[小石片]
~ rock flour 硬岩粉末
~ rock slab 硬岩石板
~ -rolled 冷轧的

~ roof 坚硬顶板
~ rubber board 硬橡胶板
~ rubber sheet 硬橡胶薄板
~ rubber tube 硬橡胶管
~ solder 含铜多的焊锡,硬焊料[药],硬[铜]焊
~ soldered joint 硬焊接合
~ soldering 硬[铜]焊(的)
~ spots 部分过硬,麻点
~ steel 硬钢(即高碳钢)
~ stock brick 耐火砖,普通硬砖
~ stopper （油漆打底用的）硬质填塞料
~ stopping 硬质填塞料
~ wall 石膏抹底墙
~ wall plaster 墙的水泥抹面,(墙的)水泥粉饰
~ wares 小五金
~ -wearing 耐磨的
~ wire 硬[高碳]钢丝
~ wood 硬木[材]
~ wood mosaic floor 硬木块镶嵌地板
~ -wrought 冷加工[锻](的)
hardas process *n.* 硬质氧化铝膜处理法
harden ['ha:dn] *v.* （使）变硬[硬化,凝结,凝固,坚固］淬火,增加硬度,(用水泥加固,设在地下)使不受爆炸[热辐射]的伤害;涨价,(市场)稳定
~ed and tempered steel 调质钢
~(ed) case 表面掺碳硬化
~ed concrete （已）硬化混凝土
~ed mortar 硬化砂[灰,泥]浆
~ed plate 淬硬钢板
~ed steel 硬化[淬火]钢
flow ~ 冷变形硬化
hammer ~ 锤硬,冷作硬化
work ~ 加工硬化
hardening ['ha:dniŋ] *n.* 硬化,淬火,硬化作用;凝结[固];增加强度;掺碳;锻炼
~ -accelerating admixture 硬化加速剂

~ agent 硬化剂
~ by cooling 冷却硬化
~ by hammer 锤锻硬化
~ characteristic 硬化特性[征]
~ curve 淬火曲线
~ degree 硬化度
~ energy 硬化能力
~ machine 淬火机
~ oil 硬化油
~ rule 硬化保护层
~ treatment 硬化处理
case ~ 表面硬化法
chill ~ 冷硬法
full ~ 全硬化,淬透
graded ~ 分级淬火
heat ~ 淬火
negative ~ 低温退火,低碳钢软化处理
point ~ 局部淬火
quick ~ 速凝,快速硬化
water ~ 水淬(硬化)
Hardenfast *n.* 哈登法斯特混凝土快速凝结剂
hardener ['ha:dənə] *n.* 凝固剂,硬化剂,硬化成分;硬化土;硬化机
~ dust 硬化填充料
~ filler 硬化填充料
copper ~ （炼铝用）铜合金
floor ~ 铁屑;水泥地面加硬处理
hardhat ['ha:d'hæt] *n.* 安全帽,建筑工人 *a.* 头戴安全帽的
hardie *n.* 方柄凿[锤]
~ hole 锤[凿]柄孔
hardihood ['ha:dihu:d] *n.* 结实,大胆
hardiness ['ha:dinis] *n.* 结实,抵抗力,耐劳(性),耐寒(性)
Hardinge 哈丁
~ ball mill 哈丁型圆锥球磨机
~ conical mill 哈丁圆锥球磨机
hardish *a.* 较硬的
hardness ['ha:dnis] *n.* 硬[刚,强]度,坚硬性

～ ageing 加工[硬化]时效
～ class 硬度等级
～ determination 硬度测定(法)
～ gauge 硬度计
abrasion ～ 磨耗硬度,耐磨硬度
abrasive ～ 磨耗[磨蚀]硬度
diamond ～ 金刚钻石角锥硬度,维克斯硬度
dynamic ～ 刮刻硬度,冲击硬度,马尔特氏硬度
file ～ 锉刀硬度(用锉刀定级的硬度)
flinty ～ (＝glass ～) 燧石(火石)玻璃硬度
heating ～ 加热硬性
hot ～ 热硬度
impact ～ 冲击韧性
indentation ～ 压痕硬度
passive ～ 纯态硬度,耐磨性
rebound ～ 反弹硬度
strain ～ 冷加工硬度
wear ～ 抗磨强度,抗磨力
Hardnester n. 锉式硬度试验器
hardometer n. 硬度计
hardpan ['hɑːdpæn] n. 硬土层,硬盘,不透水层,硬质地层,坚固的基础;最低点,底价
hards [hɑːdz] n. (pl.)麻屑,亚麻粗纤维;硬质煤
flocks and ～ 纤维屑(塞缝隙用)
hardsite 地下场,地下设施
hardstand ['hɑːdstænd] n. 硬面层,硬路面,可停放车辆飞机等的坚硬地面
hardware ['hɑːdweə] n. 硬件,硬设备;(建筑用)小五金
～ cloth 钢丝网
～ pole line 架空明线的金具[金属附件]
aluminum ～ 铝制零件
hardwood ['hɑːdwud] n. 硬木[材],阔叶树木材 a. 硬木(制成)的
～ furniture 硬木家具
～ strip floor 硬木板条地板
～ tar 硬木焦油沥青
～ -tar pitch 硬木焦油脂
hardy ['hɑːdi] n. (锻工用)方柄凿 a. 坚固的,耐劳[寒]的
～ hole 锤[凿]柄孔
hare(e)m ['hɛərem] n. = harim, haram (伊斯兰教徒家中的)闺阁,后宫
harl(e) ['hɑːl] n. (由石灰,石子等混合成的,涂在建筑物外墙上的)粗灰泥;(墙的)粗灰泥表面;毛坯 vt. 用粗灰泥涂(墙等),制……的毛坯 a. (墙等)涂粗灰泥的,(计划等)草草作成的
harm ['hɑːm] n. 伤[危]害
harmful ['hɑːmful] a. 有害的
～ impurities 有害杂质
～ substance 有害物质
～ wastes 有害废物
harmless ['hɑːmlis] a. 无害的,无恶意的
harmonic [hɑːˈmɔnik] n. 谐波[音],调和, (pl.)谐[调和]函数 a. 调和的,和谐的
～ analysis 调和[谐波]分析
～ curve 调和曲线
～ echo 和谐回音
～ fold 谐和褶皱
～ proportion 调和比例
～ ratio 调和比
～ ringing 调谐信号
harmonical [hɑːˈmɔnikəl] a. = harmonic 谐波的,调和的,悦耳的
～ vase 调和的(花)瓶饰
harmonize ['hɑːmənaiz] v. 调准,协调,(使)调和,(使)一致
harmony ['hɑːməni] n. (pl. -nies)谐[调]和,一致,融洽,和声[学]
colour ～ 色彩调和
harness ['hɑːnis] n. 导线,导线系统;装具;背带 vt. 利用(水、风等)作动力;(河流等的)开发,治理
safety ～ 安全吊带
shielding ～ 屏蔽系统
wiring ～ 装配电路
harp [hɑːp] n. 装煤铲;集电器滑轴夹

筛;形似竖琴的物件
~ mesh 平行啮合
~ (-type) screen 竖琴[平行弦]式筛
harrow ['hærəu] *n*. 耙,路耙;耙路机 *vt*. 耙平
acme ~ 刀齿耙,弯刀齿耙
jointed ~ 组合[分节]耙
harsh [ha:ʃ] *a*. 粗的,粗糙的;刚性的;严厉的,生硬的;崎驱不平的
~ aggregate 粗骨料
~ concrete 干硬性混凝土
~ mix 干硬性拌合料,粗颗粒混凝土
~ sand 粗砂,有棱角砂
~ concrete (粗骨料过多的)粗糙混凝土,干硬[粗粒]混凝土
~ mix 粗糙搅拌;干硬性混合料粗颗粒混凝土
~ mixture 干硬性混合料;粗糙搅拌
harshness ['ha:ʃnis] *n*. 轮胎耐(路面)粗糙性;(手感)粗糙度
harstigite *n*. 镁[铍]柱石
harvest ['ha:vist] *n*. 收成[获],产量;收获时节 *v*. 收获[割]
Harvey steel 固体渗碳硬化钢
hash [hæʃ] *n*. 噪声,杂乱信号,杂乱[无用]数据;复述,重申;传闻 *vt*. 把……弄乱;仔细考虑
~ -house 经济餐馆
hasp ['ha:sp] *n*. 铁扣,(门窗,箱子等的)搭扣,钩;纺锭 *vt*. 用搭[铁]扣扣上
~ and staple 铁钩和钩环,搭扣
~ iron 铁钩
~ lock 搭扣锁
haste ['heist] *n*. 匆[急]忙,紧迫;轻率;催促,急速,赶快
hastelloy *n*. 耐盐酸[耐蚀,耐热]镍基合金
ceramic ~ 陶瓷耐蚀耐高温镍基合金
hasten ['heisn] *vt*. 催促,促进 *vi*. 赶快,急行
hastings sand *n*. 哈斯丁斯(英格兰南部一城市)沙滩;铁矿砂[石],菱铁矿
hasty ['heisti] *a*. 急速的,短暂的,草率

的,赶制成的
hat [hæt] *n*. (有边的)帽子;(采矿的)顶板;随机编码 *vt*. 戴帽子
~ flange 带帽法兰
~ orifice 圆柱形锐孔
~ rack 帽架
box ~ 钢壳,靴筒(钢锭浇铸后因收缩而产生的缺陷)
hard ~ 安全帽
top ~ 帽形钢锭缺陷
hatch ['hætʃ] *n*. 闸门,沉箱的水闸室,舱口[盖];升降器口;天窗;人孔铁口;阴影线 *v*. 书阴影线;图谋,策划,计划,安排
~ grating 人孔格栅
escape ~ 应急出口,太平门
turret ~ 转塔顶门
hatched *a*. 书有阴影线的;孵出的
~ area 书有阴影线的面积,阴影部分
~ grafting 镶接
hatchet ['hætʃit] *n*. 短柄小斧,手斧
hatchety ['hætʃəti] *a*. 斧状的
hatching ['hætʃiŋ] *n*. (细平行线形成的)阴影,影线法
cross ~ 断面线
hathoric column (= Hathoric type of capital)爱神柱
hathpace ['hæθpeis] *n*. 讲[平]台
haul [ho:l] *n*. 用力拖拉;运距[输],拖拉的距离;所拖拉的量(尤指一网所打的鱼) *v*. 拖,拉,曳;运输;改变航向
~ up 木踏板绳梯
haulage [ho:lidʒ] *n*. 运输,搬运(量);运费,货车使用费;运输方式
~ appliance 拖运工具
~ cable 牵引索,拖缆
hauler ['ho:lə] *n*. 运输机,起重机;承重人,承办陆路货运者;运输工;拖曳物[者]
haulier ['ho:ljə] *n*. = hauler(英)(货运)承运人,运输工,拖曳者
hauling 拖,拉,牵引;运输,搬运,拖运;运费;运法;提升;牵引集材;拉[赶]钢

~ capstan 绞车,卷扬机
~ chain 拖链
~ container of concrete 混凝土运送容器
~ engine 牵引机[车]
~ equipment 运输设备
~ machine 拖运[牵引]机
~ plant 运输设备[装置]
~ unit 运输[工具]
~ winch 升降绞车

haunch ['hɔ:ntʃ] n. 腰,腰部;拱腰,梁[拱]腋;(pl.)后部,臀部 vt. 加腋,加托部

haunchs

~ of arch 梁[拱]腋,拱的托部,拱背圈
~ -up 拱起
~ed arch 加腋拱
~ed beam 加腋梁,变截面梁
~ed member 加腋构件,突起构件,托梁构件
~ed tenon 加腋榫,腋脚凸榫
~ing 加托臂,(道路上)混凝土镶边

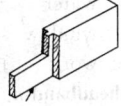
haunched tenon

~ing coefficient 加腋系数
tapered ~ 斜腋

haunching 护拱(填土层);(梁等)加腋,端部加厚,加托臂;(道路上)水泥混凝土镶边

Hauser alloy 郝氏易熔合金(铅50%,铋33.3%,其余为镉)

Hausma(n)nite n. 黑锰矿

havoc ['hævək] n. (自然力造成的)大破坏,毁坏,严重灾害,浩劫,混乱,哈佛克硅钼钒钢(0.5%碳,1.0%硅,0.2%钒,0.5%钼,其余为铁) v. 破坏,毁灭

hawk [hɔ:k] n. 镘灰板,托灰板(泥工用) v. 散布(消息);兜售

hawkbill ['hɔ:kbil] n. 坩锅钳,铁钳

hawser ['hɔ:zə] n. (系船,下锚用)粗绳,钢丝索,缆索,锚链
~ bend 缆索捆绑结头

hay [hei] n. 篱笆,围篱;干草 vt. 供给干草 vi. 制干草
~ -band 草绳(缠铸铁管用)

haydite ['heidait] n. 陶粒

haywire ['heiwaiə] n. 临时电线 a. 失去控制的

hazard ['hæzəd] n. 危险,公害,冒险;(意外)事故;未知数,不可预测 vt. 冒……的危险,危害;赌博
~ warning 危险警告灯
fire ~ 易燃性,易引起火灾的物

hazardous ['hæzədəs] a. 危[冒]险的
~ condition 危险情况
~ noise 危害性噪声
~ operation 危险作业

haze [heiz] n. 薄雾,朦胧;(思想等的)模糊不清 v. 使雾笼罩;混浊,变浊;起雾

hazel ['heizl] n. 榛,榛木;淡褐色(尤指眼睛)
~ pine 苏合香木材

haziness n. 模糊,朦胧,浊度

hazy ['heizi] a. 模糊的,朦胧的;烟雾弥漫的

H-bar 工字[H型]钢

H-beam 工字梁

H-beam bunton 工字钢罐道梁,工字梁横撑

H-beam cap 工字钢顶梁

H-beam Pile 工字(钢)桩

H-beam steel 工字梁钢

H-bit H型钻头

H-block 调整间隙垫板

H-columm 工字[H]型柱

H-columm with covers 宽翼缘加板工字型钢柱

H-display H型显示,分叉点显示

head [hed] n. (pl. heads or head) 人[动物]的头,头部,顶端[部],前部[端];上部[端];压头,(静)水头,水压;落差,扬程,蓄水高度;渠首,闸首;导

洞,水平巷道;标题,项[条,题]目,要点,方面;顶砖,拱心(石);装置,设备;首长,领导,首席[位],主任;危机,极点;才智,头脑 v. 在……的前头[顶部],为……之首,率领,使……面向(而行);主持,领导;构成……顶部,在……上加标题;遮挡,妨碍 a. 头(部)的;主要的,首席的

~ attachment 头部连接(法),顶部焊接;头部附件
~ band 拱门饰
~ bay wall 上游墙
~ beam 上横梁,顶横杆,顶梁,露梁,扶手
~ block 垫块,辙尖枕木,尖轨长枕
~ board 护顶木板,(卧车中将卧铺隔开的)轻便隔板,端面板
~ bond 丁砖砌合,全丁砖砌合法
~ casing 门窗头框
~ clearance 顶部净空
~ construction 头部连接(法),顶部焊接;头部结构
~ counterbore 顶部锥口孔
~ course 丁砖层
~ cover 顶盖
~ driller 顶部钻孔机
~ end 起点[初步,预备]的
~ face 端面
~ flashing 头顶防水板
~ -house 井口建筑物;候车室;井楼;温室控制闸;突堤码头(堤头)
~ jamb 门顶框
~ joint 端部接缝,横缝
~ log 设置在滑道前端底部的圆木,梢头材,前缘木
~ mast 吊塔,前桅,前柱
~ metal 冒口,切头
~ mo(u)lding 门窗头线(孔口的)顶部装饰线条
~ of casing 窗框上槛
~ of column(= head piece of column) 柱头
~ of pile 桩顶[头]

~ of rivet 铆钉头
~ screen 焊工面罩
~ shaft 主动轴,驱动轴
~ slab 盖[顶]板
~ space 顶空
~ wall 端墙,山墙
~ wear 头饰,帽子
~ed bolt 露头螺栓
~ed test specimen 突头试件
bolt ~ 螺栓头
bottom ~ 底盖
cable ~ 电缆分线盒,电缆接头
closure ~ 外壳顶盖
conduit ~ 贮箱
cross ~ 十字头
door ~ 门楣
gib ~ 螺栓头,钩头,销子头
joist ~ 搁栅忱
split ~ 裂口
temperature ~ 热位差,温度差
water ~ 水头,水位差,水柱高度
window ~ 窗楣
work ~ 工作台

headband n. 装饰顶带,书页[章首]花饰
headbeam ['hedbi:m] n. 顶梁
headbox n. (百叶窗)操纵机构罩壳;(造纸机的)料箱
headcut n. 沟头切割
header ['hedə] n. 半端梁搁栅;露头砖[石];首部结构;封头;头沟;水箱,蓄水池,锻造机;集水[氧,流]管
~ bland 镦锻坯料
~ bolt 冷锻[墩]螺栓
~ bond 丁砖砌合
~ brick 丁砖
~ course 露头层,丁砖层[行]
~ face 端面
~ joint 丁砖砌合,丁砖缝
~ joist 丁头搁栅
~ pipe 集[总]管
~ stone 露[丁]头石
discharge ~ (增压)集气管

blind ~ 假顶砖,两墙间的空心墙加固顶砖
bull ~ 露丁侧砖
false ~ 假丁砖
joist ~ 格栅枕
outlet ~ 泄[排]水干管
snapped ~ 半顶砖
spray ~ 喷嘴集管
stone ~ 石丁头
timber ~ 木丁头

heading ['hediŋ] *n.* 露头;导坑[洞],水平巷道,地下通道,掌子面,风道;标题,题目;(飞行)方向,方位;顶锻,镦头
~ beam 顶梁
~ bond 丁砖砌合
~ course 露丁砖层,丁砖砌层
~ die 镦粗[锻]模
~ joint 端接,直角接合

headlamp ['hedlæmp] *n.* 头灯

headlap *n.* 搭接重叠(部分)

headless ['hedlis] *a.* 无头的,没有领导的
~ nail 无头钉
~ package 无边盘卷装
~ rivet 无头铆钉
~ set screw 无头止动螺钉

headlike ['hedlaik] *a.* 形状或功能像头的,头状的

headlong ['hedlɔŋ] *a.*, *ad.* 急速,匆促;头向前

headmost ['hedməust] *a.* 领头[最先]的,最前面的

headpiece ['hedpi:s] *n.* 顶[横]梁;上[顶]部;床头板;门[窗]的楣;耳机,头戴受话器

headpin *n.* 头柱

headpost *n.* 前桅,前[端]柱

headquarters ['hed'kwɔ:təz] *n.* (*pl.*) 总部[店],司令[指挥]部

headrace ['hedreis] *n.* 进水渠,引水道,上游水道;门框上帽头

headrig *n.* 主装备

headroom ['hedrum] *n.* 上游闸室,净空,净空[开采]高度

headsheave ['hedʃi:v] *n.* 主滑轮

headsill *n.* 上槛(门或窗的框顶上的横梁);锯木坑中支撑圆木两端的垫块

headspace ['hedspeis] *n.* 预留空间;顶部空间

headstay ['hedstei] *n.* 前支架索

headstone ['hedstəun] *n.* 墙基石,拱心石

headwall ['hedwɔ:l] *n.* 山墙,拱面墙;斗壁

headward ['hedwəd] *a.*; *ad.* 朝向头部,溯源的

headway ['hedwei] *n.* 净空高度(由地板至顶棚或由地面至拱顶的垂直距离);横巷;前后两车时距

heady ['hedi] *a.* 上头的;顽固的;迷惑人的;猛烈的

heal [hi:l] *v.* 修复,使复原,治愈,愈合,(裂缝)合拢;用土覆盖,用石板或瓦片覆盖
~ing agent 修补剂
~ing property 复合[恢复]性

healant ['hi:lænt] *n.* 修补剂

healing ['hi:liŋ] 治疗,痊愈,(裂缝)合拢,使恢复;养护;(裂缝)焊合
~ stone 屋顶岩板

health [helθ] *n.* 健康(状况);卫生;健全
~ insurance 健康保险
~ resort 休养地,疗养地

healthy ['helθi] *a.* 健康[卫生]的,有益健康的;(道德或精神上)健全的;安全的,大量的

heap [hi:p] *n.* 堆,块,土堆,矸石堆;堆积;大量,群(众) *vt.* 堆积,积聚;装载[满];倾泻 *vi.* 形成堆
~ capacity 装载容量
~ of rubble 块[毛]石堆
~ of tripod 三脚架头
~ sand 填充砂,铸造用砂
~ together 堆积[起]
~ type 堆叠式
~ up 堆积[起]

~ed concrete 成堆[未摊铺]混凝土
heart [hɑːt] *n.* 心[脏],内心,中[核]心;精华,要点,本[实]质;心形物;(土地的)生产力 *vt.* 把……安放在中心部
~ and square 平勺,心形馒刀
~ board 带心板材,髓心板
~ bond 心砌合(法)中心结合
~ check (木)心(辐)裂,内部裂缝
~ plank 厚心板(厚 5～15cm,宽大于 23cm)
~ shake 木心环裂
~ wall 心墙
~ wood (=duramen) 心材[木]
~ wood-rot 心材腐蚀
~ wood tree 心材树[木]
~ing concrete 填心混凝土
hearth [hɑːθ] *n.* 炉,炉(火)床;坩埚;壁炉地面
~ bottom 炉底
~ melting furnance 床[镗]式熔炉
~ stone 炉石;磨石
heat [hiːt] *n.* 热量[度],热辐射;热学;暖气,保温;热(激)烈;炉子的容量,装炉(量);一次熔炼,一炉(钢水) *v.* 热,热处理,加[供,变,发]热;交换[传处,转化,吸收]热
~ absorbing glass 保温[吸热]玻璃
~ absorption capacity 吸热能力
~ actuated fire door 热动防火门
~ -affected zone 温度影响区域
~ balance 热平衡
~ balancer 热平衡器
~ booster 增热器,加热器
~ circulation 热循环
~ crack 热致裂缝
~ -cured 加热养护的
~ dissipating capacity 散热量
~ distortion 热致扭曲
~ drying (污泥)加热干燥

~ endurance 耐热性[度]
~ exchanger 热交换器
~ expansion 热膨胀
~ guard 绝热体
~ insulating block 绝热块体
~ insulating concrete 绝[隔]热混凝土
~ insulating layer 绝热层
~ insulating meterial 绝热材料
~ insulating slab 绝热板
~ insulation 绝热,热绝缘
~ insulator 绝热体,隔温层,热绝缘体
~ pollutipn 热污染
~ proof 耐热(的),不透热(的),防热的,保温
~ resistance index (沥青混凝土的)抗热指数
~ resistance cement plate 耐热水泥板
~ resisting paint 耐热油漆
~ resisting steel 耐热钢
~ set 热定形
~ sink 散热片,吸热器(装置)
~ storage quilt 保暖垫
~ supply pipeline 供热管道
~ supply system 供热[暖]系统
~ switch 供热开关
~ tinting 烘染
~ed air intake system 冷热空气调温进气系统
heater ['hiːtə] *n.* 加[预,发,放]热器,加热丝;暖气[保暖]设备;热源;加热元件;灯丝;加热工;燃烧室
~ casing 加热炉罩,暖气设备罩
~ coil 加热线圈(盘管)
~ housing 加热炉罩,暖气设备罩
~ planer 加热整平机(黑色路面加热整平用);烫平机
air intake ~ 进气加热器
direct contact feed ~ 给水直接加热器
electric ~ 电炉,电热器

gas ~ 煤气炉
primary ~ 预热器
space ~ 空间对流加热器
surface ~ 暖面器
tyre ~ 轮胎热压器,热箍器

heating ['hi:tiŋ] *n.*; *a.* 加[受]热(的),加温;自热;暖气(装置);电热元件
~ alloy 合金电热丝
~ and ventilation system 供暖与通风系统
~ appliance 供热设备,暖气装置
~ apparatus 加热器
~ bath 热浴,加热池
~ by town gas 城市(家用)煤气供热
~ by waste heat 废气供热
~ capacity 发热量,供热[暖]能力
~ chamber 暖气室
~ circuit 供热[暖]系统[范围]
~ coil 暖管,供暖盘管
~ device 加热器[装置],发热器
~ grid 供热管道
~ hood 暖气装置外壳[防护套,挡板]
~ installation 加热[供暖]装置
~ jacket 暖气装置外套
~ line 供热[暖]管道,暖气管道
~ main 暖气总管,供暖干管
~ mantle 暖气罩,加热外罩
~ pipe 暖气[供暖]管
~ pot 加热壶[锅]
~ radiator 暖气片
~ season 供热[暖]季节
~ spiral (加)热盘管
~ surface 受[加]热面
~ system 供热[暖]系统
~ tape 绝热带,绝热胶布
~ -ventilating assembly 供暖通风(两用)机组

heatronic welding 高频(率)电(解质加)热焊接

heave [hi:v] *n.*; *v.* 举[拉,抬]起,鼓[胀,隆,挺]起,(道路)冻胀;蠕[徐]变,潜移[伸],塑性变形;水平移动;平错(地质学名词)

heaven ['hevn] *n.* 天,天空,大气;天国,(基督教)天堂;极其赏心乐意的地方[条件,时期];超凡的宇宙或区域;宇宙法则;宇宙伦理原则
~ly bamboo 南天竹(天竹子)

heaver ['hi:və] *n.* 举者,挑夫,举起重物的装置;杠杆,小铁链,大秤

heavily ['hevili] *ad.* 重[沉]重地,缓慢地,大量地,密集地
~ coated electrode 厚涂层焊条
~ cracked 深度裂化的
~ polluted area 严重污染区域
~ reinforced 大量配筋的,重配筋的,(钢筋混凝土)超配筋的;大大加强的,重重加强的
~ supported 坚牢固定的,安全支架的
~ wooded area 树木很密的地区

heaviness ['hevinəs] *n.* 可称性,有重量性,有质性;沉重,重量

heavy ['hevi] *n.* (*pl.* heavies)重物 *a.* 重[载,型]的,负重的,浓的,大量的,难行的(道路),猛烈的 *ad.* 沉[笨]重地,大量地
~ aggregate 重集[骨]料
~ aggregate concrete 重[骨料]混凝土
~ alloy 重[高密度]合金
~ asphalt 稠[重]地沥青
~ asphaltic (base) oil 重沥青(基)油
~ -bedded 厚铺的
~ bitumen 重质沥青,稠沥青
~ bodied 粘滞的,粘的
~ building 构造坚固,结构结实
~ casting 大型铸件
~ ceramics 重质陶瓷
~ clay 重黏土
~ clay article 重黏土制品
~ clay flooring tile 重质黏土铺地瓷砖
~ concrete 重混凝土,重集料(如重晶石)制成的混凝土(防辐射用)
~ construction 重型建筑,大型工程

~ duty floor 使用频繁的楼板
~ duty frame 重型构架
~ duty scaffold 重型脚手架
~ earth 重晶石,重土
~ -edge reinforcement 边缘加强钢筋
~ enamel single-glass 厚漆单层玻璃(的)
~ ga(u)ge metal (按英国标准线规规定的)厚级板材,厚(金属)板
~ ga(u)ge roofing sheet 加厚屋面板,重型屋面板
~ ga(u)ge wire 粗导线
~ iron 厚镀层热浸镀锌铁皮
~ joist 重型搁栅;料板(厚4英寸宽8英寸以上的制材)
~ jute mats 厚麻布,重麻席
~ line (图表中的)粗线,黑线
~ mallet 重锤
~ mortar 稠灰浆,稠(水泥)砂浆
~ pine (pinus pondarosa) 美国西部黄松
~ plate 厚钢板,厚板
~ reinforcement 大量配(钢)筋
~ section prop 重型支柱
~ -section stringer 大截面桁条,粗桁条
~ shade 饱和色
~ sheet glass 厚平板玻璃
~ soil 重质土,黏质土
~ spar 重晶石
~ surface treatment 加厚表面处理(层)
~ tar 重质柏油,厚柏油,重质焦油沥青
~ textured soil 黏性土
~ timber construction 重木结构,大木建筑
~ walled 厚壁的,厚墙的
~ weight concrete 重混凝土

heavy timber construction

~ wood 重木材
hectare ['hekta:] *n.* (简 ha.)公顷(等于一百公亩,即一万平方米,或等于15市亩)
hectogram (me) ['hektəugræm] *n.* 百克(＝1/10公斤)
hectoliter ['hektəuli:tə] *n.* 100公升
hectometer ['hektəumi:tə] *n.* 百米
hectostere 百立方米(法国度量名)
hedge ['hedʒ] *n.* 树篱,栅栏,障碍物 *v.* 用篱笆[栅栏]围住[分开],包围;设障碍于……;推诿,躲闪
~ trimmer 修边[剪,整]机
heel [hi:l] *n.* 拱座,柱脚,坝踵,挡土墙(迎土面)底部,上游[迎水]坡脚,根部;钻井口;任何器具(特指平地机刮刀)的近柄处;倾向一边的动作 *v.* 加(后)跟;紧随;(使船)倾侧
~ block 垫块[板]
~ boom 底脚吊杆
~ contact 齿跟部啮合(印痕),踵形接触,锥齿轮的大端接触
~ of metal 熔金属面
~ post 门轴柱,拄脚;承重柱,带加固环的支柱,(闸门的)侧立柱
~ rope 杆底绳,绊马索
~s seat 跟座
~s slab 柱脚板,后部底板
~s stay 防滑垫
height [hait] *n.* 高度,海拔,高程,顶点;(*pl.*)高地[处],丘
~ between stories 楼层高
~ cut-out 开孔高度
~ -diameter ratio 高(度)-(直)径比
~ finder 测高仪[计]
~ gauge 测高仪
~ indicator 高度计[指示器],测高仪
~ mark 高度标记,标高
~ measurement 高程测量,测高仪
~ of arc 矢高
~ of arch 拱高
~ of camber 路拱高度,拱矢高
~ of course 层高

~ of deposit 沉[淤]积高度
~ of drop 落差
~ of layer 料层厚度
~ of level 水平面标高,水准仪高
~ of life 提升高度;一次铺筑的厚度,混凝土浇筑块高度
~ overall 全高,净高,总高度
~ scale 高度比例尺
~ under hook 钩下高度
clear ~ 净高[空]
effective ~ 有效高度
operational ~ 工作[额定]高度
true ~ 真(实)高度,实际[几何]高度
heighten ['haitn] v. 增[提,升,加]高,增大[加],加强;使出色,(使)变显著
held [held] n. (工具的)柄榫头
heldwater n. 吸着[粘滞]水
helenite n. 弹性地腊
heliarc ['hi:lia:k] n. 氩弧
~ cutting 氩弧切割(法)
~ welder 氩(弧)焊机
helical ['helikəl] n. 螺旋,螺旋面[线,形],螺线 a. 螺旋(形)的,螺纹的
~ architecture 螺旋形建筑
~ auger 螺纹钻
~ bevel gear 螺旋伞齿轮
~ burr 螺纹
~ compression spring 压力盘簧,螺旋压(力弹)簧
~ convolute 螺旋面
~ curve 螺旋曲线
~ flash lamp 螺旋闪灯
~ (out) lobe 螺旋(外)叶

helical flash lamp

~ reinforcement 螺旋钢筋
~ screw spreader 螺旋形(粒料)撒布机
~ -shaped bit 螺旋形钻

~ sounding bore 螺旋形触探钻
~ spring 螺旋弹簧,盘簧
~ stair 盘梯
~ welding 螺旋焊接
helically ad. 成螺旋形
~ grilled tube 伸缩管,连接弯管
~ welded tube 螺旋缝焊接管
helicline ['heliklain] n. 盘旋斜坡道
helicogyre = helicogyro n. 直升飞机
helicoid ['helikɔid] n. 螺圈,螺旋面,螺(旋)状体;a. 螺(旋)纹的;helicoidal, a. helicoidally ad.
helicoidal [heli'kɔidl] a. 螺旋形的
~ flow 螺旋流,螺旋状水流
~ saw 石工锯
helidrome ['helidrəum] n. 直升飞机降落场[机场]
heliocentric(al) [hi:liəu'sentrik (əl)] a. 以太阳为中心的,以日心测量的;heliocentrically, ad. heliocentricity, n.
heliochrome ['hi:liəukrəum] n. 彩色照片,天然色照片;heliochromic, a.
heliodon n. 日影仪
heliogramma n. 日照纸
heliograph ['hi:liəugra:f] n. 日照计,日光仪,(拍太阳用的)太阳摄影[照相]机,日光反射信号器;v. 以日光反射信号机通讯;heliographic, heliographical, a.; heliographer, n.; heliographically, ad.
heliogravure ['hi:liəugrə'vjuə] n. 凹版摄影
heliolamp ['hel'ə'æmp] n. 日光灯
heliolatitude ['heliəlætitude] n. 日面纬度
helioplant n. 太阳能利用装置
heliosphere n. 日光层
heliostat ['hi:liəustæt] n. 定日镜,回照器
heliothermometer n. 日温量测计
heliotrope ['heljətrəup] n. 回照器,回

光仪;淡紫色,紫红色
heliotropic *a.* 趋[向]日性的
　～ wind 日转[成]风
heliotropism [ˌhi:li:'ɔtrəpizəm] *n.* 向日[光]性
helium ['hi:ljəm] *n.* 氦(He)
helix ['hi:liks] *n. pl.* helixes, helices;螺旋线(结构),螺管(旋,弹簧),螺旋形(之物);(柱头的)涡卷
　～ angle 螺旋角

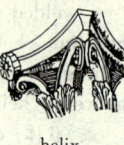
helix

helm [helm] *n.* 驾驶盘,舵轮。枢机,*vt.* 掌舵,掌握(枢机),指挥;驾驶;helmless, *a.*
　～ roof 四坡陡屋顶,尖屋顶,顶层
helmet [helmit] *n.* (安全,防护)帽,盔,压力盔形图;箍,环; helmeted, *a.*; helmetlike, *a.*
　～ liner 钢盔帽
　～ shaped 盔形的(一般多指应力分布图)
　～ed 戴头盔[安全帽,护面罩]的
　diving ～ 潜水帽
　smoking ～ 防毒面具,烟罩
　welding ～ 焊工帽
helobious = helodious *n.* 沼泽地
helodes *n.* 沼地的
helophyte ['heləfait] *n.* 沼生植物
help [help] *n.; v.* 帮助,资助,救助(济);促进,助长;治疗;阻止,避免(通常与 can, can't 连用); helpable, *a.*; helper, *n.*
　～ing wind 顺风
helper ['helpə] *n.* 助手,救助者,辅助机构;辅助炮眼
　～ grade 辅助坡度
　～ spring 辅助弹簧
　～ stringer 助纵梁
　～ -up 推车工助手
helpmate ['helpmeit] *n.* 助理人员,助手

helve [helv] *n.; v.* helved, helving; *n.* 镐柄,斧柄; *vt.* 装柄于…… helver, *n.*
　～ hammer 杠杆锤,摇锤
hematite ['hemətait] = haematite *n.* 赤铁矿(Fe_2O_3); hematitic, *a.*
hemerocology *n.* 人工环境生态学,人为地面形态
hemicolloid *n.* 半胶质[体]
hemicrystalline *a.* 半结晶的,半晶状的,半晶质
hemicycle ['hemisaikl] *n.* 半园形,半圆结构[建筑物]
hemidome *n.* 半圆屋顶,半坡面
hemihydrate *n.* 半水化合物
　～ plaster 熟石膏灰泥
hemiprism *n.* 半棱柱
hemipyramid *n.* 半棱锥体
hemisphere ['hemisfiə] *n.* 半球;半球地图[模型];(知识活动的)范围[领域]
hemispherical [hemi'sferikəl] *a.* 半球状的,半球形的
　～ coverage 半球形视野,半球形射程
　～ dome 半球形圆顶,半圆体屋顶
　～ dome structure 半球形圆顶结构
　～ projection 半球投影
hemlock ['hemlək] *n.* 铁杉木
hemp [hemp] *n.* (大),麻,苎麻,麻絮[屑]; hemplike, *a.*
　～ core 麻心
　～ cut 麻筋,麻刀
　～ hose 麻织水龙带
　～ packing 麻垫料,麻(丝封)填
　～ palm 棕榈
　～ rope 麻绳
　～ -twist 麻绳
　Manilla ～ 马尼拉麻
henry ['henri] *n.* (*pl.* -ries,-rys);亨,亨利(电感单位)
　～ meter 电感计,亨利计
hepatic [hi'pætik] *a.* 肝的,肝褐色的
　～ cinnabar 朱砂-硫化汞,硫化汞矿
　～ gas 硫化氢
heptagonus *n.* 七角形

heptahedron ['heptə'hedrən] *n.* 七面体

heptastylos [ˌheptə'stailɔs] *n.* 有七根柱子的建筑物

heptastylos

herbicide ['hə:bisaid] *n.* 除莠[草]剂;herbicidal *a.*

herbosa *n.* 本植被,草丛

herculite *n.* 钢化玻璃

heritable ['heritəbl] *a.* 可转让[继承]的,遗传性的

heritage ['heritidʒ] *n.* 传统,遗传,继承物

heritor ['heritə] *n.* 继承人

herm [hə:m] *n.* 古希腊刻有汉密士神象的石碑

hermetic(al) [hə:'metik(əl)] *a.* 密封的,不透气的
~ material 不透气材料
~ seal (真空)密封接头,气密封接头

hermetization [hə:meti'zeiʃən] *n.* 密封,封闭

hermitage ['hə:mitidʒ] *n.* 僻静的住处,隐居之处

herringbone ['heriŋbəun] *n.* 人字形,鲱骨(式); *a.* 人字形的,鲱骨状的; *v.* 作人字形,作交叉缝式
~ bond 人字形(式)砌合
~ brickwork 错缝砌体,人字形砌砖,交叉缝砌砖
~ bridging 人字撑
~ drainage system 人字形排水系统
~ dressing 人字纹修琢(石面)
~ earth 鱼骨形接地
~ fashion 人字式[形]
~ joint 人字榫
~ masonry 人字铺砌圬工
~ mesh opening 人字形钢孔
~ parquetry 人字拼木地板,席纹地面
~ pavement 人字式(铺砌)路面
~ paving 人字式铺砌法

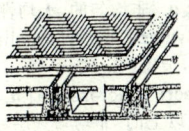
herringbone parquetry

~ strut 人字形支撑
~ strutting 搁栅撑,人字撑
~ system 人字形排水系统,鲱鱼骨形排水系统
~ wharf layout 人字式码头布置

hertz [hə:ts] *n. pl.* hertz, hertzes; (简 Hz)赫兹(频率单位,以每秒周数c/s计)

hesitation [ˌhezi'teiʃən] *n.* 犹豫,暂[短]停
~ set (水泥[混凝土])假凝

hessian ['hesiən] *n.* 浸沥青的麻绳,打包麻布,粗麻屑;砂坩锅
~ -based bitumen sheeting 粗麻布底的油毛毡
~ rope 麻绳

heterism *n.* 殊异性

heterization *n.* 异化

heterobrochate *n.* 异形网状

heterogeneity [ˌhetərəudʒi'ni:iti] *n.* 不均匀(性),多相(性),不同性质,不均质

heterogeneous ['hetərəu'dʒi:niəs] *a.* 不均匀的,非均质的,掺杂的,异种的,多相的,异次的
~ anisotropic medium 非均匀各向异性介质
~ body 非均质体
~ foundation 非均质地基
~ ground 非均质基础
~ mixture 不均匀混合料
~ sand 非均质砂
~ shear layer 非均匀剪力层
~ soil 非均质土
~ state 非均匀态
~ strain 非均匀应变
~ structure 多相组织
~ system 非均匀系统,多相系

heterophase *n.* 多相

heteropic *a.* 非均匀的,无均性的

heteropolymer [ˌhetərəu'pɔlimə] *n.* 异形[态],杂聚(合)物

heterosphere ['hetərəsfiə] *n.* (在72～80公里以上的)非均匀气层

heterostatic *a.* 异位[势]差的

heterostrophy *n.* 以与常规方向相反的方向卷绕所得的品质或状态

heterotaxy ['hetərətæksi] *n.* 地层变位
heterotaxic, *a.*

heterotopic *a.* 斜交的

heterotype *n.* 同类[型]异性物

heuristic [hjuə'ristik] *n.* 启发式的论据,启发法的科学与艺术; *a.* 启发式的; **heuristically**, *ad.*

heuristics *n.* 启发法,直观推断

hew [hju:] *v.* hewed, hewed 或 hewn, hewing; *vt.* 砍劈,斩[砍[削]成;人工采掘;方材,坡方; *vi.* 砍,削;奉行,遵守; **hewable**, *a.*

～ away 砍去,斩去

～ down 砍倒

hewn [hju:n] *a.* 粗削的,砍劈过的,(石料)粗琢的

～ squares 劈枋[材],粗角[材]

～ stone 粗削石,毛石

～ timber 披材[方],砍平的原木,方木

hexabolite *n.* 玄闪石

hexagon ['heksəgən] *n.* 六角[边]形; **hexagonal**, *a.* **hexagonally**, *ad.*

～ bar iron 六角钢

～ bolt 六角形螺栓

～ headed bolt 六角头螺栓

～ steel 六角形钻钢

hexagonal *a.* 六角[边]形的

～ bar 六角(形)铁杆

～ head screw 六角头螺钉

～ pile 六角桩

～ steel bar 六角钢条

hexagram ['heksəgræm] *n.* 六角星形,六线形;
-hexagrammoid, *a.* ;*n.*

hexahedron ['heksə'hedrən] *n. pl.* -drons,-dra;六面体;
-hexahedral, *a.*

regular ～ 正六面体,立方体

hexastyle ['heksəstail] *a.* 有六柱的,六柱式的

hexastyle

hexastylos [heksə'stailəs] *n.* (古代寺庙)有六根柱子的建筑物

hexyl ['heksil] *n.* 乙基,六硝炸药;
-hexylic, *a.*

～ alcohol 乙醇($C_6H_{13}OH$)

H-girder 工字[H]型梁

H-hinge 工字铰链

hiatus [hai'eitəs] *n. pl.* -tuses,-tus; *n.* 裂缝,间断[隙],缺失,罅隙;脱句[字]

hibernal [hai'bə:nl] *a.* 冬季的,寒冷的

hickey ['hiki] *n. pl.* -eys;(电器上的)螺纹接合器,弯管器

hickory ['hikəri] *n. pl.* -ries;胡桃木;一种坚固的棉织物

～ shirt 工作服[衫]

hicktown' [hiktaun] *n.* 远离大都市的乡镇

hicore ['hikɔ:] *n.* 不锈钼铬钢

hide [haid] *n.* ; *v.* hid, hidden 或 hid, hiding; *n.* 隐匿处; *v.* 隐匿[藏,瞒],遮掩,庇护,潜伏,守秘密; **hidable**, *a.* **hider**, *n.*

～ -out 掩蔽所

～ -den danger 隐患

～ -den line 隐线,虚线

～ -den microphone 窃听器

～ing place 躲[储]藏处

～ing power (油漆等的)遮盖[被覆]力

hidebound [haidbaund] *a.* 墨守成规的,死板的,偏狭的;紧皮的(树木); hideboundness, *n.*

hiduminium [haidju'miniəm] *n.* 铝铜镍合金,RR合金

hierarchy ['haiəra:ki] *n. pl.* -chies;体系[制],系统,谱系;分层,阶层;等级制度;科学的分类(如门、纲、目、科、属、种等)

~ of road 道路网主干线

analytic ~ process 层次分析法

high [hai] *n.* 高(气)压,高压圈[带];高峰[潮];高水准;高处[地];天空;大数字; *a.* 高的,高度[级,等,价,地,纬度]的;强烈的[非常,重大]的; *ad.* 高,大,强,高度[价]地,显著地

~ alkali cement 高碱水泥
~ alkalinity 高碱度
~ alloy steel 高合金钢
~ -alumina brick 高铝砖
~ -alumina cement 高矾土水泥
~ -alumina refractories 高铝耐火材料
~ and low water alarm 水位报警器
~ angle fault 陡角断层
~ back 高靠背
~ barometric maxima 高气压的最高值
~ bed 沙洲,浅滩
~ bond reinforcing bars 高握裹力钢筋
~ brass 高锌黄铜
~ -brown 鲜明棕色的
~ calcium lime 高钙石灰
~ -capacity 大容量
~ carbon steel 高碳钢
~ carbon tar 高碳焦油,高碳焦油沥青
~ -cement concrete 富灰混凝土
~ confining-pressure test 高侧限压力试验
~ content alloy 高合金
~ correlation 高度相关

~ -crowned arch 高顶拱
~ current 大电流,强电流
~ cut 深挖(土)
~ -cycle 高频的
~ definition 高清晰度
~ density concrete 高密度混凝土,(防辐射)重混凝土
~ density grading 高密(实)度级配
~ density polyethylene 高密度聚乙烯
~ discharge mixer 斜轴拌和机
~ -ductility steel 高塑性钢
~ duty fireclay 高耐火度黏土
~ duty pig iron 优质生铁
~ duty steel 高强度钢
~ -early (strength) cement 早强水泥
~ -early strength 早强(的),速凝(的)
~ early strength concrete 早强快凝混凝土
~ efficiency 高效率
~ elastic deformation 高弹形变
~ -elastic limit steel 高弹性限度钢
~ elasticity 高弹性
~ energy 高能的
~ exchange 高汇率,汇兑价高
~ expansion cement 高膨胀(性)水泥
~ -fidelity 高保真度(的),易感(的),高灵敏度(的)
~ finish 高度光泽,非常光泽
~ finished 精制的
~ (low) flush tank 高(低)冲洗水箱
~ forest 乔木林
~ -geared operation 高速操纵,快速施工
~ gloss enamel 高光瓷漆
~ gloss paint 强光泽涂料
~ grade 高[陡]坡;高质量(的)
~ grade cement 高标号水泥,优质水泥
~ grade concrete 高标号混凝土
~ humidity treatment 高湿度处理[养护]

~ -initial strength portland cement 早强(波特兰)水泥
~ investment 高额投资
~ iron portland cement 高铁波特兰水泥
~ joint 凸缝;(混凝土路面因冻胀而形成的)冻胀缝
~ -level 高标准,高标高
~ level cistern 高位水槽[池]
~ level intake 浅(孔)式进水口
~ -lift 高扬程,高举;高块(混凝土浇筑层)
~ -lift concrete construction method 混凝土高块浇筑法
~ -lift fork stacking truck 铲车,叉式(万能)装卸[升降]车,叉式升降机,升降叉车
~ line 高压电线,公用动力线
~ living 奢侈的生活
~ -low graph 顶端—深底的图形
~ -low lamp 明暗电灯
~ -low method 高低点法
~ -magnesia cement 高镁水泥
~ magnesium lime 高镁石灰
~ mechanization 高度机械化
~ milling 精磨
~ melting wax 高熔点(黄)腊
~ molecular polymer 高分子聚合物
~ -octane gasoline 高辛烷值的汽油
~ ohmic resistance 高欧姆律电阻
~ order 高位[阶,等级]
~ overcast 高云密布(天空),阴天
~ penetration bitumen 高渗透性沥青
~ -pitched 急倾斜的;(屋顶)高坡的,高音调的
~ -pitched roof 高坡屋顶,陡坡屋顶
~ place 高殿
~ -pole 大圆材
~ polymer 高(分子)聚合物
~ portland 多水泥的
~ powered money 强力货币
~ -precision leveling 高精度水准测量
~ -pressure 高(气)压的;倾力[强行]推销的
~ -pressure cementation 高压胶[粘]结
~ -pressure cutout 高压开关,高压断流器
~ -pressure fire system 高压消防系统
~ -pressure gate 高压闸门
~ -pressure grouting 高压灌浆
~ pressure heating 高压供暖
~ -pressure hose line 高压蛇形管[软管]线
~ -pressure lubrication grease 高压润滑油[油脂]
~ -pressure main 高压(水;电、煤气、下水道等的)总[干]管线
~ -pressure mercury (discharge) lamp 高压汞灯,高压人工太阳灯
~ -pressure oil burner 高压燃油炉
~ -pressure pipeline 高压管道
~ -pressure pump 高压泵
~ -pressure sluice valve 高压闸阀
~ -pressure spraying set-up 高压喷射[涂,洒]装置
~ -quality cement 高标号水泥,优质水泥
~ -quality grey iron 高级灰口铁
~ -quality measuring 高品位测量,高精度测量
~ -quality steel 高质量钢
~ quartz gravel 高石英(质)砾石
~ rainfall area 多雨区
~ range 大比例,大刻度,高灵敏度量程
~ rate activated sludge process 高速活性污泥处理法
~ refractory brick 高级耐火砖
~ relief 高凸浮雕

high relief

~ rented 高租金的
~ -rise building 高层房屋[建筑物],多层并备有电梯的建筑
~ -rise(parking)garage 多层停车库
~ -rise structure 多层建筑
~ rupturing capacity 高断裂强度
~ safety glazing 高安全度窗玻璃
~ -side pack 上部充填带
~ silica cement 高硅石[火山灰]水泥
~ silica brick 高硅砖
~ slope 陡坡
~ -slump concrete 高塌落度混凝土
~ -speed steel 高速钢,锋钢,硬钢
~ -speed underground railway【英】高速地下铁道
~ -speed underground railroad（美）高速地下铁道
~ -speed vibrator 高频振动器
~ spot （路面）凸起处,路障,要[特]点,名胜古迹
~ street 大街,正街,干道
~ strength bars 高强度粗钢筋
~ -strength cement 高强度水泥
~ strength concrete 高强度混凝土
~ strength friction grip bolt 高强夹紧螺栓(用于钢梁拼装)
~ symmetry 高对称
~ table 大宴会中较其他桌子高的餐桌,英国大学中院长、导师等的餐桌(常有美酒佳肴)
~ -temperature material 高温材料
~ -temperature insulation 高温绝缘
~ -temperature (coal) tar 高温(煤)焦油,高温焦油沥青(或煤沥青)
~ temperature steel 耐热钢
~ -temperature treatment 高温处理
~ temperature water 高温水
~ -tensile 高强度
~ tensile alloy 高抗拉合金
~ -tensile bolt 高强螺栓[插销]
~ -tensile reinforcing steel 高强（度）钢筋
~ -tensile steel 高级钢,耐拉钢

~ tensile wire 高强力钢丝
~ -tension 高（电）压,高张力,强拉力;高压的
~ tension apparatus 高压设备
~ tension cable 高压电缆,高压线
~ tension cell 高压间隔
~ tension circuit 高压电路
~ tension rods 高强度粗钢筋
~ -test 优质的[高级]的,适应高度需要的,经过严格试验的,高挥发性的
~ test cement 高级水泥
~ valley 高屋谷,高屋顶排水沟
~ -velocity duct 高速(水流)管道
~ -velocity road 高速公路
~ -viscosity tar 高粘性焦油
~ -voltage 高(电)压
~ wall slope 边坡,边坡坡度
~ -water table 高地下水位,高[潜]水位
~ webbed tee iron 宽腰 T 型钢
~ weight mud 超重泥土[浆]
~ wind 大风,疾风
~ yield stress steel 高屈服点强度钢
higher [haiə] a. (high的比较级),较高的,高等[级、度、次]的
~ algebra 高等代数
~ bronze 铝铁镍锰高级青铜
~ education 高等教育(大专以上教育)
~ geodesy 高等大地测量学
~ mathematic(s) 高等数学
highlight ['hailait] n.; v. , -lighted, -lighting; n. 照明效果,(图,影象中)最明亮的部分;重[要]点,显著部分,最精彩的地方; vt. 强调,着重,以强烈光线照射
~ signal 最亮信号
highly ['haili] ad. 高度地,有利地,高贵[价]地,很,极
~ plastic clay 高塑性黏土
highrise n. 多层高楼[建筑物]; a. 高耸[摩天,高层]的
highway ['haiwei] n. 公路,大路[道],

（水陆）交通干线；信息通路；达到目的的途径
~ binder 公路铺路沥青（粘结料）
~ engineering 道路工程（学）
~ structure 公路建筑物
Hiley's formula 黑莱（打桩）公式
hilt [hilt] *n.* （刀,剑或匕首的）手柄,把；*v.* 装柄[把]于……hilted, *a.*; hiltless, *a.*
hind [haind] *a.*, hinder, hindmost 或 hindermost；(指前后对称的)后面[边,部]的,在后的
hinelight *n.* 高强度荧光灯
hinge [hindʒ] *n.*;*v.*, hinged, hinging；*n.* 铰链,铰接,枢纽,门枢,活动关节,重要,重点,主旨,关键；转折点；透明胶水纸；*v.* 给……装铰链,用铰链转动[结合],以……为转移；hinged, *a.*; hingeless, *a.*; hingelike, *a.*, hinger, *n.*
~ armature 枢轴衔铁
~ fault 捩转断层[错位]

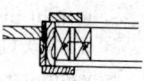

hinge jamb

~ jamb 门框枢
~ joint 铰接,铰式接缝,铰链接合
~ pillar 铰链柱
~ pin 铰链的销
~ plate 铰合板
band ~ 扁担式铰链
door ~ 门铰链
gate ~ 大门铰链
piano ~ 长铰链
window ~ 窗铰链
hinging post 带铰门柱
hinged ['hindʒd] *a.* 铰接的
~ arch 铰接拱,有铰拱
~ bar 铰接的型钢
~ barrier 铰式栅门
~ door 铰接门
~ frame 铰接框架
~ gate 回转[旋开]式闸门
~ radiator guard 铰式散热器[暖气片]防护罩
~ (window) shutter 铰式百叶窗
~ truss 铰接桁架
~ window 铰链窗
hingeless *a.* 无铰的
~ arch 无铰拱
hingle *n.* 门铰链,锅[壶]把
hip [hip] *n.*;*v.*, hipped, hipping, *n.* 屋脊,斜(屋)脊；堆尖；臀部；*a.* 内行的,熟悉内情的；超时新的；*v.* 给……造屋脊；hipless, *a.*；hiplike, *a.*
~ -and-valley roof 有交叉的四坡屋顶
~ bath (tub) 坐浴(浴盆)
~ capping 屋脊压盖

hip-and-valley roof   hip capping

~ dormer 屋脊窗
~ hook 脊瓦挂钩
~ jack (rafter) (屋顶)支(顶)木,脊椽,（四坡屋顶的）端坡椽
~ joint 上弦与斜端杆结点
~ knob 脊端饰
~ mold 脊线脚
~ point (桁架)上弦与斜端杆接点
~ rafter 脊小椽,屋脊椽木[支脊木],四坡屋顶面坡椽
~ roll 四坡屋顶中脊上的瓦或盖条
~ roof 四坡屋顶
~ tile 脊瓦,屋脊盖瓦
~ -type dormer (window) 屋顶天窗,老虎窗
~ vertical 肩竖杆(承受局部横梁荷载的)竖吊杆,后部的直立杆件
~ -ped end (屋顶的)斜坡端,肩脊端
~ -ped-gable roof 人字[三角]屋顶
~ -ped mansard roof (英)折线[复折]形斜坡屋顶
~ -(ped) roof 四坡屋顶,斜截头屋顶

conical ~ 圆锥屋肩
half ~ 半斜脊
inner ~ （平旋桥）内肩节

hip(ped)roof

hire ['haiə] n.；v.，hired,hiring；n. 租[雇]用；租金，报酬，工资 v. 租借，出租；租用，雇用
~ purchase 分期付款，租购
~ system 分期付款的购买法，租购制
~d plant 租用（的）设备
~ing hall 职业介绍所
hireling ['haiəliŋ] n. 被雇者，佣工 a. 被雇用的
hirer ['haiərə] n. 雇主，租借者
H-iron 工字铁[钢]
hirst n. 砂堆[滩]
histogram ['histəɡræm] n. （统计学的）直方[柱状，矩形]图，频率曲线
historic [his'tɔrik] a. 历史上著名的，有历史性的
historiated [his'tɔ:rieitid] a. 有图案的，用人物像装饰的
historical [his'tɔrikəl] a. 历史上的；真实的；historically ad. historicalness n.
~ city 历史名城
~ geology 地史学，历史地质学
~ profit 实际利润
~ relic preservation area 历史文物保护区
~ spot 古迹
~ time 有史时期
histosol n. 有机土（沼泽土，泥炭土）
hi-strength n. 高强度
hit [hit] n. 打击，命中，碰撞；v. hit, hitting；打击，击中，碰撞，冲击；符合，成功；hitless, hittable a.；hitter n.
~ -and-miss effect 时现时隐效应
hitch [hitʃ] n. 连接（装置），接头，栓子；套上，钩住；障碍；v. 系（绳索），绑，钩住；妥协
H-layer 腐殖质层
hoar [hɔ:] n. 灰白（色）的涂层；白霜；a. 灰白的；铺满白霜的
~ frost （白）霜，雾松
hoard [hɔ:d] n. 贮藏（物），宝库；v. 贮藏，积蓄，囤积
hoarding ['hɔ:diŋ] n. 贮藏；（pl.）贮藏物；招贴板，广告牌；（建筑过程中围在房屋或空地四周的）板垣，围篱；囤积，积蓄
hoarstone ['hɔ:stəun] n. （自太古存在的作界标用的）弧石柱，界[标]石
hob [hɔb] n.；v. 滚刀，铣刀，滚铣，切压；螺旋杆；hoblike, a.
~ nail 平头钉
~ tap 标准螺纹攻，板牙丝锥
hod [hɔd] n. 灰砂斗，砂[灰]浆桶，化灰池；煤斗，砖斗
~ carrier 搬运灰泥砖瓦工，建筑小工
~ man 辅助工，小工，送灰工
hodoscope ['hɔdəskəup] n. 描迹仪，辐射计数器
Hoffmann kiln 霍夫曼（砖,瓦）窑
hog [hɔɡ] n.；v., hogged, hogging; n. 拱，弯拱；土工，挖土工人；软管；碎木机；变形；v.（使）变[扭，拱]曲，变形，切成木片；hoglike, a.
~ back 拱背，拱起物
~ -back truss 弓形桁架
~ backed girder 弓形大梁（上缘变曲的梁）

hog backed girder

~ chain 倒拱式悬链[杆]（用于吊桥）
~ chain truss 链式桁架
~ -frame 弓背构架
~ wire 粗铁丝围栏，倒钩钢丝
hoggin ['hɔɡin] n. （英）夹砂砾石，级配碎[砾]石，筛过的碎石；滤水层，滤层
hogging ['hɔɡiŋ] n. 弯[扭，翘]曲；挠度
~ moment 挠曲力矩，负弯矩，使梁顶面受拉的弯矩，中拱弯矩，中拱力矩
~ stress 中拱应力
hogshead ['hɔɡzhed] n. 液量单位（等于

238.5升);大桶(其体积为238.5升至530升者)(英国=521/2英加仑,美国=63加仑)

hoist [hɔist] *n.* 升起,绞起;起重[卷扬,升降,启门]机;一面旗; *v.* 升[绞]起,提高; hoister, *n.*
~ elevator (美)家用电梯
~ equipment 国产设备

hold [həuld] *n.*; *v.*, held 或[古]holden, holding *n.* 抓,捉,控制,把柄;船舱 *v.* 占有,保持,抓住,固定,安装,支撑,容纳,存储;有效;抑制,阻滞,约束;同步[期];举行,开(会)
~ (-)down 夹板,压具,缩减,压制;固脚螺母
~ fast 夹(钳),支架,钩子,平头大铁钉;锚碇,地锚
~ mode 保持状态
~ of pile 桩的打入深度

holder ['həuldə] *n.* 把柄,夹;支[托]架;罐,盒,容器;持票人; holdership, *n.*

holding ['həuldiŋ] *n.* 占有物,支持物;土地

holdings *n.* 馆藏[库存]资料;拥有(土地,股票,债券等)财产
gold ~ 黄金储备额

hole [həul] *n.*; *v.*, holed, holing; *n.* 孔,穴,洞,坑洼;通路[道];频带空段;炮眼,钻孔,管道 *v.* 穿[钻,凿,冲]孔,把……放入洞中;挖井[隧道], holeless, *a.*
~ aperture 孔径
~ block 空心砖[砌块]
~ circle 孔圈
~ screen 多孔筛
~ size 孔径
~ spacig 钻孔距离
access ~ 检查孔
air ~ 风眼,通风[气]孔;砂眼
dead ~ 未穿透孔,盲孔
discharge ~ 泄放孔
man ~ 人行道,人行天井,检查孔
mortise ~ 榫眼

weep ~ 泄[排]水孔

holer *n.* 凿岩工,打眼工,挖洞者

hollow ['hɔləu] *n.* 孔,穴,凹地,(山间)盆地,谷;毛管,空心管坯,开槽刨; *v.* 挖[掏]空,弄凹; *a.* 空的,空心[洞,虚]的;凹隐的; *ad.* 完全; hollowness, *n.* hollowly *ad.*
~ -backed flooring 底(部透)心楼板
~ beam 空心梁,管梁
~ block 空心块体[砖]
~ block wall 空心砖墙
~ boring rod 空心钻杆
~ brick 空心砖
~ brick wall 空心砖墙
~ building block 空心砌块
~ buttress 空腹支墩
~ casting 空心铸塑
~ cinder block 空心煤渣砖
~ clinker block 空心熔渣[烧结]砖
~ concrete 多孔[蜂窝状]混凝土
~ column 空心柱
~ -core door 空心门

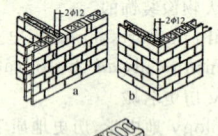

hollow masonry units

~ drill shank[steel] 空心钻杆,六角钢
~ girder 空心梁
~ masonry units 空心砌块
~ -metal door 空心金属门
~ -meta frame 空心金属框架
~ newel 空心楼梯栏杆柱,空心(螺旋梯)中柱

hollow-metal door

~ pier 空心墩
~ pile 空心桩
~ pillar 管柱
~ plaster panel 空心灰泥板
~ plywood 空心胶合板
~ quoin 空心墙角基石
~ -ribbed bridge 空心肋构桥
~ rounded skirting 凹圆踢脚
~ slab 空心板
~ slab floor 空心板楼板
~ spar 红柱石
~ spot 凹陷地点
~ square 消波混凝土块体,空方阵
~ square molding 空心方线脚

hollow square molding

~ -tile floor 空心板楼板
~ -trussed beam 花隔梁,透空梁
~ unit 空心构件[砌块]
~ wall 空心墙
~ -web girder 空心腹板梁
holohedrism [hɔləuˈhiːdrizəm] *n.* 全对称性
holometer *n.* 测高计
holoseismic method 全息地震法
holosteric barometer 固体[空盒]气压表
holy water stone 圣水石
home [həum] *n.* 家,住宅,住址;家乡,本国,产地,发源地 *a.* 家庭的;本乡[国]的;中要害的 *ad.* 在[回,到]家;*v.* homed homing;回家
~ accessories 房间内装饰品[如窗帘、地毯等等]
~ address 家庭[标识]地址
~ bank 国内银行
~ building 住宅建设
~ elevator (美)家用电梯
~ equipment 国产设备
~ -grown timber 国产的木材
~ handyman 受雇做(家庭)杂事的人
~ industry 家庭工业
~ lift (英)家用电梯
~ land scape 家庭园景
~ made 自制的,本国制的
~ office 总公司,总店,(银行)总行
~ products 本国产品
~ record 原[起]始记录
~ stead 住宅,屋基,(美)自耕农场,家园,田园
~ water softener 家用水质软化器
~ wear 便服
private ~ 私人住宅
semi-detached ~ 半独立式住宅
homeochromatic [ˈhəumiəˈkrætic] *a.* 颜色相似的
homeostasis [həumiəˈsteisis] *n.* 自动(调节)平衡,原状稳定; homeostatic, *a.*; homeostatically, *ad.*
homesite [ˈhəumsait] *n.* 住宅基址
homestead 宅基,地基,住宅;[美、加]
homoclinal [ˈhɔməclinl] *a.* 同斜层的
homogen [ˈhɔmədʒen] *n.* 均质(合金)
homogeneity [həməudʒeˈniːiti] *n.* 同种[质,性],均匀[一,质]性,一致性,齐次
~ test 同质性检验
homogeneous [ˌhɔməˈdʒiːniəs] *a.* 均匀的,单一的,同种[原,类,质,次]的,齐次的,齐(性)对等的,类似的,单色的;-homogeneously, *ad.*
~ concrete 均质混凝土
~ cover(ing) 均匀涂层,均质护壁板[屋顶]
~ deformation 均质变形
~ degree 均匀度
~ fibre wallboard 均质纤维板
~ isotropic soil 各向同性均质土
~ layer 均匀[质]层
~ light 单色光
~ mixture 均质混合物
~ polyvinylchloride cover(ing) 均质聚氯乙烯涂层[护壁板]
~ roof 匀质屋顶
~ soil 均质土

~ state 均态
~ strain 均匀应变
~ tube（非焊接的）整管
homogenizer [həˈmɔdʒənaizə] n. 均化器,均浆器
homogeny [həˈmɔdʒini] n. 同种[质,一]性;(地层)生成同一
homographic [ˈhɔməˈræfik] a. 对应的
~ solution 对应解
homolog [ˈhɔələgm] = homologue n. 同系物
homology [həˈmɔlədʒi] n. 相符[同];对应;同源,同调;透射;关系相同,(现象)对称,异体同形
homolosine projection 等面积投影
homolysis n. 均裂,均匀分解
homomorphism n. 同形,同态
homopicnal inflow = homopycnal inflow 等密[重]入流
homothermal [həumeˈθɔːmel] a. = homoiothermal 恒[同]温的
~ condition 恒[同]温条件
homothermous a. 恒[同]温的
homothetic (-al) [həuməˈθetik (əl)] a. 同形的,相似的,同位的
hondrometer [hɔnˈdrɔmitə] n. 粒度计,微粒特性测定计
hone [həun] n. ;v., honed, honing; n. 刮路器,磨孔器,磨(刀)石,油砥石,极细砂岩; vt. (在磨石上)磨,搪磨; vi. 渴望
~ stone 均密砂岩,磨刀石
honing oil 磨刀油
honeycomb
[ˈhʌnikəum] n. 蜂窝[蜂房],蜂窝(状)物;
v. 使成蜂窝状; a. 蜂窝状的

honeycomb

~ brick 蜂窝砖
~ core plywood 蜂窝夹心胶合板
~ cracks（路面)蜂窝状裂缝,龟裂,网状裂缝
~ radiator 蜂窝式散热器

~ structure 蜂窝结构
~ texture 蜂窝结构[组织]
~ weathering 蜂窝状风化
~ed wall 蜂窝式墙,花墙
hood [hud] n. 出[遮]檐;隐藏,覆盖,排风罩,(发动机)罩兜;车蓬,挡板 v. 覆盖,隐藏,加盖子,戴罩
~ mo(u)ld 门(窗)罩饰,拱檐线脚

hood mo(u)ld

hook [huk] n. (吊)钩,钩状物,镰刀;拐,箍(圈),夹[卡,制]子; v. (用钩)钩住[紧],挂上;弯成钩形;
hookless, a. ; hooklike, a.
~ accessories 吊钩附件
~ anchorage 钩形锚具
~ and butt joint 钩扣连接
~ and eye hinge 钩扣绞连
~ block 带钩滑车
~ bolt 钩头螺栓
~ face 曲面
~ feed 钩式送料
~ gauge 钩形水尺(测微水位用,读至毫米),钩形(水位)测针
~ -gaugemethod 钩形表法(加气混凝土配料法的一种)
~ -out blind 离窗遮帘
~ rebate S 形凹槽接合缝
~ spanner 钩形扳手
~ed bar 带钩钢筋
~ed bolt (带)钩螺栓
~ed nall 钩头大钉,曲钉
~ed spits 钩形沙咀
~ing iron 清缝凿
cant ~ 搬钩
lifting ~ 吊钩
hooke's law 虎克定律
hooklet [ˈhuklit] n. 小钩子
hookwrench [ˈhukrentʃ] n. 钩形扳手
hoop [hu:p] n. 环,箍,圈;箍铁,带钢,弓形小门 vt. 加箍,围绕
~ back 围圈靠背

~ cutter 拆包钳,拆铁箍钳
~ driver 紧箍器
~ fastener 紧箍钉
~ iron 箍铁
~ -iron bond 卡箍连接
~ mill 箍钢轧机,带钢压延机
~ net 袋纲,圈纲
~ steel 带钢
~ stick 车棚轻型构件
~ stress (圆)周应力,环箍应力
~ tension 环筋张力
~ wood 光滑冬青
~ed column 箍筋柱
~ed concrete 环箍混凝土
~ed penstock 加箍压力水管
~ed pile 配箍筋桩
~ed reinforecement (环)箍(钢)筋
hootch [hu:tʃ] *n.* hooch 茅屋
hootenanny [ˈhu:tənæni] *n. pl.* -nies
= hootannanny = hootanny 机器配件,小机件
hooter [ˈhu:tə] *n.* 汽笛,鸣音器
hop [hɔp] *n.; v.,* hopped, hopping;
n. 跳跃,飞行; *v.* 跳
hopper [ˈhɔpə] *n.* 戽[漏,料]斗,装料箱;有倾斜斗的手推车,底卸(式)车;泥舱;贮水槽
~ bottomed bin 底卸式料斗
~ bottom car 底卸(式手推)车
~ car 底卸式车,[漏斗式]底卸车
~ car unloader 链板卸车机
~ charging bucket 料斗,活底料罐
~ chute 漏斗式斜槽,滑槽
~ clearance 漏斗离地高度
~ closet 带贮水槽的抽水马桶
~ gritter (漏)斗式铺砂机
~ head 水落斗
~ light 外推式下旋窗扇
~ mill 斗式碾磨机
~ -on-rails 轨承式料斗
~ opening 漏斗口
~ throat 斗式卸料孔

~ trailer 料斗(式)拖车(供搬运或撒布、石等散粒料用)
~ truck 漏斗型底卸车,斗式卡车
~ vet(ilator) 漏斗式通风口[空气调节器]
~ vibrator 斗式振动筛
~ wagon (自动)倾卸斗车,漏斗车,[漏斗式]底卸车,底开门车
~ window 外推开上悬窗
hopping [ˈhɔpiŋ] *n.* 跳动; *a.* 忙碌的
horizon [həˈraizn] *n.* 地平(线),水[视]平线;地层,层位
~ colsure 水平闭合
~ glass 地[水]平镜
~ of soil 土层
~ plane 地[水]平面
~ range 水平(视线)距离
~ sample 地层试样
~ trace 地平迹线
container ~ 含油层
horizontal [hɔriˈrɔntl] *n.* 水平线[面],地平线 *a.* 水平的,地平(线)的;平放的,横的; horizontality, horizontalness, *n.*; horizontally, *ad.*
~ acceleration (建筑物在地震时的)横向加速度,水平加速度
~ adjustment 水平校正
~ alignment (英)水平定线
~ alinement (美)水平定线
~ anchorage 水平锚碇
~ angle 水平角
~ -axis mixer 横轴式拌和机
~ baffle 水平隔板
~ bandsaw 水平带锯
~ bar-chart 横条图
~ beam 水平深,横梁
~ boarding 水平木板壁
~ boom (美)水平吊杆,水平起重杆
~ boring unit 水平钻孔机
~ bracing 水平支撑
~ branch 水平支管
~ centerline 水平中线
~ circle 水平刻度盘

~ control section 水平(面)控制点
~ coordinates 水平坐标
~ coplanar 同一水平面的
~ cornice 水平挑檐
~ course （方块）平砌层,横砌层
~ crossbar 水平横梁,水平横杆
~ dampproof course 水平防潮层
~ pivot hung window 中悬式窗
~ plan 平面图,水平投影
~ plane 水平面
~ position 水平位置(即平面坐标)
~ pressure 水平压力
~ profile 水平剖[断]面
~ projection 水平投影
~ sliding door 推拉门,平推门,横向滑动门
~ silding window 水平滑动窗
~ stripe 横条
~ sunshade 水平遮阳

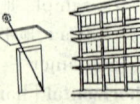

horizontal sunshade

~ thrust 水平推力

horn [hɔːn] **n.** 角,门[窗]框突角;角状物;角制品;操纵杆,机臂;垫铁;半岛;海角,岬(角);支流,(海湾)分叉 **v.** 给……装角,做成角状; hornish, **a.**; -hornlike, **a.**
~ arrester 角隙避雷器
~ balance 突角补偿
~ blende 角闪石
~ -granite 角闪花岗岩
~ stone 角岩,角石(变蛋白石)

horologe [ˈhɔrəlɔdʒ] **n.** 钟表,日晷

horse [hɔːs] **n. pl.** horses 或 horse, **v.** 马;搁架,有脚的支架;夹块[石];绳索;铁杆 **v.** horsing, horsed, 骑马 **a.** 马的,骑马的

horseshoe [ˈhɔːsʃuː] **n.** 马蹄铁,U形物; **v.** 给马钉掌
~ arch 马蹄(形)拱
~ bar 马蹄钢
~ conduit 马蹄形水管

~ holder 马蹄形夹具
~ magnet 马蹄形磁铁
~ riveter 马蹄形铆钉机
~ sewer 马蹄形(断面)阴沟,马蹄形排水管

horst [hɔːst] **n.** 地垒

horticulture [ˈhɔːtikʌltʃə] **n.** 园艺(学)

hose [həuz] **n. pl.** hoses, **n.** 皮带管,水龙带;软管,蛇(形)管; **v.** hosed, hosing; 用软管浇水[装油]; hoseless, **a.**; hoselike; **a.**
~ bib 软管龙头
~ car （消防）水管车
~ company 消防队
~ connection 软管接头,皮带式龙头

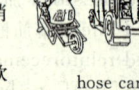

hose car

~ coupler 软管接头
~ coupling nipple 软管连接器,软管接头
~ down 用水龙带冲洗,用软管洗涤
~ flushing 软管冲洗
~ guard 软管防护层
~ -man 消防人员
~ nipple 软管接头
~ nozzle 软管喷嘴,水龙带射口
~ pipe 水龙管,蛇管
~ -proof 防水的
~ protector 软管防护装置
~ scaffold 木马架,轻便脚手架
~ siphon 虹吸软管
~ union 水龙带接头
asbestos ~ 石棉管
flexible shaft protecting ~ 软轴套

hospice [ˈhɔspis] **n.** (教会办的)旅客招待所,供学生、青年工人或贫困者寄宿的房屋

hospital [ˈhɔspitl] **n.** 医院;旅客招待所;修理各种小东西的店铺
~ service contract 医疗合同

~ switch 故障转换开关
host [həust] n. 男主人;基质材料 vt. 招待; hostless, a.; hostly, a.; hostship, n.
~ bakery 主面包房
~ -crystal 主晶
~ processor 主处理机
~ rock 主岩
hostage ['hɔstidʒ] n.; v., -taged, -taging; n. 抵押品,人质; v. 作为人质; hostageship, n.
hostel ['hɔstəl] n. 寄宿舍,旅店; v. 寄宿在招待所或旅店;开旅店,开招待所
hot [hɔt] v.; hotted, hotting; a. hotter, hottest, ad.; v. 热,把……加温,使恢复; a. 热的;热[激,强]烈的;最近的(消息等) ad. 热[猛]烈地
~ -aggregate conveyer 热集料输送机,热斗
~ -air heater 热空气[风]供暖器
~ air heater-defroster 暖气除霜器
~ -air seasoning (木材)热气烘干
~ and cold steeping (木材防腐处理用)热冷(槽)浸渍
~ and dry rolling 热干碾压法
~ application tar 热用焦油(沥青)
~ -asphalt plant 沥青加热设备
~ -banded steel pipe 热箍式钢管
~ (-)bed 温床
~ blast 热(鼓)风
~ brittle iron 热脆钢
~ bulb engine 热球柴油机
~ -cast 热铸
~ charging 热装料
~ cure 热处理[热养护]
~ cycle 热循环
~ -die 热压模
~ -dip galvanizing 热浸镀锌
~ drawn 热拉
~ driving= ~ riveting 热铆
~ ductility (混合料的)热延性
~ -film anemometer 热膜风速表

~ -finished tubing 热轧钢管
~ forming [sizing] 热压成形[冲压]
~ gas line 热气管线
~ gas welding 热风焊接,气焊
~ house 温[暖]室,窑器干燥室,烘干室,浴室
~ ice 高融点冰
~ integration 横向结合
~ investment casting 精密(腊模)铸造
~ iron 铁水
~ laboratory 强放射性物质实验室,"热"实验室
~ -laid asphaltic concrete 热铺沥青混凝土
~ laid (bituminous) macadam 热铺(沥青)碎石路
~ light 热光,电视演播厅内最主要的灯光
~ line job [work] 带电操作[作业]
~ -logging (木材的)流水式采运
~ machining 高温切削
~ metal 液态金属
~ mill 热轧设备
~ mixture 致热混合气;热拌(沥青混凝土)混合料
~ mix cold-lay type 热拌冷铺式(沥青混合料)
~ -mix plant (沥青混凝土)热拌厂
~ money 游资,流动资金
~ oven slay 高炉渣,鼓风炉渣
~ patch 热补,火补钉
~ -patch outfit 热补设备
~ -penetration (bituminous) macadam 热灌沥青碎石路
~ plasticity (混合料的)热塑性
~ plate (加)热板,电炉,煤气灶
~ pool 温泉湖
~ -poured crack filler 热灌填缝料
~ pressing 热压
~ process (筑沥青路面用的)热铺法
~ quenching 热淬,分级淬火
~ reducing 热压缩,热轧

~ repair 热修补
~ reserve 暖机预备
~ riveting 热铆
~ rolled 热轧[压]的
~ -rolled sand carpet 热压沥青砂毡层
~ rolled sheet 热轧钢板
~ rolled steel bar 热轧钢筋,热轧棒钢
~ run table 热金属辊道
~ screen 热料筛
~ shaping 加热成形
~ shoe 闪光灯插座
~ shortness 热脆(性)
~ shot 过热,淬沥过度
~ shrinkage 热缩
~ spot 热斑,潜在的危险地区,腐蚀点
~ spotting 热点
~ spraying 热喷涂
~ spring 温泉
~ storage （材料)热贮存
~ strip mill 扁[带]钢热轧机
~ tarring 热灌柏[焦]油
~ tension test 高温拉力试验
~ top 低温帽,冒口,热顶
~ wall 热损墙面,火墙
~ water 热水,困境
~ water circulating 热水循环
~ -water heating 热水供暖,水暖
~ -water heating system 热水供暖系统

hot waterline heating system

~ water line 热水管
~ water return 热水回路
~ wave 热浪
~ well 天然温井,温泉
~ wire 热线,热电阻线
~ wire pressure gauge 热线压力计
~ working 热作,热加工
~ -zone 热带

hotch-potch ['hɔtʃ pɔtʃ] n. (地层)混合物,混合岩层;财产混同
hotel [həu'tel] n. 旅馆(社);宅邸,要人的公馆;通讯系统用以代表字母 h 的电码
~ bedroom 旅馆卧室
~ car 餐卧车厢
~ construction 旅馆建筑(物)
~ entrance 旅馆进口
hour ['auə] n. 时(刻),小时,钟点,时机;时间(指规定的或某一段而言的);目前,现在；a. 时间的;hourless, a.
~ angle 时角(度)
~ curve 子午线
~ cost 计[小]时成本
~ glass (计时用)沙[水]漏
~ hand 时针
~ meter 计时仪(测量机器每小时转数的仪器)
~ zone 时区
business ~ 营业时间
idle ~ 窝工[停机]时间
off-peak ~ 轻负[非峰]荷时
peak ~ 高峰负荷时
working ~ 工作时间
hourly ['auəli] a. 每小时的,时[逐]时的,经常的；ad. 每小时地,常常
~ capacity 逐时量,小时生产能力
~ consumption 每小时消耗量
~ earnings 每小时工资
~ load 逐时负荷
~ observation 逐时观测
~ output 时出力,小时产量
~ precipitation 逐时降水量
~ traffic volume 每小时交通量
~ varation factor 逐时变化系数
~ wage(s) 计时工资
~ water consumption 每小时用水量
house [n. haus; v. hauz] n. pl. houses; v. housed, housing; a.；n. 房屋[间],宿舍,住宅,建筑物,厂[车]房;家庭[族];机构,商店; v. 收藏,容纳,收

留,供宿,居住,装[嵌]入 *a.* 有关房屋的,适合于房屋的
~ agent 房屋经纪人
~ alteration 房屋改建
~ breaking 房屋拆除
~ cable 室内电缆
~ car 棚车,冷藏车(的总称)
~ cistern 家用贮水器
~ connecting box 用户分线箱
~ connections 家庭污水管道
~ construction 房屋建筑[结构,施工]
~ decorator 室内装潢设计者
~ designer 室内设计者
~ document 内部文件
~ door 宅门
~ drain 房屋排水道,房基排水管
~ drainage 房屋排水装置[系统]
~ dust 室内尘埃
~ duty 房产税
~ dweller 房屋居住者
~ famine 房荒
~ furnishing 家具
~ heating 住房供暖,住所的集中供暖
~ hold furniture 家具
~ holder 户主
~ lead-in 进户线
~ lease 房契
~ let 小屋
~ lights 观众席座灯光
~ line 室内线,三股细油麻绳
~ maid's sink 民用洗涤[污水]槽
~ main [配电]干线
~ moss 室内尘埃
~ paint 房屋[民用,建筑]漆
~ painter 以油漆房屋为职业的人,房屋油漆工[匠]
~ phone 内线电话
~ plumbing 住房卫生设备安装
~ preservative 房屋防腐剂
~ refuse 生活垃圾
~ refuse disposal 生活垃圾处理(法)
~ sewage 家庭[生活]污水

~ sewer 家庭污水管
~ silhouette 房屋轮廓
~ stuff 家具和陈设品
~ substructure 房屋基础[下层结构]
~ top 屋顶
~ trap 室内排水管存水弯
~ voucher system 单页凭单制
~ wastewater 生活废水
~ wiring 室内配电线(路)
~ wright 建房工人
~d joint 封装接头,套入接合
~d stair 封闭式楼梯(在二道隔墙内)
housed joint
~d string 嵌入式楼梯斜梁
boiler ~ 锅炉房
duplex ~ 两套住房合成的住宅
pent ~ 屋顶小楼
split level ~ 错层式住宅
stepped hillside ~ 坡地住宅

houseboat ['hausbəut] *n.* 可供住家的船,水上住家;有铺位的游艇

housebote 修房木料

housebreaker ['hausbreikə] *n.* 承包拆屋者

housebuilder *n.* 住房营造业者,施工人员

housefoundation *n.* 房屋基础

housefront 房屋的正面

household ['haushəuld] *n.* 住房;家庭[族,属],家务消费单位;*a.* 家族[庭]的;家用的,普通的
~ consumption 住户[家庭]消费
~ disposal system 生活垃圾处置系统
~ effects 家用[住户]用品
~ finance corporation 家庭金融中心
~ store room 家用贮藏室

housekeeping ['haus,ki:piŋ] *n.* 整理[服务性]工作;家务;(工商机构)财产及设备的管理及保养

housing ['hauziŋ] *n.* 房屋,住房[宅],

公寓;外壳,护罩;轴套;盖,(包装)箱; (*pl.*)马饰物
~ allowance 住房津贴
~ and land tax 房地产税
~ area 住宅[居住]区
~ authority 住房管理局[部门]
~ betterment 房屋修缮和扩建
~ centre 居住中心,居民点
~ colony 住宅群(体)
~ code 使房建筑规范
~ density 房屋密度
~ development 住宅群
~ estate 住宅区,居民点
~ estate road 住宅区[居民点]道路
~ financing 住宅资金供应
~ funds 住宅公债[存款]
~ joint 藏纳接头
~ law 居住法(律)
~ legislation 住房立法
~ policy 住房[居住]政策
~ project;(scheme)住宅建筑方案
~ standards 住房建筑标准
~ stock 住房总量

hovel ['hɔvəl] *n.* 小屋,茅舍,杂物间; 棚,遮蔽物;哥特式教堂内代替小尖塔保护圣像的壁龛; *vt.* -el(l)ed,-el(l)-ing;放入棚舍内
~ shelter 临时暖房(如育雏用的棚)

hover ['hɔvə] *n.* 顶棚; *v.* 盘旋,守在近旁,摇曳不定; hoverer, *n.*; hoveringly, *ad.*
~ car 气垫车,悬浮运载工具
~ dredger 气垫式挖泥船(挖泥装置在汽垫船上)
~ ferry 气垫渡船
~ ground 松散地,不坚硬土壤
~ marine 气垫船
~ plane 直升飞机
~ platform 气垫平台
~ train 气垫列车

hoverbarge [英]气垫驳船,大型气垫游艇

hovercraft ['hʌvəkra:ft] *n.* 气垫飞行器,气垫车,气垫船,气垫艇,腾空船;悬浮运载工具;水陆两用垫式航行器

howe [hau] *n.* 洞,孔;低地,洼地[苏格兰]; *a.* 空的;深的;小山谷

Howe 豪威
~ truss 豪威(氏)桁架,平弦再分桁架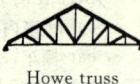
Howe truss

howf = howff [hauf] *n.* 住所,寓所;经常去的地方; *vi.* 住,常去

hoya [西]河床[谷];崎岖山区盆地

Hoyt alloy 一种锡锑铜合金

H-pile H型钢柱(宽缘)

H-pole 工字钢桩,H型(电)杆

H-post 工字杆

H-scope H型显示器

H-section H型截面

H-shaped iron 工字铁[钢]

Huaqing pool(Hot springs) 西安华清池

Huanghe Tower 武汉黄鹤楼

Huber 休伯
~'s formula 中央断面积公式(求原木体积公式)

hub [hʌb] *n.* (轮)毂;(电线)插座,套节,(衬)套;车辙;测站木桩,标柱[桩];中心[枢]
~ and spigot joint 中心向连轴节,套塞接头
~ cap 毂盖,轴端盖
~ docking 毂套对接
~ flange 毂缘
~ ratio 轮毂比(水轮机轮毂直径与转轮直径之比)
~ stake 中心桩

Hubbard tank 练力浴池

hubber *n.* 冲压机[工]

hubbing *n.* 压制阴模法(用阳模压制阴模),切压制模(法)

hudge *n.* 吊桶

hudsonite 黑铁辉石,角闪橄榄岩

hue [hju:] *n.* 色彩[调,泽];混合;形状,外观;hueless, *a.*
colour ~ 色调

dark-hued 暗黑色的
many-hued 有许多颜色的
primary ~ 基色调
spectral ~ 色调
huerta 灌溉冲积平原
huge ['hju:dʒ] *a.*, huger, hugest; 巨大的,庞大的,非常的,无限的;hugely, *ad.*; -ness, *n.*
~ concrete block 大型混凝土砌块
~ blast 大爆破
hull [hʌl] *n.* 外壳;机身,船体,骨架,车盘,薄膜;*vt.* 去壳[皮];*vi.* 漂流;huller, *n.*
~ armor plate 船身(装)甲板
~ -borne (气垫船等的)排水状
~ cell 薄膜电池
human ['hju:mən] *n.* 人,人类 *a.* 人类的;人物的;与人类有关的 human-like, *a.*; humanness, *n.*
~ activities 人类活动
~ affairs 人事
~ asset accounting 人力资产会计
~ behaviour 人类行为
~ capital 人力资本
~ ecology 人类生态学
~ engineering 人类工程学,工程心理学,工效学,人事管理,机械设备利用学,人机(工程)学
~ environment 人类环境
~ error 人为误差
(~)excreta 人类排泄物[粪便,尿]
~ factors 人类工程学,劳动环境学,劳动条件学
~ history 人类历史
~ geography 人文地理学
~ potential 人力资源的潜力
~ power 人力
~ race 人类
~ resource 人力资源
humanism ['hju:mənizm] *n.* 人道;人文主义;人文学;古典文化研究
humanoid ['hju:mənɔid] *a.* 具有人类特点的,有人形的;*n.* 类人动物,猿人

hume duct 或[**hume pipe**] 钢筋混凝土管,混凝土冷却管道
humic ['hju:mik] *a.* 腐殖的,与有机物有关的或至少部分是由有机物组成的
~ acid 腐殖酸
~ coal 腐殖煤
~ matter 腐殖物质
~ mulch 腐殖覆盖物
humid ['hju:mid] *a.* (潮)湿的,湿润的;humidly *ad.* humidness *n.*
~ air 潮湿空气
~ climate 湿润气候
~ mesothermal climate 湿温气候
~ process 加湿法
~ region 湿润区
~ room 雾室,保湿室
~ zone 湿润带,湿润(地)区
humidification [hju:midifi'keiʃən] *n.* 湿润,增湿作用
humidifier [hju:'midi‚faiə] *n.* 加湿器,湿润机(使室内空气增加湿度的装置)
humidity [hju:'miditi] *n.* 湿度,水分含量,湿气
~ barrier 防潮[隔湿]层
~ chart 湿度图
~ control 湿度控制
~ correction factor 湿度校正系数
~ density meter 湿度密度计
~ factor 湿度因子,湿润系数
~ -free material 防潮材料
~ migration 湿气迁移

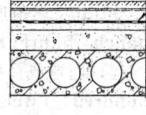

humidity barrier

humidizer *n.* 增湿剂
humidor ['hju:midɔ:] *n.* 保湿装置;恒湿室;蒸气饱和室
humidostat *n.* 湿度调节仪,恒湿器
humification *n.* 腐殖作用[化]
humify ['hju:mifai] *v.*, -fied,-fying;使潮湿;变成沃土,腐殖化,成为土壤
humin *n.* 腐殖[黑]物
humit 温湿
humiture ['hju:mitʃə] *n.* 温湿度(华氏

度数与相对湿度之和的 1/2)
hummer ['hʌmə] *n.* 蜂鸣器；= hummingbird；
~ screen 电磁簸动
humming ['hʌmiŋ] *n.* ; *a.* 蜂鸣音(的); hummingly, *a.*
hummock ['hʌmək] *n.* (小)圆丘；冰脊或冰堆；波状地
hump [hʌmp] *n.* 驼峰；山岗，丘陵，凸起；山脉，圆丘；(曲线)顶点，(巅)峰值；*v.* (使)隆起(成圆形)；急速移动，飞跑；努力，奋发；humpless *a.*
~ speed 界限速度
~ -type stilling basin 驼峰式消力[静水]池
humulite ['hju:mjulait] *n.* 腐殖岩
humulith ['hju:mjuliθ] *n.* 腐殖岩
humus ['hju:məs] *n.* 腐殖质[土]，沃土；humuslike, *a.*
~ sludge 腐殖污泥
~ soil 腐殖(质)土
~ tank 腐殖质沉积池
hunch [hʌntʃ] *n.* 厚片，块，瘤；*v.* 使隆起；推；弯成弓状
hundred(-) weight [hʌndrədweit] *n. pl.* -weights,-weight；度量衡名(等于 1/20 吨，英国为 112 磅，美国为 100 磅)，英担
hung [hʌŋ] *v.* (hang 的过去时和过去分词)悬挂
~ acoustic(al) ceiling 悬式吸音天花(板)
~ fender 悬式护弦木
~ floor 悬式楼板
~ glazing 悬式玻璃窗
~ partition 悬式隔墙
~ scaffold 悬式脚手架
~ shell 悬式骨[框]架
~ story 悬式楼板
~ tilework 悬式排水瓦管
~ window 悬窗
~ wood fender 悬木防护器

bottom ~ 下悬式(窗)
side ~ 平开式(窗)
top ~ 上悬式(窗)
Hungarian architecture 匈牙利式建筑
hunk [hʌŋk] *n.* 大片[块]；岩块
~ cable 电缆盘
hunt [hʌnt] *n.* 狩猎，搜索；*v.* 寻觅，搜索，狩[打]猎，追踪[逐]，探求；振动，摆动；huntable, *a.* ; huntedly, *ad.*
~ effect 摆动效应，振动现象
~ table 半圆形矮桌
leak~ing 测[探]漏，泄漏点寻觅
synchronous motor ~ ing 同步电动机振荡
hurdle ['hə:dl] *n.* 篱笆，栅栏；树枝编的格子；*v.* -dled,-dling；用篱围住；跳栏；hurdler, *n.*
~ dike 栅栏透水堤
~ d ore 粗筛过的矿石
~ groin 栅栏透水坝
~ groynes 栅栏透水丁坝
~ work 编[柳]条制品
hurl [hə:l] *n.* 投，掷 *v.* 猛掷，猛烈发出
~ barrow 双轮手推车
~ out 释出(粒子)
~ ing pump 旋转泵
hurricane ['hʌrikən] *n.* 飓风(十二级以上)，龙卷风
~ damage 飓风造成的损失
~ deck 最小层甲板，飓风甲板
~ globe 防风罩
~ lamp 防风灯
~ -proof 能抗飓风的
~ rain 飓风雨
~ surge 飓风涌浪[暴潮]
~ tide 飓风潮
hurrock 一堆石子[垃圾]
hurry ['hʌri] *n. pl.* -ries；*v.* -ried, -rying；*n.* 急促，慌忙；推[车]，拉[车]；筛子，溜子，漏斗；放矿口；装煤便桥；煤的装卸码头；*v.* 催促，赶紧；

慌忙 hurryingly, *ad.*
~ -durry 多风多雨的
~ gum 筛过的细料
~ -up wagon 抢修车

hurst [həːst] *n.* (有树的)沙洲,河中沙丘

hurt [həːt] *n.*; *v.* 损害[伤],创伤; *v.* hurt, hurting; 损伤, 伤[危]害, (使)受伤, (对)……有不良影响 *a.* 受伤的, 损坏的; hurtable, *a.*

hurtle ['həːtl] *n.*; *v.*, -tled, -tling; *n.* 碰撞(声); *v.* 碰[冲]撞

husbandry ['hʌzbəndri] *n.* 农业, 耕作, 饲养; 家政, 管理

husk [hʌsk] *n.* (外)壳, 外皮; 支架, 无价值的部分; *vt.* 剥去外皮; husker, *n.*; husklike, *a.*

hustiement 家庭用具

hut [hʌt] *n.* 棚屋, 小屋, 茅舍 *vt.* 供应小屋 *vi.* 住小屋; hutlike, *a.*
cable[test] ~ 电缆分线箱[配线房]
gas bomb ~ 储气瓶存放站

hutch [hʌtʃ] *n.* 棚屋, 茅屋; 箱, 容器; 矿车, 洗矿槽 *vt.* 把……装箱; 洗矿; 囤积
~ table 柜桌

Hutchinson tar tester 赫金森氏焦油黏度计

Hutchinson viscosity 赫金森氏黏(滞)度

Hveem cohesiometer 维姆粘聚力仪(测定沥青混合料及稳定土等粘聚力用)

Hveem design method 维姆(柔性路面)设计法

hyacinth ['haiəsinθ] *n.* 风信子; 红锆石, 一种(作宝石用的红色或带褐色的)钙铝榴石, 紫兰色

hyacinthine [haiə'sinθain] *a.* 风信子的, 用风信子装饰或布置的

hyaline ['haiəlin] *n.* 玻璃质, 透明物, 碧空; = hyalin *a.* 透明的, 玻璃状的
~ -quartz 玻璃石英

hyalite ['haiəlait] *n.* 玻璃蛋白石, 玉滴石

hyaloid ['haiəlɔid] *a.* 透明的, 玻璃状的; *n.* (眼球的)玻璃状膜

hyaloplasm ['haiələˌplæzəm] *n.* 透明质; -hyaloplasmic, *a.*

hybrid ['haibrid] *n.* 混合(物), 混合电路; 间生; 杂交种 *a.* 混合(式)的, 间生的, 杂化[种]的
~ computer (模拟一数字)混合式计算机
~ circuit 混合电路
~ drive 双动力驱动, 混合动力驱动, 复式动力驱动
~ -element method 混合单元法
~ integrated circuit 混合式集成电路
~ library 混合程序库
~ metal 石墨化钢
~ model (数字和模拟)混合模型
~ multivibrator 复合式多谐振荡器
~ parameter 混合参量
~ rock 混杂(浆)岩
~ vehicle 双(多)动力型汽车
~ -type 混合型[式]的, 差动式, 桥接岔路型

hybridism ['haibridizəm] *n.* = hybridity 混合性, 混染[杂]作用, 混成

hycar *n.* 合成橡胶, 丁二烯-丙烯腈共聚物

hydatogenesis [ˌhaidætə'dʒenəsis] *n.* 热液矿作用, 水成作用

hydatogenous [ˌhaidə'tɔdʒinəs] *a.* 液成的, 水成的
~ rock 水成岩

hydralime *n.* 熟石灰

hydrant ['haidrənt] *n.* 消防[配水]龙头, 给水栓, 取水管
~ barrel 消防龙头套筒
~ bonnet 消防龙头护罩
~ cap 消火栓帽
~ pipe 消防水管

hydrate ['haidreit] *n.* 水化[合]物; 氢氧化物; *v.* -drated, -drating; (使)成水化物; hydrator, *n.*
~d form 水合式

hydration

~d lime 熟石灰,消石灰($Ca(OH)_2$)
hydration [haiˈdreiʃən] *n.* 水合[化]作用
   ~ heat 水化[合]热
   ~ retarder 水化阻滞剂
   ~ water 结合水
hydraucone 喇叭形尾水管,锥形尾水管,水力圆锥体,水力喇叭口,喇叭口
   ~ (draft) tube 水力锥式尾水管,喇叭形尾水管
hydrauger 水力螺旋钻
   ~ hole 水冲钻[探]孔
   ~ method 水冲钻孔法
hydraulic [haiˈdrɔːlik] *n.* 水力、液压传动装置; *a.* 水力[工]的,液力[压]的;水硬的;-hydraulically, *ad.*; -hydraulicity, *a.*
   ~ admixture 水硬性掺料[添加剂]
   ~ accumulator 蓄水池,液压蓄能器
   ~ activity 水凝性,水硬性(水泥)
   ~ agitator 液力搅拌器
   ~ ally operated equipment 液压设备,液压[联合操作]设备
   ~ analogy 水力模拟
   ~ analysis 水力分析
   ~ architecture 水工建筑学
   ~ backhoe 液压反向铲
   ~ back-pressure valve 水压逆止阀
   ~ balancing disc 液压平衡盘
   ~ bender 液压弯管[曲]机
   ~ binding agent 水硬性胶结剂
   ~ bore 涌潮[浪]
   ~ borehole gauge 液压钻孔应变计
   ~ boring lathe 水力钻床
   ~ boundary condition 水力边界条件
   ~ brake 液压制动器,液压闸
   ~ buffer 水力缓冲[消能]器
   ~ calculation 水力计算
   ~ capstan 水力绞盘,水力起锚机
   ~ cell 液压传感器
   ~ cement 水硬性水泥
   ~ cement concrete 水硬水泥混凝土
   ~ chamber 水压室

   ~ characteristics 水力特性[征]
   ~ classifier 水力分级机
   ~ clutch 液压离合器
   ~ conductivity 导[透]水性
   ~ controlled scraper 液压式铲运机
   ~ crane 液压起重机
   ~ cylinder hoist 液压圆筒式启门机
   ~ design 水力[液压]设计
   ~ development 水力开发[枢纽]
   ~ drag 水阻力,水力牵引
   ~ dredge 水力吸泥[疏浚]机
   ~ drop 跌水
   ~ dynamometer 液压测力计
   ~ efficiency 水力效率(净水头与总水头之比率)
   ~ elements (河流的)水力因[要]素
   ~ elevator 水力提升机
   ~ emptying system (船闸的)泄水系统,排水系统
   ~ engine 水力发动机
   ~ engineering 水利工程,水力工程(学),水工学
   ~ excavation 水力挖掘
   ~ feed 液压进料,水力传送
   ~ felling wedge 液压伐木楔
   ~ fill dam 水力冲填坝
   ~ fluid 液压流体
   ~ flume 水力试验槽
   ~ flute transport 溜槽水力输送
   ~ friction 水力摩阻,水头损失
   ~ ga(u)ge 水压计
   ~ generator 水轮发电机
   ~ governing 液压调节
   ~ grade 水力坡降
   ~ grade line 水面线,水位降落曲线,水力坡降线
   ~ gradient 水力坡降[梯度]
   ~ gun 水枪,水力冲射器
   ~ hammer 液压锤
   ~ haulage 水力运输
   ~ head 水头
   ~ hoist 液压启门机,水力提升机
   ~ hook 液压挂钩

hydraulics

~ impact 水力冲击
~ inclination 水力[面]坡降
~ index 水硬率
~ indicator 水压计
~ intensifier 液压增强器
~ jack 液压[水力]千斤顶
~ jump 水跃
~ laboratory 水力实验室,水工试验所
~ levee 水力冲积堤
~ lift 水压升降机
~ lime 水硬性石灰
~ limestone 水硬性石灰石
~ loss 水力损失
~ machine 水[液]压机
~ machinery 水力机械
~ main 总水管,水管干线
~ mean radius 平均水力半径
~ mining 水力采矿
~ model 水工模型
~ modulus 水硬率
~ mortar 水凝灰浆
~ motor 水力发动机
~ navvy 水力挖泥机
~ oil 液压油
~ operated 液压操纵的
~ parameter 水力参数
~ permeability 水力渗透性
~ pile driving 液压打桩
~ pipeline 水力管道
~ plunger elevator 液压柱塞升降机
~ potential 水力位[势]能
~ power 水力[能]
~ power plant 水力发电厂
~ power project 水力发电工程
~ press 液[水]压机

hydraulic jack

hydraulic pressure elevator

~ pressure 水[液]压(力)
~ pressure elevator 液压提升装置
~ pressure regulator 液压调节器
~ pressure test 水[液]压试验
~ prime mover 水力原动机
~ propulsion system 液压推进系统
~ puller 液压张拉器
~ pump 水泵
~ punching machine 水力[液压]冲孔机
~ radius 水力半径
~ ram 压力扬吸机, 液压夯锤
~ ram pump 水锤泵
~ regulator 水力调节器
~ retaining structure 挡水水工建筑物
~ riveting 液压铆接
~ seal 防水层
~ shock 水力冲击
~ shock absorber 液压减振器
~ similarity 水力相似性
~ slip coupling 液压滑动连轴节
~ slope 水力坡度[降]
~ slushing 水力充填,水砂充填
~ sprayer 水力喷雾器
~ stability 水力稳定性
~ strength 水硬强度
~ structure 水工建筑物,水工结构
~ suction dredger 吸扬式挖泥船
~ suspension 水力悬浮
~ system 液压系统
~ transport 水力输送
~ tripping device 液压脱钩装置
~ turbine 水轮机
~ uplift pressure (水)扬压力
~ valve 液压(操作)阀
~ vane pump 水力叶轮泵
~ variable-speed gear 液压变速齿轮

hydraulics [haiˈdrɔːliks] *n.* 水力学
~ of open channel 明渠水力学

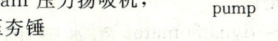
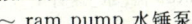

hydraulic ram pump

~ of well 管井水力学

hydro ['haidrəu] *n.*; *pl.* -dros；水，液体，氢；水力发电厂；水疗处；*a.* 氢(化)的，水的；水力发电的
  ~ alternator 水力发电机
  ~ chart 水文图
  ~ chemistry 水(质)化学
  ~ chloric acid 盐酸
  ~ cone type 虹吸式，吸管式
  ~ consolidation 水固结(作用)
  ~ cooling 水冷处理
  ~ crane 水力起重机
  ~ -densimeter (土的)含水密实度测定仪
  ~ dynamometer 流[水]速计
  ~ energy 水能
  ~ foil 水叶，水翼
  ~ foil cascade 水力翼栅
  ~ generator 水力发电机
  ~ geochemistry 水文地质化学
  ~ geography 水文地理学
  ~ -isobaric line 等水压线
  ~ isobaths 地下水等高线，等(水)深线
  ~ isohypse 等(水)深线，地下水等高线
  ~ isopleth map 地下水位涨落图，水文等值线图
  ~ -jet dredge 水力采砂船
  ~ -junction 水利枢纽
  ~ manometer 流体压力计，测压计
  ~ mechanics 流体力学
  ~ melioration 水利土壤改良
  ~ meteor 水文气象，水气凝结体，降水
  ~ meter (液体)比重计，流速计
  ~ modulus 流量模数
  ~ morphous process 水成[溃]过程
  ~ motor 液压马达，水力发动机
  ~ motor propeller 水力推进器
  ~ period 水文周期，(洪水)淹没期
  ~ phobic portland cement 憎水(硅酸盐)水泥
  ~ phone 水下测听器，水下地震检波器，漏水检查器

~ photometer 水下光度计
~ power tunnel 水力发电隧洞
~ project 水力工程
~ science 水科学，水文科学
~ scopic coefficient (of soil) (土的)吸水[湿]系数
~ scopic water 吸着水，吸湿水
~ seal 液[水]封，液压密封
~ sizer 水力分级机
~ -vacuum 油压真空制动器
~ valve 水龙头，液压开关[阀，活门]

hydrocooler *n.* 水冷却器
hydrocooling *v.* 用水冷却
hydrodynamic(al) ['haidrəudai'næmik(əl)] *a.* 水动力(学)的，流体动力(学)的
  ~ analogy 水动力模拟
  ~ form 流线形
  ~ gauge 动水压力计
  ~ head 动力水头
  ~ lag 水力延滞
  ~ phenomena 水动力现象
  ~ pressure 动水压力

hydroelectric [ˌhaidrəui'lektrik] *a.* 水电的，水力发电的
  ~ engineering 水力发电工程
  ~ plant 水力发电厂[站]
  ~ potentiality 水力蕴藏量
  ~ power plant 水力发电厂
  ~ resource 水力资源
  ~ schemes 水电开发计划，水电站系统，水电枢纽
  ~ station 水力发电厂

hydroelectricity [haidrəuilek'trisiti] *n.* 水电
  ~ generation 水力发电

hydroextractor *n.* 脱水机
hydrogel *n.* 水凝胶
hydrogen ['haidridʒən] *n.* 氢(H)
  ~ bomb 氢弹
  ~ nitrate 硝酸
  ~ nitride 氨(NH_3)

~ peroxide 过氧化氢
~ sulphide 硫化氢
~ test 测氢试验
hydrogenic rocks 水成[生]岩
hydrogeological *a.* 水文地质的
~ chart 水文地质图
~ condition 水文地质条件
~ map 水文地质图
~ nature 水文地质特性
hydrogeology [ˌhaidrəudʒi-'ɔlədʒi] *n.* 水文地质(学); hydrogeologist, *n.*; hydrogeologic, hydrogeological, *a.*
hydrograph ['haidrəɡrɑːf] *n.* (水文)过程线, 水文曲线; 自记水位计

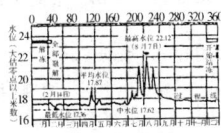

hydrograph

~ analysis 水文分析
~ assending limb (水文)过程线上升段
~ recession limb (水文)过程线下降段
hydrographic(al) [ˌhaidrə'ɡræfik(əl)] *a.* 水文地理的; 水道测量的
~ cast 水文施测, 测深
~ chart 水文[水系]图
~ curve 水文曲线
~ data 水文资料
~ datum 水文基准面
~ features 水文特征
~ map 水文测验图, 水道图
~ measurement 水文测验
~ net 水系, 河网
~ ship 水文测验船
~ station 水文站
~ survey 水道[水文]测量
~ vessel 水文测验船
~ wire 水文测验缆索
~ year 水文年(度)
hydrography [hai'drɔɡrəfi] *n.* 水文学,

水文测验学, 水道测量学; hydrographic, hydrographical, *a.*; hydrographically; *ad.*
hydrokinetics *n.* 流体动力学
hydrolith *n.* 水生岩, 水成岩
hydrologic(al) [ˌhaidrə'lɔdʒik(əl)] *a.* 水文的
~ analogy 水文情态模拟, 水文比拟
~ and meteorologic survey 水文与气象测验
~ apparatus 水文仪器
~ atlas 水文图集
~ basin 水文流域
~ benchmark 水文用水准点
~ budget 水量[分]平衡(表)
~ calculation 水文计算
~ computation 水文计算
~ condition 水文条件
~ cycle 水文循环, 水循环
~ data 水文数据
~ divide 水文分水界
~ experiment 水文实验
~ exploration 水文查勘
~ forecast 水文预报
~ front 水文锋(面)
~ gage 水位计
~ investigation 水文研究
~ manual 水文手册
~ map 水文图
~ model 水文模型
~ network 水文站网
~ process 水文过程
~ prognosis 水文预测
~ record 水文记录
~ region 水文区域
~ routing 水文演算(洪水)
~ section 水文剖面
~ sensor 水文传感器, 水文探测设备
~ services 水文服务事业
~ station 水文站
~ warning 水文警报
hydrology [hai'drɔlədʒi] *n.* 水文学;

hydromechanics

-logical, *a*.; hydrologically, *ad*.
hydromechanics *n.* 流体力学
hydrometeorology *n.* 水文气象学
hydrometric(al) [haidrəu'metrik(əl)] *a.* 水文测验的
~ float 水文测验浮标,测水浮标
~ pendulum 流速摆(测量河道流速用)
~ section 水文测验断面
~ station control 水文测站控制
hydrometry [haidrɔmitri] *n.* 水文测验(学),(液体)比重测定(法),测湿法
hydromica *n.* 水云母
hydropneumatic [haidrəunju:'mætik] *a.* 液压气动的,水气并用的
~ system 液压气动(供水)系统
hydropneumatics *n.* 液压气动学
hydrous ['haidrəs] *a.* 含水的,水化的
~ micas 含水[水化]云母
hydrostatic(al) [haidrɔ'stætik(əl)] *a.* 静水力学的,流体静力(学)的,静水压力的;-hydrostatically, *ad*.
~ accelerometer 流体静力加速仪
~ balance 比重秤
~ catenary 静水压悬链线
~ compression 静水压缩
~ curve 静水压(力)曲线
~ drive 液压静力传动
~ excess pressure 超静水压力,超流体静压力
~ equation 流体静力方程
~ gauge 静水压力计
~ head 静水头
~ level 静水位,静水平面
~ levelling apparatus 流体静力水准测量仪
~ overpressure test (钢管的)超静水压试验
~ press 水压机
~ pressure 静水压力
~ pressure ratio 静水压力比,(土压力的)静水压力系数
~ pull 静水拉[张]力
~ stress 流体静胁强[应力]

~ -structure 水工结构
~ sludge removal 静水排泥
~ tension 静水张力
~ test (静)水压试验
~ transmission 液压传动[递]
~ uplift 静水上托[扬压]力
hydrostatics [haidrəu'stætiks] *n.* 流体静力学
hydrotechnology *n.* 供水技术,水的贮存与分配工程学
hydrothermal [ˌhaidrə'θə:məl] *a.* 地上[下]水垫的,热水[液]的,水热作用的
~ alteration 热液蚀变
~ metamorphism 热液变质
~ method of curing (混凝土)湿热养护(法)
hydrotimeter *n.* 水硬度计
hyetal ['haiitəl] *a.* 雨的,降雨的
~ coefficient 雨量系数
~ interval 雨量差(等雨量线差值)
~ region 多雨地区
hyetograph ['haietəgra:f] *n.* 雨量计,雨量分布图,年平均雨量分布图表
hyetographic *a.* 雨量过程线的,雨量图[计]的
~ curve 降水量曲线
hyetometer [haii'tɔmitə] *n.* 雨量表
hygiene ['haidʒi:n] *n.* 卫生学,保健学
hygienics [hai'dʒi:niks] *n.* 卫生学
hygroautometer *n.* 自记湿度计
hygrograph ['haigrəgræf] *n.* (自记)湿度计[仪]
hygrology [hai'grɔlədʒi] *n.* 湿度学
hygrometer [hai'grɔmitə] *n.* 湿度计
dew point ~ 露点湿度计
hair ~ 毛发湿度计
hygrometry [hai'grɔmitri] *n.* 测湿法,湿度测定法
hygroscope ['haigrəskəup] *n.* 验[测]湿器,湿度计

hair hygrometer

hygroscopic [ˌhaigrə'skɔpik] *a.* 验湿器的;易湿的,吸湿性的; hygroscopically *a.*; hygroscopicity, *n.*
  ~ absorption 吸湿(作用)
  ~ coefficient 吸湿系数
  ~ effect 吸湿作用
  ~ moisture 吸湿水分
  ~ water coefficient 吸湿(水)系数

hygroscopicity *n.* 吸湿性,吸湿性,润湿度
  ~ material 吸湿性材料
  ~ moisture 吸湿含水量(土壤在自然干燥状态时的含水量),吸湿水

hygroscopy *n.* 湿度测定法,潮解性,吸水性

hygrostat ['haigrəstæt] *a.* 湿度测量控制仪;湿度检定箱

hygrothermograph [ˌhaigrə'θə:məgra:f] *n.* (自记)湿温计,温湿仪[计]

hykea (**hylaea**, **hileia**) *n.* 热带雨林

hylean *vt.* 被森林覆盖

hypabyssal rock 浅成岩,半深成岩

hypaethral [hi'pi:θrəl] *a.* = hypethral;天井的,(古典建筑)全部或部分露天的,中央无瓦顶的

hyperacid [haipə(:)'ræsid] *a.* 多酸的

hyperacoustic *a.* (特)超声(波)的

hyperbar *n.* 高气压

hyperbola [hai'pə:bələ] *n.* 双曲线

hyperbolic(al) [ˌhaipə'bɔlik(əl)] *a.* 双曲线的;夸张的
  ~ function 双曲(线)函数
  ~ plane 双曲平面
  ~ space 双曲空间
  confocal ~ 共焦双曲线

hyperboloid [hai'pə:bələid] *n.* 双曲面; hyperboloidal, *a.*

hyperborean [ˌhaipə'bɔ:riən] *a.* 极北(人)的,酷寒的; *n.* 北极人

hypercharge 超(负)荷

hypercritical ['haipə:kritikəl] *a.* 超临界的;苛求的,吹毛求疵的
  ~ flow 超临界(水)流

hyperelastic [haipəri'læstik] *a.* 超弹性的
  ~ law 超弹性定律

hyper inflation 恶性通货膨胀

hypermatic *a.* 过粘的

hypernormal 超常态
  ~ dispersion 超常态离差[离中趋势]

hyperphysical [haipə(:)'fizikəl] *a.* 超物质的,超自然的

hyperplane ['haipəˌplein] *n.* 超平面

hyperpressure *n.* 超压

hyperpycnal inflow (超)重入流

hyperpyrexia [ˌhaipəpai'reksiə] *n.* 温度过高,过热

hyperquadric *a.* 超二次曲面(的)

hyperscope ['haipəskəup] *n.* 潜望镜

hypersonic [ˌhaipə'sɔnik] *a.* 超音速的
  ~ flow 超声速流
  ~ speed 超音速

hyperspecilization 高度专门化

hypersphere ['haipəsfiə] *n.* 多维[超]球面;-hyperspherical, *a.*

hyperstatic *a.* 超静定的
  ~ structure 超静定结构
  ~ system 超静定体系

hyperstaticity *n.* 超静定性

hypersurface *n.* 超曲面,多维曲面

hyperthermic (土湿)超热状况

hypervelocity *n.* 超高速
  ~ free stream 超高速自由流

hypervisor *n.* 管理程序

hypethral [hi'pi:θrəl] *a.* 露天的,无屋顶的

hypethron *n.* 院子,天井

hyphen ['haifən] *n.* 连字符,短划[横]

hypobaric [ˌhaipə'bærik] 低(气)压(环境)

hypobatholitic zone 深成岩基带

hypocenter = hypocentre = hypocentrum *n*. 震源,地面零点
hypocycloid [ˌhaipəuˈsaikləid] *n*. 内摆线,圆内旋轮线
hypodispersion *n*. 平均分布
hypoelastic law 准弹性定律
hypofiltration 深成(的)渗透作用
hypofunction [ˌhaipəuˈfʌŋkʃən] *n*. 机能减[衰]退
hypogeal [ˌhaipəˈdʒi:əl] *a*. 地下的,地下生长的
hypogee *n*. 岩洞建筑,地下[山边]建筑
hypogene [ˌhipədʒi:n] *a*.; 地下(生成)的,深生的;上升的; *n*. (地壳)内力,深成
~ action 内力[深成]作用
~ rock 深成岩
~ water 上升水
hypogeum [ˌhipəˈdʒi:əm] *n*. 地窖,地下室;古代地下墓室,古代房屋的地下部分
hypopycnal inflow 轻入流,低重入流
hypostyle hall 多柱式走廊
hypotenuse [haiˈpɔtinju:z] *n*. (直角三角形的)斜边,弦

hypostyle hall

hypothec [haiˈpɔθek] *n*. 抵押权,担保权
hypothecate [haiˈpɔθikeit] *v*., -cated, -cating; 抵押; hypothecator, *n*.; hypothecation, *n*.
hypothermia [haipəˈθə:miə] *n*. 低温,体温过低
hypothesis [haiˈpɔθisis] *n*. *pl*. -ses;假说,假设,学说
hypothesize [haiˈpɔθisaiz] *v*., -sized, -sizing;假设,假定; hypothesizer, *n*.
hypothetic(al) [ˌhaipəuˈθetik(əl)] *a*. 假定[说]的

~ hydrograph 设计水文过程线
~ structure 假设结构
hypotonic [ˌhaipəˈtɔnik] *a*. 低渗(性)的;-hypotonicity, *n*.
hypsographic *a*. 测高(学)的
~ curve 等高[深]线,湖海等深线,高程面积曲线
~ feature 高度地势
~ names 图上地形要素名称
hypsography [hipˈsɔgrəfi] *n*. 测高学,等高线法;比较地势学,地貌表示法
hypsometer [hipˈsɔmitə] *n*. 测高计,沸点气压[测高]计
hypsometric(al) [hipsəˈmetrik(əl)] *a*. 测高(学,术)的
~ curve 等高线,等深线
~ formula 测高公式
~ map 等高线地图
~ method (地图)分层设色法
hypsometry [hipˈsɔmitri] *n*. 测高法[学,术]
Hy-rib 一种钢丝网(作轻型混凝土板中钢筋用)
hysteresis [ˌhistəˈri:sis] *n*. 滞后现象[作用],迟滞(性),磁滞(现象)
~ coefficient 磁滞系数
~ damping 磁滞阻尼
~ loop 磁带回线,滞后环
~ loss 磁滞损失
~ meter 磁滞计
~ modules 滞后模量
~ set 滞后变形
hythergraph [ˈhaiθəgra:f] *n*. 温度与湿度关系图,温湿图
hy-therm 耐[抗]热的
hy-Tuf steel 高强度低合金钢
hyzone *n*. 氚(即三原子的氢)
Hz = hertz 赫兹

I

ice [ais] ***n.*** ; ***v.*** , iced, icing; ***a.*** ; ***n.*** 冰(块); ***v.*** 结冰,冻结,冰冻,冻凝; ***a.*** 冰的
~ admixture 冰掺合料
~ age 冰期,冰河时代,更新世冰期
~ apron (桥墩的)挡冰板
~ atlas 冰图
~ auger 凿冰器,冰钻
~ ax(e) 破冰斧
~ barchan 新月形冰丘
~ -barrage 冰坝
~ barrier 冰障,冰堰
~ -bearing current 挟冰水流
~ belt 浮冰带,破冰船侧装的厚甲
~ boom 防冰栅
~ bore 冰钻孔
~ box 冰箱,冷冻机
~ bridge 冰川坚冰,冰桥,冰坝
~ breaker 破冰船
~ cap 冰帽[盖]
~ cellar 冰窖
~ chart 冰情(分布)图
~ -chest 冷箱,冰库
~ chisel 冰錾
~ chute 泄冰槽
~ coating 结冰,冰层
~ cofferdam 冰围堰
~ condition observation 冰情观测
~ content 含冰量
~ control 冰流控制,冰冻防护措施
~ control gate 冰流控制闸门
~ cover 冰盖,冰层
~ cover water rating curve 结冰影响的水位流量(关系)曲线
~ crusher 碎冰机
~ dam 冰坝
~ deflecting boom 拦冰栅
~ designation 冰情图例
~ dike 冰墙,冰脉

~ doubling 抗冰衬板
~ drift(ing) 浮冰,流冰,漂冰
~ drill 冰钻,钻冰机
~ field 冰原
~ float(ing) 浮冰,冰盘
~ -floe 大浮冰,浮冰块
~ flood 凌汛
~ fog 冰雾
~ forecast 冰情预报
~ -free harbor 不冻港
~ freezer 制冰淇淋机
~ -free period 无冰期
~ -free port 不冻港
~ gauge 量冰尺,测冰仪
~ gate 泄冰闸门
~ glass 冰花状玻璃
~ gorge 冰峡,冰坝,冰谷
~ guard 挡冰栅,桥墩的破冰构造,冰挡
~ hail 小雹
~ (-)house 冷藏库,冰室,冰屋,贮冰库,制冰厂
~ jam 冰阻,冰块拥塞,河槽由于冰块堆积而阻塞
~ jam flood 冰塞凌汛
~ jam stage 冰壅[冰塞,冰坝]水位
~ layer 冰层
~ ledge 沿岸冰,冰栅
~ load 冰荷载
~ machine 制冰机,冷冻机
~ needle 冰针(晴冷天气中飘荡的一些细长冰粒)
~ paper 透明纸(制图用),冰纹纸
~ patterned glass [window] 冰花

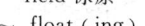

ice dam

玻璃[窗]
~ pellets 小冰球,小冰雹
~ period 冰冻期
~ pick 冰凿,冰镐
~ plant 制冰厂
~ point 冰点
~ pressure 冰压力,冰凌压力
~ prevention measure 防冰措施
~ protection 流冰保护
~ push 冰推力,冰壅
~ rain 冻雨,冰雨
~ rampart 冰(河)堤,冰成砂砾堆
~ regime forecast 冰情预报
~ removal 除冻器
~ removal agent 防冻剂
~ rink 溜冰场
~ road 冰道
~ run 淌凌,流冰
~ runoff 冰水径流
~ scouring 冰冲刷作用
~ sheet 冰层,冰原,陆冰(陆上的冰层,冰盖)
~ sluice 泄冰闸
~ stream 冰河,冰流
~ structure 冰花结构
~ thrust 冰压力,冰推力
~ undercurrent 冰川底流
~ water 冰水
blue ~ 纯洁冰(冰愈纯洁色愈蓝)
electric ~ chest 电冰箱
ground ~ 底冰

iceberg [ˈaisbə:g] *n.* 冰山,流冰
icebound [ˈaisbaund] *a.* 冰封住的,(海港,河流等)封冻的
Iceland [ˈaislənd] *n.* 冰岛
~ spar 冰洲石,双折射透明方解石
icestone *n.* (天然)冰晶石
ichnography [ˈiknəˌgrɑ:fi] *n.* 平面图[法]
ichthyoid [ˈikθiɔid] *n.* 鱼形体,流线形体; *a.* 鱼(状)的,流线形的
ICI (International Commission on Illumination) 国际照明委员会

I-column *n.* 工字柱
icon [ˈaikən] *n. pl.* icons, icones; 标志或标记(借以表示与它所代表的事物具有共同特性); 肖像, 图像, 插画; 彩色图谱, 画像; 偶像
iconography [ˌaikəˈnɔgrəfi] *n.* 插画, 插图; 图解; 影像塑造术; 影像学, 肖像学[研究]; iconograph, *n.*; iconographer, *n.*
iconolog [ˌaikəˈnɔləg] *n.* 光电读像仪; 图像学
iconometer [ˌaikəˈnɔmitə] *n.* 光像测定器; 返光镜; 测距镜, 量影仪
icy [ˈaisi] *a.* icier, iciest; 覆盖着冰的, 多冰的; 冰构成的; 极冷的, 冰冷的, 似冰的
~ shower 冰(阵)雨
~ snow 冻结雪, 冰状雪
~ wind 冰冷风
ID (industrial design) 工业制品设计, 工艺美术设计
idea-finding 研讨方案, 探索方案
Ideal *n.* 铜镍合金(铜 55% ~ 60%, 镍 45% ~ 40%)
ideal [aiˈdiəl] *n.* 理想, 设想, 典型, 概念; *a.* 理想的; 标准的; 概念的, 虚构的; 唯心论的; idealness, *n.*
~ boundary 理想边界
~ building 标准建筑物
~ climate 理想气候
~ constraint 理想约束
~ cupola 理想圆顶, 典型圆顶
~ dome 典型圆(屋)顶

ideal dome

~ energy gradient 理想能(坡)线
~ engine 理想发动机
~ fluid 理想流体
~ function 理想函数, 广义函数
~ gas equation 理想气体方程
~ grading curve 理想的颗粒级配曲线

~ grain size 理想粒径
~ liquid 理想液体
~ middle-sized grain 理想中值[等]粒径
~ partical size 理想粒径
~ plan 理想方案
~ plastic theory 理想塑性理论
~ sea level 理想海平面,海洋等势面
~ section 理想剖面,理想断面
~ solution 理想溶液,理论解
~ value 理想值
~ wave 理想波
idealine *n.* 糊状粘结剂
ideality [aidi'æliti] *n. pl.* -ties;理想(状态,性质);理想化;想像力,观念性;(*pl.*)理想的事物
idealize [ai'diəlaiz] *v.*, -ized,-izing;……理想化;形成理想;idealizer, *n.*
~d elastic continual mass 理想弹性连续体
~d fragmental mass 理想碎屑体,理想散体
idemfactor ['aidəmfæktə] *n.* 等幂因子,等幂矩阵,归本因素
identical [ai'dentikəl] *n.* 恒等式;*a.* 恒等的,恒同的;相等的,相同的
~ element 单位元素,幺元
~ equation 恒等方程
~ figures 恒等形
~ map 等角投影地图
~ points 对合点
~ relation 恒等关系,恒等式,全等式
identification [aidentifi'keiʃən] *n.* 证明,识别,鉴定,辨别;号码装定(计算技术);标志,标识符;恒等,相同
~ beacon 航标,灯塔标志;岸标
~ card 身份证
~ code 标识符,识别码
~ mark 识别符号,商标,标志
~ markings (商标)打印标号,认识标记
~ of position 位置的确定
~ system 识别系统

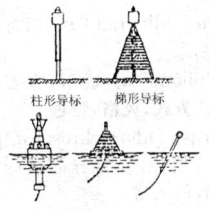

柱形导标　梯形导标

identification beacon

~ test 鉴别试验
coded ~ 编码符号,译码表示法
file ~ 文件识别
instrument ~ 仪器标记
identifier [ai'dentifaiə] *n.* 标识[标志,识别]符,文献编号
identify [ai'dentifai] *v.* -fied,-fying;识别,辨别,区别,鉴定;等同;标志;联合,参与;identifier, *n.*
~ -disc 证明牌;辨别牌
~ing plate 标号牌
identity [ai'dentiti] *n. pl.* -ties;同一性,完全相同,一致;恒等(式);本性[身];国藉,身份
~ card 身份证
~ element 单位[恒等]元素,幺元(素)
~ equation 恒等方程,认定方程
~ law 同一律
~ matrix 单位矩阵
~ unit (全)同门
identometer *n.* 材料鉴别仪
ideograph ['idiəugra:f] *n.* =ideogram;表意文字;记号,符号
Chinese ~ 汉字
ideotype *n.* 非典型标本,表意标本;外型,续型
idigbo *n.* 非洲伊地泡木,非洲伊地泡木的木材
idioadaptation *n.* 个别适应,特殊适应
idiocrasy [idi'ɔkrəsi] *n. pl.* -sies; = idiosyncrasy;敏感,特异体质,特异(反)应性;idiocratically, *ad.*; idiocratic, idiocratical, *a.*
idioelectric *n.*; *a.* 非导体,能摩擦起电

的(物体)
idiograph ['idiəgra:f] *n.* 商标;(个人)签名,私章
idiom ['idiəm] *n.* 成语,方言;习惯语法;表达方式,风格,特色
idiosyncrasy [idiə'siŋkrəsi] *n. pl.* -sies;个性,特质[性];特异品质;特异反应性;特有的风格
idle ['aidl] *n.*; *v.*, idled, idling; *a.* idler, idlest; *n.* 空转,空载,慢车[速];无功[效],停机; *v.* 空转,开慢车;浪费,虚掷; *a.* 空载的,空转的;闲散的,停顿的;无功的; idly, *ad.*
~ asset 闲置资产
~ balance 闲置余额
~ capacity 备用容量,空转功率,闲置能量
~ capital 闲置资本
~ cash 闲散现金
~ current 无功[效]电流
~ equipment 闲置设备
~ fund 游资,闲散资金
~ gear(＝idler gear) 不作用齿轮;惰轮
~ motion 空转
~ time 停工时间,空载时间
~ unit 备用机组,闲置机组
~ wheel 惰轮,空转轮,滚轮
run ~ 窝工,空转
synchronous ~ 同步(信号)空载;空载同步
idler ['aidlə] *n.* 惰轮,空转轮,中间轮;导向辊;托辊,空车,空载
~ gear 空转齿轮,惰轮
~ pulley 惰轮
~ roller 导[托张力,惰]辊
~ shaft 导轮组,惰轮轴
~ track 导轮轨
~ travel 空转行程
impact ~ 防冲托辊
running ~ 空转轮
self-aligning ~ 自调位托辊
trailing ~ 从动空转轮

idocrase ['aidəukreis] *n.* 符山石
idrialite *n.* 辰砂地蜡,绿地蜡
igdanit *n.* 硝酸铵与柴油燃料混合物
igdantin *n.* 依格达胶(光弹模型材料)
igelite *n.* 聚氯乙烯塑料
igepons *n.* 依格泡(表面活性剂)
I-girder *n.* 工字大梁
igloo ['iglu:] = iglu *n. pl.* igloos;圆顶建筑,园顶弹药库(工事),(爱斯基摩人用雪块砌成的)圆顶小舍
igneous ['igniəs] *a.* 火的,火成的,含有火的;靠火力的,熔融的;引发火山活动的; *n.* 火成岩
~ activity 火成活动(与火成岩侵入和形成有关的所有作用)
~ fusion 煅烧熔融
~ magma 岩浆
~ rock 火成岩,岩浆岩
(~)volcanic rock (火成)火山岩

igneous rock

ignescent [ig'nesnt] *n.* (碰击后)发出火花的物质; *a.* 发出火花的
ignimbrite [ig'nimbrait] *n.* 熔结凝灰岩
ignitability [ignaitə'biliti] *n.* 易燃性,可燃性
ignitable [ig'naitəbl] = ignitible *a.* 可着火的,可燃的,易燃的;易着火的
ignite [ig'nait] *v.*, -nited, -niting;点火,发火,着火,(使)燃烧;-igniter, *n.*
~ ting circuit 点火电路
~ ting fuse 传爆信管
~ ting primer 雷管,起爆药包
igniter [ig'naitə] *n.* 点火剂;点火器,点火装置;引爆装置,放炮器
booster ~ 助推器[火药],点火剂
cartridge ~ 装药点火管
delay ~ 延迟点火装置
flame ~ 点火装置,点火具;点火药
hypergolic ~ 自燃点火器,自燃发火装置
integral ~ 复合[二组元]喷嘴

pyrotechnic ~ 烟火发火剂,烟火发火器
squib-initiated ~ 火花塞式点火器
torch ~ 火舌式点火器

ignition [ig'niʃən] *n.* 点火,发火,引燃,起爆,发火装置
~ hazard 着火危险
~ charge 点火药,导火炸药
~ circuit 发火电路
~ coils 发火线圈
~ delay 发火延迟,延迟点火
~ heat 燃烧热
~ lag 延迟发火(装置)
~ pad 点火药包
~ plug 火花塞
~ point 着火点,发火点,燃点,点火瞬间
~ powder (焊接用)点火药
~ system 发光系统,发火设备
~ temperature 燃点,着火温度
~ timing 发火定时,发火正时
arc ~ 电弧点火
auxiliary ~ 利用起动燃料点火
correct timed ~ 正时点火
electro-magnetic ~ 电磁点火
high tension ~ 高(电)压点火
hot-wire ~ 电阻丝点火
multi-stage ~ 多段点火,多级点火
retard(ing) ~ 延迟点火
spontaneous ~ 自发着火,自燃
submersion-proof ~ 防水的点火系统,潜水点火系统

ignitor =igniter *n.* 点火器,发火器;引燃电极,引燃管
~ pad 点火药包
~ train 导火索[线]

ignitron [ig'naitrɔn] *n.* 引燃管,点火管,放电管,引燃电极;水银半波整流管
air-cooled ~ 气冷式引燃管
water-cooled ~ 水冷式引燃管
water-jacked metal ~ 水套冷却金属点燃管

ignitus [ig'nitəs] *n.*; *a.* 火红色(的)

ignore [ig'nɔ:] *n.*,*v.*,-nored,-noring; *n.* 无作用(符号);(电极)空点(子); *vt.* 忽略,不问,否定,抹煞,无视;-ignorable,*a.*;-ignorer,*n.*
~ character 无作用字符,非法符
~ instruction 无效指令,否定指令

ihleite *n.* 黄铁矾

ihrigizing *n.* 硅化法,固体渗硅

I-iron *n.* 工字铁,工字钢

ill [il] *n.*;*a.*, worse,worst; *ad.*; *n.* 疾病,伤害;灾难,不幸; *a.* 生病的,有害的,有缺点的; *ad.* 不善,不完善,不充分,无效地;几乎不能地
~ -condition 病态
~ -conditioned equation 病态方程
~ -conditioned matrix 病态矩阵
~ -mangement 管理不善

illation [i'leiʃən] *n.* 推理,推论,推断;演绎[法]

illegible [i'ledʒəbl] *a.* 不明了的,难阅读的;难以辨认的,字迹模糊的;illegibleness,*n.*; illegibly,*ad.*

illimitability [i,limitə'biliti] *n.* 无限,不可计量,不可限制

illiquid [i'likwid] *a.* 非现金的,无流动资金的,不动的;(债务等)不明确的
~ holding 不动产

illite ['ilait] *n.* 伊利石,伊水云母

illuminance [i'l(j)u:minəns] *n.* 照度,照明度,施照度

illuminant [i'lju:minənt] *n.* 施照体,发光体;光源;照明剂;照明装置; *a.* 发光的,照明的,照耀的

illuminate [i'lju:mineit] *vt.*,-nated,-nating;照明,照射,阐明,说明,启发
~d barrier 照明式围拦
~d body 受照体
~d bollard 照明系缆柱;照明护柱
~d ceiling 照明式顶棚,反光顶棚

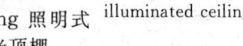
illuminated ceiling

~d circuit diagram 照明电[线]路图

~d contours 明暗等高线
~d guard-post 照明护柱
~d indoor fountain 照明式的室内喷泉
~d mirror （经纬仪上的）反光镜；照明镜
~d sign 照明标志（用外部光源照明）
~d system 照明系统
~d wall 照明墙，反光墙

illuminating [iˈljuːmineitiŋ] *a*. 照明的，照耀的；启发的；illuminatingly, *ad*.
~ device 照明装置
~ effect 照明效应
~ engineering 照明工程
~ glassware 照明器皿[仪器]
~ line 照明线路
~ mark 光标
~ mirror （经纬仪上的）反光镜，照明镜
~ power 亮度，照明能力

illumination [iˌljuːmineiʃən] *n*. 照明，照明度；照明学；发光，亮度；照明设备；涂饰，花彩；说明，解说，启发；(*pl*.) 灯饰，装饰
~ cable 照明电缆
~ calculation 照度计算[预测]
~ circuit 照明线路
~ component 照明元件
~ current 照明电流
~ device 照明设备
~ engineer 照明工程师
~ engineering 照明工程[技术]
~ from skylight(s) 天窗照明，天棚照明
~ installation 照明设备
~ intensity 照度
~ level 照度级
~ load 照明负荷
~ system 照明系统
ambient light ~ 环境光照明
background ~ 背景照明
critical ~ 中肯照明，临界照明，要求高的照明
daylight ~ 日光照明
diffused ~ 漫射照明，扩散照明
even ~ 均匀照明
highlight ~ 强光照明
indirect ~ 间接照明，反射照明
intensity of ~ 照度
laser beam ~ 激光束照明
monochromat(o) ~ 单色光照明
overhead ~ 顶部照明
stage ~ 舞台照明
threshold ~ 临界照明度

illuminator [iˈljuːmineitə] *n*. 照明器，施照器，发光器，照明设备；发光体；反光镜，反光板；（底片观察用）光源；启发者
~ level 照明高度
laser ~ 激光照明
Quasi-secretive ~ 半隐蔽照明器

illuminometer [iˌluːmiˈnɔmitə] *n*. 照度计
Macbeth ~ 麦克贝斯照度计

illustrate [ˈiləstreit] *v*., -trated, -trating；说明，插图，图解；举例说明
~d book 有插图的书
~d catalogue 插图目录表

illustration [ˌiləsˈtreiʃən] *n*. 说明，图解，注解；实例，例证；图例；插图，图表

illustrious [iˈlʌstriəs] *a*. 著名的，显赫的，杰出的，光辉的，明亮的；illustriously, *ad*.; illustriousness, *n*.

illuvial [iˈluːviəl] *a*. 淀积的
~ horizon 淀积层

illuviation [ilˌ(j)uːviˈeiʃən] *n*. 淀积作用，淋积作用

illuvium [iˈluːviəm] *n*. 淀积层，淋积层，淀积物

image [ˈimidʒ] *n*.; *v*., -aged, -aging；*n*. 图像，形像，映像；塑像；典型；概念，想像，印象；比喻，隐喻，*vt*. 描绘，想像出，使……成像，反映，映射；

image

象征
~ dissector 析像器[管]
~ scale 图像比例尺,象标
acoustic ~ 声像
circuit ~ 电路图像
developed ~ 放大照片
diminished ~ 缩小像
fault ~ 失真图像
full-colour ~ 全彩色影像
hard ~ 黑白鲜明图像
harsh ~ 色调鲜明图像
high-contrast ~ 对比度强的图像
hologram ~ 全息照相图像
low-luminosity ~ 低亮度图像
map ~ 地图照相图
out-of-focus 散焦图像,不清晰图像
real ~ 实像
secondary ~ 重像
sharp ~ 清晰影像
soft ~ 色调不鲜明图像
stereoscopic ~ 立体像
superimposed ~ 重叠影像
television ~s 电视图像
virtual ~ 虚像
visual ~ 可见图像,目视图像
imaginal [i'mædʒinəl] *a.* 形像的,想像力的,(有关)想像的
~ type 表象类型
imaginary [i'mædʒinəri] *n.* 虚数; *a.* 假设的,幻想的,虚(构)的;虚数的; imaginarily, *ad.*; imaginariness, *n.*
~ axis 虚轴
~ component 虚数部分
~ hinge 假铰,虚铰
~ intersection 虚相交
~ line 虚线
~ load factor 虚荷载因素
~ number 虚数
~ part (复数)虚部
~ point 虚点
~ (pre)stressing 虚预加应力
~ quantity 虚量
~ root 虚根

~ tensioning 虚张拉
~ stressing 虚加力
~ stretching 虚拉伸
imagination [imædʒi'neiʃən] *n.* 假想,空想;想像力,创造力
imbank [im'bæŋk] =embank *v.* 筑堤(防护),用(土)堤围起
imbankment [im'bæŋkmənt] *n.* 堤防,路堤,堤坝;筑堤(工程),填土
imbed [im'bed] *v.*, -bedded -bedding =embed 嵌入,埋置,灌封
~ ded chips 嵌入(有机结合料层中的)石屑
imbibant [im'baibənt] *n.* 吸涨体
imbibe [im'baib] *n.* 吸液膨润
imbibition [ˌimbi'biʃən] *n.* 吸胀作用,吸液,吸收;(土壤)水分;透入,浸润,浸染
~ pressure 吸胀压
~ process 吸液印相法
~ moisture 吸入水
~ water 渗吸水,吸胀水
imbrex ['imbreks] *n.* 槽瓦,筒瓦;鳞状装饰中的一片或一部分
imbricate ['imbrikeit] *v.* ['imbrikit] *a.*; *v.*, -cated -cating; *v.* 使成鳞状,(使)叠盖; *a.* 覆[叠]瓦状的,鳞状的;重叠的; imbricately, *ad.*
~ stucture 叠瓦结构
~d dune 叠置沙丘
~d ornament 鳞状装饰,鳞状饰物
~d plate 搭盖板
~d roof 叠瓦屋顶
~d texture 鳞状结构
imbrication [ˌimbri'keiʃən] *n.* 叠覆,叠盖,鳞状叠覆;鳞形的装饰或图案;叠瓦作用

imbrication

imbue [im'bju:] *vt.*, -bued -buing; 使吸入(水分等),浸染;灌注,充满;使蒙受,感染,影响; imbuement, *n.*

imerina stone *n.* 散光闪石
imernite *n.* 钠透闪石,散光闪石
Imhoff(-) cone *n.* 殷氏圆锥管,英霍夫锥形管(测定沉淀性物质用)
　～ test 殷氏圆锥管试验(污水沉淀)
Imhoff tank *n.* 隐化池;双层沉淀池,英霍夫沉淀池
imine [i'mi:n] *n.* 亚胺
imitant ['imitənt] *n.* 赝品或代用品
imitate ['imteit] *vt.*, -tated,-tating;模仿,仿造,模拟;伪造
imitation [imi'teiʃən] *n.* 仿造品,赝品,复制品;模仿,模拟,效法;*a.* 冒充的,假造的; imitational, *a.*
　～ art paper 仿美术纸
　～ brick 假砖
　～ gold 装饰用的铜铝合金(铝3%～5%,其余铜)
　～ leather 假皮,人造革
　～ marble 仿大理石,人造大理石
　～ stone 假石,人造石
　～ wood 人造材
　signal ～ 信号模拟
imitator ['imiteitə] *n.* 仿真品,模拟器;模仿者
immalleable [i'mæliəbl] *a.* 无韧性的,无延展性的
immaterial [imə'tiəriəl] *a.* 非物质的,无形的;不足道的,不重要的; immaterially, *ad.*; immaterialness, *n.*
immature [imə'tʃuə] *v.* 使无实体,使无形; *a.* 未成熟的,幼年的;轻度侵蚀; immaturely, *ad.*; immatureness, *n.*
　～ concrete 未凝结的混凝土
　～ residual soil 新残积土
　～ soil 生土,未成熟土
immeasurable [i'meʒərəbl] *a.* 不能测量的,无限的,(广大)无边的
immediate [i'mi:djət] *a.* 即时的,立即的;直接的,极近的
　～ access 即时[快速]存取
　～ acidity 直接酸度
　～ compression 瞬时压缩
　～ delivery 即交
　～ inference 直接推理
　～ interests 眼前利益
　～ movement 瞬时位移
　～ processing 即时加工
　～ runoff 地表径流,直接径流
　～ settlement 瞬时沉降
immense [i'mens] *a.* 无限的,无边的,极广大的; *n.* 无际,无量,无限,无边; immensely, *ad.*; immenseness, *n.*
immeasurability *n.* 无法计量的性质或状态
immerge [i'mə:dʒ] *v.*, -merged,-merging;浸入[没],沉入,专心,埋头于
immergence [i'mə:dʒəns] *n.* 沉浸
immerse [i'mə:s] *vt.* -mersed,-mersing;浸入,沉没,专心,埋头于;陷于
　～ body 潜没体
　～d gas cutting 水下气割
　～d method 水浸法(探伤)
　～d nozzle 潜入式喷嘴
　～d section 浸水[水下]断面
　～d tube 沉管
　～d tunnel 浸埋式隧道
immersion [i'mə:ʃən] *n.* 浸没,沉没;专心;沉溺
　～ compression test 浸水抗压试验(测定沉淀沥青混凝土水稳定性用)
　～ cooling 浸冷
　～ cured 浸水养护
　～ depth 浸没深度
　～ heater 浸入式加热器
　～ pipe 淹没管
　～ scanning 水中扫描法,浸液扫描法
　～ stability test 水浸安定度试验(评定沥青混合物耐水性的试验方法)
　～ technique 水中扫描术,浸液扫描术
　～ test 浸渍试验
　～ -type vibrator 插入式振动器
　～ unconfined compression test 浸水(无侧限)抗压试验
　～(immersium) vibrator 插入式振捣器,振捣棒

~ weighing 水中称重
immigrant ['imigrənt] **n.**; **a.**（自国外移入）侨民（的）
~ remittance 侨汇

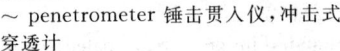
immersion type vibrator

immiscibility [imisi'biliti] **n.** 不溶混性,难混溶性,不可混合性

immittance [imitəns] **n.** 导抗,阻纳

immixture [i'mikstʃə] **n.** 混合物[作用]

immovable [i'mu:vəbl] **a.** 不可移动的,固定的;坚定的;静止的; **n. pl.** 不动产; immovableness, **n.**
~ bed model 定床模型
~ bed model testing 定床模型试验
~ property 不动产
~ restraint 固定约束

immure [i'mjuə] **vt.** 镶在墙上,埋在墙里

impact [n. 'impækt v. im'pækt] **n.** 冲击,碰撞;冲击力[量],动能;弹着,命中;着陆;脉冲;影响; **vt.** 击中,碰撞;装填,压紧; **vi.** 撞击;发生影响
~ allowance load 容许冲击荷载
~ attenuation device 减冲设施
~ bend(ing) test 冲弯试验
~ bending strength 冲击挠曲强度
~ breaker 冲击(式)破碎机
~ brittleness 冲击脆性
~ (cast-)in-situ pile 现场打桩
~ clam 冲击式抓斗
~ coefficient 冲击系数
~ compaction 夯击压实
~ compactor 夯实器,冲击(式)夯(具)
~ crusher 冲击式破碎机
~ damper 缓冲器,减震器
~ device 冲击装置,击打装置
~ ductility 冲击延性,冲击韧性
~ -endurance test 耐冲试验,反复冲击试验
~ energy 冲击动能,冲击能量
~ extrusion 冲击挤压
~ fatigue 冲击疲劳
~ formula 冲击公式
~ -grab boring machine 冲击式钻孔机
~ grinding 冲击研磨
~ hammer 冲(击)锤
~ hardness 冲击硬度
~ load 冲击荷载
~ load stress 冲击荷载应力
~ loss 冲击损失
~ (testing) machine 冲击试验机
~ method （夯实土用）冲击法
~ noise 冲击声响 [噪声]

impact noise

~ penetrometer 锤击贯入仪,冲击式穿透计
~ pile 冲击桩
~ pile driver 冲击式打桩机
~ pressure tube 冲击压力管,毕托管
~ -pressure velocity meter 冲击式流速仪
~ pulverizing 冲击粉碎
~ rammer 冲击锤
~ resilience 冲击回弹性
~ resistance 冲击阻力
~ roller 冲击式压路机
~ screen 振动筛
~ shell pile 冲击式套管桩
~ shock 冲击振动
~ spot welding 冲击点焊(法)
~ strength 耐[冲击]强度
~ stress 冲击应力
~ test 冲[撞]击试验
~ testing machine 冲击试验机
~ type energy dissipator 冲击式消能工
~ wheel 冲击式水轮(机)
~ wrench 机动(套筒)扳手;冲击扳手
~ zone （波浪）冲击区
alternating ~ 交变冲击

central ~ 对[中]心碰撞
direct ~ 直接碰撞,直冲
eccentric ~ 偏心碰撞
elastic ~ 弹性碰撞
hydraulic ~ 水力冲击
semi-elastic ~ 半弹性碰撞

impaction [im'pækʃən] *n.* 嵌入,嵌塞,压紧,撞击,碰撞
inertial ~ 惯性撞击

impactor =impacter *n.* 冲击机,冲击式打桩机;装填机;卧式锻造机,锤碎机

impages *n.* 门栏杆

impair [im'pɛə] *n.*; *vt.*, 削弱,减少;损害,障碍;奇数; *a.* 不成对的,奇数的
~ed area 损坏面积

impairment [im'pɛəmənt] *n.* 损害,毁[减]损

impale [im'peil] *vt.*, -paled,-paling;刺穿;钉住;使绝望;围以木桩;impalement, *n.*

impalement *n.* 刺穿;围栏或栅栏

impalpable [im'pælpəbl] *a.* 细微的,微粒的,无形的;难以理解的;impalpably, *ad.*

impansion [im'pænʃən] (面积,经营范围或人员)缩减

impassability ['impa:sə'biliti] *n.* 阻塞,堵塞,不可通性;无感觉,麻木

impassible [impa:səbl] *a.* 不可通行的,无路可通的;麻木的,无动于衷的;impassably, *ad.*

impasse [æm'pɑ:s]【法】*n.* 尽头路,绝境,死胡同,僵局

impaste [im'peist] *vt.* impasted, impasting;用浆糊封,用糊状物涂;使成糊状

impasto [im'pɑ:stəu] *n.* 厚涂颜料

impedance [im'pi:dəns] *n.* 阻抗;(全)电阻,电阻抗,表现电阻
~ bond 阻抗结合
~ bridge 阻抗电桥
~ matching 阻抗匹配
~ screw 节流螺钉
capacitive ~ 电容性电抗,容抗
load ~ 负载阻抗
nominal ~ 标称阻抗
short circuit ~ 短路阻抗

impede [im'pi:d] *vt.*, -peded,-peding;阻碍,妨碍,阻止; impeder, *n.*
~d drainage 不良排水

impediment [im'pedimənt] *n.* 阻碍,障碍,障碍物;(*pl.*)行李,辎重

impel [im'pel] *vt.* -pelled,-pelling;推进,行动,刺激;迫使,强迫,抛,投

impellent [im'pelənt] *a.* 推的,推进的; *n.* 推动力;发动机

impeller [im'pelə] *n.* 叶轮,涡轮;压缩器,推进器;刀盘;推动者
~ blade 叶轮叶片
~ breaker 叶轮式破碎器
~ pump 叶轮泵
~ shaft 叶轮轴
~ suction 叶轮吸水口
~ turbine 叶轮式水轮机
~ vane 叶轮叶片
centrifugal ~ (离心式)叶轮
radial-inlet ~ 径向进口式泵
screw ~ 螺旋式叶轮
shrouded ~ 闭式叶轮

impeller blade

impending [im'pendiŋ] *a.* 即将发生的;悬在上面的
~ cliff 悬崖
~ skidding 急刹车滑行,紧急滑行

impending cliff

impenetrable [im'penitrəbl] *a.* 不可贯入的,不能穿过的;不可测知的;密封的,不透……的,防……的;费解的,顽固的;不接受的

imperative [im'perətiv] *n.* 命令,规则;

a. 强制的,不可避免的,必须的,紧急的,迫切的; imperatively, *ad.*; imperativeness, *n.*
~ duty 紧急任务
~ necessity 迫切需要

imperceptible [impə'septəbl] *a.* 看不见的,觉察不到的,细微的; imperceptibleness, *n.*; imperceptibly, *ad.*

imperfect [im'pə:fikt] *a.* 不完全的,未完成的,有缺点的,不标准的; imperfectly, *ad.*; imperfectness, *n.*
~ arch 斜圆拱
~ combustion 不完全燃烧
~ earth 接地不良
~ equilibrium 不完全平衡
~ frame 不完全框架
~ gas 非理想气体
~ sparking 不完全发火
~ stage 不完全阶段
~ tape 缺陷带

imperfection [impə'fekʃən] *n.* 不完全,不健全,不完整性,缺陷

imperforation [impə:fə'reiʃən] *n.* 无孔的性质[状态,物]

imperial [im'piəriəl] *n.* 特大,特等品; *a.* 帝国的;壮丽的,堂皇的;特大的,质地最优的;英国度量衡制的; imperially, *ad.*; imperialness, *n.*
~ apartments 帝国公寓
~ basilica 皇宫,古罗马长方形会堂(建筑)
~ cathedral 帝国大教堂
~ cupola 大圆顶
~ dome 葱头形圆顶
~ gallon 英国标准加仑(4.546 升)
Imperial Garden 北京故宫御花园
Imperial Heavenly Vault 北京故宫皇穹宇
Imperial Mansion 帝国大厦(美国)

Imperial Mansion

~ palace 皇宫,宫殿
~ purple 深紫红色
~ red 朱红色,朱砂色
Imperial Rome style 罗马帝国的(建筑)风格
~ roof 大屋顶
~ scales 英制
~ sizing 英制尺寸

imperious [im'piəriəs] *a.* 专横的,傲慢的;紧急的,迫切的;不透(水)的; imperiousness, *n.*

impermeability [im,pə:mjə'biliti] *n.* 不透水性,抗渗性,防水性,气密性
~ factor 不透水系数(流域面积上的径流量与降水量之比)
~ test (混凝土)抗渗试验

impermeable [im'pə:mjəbl] *a.* 不渗透的,不能透过的,密封的,不透水的; impermeably, *ad.*; impermeability, *n.*; impermeableness, *n.*
~ base 防渗地基
~ barrier 不渗水层,不透水隔层
~ clay 防渗黏土
~ confining bed 不透水隔层
~ course 不透水层
~ diaphragm 阻水层,防渗心墙[斜墙]
~ foundation 防渗地基
~ groin 不透水丁坝[挑水坝]
~ groyne 不透水析流坝
~ layer 不透水层
~ material 不透水材料,防渗材料
~ rock bed 不透水岩层
~ seam 不透水接缝
~ soil 不渗透土
~ stratum 不透水地层
~ wall 阻声[水]墙
~ zone 不透水区

impermephane *n.* 透明保护敷料

impervious [im'pə:vjəs] *a.* 不透水的;不能透过的,密封的,抗渗的; imperviously, *ad.*; imperviousness, *n.*
~ area runoff 不透水区径流

~ asphalt(ic) concrete layer 防水沥青混凝土层
~ base 不透水基础
~ bed 不透水层,不透水河床
~ blanket 防渗铺盖,不透水覆盖层,不透水面层
~ bottom 不透水基底,不透水底板
~ boundary 不透水边界
~ clay blanket 不透水黏土覆盖层
~ concrete diaphragm 混凝土防渗墙
~ core 不透水[防渗]心墙
~ course 不透水层
~ cut-off 不透水[防渗]齿槽
~ dam 不透水坝
~ diaphragm 不透水心墙[隔层]
~ element 防渗设施
~ foundation 不透水地基
~ layer 不透水层
~ material 不透水材料
~ rock 不透水岩石
~ rolled fill 不透水的碾压填土
~ section 防渗截面
~ soil (空气、水、植物根系等不能透过的)不透性土
~ stratum 不透水地层
~ surface 不透水地面[面层]
~ wall 防渗墙
~ zone 不透水区

imperviousness *n.* 不透[过]水性
~ coefficient 不透水系数
~ factor 不透水系数

impetus ['impitəs] *n.*; *pl.* -tuses;冲量,动量;原动力;刺激,促进,推动;冲击

impinge [im'pindʒ] *vi.*, -pinged, -pinging;撞[冲]击;侵犯

impingement [im'pindʒmənt] *n.* 碰撞,冲[撞]击;弹[回,反]跳;冲突;水锤
~ aerator 冲击曝气器
~ attack 浸蚀,腐蚀,滴蚀
~ black 烟道炭黑
~ separator 冲击式分离器

impinger [im'pindʒ] *n.* 冲击滤尘器,撞击器

implant [im'pla:nt] *n.* 插入物,移植物; *v.* 注入,插入;掺杂

implastic *a.* 不可塑的;不易放入模子里的

implement [*n.*'implimənt *v.*'impliment] *n.* 器械,工具,用具,仪器; *v.* 供应器具,提供方法;执行,实现;填补
~ shed 工具仓库,工具堆栈
building
inserted ~ (拖拉机上的)悬挂式机具
pruning ~ 树木修剪工具

implementation [implimen'teiʃən] *n.* 供给器具;实现,履行;仪器,工具

implicit [im'plisit] *a.* 含蓄的,隐含的,暗示的;绝对的;无疑的,不明显的
~ computation 隐函数法计算
~ declaration 隐式说明
~ difference equation 隐式差分方程
~ differentiation 隐(式)微分法
~ function 隐函数
~ iterative method 隐式迭代法
~ scheme 隐式(差分)格式
~ smoothing 隐式平滑,隐式修匀

implode [im'pləud] *v.*, imploded, imploding;爆聚,(向)内(破)裂,内向爆炸;压破

implosion [im'pləuʒen] *n.* 爆聚,内爆,内破裂,压破,从外向内的压力作用,挤压

impluvium [im'plu:viəm] *n. pl.* -via;(古罗马)住宅前室(atrium)或中庭(peristyle),存积雨水的方形贮水池

imponderable [im'pɔndərəbl] *n.* 不可量物(如热、光等);(*pl.*)无法估量的事物[影响作用]; *a.* 极轻的,无重量的;不可称量的; imponderableness, *n.*; imponderably, *ad.*

imporosity [impɔ:'rɔsiti] *n.* 无孔性,结构紧密性,不透气性

imporous [im'pɔ:rəs] *a.* 无孔隙的

import [*v.*im'pɔ:t *n.*'impɔ:t] *n.* 输入,

引入,移入;输入功率;(*pl.*)进口(货,商品);含义,意义;重要(性);*v.* 含有……的意思,意味着;输入,引入
- ~ and export duties 进出口税
- ~ commission 进口代办行
- ~ commission house 进口代理商
- ~ declaration 进口报关单
- ~ deposit 进口保证金
- ~ duty 进口税
- ~ entry 进口报关
- ~ quantum 进口量
- ~ quota system 进口配额制
- ~ surplus 入超
- ~ trade 进口贸易

importance [im'pɔːtəns] *n.* 重要的性质或情况,重要(性),重大,价值

imported *a.* 导入的,输入的
- ~ labo(u)r 输入劳动
- ~ water 输入水,超流水(来自所研究的地下水盆地范围以外的水)

impose [im'pəuz] *v.*, -posed,-posing; 征(税);将……强加于,使……负担;利用发生影响; imposable, *a.*; imposer, *n.*
- ~ed load 作用荷载,外施荷载

imposite *n.* 一种不溶于松节油的天然沥青

impossible [im'pɔsəbl] *a.* 不可能的,不会发生的,办不到的
- ~ event 不可能事件

impost ['impəust] *n.* 拱墩,支撑拱门的柱头,起拱点;关[进口]税;*v.* 分类估税
- ~ block 拱墩块体
- ~ capital 拱墩柱头
- ~ hinge 拱墩柱头铰
- ~ joint 拱墩柱头接缝
- ~ moulding 拱墩柱头造型
- ~ pressure 拱墩压力
- ~ section 拱墩断面
- ~ stringer 拱脚石

impotable [im'pəutəbəl] *a.* 不可饮的,不适合饮用的
- ~ water 非饮用水

impound [im'paund] *v.* 筑堤,拦[蓄]水;修圩;扣押;没收; impoundable, *a.* impoundage, *n.*; impounder *n.*
- ~ed area 蓄[集]水面积
- ~ed basin 蓄水池
- ~ed harbor 封闭式港口,有闸港
- ~ed level 蓄水位[高程]
- ~ed surface water 地面积水
- ~ed water 积滞水,拦蓄水
- ~ing dam 拦[蓄]水坝
- ~ing dock 通潮闸坞,有闸港池
- ~ing reservoir 水库,蓄水库
- ~ing schemes 蓄水方案

impoundment [im'paundmənt] *n.* 蓄水区,蓄水池;集水;扣押,保管

impoverishment [im'pɔvəriʃmənt] *n.* 贫脊,贫乏,缺乏;合金成分的损失

imprecision [impri'siʒən] *n.* 不精确性[状态],精密度不够

impreg ['impreg] *n.* 浸渍木,树脂浸渍木材,浸渍处理材

impregnability [impregnə'biliti] *n.* 浸透能力,浸透性能;坚固

impregnable [im'pregnəbl] *a.* 坚固的,填塞坚实的,充满的;可渗透的; impregnably, *ad.*; impregnability, *n.* impregnableness, *n.*

impregnant [im'pregnənt] *n.* 浸渍剂,浸渍物
wet-strengthening ~ 防湿浸渍剂

impregnate [*v.* 'impregneit; *a.* im'pregnit] *v.* -nated,-nating;注入,灌注;浸透,使充满;*n.* 浸渍树脂;*a.* 浸透的,饱和的; impregnator, *n.*
- ~ tie 浸油轨枕
- ~ with ereosote 用杂酚油灌注
- ~d cement 饱和水泥
- ~d felt 浸渍毛毡
- ~d paper 浸染纸

~d rock 浸染岩
~d wood 浸渍防腐木材
impregnating n. 浸渍,浸染,浸透
~ agent 浸渍(防腐)剂
~ compound 浸渍(防腐)剂
~ fluid 防腐液
~ installation 浸渍(防腐)设施
~ liquid 防腐液
~ oil 浸渍防腐油
~ scumble 防腐涂料
~ test 浸渍(防腐)试验
~ varnish 浸渍清漆
vacuum ~ 真空浸注
impregnation [impreg'neiʃən] n. 注入,浸染,浸透;充满,饱和;注入物
~ bath 浸泡池(槽)
~ degree 浸渍度
~ installation 浸渍(防腐)设施
~ period 浸渍(防腐)时期
~ speed 浸渍速度
~ time 浸渍时间
aluminium ~ 铝化,渗铝
vacuum ~ 真空浸透
waxed ~ 蜡浸透
impregnator n. 浸渍机
impress [n. 'impres v. im'pres] n. 印象;记号;特征;v. 盖印,刻记号,使记住;压入,压痕;影响;标志;引用,利用
~ed distortion 强迫扭曲
~ed voltage 外加电压
~ed watermark 压纹,外压水印
impression [im'preʃən] n. 印象,概念;感应,影响;痕迹,印记;凹槽
~ block 压印模板
~ cylinder 压印滚筒
ball ~ (硬度试验的)球凹,球印
blind ~ 平压印
die ~ 刻印模,模槽
wax ~ 蜡版
imprest [im'prest] n. 预付款; a. 预付的,借予的

imprint [n. 'imprint, v. im'print] n. 印(记),痕迹;特征,印象; vt. 刻上记号,加特征,印(刷),盖(印); imprintability, n.; imprintable, a.
imprinter n. 刻印机,戳印机,印码机
impromptu [im'prɔmptju:] n. 即席讲话; a. 临时的,无准备的,即席的; ad. 临时地,即席地
improper [im'prɔpə] a. 不正确的,不适当的,不合理的,不规则的,非正常的; improperly, ad.; improperness, n.
~ fraction 假分数,可约分数
~ integral 广义积分,奇异积分
~ rotation 非正常旋转
~ timing 不正确的点火
~ ventilation 通风不良
improvability [imˌpru:və'biliti] n. 改良,可改进性
improve [im'pru:v] v. -proved,-proving;提高,改进,改良,增进;好转,进步,矫正;利用;涨价
~d design 改进设计
~d road 改善道路
~d soild wood 改良实心木
~d subgrade 改善路基
~d Venturi flume 改良文德里槽
~d wood 压缩木材
~ing dirt roads 泥路强固
improvement [im'pru:vmənt] n. 改良,改善,改进(之处);进步
~ district 改建区
~ line 道路扩建用地线,道路改建路界线
~ works 整治[技术改造]工程
highway ~ 公路改良
operation ~ 操作改善
river ~ 河道整治
improver [im'pru:və] n. 改良者,改善者;添加剂;实习生,学徒
~ color 颜色改良剂
fluorescene ~ 荧光促进剂
viscosity index ~ 黏度指数改进剂
improvise ['imprəvaiz] v. 即席[临时]

创作,临时准备,凑合,现凑
~d makeshift 临时凑合的办法
impsonite *n.* 一种焦油沥青;脆沥青岩
impulsator [im'pʌlseitə] *n.* 脉冲发生器
impulse ['impʌls] *n.* 冲动,冲击;冲量;脉冲,脉动;*v.* 冲击,推动,激励;发出脉冲
~ blade 冲击式叶片
~ charge 推力炸药
~ condensing turbine 冲击式凝气透平
~ frequency 脉冲频率,冲击频率
~ generator 脉冲发生器
~ loading 冲击加荷
~ momentum equation 冲量-动量方程
~ pallet 推力钻
~ pump 冲击式泵
~ -reaction turbine 冲击-反击式透平,冲击-反击式水轮机
~ response 脉冲响应
~ testing 冲击试验
~ testing apparatus 冲击试验设备[装置]
~ turbine 冲击式涡[水]轮机
~ water turbine 冲击式水轮机
~ wheel 冲击式水(叶)轮
actual specific ~ 实际比推力
boosting ~ 助推器冲量
ideal specific ~ 理想比冲量,理想单位冲量
inceptive ~ 初始脉冲,占线脉冲
initial ~ 起始冲量,初始冲量
instantaneous ~ 瞬时脉冲
oscillating ~ 振荡脉冲
impulsion [im'pʌlʃən] *n.* 冲击,冲动,

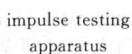
impulse testing apparatus

脉冲;冲量,冲力,推动力
~ fan 冲击式通风机
~ wave 脉冲波
impulsive [im'pʌlsiv] *a.* 冲击的,冲动的,撞击的; impulsiveness, *n.*
~ force 冲力
~ load 冲击荷载
~ warming 爆发性加温
impurity [im'pjuəriti] *n. pl.* -ties;不纯,不洁,不纯洁性;污染,沾染;污垢;(*pl.*)杂质,夹杂物
absorption ~ 吸收沾染
ash-forming ~ 成灰杂质
detrimental ~ 有害杂质
gas ~ 气体杂质
mechanical ~ 机械杂质
metal ~ 金属杂质
imputrescibility *n.* 不易腐烂性
imputrescible [ˌimpjuː'tresibl] *a.* 不易腐烂的,不腐朽的; -imputrescibility, *n.*
in [in] *prep.* 在……方面,在……内;进入
~ -and-out bond 内外砌合,丁砖与顺砖交替砌合
~ bond 货物在关栈中尚未完税
~ bridge 并联,跨接
~ bulk 散装;成堆,大批,大量
~ charge 主管,掌握
~ constant dollar value 按不变价值美元计算
~ current price 按现价计算,按时价计算
~ duplicate 一式二份
~ demand 顾客需要
~ dollar 按美元,按金额
~ echelon (排)成梯形
~ economy 按经济原则
~ funds 有资本
~ -group rating 分组指标,分组指数,类集指数
~ hauling 在搬运(拖曳)中
~ hole perforator 潜孔钻机
~ kind 以货代款

~ lay 镶入(物)
~ line production 流水作业
~ line pump 管线泵,直列式柱塞泵
~ metric 用公制
~ -phase 同相(位)
~ -place bedrock 原状基岩
"in place" conditions (土的)原状,自然层理状况
~ -place mixing 现场拌和
"in place" soil shearing test 原状土剪力试验
"in-place" test 现场试验,实地试验
~ red ink 有赤字
~ round number 约计,以整数计,以整数表示,以约数表示
~ rush of water (钻孔中)进水,决口
~ series 串联
~ -serve behavior 使用品质,使用情况;工作状态,功效,作用
~ service life 使用寿命
~ situ 就地,在原处;在(施工)现场;在天然岩层中
~ situ concrete 就地浇筑混凝土,现浇混凝土
~ situ porosity 原始孔隙率
~ -situ pressiomenter 现场压力计
~ -situ soil 原地土
~ -situ strength 自然条件下的强度,现场强度
~ -situ terrazzo 现浇水磨石
~ -situ test 现场试验
~ the clear 净空,净高
~ the log 未锯开的状况,原木的形式
~ the open air 露天
~ the rough 未经修琢的,生料,未加工的
~ the substance 全重,毛重,大意,要点
~ the wet (基坑的)湿挖,湿施工
~ -time operation 定时操作
~ transit 在运输中

inaccessible [inæk'sesəbl] *a.* 难达到的,难接近的

~ area 荒野,偏僻之处
~ roof 高顶板
~ value 不可达值

inactive [in'æktiv] *a.* 钝态的,钝性的,不活动的,不活泼的,反应缓慢的
~ file 待用文件
~ glacier 固定冰川,停滞冰川
~ line 虚描线
~ optically 不旋光的,非旋光的
~ storage 无效库容
~ volcano 死火山,不活动火山
~ well 不出油井
~ workings 废巷道,停工区

inadhesion *n.* 不粘着

inalium *n.* 因阿铝合金(0.5%Si,1.2%Mg1.7%Cd,余量 Al)

in-and-in ['inənd'in] *a.;ad.* 互锁,互卷入;近亲交配的(地)

in-and-out 进进出出的;彻底地
~ bolt 贯通螺栓
~ bond 丁顺砖逐层交替的砌合法
~ plating 内外搭接板;内外搭接制
~ strakes 内外叠板

inaperture *a.* 无孔隙的

inartistic [ina:'tistik] *a.* 非艺术的,缺乏艺术性的

inauguration [inɔ:gju'reiʃən] *n.* 开始,开幕(仪)式,落成[成立]典礼

inaugurator [i'nɔ:gjureitə] *n.* 创始者,主持就职[开幕]仪式者

inaurate *n.* 黄金光泽

inband *n.* 露头石,丁头

inbond *n.* 丁头砌合
~ brick 丁砖,内砌砖

inbound ['inbaund] *a.* 归航的,入境的,入站的
~ line 入站线路
~ ship 归航船只
~ traffic 入境交通

in-built ['inbilt] *a.* 埋没的,固定的,装入的,嵌入的
~ ashtray 嵌入式灰缸
~ antisplash 固定挡泥(水)板

~ (bath) tub 埋设式浴盆
~ cabinet 嵌入式柜橱
~ cupboard 嵌入式碗橱
~ furniture 固定式家具
~ garage 内部车库
~ kitchen 内部厨房
~ shower stall 内部淋浴间
~ wardrobe 埋入式衣橱
inbye ['in'bai] *ad.* 近旁的;向工作区的;地下的,巷道内的
~ side 地下的,巷道内的
incandescent lamp 白炽灯
incanus *n.* 灰白色
incase [in'keis] *vt.* -cased,-casing;装在箱内,镶在框子内,包在……内
inca stone *n.* 黄铁矿
incavation *n.* 空心的东西或洼地
incense ['insens] *n.* 香味;*v.* 加香味,用香水;incensement, *n.*
~ cedar 翠柏
~ juniper 香刺柏
~ wood 香木
incenter ['insentə] *n.* 内(切圆)心
incentive [in'sentiv] *n.* 刺激;鼓励;诱因,动机;*a.* 刺激的,诱发的
~ force 鼓动力
~ scheme 奖励计划
~ system 奖励制度
~ wage payment 奖励工资
material~s 物质刺激
inception [in'sepʃən] *n.* 开始,发端,初始位置
fruit ~ 开始奏效,开始收益
knock ~ 起爆
inceptiso(i)l *n.* 始成土
inch [intʃ] *n.* 英寸,吋;少量,少许;小岛;*v.* 慢慢移动,渐动
~ plank 一时板
~ scales 英制比例
~ screw thread 英制螺纹
~ size 英制尺寸
circular ~ 圆(直径为一吋,面积为 5.06平方厘米)
incher ['intʃə] *n.* 小管;以英寸数作为量纲的东西
inching ['intʃiŋ] *n.* 低速运转;微调整,微动
~ control 微动控制
incidence ['insidəns] *n.* 入射,入射角,倾角;落下;影响(范围);关联
blade ~ 叶片安置角
effective ~ 有效迎角
normal ~ 正入射,法线入射,垂直入射
oblique ~ 斜入射
wing ~ 翼倾角,翼安装角,翼迎角
incident ['insidənt] *n.* 偶发事件,事变[故];*a.* 入射的;附带的,外来的;易发生的; incidentless, *a.*
~ angle 入射角
~ beam 入射光束
~ detection 事故检查
~ flow 入射流
~ light 入射光
~ normal 入射法线
~ ray 入射线
~ wave 入射波
incidental [insi'dentl] *a.* 偶然的,偶发的,临时的;附带的
~ charge 杂费
~ losses 附带损失
~ revenue 临时性收入
incinerate [in'sinəreit] *v.* -ated,-ating;烧成灰,焚化;-incineration, *n.*
incineration [insinə'reiʃən] *n.* 焚化,烧尽
~ house 垃圾焚化站
~ of garbage 垃圾焚化
~ plant (垃圾)焚化厂
~ unit of garbage 垃圾焚化装置
incinerator [in'sinəreitə] *n.* 焚化炉,焚秽炉,燃烧炉,(垃圾)化灰炉
~ plant (垃圾)焚化厂
sanitary ~ 卫生焚化炉
incipient [in'sipiənt] *a.* 起始的,初

期的
- ~ crack 初始裂纹
- ~ decay （木材的）初期腐烂
- ~ erosion 早期冲蚀
- ~ failure 初期破坏,将临故障
- ~ fault 潜在事故,隐患
- ~ piping 初期管涌
- ~ rupture 初期破坏
- ~ scour 初始冲刷
- ~ sediment motion 初始泥沙运动

incircle [in'sə:kl] *n.* 内切圆

incise [in'saiz] *vt.* -cised,-cising;刻,雕刻,切,切开
- ~d decoration in bronze 铜雕饰
- ~d inscription 阳刻
- ~d meander 深切河湾[曲流]
- ~d ornament 雕刻饰物
- ~d river 深切河,深槽河
- ~d slab 雕刻石板[木板]
- ~ing and filling with colored lacquer 雕填
- ~ing and filling with gold dust 戗金

incision [in'siʒən] *n.* 切口,缺口,切割,（雕）刻,刀痕
- V-shaped ~ V形切口法

incitation [insai'teiʃən] *n.* 激励,刺激,激发

incitement [in'saitmənt] *n.* 鼓励,激发,煽动(物)

inclination [inkli'neiʃən] *n.* 倾斜[角,向];斜度;趋势;-inclinational, *a.*; -inclinatory, *a.*
- ~ compass 倾[俯]角罗盘
- ~ correction 倾斜改正
- ~ limit 极限倾角
- ~ line 比降[坡度]线
- ~ method 倾度计算法
- ~ of ramp （高速公路）进出车道坡度角
- ~ of roof 屋顶斜度
- air base ~ 航测基线倾斜
- magnetic ~ 磁倾角,磁力线偏转
- orbit ~ 轨道倾角,轨道平面对赤道的

倾角（卫星）

inclemency [in'klemənsi] *n. pl.* -cies;寒冷,凛冽的天气;残酷

inclement [in'klemənt] *a.* 残酷的,恶劣的;（天气）寒冷的;-inclemently, *ad.*; inclementness, *n.*
- ~ weather 狂风暴雨的天气

inclinable [in'klainəbl] *a.* 倾向于……的,可倾斜的

inclinator = inclinometer, *n.* 倾斜仪,倾角仪,磁倾计;倾倒器;(沿楼梯的)自动升降椅

inclinatorium [in,klainə'tɔ:riəm] *n.* 倾斜仪,磁倾仪,测斜器

incline [n. 'inklain v. in'klain] *n.*; *v.*, -clined,-clining; *n.* 斜面[坡];斜（车）道;斜井; *v.* 倾斜[向],赞同;-incliner, *n.*
- ~ hoist 斜井提升机
- ~ level 倾斜计
- gravity ~ 重力斜坡道,轮子坡
- self-acting ~ 自动[重]斜坡道,轮子坡

inclined [in'klaind] *a.* 倾斜的,有坡度的;倾向于……的
- ~ anchor 斜锚

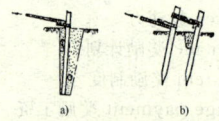

inclined anchor

- ~ apron 斜护坦
- ~ approach 斜坡道
- ~ arch 斜拱
- ~ axis mixer 斜轴拌和机
- ~ bar 斜杆[钢筋]
- ~ barrel arch （连拱坝的）斜拱筒
- ~ barrel vault 斜筒穹顶,倾斜半圆形拱顶
- ~ boom 倾斜吊杆

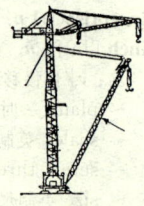

inclined boom

(起重机)
~ borehole 斜钻孔
~ breakwater 斜坡式防波堤
~ bridge 坡桥,有纵坡的桥
~ cableway 斜缆道(坡度常采用1∶4,料斗可自行滑下)
~ catenary construction 斜链线结构
~ chord (桁架)斜弦杆
~ clerestory 倾斜天窗,斜联窗假楼
~ conveyer 倾斜运输机
~ cut and fill 倾斜分层充填(法)
~ drop 陡[斜]槽跌水
~ elevator 倾斜提升机,升降机
~ escalator shaft 斜自动楼梯通道
~ fault 倾斜断层
~ force 斜向作用力
~ grate 倾斜的壁炉铁栏[炉架],斜的门窗铁栅栏
~ haunched beam 斜变截面[加腋]梁
~ hole cut 斜孔掏槽
~ intake 倾斜进水口[入口]
~ jib (=gib) 斜起重臂,斜突梁
~ ladder 倾斜梯
~ motorstair shaft 倾斜汽车螺旋形楼梯通道
~ moving stair (case) shaft 倾斜的正面大楼梯通道
~ parking 斜列停车
~ piece 斜撑
~ pile 斜桩
~ plane 倾斜面,倾斜轨道
~ reaction 斜向反力
~ ropeway (运重物的)斜坡的空中索道
~ screen 倾斜栅网,斜筛
~ sewing screen 倾斜的排水(拦污)栅
~ shaft 斜井,斜(式)导井
~ ship lift 斜面升船机
~ shoot 斜槽
~ shore 斜(海)岸,斜(支)撑

~ staff ga(u)ge 倾斜水尺,斜坡水尺
~ steel 斜钢筋
~ stirrup 斜钢箍[箍筋]
~ subway 斜坡道
~ trough chute 斜槽
~ tube manometer 斜管式测压计
~ web (板梁的)斜腹板
~ weir 斜背堰
~ wharf 斜码头
inclinometer [ˌinkliˈnɔmitə] ***n.*** 测斜机,倾斜仪,倾角计,磁倾计
bubble ~ 气泡测斜仪
gyroscopic ~ 回转倾斜计,陀螺式倾斜计
pendulum ~ 摆式倾斜仪,单摆倾斜计
universal ~ 万向倾斜计
inclose [inˈkləuz] ***v.*** = enclose;围绕,包围;包装,(随函)封入,附寄
~d meander 环形河湾,环形河曲
inclosure [inˈkləuʒə] ***n.*** = enclosure;包围,围绕;围墙,围栏;(信中的)附件
~ act 圈地法
~ wall 围墙
include [inˈkluːd] ***vt.*** -cluded,-cluding;包括,包含,把……包括在内;算入,计入
~d angle of arch 拱圈中心角[包角]
~d angle of crest 堰顶(圆弧)包角
~d sapwood 内涵边材
inclusion [inˈkluːʒən] ***n.*** 包括[含];夹杂;包含关系;包体,内含物,杂质
~ fee 土地增值税
~ of peat 泥碳夹层,泥碳包体
~ texture 包体结构
dross ~ 氧化夹杂物
metal ~ 金属夹杂物
non-metallic ~ 非金属夹杂物
slag ~ (焊缝)夹渣
inclusive [inˈkluːsiv] ***a.*** 包含的,的,可兼的;包括一切的
~ -NOR-gate "或非"门
~ -OR-gate "或"门

incoagulable [inkəu'ægjuləbəl] *a.* 不凝聚的,不能混凝的,不可凝结的

incochrome nickel *n.* 镍铬耐热(因科镍)合金

incoercibility *n.* 不可压缩性

incoercible [ˌinkəu'əːsibəl] *a.* 不可压缩的,不可强迫的

incoherence [inkəu'hiərəns] *n.* 松散性,不连贯性;不相干性

incoherent [ˌinkəu'hiərənt] *a.* 无粘性的,无内聚力的,松散的,不连贯的;无条理的,不相干的
~ alluvium 无粘性冲积层,松散冲积层
~ material 无粘性材料,松散材料,非固结材料
~ noise 不相干噪声
~ rock 不粘结[松散]岩石

incohesion *n.* 不粘结性,无内聚性

incohesive [inkəu'hiːsiv] *a.* 无粘聚力的

Incoloy *n.* 因科洛伊(一种耐高温的镍铬铁合金)

incombustibility ['inkəmˌbʌstə'biliti] *n.* 不可燃性,不燃性

incombustible [ˌinkəm'bʌstəbl] *n.* 不能燃烧之物; *a.* 不能燃烧的,防火的; incombustibleness, *n.*; incombustibly, *ad.*
~ material 不燃烧[防火]材料
~ paper 耐火纸

income ['inkəm] *n.* 进款,收益,收入,所得
~ account 收益[进款]账
~ apportionment 收益分配
~ bone 收益债券,营业收入债券,不担保有利息债券
~ consumption curve 收入消费曲线
~ sheet 收益表
~ statement 收[损]益表,损益计算书
~ tax 所得税
gross ~ 总收入
net ~ 净收入

incoming ['inˌkʌmiŋ] *n.* 进来,进料;(*pl.*)收入; *a.* 进来的,接任的,移民的,增殖的,引进的,输入的
~ flow 入流,来水,进水(量)
~ line 进线,输入线路
~ panel 进线配电盘
~s and outgoings 收支
~s and outgoings balance 收支平衡
~ sediment 来沙(量)
~ solar radiation 日照[射]
~ stone (轧石机的)进给石料
~ teletype 输入电传打字机
~ tide 涨潮
~ waste 进入的废水,流入的废水
~ wave 入波,来波

incompact [ˌinkəm'pækt] *a.* 松散的,不紧凑的,不结实的

incomparable [in'kɔmpərəbl] *a.* 无比的,无双的,不能比较的; incomparableness, *n.*; incomparability, *n.*; incomparably, *ad.*

incompatibility ['inkəmˌpætə'biliti] *n. pl.* -ties;不相容[协调,亲和](性),不能和谐共存,(*pl.*)不能相容的特点

incompetent [in'kɔmpitənt] *a.* 无能力的,不能胜任的,不适当的,不合格的,
~ bed 软岩层,弱胶结地层

incomplete ['inkəm'pliːt] *a.* 不完全的,不足的
~ combustion 不完全燃烧
~ compaction 不完全捣实,(混凝土)捣实不足
~ contraction 不完全收缩
~ fusion 不完全熔接
~ mixing 不均匀拌和
~ overflow 不完全溢流
~ reaction 不完反应,未完反应
~ shear crack 不完全剪切裂隙
~ task log 未完成任务记录

incompressibility *n.* 不可压缩性

incompressible [ˌinkəm'presəbl] *a.* 不可压缩的, 坚硬的; incompressibility, *n.*
～ fluid 不可压缩性流体
～ material 不可压缩材料
～ stratum 不可压缩层

inconel *n.* 铬镍铁合金, 因康镍合金(80% Ni; 14% Cr; 6% Fe)

inconsequent [in'kɔnsikwənt] *a.* 不连贯的; 不合理[逻辑]的; inconsequently, *ad.*
～ drainage (system) 不顺向水系
～ river[stream] 非顺向河

inconvertible [ˌinkən'və:təbl] *a.* 不能兑换的纸币; 不能交换[兑现]的; inconvertibility, *n.*; inconvertibly, *ad.*
～ currency 不能自由兑换的通货
～ paper 不兑现契约

incorporate [*v.* in'kɔ:pəreit *a.* in'kɔ:pərit] *v.*, -rated, -rating; 联[结]合; (使)加入, (使)合并, 组成公司[法人]; *a.* 联合的, 合并的, 组成的; incorporative, *a.*
～d company 股份有限公司

incorporation [inˌkɔ:pə'reiʃən] *n.* 结合, 合并; 团体, 社团, 公司; 掺合, 掺人, 混合

incorporeal [inkɔ:'pɔ:riəl] *a.* 无形的, 精神的, 非物质的; incorporeity, *n.*; incorporeality, *n.*; incorporeally, *ad.*
～ capital 无形资本

incorrodible *a.* 不受腐蚀影响的, 抗腐蚀的

incorrupt [ˌinkə'rʌpt] *a.* 无差错的, 纯粹的, 清廉的; 无改动的; incorruptly, *ad.*; incorruptness, *n.*

increase [*n.* 'inkri:s *v.* in'kri:s] *v.* -creased, -creasing; *n.* 增长, 增加, 增殖, 增进; 增加量, 增大额; 生产物, 利益; *v.* 增加, 增大, 增进, 增殖; 提高, 上升
～ accumulation 增加积累
～ amount 增加金额

～d demand 需求增加
～d units 增产量
～ing annuity 递增年金
～ing cost 成本递增
～ing function (递)增函数
～ing return 递增报酬[利润]
～ of budget 增加预算
～ of production 增加生产
～ of risk 增加风险

increaser [in'kri:sə] *n.* 扩径水管, 异径接头

increment ['inkrimənt] *n.* 生长, 增长, 增大; 增值, 增量, 余差, 盈余; 药包; incremental, *a.*
～ budgeting 增量预算法
～ cost 增值成本
～ -drop test 落锤增量(冲击弯曲)试验
～ of load 荷载增量
～ tax 增值税
～ value duty 增值税

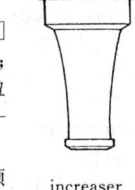
increaser

incremental [inkri'məntəl] *a.* 递增的
～ analyses 增量分析, 边际分析, 限界分析
～ benfit 边际利益
～ benefit-cost ratio 边际利益-费用比
～ cash flow 增加的现金流动
～ connector 加长连接件
～ construction 建筑增量
～ launching 顶推法
～ load method 渐增荷载法
～ theory 增量理论

increscent [in'kresənt] *a.* 渐增的; (月亮)变圆的; -increscence, *n.*

incrust [in'krʌst] *v.* 覆以硬壳, 生壳; 镶嵌, 用皮(或壳)包裹

incrustant *n.* 水垢, 垢壳

incrustation [ˌinkrʌs'teiʃən] *n.* 结壳(作用), 结锅垢; 水垢[锈]; (建筑物)表

面装饰,镶嵌细工
    sediment ～ 积垢
incubator [ˈinkjubeitə] *n.* 恒温箱,培养箱,孵卵器
    ～ test 稳定度试验
incumbent [inˈkʌmbənt] *n.* 在职者;居住者,房客; *a.* 现任的,在职的;上覆的,上层的
incursion [inˈkəːʃən] *n.* 侵入(作用),侵犯,袭击;进入,流入
incurvate [ˈinkəːveit] *v.*, -vated,-vating; 凹入的,(使向内)弯曲; *a.* 弯曲的,内曲的
incurvation [ˌinkəːˈveiʃən] *n.* 内曲(现象),弯曲;挠度
incurvature [inˈkəːvətʃə] *n.* 内曲,向内弯
incurve [ˈinkəːv] *n.*; *v.*, -curved, -curving; *n.* 内曲,弯曲; *v.* (使)内弯
incus [ˈinkəs] *n. pl.* -cudes. 砧骨;砧状云;音卡斯合金钢(0.55% C, 0.7% Mn, 0.7% Cr, 1.75% Ni, 0.7% Mo, 余量 Fe)
indalloy *n.* 钢银焊料(钢 90%,银 10% 合金)
indanthrene [inˈdænθriːn] *n.* 阴丹士林,靛茵醌,标准还原蓝
    ～ blue 阴丹士林蓝
indanthrone [inˈdænθrəun] *n.* 靛蒽醌(阴丹士林的正式名称),阴丹酮
    ～ blue 靛蓝
indate *n.* 有效期
indebted [inˈdetid] *a.* 负债的,受惠的
indefinite [inˈdefinit] *a.* 不明确的,未确定的,无定限的;模糊的,无限期的
    ～ equation 不定方程
    ～ integral 不定积分
indefinitely *ad.* 无穷地,无限(期)地
    ～ small 无穷小
indelta *n.* 河流分流内陆区
indemnification [indemnifiˈkeiʃən] *n.* 赔偿,赔偿金[物]
indemnity [inˈdemniti] *n. pl.* -ties; 赔偿金,损失赔偿,补偿,保险,赦免
    ～ agreement 补偿协仪
indent [*v.* inˈdent *n.* ˈindent] *n.* 穴,锯齿形;刻痕,凹槽;双联订单(印刷或书写中的)缩进,空格; *v.* 刻成锯齿形,刻凹槽,使凹进;订合同,一式数份地起草(文件,合同等); indenter, *n.*
    ～ed bar 刻痕[螺纹]钢筋
    ～ed beam 错口[锯齿]式组合梁
    ～ed bolt 带纹螺栓
    ～ed chisel 锯齿鏨[凿子]
    ～ed coast-line 曲折[锯齿形]海岸线
    ～ed joint 齿接合,齿合接健
    ～ed molding 锯齿形线脚

indented moding

    ～ed ribbed bar 锯齿形竹节钢(筋)
    ～ed roller 凹纹压路机[路碾]
    ～ed steel bar 齿纹[刻痕]钢丝
    ～ing apparatus 凹进仪
    ～ing ball 硬度计球
    ～ing course 锯齿形水道
    ～ing hammer 锤[冲]头

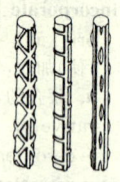
indented steel bar

indentation [indenˈteiʃən] *n.* 呈锯齿形;海岸线凹入处,缺口;压痕,刻痕
    ～ hardness 刻痕硬度
    ～ machine 硬度机

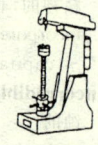
indentation machine

indenter [inˈdentə] *n.* (硬度试验的)压头
indenture [inˈdentʃə] *n.*; *v.*, -tured, -turing; *n.* 合同,契约,清单;成犬牙交错状,凹凸不平
    ～ labour 合同工
independence [indiˈpendəns] *n.* 独立[性],自立,不依靠,无关[性]

~ accountant 独立会计师
~ accounting unit 独立核算单位
~ audit 独立审计
~ contractor 独立承包商
~ test 独立性检验
~ variable 自变量

independent [indi'pendənt] *a.* 独立的,线性无关的,自主的
~ axle 自动轮[轴]
~ component 独立成分
~ event 独立事件
~ footing 独立底脚
~ foundation 独立地基,独立基础
~ function 独立函数
~ girder 独立主梁[桁架]
~ medium 独立式中央分隔带
~ observation 独立观测
~ partial tide 独立分潮
~ sample 独立样本
~ scaffold 独立脚手架
~ stairs 不靠墙楼梯
~ suspension 独立悬架
~ trailer 全挂车
~ variable 自变量,独立变量
~ ventilation 独立通风
device ~ 装置独立性,设备独立性
linearly ~ 线性无关,线性独立

indestructibility ['indistrʌktə'biliti] *n.* 不灭性,不可损坏性
~ of matter 物质不灭(定律)

indestructible [indi'strʌktəbl] *a.* 不灭的,耐久的,不能破坏的; indestructibly, *ad.* ; indestructibleness, *n.*

indeterminacy *n.* 不确定度,不明确,不确定(性),含糊不清,测不准

indeterminable [indi'tə:minəbl] *a.* 不能确定[解决]的,不定数; *n.* 难以解决的问题;indeterminably, *ad.*

indeterminate [indi'tə:minit] *a.* 不(确)定的,超静定的; indeterminately, *ad.* ; indeterminateness, indeterminacy, *n.*

~ analysis 不定解分析
~ coefficient 不定系数,未定系数
~ equations 不定方程组,不定方程式
~ error 不定误差
~ form 不定式
~ frame 超静定构[框]架
~ principle 测不准原理
~ structure 超静定结构

index ['indeks] *n. pl.* indexes, indices; *v.* ; *n.* ,指数,系数;索引,目录,指标,指针;指示器;指南;下标,变址; *vt.* 给……编索引,编入索引中
~ -area method（径流预报的）指标面积法
~ arm (of sextant)（六分仪的）指臂
~ bar 标杆
~ basin 参证[指标]流域
~ bed 标准层,指示层
~ book 目录书
~ catalog (ue) 索引表
~ chart (map, diagram) 索引[指示,接合]图
~ contour 标注[注数字]等高线,指标等值线
~ correction 指标改正,仪器订正
~ crant 分度头曲柄
~ edge 索引栏
~ed loan 指数化贷款
~ed plane 标高平面
~ error 指标(误)差,分度误差
~ glass (测角器)指示镜,(六分仪的)动镜
~ guage 指示计[表]
~ hand 指针
~ head 分度头,分度器
~ hole 定位孔
~ intensity 标记亮度
~ law 指数律
~ line 等值线,指标线
~ liquid 标记液体,折射率液
~ manual 牵引手册
~ map 索引图,(建筑物)位置图

~ mark （读数）指标，目录，索引；商标，品号；枪管上的瞄准线
~ mineral 标准[指标]矿物
~ movement 指数变化
~ number 指数，索引号
~ of blasting action 爆破作用指数
~ of correlation 相关指数
~ of inertia 惰性指数，惯性指数
~ of liquidity 液性指数，液化指数
~ of living standard 生活水平指数
~ of performance 效能[性能]指数
~ of plasticity 塑性指数
~ of pollution 污染指数[标]
~ of precision 精(确)度指数
~ of quality 质量指标
~ of radicals 根式的指数
~ of refraction 折射[光]率
~ of reliability 可靠性指数
~ of sensitivity 灵敏度指数
~ of speciality 特性指数
~ of stability 稳定指数[率]
~ of turbulence 紊流指数
~ of wetness 湿润[年水量]指数
~ organism 指标生物
~ part 变址部分
~ percent 指数率
~ plan 索引图
~ plane 标志[标准]面
~ plate 分度[标度]盘
~ register 变址[指数]寄存器
~ ring 分度圈，刻度环
~ station 参证[参考]站
~ storage 编址存储器
~ stress 指示应力
~ table 水平分度头
~ tag 标牌，标签，记号，（测深索上的）标记
~ time 转位时间
~ unit 指示装置
~ variable 下标变量
~ word 变址字，下标字，索引字
~ed plane 标高平面
~ed system 加标系

absorption ~ 吸收指数
acidity ~ 酸度指数
adiabatic ~ 绝热指数
asphalt penetration ~ 沥青针入度指数
bitumen ~ 纯沥青指数
cementation ~ 粘结性指数
cooperation ~ 协同索引[记号]
covariant ~ 协变指标
cycle ~ 循环指数
decontamination ~ 净化指标
fixed ~ 固定式照准器
group ~ （路基土壤分类）组指数[标]
liquidity ~ 流性指数
penetration ~ 渗透指数
performance ~ 特性[性能]指数
plasticity ~ 塑性指数
pollutional ~ 污染[污浊]指数
production ~ 生产指标
quality ~ 质量指标
refraction ~ 折光指数
sensitivity ~ 灵敏度指数
sliding ~ 游标
spectral ~ 光谱指数
time ~ 时标，记时
viscosity ~ 粘度指数

indexing ['indeksiŋ] *n.* 标刻度，索引，加下标，变地址
~ plate 刻度盘
~ system 指数调整制度
aspect ~ 方面[向]性检索
compound ~ 复式分度法
coordinate ~ 坐标检索（法），对等检索
correlative ~ 相关检索
datacode ~ 数据码检索
differential ~ 差动分度法

India ['indjə] *n.* 印度
~ rubber 天然橡胶，（擦字用）橡皮，印度橡胶
~ rubber wire 橡胶绝缘线

Indian ['indjən] *n.; a.* 印度人（的），印度的

~ architecture 印度式建筑
~ ink 黑墨,墨汁

Indian architecture

~ monsoon 印度季风
~ Ocean 印度洋
~ paper 凸版纸,字典纸
~ red 印度红,三氧化铁
~ red pigment 黄红色铁矿矿砂颜料
~ redwood 印度红木
~ rock-fill dam 印度式堆石坝
~ rubber 弹性橡皮
~ summer 秋老虎,(深秋,初冬之)小阳春
~ tide plane 印度洋基准面

indicate ['indikeit] *v.*, -cated, -cating;指示,显示,表示,指出; indicatable, *a.*
~ed altitude 计示高度
~ed efficiency 指示功率[效率]
~ed horsepower 指示马力
~ed number 指示数
~ed pressure 指示压力
~ed reserve 推论储量
~ed thrust 指示推力
~ing accuracy 指示精度
~ing arm 指示杆
~ing floor stand 指示地轴架
~ing ga(u)ge 指示计
~ing (measuring) instrument 指示式(量测)仪器
~ing light 指示灯
~ing needle 指针

indication [indi'keiʃən] *n.* 指示,显示,表示,暗示,指出;象征,迹象;读数
~ error 读数误差
~ sign 指示标志
brief ~s 简述,概述
group ~ 组号,分类标志
remote ~ 远距离指示,遥测

indicative [in'dikətiv] *a.* 指示的,表示的,表示特征的
~ figure 指示性数字
~ growth rate 指示性增长率
~ mark 指示性标志
~ planning 指示性计划
~ price 指示性价格

indicator ['indikeitə] *n.* 指示器,计数器,转速器,记录盘;指示剂;经济指标
~ card 示功图,指示卡(片)
~ diagram 示功图,指示压容图
~ gage 指示表[针]
~ lamp 指示灯
~ paper 试纸
~ post 开关指示柱
dial ~ 千分表,标度盘指示器
load ~ 荷载指示器
remote plan ~ 平面位置遥测器
revolution ~ 转数计[表]
strain-gauge ~ 应变仪
surface ~ 表面规,表面找正器
wind (direction) ~ 风向指示器
wind velocity ~ 风速仪

wind velocity indicator

indices ['indisi:z] *n. pl.* of index;指出,记号,标志;标高,高程;分数,指数

indiction [in'dikʃən] *n.* 小纪(15 年)

indifference [in'difərəns] *n.* 不重要[视],琐事,小事;不关心;无差别;中性,中立
~ curve 无差异曲线
~ level 无差异水平

indifferent [in'difərənt] *a.* 无差异的;中[惰]性的;无关紧要的,不重要的;质量不高的;-indifferently, *ad.*

~ equilibrium 随遇[中性]平衡
~ gas 惰性气体

indigenous [in'didʒinəs] *a.* 本地产的,国产的；天生的,固有的；indigenously, *ad.*; indigenity, indigenousness, *n.*
~ capital 本国资本
~ forest 原始森林,处女林
~ soil 原生土,定积土
~ stream 本地河
~ value 固有价值

indigestible [indi'dʒestəbl] *a.* 不能消化的,难理解的,难接受的；indigestibly, *ad.*; indigestibleness, *n.*

indigestion [indi'dʒestʃən] *n.* 消化不良,难理解,难领会

indigitation [indidʒi'teiʃən] *n.* 套迭

indigo ['indigəu] (= indigotin) 靛蓝[色]

indirect [indi'rekt] *a.* 迂回的,间接的,次要的；indirectly, *ad.*; indirectness, *n.*
~ action 间接作用
~ activities 辅助业务
~ addressing 间接寻址
~ aerological analysis 间接高空分析
~ bank protection 间接护岸工程
~ benefit 间接效益[受益]
~ catchment 间接集水面积
~ charges 间接费用
~ construction costs 间接工程费用
~ cost 间接成本
~ damage 间接损失[害]
~ department 辅助部门,服务部门
~ discharge measurement 间接测流法
~ drainage 间接排水
~ eruption 间接喷发
~ exchange 间接汇兑[交换]
~ expense 间接费用
~ factor 间接因素
~ flood damage 间接洪水损害
~ heating (system) 间接供暖(系统)
~ illumination 间接[反射]照明
~ influence 间接影响[作用]
~ investment 间接投资
~ labo(u)r 间接劳动
~ laying 间接瞄准
~ leveling 间接高程测量
~ liability 间接责任[负债]
~ lighting 间接照明
~ load 间接荷载
~ loss 间接损失
~ measurement 间接量度
~ mouth 河口(河口与海湾相接之处)
~ national tax 间接国家税
~ observation 间接观测

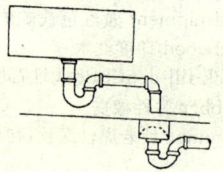

indirect waste pipe

~ radiator 间接散热器
~ stress 间接应力,合成应力
~ waste pipe 迂回排泄管

indirection [indi'rekʃən] *n.* 间接的行动[方法、步骤]；迂回

indiscriminate [indis'kriminit] *a.* 紊乱的,无差别的,混淆的；不辨优劣的,不加选择的；indiscriminateness, *n.*; indiscriminately, *ad.*

indispensable obligation 不可推卸的责任

indissoluble [indi'sɔljubl] *a.* (indissoluable)不溶的,不溶解的,不能分解的,不能分离的

inditron ['inditrɔn] *n.* 字码管,氖灯,指示管,示数管

indium ['indiəm] *n.* 铟(In)

individual [indi'vidjuəl] *n.* 个人,个体；*a.* 个人的,个体别的,单一[独]的；独特的,特殊的,有特性的

~ bargaining 个别买卖,个别交易
(~)base 独立基础
~ batcher 分批计量器;单个混凝土拌和机
~ capital 个人资本
~ cellar 独立地下室[地窖]
~ coat 特殊涂层
~ consumer 特殊用户
~ earnings 个人收入
~ enterprise 私人企业
~ elevation 高程注记点
~ equilibrium 单个均衡
~ fabrication 特殊装潢
~ footing 单独基础,独立基础
~ footing foundation 单底脚基础
~ form of art 艺术的独特风格
~ heating 单独供暖
~ item 个别项目
~ labour 个体劳动(者)
~ management 个别管理
~ mounting 个别装饰,特殊装饰
~ negotiation 个人转让
~ office 个人办公室
~ packing 单件包装
~ personality 个人人格
~ prepared-roofing shingle 个别预制的屋顶盖板
~ project 单项工程
~ proprietorship 独资,个人企业
~ sewage disposal system 专用污水处理系统
~ shot 单孔爆破
~ style 独特风格
~ tile 专用瓦[砖]
~ vent 单独通气管

individual vent

individuality [individju'æliti] *n. pl.* -ties;个体,个性;(*pl.*)特质,特征;独立存在

individualization [individjuəlai'zeiʃən] *n.* (使具)个性化

individuation [individju'eiʃən] *n.* 个体化,个性化,区域分化

indivisibility [indivizə'biliti] *n.* 不可分性,不能整除

indivisible [indi'vizəbl] *a.* 不可分割的(东西),不能整除的,极微的

indolence ['indələns] *n.* 进展缓慢的情况,懒惰,怠惰

indoor ['indɔː] *a.* 室内的,户内的,内部
~ air cooler 室内冷气机
~ arena 室内斗技场
~ building board 室内建筑板
~ building sheet 室内建筑钢板
~ chlorinated rubber paint 室内氯化橡胶涂料
~ (clear) varnish 室内清漆
~ climate 室内环境
~ coating 室内涂料[涂层]
~ decor 室内装饰
~ decorating 室内装饰[布置]
~ design temperature 室内设计温度
~ emulsion paint 室内(感光)乳剂油漆,室内乳化油漆
~ finish(ing) paint 室内饰面油漆,室内终饰油漆
~ fixture 室内(附属)装置(如电线,自来水管等)
~ fountain 室内喷泉,室内喷水池
~ glass door 室内玻璃门
~ gloss (clear) varnish 室内抛光油漆
~ humidity 室内湿度
~ ice rink 室内溜冰场
~ illumination 室内照明
~ installation 室内装置
~ lacquer 室内漆器
~ learner's pool 室内初学者泳池
~ light fitting 室内照明设备
~ luminaire (fixture) 室内照明装置
~ marble 室内大理石
~ market 室内市场
~ masonry (dividing) wall 室内污工[砖石](分隔)墙
~ moisture 室内湿度,室内保湿

~ noise 室内噪音
~ partition 室内分隔

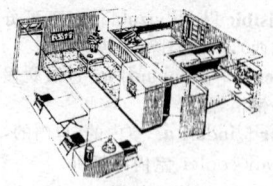

indoor partition

~ pigmented varnish 室内有色清漆
~ pipe system 室内管网
~ plant 室内植物
~ potted landscape 室内盆景
~ relative humidity 室内相对湿度
~ setting (scene) 内景
~ slatted blind 室内的木条百叶窗
~ swimming bath(s) 室内游泳池
~ swimming pool 室内游泳池
~ teaching pool 室内教练池
~ temperature 室内温度
~ temperature control 室内温度控制
~ tile 室内铺地砖
~ track 室内跑道
~ whitewash 室内粉刷[饰]
~ window cill [sill] 室内窗台

indorse [in'dɔ:s] *v.* =endorse; 保证,担保;认可,支持;支票(等的)背书

indorsee [indɔ:'si:] *n.* = endorsee; 受让人,被背书人

indraft ['indrɑ:ft] *n.* 吸入(物),引入,流入,向内流或向内的(气,水)流

indraught ['indrɑ:ft] *n.* 吸入,引入,吸入物,流入,向岸流;海的入口

indrawing ['indrɔ:iŋ] *n.* 吸入,向内,吸进; *a.* 引入的,吸入的

induce [in'dju:s] *v.* -duced,-ducing; 诱导,惹起,感应
~ business 招徕生意
~d charge 感应电荷,感生电荷
~d cleavage 诱生劈理

~d cracking 感应裂缝
~d current 感应电流
~d decomposition 诱导分解
~d draft 诱导通风,强制通风
~d draft fan 吸风风扇
~d drag 诱导阻力,感应阻力
~d draught 诱道通风,强制通风
~d draught fan 吸风机,引风机
~d draught water cooler 吸风水冷却器
~d earthquake 诱发地震
~d flow 诱导水流
~d infiltration 诱导入渗[下渗],次生下渗
~d item 诱发项目
~d porosity 次生孔隙
~d recharge (of an aquifer) 含水层的,诱导回灌,诱导补给,人工回灌,人工补给,从河流或其他水体向含水层中泄水
~d strees 诱导应力
~d surface runoff 诱导地面径流
~d traffic 诱增交通量(因道路交通设施改进所增加的交通量,包括新增交通量,导增交通和变增交通量)
~d voltage 感应电压

inducement [in'dju:sment] *n.* 诱导,刺激;引诱物,动机

inducer [in'dju:sə] *n.* 诱导者,劝说者,诱导物,诱因;叶轮
impeller ~ 导流叶轮,导风叶轮
knock ~ 爆震诱导物

induct [in'dʌkt] *v.* 导入,吸入,引导,引进;介绍,传授,归纳;使正式就任[职]; inducter, *n.*

inductance [in'dʌktəns] *n.* 电感,电感器,感应系数
~ bridge 电感电桥
~ effect 电感效应
~ strain gage 电感应变计

induction [in'dʌkʃən] *n.* 引导,诱导作用;感应;归纳(法);序言,前言,序论,绪言;就职[任]式

~ coil 感应线圈
~ compass 感应式罗盘
~ furnace 感应炉,感应电炉
~ generator 感应发电机
~ hardering 感应淬火
~ heating furnace 感应加热炉
~ manifold 吸水管,进油[气]管
~ method 归纳法
~ motor 感应电动机,异步电动机
~ period 诱导期,感光胶卷显影时间
~ pipe 吸管,吸入管,送水[气]管
~ stroke 进气冲程
~ valve 吸入阀,进气门
~ ventilation 诱导通风,导入式通风

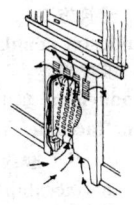
induction ventilation

finite ~ 有限归纳法
mathematical ~ 数学归纳法
successive ~ 逐次归纳法
transfinite ~ 超穷归纳法

inductive [in'dʌktiv] *a.* 感应的;归纳的;吸入的,诱导的;序论的,入门的
~ coil 电感线圈,有感线圈
~ method 归纳法
~ sailnometer 感应式盐度仪

inductivity [indʌk'tiviti] *n.* (绝对)电容率,介电常数,诱导率,感应性
magnetic ~ 导磁率

inductor [in'dʌktə] *n.* 感应器,诱导物;电感线圈;擅加化学反应速率的物质
~ form 线圈架,线圈管

indurate ['indjuəreit] *v.* 使坚固,(使)变硬,(使)硬化,固结,-rated,-rating. ['indjuərit] *a.* 硬化的
~d clay 硬化黏土
~d talc 硬滑石,滑石板岩
~d rock 硬化的岩石
~d soil 硬盘土,硬结土壤,固结土壤

induration [indju'reiʃən] *n.* 硬化,硬结,固结

industrial [in'dʌstriəl] *n.* 工业[产业]工人;工业家;工业公司,工业股票; *a.* 实业的,产业的,工业上的;有关产业工人的
~ architect 工业建筑师
~ architecture 工业建筑(学)
~ area 工业区
~ art 工艺;工艺美术,工业美术
~ asphalt(ic) tile 工业沥青瓦
~ base 工业基地
~ bond 工业债券
~ building 工业建筑物,厂房
~ capital 工业资本
~ center 工业中心
~ chimney 工业烟囱
~ city 工业城市
~ constution 工业建筑(物)
~ construction project 工业建筑工程
~ construction site 工业施工工地[现场]
~ consumption 工业耗水量
~ control 生产过程控制
~ curtain wall 工业用帷幕墙
~ design 工业设计
~ detergent 工业洗涤剂
~ diamond 人造金刚石,工业金刚石
~ district 工业区
~ dust 工业垃圾
~ engineering 工业管理学,企业管理学
~ exhaust system 工业废气排放系统
~ extract ventilation unit 工业吸风装置,工业抽气通风装置
~ floor 工业楼板[面]
~ floor finish 工业楼面修饰
~ floor(ing) tile 工业用铺楼面砖
~ fund 工业资金
~ garage 工业停车房[汽车间]
~ gas 工业煤气
~ glass 工业玻璃
~ glazing 工业磨光(机)

~ hazard 工业事故
~ heating facility 工业供暖装置
~ housing 工业房屋(建筑)
~ hygiene 工业卫生(学)
~ investment 工业投资
~ light fitting 工业照明装置
~ light(ing) fixture 工业照明设备
~ loader 工业装载机;工业装卸工
~ luminaire (fixture) 工业照明设备
~ mastic floor(ing) 工业胶脂地板
~ noise 工业噪声
~ nuisance 工业公害
~ occupancy 工业区
~ park 工业区
~ parquet(ry) block 工业镶木[拼花]地板块料
~ partition (wall) 工业隔墙
~ plant 工厂;工业设备
~ pollution 工业污染
~ population 工业人口
~ port 工业港(口)
~ premises 工业房产连地基
~ product 工业生产
~ property 产业所有权
~ reservoir 工业蓄水池[水箱]
~ security manual 工业安全手册
~ sewage 工业污水,工业废水
~ shutter door 工业百叶门,工业遮门
~ space-heating 工业小暖炉
~ stact 工业烟囱
~ stair (case) 工业楼梯(间)
~ standard 工业标准
~ storied building 工业高层建筑
~ system 工业体系
~ tile 工业砖[瓦]
~ town 工业城市
~ union 产业工会
~ vapours 工业蒸汽[废气]
~ varnish 工业油漆
~ wall tile 工业墙砖
~ waste(s) 工业废水[废料,废液]
~ waste gas 工业废气
~ waste water 工业废水
~ water service 工业用水设施
~ water supply 工业供水
~ water tower 工业水塔
~ zone 工业区

industrialize [in'dʌstriəlaiz] *v.* (使)工业化
~d building 工业化建筑
~d (building) construction 工业化建筑施工

industry standard specification 工业标准规格

induvia *n.* 集中层体

ineffaceable [ini'feisəbl] *a.* 消除不掉的,不可磨灭的; ineffaceably, *ad.*; ineffaceability, *n.*

ineffective [ini'fektiv] *a.* 无效的,无益的,无能的;缺乏艺术效果的; ineffectively, *ad.*; ineffectiveness, *n.*

inefficiency [ini'fiʃənsi] *n.* 无效率,无能
discharged for ~ 因无能被解职

inefficient [ini'fiʃənt] *a.* 效率很低的,无能力的; inefficiently, *ad.*
~ pump 低能泵

inelastic [ini'læstik] *a.* 无弹性的,无伸缩性的,无适应性的
~ behaviour 非弹性状态
~ bending (梁)的非弹性弯曲(即塑性弯曲)
~ buckling 非弹性压曲(即塑性压曲)
~ consolidation 非弹性固结
~ deflection 非弹性挠曲
~ deflexion 非弹性挠曲
~ deformation 非弹性变形
~ demand 无弹性需求
~ lateral buckling 非弹性侧向压屈
~ property 非弹性性质
~ range 非弹性阶段,非弹性范围
~ region 非弹性区域
~ scattering 非弹性散射

~ strain 非弹性应变
~ struction 非弹性结构
~ supply 无弹性供给
inelasticity [inilæs'tisiti] *n.* 无弹性,无适应性
inelegance [in'eligəns] *n.* 不精致,粗糙
inelegant *a.* 不精致的,粗糙的
ineligible [in'elidʒəbl] *a.* 不合格的,不适当的,不能入选的; *n.* 不合格者; ineligibleness, *n.*; ineligibly, *ad.*
inequality [ini(:)'kwɔliti] *n.* 不等式,不平等,不平均;均差;不平度,地形崎岖度
~ constraint 不等式约束
integral ~ 积分不等式
inequation [ini'kweiʃən] *n.* 不等(方程)式
inequiaxial [ˌini:kwə'æksiəl] *a.* 不等轴的
inequigranular *n.* 不等粒状
inequilateral [ini:kwi'lætərəl] *a.* 不等边的
inequitable [in'ekwitəbl] *a.* 不公平的,不公正的
~ exchange 不公平交换,不等价交换
~ taxation 不公平课税
inert [i'nə:t] *a.* 惰性的,不活泼的,无自动力的,缺乏活动性的,迟钝的,不起化学作用的
~ aggregate 惰性骨料
~ dust 岩粉
~ filler 惰性填料
~ force 惯性力
~ gas 惰性气体
~ material 惰性材料,惰性物质
~ matter 惰性物质
~ pigmen 惰性颜料
~ solid 惰性固体(即矿料)
inertance *n.* 惰性,惯性,迟滞
acoustic ~ 声惯量,声感抗
inertia [i'nə:ʃiə] *n.* 惯性,惰性;惯性,惰性值;迟钝,缺乏活动性; inertial, *a.*
~ block 惰性块

~ coefficient 惯性系数
~ couple 惯性力偶
~ current 惯性流
~ damping 惯性阻尼
~ effect 惯性效应
~ flow 惯性流
~ force 惯性力
~ governor 惯性调速器
~ grade (公路等的)惯性坡度
~ head 惯性水头
~ load 惯性荷载
~ mass 惯性质量
~ modulus 惯性模量,惯性系数
~ oscillation 惯性振动
~ resistance 惯性阻力
~ test 惯性试验
~ wave 惯性波
bending ~ 弯曲惯性
electromagnetic ~ 电磁惯性[惰性]
fuel ~ 燃料惯性
gyroscopic ~ 陀螺惯性
magnetic ~ 磁惯性
moment of ~ 惯性矩,转动惯量
rolling-moment ~ 滚动惯性矩,滚转转动惯量
rotary ~ 转动惯量
thermal ~ 热惯性
inertial [i'nə:ʃəl] *a.* 惯性的,惰性的,不活泼的
~ boundary layer 惯性边界层
~ impact force 惯性冲击力
~ reactance 惯性抗力
~ settling 惯性沉降
~ system 惯性系
inertio-gravitational wave = inertia-gravity wave 惯性重力波
inessential [ini'senʃəl] *a.* 无关紧要的,不重要的,非必要的;非本质的,非物质的
inestimable [in'estiməbl] *a.* 难估量的,极贵重的,无价的; inestimably, *ad.*; inestimableness, *n.*

inevitability [inevitə'biliti] *n.* 不可避免,必然性

inevitable [in'evitəbl] *a.* 不可避免的,必然的; inevitableness, *n.*; inevitably, *ad.*

inexact [inig'zækt] *a.* 不精确的,不精密的,不正确的,不准确的; inexactly, *ad.*; inexactness, *n.*
~ reasoning 不精确的推理
~ statement 不正确的断语

inexhaustible [inig'zɔːstəbl] *a.* 不会耗尽的,无穷无尽的,源源不绝的; inexhaustibility, inexhaustibleness, *n.*; inexhaustibly, *ad*.

inexpansibility *n.* 不可膨胀性

inexperience [iniks'piərjəns] *n.*; *v.* 缺乏实际经验,不熟练,外行
~d operator 生手

inexplosive [iniks'pləusiv] *a.* 不爆炸的,不破裂的,不爆发(性)的

inextensibility [inikstensə'biliti] *n.* 不可延展性

in extenso [in eks'tensəu] *ad.*
【拉】全部,不省略,不删节,详尽

infall ['infɔːl] *n.* 降落,塌陷,崩陷;入侵;(水库,运河等的)进水口

infallibility [infælə'biliti] *n.* 无拒绝性,无(失)误性,绝对可靠性

infan *n.* 输入(端),扇入

infant ['infənt] *n.* 婴儿,幼儿,法定未成年者; *a.* 婴儿的,幼年的,初期的,未成年的
~ industry 新建工业,新生工业
~ marketing 外销初始期

infantile ['infəntail] *a.* 婴儿的,幼稚的,初期的,发端的
~ landforms 幼年地形

infection [in'fekʃən] *n.* 传染,感染,传染病;(流行于社会的)影响[思想]
~ period (~ stage) 感染期
art ~ 艺术感染

infeed [infiːd] *n.* 横进给,横向进磨,切入磨法,横切;馈电

infer [in'fəː] *v.* -ferred,-ferring;推论,推断;表示,意思就是
~ red 推断的,推论的
~ red reserves 推断储备(量),推定储量

inference ['infərəns] *n.* 推理,推断,推论,演绎,含意
~ technique 推论技术

inferential [infə'renʃəl] *a.* 推理(上)的,推论(上)的; inferentially, *ad.*

inferior [in'fiəriə] *a.* 劣等的,下级的,初等的; *n.* 下级,下辈,部下
~ advantage 较小利益
~ field 无穷域
~ figures 下附数字
~ goods 次等商品,低档货物
~ limit 下限
~ purlin 下桁条
~ tide 下位潮

infill [in'fil] *v.* 填充
~ block 填充块
~ brick 填充砖(块)
~ panel 填充板
~ slab 填充板
~ tile 填充砖[瓦]
~ wall(ing) 填充墙(砌体)

inferior tide

infilling [in'filiŋ] *n.* 填充(物),填充空隙
~ concrete 填充混凝土
~ material 填充料

infiltrate [in'filtreit] *n.*, *v.*, -trated,-trating; *n.* 渗透物,渗入物,渗滤水; *v.* 渗透,渗入;浸润,浸透;突破
~ d course 渗水层

infiltrated course

~d dampproof course 渗排水防潮层
~d flow 渗流
~d water 下渗水,入渗水

infiltration [infil'treiʃən] *n.* 下渗,入渗,渗透,渗滤,浸润
~ area [zone] 下渗区,地下水补给区
~ basin 渗入盆地
~ capacity 下渗容量,渗入能力,渗透量
~ channel 渗水渠,渗水沟
~ coefficient 入渗系数,渗透系数
~ ditch 渗水沟,盲沟
~ diversion 渗透取水,(河底)渗水引水管
~ experiment 渗透试验
~ flow 渗入流量,渗流
~ gallery 渗水渠,渗水廊道,集水管道(埋设在地下的有孔管道)
~ head 渗水水头
~ index 下渗指数
~ intensity 下渗强度
~ loss 渗流损失
~ pond 渗水池

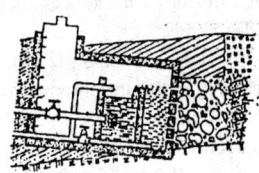

infiltration pond

~ rate 渗透率,入渗率
~ rate curve 渗透率曲线
~ slit 渗水缝
~ stress 渗透应力
~ theory 下渗理论,入渗理论
~ tunnel 渗水隧道
~ value 下渗值
~ volume 渗入容量
~ water 下渗水,入渗水,过滤水
~ well 入渗井

infiltrometer *n.* 测渗仪,渗透计
~ plot 下渗实验区

infimum [in'faiməm] *n.* 最大下界,下确界

infinite ['infinit] *a.* 无限的,无穷的,极大的,无穷大的; *n.* 无穷数;infiniteness, *n.*
~ aggregate 无穷[限,尽]集料
~ aquifer 无限含水层
~ convergence 无限收敛
~ decimal 无穷小数
~ degree of stability 无限稳定度
~ elasticity 无限弹性
~ integral 无穷积分
~ medium 无限介质
~ number 无穷大数
~ population 无限总体
~ product 无穷乘积
~ regress 无穷回归
~ sequence 无限序列,无穷序列
~ series 无穷级数,无限级数
~ set 无穷集,无限集
~ span 无限跨度,无限翼展
~ strip 无限(地)带,无限露天采掘
~ static stability 无限静稳定性

infinitely ['infinitli] *ad.* 无限地,无穷地
~ great 无穷大,无限大
~ rigid pile 无限刚性桩

infinitesimal [infini'tesiməl] *a.* 无穷小的,无限小的,无限小(量),极微(量)
~ calculus 微积分(学),微分运算,积分运算
~ change 无限小变化
~ deformation 无限小形变
~ displacement 无限小位移
~ element 无穷小元素,微量元素
~ linear trasformations 无穷小线性变换
~ radius 极小半径
~ value 无穷小数

infinitude [in'finitju:d] *n.* 无限,无穷,无限量,无限小的数量[范围],一个无穷数

infinity [in'finiti] *n. pl.* -ties;无穷,限,无数,无穷大,无限大;超限数;无边,无量
~ bar (测距仪的)校正杆

actual ~ 实无穷,真无穷
infix [v. in'fiks n. 'infiks] v. 固定,插入;中缀,中加成分,镶嵌; n. 字腰; infixion n.
inflame [in'fleim] v. -flamed,-flaming;燃烧,着火; inflamer, n.; inflamingly, ad.
inflamer n. 燃烧器,燃烧物
inflammability [inflæmə'biliti] n. 易燃性,可燃性
～ limit 可燃极限
～ point 燃点,着火温度
～ test 易燃性试验
inflammable [in'flæməbl] n. 易燃物,可燃物; a. 易燃的,易激动的; inflammably ad.; inflammableness, n.
～ gas detector 可燃气体探测器
～ liquid 可燃液体,易燃液体
～ material 易燃材料
～ store 易燃品贮藏库
～ substance 易燃物,易燃料
inflammation [inflə'meiʃən] n. 燃烧,着火,点燃;发光;起爆
～ temperature 着火温度
inflatable [in'fleitəbl] a. 充气的,可充气的物品;可膨胀的
～ building 充气建筑

inflatable building

～ cushion 气垫
～ dam 充气坝
～ mattress 充气垫
～ packer 充气垫,充气止浆塞
～ skeleton 充气骨架,充气房屋
～ structure 充气结构
inflate [in'fleit] v. ,-flated,-flating;充气,(轮胎)打气,膨胀;通货膨胀;inflater,inflator, n.
～d 膨大的,(通货,物价等)膨胀的
～d bridge 充气桥
～ d-hose 充气(加劲)软管
～d slag 多孔熔渣
～d structure 多孔结构,充气结构
～d tyre 充气轮胎
inflation [in'fleiʃən] n. 充气,膨胀,通货膨胀;(气体)补给,充气
～ pressure 充气压力,轮胎压力,气胀压力
～ tax 通货膨胀税
air ～ 充气
structural ～ 结构性通货膨胀
inflationary [in'fleiʃənəri] a. 使膨胀的,有膨胀性的;通货膨胀的
～ gap 通货膨胀差距,通货膨胀缺口
～ spiral 膨胀螺旋
spiral ～ 通货,物价;工资的螺旋形膨胀,恶性通货膨胀
inflator [in'fleitə] n. 增压器,压送泵,打气筒;充气机
inflect [in'flekt] v. 使弯[反]曲,使屈折,使向内弯曲; inflectedness, n.; inflector, n.
～ed arch 反曲拱
inflection [in'flekʃən] n. 反弯[曲],反挠,拐折;凹陷,向内弯曲;偏差,偏转,偏移,偏斜; inflectionless, a.
～ angle 拐角,偏转角
～ point 拐点,转折点,反曲点
inflexibility [infleksə'biliti] n. 不挠性,不弯曲性,不屈性,硬性,刚性,不可压缩性;劲度,刚度
inflexible [in'fleksəbl] a. 不曲的,刚性的,刚直的; inflexibly, ad.; inflexibleness, n.
～ burden 固定负荷
～ steel wire rope 硬钢丝绳
～ tile 刚性砖
inflexion [in'flekʃən] n. =inflection;反曲,反挠;偏差,偏移;拐折,回折;inflexional, a.; inflexionless, a.

~ point 反曲点,拐点,转折点
inflow ['infləu] *n.* 入流,来水,流入;入流量,进水量
~ channel 进水渠(道)
~ current 进流,正极电流
~ hydrograph 来水过程线,入流过程线
~ of capital 资本流入
~ of ground water 地下水来水量
~ of phreatic water 地下水[潜水]来水量
~ -storage-discharge curve 进水-蓄水,泄水曲线
~ storage-outflow method (洪水演算的)入流-蓄流-出流法
~ wheel (液力传动的)向心式叶轮
~ing T-piece (支管)入流三通,汇流三通
influence ['influens] *n.,v.*, -enced, -encing; *n.* 影响,效应,作用,势力,感应;*v.* 感化,影响,对……有作用;influenceable, *a.*; influencer, *n.*
~ area 影响范围,影响面积,影响圈
~ basin (area) (水井抽水漏斗)影响范围,浸没面积
~ characteristic 感应特征
~ chart 影响图
~ circle (水井抽水漏斗)影响圈
~ coefficient 影响系数
~ diagram 影响(线)图
~ field 影响场
~ line 影响线,感应线
~ line of moments 弯矩影响线
~ of heat 热效应
~ surface 影响面
~ value 影响值
~ zone 影响带
influencing factor 作用因素,影响因素
atmospheric ~ 大气影响
disturbing ~ 扰动影响
frequency ~ (对仪表读数准确度)频率影响

friction ~ 摩擦影响
temperature ~ 温度影响
influent ['influənt] *a.* 流入的,汇入的,进水的;*n.* 入流渗水;支流;渗流
~ action 渗水作用
~ channel 渗水渠
~ flow 入流量
~ pipe 渗流管
~ seepage 渗漏,渗透
~ stream 渗水河,(补给地下水的)亏水河
~ water 渗漏水

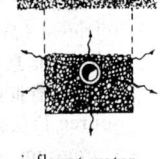
influent water

influential [influ'enʃəl] *a.* 有影响的,有力的,有势力的;*n.* 有影响力的人物(常用复数);influentially *ad.*
influx ['inflʌks] *n.* 流入(量),灌入,汇集,涌入;河口,河流的汇合处,注入口
~ rate 涌入速度
~ of order 定货单涌至
~ of traffic 交通汇合处
inform [in'fɔ:m] *v.* 通知,通告,传达
informal [in'fɔ:məl] *a.* 非正式的,通俗的
~ agreement 非正式契约
~ English 通俗英语
~ group 非正式组织
~ record 非正式记录
informatics [infə'mætiks] *n.* 信息学,信息控制论
information [infə'meiʃən] *n.* 信息,情报,资料,消息,数据;新闻,报导,知识,通知,通报
~ audit 情报审计
~ bank 资料库,情报所,信息所
~ bits 信息位,信息比特
~ channel 信道
~ circular 资料通报
~ counter 服务台,问讯台
~ desk 问讯处,服务台
~ engineering 信息工程学
~ feedback 信息反馈

~ flow 信息流
~ generator 信息源
~ management system 信息管理系统
~ office 问讯处,情报处[室]
~ processing 情报处理
~ processing language 信息处理语言
~ retrieval 信息检索
~ science 信息科学
~ service 查询业务,查询台
~ source 信息源
~ storage means 信息存储方式
~ theory 信息论
~ transfer 信息传递
~ transmission 信息传输
commodity ~ 商品信息
financial ~ 金融信息
market ~ 市场信息
topographical ~ 地形资料

informational [infəˈmeiʃənəl] *a.* 信息的,新闻的,情报的,供资料的
~ sign 导向标志,指示标志

infra [ˈinfrə] *ad.* 在下,以下(指书的前后);[字首]表示"在下,在外"的意思
~ -acoustic 次声的,闻域以下的,亚声的
~ -audible sound 次声
~ -gravity wave 长周期重力波
~ -littoral 远离岸的
~ -littoral deposit 远岸沉积

infrabar *n.* 低气压
infragranitic *a.* 花岗岩层下的
infralittoral [ˌinfrəˈlitərəl] *a.* 远岸的
~ deposit 远岸沉积

inframedian zone 水深在 300～600 英尺间的海底

infraneritic *a.* 浅海的(水深36～183米)
~ environment 浅海环境

infrared [ˈinfrəˈred] *a.* 红外线的,红外辐射,红外区,电磁波谱的红外段
~ detection technique 红外探测技术
~ drying 红外线干燥
~ fire 红外炉
~ heater 红外线取暖器
~ lamp 红外线灯
~ oven 红外线烤炉
~ pavement heater 红外铺路(用)加热器
~ photograph(y) 红外线摄影(术)
~ radistion 红外无线电导航系统,红外无线电测距
~ rays (or light) 红外线
~ spectrum analysis 红外光谱分析
~ stove 红外炉
~ warmer 红外取暖器
~ welding machine 红外线焊接机

infrasil *n.* 一种红外硅材料
infrastructure [ˈinfrəstrʌktʃə] *n.* 下部结构,底层结构,下层构造,基底,基础设施;公共设施;永久性防御设施
infundibular [infʌnˈdibjulə] *a.* 漏斗形的
infusion [inˈfju:ʒən] *n.* 浸入,注入;输液;浸剂
~ process 浸渍法

infusorial earth *n.* (板状)硅藻土
infusum *n.* 浸剂
in(-)gate [ˈingeit] *n.* 入口孔,马头门,内门,输入门,内浇口
ingather [inˈgæðə] *v.* 收集,聚集,收获
ingathering [ˈingæðəriŋ] *n.* 收集,收获[集]物;*a.* 收集的,聚集的
ingenuity [indʒiˈnju:iti] *n. pl.* -ties;巧妙,精巧;创造力[性];才能,设计新颖,独出心裁
ingestion [inˈdʒestʃən] *n.* 把空气,气体或液体注入引擎里;摄取,取入,吸入,吸收; in-gestive, *a.*

ingle [ˈiŋgl] *n.* 火炉,壁炉

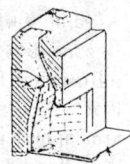

ingle

inglenook ['iŋglnuk] *n.* 炉边,炉隅
ingleside ['iŋglsaid] *n.* =fireside 炉边
ingoing ['ingəuiŋ] *n.*; *a.* 进来(的),注[深]的;就任的
~ flood 进潮流
~ stream 进潮流,进入水流
~ tenant 新租户
~ vessel 驶进来的船
ingoldsby car *n.* 重型卸料车
ingot ['iŋgət] *n.* (金,银,钢等的)锭[块],铸锭[模,块],浇锭,坯料
~ bar 铸块
~ blank 锭坯
~ breaker 碎锭机
~ buggy 运金属的卡车
~ car 运锭块的车
~ charger 装锭机
~ crane 锭块起重机
~ crusher 锭块压碎机
~ drawing machine 锭块牵引机
~ ejector 排锭器
~ iron 锭铁,工业纯铁,低碳钢
~ iron pipe 铸铁管
~ lathe 铸锭车床
~ manipulator 翻砂工(人),铸造用器具
~ mold 铁模
~ piler 垛锭机
~ slab 扁钢锭
~ steel 钢锭,铸钢
~ (transfer) car 运锭车
flat-shaped ~ 扁钢锭
thin-skinned ~ 薄皮钢锭
ingotism [iŋgətizəm] *n.* 树枝状结晶(钢锭结构缺陷)
ingrain ['in'grein] *v.* 原纱染色,深染; *a.* 深染的,固有的; *n.* 原料染色的产品
~ carpet 双面提花(无绒头)地毯
~ dye 显色染料
~ jute carpeting 双面提花黄麻地毯
~ wallpaper 双面提花墙纸
~ wallpaper coat 双面提花墙纸覆盖层

ingrain jute carpeting

~ed 纱染色的,包晶的,浸染的
ingredient [in'gri:djənt] *n.* (混合物的)成分,拼合料,配料
~ of concrete 混凝土成分
accessory ~ 副成分
ingress ['ingres] *n.* 进入,入口,进路,通道;初切,初亏;入场权;ingression, *n.*; ingressive, *a.*; ingressiveness, *n.*
~ of groundwater 地下水侵入
~ pipe 导入管
~ reflector 入射反射镜
~ transition 侵入过渡层
ingression [in'greʃən] *n.* 内移,进[侵]入,海侵
~ sea 进侵海
ingrown ['ingrəun] *a.* 长入内部的,向内生长的
~ bark (bark pocket) 树穴
~ valley meander 深切河谷曲流
ingrowth ['ingrəuθ] *n.* 向内生长(物)
inhabitancy [in'hæbitənsi] *n. pl.* -cies; 居住;住所,家
inhabitant [in'hæbitənt] *n.* 居民,居住者,栖息的动物
inhabitation [inhæbi'teiʃən] *n.* 居留,居住;住处[宅]
inhalant [in'heilənt] *n.* 吸入孔,吸入器,吸入剂; *a.* 吸入的
~ siphon 进水管
inhalation [inhə'leiʃən] *n.* 吸入,吸入剂[物]
~ tube 吸气管
~ valve 吸气阀

inhalator ['inhəleitə] *n.* 气雾吸入器,人工呼吸器
inhale [in'heil] *n.*；*v.*, -heled, -haling；*n.* 吸(人), 喝；*v.* 吸入；inhalement, *n.*
inhaler [in'heilə] *n.* 吸入器, 吸气泵, 滤气器, 空气滤过器
inhaul ['inhɔ:l] *n.* 牵收索, 拖铲索
　~ cable 卸载拉绳
inherence [in'hiərəns] *n.* 固有, 具有, 内在；基本属性
inherent [in'hiərənt] *a.* 内在的, 固有的, 本来的, 先天的；inherently, *ad.*
　~ ash 固有[原生]灰分
　~ austenitic grain size 奥氏体本质晶粒度
　~ balance (发动机等本身结构决定的)固有平衡
　~ burst 固有岩爆
　~ concrete heat 混凝土发热
　~ defect 固有缺陷
　~ error 固有误差
　~ fine grain 本质细晶粒
　~ grain size 本质晶粒度
　~ impurity 固有[内在]杂质
　~ law 内在规律
　~ moisture 结构水, 固有水分
　~ noise level 固有噪声级
　~ oscillation 内部振荡
　~ porosity 内在[固有]孔隙
　~ reliability 固有[内在]可靠性
　~ road right 固有路权
　~ stability 固有稳定性
　~ stress 内在应力
　~ value 原本价值
　~ vice 内部缺陷
　~ viscosity 特性黏度
inherit [in'herit] *v.* 继承, 遗传
　~ed error 继承误差
　~ed river 遗留河
inheritance [in'heritəns] *n.* 继承, 遗传, 承受, 继承权[物]
inhesion [in'hi:ʒən] *n.* =inherence；先天就有的情况, 固有(性), 内在(性)

inhibit [in'hibit] *v.* 防止, 制止, 抑制, 禁止；否定；inhibitable, *a.*；inhibiter, *n.*
　~ing 铠装(指火药柱)；加抑制剂
　~ing agent 缓凝剂
　~ing oil 防锈油
　~ing pigment 缓凝颜料
　~ing value 安定性增高值
inhibition [inhi'biʃən] *n.* 抑制, 制止, 阻止, 阻化, 延缓, 反催化
inhibitor [in'hibitə] *n.* 抑制剂, 阻化剂, 防锈蚀剂, 抗氧化剂, 防腐蚀剂
　~ coating 保护层
　~ valve 限制[止]阀
　antigum ~ 防胶剂
　corrosion ~ 抗腐蚀剂
　detonation ~ 防爆剂
　emulsion ~ 乳胶阻化剂
　flash ~ 消焰剂(硫化钾)
　oxidation ~ 抗氧化剂
　rust ~ 抗腐蚀添加剂, 防锈剂
inhomogeneity ['inhəumədʒi'ni:iti] *n.* 非均质, 不均匀性, 多相性, 杂色性
inhomogeneous ['inhəumə'dʒi:niəs] *a.* 不均匀的, 不均值的, 不均等的；不均质的；不同源的；非齐次的
　~ anisotropic medium 不均匀各向异性介质
　~ coordinates 非齐次坐标
　~ deformation 不均匀变形
　~ difference equation 非齐次差分方程
　~ field 非均匀场
　~ force system 不均匀力系
　~ mesh calculation 非均匀网格计算
　~ plastic flow 不均匀塑性流变
　~ soil 非均质土
　~ turbulence 非均匀(性)紊流
inhour =inverse hour *n.* 倒时数；反时针；逆时针；核反应(速度)单位
inhouse *a.* 内部的, 固有的, 自身的, 室内的
　~ colour 初色

~ design 自办设计
~ facility （企业）内部的设备
~ system （=in plant system）近距离（控制）系统
~ test 实验室试验

initial [i'niʃəl] *n.*; *v.*, -tial(l)*ed*, -tial(l)ing; *n.*; *a.* 最初（的），初始（的），原始（的），初期（的），固有（的）；字首，字头；*v.* 签姓名的首字母于……草签

~ address 起始地址
~ air content 初始空气含量
~ allowance 机械加工留量
~ amplitude 初（振）幅
~ an agreement 草签协定
~ azimuth 起始方位角
~ angle 初始角
~ appropriations 初步拨款
~ audit 初步审计
~ bed 底料层
~ bending 初始弯曲
~ blasting 初次爆破
~ breakdown 初次压轧，毛坯
~ budget estimates 初步概算
~ capacity 初始容量
~ compression 初（始）压力，预压力
~ condition(s) 初始条件，原始数据
~ configuration 初始构形
~ consolidation 初始固结
~ contraction crack 初期[原始]收缩裂缝
~ convergence 初始下沉[闭合]
~ cooling 预冷却
~ cost 原价，基本建设费，创办成本[费]
~ crack 初（期）裂缝
~ curing 初期养护
~ cut 切槽，掏槽
~ data 原始数据
~ database 初始数据库
~ day 起算日
~ deflection 初始弯沉，初期弯沉
~ density 初始密度
~ design 原设计，初步设计

~ detention 初始滞流
~ displacement 初位移，初始排水量
~ dredging 基建挖泥
~ equation 初始方程
~ error 原始误差，初始误差
~ estimates 初步概算，初始估计(值)
~ expenditure 创办费，开办费
~ expenses 创办费，开办费
~ face 爆破后的边坡面
~ filling 初期蓄水
~ flaw （混凝土骨料周围的）原始裂纹
~ flow 初始流量，起涨流量（指过程线上洪水开始上涨时刻的流量）
~ fund 开办基金
~ gradient 初始梯度（最小的水力梯度，小于此值时土壤中无水流动）
~ installaion 初期装机
~ interface velocity 初始沉降速度
~ investment 开始投资
~ landmass 原始陆体
~ line 始边，起[始]线，极轴
~ load 初始荷载
~ load-bearing capacity 荷载承受能力
~ loss 初损，初始（径流）损失
~ maintenance 初期养护
~ meridian (= Greenwich meridian) 首子午线，格林威治子午线
~ moist curing 初期湿养护
~ moisture content 初始含水量，原始含水量
~ operation 初运转，初始运行
~ order 期初定货
~ ornamental 饰花字头，外文词头大写字母的美化
~ overburden pressure 起始超载压力
~ outlay 初期投资，工程施工投资
~ payment 首期付款
~ period 初期
~ point 原点，起点
~ porosity 初始孔隙度

~ percipitation 初期降水（填注量满足前的降水）
~ pressure 起始压力，初期压力
~ prestress 初施预应力，初始预应力
~ price 牌价，开价
~ reading 初读数，起始读数
~ reserve 初期储备
~ rolling 初压，初次碾压
~ sett(ing)（水泥混凝土或水泥浆的）初凝，初始沉陷
~ setting energy 初凝能量
~ setting heat 初凝热
~ setting period 初凝期
~ settlement 初始沉陷，初期沉陷
~ shear stiffness 初始剪切劲度
~ shrinkage 初始收缩量
~ side 起算边
~ speed 初速度
~ stage 初始阶段
~ state 初始阶段，初始状态
~ strength 初期强度
~ stress 预[初]应力
~ stress tensioning 预应力张拉

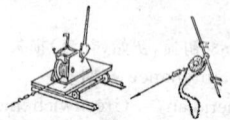

initial stress tensioning

~ surplus 初期盈余
~ task index 起始任务指标
~ temperature 初始温度
~ tension 初始张力
~ (thermal) strain 初始（热）应变
~ treaty 初期协议
~ value 初始值
~ velocity 初速（度），起始速度
~ viscosity 初始黏度
~ void ratio 初始孔隙比
~ water deficiency 初始亏水量
~ yield load 初始屈服荷载
~ zone 零时时区

initialization n. 预置，恢复，清除；初始化；初始准备（工作）

initialize [iˈniʃəlaiz] v. 预置，恢复，清除；设定初值

initialized a. 起始的，预备的，准备工作的，预备步骤的

initiate [iˈniʃieit] v. 启动，创始，发起，开始，着手，发动，激发[磁]，促使
~ key 启动键
~ing additive 引发添加剂
~ing pulse 角发脉冲
~ing signal 启动信号
~ tigger 启动触发器[脉冲]

initiation [iniʃiˈeiʃən] n. 创始，发起；正式加入（社团等），入门；起爆，起燃，激磁
~ system 起爆系统

initiative [iˈniʃiətiv] n. 发端初步，创始，着手，首创精神，主动力，积极性，创造性；a. 起始的，创始的，初步的，自发的；initiatively, ad.

initiator [iˈniʃieitə] n. 起爆器[剂]，励磁机，首创者，指引者，传授者

inject [inˈdʒekt] vt. 喷射，射水，灌入，注入，注射；injectable, a.；injected, a.
~ed body 贯入体
~ed dike 侵入岩脉
~ed hole 注浆孔，灌浆[凝固钻]孔
~ed rock 贯入岩（层），受压的岩石

injection [inˈdʒekʃən] n. 喷射，注射，注入，滚入，贯入，进入轨道；注满，灌浆，发射；加压
~ agent 灌浆剂，注入剂
~ column 灌浆管柱
~ cup 喷头，喷嘴头
~ drill-bit 湿式凿岩钻头
~ gallery 灌浆廊道
~ grouting 灌浆
~ gun 喷枪
~ head 注入头（使钻液注入钻杆

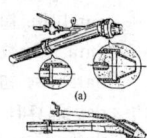

injection gun

的水龙头接箍)
~ installation 喷射装置
~ lance 喷枪,(喷雾机的)喷杆
~ method 喷射法,贯注法(灌注土稳定剂),注入法
~ method in tunnel construction 注射法隧道施工
~ molding 喷射铸造(模型)法
~ mortar 喷射灰浆
~ nozzle 喷射管嘴,喷嘴
~ pipe 喷射管
~ plant 喷射设备
~ pressure 喷射压力
~ process 注液法用于壤加固
~ pump 喷射泵,喷油泵
~ rod 喷杆
~ well 注入井,(油田的)灌水井
cement ~ 水泥灌注
leaf by leaf ~ 层层贯入
lit par lit ~ 间层注入,层叠贯入

injectivity *n.* 井孔或岩层容纳注入液的能力,注入性
~ -index test 注水系数测试

injector [in'dʒektə] *n.* 喷射器,注射器,灌浆机;喷嘴,喷头
~ blowpipe 低压喷焊器
~ condenser 喷射冷凝器
~ stream 射流
air ~ 空气喷射器
spray-type ~ 喷雾器,喷注器

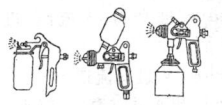

spray-type injector

injectron [in'dʒektrɔn] *n.* 高压转换管

injunction [in'dʒʌŋkʃən] *n.* 命令,指令,禁令,责成

injurant [in'dʒərənt] *n.* 伤害物,伤害剂

injure ['indʒə] *vt.* -jured,-juring;毁坏,伤害,损害

~ -accident rate 伤害事故率(交通)

injurious [in'dʒuəriəs] *a.* 有害的,不公平的,诽谤的; injuriously, *ad.*; injuriousness, *n.*
~ amount 有害含量

injury ['indʒəri] *n. pl.* -ries;伤害,损害,诽谤;不法行为
industrial ~ 工业性伤害
irradiation ~ 射线杀伤,射线病

ink [iŋk] *n.* 墨水,墨汁,印色,油墨; *vt.* 涂墨水于;签(合约); *vi.* 加墨水
~ black 墨黑色
~ bottle 墨水瓶
~ drawing 墨线图
~ film 彩色胶卷
~ fountain 墨斗
~ knife 油墨刮刀
~ed 修补着色,墨水涂染
~ed drawing 上墨图
~ing 上墨水线,涂油墨,着墨,印色
~ print 油墨印件
copying ~ 复写墨水
drafting ~ 绘图墨水
etching ~ 耐酸漆
fast-setting ~ 快干漆
glossy ~ 亮光漆
heat-set ~ 热干漆,热度定印墨
indelible ~ 不变墨水,不灭墨水
invisible ~ 隐显墨水(用于秘密通信)
moulding ~ 造型用涂料
permanent ~ 不变墨水

ink-lab *n.* 砚

inkpad *n.* 印泥[台]

inky ['iŋki] *n.*;*a.*, inkier, inkiest; *n.* 小型聚光灯; *a.* 漆墨的;给墨水弄脏的,墨水似的; inkily, *ad.*

inlaid ['in'leid] *v.* inlay 的过去式和过去分词 *a.* 嵌花的,镶嵌的
~ brick 平埋砖,路面标示砖
~ flooring 镶嵌地板
~ flooring brick 镶嵌铺面砖
~ lino (leum) 嵌花油地毯,嵌花漆布

~ -strip 嵌花板条
~ vinyl goods 乙烯嵌花（装饰）品
~ wood 镶嵌木
~ work 镶嵌细工
inland [a.ˈinlənd, n.ˈinlənd, ad. inˈlænd] *n.*; *a.* 内陆(的), 内地(的); *ad.* 在内地, 去内地
~ basin 内陆盆地
~ bill 国内汇票
~ canal 内陆运河
~ customs dues 国内关税
~ delta 内陆三角洲
~ desert 内陆沙漠, 内地沙漠
~ dike 内堤
~ dunes 内陆沙丘
~ forwarding expenses 内地运费
~ freight haulage （港口货物的）陆上疏运, 内地货运
~ harbour 内陆港, 内地港, 内河港
~ inundation 内涝
~ lake 内陆湖
~ navigation 内河航行[运]
~ port 内河港口
~ river 内河
~ sand 内陆砂
~ sea dike 内陆海堤
~ terminal depot 内河码头, 内陆枢纽站
~ waterway 国内水道, 可通航的内河水路系统
~ waterway network 内河水道网
inlay [v. ˈinˈlei, n. ˈinlei] *n. pl.* -lays; *v.*, -laid, -laying; *n.* 嵌体, 内置, 插入; 嵌入法, 镶嵌工艺; *vt.* 嵌入, 镶嵌
~ pattern 镶嵌花纹
~ window 窗嵌体

inlay

inlayer [ˈinleiə] *n.* （镶嵌）工匠; 内层, 人鞘
inleakage [ˈinliːkidʒ] *n.* 泄漏量, 渗入, 吸入; 贯穿内部; 漏电
inlet [n. ˈinlet, v. inˈlet] *n.*; *v.*, -let, -letting; *n.* 海口, 港, 湾; 进水口, 雨水口; 镶嵌物; *vt.* 放进, 插入
~ chamber 进气室
~ channel 进水渠
~ connection 进口连接(件), 进气管接头
~ culvert 进水涵洞, 进水暗渠
~ duct 进水道, 进风道
~ elbow 喷管
~ fan 进气风扇, 通风进气口
~ grate 进水口帘格, 进水口笼
~ grating 进水口帘格, 进水口笼
~ grid 进水口栅格
~ hole 进水孔
~ manhole 人孔入口
~ manifold 进气歧管, 进气多头管
~ of sewer 下水道入口
~ of ventilating system 通风系统
~ outlet 进出口, 吸排气口
~ pipe 进水管
~ pipe gallery 进水管廊道
~ relief valve 进口安全释放阀
~ shaft 进水口竖井
~ sill 进水口底槛
~ sluice 进水闸
~ strainer 进水口滤网
~ stroke 吸入冲程
~ structure 进口建筑物
~ submerged culvert 半压力式涵洞
~ transition 进口渐变段
~ valve 进水阀, 入流阀, 进气门, 进给阀
~ velocity 进口流速
~ well 集水井, （泵站）吸水井
air ~ 进气口
air suction ~ 空气吸入口
cable ~ 电缆入口
duct ~ 进水道, 进风道, 导管入口, 气道入口
evaporated nitrogen ~ 氮蒸汽入口（冷却装置）
gutter ~ 檐[街]沟进水口

pressure ~ 增压管,接管嘴
street ~ 街道雨水口
trumped-shaped ~ 喇叭状入口,漏斗状入口
weir ~ 溢流引水

inlier ['inlaiə] *n.* 内围体,内围层,内围区,内露层;内窗层;岩石的内层

inmost ['inməust] *a.* 最深(部)的,最内部的

inn [in] *n.* (小)旅馆,客栈;酒馆,酒店

innards ['inədz] *n.* 内部(结构)

innate ['i'neit] *a.* 固有的,内在的,天赋的,先天的;innately, *ad.*; innateness, *n.*

inner ['inə] *a.* 内部的,里面的,靶心与外圈间的部分; innerly, *ad.*; innerness, *n.*

~ air cooler 内部空气冷却器
~ arch 内拱
~ bank (河流)内岸
~ bar 内滩
~ bearing 内轴承
~ belt 内环(路)
~ bending moment 内弯矩
~ berm(e) 内戗道
~ brake 内制动器,内闸
~ break water 内防波堤
~ building board 内部建筑板
~ circumference highway 内环路
~ coat 内涂层,内保护层
~ column 内柱,内墩
~ contour of tunnel lining 隧道衬砌内轮廓
~ corridor 内(走)廊
~ court 内场,内院,内宫
~ curtain (wall) 内隔板,内隔墙
~ decor 内部装饰
~ decorating 内装修,内部布置
~ diameter 内径
~ dimension 内部尺寸
~ door 内(部)门
~ edge 内缘
~ emulsion paint 内(装饰)乳胶漆

~ enceinte (wall) 内围墙
~ facing 内衬,内饰,内工作面
~ fibre board finish 内部纤维板装饰
~ finish(ing) paint 内装饰涂料
~ fixtures 室内装置
~ flue 内烟道
~ formwork 内模板
~ friction 内摩擦
~ gallery 内走廊
~ gallery apartment building 内廊式公寓建筑
~ garden 内花园
~ gates 内大门

inner garden

~ glass door 内部玻璃门
~ glazing 内部门、窗镶嵌用玻璃
~ gloss (clear) varnish 内部上光清漆
~ handrail 内扶手栏杆
~ harbo(u)r 内港
~ hardboard finish 内侧[部]硬质板饰面
~ heat 内部供暖
~ humidity 内部湿度
~ illumination 内部照明(度)
~ insulation 内部隔声,隔层,内部保温
~ joinery 室内细木工
~ lane 内侧车道
~ layout 室内布置
~ lighting 室内采光
~ lining 内部衬砌
~ loop (立体交叉中的)内转(插入)匝道;内环路,内侧道
~ masonry dividing wall 内分隔墙
~ partitioning 内部分隔(墙)

~ piping 内部管道系统
~ primer （内部）底漆
~ proscenium 内部舞台
~ rail 内围栏
~ redecoration 内部重新装饰[油漆]
~ reveal 内部的窗[门]侧（指墙与门或窗之间），内部外抱，内侧壁
~ seal(ing) 内部封闭
~ side 内(侧)壁
~ (side) aisle 内部侧廊[耳房]
~ skin 内部罩面层
~ slatted blind 内部木条百叶窗
~ slope 内坡
~ stairrail 内楼梯栏杆
~ string 内楼梯斜梁
~ surface 内壁
~ stud 内门窗梃,（内部）房间净空高度,内部壁骨
~ temperature control 内部温度控制
~ tile 内瓦[砖]
~ tracery 内窗（花）格
~ tube 管道内层,内胎
~ viscosity 结构黏度
~ wall(ing) 内墙（砌体）
~ width 净宽,内宽
~ window 内窗
~ window cill [sill] 内窗槛
~ window frame 内窗框架
~ work 内部工作[工程]

inner(-)most ['inəməust] *a.* 最内部的,最深的；*n.* 最深处,最内部
~ part 最里面部分,最深部
~ suburbs 近郊

inning ['iniŋ] *n.* 围垦,围垦地,冲积土,涨出地；(*pl.*)在海中填筑的陆地,海埔新生地

innocuity [i'nɔkjuiti] *n.* 无害,无毒

innocuous [i'nɔkjuəs] *a.* = harmless, 无害的,无毒的; innocuously, *ad.*; innocuousness, *n.*
~ effluent 无害废水[排出物]

innovation [inəu'veiʃən] *n.* 改革,革新,改善,改进；新设施,新制度；新发明,技术革新,合理化建议
~ cost 革新成本
~ technological 工艺革新,技术革新
~ theory 革新理论,创新理论

innovator ['inəveitə] *n.* 改革者,革新者

innoxious [i'nɔkʃəs] *a.* 无害的,无毒的; innoxiously, *ad.*; innoxiousness, *n.*

innumerable [i'nju:mərəbl] *a.* 无数的,数不清的; innumerableness, *n.*; innumerably, *ad.*

innumerous [i'nju:mərəs] *a.* = innumerable 无数的,数不清的

innyard *n.* 客栈的庭院

inodo(u)rous [in'əudərəs] *a.* 无臭气的,无气味的; inodo(u)rousness, *n.*

inorfil [inɔ:fil] *n.* 无机纤维

inorganic [inɔ:'gænik] *a.* 无机的,无机物的,无组织体系的; inorganically, *ad.*
~ acid 无机酸
~ binder 无机结合料
~ building material 无机建筑材料
~ cementing material 无机胶凝材料
~ clay 无机黏土
~ chemistry 无机化学
~ colloid 无机胶体
~ compound 无机化合物
~ (concrete) aggregate 无机（混凝土）骨料
~ constructional material 无机建筑材料
~ fibre board 无机纤维板
~ heat insulating material 无机隔热材料
~ insulation (grade) material 无机绝缘（等级）材料
~ matter 无机物
~ structural material 无机建筑材料
~ synthetic dye (stuff) 无机合成颜

[染]料
inorganization [inɔːgənaiˈzeiʃən] *n*. 无组织
inornate [inɔːˈneit] *a*. 无华丽装饰的
in(-)phase [ˈinfeiz] *a*. 同(位)相的
　~ component 同相分量,同相部分
in-place construction 现场施工
in-place grave 墓地
in-place measurement 现场测量[定]
inpolar [inˈpəulə] *n*. 内极点
inpolygon *n*. 内接多边形
inpolyhedron *n*. 内接多面体
input [ˈinput] *n*. 输入,输入额,输入端,输入量;生产的要素;*v*. 投放,资料输入
　~ block 输入部件
　~ circuit 输入电路
　~ combination 成本投入组合
　~ data 输入数据
　~ device 输入设备
　~ equipment 输入装置,输入设备
　~ in cash 现金投入,现金投放
　~ information 输入信息
　~ in kind 实物投入,实物投放
　~ output accounting 投入产出会计
　~ output price 投入输出成本
　~ power 输入功率,功率输入
　~ed income 估算收入
　analog ~ 模拟输入
　automatic ~ 自动输入数据,自动装入数据
　card ~ 卡片输入
　erase ~ 消去信号输入
　heat ~ 供热
　on-line ~ 联机输入
　work ~ 消耗功,机器的总功(包括摩擦功),指示功
input/output 输入/输出(装置),输入/输出数据,输入/输出方式
　buffered ~ 输入/输出缓冲装置
　concurrent ~ 同时输入/输出
　programmed ~ 程序控制输入输出
　real-time control ~ 实时控制输入

输出

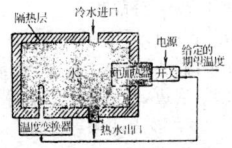

input/output

　remote message ~ 远距离信息输入输出,远端信息输入输出
　simultaneous ~ 同时输入输出
　storage ~ 存储器输入输出
inquiry [inˈkwaiəri] *n*. 询问,打听,调查,查询;探[研]究
　~ office 问讯处
　~ trunk directory ~ 长途电话局查询台
inrush [ˈinrʌʃ] *n*. 突然崩坍,流入或涌入的动作或情况
inscription [inˈskripʃən] *n*. 碑文,题字; inscriptional, *a*.
insculptate *a*. 雕空的,挖空了的,具有凹点的
insecticidal [inˈsektisaidəl] *a*. 杀虫的
　~ paint 杀虫涂料
insecticide *n*. 杀虫剂
insection *n*. 切开,切断,切口;齿纹,锉纹
insecure [insiˈkjuə] *a*. 不可靠的,不安全的,易塌的
insert [ˈinsəːt] *n*. 插入物,镶嵌物,镶块,垫片;嵌入路面的标志;插页;[inˈsəːt] *vt*. 嵌入,插入;刊登;insertable, *a*.; inserter, *n*.
　~ chip 镶装刀片
　~ map 插图,附图
　~ moulding 镶嵌造型,镶嵌饰
　~ nozzle 喷管衬套
　~ pattern 组合模(板)
　~ pin 插销
　~ed piece 嵌入加强块
　~ed tenon 嵌入榫头
　~ed tool 硬质(合金)刀具

lock drill ~ 凿岩机合金衬片
separator ~ 分隔板
inset [*n.* ˈinset, *v.* inˈset] *n.*; *v.*, -set, -setting;插入,嵌入;插入物,插图,插书,插页;镶边,镶嵌物;流入,水道
~ balcony 嵌入式阳台
~ grate 嵌入式帘格
~ -type 嵌入式(的)
horizon ~ 井底车场
incline ~ 斜井井底车场
shaft ~ 井底车场
inshore [ˈinˈʃɔː] *a.* 沿海的,近海的,向陆的; *ad.* 靠近海岸,向着海岸
~ current 近海流,近岸流
~ longitudinal girder (减荷台式码头的)内岸纵主梁
~ wind 向岸风,向陆海风
~ zone 近海带
inshot [inˈʃɔt] *n.* 跃进装置,跃升
inside [ˈinˈsaid] *n.* 内部,里面,内侧,内容; *a.* 内部的,内侧的,里面的,户内的; *ad.* 在里面[内部]
~ air temperature 内部(空)气温(度)
~ arch 内拱
~ back 里封底
~ bath(room) 内浴室
~ brake 内制动,内闸
~ building board 内部建筑板
~ calliper [caliper] 内径规,内卡钳
~ casing (门窗)内框,内压条
~ cellar wall 内部隔墙
~ clinch 内绳扣,内活络圈套
~ coat 里层
~ column 内柱
~ core 内芯
~ corner moulding 内角线
~ court 内院[场]
~ cover 里封底
~ curve 内曲线
~ decorating 内部装修
~ diameter 内径
~ dimension 内尺寸

~ door 内(部)门,里门
~ dozer (blade) 内侧推土机(刮铲)
~ -emptied brick 中空砖
~ emulsion paint 室内乳胶漆
~ facing 内衬,内饰,内工作面
~ fibreboard finish 内部纤维板装饰[修]
~ finish 内粉刷,内部装修
~ fixtures 室内装置
~ -frosted lamp 乳白灯泡
~ gallery 内走廊
~ gallery apartment building 内走廊式公寓建筑
~ glass door 室内玻璃门
~ glazing 内部门[窗]镶嵌用玻璃
~ gloss (clear) varnish 内部上光清漆
~ handle 内把手
~ handrail 内扶手栏杆
~ hardboard finish 内侧[部]硬质板饰面
~ heat 内部供暖
~ height 内高
~ humidity 室内湿度
~ illumination 室内照明(度)
~ insulation 内部隔声,内隔层,内部保温
~ joinery 室内细木工
~ kitchen 内厨房
~ knowledge 内幕
~ layout 室内布置
~ lighting 室内采光
~ lining 内衬,内部衬砌
~ masonry dividing wall 内部分隔墙
~ micrometer 内径千分尺
~ moisture 室内湿度
~ noise 室内噪声
~ -out filter 外流式过滤器
~ paint 底漆
~ paint coat 底漆层
~ partitioning 内部分隔
~ plant 室内设施(线缆)

~ radius 内半径
~ redecoration 内部重新装饰[油漆]
~ reveal 内部的窗[门]侧(指墙与门或窗之间),内部外抱
~ seal(ing) 内(部)封闭
~ skin 内部罩面层
~ slatted blind 内部木条百叶窗
~ slope 内坡
~ stairrail 内楼梯栏杆
~ stud 内门窗梃,房间净空高度,内部壁骨
~ (sur)facing 内部表面加工[修整]
~ tile 内砖[瓦]
~ tracery 内窗(花)格
~ track 里圈
~ turn 内转弯
~ turning radius 内转弯半径
~ wall 内墙
~ wall frame 内墙框架
~ wall sill 内墙基石[木]
~ wall surface 内墙(表)面
~ width 内宽,净宽
~ window 内窗
~ window cill 内窗槛
~ wiring 室内线路[布线]
~ work 室内工作

insignia [in'signiə] *n.* 国徽,证章,勋章,标志

insition [in'sifən] *n.* 添加物,移入物,接种物

in(-)situ [in'saitju:]【拉丁】现场,就地,工地
~ aerated concrete 现场浇注加气混凝土
~ architectural concrete 现场浇筑混凝土

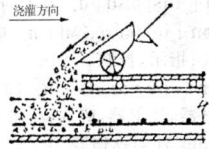

in(-)situ aerated concrete

~ brickwork 就地砌砖
~ cable duct 工地电缆槽
~ concrete 现场混凝土
~ cast 就地浇注
~ cast concrete floor 就地浇注混凝土楼板
~ (cast) staircase 现场浇注楼梯间
~ (cast) pile 就地浇注桩
~ (cast) rib 就地浇注肋
~ construction 现场施工
~ curb beam 就地(浇筑)路缘梁
~ flooring 现浇地板
~ filling 就地装填,就地灌注
~ landing 就地着陆;现场卸货处,工地码头
~ loading test 现场荷载试验
~ mortar 就地(制)灰浆
~ pile 现浇[混凝土]桩
~ pipelining 就地敷设管道
~ sand 现场(制)沙
~ soil test 现场土壤试验
~ topping 就地封顶

insoak *n.* 未饱和土壤对地表自由水的吸收

insolation [insəu'leifən] *n.* 暴晒,日照,日射;日射率;日光浴
~ area 日照面积
~ duration 日照时间
~ gauge 日照仪

insolubilize [in'sɔljubəlaiz] *vt.*, -ized, -izing; *v.* 不溶解;insolubilization, *n.*

insolubilizer *n.* 不溶粘料

insoluble [in'sɔljubl] *a.* 不溶解的,难溶解的,不能解决的;insolubility, *n.*; insolubly, *ad.*
~ compound 不溶化的物质
~ matter 不溶物质

insolvancy [in'sɔlvənsi] *n.* 破产,无偿还能力
~ clause 破产条款

insolvent [in'sɔlvənt] *n.*; *a.* 无偿还能力者(的)
~ debtor 破产者

~ laws 破产法

insonify [in'sɔnifai] *v.* 声照射,声穿透
~ (fied) zone 声音传播区,有声区

inspect [in'spekt] *vt.* 检查,考查,审查;探伤;检阅,视察; inspectability, *n.*; inspectable, *a.*; inspectingly, *ad.*
~ing hole 检查孔,观察孔
~ing pit 检查[修]坑,探坑

inspection [in'spekʃən] *n.* 检查,检验;验证,检修,观察,调查;探伤; inspectional, *a.*
~ and checkout 检查与测试
~ and claim 检查和索赔
~ bureau 检查所(局)
~ car 铁路轨道检查车
~ certificate 检验证明书,检查证明;技术检查报告
~ chamber 检查室[井]
~ chart 检查表
~ clause 检验条款
~ cover 检查孔盖
~ door 检查孔[门]
~ eye 检查孔[口]
~ declined! 谢绝参观!
~ gallery 检查廊道
~ gauge 检查规
~ hole 检查口,检验孔,检查孔
~ hole cap 窥视孔盖
~ lamp 检查灯,修车用灯
~ manual 检查手册
~ memorandum 检查记录
~ pane 窥视玻璃孔
~ pit 检查坑,检验井;探坑,探井
~ plate 检查孔盖板
~ plug 观察孔塞
~ rack (汽车)检修台
~ report 检验报告,视察报告
~ shaft 检查[竖]井
~ sheet 检验单
~ trolley 检查用吊车
~ tube 检查管道
~ tunnel 检查隧道
~ well 检查井
~ window 检验窗
annual ~ 年度检查,年度检修
building ~ 施工监督,房屋检查
casual ~ 不定期检查,临时检查
China Commodity Inspection Bureau (CCIB) 中国商品检验局
construction ~ 施工检查
curtailed ~ 抽样检查,抽查
daily ~ 日检查,小检查,日常查看
electronic ~ 电子检测
infrared ~ 红外检验,红外探伤
magnetic ~ 磁性探伤
non-destructive ~ 非破坏性检查
outer ~ 外观检查
periodic ~ 定期检查
photoelectric ~ 光电检查
quality ~ 质量检查
radiographic ~ 射线照相、探伤法,射线故障检验法
random ~ 抽查,随机检查
regular ~ 定期检查
routine ~ 常规检查
safety ~ 安全检查
sampling ~ 抽样检查,采样检查
shaft ~ 检查井筒,探坑,探井,检查坑
shut-down ~ 停工检查
supersonic ~ 超声(波)检查
visual ~ 外部检查,直观检查

inspector [in'spektə] *n.* 检验[检查,监工]员
~ -general 总检查长,监察长
~ -in chief 总检查员,监察长
~ of material 材料检验员

inspectoscope [in'spektəskəup] *n.* 金属裂缝探伤器,探伤器,探伤镜

inspersed ['inspə:sd] *a.* 渗入的

inspiration [inspə'reiʃən] *n.* 吸气,吸入,进气;指示,授意

inspirator ['inspəreitə] *n.* 喷气注水器,喷射器,注射器,吸入器[者]
~ burner 注射燃烧器

inspire [in'spaiə] *v.*, -spired,-spi-

ring；*v.* 灌输,注入,吸入；产生,引起；鼓舞,启发；inspirer, *n.*
inspirometer *n.* 吸气测量计
inspissate [in'spiseit] *v.*, -sated,-sating；*a.* 浓厚的,浓缩了的,强烈的；*v.* 蒸浓,使浓缩；inspissator, *n.*
～doil 浓缩石油
inspissation [inspi'seiʃən] *n.* 蒸浓(法),浓缩作用,增稠
inspissator *n.* 蒸浓器
instability [instə'biliti] *n.* 不稳定(性),不稳定(度),不安定(性)
～ at resonance 共振不稳定性
～ coefficient 不稳定(度)系数
～ in pitch 纵向不稳定性
amplitude ～ 振幅不稳定度
dynamic ～ 动(力)不稳定,动(力)不安定性
elastic ～ 弹性不稳定(性)
longitudinal ～ 纵向不稳定性
plastic ～ 塑性不稳定性
static(al) ～ 静(力)不稳定性,静(力)不安定性
instable [in'steibəl] *a.* 不稳定的,易变的；不安定的
～ buoyancy equilibrium 不稳定(静)浮力平衡(状态)
～ frame 不稳定系统
instal(l) [in'stɔ:l] *vt.* 装置,装设,安装,装配(机器)；任命；installer, *n.*
～ed capacity (发电厂)设备容量
～ed (name plate) capacity 装机(铭牌)容量
～ed wheel capacity 装机轮容量
installation [instə'leiʃən] *n.* 装置,设备；安装,装配,调整
～ and checkout 安装与测试
～ and maintenance 安装维修的,安装与维修
～ and overhaul specification 安装与检修规范
～ charge 安装费
～ cost 安装成本[费用]
～ diagram 安装图
～ dimension 安装尺寸
～ drawing 安装图
～ exercise 安装作业
～ equipment 安装设备
～ fee 安装费
～ floater 安装临时工
～ instructions 安装工程
～ of track 轨道敷设
～ parts list 安装零件清单
～ squadron 安装中队
～ work 安装工程
accelerating ～ 加速装置
bottom ～ 井底车场设备
computer ～ 计算机安装[调整]
concealed ～ 隐式装置(法),隐531敷设
cooling-fan ～ 冷却通风装置
dial system ～ 自动电话装置
electrical ～ 电气工程
emergency ～ 备用装置,应急设备
engine ～ 动力装置
exposed ～ 露明式装置(法),明管布设法
ground ～ 地面设备,地面装置
heating installation
heating ～ 供暖设备,暖气装置
hydroelectric ～ 水电站,水力发电装置
laser satellite tracking ～ 激光卫星跟踪装置
multilevel hoisting ～ 多级提升设备
multipe ～ 复式装置,组合吊架
pipe ～ 管道敷设,管道安装,管道装置
pneumatic water supply ～ 气压给水设备
refrigerating ～ 冷冻设备
sanitary ～ 卫生设备[装置]
site ～ 工地设施
subscriber's intercommunication ～ 用户内部通话装置

turbine ～ 涡轮机
ventilation ～ 通风设备
wind tunnel ～ 风洞
winding ～ 提升设备

installer *n.* 支座,安装者

instal(l)ment [in'stɔːlmənt] *n.* 安装,装配;分批,连载,分期付款
～ and interest charge 分期还本付息
～ and margin purchase 分期付款和保证金购买
～ bond 分期偿还公司债
～ buying 分期付款购货
～ credit 分期付款信贷
～ delivery 分期交货
～ long 分期付款贷款
～ payment 分期付款
～ plan 分期付款销售方式

instance ['instəns] *n.*; *v.*, -stanced, -stancing; *n.* 实例,例子,范例,样本,阶段,步骤;建议,请求,诉讼程序; *v.* 举例,引证,举……为例

instancy ['instənsi] *n.* 紧急,急迫,迫切,瞬间,即时,立即,顷刻;坚持,强迫

instant ['instənt] *n.* 瞬间,瞬时,顷刻,即时,此刻; *a.* 即刻的,急迫的,当月的,本月的
～ acting 有效瞬间
～ bathroom 快速浴室
～ bridge 瞬时建成桥(一种快速建成的新型预应力装配式混凝土桥)
～ lock 碰锁,暗锁
～ of failure 故障瞬间,中断瞬间
～ reply 可即时放送的录像,慢镜头重演
significant ～ 有效瞬间

instantaneity [ˌinstæntə'niːiti] *n.* 瞬时,即时

instantaneous [ˌinstən'teinjəs] *a.* 即时的,瞬时的,同时的,瞬时作用的; instantaneously, *ad.*; instantaneousness, *n.*
～ acceleration 瞬时加速度
～ cap 瞬时或同步起爆雷管,瞬发雷管
～ centre 瞬时中心,瞬心
～ complete rejection 瞬时全部弃荷
～ corrosion rate 瞬时腐蚀(速)率
～ deflection 瞬时变位
～ deformation 瞬时变形
～ drawdown 瞬时(水位)下降,瞬时消耗
～ electric detonator 瞬时电子引信,瞬时电子引爆剂

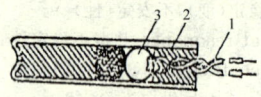

1.脚线; 2.硫磺绝缘涂料;
3.球形发火剂
instantaneous electric detonator

～ erection 瞬时安装(法)
～ exposure 自动快速曝光
～ firing 瞬时引爆
～ flow rate 瞬时流量
～ fusing 瞬熔,瞬发引线(装置)
～ gas water heater 快速煤气热水器
～ linear velocity 瞬时线速度
～ load(ing) 瞬时荷载,瞬时载重
～ maximum wind 瞬时最大风(速)
～ modulus of elasticity 瞬时弹性模量
～ power 瞬时功率
～ pressure 瞬时压力
～ pulsation 瞬时脉动
～ sound pressure 瞬(时)声压
～ speed 瞬时车速
～ state 瞬时状态
～ strain 瞬时应变
～ strength 瞬时强度
～ surge 瞬时涌浪
～ unit hydrograph 瞬时单位(过程)线
～ value 瞬时读数,瞬时值
～ velocity 瞬时速度
～ warm water 快速暖水器

~ weight 瞬时重量
~ wind speed 瞬时(最大)风速
instaseal *n.* 防漏粗黏粉
instate [in'steit] *vt.* -stated,-stating; 安置;任命,授予资格[职位];instatement, *n.*
instauration [ˌinstɔː'reiʃən] *n.* 修复,恢复,重建
instead [in'sted] *ad.* 代替,当作;更换
~ of 代替
in-step ['in-step] *a.* 同步的,同相[位]的
instinct ['instiŋkt] *n.* 本能,本性;直觉,天才; *a.* 生动的,充满……的
institute ['institjuːt] *n.*; *v.*, -tuted, -tuting; *n.* 学院,学会,协会,学术团体或福利组织,研究所[院];会址,院址,校址;讲座;民法;(*pl.*)(基本)原理[则]; *v.* 设立,制定;开始,着手,实行
research ~ 研究所[院]
institution [ˌinsti'tjuːʃən] *n.* 建立,制定;规定,制度,惯例;协会,学会,机构,机关,公共设施;学院,研究所
Institution of Civil Engineers (I.C.E) 土木工程师学会
institutional [ˌinsti'tjuːʃənl] *a.* 设立的,制定的;规定的,制度的;研究所的,学会的,学校的,公共机构的;原理的;institutionalism, *n.* institutionally, *ad.*
~ accounting 机关会计,事业会计
~ approach 惯例方法
~ property 社团地产,社团财产
~ waste water (公共)机关废水,单位废水
in(-) streaming ['instriːmiŋ] *n.* 内流,流入,涌进; *a.* 内流的,涌进的
instron *n.* 拉伸强度试验机
instruct [in'strʌkt] *v.* 教育,教导,指示;说明,通知;命令

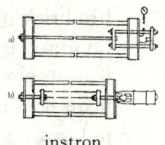

instron

instruction [in'strʌkʃən] *n.* 规程,说明,指南,须知,指示,程序;讲授,教导;指令书,说明书;指令,码
~ address 指令地址
~ book 说明书,指南
~ clerk 工作指导,管理人员
~ for erection (work) 装配[安装](工作)技术规程
~ in draughtmanship 制图质量技术规程
~ manual 说明书,使用手册
~ of technical operation 技术操作规程
~ repertoire 指令系统
~ sheet 说明书,说明片,样本
accident prevention ~ 技术安全规程
engineering ~s 工程说明书
fixing ~ 安装说明
loading ~ 载重定额[规定]
operating ~ 工作细则,使用说明书
output ~ 输出指令
overhaul ~ 大修指导
service ~ 操作规程,使用规程,维护规程
working ~ 操作规程,工作细则
instructive [in'strʌktiv] *a.* 指导性的,有(教)益的,教训的;instructively, *ad.*; instrctiveness, *n.*
~ card 工作指导卡
instructor [in'strʌktə] *n.* 教师,讲师,指导员;-instructorship, *n.*
instrument ['instrumənt] *n.* 仪器,仪表,器械,工具;方法;合同,法定文件,证券,公证状; *vt.* 给……装备仪表,提交法律文件给……
~ board 仪表板[盘]
~ case 仪器柜,仪表箱,仪表盒
~ constant 仪器常数
~ contour 实测等高线
~ drawing 仪器图
~ light 仪表(操纵)板照明指示灯
~ multiplier 仪表扩(量)程器

~ of credit 商业证券,信用证书
~ of production 生产工具
~ of retification 批准书
~ panel 仪表板
~ room 仪表室
~ shelter 百叶箱(气象观测用)

instrument shelter

~ shunt 仪表分流器
~ station 测站
all-purpose ~ 万能仪表
altazimuth ~ 地平经纬仪
angle ~ 测角仪
automatic recording ~ 自动记录装置
azimuth ~ 方位仪,测角仪
contour ~ 等高线笔,回转笔
current measuring ~s 测流计,测流仪器
curve-drawing ~ 曲线板
echo-sounding ~ 回声测深仪
first-order ~ 精密仪器
flow ~ 流量表,液[气]体消耗量测仪
height-finding ~ 高度计
knock indication ~ 爆振指示器
level(ing) ~ 水准仪
micrometer ~ 微米经纬仪,光学经纬仪
mud-loss ~ 泥浆漏失测定器
multi-purpose ~ 通用工具,万用[能]工具
sampling ~ 取样器
self-levelling ~ 自动定平水准仪,自动定平仪器
stereometer-type ~ 立体测量仪
stereoscopic plotting ~ 立体测图仪
strain measuring ~ 应变仪
totalizing ~ 求积仪,积分仪
transit ~ 经纬仪

universal ~ 全能经纬仪,通用测量仪表
wind ~ 风速计,风力表,气流计
insula [insjulə] *n.* 建筑群
insulate ['insjuleit] *vt.*, -lated,-lating;保温,绝热,绝缘,隔离,隔声
~d column 隔离柱,独立柱
~d conductor 绝缘线
~d neutral (对地)绝缘中线
~d paint 绝缘漆
~d wire 绝缘线
insulater *n.* 保温工
insulating *a.* 绝缘的,绝热的,隔离的
~ ability 绝缘能力
~ article 绝缘制品
~ asbestos board 绝缘石棉板
~ base 绝缘地基
~ block 绝缘(砌)块
~ board (=fiber board) 隔热板(纤维板),绝缘板
~ brick 保温砖,隔热砖
~ building material 绝热建筑材料
~ celling 隔热天花板[顶棚]
~ coating 绝热涂料[油漆,涂层]
~ concrete 隔热混凝土
~ cork sheet(ing) 绝缘软木(薄)板,绝缘软木纸
~ cork slab 绝缘软木(厚)板
~ corrugated cardboard 绝缘波纹卡(片)纸板,绝缘波纹(厚硬)纸板
~ course 隔热层,绝缘层
~ door 隔声[热]门
~ efficiency 绝缘性能[功能],绝缘有效系数
~ facing 绝缘涂料
~ felt 绝缘油毛毡
~ fibre board 绝热纤维[硬纸]板
~ fire brick 绝热耐火砖
~ floor 绝缘楼板
~ foam 保温泡沫材料[塑料,橡胶]
~ foam board 保温泡沫塑料板
~ foil 绝缘(金属)薄片

~ glass 绝缘[热]玻璃
~ glaze 绝缘瓷釉[珐琅(质)]
~ glazing 窗用玻璃,隔热玻璃
~ glazing unit 绝缘釉料元件
~ gypsum 隔热石膏
~ gypsum board 隔热石膏板,隔热灰泥板
~ (gypsum) plasterboard 绝热灰泥板
~ insert(ion) 绝缘垫圈[片],绝缘衬垫
~ jacket 隔热挡板,绝缘外壳
~ layer 隔热层,保温层,隔离层,绝缘层
~ lining 绝缘里衬
~ masonry (work) 绝缘[热]圬工工程
~ material 绝缘材料
~ medium 绝缘介质
~ (mixed) plaster 绝缘(混合)灰泥
~ oil 绝缘油
~ panel 绝缘板
~ paper 绝缘纸
~ partition (wall) 绝热隔板
~ paste 绝缘胶
~ plank 绝缘厚板
~ plaster 绝缘涂层,保温灰泥
~ plastic foam board 保温泡沫塑料板
~ powder 泡沫粉末
~ pumice 保温浮[轻]石
~ pumice gravel 保温浮[轻]砾石
~ refractory 隔热耐火材料
~ roof fill 隔热保温屋面充填(材)料
~ roof(ing) tile 隔热屋瓦
~ sheet 绝缘(薄)板
~ skin 绝缘外壳
~ slab 绝缘(厚)板,隔热保温板
~ sleeve 绝缘套筒
~ straw board 绝缘稻草[麦秆]板
~ structural panel 隔热保温建筑板
~ tape 绝缘胶带
~ tile 隔热瓦[砖]

~ trim 绝缘贴面[镶边]
~ tube 绝缘管
~ value 绝缘值
~ varnish 绝缘清漆
~ wall 绝热墙,保温墙
~ wallboard 绝缘墙板
~ wallpaper 绝缘墙纸
~ window 隔热保温窗
~ work 隔热保温工程

insulation [insju'leiʃən] *n.* 隔离,绝缘,绝热,保温;绝缘体,绝缘材料,隔层
~ against vibration 隔振
~ (-grade) block 绝缘(等级)块体
~ material 隔热保温材料
abiabatic ~ 绝热绝缘
air ~ 空气绝缘
air space ~ 空隙绝缘
anti-corrosion ~ 防蚀层
cambric ~ 黄蜡布绝缘
ceramic ~ 陶瓷绝缘
cork ~ 软木绝缘
cotton cover ~ 纱包绝缘
damp-proof ~ 防湿绝缘
fibreglass ~ 玻璃纤维绝缘
heat ~ 绝热保温层
roof ~ 屋面绝热
rubber ~ 橡胶绝缘,橡皮布
sound ~ 隔声法,隔声,隔声层
thermal ~ 绝热,热绝缘,绝热物

insulator ['insjuleitə] *n.* 绝缘子[体],隔电子;绝缘材料,电介质,电介质,隔离物,非导体
antenna lead-in ~ 天线引入绝缘子
sound ~ 隔声材料,隔声器
thermal ~ 绝热器,绝热体
wall tube ~ 穿墙绝缘管

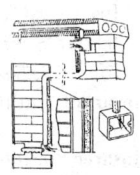

wall tube insulator

insulcrete *n.* 绝缘(混凝土)板

insulite floor 胶料地面

insullac *n.* 绝缘漆
insult [*n.* ˈinsʌlt, *v.* inˈsʌlt] *n.* 损伤,伤害,危害; *v.* 伤害,蔑视
 environmental ~ 环境对人体的危害
 radiation ~ 辐射损伤,辐射伤害
insurable [inˈʃuərəbl] *a.* 可以保险的; -insurability, *n.*
 ~ interest 可保(险)的权益
 ~ property 可接受保险的财产
insurance [inˈʃuərəns] *n.* 保险(业),保险费[金额],保证,安全措施; insurance【美】; assurance【英】
 ~ against all risks (=all-risk ~) 全能保险
 ~ against earthquake damage (= earthquake damage ~) 地震(损害)保险
 ~ against fire (=fire ~) 火灾保险
 ~ against litigation 诉讼保险
 ~ against theft 保失窃险,保窃盗险
 ~ agent 保险经纪人
 ~ amount 保险总额
 ~ applicant 要保人,保险申请人
 ~ company 保险公司
 ~ contract 保险契约
 ~ expense 保险费用
 ~ fund 保险基金
 ~ indemnity (= ~ compensation) 保险赔偿金
 ~ instruction 授保通知
 ~ policy 保险单
 compulsary ~ 强制保险
 fire ~ 火(灾)保险
 labour ~ 劳动保险
 life ~ 人寿保险
insurant [inˈʃuərənt] *n.* 投保者,受保人
insurer [inˈʃuərə] *n.* 承保人,保险商
inswept [ˈinswept] *a.* 流线形的,前端缩窄的;流过,扫过
 ~ frame 窄式车架
in-swinging window *n.* 活动窗,旋转窗
intact [inˈtækt] *a.* 未扰动的,未损的;完整的;未受影响的
 ~ clay 原状黏土,未扰动黏土
intagliated [inˈtæljeitid] *a.* 凹刻的,凹雕的,阴雕的
intaglio [inˈtɑːliəu] *n.*, -taglios,-tagli; *v.*; *n.* 凹板,凹雕; *vt.* 阴雕,凹雕,用凹印技术或方法操作
 ~ printing 凹板印刷
 ~ rilevate 凹浮雕
intake [ˈinteik] *n.* 取水口,进口,进气[水]装置,进风巷道;进风量,需求功率
 ~ area 受水区,含水层区,补给区,进水面积,回灌区
 ~ basin 取水池,进水池
 ~ canal 进水渠
 ~ chamber 进水室,进气室
 ~ channel 进水槽
 ~ check valve 进水止回阀

intake check valve

 ~ conduit 进水管道
 ~ dam 引水坝,取水坝
 ~ duct 进水渠,进气管
 ~ fish baffle 进水口挡鱼结构
 ~ gallery 进水廊道
 ~ gate 进水闸门
 ~ grate 进水(口护)栅
 ~ header 进水渠首
 ~ loss 进水损耗[失]
 ~ main 进水干管,取水干管
 ~ manifold 进气歧管
 ~ of water works 水厂取水口
 ~ of well 窨井的取水设备,取水戽斗
 ~ pipe 进水管,取水管
 ~ pipe gallery 进水管廊道
 ~ pipe line 进水管线
 ~ rate 进水速度,(土壤的)吸入速度,入流率
 ~ recharge (含水层的)人引回灌量
 ~ screen 进料滤网,进水口拦污栅

~ shaft 进水竖井
~ silencer 消声[音]器
~ stroke 进气冲程
~ structure 取水建筑物
~ tower 进水塔
~ tunnel 取水隧洞
~ valve 进[气]阀(门),进水阀
~ velocity 进水速度
~ weight 运入重量
~ with automatic flushing 自动冲沙进水口
~ works 进水工程,取水工程,进水建筑物
air ~ 进气口,入风巷道
chimney ~ 烟道进口
controllable nose ~ 头部可调进气口
flush air ~ 戽斗形进气口,平直进气口
fresh ~ 新鲜入风
main ~ 总入风(道)
mountain ~ 山麓取水口,底部取水口
power-tunnel ~ 发电隧洞进水口
river ~ 岸边取水口,无压取水口
screened ~ 带拦污栅的进水口
siphon ~ 虹吸式取水口
stream ~ 山溪取水口
submerged ~ 浸没式进水口,深水隧洞进水口
subsidiary ~ 山麓取水口
water ~ with filter gallery 渗滤式进水结构

intangible [in'tændʒəbl] *a.* 不现实的;不能触摸的,无实体的,无形的,空虚的;模糊的; *n.* 不可能摸的东西,无实体的东西; intangibility, *n.*; intangibleness, *n.*; intangibly, *ad.*
~ assets 无形资产,名义资产
~ benefit 不可计算的利益,无形效益
~ factor 难以确定的因素
~ flood damage 难确定的洪水灾害
~ loss 无形损失

intarsia [in'tɑːsiə] *n.* 嵌花,细木镶嵌装饰,细木镶嵌的制作艺术,木工镶嵌术

marble ~ 云石装饰
intarsist [in'tɑːsist] *n.* 细木镶嵌工
in tato 完全
integer ['intidʒə] = integral *n.* 总体,整数,整体,整型,完整的东西
~ programming 整体规划
integrability [intigrə'biliti] *n.* 可积性,可积分性
integrable ['intigrəbl] *a.* 可积(分)的,可积的
~ function 可积函数
integral ['intigrəl] *a.* 可积(分)的,总体的,整型的,综合的,组成的,主要的,固定的; *n.* 完整的东西;总体,整体;积分
~ basis 整体基础
~ bumper 整体挡板,整体缓冲垫
~ calculus 积分学
~ cast 整体浇铸
~ cast handle 固定把手
~ curb 整体路缘(和路面结合在一起的路缘石)
~ -differential equation 积分微分方程
~ distribution curve 累积分布曲线
~ domain 整域[环]
~ equation 积分方程
~ hardener 整体硬化剂,整体母合金,整体中间合金
~ hardening agent 整体硬化剂
~ linear programming 整数线性规划
~ lock 固定锁
~ method of water proofing 总体防水法
~ multiple 整数倍
~ point 整体式探头
~ process 整数法,整体过程
~ steel casting 整体钢铸件
~ test system 综合测试设备

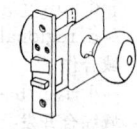

integral cast handle

integral lock

~ waterproof(ing) agent （整体）防水剂
~ water repelling agent （整体）防水剂
space ~ of force 力的空间积分（即功）
time ~ of force 力的时间积分（即冲量）

integraph ['intigrɑ:f] *n.* 积分仪,积分器

integrate ['intigreit] *v.*,-grated,-grating; *vt.* 整化,(求)积分,集成,综合; *vi.* 同化; integrative, *a.*
~d appraisal (evaluation) 综合评价
~d belt system 整套的输送带系统
~d case studies 综合个例研究
~d circuit 集成电路
~d contractor 综合承包商
~d control system 综合控制系统
~d curb 整体路缘
~d data processing (IDP) system 集总数据处理制度,综合数据处理系统
~d drainage 合成水系
~d flow curve 累积流量曲线
~d fund 综合基金
~d iron and steel plant 综合钢铁厂
~d logging 集成伐木,木材综合利用采运
~d management approach 综合管理方法
~d network 总体网络
~d plan 整体规划
~d power grid 综合电力网
~d regional cooperation 一体化区域合作
~d river basin development 江河流域综合开发
~d rural development 农村综合发展
~d survey 综合考察
~d task index 综合任务指标
~d test requirement outline 综合测试要求大纲
~d utilization 综合利用
~ing (integrating) float 积深浮标
~ing flow meter 积分流量计,累积流量计
~ing meter 积分仪
~ing water sampler 积深采（水）样器

integration [inti'greiʃən] *n.* 集成,积分（法）,综合,整体化,累积（信号）
~ method 积分法（测流）
~ of operation 联合作业

integrator ['intigreitə] *n.* 积分器,累计器,积分仪[装置],积分文件[电路]
~ circuit 积分电路

integrity [in'tegriti] *n.* 完整（性）,完全（性）,不间断性,牢靠性
~ basis 整基

integro-difference equation 积分差分方程

integrometer *n.* 矩求积仪,惯性矩面积仪

intellect ['intilekt] *n.* 智力,智士,智者；知识界

intellection [inti'lekʃən] *n.* 智力[思维]活动,思考；推理,理解概念

intellectual [inti'lektjuəl] *n.* 知识分子; *a.* 智力的,知（识）的,理智的,聪明的; intellectually, *ad.*; intellectualness, *n.*
~ faculties 智能
~ work 脑力劳动[工作]

intelligence [in'telidʒəns] *n.* 情报,消息；知识,智力；引导信号,信息,指令
~ bureau 情报局[部门,处]
~ platform （航天）侦察站
~ quotient 智商,智力商数
~ signal 载波信号,情报数报信号
artificial ~ 人工智能
current ~ 动态情报
guidance ~ 制导信息
homing ~ 自动引导信息
photographic ~ 照相情报

intelligent [in'telidʒənt] *a.* 有才智的,

聪明的,有理解力的,有理智的;intelligently, *ad.*
~ terminal 智能终端,灵活终端
intelligibility [inteˌlidʒə'biliti] *n. pl.* -ties;可理解性,清晰度
INTELSAT ['intelˌsæt] *n.* 国际通信卫星(组织)
intemperate [in'tempərit] *a.* 激烈的,过度的
~ wind 烈风
~ zone 热[寒]带
intense [in'tens] *a.* 强(烈)的,激烈的,热烈的;intensely, *ad.*; intenseness, *n.*
~ colour 浓色
~ cooling plant 深度冷冻厂
~ depression 强低压
~ fall 暴雨
~ heat 高温,酷暑;急剧加热
intensifier [in'tensifaiə] *n.* 加强物,加厚剂;照明装置;扩大器,增强剂
sweep ~ 扫描照明装置
intensimeter *n.* 声强计
intension [in'tenʃən] *n.* 强度,紧张,内涵; intensional, *a.*; intensionally, *ad.*
intensity [in'tensiti] *n. pl.* -ties;剧烈,强烈,紧张,能通量密度,强度,亮度;厚度;应力
~ of breaking 拉(伸)应力,断裂应力
~ of compression 压强度,压强
~ of draught 牵引强度
~ of gravity 重力强度
~ of illumination 照度
~ of labor 劳动强度
~ of light 光度,照度
~ of load(ing) 荷载强度
~ of oscillation 振动强度
~ of precipitation 降水强度
~ of pressure 压力强度
~ of radiation 辐射强度
~ of rainstorm 暴雨强度
~ of shear 剪力强度
~ of stress 应力强度
~ of tension 拉力强度
~ of vibration 振动强度
~ of washwater (快滤池的)冲水强度
beam ~ 光束强度
calorific ~ 热卡强度
colour ~ 彩色亮度
earthquake ~ 地震烈度
illumination ~ 照度
incident ~ 入射强度
jamming ~ 干扰强度
noise-field ~ 噪声场强
radiant ~ 辐射强度
threshold ~ 临界强度,临阈强度
traffic ~ 话务量,业务量(强度),通信密度,交通密度
turbulence ~ 紊流强度,紊度
vortex ~ 旋涡强度,涡流强度
zero ~ 零强度,零位强度
inter [in'tə:] *vt.* , -terred,-terring;埋葬;['in'tə:]【拉丁】*prep.* 在……中间,在内
~ -bank rate 银行同业汇价
~ -combination 相互组合
~ -element traction (有限单元法中)单元间的牵引
~ -pile-sheeting 桩间插板[板墙]
~ -zone trip 区间行程
interact [intər'ækt] *v.* 相互作用,反应,相互影响; *n.* =entracte
interaction [intər'ækʃən] *n.* 相互作用,相互影响,交互作用,交相感应,干扰,干涉; interactional, *a.*

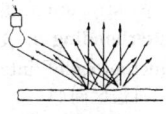
interaction

~ formula 交接公式
~ space 互作用空间
~ time 互作用时间
attractive ~ 相互吸引,互相吸力
man-machine ~ 人机对话

parametric ～ 参量互作用
repulsive ～ 互斥力,互相推斥作用
scalar ～ 无向量的相互作用;标量相互作用
shock (-wave) boundary layer ～ 激波与边界层相互作用
shock turbulence ～ 激波与紊流相互作用
spin-orbit ～ 自旋与轨道相互作用
structural ～ 结构相互作用
tensor ～ 张量相互作用
vector ～ 矢量相互作用,向量相互作用
interactive [intər'æktiv] *a.* 相互作用的,相互影响的,相互干扰的
～ model 相互作用模型
～ query 对话式(相互)查询
interadaptation *n.* 相互适应[顺应]
interagency *a.* 跨部门的,部门之间的
interalloy ['intərəlɔi] *n.* 中间合金
interaquifer flow 含水层间水流,层间流
interarea = cardinal area *n.* 主面,基面
interattraction *n.* 相互吸引
interaxis *n.* 两轴线之间的空间
interbaluster *n.* 栏杆(空)档
interband *n.* 中间带
interbasinal development planning 跨流域开发计划
interbedded [intə'bedid] *a.* 位于不同地层之间的(岩石、矿物等),互层的,层间的,镶嵌的,混合的
～ sands 互层砂土
～ stratum 夹层
interbonding *n.* 交织砌合
interborough ['intə'bʌrə] *a.* 市镇间的,自治区间的
intercalary [in'tə:kələri] *a.* 闰的,插入的,添入的,添加的,夹层的
～ day 闰日(即2月29日)
～ month 闰月(即2月)
～ strata 夹层
～ year 闰年
intercalate [in'tə:kəleit] *vt.* , -lated,

-lating; 闰,插入,深入,成夹层
-intercalative, *a.*
～d bed 夹层,间层
～d beds of sandstone 砂岩夹层
～ tapes 叠式[夹层]绝缘带
intercalation [intə:kə'leiʃən] *n.* 置闰,插入,夹层;隔行扫描
inter(-)call telephone 内部电话
intercardinal ['intə'ka:dinəl] *a.* (方位)基点间的,罗盘上的(方位)点间的点
～ point 象限点,隅点
intercept [v. intə'sept, n. 'intəsept] *vt.* 拦截,截断,截声;侦听,窃听,监听;*n.* 截距;交点;窃听,监听;interception, *n.*
～ equation 截段方程
～ form 截距式
～ receiver 截听接收机
～ tape 暂录带
～ valve 截流阀
～ed crossroad 横路交通截断型交叉路
～ed drain system 截流式排水系统
～ed stream 间断性河流
～ed water 截留水,拦蓄水
～ing chamber 截流井
～ing channel 截水沟
～ing dike 截土堤(公路上保护路基的挡土墙)
～ing drain(s) 截流排水道,截水沟
～ing layout 截流式布置
～ing sewer 截流下水道
～ing system 截流式系统(陶瓦排水管布置方法之一)
～ing trap 汽水阀
interception [intə'sepʃən] *n.* 拦截,截留,植物截留;交叉,相交
～ channel 截水沟
～ loss 截留损失(植物对降水的截留)
～ by crope 作物截留(量)

intercalated bed

~ by forest 森林截留(量)
interceptometer *n.* 截留仪,截留测定计
interceptor [intəˈseptə] *n.* 截留管,截水沟,截流井,截流器,窃听器,隔断,遮断器
~ drain 截水沟
~ valve 阻止[截断]阀
interchange [*n.* ˈintətʃeindʒ, *v.* intəˈtʃeindʒ] *n.*; *v.*, -changed, -changing; 道路立体枢纽,互通式立交,高速道路入口处;互换,转运; *v.* 交换,交替,转接; **interchanger**, *n.*
~ ability and replacement 可互换性与置换

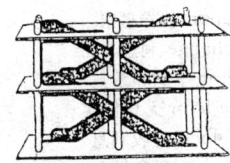

interchange cross autostairs

~ cross-platform 互通式立体交叉平台
~ cross autostairs 互通式立体交叉自动楼梯
~ method (桁架分析用)替换法
~ of air 空气交换
~ of energy 能量交换
~ of heat 热交换
~ point 货物转运站
~ ramp 互通式立交坡道
~ station 交换站,枢纽站
~ track 在转运时货车的运行轨道
~ with weaving section 交织型互通式立体交叉
all-direction ~ 全定向道路立体枢纽
cloverleaf ~ 四叶式立体交叉
cross ~ 十字立体交叉
diamond ~ 菱形立体交叉
direct-connection ~ 直连式立体交叉
directional ~ 定向立体交叉
grade separated ~ 立体交叉
rotary ~ 环形立体交叉
successive ~ 连续式道路立体枢纽
terminal ~ 道路终点枢纽
trumpet-type ~ 喇叭式立体交叉
interchangeability [ˈintətʃeindʒəˈbiliti] *n.* 互换性,可交换性,可交替性
interchangeable [intəˈtʃeindʒəbl] *a.* 可互换的,可替代的,通用的; **interchangeably**, *ad.*
~ anti-siphonage device 可互换的反虹吸装置
~ design 通用设计
~ equipment (item) 通用装备[仪器,铁道车辆]
~ parts 可互换零件,通用配件
~ program tape 可互换程序带
~ term 通用条款
interchanger *n.* 交换器,交换机
heat ~ 热交换器
intercity [intəˈsiti] *a.* 城市之间的,市际的
~ communication 市际通信
~ network 城市间电网
~ relay system 城市间中继系统,城市间电视转播系统
~ road 市际(道)路
~ traffic 市际交通
~ trucking 市际汽车货运,长途货运
intercolumnar [intəkəˈlʌmnə] *a.* 柱间的,塔间的
intercolumniation [intəkəlʌmniˈeiʃən] *n.* 柱间空地,柱间距;分柱法

intercolumniation

incercom [ˈintəkəm] = intercommunication *n.* 对讲电话,内部通信联络系统
~ master set 内部通话主机
~ set 内部通信设备
intercommunication [intəkəmjuːniˈkeiʃən] *n.* 连通,多向

通信,相互往来[通信,连络];对讲电话装置
intercommunication system 内部通信系统,内部通话制
intercommunicator n. 内部通信机
interconnect ['intəkə'nekt] v. (相)互连(接),横向连接,内连;interconnection, n.; interconnectedness, n.
　~ wiring diagram 接线[布线]图
　~ed electric power system 互联电力系统
　~ed estuarine channels 河口河网
　~ed grouting ducts 互连灌浆导管
　~ed pipe system 互连管道系统
　~ed voids 相互连通的孔隙
　~ing road (way) 联络道路
　~ing taxiway 互联汽车道
interconnection [intəkə'nekʃən] n. 内连,相互连接[联系],联络线,横向通[过]道
　~ diagram 接线图
　~ grouting ducts 互连灌浆导管
　multi-layer ~ 多层互连
intercontinental [intəkəntə'nentəl] a. 洲际的
　~ airport 洲际航空港,洲际机场
　~ circuit 洲际电路
　~ through-shipment 洲际直达货物[运输]
　~ traffic 洲际报务[话务]
intercostal [intə'kɔstl] n. 加强肋,肋际;a. 肋间的
　~ angle 间断角钢
　~ floor 间断肋板
　~ girder 间断纵桁
　~ keelson 间断内龙骨[筋]
　~ member 间断构件
　~ plate 间断板
interdependence n. 相互依赖性
interdetermination n. 互为因果的关系
interdisciplinary [intə'disəplinəri] a. 各学科间的,交叉学科的,边缘学科的,多学科的

　~ approach 多科性[跨学科]研究法
interdome
['intədəum] n. 圆屋顶间,穹隆顶或圆屋顶内壳与外表之间的空隙

interdome

interest ['intrist] n. 利益,利息;股份,财产;兴趣,开心,注意;所有权,权力 vt. 使发生兴趣,使发生关系,使注意
　~ bill of exchange 附息汇票
　~ clause 利息条款
　~ coupon 债券息票
　~ exchange 利息支出
　~ factor 本利和系数
　~ free loan 无息贷款
　~ in block 应收利息
　~ in come 利息收入
　~ in red 应付利息
　~ period 利息期
　~ policy 利息政策
　~ rate 利率
　~ receivable 应收利息
　~ restriction 利息限制
　~ subsidization 利息补贴
　~ed administration 有关主管机关
　~ed parties 有关方面
　business ~s 商业界
　compound ~ 复利
　harmony of ~ 利益协调
　places of ~ 名胜
　simple ~ 单利
interface ['intəfeis] n. 界面,边界,国际间[线],学科间共有的问题[理论,事实等];v. 对接,联系,连接
　~ between horizons 层间接触面,层间面
　~ condition 交界条件
　~ event 交界事项
　~ relative moisture content 分界相

对含水量
~ strength 接触面强度(粘结强度)
depositional ~ 沉积间面
interfacial [intəˈfeiʃəl] *a.* 分界面的,边界的,面际[间]的,层间的
~ friction (路面结构的)层间摩擦,接触面摩擦
~ tension 界面张力,面际张力
~ wave 界面波,内波
~ weld 摩擦面焊合(点)
interfacing *n.* 衬布,衬头,分界面
interfenestral *a.* 窗间的
interfenestration [intəfenəsˈtreiʃən] *n.* 窗间墙墩,窗间墙宽度,窗距布置
interference [intəˈfiərəns] *n.* 干扰,干涉,冲突,相互影响,过盈; interferential, *a.*
~ factor 干扰因素[系数]
~ settlement 干扰沉陷
~ zone 干扰区[范围]
interfile *a.* 文件[资料]间的; *v.* 把……归档,把……编入档案
interfinger [intəˈfiŋgə] *vi.* (地)层间的交错,指状交错,相互贯穿; *n.* 楔形夹层
~ ing 相互贯穿
interflood period 洪水间隔期,间洪期
inter-floor stair(case) 楼面[层]间楼梯
inter-floor traffic 楼面间交通
inter-floor travel 不同层次(桥面或楼面)间的交通

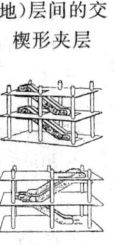

inter-floor stair(case)

interflow [ˈintəfləu] *n.* 壤中流,土内水流,表层流,层间流; *v.* 汇合,混合,交流,互通
interfluent [inˈtəːfluənt] *n.* 内流熔岩; *a.* 交错的,混淆的,汇合的
interfluve [ˈintəfluːv] *n.* 河间地,相邻两流域间的分界地
interfold [intəˈfəuld] *vt.* 交互折叠,叠合(将另一个折在里面)

interformational sill 内部构造(底)槛
interglyphe *n.* 柱槽间
intergrade *n.* 中间级,中间期,中间的或过渡的形式
intergranular *a.* (颗)粒间的,晶(粒)间的,内在(晶)粒状的
~ martensite (颗)粒状(结构)马氏体钢
~ slip-plane 颗粒间滑动面
~ space 粒间孔隙
~ stress 粒间应力
intergrind *v.* 相互研磨
interground 相互研磨
~ addition 研磨时的添加料
intergrout *n.* 中速凝结浆
interim [ˈintərim] *n.* 间歇,期中,暂时,临时; *a.* 暂时的,过渡时期的
~ andit 期中审计
~ charges 临时开支
~ closing 期中结账
~ engineering order 临时工程指标
~ income statement 期间损益表
~ spare parts list 临时备件单
~ specification 暂行规范
interior [inˈtiəriə] *n.* 中间段,内部[地],内陆部分;内景,内政;室[屋]内图; *a.* 国内的,室内的,内部的;内地的; interiority, *n.*; interiorly, *ad.*
~ adhesive 内部粘着力
~ air cooler 室内冷空气
~ angle 内角
~ arch 内切拱
~ architecture 室内建筑
~ basin 内陆盆地
~ bending moment 内部弯矩
~ bonding agent 内部粘结剂
~ (building) board 室内(建筑)板
~ (building) panel 室内(建筑)板
~ cement(ing agent) 室内粘结剂
~ chlorinated rubber paint 内部氯化处理的橡胶漆
~ column 内柱

~ continent 内陆
~ corner 内墙角
~ corner reinforcement 内墙角加强肋
~ corridor 室内走廊
~ -corridor type building 内走廊型建筑物
~ corrosion 内部腐蚀
~ court 内庭院
~ decor （室内）装饰,全部陈设
~ decorating 内部装饰
~ decoration 室内装饰
~ design(s) 内部设计,室内设计
~ designer 室内设计师,内部设计师
~ delta 内陆三角洲
~ designer 内部设计师,室内设计师
~ door 内门
(~)doorjamb block 内部门窗侧壁砌块[体]
(~)doorjamb tile 室内门窗侧壁砖
~ drainage 内陆水系,内部排水
~ fiberboard finish 内部纤维板装修
~ finishing 内装
~ fire protection 内部防火
~ fixtures 内部装饰物[品]
~ flooring 内楼面
~ focusing telescope 内对光望远镜
~ gallery 内部长廊
~ gallery apartment building 内廊道式公寓建筑

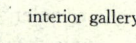

interior gallery

~ garden 室内花园
~ garage sign 内部车库标志
~ glass door 室内玻璃门
~ gloss (clear) varnish 室内上光清漆
~ handrail 内（部）扶手
~ heat 室内（供）热
~ heat gain 室内增热
~ humidity 室内湿度
~ illumination 室内照明（度）
~ insulation 内部绝缘

~ joinery 内部细木工作
~ lake 内陆湖
~ layout 室内布置[规划]
~ liabilities 对内负责
~ lighting 内部照明
~ lining 内部衬砌
~ lining panel 室内护板
~ (load) stress 中载应力
~ lot 内地段
~ masonry (dividing) wall 室内砌筑（分隔）墙
~ measure 内测度
~ noise 室内噪声
~ opposite angles 对（顶）角
~ orientation 内方位,内定向
~ paint coat 内部涂料
~ panel 室内护墙板
~ partitioning 室内间隔
~ picture 室内画,内景画
~ piping 内部管线
~ plain 内陆平原
~ primer 核心雷管
~ relative humidity 室内相对湿度
~ relative moisture 内部相对含水量
~ reveal 室内外抱,内部的窗[门]侧（指墙与门或窗之间）
~ span 内跨度,中间跨
~ stain 内部色彩
~ stair (case) 室内楼梯间

interior stair(case)

~ stairrail 室内楼梯栏杆
~ stanchion 室内支柱
~ structure 内部结构
~ stud 室内壁骨
~ support 内(部)支承
~ sur (facing) 室内表面修饰

~ temperature 室内温度
~ temperature control 室内温度调节[控制]
~ tile 室内砖
~ tracery 内部窗(花)格
~ trim 内部装修
~ valley 灰岩盆地,岩溶盆地
~ valve 内阀门
~ view 内景,内视图,内部图
~ wall 内墙
~ waterway 内陆水道,内河水道,内河航线
~ window frame 内窗框
~ window sill 内窗槛
~ work 室内工作[程]
~ yard 里院,内院

interjoist *n.* 跨度,跨距;搁栅间,搁栅档
interknot [intə'nɔt] *v.* -knoted,-knoting;连接在一起,纠缠在一起
interlace [intə'leis] *n.*; *v.*, -laced, -lacing;交错,交织,交替,夹层;组合;隔行(扫描),交错(存储); interlacedly, *ad.*; interlacement, *n.*

interlacing arcade 内叉拱廊
interlacing arches 交错拱
interlacing diagram (交通)交叉图
interlacing ornament 缨带饰

interlacing ornament

interlayer ['intəleiə] *n.* 夹层,间层; *a.* 层间的
~ of spun glass (玻璃纤维)夹层
~ temperature 层间温度
~ water 层间水

interline [intə'lain] *n.* (铁路)线之间连系,(两条)线之间虚线
~ freight 铁路联运货物
~ waybill 联运货物路程单

interlink [*v.* intə'link, *n.* 'intə(:) link] *n.* 连环,连锁; *v.* 使连锁,环接
interlock ['intəlɔk] *n.* 连锁装置,连锁转车充辙器;保险设备; *v.* 连锁,嵌锁,锁结,连接,结合;-interlocker, *n.*
~ assembly 连锁装置系统
~ed grain 交错纹理
~ing board 嵌锁(式)板
~ing (building) panel 锁结式(建筑)板
~ing concrete pipe 嵌锁式混凝土管
~ing device 连锁装置
~ing grain 斜交木纹
~ing revetment 连锁护坡
~ing roof tile (= ~ing tile) 连锁屋瓦
~ sheeting 搭扣板桩
~ing surface 嵌锁式路面,嵌挤式面层
~ing tile 咬接瓦,连锁屋瓦
~ing tile roofing 连锁瓦(片)屋面
~ing tooth 交错齿

interlocker *n.* 连锁装置,连续者,连锁器
intermediate [intə'mi:djət] *n.* 中间体;中间人,调停人;中间阶段;中间产品; *a.* 中间的,层间的;中级的; intermediately, *ad.*; intermediateness, *n.*
~ axle 中轴
~ bay (桥)的中间跨;(混凝土路的)中间板
~ bearer 中间承木,桩间板壁;中间托梁
~ belt 中间带,中环(路)
~ column 中型柱
~ course 中间层
~ cracking (混凝土路面的)中间裂缝,板中开裂
~ (crushing) chamber (碎石机的)中碎室,(轧石机的)中轧室
~ deck 中承式桥面,中层桥面,中层
~ (door) rail 中门横档
~ floor 中间楼层
~ floor beam 中间楼面间梁
(~)(floor) joist 中层[间]楼板搁栅
~ floor slab 中间楼板
~ foam layer 中间泡沫层

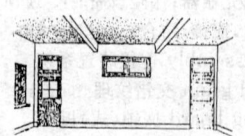
intermediate floor beam

~ frame 中间构架[肋骨]
~ gate 中门
~ grain 中间颗粒
~ grinding 中间研磨
~ ground water 过渡层地下水
~ igneous rocks 中性火成岩
~ inflow 区间入流,区间来水
~ jack rafter 中间小椽
~ lava 中性熔岩
~ layer 中间层,夹层
~ panel 中间镶板
~ pier 中窗间壁,中间墩
~ platform 中间平台
~ post 中间(支)柱
~ pressure 中等压力
~ principal stress 中主应力
~ pulley 中滑轮
~ pumping station 中间泵站
~ purlin 中间檩
~ rafters 中椽
~ rail 中间(门窗的)横档,中档
~ rib (哥特式弯的)居间肋,中间肋条
~ ring road 中环路
~ rock 中性岩,岩石夹层
~ scale 中间尺度,中尺度
~ section 中截面
~ sedimentation tank 中间沉淀池
~ settling tank 中间沉淀池
~ sill 中层帽木
~ span 中间跨度[桥孔]
~ stair (case) 中间楼梯
~ stiffener 中间加强杆,中间加劲杆
~ stiffness 中级刚度
~ stock 居间的砧木
~ switch 中间开关

~ tie 中间系杆
~ truss 中间桁架
~ tubular column 中型管柱
~ -type pavement 中级路面
~ value 中值,介值
~ value theorem 中值定理,介值定理
~ variable 中间变量,层间变量
~ water 过滤带水,中间水,中层水,通气层水
~ water zone [belt] 隔水层[带]
~ zone 中间带
intermittence [intə'mitəns] n. 中间,中断,间歇(性),周期(性)
intermittent [intə'mitənt] a. 间歇的,断续的,周期的,脉动的,急撞的;-intermittently, intermittingly, ad.
~ bucket (type) elevator 间歇吊斗式升降机
~ disconnection 时断时续
~ drainage way 间歇式排水道
~ dredger 间歇挖泥机
~ dryer 间歇干燥器
~ excarator 间歇性挖土机
~ filter 间歇式滤池,间歇滤池
~ flush 间歇冲洗
~ fountain 间歇式喷泉

intermittent fountain

~ grading 间断级配,升降交替的坡度
~ heating 间歇加热
~ interrupt stream 间歇性河流
~ light 断续光
~ load 间歇荷载,间歇负载
~ mixer 间歇式拌和机
~ mixing 间歇抖和(法)
~ operation 间歇运行[操作]
~ rain 间歇雨

~ recorder 打点式记录,断续记录
~ sand filter 间歇砂滤池
~ sedimentation 断续沉淀(法),间歇沉降(法)
~ shoulder 间断式路肩,间断没有避车道的路肩
~ spring 间歇泉
~ stream 间歇河(流)
~ (type) track shifting machine 间断性移轨机
~ welding 间断焊

intermix [ˌintəˈmiks] *v.* 搅拌,拌和,(使)混合[杂]

intermixture [ˌintəˈmikstʃə] *n.* 混合,混合料,混合液[剂],混合物

intermodal *a.* 综合运输的,联运的,用于联运上的
~ cargo 联运货物
~ container 联运集装箱
~ transportation 协同联运,多式联运

intermodillion [ˌintəmouˈdiljən] *n.* 檐托座间,斗宫档,斗栱间

intermontane [ˌintəˈmɔntein] *a.* 山间的
~ basin 山间盆地
~ plain 山间平原
~ trough 山间槽地

intermountain [ˈintəˈmauntən] *a.* = **intermontane**;山间的
~ basin 山间盆地
~ seismic belt 山间地震带

intermural [ˌintəˈmjuərəl] *a.* 墙壁间[距]的,校际的,埠际的

internal [inˈtə:nl] *n.* (*pl.*)内脏,内部零件,本质; *a.* 内部的,内在的,内政的,国内的,固有的
~ air cooler 内部空气冷却
~ angle bead 阴条砖(釉面砖的配件砖)
~ annular shake 木心坏裂
~ arch 内拱
~ behavior 内部性状

~ beding moment 内弯矩
~ bird's beak 阴条三角(釉面砖的配件砖)
~ boundary condition 内边界条件
~ (building) board 内部建筑板
~ (building) sheet 内部建筑(薄)板
~ bull-nose brick 内凹砖
~ check 内部核查
~ cohesion (土等的)内聚力
~ column 内柱
~ combustion compactor 内燃夯实机
~ combustion engine 内燃机
~ corridor 内走廊
~ corrosion 内部锈蚀
~ court 内部庭园[天井]
~ decorating 内部装饰
~ deformation 内部变形
~ designer 内部设计师
~ diameter 内径
~ diffusion 内扩散
~ dimensions 内部尺寸
~ door 内门
~ dormer 内凹式老虎窗(斜屋顶的)
~ drain 内部排水(管)
~ drainage 内陆水系,内部排水(土壤)
~ elevation 内部标高
~ equipment 内部设备
~ facing 内部面饰
~ fibre board finish 内部纤维板装修
~ field distribution 田间配水
~ financing 内部财务
~ finish(ing) paint 内部装修用漆
~ fire-extingushing system 内部灭火[消防]系统
~ fittings 内部配件
~ fixtures 内部装置
~ flow 内部水流;(金属轧断时的)内变形
~ force 内力
~ focusing 内调焦
~ focusing telescope 内对光望远镜

~ friction 内摩擦(力),内摩阻力
~ gage 内径规
~ gallery 内部廊道
~ gallery apartment building 内廊道式公寓建筑
~ glass door 内部玻璃门
~ gloss (clear) varnish 内部上光清漆
~ handrail 内扶手
~ hardboard finish 内部硬质纤维板装修
~ heat gain 内部取暖
~ humidity (室)内湿度
~ illumination 内部照明(度)
~ installation 内部设备[安装]
~ installation work 内部安装工作
~ joinery 内部细木工作
~ layout 内部规划[布置]图
~ leaf 内部门扉
~ lighting 内部采光
~ lining 内部衬砌
~ masonry (dividing) wall 内部坊工(分隔)墙
~ member 内套件,被包容件
~ memory 内存储器
~ mitre 内斜削
~ mixing blade 内拌和翼
~ moisture 内部湿度
~ moment 内(弯)矩,内力矩
~ navigation 内河航行
~ noise 内部噪声
~ orientation 内部定位
~ (origin) water 内(源)水,内部水
~ paint 内部涂料
~ panel 内部护墙板[镶板]
~ partitioning 内部分配,内部分隔
~ pipe size 管的内径
~ pipe thread 管的内螺纹
~ plastering 内部抹灰
~ plumber (= plumber) 装管工人
~ plumbing 内部卫生管道工程
~ pressure 内压力,内压强
~ pressure current 内压流,密度流
~ -prestress 内部预应力

internal partitioning

~ primer 内底漆
~ ramp (立体交叉的)内匝道,内坡道
~ rate of return 内部收益率
~ rational separation 室内合理分隔
~ redecoration 内部重装修
~ regression 内回归
~ resistance 内阻力
~ reveal 内部外抱,内部门[窗]侧(指墙与门或窗之间)
~ revenue 国家税收
~ ring 内环
~ scaffold(ing) 内脚手
~ scour 内部淘洗,内部冲刷
~ seiche 内假潮,内静震(波)
~ settlement measurement 内部沉陷量测
~ shake 内环裂(木)
~ shield 内部屏蔽
~ shuttering 内模壳
~ skin 内层墙
~ skip-hoist 内吊斗提升机
~ slatted blind 内石板百叶窗
~ soil drainage 土壤内部排水
~ span 内跨度
~ (spun) vibrator 内插式振捣器
~ stagnant water 内(部积)滞水
~ stair (case) 内楼梯
~ stairrail 内楼梯拦杆
~ storage 内存储器
~ strain 内应变,内变形
~ stress 内应力
~ stud 内部壁骨
~ surface 内表面
~ (sur) facing 内表面修装
~ telephone system 内部电话系统
~ temperature control 内部温控

~ terrestial heat 地热
~ threads 内螺丝
~ tide 内潮
~ tile 内砖
~ tracery 内窗(花)格
~ trip 市内乘车出行, 境内出行
~ vibrating machine 插入式振捣机 internal vibrator
~ vibrator 插入式(混凝土)振捣器
~ wall 内墙
~ wall surface 内墙面
~ wall(ing) panel 内墙板
~ water 内脏水, 内部水
~ water circulation 水文内循环, 内部(水)循环
~ waters 内部水域, 内地水域
~ welding 内部焊接
~ window 内窗
~ window cill 内窗槛[台]
~ window frame 内窗框
~ window sill 内窗槛[台]
~ wiring 内部线路, 暗线

internality [ˌintəˈnæliti] *n.* 内在(性)
international [ˌintəˈnæʃənl] *n.* 国际性组织, 国际比赛; 侨居国外者; *a.* 国际(上)的, 世界的
~ accounting standards 国际会计标准
~ aerial navigation 国际航运
~ airway 国际航空线
~ balance of payments 国际收支平衡
~ business 国际企业
~ commercial exchange 国际商品交易所
~ commodity 国际商品
~ convention 国际惯例
~ cooperation 国际合作
~ debt 国际债务
~ de luxe type hotel 国际豪华型旅馆
~ exchange 国际汇兑
~ investment 国际投资
International Investment Bank (IIB) 国际投资银行
~ labour compensation facility 国际劳动补偿办法
~ loan 国际借款
~ knot 国际节(速度单位＝1.852公里/小时)
~ money market 国际金融市场
~ money orders 国际汇票
~ monopoly 国际垄断
~ patent 国际专利
~ rated horsepower 国际额定马力
~ sea-borne shipping 国际海运船舶吨数
~ security 国际证券
~ standard 国际标准
~ standard atmosphere 国际标准大气压
~ syndicated bank loans 国际银行贷款
~ (passenger) terminal building 国际(旅客)度假村, 国际旅客车站[港口, 码头, 终点站]房屋
~ test sieve series 国际成套试验筛(包括 19 种筛号)
~ trade 国际贸易
~ trading partner 国际贸易伙伴
~ trade policy 国际贸易政策
International Trade Organization (ITO) 国际贸易组织
~ understanding 国际谅解
~ unit 国际单位
~ waters 国际水域
~ waterway 国际航道

internode [ˈintənəud] *n.* 节间部, 波腹; internodal, *a.*
internship [ˌintəːnʃip] *n.* 实习, 实习期, 实习补助金
interpenetrate [ˌintəˈpenitreit] *v.*, -trated, -trating; 渗透; 贯通; interpene-

tration, *n*. ; interpenetrable, *a*. ; interpenetrant, *a*. ; interpenetrative, *a*. ; interpenetratively, *ad*.
interpenetrating molding 交截线脚
interpenetration ['intəpeni'treiʃən] *n*. 互相渗透,互相贯通,互相钩接;交截细工
interpilaster [ˌintə(ː)pi'læstə] *n*. 壁柱空间[间距]
interpile sheeting 桩间水平支撑
interpolate [in'təːpəuleit] *v*., -lated, -lating;窜改,添进,插入,内插;写进
  ~d contour 内插等高线
  ~d section 插补断面
interpolating function 插值函数
interpolation [intəːpəu'leiʃən] *n*. 内插法,插值法,插入物
  ~ error 内插[插补]误差
  ~ function 插值函数
  ~ of contours 等高线内插
interpretation [intəːpri'teiʃən] *n*. 解释,说明,描述;翻译,译码;(实验数据)整理[分析]
  depth ~ 深度推断
interprovincial [intəprə'vinʃəl] *a*. 省际的
  ~ highway 省际公路
interradium ['intə(ː)'reidjəm] *n*. 间隔区
interrupt [intə'rʌpt] *v*. 中断,切断,阻止,妨碍,打扰
  ~ arch moulding 间断拱线脚
  ~ed discharge of traffic 交通[车流]中断
  ~ed loading 断续加载
  ~ed oscillation 间歇振荡
  ~ed screw 分瓣螺钉
  ~ed stream 断续河流,中断河,间断河
  ~ed water table 间断潜水面,间断地下水面
  ~ing drainage well 间歇排水井
interruption [intə'rʌpʃən] *n*. 中断,遮断,断路,停歇,中止,阻止,断续;打扰
  ~ loss (植物)截留(雨量)损失
intersect [intə'sekt]
v. 横断[切],交叉[切];贯穿
  ~ed country 地形起伏地区,丘陵地区
  ~ed terrace 交切阶地

intersecting arcade

  ~ing arcade 交叉拱廊
  ~ing fault 交切断层
  ~ing line 交线
  ~ing point 交(会)点,转角点
  ~ing roads 相交道路
  ~ing roof 交叉屋顶
  ~ing tracery 交叉花格窗
  ~ing vault 交叉拱

intersecting vault

intersection [intə'sekʃən] *n*. 交会,相交,交叉;交集,逻辑乘法[积];前方交会
  ~ at grade 平面交叉
  ~ census 交叉口交通调查
  ~ chart 网络图,交织图
  ~ crosswalk 交叉口人行横道
  ~ diagram 交叉口略图
  ~ entrance 交叉口入口
  ~ gate 与门交叉口
  ~ leg(s) 岔道,(交叉口的)相交路段
  multiway ~ 多路交叉口
  roof ~s 屋顶的交接
  T-type ~ T 字形交叉道
interspace [*n*. 'intə'speis, *v*. intə'speis] *n*.;*v*.、-spaced,-spacing;*n*. 空间[隙],间隔
intersperse [intə'spəːs] *v*. -spersed, -spersing;点缀;撒布,分散;交替,更迭;引入; spersedly, *ad*; interspersal, interspersion, *n*.

interstice [inˈtəːstis] *n. pl.* -stices；空隙，裂缝
interstitial [intəˈstiʃəl] *n.* 填隙(原子)，结点间，节间；*a.* 空隙的，间隙的，中间的，成裂缝的；-interstitially，*ad.*
　～ flow 间隙渗流
　～ ice 缝隙冰，间隙冰，表层以下的冰
　～ gas 间隙气体
　～ (hydraulic) pressure 孔隙(水)压力
　～ surface area 孔隙面面积
　～ surface (of soil) (土的)孔隙面
　～ water 间隙水，缝隙水，(岩石)缝间水
interstitialcy *n.* 结点间，节间，堆填子
interstock *n.* 中间钻木
interstratification *n.* 间层(作用)，层间(排列)
interstratified *a.* 间隔的，层间的
interstream *n.* 分水岭，分水界
　～ area 河间地区
intertexture [intəˈtekstʃə] *n.* 交织(物)；组合，编织
intertidal [intəˈtaidəl] *a.* 潮间的，(位于)潮区内的；*n.* 涨潮与落潮线之间的地段
　～ natural history 潮区自然特性
　～ region 潮间带，潮浸区，潮浦
　～ zone 高低潮位之间的竖直区，潮间带
intertie 交接横木，交叉拉杆，中间系杆
intertown *a.* 市际的，长途的
　～ bus 长途公共汽车
intertrabecula *n.* 间小梁
intertropical [intəˈtrɔpikəl] *a.* 在南北回归线之间的，在赤道地带的，热带的
　～ confluence zone 热带汇流区[带]
　～ convergence zone [belt] 热带辐合带
　～ front 热带锋
intertropics *n.* 南北回归线之间地带内的任何地区
intertube *a.* 管间的，偏平流的

interturn *a.* 匝间的
intertwine [intəˈtwain] *v.* -twined,-twining；编结，缠绕，交纽，编合；-intertwinement, *n.*；-intertwiningly, *ad.*
intertwining 编织，缠结
interurban [intəˈəːbən] *n.*；*a.* 城市间(的)，市际(的)；城市间的(交通路线，交通车辆)
　～ railroad 市际铁路，穿城铁道
interval [ˈintəvəl] *n.* 间距，区间，节间，间隔，中断期间；周期，间歇，范围，行程
　～ linear programming 区间线性规划
　～ measurement 区间计量法
　～ of columns 柱间距离
　mist ～ 生雾温差
　sampling ～ 测量间隔，测量区间，采样间隔，抽样区间
　time-space ～ 时空间隔
　temperature ～ 温度范围
　winding ～ 提升间歇时间，装卸载时间
intervallum *n.* 内外墙间隔带，间隔带
intervalometer [intəvəˈlɔmitə] *n.* 时间间隔测量器，间隔定时器，曝光节制器，定时器
intimate [*a.* ˈintimit *v.* ˈintimeit] *v.*, -mated,-mating；*a.* 直接的，相近的，紧密的，亲密的；精湛的；熟悉的，秘密的；*v.* 暗示，暗指；宣布，通知；-intimately, *ad*；-intimateness, *n.*；-intimater, *n.*
　～ contact (颗粒的)紧密接触
　～ mixing 均匀拌和
　～ mixture 均匀混合物
intracoastal waterway 内陆水道
intracontinental *a.* 陆内的
　～ sea 陆内海
intrados [inˈtreidɔs] *n.* 拱腹线，拱内圈；内弧面
　～ springing line 拱内圈起拱线，拱腹起

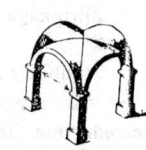
intrados springing line

拱线
intraformational bed 层内夹层
intra-vane pump 内翼泵
intravibrator *n.* 插入式振捣器
intra-zone traffic 区内交通
intrench [in'trentʃ] *v.* = entrench 掘工事, 构筑工事
　～ed meander 峡谷蜿蜒河道, 嵌入曲流
　～ed stream 嵌入河
　～ed valley 嵌入河谷
　～ing tools 土工器具
intrenchment [in'trentʃmənt] *n.* = entrenchment; 掘壕沟, 筑垒; *n.* 壕沟, 堡垒
intricate ['intrikit] *a.* 错综的, 交错的, 复杂的; -intricately, *ad.*; -intricateness, *n.*
　～ (cross) section 交错截面
intrinsic(al) [in'trinsik(əl)] *a.* = intrimsic; 内在的, 内部的, 本质的, 真正的, 实在的, 本征的, 固有的; -intrinsically, *ad.*
　～ accuracy 内在精度
　～ acidity 固有酸度
　～ basicity 固有碱度
　～ curve (= Mohr's envelope) (摩尔圆) 包络曲线
　～ energy 内能
　～ error 固有误差, 基本误差, 特征误差
　～ equation 本征方程, 内禀方程
　～ nature 固有性质
　～ permeability 内渗透性
　～ pressure 内压力
　～ shrinkage (水泥硬化时的) 内在收缩
　～ solubility 固有溶(解)度
　～ viscosity 固有粘度
introduction [ˌintrə'dʌkʃən] *n.* 引言, 绪言, 序论, 介绍, 绪论, 前言; 推广, 放入, 插入; 起引进作用的事物, 引入的事物

　～ course 插入层
introflexion *n.* = introflection 向内弯曲
intrusion [in'tru:ʒən] *n.* 侵入, 干涉, 侵入岩; 材料的下沉; -intrusional, *a.*
　～ agent 外加剂, 灌入剂
　～ of saltwater 盐水侵入
　～ pipe (水泥砂浆) 灌注管
　～ spring 侵入泉, 障碍泉
intrusive [in'tru:siv] *a.* 插入的, 侵入的, 侵入岩形成的; 妨碍的, 干涉的; -intrusively, *ad.*; -intrusiveness, *n.*
　～ body 侵入岩体
　～ growth 侵入生长
　～ rock 侵入岩
　～ sheet (sill) 侵入岩床
　～ vein 侵入脉
inundate ['inʌndeit] *v.* -dated, -dating; 淹没, 泛滥; -inundator, *n.*
　～d area [district] 淹没面积, 洪泛区, 泛滥区

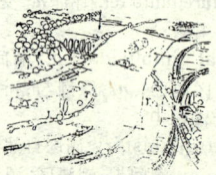

inundated area [district]

　～d district (flood perihyery) 泛滥区域, 洪水区域, 淹水地区
　～d land 淹没地, 洪泛地, 泛区, 洪泛区
　～d plain 洪泛平原, 泛滥平原
　～ing flood 泛滥洪水, 大洪水
inundation [ˌinʌn'deiʃən] *n.* 泛滥, 水灾, 淹没, 洪水, 大水, 充满, (水泥) 浸水
　～ area 洪水淹没区域
　～ bank 泛滥河岸
　～ canal 泛灌渠, 洪灌渠
　～ map 泛滥区地图, 淹没地区图
　～ method 泡水法, 淹灌法
　～ protection 防洪, 防淹没

inundative irrigation 淹灌
invade [in'veid] *v.*, -vaded,-vading；侵入，侵略，拥入；充满，遍布
~d zone 侵入带
~ing air 侵入空气
~ing sea 侵入海
~ing shoreline 侵入海岸线
~ing water 侵入水
~ing waves in estuary 河口入侵波
invar [in'vɑ:] *n.* 因钢，不变钢，不胀钢（含镍的合金钢），因瓦合金，因瓦铁镍合金，微胀合金
~ leveling staff 因瓦水准标尺
~ (metal) 因钢
~ (plotting) scale 殷钢(绘图)尺
~ steel 殷钢，不胀钢
~ tape 殷钢卷尺，殷钢带尺
~ wire 因瓦基线尺
invariant [in'vɛəriənt] *n.* 不变式，不变量，音瓦里铁镍合金；常数因子；标量，纯量，无向量，标量张量；*a.* 不变(形)的，恒定的，无变度的；-invariantive, *a.*；-invariantly, *ad.*
~ equilibrium 不变[无变度]平衡
~ factor 不变因子
~ in space-time 时空不变式
~ of stress 应力分量不变式，分应力不变式
unrestricted ~ 无约束不变式，无制不变式
invaro *n.* 因瓦劳合金钢，因瓦劳锰铬钨钒钢
invention [in'venʃən] *n.* 发明，创造；创造性的想像力，独创设计
inventory ['invəntri] *n. pl.* -tories；*v.*, -toried,-torying；清单，清册，目录，报表；装(投)料量，库存量，总量，存货；开清单，编(制)目录；清点，存储，盘存，财产注册；*vt.* 清点存货；编列详细目录；inventorial, *a.*；-inventorially, *ad.*
~ assets 库存资产
~ carrying cost 存货储存成本
~ cut off date 盘存截止期
~ date 盘存日期
~ equation 盘存公式(期初存货＋本期增加-本期提取＝期末存货)
~ index 存货指数
~ list 盘存清单
~ profit 盘存盈余
~ sheet 盘存报表
~ system 盘存制度
~ turnover ratio 存货周转率
inverarite 镍流铁矿
inverse ['in'və:s] *n.*；*v.*, -versed,-versing；*a.*；(相)反的，反向的，倒置的，逆的；倒数；*n.* 倒转状态[物]；*vt.* 使倒转；-inversely, *ad.*
~ convolution 逆褶积，逆卷积
~ correlation 负相关，逆相关
~ distance 反向距离，距离倒数
~ estuary 逆向河口
~ flow 逆流，倒流，反流
~ graded bedding 逆序粒层，逆粒级层
~ Fourier transform 傅里哀逆变换
~ function 反目数
~ grading 逆级配，逆程序
~ Lapalace transform 拉普拉斯逆变换(式)，逆拉普拉斯变换(式)
~ limit 逆向极限
~ logarithm 反对数
~ matrix table 逆矩阵系数表
~ operation 逆运算
~ plummet 倒锤线
~ proportion 反比例
~ ratio 反比
~ sampling 颠倒采样
~ -square law 平方反比律
~ volume 容积的倒数
inversion [in'və:ʃən] *n.* 颠倒，倒置，倒转；反向，转换，转化，转变
~ fog 逆温雾
~ layer 逆温层，逆转层
~ of water temperature 海水逆温

invert [v. in'və:t, n. 'invə:t] *n.* 仰拱,管道内底(管道内壁最低点),内底,底板; *a.* 反转的,转化的,颠倒的,逆的, *vt.* 反转,转化,倒置,颠倒; inverted, *a.*; inverter, *n.*; invertible, *a.*

inverted arch

~ed arch 仰拱,倒拱
~ed block 倒拱砌块
~ed bow and chain girder 倒置拱式大梁,鱼肢形大梁
~ed form 仰拱模板
~ed 倒立的,倒置的,内外翻转的,与既定格式相反的
~ed asphalt emulsion 沥青倒乳液(水分散在沥青中的乳液)
~ed beam 上翻梁
~ed camber 倒路拱,反拱,反拱度
~ed capacity 吸收容量
~ed carburetor 倒置汽化器,倒置化油器
~ed cavetto 倒拱修圆
~ed choke 倒滤层(铺在碎石路面下的),反嵌料(防止路基土渗入碎石层)
~ed curb 倒置缘石,倒置式路缘石
~ed dipper 反向铲
~ed elevation 管道内底高程(管道内壁最低点的高程)
~ed filter 倒滤层
~ed kerb 倒拱路边石
~ed king (post) truss 倒单柱桁架
~ed lining 逆砌法(隧道施工时用的方法)
~ed order 倒序排列
~ed penetration 倒贯入,反绩入(一种先洗沥青后撒石料的路面施工方法,如表面处治等)
~ed penetration macadam 倒贯入式碎石(路)
~ed pleat 褶叠
~ed queen (post) truss 倒双柱桁架
~ed roof 倒置屋顶(防水层在隔热层下)
~ed siphon 倒虹吸管
~ed siphon culvert 倒虹吸涵洞
~ed siphon sewer 倒虹吸下水道
~ed stream 倒流河,逆流河
~ed talon 反爪饰
~ed T beam 倒 T 形梁
~ed tide 倒潮,逆潮
~ed triangular truss 倒[置]三角桁架
~ed T section 倒 T 形截面
~ed U yoke 倒 U 形轭架
~ed valve 止回阀,逆止阀
~ed vault 倒拱券顶
~ed welding 仰焊接
~ed well 吸水井,回灌井,渗水井,补给井
~ grade 管道内底坡度
~ level 管道内底标高,(水)池底标高
~ range finder 倒像测距仪

invest [in'vest] *v.* 投资,使带有(性质等),授与,投入;(包)围,笼罩;购买; -investable, investible, *a.*
~ material 覆盖材料
~ed capital 投资资本
~ed firm 投资公司

investigate [in'vestigeit] *v.*, -gated, -gating;调查(研究),勘测,试验,审查;investigator, *n.*
~ by digging 钻探

investigation [investi'geiʃən] *n.* 调查,审查,勘测,试验,研究,调查报告,研究论文;-investigational, *a.*
~ area 勘察区,调查区
~ cost 勘察费用,调查研究费
~ hole 探孔
~ pit 探孔,探槽
~ shaft 勘探井
field ~ 野外试验,现场试验
full-scale ~ 原型试验研究

model ～ 模型(实验)研究
wind tunnel ～ 风洞试验,风洞中的研究
investment [in'vestmənt] *n.* 投资,投入资本,花费,授与,包埋料,围模法,围模料,被覆物
～ abroad 对外投资额
～ account 投资账
～ advisor 投资顾问
～ advisers act 投资顾问法
～ certificate 投资证书
～ climate 投资环境,投资气候
～ credit 投资信贷
～ market 投资市场
～ policy 投资政策
～ project 投资方案,投资项目
～ securities 投资性证券
～ tax credit 投资宽减税额
～ trust 投资信托
～ turn over 投资周转(率)
goverment ～ 国家投资
gross ～ 投资总额
investor [in'vestə] *n.* 投资者
inviscid [in'visid] *a.* 非粘性的,无韧性的,不能展延的,流动的
～ flow 非粘性流
～ fluid 非粘性流体
invitation [invi'teiʃən] *n.* 请柬,招待[券];邀请; *a.* 邀请的;
～ card 招待券
～ issuing date 招标日期
～ to bid 招标
～ to tender 招标
invite [*v.*in'vait, *n.*'invait] *n.* ; *v.*, -vited,-viting; 邀请, 招待; 吸引, 招致, 要求, 请求; inviter, *n.*
～ to tender 招标
invoice ['invɔis] *n.* ; *v.*, -voiced,-voicing; *n.* 发票,发货票;货物托运单; *v.* 开发票,列清单
～ amount 发票金额
～ book 发票簿

～ book inwards 进货簿
～ number 发票号
～ book outwards 出货簿
～ requisition 请购单
～ specification 发票明细单
involute ['invəlu:t] *n.* 渐开线,渐伸线,切展线,包旋式(内旋); *a.* 渐开的,渐伸的;纷乱的,错综的;内卷的; *v.* 消失,恢复原状,卷起
inwall [*v.*in'wɔ:l, *n.*'inwɔ:l] *n.* 内壁, *vt.* 围以墙
inward ['inwəd] *n.* 内部(物),里面,实质;(*pl.*)进口商品,进口税; *a.* 里面的,内部的,向内的,在内的;固有的;进口的,输入的; *ad.* =inwards;向内,在内,向中心
～ bound 向内行驶
～ cargo 进口货物
～ charge 入港费
～ collection 进口托收
～ -flow turbine 内流式透平机
～ freight 进口运费
～ freight and cartage 运入运费
～ manifest 进口货单
～ opening 向内开(窗)
～ osmosis 内向渗流
～ pilotage 进港引水费
～ pressure 内向压
～ remittance 汇入款项
～ seepage 向内渗漏
～ tipping 向内倾斜
in（ward）window 向内开的窗,里面的窗
inwash 水力冲填,冰川边缘沉积,巨原冲积层,岸边淤积
inwelling 海水倒灌
iochroite 电气石
iodate ['aiədeit] *n.* ; *v.*, -dated,-dating; *n.* 碘酸盐; *v.* 用碘处理,碘化; iodation, *n.*
iodation *n.* 碘化作用,加碘
iodide ['aiədid] *n.* =iodide;碘化物

iodinate

~ flux 碘熔剂
~ -process 碘化物法,碘化物热离解法
iodinate *v.* 碘化,碘处理
iodine [ˈaiədi:n] *n.* 碘,碘酊
~ consumed 碘消耗量(水质污染指标)
~ number [~ value] 碘值
sublimed ~ 升华碘
iodism [aiədizm] *n.* 碘中毒
ion [ˈaiən] *n.* 离子
~ exchange 离子交换
~ exchanger 离子交换器
~ exchange resin 离子交换树脂
~ exlusion 离子排斥
liquid ion-exchange 液体离子交换,溶剂萃取
ionic [aiˈɔnik] *a.* 爱奥尼亚(Ionia)(人)的;爱奥尼亚式的(其特征为柱头带涡卷形装饰);爱奥尼亚语的;*n.* 爱奥尼亚语
~ architectural order 爱奥尼亚建筑柱型
~ architecture 爱奥尼亚建筑
~ base 爱奥尼亚底座[基础]
~ colonnade 爱奥尼亚柱廊
~ column 爱奥尼亚柱
~ cyma (tium) reverse 爱奥尼亚里反曲线(状)
~ entablature 爱奥尼亚柱顶盘
~ order (of architecture) 爱奥尼亚(建筑)柱型
~ pilaster 爱奥尼亚式壁柱[半露柱]
~ portico 爱奥尼亚式门廊
~ style 爱奥尼亚格式[字体]

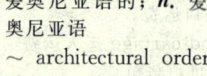

ionic order

ionic pilaster

~ temple 爱奥尼亚式寺院[神殿],爱奥尼亚式庙宇
~ volute (= Ionic scroll) 爱奥尼亚式盘涡
iozite *n.* 方铁矿
ipsojure [ˈipsəuˈdʒu:ri:] 根据法律本身
ipsonite *n.* 残沥青
Ir(**Iridium**) *n.* 铱
Iraurita(e) *n.* 铱金
iridescence [iriˈdesns] *n.* 虹彩,晕虹,虹色
iridescent [iriˈdesnt] *a.* 闪光的织物,服装或小配件;晕色的,虹色的,虹色的;*n.* 虹色的衣料或其他物质; iridescently, *ad.*
~ cloud 虹彩云
Iridium [aiˈridiəm] *n.* 铱(Ir)
Iridoplatinum *n.* 铱铂合金
Iridosmine [irəˈdɔzmin] *n.* 铱锇矿
iris [ˈaiəris] *n. pl.* irises, irides;膜片,隔板,档板;光圈,可变光阑[圈];虹彩,晕彩,虹状物;窗孔
~ diaphragm 可变光阑
~ mauve 彩虹紫色,灰黄粉红色
Irish [ˈaiərif] *n.* ; *a.* 爱尔兰,爱尔兰的
~ architecture 爱尔兰建筑
~ bridge 爱尔兰桥
~ confetti 一块岩石[砖,砖石碎片]
~ diamond 水晶
~ ivy 爱尔兰常春藤
iron [ˈaiən] *n.* 铁,烙铁,熨斗;铁器,铁制品,刑具; *a.* 铁(制)的,铁似的
~ aggregate 铁屑集料
~ aggregate concrete 铁屑(集料)混凝土
~ (air) trap 铁阻气
~ alum 铁矾,铁明矾
~ angle 角铁
~ -arc 铁弧拱
~ arch 铁拱
~ arch bridge 铁拱桥
~ arch(ed) girder) 铁拱主梁
~ architecture 铁建筑

~ article 铁制品
~ -back 壁炉背部的铁板,护铁(刨刀的)
~ ballast 铁碇,铁道碴
~ band 铁箍,扁铁条,带铁
~ bar 铁条,钢筋
~ bars 铁栅拦
~ -basalt 铁玄武岩
~ based material 铁基材料
~ beam 铁梁
~ -bearing 铁支承,铁轴承
~ black 铁黑,黑锑粉
~ (blastfurnace) slag (高炉)(炼铁)炉渣
~ (blastfurnace) slag aggregate (高炉)(炼铁)炉渣粒料[团粒]
~ (blastfurnace) slag chip(ping)s (高炉)(炼铁)炉渣板
~ (blastfurnace) slag concrete (高炉)(炼铁)炉渣混凝土
~ (blastfurnace) slag filler (炼铁)(高炉)炉渣填料
~ (blastfurnace) slag sand (炼铁)(高炉)炉碴砂
~ (blastfurnace) slag sand concrete (炼铁)(高炉)炉渣砂混凝土
~ block pavement 铁块路面
~ blue 铁青色
~ board 烫衣台,烫衣板
~ bolt 铁(螺)栓
~ -bound 包铁的
~ box switch 铁壳开关
~ brick 铁板料,铁砖
~ brick paving 铁砖铺砌
~ bridge 钢桥,铁桥
~ brown 铁褐色,铁深黄
~ buff 铁浅黄
~ cake 铁渣,含铁滤饼
~ carbon alloy 铁碳合金
~ carrier 铁件(自动)装卸机
~ cement 铁腻子,铁胶结料,含铁水泥
~ chain bridge 铁链桥

~ chill 冷铁
~ claw 铁爪,撬棍
~ clay 铁黏土
~ compound 铁化合物
~ (aggregate) concrete 铁屑混凝土
~ connector 铁制连接件
~ covering 铁盖
~ dog 狗头钉,两爪铁钩
~ dowel 铁暗壁,合缝钢条,外伸的短钢筋
~ dust 铁屑,铁粉
~ firedog (铁)柴架壁炉
~ for concared planes 凹刨刨刀
~ form (work) 铁模板(工程)
~ frame work 铁框[构]架
~ framing construction 铁框架结构
~ front 铸铁制的建筑物正面
~ gate 铁门
~ gauze 铁丝网
~ girder 钢梁
~ girder bridge 钢梁桥
~ glazing bar 上釉铁条,铁制嵌玻璃条
~ gray 铁灰色,灰白色
~ hat 钢盔,铁帽,安全帽
~ hoop 带钢
~ intecepting (air) trap 铁制阻截(空气)用存水弯
~ intercepter 铁隔断
~ lath(ing) 铁板条
~ leg 段铁
~ lintel 铁楣
~ mica 铁云母
~ mica paint 铁云母涂料
~ mold 锈痕,墨水迹
~ muntin(g) 铁窗格条,铁门中梃
~ ore cement 矿渣水泥,铁矿水泥
~ ore road surface 铁矿石路面
~ oxide 氧化铁
~ oxid(e) coating 氧化铁涂料
~ oxid(e) paint 氧化铁涂料
~ pavement 铸铁块(铺砌)路面

~ picture 铁画
~ pike 铁镐
~ pipe 铁管
~ pipe culvert 铁管涵洞
~ pile 铁桩,钢桩
~ plate 铁板,钢板,板钢,板铁
~ portland cement 铁硅酸盐水泥,矿渣硅酸盐水泥
~ powder cement 铁粉水泥(防水用)
~ product 铁制品
~ putty 含铁,油灰
~ pyrite 黄铁矿,二硫化铁
~ red (pigment) 铁红,红色氧化铁(颜料)
~ ridging 铁(屋)脊
~ rod 铁杆[棒]
~ roller 铁滚筒,铁碾
~ roof cladding 屋顶(铁)金属覆盖层
~ roof cover(ing) 屋顶铁覆盖层
~ rope bridge 钢索桥

iron rope bridge

~ rust 铁锈
~ scaffold 铁脚手架,金属脚手架
~ section (=setion ~) 型钢
~ sheet 铁片,铁皮,黑铁皮
~ (sheet) roof cover(ing) 铁皮屋顶覆盖层
~ (sheet) roof sheathing 铁皮屋顶夹衬板
~ sheetpile 铁制薄板桩,钢板桩
~ slag aggregate 铁矿炉渣粒料
~ slag concrete 炼铁炉渣混凝土
~ stairs 铁制楼梯
~ tic 铁系杆
~ trap 铁存水弯
~ weight 压铁
~ window bar 铁窗条
~ wood 铁木,坚硬的木料

~ wire 铁丝
~ wire sleeve 铁丝筛
~ wood 硬木,坚硬的木料
~ work 铁工,锤铁,锻铁或磨光;铁制零件,铁制品,钢铁厂
angle ~ 角钢,角铁
beak ~ 鸟嘴铁(丁字钻)
black ~ 黑铁皮;黑铁矿
black sheet ~ 黑钢皮
bloomery ~ 熟铁块
box ~ 槽钢,槽铁
break ~ 护铁(刨刀的)
building ~ 建筑钢
calking ~ 密缝錾
cast ~ 铸铁,铸造生铁
channel ~ 槽钢,槽铁
checkered ~ 网纹铁板
chromium ~s 铬铁合金
climbing ~ 金属踏步
construction ~ 建筑用铁
corner ~ 角钢,角铁
corrugated (sheet) ~ (= roofing ~) 瓦楞铁(片),瓦楞钢皮
corrugated sheet ~ 波纹铁皮
cramp ~ 两爪钉,钉钩,把钩
double ~ (= H-shaped ~) 工字铁,工字钢
dry ~ 低硅生铁
ductile cast ~ 球墨铸铁
dug ~ 熟铁
edge ~ 角铁
electric ~ 电烫头
embossing ~ 压花铁
equal angle ~ 等边角铁,等边角钢
fashioned ~ 型钢
figured ~ 型钢
flange angle ~ 翼角铁
flat ~ 扁钢
flat hoop ~ 平箍钢
floor ~ 底铁
galvanized ~ 白铁,镀锌钢皮,镀锌铁皮,白铁皮

grab ~ 打捞工具
graphitic pig ~ 灰口铁
groy cast ~ (greg cast ~) 灰口铁
grozing ~ 烙铁
haematite ~ 赤铁矿
heavy ~ 厚镀层热浸镀锌铁皮
hexagen bar ~ 六角钢
hoop ~ 箍钢,箍铁
horseshoe ~ 马蹄铁
knotted bar ~ 竹节钢
laminated sheet ~ 成层薄钢板
light-weight ~ 轻便式,熨烫斗
magnetic ~ 磁铁
malleable cast ~ 可锻铸铁,韧性铸铁
mangance ~ 锰铁
melleable cast ~ 可锻铸铁
mottled (pig) ~ 麻口(生)铁
nickel cast ~ 镍铸铁
nickel white ~ 镍白口铁
niresist cast ~ 不锈镍铸铁
pig ~ 生铁
pig ~ alloy 合金生铁
plain sheet ~ 无楞铁皮
plane ~ 刨刀,刨铁
profiled ~ (= section ~) 型钢
profiled sheet ~ 成型薄钢板
quadrant ~ (square bar ~) 方钢,方铁
rail ~ 轨钢
refined (cast) ~ 精炼生铁
ribbon ~ 带钢,带铁
rivet ~ 铆钉钢
rod ~ 圆钢,圆铁
rolled sheet ~ 轧制钢板,轧制铁皮
rustless ~ 不锈钢,不锈铁
scrap ~ 废铁[钢]
section ~ 型钢
sheet ~ 薄铁皮
silicon ~ 硅钢
sow ~ 沟铁(高炉)

profiled sheet iron

stainless ~ 不锈钢
stepping ~ 金属踏步
supporting angle ~ 支撑角铁
sweat ~ 焊铁
unequal angle ~ 不等边角钢
white (cast) ~ (white pig ~) 白口铁
wrought ~ 熟铁,锻铁
ironing board *n.* 熨烫板[台]
ironingroom *n.* 熨衣室,熨烫室
ironing-screed *n.* 熨烫样板
ironmonger [ˈaiənmʌŋgə] *n.* 小五金商,五金安装工
ironmongery [ˈaiənmʌŋgəri] *n.* 五金店,铁器,五金器具
 ~ materials 小五金材料
 ~ finishes 小五金饰面
iron-sandstone *n.* 铁砂岩
iron-stone [ˈaiənstəun] *n.* 铁矿,菱铁矿,含铁矿石
 ~ china 硬质陶器
 brown ~ 褐铁矿
 clay ~ 泥铁矿,泥铁岩
iron(-)ware [ˈaiənwɛə] *n.* 铁器,铁制物品,家用铁器,五金店
irradiation [ireidiˈeiʃən] *n.* 辐射,照射,光渗,光照,发光,(用紫外线)照射
irregular [iˈregjulə] *a.* 不规则的,不整齐的,非正规的,无规律的,不均匀的; *n.* 不合规格的物品; **irregularly**, *ad.*
 ~ bedding 不均匀基床
 ~ channel 不规则渠道
 ~ course 乱砌层,不规则层
 ~ coursed rubble 不成层乱砌毛石
 ~ French curve 不规则曲线规
 ~ (paving) sett 不整齐(铺砌的)小方石
 ~ profile 不规则纵断面

irregular rubble gable

~ rubble gable 不规则毛石山墙
~ sett 不均匀小方石
~ weir 不规则小堰
irregularity [iregju'læriti] ***n.*** 不符,不合常规,不均匀性,不一致性,不平衡性,奇异性,不合规定;不平整,凹凸不平
~ distributed load 不规则分布荷载
~ index (路面的)不平整指数
flow ~ 流动的不均匀性,气流不均匀性
operation ~ 工作事故
surface ~ 路面不平,表面不平度,表面上奇异性,表面上的奇点
irreversible [iri'və:səbl] ***a.*** 不可逆的,不能翻转的,不能倒置的,不可改变的; irreversibility, ***n.***, irreversibleness, ***n.***; irreversibly, ***ad.***
~ creep 不可逆转的徐变,不可逆转的[塑性变形]
~ deformation 不可逆变形
~ displacement 不可恢复的位移
~ operation 不可逆运算
~ shrinkage 不可逆收缩
irrigable ['irigəbl] ***a.*** 可灌溉的; irrigably, ***ad.***
~ area 可灌溉区,可灌溉面积
~ land 可灌溉地
irrigate ['irigeit] ***v.***, -gated, -gating; 灌溉,浇,冲洗 ~d 已灌溉的
~d area 灌溉面积
~d farms 灌溉农田
~d land 灌溉地,水田
~d region 灌溉区域
~ing head 灌溉水头
~ing net 灌溉网
~ing stream 灌溉水流
~ing water 灌溉用水
~ water quota 灌水定额
irrigation [iri'geiʃən] ***n.*** 灌溉,冲洗,灌注; irrigational, ***a.***
~ area 灌溉面积
~ by gravity 自流灌溉

~ by pumps 泵灌
~ canal 灌溉渠
~ channel 灌溉渠
~ dam 灌溉用坝
~ demand 灌溉需水量
~ design 灌溉设计
~ district 灌溉区
~ ditch 灌溉明沟,灌溉支渠
~ draft 灌溉取水
~ efficiency 灌溉效率,水的有效利用系数
~ engineering 灌溉工程
~ equipment 灌溉设备
~ field 灌溉场,灌溉地
~ furrow 灌溉水沟
~ hydrant 灌溉水栓
~ installation 灌溉设备
~ intensity 灌溉强度,灌溉面积(与总面积之)比
~ lateral (canal) 灌溉支渠
~ line 灌溉线
~ main 灌溉干管,灌溉干渠
~ method 灌溉方法
~ mouth 灌水口
~ network 灌溉网
~ period 灌水一次的时间(作物耗水高峰阶段)
~ pipe 灌溉水管
~ pool 灌溉水池
~ practice 灌溉制度,灌溉措施
~ project 灌溉工程,灌溉计划
~ pump 灌水泵
~ pumping 灌溉抽水[扬水]
~ pumping station 灌溉泵站
~ rate 灌溉[用水]率,灌水率
~ requirement 灌溉需水量,灌溉定额
~ rotation 轮灌,灌溉轮换
~ scheme 灌溉方案[流程图]
~ season 灌溉期,农田用水季节
~ storage 灌溉库容
~ structure 灌溉建筑物

~ system 灌溉系统
~ system tailwater recovery 尾水回收灌溉系统
~ technique 灌溉水技术
~ trench 灌溉沟渠
~ tunnel 灌溉隧洞,灌溉隧道
~ water 灌溉(用)水
~ water management 灌溉用水管理
~ water quality 灌溉水质
~ -water requirements 灌溉需水量
~ weir 灌溉用堰
~ well 灌溉水井。

irrigationist *n.* 灌溉用户,水利专家,提倡灌溉者

irrigator ['irigeitə] *n.* 灌溉设备,冲洗器,灌喷机

irrotational [irəu'teiʃənəl] *a.* 无漩涡的
~ deformation 无漩形变
~ field 无旋场
~ flow 无旋流,无涡流
~ motion 非旋(转)运动,无涡运动
~ wave 无旋波,非旋转波

irrotationality *n.* 无旋性,无涡性,无漩涡现象,有势性

irruption [i'rʌpʃən] *n.* 侵袭,闯入
~ vein 侵入(岩)脉
IR-system ~ 红外系统
mepping ~ 红外测绘系统
ranging ~ 红外测距系统

isacoustic lines 同强震声线

isalea *n.* 等日射[量]线

isallotherm [ais'æləθə:m] *n.* (等)变温线

isanemone *n.* 等风速线

isarithm *n.* 等值线

isba [iz'ba] *n.* 俄国式木屋

iscorex 艾斯科勒克(装饰用金银线)

isentropic [aisen'trɔpik] *n.*; *a.* 等熵线,等熵的
~ analysis 等熵面分析
~ change 等熵变化,绝热变化
~ chart 等熵线图
~ flow 等熵流
~ graph 等熵图

I-shaped *n.* 工字形的

I-shaped cross-section 工字形截面

isinglass ['aiziŋglɑːs] *n.* 鱼胶,云母,透明薄云母片,白明胶

iskmeter *n.* 现场土壤剪切仪,现场剪切触探仪

Islam ['izlɑːm] *n.* 伊斯兰教,Islamic *a.* 伊斯兰的
~ archs 伊斯兰教拱

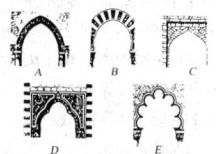

Islam archs

island ['ailənd] *n.* 岛,岛屿;路岛,安全岛,船上的甲板,舰台,航空母舰右舷的甲板室;*vt.* 使变成岛置于岛上;孤立; -islandish, islandlike, *a.*; -islandless, *a.*
~ breakwater 岛式防波堤,独立式防波堤
~ chain 岛链,列岛
~ of resistance 孤立支撑点
~ platform 岛式站台
~ platform roof 岛式平台屋顶
~ station roof 单柱屋顶
~ type wharf 岛式码头
floating ~ 浮岛,浮站,水上飞机场
loading ~ 装peds, 候车岛

isle [ail] *n.* 岛(尤指小岛);*vt.* 使成为岛屿,置于岛屿上; isleless, *a.*
~s 群岛,列岛

islet ['ailit] *n.* 小岛,小岛状物,岛状地带,孤岛; isleted, *a.*

isobar ['aisəubɑː] *n.* 等压线,同量异位素,同量异序元素,异序素,同质异位素,核同质异位素; isobarism, *n.*

isobath ['aisəubɑːθ] *n.* 等(水)深线

isocatabase *n.* 等降线
isochronal [ai'sɔkrənəl] *a.* 等时的；**isochronally**, *ad.*
　～ map 等时线图
isochronograph *n.* 等时图
isoclimate line 气候等值线，等气候线
isoclimatic zone 同气候带，气候相同地带
isoclinal [aisə'klainl] *n.*；*a.* 等倾[斜]的，等(磷)倾线
　～ fold 等斜褶皱
　～ valley 同斜屋谷，等斜谷
isocline *n.* 等倾线，等斜线，等向线
isoclinic [aisə'klinik] *n.*；*a.* = isoclinal；等(磁)倾线，等倾的
　～ line (主应力)等倾线
isodmon = isodomun 整块(石)端砌，整块(石)面砌
isodomun 整块(石)端砌

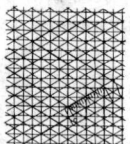
isodomun

isodyne 等力线
isogram(s) ['aisəgræm(z)] *n.* 等值线图
isohyet [aisə'haiət] *n.* (= isohyetal, isohyetal line)等雨量线
　～ grandient 等雨量线梯度
isohyetal *a.* 等雨量的
　～ line 等雨量线，等沉陷线
　～ map 等雨量线图
　～ method 等雨量线法(求流域平均雨深)
isohyetose 等雨量线
isohygromen 等湿度线
isohygrometric line 等湿度线
isohypse *n.* (= contour) 等高线
　～ -line 等高线(水准面)，水平线
isolate ['aisəleit] *v.*, -lated,-lating；隔离,孤立,断开,使绝缘,使隔振；*a.* 孤立的,隔离的;-isolator, *n.*
　～d (bath) tub 独立浴盆
　～d beam 独立梁
　～d bell-tower 孤立钟楼
　～d block 独立区段
　～d building 独立建筑物
　～d chapel 独立小教堂
　～d chimney 孤立烟囱
　～d column 独立柱
　～d footing (= ～ foundation) 独立底脚
　～d foundation 防振基础,隔振基础,独立基础
　～ hall 独立大厦
　～ house 独立住宅[建筑物]
　～ masonry wall 独立坊工墙
　～ pier 独立桥墩
isolation [aisə'leiʃən] *n.* 隔离,孤立,单独,绝缘,隔音
　～ booth 隔声间
isolator ['aisəleitə] *n.* 绝缘子,绝热体,隔离器
isoline ['aisəlain] *n.* = isogram；等斜褶皱,等值线,等深线
isomenal *n.* 月平均等值线(特指气温的)
isometric [aisə'metrik] *n.* 等容线；*a.* 等量的,等角的,等距的,等轴的,等容的,等体积的,同质异能的
　～ diagram 等轴测图
　～ drawing 等角图,等轴测图
　～ line 等值线,等容线
　～ projection 等角投影,等距投影
　～ sketch 等轴测图
　～ view 等轴测图

isometric drawing

isometrography *n.* 等角线规
isonif 等雪量线
isopectic (line) 等冰冻线,冰冻等时线
isopiestic(s) [aisəupai'estik(s)] *n.* 等压线；*a.* 等压的
isopluvial *a.* 等雨量线
isopotential *n.* (= equipotential line)等位[势]线
　～ level 等压面

isosceles [ai'sɔsiliːz] *a.* 二等边的,等腰的
isoseismal *n.* ;*a.* 等振线,等振线的
ISO sieves 国际标准(试验)筛
isostatic [aisə'stætik] *a.* (地壳)均衡的,等压的; isostatically, *ad.*
　～ line 等静力线
　～ structure 等压结构
isotherm ['aisəuθəːm] *n.* 等温线,恒温线
isotope ['aisəutəup] *n.* 同位素
　～ tracer 同位素示踪剂
　～ tracer technique 同位素示踪技术
isotropic [aisəu'trɔpik] *n.* ;*a.* 各向同性(的),无向(的),均质(的)
　～ body 各向同性体
　～ linear elasticity 各向同性线性弹性
　～ mass (= ～ body) 各向同性体,均质体
　～ material 各向同性材料,均质材料
　～ soil 均质土壤
　～ tracer 同位素示踪剂,示踪原子
isovelocity *n.* 等速线
isovent *n.* 等风速线
issue ['isjuː] *n.* ;*v.* , -sued, -suing;*n.* 河口,出口;结果,结局;发行;期号(报刊);问题,论点;流出;*v.* 流出,放出;发行,颁发,出版; issuable, *a.* ; issuably, *ad.* ; issuer, *n.*
　～ house 证券发行公司
I-strut *n.* 工字铁支撑
Italian [i'tæljən] *n.* ;*a.* 意大利人(的),意大利语(的),意大利的
　～ architecture 意大利式建筑
　～ asbestos 意大利石棉
　～ gothic 意大利哥特式(建筑),意大利尖拱式建筑
　～ modern style 意大利现代式
　～ mosaic 意大利镶嵌花砖
　～ ochre 意大利赭色,意大利黄褐色
　～ Renaissance 文艺复兴式(建筑)

Italian Renaissance

　～ Romanesque 意大利罗马式(建筑)
　～ roof tile 意大利屋瓦(同时使用平、弧两种瓦)
italsi *n.* 意大利硅铝合金
item ['aitəm] *n.* 项目,科目,条目,条款;节,段; *ad.* 又,同样地; *vt.* 记下,备忘
　～ advance 项目转移
　～ design 项目设计
　～ forecast 单项预测
　～ received in advance 预收项目,预收款
　spanned ～ 跨项,跨组,共存项,共轭项
itemize ['aitemaiz] *vt.* -ized, -izing;分条,分项,列举,列单; itemization, *n.* ; itemizer, *n.*
　～d schedule 项目一览表
iteration [itə'reiʃən] *n.* 反复,重复;迭代(法),逐步逼近法
　～ method 迭代法
　～ process 迭代过程
iterative ['itərətiv] *a.* 反复的,迭代的
　～ design 迭代设计
I-terminals Ⅰ信号输出端
　～ structure 迭合结构
itinerary [ai'tinərəri] *n. pl.* -aries; *a.* ; *n.* 旅行日志,旅行计划,旅行路线
　～ map 旅行路线图
It-plate 石棉橡胶板
ivory ['aivəri] *n. pl.* -ries, *a.* 象牙(制品);乳白色,象牙色;厚光纸; *a.* 象牙色的,象牙制的; ivorylike, *a.*
　～ board 象牙白纸板
　～ glazed coat(ing) 象牙色表层,象牙釉表层,镶牙表层
　～ glazo(d finish) 象牙色表层
　～ sculpture 象牙雕刻品
　～ shell 象牙螺
　～ white 象牙白,乳白(釉色)
ivy ['aivi] *n. pl.* -vies;常春藤;学院的,学究式的,纯理论的,抽象的

~ -leaf 常春藤叶
Ivy League 大学的学生或校友
Izod 悬臂式(的)
~ impact machine 悬臂梁式冲击试验机
~ notch V 型缺口
~ test 悬臂梁式试验

J

jack [dʒæk] *n*. 千斤顶,起重器;插座[孔];支撑,支柱,弹簧开关;传动装置; *vt*. (用千斤顶)顶起,张拉;抬高
~ arch 平拱,单砖拱,等厚拱
~ arm 起重平衡臂
~ base 插孔板
~ bit 钻头
~ bolt 起重螺栓
~ boom (挖掘机)辅助支架
~ box 配电箱
~ -carrying frame 千斤顶载架
~ chain 8 字形链,起重链
~ cylinder 千斤顶油缸
~ down (用千斤顶)降下
~ engine 辅助引擎,小型发动机
~ fastener 插口线夹
~ frame 绞车架,千斤顶架
~ hammer 风镐,凿岩机[锤]
~ hammer drill 撞钻,冲钻
~ -in method 顶入法
~ -knife door 折叠(刀)式门
~ ladder 索梯
~ lagging 筑拱模板用支承板,砌筑壳体用的模板
~ lamp 安全灯
~ lever 顶重杠杆
~ loading method 千斤顶加载法
~ pad (汽车起重机两侧支承下的)承压盘
~ pair 双千斤顶,千斤顶对
~ panel 接线板,转插板
~ pile 压入桩
~ pine 黑松,短叶松
~ plane 粗[大]刨,台车

~ post 轴柱,(机轴的)撑杆
~ rafter 支[顶]木,椽子
~ rib 短肋,短椽
~ ring 支撑环
~ rod 滑[导]杆
~ rod sleeve 滑杆套

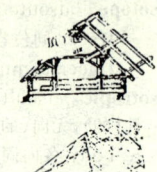

jack rafter

~ roll 手动绞车
~ screw 起重螺旋
~ shaft 传动[中间]轴,副轴
~ stringer 小纵梁,衍条
~ switch 插接开关
~ timber 支[顶]木,椽子
~ truss 次桁架,[四面坡屋顶]半桁架,托架
~ up (用千斤顶)顶起
~ -up partition 顶升隔墙
~ up the price 抬高价格
~ vault 平拱顶,单砖穹顶,等厚穹窿
~ with handle 手摇千斤顶
~ wood 细纹木
air ~ 气力起重器
anchor ~ 锚定柱
differential-screw ~ 差动千斤顶
glassing ~ 磨光机
hydraulic ~ 液压千斤顶
lagging ~ 拱架

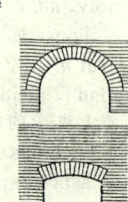

jack vault

lifting ~ 千斤顶,起重器
mud ~ 泥浆泵
pipe ~ 管子拉紧装置
pneumatic ~ 气动起重器
pulling ~ 手动葫芦
retraction ~ 升降机,收放机构
riveting ~ 铆接托架
screw ~ 螺旋千斤顶
sliding (screw) ~ 滑座螺旋起重器
track ~ 起轨器
jackbit ['dʒækbit] *n.* 钻头
~ insert 刃具,刃口
jackdrill *n.* 凿岩机
jacket ['dʒækit] *n.* 套,盖,罩,盒,外壳；*vt.* 给……装上套,用外壳遮盖；jacketed, *a.*; jacketless, *a.*; jacketlike, *a.*
~ cylinder 有套汽缸
~ pipe 套管
~ space 护套空间
~ valve 套层阀
boiler ~ 锅炉外套
column ~ 柱管,外柱
concrete ~ 混凝土壳
cooling air ~ 气冷式套筒
cooling water ~ 水冷式套筒
cylinder ~ 汽缸套
life ~ 救生衣
protecting ~ 保护罩
splash ~ 水套
steam ~ 汽套,蒸汽加热套
valve ~ 阀盒
jacketing ['dʒækitiŋ] *n.* 外套；套式冷却；套式加温
jacking *n.* (预应力混凝土)张拉；(用千斤顶)起重,顶进,压入,压紧；顶管法
~ accessible cableconnections 张拉斜缆接合
~ anchorage 张拉锚碇
~ base 顶推基座
~ beam 顶梁,反作用梁
~ block 千斤顶垫块
~ device 张拉装置,顶升装置,顶进设备
~ dice 千斤顶垫块
~ force 顶推力,张拉力
~ frame 顶推托[构]架
~ -jetting process 压入冲洗法
~ of foundation 基础抬升
~ plate 千斤顶垫板,(压桩时置于桩顶传递压力的)垫板
~ pocket 千斤顶座孔(放置千斤顶的预留孔)
~ stress 顶拉应力
~ strut 顶进支杆,顶柱
jackleg ['dʒækleg] *n.* 风动钻架
jacklift *n.* 起重托架
jackstay *n.* 撑杆；分隔索,支索
Jacob ['dʒeikəb] *n.* 木踏板绳梯,索梯
Jacobean style (17世纪英国的)雅各宾式(建筑),英王詹姆斯一世时代的哥特式建筑风格
Jacobian [dʒə'kəubjən] *n.* 函数[雅各比]行列式
~ matrix 雅各比矩阵
jade [dʒeid] *n.* 玉,翡翠；*a.* 玉制的,绿玉色的；jadelike, *a.*
Jade Belt Bridge 颐和园玉带桥
jag [dʒæg] *n.*；*v.*, jagged, jagging；*n.* 锯齿状缺口,*v.* 字形凹口；*v.* 使成锯齿状；jagger；jagless, *a.*
~ bolt 地脚螺丝,棘螺丝
jagged ['dʒægid] *a.* 锯齿状的,有缺口的,参差不齐的；jaggedly, *ad.*; jaggedness, *n.*
~ bit 齿形钻头
~ casing 齿缘套管
~ core bit 凿孔空心钻头,齿形钻头
~ edge 不平坦边缘
~ rocks 巉岩
~ terrain 多裂缝地带
jail [dʒeil] *n.* 监狱；*vt.* 下狱；jailless, *a.*; jaillike, *a.*
Jain architecture *n.* 耆那式建筑(耆那教是起于印度的宗教)
jalousie ['ʒæluːziː] *n.* (固定)百叶窗,遮

窗;-jalousied, a.
jam [dʒæm] v., jammed, jamming, n.;v. 压[挤]紧,卡住,堵塞;干扰,失真;n. 障碍,拥挤
~ density 堵塞密度
~ nut 锁紧[止动]螺母
~ weld 对头[挤压]焊,塞焊
card ~ 卡片阻塞
traffic ~ 交通阻塞
jamb [dʒæm] n. 门窗边框,门窗梃,门档;壁炉侧墙
~ block 门窗边框圆角砌块
~ brick 门窗边框圆角砖
~ brush (刷漆前用)掸灰刷
~ casing 门窗边框厢
~ duster 扫尘刷
~ extension 侧柱盖板
~ guard 侧柱护铁
~ liner 窗边框筒子板
~ moulding 门窗边框线条
~ post 门窗边框柱
~ shaft 门柱
~ stone pillar 门窗边框石柱
~ wall 门窗边框墙
door ~ 门侧壁,炉门侧柱
post ~ 柱壁
splayed ~s 八字侧墙
squared ~ 方角侧墙
window ~ 窗侧墙

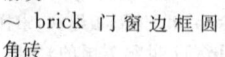
jambs

jami n. 伊斯兰大教堂
jammer ['dʒæmə] n. 干扰发射台;支柱[座]
janitor ['dʒænitə] n. 房屋管理员,看门人;
janitorial, a.; janitorship, n.
~ closet 房屋管理室
~ room 传达室
japan [dʒə'pæn] n.;vt. japanned, japanning; n. 亮漆,假漆;漆器;日本瓷

器;vt. 涂以假漆;japanner, n.
~ black 黑(亮)漆
~ wax 漆蜡
~ ning 涂[上]漆
~ ning oven 涂漆用炉
Japanese [ˌdʒæpə'niːz] a.;n., pl. Japanese; a. 日本的,日本人的;n. 日本人,日语
~ architecture 日本式建筑
~ bell tower 日本钟楼
~ flowering cherry 樱花
~ lacquer 深黑漆
~ roof truss 日本式屋架
~ temple 日本神殿

Japanese temple

~ tung oil 日本桐油
~ wood oil 日本桐油
Japonic style n. 日本式(建筑)
jar [dʒɑː] n.;v., jarred, jarring; n. (大口)瓶,罐,缸,容器,罩;电瓶,电瓶外壳;震动,冲击;噪[杂]音;v. 使振动;震惊,刺激;冲击打钻;发出噪声 jarringly, ad.; jarless, a
~ -proof 防震的
~ ram moulding machine 振动制模机
~ red loose ground 震松地
~ ring 振[抖]动;震声;不调和
~ ring effect 振动效应
~ ring machine 振动机
~ ring motion 振动
accumulator ~ 蓄电瓶
battery ~ 蓄电池容器,电池槽
bell ~ 钟,钟罩;钟形烧结炉
filtering ~ 滤缸
hydrometer ~ 比重计瓶
syntonic~s 谐振瓶
jardiniere [ˌʒɑːdiː'njɛə] n. (装饰用)花盆,花架
jarrah n. (铁路枕木用)赤桉木
jaspe n. 一种有细纹装饰的亚麻油毡,铺

地用漆布
jatex *n.* 浓缩橡浆
jaw [dʒɔ:] *n.* 破碎机颚板;爪;钳子,虎钳;销键;滑块;凸轮;jawless, *a.*
~ brake 爪闸
~ breaker 颚式轧碎[碎石]机
~ clutch 颚式离合器
~ coupling 爪盘联轴节
~ crusher 颚式轧碎机
~ nut 锁紧螺帽
~ of pile 桩靴
~ of spanner 扳手钳口
~ opening (破碎机入口)颚板间距
~ plate 破碎机颚板
~ setting (破碎机出口)颚板间距
~ vice 虎钳,夹具
alligator ~ 鳄鱼牙式桩靴
centering ~ 中心凸轮
grab ~ 抓铲
gripping ~ 夹爪
roller ~ 滚压颚板
safety ~ 安全吊
universal joint ~ 万向节爪
vice ~ 钳口
jawab *n.* 对称房屋,配称建筑
jedding axe *n.* (采石场用)鹤嘴斧
jeep [dʒi:p] *n.* 吉普车,小型越野车;一种小型侦察联络飞机;小型航空母舰
jell [dʒel] *n.* 凝胶,冻胶
jellification [dʒelifi'keiʃən] *n.* 冻结,凝结,胶凝作用
jelly ['dʒeli] *n. pl.* jellies, *v.*, jellied, jellying; *n.* 胶体,胶状物,(透明)冻胶; *v.* 使成胶状,凝结;-jellylike, *a.*
~ consistency 冻胶稠度
~ mould 胶模
lubricating ~ 凝胶润滑剂
mineral ~ 矿物胶冻
petroleum ~ 矿脂,凡士林
jemmy ['dʒemi] *n. pl.* jemmies, *v.*, jemmied, jemmying; *n.* 短铁棍, *vt.*

撬开
Jena planetarium ['jeinəplæni'tɛəriəm] *n.* 耶命天象馆(1922年设计的第一个薄壳混凝土圆屋顶)
jenny ['dʒeni] *n. pl.* jennies; *n.* 移动起重机,卷扬机,划线规,纺纱机
~ scaffold 活动脚手架
~ wheel 单滑轮起重机
jerk [dʒə:k] *n., v.* 急摔[拉,撞,扭];突然跳动,猛一冲撞;加速度变化率;-jerker, *n.*; -jerkingly, *ad.*
~ line (一种钻机)启动绳
~ pump 高压燃油喷射泵
jerkinhead 山墙尖呈斜坡的(屋顶)
~ roof 两坡式屋顶
jerry ['dʒeri] *n. pl.* jerries; *n.* 碳质页岩; *a.* 偷工减料的,工程草率的
~ building 偷工减料的建筑工程
jerrycan *n.* 五加仑装的汽油罐
Jerusalem pine 耶路撒冷松
Jesse window *n.* 教堂用窗(树杆状窗格,用彩色玻璃嵌成圣象)
Jesuitical style *n.* (拉丁美洲的)耶稣教会式(建筑)
jet [dʒet] *n.*; *v.*, jetted, jetting; *n.* 射[水]流,喷气,喷射;喷嘴,喷射器,喷气发动机,套管[筒],煤玉,黑玉; *v.* 喷[射]出;用喷气机运送; *a.* 喷气式推进的; jettingly, *ad.*
~ and fuel resisting joint sealing compound 抗喷气和燃油的(路面)填缝料
~ area 射流截面,喷嘴面积
~ bit 喷射钻头
~ black 炭黑
~ blower 喷气鼓风机
~ boundary 射流边界
~ carburetor 喷雾式汽化器
~ centrifugal-pump 喷射式离心泵

~ coal 长焰煤
~ condenser 喷水冷凝器
~ condenser pump 喷水冷凝泵
~ cutter method 喷射挖掘法(冲水打井,用泵吸泥)
~ cutting 水力开挖,射流切割
~ deep-well pump 喷射式深井泵
~ deflector 射流转向器,挑流装置
~ diffuser 射流扩散器
~ disperser 射流消能[扩散]装置
~ dredge 射流式挖泥船
~ drilling rig 喷焰钻孔机
~ elevator 水力提升器
~ engine 喷气发动机
~ exhauster 喷射抽气机[真空泵]
~ fan 射流风机
~ flow gate 射流式闸门
~ flame process 热焰喷射法
~ generator 喷注式(超声波)发生器
~ grouting 加固土体的旋喷注浆法
~ head 喷头
~ height 喷水高度,射流高度
~ hole 喷水孔
~ humidifier 喷射加湿器
~ lifter method 提升喷射法(用电动泵将钻井中泥砂喷出)
~ mill 喷射碾磨机,风动冲压粉碎机
~ noise 喷射气流噪声
~ nozzle 喷嘴,尾喷管
~ passage 喷气通道
~ piercing method 热焰喷射钻孔法(开石方)
~ pipe 喷管
~ pipe temperature 喷管温度
~ pressure 喷射压力
~ priming 射流起动
~ probing 射水杆(土壤)探测
~ process 水冲(成粒)法,水淬法
~ pulverizer 喷射碾磨机
~ pump 喷射泵
~ ratio 射流比
~ regulator 喷射调节器
~ resistant pavement 抗喷气路面
~ shale 煤玉页岩
~ steam 射流;喷气引擎的废气
~ trajectory length 射流挑出长度
~ valve 喷射阀
~ washer 喷射洗涤机
~ wave 射流波
~ ted 用水冲的;用黑玉装饰的
~ ted pile 射水沉桩,水冲桩
~ ted screw pile 射水螺旋桩
adjustable ~ 可调节喷管
annular ~ 环状喷射流
axial flow ~ 轴流式喷气发动机
carburetter ~ 汽化器喷口
cement throwing ~ 水泥喷枪
fishtail ~ 鱼尾式喷嘴
free overflow ~ 自由降落水舌
gasturbine ~ 燃气涡轮喷气发动机
power ~ 动力喷口
sand ~ 喷砂器

jetburner *n.* 喷灯;喷射口
jetcrete *n.* 喷枪喷射水泥浆,喷浆
jetter *n.* 喷洗器,喷洗装置
jettied construction *n.* 悬挑式建筑
jetting *n.* 喷射,喷注;土方的水力冲填,土方工程水力机械化;水力沉桩
~ angle 抛离角
~ bit 高压射流钻头
~ cutter rod 喷枪杆
~ device 射水装置
~ distance (喷射运输机)抛射距离
~ drill 热焰钻孔装置
~ drilling 高压射流钻进
~ fill 水力冲填
~ gear 水枪;高压射流冲矿机,高压射流挖掘机
~ hose 喷注水管
~ lance 喷枪
~ method (井点的)射水沉没法
~ nozzle 高压喷嘴
~ orifice (井点的)射水口

~ out 挑出,射出
~ piling 水冲打桩法
~ pipe 射水水管
~ process 水冲法,喷射打井法
~ pump 射水泵
~ rod 喷枪杆
~ sled (汽车式)高压射流挖掘机
~ tool 喷砂工具
~ type well-point 喷射井点
~ water 喷射水

jettison [dʒetisn] *n.*; *vt.* (紧急情况下)抛弃货物等,投掷,放出;jettisonable, *a.*
~ device 弹射装置
~ gear 投弃装置,放油装置
fuel ~ 燃料放出

jetty ['dʒeti] *n. pl.* jetties, *v.*, jettied, jettying; *n.* 突码头,突防波堤,导堤,建筑物突出部分; *vt.* 伸出,突出; *a.* 煤玉似的,乌黑发亮的; jettiness, *n.*
~ bent pile 码头排架桩
~ dock 堤式船坞
~ harbour 突堤港
~ head 堤头
~ wharf 突堤码头
floating ~ 浮码头
projecting ~ 丁坝

jewel ['dʒu:əl] *n.*; *v.*, jewel(l)ed, jewel(l)ing; *n.* 珠宝,宝石;(钟表的)宝石轴承; *vt.* 镶以宝石;jewellike, *a.*
~ bearing 宝石轴承
~ block 球滑车
~ post 钻石柱
V ~ V 型支承面宝石轴承

jewel(l)ery = jewelry *n.* 珍宝,宝石,镶嵌有宝石的饰物
~ alloy 装饰用合金

Jewish ['dʒu:iʃ] *a.* 犹太的,犹太人的;Jewishly, *ad.*; Jewishness, *n.*
~ architecture 犹太式建筑
~ bema 犹太教堂中的讲坛

Jew's pitch 沥青,死海沥青
jewstone *n.* 白铁矿
jib [dʒib] *n.*; *v.*, jibbed, jibbing; *n.* 起重臂,吊杆,突梁,准,扁栓;夹具;绞辘;(截煤机)截盘; *v.* 改变方向
~ and cotter 合楔
~ angle indicator 起重臂角度指示器
~ arm 起重机机臂
~ barrow 支梁式手推车
~ boom 辅助延伸臂,顶部附加臂,飞臂
~ brake 起重臂制动闸
~ cable 起重臂缆
~ component 起重臂部件
~ crab 吊杆车
~ crane 旋臂起重机
~ door 与墙齐平的隐门
~ drum 吊臂索滚筒
~ extension 辅助伸臂,顶部附加臂
~ extension ram 吊臂外伸作动筒
~ foot spool 起重臂脚鼓
~ head 吊臂上端
~ height 吊臂高度
~ hoist 吊臂卷扬机
~ hoist limit switch 吊臂提升止限开关
~ inclination 吊臂倾斜度
~ inclination angle 吊臂角
~ length 吊臂长度
~ lift cable 起重臂提升缆
~ lift valve 吊臂提升阀
~ loader 旋臂装料机
~ -mounted drill 转臂式钻机
~ pin 吊臂脚销
~ point 吊臂顶端
~ point pin 吊臂顶销
~ point sheave 吊臂顶滑轮
~ section 吊臂接长段
~ shipper (shaft) 吊臂移动轴
~ side sheave 吊臂侧缆滑轮
~ swing angle 吊臂旋转角
~ tooth 截盘齿

~ winch 吊臂卷扬机
adjustable ~ 可调梃杆
chain ~ 链式截盘
conveyer ~ 运输机壁
cutter ~ 截煤机截盘
double lever ~ 象鼻架伸臂
drill ~ 斜杆钻架
elevator ~ 升降机臂
fly ~ 飞臂(起重臂顶端铰接短臂)
hammerhead ~ 平衡锤臂
lattice ~ 格构吊杆
luffing ~ 鹅头伸臂(伸臂可垂直升降保持重物水平移动)
overcutting ~ 上截盘
roofing ~ 截顶槽截盘
undercutting ~ 下截盘
underslung ~ 悬吊截盘
jig [dʒig] *n.*；*v.*，jigged，jigging；*n.* 夹具，夹紧装置；模具，钻模；样板，规尺；筛，筛选机，淘汰机；衰减波群；*v.* 簸动，筛选，分类；用夹具加工；jiglike，jigish，*a.*
~ borer 坐标镗床
~ bush 钻套
~ drill 钻模钻床
~ grinder 坐标磨床
~ key 钻模键
~ plate 夹具[钻模]板
~ point 基点
~ saw 细锯，线锯
~ washer 淘汰洗矿机
air ~ 风力淘汰机
assembling ~ 装配夹具
basket ~ 动筛式淘汰机
box type ~ 箱式钻模
buddle ~ 定筛式淘汰机
channel type ~ 槽式钻模
finishing ~ 精选淘汰机
mounting ~ 安装夹具；装配架
rotary ~ 旋转式固定架
standard ~ 标准样板
testing ~ 试验架[台]
welding ~ 焊接夹具

jigger [ˈdʒigə] *n.* 滑车，小绞车；淘汰机，筛矿机，振动器；(制陶器的)盘车；减幅振荡变压器，耦合器
~ bars (路面)搓板带
~ conveyor 振动输送机
~ saw 往复式线锯
hand ~ 手工磨光器
pendulum ~ 摆锤打光机

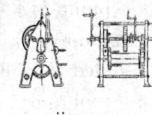
jigger

jiggering *n.* (制陶器)盘车拉坯；(石料加工)镟削外径

jigging *n.* 筛；振动；簸选
~ block (机械加工用)样板
~ conveyer 振动输送机
~ grate 簸动炉箅
~ motion 颠簸运动
~ platform 振动台
~ screen 振动筛
~ sieve 振动筛
hand ~ 人力淘汰
hydraulic ~ 水力淘选
machine ~ 机械淘选
pneumatic ~ 风力淘选

jiggle [ˈdʒigl] *v.*，jiggled，jiggling，*n.*；*v.*；*n.* 轻摇，摇动，摆动；jiggly，*ad.*
~ bar 摇手柄，摇杆
~d frame 摇动架

jim crow *n.* 弯轨器，轨条挠曲器

jimmy [ˈdʒimi] *n. pl.* jimmies，*v.*，jimmied，jimmying；*n.* 短撬棍；运煤车，料车；*vt.* 撬开

jink [dʒiŋk] *n.* 矿车连接器，车钩

jinny *n.* 自重滑行坡；固定绞车

jinrik(i)sha [dʒinˈrikʃə] *n.* 人力车，黄包车

Jinshan Temple 镇江金山寺

jitter [ˈdʒitə] *n.*，*v.* 振动，跳动；图像跳动，信号不稳；速度误差，偏差；散开，(扫描点错误移动的)图像失真
angular ~ 角度不稳定性

frequency ~ 频率波动
time ~ 时间起伏
jitty *n.* 联络小巷道,短通道
job [dʒɔb] *n.*; *v.*, jobbed, jobbing; *n.* 施工,工程,作业;职务;加工件,零部件;工地;事件;成果; *v.* 做临工,打杂;加工;承包;雇用
~ breakdown 工作细目分类
~ captain 工地负责人
~ change analysis 工作变更分析
~ change request 工作变更要求[申请]
~ class 作业类别
~ cleanup 建筑工地清理
~ -cured concrete 现场养护混凝土,工地养护混凝土
~ curing 现场养护
~ evaluation 工作评价
~ experience training 工作经验训练
~ foreman 工作队长
~ history card 记事卡片
~ initiation 作业起始
~ instruction sheet 工作说明单
~ laboratory 工地实验室
~ location 施工现场
~ made 现场制作
~ management 作业管理
~ mechanic 工地钳工
~ mix 现场拌合
~ mix design 现场配合比设计
~ -mixed paint 现场调制涂料
~ mixer 现场搅拌机
~ -mixed paint 现场调制涂料
~ number 工号
~ operation 加工方法
~ order 工作通知单,工作程序
~ out 分包出去
~ placed concrete 现浇混凝土
~ plan 施工进度表

job mixer

~ planning 施工计划
~ plaster 现场拌制灰泥
~ -poured concrete 现场浇筑混凝土
~ practice 施工方法
~ processing 作业处理,作业加工
~ production 分批生产
~ program 工作[加工]程序
~ rates 生产定额
~ report 施工报告
~ schedule 工程进度
~ sequence 加工序列[程序]
~ site 工地,施工现场
~ specification 施工[工作]规范
~ -splitting 全日工改为两个半日工
~ -stored 工地存储的
~ stretching 现场预应力张拉
~ step 工作步骤
~ task analysis 工作任务分析
~ tensioning 现场预应力张拉
~ throughput 作业处理能力[吞吐量]
~ ticket 工作通知单
~ time report 工时报表
~ -to- hauling 工地搬迁,施工现场变更
~ training standards 工作训练标准
~ -work 包工,散工,临时工
~ (work) shop 现场加工车间
casing cementing ~ 套管注水泥施工
odd ~s 零碎工作,散工
part-time ~ 零星工作
piece work ~ 计件工作
shaft ~ 开凿井筒
stacked ~ 叠式工作;汇总工作
jobber [dʒɔbə] *n.* 临时工;批发商;股票经纪人
jobbing *n.* 做临工;重复次数很少的工作
~ casting 商业铸造
~ foundry 中心铸造车间
~ mill 中小型钢轧机
~ sheet 中[轻]型板
~ shop 修理车间

~ work 临工,散工
jockey ['dʒɔki] *n.*; *pl.* -eys, *v.*, -eyed, -eying; *n.* 膜片,薄膜,导轮,驾驶员,操作者; *v.* 操作,驾驶,移动; jockeylike, jockeyish, *a.*; jockeyism, jockeyship, *n.*
~ pulley 导轮,(皮带的)支持轮,张紧辊
~ weight 活动砝码,调节衡重
jog [dʒɔg] *v.*, jogged, jogging; *n.* 轻推[摇]粗糙面,墙的凹凸面;拼[啃]合;精密进料,缓慢走刀; *v.* 轻推[摇];提醒,唤起;接[拼]合;-jogger, *n.*
jogging *n.* 轻摇,微动,平稳移动;瞬时断续接电;缓慢加压
joggle ['dʒɔgl] *v.* -gled, -gling *v.*; *n.* 轻摇[推],榫接,啃合;折曲,偏斜;joggler, *n.*
~ beam 拼接梁,榫接梁
~ joint 啃合接头,榫接
~ piece 啃合件
~ post 啃合木柱
~ truss 拼接桁架
~ work 拼接工作,啃合接榫作业

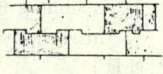
joggle joint

joggled *a.* 啃合的,榫接的
~ beam 拼接[镶合]梁
~ butt joint 啃合对接
~ frame 错折肋骨(造船)
~ joint 啃合接,榫接
~ lap joint 啃合搭接
~ lintel 啃合小过梁
~ single lap joint 啃合单面搭接
~ stone 企口石块
joggling *n.* 抖动;卷边
~ machine 折曲机
~ table 振动台
~ test 折曲试验
johnboat ['dʒɔnbəut] *n.* 驳船,平底船
Johnson ['dʒɔnsn] *n.* 约翰逊(人名)
~ bronze 轴承青铜
~ noise voltage 热噪声电压

~ powermeter 微分功率表
~ regulator 差动式调压箱[井,池]
~ valve 针型阀门,高落差水轮机阀
join [dʒɔin] *v.*; *n.* 联合,接合,连接,接缝;参加,加入;-joinable, *a.*
~ monolithically 无缝连接
~ on skew 斜向接
~ operation 联合运算
~ up 连接起来,咬合
joiner ['dʒɔinə] *n.* 接合物;细木工;刨床;装配工
~'s belt 木工工作台
~'s bench 木工工作台
~'s cross-cut saw bench 细木横割锯台
~'s glue 木工胶,水胶
~'s hammer 木工锤
~'s kit 细木工工具箱
~'s shop 细木工场
general ~ 万能木材加工机
wood ~ 木工接合器

joiner's bench

joinery ['dʒɔinəri] *n.* 细木工作,细木工,细木工制品
~ component 细木工件
~ member 细木工件
~ timber 细木工用材
~ unit 细木工件
~ wood 细木工用材
joining *n.* 接合,连接[拼]接
~ balk 联系木梁
~ beam 联系梁,联结杆
~ by mortise and tenon 榫眼接合
~ by rabbets 企口接合,槽舌接合
~ nipple 接合螺管
~ on butt 对头接
~ shop 拼接工场
~ up 连接,接线
~ with keypiece 键连接
~ with passing tenon 穿榫接头

~ with peg-shoulder 直榫接头
~ with swelled tenon 扩榫接头
~ yard 拼接场

joining-up *n.* 连接,咬合
~ differentially 差接
~ in parallel 并接
~ in series 串接

joint [dʒɔint] *n.*；*v.* 接合,连接；接合点,接缝[头]；连轴节,铰接头；关节；节点；节理；*a.* 连接的,共同的
~ acceptance plan 联合接收计划
~ action 接头作用
~ aging time 粘合期
~ agreement 共同契约
~ allowance 接缝允许间隙
~ applicator 填缝机
~ architect 联合建筑
~ area 接缝面积
~ assembly 接缝组装件
~ audit 联合审计
~ bar 鱼尾板；接头销钉
~ block 接头块
~ bolt 接合[插销]螺栓
~ box 电缆接线箱；连接套筒
~ cap 密封盖
~ capital 联合资本
~ cement (管道)填缝料
~ chair 钢轨接座,接轨垫板
~ charge 联合运费
~ cleaning 清缝
~ close 合缝
~ closure 路面填缝
~ compound 填缝混合料,密封剂
~ construction 合作施工；接缝制作
~ cost 联合费用
~ coupling 万向接头；电缆接头套管
~ cover(ing) 接缝搭盖
~ cross (万向接头的)十字头
~ current 总电流
~ cutter 切缝机
~ demand 共同需求

joint cutter

~ depth 接缝深度
~ design 接缝[节点]设计
~ dimension 接缝尺寸
~ displacement 节点位移
~ disposal 联合排放(污水)
~ distribution 联合分布
~ dowel 定缝销
~ efforts 共同努力
~ element (有限元分析中的)连接单元
~ entropy 相关平均信息量,相关熵
~ equilibrium 节点平衡
~ estate 共有财产
~ expense 共同营业费
~ face 接缝面；节理面
~ failure (深钻技术中的)脱缝
~ fastener 接合片
~ faulting 错缝,断缝
~ filler 膨胀缝填料,嵌缝料
~ fillet 填缝板,嵌缝条
~ filling 填缝,嵌缝
~ filling compound 填缝料
~ fixing 接缝固定
~ flange 接合法兰
~ floor(ing) 拼缝地板
~ free 无缝的
~ gap 接缝间隙
~ gauge 测缝仪
~ grid 接缝网
~ grinding (花岗岩和沥青)联同辗磨
~ grouting work 接缝灌浆
~ hinge 接合铰链
~ holding moment 节点钳固弯矩
~ imperfection 接缝缺陷
~ income 共同收益
~ installation plan 联合安装计划
~ installing machine (混凝土路面)接缝安装机
~ iron 接缝铁
~ leakage 接缝渗漏
~ load 节点荷载
~ly operated 合营
~ masking tape 接缝覆盖带

~ measurement 测缝
~ mechanism 节点原理
~ meter 测缝计
~ mobility 节点位移
~ moment 节点弯矩
~ mortar 接缝灰浆
~ movement 节点[接缝]移动
~ observation 联合观测
~ occupancy date 联合占用日期
~ of framework 框架节点
~ of rupture 断裂缝
~ on square 直角接合
~ opening 缝口,缝隙
~ operating procedure 联合操作程序
~ packing 接合填密,填充垫圈
~ parts 接合配件
~ pattern 节理型
~ permeability 接缝渗透性
~ pin 连接销
~ pipe 连接管,接合管
~ placing machine 接缝机
~ plane 节理面,接缝面
~ plate 接缝[合]板
~ plug 接缝堵塞物
~ pointing 勾缝,嵌缝
~ pouring 接缝灌浆
~ pressing-in method 压缝法
~ probability 联合概率
~ profile 接缝断面
~ quotas 共同配额
~ raker 清缝耙
~ random variable 联合随机变量
~ reinforcement 接缝钢筋
~ resealing 重新封缝
~ resistance 合成电阻
~ rigidity 节点刚度
~ rosette 节理玫瑰图
~ rotation angle 节点转动角
~ rule 接榫量规

joint runner

~ runner 填缝索,索状(铅)充填料
~ sawing 接缝磨光
~ seal 接缝止水,封缝
~ sealant 接缝止水剂
~ sealer 填缝料[机]
~ sealing compound 填缝混合料
~ section 合页铰页片
~ sheet 接缝垫片
~ shield 接缝(防漏)盖面
~ sleeper 接缝垫枕
~ spacer 接缝隔片
~ spacing 接缝间距
~ spalling 缝口碎裂,缝口低陷
~ spider (万向接头的)十字叉
~ spring 节理泉
~ statement 联合声明
~ state-private enterprise 公私合营企业
~ -stock company 股份公司
~ stool 折叠椅子
~ strength 接缝[节点]强度
~ strap 带状接头
~ strip 压缝条
~ system 节理系统
~ surface 接缝[节理]面
~ tenancy 共有不动产权
~ tie 钢轨接头轨枕;拼接板
~ tongue 榫舌
~ trajectory 节点轨迹
~ translation 节点位移
~ undertaking 共同事业
~ venture (firm) 联合企业[投资],合资企业
~ with dovetail groove 燕尾榫接合
~ zone 接缝区
abutment ~ 平接缝,对接头
air-cushion ~ 气垫缝
articulated ~ 活络接头
ball ~ 球节点
beams ~ 梁的[连]接合
bolted ~ 螺栓接合
brazed ~ 铜焊接头

break ～ 间缝,错缝接合;间砌法
brickwork ～ 砖缝
build ～ 构造缝
butt ～ 对接,平缝
caulked ～ 嵌缝
clip ～ 砖砌厚缝
closed ～ 密缝
dilation ～ 膨胀缝
dovetailed ～ 燕尾榫接
dowel ～ 暗榫接合
dowelled tongue and groove ～ 有榫钉的企口缝
expansion ～ 伸缩缝
fillet and groove ～ 企口接合
fillistered ～ 凹槽接
finger ～ 马牙榫(木工)
fish plate ～ 鱼尾板接合
flush ～ 平灰缝
grouted ～ 浆缝
header ～ 砖砌合
jump ～ 对(头)接合
knuckle ～ 铰接,肘(形)接
longitudinal ～ (砌体)横缝
match ～ 舌槽接,企口接
metal cavity ～ 埋铁接缝
mortar (bed) ～ 灰缝
mortise ～ 榫缝
off-set ～ 错缝
open ～ 凹缝,开口缝
overlap ～ 搭接
panel ～ 格接
perpend ～ (砖砌的)竖缝
pin ～ 枢接合
ploughed and tongued ～ 企口接合
pointed ～ 勾缝
preformed ～ 预制缝
pumping ～ 汲泥接缝
push ～ 挤接(砖工)
rabbet ～ 嵌接
raked ～ 剔缝
recessed ～ 方槽凹缝
ribbed ～ 肋接

rivet ～ 铆(钉接)合
riveted butt ～ 对接[头]铆接
rusticated ～ 粗糙缝,显著缝
sawed ～ 锯成缝(混凝土路面)
scarf ～ 嵌接,斜口接合
seal ～ 密封接头[焊缝]
shove ～ in brickwork 挤浆砌砖法
spade type ～ (水泥路面)铲式接缝
spigot ～ 插管接头,套管接合
splayed ～ V形缝
spline ～ 填实缝
staggered ～ (砌体)错缝
T ～ T型接合[头]
table ～ 嵌接
tongue and groove ～ 企口[凸凹]接头
transversal ～ 直[竖]缝
twist ～ 绞[扭]接
union ～ 管子接头
universal ～ 万向节
voussoir ～ 拱块缝
welded ～ 焊缝

voussoir joint

jointer ['dʒɔintə] *n.* 管子工;连接器,接缝器;(泥工用)涂缝镘
～ plane 木工长刨
saw ～ 连锯器

jointing *n.* 接合榫[缝];连接;填缝;垫片,封泥;砌体勾缝;节理
～ arrangement 接缝排列布置
～ cast units by prestressing 用预应力结合的预制件
～ clamp 接线夹
～ compound 密封剂,接缝混合料
～ element 连接件
～ material 接合密封材料,填料
～ of pipes 管封
～ paste 灌缝浆
～ piece 连接件
～ plane 修边刨
～ press 板材压合机
～ refractory cement 耐火填缝灰浆

~ rivet 连接铆钉
~ rule 接榫规
~ tool 接榫工具,接缝器
~ washer 接头垫圈
asbestos ~ 石棉填料
hemp ~ 麻填料
metal(lic) ~ 金属填料
jointless ['dʒɔintlis] *a.* 无缝的,无关节的
~ construction 无缝(路面)构造
~ floor 无缝地板
~ structure 无缝结构
jointly ['dʒɔintli] *ad.* 联合地,共同地
~ ergodic random process 联合遍压随机过程
~ stationary random process 联合平稳随机过程
joints *n.* (筛)网线的交织
jointy *a.* 成层的,多接缝的
joist [dʒɔist] *n.* 小梁,托梁,工字钢梁,搁栅,桁条; *vt.* 给……架搁栅,给……安托梁; joistless, *a.*
~s and planks 搁栅与厚板
~ bearing 托梁支座
~ ceiling 搁栅平顶
~ floor 搁栅楼板
~ grillage 搁栅排格
~ hanger 搁栅锚件 [栅吊钩]
~ head 搁栅枕
~ nail 搁栅钉
~ shears 型钢剪切机
~ steel 工字钢,梁钢
attic ~ 屋顶层搁栅
binding ~ 联结搁栅
boarding ~ 裸搁栅
bridging ~ 加斜撑的搁栅
common ~ 地板托梁
exposed ~ 明搁栅
ground ~ 地搁栅
landing ~ (楼梯)平台搁栅
open-web ~ 空腹搁栅
plate girder ~ 工字板梁

joist hangers

strip(ped) ~ 板条搁栅
tail ~ 挑出的搁栅尾端
track ~ 托轨梁
trim(mer) ~ 托梁,承接梁
trimming ~ 整饰横梁,过梁,楼梯搁栅
Jolly *n.* 耐火砖成型机
~ balance (以 Jolly 命名的)测密实度天秤,乔利秤
jolt [dʒəult] *v.*; *n.* 振动,摇动,颠簸; 震惊; jolter, *n.*; joitingly, *ad.*; joltless, *a.*
~ capacity 振击能力
~ moulding 振动成型
~ packing 振动填料
~ ramming 振捣,振动夯击
~ squeeze moulding machine 振压造型机
~ knock-out grid 振击落砂架
~ vibrator 振动式振捣器
~ing 振动,振实
~ing table 振动台
jolter *n.* 振实制型机,振筑器
Joosten process *n.* 周斯坦氏化学加固土壤法
joule [dʒuːl] (简 J)焦耳(能量单位,等于 10^7 尔格)
~ effect 焦耳效应
~ equivalent 焦耳当量
journal ['dʒəːnl] *n.* 轴颈,枢轴; 杂志; journalary, journalish, *a.*
~ bearing 轴颈轴承
~ box 轴颈箱
~ bronze 轴颈轴承青铜
~ packing 轴颈填料
~ rest 轴颈支承
~ with collar 有环轴颈
block ~ 止推轴颈
crank ~ 曲轴轴颈
pointed ~ 锥形轴颈
spherical ~ 球轴颈
thrust ~ 止推轴颈

vertical ~ 枢轴颈
Journal of the Construction Division
(美国土木工程师学会)建筑杂志
journey ['dʒə:ni] *n. pl.* -neys, *v.*,
-neyed,-neying; *n.*; *v.* 旅行,旅程;
流动,移动; journeyer, *n.*
~ man 计日工,雇工,散工,满师的熟练工
~ speed 行程速度
~ time 行程时间
~ -work 临时工,短工
joy [dʒɔi] *n.*; *v.* 快乐,喜悦
~ stick 操纵[杆]手柄
jube *n.* 屏障；教堂中殿与圣台之间的隔墙(通常装饰有雕刻品等)
jubilee ['dʒu:bili] *n.* 纪念喜庆,庆祝

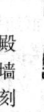
jube

~ skip 小型轻轨翻斗车
~ track system 窄轨铁路系统
~ truck (一种车身狭而深的侧卸式)小型货车,轻轨料车
~ wagon 小型倾卸车
judas ['dʒu:dəs] *n.* 监视孔
~ window 监[窥]视窗[孔]
judder ['dʒʌdə] *n.* 强振动声,颤振；位移
judge [dʒʌdʒ] *n.*; *v.*, judged,judging; *n.*; *v.* 判断,鉴定,评价; judgeable, judgeless, judgelike, *a.*; judger, *n.*; judgingly, *ad.*
judg(e)ment ['dʒʌdʒmənt] *n.* 判断力,评定[价],鉴定；意见,看法,见解；judgmental, *a.*
~ time (行车)判断时间
judicial [dʒu(:)'diʃəl] *a.* 司法的
~ authority 司法机关
~ decision 司法裁决
~ powers 司法权
~ practice 司法程序[惯例]
~ precedent 司法制例
~ settlement 司法解决
jug [dʒʌg] *n.*; *v.*, jugged, jugging; *n.* 有柄水壶；气缸; *v.* 装入壶中；挤成一群
~ ged 齿形的,锯齿形
~ handle 壶柄；(蒸馏塔中引出的)操作用管线
juggler *n.* 斜柱
jumbo ['dʒʌmbəu] *n. pl.* -bos, *a.* ; *n.* 庞然大物；运木滑板；隧道盾构；钻车；隧洞运碴车; *a.* 巨大的
~ base 大号(管)基
~ boom 重型吊杆
~ brick 大型砖
~ loader 凿石装车联合机
~ windmill 巨型风力发动机
air-motor driven ~ 风动钻车
hgdro-boom ~ 液压钻架钻车
long-feed drill ~ 深进钻车
roof-bolting ~ 锚杆孔钻车
sinking ~ 凿井用钻车
tunnel ~ 巷道掘进钻车
jumbolter *n.* 杆柱[锚杆]钻车
jumbos *n.* 橛
jump [dʒʌmp] *v.*; *n.* 跳跃,跳动；突变,跃迁；水跃；阶差,一跳的距离；(指令)转移；矿脉的断层；薄板叠轧时的折皱; jumpable, *a.*; jumpingly, *ad.*
~ counter 跳进式脉冲计数器
~ coupling 跳合联轴节
~ dam 跳越坝
~ distance 水跃距离
~ -down 下落断层
~ drilling 冲击钻孔,撞钻
~ function 跃变函数
~ grading 间断级配
~ height 水跃高度
~ in brightness 亮度落差[跃变]
~ joint 对(头)接
~ order 跳越指令
~ operation 转移操作
~ saw 升降圆截锯

~ seat 活动[折叠]坐位
~ spark 跳跃火花机
~ the rails 出轨
~ weld 平头焊接
energy ~ 能量跃迁
hydraulic ~ 水跃
phase ~ 相位突变
pressure ~ 压力剧变
jumper [ˈdʒʌmpə] *n.* 搭接片,跨接线；长钻,穿孔凿；桩锤；冲击钻杆；鹤嘴
~ bar 撞钻,冲钻
~ bit 冲击锤[钻头]
~ boring bar 冲钻杆
~ cable 跨接电缆
~ hammer 撞击锤,冲孔器
~ stay 制动支架
~ tube 跨接管
~ wire 跨接线
flat ~ 扁凿
flexible ~ 挠性接合器
percussion ~ 冲击式钎子
saw ~ 锯齿器

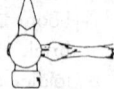

jumper hammer

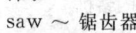

flat jumper

jumpforming *n.* 升模施工
jumping *n.*; *a.* 突变现象；跳动的
~ drill 手钻,跳动钻
~ form 升[滑]模
~ frog 蛙式打夯机
~ matrix 跳跃矩阵
~ shuttering 升模施工
~ vault 跃式拱顶(一种具有西里西亚地方特色的建筑形式)
junction [ˈdʒʌŋkʃən] *n.* 接合,连接；接合处,接点,交会点；道路交叉口,枢纽,河流合流处；中继线；junctional, *a.*
~ arm 交叉臂
~ block 接线块；(排水管道的)连接段
~ box 接[分]线盒
~ canal 连接运河
~ chamber 连接室,汇流井
~ closure 接点闭合
~ gallery 连接廊
~ gusset (plate) 接点板
~ manhole 管路交汇处人孔,检查井
~ marker 路口指示标
~ of the star type (道路)星形枢纽
~ piece without saddle 无法兰盘接合件
~ plate 接合[接点]板
~ point 接合点
~ pole 分线柱
~ port 转口港
~ roundabouts 道路枢纽环道
~ signal 枢纽信号
~ station 枢纽[联轨]站
butt ~ 对接接头
fly-over ~ 立体交叉
tumbling bay ~ 陡坡排水管与检查井的连接
junctor *n.* 联络线；连接机
juncture [ˈdʒʌŋktʃə] *n.* 接合[连接]点,接缝；时机,关键
jungle [ˈdʒʌŋgl] *n.* 丛林,密林；稠密居住区；-jungled, *a.*
junior [ˈdʒuːnjə] *a.* 次的,低级的,初级的；*n.* 年幼者
~ beam 次梁
~ channel 小型槽钢
~ engineer 初级工程师
~ machine 新式机床
juniper [ˈdʒuːnipə] *n.* 桧木；黄绿色
junk [dʒʌŋk] *n.* 碎片；废料,小块废铁；填缝用绳屑；舢板船；大(尺寸)材
~ -bottle 深绿色厚玻璃瓶
~ pipe 非标准化管道
~ ring 填料盒压环,压盖,衬圈
~ value 废品值
~ed tire 废旧轮胎
Jupiter [ˈdʒuːpitə] *n.* 木星,孤光灯
Jurassic [dʒuəˈræsik] *n.*; *a.* 侏罗纪(的)
~ limestone 侏罗纪石灰岩
~ period 侏罗纪
~ reef 侏罗礁

～ sandstone 侏罗纪砂岩
～ type of folding 侏罗式褶皱
jurupaite *n.* 硅钙镁石
jury ['dʒuəri] *n. pl.* -ries；*n.* 应急[临时]构件；*a.* 应急[临时]的；jury-less, *a.*
～ mast 临时桅杆
～ pump 备用泵
～ repairs 应急修理
～ rig 应急索具
～ rudder 应急舵
～ strut 应急支柱
just [dʒʌst] *a.* 正确的，公正的
～ compensation 正确[公正]补偿
～ size 正确尺寸
jut [dʒʌt] *v.*, jutted, jutting, *n.*；悬臂，突出部；*vi.* 突[伸]出；juttingly, *ad.*
～ window 突出窗
jute [dʒu:t] *n.* 黄麻；jutelike, *a.*
～ backing 麻靠垫，麻衬里
～ bag 麻袋
～ board (止水用)麻片
～ burlap 麻布，粗麻帆布
～ burlap mat 麻布垫
～ canvas 棉麻帆布
～ factory 麻加工厂
～ fiber 黄麻纤维
～ hessian drag finish (混凝土路面的)麻布刮平抹光
～ insertion 麻衬垫
～ lamination 麻叠层片
～ packing 黄麻填缝
～ rope 麻绳
～ wrap(ping) 麻包装皮
jutter ['dʒʌtə] *n.*；*v.* 突[伸]出，摇动
jutting-off-pier 悬臂桥墩
jutty ['dʒʌti] *n. pl.* -ties, *v.*, -tied, -tying；*n.* 建筑物上层突出部分；突码头；*v.* 突出
juvenile ['dʒu:vinail] *a.* 岩浆源的，童期的；juvenilely, *ad.*；juvenileness, *n.*
～ motion 岩浆活动
～ relief 幼年地形
～ water 岩浆水，初生水
～ wood 幼龄木材
juxtaposition [ˌdʒʌkstəpə'ziʃən] *n.* 斜接；并置，并列；毗连；交叉重叠法，juxtapositional, *a.*

K

K = cellophane 赛璐粉，玻璃纸；= constant 常数；= (Kelvin degree(°K)) 开氏绝对温度；= kilo 千；= knot 海里，节
～ -bracing K 形撑架，K 形连接杆
～ -crossing K 形路口
～ -display K 形显示，位移距离显示
～ factor 热通量，增殖系数
～ -scope K 形位移显示器
～-s softening point 克一沙氏软化点
～ -support 德国一种 K 形门柱
" ～ "test 平板承载[压力]试验
～ -truss K 形桁架
～ -type anchorage K 式锚

K-truss
～ -value K 值(表示路基或基层承载力的指标)，粘度值，增殖系数
Kainozoic era = Cenozoic era 新生代
Kaiser ['kaizə] *n.* (k-)皇帝；凯撒
～ dome (凯撒屋面板股份有限公司生产的)半球形圆屋顶
～ floor 上述公司生产的楼板
Kalamein door 金属包门，铁皮包门
kalamein fire door 金属包皮防火门
kalamein sheathing 金属包皮

kalasa 印度式建筑的瓶形顶部装饰
kalema *n.* 海滨激浪
kalk *n.* 石灰
kalsomine = calcimine *n.* 刷墙粉
kalzium metal 铝钙合金
kamash alloy 锡基合金
kame [keim] 冰碛阜,冰砾阜(呈小山状的成层的冰川砂及砾石沉积)
　～ terrace 冰碛阶地
kampometer *n.* 视场测量仪,热辐射计
kamptulicon [kæmp'tju:likən] *n.* 橡皮地毯
kanat 暗渠,地下暗渠;坎儿井;帷幕墙(印度)
Kansas Triaxial Method (美国)堪萨斯州三轴试验法(设计柔性路面厚度的方法)
kaolin(e) ['keiəlin] *n.* 高岭土,瓷土; kaolinic, *a.*
　～ clay 高岭黏土
　～ porcelain 瓷器
　～ powder 高岭土粉
　～ sand 高岭砂
kaolinisation *n.* 高岭土化作用(使含陶土的矿物变为纯高岭土)
kaolinite ['keiəlinait] *n.* 高岭石,纯粹高岭土
　～ flakes 纯高岭石薄片
　～ group 高岭石类
kapnometer 烟密度计
kapok ['keipɔk] *n.* 木棉
kapur = kapor 龙脑香樟树
kar *n.* 冰坑,凹地
Karaoke hall 卡拉OK厅
karat ['kærət] *n.* 开(纯金含量单位,纯金为24开);克拉(宝石的重量单位,等于0.2克)
karez = kariz (巴基斯坦)坎儿井,灌溉暗渠
Karman 卡门
　～ street 卡门涡列
　～ vortex flow 卡门涡流
　～ vortex trail 卡门涡迹

karri 卡里桉树,考里木
karst [ka:st]【德】*n.* 喀斯特,岩溶,石灰岩溶洞
　～ base level 岩溶[喀斯特]基准
　～ basin 喀斯特盆地
　～ erosion 岩溶[喀斯特]侵蚀
　～ funnel 喀斯特漏斗
　～ lake 岩溶湖
　～ landscape 喀斯特景观
　～ pit 岩溶[喀斯特]井
　～ source 喀斯特泉
　～ spring 喀斯特泉
　～ topography 喀斯特地形
　～ water 喀斯特水(含碳酸盐高),岩溶水
karstenite ['ka:stənait] *n.* 硬石膏
karstification 喀斯特作用
kata factor 降幂因数
katabatic wind 下降气流
katabolism [kə'tæbəlizm] *n.* = catabolism 分解代谢,分解作用; katabolic, *a.* ; katabolically, *ad.*
kataklastic rock 碎裂岩
katamorphic *a.* 破碎裂变质的
　～ zone 碎裂变质带
katamorphism *n.* 碎裂变质现象
katathermometer [kætəθə'mɔmitə] *n.* 冷却温度计,低温温度计
kathode ['kæθəud] *n.* = cathode 阴极
katogene *n.* 破坏作用
kauri ['kauəri] *n. pl.* -ris; *n.* 新西兰产的一种高的松树,贝壳松脂
　～ resin 贝壳松树脂
kay = cay *n.* 小礁岛,礁沙丘
kedge [kedʒ] *n.* 小锚,牵引锚
keel [ki:l] *n.* 龙骨,船脊骨;平底船; *v.* 装龙骨;将船翻转; keelless, *a.*
　～ -and-bilge block 垫船木块
　～ arch 波斯[桃尖,龙骨]拱
　～ block 龙骨墩木

keel arch

~ block tier 龙骨垫木台
~ cooler 龙骨冷却机
~ laid 开工,动工造船
~ line 龙骨线
~ molding 盘木线脚
ballast ~ 压载龙骨(潜艇)
bilge ~ 船底龙骨
docking ~ 坞龙骨

keel molding

keelson [kelsn] *n.* 内龙骨;底材
center ~ 中心内龙骨
engine ~ 承受发动机的内龙骨

keen [ki:n] *a.* 锋利[尖锐]的;强烈的; **keenly**, *ad.* **keenness**, *n.*
~ alloy 铜基合金

Keene's cement 金氏水泥,干固白水泥,D级硬膏泥

keep [ki:p] *v.*, kept, keeping, *n.*; *v.* 保持,维持,持续;保存,贮备;经营,管理;照料;制止,抑制;遵守,履行;拿着; *n.* 保持,保条;管理;(下承轴)盒底;支持零件;城寨
~ a contract 遵守合同
~ and bailey castle 带主塔楼城堡
~ -like tower 城堡主楼形塔
" ~ off median" sign 中央分隔带禁入标志
~ relay 保护继电器
~ right sign 右行标志
~ tower 城堡主塔
~ upright 切勿倒置
~ one's words 遵守诺言

keep-like tower

keeper ['ki:pə] *n.* 保管员,记录员;保持器,定位件,夹头,卡箍;锁紧螺母,定位螺钉;门栓,带扣;永磁铁衔铁;吸气剂;竖向导板; **keeperless**, *a.*; **keepership**, *n.*
~ hook 扣牢门窗用的S形钩子
~ of magnet 保磁用衔铁
dust ~ 防尘装置
elevator wrist pin ~ 升降机肘节销定位螺钉

oil ~ 油承
time ~ 计时机构,测时计,记时员

keeping ['ki:piŋ] *n.* 保存,看守,遵守; 供养;一致,协调
~ dry 保持干燥
~ gauge 砌墙高度规准尺,砌墙用标杆
~ room 起居室
~ (spare) parts 备用件
~ the gauge 按层砌筑

keeps 罐[笼]座,罐托

keeve [ki:v] *n.* 大桶,缸,盆

keg [keg] *n.* 小桶(1/2keeve)
~ buoy 木桶浮标
~ float 浮桶,桶浮标

Keller machine 自动机械雕刻机

kelly ['keli] *n. pl.* -lies,-lys; *n.* (旋转钻孔)钻杆,卡杆;凯氏方钻杆

Kelly ball test 凯氏球体贯入试验(测混凝土稠度)

Kelly bar 凯氏方钻杆
~ bushing 钻杆衬套[套管]
~ drive 钻杆驱动

Kelvin ['kelvin] *n.* 绝对温度,热力学温度单位(开,K),(开尔文)
~ absolute scale 开氏绝对温度
~ and Hughes echo sounder 开尔文—休斯式回声测探仪
~ degree 开氏[绝对]温度
~ model 开尔文模型
~ scale of temperature 开氏绝对温标
~ solid 开尔文固体
~ sounder 开尔文测探仪
~'s visco-elastic stress-strain relation 开尔文粘—弹性应力—应变关系
~ tide gauge 开尔文验潮仪
~ wave 开尔文波

Kemidol *n.* 凯米多尔石灰,细石灰粉

Kennedy Airport

Kennedy Airport 肯尼迪航空楼
Kennedy Key 方形切向键
Kennedy's cirtical velocity 肯尼迪临界速度(明渠中不淤不冲速度)
kennel ['kenl] *n.*；*v.*，-nel(l)ed,-nel(l)ing；*n.* 沟渠,阴沟；下水道；*v.* 钻进
　～ coal 长焰煤
kenotron ['kenətrɔn] *n.* 高压整流二极管
kent [kɛnt] *n.* 制[绘]图纸；长撑竿
kentite *n.* 肯太炸药；铵硝、钾硝、三硝甲苯炸药
kentledge ['kentledʒ] *n.* 压重物,压载铁,平衡重
Kentucky-rock asphalt 肯塔基岩与地沥青混合料(一种特制的薄层防滑面层)
kep *v.*，kepped, keppen or kippen, kepping，*n.*，门扣,拉手,窗钩；夹子；*vt.* 挡住,抓住
　cage ～ 罐[笼]座
keps 罐笼座,罐托
keramics [ki'ræmiks] *n.* (*pl.*)＝ceramics 陶器
keramzite concrete 苏联的一种膨胀土骨料混凝土
kerargyrite *n.* ＝cerargyrite 角银矿
keratin ['kerətin] *n.* 角质素
keratinize ['kerətinaiz] *v.*，-ized,-izing；角质化；keratinization，*n.*
Keratophyre *n.* 角斑岩
kerb [kə:b] *n.* ＝curb 路边石,路缘,路缘石；井栏石
　～ alignment 路边线
　～ below ground 埋入式路缘,深路缘
　～ brick 路缘砖
　～ grab(ing) equipment 路缘开挖机
　～ inlet 路边窨井(下水道的路边进口)
　～ lane 外侧车道,路边线
　～ machine 路缘压路机
　～ mould 路缘模板
　～ parking 路边停车
　～ press 路缘压路机
　～ shoe (沉井带刃口的)边脚,沉井刃脚
　～ed footway 带路缘石之人行道
　kerbside loading 靠人行道上车或装货,路边装卸
kerbstone ['kə:bstəun] *n.* 边石,栏石,路缘石
　～ market 交易所外的有价证券市场,场外交易
kerf [kə:f] *n.* 截口,切口,劈痕,割缝,切断沟；*v.* 截断,切开,锯断；采,掘
　～ed beam 有截口的梁
　bottom ～ 底槽(采矿)
　thin ～ 窄截槽(采矿)
　top ～ 顶槽(采矿)
kerites *n.* 煤油沥青
kermesinus *n.* 暗红色
kermesite *n.* 黄锑矿
kern [kə:n] *n.* 核心,核部；截面核心；颗粒；型芯撑
　～ area (截面)核心面积
　～ concrete 柱芯混凝土
　～ counter 计核器
　～ cross-section 柱芯(混凝土)断面
　～ distance 核心距
　～ limit 截面核心周界
　～ line 截面核心线
Kern method 克恩涂料粘附性试验法
　～ point 截面核心点(截面主轴上的截面核心周界点)
　～ point moment 截面核心点矩
　～ stone 粗粒砂岩
　～ strength 柱芯混凝土强度
Kernbut *n.* 断层外侧丘
kernel ['kə:nl] *n.*；*v.*，-nel(l)ed,-nel(l)ing；*n.* 核[中]心,模芯；原子核；颗粒,种子；要点；影响函数核；零磁场强度线；*vt.* 包在核内；kerneled, *a.*；kernelly, *ad.*
　～ (test) program 核心(检测)程序
　atomic ～ 原子核
　closed ～ 闭核

diffusion ～ 扩散影响函数,扩散核
displacement ～ 位移影响函数,位移核
energy ～ 能量核
flame ～ 火焰中心
point ～ 点源影响函数,点源核
transport ～ 迁移影响函数,迁移核
kernite *n.* 斜方硼砂
kerogen ['kerədʒən] *n.* 油母岩质
～ shale 油页岩
kerosene ['kerəsi:n] = kerosine *n.* 煤[火]油
～ cutback 用煤油轻制的沥青
～ emulsion 石油乳制
～ engine 煤油发动机
～ fuel burner 煤油燃烧炉
～ lamp 煤油灯
～ method (测密度用)煤油法
～ number 煤油号数
～ oil 煤[火]油
～ paint 煤油涂料
～ resistance 石油稳定性[耐久性]
aviation ～ 航空煤油
burning ～ 照明煤油
power ～ 动力煤油
kerosene-fluxed 用煤油助溶的,用煤油软制或稀释的
kerotenes *n.* 焦化沥青质
kerrite *n.* 黄绿蛭石
kerve 掏槽,截槽,切缝
Kessener revolving brush 凯氏旋转刷(活性污泥中用以维持循环和供氧的圆筒形金属刷)
Kesternich test 耐蚀试验
ket *n.* 刃
keto ['ki:təu] *n.* 酮基; *a.* 酮类的
kettle ['ketl] *n.* 水壶,锅,釜;白铁桶;溪水冲成的凹处
～ boiled linseed oil 炼制的亚麻油,亚麻子油清漆
～ depression 锅形陷落
～ fault 锅形断层
～ hole 锅形陷洞,坑洞,壶穴

～ lake 锅形湖
～ moraine 锅形冰碛,多穴碛
asphalt ～ (铺路)柏油壶,地沥青锅
asphalt heating ～ 沥青加热锅
dross ～ 除渣锅
pouring ～ 浇注勺
softening ～ 软化锅
still ～ 蒸馏釜
vacuum ～ 真空釜
kettledrum ['ketldrʌm] *n.* (釜状)铜鼓,定音鼓
kettleholder ['ketl,həuldə] *n.* 水壶柄,勺柄
kevel ['kevəl] *n.* 鹤嘴锤(用于砍石);石工锤,缆耳
kewal *n.* 冲积土
key [ki:] *n. pl.* keys, *a.*; *v.*, keyed, keying; *n.* 钥匙,键,按钮,开关;销,栓,双头螺栓;拱心石,楔形砖;关键,要害;顶,冠;索引,检索表,解答,图例;沙洲,暗礁;混凝土中的槽缝;板条间的灰泥;表面附着性; *v.* 销上,插上,用键固定;按键钮,键控;发报;自动开关;提供线索; *a.* 主要的,关键的
～ aggregate 嵌缝集料
～ bar 桁架封闭杆(非受力杆)
～ beam 楔形梁
～ bed 键座;标准[分界]层
～ bit 钥匙齿[头]
～ block (=crown block) 拱顶石
～ board 键盘,开关板
～ board perforator 键盘穿孔机
～ bolt 键螺栓,螺杆销
～ brick 拱顶砖
～ cabinet 电话控制盒
～ card puncher 卡片穿孔机
～ chuck 钥匙式卡盘
～ city 中心城市
～ colour 基本色
～ columns 关键列
～ component 主要部分
～ count 主要交通站点数;控制点计数

~ course 拱顶石层
~ cut 基本挖掘
~ diagram 索引[解说]图
~ drawing 索引[解释]图
~ driver 销[键]起子
~ drop 钥匙孔盖片
~ factor 关键因素
~ floating 抹灰的底层；基础的平整层
~ function 主要功能
~ groove 键槽
~ group 指示组
~ hinge 拱顶铰
~ hog 拱顶加高量
~ hole 钥匙孔；键孔；基准井
~ hole escutcheon 钥匙孔罩
~ -hole saw 钢丝锯，栓孔锯
~ -hole type notch 钥匙孔式槽口，（冲击试件用）锁眼式刻槽
~ horizon 基准标志层
~ -in 插上，嵌上；键盘输入
~ industry 关键工业
~ -in-knob lock 门把锁
~ instruction 引导指令
~ job 关键工作
~ joint 键接合
~ joint pointing 凹形勾缝
~ lever 主控手柄
~ light 主光
~ line 主线，母线
~ link 主要环节
~ machine 主导机械
~ map 索引图，总图
~ metal 母合金
~ pattern 回纹饰；形花样
~ pile 枢桩，主桩
~ plan 平面布置总图
~ plate 键板；钥匙孔板
~ point 关键，要点
~ position 枢纽位置
~ post 主要岗位
~ protrusion 榫舌

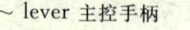
key joint pointing

~ punch machine 键控穿孔机
~ rack 钥匙挂板
~ relay 键控继电器
~ rendering （粉刷）底涂
~ sag 拱顶下垂度
~ search 关键字检索
~ seat 键槽，电键座
~ segment 刹尖石[砌块]
~ set 电键
~ slot 键槽；钥匙槽
~ socket 旋钮灯口
~ station 主台站，中心站
~ stone 拱顶[心]石，填缝石，嵌缝石
~ switch 钥匙[琴键]开关
~ trench 坝基截水墙槽
~ valve 扳手阀门
~ wall 齿（刺）墙
~ way 键槽，销座
~ well 基准钻孔
~ word 关键字
~ wrench 套筒扳手
activate ~ 起动键
arch ~ 拱石
box ~ 套筒扳手
brake ~ 制动键
coiled ~ 旋簧键
drift ~ 锥柄工具；卸出斜铁
flange ~ 法兰扳手
hook ~ 钩形扳手
ignition switch ~ 发火开关钥匙
make-and break ~ 开关
master ~ 万能钥匙
rail ~ 轨条固定楔
screw ~ 螺丝扳手
spoke ~ 条幅扳手
taper ~ 斜键
tightening ~ 斜扁销
tubular ~ 管形螺丝扳手
voussoir ~ 拱楔石
wood ~ 木楔

keyed *a.* 有键的，键控的，用键结合的；以石楔咬紧的，锁着的
~ amplifier 键控放大器

～ beam 键接梁,键迭合梁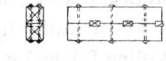
keyed beam
～ brick 楔形砖
～ compound beam 键接组合梁
～ construction joint 键接施工缝
～ dowel 槽榫
～ girder 键合梁
～ into 嵌进
～ joint 嵌缝;楔形缝
～ miter joint 嵌键斜角缝
～ mortise and tenon 加键镶榫接合
～ pointing 嵌键勾缝

keyer ['ki:ə] ***n.*** 键控器,电键电路;定时器
　　～ tube 键控管

keyhole ***n.*** 锁眼,栓[钥匙]孔
　　～ charpy impact test specimen 钥匙形缺口冲击样
　　～ specimen 有刻痕的冲击试块

keyholed back plate 带锁眼底板

keying ['ki:iŋ] ***n.*** 锁入,用键固定;键控法,发报;按键,自动开关
　　～ action 连接键作用;楔住
　　～ action of aggregate 骨料的锁结作用
　　～ circuit 键控电路
　　～ strength 咬合强度

keyless ['ki:lis] ***a.*** 无键的,无钥匙的
　　～ lock 无钥匙锁
　　～ socket 无开关灯口

keynote ['ki:nəut] ***n.*** 重点,要旨,基本方针,基调

keyseat ***n.*** 键槽,销槽

keyshelf ***n.*** 键座,电键盘

keystone ['ki:stəun]
n. 拱心石,拱顶嵌缝石;关键,要旨;***v.*** 用拱顶石支承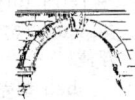
hkeystone
　　～ layer 拱顶石层
　　～ plate 瓦垅[波]纹钢板

kibble ['kibl] ***n.***;***v.***,-bled,-bling;***n.*** 木桶,竖井吊桶;***vt.*** 把……碾成碎块,粗磨

kibbler ***n.*** 粉碎机

kick [kik] ***v.*** 踢;反冲;(仪表指针的)急冲,突跳,跳动;逆转,反抗;***n.*** 反冲[后座]力,弹力,反应力;(发动机)起动;凹槽模,制槽砖模条;kickable,***a.***;kickless,***a.***
　　～-atomizing 冲击喷射[雾化,粉化]
　　～-atomizing pile driver 冲击喷射打桩机
　　～-back 逆转,回程,反冲,后坐力;回扣
　　～-back dump 逆转翻笼
　　～ circuit 脉冲电路
　　～-off (卫星与火箭)分离,断开;拔料机,推出机
　　～-off arm 出料推杆
　　～-out 反冲击;断路
　　～-out arm 出料推杆
　　～-out lever 断路器杠杆
　　～-out master cylinder 断路器主活塞缸
　　～-pedal 脚蹬起动踏板
　　～ plate 蹴板
　　～ starter 反冲式起动器,脚踏起动器
　　～ strip 门脚护条
　　～ transformer 脉冲变压器
　　～ing piece 垫木
　　～(ing) plate 门脚护板
　　～ing recess 柜下防踢的凹进处

kicker ['kikə] ***n.*** 喷射器,抛掷器;艇外推进器;落后反弹的物体;轧条;加强板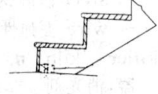
kicker plate
　　～ light 强聚光
　　～ port 排气孔
　　～ plate 加强横木

kickpoint ***n.*** 转折点

kicksort ***v.*** 振幅分析

kicksorter ***n.*** 振幅分析器

kickstand ***n.*** 撑脚架

kick-up ***n.*** 向上弯曲;翻车机,翻罐笼
　　～ frame 降低车辆重心的车架

kiddle ['kidl] *n.* （河中捕鱼用的）拦河栅，鱼梁

kidney ['kidni] *n. pl.* -neys；*n.* 小圆石，小卵石；矿肾；（转炉）结块 *a.* 肾状的，卵形的
~ joint 挠性接头，气隙耦合器

kier [kiə] *n.* 漂煮锅；精炼锅

kieselguhr ['ki:zlguə] *n.* 硅藻土
~ brick 硅藻土（烧制的）砖
~ concrete 硅藻土混凝土
~ covering cord 硅藻土包线[塞绳]（绝缘，隔热用）
~ slab 硅藻土板

kieserite ['ki:zərait] *n.* 硫[水]镁矾

kietyoite *n.* 磷灰石

kikekunemalo *n.* 漆用树胶

kil *n.* 黑海白黏土

kilderkin ['kildəkin] *n.* 英国容量单位（=1/2 barrel），小桶

kilfoam *n.* 抗泡剂

kilhig *n.* 推杆

kilkenny coal 无烟煤

kill [kil] *v.* 杀死；摧毁，击落；停住（机器），切断（电流；抑制（振荡），衰减；中和；脱氧；沉积[淀]；使失效，删除，放弃；*n.* 杀死；破坏；摧毁；脱氧；沉积[淀]；水道，小河；爆炸性气体，瓦斯

killagh *n.* 小艇锚，石锚

killas *n.* 板岩，片岩

killed [kild] *a.* 饱和了的，镇静的，已断路的，已停电的
~ alloy 安定合金
~ lime 失效石灰
~ spirit 焊酸，焊接用药水
~ steel 镇静钢
~ wire 去弹性钢丝

killer ['kilə] *n.* 瞄准器，抑止器，断路器，消光剂
~ circuit 抑制电路
~ switch 断路器开关
echo ~ 回声抑制器
noise ~ 噪声抑制器
spark ~ 减弧器

killick ['kilik] *n.* =kelleg, kellick, kellock, killoch, 小锚, 石锚

killing ['kiliŋ] *a.*；*n.* 致死（的），破坏的；切断电流，镇静的，脱氧（的）；沉积；killingly, *ad.*
~ agent 镇静剂，脱氧剂
~ freeze 严冬
~ frost 严霜
~ knot 除掉木节
~ line 停泵线路
~ period 镇静期，脱氧期
~ substance 毁灭性物质
~ wells 压制井喷

kiln [kiln] *n.* 窑，炉，干燥器；*vt.* 窑内烘干或焙烧
~ brick 窑烘砖
~ building 炉窑车间
~ cart 窑车
~ chamber 窑室
~ discharge 出窑
~ dried 窑室烘干的
~ dried wood 窑[烘]干木材
~ drying 窑内烘干
~ evaporator 窑式脱水器
~ feeder 装窑机
~ -fired 窑烘的，用窑烧制的
~ -fresh brick 刚出窑的砖
~ furniture 烧窑辅助设备
~ insulation 窑隔热层
~ liner 窑内衬
~ lining 窑内衬砌
~ operation 炉窑的运转
~ pier （转炉）基座
~ placing 装窑
~ refractory brick 窑烘砖
~ scum （窑制品）吐渣（表面析出白色浮渣）
~ shell 炉壳
~ size 炉窑尺寸
~ stone 炉料
~ transformation 窑变
~ waste heat 烧窑排出的余热

archless ～ 无拱炉
bee-hive ～ 蜂窝窑
brick ～ 砖窑
calcining ～ 焙烧窑
dolomite calcining ～ 白云石煅烧炉
double-chamber type tunnel ～ 双室式隧道窑
dry ～ 烘干窑[器]
gas-fired calcining ～ 煤气煅烧窑
glaze ～ 釉瓷窑
lime ～ 石灰窑
pit ～ 炼焦炉
plaster ～ 石膏窑
pottery ～ 陶瓷窑
rotary ～ 回转炉
rotating cylindrical ～ 回转圆筒窑
shaft ～ 竖窑
wet process rotary ～ 湿法烧制转窑
kilnboy *n.* 干燥程序自动控制记录器
kilneye *n.* 窑孔,出灰孔
kilning *n.* 烧窑,窑烘
kiloampere ['kiːləuˌæmpiə] *n.* 千安培
kilobar [ˌkiləbaː] *n.* 千巴
kilobit *n.* 一千位,一千个二进制数,千比特
kilocalorie ['kiləˌkæləri] *n.* 千卡
kilocurie ['kiləˌkjuri] *n.* 千居里
kilocycle ['kiləˌsaikl] *n.* 千周,千赫
kilodyne ['kilədain] *n.* 千达因
kilogauss ['kiləˌgaus] *n.* 千高斯
kilogram(me) ['kiləgræm] *n.* 千克,公斤
kilogram-meter = kilogram-metre ['kiləˌgræm'mitə] *n.* 公斤-米,千克-米
kilohertz ['kiləhəːts] *n.* 千赫
kilohm ['kiləum] *n.* 千欧姆
kilojoule ['kilədʒuːl] *n.* 千焦耳
kiloliter = kilolitre ['kiləˌliːtə] *n.* 千(公)升,一立方米
kilolumen ['kiləˌluːmən] *n.* 千流明
～ -hour 千流明(小)时

kilomegacycle [ˌkiləˈmegəˌsaikəl] *n.* 千兆周
kilometer = kilometre ['kiləˌmiːtə] *n.* 千米,公里
～ post 里程标
～ stone 里程碑[石]
kilometrage *n.* 在公路、航道竖立公里里程碑
kiloroentgen *n.* 千伦琴
kilostere ['kiləˌstiə] *n.* 千立方米,千立方公尺
kilovolt ['kiləˌvəult] *n.* 千伏
～ -ampere 千伏安
～ meter 千伏表
kilowatt ['kiləwɔt] *n.* 千瓦
～ -hour 千瓦小时,度
～ -hour meter 电度表
kilsyth basalt *n.* 次辉绿玄武岩
kimberlite ['kimbəlait] *n.* 角砾云橄岩,金伯利岩
kimzeyite *n.* 锆榴石
kin [kin] *n.*;*a.* 同类(的),同性(的); kinless, *a.*
kind [kaind] *n.* 种类,型式;性质,特性; *a.* (矿石)易采的;和气的
in ～ 以货代款
Kind Chaudron method of shaft sinking 一种竖井开挖[钻进]法
kindergarten ['kindəˌgaːtn] *n.* 幼儿园
kindle ['kindl] *v.*, -dled, -dling; *v.* 点火,着火,燃烧;引起,激起; kindler, *n.*
kindling ['kindliŋ] *n.* 点火,燃烧,发亮;易燃之引火物
～ point 燃点,着火点
～ temperature 着火温度
～ wood 引火木,生火柴
kindred ['kindrid] *n.*;*a.* 相似(的),同种类(的),同源(的);亲缘族,亲戚; kindredless, *a.*; kindredly, *ad.*; kindredness, *n.*
～ effect 邻近效应
～ type 类似型式
kinegraphic control 远距离控制

kinema ['kinimə] = cinema *n.* 电影；电影院
　~ camera 电影摄影机
　~ colour 彩色电影,彩色片
kinematic [ˌkaini'mætik] *a.* 运动的,运动学的
　~ boundary condition 运动边界条件
　~ design 机动设计
　~ eddy viscosity 涡动运动粘滞度
　~ indeterminacy 运动不确定性[测不准性]
　~ similarity 运动相似(性)
　~ similitude 运动相似(性)
　~ theory of framework 结构运动理论
　~ viscosity 运动粘滞度
　~ viscosity coefficient 运动粘滞系数
kinematically [ˌkaini'nætikəli] *ad.* 运动学上
　~ acceptable solution 满足运动条件的解
　~ admissibe multiplier 动性许可乘数(结构产生塑性铰活动时容许的安全系数)
　~ determinate 镇定的
　~ indeterminate structure 超静定结构
　~ nonlinear analysis 运动学非线性分析
kinematics [ˌkaini'mætiks] *n.* 运动学
kinemometer [ˌkaini'mɔmitə] *n.* 流速计；感应式转速表
kinescope ['kainiskəup] *n.* 显像管；显像管录像
　~ recorder 屏幕录像机
　colour ~ 彩色显像管
kinesis *n.* 运动,动态
kinestate *n.* 运动状态
kinetic [kai'netik] *a.* 动的,动力的,动力学的；能动的；kinetically, *ad.*
　~ analogy 动力模拟
　~ art 活动艺术
　~ boundary condition 动力边界条件
　~ characteristic 动态特性
　~ coefficient of friction 动摩擦系数
　~ -control 动态控制
　~ eddy viscosity 涡动动力粘滞度
　~ energy 动能
　~ equation 动力方程
　~ equilibrium 动态[力]平衡
　~ friction 动摩擦
　~ head 动力[流速]水头
　~ metamorphism 动力变质作用
　~ moment 动力矩
　~ pressure 动压力
　~ property 动力学性质
　~ stability 动力稳定
　~ tank 活动油箱
　~ viscosity 动力粘滞度
kineticist *n.* 动力学家,活动艺术家
kinetics [kai'netiks] *n.* 动力学
　~ of comminution 破碎动力学
　~ of flotation 漂浮动力学
　fluid ~ 流体动力学
kinetostatics *n.* 运动静力学
king [kiŋ] *n.* 国王；*a.* 主……特大的；kingless, *a.*；kinglessness, *n.*；kinglike, *a.*
　~ and queen post truss 立字桁架
　~ bolt 主销,中枢销,主螺栓；中心立轴；主吊杆
　~ bolt ball and socket 主销球窝
　~ bolt truss 钢吊杆桁架
　~ bulkhead 主舱壁
　~ closer 超半砖丁接砖
　~ head 桁架中柱的大头

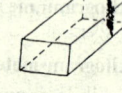

king closer

　~ lever 主握柄
　~ piece 主柱,桁架中柱
　~ pile 主桩
　~ pin 中心销,主销；中心立轴
　~ pin angle 主销倾角
　~ pin bearing 主销座,中心轴支座
　~ plank 甲板中心纵列板
　~ post 桁架中柱,主杆,吊杆柱；单柱

屋架
~ post girder 单柱大梁
~ post roof 单立柱桁架屋顶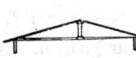
king post roof
~ post truss 单立柱桁架
~ rod 桁架中柱，桁架中吊杆
~'s blue 钴兰色，深兰色
~'s chamber (金字塔中)国王之墓室
~'s highway 主要道路；公路
~'s master mason 王宫建筑师
~'s tomb 王陵
~ -tower 塔式起重机的主塔
~ truss 有中柱的桁架
~ wood 紫木，王木

Kinghoren metal *n.* 铜锌合金
kingsbury bearing *n.* 金斯布里式轴承
kink [kiŋk] *n.* 扭结，绞结，结点，死扣，疙瘩；套索，绞链，环；弯折，转折点；*v.* 打结，拧绞，纠缠
~ mark 折痕
~ing 扭结；缠线

kinker *n.* (打结器的)扭结轴
kinks *n.* 操作指南
kino ['ki:nəu] *n.* 电影院；胺树胶
~ lamp 显像管
kinscord 花式沟纹
kinsh *n.* 撬棍
kintal ['kintl] *n.* (一)百公斤
kiosk ['ki:ɔsk] *n.* 凉亭；小配电亭；公用电话亭；售货亭
kip [kip] *n.* 旅店，客栈；千磅
kipp phenomenon 跳跃现象
kir *n.* 岩沥青，含沥青岩
kirksite *n.* (模具用)锌合金
kirn [kə:n] *v.*；*n.* 人工打眼；剧烈搅拌
kirner *n.* 手工冲击钻
kirunavaarite *n.* 磁铁岩
kirve *n.* 掏槽，截槽
kish *n.* 大柳条方框；铁水上的漂浮石墨，石墨泡；碴壳，残留金属
~ lock 石墨[漂浮]集结
~ slag 石墨渣

kiss [kis] *n.* 接缝，缝口
~ coating 单面给胶涂层法
~ core 预埋型蕊
kissing gate *n.* 单人转门
kit [kit] *n.* 工具，工具箱；元件，器材；成套零件；背囊，小桶；一整套，一组
adapter ~ 成套附件
first aid ~ 急救药箱
flyaway ~ 随机器材包
generator ~ [装配]发电机的整套零件
installation ~ 装置工具包
repair ~ 修理工具包
spare parts ~ 零备件箱
kitchen ['kitʃin] *n.* 厨房，全套炊具；kitchenless, *a.*; kitcheny, *a.*
~ buffet car (铁路)餐车
~ building block module 装配[组合]式厨房
~ burning appliance 厨房灶具
~ cabinet 厨房橱柜
~ car 餐车
~ central heating 厨房集中供暖
~ closet 厨房壁柜
~ dining room 连餐厅一起的厨房
~ fire 厨房炉灶
~ fitments 厨房设备
~ fittings 厨房设备
~ floor 布置厨房的楼层
~ furnace 厨房炉灶
~ furniture 厨房家俱
~ garden 家庭菜园
~ installation 厨房设备
~ range 炉灶
~ sink 厨房洗碗槽
~ storey (英)布置厨房的楼层
~ system 厨房设备
~ tower 厨房水塔[箱]
~ -ware 炊具
~ waste 厨房垃圾[污水]
~ waste disposal pipe 厨房下水管
~ waste disposal shaft 厨房下水道

窨井
~ yard 厨房院子
kitchener ['kitʃinə] *n.* 厨师;厨灶
kitchenet(te) [ˌkitʃi'net] *n.* 小厨房,室内厨房
kite [kait] *n.*;*v.*, kited, kiting;*n.* 风筝;轻型飞机;*vi.* 像风筝一样飞翔;-kiter, *n.*,-kitelike, *a.*
~ -airship 系留气艇
~ balloon 系留气球
~ -camera 俯瞰图照相机
~ float 鸢式浮子
~ winder 转向斜踏步
klaxon ['klæksn] *n.* 电喇叭,电警笛
~ device 喇叭
Kleine(hollow-brick) floor（由空心砖缝间置钢筋然后用混凝土浇筑成整体的）楼板
kleine (houow) floor (clay) brick 克赖恩纳空心楼板砖
Kleinpflaster *n.* 嵌花式砌块,小方石块
kleit *n.* 高岭土
klinker brick *n.* 缸砖
klinkstone=clinkstone *n.* 响岩
klint *n.* 陡崖;硬岩礁
klintite *n.* 硬礁岩
klip *n.* (岸边)悬崖
klong [klɔŋ]【泰】*n.* 水道,运河
kloof [klu:f]【南非】*n.* 峡谷,冲沟
kluf *n.* 沟,峡谷
klystron ['klaistrən] *n.* 调速管,速度调制电子管
kmaite *n.* 绿云母
knack [næk] *n.* 窍门,技巧,妙诀
knag [næg] *n.* 木节,木瘤;木钉
knap [næp] *v.* knapped, knapping;*n.* 小山,丘顶;*v.* 敲打
~ ping machine 碎石机
Knapp bottom pressure gauge 纳普底压计
knapper ['næpə] *n.* 碎石工;碎石锤
knapsack ['næpsæk] *n.* 帆布背包;背罐;-knapsacked, *a.*

~ pump 背负式喷雾器
~ sprayer 背负式喷雾器
knar [na:] *n.* 木瘤,木节;-knarred, knarry, *a.*
knead [ni:d] *vt.* 揉,捏,按摩,搓;混合,搅拌;kneadability, *n.*;kneadable, *a.*;kneader, *n.*;kneadingly, *ad.*
~ ed 被揉搓的
~ ed eraser (rubber) 橡皮擦,橡皮
~ ed graval 泥流砾
~ ed structure 捏合结构
~ ing action 揉搓[混合]作用
~ ing action of traffic 交通对路面的揉搓作用
~ ing compactor 揉压机
~ ing machine 搅拌机,搅搓混砂机
~ ing mill 捏和机,搅拌机
~ ing trough 揉合槽[钵]
kneader *n.* 捏和机;黏土拌合器
~ type mixer 搅拌机;搓揉式混砂机
knee [ni:] *n.* 膝盖,膝形杆;弯头,弯管;肘管;曲线最大曲率处;合角铁;扶手弯头;铣床的升降台;木船的肋材 *a.* 膝形的 *v.* 用合角铁接合
~ -and-column 升降台
~ batten 角撑板
~ bend 肘形弯管,弯管接头
~ brace 斜[角,隅]撑
~ braced bracket 带斜撑托座[架]

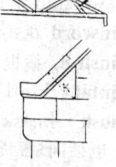
knee brace

~ -brace roof truss 斜撑屋架
~ bracket plate 弯角连接板,节点板
~ colter 膝形犁刀
~ frame 拐弯式构架
~ -girder 肘状梁
~ -girder bascule 肘状梁式竖旋桥
~ hole (写字台下)容膝空档
~ iron 角补强铁,隅铁
~ joint 膝关节,肘接,弯头接合
~ -lever 直角形杠杆,角杆

~ loss 弯管水头损失
~ of head 船头悬伸木
~ piece 曲橡;曲形部件
~ pipe 弯管,弯头
~ plate 肘板,角撑板
~ rafter 曲橡,弯橡
~ roof 复斜屋顶
~ stone 山墙角石
~ table 三角桌
~ timber 曲肘形木材
~ wall 顶层间壁墙,支撑墙
beam ~ 梁尾接铁
hanging ~ 船的吊板,倒挂龙骨
soft ~ 曲线缓变弯折处
kneeboard *n.* (赛马场周围)矮墙
kneed [ni:d] *a.* 关节状的,弯的;多瘤的
kneeler ['ni:lə] *n.* (支承拱脚的)斜交石
kneeling figure *n.* 祈祷雕像
knick point *n.* 裂点
knife [naif] *n. pl.* knives, knifed, knifing; *n.* 小刀,刀口[刃],刀具;切割器;*v.* 用刀切,劈开;knifeless, *a.*; knifelike, *a.*; knifer, *n.*
~ arbor 圆盘刀片的心轴
~ -cutter (制砖)切割机
~ -cut veneer 切成层板
~ edge 刀口;刀刃支承;刃形边缘
~ edge bearing 刀口支承
~ edge load (作用于狭长构件上的)线荷载,条形荷载;刃口荷载
~ edge pivot 刀形枢轴
~ edge support 刃形支承
~ -edge weir 刃刃堰
~ grinder 磨刀装置,磨刀石,砂轮
~ holder 刀夹,刀架
~ machine 磨刀机
~ mill 切碎机
~ rest 拒马;餐桌上的刀架
~ -scratch test 刀片划痕(硬度)试验
~ stone 磨刀石
~ switch 闸刀开关

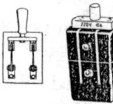

knife switch

~ test for plywood 胶合板刀齿试验
~ tool 修边刀具
knifing (切深孔型中的)切深;刮膜
knifing filler 适于用灰刀填塞的填缝物(有别于用刷的填缝物)
breaking ~ 轧刀
bush ~ 刮轴衬用的刀
dead ~ 固定刀片
erasing ~ 刮刀
group ~ 组合闸刀
hollowing ~ 弧刮刨
putty ~ 刮铲,油灰刀
spreading ~ 刮刀,涂胶刀
striking ~ 三棱刮刀
knight [nait] *n.* 系缆柱
~ engine 套阀发动机
knit [nit] *v.*, knitted or knit, knitting; *v.* 编织;结[粘,联]合;弄紧,使紧凑;knittable, *a.*
~ goods 针织品
knitting ['nitiŋ] *n.* 编织物,编织法;接合
~ action (路面上层混合料的)交织作用,网结作用
~ frame 针织机架
~ layer (新旧混凝土之间的)接合层
~ machine 针织机
knob [nɔb] *n.*; *v.*, knobbed, knobbing; *n.* 节,瘤,疙瘩;旋钮,圆球形把手;球形柄;鼓形绝缘子;小丘;*vt.* 将(多余石块)敲掉;装球形把手;突出,鼓起;-knobbed, knoblike, *a.*
~ -and-basin topography 凸凹地形
~ -and-tube wiring 穿墙布线法
~ (door) fitting 门把配件
~ (door) furniture 门把配件,门钮装置
~ (door) hardware 门把五金配件
~ hole 定位锁孔
~ latch 球把框锁
~ lock 球把门锁
~ -operated control 旋钮控制

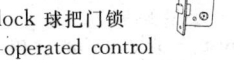

knob-lock

control ~ 控制手柄
door ~ 门把
fuse ~ 保险丝端帽
gear shift lever ~ 变速杆捏手
riveting ~ 铆工模
shift ~ 翻转开关,开关按钮

knobble ['nɔbl] *n.* 节瘤,小节,小圆块; *v.* 开坯;压平隆起;削节

knobb(l)ing *n.* 石块粗加工(琢平);木材削节;压平表面的隆起;小延压量轧制(除去氧化皮)

knobboss *n.* 球形捏手

kobstone *n.* 瘤石

knock [nɔk] *v.*;*n.* 敲打,打击,碰撞;爆震,震动;破坏;knockless, *a.*
~ -boring machine 冲击式钻机
~ characteris tic 抗震性
~ -compound 抗震剂
~ down test 可镦性试验
~ hole 定位销孔
~ meter 爆震[测震]计
~ -off 敲落;停止,中止;自动停机
~ off joint 脱钩装置
~ -on 撞出,反跳,弹跳
~ pin 定位销,顶销
~ rating 防爆率
~ -reducer 抗爆[震]剂
~ -test 抗爆性试验
~ value 抗震值
~ wave 冲击波
~ed down 船在大风下的横倾状态;建筑构件或机械部位错位;散装
~ing 敲击信号;水锤;爆震;锤破的大矿石
~ing-bucker 采石器
~ing-out grid 落砂栅[架]
diesel ~ 柴油机爆震
engine ~ 发动机爆震
piston ~ 活塞爆震,敲缸
side ~ 活塞松动,横向冲击
taper ~ 圆锥顶销

knockdown ['nɔk'daun] *n.* 击倒,撞倒;易于拆卸的东西 *a.* 能拆卸的;压倒的,锐不可挡的;折叠式的
~ tower 拆卸式舰标

knocker ['nɔkə] *n.* 门环,门锤;敲门者;爆震物质;脱圈器

knockout ['nɔkaut] *n.* 打出,敲出;脱模,出坯;分离;落砂;拆卸工具;脱模装置;分液器;喷射器;抛掷器
~ box 气体分离箱
~ coil 一种水冷分凝管
~ cylinder 顶件油缸
~ drum 分离鼓
~ grid 落砂栅
~ machine 脱模机,落砂机
~ pin 顶出杆
~ plate 脱模板
~ press 脱模机
~ stroke 出坯冲程
~ tower 分离塔
core ~ 泥蕊打出机
spring ~ 弹簧顶件器

knockrating *n.* 防爆率,抗爆值

knoll [nəul] *n.* 圆丘,小山;海丘(高出海底 500 ~ 1000 米);土墩;knolly, *a.*
~ spring 丘泉

Knoop (indentation) microhardness test 诺氏(刻痕)微硬度试验

Knoop number 诺氏硬度值

Knoop scale *n.* 诺氏硬度标度

knop [nɔp] *n.* 圆形把手,捏手;节,瘤;顶华,蕾形装饰,突出的雕饰

knot [nɔt] *n.*;*v.*, knotted, knotting; *n.* 结,结点;木节;波节点;节疤;(航速)节,海里/小时;关键,要点;*v.* 打结;捆扎;聚集;knotlike, *a.*
~ cluster 节群(木材)
~ hole 木节孔
~ strength (钢丝绳心的)打结强度
~ wood 有节木料
~ work 交错线饰,编织细工
~ ted pillar 打结柱
~ ted schist 瘤斑片岩

knot work

~ ting 塞木节孔,节疤封闭;打结;虫胶在酒精中的溶液(用以覆盖木材上的节)
~ ting strength 打结强度
builder's ~ 死结
ebonite ~ 硬橡皮扣
figure-8 ~ 8 字形结
knothol mixer *n.* 隔膜混合器
knothole ['nɔthəul] *n.* 木节孔
knotter ['nɔtə] *n.* 打结器;除节机
knotty ['nɔti] *a.*, -tier,-tiest; *a.* 有[多]节的;难解决的,死结似的; knottily, *ad.* knottiness, *n.*
~ and knaggy wood 多节木
~ pine 多节松木
~ wood 多节木
know [nəu] *v.*, knew, known, knowing; *v.* 知道,了解,认识,识别; knower, *n.*
~ how 专门技能,生产经验,技术情报,窍门
knowledge ['nɔlidʒ] *n.* 知道,了解,通晓;知识,学问,认识,经验; knowledgeless, *a.*
general ~ 普通知识,常识
perceptual ~ 感性知识
rational ~ 理性认识
known number 已知数
known reserves 已知储量,探明储量
knuckle ['nʌkl] *n.; v.*, -led,-ling; *n.* 关节,肘,万向接头,铰接;钩爪,钩舌;(屋顶的)脊;棱缘,卷边; *vt.* 弯成肘节; knuckly, *ad.*
~ arm 关节臂,(汽车)转向节臂
~ bearing 铰式支座
~ buster 板钳,扳手
~ centre 万向接头十字轴
~-gear 圆齿齿轮
~ joint 肘形关节,叉形接头,铰接
~ joint press 肘杆式压力机
~ length 铰轴长度
~ line 棱角线
~ man 挂钩工

~pin 转向节销,(万向)接头插销
~ post 转向节柱
~ press 曲柄连杆式压力机
~ (screw) thread 圆螺纹
~ spindle 转向节销
~ thrust bearing 转向节止推轴承
~ tooth 圆顶齿
~ trunnion 转向节枢销
coupler ~ 车钩关节
steering ~ 转向节

knurl [nə:l] *n.; v.* 节,瘤,隆起;球饰;圆形按钮;滚花;压花纹
~ wheel 滚花轮
~ed 滚花,压花
~ed roll 滚花辊[筒]
~ing 滚压花纹;断圆线条饰;滚花(刀);刻痕
~ing tool 滚花工具

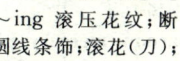
knurled roll

koalmobile *n.* 自行式井下无轨矿车
kochenite *n.* 琥珀树脂
kodak ['kəudæk] *n.; v.*, -daked,-daking; *n.* (K-)柯达照相机; *v.* 用上述照相机照相; kodaker, *n.*
koenlite *n.* 重碳地蜡
kogel process *n.* (路面防滑)热处理
Kollag *n.* 固体润滑油
kollergang【德】*n.* 轮碾机,碾砂机
kollermill *n.* 轮碾机,碾砂机

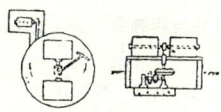
kollermill

Konal *n.* 考纳尔镍钴合金
konel *n.* 考涅尔代用白金
kong *n.* 缸;(含锡砾石下的)无矿基岩
Konik(e) *n.* 镍锰钢
konimeter [kəu'nimitə] *n.* 尘度计,计尘器
konstantan ['kɔnstəntæn] *n.* 康铜,镍铜合金

koplon *n.* 高湿模量粘胶纤维
Kopol *n.* 化石树脂
Korduct *n.* 库尔管（石棉水泥电线导管的一种）
Korean pine *n.* 红松
koris *n.* 干河谷（北非）
kornish boiler 水平单火管锅炉
koron *n.* 聚氯乙烯
koronit *n.* 柯罗炸药
koroseal [ˈkɔːrəsiːl] *n.* 氯乙烯树脂［塑料］,柯罗封料；(K-) 上述材料的商标名
Koval *n.* 考瓦尔铝合金
Kovar *n.* 柯伐铁镍钴合金
kraal [krɑːl] *n.* 茅屋；南非土人的小村庄；畜栏；浅海围场
Kraemer and Sarnow mercury pipeline *n.* 克-沙氏水银吸移管（确定沥青材料软化点用）
kraft [kræft] *n.* 牛皮纸
  ~ paper 牛皮纸
  union ~ 双层夹沥青包装纸
Kranenburg method *n.* 直接水压成形法
Kranz Triplex method 可锻铸铁制造法
Kratogen 克拉通区，台地区
Kraus process 克劳斯法（一种活性污泥法的改型）
kremastic water 含气带水,渗流水,通气层的水
Kremlin 克里姆林宫

Kremnitz white 铅白（颜料）
kreosote [ˈkriːəsəut] *n.* 木馏油,杂酚油
Kroll corrosive liquid 氢氟酸腐蚀液
Kromarc *n.* 可焊不锈钢
Krupp austenite steel 奥氏体铬镍合金钢
Krupp triple steel 克房伯高速钢
Krupp section 克房伯型材
kryogenin *n.* 冷却剂
kryometer *n.* 低温计
kryoscope *n.* 凝固点测定计
kryptomere *n.* 隐晶岩
kuchersite *n.* 油页岩
Kuhl cement 低热水泥,铝土水泥
Kunial *n.* 含铝铜镍弹簧合金
kuppe *n.* 导流罩,透声穹室
kurhaus *n.* 疗养所
kurkar *n.* 凝砂块（砂丘砂和碳酸钙胶凝砂的混合料）
kurtosis [kəˈtəusis] *n.* (曲线的)峰态,突出度,峭度,峰度(统计图)
kyanite [ˈkaiənait] *n.* 兰晶石
kyanize [ˈkaiənaiz] *vt.*, -nized, -nizing; 用升汞浸渍木材,用升汞注入木材防腐; kyanization, *n.*
kyanizing [ˈkaiənaiziŋ] *n.* 升汞防腐法
kybernetics *n.* 控制论
kymatology *n.* 波浪学；脉波学
kymograph [ˈkaiməgrɑːf] *n.* 波形自动记录器；转动记录器 kymographic, *a.*
kyrtometer *n.* 曲度计,曲面量测计

L

L 或 [el] *n. pl.* L's, Ls, L's, Ls [elz]; L 形物；建筑物直角侧翼; *pl.* 高架铁路(=elevated railroad)
  ~ -shaped retaining wall L 形挡土墙
  ~ square 直角尺,角直板
  ~ type curb L 形路缘石

label [ˈleibl] *n.*; *v.*, -beled, -beling; *n.* 标签,名牌,牌号[商标],符号；披水石；门楣；示踪; *vt.* 贴标签,做记号; 示踪;分类; labele, *n.*
  ~ check 标记检验
  ~ coding 标记编码
  ~ -corbel table 托臂滴水石

~ course 门窗披水线条
~ processing 标记处理
~ stop 滴水石浮雕,披水端饰
~ing 加标签
~ ing machine 贴标签机
file ~ 文件标记
operational ~ 操作标记
tape ~ 磁带标记

label stop

label(l)er *n.* 贴标签机
labile ['leibail] *a.* 不稳定的,活泼的;不坚固的;滑动的; lability, *n.*
~ flow 不稳定流
~ state 易变态,不稳定态
lability *n.* 不安稳定性,易滑性
~ coefficient 不稳定系数
~ effect 不稳定效应
~ number 不稳定系数
labo(u)r ['leibə] *n.* 劳动,工作;劳动者;劳动力; *v.* 劳动,工作,努力;仔细分析解释; laboringly, *ad.*; laborless, *a.*
~ agreement 劳资协议
~ camp 劳工驻地
~ capacity 劳动生产率
~ condition 劳动条件
~ constant 劳动定额,劳力常数
~ contract system 包工制
~ cost 劳动成本,人工费
~ day 劳动[工作]日
~ discipline 劳动纪律
~ economics 劳动经济学
~ efficiency 人工效率
~ emulation 劳动竞赛
~ exchange 职业介绍所;劳力交换
~ famine 劳动力缺乏
~ feather bedding 强迫雇佣劳动
~ flow table 劳动流程表
~ flux 劳动力流出
~ force 劳动力
~ grade 劳动等级
~ hour 人工小时,工时
~ hour rate 工时率[费]
~ liquidity 劳动流动性
~ insurance 劳工保险
~ insurance regulation 劳保条例
~ market 劳力市场
~ measurement 人工测定
~ organization 劳动组织
~ productivity 劳动生产率
~ protection 劳动保护
~ quantity variance 人工工时差异
~ quotas 劳动份额
~ rate standards 工资率标准
~ reward 劳动报酬
~ saving 省工,节省劳力
~ service company 劳动服务公司
~ slowdown 怠工
~ turnover 劳工周转率
~ union 工会
~ value 劳动价值
auxiliary face ~ 工作面辅助工
day ~ 计日工作
manual ~ 体力劳动
nonproductive ~ 非生产劳动,辅助劳动
on-the-site ~ 工地劳动力
season ~ 季节工
skilled ~ 熟练工
laborage ['leibəridʒ] *n.* 工资
laboratory [lə'bɔrətəri] *n. pl.* -ries;实[试]验室,研究室;药厂,化学厂;熔炼室; laboratorial, *a.*; laboratorially, *ad.*; laboratorian *n.*
~ apparatus 实验室设备
~ bench tile 实验室桌面砖
~ equipment 实验室设备
~ flume 试验水槽
~ mixer 实验室拌合器
~ procedure 实验程序
~ proportioning 试验室配合比
~ report 实验报告

~ -scale 实验室规模的,小型的
~ sifter 震动筛分机
~ sink 化验室用洗盆
~ sole 炉底,炉床
~ stirring device 实验室搅拌器
~ technician 实验员
~ technique 实验室技术
~ test 室内试验
~ tile 实验室铺面砖
~ tower 实验塔
~ trailer 实验室拖车
~ vehicle 实验室用车
cold ~ "冷"实验室(无放射性实验室)
high-altitude ~ 高空实验室
hot ~ "热"实验室(强放射性物质实验室)
testing ~ 试验室
wind tunnel ~ 风洞实验室

labo(u)rer ['leibərə] *n.* 工人,劳动者

labradorite ['læbrədɔːrait] *n.* 拉长岩,富拉玄武岩,曹灰长石

labyrinth ['læbərinθ] *n.* 迷宫,曲径;密封圈
~ box 迷宫式密封盒
~ fret 曲折回纹饰
~ gland 迷宫式压盖
~ of Crete 克里底王的迷宫
~ packing 迷宫式密封,迷宫式填充材料
~ seal 迷宫式止水[密封]
~ seal ring 迷宫止水环
~ waterstop 迷宫式填缝亲

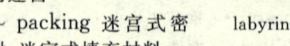

labyrinth

lac [læk] *n.* 虫胶,紫胶,假漆,虫漆;十万,无数
~ -la(c)ker 虫漆染料
~ resin 虫胶树脂
~ varnish 虫胶清漆
asphalt ~ 沥青漆
needle ~ 针状虫胶
Ningpo ~ 金漆
seed ~ 粗虫胶
stick ~ 树枝虫胶
white ~ 漂白虫胶

lacca *n.* 虫胶,虫漆
~ coerula 石蕊

laccolith ['lækəliθ] = laccolite *n.* 岩盖

lace [leis] *n.*;*v.*, laced, lacing; *n.* 束带饰,花边; *v.* 束紧,编织,缀合;(在卡片上)穿花边; laceless *a.*; lacelike, *a.*; lacer *n.*
~ bar 缀条
~ -like spire 花边状塔顶
~d 用带结紧的,有饰带的,以带镶边的
~d beam 缀合[空腹,花格]梁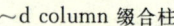
~d column 缀合柱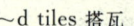
~d tiles 搭瓦
~d valley 搭瓦天沟

laced valley

lacer *n.* 系紧的用具
belt ~ 皮带扣,皮带卡子

lacerate ['læsəreit] *v.*;*a.*,-ated,-ating,*a.*; *vt.* 扯[划]破,撕[劈]裂; *a.* 扯破[撕裂]的

lacerating machine = tensile testing machine *n.* 拉力试验机

laceration [ˌlæsə'reiʃən] *n.* 划破,裂口,削切,撕[劈]裂

lacet [lə'set] *n.* 盘山[回旋]道路;带子
~ road 盘山道路

lacework ['leisˌwəːk] *n.* 网络,网状结构;花边

lacing ['leisiŋ] *n.* 束带,纽带;单缀,联缀;导线;花边装饰;钢筋分布;拱圈结合层
~ bar 缀条
~ board 系紧板,布缆板
~ course 带层,拉结层
~ stand 绕丝架
belt ~ 皮带结合,皮带扣

lack [læk] *v.*;*n.* 缺乏,不足
~ of alignment 中心线偏斜
~ of penetration 未焊透,假焊

lacker ['lækə] *n.*;*v.*;= lacquer 真漆,亮漆,天然漆

laconicum [ləˈkɔnikəm] *n. pl.* -ca；古罗马浴室中的热气浴室

lacquer [ˈlækə] *n.* 清漆，亮漆，喷漆，蜡克；漆膜,胶膜；涂漆镀锡薄钢板；*vt.* 上漆，喷漆；lacquerer, *n.*
- ~ brush 清漆刷
- ~ coat 漆涂层
- ~-coated steel sheet 涂漆薄钢板
- ~ curtain coating (method) 喷漆法
- ~ diluent 稀漆剂
- ~ disk 蜡克盘
- ~ enamel 珐琅,磁漆
- ~ film 漆膜
- ~ master 蜡克主盘
- ~ oil 喷漆用油
- ~ original 蜡克原盘
- ~(ed) plate 涂漆镀锡薄钢板
- ~ putty 腻子,油灰
- ~ solution 漆溶液
- ~ solvent 溶漆剂
- ~ thinner 漆冲淡[稀释]剂
- ~ varnish 亮漆
- ~ vehicle 亮漆媒液
- ~ed hardboard 涂漆硬质木纤维板
- antirust ~ 防锈漆
- clear ~ 透明漆
- moisture-resisting ~ 防潮漆
- priming ~ 底漆
- stop-off ~ 护皮漆,罩漆

lactic [ˈlæktik] *a.* 乳的,来自乳汁的
- ~ acid 乳酸

lactolite *n.* 乳酪塑料

lactoprene *n.* 乳胶,聚丙烯酸酯橡胶,人造橡胶

lacuna [ləˈkjuːnə] *n. pl.* -nae, -nas；脱漏，缺项；空[间]隙,凹穴,小孔

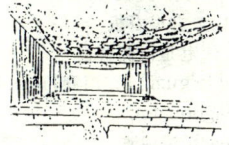

lacunar

lacunar [ləˈkjuːnə] *n. pl.* lacunars, lacunaria；花格平顶,凹格天花板

lacustrine [ləˈkʌstrin] *a.* 湖积[成]的；生在湖中的
- ~ clay 湖成黏土
- ~ deposit 湖沉积
- ~ limestone 介壳灰岩
- ~ plain 湖成平原
- ~ soil 湖积土

ladder [ˈlædə] *n.* 梯子,梯形物,梯形裂缝,阶梯；多斗挖土机或链斗挖泥船的斗架
- ~ access 梯道
- ~ bracing 梯形拉条
- ~ bucket dredge 链斗式挖泥机[船]
- ~ chain 梯状链
- ~ ditcher 链斗挖沟机
- ~ dredge 链斗挖泥机[船]
- ~ drilling 梯架式钻进(法)
- ~ escape 安全梯
- ~ excavator 多斗挖土机
- ~ frame 链斗框架
- ~ hoist 多斗升降机
- ~ hook 梯钩
- ~-mounted drill 链斗式钻机
- ~ of cascades 梯级跌水
- ~ rung 梯子的横档
- ~ scaffold 梯式脚手架,梯台架
- ~ stair 梯级
- ~ stay 梯子架
- ~ trencher 多斗挖沟机
- ~ truck 带梯卡车,有云梯的消防车
- ~ type network 梯形网络
- ~ way 带梯子的垂直矿井通道
- ~ web (百叶窗)梯形带
- ~ well 挖泥槽
- ~ work 梯工
- bucket ~ 挖泥船斗架
- companion ~ 升降梯
- fire ~ 消防梯
- folding ~ 折叠梯
- jack ~ 索梯
- Jacob's ~ 软梯,绳梯

scaling ~ 云梯
laddertron *n.* 梯形管
lade [leid] *v.*, laded, laden or laded, lading; *v.* 装载,汲取,获得,塞满; *n.* 河口,水道;支墩; lader *n.*
laden ['leidn] *a.* 装满[载]的,装着货的,充满了的,荷载的
 ~ in bulk 散装
 ~ weight 装载重量
ladies *n.* (*pl.*) 女士,夫人;女厕所,女盥洗室
 ~'changing room 女更衣室
 ~'drawing room 女客厅
 ~'hairdressing shop 女子理发室
 ~'room 女厕所
 ~'toilet 女厕所
lading ['leidiŋ] *n.* 装载,装货,荷重
 ~ door 装料门
ladle ['leidl] *n.*; *v.*, -dled,-dling; *n.* 杓,戽斗;钢水包; *vt.* 用勺子舀,戽出; ladler, *n.*
 ~ brick 盛钢桶[铁水包]衬砖
 ~ clay 造型黏土
 ~ crane trolley 钢包吊车
 ~ lining 浇铸构内衬
 ~ lip 浇包嘴
 ~ pit 出钢坑
 charging ~ 装料桶
 slag truck ~ 运渣桶
ladleful *n.* 满包量,满勺
Ladoga precision depth recorder 拉多加型精密回声测深仪
lady ['leidi] *n. pl.* -dies;(屋面)小石板;探照灯控制设备;女士; ladyish *a.*; ladyishly *ad.*; ladyishness *n.*; ladyless *a.*
Lady Chapel 大教堂中的圣母堂
lag [læg] *n.*; *v.* lagged, lagging; 滞[落]后,拖延,耽搁;错开,平移;惯[惰]性;套板,外套,防护罩;木材裂缝
 ~ bolt 方头螺栓
 ~ coefficient 滞后系数
 ~ deposit 残留沉积,

lag bolt

粗化沉积
 ~ fault 滞后断层
 ~ gravel 残积砾石;河底推移较慢的砾石
 ~ intake 迟关进汽门
 ~ phase 停滞阶段
 ~ pile 加套桩
 ~ screw 方头木螺钉
 ~ spike 螺钉
 altimeter ~ 高度差
 dynamic ~ 动态时滞
 magntic ~ 磁滞
 pickle ~ 酸浸时滞性试验
 signal ~ 信号滞后
 thermal ~ 热惯性,缓慢加热
 time ~ 时滞,时间延迟
 time ~ fuse 延时熔丝
lagengneiss *n.* 层状片麻岩
lagging ['lægiŋ] *n.* 落[滞]后,延迟;套[挡]板,板条;外套,防护[隔热]套;矿道顶木;桥面铺板;砌拱用弧形型板,拱模撑
 ~ cover 粗涂,粗镀
 ~ current 滞后电流
 ~ edge 脉冲后沿
 ~ indicators 滞后指标
 ~ jack 拱鹰架
 ~ jacket 汽缸保温套
 ~ of pile 桩箍
 ~ section 预制管道隔热块
 ~ water 缓流水
 boiler ~ 锅炉隔热外套
 gob ~ 充填隔板
 head ~ 顶板背板
 jack ~ 砌筑壳体用模板
 roof ~ 顶板背板
 shaft ~ 井壁背板
 side ~ 巷道两侧背板
lagoon [lə'gu:n] *n.* 泻湖,咸水湖;污水池; lagoonal *a.*
 ~ beach 泻湖滩
 ~ facies 泻湖相

~ harbour 河口浅水港,近海湖港,泻湖港
~ -side 泻湖岸边土地
~ -island 环礁,泻湖礁
~ mouth 泻湖口
~ mud 泻湖淤泥
~ sand 泻湖砂
~ slope 泻湖坡

lagtime *n.* 滞时;汇流时间;滞后时间
laguna, lagune = lagoon
lahar *n.* 泥石流,火山泥流
~ deposit 泥石流沉积

laid [leid] (lay 的过去分词)放置,铺砌,敷设
~ bare 露出,露明
~ cold 冷铺
~ contrary to the stratum 递层铺砌
~ dry 干砌
~ hot 热铺
~ -in moulding 镶入的线条
~ in panels 按格铺砌
~ length 敷设长度
~ line 直纹线条
~ -off work 停工,解雇
~ on edge course (砖的)侧砌
~ -on moulding 安装线脚
~ -on thread 外嵌[贴]线条(装饰用)
~ paper 直纹纸
~ paving mix(ture) 嵌置的铺面混合料
~ touching (砌石的)咬接
~ -up 拆卸修理
~ wire 直纹网
hand ~ 手砌

laitance [ˈleitəns] *n.* 水泥浮浆,翻沫,混凝土面上的浆沫
~ coating 水泥浮浆层
~ layer 水泥翻沫层

laitier [ˈleiitiə] *n.* 浮渣
lake [leik] *n.* 湖泊,湖水;深红色(颜料); lakelike, *a.*
~ asphalt 湖沥青
~ bank 湖堤
~ basin 湖盆地,湖流域
~ basin bog 湖沼
~ bed 湖泊沉积矿;湖床
~ -bottom mud 湖底淤泥
~ bottom reclamation 开垦[围垦]湖地
~ circulation 湖泊湖流
~ clay 湖泥
~ colours 色淀染料
~ conservation 湖泊调蓄
~ delta 湖泊三角洲
~ deposit 湖相沉积
~ drainage 湖泊集水区域
~ due to landslide 山崩湖
~ dwelling 史前湖上的木屋;水上房屋
~ facies 湖相
~ -floor plain 湖底平原
~ intake 湖水引水口
~ loess 湖积黄土
~ marl 湖成泥灰岩,沼灰土
~ moor 湖沼
~ oil 琥珀油
~ -outlet channel 湖泊出口水道
~ pigment 色淀染料;颜料漆
~ pitch 湖沥青
~ plain 湖平原
~ sand 湖砂
~ side 湖滨
~ -village 水上住户
~ wall 湖墙(受冻结湖水平面上形成的低的砂石脊)
artificial ~ 人工湖

lakefront [ˈleikfrʌnt] *n.* 湖边平地
lakelet [ˈleiklit] *n.* 小湖
lakeshore [ˈleikʃɔː] *n.* 湖岸
Lala *n.* 康铜
Lalique glass 浮雕图案玻璃
lally column 混凝土填心圆筒形钢结构柱
lam [læm] *n.* 砂质黏土,亚黏土

lamagal *n.* 镁铝耐火材料
lamasery [lɑːməsəri] *n. pl.* -series; 喇嘛寺院
lamb [læm] *n.* 翼形螺帽；操舵盘；羔羊
~'s tongue 羊舌饰
lambert ['læmbəːt] *n.* 朗伯(亮度单位)
lambrequin ['læmbəkin] *n.* 门窗垂饰, 扇形花纹
lame [leim] *a.*, lamer, lamest; 跛的, 不完美的, 损坏了的；*v.*, lamed, laming; 使损坏, 使不完美；*n.* 金属薄板；lamely, *ad.*; lameness, *n.*
lamel *n.* 薄板
lamella [ləˈmelə] *n. pl.* -lae,-las; 薄片, 薄壳层, 薄板；lamellar, *a.*; lamellarly, *ad.*
~ cupola 肋格半球顶
~ dome 肋格穹顶

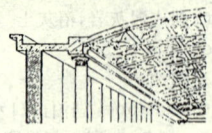
lamella dome

~ mat 叠层垫(石棉制品)
~ roof 叠层式屋顶
~ system 格网结构系统
~ truss 叠层屋架
glide ~ 滑动夹层
lamellar *a.* 层状的, 薄板状的, 成层的, 页片状的
~ flow 层流, 片流
~ (pit) prop 格状坑道支撑
~ structure 层状[叠层]结构
lamellation *n.* 纹[页]理
lamina ['læminə] *n. pl.* -nae,-nas; 薄板[片,层]；层状体；叶片
~ cribrosa 筛板
laminac *n.* 聚酯树脂, 泡沫塑料
laminal ['læminəl] *a.* 由薄板组成的, 成层的, 鳞片状的, 层理的
laminar ['læminə] *a.* 成薄层的, 薄片的, 层理的, 流线的
~ boundary layer 层流边界层
~ damping 层流衰减
~ displacement 薄层位移
~ film 层流薄模
~ flow 层流, 片流
~ fracture 成层断裂
~ layer 层流层
~ motion 层流, 片流
~ sand-flood 片状积沙
~ separation 层流分离现象
~ skin friction 层流表面摩擦
~ sublayer 层流底层
~ surface flow 水面层流
~ -turbulent transition 层流—紊流过渡段
~ velocity 层流速度
laminate ['læminit] ['læmineit] *a.*; *v.* -nated,-nating; 分层, 分成薄片；层压, 叠层；制成薄板, 包以薄片；*a.* 薄板状的, 片状的, 成层的；*n.* 层压板, 叠层制品, 层压材料；lamination, *n.*; laminator, *n.*
~ insulating board 成层绝缘板
~ moulding 层压模法
~ structure 片[层]状结构
phenolic ~ 酚醛树脂层压板
synthetic resin bonded ~ 合成树脂层压板
laminated ['læmineitid] *a.* 薄板状的, 由薄片组成的, 叠层的, 层压的
~ arch 层板[叠层]拱
~ article 叠层制品
~ basalt 板状玄武岩
~ beam 叠层梁
~ bearing 层压材料支承
~ board 层压板
~ clay 层状黏土, 纹泥
~ coal 泥煤, 页煤
~ composite structure 层叠式组合结构

~ construction 叠层结构
~ cover 叠层盖板
~ curved beam 弧形叠层梁
~ fabric 层压网纸板
~ fiber wallboard 层压纤维墙板
~ filter 叠层滤片
~ floor 叠层地板
~ form(work) board 叠层模板

laminated floor
~ frame 胶合叠层框架
~ girder 胶合叠层梁
~ glass 夹层玻璃,层压玻璃
~ glued timber arch 胶合叠层板拱
~ insulating board 叠层绝缘板
~ insulating glazing 多层绝缘玻璃片
~ insulating sheet 叠层绝缘板
~ jointing 板状节理
~ lattice girder 叠层格构梁
~ lead sheet 馏铅薄板
~ leaf spring 叠板弹簧
~ light (weight) building slab 建筑用轻型叠层板
~ lime 叠层石灰
~ limestone 叠层石灰岩
~ material 层压材料
~ melamine resin board 蜜胺树脂层压板
~ panel 夹层胶压镶板
~ paper 层压纸板
~ plastic 层压塑料,塑料贴面板
~ plastic panel with decorative surface 层压塑料装饰面板
~ polyvinyl chloride (PVC) cover(ing) 聚氯乙烯层罩面板
~ portal frame 叠层门架
~ rib 叠层肋
~ rock 纹层岩
~ safety glass 叠片胶合安全玻璃
~ shell roof 叠层薄壳屋顶
~ shim 叠层薄垫[填隙片]

~ shuttering board 叠层模壳板
~ soil 层状土
~ spring 叠板弹簧
~ structure 层状结构
~ system 叠层结构
~ timber 叠层木材
~ wire(d) glass 钢丝叠层保险玻璃
~ wood 层压[胶合]板
laminating *a.* 层压[叠合]法
~ composition 胶合料
~ film 层压薄膜
~ resin 层压树脂
~ sheeting 层压薄膜
contact ~ 接触层压
continuous ~ 连续层压
roll ~ 辊压层压
solvent ~ 溶剂法制塑料薄片
lamination [læmi'neiʃən] *n.* 层压,叠合,分层;成层状,叠层结构;层理,纹理
~ factor 层叠系数
cross ~ 交错纹理
parallel ~ 同向纹理层压
random ~ 非同向纹理层压
spring ~ 弹簧板,叠层板弹簧
laminator *n.* 层压机
laminboard *n.* 薄木片层夹板
laminwood *n.* 叠层[胶合]木
Lamipore *n.* 层压多孔金属材料
lamp [læmp] *n.* 灯,灯泡;电子管;*v.* 照亮
~ adapter 灯座适配器
~s and lanterns 灯和信号灯
~ bank 白炽灯组
~ base 灯座
~ black 灯烟,烟墨,灯黑
~ bracket 壁灯托架
~ cap 灯头
~ check 灯光检验
~ compartment 灯房
~ cord 柔性连接线
~ dimmer 灯光控制器
~ efficiency 发光效率

~ globe 圆灯罩
~ holder 灯头
~ hole （从地面伸向地下的）灯孔，灯管
~ house 灯房；灯罩，（仪器上的）光源
~ housing 灯光屏蔽罩
~ lens 灯玻璃
~ light 灯光
~ pendant 灯垂饰
~ plug 灯塞线，灯插头
~ pole 路灯柱
~ post 灯柱，路灯柱
~ -shade 灯罩
~ stand 灯台
~ socket 灯座，灯插头
~ switch 灯开关
approved ~ 安全灯
arc ~ 弧光灯
astral ~ 无影灯
blast ~ 喷灯
built-in ~ 墙顶内灯
ceiling ~ 吸顶灯
daylight ~ 日光灯
filament ~ 白炽灯
fluorescent ~ 荧光灯
incandescent ~ 白炽灯
mercury discharge ~ 汞灯
neon ~ 霓虹灯，氖灯
pendent ~ 吊灯
position ~ 座席灯
quartz mercury ~ 石英汞灯
strip-light ~ 顶灯，天幕灯
table ~ 台灯
wall ~ 壁灯
lamphole *n*. 灯井
lamproom *n*. 灯房
lamprophyre ['læmprəˌfaiə] *n*. 煌斑岩
lanai [lɑː'nɑːi] *n*. 门廊
Lancashire ['læŋkəʃiə] *n*. 英国的兰开郡

lamp house

~ boiler 兰开夏锅炉
~ brick 兰开夏砖
~ pattern distempering brush 兰开夏式平刷
~ two-flue boiler 兰开夏双烟道锅炉
lance [lɑːns] *n*.；*v*., lanced, lancing, 枪，矛；(前膛炮的)撞杆；喷杆，喷枪；*v*. 冲进；刺破
~ door 吹灰门
~ pipe 钻管
~ point 枪点；钻点
oxygen ~ 氧气切割枪
steam ~ 低压蒸汽除冰器
lancet ['lɑːnsit] *n*. 矢状饰；长窄尖头窗；小刀，小枪；折角条
~ arch 尖拱

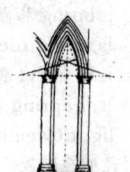

lancet   lancet arch

~ architecture 尖拱式建筑
~ window 尖头窗
lancing *n*. 气割；切缝
land [lænd] *n*. 土地，陆地，地面，国土；岸；国家；刀刃的厚度，刃背；活塞环槽脊；焊接区；*v*. 登陆，到达，降落，接地，卸船
~ abutment 桥台，岸墩
~ access benefit 田庄通道受益（道路筑成后田庄之得益）
~ accretion 土地围垦，土地填筑
~ agency 土地管理处
~ allotment 土地核记
~ annexation 土地兼并
~ area 陆域；土地面积
~ asphalt 地沥青
~ batture 河滩地（洪水期淹没）

~ bedding 土地分层耕作
~ betterment 土壤改良
~ bond 土地债券
~ boundary 地界
~ breeze 陆地风
~ capability 地力,土地生产力
~ capability map 地力图,地力分类规划图
~ carriage 陆上运输
~ certificate 土地证
~ chain 测链(66英尺)
~ clearing 清理场地
~ clearing blade 灌木清除机
~ contracts 土地契约
~ -controlled climate 大陆性气候
~ cost 地价
~ depot 贮木场
~ depression plant 内陆洼地电站
~ disposal 农田灌溉污水处理;土地处置
~ drain 地面排水
~ drainage 地面排水
~ dredger 挖土[泥]机
~ driver 陆上打桩机
~ed estate 地产
~ evaporation 陆面蒸发
~ expropriation 土地征用
~ expropriation right 土地使用权
~ facies 陆相
~ fall 山崩;滑坡
~ feature 地貌
~ fill 土地填筑
~ filtration 土地过滤,滤床灌溉
~ floe 大陆冰,厚浮冰块
~ form 地形
~ grader 灌木清除机
~ grading 土地平整
~ hemisphere 陆半球
~ ice 岸冰
~ improvement 土地改良
~ jobber 土地投机商
~ law(s) 土地法

~ leveler 平土机
~ leveling 土地平整
~ leveling project 土地平整工程方案
~ liable to floods 洪泛区
~ line 岸边线,道路用地边线
~ management 土地[农田]管理
~ map 地形图
~ mark 岸标,界标[桩]
~ melioration 土地改良
~ movement 土体位移[滑动]
~ ownership 土地所有权
~ patent 土地专利
~ pier 岸墩,桥台
~ pitch 地沥青
~ plane 土地整平机
~ planning 土地规划,建筑用地规划
~ plaster 石膏粉
~ pollution 大地污染
~ preparation 整地
~ price 抵岸价格
~ readjustment work 土地调整工作
~ reclamation 土地围垦,土地填筑
~ registration 土地登记
~ rentals 土地租费
~ resource 土地资源
~ resting 土地休耕
~ retirement 土地退化
~ revenue 土地收入
~ right 土地权
~ -service road 地方道路
~ shaping 土地整形
~ shield 陆上隧道建筑防护板
~ side 背水面
~ slide 坍方,滑坡
~ slip 坍方,滑坡
~ slope 土地坡度
~ smoothing 土地平整
~ spout 陆龙卷风
~ spring 表层泉
~ storage tank 地上油罐
~ subsidence 地面沉降

~ surface 地面
~ surveying 土地测量
~ swell 拍岸浪
~ tax 土地税
~ tenancy 土地租佃
~ tenure 土地使用, 土地使用权
~ tie 着地拉杆
~ topography 地形, 地貌
~ transport 陆上运输
~ treatment (污水)田间处理
~ trust 土地信托
~ trust certificate 土地信托证
~ upheaval 地面隆起
~ use 土地利用
~ use map 土地利用图
~ use planning 土地使用规划
~ value 土地价值
~ wall 岸壁
~ wash 海滨高潮线, 满潮海岸线
~ waste 岸屑, 风化石; 砂砾
~ water 陆地水
~ wave 地形起伏
back ~ 后方, 腹地
bottom ~ 河滩, 洼地
flood ~ 河滩, 洪泛区
made ~ 填土
marsh ~ 沼泽地
overflow ~ 漫流滩地
table ~ 高原, 台地
lander ['lændə] *n*. 罐座, 罐托; 登陆者
landfill waste site *n*. 垃圾填筑地
landing ['lændiŋ] *n*.
登陆, 着陆, 上岸; 码头, 月台; 楼梯平台; 沉淀
~ aid 着陆辅助设备
~ area 降落场
~ beam 楼梯平台梁
~ bearer 楼梯平台托架
~ (binding) joist 平台联结搁栅
~ bridge 登岸桥
~ carriage 平台搁栅; 登岸跳板

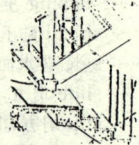
landing beam

~ ceiling joist 平台顶板搁栅
~ door (住宅楼的)楼门
~ entrance 电梯口
~ girder 平台梁
~ joist 楼梯平台搁栅
~ mat 金属路面板, 降落垫板
~ mats 装拆式飞机起降跑道
~ opening 电梯口
~ pier 登岸码头, 栈桥
~ place (飞机)降落场, 卸货地点
~ platform 楼梯平台
~ quay 靠岸横码头
~ runway 降落跑道
~ slab 楼梯平台板
~ stage 栈桥, 浮平台
~ step 平台踏步
~ strip 跑道
~ weight 卸货重量
floating passenger ~ 浮栈桥
half pace ~ 半中平台
incline ~ 下山车场
panel ~ 盘区车场
quarter space ~ (直角转弯的)梯台
stair ~ 楼梯平台
landlocked *a*. 陆地围绕的
~ body of water 闭塞水体
~ harbour 天然港, 陆封港
landmark *n*. 界标; 目标; 陆标; 里程碑
landscape ['lænskeip] *n*.; *v*., -scaped, -scaping; *n*. 风景, 景观; *v*. 使美化
~ architect 园林建筑师
~ architecture 园林建筑学
~ conservation 园林保护
~ design 园林设计
~ development 园林设计, 环境美化
~ engineering 景观工程
~ gardening 园艺学
~ painting 风景画
~ planning 环境规划
~ preservation 环境维护
~ protection 环境保护

~ treatment 园林设计,环境美化
~ work 环境美化工作
landside slope 内坡,背水坡
landslide ['lændslaid] *n.* 山崩,滑坡
~ flow 山崩流,泥石流
~ spring 山崩泉
landwash *n.* 高潮线
landwaste *n.* 岩屑,风化石,砂砾
lane [lein] *n.* 小路,小巷;车道;跑道,通道
~ -at-a-time construction 路面按车道施工法
~ capacity 车道通过能力
~ change (道路建筑时的)车道转换
~ divider 分道线,分车岛
~ flow 车道车流
~ gate 巷门
~ guide 车道指示标志
~ indicating signal 车道指示信号
~ joint 道路纵向缝
~ line 路面分道线
~ loading 车道荷载
~ marker 车道划线机
~ marking 车道标线
~ moving 车道转换
~ separator 分车带,车道分隔带
~ tortuous line 车道折线
~ -use control 车道使用控制
~ width 车道宽度
crossover ~ 转换车道
diagonal parking ~ 斜列式停车道
entrance ~ 进入快速干道的车道
exit ~ 驶出车道
median ~ 分隔带车道
turning ~ 回车道
waiting ~ 短时停车道
Lange lay *n.* 兰格捻(钢绞线捻向与钢丝绳捻向相同),顺捻
language ['læŋgwidʒ] *n.* 语言;机器代码;符号组; languageless *a.*
algorithmic ~ (ALGOL) 算法语言
assembly ~ 汇编语言
Common Business Oriented ~ (CO-BOL) 通用商务语言
computer ~ 计算机语言
conversational ~ 会话语言
FORTRAN ~ 公式变换语言
general purpose ~ 通用语言
object ~ 目标语言
original ~ 源语言
simulation ~ 模拟语言
lantern ['læntən] *n.* 灯,信号灯,提灯;罩,外壳;离心式泵外壳;灯笼式屋顶;塔式天窗
~ glass 灯用玻璃
~ light 提灯,挂灯;灯笼式天窗
~ light roof 灯笼式屋顶
~ opening 灯孔
~ pane 灯用玻璃
~ ring 套环;圆顶天窗圈梁
~ slide 幻灯片
~ tower 灯塔

lantern

lanthanum ['lænθənəm] *n.* 镧(La)
~ flint glass 镧火石玻璃
lanyard ['lænjəd] *n.* 牵索,短索;拉火绳
lap [læp] *v.*, lapped, lapping, *n.*;搭接,重叠,摺叠,研磨,磨光;包住,使成卷,使包住;盖板
~ belt (座位上的)安全带
~ cement 搭接胶合料
~ dovetail 搭接鸠尾榫
~ drag 叠板刮路机
~ fillet welding 搭接角焊
~ former 成卷[条卷]机
~ guide 导卷架
~ joint 互搭接头,搭接,叠接
~ -jointed sheeting 搭接板
~ length 搭接长度
~ link 连接两段链条的接头

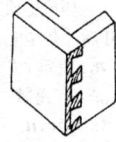

lap dovetail

~ mark 摺皱
~ of splice 搭接长度
~ -over seam 搭接缝
~ position 重叠位置
~ riveting 互搭铆接,叠式铆接
~ scarf 互搭楔接
~ seam 搭接缝
~ shaver 削匀机,刮皮刀
~ siding 互搭板壁
~ splice 互搭接头
~ welding 搭头焊,叠式焊
~ width (屋面瓦)搭接宽度
~ped butt 端搭接
~ped tenons 搭接榫
~ped type 重叠式(混凝土沉排型式)
forward ~ 前后重叠
side ~ of corrugation 互楞边搭
under ~ 重叠不足

lapicide ['læpisaid] *n.* 石工
lapidary ['læpidəri] *n. pl.* -daries; 宝石工艺,刻石工艺;宝石匠
lapie *n.* 岩沟,石灰岩沟
lapillus [lə'piləs] *n. pl.* lapilli;火山砾
lapie lazuli ['læpi'læzjulai] *n.* 天青石,琉璃璧;天蓝色
lapper ['læpə] *n.* 磨床,研磨机
lapping ['læpiŋ] *n.* 搭接,重叠;研磨,抛光
~ machine 精研机
~ plate 研磨板;搭接板
lapse [læps] *n.*, *v.*, lapsed, lapsing; *n.* 差错,失误;时间流逝,经过;废止,消失,递减,跌落; *v.* 失误,消失,失效; lapser, *n.*
~ limit 对流层顶
~ line 递减线
~ rate (温度)垂直梯度,递减率
lapwork *n.* 搭接工
laquear *n.* =lacunar 凹穴;凹格天花板
larch [lɑ:tʃ] *n.* 落叶松,落叶松木
~ fir 松杉木材
~ pine 南欧黑松
-tree 日本落叶松
lardalite *n.* =laurdalite 歪霞正长岩
larder ['lɑ:də] *n.* 食品室,贮藏间,食橱
large [lɑ:dʒ] *a.*, larger, largest;大的,大规模的,大量的,广泛的
~ aggregate concrete 大粒料混凝土
~ annual rings 大年轮
~ -area flood 大面积洪水
~ ballast 大块石渣
~ bin 大型储仓
~ blocked structure 大型砌块结构
~ bore 大孔径钻孔
~ brickblock 大型砖块
~ bubble aeration 大气泡曝气
~ calorie 大卡,千卡
~ -capacity air raid shelter 大型防空掩护所
~ cast (concrete) panel 大型预制混凝土板
~ core memory 大容量磁心存储器
~ deflection theory 大挠度理论
~ -diameter drilling machine 大孔径钻机
~ displacement matrix 大位移矩阵
~ electronic display 大型电子显示器
~ electronic panel 大型电子设备控制板
~ end bell 大头承口
~ finisher 大型路面抹面机
~ flat head nail 大头钉
~ format board 大规格板
~ grain 大颗粒
~ housing estate 大型居民点
~ injection 大量注入
~ knot 大树节(直径1.5英寸以上)
~ mallet 大锤
~ panel structure 大板结构
~ pole 大桶子木,大圆材
~ praying chamber 祈祷室
~ project 大规模工程
~ riprap 大块石,大块抛石护板
~ sample 大样本

~ scale 大规模的,大比例的
~ -scale construction 大规模施工
~ -scale convection 大规模对流
~ -scale corporation 大规模企业
~ -scale integration 大规模集成（电路）
~ scale production 大规模生产
~ -scale map 大比例尺地图
~ -scale photograph 大比例尺像片
~ scale site 大型建筑工地
~ -scale setting out 定线,放样
~ -scale survey 大比例尺测量
~ -span 大跨度的
~ -span beam 大跨度梁
~ -span floor slab 大跨度楼板
~ -span frame 大跨度构架
~ -span (load) bearing system 大跨度承重结构
~ -span roof 大跨度屋顶

large span roof

~ -span shell 大跨度薄壳结构
~ -span shell vault 大跨度薄壳拱顶

largest peak discharge 最大洪峰流量
Larimer column 工字钢组合柱
larmier *n.* 滴水槽,飞檐
larnite *n.* =belite 硅酸二钙,斜硅酸石
larry ['læri] *n. pl.* -ries;薄浆;拌浆锄;小车;铺层;手推车;翻底小车
~-ing 薄浆砌筑法,灌浆
Larssen section *n.* 拉森型材;拉森厚木板
Larssen steel sheet pile wall 拉森钢板桩墙
lase [leiz] *vi.* lassed, lassing;激光辐射,光激射
laser ['leizə] *n.* 激光;激光器,光激射发生器;*v.* 光激射
~ array 激光阵列
~ beam 激光束
~ beam projector 激光束投射器
~ collimator 激光准直仪
~ deflector 激光偏转器

~ distance measuring instrument 激光测距仪
~ flash tube 激光闪光管
~ hole drilling system 激光钻孔系统
~ level 激光水准仪
~ light 激光
~ processing 激光加工
~ range finder 激光测距仪
~ theodolite 激光经纬仪
~ velocimeter 激光测速仪
~ welding 激光焊接

lash [læʃ] *n.* 冲击,打击;鞭梢;空隙; *v.* 冲击,鞭打;用绳系住;清除岩石;lashless *a.*; lashingly *ad.*
~ method 振动法
~ rope 绑扎绳[子]
lasher ['læʃə] *n.* 拦河坝,溢流堰,蓄水池;清[装]石工;系索,捆绑绳索
lashing ['læʃiŋ] *n.* 捆索;耦合,连接;楔住,扭紧;清除岩石
~ wire 捆绑线
last [lɑ:st] *a.* 最后的,最近的;结论性的,极端的; *n.* 最终,结局; *v.* 持续,延长;经受住
~ ice date 终冰期
~ in 逆序换算
~ in first out 后进先出法
~ in on hand 后进现存法
~ in last out 后进后出法
~ quarter 下弦
~ing 持久的,坚固的,稳定的,最后的;持久,耐久
~ing properties 耐久性
lastics *n.* 塑料;弹料
latch [lætʃ] *n.* 门闩,插销;碰锁,弹簧锁,弹键;掣子,卡铁;凸轮; *v.* 闩上,上插销,上碰销;锁住,系紧,抓住
~ and lock 弹簧锁,碰锁
~ bolt 弹簧闩,碰簧销

latch

~ drive 闭锁驱动
~ hook 挂钩,锁钩
~ lock 弹键闭锁
~ nut 防松[锁紧]螺母
~ retainer 吸持塞孔,控制掣子
~ screw 弹键螺钉
~ spring 碰锁弹簧
~ stop 扣铁挡
~ system 译码装置
brake ~ 制动掣子
fire door ~ 炉门闩
holding ~ 卡子;定位销
quadrant ~ 掣子弧形板
thumb ~ 门窗插销
late [leit] *a.*, later or latter, latest or last;迟的,晚的,后的,最近的,滞后的
~ Art Nouveau 后期新艺术风格
~ Baroque 后期巴洛克艺术风格
~ Baroque architect 后期巴洛克建筑形式
~ -Frankish architecture 法兰克晚期建筑艺术
~ gate 后闸门
~ geometrical 晚期几何形窗花格,英国式几何窗花格
~ glacial deposit 后期冰川沉积物
Late Gothic 后期哥特式(装饰)风格
Late Gothic hallchurch 后期哥特式大厅教堂
Late Gothic tracery 后期哥特式几何窗格
Late Gothic vault 后期哥特式尖拱
~ maturity (river) 晚成熟期(河流)
~ (-)mediaeval architecture 中世纪晚期建筑艺术
~ model 新型的
Late Modern Style 近代艺术风格
~ opus 晚期作品
Late Pointed Style 英国哥特式建筑风格之后期,垂直线建筑风格
~ Renaissance architecture 文艺复兴后期建筑艺术,巴洛克建筑艺术
~ Romanesque (style) 仿罗马式建筑(风格)
~ Romanesque church 仿罗马式教堂
~ strength 后期强度
~ wood 秋材,晚材
latent [ˈleitənt] *a.* 潜在的,潜伏的
~ defect 潜在瑕疵,内在缺陷
~ clause 潜在瑕疵条款
~ energy 潜能
~ fusion 潜熔
~ heat 潜热
~ hydraulic binding agent 潜伏水硬性胶结剂
~ reactivity 潜在活性
~ solvent 惰性溶剂
~ stress 潜在应力
lateral [ˈlætərəl] *a.* 横[侧]向的,旁边的,支线的; *n.* 横向排水沟;支线线路;侧面;水平[抗风]支撑
~ abrasion 侧向侵蚀
~ acceleration 横向加速度
~ arch 侧向拱
~ area 侧面积
~ bars tread 横向条纹
~ bending 侧向弯曲
~ bracing 横向支撑,侧向水平连接系
~ branch 支流
~ buckling 侧向压曲
~ bulging 侧向膨胀
~ canal 旁支运河,支渠
~ clearance 侧向净空
~ cofferdam 侧向围堰
~ colonnade 侧柱廊
~ compression 侧向压缩
~ confinement 侧限
~ contraction 侧向收缩
~ corrosion 侧向磨蚀[冲刷]
~ crevasse 侧向裂缝
~ deflection 侧挠度
~ deformation 侧变形,横向变形
~ deviation 侧向偏差
~ diagonal 横向斜材

~ dimension 横向尺寸
~ displacement 侧位移
~ ditch 边沟,支沟
~ drainage 侧向排水
~ duct 支管道
~ earth pressure 侧向土压力
~ elevation 侧视图
~ entrance 侧门
~ erosion 旁蚀,侧向侵蚀
~ escape 侧向出口
~ expansion 侧膨胀
~ extensometer 横向伸长仪
~ facade 侧立面
~ face 侧面
~ fault 横错断层
~ flexure 纵向弯曲
~ flow 侧[横]向流
~ force 侧向力
~ friction 侧摩阻
~ gradient 横向坡降
~ headgate 支渠进水闸门
~ illumination 侧向照明
~ impact 侧向冲击
~ inflow 横向入流
~ lap 侧向重叠
~ leaching 侧向浸出
~ load 侧向荷载
~ moraine 侧碛,冰川侧翼的冰碛
~ movement 横向水平运动;沿层横向错动
~ overlap 侧向重叠
~ pipe 支管
~ placement 路边布局,侧向构筑物
~ pressure 侧向压力
~ pressure coefficient 侧压力系数
~ reaction 侧向反力
~ refraction 旁折射
~ reinforcement 横向钢筋,箍筋
~ resistance 横向阻力
~ restraint 侧向约束
~ ridge 侧面边缘
~ rigidity 侧向刚度

~ section 横截面
~ seepage 侧向渗流
~ service piping 消费管线,供水支管
~ sewer 侧向暗沟,污水支管
~ shear 横向剪切
~ sight distance 侧向视距
~ spillway channel 旁侧溢洪道
~ stability 侧向稳定性
~ storage 沿岸调蓄,支流蓄水
~ strain 侧应变
~ strength 横向强度
~ strut 侧向支撑
~ support 横向支承,侧面支撑
~ system 横向支撑系统;支渠系统
~ terrace 侧向阶地
~ thrust 横向推力
~ tie 横栏杆
~ ties of column 钢筋混凝土柱的横箍
~ tilt 侧向倾斜
~ truss 横向[抗风]桁架
~ variation 横向偏差
~ vibration 横振动
~ view 侧视图
~ wedge 侧向楔体
~ weld 侧焊,侧焊缝
~ wing 侧翼

laterally *ad.* 横向地,在[由]侧面
~ confined compression test 受侧限压缩试验
~ confined specimen 受侧限的试样

Lateran [ˈlætərən] *n.* (罗马)拉特兰大教堂;拉特兰宫

laterite [ˈlætərait] *n.* 红土,铁矾土
~ chip(ping)s 红土碎片
~ deposit 红土沉积
~ gel 红土凝胶
~ soil 红土,砖红壤

laterization [ˌlætəriˈzeiʃən] *n.* 红土化作用

laterolog *n.* 侧向测井

latex [ˈleiteks] *n. pl.* latices, latexes;

乳胶,橡胶液,乳液
~ adhesive 乳胶粘合剂
~ base 乳胶基
~ bitumen mixture 橡胶沥青混合物
~ emulsion 橡乳胶
~ floor(ing) (finish) 乳胶地面涂层
~ foam 泡沫乳胶
~ glue 乳胶粘合料
~ liquid 橡胶乳液
~ mastic 乳胶粘结剂
~ -modified cement paste 掺橡胶水泥浆
~ paint 乳胶漆

lath [lɑːθ] *n. pl.* laths;板条,灰板条;*v.* 钉条板,挂瓦条
~ and a half 1/4 英寸厚板条
~ and plaster 板条抹灰
~ brick 条形砖
~ ceiling 板条抹灰泥顶棚
~ fence 板条围栏
~ hammer 石膏工手槌;板条槌
~ insulating mat 板条抹灰泥绝热层
~ mesh 板条抹灰网
~ nail 板条钉
~ screen 板条帘
back ~ 板桩;顶板背板
eaves ~ 屋架端部的支梁,檐梁
espalier ~ 棚架板条
gauge ~ 挂瓦条
metal ~ 钢丝网
plaster's ~ 灰板条
rendering ~ 灰板条
rib ~ 肋条钢丝网
stiffened wire ~ 加劲钢丝灰网
wood ~ 木灰条

lathe [leið] *n.*;*v.*, lathed, lathing; *n.* 车床,旋床;*v.* 用车床加工
~ bed 车床床身
~ carriage 车床刀架[夹头]
~ chuck 车床夹盘
~ dynamometer 车床测力计
boring ~ 镗床
burnishing ~ 抛光车床

drill ~ 钻床
universal ~ 万能车床

lathing [ˈlɑːθiŋ] *n.* 钉板条;板条
~ ceiling 板条抹灰泥顶棚
~ hatchet 板条斧

lathwork [ˈlɑːθwəːk] *n.* 钉板条,板条涂灰泥;板条工作

Latin [ˈlætin] *n.* 拉丁语;拉丁人;*a.* 拉丁语的
~ architecture 拉丁建筑
~ square method 拉丁方格法

latite *n.* 熔岩,二长安山岩

latitude [ˈlætitjuːd] *n.* 纬度,黄纬;幅度,范围;纵距
~ determination 纬度测定
~ difference 纬差
~ line 纬线

latrine [ləˈtriːn] *n.* 沟厕,坑厕

latten [ˈlætən] *n.* 黄铜片;金属薄板

lattice [ˈlætis] *n.*;*v.*, -ticed, -ticing; *n.* 格子,格构,网[栅]格;点阵;晶格;支承桁架;*vt.* 做成格子
~ bar 格条
~ beam 格构梁
~ boom 格构梃杆
~ bracing 格构撑架
~ bridge 格构[梁]桥

lattice bridge

~ column 格构柱
~ derrick crane 格梁动臂起重机
~ door 格构门
~ drainage 格状排水系
~ fence 格构围栏
~ frame 格式构架
~ girder 格构大梁
~ jib 格构梃杆
~ mast 格构杆,桁架杆
~ network 格状网络
~ pole 格构桅杆

~ purlin 格构檩条
~ pylon 格构牌楼
~ rod 格条
~ roof 格构屋顶
~ screen 格构式栏污栅
~ shell 格构薄壳
~ square 方格形
~ structure 网格结构,格状构造
~ strut 格构支撑
~ tower 格架塔
~ truss 格构桁架
~ -type portal frame 格构型框架
~ wall 花格墙
~ web 花腹板
~ window 斜条格构窗,花格窗
~ work 格构工作
~d 结构的,缀合的,有格子的
~d bar 缀合杆
~d column 格构柱
~d cupola 格构穹顶
~d partition wall 缀合隔墙
~d stanchion 缀合支柱
~d structure 格构结构
~d strut 格构支撑
space ~ 立体格子
lauan [ləˈwæn] *n.* 柳安木
launch [lɔːntʃ] *v.* 下水;发射,升空;滑曳; *n.* 下水;汽船; launchable, *a.*; launcher, *n.*
~ing 入水,下水;升空;(桥梁架设)滑曳
~ing cable 牵索
~ of caisson 沉箱下水
~ing nose 滑曳导梁
~ing slide 下水滑道
~ing trolley 滑曳空中吊车
~ing way 下水滑道
launder [ˈlɔːndə] *n.* 槽,洗矿槽,出钢槽,洗涤槽; launderable *a.*; launderability *n.*; launderer *n.*
discharge ~ 卸载槽
primary ~ 初洗槽
rewash ~ 再洗槽
tailing ~ 尾矿槽
laundry [ˈlɔːndri] *n. pl.* -dries;洗衣房[店]
~ chute 脏衣物滑槽
~ club 公共洗衣店
~ drying ground 晾衣场
~ machine 洗衣机
~ room 洗衣房
~ tray 洗衣槽
~ tub 洗衣盆
~ wastes 洗衣废水
lava [ˈlɑːvə] *n.* 火山熔岩
~ ash 火山灰,熔岩灰
~ basalt 玄武岩
~ bed 熔岩层,熔岩床
~ flow 熔岩流
~ plain 熔岩平原
~ slag 熔岩渣
~ soil 熔岩土壤
~ tub 熔岩管(熔岩流空的管状孔道)
lavabo [ləˈveibəu] *n. pl.* -boes,-bos;盥洗器;洗手盆,壁盆,水箱
lavatory [ˈlævətəri] *n. pl.* -ries;盥洗室,厕所
~ basin 盥洗盆
~ fan 盥洗室排气扇

lavatory waste pipe

~ flush(ing) 厕所冲水
~ pan 大便器
~ seat 抽水大便器座
~ waste pipe 盥洗排水管
lavender [ˈlævində] *n.;a.* 淡紫色(的)
law [lɔː] *n.* 法律;定律;守则,规则
~ of conservation of energy 能量守恒定律

~ of conservation of matter 物质不灭定律
~ of inertia 惯性定律
~ of similitude 相似律
~ of universal gravitation 万有引力定律

lawful [ˈlɔːful] *a.* 合法的,法定的,法律上的
~ operations 合法经营
~ property 合法财产
~ rights and interests 合法权益

lawn [lɔːn] *n.* 草地[坪,场]
~ mower 割草机
~ sprinkler 草地喷灌器

lawn mowers

lawrencium [lɔːˈrensiəm] *n.* 铹(Lw)

lay [lei] *v.* laid, laying; 放置,安排.铺砌,敷设;浇筑;提出,拟定;绞,捻; *n.* 位置,地形,情况,状态;绳索的股数及拧法,绞距
~ -aside 路侧停车处;超车车道;备用车道
~ barge 敷设管线驳船
~ by 保留;延展;路旁停车处;路支叉及其加宽部;避车道
~ -by material 放于路侧的备用材料
~ day 停工日
~ -days 停泊[装卸]时间
~ -down thickness 松铺厚度
~ edges 空白边
~ -in ceiling 嵌入式顶棚
~ land 生荒地,处女地
~ light 平窗采光
~ of land 地形,地貌
~ on 安装(水,煤气等);涂漆,涂水泥
~ out 绘样,设计;镶砌;陈列
~ panel 平纹镶板
~ ratio 绞距系数
~ shaft 副轴
~ the grain 整光,砑光;上浆
~ tongs 长柄管钳

Lang ~ 同向捻,兰氏捻
ordinary ~ 异向捻,交叉捻
simple ~ 单绞

laydown *n.* 沉积[淀]作用;搁置,放下;铺设,建造
~ machine (沥青混凝土的)铺设机

layer [ˈleɪə] *n.* 层,层数,层次;焊层,涂层,铺层工,砌砖工
~ board 天沟托板
~ -built embankment 分层铺筑的(路)堤
~ construction 分层铺筑[施工]
~ equivalency 当量层厚
~ insulation 层间绝缘
~ silicate 黏土矿
~ed 成层的,分层的
~ed clay 分层黏土
~ed map 分层着色地图
~ed pavement 成层路面
~ed permafrost 层隔永冻层
~ed soil 成层土
~ed structure 成层结构
~ed system 层状体系
absorbed ~ 吸收[附]层
barrier ~ 阻挡层
bearing ~ 含水层
boundary ~ 边界层,附面层
brick ~ 砌砖工
finishing ~ 上层,饰面层
laitance ~ 水泥翻沫层
pipe ~ 铺管工
plate ~ 铺路工
rail ~ 铺轨机
supporting ~ 承托层
surface accumulation ~ 表面增强层

laying [ˈleɪɪŋ] *n.* 铺设,浇筑,砌筑;绳的搓法;初涂底层
~ and finishing machine 摊铺整修机
~ area 砌筑面积
~ attachment 管道铺设附属设备
~ barge 铺管驳船

~ blocks 砌石
~ bricks 砌砖
~ depth 埋置深度;铺筑厚度
~ drawing 铺设图
~ guide (屋面卷材的)铺设线
~ length 铺设长度
~ line 铺设线
~ machine 铺路机
~ of pipe 埋管,铺管
~ -off 停工,下料
~ out 定线,放样,敷设
~ pattern 铺砌用模型板
~ pipes 铺设管道
~ point 照准点;工作面
~ rate 铺砌速度
~ ship 管道铺设船
~ site 工作面
~ technique 铺砌技术
~ temperature (道路涂面的)铺筑温度
~ tiles 砌石
~ trowel 大灰抹子
~ winch (管道)铺设绞车
brick ~ 砌砖
cable ~ 电缆铺设
final ~ 最终瞄准
grout ~ 灌浆铺砌
optical ~ 光学瞄准
track ~ 轨道铺设
tunnelless underground ~ 无沟地下铺设

layman ['leimən] **n. pl.** -men;非专业人员,外行;laymanship,**n.**

layout ['leiaut] **n.** 计划,设计,方案,布置,规划,定线,放样,草[略]图,轮廓,草案,事态,形势
~ chart 布置图,线路图,施工流程图
~ design 图纸设计
~ location 定线
~ machine 测绘缩放仪
~ of construction work 施工场地布置
~ of reinforcement 加固方案
~ of round 一组炮眼分布
~ plan 规划,平面布置图
~ survey 定线测量
alternate ~ 比较设计
fan ~ of sewers 扇形排水管系
gridiron fashion street ~ 格状道路布置
service ~ 屋内管线布置
wiring ~ 装配图,布线图

layover ['leiəuvə] **n.** 中途停留;(公共交通)终点停车处

layshaft ['leiʃɑːft] **n.** 副轴,中间轴

layup ['leiʌp] **n.** 扭绞,绞合;铺叠;敷层;树脂浸渍增强材料
RR ~ 聚脂树脂增强层压材料

lazulite ['læzjulait] **n.** lazulitic **a.** 天蓝石(的)

lazurite ['læzjurait] **n.** 天青石,青金石

lazy ['leizi] **a.** , -zier,-ziest;懒惰的,迟钝的
~ arm 吊臂
~ bar 火炉工具挂杆
~ board 木制支架
~ guy 吊杆稳索
~ -jack 屈伸起重机;胀缩补偿装置
~ stream 缓流
~ thermometer 惰性温度表
~ -tongs 同步机构;长夹具

L-bar (不等边)角钢

L-beam (不等边)角钢

leach [liːtʃ] **v.** 沥滤,浸出,滤取;淋溶[洗],漂洗 **n.** 沥滤器,滤灰池
~ed brown earth 强淋溶土,灰棕土
~ed horizon 淋溶层
~ed layer 淋溶层
~ed soil 淋溶土
~ed surface 渗漏面,淋溶面
~ed zone 淋滤带
~ing 滤出,过滤,淋滤,浸出;混凝土渗溶作用;浸析物
~ing agent 助滤剂
~ing basin 滤水池

~ing cesspool 污水渗井
~ing factor 淋洗因数
~ing operation 淋洗作业
~ing pit 渗水井坑
~ing rate 渗出率
~ing test 淋溶试验
~ing water 淋洗水
~ing well 渗水井
agitation ~ 搅动沥滤
confined ~ing 槽内浸滤
heap ~ing 堆摊浸滤
neutral ~ing 中性浸出
sand ~ing 砂滤
vat ~ing 槽浸

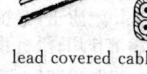
leaching cesspool

lead [li:d] *n.*; *v.*, led, leading; *n.* 导线,引线;导管[柱],导程;螺距;通路,引水沟; *v.* 领导,导向;移前,输入; [led] *n.* 铅,铅锤;枪弹
~ alloy 铅合金(焊锡)
~ apron 铅护裙,铅护檐板
~ attachment 引线焊接
~ (base)(priming) paint 铅基漆
~ biological shield 生物防护铅屏
~ block 导向滑轮;铅砖(用于原子反应堆)
~ bonding 引线接合
~ bronze 铅青铜
~ burned joint 铅焊接头
~ burning 铅焊
~ cable 铅皮电缆
~ cable sheath 电缆铅包皮
~ capping 铅帽盖;铅排水口
~ caulking 铅封,铅填缝
~ chromate pigment 铬酸铅颜料
~ coating 包铅,铅皮
~ coping 铅帽盖
~ covered cable 包铅电缆

lead covered cable

~ cupola 包铅圆屋顶

~ curve 导向曲线
~ damp course 铅皮防水层
~ dome 包铅圆屋顶
~ dowel 铅销钉
~ draining pipe 排水铅管
~ drier 铅催干剂
~ dust 铅灰
~ elbow 铅管弯头
~ facing 铅涂面层
~ filler 铅衬垫[垫片]
~ flashing (piece) 铅泛水条,铅挡条
~ flat 铅皮平屋顶
~ flushing pipe (厕所)冲水铅管
~ foil 铅箔
~ foundry 铸铅
~ fuse wire 铅保险丝
~ glance 方铅矿
~ glass 铅玻璃
~ glaze 铅釉
~ glazing 铅条镶嵌玻璃
~ grey 铅灰色
~ gutter 铅檐沟
~ hammer 铅锤
~ hip 铅斜屋脊
~ insert 铅垫片
~ joint 填铅接合缝
~ limit switch 行程限位开关

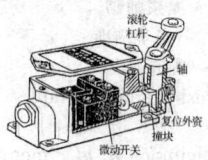
lead limit switch

~ line 测深绳
~ lined pipe 衬铅管
~ lining 铅衬,镀铅
~ manganese drier 铅锰干燥剂
~ matte 铅锍
~ monoxide 氧化铅
~ ore 铅矿石
~ oxide 氧化铅
~ pad 铅垫

~ paint 含铅漆
~ patenting 铅浴淬火
~ pigment 铅颜料
~ pipe 铅管
~ plate 铅板
~ -plating 镀铅
~ plug 铅塞,铅销
~ plumb 铅垂线
~ plummet 铅垂球
~ poisoning 铅中毒
~ primer 铅底[基]漆
~ rainwater gutter (屋顶)铅檐沟
~ red 红丹
~ regulus 铅块
~ riser 引线头
~ roof gutter 铅屋檐沟
~ roofing 铅皮屋面
~ screw 导杆,丝杠
~ seal 铅封;灌铅
~ sheet 铅皮
~ sinker 铅锤
~ soaker 铅皮泛水
~ time 准备时间;订货至交货的时间
~ -tin solder 铅锡焊料
~ tower 指向塔;打桩架
~ trough 铅屋檐沟
~ washer 铅垫片
~ weight 铅锤
~ wire 铅丝
~ wool 铅毛
~ zine primer 铅锌底漆
~ed 加铅的;有引线的
~ed bevelled plate (镶铅条的)斜边彩画玻璃
~ed bronze 铅青铜
~ed glass 镶铅玻璃
~ed joint 铅封接
~ed light 花饰铅条窗
~ing 领导的,导向的;超前的,主要的;铅制品,加铅
~ing block 导向块
~ing diagonal 主对角线

~ing edge 前缘[沿]
~ing end 前端,引导端
~ing frame 主构架
~ing-in girder 导梁(干船坞中用)
~ing-in wire 引入线
~ing mark 标志
~ ing marks method 导标断面测深法
~ing pile 导桩,定位桩
~ing screw 导杆,丝杠[攻]
~ing truck 前导转向架
~ing wheel 导轮,主动轮
~ing wire 导线
black ~ 石墨
flat ~ 铅皮,铅薄板
hand ~ 测深铅锤
ground ~ 接地线
main ~ 母线,电源线
pile driver ~ 打桩机导柱
leader [li:də] *n.* 导管[杆];引出线,落水管,排水沟;领导者
~ head 水落斗
~ hook 水落管箍
~ shoe 水落管斜口
~ pipe 水落管
~ strap 水落管固定带
leads *n.* 打桩机导柱
leadwork *n.* 铅衬,铅制品;铅熔炼厂
leaf [li:f] *n. pl.* leaves,*v.*; *n.* 叶片,薄片;门扉;页,箔;开合桥的翼;*v.* 长叶;翻页
~ and dart 叶饰与箭头饰
~ and tongue 叶饰与舌饰
~ bridge 开合桥
~ dam 活瓣式闸门坝,活板坝
~ door 摺门,双扇门
~ -gilding 贴金箔,包金
~ -like decoration 叶饰
~ mould 腐叶土
~ of blocks (空心墙)砖外壳
~ of hinge 合页片
~ -shaped curve (哥特式)叶状窗

花格
~ spring 片弹簧,钢板弹簧
~ -valve 舌阀,簧片阀
~ wood 阔叶林木,硬材
~ed pigment 片状颜料
acanthus ~ 叶形板
door ~ 门扇
gold ~ 金箔
tin ~ 锡箔
lenfage ['li:fidʒ] *n.* (树)叶,叶饰
league [li:g] *n.* 里格(长度单位,约等于三英里);联盟(联合会);种类,范畴
leak [li:k] *v.* 泄[渗]漏,渗流,泄露; *n.* 渗漏处,漏缝;漏气[水]漏失量
~ clamp 防漏管箍
~ detector 检漏器
~ hunting 检漏,测漏
~ indicator 渗漏显示器
~ locator 测漏器
~ -off chute 漏水槽
~ preventive 防漏剂
~ test 渗漏试验
~ -tight 不漏的,密封的,无漏损的
leakage ['li:kidʒ] *n.* 漏,泄漏;渗透,渗流;漏泄物;漏泄量
~ conductor 线路避雷器
~ factor 渗漏系数
~ loss 渗漏损失
~ path 渗径
~ water 渗漏水
leakance ['li:kəns] *n.* 泄漏,漏电,泄漏系数
leakless ['li:klis] *a.* 不漏的,防漏的
leakproof ['li:kpru:f] *a.* 防漏的,不漏的
leaky ['li:ki] *a.* leakier, leakiest;漏的,易泄漏的,有漏洞的;松的;-leakiness, *n.*
~ aquifer 漏水含水层,渗漏蓄水层
~ foundation 漏水地基
~ joint 渗漏的接缝,不密闭的接缝
~ riveting 松铆
leam [li:m] *n.*; *v.* 沼泽地排水
lean [li:n] *n.*; *a.*; *v.*, leaned or leant,

leaning; *v.* 倾[偏]斜;歪曲;倾向于,依靠;使贫困化; *a.* 贫瘠[乏]的,瘦的,质劣的; *n.* 倾[偏]斜
~ clay 贫黏土
~ coal 低级煤
~ concrete 贫混凝土,少灰混凝土
~ concrete base 贫混凝土底层[基底]
~ gas 贫(煤)气
~ lime 贫石灰
~ material 选矿后的废石
~ mix 贫拌合,少灰混合,少灰混合料
~ -mixed concrete 贫混凝土,少灰混凝土
~ mixture 贫拌合,少灰混凝料
~ mortar 贫砂[灰]浆
~ ore 贫矿
~ rolled concrete 碾压贫混凝土
~ -to 单坡棚,一面坡屋
~ -to dormer 单坡屋顶老虎窗
~ -to mansard roof 单面折线形屋顶
~ -to roof 单坡屋顶
~ -to roof purlin(e) 单坡屋架檩条
~ -to trussed strut 单坡桁架支撑
~ing tower 斜塔

lean-to

leap [li:p] *n.*; *v.*, leaped or leatp, leaping; *n.* 跳跃,跳跃的距离[高度]; *v.* 跳跃,跃过
~ -frog fashion 蛙跳式排水,多级明坑排水
~ -frog method 跳步[跳点]法
~ -frogging 蛙跳式
~ing formwork 升模
~ing frog rammer 蛙式打夯机
~ing shuttering 滑升模板
~ing weir 下水道溢流堰
leapfrog ['li:pfrɔg] *n.* 蛙式夯,动力夯;蛙跳式前进
lear board 排水槽支承板,天沟托板
lease [li:s] *n.*; *v.*, leased, leasing; *n.*

租约[契];租期; v. 租借,出租; lease-less v. ; leaser n.
~ed circuit 专用路线
least [li:st] *a.* 最小的,最少的; *n.* 最小,最少
~ cost 最小成本
~ cost estimating and scheduling 最低成本估计与预定
~ count 最小读数
~ dimension 最小尺度
~ disadvantage 最小损失
~ -energy principle 最小能量原理
~ error 最小误差
~ load 最小荷载
~ moment 最小弯矩
~ perceptible chromaticity difference 最低可见(色度)差
~ price 最小价格
~ radius of gyration 最小回转半径
~ reading 最小读数
~ square 最小二乘法
~ work principle 最小功原理
leastone *n.* 层状砂岩
leather ['leðə] *n.* 皮革,革制品
~ bellows 皮风箱
~ belt 皮带
~ cloth 人造革,漆布
~ collar 皮垫圈
~ cup 皮碗
~ hard (毛坯的)半干状态
~ gasket 皮垫圈
~ packing 皮革填充物,皮垫
~ seal(ing) ring 皮带止水密封圈
~ washer 皮垫圈
fancy ~ 装饰革
lining ~ 衬里革
patent ~ 漆皮
leatheret(te) [ˌleðə'ret] *n.* 人造革,假皮
leave [li:v] *v.* , left, leaving; 离开,动身,出发;留下,放置
lebbek-tree 山槐,大叶合欢
lech *n.* 拱顶石

leck *n.* 硬黏土,致密黏土,石状黏土
lectern ['lektə:n] *n.* 桌面倾斜的讲台
lecture ['lektʃə] *n.* 讲演,讲座
~ experiment 演示实验
~ hall 大讲堂,演讲厅
~ room 讲堂,教室
~ theater 大讲[礼]堂,阶梯教室
ledbit *n.* 有铅垫板的沥青防水层
ledge [ledʒ] *n.* 横档,壁架,窗台;突出部分;急流,浅滩,矿脉,地层; ledged, *a.* ; ledgeless, *a.*
~ excavation 岩石开挖
~ joint 搭接接合
~ rock 礁石,岩床
~d and braced door 直拼斜撑门

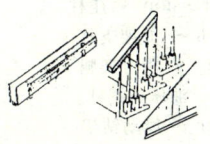

ledged and braced door

~d door 直拼撑门
epitaxial ~ 外延台阶
ledger ['ledʒə] *n.* 总账;(脚手架的)横杆,卧木,底板
~ account 分类账
~ assets 账面资产
~ board 脚手架,栏杆扶手,木架隔层横木,栏顶板
lee [li:] *n.* 下风,背风面 *a.* 背风的
~ breakwater 纵向防波堤
~ shore 背风岸
~ -side 背风面
~ tide 下[顺]风潮
leeway ['li:wei] *n.* 活动余地;容许偏差;时间损失;风压差
lefkoweld *n.* 环氧树脂类粘合剂
left [left] *a.* 左面[侧,翼]的 *n.* 左侧
~ elevation 左视图
~ hand door 左开门
~ -handed twist 左扭转
~ -hand screw thread 方旋螺纹

~ lane 左边车道
~ lay 左捻
~ marginal bank 左边岸
~ shift 左移位
~ view 左视图

leg [leg] *n.*; *v.*, legged, legging; *n.* 腿,柱,支架;支线,支路;(三角形的)股,边;(三相系统的)相位
~ bridge 以立柱作支承的梁式桥
~ of frame 构架支柱
~ piece 立柱,撑杆
~ pipe 冷凝器气压管
~ room （座位前）伸脚的空间
~ type 支腿式
~ vice 台虎钳
compression ~ 压柱
derrick ~ 井架支柱
folding ~ 折合架腿
oleo ~ 油压减震柱
pusher ~ 风动钻架
shear ~ crane 动臂起重机
spacer ~ 隔离支柱
stiff ~ derrick 刚性柱架,斜拉杆式起重机
telescope ~ 伸缩腿架

legal ['li:gəl] *a.* 法律上的,法定的,合法的
~ effect 法律效力
~ entity 法律实体
~ framework 法律体制
~ heir 法定继承人
~ hour 法定时
~ guardian 法定监护人
~ limit （汽车的）法定速率限制
~ instrument 法律文件
~ interest 法定利息
~ investment 合法投资
~ reserves 法定发行储备
~ securities 法定担保
~ standard 法定标准
~ surplus 法定公积金
~ system 法制

~ value 法定价值
~ water level 正常水位

legend ['ledʒənd] *n.* 图例,符号表;传说

legibility [ˌledʒə'biliti] *n.* 易读性,清晰度
~ distance （标志）可读距离

lehr [liə] *n.* 玻璃锻烧炉;长隧道窑

leisure ['leʒə] *n.* 休养,游览
~ area 游览区

Lemac *n.* 聚醋酸乙烯酯系列

lemon ['lemən] *n.* 柠檬; *a.* 柠檬的;淡黄色的
~ yellow 柠檬黄

length [leŋθ] *n.* 长度,距离;持续时间
~ average 平均长度
~ breadth ratio 长宽比
~ change 长度变化
~ coefficient 长度系数
~ diameter ratio 长度直径比
~ measurement 长度量测
~ of cantilever 悬臂[梁]长度
~ of embedment 埋入长度
~ of penetration 贯入度
~ of perpendiculars 垂直距离
~ of pipe 管子长度
~ of stroke 冲程
~ of run 运程
~ of side 边长
~ rod 测杆
crest ~ （坝、堰的）顶长
critical ~ 临界长度
gap ~ 隙宽
laid ~ （管道的）铺设长度
overhanging ~ 自由长度,梁的伸长
space ~ 空间距离,间距
unsupported ~ 自由长度,无支承长度
wall ~ 工作面长度

lengthen ['leŋθən] *v.* 加[接]长,延伸
lengthener, *n.*
~ing bar 延伸杆
~ing piece 接长杆

lens [lenz] ***n. pl.*** lenses;透镜,镜头,镜片;(汽车)灯玻璃; lensless, ***a.***; lenslike, ***a.***
　~ aperture 透镜孔径
　~ of clay 黏土透镜体
　~ truss 叶[梭,鱼]形桁架
lenticular [len'tikjulə] ***a.*** 透镜状的,双凸的,扁豆状的; lenticularly, ***ad.***
　~ arch 双叶拱
　~ beam 鱼腹式梁,组合梁
　~ girder bridge 鱼腹式梁桥
　~ truss 鱼腹式桁架,叶形桁架
lentiform ['lentifɔːm] ***a.*** 透镜状的,双凸的,扁豆状的
　~ beam 鱼形梁
lepol kiln 列波尔水泥回转窑
lesbian cyma（tium） 古希腊波状花边,反曲线状花边
lesche 古希腊宴会厅
lessee [leˈsiː] ***n.*** 租户,承租人; lesseeship, ***n.***
lessor [leˈsɔː] ***n.*** 出租者
let-in brace 嵌[插]入式斜撑
letter ['letə] ***n.*** 字母;文字;书信; ***v.*** 写上字;加标题
　~ case 分信箱,文件夹
　~ character 字母符号
　~ constract 书面合同
　~ decoration 字母装饰
　~ of authority 授权书
　~ of authorization 委托书
　~ of intent (合同)意向书
　~ of introduction 介绍信
　~ of offer and acceptance 交货验收单
　~ of undertaking 承诺书
　~ ornament 字母装饰
　~ plate 信箱投信口开缝板
　~s patent 专利[特许]证
　~ (-type) combination lock 字母锁
　~ed message (标志上的)文字通告
leucite ['ljuːsait] ***n.*** 白榴石
leucocrate ***n.*** 淡色岩

levee ['levi] ***n.***; ***v.***, -eed, -eeing; ***n.*** 防洪堤,天然冲积堤;码头; ***v.*** 筑防洪堤
　~ crown 堤顶
　~ grade 堤顶纵坡;堤顶高
　~ maintenance 堤防维护
　~ muck ditch 堤脚泥沟
　~ ramp 堤上坡道
　~ revetment 堤面护坡
　~ ridge 冲积堤脊线
　~ sluice 堤上泄水闸
　~d bank 堤岸,淤填沙滩
　~d channel 有堤渠道
　~d pond 堤成池
level ['levl] ***n.***; ***a.***; ***v.***, -el(l)ed, -el(l)ing; ***n.*** 水平[准],标高;水平线[面];水准仪,级别,等级; ***a.*** 水平的;相等的 ***v.*** 使成水平; levelness, ***n.***
　~ bar 水准[平]尺
　~ book 水准手册
　~ bridge 平桥
　~ controller 水位控制器
　~ correction 水平校正
　~ crossing 平面交叉,水平交叉
　~ crossing gate 平交道拦,路栅
　~ cutting 水平开挖[路堑]
　~ detector (贮仓)仓面指示器
　~ gauge 水准仪,液面指示器
　~ gutter 平檐沟
　~ hydrograph 水位过程线
　~ instrument 水准仪
　~ line 等高[水平]线
　~ -luffing crane 鹅头伸臂起重机
　~ net 水准网
　~ off flush 整平,使平整
　~ of illumination 照度等级
　~ of management 管理水平
　~ of service 路况等级;服务水平
　~ of zero of gauge 水尺零点高程
　~ plane 水准面
　~ surface 水准面

level controller

~ survey 水准测量
~ terrain 平地,平坦地区
~ theodolite 水准经纬仪
~ track 水平轨道,平路
~ -up course 平整层
~(l)ing 水准测量;整平
~ing adjustment 水准测量平差
~ing base 水准基点
~ing blanket 找平层
~ing course 整平层
~ing instrument 水准仪
~ing layer 整平层
~ing machine 平土机,平整机
~ing mass 整平充填料
~ing mortar 找平灰浆
~ing network 水准网
~ing peg 水准测量桩
~ing pole 水平标杆
~ing process 水准测量;整平横断面
~ing ring (道路排水口的)整平圈
~ing rod 水准尺,水平标杆
~ing screw 校平螺旋
~ing staff 水准标尺
~ing support 校平[水准]架
~ing survey 水准测量
~ing work 整平工作
acceptable quality ~ 允许质量指标,质量合格标准
addressing ~ 选址级数
adit ~ 平硐水平
bottom ~ 底石,座板
extracting ~ 回采水平
flood ~ 洪水位
formation ~ 施工基面
mean sea ~ 平均海面
rail ~ 轨面
still water ~ 静水位
stop ~ 汽车停车场标高
trigonometric ~ing 三角高程[水准]测量
zone ~ing 区域平均法,区域整平

leveling rod

leveler ['levlə] n. 水准测量员;整平器;平地机,矫平机
rail ~ 钢轨整平机,平轨机
roll ~ 辊式矫直机
stretcher ~ 拉伸矫直机
lever ['li:və] n. 横杆,手柄
~ amplification strain gauge 杠杆放大式应变仪
~ arm 杠杆臂
~ boards 活动百叶窗,活动气窗
~ bolt 杠杆式插销
~ brake 杆闸,手刹杆
~ change 变速杆
~ clamp 杠杆夹具
~ deflectometer 杠杆弯沉仪(贝克曼弯沉仪)
~ drawbridge 杠杆操纵活动桥
~ fulcrum 杠杆支点
~ gear door 提升门
~ handle 搬把式门拉手
~ jack 杠杆式千斤顶
~ latch 杠杆锁
~ principle 杠杆定理
~ shears 械杆式剪切机
~ tongs 杠杆式夹具
~ -wood 美洲铁木
leverage ['li:vəridʒ] n. 杠杆作用;杠杆装置;杠杆效率
levigate ['levigeit] v., -gated,-gating;粉碎,研末,磨光,淘选;levigable, a.; levigation, n.
~d abrasive 细磨磨料
levy ['levi] v., levied, levying, n. pl. levies; v. 征税,抽税; n. 征税,征收; levier, n.
~ tax 征税
levy's criterion (重力坝设计的)利维原理
lew n. 避雨棚
lewis ['lu:is] n. 起重爪,吊楔
~ bolt 地脚螺栓;吊楔螺栓

lewis

library ['laibrəri] **n. pl.** - braries; 图书馆; 库程序库
~ routine 库存程序
licence ['laisəns] **n.** 许可证, 执照, 牌照; **v.** 许可证, 批准
~ fee 执照费
~ plate 汽车牌照
lich house n. 殡舍, 停尸房
lid [lid] **n.** 帽, 盖, 罩
lierne [li'ə:n] **n.** 枝肋
~ vault 扇形肋的穹顶
life [laif] **n. pl.** lives; 生[寿]命; 使用期限, 耐久性
~ belt 安全带
~ boat 救生船
~ -buoy 救生圈
~ expectancy 预期使用寿命
~ line 安全线, 保险带
~ preserver 救生器
~ size 和实物一样大小, 原大小[尺寸]
~ span 使用年限
flex(ing) ~ 挠曲寿命
overhaul ~ 大修周期
rupture ~ 持久强度
lift [lift] **n.** 举升, 提起, 上升; 升力; 上升高度; 楼层; 升船机, 升降机, 电梯, 卷扬机; **v.** 提升, 举起; liftable, **a.**
~ arm 提升臂
~ attendant 电梯操纵员
~ away shutter door 提升门
~ block 起重滑轮; 电梯间
~ bridge 升降桥, 吊桥
~ capacity 提升能力
~ car 电梯厢
~ cargo 承运货物
~ dock 浮船坞
~ gate 提升式闸门
~ guide rail 电梯导轨
~ hook 吊钩
~ irrigation 提水灌溉
~ latch 弹簧锁

lift bridge

~ lock 单级船闸
~ of lock 船闸水头
~ off 卸下
~ of pump 水泵扬程
~ -off pressure (预应力混凝土构件)离地张力
~ -on/ ~ -off (LO/LO) system 吊装法
~ pot 提升罐
~ shaft 升降机井
~ -slab 顶升楼板
~ -slab column 升板支柱
~ -slab construction 升板法施工
~ span 提升式桥孔
~ station 提升站, 扬水站
~ truck 叉车, 自动装卸车
~ -type car park 升降式车库
~ valve 提升阀
~ van [货物]装卸箱
~ wall 闸室墙
~ well 升降机井, 电梯井
~ with axial floater 轴向浮筒式升船机
~ with counter weight 平衡重式升船机
~ing appliance 提升装置, 起重设备
~ing barrier 升降式栏木
~ing beam 起重梁, 启门梁
~ing block 起重滑轮
~ing bolt 吊环
~ing chain 起重链
~ing device 提升设备
~ing dog 抓钩
~ing eye 吊眼, 吊环
~ing force 提升力
~ frame 吊运架
~ing gate 提升式闸门; 栏木
~ing-gate feeder 升板式装料机
~ing gear 提升传动装置
~ing guard 井口防护栏
~ing height 提升高度
~ing hook 吊钩

~ing hook-type gate 钩式提升闸门（双层平板闸门的上门扉）
~ing jack 千斤顶；起重器
~ing magnet 起重磁铁，磁铁起重机
~ing moment 上托［倾覆］力矩
~ing platform 升降平台
~ing power 起重力，举升力
~ing rope 吊索
~ing shutter 提升式百叶门
~ing sliding door 提升滑动门
~ing tackle 提升滑轮
~ing truck 起重卡车
~ing winch 卷扬机
~ing window 上拉窗，提升滑动窗
dual action ~ 双向升降机
hydraulic ~ 液压起重机
window ~ 吊窗提手

lifter ['liftə] *n.* 启门机，升降机
fork ~ 叉车，叉式升降机
valve ~ 阀［气门］挺杆，起阀器

lifterloader *n.* 升运装载机

light [lait] *n.* 光，光线，灯光；*a.* 亮的；淡色的；轻质的；少量的
~ activated switch 光敏开关
~ adobe brick 轻质黏土砖
~ aggregate 轻质骨料
~ alloy 轻质合金
~ -alloy girder 轻合金梁
~ angle 光入射角
~ beacon 灯标，灯光信号
~ beam 光束，光线
~ blading 轻刮，整修路型
~ block masonry 轻型圬工结构
~ bracket 轻型托架
~ brick 轻型砖
~ building board 轻型建筑板
~ cast member 轻质（混凝土）预制件
~ chamotte brick 轻质耐火砖
~ clay 轻质黏土
~ coating 薄涂层
~ -coloured 轻度着色的
~ column 轻型柱
~ component 轻型建筑构件
~ concrete 轻质混凝土
~ concrete structure 轻质混凝土结构
~ construction 轻型建筑
~ continuance welding 轻型连续焊
~ control 灯火管制
~ cutting 浅挖
~ -cupola 上部采光的圆屋顶
~ delivery truck 轻型送货车
~ -diffusing ceiling 发光顶棚
~ displacement 空载排水量
~ distribution box 轻配电箱
~ -duty test 轻负荷试验
~ equipment 轻型设备
~ expanded clay aggregate 轻质膨胀黏土骨料
~ fill 低填土
~ fillet 浅角焊缝
~ fitting 照明装置，灯具
~ float 灯标船，灯浮标
~ fuel 轻质燃料
~ gauge plate 薄钢板
~ gauge sheet 薄钢板
~ gauge railway 轻轨铁道
~ -gauge steel structure 轻型钢结构
~ (weight) grader 轻型平地机
~ hopper wagon 轻便漏斗车
~ house 灯塔
~ industry 轻工业
~ intensity 光强
~ load 轻荷载
~ loam 轻质亚黏土
~ locomotive 轻型机车
~ mast crane 桅杆转臂起重机
~ material bucket 轻型料斗
~ metal alloy 轻合金
~ metal construction 轻金属建筑
~ metal lattice(d) girder 轻金属格构桁架
~ mineral aggregate 轻质矿物骨料

~ -panel 轻型方格
~ partition wall 轻质隔墙
~ -passing board 透光板
~ -passing plastic 透光塑料,有机玻璃
~ permeability 透光性
~ point 照明插座
~ pole (路)灯杆
~ porous clay brick 轻质多孔黏土砖
~ -proof louver 不透光(的)百叶窗板
~ proof vent 不透光通气孔
~ -proofness 遮光性
~ railway 轻便铁道
~ reflecting characteristics (路面的)反光持性
~ -reflecting curb 反光路缘
~ road oil 轻质铺路油
~ roof 轻型屋顶
~ sandy loam 轻亚砂土
~ (weight) section(al) steel 轻型钢材
~ shade 淡色调
~ sheet 薄钢板
~ shot 小爆破
~ silty loam 轻粉壤土
~ slag 轻质炉渣
~ slag concrete 轻质炉渣混凝土
~ soil 轻质土
~ source 光源
~ staff 灯柱
~ stone 浅石色
~ surface mulch 路面细覆盖料
~ surfacing 简易路面
~ switch 照明开关
~ tar 轻质焦油沥青
~ -textured soil 砂性土
~ -tight jalousie 遮光百叶窗
~ tower crane 轻型塔式起重机
~ traffic 轻量交通
~ type (pile) frame 轻型桩架
~ vault 轻型拱

~ vehicle 轻型车辆
~ weight concrete 轻量混凝土
~ weight insulating concrete 轻量绝热混凝土
~ weight motorcycle 轻型摩托车
~ well 采光井
~ wood 易燃干木料
~ing cable 照明线
~ing column (路灯)灯柱
~ing current 照明电流
~ing design 照明设计
~ing diffuser 漫射光照明装置
~ing engineering 照明工程
~ing facilities 照明设备
~ing glass 漫射光玻璃
~ing level 照明等级
~ing panel 照明配电板[箱]
~ing peak 照明峰荷
~ing standard 灯柱
~ing system 照明系统
bracket ~ 壁灯
casement ~ 竖铰链气窗
ceiling ~ 吸顶灯
drop ~ 吊灯;吊窗
floor ~ 落地灯
hopper ~ 倒开气窗
indirect ~ 间接采光
lantern ~ 天窗,圆顶窗
leaded ~ 花饰铅条窗
pavement ~ 路面灯光
pivot ~ 摇窗
transom ~ 气窗,楣窗
lighter [ˈlaitə] *n.* 打火机;驳船
~ aboard ship 载驳母船
~ berth 驳船泊位
~'s wharf 驳船码头
lightning [ˈlaitniŋ] *n.* 闪电,雷电
~ arrester 避雷针,避雷装置
~ conductor 避雷导线
~ guard 避雷装置

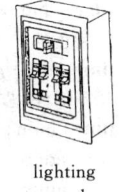

lighting panel

lightning arrester

~ protector plate 避雷板
~ rod 避雷针
~ spike 避雷针
~ switch 避雷开关

lignin(e) ['lignin] *n.* 木质素, 木纤维; 木素制塑胶绝缘料
~ plastics 木质纤维塑料
~ tar pitch 木质素硬沥青

lignite ['lignait] *n.* 褐煤
~ resin 褐煤树脂
~ tar pitch 褐煤焦油沥青

like-grained *a.* 单一粒径的, 颗粒均匀的
~ concrete 单一粒径骨料混凝土
~ gravel 等粒径卵石

lilac ['lailək] *n.* 紫丁香; 淡紫色; *a.* 淡紫色的

limb [lim] *n.* 肢, 臂, 边缘, 山侧; 支流, 零部件; 芯柱, 电磁铁芯

lime [laim] *n.* 石灰, 氧化钙
~ addition 石灰添加量
~ agitator 石灰搅拌机
~ and cement mortar 白灰水泥(砂)浆
~ -ash flooring 灰渣石灰铺面[屋顶]
~ bin 石灰库[仓]
~ -binding capacity 石灰浆粘结能力
~ -bound aerated concrete 石灰胶结多孔混凝土
~ bound macadam 石灰粘结碎石路
~ brick 石灰砖
~ brush 石灰浆刷
~ burner 石灰窑, 石灰锻烧工
~ burning plant 石灰厂
~ carbonate 碳酸钙
~ cement 石灰水泥, 石灰胶结料
~ -cement finish 石灰水泥灰浆抹面
~ -cement-flyash concrete 石灰水泥粉煤灰三合土
~ concrete 石灰混凝土
~ chloride 氯化钙
~ deposit 石灰质沉积物
~ dust 石灰粉
~ encrusted 钙化
~ emulsion 石灰乳浊液
~ feldspar 钙长石
~ finish coat 石灰抹面层
~ glass 钙玻璃
~ grout 石灰灌浆
~ -gypsum mortar 石灰石膏灰浆
~ hydrating machine 石灰水化机
~ injection 石灰粉喷射; 打石灰桩
~ insulating plaster 石灰隔热抹灰层, 石灰隔热灰泥
~ kiln 石灰窑
~ marl 石灰泥灰岩
~ milk 石灰浆
~ mill 石灰(石)粉碎机
~ mixer 石灰拌和机
~ mortar 石灰砂浆
~ paste 石灰膏
~ pit 石灰池[坑]
~ plaster 石灰砂浆
~ -pozzolanic cement 石灰火山灰水泥
~ putty 石灰膏
~ raker 石灰搅拌器
~ ratio 石灰比
~ -rock 石灰岩
~ sand 石灰砂
~ -sand brick 灰砂砖, 硅石砖
~ saturation factor 石灰饱和系数
~ slaking 石灰熟[消]化
~ slaking pit 石灰消化坑
~ -soda glass 钙钠玻璃
~ soda softening method 石灰苏打软化法
~ soil 钙质土
~ stabilization 石灰稳定法
~ -stabilized soil 石灰加固土
~ sulphur 石灰硫磺合剂
~ suspension 石灰悬浊液
~ -trass mortar 石灰火山灰浆
~ -treated aggregate 石灰处理骨料
~ tree 菩提树

~ wash 刷石灰水,粉刷石灰
~ water 石灰水
~ white 熟石灰
anhydrous ~ 干[生]石灰
calcium ~ 生石灰
carbonate-free ~ 纯石灰,纯氧化钙
drowned ~ 熟石灰
quick slaking ~ 快熟石灰
refractory ~ 耐火石灰
limestone [ˈlaimstəun] *n.* 石灰石
~ addition 石灰石掺合料
~ block 石灰石块
~ cave 石灰岩溶洞
~ coarse aggregate 石灰石粗骨料
~ filler 石灰石屑填料
~ rock asphalt 石灰岩地沥青
~ rubble 石灰岩碎石
limit [ˈlimit] *n.* 极限,范围;公差,极限值
~ analysis 极限分析
~ deformation 极限变形
~ design 极限状态设计
~ equilibrium 极限平衡
~ -load approach 极限荷载法
~ of accuracy 精度极限
~ of bearing capacity 承载能力极限
~ of consistency 稠度限度
~ elasticity 弹性极限
~ of error 误差限度
~ of stability 稳定限度
~ of visibility 可见度极限
~ of yielding 屈服极限
~ state 极限状态
~ profitable haul 最大有利运距
~ strength 极限强度
~ switch 限位开关,极限开关
~ed-access road 限制进入的道路
~ed prestressing 有限预加应力
~ed range 限界,限度
~ed speed 限制速度

~ing angle of friction 极限摩擦角
~ing condition 极限状态[条件]
~ing curve 极限曲线
~ing depth 极限深度
~ing gradient 限制坡度
~ing head 极限水头
~ing intensity 极限强度
~ing pressure 临界压力
~ing stress 极限应力
~ing surface 界面
~ing value 极限值
limmer [ˈlimə] *n.* 沥青质石灰岩
limnology [limˈnɔlədʒi] *n.* 湖沼学
limonite [ˈlaimənait] *n.* 褐铁矿
~ rock 褐铁岩
limy *a.* 含石灰的
linden *n.* 椴木
line [lain] *n.*, *v.*, lined, lining; *n.* 线条,线路,管线,交通线;排,行;绳索;轮廓,外形; *v.* 划线,排齐,做衬里,衬砌
~ -and-grade stakes 放样桩
~ building 行列式建筑
~ drawing 画线
~ drilling 成行钻孔
~ flow assembly 流水作业线装配
~ level (砌砖)拉线找平
~ load 线荷载
~ manhole 管道人孔
~ map 线路图
~ of balance 平衡线,对称线
~ of communication 通信线路
~ of position 定位线,位置线
~ pipe 干线管,总管
~ pull 绳索拉力
~ relay 线路继电器
~ stake 路线桩
~ survey 线路测量
~ welding 线焊
~d canal 有衬砌渠道
arch springing ~ 起拱线,拱脚
base ~ 基[底]线
border ~ 边线

limit switch

branch ～ 支线[路]
broken ～ 虚线;折线
camber ～ 脊线
contour ～ 轮廓线;等高线
crown ～ 路拱线
crown ～ of vault 穹顶线
curb ～ 路缘线
floor ～ 楼面线
formation ～ 施工线
high ～ 索道;高压线
house ～ 室内线
neat ～ 边线;墙面交接线
ridge ～ 分水[山脊]线
stream ～ 流线
trunk ～ 干线
linear ['liniə] *a.* 直线的,线性的;纵的,沿线的
～ acceleration 线加速度
～ analysis 线性分析
～ changes (混凝土路面的)线变形
～ city 带形城市
～ correlation 线性相关
～ displacement 线位移
～ equation 线性方程
～ expansion 线膨胀
～ expansion coefficient 线性膨胀系数
～ fracture mechanics 线性断裂力学
～ load 单位长度线荷载
～ meter 延米
～ programming 线性规划
～ regression 线性回归
～ shrinkage 线性收缩率
～ strain 线应变
～ structure 直线型结构
～ transformation 线位移
～ velocity 线速度
liner ['lainə] *n.* 衬砌,衬里,轴瓦,垫片;混凝土模板;两梁间的横梁;班机[船]
～ plate 衬砌板,垫板
～ tube 衬管
chimney ～ 烟囱内套
combustion ～ 燃烧室耐火层
die ～ 压模衬里
end ～ 底衬
joint ～ 接合衬垫
jolt-packed ～ 振筑衬里
outer ～ 外套
taper ～ 楔形塞垫,斜垫
lining ['lainiŋ] *n.* 衬砌,衬垫,垫板,镶衬;面料,涂层;套筒;挡板;涂底
～ arch 衬砌拱
～ board 衬板
～ brick 衬里砖
～ concrete 衬里混凝土
～ machine 衬里铺砌机
～ material 衬垫料,衬砌材料
～ of tunnel 隧道衬砌
～ paper 衬纸
～ plank 衬垫板
～ pole 花[标]杆
～ section 衬砌剖面
～ sheeting 衬板,封檐板
～ slab 外墙加劲板
～ (stone) ware 衬面陶瓷
～ strip 内墙板条
～ tube 衬砌管
～ wall 衬砌墙
block-stone ～ 块石衬砌
brick ～ 砖内衬
gunite ～ 喷混凝土衬砌
jamb ～ 门窗侧板
sinking ～ 沉井
tile ～ 片瓦内衬
timber ～ 木支架
link [liŋk] *n.* 链环[节];钢箍;关节;连杆,连接物 *v.* 连接;耦合; linkable, *a.*; linker, *n.*
～ bar 连接钢筋;铰接杆件
～ belt 链带
～ bolt 铰链螺栓
～ bridge 链式悬桥
～ dormer 连接的老虎窗
～ lever 连杆

link dormer

~ed arch 铰接拱
~ed connection 铰接
~ed system 联动系统
~ed switch 联动开关
~ing member 联系杆件
~ing route 连接路线
linkage [ˈliŋkidʒ] *n.* 链合,成链;联锁,联动装置
　~ drawbar 牵引板
　~ winch 悬挂式绞车
links [liŋks] *n.* 海岸草原,海岸沙滩;高尔夫球场
linn [lin] *n.* 瀑布;悬崖;溪谷
linoleum [liˈnəuljəm] *n.* 油毡,油布,漆布
　~ (bonding) adhesive 油毡粘结剂
　~ base 油毡底层
　~ cover 油毡护面,油毡覆盖层
　~ floor cover(ing) 油毡铺地面层
　~ tile 油毡块,油毡片
linseed [ˈlinsi:d] *n.* 亚麻子
　~ oil 亚麻子油
　~ oil putty 亚麻子油灰
　~ oil varnish 亚麻子油漆
lintel [ˈlintl] *n.* 过梁,楣石
　~ brick 过梁砖
　~ course (门窗)过梁层
　~ of a door 门楣
　chase ~ 槽楣
　concrete ~ 混凝土过梁
　steel ~ 钢过梁
　stone ~ 石过梁
　　　　　　　concrete lintel
lintol *n.* 过梁
Linville truss 林维尔桁架(双重 N 形腹杆系)
lip [lip] *n.* 唇,凸缘,突起;悬臂,支架;挖土机舌瓣,挖斗前缘;翼缘,边,端;鱼鳞板,百叶窗片
　~ block 支撑垫块
　~ crub 唇状路缘
　~ mould 唇瓣装饰
　~ped joint 唇状接合,半搭接

liquefaction [likwiˈfækʃən] *n.* 液化作用,熔解;稀释,冲淡
　~ failure 液化破坏
liquefied *a.* 液化的
　~ natural gas 液化天然气
　~ petroleum gas 液化石油气
liquid [ˈlikwid] *n.* 液[流]体; *a.* 液态的,流动的
　~ admix(ture) 液态(混凝土)附加剂
　~ agent 液态(混凝土)附加剂
　~ asphalt 液体地沥青
　~ bitumen 液体沥青
　~ coating (material) 液态涂面料
　~ concrete 液状混凝土
　~ crystal 液晶
　~ distributor 喷液机
　~ explosive 液态炸药
　~ fuel 液体燃料
　~ gas 液化气体
　~ glue 液胶
　~ level recorder (自动)液面记录器
　~ limit apparatus 流限仪,(土的)液限仪
　~ membrane curing (混凝土)液膜养护
　~ mortar densifier 液态灰浆止水剂
　~ piezometer 液体测压计
　~ polishing agent 抛光液
　~ seal 水封
　~ sludge 液态污泥
　~ state 液态
　~ waste 废液[水]
　preservative ~ 防腐液
liquidation [ˌlikwiˈdeiʃən] *n.* 清算[理];偿还,破产;液化,熔解[析]
liquidity index (土的)液性[化]指数,流性指数
liquor [ˈlikə] *n.* 液[流]体;水溶液,酒
L-iron 角铁,角钢
list [list] *n.* 表,册;目录,清单,价目单;镶边;狭条,边饰; *v.* 列表,编目;镶边
　~ edge 毛翅,板材边缘瘤

~ of architects 建筑师目录[索引]
~ of constructions 建筑结构索引
~ of drawing 图纸清单
~ of errata 勘误表
~ of modification 更改清单
~ of parts 零件目录
~ing 列表,排列;边材,镶边
authorized data ~ 技术数据资料一览表
drawing ~ 图纸
equipment component ~ 全套设备部件零件表
general ~ 总清单
parts ~ 零件单
proof ~ 校对表,检验目录
spare parts ~ 备(用零)件表

listel *n.* 平线脚,扁带饰
listening ['lisniŋ] *n.* 收听,监听
~ method (检查地下管道漏水)听声法
lister ['listə] *n.* 制表人;双壁开沟犁
~ furrow 犁沟
liter ['li:tə] *n.* =litre 公升,升
litharge ['liθa:dʒ] *n.* 氧化铅,密陀僧,铅黄,黄丹
lithe [laið] *a.*, liter,lithest;易弯的,柔软的; lithely,*ad.*; litheness,*n.*
~ board 易变形板,柔性板
lithic ['liðik] *a.* 石的,石制的; lithically,*ad.*
~ contact 母质层(土壤岩石接触层)
lithification [ˌliθifi'keiʃən] *n.* 成岩[岩化]作用,矿化,石化
lithistid *n.* 压缝石条
lithium ['liθiəm] *n.* 锂(Li)
lithofacies ['liθəˌfeiʃiz] *n.* 岩相
~ map 岩相图
lithology [li'θɔlədʒi] *n.* 岩石[性]学; lithologic,lithological,*a.*; lithologically,*ad.*; lithologist,*n.*
lithometeor *n.* 大气中浮悬尘土
lithopone ['liθəpəun] *n.* 锌钡白
Lithosite *n.* 利索赛特(一种砂浆料)

lithosol *n.* 石质土,岩石状土壤
lithostrome *n.* 均质岩层
little ['litl] *a.*, less or lesser,least;or littler,littlest; *ad.*, less, least; 小的,少的,短的
~ hydraulic lime 低水凝性石灰
littoral ['litərəl] *a.* 沿岸的,沿海的,海滨的;-littorally,*ad.*
~ area 海岸区,潮汐区,滨海区
~ current 沿岸流
~ deposit 沿岸沉积
~ drift 沿岸漂流[漂砂]
~ dune 海岸砂丘
~ shelf 海岸浅滩,浪成滨海台地
~ transport 沿岸轮砂
live [laiv] *a.*, [liv] *v.*, lived,living *v.* 生活,居住,活着; *a.* 活的,有生命的;有效的,起作用的;有电的;liveness,*n.*
~ axle 传动轴,主动轴
~ enclosure 树篱,矮树栏
~ fence 绿篱,树篱
~ guy 活动牵索
~ knot 活疖疤
~ load 活[动]荷载
~ pickup 室内摄影,播送实况
~ rail 导电轨,汇流排
~ room 混响室
~ silo 卸料仓
~ storage 活库容,有效库容
~ wire 火线,载电线
~ wood 新伐木材
~ zone (拌和机拌缸的)活动区,净体积
lividity [li'viditi] *n.* 铅[青]灰色
living ['liviŋ] *a.* 活着的,有作用的; *n.* 生活,活动
~ area 住宅区

living area

\~ hut 临时居住木板房
\~ quarter 宿舍,住宅区
\~ room 起居室
\~ unit 居住单元

load [ləud] *n.* 荷载[重]装载; *v.* 加荷[载]; loadless, *a.*
\~ applying unit 加载设备
\~ at failure 破坏荷载
\~(ing) bay 装料间
\~ bearing 承重,承载
\~ bearing characteristic 荷载[负荷]特性
\~ bearing frame 承重框架
\~ -bearing panel construction 承重板墙结构
\~ bearing partition 承重隔墙
\~ -bearing skeleton 承重骨[构]架
\~ -bearing skeleton rigidity 承重骨架刚度
\~ -bearing structural tile 承重瓦,预应力槽瓦
\~ -bearing structure 承重结构
\~ bearing wall 承重墙
\~ -carrying capacity 承载能力,载重量
\~ -carrying member 承重构件
\~ cell 压力盒;(石料压力试验用)加载筒
\~ classification number 荷载分类指数(用以设计水泥混凝土路面)
\~ coefficient 荷载系数
\~ -consolidation curve 荷载—固结曲线
\~ constant 荷载常数
\~ curve 荷载曲线(表示荷载与时间变化的曲线)
\~ -deflection diagram 荷载-挠度图
\~ deformation diagram 荷载变形图
\~ diagram 荷载图
\~ -dispatching office 配电所
\~ distribution line 荷载分布线
\~ distribution plate 承压板
\~ duration curve 荷载历时曲线

\~ equivalent 荷载当量
\~ factor 荷载[负荷]系数
\~ -factor method (钢筋混凝土的)荷载系数法,破损阶段法
\~ gauge 负荷计
\~ hopper 装料漏斗
\~ impact allowance 容许冲击荷载
\~ indicator 测力[载]计
\~ limitation 荷载限度
\~ line 负载线,载重吃水线
\~ moment 负载力矩
\~ -off 卸荷
\~ peak 负荷峰值
\~ -penetration test 荷载-贯入试验
\~ piston (土壤承载力试验用)加压柱塞
\~ point 荷载作用点
\~ range 荷载幅度
\~ rate 荷载率,单位荷载
\~ removal 卸荷
\~ rig 加荷设备
\~ -settlement curve 荷载-沉陷曲线
\~ spreading property 荷载分布特性
\~ -strain diagram 荷载应变图
\~ stress 荷载应力
\~ transfer 荷载传递
\~ -transfer joint (混凝土路面的)传力式接缝
\~ water line 载重吃水线
\~ed area 承载面积
\~ed brick 承载砖
\~ed chord 承载弦(有节点荷载的弦杆)
\~ed filter (坝址)承载反滤层
\~ed masonry (work) 承载圬工结构
\~ed plane 承载平面
\~ed rubber 填料橡胶
\~ed stream 含沙河流
actual \~ 有效荷载
appliance \~ 民用[生活]负载
bed \~ 河床质,推移质
dead (weight) \~ 死荷载,恒重
dummy \~ 虚荷载,等效荷载
eccentric \~ 偏心荷载

gross ~ 总重,毛重
panel ~ 节点荷载
specified ~ 规定荷载
loader [ˈləudə] *n.* 装料[载]机;装卸工人;输入程序
~ -dozer 装载推土两用机
chain ~ 链式装车机
grab ~ 抓斗式装岩机
shovel ~ 铲装机

loader

loading [ˈləudiŋ] *n.* 装载,加料;荷[载]重
~ and unloading track 装卸线
~ area 加载面积
~ bin 装载[料]仓
~ boom 吊货杆,装卸臂
~ bridge 桥式装料机
~ chute 供料滑槽
~ crane 装载起重机
~ cycle 加荷周期,重复加荷
~ density 加载密度
~ frame 加荷架
~ index 荷载指数
~ intensity 加载强度
~ limitation 载重限制
~ mechanism 加荷机构
~ pipe 填料管(弯管时管内填料以防变形)
~ plate 荷载板
~ platform 装卸站台
~ point 加载点
~ ram 加荷活塞
~ ramp 装载坡道
~ rate 加荷速率
~ ratio 装药率(单位体积岩石的炸药量)
~ routine 输入程序
~ shock 冲击加荷,负荷态变
~ shovel 装料铲
~ stage 装载台
~ state 负荷状态
~ stress 荷载应力

~ system (试桩的)加荷装置
~ test 加载试验
~ tray 装料盘
~ tinch 起重绞车
~ yard 装卸场
~ zone 装卸区
clamshell ~ 抓岩机装岩
scraper ~ 铲运机装载

loam [ləum] *n.* 壤土,亚黏[砂]土,钙质黏土
~ concrete (由软黏土和水泥组成的)黏土混凝土
~ core 亚黏土芯墙
~ fill(ing) 回填亚黏土
~ seal(ing) 亚黏土铺盖
~ wall 亚黏土墙,夯土墙

loamy [ˈləumi] *a.*, loamier, loamiest; 壤土质的,亚黏[砂]土质的; loaminess, *n.*
~ clay 壤质黏土
~ coarse sand 壤质粗砂
~ fine sand 壤质细砂
~ gravel 带壤土砾石
~ soil 壤土
~ texture 壤质构造

loan [ləun] *n.* 贷款,借出物,公债
~ account 贷款账户
~ agreement 贷款协定
~ application 贷款申请书
~ bill 贷款汇单
~ contract 贷款合同
~ delinquency 贷款拖欠
~ holder 债券持有者
~ interest 借款利息
~ restriction 贷款限制
~ syndicate 贷款财团

lobby

lobby [ˈlɔbi] *n.*, -bies 前厅,门廊,门厅,穿堂;休息室;接待室
lobe [ləub] *n.* 叶,瓣,凸角;叶形轮-lobeless, *a.*
local [ˈləukəl] *a.* 当地的,局部的,地方的;-localness, *n.*
～ action 局部作用
～ adjustment 局部平差
～ base level 当地基准面
～ borrow 就地取土坑
～ branch road 地方支路
～ break 局部开裂
～ buckling 局部压弯
～ contraction 局部收缩
～ deflection 局部挠曲
～ distortion 局部变形
～ disturbance 局部扰动
～ hardening 局部硬化
～ inflow 区间流入
～ material 当地材料
～ shear failure (土的)局部剪力破坏
～ strain 局部应变
～ stress 局部应力
～ time 当地时间
～ view 局部视图
locality [ləuˈkæliti] *n. pl.* -ties;位置,场所;地区,区域
location [ləuˈkeiʃən] *n.* 定线,定位;位置,地点;储存单元; locationally, *ad.*
～ centre line 定位中心线
～ layout 定位[场]地布置
～ map 定位图,位置图
～ survey 定线测量
run ～ 运行操作位置
lock [lɔk] *n.* 锁,闩;闸门,船闸,;联锁,同步;*v.* 锁住,关闭;联锁,同步;制动;过闸;lockable, *a.*
～ and bolt 锁加闩
～ bar steel pipe 加箍钢管
～ bolt 锁紧螺栓
～ bolt support with shotcrete 喷锚支护
～ check gate 止逆闸门

～ joint 锁口接头
～ miter 斜接头
～ pin 保险销
～ rail 装锁横档
～ seam 锁口缝
～ stile 装锁竖框
～ valve 保险阀
～ing device 锁闭装置
～ing grip 夹具
～ing material 联锁件
～ing nut 锁紧[防松]螺母
～ing ring 锁环
～ing screw 锁紧螺丝
～ing washer 锁紧垫圈
cylinder ～ 弹子锁
man ～ 人孔闸
mortice ～ (门上)插锁

lock rail and lock stile

lockage [ˈlɔkidʒ] *n.* 水闸高差;船闸通行费;过闸
locker [ˈlɔkə] *n.* 有锁的抽屉,衣帽柜,橱柜
～ plant 抽屉式冷柜,小冷库
～ room 衣帽间,更衣室
locomobile [ˌləukəˈməubiːl] *n.* 自动机车,锅驼机; *a.* 自动推进的; locomobility, *n.*
locomotive [ˈləukəˌməutiv] *n.* 火车头,牵引机车; *a.* 自行移动的,运行[转]的; locomotively, *ad.*; locomotiveness, *n.*
～ shed 机车库
locus [ˈləukəs] *n. pl.* loci; loca;场所,位置,所在地;轨迹[线]
～ of centres 拱圈的中心轨迹线
lode [ləud] *n.* 矿脉,岩脉;排水沟,水路
lodge [lɔdʒ] *n.*; *v.*, lodged, lodging; *n.* 传达室,小屋; *v.* 住宿;堆积
lodging [ˈlɔdʒiŋ] *n.* 临时住宿处,寓所;(*pl.*)公寓
～ -house 宿舍
loess [ˈləuis] *n.* 黄土
～ formation 黄土层

~ soil 黄土性土壤,大孔性土

loft [lɔːft] *n.* 阁楼,顶楼;(教堂或讲堂的)楼厢;货仓的楼上一层;鸽房
~ drier 箱式干燥器,干燥箱

log [lɔg] *n.*; *v.*, logged, logging; *n.* 圆[巨,原]木;测程仪,测井; *v.* 伐木,采伐;记录; loglike, *a.*
~ cabin 圆木小屋
~ chute 放木滑槽,滑木道
~ conveyer 木料输送机
~ -crib 木笼,圆木叠框

log conveyer

~ -crib abutment 木笼式桥墩
~ -crib revetment 木笼护岸
~ dam 木坝
~ frame 垂直锯(木)架,多锯机
~ loader 圆木装载机
~ pole 圆木桩
~ pond 原木池
~ rule 材积表
~ sheet 对数纸,记录表(卡片)
~ stop 木闸,叠梁闸板
~ training wall 原木导墙
~ weir 木堰
building ~ 建筑材
saw ~ 锯架
waney ~ 不等径圆木

logarithm [ˈlɔgəriθəm] *n.* 对数
logarithmic [ˌlɔgəˈriθmik] *a.* 对数的; logarithmicaly, *ad.*
~ scale 对数比例尺,计算尺
~ table 对数表
~ viscosity number 比浓对数黏度

loggia [ˈlɔdʒə] *n. pl.* -gias, loggie ; 敞廊,凉廊

logic [ˈlɔdʒik] *n.* 逻辑学,论理学,逻辑; *a.* 合逻辑的,逻辑上的
~ design 逻辑设计
~ flow chart 逻辑流程图
~ order 逻辑指令

Lombard architecture 朗伯特建筑

London smoke 灰黑色

long [lɔg] *a.* 长的,长久的; *n.* 全长
~ -and-short work 长短砌合
~ -boom(ed) dragline 长臂索铲
~ church 长列式教堂
~ column 长柱
~ corridor 长廊
~ continued load 长期负荷

Long Corridor (颐和园的)长廊
~ distance recorder 远程记录仪

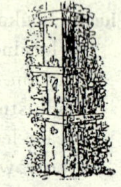
long-and-short work

~ dolly (打桩时用的)送桩,垫桩
~ -duration static test 长期静力试验
~ face (顺砖的)长边面
~ fascine 长梢捆,梢龙
~ -handled float 长柄抹子
~ header (沿墙面的)连续丁砖
~ -hole method 深孔爆破法
~ -line method (预应力混凝土的)长线法
~ -line prestressed concrete 长线张拉法预应力混凝土
~ -line pretensioning banks 长线预应力先张法台座
~ nipple 长螺纹接头
~ -period force 长周期引潮力
~ -period wave 长周期波
~ radius elbow 大半径弯头
~ -range forecast 长期预报
~ -range planning 远景规划
~ -ski system 长橇方式(沥青混合料摊铺机控制高程的一种装置)
~ span bridge 长跨桥,大跨度桥
~ strip footing 长条形基脚
~ taper 长锥体
~ -term creep 长期蠕变
~ -term plan 长期计划
~ -term strength 后期强度
~ -time load 持久荷载

~ -time strength 持久强度
~ ton 长吨(合 1.016 公吨)
longeron ['lɔndʒərən] *n.* 纵梁,大梁
longitude ['lɔndʒitju:d] *n.* 经度,经线
longitudinal [ˌlɔndʒi'tju:dinl] *a.* 纵向的,轴向的; *n.* 框架中的纵柱;longitudinally, *ad.*
~ alignment 纵向定线
~ arch 纵向拱
~ axis 纵轴
~ bar 纵向钢筋
~ beam 纵梁
~ bending 纵向弯曲
~ bent 纵向构架
~ bond 纵向多层砌体
~ bracing 纵向支撑;水平联结系
~ cable (预应力配筋的)纵向钢丝束
~ cavity (黏土带孔板的)孔洞
~ center joint 纵向中缝
~ construction joint 纵向工作缝
~ crack 纵向裂缝[隙]
~ culvert 纵向输水涵洞
~ dam 顺坝
~ deformation 纵向变形
~ drainage 纵向排水沟
~ expansion joint 纵向膨胀缝
~ fault 纵向断层
~ fiber 纵向纤维
~ fillet weld 纵向贴角焊
~ fish plate 纵向接合板
~ force 纵向力
~ frame 纵构架
~ girder 纵向大梁
~ girt 纵向围梁
~ grade 纵坡
~ grain 纵向木纹
~ joint 坝的纵缝;砖墙的横缝
~ manifold 纵向集合管道
~ member 纵向杆件
~ pressure 纵向压力
~ profile 纵剖面
~ reinforcement 纵向钢筋
~ rod 纵向钢筋;纵向杆
~ seam 纵向焊缝
~ separation 纵向间隔
~ shear 纵向剪切
~ slope 纵向坡度
~ stiffener 纵向(腹板)加劲杆
~ strut 纵向支撑杆
~ tie 纵向联结杆
~ truss 纵向桁架
~ underdrain (路面下的)纵向暗沟
~ vertical bracing 纵向垂直支撑
~ vibration 纵向振动
~ view 纵视图
~ wall 导墙
~ warping 纵向翘曲,纵向弯翘
~ web stiffener 纵向腹板加劲肋
~ weld 纵向焊缝
longleaf pine 长叶松
longshore ['lɔŋʃɔ:] *a.* 海岸的,沿岸的
~ bar 沿岸沙洲
~ current 沿岸流
~ drift 沿岸漂流[漂沙]
~ transport 沿岸输运
longwall *n.* 长工作面;连续工作面,长壁法回采
~ advancing 前进式回采工作面
~ face 长壁工作面
~ retreating 后退式回采工作面
full ~ 全面开采法
lookout ['luk'aut] *n.* 了望台
~ deck 了望平台
~ gondola (观光塔的)了望舱
~ tower 了望塔

lookout tower

lookum ['lu:kəm] *n.* 遮盖起重机的小屋,披屋
loop [lu:p] *n.* 环,圈,匝;孔眼;环[回]路;波腹; *v.* 成圈[卷];循环
~ bar 环头杆,套杆
~ -hole door 带狭孔的门
~ing channel 卷取[料]槽
~ing floor 转环地面

~ lake 牛轭[环形]湖
~ line 环形线
~ strength 互扣强度
~ window 竖直狭通气孔,高狭窗
~ed barrier 环形沙洲
inner ~ (立体交叉的)内转车道,内环道
passing ~ 让车道,错车环道

loophole [ˈluːphəul]
n. 射击孔；了望孔；小窗；漏洞

loophole

loose [luːs] *a.*, looser, loosest, *v.*, loosed, loosing; *a.* 松散的,宽松的；不牢固的；无束缚的,自由的；*v.* 松', 放开, 释放
~ aggregate 疏松集料,松散骨料
~ alluvium 松散冲积层
~ and open surface 松散透水的路面
~ apron 松散石料护坦
~ base 松底层
~ bush 可换[活动]衬套
~ cement 散装水泥
~ change gear 可互换变速齿轮
~ core 松散夹心(混凝土的水泥浆流失及骨料离散)
~ -coupled type pipe 松接式管道
~ depth 松散厚度
~ earth 松土
~ fill 坝料干填
~ fit 松配合[装配]
~ flange 松套法兰
~ ground 松散地面
~ joint 松接头
~ -joint butt 可拆铰链
~ -joint hinge 插销铰链
~ knot (木材)疏松节疤
~iy spread concrete 未捣实混凝土
~ masonry 干砌圬工
~ -measure volume 松方体积
~ mix 松级配混合料
~ packed 散装
~ penstock 松接式压力钢管

~ -pin hinge 枢轴铰链
~ rivet 松铆钉,未铆紧的铆钉
~ rock 松散岩石
~ -rock dam 堆石坝
~ rock dump (坝体下游部位的)堆石体
~ rock fill 抛石充填
~ sand 散沙
~ soil 松散土
~ -stone 干砌石
~ tongue 嵌入榫；合板钉
~ tongue joint 抽心接头
~ unit weight 干松容量
~ volume 松体积,松方
~ yards (土)松方量

loosen [ˈluːsn] *v.* 弄(放,翻)松；解除,放宽；-loosener, *n.*
~ed concrete 已损坏的混凝土
~ing depth 松土深度
~ing earthwork 松土工程
~ing strength 卸荷强度

lopwood *n.* 枝材
Lorenz bored pile 劳伦茨式钻桩
Lorenz/Fehlmann method 劳费氏沉箱沉降法(沉箱做成上小下大,上部灌膨润土稀浆)

lorry [ˈlɔri] *n. pl.* -ries；手推车,运料车,运货汽车
~ entrance 卡车入口
~ haul(ing) 卡车运输
~ -mounted auger 汽车式钻机
~ -mounted crane 汽车式起重机
~ rail 手推运料车轨道
~ scale 汽车秤
breakdown ~ 抢险工程车
tip ~ 自卸汽车

lose [luːz] *v.* 失(去,落),错过,漏过,失败
~ time 延误,失去时机

loss [lɔːs] *n.* 损失,遗失；损耗
~ and gain 损益
~ angle 损耗角
~ by solution 溶失量

~ in weight 重量损失
~ on heating 加热损失(测定沥青材料用)
~ rate 损失率
counting ~ 计数损失
energy ~ 能量损失
filter ~ （泥浆的）失水量
mud ~ 泥浆漏失
lost [lɔst] *a.* 失去的，遗失的；浪费的，分离开的
~ -colour process 落色工艺
~ head 水头损失
~ material 损耗的材料
~ surfacing 磨耗的面层
lot [lɔt] *n.* 组[批，群]；地块[段，皮，区] *vt.* 分[给，到]，划分，分堆
~ plan 地段图
~ production 大批[批量]生产
~ size 批量
lotus [ˈləutəs] *n.* 荷花饰
~ capital 荷花饰柱头
~ column 荷花饰柱子
lounge [laundʒ] *n.* 卧榻，沙发；休息室，吸烟室
louver [ˈluːvə] *n.* 天窗，气窗；百叶窗；汽车的散热孔
~ board 百叶窗[门]板
~ door 百叶门
~ shutter 活动百叶窗
~ window 百叶窗；气窗
~ed air outlet 百叶窗式出气口

louvered overhand

~ed overhand 鱼鳞板，百叶吊帘
Louvre [luːvr] *n.* （法国巴黎）罗浮宫
low [ləu] *a.* 低的，矮的；弱的，小的；-lowness, *n.*
~ accuracy 低精确度
~ -alkali cement 低碱水泥
~ -alloy steel 低合金钢
~ -bake finish 低温烤漆

~ block 矮平房
~ boom （桁架）下弦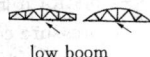
low boom
~ boiling point 低沸点
~ carbon steel 低碳钢
~ -carbon tar 低碳焦油沥青
~ -ceilinged room 低顶棚房间
~ crown 低路拱，横坡小的路拱
~ deck of bridge 双层桥的下层桥面
~ -duty fireclay brick 轻荷耐火砖
~ earth barrier 矮土围堰；矮土墙
~ efficiency 低效率
~ -fall installation 低水头电站
~ flow 枯水流量
~ -flow frequency curve 枯水流量频率曲线
~ frequency 低频
~ gear 低排档[速率]
~ -grade cement 低标号水泥
~ -grade concrete 低标号混凝土
~ gradient 平缓坡度
~ ground 低洼地
~ hardness water 低硬度水
~ -head hydroelectric plant 低水头电站
~ -head cement 低热水泥
~ -hedge 矮树篱
~ -income block 低租金住宅
~ -intensity 低强度
~ -level groin 潜丁坝
~ level intake 深式进水口
~ level outlet 泄水底孔
~ -level water tank 底部[低位]水箱
~ -lift blade grader 低举式平地机
~ -lift construction （大体积混凝土的）薄层施工
~ -lift pump 低扬程水泵
~ -limed cement 低石灰水泥
~ manganese steel 低锰钢
~ -penetration asphalt 低针[贯]入度地沥青
~ pitched roof 缓坡层顶

~ platform 低站台
~ pressure compressor 低压通气机
~ pressure heating boiler 低压加热锅炉
~ -pressure grouting 低压灌浆
~ pressure steam heating 低压蒸汽供暖
~ -profile plant 自行式拌制混凝土装置
~ relief 浅浮雕
~ -relief terrain 丘陵区
~ -rent apartment (unit) 低租金公寓,社会资助公寓
~ -rise building 低层建筑
~ -rise spherical shell 低曲率球壳结构
~ -shrinkage concrete 低缩性混凝土
~ -slump concrete 低坍落度混凝土,干硬性混凝土
~ -speed wind tunnel 低速风洞
~ -strength cement 低强水泥,低标号水泥
~ -temperature (coal) tar 低温煤焦油,低温焦油沥青
~ -temperature setting 低温固化
~ truss bridge 下承式低桁架桥
~ -type pavement 低级路面
~ -velocity material 低速传震材料
~ volatile tar 低挥发性焦油
~ voltage switchgear room 低压开关室
~ water 低潮,低水位
~ -water discharge 低水位流量
~ -water-loss cement 低析水性水泥
~ water mark 低水位标志
lower ['ləuə] *a.*;(low 的比较级); *v.* 降低,降下,减低; lowerable, *a.*
~ basement 第二层地下室
~ bearing spider 下轴承架
~ boom (桁架的)下弦杆
~ border (图廊之)下边线
~ boundary 下边界,下限
~ chord 下弦杆
~ coat (面层)下层
~ course 下游(段)
~ desk 下层桥面
~ development (水电站)下游梯级
~ discharge tunnel 流底涵[孔]
~ expansion chamber (调压井)下扩大室
~ falsework 下部脚手架
~ fascia 柱顶过梁的最下一条
~ flange of girder 梁的下翼缘
~ floor 下层楼
~ guide track (门的)底导轨
~ lateral bracing 下弦横撑
~ lip (虹吸道)出口反弧段
~ low water 低低潮
~ plastic limit 塑限下限(土壤含水量指标)
~ pool 尾水池
~ surge basin 下调压池
~ triassic sandstone 杂色砂岩
lozenge ['lɔzindʒ] *n.* 菱形;lozenged, *a.*;lozenge-shaped, *a.*
~ moulding 菱形装饰
~ riveting 错列[菱形]铆接
~ roof 菱形(石板)屋顶
L-roof plate L 形屋盖板

L-roof plate

L-shaped retaining wall L 形挡土墙
L-square 直角板,角直板
L-type curb L 形路缘石
lube [lu:b] *n.* 润滑油
~ chart 润滑系统图
~ grease 润滑脂
~ oil 润滑油
~ point 加润滑油位置
lubricant [lu:brikənt] *n.* 润滑油[剂]
~ grease 润滑脂
~ groove 润滑油槽
lubricate ['lu:brikeit] *v.*, -cated,-cating 使润滑,加润滑油

lubricating *a.* 上润滑油的
~ can 注油壶
~ gun 注油枪
~ hole 注油孔,润滑油孔
~ oil 润滑油
~ oil circulation 润滑油循环
~ oil pump 润滑油泵
~ water 润滑性水分(降低土粘聚力的水分)

lubrication [ˌluːbriˈkeiʃən] *n.* 润滑;lubricational, *a.*

lucarne [luːˈkɑːn] *n.* (屋顶)天窗,老虎窗

lucite [ˈluːsait] *n.* 有机玻璃,人造荧光树脂

Ludowici tile *n.* (法国马赛的一种)多槽瓦

luff [lʌf] *v.* 起重臂起落摆动,起重臂调整
~ing 起吊;起重杆升降
~ing cableway 塔架可倾式缆索道
~ing crane 鹅头式伸臂起重机

lug [lʌg] *n.*;*v.*, lugged, lugging; *n.* 突缘,突出部,凸耳;把手,手柄;肋,突纹;*v.* 使劲拉;强拖
~ bolt 长平头螺栓
~ brick 企口砖
~ connector 搭接头
~ sill 突缘窗台板

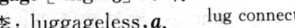
lug connector

luggage [ˈlʌgidʒ] *n.* 行李; luggageless, *a.*
~ carrier 行李架
~ compartment 行李间
~ elevator 运行李升降机
~ room 行李间
~ store 贮藏室

lugless brick *n.* 光面砖

lumber [ˈlʌmbə] *n.* 木材,木料
~ drying 木材干燥(法)
~ gauge 板材厚度规
~ kiln 木材干燥窑
~ mattress 木(沉)排
~ mill 锯木厂
~ yard 堆木场
dimension ~ 标准尺寸木材
end matched ~ 端边企口板

lumen [ˈluːmin] *n. pl.* -mens,-mina;流明(光通量单位)
~ meter 流明计

luminaire [l(j)uːmiˈnɛə] *n.* 灯具,照明装置
~ grid suspension ceiling 光格吊顶

luminance [ˈluːminəns] *n.* 亮度
~ factor 亮度因数

luminous [ˈluːminəs] *a.* 发光的,光亮的; luminously, *ad.* luminousness, *n.*
~ beacon 灯标,灯塔
~ body 发光体
~ ceiling 发光顶棚
~ efficiency 发光效率
~ intensity 发光强度
~ -intensity distribution curve 配光曲线,光强分布曲线

luminous-intensity distribution curve

~ paint 发光漆
~ pigment 发光颜料
~ wall 发光墙面

lump [lʌmp] *n.* 小块,堆,团;瘤,节;粒度
~ lime 块石灰
~ material 块料
~ -sum contract 包干合同
~ work 包工

lunar [ˈluːnə] *a.* 月的,太阴的,阴历的
~ calendar 阴历
~ time 太阴时

lunch [lʌntʃ] *n.* 午餐
~ counter 便餐用长柜台
~ room 小食堂,便餐馆

lunette [luːˈnet] *n.* 新月形之物;弧面窗;墙顶半圆形转角

lute [luːt] *n.*;*v.*, luted, luting; *n.* 水泥封涂,封泥;(水泥及沥青混凝土路面

的)修整样板；v. 用泥封涂
luting ['lu:tiŋ] n. 油灰,封泥,腻子
lux [lʌks] n. pl. luxes, luces ,勒(克斯)(照度单位,等于1流明/米²)
~ candle 米烛光
~ gauge 照度计,勒克司计
luxury ['lʌkʃəri] a. 豪华的
~ apartment 豪华公寓
lyceum [lai'siəm] n. 演讲厅,大教室,讲堂
lych-gate 教堂墓地前面有顶盖的门

lych-gate

lye [lai] n. 灰汁,碱液
lying ['laiiŋ] n. 天窗,横卧；a. 躺[凹]着的
~ light 天窗
~ panel 平纹镶板

lying panel

lysimeter [lai'simitə] n. 测渗仪,土壤渗透仪；溶度计(测定物质溶解度的仪器)
lysis n. 溶解,溶化,逐渐消退
lytag n. 粉煤灰陶粒,烧结粉煤灰轻质骨料
Lytoy n. 烧结粉煤灰陶粒

M

m ＝metre 米(长度单位)
macadam [mə'kædəm] n. 碎石路(面)；碎石
~ aggregate 锁结式集料,粒径相近的粗集料
~ aggregate type 碎石路面,嵌锁式路面
~ base 碎石基[底]
~ foundation 碎石基础
~ mixing plant 碎石混合料工厂
~ pavement 碎石路面
~ road 碎石路
~ roller 碎石压路机,碎石路碾
~ sub-structure 碎石下部结构 macadam roller
~ surface 碎石路面[面层]
asphalt ~ 沥青碎石路
bituminous ~ 沥青碎石路
cement-bound ~ 水泥结碎石路
clay-bound ~ 泥结碎石路
coated ~ 表面处理的碎石路
cold penetration bituminous ~ 冷灌沥青碎石路
composite ~ 复合碎石路
dry-bound ~ 干结碎石路
gravel-topped ~ 砾石铺面碎石路
grouted ~ 灌浆碎石路
hot penetration bituminous ~ 热灌沥青碎石路
lime-bound ~ 石灰结碎石路
oil ~ 用胶结材料处理的碎石路
penetration ~ 灌入法碎石路
plain ~ 纯石子路
silicated ~ 以水玻璃处理的路面,混合硅酸处理的碎石路面
tar ~ 焦油混凝土；焦油处理的碎石路
tarviated ~ 柏油碎石路
water-bound ~ 水结碎石路,水结路面
macadamization [məˌkædəmai'zeiʃən] n. 建筑碎石路,碎石筑路法
macadamize [mə'kædəmaiz] vt. ,-ized, -izing；铺碎石；建筑碎石路
macadmix n. 拌有沥青或其他粘结料的碎石混合料
Macasphalt n. 马克他沥青混合料(一种冷铺沥青混合料)
~ type pavement 碎石柏油路面
macerate ['mæsəreit] v. ,-ated,-ating；浸软；浸解；浸渍；浸化

maceration [ˌmæsəˈreiʃən] *n.* 浸软；浸解；浸渍作用

machicolate [məˈtʃikəuleit] *vt.*, -lated, -lating；作突出的堞眼于胸墙

machicolation [məˌtʃikuəˈleiʃən] *n.* 堞眼；枪眼

machinability [məˌʃinəˈbiliti] *n.* (可)切削性，机械加工性能

machinable [məˈʃiːnəbl] *a.* 可切削的，可机械加工的，可用机械的

machine [məˈʃiːn] *n.* 机器；机械；机制，机器加工
～ attendance 机器保养
～ -banded pipe 机制加箍木管
～ bolt 机械螺丝
～ -broken stone 机器粉碎石块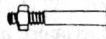
machine bolt
～ casting 机铸，机器铸造
～ -cleaning method 机械清洁[除]法
～ code 机器代码
～ -cut peat 机切泥
～ drawing 机器作图
～ drill 机械钻井设备，钻机
～ -extruded terra-cotta 机制琉璃砖
～ finish(ing) 机械修理，机器最后加工
～ -finish concrete 机器磨面混凝土
～ -foaming process （膨胀渣的）机械成形法，蒸汽转筒法
～ forging 机器锻造
～ foundation 机器基础[座]
～ framing 机架
～ hour method 机时法
～ hour rate 机时(生产)率
～ instruction code 机器指令码
～ language programme 机器语言程序
～ -made brick 机制砖
～ -made mortar 机拌灰浆
～ -made nail 机制钉
～ -made plaster 机拌[制]石膏
～ -mixed concrete 机拌混凝土
～ mixer 搅拌机
～ mixing 机拌(法)

～ operation 机器操作
～ operator 机工，机械操作工
～ parts 机械零件
～ -printed paper 机印墙纸
～ rammer 机械夯具，夯击机
～ repair shop 机器修理车间
～ rivetting 机铆，机械铆接
～ room 机器间，机械设备间
～ -sized 机测的，机器度量的
～ sweeping 机器扫路
～ -tamped concrete 机夯混凝土
～ tamper 机夯，机械捣固器
～ tools 工具机，机床
air drying ～ 空气干燥机
air refrigerating ～ 空气制冷机
ash crusher ～ 碾灰机
bending ～ 弯折机，弯钢筋机
brick ～ 制砖机
brush ～ 清扫机，扫街机；沥青涂抹机
brushing ～ 刷光机
bubble ～ 气泡式浮选机
bucked trenching ～ 斗式挖沟机
bump-cutter ～ 整平机
cable dredging ～ 绳索式挖泥机
cascade ～ 高落式浮选机
caterpillar excavating ～ 履带式挖掘机
cement testing ～ 水泥试验机
cementing ～ 擦胶机
centrifugal ～ 离心机
clay-cutting ～ 黏土切割机
clay-working ～ 黏土加工机
concrete vibratory ～ 混凝土振捣机
crushing ～ 碎石机
ditch cleaning ～ 清沟机
ditch(ing) ～ 挖沟机
flexing ～ 挠曲机
hydraulic ～ 液[水]压机
joint cleaning ～ 清缝机
joint ～ 接合[缝]机
knapping ～ 碎石机
laser-beam drilling ～ 激光钻孔机

lining ~ 渠道砌筑机
material testing ~ 材料试验机
model ~ 样机
paving ~ 铺路机
pile drawing ~ 拔桩机
pile driving ~ 打桩机
pneumatic ~ 风[气]动机械
pumpcrete ~ 混凝土泵
road ~ 筑路机
spinning ~ (制混凝土用的)离心机;纺纱机
universal testing ~ 万能试验机
weighing ~ 称桥,称重机
machinery [məˈʃiːnəri] *n*. 机器,机械;工具;手段,方法
~ component 机器零[部]件
~ equipment 机器设备
~ foundation 机器基础[座]
~ noise 机械噪声
~ oil 机油
~ vibration 机械振动
auxiliary ~ 辅助机械
construction ~ 施工机械
conveying ~ 输送机械
excavating ~ 挖掘[泥]机,掘土机
handling ~ 装卸机械
hoisting ~ 起重机械
hydraulic ~ 液压[水力]机械
mixing ~ 拌合机械
road ~ 道路机械
screening ~ 筛分设备
sizing ~ 筛分设备
macro [ˈmækrəu] [词头]宏(观),大(量),常量粗视,长
macroanalysis *n*. 常量分析
macroassembler *n*. 宏汇编程序
macro-axis *n*. 长轴,[晶体的]长对角轴
macrocode *n*. 宏代码
macrocorrosion *n*. 宏观腐蚀,大量腐蚀
macro-cracks *n*. 宽裂缝
macrocrystalline [ˌmækrəuˈkristəlain] *a*. 粗(粒结)晶的,宏晶的,大(块)结晶的; *n*. 宏[粗]晶,粗晶体
macroeconomic 宏观经济的
~ model 宏观经济模型
~ theory 宏观经济理论
macrographic *a*. 宏观的
macrography [məˈkrɔgrəfi] *n*. 肉眼检查;宏观图
macro-instruction [ˌmækruinˈstrʌkʃən] *n*. 宏指令
macrometer [məˈkrɔmitə] *n*. (光学)测远[距]器
macro-modular computer 宏模组件计算机
macromolecule [ˌmækrəuˈmɔlikjuːl] *n*. 大[高]分子
macropore *n*. 大孔隙
~ coefficient 大孔隙系数
~ volume 大孔隙体积
macroporosity *n*. 大孔隙率
macroporous *a*. 大孔性的
~ soil 大孔性土,黄土
macroprecipitation *n*. 大降雨,大范围降水
macro-relief *n*. 广域地形;大区地形
macro-rheology *n*. 宏观流变学
macroscale *n*. 宏观尺寸
macroscopic [ˌmækrəˈskɔpik] *a*. 粗量的;粗大的;肉眼可见的;宏量的; macroscopically *ad*.
~ cavitation 大空穴
~ model 宏观模型
~ scale 宏观规模
~ void 大孔,大空洞
macroseism *n*. 强震
macrostrain *n*. 宏应变[胁变]
macrostress *n*. 宏应力[胁强]
macrostructure [ˌmækruˈstrʌkʃə] *n*. 宏观结构[构造],大型构造
macro-turbulence *n*. 宏观紊流,大尺度紊流
macroviscosity *n*. 宏观黏性
macrovoid ratio *n*. 大孔隙比

maculose *a.* 斑结状的
  ～ rock 斑结状岩
made [meid] *v.* make 的过去式和过去分词；*a.* 特制的，人工造的；完成的，制成的；捏造的
  ～ block 组成滑车
  ～ ground 填土[地]，现代沉积
  ～ land 填土[地]
magdolite *n.* 两次煅烧白云石
magic ['mædʒik] *n.* 魔术，戏法 *a.* 魔术的；神奇的
  ～ hand 机械手
  ～ ink 万能笔
  ～ lantern 幻灯，映画器
  ～ scanner 光笔显示器

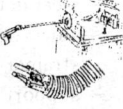
magic hand

  ～ square 魔方，纵横图
magma ['mægmə] *n. pl.* -mate, -mas 岩浆；稠液 magmatic, *a.* magmatism, *n.*
  ～ pump 糊泵，稠液唧筒
magmatic [mæg'mætik] *a.* 岩浆的
  ～ assimilation 岩浆同化作用
  ～ corrosion 岩浆熔蚀
  ～ eruption 岩浆喷溢
  ～ explosion 岩浆爆发
  ～ intrusion 岩浆侵入
  ～ minerals 岩浆矿物
  ～ origin 岩浆成因[起源]
  ～ rock 岩浆岩
  ～ water 岩浆水
magnaflux ['mægnəflʌks] *n.* 磁粉检查法，磁力探伤法，电磁探矿法；磁通量；*vt.* 用磁粉检查法检验，磁力探伤
  ～ method 磁通量检测法
  ～ test 磁力线检验，磁流试验
Magnel anchoring system 马氏预应力张拉系统
magnesia [mæg'ni:ʃə] *n.* 氧化镁，镁氧 (MgO)；magnesial *a.*
  ～ brick 镁(氧)砖
  ～ cement 镁氧水泥，菱镁土水泥
  ～ ceramics 镁氧陶瓷
  ～ hardness 菱镁硬度
  ～ lime 镁氧石灰
  ～ refractory 镁氧耐火砖
magnesian [mæg'ni:ʃən] *a.* 镁质的，(含)镁氧的，氧化镁的
  ～ chalk 镁质白垩
  ～ lime 镁质石灰
  ～ lime paste 镁质石灰岩浆
  ～ lime putty 镁质石灰油灰
  ～ limestone 镁质石灰岩，白云石
  ～ marble 镁质大理石
  ～ quicklime 镁质生石灰
  ～ semi-hydraulic lime 镁质半水石灰
magnesite ['mægnisait] *n.* 菱镁矿，菱苦土，菱镁土
  ～ (building) board 镁氧建筑板
  ～ brick 镁砖
  ～ cement 镁氧水泥，菱镁土水泥
  ～ -chrome brick 镁铬合金砖
  ～ flooring 氧化镁铺面，氧化镁地面材料
  ～ flooring tile 氧化镁铺面砖
  ～ refractory 镁氧耐火砖
  ～ refractory product 镁氧耐火制品
  ～ rock 菱镁岩
  ～ sheet 镁氧建筑板
magnesium [mæg'ni:zjəm] *n.* 镁 (Mg)
  ～ carbonate 碳酸镁
  ～ casting alloy 镁铸合金
  ～ chloride 氯化镁
  ～ chrome brick 铬镁砖
  ～ fluo-silicate 硅氟酸镁
  ～ light 镁光
  ～ limestone 镁质石灰石，白云石
  ～ oxide 氧化镁
  ～ oxide cement 镁氯氧水泥
  ～ oxychloride cement 镁氯氧水泥，菱镁土水泥
  ～ silicate 镁硅砂
magnet ['mægnit] *n.* 磁铁，磁石；磁体
  ～ crane 磁力起重机
  ～ current meter 磁力流速仪

~ road sweeper 磁力扫路机
lift(ing) ~ 起重磁铁
magnetic [mæg'netik] *a.* 磁性的,磁石的
~ anomaly 地磁异常
~ attraction 磁(引)力
~ azimuth 磁方位角
~ bearing 磁方向角,磁象限角
~ body 磁体
~ brake 磁闸
~ catch 磁性掣子
~ cell 磁元件,磁存储单元
~ circuit 磁路
~ clutch 磁性离合器
~ coil 电磁线圈
~ compass 罗盘仪
~ controller 磁力控制器
~ core 磁芯
~ core memory 磁芯存储器
~ core storage 磁芯存储器
~ coupling 磁耦合
~ course 磁针方向
~ crack detection 磁化裂缝检测法
~ declination 磁偏角
~ deflection 磁偏转
~ disc 磁盘
~ disk memory 磁盘存储器
~ disturbance 磁扰
~ domain 磁畴
~ drag 磁性曳力
~ drum 磁鼓
~ drum storage 磁鼓存储器
~ field 磁场
~ field intensity 磁场强度
~ flux 磁通量
~ force 磁力
~ inclination 磁倾角
~ induction 磁感应[强度]
~ iron 磁铁
~ iron ore 磁铁矿
~ map 地磁图

magnetic catch

~ meridian 磁子午线
~ moment 磁矩
~ needle 磁针
~ north 磁北
~ paper 磁性纸带
~ pole 磁极
~ pulley 磁性滑轮
~ sand 磁铁矿砂
~ separation 磁选,磁力分离
~ separator 磁选机,磁力分选机
~ shield 磁屏
~ steel detector 磁性钢筋探测仪
~ storm 磁暴
~ strainer 磁性筛[滤]网
~ support 磁垫,磁座
~ surveying 磁性测量
~ sweeper (清除铁屑用的)磁力扫路机
~ switch 磁开关
~ tape 磁带
~ test(ing) 磁力探伤
~ torque 磁扭转
~ type speedometer 磁力式速率计
magnetite ['mægnitait] *n.* 磁铁矿
~ aggregate 磁铁矿骨料
~ concrete 磁铁矿混凝土
magnification [ˌmægnifi'keiʃən] *n.* 放大率;放大;伸缩
~ factor 放大系数
~ ratio 伸缩比,放大率
magnifier ['mægnifaiə] *n.* 放大镜;放大器
~ for reading 读数放大镜
magnify ['mægnifai] *v.*, -fied,-fying;放大;增加,强调
magnitude ['mægnitjuːd] *n.* 大小;数量;等级;震级
~ of stress 应力值
stellar ~ 星等级
mahoganize [mə'hɔgənaiz] *vt.* -ized, -izing;将(木料)油漆成桃花心木状
mahogany [mə'hɔgəni] *n.* 桃花心木,红

木;桃花心木制成的家具;红褐色
~ brown 红木褐色,赤褐色
~ chair 红木椅
~ -faced 红木铺面
~ furniture 红木家具
~ -lined 红木装饰
mail [meil] **n.** 邮件,信件[汇],邮递工具[员]
~ -box 邮箱,邮筒
~ drop 邮箱,邮筒
main [mein] **a.** 主要[基本]的;总的;充分的;强力的 **n.** 干道[线];总线;总管;主要部分,要点
~ air duct 主风道
~ aisle 主走廊
~ anchor 主锚
~ and auxiliaries 主机及辅机
~ archway 主拱道
~ assembly hall 主装配间;主会议厅
~ assembly line 主装配线
~ axis 主轴
~ bar 主钢筋
~ beam 主梁
~ bearing structure 主要承重结构
~ block 主街区
~ body of road 路基
~ bottom 基座[岩]
~ brace 主转桁索;主斜撑
~ budget 总预算
~ building 主建筑,正屋
~ cable 载重索,主索[缆]
~ canal 干[主]渠
~ carriageway 主要车道
~ channel 主水道,主河槽
~ conduit 综合(地下)管道
~ control board 主控盘
~ control room 主控制室
~ couple 主屋架
~ course 主航道,干道
~ crossing 主要交叉口
~ diagonal 主对角线;主斜撑;主斜拉杆
~ diagonal rib 主斜肋

~ dike 主堤
~ drain 排水总管,干线沟渠
~ drive gear 主动齿轮
~ drive shaft 主动轴
~ exchange 电话总局,交换总机
~ fault 主断层
~ floor 主层
~ gallery 主廊道
~ girder 主大梁,主梁
~ highway 主要公路,干路
~ hook 主钩
~ jet 主喷嘴,(汽化器的)高速用喷嘴
~ job 主体工程;主要工作
~ joint 主节理
~ mast 主桅
~ pipe 主管
~ plane 主平面
~ program 主程序
~ pump 主泵
~ rafters 主椽
~ reinforcement 主钢筋
~ riser 主立管
~ rod 主杆
~ sewer 污水干管,排水总管
~ shaft 主轴;主井
~ span 主跨
~ specification 主要规格
~ station peg 主测站桩
~ stem 干线
~ strain 主应变
~ stress 主应力
~ stress line 主应力迹线
~ supply 供电干线,主供油管
~ thoroughfare 主要通道,主要大街
~ tie 主拉杆
~ traffic flow 总交通流量
~ transformer 主变压器
~ truss 主桁架
~ -water 自来水
~ wiring diagram 主接线图

main truss

blast ~ 总风管

collecting ~ 总管;集流管;汇流排
force ~ 承压干管
gas ~ 煤气总管,总气管
heating ~ 供暖干管,暖气总管
public supply ~s 城市配水网,公用配水干管网
supply ~ 总供水(气)管;输电干线
mainland ['meinlənd] *n.* 大陆;本土;mainlander, *n.*
~ climate 大陆性气候
maintain [men'tein] *v.* 养护,保养;维持,支持;继续;主张,坚持;maintainer, *n.*
maintenance ['meintinəns] *n.* 养护,维修,保养,养路;维持,支持;保存
~ and repair budget 维修预算
~ apron 机修坪
~ building 养路工房,道班房
~ charge 修缮费
~ contracts 维修合同
~ cost 养护[维修,保养]费
~ crew 养护工作队
~ depot 养护补给站;修配厂,保养厂
~ division 养路段,养路工区,维修队
~ dredging 维护性挖泥,例行清淤
~ engineer 养护[路]工程师
~ equipment 养路[护]设备
~ factor 保养系数
~ footway 维修用便道
~ force 养护工队[力量]
~ foreman 养路领班;养护力量
~ -free 无需维护
~ -free life (路面)耐久年限,无养护年限
~ gang 养路道班,养护工作队
~ hangar (飞机)修理棚
~ instruction 维修说明书
~ job 维修工作
~ load test 维持载荷法(试桩)
~ log 保养时程表
~ machinery 养路机械

maintenance machinery

~ man 维修工
~ manhole 检修人孔
~ manipulation 养路[维修]工作,养路操作
~ manual 养护手册
~ organization 养护机构
~ outlay 维修费用,维护费
~ overhaul 大修
~ overhead 经常性维修
~ painting 维修性油漆
~ period 养护周期,保修期限
~ point 维修点[站]
~ prevention 维修设施
~ process 养路工程;养护方法[程序]
~ quota 养护定额
~ shop 维修车间
~ sidewalk 维修用便道
~ superintendent 养路段长,养路主管工程师
on-line ~ 不停产检修
operating ~ 小修,日常修理
periodic ~ 定期维修
routine ~ 日常维修,经常性养护
scheduled ~ 定期检修
maisonette [ˌmeizə'net] *n.* 小房屋,出租房间;二层公寓
majolica [mə'jɔlikə] *n.* (石灰质)陶器,涂不透明釉的陶器
~ mosaic 彩陶镶嵌
major ['meidʒə] *a.* 较大的;较多的;较长的;多数的;主要的 *n.* 大分类
~ axis (椭圆的)长轴
~ bed 主河槽
~ calorie 大卡
~ cycle 大[主]循环
~ dam 主坝
~ fold 主褶皱
~ highway 主要公路
~ -minor rotary 主要路与次要路的环形交叉
~ -orthotropic bridge 立正交桥
~ overhaul 大修,总检修

~ parts 主要零部件
~ principal plane 主平面
~ principal stress 第一主应力
~ project 大型[规模]工程
~ repair 大修
~ total 总计,主要统计量
~ wave 主波
make [meik] v. made, making; 制造, 做成,使……成为 n. 形状,样式;构造,组成;商标;制造
~ -contact 闭合触点,接通触点
~ good 修缮,修复
~ reservoir 调节水库
~ -up cathead 备用锚栓
~ valve 补偿阀;调压阀
~ -up water 补充水
quick ~ 快速闭合
total ~ 总出[产]量
malachite ['mæləkait] n. 孔雀石
~ green 孔雀绿
blue ~ 蓝铜矿
maladjustment [ˌmælə'dʒʌstmənt] a. 失调,不匹配,调整不良
malaxation n. 捏,揉(混),拌和
malaxerman n. 拌料工
malchite n. 微闪长岩
male [meil] a. 阳性的,雄的; maleness, n.
~ cone 预应力锚塞
~ die 内螺模;阳模
~ dovetail 榫舌
~ fitting 外螺纹配件
~ gauge 内径规
~ nipple 有外螺纹的管接头
~ plug 插头
~ screw 阳螺钉,外螺纹
~ T 外螺纹三通管
maleic acid n. 马来酸,缩苹果酸,顺丁烯二酸
~ ester resin 马来酸脂树脂,顺丁烯二酸酯树脂
malfunction [mæl'fʌŋkʃən] n. 故障;

失误;失灵;误操作
mall [mɔːl] n. 林荫路;商业广场;手用大锤
~ hammer 大锤
malleability [ˌmæliə'biliti] n. 延展性,韧性
malleable ['mæliəbl] a. 有延展性的;可锻的; malleableness, n.
~ cast-iron 韧性铸铁
~ iron 韧性铁,可锻铁
~ steel 韧性钢,软钢
~ wrought iron 韧性锻铁
mallet ['mælit] n. 木锤
~ perforator 锤式冲孔机
calking ~ 填隙锤
malm [mɑːm] n. 石灰质砂,泥灰岩,灰泥;白垩土
~ brick 灰砂砖,白垩砖
~ rock 黏土砂岩
~ rubber 软白垩砖
~ stone 砂岩
maltha ['mælθə] n. 软沥青,半液质沥青,沥青焦油胶
malthene n. 软沥青质
malthoid n. 油毛毡
~ roofing 油毡屋面
mamelon =(maneron) n.(小)圆丘
mammock ['mæmək] n. 岩块;碎片,破片; vt. 破[撕,切]成碎片
mammoth ['mæməθ] a. 巨大的
~ blast 大暴破
~ pump 气动泵
~ structure 巨型结构
man [mæn] n. pl. men, v., manned, manning; n. 男人; v. 配备人员;使载人; manless, a. manlessly, ad.
~ -assignment 人力分配[配置]
~ cage 提人罐笼
~ -hole cover 人孔盖
~ -hour 工时
~ -hour cost 人时成本
~ less face 无人工作间
~ lid 进人孔盖

~ lift elevator 手动升降机
~ lock （沉箱）进人闸
~ -machine communication 人机联系
~ -machine engineering 人机工程学
~ -machine interaction 人机联系
~ -machine interface 人机接口
~ power 人力，劳动力
~ power management 劳动力管理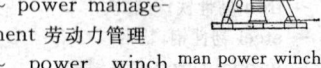
~ power winch man power winch 人力[工]绞盘

management ['mænidʒmənt] n. 管[办，处]理；经营；管理处，经理处 managemental, a.
~ activities 管理活动
~ advisory service 管理咨询服务
~ analysis center 管理分析中心
~ by objectives 目标管理
~ contract 管理合同
~ information system 管理信息系统
~ performance 管理成效
~ planning 管理计划
~ procedure 管理程序
~ process 管理过程
~ science 管理科学
~ technique 管理技术
construction ~ 施工管理
data ~ 数据管理
data base ~ 数据库管理
facility ~ 设备管理
factory [plant] ~ 生产组织，工厂管理
objective ~ 目标管理
personnel ~ 人事管理
production ~ 生产管理
project ~ 项目管理
storage ~ 库存管理
task ~ 任务管理
technical ~ 技术管理

mandatory ['mændətəri] ad.；n. 命令的，无选择的，强制的

~ control 强制性控制
~ expenditure 强制性支出
~ provision 强制性规定
~ rule 强制性守则
~ sanction 强制性制裁
~ sign 指示标志

mandrel ['mændril] n. 鹤嘴锄；丁字镐，(圆形)心轴；圆棒；芯棒
~ press 心轴压床

maneuver [mə'nu:və] =manoeuver n.；v.，-vered,-vering. n.；v. 机动；操纵；调遣；maneuverer, n.
~ room 操纵室
~ing chain 操纵链

manganese [ˌmæŋɡə'ni:z] n. 化学元素锰(Mn)
~ alloy 锰合金
~ bronze 锰青铜
~ cast-iron 锰铸铁
~ cement 锰水泥
~ silicon steel 锰硅钢
~ steel 锰钢

mangle ['mæŋɡl] n.；v.，-gled,-gling；n. 碾压[滚轴，光泽]机，轧板机；v. 刮切；割碎；破坏

mangler ['mæŋɡlə] n. 压延[砑光]机，切碎机，绞肉机

manhole ['mænhəul] n. 探井，检查井；检查孔，进入孔；升降口；避险洞
~ chamber 进入孔间
~ cover 探井盖；人孔盖；升降口盖
drop ~ 跌落井
flushing ~ 冲洗井
inspection ~ 检修人孔
sewer ~ 污水检查井
street ~ 街道进人孔
water pipe ~ 水管井

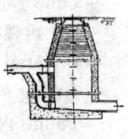
manhole

manifold ['mænifəuld] a. 多样的；多方面的，许多的；n. 多支管；复式接头；复写本；簇；v. 复写成数份；装成数份；manifoldly ad. manifoldness n.

~ condenser 多管冷凝器
~ crane 多用起重机
~ flow 簇流；分叉管水流
~ heater 多管加热器[蒸汽炉]
~ penstock 多叉压力水管
~ pipe 多支路管，多叉管
~ pressure 歧管压力
~ tunnel 多叉隧洞
delivery ~ 进水干管
distribution ~ 分配的干线
exhaust ~ 排气歧管
induction ~ 进气管；吸水管；进油管
inlet ~ 进气歧管
main ~ 主导管

manipulate [məˈnipjuleit] *vt.*, -lated,-lating；(用手)操作；操纵；使用；处理；manipulatory, *a.* manipulatable, manipulable, *a.*
~d soil 重塑土

manipulation [mənipjuˈleiʃən] *n.* (用手)操作；操纵；手术
bit ~ 二进制处理，数字处理
byte ~ 字组[字节]处理
charactor ~ 字符处理
date ~ 数据处理
manual ~ 手工[动]操作，人工控制
remote ~ 远距离操纵[遥控]

manipulative device 手工器[工]具
manipulator [məˈnipjuleitə] *n.* 操作器，机械手，键控器，操纵者
manjak *n.* 纯沥青，硬化沥青
man-made [ˈmæn,meid] *a.* 人制造的；人造的(别于天然的)，人工的
~ adhesive 人造胶粘剂
~ aggregate 人造(混凝土)骨料
~ asbestos 人造石棉
~ bending medium 人造胶粘剂
~ bonding adhesive 人造胶粘剂
~ bonding agent 人造胶粘剂
~ building material 人造建筑材料
~ building product 人造建筑制品
~ cementing agent 人造胶粘剂
~ construction material 人造建筑材料
~ construction product 人造建筑制品
~ diamond 人造钻石
~ earthwork 人工土方工程
~ fiber [美]人造纤维
~ fibre [英]人造纤维
~ fibre panel 人造纤维板
~ glue 人造胶水
~ ground 填土
~ lake 人工湖，(人造)水库
~ marble 人造大理石
~ masonry unit 人造圬工体
~ noise 人造[为]噪声
~ obstacle 人造[为]障碍
~ plastic 人造塑料[胶]
~ pollution 人为污染
~ pozzolana 人造火山灰材料
~ quartz 人造石英
~ resin 人造[合成]树脂
~ stone 人造石
~ resin-based adhesive 人造树脂类胶粘剂
~ rubber 人造[合成]橡胶
~ structural material 人造建筑材料
~ structural product 人造建筑制品
~ travertine 人造凝灰石

manner [ˈmænə] *n.* 方式[法]，样式，风格，手法；种类
mannerist architecture 程式化建筑，守旧派建筑
manograph [ˈmænəgrɑːf] *n.* (流体)压力记录器，测压计
manometer [məˈnɔmitə] *n.* 压力计，测压管，风压表，血压计；-manometric, *a.* -manometrical *a.* -manometrically, *ad.* -manometry, *n.*
~ method 测压管法
~ tube 测压管
absolute ~ 绝对压力表
air ~ 空气压力表
alarm ~ 气压报警计

manometric(al)

back pressure ~ 后压测压计
bell ~ 浮钟压力计
contact ~ 接触压力表
diaphragm ~ 膜片压力表
differential ~ 差示压力表
electrical ~ 电动压力表
hot-wire ~ 热效压力表
liquid ~ 液拉压力表
liquid tubular ~ 液体管式压力表
mechanical type ~ 机械式压力表
mercury ~ 水银压力表
spring ~ 弹簧压力表
well-type ~ 杯式压力表

manometric(al) [mænə'metrik(al)]
a. 测压的,压力计的,压差的
~ bomb 测压弹[器]
~ head 测压水头
~ thermometer 测压[压差]温度计

manor ['mænə] *n.* 庄园,大宅邸
~ house 庄园主住宅
~ place 庄园住宅

manostat ['mænəstæt] *n.* 恒[稳]压器;-manostatic,*a.*

manpower ['mæn,pauə] *n.* 人力,劳动力,人工
~ shortage 人[劳动]力短缺

mansard ['mænsɑːd] *n.* 复摺屋顶
~ cornice moulding 复摺式檐口装饰线脚
~ dormer window 复摺式屋顶(老虎)窗
~ flat roof 复摺式平板屋顶
~ roof 复摺屋顶
~ (roof) truss 复摺屋顶桁架

Mansion of World Trade Center

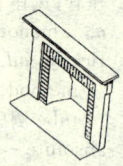
mantel

Mansion of World Trade Center 世界贸易中心大厦(美国)

mantel ['mæntl] *n.* 壁炉面饰,壁炉架
~ board 壁炉架板
~ piece 壁炉台
~ shelf 壁炉架
~ tree 壁炉过梁

mantle ['mæntl] *n.*;*v.*,-tled,-tling;
n. 外罩,机壳;覆盖物;地幔,风化覆盖层;*v.* 罩,覆;掩藏,遮住
~ of lamp 灯罩
~ of rock 风化层,表层岩
~ of soil 土被,土壤覆盖层
~ of vegetation 植被
~ pipe 套管

manual ['mænjuəl] *a.* 用手的,手动[工,制]的;*n.* 手册,说明书;键盘
~ acting 人工操作的
~ adjustment 人工[手动]调节
~ blending mixer 手动拌和机
~ boring 人力钻探
~ brake 手刹
~ computation 手算
~ consolidation 手工捣实,人工固结
~ control 手控
~ counting 手工计数
~ crane 手动起重机
~ cutting 手[人]工切割
~ excavation 人工开挖
~ extinguisher 手提式灭火机
~ fire alarm 手动火灾报警器
~ fitting 手工配合
~ hoist 手动提升机
~ input 人工输入
~ loading 手动加荷
~ labour 手工,体力劳动
~ of accounting 会计手册
~ operation 手工[动]操作
~ placement 人工填筑
~ pump 手动泵
~ riveting machine 手工铆接机
~ safety switch 手控安全开关

~ sampler 手动取样器
~ setting 手工调整
~ signal 人工信号
~ shift 手动转换
~ switch 手动开关
~ work 手工
design ~ 设计手册
service ~ 使用说明书
manufactory [ˌmænjuˈfæktəri] ***n. pl.***
-ries ;工厂
hardware ~ 五金工厂
manufacture [ˌmænjuˈfæktʃə] ***n. ;v.*** ,
-tured,-turing; ***v.*** ; ***n.*** 制造,加工;产品; manufacturable, ***a.*** manufactural, ***a.***
~d adhesive 人造胶粘剂
~d aggregate 人造骨料
~d article 人工制品,工艺器
~d brick chip(ping)s 加工砖片
~d building 预制构件建筑物
~d goods 制成品
~d lumber 加工木材
~d marble 人造大理石
~d products 工业制品
~d sand 人造砂
~'s standard 制造厂标准
extensive ~ 大批生产
interchangeable ~ 零部件可互换的生产
large scale ~ 大批生产
quantity ~ 大量生产
series ~ 成批[系列]生产
small lot ~ 小批生产
wholesale ~ 大批量生产
manufacturer [ˌmænjuˈfæktʃərə] ***n.***
制造者[厂],厂商,工厂主
~'s agent 经销商
~'s certificate 厂商证明书
~'s instruction 厂家说明书
~'s sales branches 制造商的分销机构
~'s test 厂方试验
manufacturing [ˌmænjuˈfæktʃəriŋ] ***a.***
制造(业)的,生产的; ***n.*** 制造(业),生产
~ cost 制造[生产]成本
~ cycle 生产循环[周期]
~ department 制造部门
~ district 工业区
~ extablishment 制造公司
~ expense 制造费用
~ expense ledger 制造费用分类账
~ expense order 制造费用通知单
~ facilities 生产设施
~ industry 制造工业
~ management 生产管理
~ method 生产方法
~ order 生产命令
~ operation 生产(制造)过程
~ overhead 制造间接费用
~ statisics 生产统计
~ process 加工程序,制造工艺
~ technology 制造技术
~ tolerance 制造公差
manway [ˈmænwei] ***n.*** 进入道[孔],检查孔(井)
~ cover 进入孔盖
~ frame 进入道框架
many-columned ***a.*** 多柱的,多柱式(建筑)

many-columned

many-sided ***a.*** 多边的
map [mæp] ***n.*** ; ***v.*** , mapped, mapping; ***n.*** 图,地图;映像,变换; ***v.*** 绘制(图,地图) mapable, ***a.*** maplike, ***a.***
~ accuracy 地图精度
~ board 图板
~ border 图廊,图框
~ code 地图符号
~ coordinates system 地图坐标系
~ cracking 网状裂缝,龟裂
~ grid 地图方格网

~ interpretation 地图判读
~ measurer 量图仪
~ projection 地图投影
~ relationship 图幅接合表
~ revision 地图修订
~ scale 地图比例尺
~ symbols 图例
accident ~ 事故图
aerial navigation ~ 航空图
conformal ~ 保角映射图
construction ~ 建筑图,施工图
contour ~ 等高[值]线图
exploration ~ 勘探图
flow ~ 流(线)图
geographic ~ 地形图
hydrogeological ~ 水文地质图
hydrographic ~ 水系图,水文测验图
planimetric ~ 平面图,测面图
progress ~ 进度图
sketch ~ 草图,示意图
topographical ~ 地形图
weather ~ 天气图,气候图
mapping ['mæpiŋ] *n.* 绘制地图,制图;
映射,映像
~ control 测图控制
~ degree 映像度
~ function 映像函数
~ pen 绘图笔
~ space 映像空间
aerial ~ 航空摄影
conformal ~ 保角变换[映射]
laser ~ 激光测绘
photographic ~ 像法测图
topographic ~ 地形测量
marble ['mɑ:bl] *n.*; *a.*; *v.*,-bled,
-bling; *n.* 大理石,大理岩;石弹; *a.*
大理石的;似大理石的; *pl.* 大理石雕
刻[艺术]品; *vt.* 使有大理石样色纹.
marblelike, *a.* marbler, *n.*
~ aggregate 大理石骨料
~ building 大理石建筑
~ cutter 大理石采石工;大理石切
割机
~ decoration 大理石装饰
~ dressing 大理石面修整
~d glass 大理石纹玻璃,斑纹玻璃
~d glazed paper 玳瑁蜡光纸
~ facing 大理石贴面
~ inlay 大理石镶嵌
~ lining 大理石贴面
~ intarsia 大理石镶嵌装饰
~ mosaic 大理石镶嵌艺术
~ pavement 大理石铺面
~ polishing machine 大理石上光机
~ polishing material 大理石上光
材料
~ quarry 大理石开采场
~ slab 大理石板
~ test 大理石腐蚀试验
~ texture 类似大理石花纹
imitation ~ 假[人造]大理石
white ~ 白大理石
marcasite ['mɑ:kəsait] *n.* 白铁矿
margin ['mɑ:dʒin] *n.* 边缘;极限;页边
之空白;余量,富裕;保证金; *vt.* 页边
留空白;加注;交保证金
~ and guaranty 保证金,押金
~ calls 追加保证金
~ capacity 备用容量
~ draft 石缘琢边
~ of continental shelf 大陆架边缘
~ of drill 钻锋圆边
~ of energy 能量储备
~ of error 误差限
~ of safety 安全限度
~ stop 极限挡块
~ tolerance 公差
~ trading 保证金信用交易
~ trowel 卷边线脚用抹子
current ~ 工作电流范围
noise ~ 噪声边限
overload ~ 过载定额
marginal ['mɑ:dʒinl] *a.* 边缘的;有旁
注的;界限的,边境的; marginality,

n. marginally, *ad.*
~ access 侧面通道
~ aggregate 边次集料
~ analysis 边际分析
~ balance 余额，赚头
~ bank 边岸，路堤岸
~ bar 边部钢筋，护栏
~ beam 边(橡)梁
~ bore 边孔[井]
~ bund 边缘堤岸
~ business 限界营业
~ condition 边际条件
~ convolution 边缘叶，边缘回卷
~ cost 边际成本
~ cost curve 边际成本曲线
~ cost equation 边际成本方程式
~ credit 定额信用证
~ crevasse 边缘裂缝
~ ditch 边缘沟
~ expenditure 限界消费
~ friction 路边阻力
~ ground water 边缘地下水
~ investment 边际[界限]投资
~ land (道路)边缘用地
~ land acquisition (路边)界地申请(取得路边地权)
~ lappet 缘垂
~ mold 边缘饰条
~ note 边注
~ obstruction 路边障碍物
~ profit 边际利润
~ quay 沿岸码头
~ reinforcement 边缘钢筋
~ sea 边缘海
~ strip (车行道)路缘带
~ swell 沿岸涌浪
~ texture 边缘结构
~ tile (for gables) (山墙)边瓦
~ torque moment 极限扭矩
~ track 最外股道
~ torsional moment 极限扭矩
~ trough 边缘凹陷

~ utility 边际效用
~ value 临界值
~ wharf 堤[顺]岸码头
~ zone 边缘地带
marguise *n.* 挑棚(大门雨罩)
marigram ['mærigræm] *n.* 潮汐(涨落)自记曲线
marigraph ['mærigra:f] *n.* 自记潮位仪
-marigraphic, *a.*
marine [mə'ri:n] *a.* 海的，海中的，海中产的；渔船的；*n.* 船舶，船队；海运业
~ abrasion 海蚀
~ aircraft 海上飞机
~ aggregate 海成骨料
~ anemometer 船用风速表
~ atmosphere 海洋大气层
~ barometer 船用气压计
~ bed 海床
~ belt 领海
~ borer 海洋钻蛀虫
~ bottom sediments 海底沉积
~ cable 海底电缆
~ cargo insurance 海上运输保险
~ chart 海图
~ chemistry 海洋化学
~ clay 海成黏土
~ climate 海洋(性)气候
~ construction 海洋[岸]工程
~ corrosion 海水腐蚀
~ current 海洋流
~ cut terrace 海蚀阶地
~ cycle 海蚀周期
~ denudation 海水剥蚀
~ deposit 海洋[相]沉积
~ dredger 海上挖泥船
~ erosion 海蚀，浪蚀
~ foreland 海岬
~ fouling 船底污物
~ geology 海洋地质学
~ glue 船用胶，防水胶
~ gravel 海相砾石
~ grey 船用灰漆
~ hydrographic surveying 海洋水

文测量
~ installation 海岸设施
~ insurance 海上保险
~ law 海上法,海商法
~ leg 浮式驳船提升机
~ light 海上导航灯标
~ marl 海成泥灰[岩]
~ navigation 航海,海上航行
~ navigation aid 海上救护
~ oil 航海油
~ organism 海洋生物
~ paint 船用漆
~ plain 海蚀平原
~ radio 海上无线电通讯,船舶电台
~ reserve 海上禁捕区
~ salina 海边盐滩
~ sand 海砂
~ shells 海贝壳
~ soil 海积土
~ structure 海岸[上]结构
~ surveying 海洋测量[调查]
~ terminal 海运终点站
~ timber 海洋森林
~ traffic 海运,海上交通
~ transgression 海侵
~ transportation 海上运输
~ varnish 海上建筑物用清漆
~ vehicle 船舶
~ work 海事工程

marine leg

maritime ['mæritaim] n. 海的,海上的;近海的;海员的
~ affair 海运事务
~ belt 领海
~ canal 通海运河
~ climate 海洋性气候
~ casualty 海上事故
~ commerce 海上贸易
~ country 沿海国家
~ custom 海关
~ engineering 海(洋)工(程)学
~ insurance 海上保险
~ law 海洋法,海商法
~ losses 海上损失,海损
~ pine 海岸松
~ plants 海生植物
~ port 海港
~ power 制海权
~ resources 海洋资源
~ section of stream 盛潮河段
~ survey 海洋测量[调查]
~ territory 领海
~ trade 海上贸易
~ transport 海运
~ work 海洋工程,海事工程

mark [mɑ:k] n. 记号,标志,商标;印[痕]迹;界限,标准;分数,点数;型号;特征;马克(西德货币单位) v. 作记号;记分数;划界限;注意
~ buoy 标识浮标
~ -down 降价,标低价格,减成
~ -down sale 降价销售
~ mumber 标号
~ of reference 参照符号
~ -on 成本外加数
~ pin 测钎
~ post 标杆
~ pulse 标志脉冲
~ed capacity 额定生产率,额定容量
~ed route 有标志路线
~ed station 埋石测站
~ed stone 标石
~ed transfer 标明转让
adjusting ~ 调整标记,安装标记
admissible ~ 允许符号,容许数字
aiming ~s 角度标志(天线位置);瞄准点
all ~ 全标记,全穿孔
angle ~ 角度标记[符号]
assembling ~ 装配记号
auxiliary aiming ~ 辅助瞄准点
azimuth ~ 方位标
bearing ~ 方位标记

bench ～ 水[基]准点;标高标志;试射点
block ～ 字组标记[符号]
brand ～ 工厂印记,商标
calibration ～ 校准标记
cement ～ 水泥等级
collimation ～ 框[直]标[志]
core ～ 型心记号,铁心上的符号
course ～ 航向标记
datum ～ 基准点
draft ～ 吃水标志
drum ～ 磁鼓标记
enter ～ 进口轧痕
erection ～ 安装标高[高程]
erosion ～ 侵蚀痕
eyepiece ～ 目镜标记
fiducial ～s 框(直)标(志),基准标
finish ～ 加工符号
fixed timing ～ 固定记时标记
floating ～ 浮标
flood ～ 满潮标,高水标
flood level ～ 洪水痕迹
flow ～ 流线谱
focusing ～ 聚焦标记
gauge ～ 定位刻线,印痕
group ～ (信息)组标记,组号
guide ～ 标记,记号,导板划痕
high water ～ 高[洪]水痕迹
hole ～ 洞孔测标
identification ～ 识别符号,商标
illuminating ～ 光标
index ～ 指数;目录索引,商标,品号;轮管上的瞄准线
land ～ 界标,陆标
load ～ 载重标志
location ～ 装配标记
low-water ～ 低水位标志,低潮线
match ～ 配合记号
measuring ～ 测标,丈量标
meridian ～ 子午(线)标
movable ～ 可动刻度标志
optical ～ 光学测标
permanent bench ～ 永久[固定]水准点
record-storage ～ 存储器记录标志
reference ～ 基准点,参考标高,控制高程
shuttering ～ 模(板)痕
space ～ 空间测标
spatial ～ 浮标[空间]测标
station ～ 测站标石
storage ～ 存储器(位置)标号
subsurface ～ 地下标点
supplementary bench ～ 水准补点
surface ～ 地面标点
timing ～ 时间标志,定时记号
track-error ～ 磁道出错标记
underground ～ 地下标石
water ～ 水印[线,位]
word ～ 字标记[符号]
zero ～ 零点刻度,零位记号
marker [ˈmɑːkə] *n.* 登记员;指示器,路标
～ bed 标准(地)层
～ line 标志线
～ light 信号灯
～ post (反射式)向导标
boundary ～ 界标
construction ～ 施工标志
road ～ 路面划线机
roadway ～ 路标
route ～ 路线标
sea ～ 海上标志
street ～ 路牌
time ～ 时间标记

construction marker

market [ˈmɑːkit] *n.* 市场;市面;销路;买卖 *v.* 销售;买卖
～ analysis 市场分析
～ appraisal standard 市场作价标准
～ behavior 市场行为
～ counselor 市价评议员
～ cycle 销售循环
～ day 集日;集市
～ demand 市场需求

~ demand curve 市场需求曲线
~ demand price 市场需求价格
~ development 市场发展
~ directed economy 市场导向经济
~ economy 市场经济
~ efficiency 市场效率
~ finance 市场金融
~ forecast 市场预测
~ garden 供应市场的菜园
~ hall 室内市场
~ house 市场管理所,市场(指建筑物)
~ index 市场指数
~ information 市场信息
~ instinct 市场本能
~ off 市场休业
~ order 市场定购
~ -oriented production 面向市场的生产
~ place 市场商业中心区
~ pot 熔铝锅
~ potential 市场潜力
~ price 市场价格
~ quotation 市场行情
~ rate 市场利[汇]率
~ readjustment 市场再调节
~ report 市场报告
~ research 市场调查
~ risk 市场风险
~ share 市场份额
~ stead 市场
~ structure 市场结构
~ survey 市场调查
~ testing 市场测验
~ town 集镇
~ transaction 市场交易
~ value 市场价格
black ~ 黑市
buyer's ~ 买方市场
flea ~ 旧货市场
grey ~ 半黑市场
jobbing ~ 中间市场
local ~ 生产地市场
retail ~ 零售市场
seller's ~ 卖方市场
terminal ~ 终点市场
marking ['mɑ:kiŋ] *n*. 标志,记号,印痕,条纹;作记号,划线打印; *a*. 有特征的
~ board 记分牌
~ chalk 记号笔
~ ga(u)ge 划线规
~ ink 不褪色墨水
~ machine 划线机
~ material 路面划线材料
~ out [道路]立标桩定线立桩标界
~ paint 划线漆,路标漆
~ pin 标杆,测杆
marl [mɑ:l] *n*. 泥灰岩,灰泥
~ loess 泥灰质黄土
~ loam 泥灰质壤土
~ pellet 泥灰质团粒
~ pit 泥灰岩采掘场
~ slate 泥灰板岩
~ stone 泥灰岩[石]
lime ~ 灰质泥灰岩
marlite ['mɑ:lait] *n*. (抗风化的)泥灰岩
marly ['mɑ:li] *a*. 泥灰质的;石灰泥的;含石灰泥的
~ clay 泥灰质黏土
~ limestone 泥灰质石灰岩
~ sandstone 泥灰质砂岩
~ sandy loam 泥灰砂质壤土
~ shale 泥灰质页岩
~ soil 泥灰质土
marmorate(d) *a*. 具有大理石纹的
marmoration *n*. 用大理石贴面[表面装饰]
marmoratum *n*. 大理石粉胶泥
marquetry ['mɑ:kitri] *n*. 镶嵌细工
Marseilles pattern tile 马塞式样的瓦
marsh [mɑ:ʃ] *n*. 沼泽,沼地,湿地

marquetry

~ deposit 沼泽沉积
~ gas 沼气
~ muck 湿地腐殖土
~ land 沼泽地
~ podzol 湿地灰壤
~ peat 沼泽泥煤
~ soil 沼泽土,湿地土

Marshall ['mɑːʃəl] *n.* 马歇尔(男子名)
~ (cylindrical) test (沥青混凝土)马歇尔(圆柱形)试验
~ flow value (沥青混凝土的)马歇尔流动值
~ mix design method 马歇尔混合设计法
~ properties 用马歇尔法测定的(沥青混凝土)性质
~ stability 马歇尔稳定度
~ stability test 马歇尔稳定度试验
~ stability value 马歇尔稳定度值

marshalling ['mɑːʃəliŋ] *n.* 排列,整理;调度,编组;砌筑
~ area 编组区,备用月台
~ equipment 编组设备
~ masonry 砌筑圬工
~ track (列车)编组线
~ tractor 编组车头
~ vehicle 编组列车
~ yard 调车场

marshy ['mɑːʃi] *a.* marshier, marshiest;沼泽的,湿地的 marshiness, *n.*
~ area 沼泽地,沼沼地,湿地
~ district 沼泽地区
~ ground 沼泽地[土],湿地
~ land 沼泽地,泥沼地,湿地

mascaron 怪状头饰(在拱顶石上)

mason ['meisn] *n.* 泥瓦工,石工,圬工,砖石工;masonic, *a.*
~ architecture 圬工建筑[结构]
~ flagger 砖石铺路
~ jar 陶瓷瓶

mascaron

~ sand 圬工砂
~ 's float 圬工镘刀[板]
~ 's hammer 石工锤
~ 's level 圬工(用)水平尺,泥瓦工(用)水准器
~ 's mold 泥瓦工用型板
~ 's rule 泥瓦工用尺
~ 's runway 圬工跳板
brick ~ 砌砖工
master ~ 熟练泥水工
stone ~ 石工

masonry ['meisnri] *n. pl.* -ries;(砖)石工程,圬工,砖石建筑,砌筑体
~ anchor 圬工锚碇
~ arch 圬工拱
~ block 圬工块
~ bond 圬工粘结剂
~ brick 圬工砖
~ bridge 圬工桥
~ building 圬工建筑物
~ buttress 圬工支墩
~ cement 圬工水泥
~ cleaner 圬工清洁剂
~ component 圬工建筑构件
~ construction 圬工建筑
~ course 圬工砌层
~ cutting blade 圬工用切刀
~ dam 圬工坝
~ diaphragm 圬工隔膜
~ envelop 砌体
~ failure test 圬工破坏试验
~ fastener 圬工紧固件
~ fill 圬工填充(料)
~ footing wall 圬工建基脚墙
~ foundation wall 圬工基础墙
~ gallery 圬工廊道
~ grout(ing) 圬工灌浆
~ joint 圬工砌缝
~ lining 圬工衬砌
~ material 圬工材料
~ member 圬工建筑构件
~ nail 水泥钉

~ paint 圬工涂料
~ pier 圬工墩,砌石墩
~ point 圬工灰缝
~ reinforcing 圬工加筋
~ revetment 圬工护岸
~ ruins 圬工碎片
~ saw 石工用锯
~ sewer 烧砌沟管
~ structure 圬[石]工结构
~ tool 圬[石]工工具
~ unit 圬工单位,圬工体
~ vault 圬工穹窿
~ veneer 砖石建筑饰面
adobe ~ 土砖圬工
ashlar ~ 琢石圬工
bastard ~ 乱石圬工
block-in-course ~ 成层砌石圬工
brick ~ 砖石圬工
brick-lined ~ 砖砌[衬]工
broken ashlar ~ 不等形琢石圬工
broken range ~ 不分层砌石圬工
cement-rubble ~ 浆砌毛石圬工
clinker ~ 熔渣块圬工,炼砖圬工
cobweb rubble ~ 蛛网形缝毛石圬工,虎皮形石圬工
coursed rubble ~ 成层毛石圬工
cyclopean ~ 蛮石圬工
dressed ~ 敷面圬工
dry rubble ~ 干砌毛石圬工
free-stone ~ 毛石圬工
green ~ 新筑圬工
hammer-dressed (ashlar) ~ 锤琢石(圬)工
loose ~ 干堆圬工
mortar rubble ~ 浆砌毛石圬工
pointing ~ 勾缝[嵌灰]圬工
polygonal ~ (砌)多角石圬工
quarry-stone ~ 毛石砌体[圬工]
random (rubble) ~ 乱石圬工
range ~ 成层圬工
reinforced brick ~ 加筋砖砌圬工

masonry veneer

repointed ~ 重嵌(灰缝)圬工
rubble ~ 毛石圬工
rustic ~ 粗面圬工,粗石工
tile ~ 砖瓦圬工
tool-faced ~ 锤琢石圬工
uncoursed rubble ~ 不成层毛石圬工

mass [mæs] *n.* 团,块;大量,大多数;容[体]积;质量; *v.* 集合; *a.* 大规模的,群众的
~ acceleration 质量加速度
~ accounting 群众核算,容[体]积核算
~ action law 质量作用定律
~ avalanche 大规模崩坍(冰雪,土石等)
~ balance 质量[物料]平衡
~ calculation 土方计算
~ center 质心
~ centroid 质量中心
~ concrete 大体积[大块]混凝土
~ concreting 大体积混凝土浇筑
~ conservation 质量守恒
~ control 质量控制
~ curing 密封绝热养护
~ curve 累积曲线
~ data 大量数据
~ density 质量密度
~ diagram 累积图,土方累积曲线图
~ diagram of runoff 径流累积图
~ distribution 质量分布
~ element 质点
~ energy curve 质能曲线
~ float 集束浮标
~ flowmeter 质量流量计
~ force 体积[质量]力
~ foundation 大体积基础,大块基础
~ function 大数函数
~ haul curve 开挖运输曲线
~ matrix 质量矩阵
~ management 集体管理
~ movement 块体运动

~ of concrete 混凝土块
~ of dam 坝体
~ of masonry 圬工块
~ of sand 砂团
~ output 累计出力[产量]
~ production 大量生产
~ rainfall curve 雨量累计曲线
~ ratio 质量比
~ runoff curve 径流累计曲线
~ sacking 大批裁员
~ selling 集体销售
~ shooting 集爆
~ spectrograph 质谱仪
~ spectrometry 质谱测定法
~ spectroscopy 质谱
~ spectrum 质谱(化)
~ storage 大量存储
~ transit 公共交通,大量运输
~ transportation facilities 公共交通设施
~ transportation loading zone 公共车辆站台
~ volume 土方累计体积
ore ~ 砂坝
pilaster ~ 壁柱墩子
shrinkage ~ 收缩量

massicot [ˈmæsikɔt] *n.* 黄铅丹,黄丹(pho)

massif [ˈmæsiːf] *n.* 断层块,山丘,(古地块)台

massive [ˈmæsiv] *a.* 宽而大的,大而重的,整块的;结实的; massively, *ad.* massiveness, *n.* massivity, *n.*
~ arch dam 重力拱坝
~ board 大块型板
~ -buttress dam 大头坝
~ concrete 大体积混凝土
~ construction 整体结构
~ footing 块状底脚

massive board

~ gypsum 天然石膏
~ -head buttress 大头式支墩
~ -head dam 大头坝
~ layer 大体积浇筑层
~ quay wall 大体积岩壁
~ rock 块状岩
~ structure 整体结构,大体积[块]结构;块状构造;重型结构
~ texture 整体结构,块状结构[组织]

mast [mɑːst] *n.* 桅,柱,杆,起重架; *v.* 立柱;装桅杆; mastlike, *a.*
~ arm (照明)灯具悬臂
~ bottom 起重杆支柱底座
~ crane 桅杆式起重机,桅式吊机
~ extension 伸缩杆,套管式天线杆
~ timber 桅木
aerial ~ 天线杆
portable ~ 轻便井架
radio ~ 天线杆

mastabe 露天石头平台(伊斯兰教建筑)

master [ˈmɑːstə] *n.* 主人,雇主;教师,家长;师傅;专家;硕士;(录音)主盘,母唱片; *a.* 主人的;主要的; *vt.* 控制,征服;掌握,精通; masterhood, *n.* masterless, *a.*
~ altimeter 校正用高度计
~ antenna 共用天线
~ bedroom 主卧室
~ board 主控制(仪表)板
~ budget 总预算
~ builder 营造师,监工
~ card 原版卡片
~ carpenter 木工师傅
~ clock 母钟
~ clutch 主离合器
~ computer 主电子计算机,总站电子计算机
~ container 主集装箱
~ contract 交易主约
~ controller 主控器
~ control room 中央控制室
~ curve 通用曲线

mastic

~ cylinder 主油缸
~ designer 总设计师
~ driving program 推算程序
~ equation 主方程
~ ferquency 主频
~ gauge 总表,校对量规
~ gear 主齿轮,标准齿轮
~ gulley 集水主沟
~ joint 主要节理
~ key 万能钥匙
~ link 主联杆
~ map 原图
~ mason 熟练圬工,圬工领班
~ mechanic 技工长,熟练技师
~ of architecture 建筑(学)硕士
~ of civil engineering 土木工程硕士
~ piece (of work) 杰作
~ plan 总平面图,总图,总体规划
~ plate 模板
~ program 主程序
~ radio antenna 无线电主天线
~ river 主河流
~ routine 主程序
~ sample 标准样品
~ schedule 总日程表
~ serial number 总编号
~ sheet 原图
~ slave system 主从系统
~ station 主站
~ stream 干流,主流
~ switch 总开关
~ timer 主要定时开关
~ work 杰作

mastic ['mæstik] *n.* 砂胶,树脂,乳香,玛脂;淡黄色
~ asphalt 地沥青砂胶,沥青膏
~ asphalt surfacing 地沥青砂胶铺面
~ asphalt finisher 地沥青砂胶磨光机
~ bed 砂胶层,胶脂层
~ -bitumen rubble 沥青砂胶橡皮
~ block 砂胶块
~ cement 水泥砂胶
~ cooker 砂胶加热锅
~ cushion 砂胶垫层
~ filler 砂胶地面
~ floor 砂胶地面
~ gum 胶粘脂
~ joint 砂胶接缝
~ jointing 砂胶勾缝
~ macadam 砂胶碎石路
~ pavement 沥青砂胶路面
~ sealing 砂胶密封
bituminous ~ 沥青砂胶
granited ~ 花岗石砂胶

masthead ['mɑːsthed] *n.* 柱顶,桅顶,杆顶

mat [mæt] *n.; a.; v.,* matted, matting, *n.* 柴排,席子,坐垫,面层,钢筋网; *a.* 暗淡无光的;粗糙的,未抛光的; *v.* 铺席;缠结褪光; matless, *a.*

masthead

~ base 垫层
~ coat 面层,保护层
~ -covered pavement 有保护层的路面
~ finish 磨砂表面处理
~ finish paint 磨砂表面涂料
~ foundation 席形[底板]基础
~ glass 磨砂玻璃,毛玻璃
~ layout 织物敷层
~ surface (相纸)布纹面
~ work 编织工;席工
bar ~ 钢条网
bituminous ~ 沥青层
curing ~ 混凝土养护席
fire-proof ~ 耐火席子
glass ~ 玻璃纤维板,玻璃垫
insulating ~ 绝缘垫
reinforcement ~ 钢筋网

match [mætʃ] *v.* 比赛;匹配;与……相适应; *n.* 火柴;比赛;导火线; matcher, *n.*
~ board 横板,企口板
~ boarding machine 企口镶板机
~ casting 镶合浇制

~ joint 企口[舌槽]接合
~ lines 对口线
~ marking 配合记号
~ plane 边刨,槽刨
~ plate 模板
~ed aggregate 配好的骨料
~ed ceiling 镶配天花板
~ed floor 镶配[企口]地板
~ing 嵌缝板条

matching

material [mə'tiəriəl] *n.* 材料,原料;物质;资料;设备；*a.* 具体的;实质的;主要的; materialness, *n.*
~ account 材料账户
~ and supply 物资与供应
~ at site 工作地点存料
~ audit 原材料审计
~ balance 物料进出平衡
~ balance equation 物质平衡方程(式)
~ balance sheet 物资平衡表
~ behaviour 材料性质
~ card 材料卡
~ checker 材料检查员
~ constant 材料常数
~ consumption 原料耗费
~ control 材料管理
~ cost 原料成本
~ designation 材料标志
~ distribution map 材料分配图
~ engineer 材料工程师
~ handling 材料供应[转运]
~ handling crane 运料吊车
~ engineering 材料储运工程
~ in bulk 整批材料
~ in process 在制材料
~ inspection 材料检验
~ list 材料明细表[清单]
~ manual 材料手册
~ model 实体模型
~ pit 储料坑
~ purchasing and stock 材料购买与储存
~ quantity standard 材料用量标准
~ receipt 收料报告
~ receipt sheet 收料单
~ record 材料记录
~ requirement 原材料需要量
~ requisition 领料单
~ reserves 材料储备
~ review standards 材料审核标准
~ salvage 物资回收
~ sample 材料样品
~ schedule 材料表,材料清单
~ shaft 材料运送竖井
~ shortage 材料短缺
~ specification 材料规格
~ standard 材料标准
~ stock 材料储备
~ storage 储料场,材料储量

material storage

~ supplies 材料供应
~ supply depot 材料供应仓库
~ testing 材料试验
~ testing laboratory 材料试验室
~ testing machine 材料试验机
~ testing report 材料试验报告
~ turnover 原材料周转率
~ unit price 原材料单价
~ variances 材料差异
acid refractory ~ 酸性耐火材料
acid-proof ~ (=acid resisting ~) 耐酸材料
acoustic ~ 隔声材料
backing ~ 座层衬板,背(贴)材(料)
basic ~ 碱性材料;原料
bituminous ~ 沥青材料
building ~ 建筑材料
bulk ~ 松散[粒状]材料
cementing ~ 粘(胶)结材料

ceramic ~ 陶质材料
choke ~ 填塞材料
cladding ~ 镀层;覆盖材料
coarse ~ 粗料
cohesionless ~ 非黏性材料
composite ~ 成层[合成]材料
condensed ~ 冷凝[凝结,密度]物质
conductor ~ 导电材料
coustructional ~ 建筑材料
damping ~ 隔[吸]声材料
discrete ~ 松散[粒状]材料
divided ~ 松散[粒状]材料
dust laying ~ 防尘料
engineering ~s 工程材料
englacial ~ 冰川内碛
excavated ~ 挖掘[开挖]料
explosive ~ 爆炸物质,炸药
extinguish ~ 灭火材料
facing ~ 饰[护]面材料,涂面材料
fettling ~ 补炉材料
filling ~s 填塞料,填充材料
filter ~ 渗滤材料
fine ~ 细料
finished ~ 制品,成品
fire-proof ~ 耐火材料
fluid bituminous ~ 液体沥青材料
fluorescent ~ 荧光材料
foreign ~ 杂质,外车material,异物
gasket ~ 垫衬材料
graded ~ 级配料
grinding ~ 磨料
half-finished ~ 半制[成]品
heat-sensitive ~ 易燃剂,易燃物质
heat-transfer ~ 载[散,传]热材料
high-espansion ~ 高膨胀材料
inert ~ 惰性材料
inflammable ~ 易燃品[料]
in-process ~ 在制品
insulating ~ 绝缘材料
istropic ~ 各向同性材料,均质料
jointing ~ 接缝料
laser ~ (=lasing ~) 激光材料

light ~ 小截面材料,薄型料
line ~ 线路器材
liquefied bituminous ~ 液态沥青材料
loading ~ 填料
lump ~ 块料
morainic ~ 碛物,堆石
moulding ~ 造型材料
multilayer ~ 多层材料
optical ~ 光学材料
oxidizable ~ 易氧化材料
packing ~ 填充[密封]材料
padding ~ 填密材料
paving ~ 铺面材料
photoelectric ~ 光电材料
plastic ~ 塑性材料,塑料
raw ~ 原(材)料
refractory ~ 耐火材料
reinforcing ~ 加强(材)料
reinforced thermoplastic ~ 加强热塑材
rigid ~ 刚性材料
rolled sheet ~ 轧制钢板
section ~ 型材
semi-finished ~ 半成品
synthetic ~ 合成材料
thermalin sulating ~ 绝热[保温]材料
thermalplastic ~ 热塑材料
thermalsetting ~ 热固材料
thin-film ~ 薄膜材料
transparent ~ 透明物质[材料]
weighting ~ 填充物,填料
mathematical [ˌmæθiˈmætikəl] *a.* 数字的,数理的;精确的; matnematically, *ad.*
~ analysis 数学分析
~ check 数字检验
~ control 数字控制
~ economics 数理经济学
~ expectation 数学期望值
~ expective value 数字期望值

~ expression 数字表达式
~ forecast 数值预报
~ induction 数字归纳法
~ logic 数理逻辑
~ model 数字模型
~ physics model 数学物理模型
~ point 虚点
~ programming 数学规划
~ solutions 数学解
~ statistics 数理统计
~ treatment 数学处理

mathematics [ˌmæθiˈmætiks] ***n.*** 数学
　applied ~ 应用数学
　elementary ~ 初等数学
　fixed-point ~ 定点运算
　floating-point ~ 浮点运算
　higher ~ 高等数学
　pure ~ 纯粹数学

matrix [ˈmeitriks] ***n. pl.*** matrices, matrixes; 胶结材料, 填充料; 基质; 母岩; 矩阵
　~ addition 矩阵加法
　~ algebra 矩阵代数
　~ analysis 矩阵分析
　~ differential equation 矩阵微分方程
　~ equation 矩阵方程
　~ formulation 矩阵形成
　~ inversion 矩阵求逆
　~ language 矩阵语言
　~ management 矩阵管理
　~ material 胶结材料
　~ method 矩阵方法
　~ multiplication 矩阵乘法
　~ operation 矩阵运算
　~ solution 矩阵解法
　~ store 矩阵存储器
　~ structural analysis 矩阵结构分析
　~ subtraction 矩阵减法
　~ summation 矩阵求和
　~ -theory 矩阵理论
　binder ~ 结合混合料
　coefficient ~ 系数矩阵

mattress [ˈmætris] ***n.*** 沉排; 柴排; 混凝土基础板; 床垫
　~ covering 沉排护面
　~ foundation 沉排基础
　asphalt ~ 沥青毡
　ballast ~ 沉碴垫
　brush ~ 梢捆垫, 柴排
　brush and wire envelop ~ 铁丝网柴褥沉排
　fascine ~ 柴排席, 沉排
　framed ~ 框架沉排
　inflatable ~ 充气垫
　spring ~ 弹簧垫

mature [məˈtjuə] ***a.; v.***, -tured, -turing; ***a.*** 成熟的, 壮年的; 熟思的; ***v.*** 老化, 成熟; 完成; 到期
　~d bill 到期汇票
　~d check 到期支票
　~d endowment 到期养老保险金
　~d liability 到期负债
　~d principal 到期本金
　~ repayment 到期应付款
　~d concrete 养护后硬化的混凝土

maturing ***n.*** （混凝土的）硬化

mauln 大木锤

Mausoleum 壮丽之墓

maximum [ˈmæksiməm] ***a.; n. pl.*** -mums, -ma; ***a.*** 最大的, 极大的; ***n.*** 最[极]大值; maximumly, ***ad.***
　~ aggregate size 骨料最大粒径
　~ allowable pressure 最大容许压力
　~ and minimum thermometer 最高最低温度度
　~ annual hourly volume 年度最高小时（交通）量
　~ apparent error 最大可视误差
　~ available water （土内）最大有效含水量
　~ average temperature 最高平均温度

~ axle load 最大轴荷载
~ bending moment 最大弯矩
~ capacity 最高生产能力
~ carrying capacity 最大承载能力
~ clearance 最大净空
~ compacted dry density 最大压实干容重
~ demand 最大（水，电等）需要量
~ density 最大密度
~ depth 最大水深[井深]
~ discharge 最大流量
~ draft 最大吃水深度
~ dry density 最大干容重
~ efficiency 最高效率
~ flood level 最高洪水位
~ flow 最大流量
~ flow capacity 最大过水能力
~ frequency 最大频率
~ gradient 最大梯度[坡度]
~ headwater 最高上游水位
~ high-water 最大高水位
~ hourly consumption 最大小时耗水量
~ hygroscopicity 最大吸湿度
~ intensity 最大强度
~ likelihood method 最大似然法
~ load 最大荷载
~ moment 最大力矩
~ output 最大生产量，最大出力
~ possible flood 最大可能供水
~ possible precipitation 最大可能降水
~ power output 最大输出功率
~ principle 极大值原理
~ rated load 最大额定荷载
~ safe capacity 最大安全容量
~ safe speed 最大安全速率
~ scale 最大刻[标]度
~ seismic intensity 最大地震烈度
~ service life 最大使用寿命
~ setting rate 最大沉降率
~ shear 最大剪力
~ shear stress 最大剪应力

~ size aggregate 最大粒径骨料
~ skid-resistance potential 最大抗滑势能
~ snowfall line 最大降雪线
~ stage 最高水位
~ strain theory 最大应变理论
~ stress 最大应力
~ stress theory 最大应力理论
~ superelevation 最大超高度
~ superelevation rate 最大超高率
~ surge 最高涌浪
~ swelling 最大膨胀
~ temperature 最高温度
~ value 最大值
~ velocity 最大速度
~ water capacity 最大含水量
~ water consumption 最大耗水量
~ water holding capacity （土壤）最大持水量
~ water level 最高水位
~ work 最大功
~ working pressure 最大工作压力
critical ~ 临界极大值
Maya arch 玛雅拱门
mean [mi:n] *n.* ; *a.* ; *v.* , meant, meaning; *n.* 中间；均值，平均；中值； *a.* 平均的；中间的；劣等的； *v.* 意思是，意味着
~ absolute error 平均绝对误差
~ accumulation 平均积厚
~ annual discharge 年平均流量
~ annual precipitation 年平均降水量
~ annual temperature 年平均温度
~ breadth 平均宽度
~ breakup date 平均解冻日期
~ cycles between failure 损坏周期平均值
~ daily temperature 日平均温度
~ deviation 平均偏差
~ diameter 平均直径

~ discharge 平均流量
~ effective pressure 平均有效压力
~ error 平均[标准]误差
~ freezing index 平均冰冻指数
~ grain diameter 平均粒径
~ high water neap tides 平均小潮高潮位
~ high water spring tides 平均大潮高潮位
~ level 平均水平,平均能级
~ life 平均寿命
~ line 等分线
~ low water 平均低潮位
~ low water neap tides 平均小潮低潮位
~ low water spring tides 平均大潮低潮位
~ particle diameter 平均粒径
~ proportional 比例中项
~ range 平均潮差
~ sea level 平均海平面
~ sea level datum 平均海平面基点
~ service life 平均使用寿命
~ solar day 平均太阳日
~ square deviation 均方偏差
~ square error 均方[误]差
~ square regression 均方回归
~ square value 均方值
~ terms 中间项
~ tide level 平均潮水位
~ time 平均时间
~ value 平均值
~ variation 平均偏差
~ water level 平均水位
arithmetic(al) ~ 算术平均,等差中项
cumulative ~ 累加平均
daily ~ 逐日平均
geometric ~ 几何平均数,等比中项
population ~ 总平均
proportional ~ 比例中项
quadratic ~ 均方值

residual ~ 残余平均
sample ~ 采样平均
simple ~ 简单平均
weighted ~ 加权平均
zero ~ 零平均
means [mi:nz] ***n.*** 方法,手段,工具
~ of labour 劳动手段[工具]
~ of livelihood 生活资料
auxiliary ~ 辅助设备
cooling ~ 冷却设备
electrooptical of communication ~ 光电通信工具
information storage ~ 信息存储方式
measuring ~ 测量装置[方法]
measure ['meʒə] ***n.***;***v.***,-ured,-uring;
v. 量测;***n.*** 量度;尺度,大小;方法,措施;measurer,***n.***
~ analysis 量测分析
~ expansion 体(积)膨)胀
~ of capacity 容量
~ of location 定位测量
~ of value 价值尺度,定价
~ed day work 计时日工
~ed profile 实测断面
~ing accuracy 量测精度
~ing apparatus 测量仪器
~ing appliance (测)量(用)具[设备],仪表
~ing beam 测杆
~ing bin 量斗
~ing bolt 量测用螺栓
~ing box 量料斗
~ing bridge 量测电桥
~ing by sight 目测
~ing chain 测链
~ing channel 量水槽
~ing circuit 量测电路
~ing column 水银柱
~ing condition 测量条件
~ing cylinder 量筒
~ing dam 量水堰

~ing device 量具,测量仪表
~ing equipment 量测仪器
~ing error 量测误差
~ing implement 量具,量测仪器
~ing instrument 量测仪器
~ing line 测线,测绳
~ing mark 测标
~ing peg 测桩
~ing pin (油)量(控制)针
~ing platform 观测台
~ing plug 测量标志,测桩
~ing point 测点
~ing projector 轮廓投影仪
~ing range 量程,测量范围
~ing reel 卷尺
~ing rope 测绳
~ing rule 测尺
~ing scale 量尺,标尺,比例尺
~ing section 量测断面
~ing staff 测杆
~ing stereoscope 量测立体镜
~ing tank 测箱,量箱
~ing tape 测尺,卷尺
~ing technique 测量技术
~ing weir 量水堰
~ing well 量测井
~ing wheel 测轮
absolute ~ 绝对量度
angular ~ 角度法
capacity ~ 容量
circular ~ 弧度法
common ~ 公约数,公测度
exterior ~ 外测度
interior ~ 内测度
nondestructive ~ 不破损测量法
precautionary ~ 预防措施
protective ~ 保护措施
remote ~ 遥测
soil sample ~ 土样测量
safety ~ 安全措施
shrinking ~ 收缩度;放尺
square ~ 面积单位

tape ~ 卷尺
technical ~s 技术措施
measurement [ˈmeʒəmənt] *n.* 量度,测量;测量方法[制度];体积
~ condition 测量条件
~ circuit 测量线[回]路
~ data 测量数据
~ device 量测装置
~ errors 测量误差
~ goods 体积货物
~ items 测量项目
~ of angles in all combination 全组合角度测量法
~ of quantities 计量,量的测定,量方
~ range 量测范围,量程
~ standard 测定标准
~ transducer 量测传感器
absolute ~ 绝对测量[量度]
accurate ~ 精密测量
acoustical ~ 声学测量
actual ~ 实测
angular ~ 角测
arc ~ 弧度测量
barometric height ~ 气压测高
base ~ 基线测量
bearing ~ 方位测量
bridge ~ 电桥测量法
comparative ~ 比较测定法,比较量度
continuous ~ 连续测量
direct ~ 直接观测
down-hole ~ 井下测量
full-scale ~ 全尺寸测量
height ~ 高程测量,测高法
indirect ~ 间接量度
infrared ~ 红外测量
noise ~ 噪声测定
optical ~ 光学测量法
oscillator ~ 振动测量
phase-delay ~ 相位延迟测定
remote ~ 遥测
soil sample ~ 土样测量

time average ~ 平均时间测定
ultrasonic ~ 超声波测量
mechanic [mi'kænik] *n.* 机械工人
mechanical [mi'kænikəl] *a.* 机械的；力学的； mechanically, *ad.* mechanicality, *n.* mechanicalness, *n.*
~ advantage 机械效益
~ agitation 机械搅动
~ analogy 机械模拟
~ analysis 力学分析；(土的)粒径分析
~ analysis curve 粒径分析曲线
~ aptitude test 机械性能试验
~ bond 机械结合；(砖的)机械砌合
~ brain 人工脑
~ clarification 机械澄清法
~ clarifier 机械澄清池
~ control (水土保持)工程措施
~ corrosion 机械磨损
~ cycling 机械循环
~ damage 硬伤，机械损伤
~ devices 机械装置
~ dewatering 机械脱水
~ digger 挖掘机
~ -draft water cooling tower 机动气流水冷(却)塔
~ draught 机械通气
~ drawing 机械图；机械制图
~ drier 干燥机
~ drilling 机钻
~ drying 机械烘干
~ effect 机械效能
~ efficiency 机械效率
~ elevator (矿料)升运机
~ energy 机械能
~ equipment schedule 机械设备表[清单]
~ equivalent of heat 热功当量
~ excavation 机械开挖
~ execution 机械施工
~ extractor 机械抽风机
~ haulage 机械运输方式

~ impedance 力[机械]阻抗
~ jack 机械式千斤顶
~ lift-lock 机械单级船闸
~ loader 装载机
~ loss 机械损失
~ -optical strain gage 机械-光学应变仪
~ power 机械功率
~ predictor 机械(潮汐)预报器
~ property 机械性能，力学性质
~ rammer 机械夯
~ screeding 机械整[刮]平
~ sewage treatment 污水机械处理法
~ shovel 机械铲，挖土机
~ sifter 机动筛
~ similarity 力学相似性
~ stabilization 改善级配加固法
~ stoker 机动加料机
~ strength 机械强度，力学强度
~ subgrader 路基整修机，平地机
~ sweeper 机动扫，扫路机
~ tamper 机械夯，机械路ș
~ testing 机械(性能)试验,力学(性能)试验
~ transport 机动运输，汽车交通
~ trowel 抹灰机
~ vehicle 机动车
~ ventilation 机械通风
~ vibration 机械振动
~ wear 机械磨耗
mechanics [mi'kæniks] *n.* 力学
~ of bulk materials handling 散装材料起重运输机械
~ of elasticity 弹性力学
~ of landslide 滑坡力学
~ of materials 材料力学
~ of plasticity 塑性力学
~ of structure 结构力学
air ~ 空气动力学
analytical ~ 分析力学
applied ~ 应用力学

celestial ~ 天体力学
classical ~ 经[古]典力学
continuous medium ~ 连续介质力学
engineering ~ 工程力学
fluid ~ 流体力学
fracture ~ 断裂力学
rational ~ 理性力学
rock ~ 岩石力学
soil ~ 土力学
statistical ~ 统计力学
theoretical ~ 理性力学

mechanise vt. =machanize
~d longhole drilling 机械化深孔凿岩

mechanism ['mekənizm] n. 机构；机理；-mechanismic, ad.
~ of cavitation 空蚀机理
~ of creep 徐变机理
~ of fracture 破坏[断裂]机理
~ of poisoning （树脂）中毒机理
control ~ 操纵机构
crystallization ~ 结晶机理
heat-removal ~ 排热设备
safety ~ 安全装置，保安机械装置

mechanization [mekənai'zeiʃən] n. 机械化

medallion molding 圆雕饰线脚

medallion molding

median ['mi:diən] n. 中值，中位数；中线，中点；a. 中间[央]的；中线的；medianly, ad.
~ barrier 路中护栏
~ deviation 中位差
~ diameter 中值[间]粒径
~ discharge 中（常）流量
~ island 中央分车岛
~ line 中线
~ point 中点；重心
~ size 中等大小；中值粒径
~ stream flow 河流中值流量
~ tolerance limit 中值容许限度
~ year 中常年，平均年

medical ['medikəl] a. 医学的；医术的；内科的；药的；医药的-medicall, ad.
~ certificate 健康证明，诊断书
~ examination 体格检查
~ history 病历

medium ['mi:diəm] a.; n. pl. -diums, -dia; a. 中间的, 中等的；n. 媒介；方法；装置
~-breaking （乳代沥青的）中裂
~ capacity plant 中容量电站
~ carbon steel 中碳钢
~ clay loam 中黏壤土, 中亚黏土
~ coarse sand 中粗砂
~-curing （沥青材料的）中凝，中级处理
~ diameter 中值粒径
~ dimension 中等尺寸
~ duty engine 中型发动机
~ exchange （货币、支票等）交换媒介
~ frequency 中频
~-grained 中粒度的
~ gravel 中砾石
~ heavy traffic 中等密度交通
~ maintenance 中修
~ of advertisement 广告手段
~ of payment 支付手段
~ pressure 中等压力
~ pressure tunnel 中压隧洞
~ range forecast 中期预报
~ reinforced 中等配筋的
~ sand 中砂
~ scale integration 中等规集成电路
~ section 中（型）材
~ setting 中凝的
~ silt 中粉砂
~-slaking lime 中消[化]石灰
~ steel 中碳钢, 中硬钢
~ strip 中间分隔带
~ tar 中质焦油, 中质焦油沥青
~ term forecast 中期预测

~ water 中水位
absorbing ~ 吸收媒[介]质
actuating ~ 工作介质
anisotropic ~ 各向异性介质
binding ~ 粘合剂
compressible ~ 可压缩介质
continuous ~ 连续介质
elastic ~ 弹性介质
fluid ~ 流质,流体
homogeneous ~ 均匀介质
hydraulic ~ (液压系统的)工作液体
infinite ~ 无限介质
ionization ~ 电离介质
isotropic ~ 各向同性介质
optical ~ 光学介质
meerschalminite *n.* 铝海泡石
meerschaum ['miəʃəm] *n.* 海泡石
meet [miːt] *v.*, met, meeting;碰见,迎接;符合,满足;对付,对抗
~ a bill 准备支付到期票据
~ a condition 满足条件
~ a criterion 符合标准
~ an objective 达到目的
~ demand 满足要求
~ the specification 合乎规范[规格]
~ with stresses 承受应力
meeting ['miːtiŋ] *n.* 会议;会合
~ hall 会议厅
~ house 会堂
~ point 交汇点
~ post (人字门上的)接合柱
~ rail (窗的)碰头横挡
~ room 会议室
megalith ['megəˌliθ] *n.* 巨石 megalithic, *a.*
megalithic [megə'liθik] *a.* 巨石的
~ age 巨石器时代

megalithic burial chamber

~ burial chamber 巨石墓室
~ masonry 巨石圬工
~ monument 巨石纪念碑
megarelief *n.* 大地形,大起伏
megaseism ['megəsaizəm] *n.* 大地震,剧震;megaseismic, *a.*
Mekong ['meiˈkɔŋ] *n.* 湄公河
melamine ['meləmiːn] *n.* = melamin 三聚氰胺,蜜胺
~ -formaldehyde 聚氰胺甲醛树脂
melaminoplast *n.* 三聚氰胺塑料,蜜胺塑料
melaphyre 胡斑岩
melilite ['melilait] *n.* 黄长石,方柱石
mellow ['meləu] *n.* 甜而多汁的;软熟的;熟练的,光练的;肥沃的;mellowly, *ad.* mellowness, *n.*
~ soil 松软土
melt [melt] *n.*; *v.*, melted, melted or melten, melting;熔化,软化 meltability, *n.* meltable, *a.*
~ cement 熔化水泥
~index 熔化指标
~ kettle(= ~ pot) 坩埚,熔炉
~ viscosity 熔解黏性
~ water (冰雪的)融(潺)水,熔融液
~ed asphalt 摊铺地沥青(混合料)
~ ed iron 铁水
~ed snow runoff 融雪径流
~ing chamber 熔油罐
~ing heat 熔解[熔化]热
~ ing-rock-and-earth method (建筑隧道用)融化土石法
~ing temperature 熔解[融化]温度
member ['membə] *n.* 构件;杆件;成员;(数学上的)元
~ connection 杆件连接
~ in bending 受弯构件
~ in compression 受压杆件,压杆
~ in flexure 受弯构件
~ in shear 受剪构件
~ in tension 受拉杆件,拉杆

~ in torsion 受扭构件
~ moment 杆件弯矩
~ size 构件尺寸
~ splice node 杆件拼接节点
~ stress 杆应力
~ system 构件[杆件]系统
~ transposition 杆件互[置]换
bow ~ 拱架,弓形构件
compression ~ 受[抗]压件
cross ~ 横梁
diagonal ~ 斜构件
diagonal web ~ 斜腹杆
driven ~ 从动构件
driving ~ 主动构件
external ~ 套件,包容件
frame ~ 构件
guide ~ 导引构件
interanl ~ 被套件,被包容件
latticed ~ 缀合杆
linking ~ 连杆
longitudinal ~ 从向杆[构]件
mating ~ 配合件
prismatic ~ 等截面杆
supporting ~ 承重构件
suspension ~ 吊杆,吊索
tension ~ 受[抗]拉件
tie ~ 系杆
torsion ~ 抗扭构件
transverse ~ 横梁,横杆,横向撑,横桁
truss ~ 桁架杆件
unstressed ~ （构架的)无应力杆件
web ~ 腹杆
wing ~ 翼构件
membrane ['membrein] *n.* 薄膜；振动片；表层；membraneless,*a.*
~ action 薄膜作用
~ analogy 薄膜模拟
~ analysis 薄膜分析
~ barrier 隔(水,气)膜
~ curing 薄膜养护
~ equation 薄膜方程
~ equilibrium 膜渗平衡

~ filter 薄膜滤器
~ force 薄膜力
~ grouting process 薄膜灌浆法
~ model 薄膜模型
~ sealed 薄膜止水[密封]的
~ stabilization 薄膜稳定法
~ stress 薄膜应力
~ theorem 薄膜理论
adiabatic ~ 绝热膜
bituminous ~ 沥青防水膜
water-proof ~ 防水膜
woven ~ 织物滤层

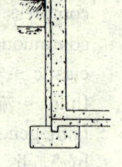
water-proof membrane

memorial [mi'mɔːriəl] *n.* 纪念物；*a.* 纪念的；memorially,*ad.*
~ arch 拱形纪念门
~ church 纪念教堂
~ column 纪念柱
~ hall 纪念堂
~ stone 纪念石,奠基石
memory ['meməri] *n.* 记忆,存储；memorless,*a.*
~ block 存储区
~ capacity 存储容量
~ core 存储磁心
~ cycle time 存储周期时间
~ drum 存储磁鼓
~ effect 记忆效应
~ location 存储单元
~ register 存储寄存器
addressable ~ 可选存储器
buffer ~ 缓冲存储器
byte ~ 二进位组存储器,字节存储器
erasable ~ 可清存储器
external ~ 外部存储器
file ~ 文件存储
magnetic ~ 磁存储器
magnetic-core ~ 磁心存储器
magnetic-disc ~ 磁盘存储器
read only ~ 只读存储器

mend [mend] *n.*; *v.* 修理[补]；改良，恢复
　～ing plate 加固板
meniscus [mi'niskəs] *n.* 新月形物，弯月形饰
mercury ['mə:kjuri] *n.* 水银；汞(Hg)
　～-arc lamp 水银弧光灯
　～ barometer 水银气压表
　～ columm 水银柱
　～ thermometer 水银温度计
　～ vapor lamp 水银灯
merlon ['mə:lən] *n.* (城)堞，城齿
mermaid ['mə:meid] *n.* 美人鱼(城市雕塑)
mesh [meʃ] *n.* 网格，筛眼；*v.* 衔接，咬合
　～ analysis 筛分析
　～ boundary 网格边界
　～ coordinate 网格坐标
　～ division (有限元法)划分网格
　～ lines 网格线
　～ screen 网筛
　～ sieve 网筛
　～ size 网眼大小，筛孔尺寸
　～ sinker (路面)钢筋网沉放器
Mesozoic era 中生代
messenger ['mesindʒə] *n.* 悬缆，吊线缆，悬索
　～ cable 吊线，悬缆线，承力吊索
　～ wire 承力吊索，吊索
messhall [messhəuse] *n.* 食堂，餐厅
metacenter ['metə,sentə] *n.* 定倾中心；metacentric, *a.* metacentricity, *n.*
　～ radius 定倾半径
　～ height 定倾中心高度
metal ['metl] *n.*, *v.*, -al(l)ed, -al(l)ing; *n.* 金属；铸铁溶液；碎石料；*v.* 用金属包；用碎石铺
　～ alloy 金属合金
　～ anchor block 金属地锚
　～ arc welding 金属(电)弧焊接
　～ awning-type window 金属遮篷式窗
　～ bladed drag 刮板式刮路机
　～ brush 钢丝刷

　～ casting 金属铸件
　～ ceiling panel 金属天花板框架
　～ coating 金属涂层
　～ conduit 金属导管
　～ corner bead 金属墙角护条
　～ covered door 金属外包门
　～ curb 金属边饰
　～ edged mirror 金属镶边镜子
　～ fabrication 金属结构
　～ framed window 金属框架窗，钢窗
　～ gauze 钢丝网
　～ grill 金属格栅
　～ lath 钢丝网板条
　～ lath and plaster 钢丝网抹灰
　～ lattice work 金属格构工
　～ lined 金属衬里
　～ partition 金属隔墙
　～ -plate guard rail 铁板护栏
　～ processing and finishing 金属加工和修饰
　～ road 碎石路
　～ rolling shutter(= ～ roller shutter) 金属辊式百叶窗；金属辊式翻板闸门
　～ roof 金属屋顶
　～ scaffolding 金属脚手架
　～ strip 金属带
　～ surface 碎石路面
　～ working machinery 金属加工机械
active ～ 活性金属
coated ～ 镀层金属
road ～ 筑路碎石
semi-finished ～ 半成品轧材
metaling ['metliŋ] *n.* 碎石料
broken stone ～ 碎石道碴
metallic [mi'tælik] *a.* 金属的；碎石的
　～ corrosion 金属腐蚀
　～ link belt 金属链带
　～ packing 金属填料
　～ paint 金属漆

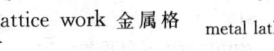
metal lath

~ tape 钢卷尺
metamorphic rock 变质岩
meteoric [ˌmiːtiˈɔrik] *a.* 陨石的,流星的;瞬息的,疾速的;大气;meteorically, *ad.*
~ iron 陨铁
~ shower 流星雨,陨石雨
~ water 雨水,大气水
meteorite [ˈmiːtjərait] *n.* 陨星,流星;meteoritic, meteoritical, meteorital, *a.*
meteorological [ˌmiːtjərəˈlɔdʒikəl] *a.* 气象的,气象学的;meteorologically, *ad.*
~ observatory 气象观测台,气象台
~ report 天气报告,天气预报
~ station 气象站
~ summary 气象报告
~ tide 气象潮
meteorology [ˌmiːtjəˈrɔlədʒi] *n.* 气象学,气象状态;meteorologist, *n.*
meter [ˈmiːtə] *n.* 米;仪器,仪表; *v.* 计量,记录,统计
~ -candle 米烛光(照度单位)
~ constant 仪表[校正]常数
~ rate 水表收费率
~ reading 表读数
~ sensitivity 仪表灵敏度
~ series 公制系列
~ stop 记录中止
air ~ 气流[风速,量气]计
angle ~ 测角器
batch ~ 分批计量器
consistency ~ (混凝土)稠度计
crack ~ 裂缝探测仪
flow ~ 流量表
orifice ~ 孔口流量计,测流[量水]孔
power ~ 功率表
pressure differential ~ 差压式流量计
strain ~ 应变计[仪]
temperature ~ 温度计
tension ~ 张力计
torque ~ 转矩计
torsion ~ 扭力计
universal ~ 万用表,通用电表
vibration ~ 振动计
viscosity ~ 黏度计
water ~ 水量计,水(量)表
methane [ˈmeθein] *n.* 沼气,甲烷
method [ˈmeθəd] *n.* 方法,方式;规律,程序;整理,顺序;methodless, *a.*
~ by trial 试算[试探]法
~ of adjustment (测量的)平差法
~ of anchoring 锚固方法
~ of approach 渐近法
~ of aproximation 近似法
~ of characteristics 特征线法
~ of column analogy 柱比法
~ of elastic center (结构力学)弹性中心法
~ of elimination 消去法
~ of extrapolation 外推[插]法
~ of fictitious loads 虚荷载法
~ of finite difference 有限差分法
~ of finite element 有限单元法
~ of finite increment 有限增量法
~ of fixed point 定点法
~ of images 镜像法,反映法
~ of induction 归纳法
~ of interpolation 内插法
~ of iteration 迭代法
~ of joints 桁架节点分析法
~ of least squares 最小二乘法
~ of least work 最小功法
~ of moment distribution 弯矩分配法
~ of operation 操作方法
~ of panel point 节点法
~ of rotation 转角法(计算刚架内力的方法)
~ of separate joint displacement 节点位移法
~ of slice (土坡圆弧滑动)条分法

~ of slope deformation 倾度变位法
~ of substitution 替代法
~ of superposition 迭加法
~ of tangent offsets 切线支距法
~ of transformed section（钢筋混凝土的）折算截面法
~ of trial and error 试算法
~ of undetermined coefficient 待定系数法
~ of weighted mean 加权平均法
~ of weighted residual 加权残数法
assembly line ~ 流水作业法
back and forth ~ 选择[尝试]法
construction ~ 施工方法
correlation ~ 相关法
cut and try ~ 试算[逐次逼近]法
discounted payback ~ 按现值计算投资回收法
frequency response ~ 频率特性法
graphical ~ 图解[示]法
image ~ 镜像法
interchange ~（桁架分析）替换法
matrix ~ 矩阵法
maximum likelihood ~ 最大似（或）然法
minimising energy ~ 最小能量法
Monte-Carlo ~ 蒙特卡洛法
mortar void ~（混凝土）灰浆孔隙法
sequential sampling ~ 按序采样法
topological ~ 拓扑方法
variational ~ 变分法
void ~ of proportioning（混凝土）孔隙配合比
zero load ~ 零荷载法
methodology [ˌmeθəˈdɔlədʒi] *n.* 方法学,方法论,分类法；methodological, *a.* methodologically, *ad.*
methylalcohol 甲醇,木精
methyltrinitrobezene [ˈmeθiltrainaitrəuˈbenziin] *n.* 三硝基甲苯(TNT)
metope 三槽板间平面（陶立克柱式雕带上的）

metric [ˈmetrik] *a.* 公制的;度量；*n.* 度量标准,量度
~ conversion 米制换算
~ demension 公制尺寸
metope
~ horse power 公制马力
~ measure 米[公]制度量
~ scale 公制比例尺,米尺
~ screw 公制螺纹
~ screw pitch gage 国际螺距规
~ series 公制系列
~ size 公制尺寸
~ system 米制,公制,十进制
~ tensor 度量张量
~ thread 公制螺纹
~ ton 公吨
~ unit 公制单位
metro [ˈmiːtrəu] *n. pl.* -ros；地下铁道；市政
metropolitan [ˌmetrəˈpɔlitən] *a.* 都市的,大城市的；母国的；大主教区；*n.* 大都市的人；母国的居民；大主教
~ area 首都地区,大都市区域
~ district 大城市管区[行政区]
~ railway 城市[市内]铁道
mica [ˈmaikə] *n.* 云母, micalike, *a.*
~ condenser 云母电容器
~ flap 云母页片
~ gneiss 云母片麻岩
~ schist 云母片岩
~ slate 云母板岩
artificial ~ 人造云母
micaceous [maiˈkeiʃəs] *a.* 云母的,含云母的;薄层的;闪亮的
~ iron ore 云母铁矿
~ sand 云母砂
~ sandstone 云母砂岩
~ shale 云母页岩
micarta *a.*（耐酸耐碱衬里用）层状酚塑料,胶纸板胶木
micro [ˈmaikrəu] *a.* ；*n.* 微小的,微观

的,显微的;微米;百万分之一;测微计,千分表
- ~ analysis 微量分析
- ~ bar (气压单位)微巴
- ~ crack 微裂缝
- ~ crystal 微晶体
- ~ film 显微胶卷
- ~ flaw 发裂纹

micro- [词头]微(量);百万分之一;显微,扩[放]大

microammeter [ˌmaikrəu'æmitə] *n.* 微安(培)计

microampere [ˌmaikrəu'æmpiə] *n.* 微安(培)

microbarogram *n.* 微气压记录图

microbarograph [maikrə'brəgræf] *n.* 微气压计[器]

microbaromter *n.* 微气压表

microbeam ['maikrəubi:m] *n.* 微光束

microcal(l)ipers ['maikrəukælipəz] *n.* 千分表,测微计
 vernier ~ 游标千分尺

microcamera *n.* 显微摄影机

microcard ['maikrəuka:d] *n.* 缩影胶片卡

microcharacter *n.* 显微划痕硬度计

microchemistry ['maikrəu'kemistri] *n.* 微量化学

microcircuit ['maikrəu'sə:kit] *n.* 微型电路

microclimate ['maikrəu'klaimit] *n.* 小气候

microclimatology *n.* 小气候学

microcline *n.* 微斜长石

microcomputer ['maikrəukəm'pju:tə] *n.* 微型电子计算机

microcopy ['maikrəkɔpi] *n.* 显微照片,缩影印刷品; *v.* 显微照相;缩影印刷,缩微复制

microdiorite *n.* 微闪长岩

microeffect *n.* 微观效应

micro-economic model 微观经济模型

microfissure *n.* 微裂隙

microgranite *n.* 微花岗岩

micrograph ['maikrəugra:f] *n.* 显微照片

microholography [maikrəuhə'lɔgra:fi] *n.* 微全息照相术

micrometer [mai'krɔmitə] *n.* 测微计;测距器;千分尺
- ~ caliper 千分卡尺;测微镜
- ~ dial 测微表;千分表
- ~ knob 测微旋钮
- ~ screw gauge 螺旋测微器

microphone ['maikrəfəun] *n.* 扩音器,麦克风
- ~ effect 颤噪效应
- ~ noise 颤噪噪声

microphysics [ˌmaikrəu'fiziks] *n.* 微观物理学,粒子物理

microprint ['maikrəprint] *n.* 缩微印刷品

microrheology *n.* 微流变学

microscope ['maikrəskəup] *n.* 显微镜

microscopic [ˌmaikrə'skɔupik] *a.* 微观的

microsecond [ˌmaikrəu'sekənd] *n.* 微秒(百万分之一秒)

microseism ['maikrəusaizəm] *n.* 微震

microseismic ['maikrəsaizmik] *a.* 微震的
- ~ forecasting 微震预报

microstructure ['maikrəuˌstrʌktʃə] *n.* 微观[显微]结构

microviscometer *n.* 微黏度计

microwave ['maikrəweiv] *n.* 微波

mid [mid] *a.* 中部的,中间的
- ~ -block 道口区间
- ~ -blue 淡蓝的
- ~ -board 中隔墙,间壁,中间纸板
- ~ -channel 中流,水路的中段
- ~ -contour 中值等高[深]线
- ~ -line 中线
- ~ -ordinate method 中距法,中央纵距法
- ~ -range forecast 中期预报

~ section 中间截面[剖视]
~ -span 中跨
~ -value 中值
~ -wall column 承墙柱,墙壁柱
~ wall shaft 承墙柱

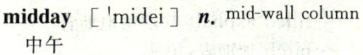
mid-wall column

midday ['midei] *n.* 中午
~ break 午间休息

middle ['midl] *n.* 中间;中间物,媒介物;中项;*a.* 中间的
~ course 河流中游
~ girder 中主梁,中间梁
~ lane 中间车道
~ man 中间人,岩层中的夹层
~ point 中点
~ post 桁架中柱
~ purline 中间桁条
~ -sized 中型的
~ term 中项,内项
~ third rule 三分点准则
~ wall 隔墙

midfeather *n.* 中间隔墙;纵隔板支承隔墙

midget ['midʒit] *n.* 小型物;矮人;*a.* 微型的,小型的;袖珍的
~ concrete mixer 微型混凝土拌合机
~ condenser 小型电容器
~ excavator 微型凿岩机
~ tractor 微型拖拉机

midline ['midlain] *n.* 中线

midmorning ['mid͵mɔ:niŋ] *n.* 上午的中段时间

midmost ['midməust] *a.*;*n.* 正中,最当中

midnight ['midnait] *n.* 午夜,半夜,子夜;漆黑;黑暗

midnoon ['mid'nu:n] *n.* 正午

midocean ['mid'əuʃən] *n.* 海中央,远洋,外洋

mid-plane *n.* 中平面

midpoint *n.*;*a.* 中点(的)

~ crossing (道路)区间交叉口

mid-position *n.* 中间位置

midrange ['midreindʒ] *n.* 中列数
~ forecasting 中期预报

midsea ['midsi:] *n.* 外海

midspan moment 中间跨弯矩

midstream ['midstri:m] *n.* 中流;水流的中心线;中点水流

midsummer ['mid͵sʌmə] *n.* 仲夏

midterm ['mid'tə:m] *a.* 中间的,期中的;*n.* 中项

mid-value *n.* 中值

midwinter ['mid'wintə] *n.* 仲冬

midwood *n.* 丛林,密林

migmatite *n.* 混合岩

migratory dune 游动沙丘

mihrab (清真寺院面向麦加的那道墙内的)壁龛

mihrab

Milan Cathedral Church 米兰主教堂

mild [maild] *a.* 柔软的,软弱的;温和的;适度的;淡味的
~ clay 软黏土,亚黏土
~ iron 软铁
~ mortar (由石灰和砂组成的)无水砂灰浆
~ steel 软钢,低碳钢

milestone ['mailstəun] *n.* 里程碑,里程石

military ['militəri] *a.* 军人的,军事的,战争的;好战的;*n.* 军人,军队;-militariness,*n.*
~ architecture 军事建筑
~ engineering 军事工程
~ geology 军事地质学
~ map 军用地图

Military Museum 军事博物馆
~ railway 军用铁道
~ road 军用道路

milk [milk] *n.* 牛奶;乳状物 *v.* 抽取;

挤奶;-miklike,a.
~ glass 乳白玻璃
~ of lime 石灰乳[浆]
~ scale 乳白度
~ -stone 乳石,白燧石
mill [mil] n. 磨坊;工场;粉碎机 v. 粉碎,磨细
~ bent 磨坊的主柱屋架
~ blank 夹层纸板
~ board （建筑用）麻丝板,马粪纸
~ building 厂房
~ floor 车间地面
~ stone 磨石
~ surface 磨碎面
ball ~ 球磨机
cement ~ 水泥厂
clay ~ 黏土拌合器,碾泥机
conglomerate ~ 砾石破碎机
crush ~ 粉碎机
grinding ~ 球磨机,磨碎机
milli- （词头）毫,千分之一
mimic ['mimik] a.; n.; v. -icked, -icking; a. 模仿的; n. 仿造物; v. 模仿;mimical,a. mimically,ad.
~ diagram 模拟现场,活动的监视屏
~ colouring 保护色
~ panel 模拟显示屏
minarets 伊斯兰教尖塔
mine [main] n.; v., mined, mining; n. 矿山;矿坑;矿;资源;地雷; v. 开矿;开坑道;设雷;-minable,a.
~ car 矿车
~ detector 探雷器
~ gallery 坑道
~ hoist 矿井提升机
~ mouth power plant 坑口（大力）发电厂
~ quarry 矿石
~ rock 矿石
~ tailing （尾）矿渣
mineral ['minərəl] n. 矿物,矿石;无机

minarets

物;矿泉水; a. 矿物的;无机的
~ acid 无机酸
~ aggregate 矿料,矿质集料,石料
~ black 石墨
~ compound 无机化合物
~ deposit 矿床
~ dust 矿粉
~ filled asphalt 掺填料地沥青
~ pitch 地沥青,柏油
~ tar 软沥青,矿沥油
~ water 矿泉水
~ white 石膏
minimum ['miniməm] a.; n. pl. -mums, -ma; n. 最小,最低,最少,最小量,最小限度; a. 最小（限度）的,最低的,最少的
~ allowable radius 最小容许半径
~ area 极小区域
~ average value 极小平均值
~ bearing capacity 最小承载力
~ curb offset 最小缘石支距
~ cost estimating 估计最低成本
~ energy line 最小能量线
~ energy principle 最小能原理
~ freight 最低水脚
~ potential energy 最小势能
~ problem 极小（值）问题
~ rate 最低率
~ retention 最低自留额
~ selling price 最低售价
~ size 最小尺寸
~ spacing 最小间隔
~ superelevation 最小超高度
~ taxable ceiling 最低应税限度
~ taxable income 最低应税所得额
~ traiff 最低税率
~ variance estimate 极小方差估计
~ -void mix 最小空隙混合料
mining ['mainiŋ] n.; a. 开矿,采矿;设雷
~ core tube 岩心管
~ effect 地下爆炸效力
~ excavation 矿山掘进

~ geology 矿山地质学
~ machine 采矿[掘]机
~ subsidence 矿坑下陷
~ survey 矿山测量
~ system 坑道工事
minor ['mainə] *a.* 较小的；较少的；次要的；*n.* 小分类；子式
~ axis 短轴
~ bed 枯水河床，小河床
~ control 低等(大地)控制
~ control point (测量的)低等控制点
~ crane 小型起重机
~ cycle 小循环
~ dam 副坝
~ fault 小[次]断层
~ karst features 幼年溶岩地形
~ coverhaul 小修
~ principal stress 最小主应力
~ shock 副震
~ triangulation 小三角测量
complement ~ 余子式
principal ~ 主子式
minus ['mainəs] *a.* 负的；减的；*prep.* 去掉，少掉，减去；*n.* 负数；负号
~ charge 负[阴]电荷
~ grade 下[降，负]坡
~ material 次品
~ sight 前视
~ sign 负号
minute ['minit] *n.* 分钟(六十秒)；记录；片刻；*a.* 细微的；详细的；微小的
~ adjustment 精密调节
~ book 会议记录簿
~ bubbles 小气泡
~ crack 发状裂缝，细裂缝
~ projections 摩擦面的粗糙度
Miocene ['maiəsi:n] *a*；*n.* 中新世
~ clay 中新世黏土
misadjustment [ˌmisə'dʒʌstmənt] *n.* 误调；失调
misalignment [misə'lainmənt] *n.* 未校准；不同轴；不同心；失调

miscibility [ˌmisi'biliti] *n.* 可混合[溶和]性
complete ~ 完全混溶性
partial ~ 部分混溶性
water ~ 水混溶性
mission tile 拱形瓦，半圆形截面瓦

mission tile

Mississippi [ˌmisi'sipi] *n.* 密西西比河
miter ['maitə] = mitre *n.* 斜接缝，斜接，45°接合
~ arch 斜接拱
~ bearing 斜接支承，人字形支承
~ bend 斜面弯管
~ box (木工用)45°角尺，轴锯箱
~ cap 楼梯扶手柱头饰
~ cutting 斜切割
~ dovetail 斜楔榫
~ gates 人字形闸门
~ aguge 定角规
~ gear 等径伞齿轮
~ joint 斜面接合；斜角联接

miter box

~ plane 斜接刨，斜接面
~ post 斜接柱
~ rod 斜角棒
~ sill 人字槛
~ square 斜角尺
~ valve 锥形阀
~ wall 人字墙
mix [miks] *n.*；*v.*，mixed or mixt，mixing；*v.* 拌合，混合；搅拌；*n.* 混合料物；mixable, mixible, *a.* mixableness, mixibleness, *n.* mixability, mixibility, *n.*
~ consistency 拌合稠度
~ design 配合比设计
~ enroute 在运输过程中拌合，路拌
~ ingredient 拌合成分
~ -in-place 就地拌合
~ muller 混砂机，混合碾砂机
~ seal 拌合式封层，混合料封层
~ selector 选[配]料器

~ed avalanche 混合崩坍
~ed base crude oil 混合基原油
~ed batch capacity 成批拌合能力
~ed bituminous macadam pavement 拌合式沥青碎石路面
~ed bituminous road 拌合式沥青路,沥青混合料路
~ed cement 混合水泥
~ed cold 冷拌合
~ed concrete 拌制好的混凝土
~ed construction 混合构造[建筑]
~ed current 混合潮流
~ed decimal 带小数
~ed flow pump 混流式水泵
~ed flow water turbine 混流式水轮机
~ed fraction 带分数
~ed garden-wall bond 三顺一丁砌墙法
~ed-grained 混粒的,多种粒径混合的
~ed-grained sand 多种粒径的混合砂
~ed gravel 混合砾石
~ed hot 热拌合
~ed layer depth 混合层厚度
~ed-powder 混合粉末
coarse ~ 粗混合物
mixer ['miksə] *n.* 混合器,搅拌[拌合]机;混合者[器]
~ car 搅拌车
cement ~ 水泥浆搅拌机
column ~ 混合柱[塔]
concrete ~ 混凝土搅拌机
handy ~ 手动式拌合机,轻便拌合机
rotary ~ 滚筒搅拌机
transit ~ 运送搅拌机
mixing ['miksiŋ] *n.* 混合,拌合
~ at site 工地[现场]拌合
~ channel 混合槽
~ coat 复合面层
~ cycle 拌合周期,一次拌合时间
~ machinery 拌合机械
~ mill 混砂机;混合碾磨机
~ platform 拌合台
~ proportion 配合比
~ ratio 混合比
mixture ['mikstʃə] *n.* 混合料;配料
~ calculation 配料计算
~ making 配料(计算)
~ placing 拌合料浇筑
~ ratio 配合比
~ specifications 配料规范
anti-freezing ~ 防冻剂
moat [məut] *n.* 水沟,壕沟,城壕,护城河
mobile ['məubail] *n.* 可动装置;汽车(美);发动机 *a.* 可动的,机动的
~ boring island 移动式钻孔平台
~ breaker 可动式粉碎机
~ moisture 游离水分
~ scaffold 活动脚手架
~ structure 活动建筑物
mobility [məu'biliti] *n.* 流[机]动性,变[活]动性;迁移率
mock [mɔk] *a.* 假的,模拟的 *v.* 嘲笑;造样板 *n.* 嘲笔;模仿;仿造品
~ arcade 模拟拱廊
~ architecture 模拟建筑
~ rafter 假椽
~ -up 大模型,足尺模型
~ window 假窗
mode [məud] *n.* 众数;型式;方法
~ of vibration 振动模式
operating ~ 工作状态,工况
model ['mɔdl] *n.* 模型;标本;模范;样式;模式 *v.* 做模型;模仿
~ number 型号[序数]
~ of structure 结构模型
~ pile 模型桩
~ pile group 模型桩群
~ scale 模型比尺
~ set 模型
architecture design ~ 建筑设计模型
optimizaion ~ 优化模型

plaster ~ 石膏模型
reconstructed ~ 复制模型
scale ~ 缩尺模型
structural ~ 结构模型
typical ~ 定型
moderate ['mɔdərit] *a.* 适度的;中等的;温和的;有限的
~ crack 中度开裂
~ erosion 中度侵蚀
~ temperature 中等温度
modern ['mɔdən] *a.* 现代的,时髦的
~ architecture 现代建筑
~ conveniences (住房内)现代设备
~ style 现代风格
modernization [ˌmɔdənai'zeiʃən] *n.* 现代化
modification [ˌmɔdifi'keiʃən] *n.* 变更;修改,修饰;变态
~ kit 附加器,成套改造器材
~ work 改造工程
modified ['mɔdifaid] *a.* 变更的;改进的;修改的
~ asphalt 改良地沥青
~ cement 改良水泥
~ epoxy 改性环氧树脂
~ portland cement 改良的硅酸盐水泥
~ scissor truss 斜撑桁架,改良剪式桁架
modify ['mɔdifai] *v.* 变更,修改,修饰;限制;减轻,缓和
modillion *n.* 托饰,飞檐托
modular ['mɔdʒjulə] *a.* 模的,模数的,比率的;按标准尺寸设计制造的

modillion

~ building unit 标准建筑单元[构件]
~ connector 组合式接插件
~ construction 部件[单元]结构
~ coordination 标准量测系统
~ design 积木式设计
~ design method 定型设计法
~ housing 积木式房屋
~ method 标准化方法
~ plane 模数平面
~ proportion 柱半径倍数比
~ ratio 模量比
~ space 模数空间
~ system 模数体系[制]
modulus ['mɔdjuləs] *n.* (*pl.* -li) 模量,系数;模数;基本单位
~ of deformation 变形模量
~ of diltation 膨胀系数
~ of elongation 伸长模量
~ of plasticity 塑性模量
~ of rupture 断裂模数,折断系数
elastic ~ (= ~ of elasticity) 弹性模量
fineness ~ 细度模数
section ~ 截面模量
Young's ~ 杨氏模量
moil [mɔil] *n.* 十字镐,鹤嘴锄
moist [mɔist] *a.* 潮湿的;多雨的
~ -adiabatic process 湿绝热过程
~ closet 保湿室
~ -cured concrete 湿养护混凝土
~ curing 湿养护,雾室养护
~ mixing 湿拌和[混合]
~ steam 湿蒸汽,饱和水蒸气
moisten ['mɔisn] *v.* 浸[沾]湿,湿润
moisture ['mɔistʃə] *n.* 湿度;水分;潮汽
~ apparatus 测湿器
~ barrier 防潮层
~ -bearing 含[持]水的
~ can 含水量盒,土盒
~ capacity (= ~ content) 含[持]水量,潮[湿]度

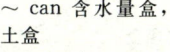
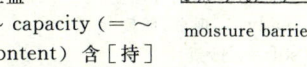
moisture barrier

~ content meter 测湿计
~ damage 潮汽损坏
~ density 湿密度
~ -density control 湿密度控制
~ -free 干的,不含水的

~ -free material 干材
~ gage 湿度计
~ laden 饱水的
~ indicator 湿度指示器
~ migration 潮汽[水分]迁移
~ penetration 水分渗入
~ -proof adhesive 防水胶粘剂
~ regain 回潮,吸湿
~ repellent 防[憎]水
~ seal (= ~ stop) 防潮密封
capillary ~ 毛细水分
hygroscopic ~ 吸湿水,吸湿含水量
molasse [məˈlɑ:s] *n.* 磨砾层
mold [məuld] *n.* 铸模;模型;性质,气质 *v.* 铸造;形成
~ lofter 放样工
~ rains 梅雨
molding [ˈməuldiŋ] *n.* 装饰嵌线,饰线,线脚
returned ~ 转延装饰嵌线
molybdenum-copper 钼铜合金
molybdenum steel 钼钢
monastery [ˈmɔnəstəri] *n.* 修道院,寺院,庙宇
monastic [məˈnæstik] *a.* 寺院的,僧侣的,修女的
~ architecture 寺院建筑
~ building 寺院[庙宇]房屋
money [ˈmʌni] *n.* 金钱,货币;金额,款项
~ down [out of hand] 现金,现款
~ on account 账上货币
~ rates 利息
pay ~ down 付现金
ready ~ 现金,现款
soft ~ 钞票
monitor [ˈmɔnitə] *n.* 通风顶,采光顶;监听器,监视器;喷水枪 *v.* 检验,检查,监测,监控;校音
~ network 监测[听]网
~ roof 采光屋顶,通风顶
monkey [ˈmʌŋki] *n.* 猴子;打桩锤;煤

returned molding

矿中之小通道
~ blood 喷在混凝土表面上的液体混合物
~ bridge 天桥
~ spanner 活[动]扳手,万能螺旋扳手
~ tail [栏杆扶手]卷尾形扶手端
~ winch 手摇小绞车
~ wrench 活动扳手
monocline [ˈmɔnəklain] *n.* 单斜层,单斜折皱[结构]
monolith [ˈmɔnəliθ] *n.* 独石柱;整料;块体混凝土,整块石料
monolithic [ˌmɔnəˈliθik] *a.* 独块的,整体的;统一的 *n.* 单块,单片
~ arch 整体式拱,独石拱
~ beam 整体式梁,独石梁
~ block 整体坝块,大型砌块
~ brick pavement 整体式砖路面
~ column 整体式柱,独石柱
~ masonry wall 整体式圬工墙
~ concrete structure 整体式混凝土结构
~ construction 整体构造[建筑,施工]
~ finish 整体修整
~ fireproof floor 整体式防火地板
~ joint (填充后形成的)整体式缝
~ pillar 整体式柱
~ terrazzo 整块水磨石
monopolist [məˈnɔpəlist] *n.* 独占[垄断],专利[者]
monopolization *n.* 获得专利权,独占,垄断
monopteron 圆形外柱廊式建筑
monospar [ˈmɔnəspɑ:] *n.* 单梁
monostyle *n.* 单柱式,独立柱式
monotint [ˈmɔnətint] *n.* 单色画 *a.* 单色的
monument [ˈmɔnjumənt] *n.* 纪念碑;界碑,标石

monopteron

~ mark 标石,标志桩
~ tie 界碑[标石]系桩
location ~ 界标
Moon [mu:n] *n.* 月亮
~ gate (中国建筑中的)月洞门
moonstone *n.* 月长石
moor [muə] *n.* 沼泽,湿土,沼泽土;停泊,下锚 *v.* 停泊,下锚
Moorish arch *n.* 摩尔式拱,马蹄形拱
~ architecture 摩尔式建筑
moorstone ['muəstəun] *n.* 花岗石,花岗石质孤石
morning-glory column 喇叭形顶的柱
more-centered arch 多心圆拱
mortar ['mɔ:tə] *n.* 砂浆,灰浆[泥] *v.* (用砂浆)涂抹,(用砂浆)接合
~ admixture 灰浆混合料
~ additive 灰浆附加剂
~ aerated with foam 泡沫[加气]砂心
~ agent 灰浆混合料
~ bed 灰浆层,化灰池
~ -board 灰浆板,镘板
~ -bound macadam 灰浆结碎石面
~ -calked joint 灰浆嵌缝
~ coating 砂浆涂层
~ course 灰浆层
~ cover 灰浆抹面
~ cylinder 灰浆圆柱块
~ densifier 灰浆稠化器
~ dropping 灰浆喷洒[射]
~ filling 灰浆勾缝
~ from trass 火山灰砂浆,浮石凝灰砂浆
~ gun 灰[砂]浆喷射枪
~ grouting method 压力灌浆法
~ injection 砂浆喷射
~ intrusion 灰浆侵入(岩层);压浆
~ joint 灰缝,砂浆接缝
~ layer 灰浆层
~ less wall 干砌墙
~ lining 灰浆衬砌,灰浆抹面

~ lump 灰浆团
~ -making property test 成浆性质试验
~ mixer 砂浆拌和机
~ mix ratio (= ~ mix proportion) 灰浆配比
~ pan 灰浆池,灰浆底座
~ paving 砂浆砌面
~ plasticizing agent 灰浆增塑剂
~ pump 灰浆泵
~ rubble masonry 浆砌毛石圬工
~ sand 灰浆用砂
~ screed 灰浆样板
~ specimen 灰浆试块
~ splashing 灰浆喷射
~ spreader 灰浆撒铺机
~ stain 灰浆斑点
~ strength 灰浆强度
~ tank 灰浆池
~ top 砂浆抹面
~ -void ratio 砂浆空隙比
~ waterproofer (用于干拌的)灰浆防水粉
~ water-reducing agent 灰浆减水剂
~ wetting agent 灰浆增塑剂
~ workability agent 灰浆减水剂
asphalt ~ 沥青砂浆
cement lime ~ 水泥石灰浆
clay ~ 黏土浆
gypsum ~ 石膏浆
mason's ~ 砌筑砂浆
refractory ~ 耐火泥浆
mortgage ['mɔ:gidʒ] *v.* ;*n.* 抵押,保证
mortise ['mɔ:tis] *n.* =mortice 榫眼 *v.* 开榫眼;用榫眼接合
~ and tenon 凹凸榫;镶榫
~ and tenon joint 凹凸榫接合
~ axe 凿榫斧
~ chisel 榫凿
~ gauge 划榫线具
~ hole 榫眼[孔]
~ joint 榫接

~ latch 插锁
~ lock 插锁
~ed astragal 榫眼圈线
chase ~ 槽榫
mosaic [mə'zeiik] *n.*; *a.*; *v.* 镶嵌［细工，砖］,锦砖（马赛克）

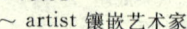
mosaic

~ artist 镶嵌艺术家
~ asphalt 预制镶嵌地沥青
~ binding 镶嵌装订
~ block 嵌镶块
~ clad 镶嵌式外壳,镶嵌装饰
~ clay tile 彩色拼装黏土瓦
~ cover 镶嵌式外包装
~ cupola (= ~ dome) 镶嵌式圆屋顶
~ decoration (= ~ decorative finish= ~ enrichment= ~ ornamental feature) 镶嵌装饰
~ external finish 马赛克外部装饰
~ exterior finish (= ~ external finish= ~ outdoor finish= ~ outer finish) 外部装饰
~ facade 镶嵌门面
~ figure 镶嵌图
~ fingers 镶嵌细木工地板
~ floor 镶嵌地板
~ glass 镶嵌玻璃,五彩玻璃
~ gold 彩色金,金箔,装饰用黄铜
~ masonry work 镶嵌石块砌筑工程
~ panel 镶嵌板,镶嵌框架
~ parquetry 镶木地板
~ pattern 镶嵌形式
~ pavement 嵌花式地面
~ paving sett 小方石拼花路面
~ piece 用于镶嵌的小块
~ screen 感光镶嵌幕
~ staircase 镶嵌式楼梯间
~ structure 晶体镶嵌结构
~ surface 马赛克面,拼花面
~ tessera 镶嵌用小块
~ tile 彩色［镶嵌］瓷砖
~ vault 镶嵌式圆屋顶
~ window 镶嵌玻璃窗
~ woodwork 镶嵌细木工
~ work 镶嵌工
~ woolwork 镶嵌式绒线刺绣
aerial ~ 空中照片嵌接地图
ceramic ~ 陶瓷锦砖（马赛克）
glass ~ 玻璃锦砖（玻璃马赛克）
Roman ~ 罗马式镶嵌玻璃,罗马式镶嵌细工
translucent ~ 半透明镶嵌
mosaicsit *n.* 镶嵌者细工师,嵌镶商
mosque [mɔsk] *n.* 清真寺
~ arch 马蹄形圆拱
~ architecture 清真寺建筑
most [məust] *a.*; *n.* 最大(的限度),最高(额,程度)
mota *n.* 黏土
mote [məut] *n.* 微尘,尘屑
motel [məu'tel] *n.* 汽车旅馆,公路两旁的旅社
mother ['mʌðə] *n.* 母亲;母体;源泉 *a.* 母亲的;本国的
motion ['məuʃən] *n.* 运动,动作;运转,运转机械
~ time analysis 工时分析
motor ['məutə] *n.* 发动机,摩托,马达,电动［拖动］机
~ drive 马达传动
~ wrench 管子钳
mould [məuld] *v.* 模制;翻砂,铸造 *n.* 模型,模板;形状;线脚
~ board 模板,样板
~ construction 模制结构
~ for precast work 预制构件模型
~ plate 模板
~ed brick 模制砖
~ed chamfer 模制倒角
~ed concrete 模制混凝土
~ed cornice 模制挑檐
~ed masonry 模制砌体
~ed piece 模制件
~ed plastic 模制塑料
~ed resin 模制树脂

bead ~ 半圆饰
beam ~ 梁拱样板
brace ~ 蒽拱样板
fillet ~ 平条线脚
hip ~ 脊饰
hoed ~ 披水饰
lip ~ 唇状饰
neck ~ 领饰
moulding ['məuldiŋ] **n.** 制模；翻砂；造型；花边，线脚，线条，嵌条
~ board 模板，样板
anthemion ~ 棕叶饰
arch ~ 拱饰
back ~ 底饰
band ~ 带饰
base ~ 底座饰
bead ~ 半圆饰
beak ~ 鸟嘴饰
bed ~ 深凹饰
billet ~ 错齿饰
bolection ~ 镜框饰
brace ~ 蒽花饰
cable ~ 卷缆纹饰
cant ~ 斜状线饰,多角线饰
chain ~ 链饰
channel ~ 槽饰
chevron ~ 波浪饰
crenellated ~ 齿饰
crown ~ 冠顶饰
dovetail ~ 鸠尾饰
egg ~ 卵圆饰
fluted ~ 凹槽饰
hip ~ 脊饰
insert ~ 镶嵌造型
keel ~ 葱形饰
laid-in ~ 粘附饰
lintel ~ 门楣饰
lozenge ~ 菱形饰
nail-headed ~ 四叶花饰
nebulous ~ 云状饰
old Greek ~ 古希腊饰
ovolo ~ 馒形饰

egg moulding

pedestal ~ 台脚[口]线
pellet ~ 小球饰
picture ~ 挂镜条
plain round ~ 纯圆饰
quirk ~ 鸟琢饰
roll ~ 卷饰
scroll ~ 旋涡饰
sprung ~ 墙顶角饰
sunk ~ 凹沉饰条
sweep ~ 刮板造型
wooden ~ 木饰线条
zigzag ~ 曲折[波浪]饰

roll moulding

mound [maund] **n.** 丘陵，小山，墩；堤 **v.** 筑堤，堆土
mount [maunt] **v.** 安装；固定 **n.** 支架，底座
mountain ['mauntin] **n.** 山,山岭,山地
~s and waters painting 山水画
~ flax 石棉
~ flour 石粉
~ green 孔雀石
mounting ['mauntiŋ] **n.** 安装；固定；装配；框架；底座
~ bracket 安装平台,安装牛腿
~ bolt 装架螺栓
~ hole 安装孔,固定孔
~ rack 安装架,工作台,机架
~ beam 梁架
~ wall 墙上安装
mouth [mauθ] **n.** 口；孔口，出口；坑口，洞口
~ of inlet 进水口
~ of pipe 管口
~ piece 管接头
movable ['mu:vəbl] **a.** = moveable 移动的,可动的
~ bearing 活动支承
~ connection 活动连接
~ dwellings 活动房屋
~ form 活动模板
~ glass wall 可动玻璃墙
~ grate 活动格栅

~ joint 伸缩缝
~ screen 活动格网
~ support 活动支座

movement ['mu:vmənt] *n.* 移动；运动；行动
~ joint 活动缝，施工缝
vibration ~ 振动

moving ['mu:viŋ] *a.*；*n.* 移动，运动，活动
~ form 活动模板
~ partition wall 活动隔墙
~ shuttering 活动模板
~ stair 自动楼梯
~ staircase 活[自]动楼梯

moyle *n.* 鹤嘴锄，十字镐

M-roof M形[双山墙]屋顶

mud [mʌd] *n.* 泥浆；淤泥 *v.* 以泥涂
~ blanket 泥毡层
~ brick 泥砖
~ daub 修补裂缝
~ mortar 泥灰[砂]浆
~ scraper 刮泥刀[机]
~ seam 泥缝
~ wall 土墙

muddy ['mʌdi] *a.* 多泥的；泥浆覆盖的；混浊的

mulch [mʌlʃ] *n.* 覆盖料；腐土，护根 *v.* 覆盖
~ method 覆盖法，覆盖养护法
~ treatment 覆盖处理
~ing 铺覆盖料
~ing paper 覆盖纸
asphalt ~ 沥青盖料

muller ['mʌlə] *n.* 碾磨[碾砂]机，(辗轮式，摆轮式)混砂机

mullion ['mʌliən] *n.* 竖框；(窗的)直梃
adjacent ~ 附加木杆
window ~ 窗梃

multi- (词头)多

multi-aisle *a.* 多走道的

multiangular [ˌmʌlti'æŋgjulə] *a.*

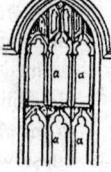
mullions

多角的
~ bar 多边形棒
~ building 多边形建筑
~ cavity block 多边形空心块
~ column 多边形柱
~ frame 多边形框架
~ ornament 多边形装饰
~ tower 多边形塔

multi-cell box girder 多孔箱梁

multi-cellular *a.* 多格的；多细胞的
~ hollow section 多格空心截面

multicentered arch 多心拱

multicoloured [ˌmʌlti'kʌləd] *a.* 多颜色的，几种颜色的
~ brick 多色砖，彩色砖
~ effect 多色效应
~ finish 多色饰面，彩色饰面

multicomponent admixture 复合料，多成分掺合料

multi-core *a.* 多芯的
~ brick 多孔砖
~ panel 多芯镶板

multideck *a.* 多层的
~ screen 多层栅格(骨料)，多级器

multi-floor *a.* 多层的
~ building 多层建筑
~ car park 多层停车场
~ factory block 多层工厂建筑
~ garage 多层车库
~ hotel 多层旅馆
~ library 多层图书馆

multifoil 多叶饰

multifoil arch 多瓣拱

multifolding door 折叠门

multifunctional *a.* 多功能的
~ room 多功能房间

multihole 多孔的

multilayer [ˌmʌltiˌleiə] *n.* 多层 *a.* 有多层的 *v.* 使多层
~ construction 多层结构，多边路面
~ monolithic wall 多层独石墙
~ structure 多层结构

multi-leaf door 多页门
multilevel [ˌmʌlti'levəl] *a.* 多层的；多平面的
multiple ['mʌltipl] *a.* 倍数的；复合的，复式的；多样的，多重的；多的，并联的 *n.* 倍数
～ arch 连拱
～ -barrel structure 多圆筒结构
～ bay 多跨
～ -bay frame 多跨框架
～ casement window 多扇窗
～ coat 多层，复层
～ -column 多柱式
～ crossing 复式交叉
～ folded plate roof 复式折板屋顶
～ frame 复式框架
～ -grid 多栅
～ -hole 多孔
～ lattice work 多重格构
～ -layer composition roofing 多层接合屋顶
～ -layer monolithic wall 多层独石墙
～ -layer panel 多层框
～ layered waterproofing 多层防水
～ lift 多层
～ line 多重线，复式线
～ -member structure 复式构件结构
～ -plate 多层板
～ portal frame 复式龙门框架
～ prismoid roof 复棱柱式屋顶
～ rib pillar 多肋柱
～ ribless shell 复式无肋壳
～ room building 多房间建筑
～ room dwelling unit 多房间居住单元
～ sample plan 多次选择方案
～ sash window 多框窗
～ sash shingle-strip 多框带状板

multiple folded plate roof

multiple rib pillar

～ size aggregate 多粒径骨料
～ -span frame 多跨[连续]构架
～ span gabled frame 多跨三角形框架
～ span girder 多跨梁
～ -story 多层
～ truss 复式桁架
～ web 多肋式
multiplex ['mʌltipleks] *n.* 多倍绘图仪 *a.* 复合的，复式的；多重的；多样的；倍数的
multiply ['mʌltiplai] *v.* 乘，繁殖
～ plywood 多层胶合板
～ wood 多层板
multipoint *a.* 多点的
multipurpose [ˌmʌli'pə:pəs] *n.*；*a.* 多目标(的)；多用途(的)，综合利用(的)
～ adhesive 多用胶粘剂
～ block 多用建筑，综合利用块
～ building 多用途建筑
～ door 多用途门
～ dryer 通用干燥剂
～ inhibitor 通用防锈剂，万能防腐蚀剂
multi-ribbed 多肋的
～ plate 多肋板
～ floor 多肋地板
multiscope *n.* 简易绘图仪
multispan 多跨；多跨的
～ beam 多跨梁
～ girder 多跨栅格
～ rigid frame 多跨刚架
～ structure 多跨结构
multistage ['mʌlisteidʒ] *a.* 多级，多阶段
multistep *a.* 多级的，阶式的
multistory [ˌmʌlti'stɔ:ri] *a.* 多层
～ building 多层房屋建筑
～ car park 多层停车场
～ factory building 多层工厂建筑
～ frame 多层框架
～ garage 多层车库，多层停车场
～ hotel 多层旅馆

~ library 多层图书馆
~ structure 多层建筑物
~ villa 多层别墅
multitubular [mʌlti'tju:bjulə] *a.* 多管道的
multitude [ˈmʌltitjuːd] *n.* 大群,大量;许多人,群众;多倍
multi-use *a.* 多用途的
~ building 多用途建筑物,多功能大楼
multiwall [ˈmʌltiwɔːl] *a.* 有多层外皮的
municipal [mjuːˈnisipəl] *a.* 城市的,市政的,内政的
~ architecture 城市建筑
~ water facilities 城市供水设施
munjack *n.* 硬化沥青
munnion [ˈmʌnjən] *n.* =mullion 窗的直棂
muntin(g) *n.* 窗格条;门中梃
mural [ˈmjuərəl] *a.* 墙壁上的;壁形的 *n.* 墙壁;壁饰,壁画
~ arch 壁拱
~ background 墙壁背景
~ decoration 壁饰
~ enrichment 壁饰
~ hanging 墙上悬吊物
~ mosaic 壁上镶嵌
~ ornamentation 壁上装饰
~ painting 壁画
muriate [ˈmjuərit] *n.* 硬石膏,氯化钾[物],盐酸盐
muriatic [mjuəriˈætik] *a.* 氯化的,盐酸的
muscovite [ˈmʌskəvait] *n.* 白[优质]云母
~ granite 白云母花岗岩
~ schist 白云母片岩
museum [ˈmjuːziəm] *n.* 博物馆
mush [mʌʃ] *n.* 烂泥,软块;噪声,干扰;废话,梦呓 *v.* 切,刻;干扰;讲废话
mushroom [ˈmʌʃrum] *n.* 蘑菇,蕈,菌
~ button 菌形按钮
~ column 蘑菇头形柱
~ construction 无梁板构造
~ floor 无梁楼盖
~ floor slab 无梁楼盖板
~ head (支撑无梁板的)锥形头
~ shell 伞形壳体
music [ˈmjuːzik] *n.* 音乐
~ hall 音乐厅
mutual [ˈmjuːtjuəl] *a.* 相互的,共同的
~ anchorage (钢筋混凝土内钢筋的)搭接
mutule [ˈmjuːtjuːl] *n.* 建筑(Doric)式檐板下的凸块[托块]
mycalex *n.* 云母玻璃,(压粘)云母块
mylar *n.* 聚酯薄膜[树脂]
~ film 聚酯[密拉]薄膜
myriorama [miriəˈrɑːmə] 万景画

N

nab [næb] *n.* 水下礁丘,门锁槽
nacarat [ˈnækəræt] *n.* 鲜艳的橘红色
nacrite *n.* 珍珠陶土
nacreous varnish 珍珠光清漆
Nada *n.* 一种铜基合金
nail [neil] *n.* 钉;指甲 *v.* 钉钉;将……钉牢,使……固定
~ claw 拔钉钳
~ clinch 弯脚钉
~ dog 勾头钉
~ drawer 拔钉钳
~ extractor 起钉机,拔钉器
~ -glued truss 胶钉结合桁架
~ hammer 羊角锤;拔钉锤
~ head 钉头饰,钉头
~ headed moulding 钉头状的线脚,四叶花装饰
~ hole 钉眼

~ puller 拔[起]钉器
blind ~ 暗钉
brad ~ 角钉,无头钉,曲头钉
bullen ~ 阔头钉
calking ~ 填隙钉,开尾钉
cut ~ 方针,切钉
double pointed ~ 双头尖钉
dog ~ 道钉
draw ~ 起模钉
finish ~ 暗钉
flooring ~ 地板钉
galvanized sheet iron ~ 镀锌钢板用钉
hob ~ 平头钉
hooked ~ 曲钉,钩头钉
joist ~ 搁栅钉
lath ~ 板条钉
machine-made ~ 机制钉
roof ~ 屋面板用钉
rose ~ 圆花钉
screw ~ 木螺钉
secret ~ 暗钉
spike ~ 长折钉,道钉
square ~ 方钉
stub ~ 短粗钉
toe ~ 斜钉
wire ~ 圆铁钉

nailable [ˈneiləbl] *a.* 可打钉的,能受钉的
~ brick 受钉砖
~ concrete 可钉钉的混凝土
~ plate 受钉板

nailcrete *n.* 受钉混凝土

nailed *a.* 钉的
~ beam 钉梁
~ connection 钉结合
~ construction 钉结构
~ joint 钉结合
~ plate girder 钉板梁
~ roof 钉接屋顶
~ truss 钉接桁架
~ wooden plategirder 钉接木板梁

nailer [ˈneilə] *n.* 制钉工人;自动敲钉机,钉钉板

nailhead [ˈneilhed] *n.* 钉头;钉头装饰
~ bonding 钉头焊,球焊
~ molding 钉头装饰线脚

nailhead

nailing [ˈneiliŋ] *a.* 敲钉用的,受钉的
~ batten 可钉板条(用来挂瓦或压缝的)
~ block 受钉块
~ brick 受钉砖
~ concrete 受钉混凝土
~ plug 受钉块,钉条
~ strip 受钉条(附着在钢板或混凝土表面以供钉钉用的木板)
~ strip bracket 可钉板条托架

nailsick [ˈneilsik] *a.* 因钉孔而产生渗漏的;因钉孔而减少强度的
~ board 因钉孔太多而失去强度的木板

naked [ˈneikid] *a.* 无保留的;裸体的
~ floor 未铺地板
~ frame 裸构架
~ light 无遮光线,无罩灯
~ radiator 无保护罩散热器
~ wire 裸线

Nalcite *n.* 离子交换树脂

named [ˈneimd] *a.* 记名的;有名的,著名的

nanometer [ˈneinəˌmiːtə] *n.* 十亿分之一米,毫微米

naos [ˈneiɔs] *n.* (*pl.* -oi)庙宇;古希腊与罗马庙宇之中央部分,中堂,主殿

naphtha [ˈnæfθə] *n.* 石脑油,挥发油,粗汽油
paint ~ 调漆油,油漆溶剂油

napkin [ˈnæpkin] *n.* 餐巾
~ ring 套餐巾用的小环

Naples yellow 那不勒斯[拿浦]黄,锑酸铅

napoleon marble 褐色带红纹的法国大

理石
nard [nɑːd] *n.* 甘松；甘松香
narrow ['næreu] *n.* 狭窄地段 *v.* 弄窄，缩小 *a.* 狭窄的；细的，有限的，严密的
 ～ -bore tube 小直径管
 ～ carpet 狭幅地毯
 ～ rule 狭规，窄尺
narthex *n.* (教堂的)前廊
nasty ['nɑːsti] *a.* 难处理的，严重的；脏的，不洁的
 ～ smell 臭味，难闻的气味
natatorium [ˌneitəˈtɔːriəm] *n.* (*pl.* -riums, -ria)(室内)游泳池
national ['næʃənl] *a.* 国家的；国民的；国立的；民族的 *n.* 国民
 ～ architecture 民族建筑(学)
 ～ customs 国民习俗，民俗
 ～ drawing 国画，民族绘画
 ～ feature 民族特色
 ～ ornament 民族装饰，民族图案
 ～ Romantic style 民族浪漫风格
 ～ standard 国家标准
 ～ style 民族风格
 ～ treasure building 国家保存的文物建筑
native ['neitiv] *a.* 本地的；天然的；出生地的 *n.* 本地人，土产
 ～ asphalt 天然沥青
 ～ bitumens 天然沥青
 ～ paraffin 天然石蜡
 ～ style 当地风格，天然形式
natrium ['neitriəm] *n.* 钠(Na)
 ～ lamp 钠蒸气灯
 ～ lead 含钠铅合金
natte 席纹柱
nattier-blue ['nætiə-'bluː] *n.* 淡青色，淡蓝色
natural ['nætʃərəl] *a.* 天然的，自然的；未加工的，粗加工的；固有的；天生的 *n.* 自然物
 ～ abradant 天然研磨料(如金刚砂)；磨蚀剂
 ～ abrasive 天然研磨料；磨蚀剂
 ～ adhesive 天然胶粘剂
 ～ aggregate 天然骨料，天然集料
 ～ aging 自然老化
 ～ anhydrite 天然硬石膏
 ～ arch 自[天]然拱
 ～ asphalt 天然沥青
 ～ bitumen 天然沥青
 ～ bristle brushes 天然鬃刷
 ～ building sand 天然建筑用砂
 ～ cement 天然水泥，天然胶结料
 ～ clay 天然黏土
 ～ cleft 自然裂缝
 ～ coarse sand 天然粗砂
 ～ color 天然色，彩色
 ～ conservation 自然保护
 ～ copal 天然硬树脂(胶)
 ～ cork 天然软木
 ～ crack 自然裂纹
 ～ desiccate 风干，自然干燥
 ～ draft ventilation 自然通风
 ～ drainage 自然排水(系统)
 ～ draught 自然风，自然通风
 ～ drying out 自然干燥
 ～ dyestuff 天然染料
 ～ emulsion 天然乳胶
 ～ erosion 自然侵蚀
 ～ evaporation 自然蒸发
 ～ feature (自然)地形，地势
 ～ feature of terrain 地貌
 ～ finish 天然涂料，天然装饰
 ～ foliage 自然叶饰
 ～ glue 天然胶
 ～ grain finish 露木纹油漆
 ～ graphite 天然石墨
 ～ gravel 天然粗砂[砾石]
 ～ gray (= ～ grey) 天然灰色
 ～ grindstone 天然磨石
 ～ gypsum 天然石膏
 ～ hardness 自然硬度
 ～ lighting 天然光照明

~ marble 天然大理石
~ mica 天然云母
~ moisture 自然湿度,天然含水量
~ pigment 天然颜料[色素]
~ plastic (material) 天然塑料性材料
~ product 原料,天然产品
~ resin 天然树脂
~ resin mastic 天然树脂胶粘水泥
~ resin varnish 天然树脂清漆
~ rock 天然岩石
~ rubber 天然橡胶
~ sand 天然砂
~ seasoning 自然风干法
~ sienna 天然赭土(一种矿物颜料)
~ slate 天然石板瓦
~ split 天然裂缝
~ stone (angle-) quoin 天然(接合墙壁用的)隅石块,(拱门的)楔形石
~ stone arch 天然石拱
~ stone column 天然石柱
~ stone coping 天然石顶
~ stone corbel 天然石托,石撑
~ stone course 天然石层
~ stone order 天然石柱型(尤指古典建筑的柱型)
~ stone polishing machine 天然石面抛光机
~ stone rubbing machine 磨石机
~ ventilation 自然通风
~ wastage 自然损耗
~ water content 天然含水量
~ wax 天然蜡
~ weathering 自然风化
~ wind 自然风
~ wood color 天然木色
naturalism ['nætʃrəlizəm] *n.* (建筑、艺术品)自然主义
naturalistic [,nætʃərə'listik] *a.* 自然主义的,博物学的
~ form 自然主义形式
naturally ['nætʃərəli] *ad.* 自然地;天然地;当然地

~ lighted 太阳光的,自然光的
nature ['neitʃə] *n.* 自然;性质,特性;种类,样子;天性,本性
naught [nɔ:t] *n.* 无;零 *a.* 无价值的,无用的
nave [neiv] *n.* 教堂的中部,中殿;(轮)毂;听众席;车站的中间广场
~ aisle 厢堂,耳堂
~ aisle bay 厢堂桁幅
~ aisle gallery 厢堂廊台
~ arcade 中堂连拱廊,大廊
navel ['neivəl] *n.* 中央,中心
navvy ['nævi] *n.* 挖掘机,掘土机;挖土工人 *v.* 挖掘
~ barrow 土车,运土手推车
~ pick (挖)土镐
navy ['neivi] *n.* 海军;藏青色
~ blue 深蓝色
~ bronze 海军青铜
near [niə] *a.* 近的,近旁的;亲近的;左方的 *ad.* 近;亲近地;几乎,差不多 *prep.* (空间,时间)接近,走近
nearby ['niəbai] *a.* 近旁的,附近的
~ material 地方材料
nearside ['niəsaid] *n.* 近侧;左边 *a.* 左边的
neat [ni:t] *a.* 净的,纯的;干净的,整洁的;精巧的
~ cement 纯水泥,净水泥
~ lime 净石灰
~ line 净(开挖)线;内图廓线;细线,准线;墙面交接线
~ plaster 纯灰浆
~ work 清水墙勾缝
neatline *n.* 准线,图表边线,墙面交接线
nebula ['nebjulə] *n.* (*pl.* -lae, -las) 星云
~ molding 星云迹脚,波边饰
nebulizer ['nebjulaizə] *n.* 喷雾器,喷散器

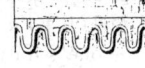

nebula molding

necessary ['nesisəri] *n.* (*pl.* -saries) 必需品 *a.* 必需[要]的
~ labour time 必要劳动时间

~ minimum profit 必要最低利润
neck [nek] *n.* 颈,颈弯饰;柱身之领卷 *v.* 缩细
chimney ~ 烟道
swan ~ 弯曲管
necked *a.* 压[收]缩的,拉细的
~ bolt 长颈螺栓
necking ['nekiŋ] *n.* 柱颈;柱颈花边装饰;颈缩

necking

needle ['ni:dl] *n.* 针;横撑木;(桥下)横梁;方尖塔 *vt.* 用横撑木支撑;用针缝
~ bath 淋浴
~ beam 小横梁,簪梁(托换基础用)
~ beam scaffold 轻型脚手架
~ lac 针状虫胶
~ leaved tree 针叶树
~ nozzle 针形喷嘴
~ scaffold 横梁支托的脚手架
~ shoring 斜撑
~ support 横撑
~ vibrator 针形振捣器
cement ~ 水泥硬固检验针
compass ~ 罗盘针
dead shore ~ 围墙用撑
penetration ~ 贯入针
needlestone ['ni:dlstəun] *n.* 针状结晶岩石
needling ['ni:dliŋ] *n.* 横撑木;柱窝
needly ['ni:dli] *a.* 如针状的
negative ['negətiv] *n.* 负数;负电,阴电;反面;照相底片 *a.* 负的,否定的
~ elongation 缩短
~ painting 负绘彩
~ reinforcement 负弯矩钢筋
~ thread 阴螺纹
neglect [ni'glekt] *vt.* 忽视,忽略;疏忽,漏做(某事) *n.* 忽视,疏忽
~ one's duty 玩忽职守
negligence ['neglidʒəns] *n.* 忽视,疏忽,粗心大意
negotiate [ni'gəuʃieit] *vt.* 商议,交涉;转让,兑现;解决,处置 *vi.* 谈判
negotiated *a.* 谈判的,协商的;转让的,兑现的;处置的,解决的
~ contract 合同
neighbo(u)rhood ['neibəhud] *n.* 邻近,附近;地区,街道;(数学的中)邻域 *a.* 邻区的
~ centre 邻里中心
~ commercial district 邻里商业区
~ density 邻里密度
~ unit area 住宅单元面积
neighbo(u)ring ['neibəriŋ] *a.* 邻[附]近的,邻界的
~ opening 相邻宽度
~ pile 邻桩
~ premises 邻屋
~ region 邻域
~ span 相邻宽度
~ unit 住宅单元
~ wall 邻墙
nemo ['ni:məu] *n.* (*pl.* -mos)室外广播
neo-Baroque *n.* 新巴洛克建筑形式
neo-Byzantine architecture 新拜占庭式建筑
neo-classic architecture 新古典主义建筑
Neodymium [ni:ə'dimiəm] *n.* 钕(Nd),稀土金属元素之一
~ glass 含钕玻璃
Neogen *n.* 一种镍黄铜
neogenesis [niəu'dʒenisis] *n.* 新生,再生
neo-Gothic [ni:əu'gɔθik] *a.* 新哥特式的(建筑)
neo-Greek [ni:əu'gri:k] *a.* 希腊复兴式的(建筑)
neoimpressionism [ni:əuim'preʃənizm] *n.* 新印象派
neo-liberty *a.* 新自由派风格的
neolithic [ni:əu'liθik] *a.* 新石器时代的
neo-loess *n.* 新黄土
neo-medi(a)eval *a.* 新中世纪式的

（建筑）

neon ['ni:ən] *n.* 氖(Ne)
~ arc lamp 氖弧灯
~ argon luminous tube 氩氖发光管
~ bulb 霓虹灯
~ lamp 霓虹灯
~ light 霓虹灯,氖光灯
~ sign 霓虹灯广告
~ -street 霓虹灯街
~ tube 氖管

Neonalium *n.* 一种铝合金

neo-plasticism [ni:əu'plæstisizm] *n.* 新塑造主义(多用直线素描的一种抽象画派)

neoprene ['ni:əpri:n] *n.* 氯丁橡胶
~ adhesive 氯丁橡胶胶粘剂
~ (compression) seal 氯丁橡胶压力填封;氯丁橡胶填缝料
~ control strips 橡胶控制板(条)
~ foam 合成泡沫橡胶
~ gasket 氯丁橡胶填塞物
~ glue 氯丁橡胶胶
~ paint 氯丁橡胶涂料
~ rubber 氯丁橡胶
~ sealing gasket 氯丁橡胶密封填料
~ seal(ing) ring 橡胶止水环
~ section expansion joint 橡胶伸缩接头
~ sleeve 氯丁[人造]橡胶套
~ washer 橡胶衬垫[垫圈]
~ water stop 橡胶止水

neorealism [ni:əu'riəlizəm] *n.* 新写实主义

neo-Renaissance *n.* 新文艺复兴式

neo-Romanesque *n.* 新罗马(建筑)风格

neo-Romanticism [ni:əurə'mæntisizəm] *n.* 新浪漫主义

neosolidizit cement 新索利迪契特水泥(一种混合硅酸盐水泥)

neoteric [ni:əu'terik] *n.*;*a.* 新式的,新发明的

nepheline ['nefəlin] (= nephelite) *n.* 霞石

~ basalt 霞石玄武岩
~ syenite 霞石正长岩

nervure ['nə:vjuə] *n.* 拱房顶的侧肋拱,交叉侧肋;叶脉

Nessler-tube *n.* 纳氏比色管

nesslerization *n.* 等浓比色法

nest [nest] *n.* 嵌套;蜂窝(混凝土缺陷);巢窝 *vi.* 筑巢
~ of saws 组合锯条
~ of screens 组合筛
~ of sieves 组合筛
~ed tables 套几

net [net] *n.* 网,网状物 *vt.* 撒网 *a.* 净的,纯的;最后的;网状的
~ absorption 净吸收量
~ apartment unit area 公寓净面积
~ area 净面积
~ benefit 净效益
~ cost 实价,成本
~ earning(s) 净赚,净收益
~ efficiency 净效率
~ floor area 净楼板面
~ income 纯收入
~ price 实价,净价
~ proceeds 净收入
~ profit 纯利润
~ residential area 居住净面积
~ section 净截面
~ structure 网状结构
~ tracery 网式窗格
~ type bed bottom 网式床屉
~ vault 网状拱顶
~ -worked dome 网架圆顶,网状穹盖
~ting lath(ing) 板条网架
crimp wire ~ 波形钢丝网
insecticide ~ 防虫网
safety ~ 安全网
wire ~ 钢丝网

net tracery

nether ['neðə] *a.* 下面的,地下的

netty ['neti] *a.* 似网的,网状的

network ['netwə:k] *n.* 网格,网状组织,网络
～ cupola 网状圆屋顶
～ dome 网状圆屋顶
survey ～ 控制点网,测图网

neutral ['nju:trəl] *a.* 中性的;中立的;中和的
～ brick 耐火砖
～ colo(u)r 灰色,中和色
～ fat 中性油
～ fibre (= ～ fiber) 中性纤维
～ fire brick 中性耐火砖
～ glass 中性玻璃
～ refractory 中性耐火材料
～ slag 中性矿渣
～ solution 中性溶液
～ tint 青灰色,不鲜明的色彩

neutralization [ˌnju:trəlaiˈzeiʃən] *n.* 中和(作用);抵消,使失效;平衡
～ level 平衡水准面
～ with lime 石灰中和处理

neutron ['nju:trɔn] *n.* 中子
～ moisture meter 中子含水量测定仪
～ sensor 中子探测器
～ shield 中子防护屏

nevadite [nəˈvɑ:dait] *n.* 斑流岩

new [nju:] *a.* 新的,新式的,新奇的;新造的;生的
～ -coined 新造的
～ developed area 新开发区

New Baroque 新巴洛克(建筑)

New England colonial 美国初期的新英格兰式建筑

newborn ['nju:bɔ:n] *a.* 新生的,再生的

newbuild *v.* 新建,重建

newel ['nju:əl] *n.* (旋梯)中柱,楼梯端柱;桥梁或高架桥竖墙终端的圆柱墩
～ cap 楼梯扶手中柱柱顶(脚)饰
～ post (楼梯转弯处的)栏杆
～ tube 旋梯轴管,轴管
～ type spiral stair (case) 有中柱的盘旋扶梯

hollow ～ 装旋梯柱的井孔
solid ～ 实砌(楼梯)旋栓

newfangled *a.* 新奇的,爱好新事物的

newfashioned *a.* 新式的,新流行的

newly ['nju:li] *ad.* 新近地;全新地
～ -laid 新铺[浇](混凝土)
～ -made panel 新嵌[镶]板
～ -made product 新产品
～ -placed concrete 新浇混凝土

news [nju:z] *n.* 新闻,消息,有新闻价值的事或人;报纸

newspaper ['nju:speipə] *n.* 报纸

next [nekst] *a.*;*n.* 下次(的),下一个紧接(的),隔壁(的)
～ -door neighbour 隔壁邻居

nexus ['neksəs] *n.* 网络;节、段;连接;连杆

N-girder N 形格构大梁

nib [nib] *n.* 尖头;突边;模孔 *vt.* 装尖头,弄尖

nibbed bolt 尖头螺栓

nibbed tile 尖头瓦

niche [nitʃ] *n.* 壁龛(放置雕像、花瓶等物的墙壁凹入处,不延伸到地板)
～ baldaquin 壁龛龛室
～ -buttressed 用壁龛支撑的
～ canopy 壁龛华盖
～ pediment 壁龛(顶上的)三角饰

niche

nichrome ['nikrəum] *n.* 镍铬合金

nick [nick] *n.* 刻痕,裂痕;凹隙[处] *vt.* 刻痕
～ action 交 咬作用
～ed column 刻槽柱
～ed tooth 刻齿痕

nickel ['nikl] *n.* 镍;美五分镍币
～ alloy 镍合金
～ bare welding filler metal 镍焊丝
～ brass 镍铜锌合金

~ bronze 镍铜
~ -chrome steel 镍铬钢
~ -chromium-iron 镍铬铁
~ -clad steel plate 镀镍钢板
~ coat 包镍,镍表皮
~ -coated 镀镍
~ -cobalt (alloy) 镍钴合金
~ -leaded bronze 镍铅铜
~ -lined 衬镍的
~ -molybdenum steel 镍钼钢
~ plating 镀镍
~ silver 德银(一种镍铜锌合金)
~ stainless steel 不锈镍钢
~ steel 镍钢
~ titanate yellow (pigment) 钛酸镍黄色颜料
nickelage *n.* 镀镍
nickelex *n.* 光泽镀镍法
nidge [nidʒ] *vt.* 以凿子修削(石块)
niello [ni'eləu] *n.* 黑金,黑金镶嵌品
nigh [nai] *a.* 近的;短的;直接的
night [nait] *n.* 夜(间)
~ bolt 弹簧[保险]插销
~ illumination 夜间照明
~ -latch 夜间安全弹簧销
~ light 夜灯(通宵不熄灯)
~ stand 床头柜
~ storage heater 夜间廉价用电的热水器
~ table 床头柜
~ works 弹子门锁机件
nigrify ['nigrifai] *vt.* 使黑,使变黑
nigritude ['nigrətju:d] *n.* 黑色,黑色之物
Nile [nail] *n.* 尼罗河
~ blue (稍带绿色的)尼罗蓝
~ green 尼罗绿(浅淡的蓝绿色)
~ mud brick 尼罗河泥砖
nimiety [ni'maiəti] *n.* 过多,过剩
ninefold ['nainfəuld] *a.;ad.* 九倍的(地),九重的(地)
nine-tenths *n.* 十分之九,差不多全部

ninety [nainti] *a.;n.(pl.* -ties)九十,第九十(的)
~ degrees bend (pipe) 90°弯管
~ degrees bend junction 90°弯头
ningpo varnish (宁波)金漆
nip [nip] *n.* 钳子,夹子;海岸低悬崖;浪蚀洞;河底切槽 *v.* 抓,咬,剪,挟,箝
~ action (钢丝绳股中的钢丝)交咬作用
~ off 剪断,摘掉
double ~ 双接口
nipper ['nipə] *n.* 各种夹具,钳子;剪钳,拔钉钳
adjustable cut ~s 活刃剪钳
cut ~s with spring 弹簧剪钳
plain-cut ~s 普通剪钳
wire ~s 钢丝钳
nipping ['nipiŋ] *n.* 搬钻,元宝榫 *a.* 夹住的
nipple ['nipl] *n.* 乳头状突起;短接管;螺纹接套;喷灯,喷嘴
~ for fittings 螺纹接口管
~ joint 螺纹接头
brazed ~ 黄铜螺纹接口
close ~ 螺纹接口[管]
cutlet ~ 出口螺纹接套
double ~ 双螺纹接套
hose ~ 软管螺纹接套
inlet ~ 进口螺纹接套
long screw ~ 长螺纹套筒
pipe ~ 管螺纹接套
reducing ~ 异径螺纹管接头
screwed ~ 螺钉接头,短接接头
nit *n.* 尼特(亮度单位,等于1新烛光/米²)
nitralising *n.* 硝酸钠溶液浸渍处理法
Nitralloy *n.* 氮化合金
nitrate ['naitreit] *n.* 硝酸盐 *v.* 硝化
~d asphalt 硝化沥青
~d cement 硝化纤维素胶泥
~d coal-tar 硝化煤焦油沥青
nitration [nai'treiʃən] *n.* 硝化作用,氮化

nitride ['naitraid] *n.* 氮化物
 ~ refractory material 含氮耐火材料
nitrification [naitrifi'keiʃən] *n.* 硝化作用
nitrile ['naitril] *n.* 腈
 ~ glue 腈胶
 ~ -phenolic rubber 腈-酚醛橡胶
 ~ resins 腈树脂
 ~ rubber 腈橡胶（一种不溶于汽油等溶剂的人造橡胶）
 ~ rubber adhesives 腈橡胶粘剂
nitrite ['naitrait] *n.* 亚硝酸盐
nitrobarite *n.* 钡硝石
nitrobenzene [ˌnaitrə'benziːn] *n.* 硝基苯
nitrocellulose [naitrəu'seljuləus] *n.* 硝化纤维素, 硝化棉, 棉花炸药
 ~ paint 硝基涂料
nitrodope *n.* 硝化涂料, 硝基清漆
nitroenamel *n.* 硝基磁漆
nivation [nai'veiʃən] *n.* 雪[霜]蚀作用
niveous ['niviəs] *a.* 雪白的, 白雪般的
no [nəu] *a.; ad.* 没有, 很少, 很小; 并非, 决非
 ~ -account (= ~ count) 无用的, 无价值的
 ~ dressing 不修整的
 ~ -fines concrete 无细骨料混凝土, 无砂大孔混凝土, 多孔混凝土
 ~ -fuse switch 无熔丝开关
 ~ -girder floor 无梁楼盖
 ~ -heat adhesive 冷粘剂
 ~ -slump concrete 无坍落度混凝土, 干硬性混凝土
nob [nɔb] *n.* 球形门柄, 雕球饰
noble ['nəubl] *a.* 壮丽的, 宏伟的; 极好的, 杰出的, 崇高的; 惰性的 *n.* 贵族
 ~ antique (style) 古风风格
 ~ fir 大冷杉
 ~ metal 贵金属
nocturne ['nɔktəːn] *n.* 夜景画
nodal ['nəudl] *a.* 有节的, 交点的
node [nəud] *n.* 节, 结; 节点, 结点; 波节

nodical ['nəudikəl] *a.* 交点的
nodose ['nəudəus] *a.* 有节的, 多结节的
nodular ['nɔdjulə] *a.* 结节的, 结节状的; 块状的, 结核状的
 ~ fireclay 块状耐火黏土
nodulizing *n.* 团矿; 烧结, 熔结 *a.* 烧结的, 粘结的
nog [nɔg] *n.* 木栓, 木钉; 木砖 *vt.* 用木钉钉牢, 砌木砖
 ~ ging piece 压墙槛, 木砖
 wood ~ 榫销, 木栓
 brick ~ ging 木架砖壁
no-hinged *a.* 无铰(链)的
 ~ arch 无铰拱
 ~ beam 固端梁
 ~ column 固端支柱
 ~ girder 无铰大梁
 ~ support 固定支座
noise ['nɔiz] *n.* 噪声, 闹声, 声音 *v.* 大声喧闹
 ~ abatement 减声, 噪声抑制
 ~ -absorbing ceiling 吸声顶板
 ~ absorption 噪声吸收
 ~ barrier 隔声栅
 ~ coefficient 噪声系数
 ~ control 噪声控制
 ~ criteria 噪声标准
 ~ dose 噪声量
 ~ elimination 消声
 ~ equivalen power 噪声等效功率
 ~ isolation class 隔声等级
 ~ immunity 抗噪声性
 ~ insulation 隔声
 ~ intensity 噪声强度
 ~ -killer 静噪器
 ~ level 噪声级, 噪声水平
 ~ pollution 噪声污染
 ~ precaution 噪声预防
 ~ -proof 隔声的
 ~ reduction 减噪, 静噪
 ~ source 噪声源
 ~ squelch 噪声消除器

~ trap 静噪器
ambient ~ 环境噪声
city ~ 城市噪声
exterior ~ 外部噪声
ground ~ 环境噪声
hazardous ~ 危害性噪声
mechanical ~ 机械噪声
thermal ~ 热噪声
white ~ 白噪声
noiseless ['nɔizlis] *a*. 无噪声的,无声的
nominal ['nɔminl] *a*. 额定的,标称的,名义上的;铭牌的
~ aperture width 标称孔宽
~ bore 标称内径
~ capacity 额定容量
~ diameter 标称直径
~ depth 额定[名义]深度
~ dimension 标称[志]尺寸
~ grain size 标称粒径
~ internal diameter 标称内径
~ internal width 标称内宽
~ length 标称长度
~ scale 标称比例尺
~ size 名义尺寸,公称尺寸
~ stress 标称应力
~ thickness 标称厚度
non-absorbent *a*. 不吸收的
non-adhesive *a*. 无黏性的
non-adiabatic *a*. 非绝热的
non-affine *n*. 非均匀的
non-ag(e)ing *a*. 不老化的,经久的,无时效的
non-aggressive *a*. 非侵蚀性的
non-articulated arch 无铰拱
non-asphaltic *a*. 非沥青质的
non-bearing *a*. 非承重的
~ partition 非承重隔墙
~ structure 非承重结构
~ wall 非承重墙
non-blended cement 纯水泥
non-bloated *a*. 无胀性的,不膨胀的
non-bonded *a*. 无黏性的,无粘结力的

non-breakable glass 刚性玻璃
non-browning glass 不变色玻璃
non-ceramic tile 非陶瓷瓦
non-chromatic *a*. 无色彩的,无颜色的
non-cohesive *a*. 无黏性的
non-combustible [ˌnɔnkəm'bʌstəbl] *a*. 非易燃的 *n*. 非易燃性物质
~ building 防火建筑
~ construction 防火结构
~ construction(al) material 耐火建筑材料
~ molded asbestoscement panel 防火石棉水泥板
non-constructive cement 杂用水泥,低强度等级水泥
non-corrodible *a*. 不锈的,抗蚀的
non-corrosive *a*. 不锈的
~ metal 不锈金属
~ pipe 不锈钢管
~ steel 不锈钢
non-criteria *a*. 非标准的
non-defective *n*. 合格品
non-destructive [nɔndis'trʌktiv] *a*. 非破损性的,无损的
non-domestic *a*. 非居住的,非住宅
~ block 非住宅区
~ building 非住宅楼
~ construction 非民用建筑
non-ductile fracture 无塑性断裂
non-ecclesiastical *a*. 非宗教的,非基督教的
~ architecture 非宗教建筑
~ building 非宗教建筑物
~ Gothic (style) 非基督教哥特式(建筑)
non-electric *a*. 不用电的,非电的
non-equilibrium *a*. ;*n*. 非平衡(的)
non-erodible *a*. 不冲刷的
nonex *n*. 铅硼玻璃
non-extruding *a*. 不凸出的
~ joint filler 不凸出的填缝
nonfading *a*. 不褪色的

nonferrous [nɔn'ferəs] *a.* 非铁的
  ~ alloy 非铁合金,有色合金
  ~ metal 有色金属
non-figurative art 非图形表示的艺术
non-fireproof construction 非[不]防火建筑
non-flammable *a.* 不可燃的,非易燃的
non-floating rail 固定栏杆
non-freezing *a.* 不冻的,耐寒的
non-glare glass 毛玻璃
non-hermetic *a.* 不密闭的
non-homogeneity *n.* 不均一性,非均质性
non-homogeneous ['nɔnhə'mɔdʒənəs] *a.* 非均质的,不均匀的
  ~ deformation 不均匀变形
  ~ medium 非均匀介质
  ~ plastic body 非均质塑性体
nonhousing building 非住宅楼
non-hygroscopic *a.* 不吸湿的
non-ignitable *a.* 耐火的,防火的
non-inert impurity 活性杂质
non-inflammable *a.*;*n.* 不可[易]燃(的)
non-loss *n.* 无损耗
non-luminous *a.* 无光的,不发光的
  ~ body 不发光体
  ~ sign 无光标志
non-lustrous finish 无光泽涂料
non-mechanical *a.* 非机械的
non-metallic *a.* 非金属的
  ~ fireproof building material 非金属防火建筑材料
  ~ lustre 非金属光泽
  ~ pipe 非金属管
nonmilitary architecture 非军事建筑
non-natural *a.* 人工[造]的
nonobjectivism *n.* (艺术的)抽象主义
non-pigmented *a.* 不变色的;不易着色的
nonplastic ['nɔn'plæstik] *a.* 非塑性的
  ~ concrete 干硬性混凝土
  ~ material 非塑性材料
nonpolishing aggregate 不易磨光(的)集料,耐磨集料
non-reflecting glass 非反光玻璃
non-reinforced *a.* 非加强的;无钢筋的
  ~ concrete 无筋混凝土,素混凝土
  ~ concrete pipe 无筋混凝土管
  ~ slab 无筋混凝土板
nonrental area 非租赁区
non-repressed brick 活动砖
nonreverberant room 无回声房间
nonrevolving door 非旋转门
nonrigid [nɔn'ridʒid] *a.* 非硬性的,非刚性的
  ~ floor 非刚性楼板
  ~ joint covering 非刚性盖缝物
  ~ plastic 软塑料
non-rust steel 不锈钢
non-rusting solution 防锈剂
non-safety lock 非安全锁
nonsettling *a.* 不沉降的,不沉淀的
non-shattering glass 不碎玻璃
nonshrinkage concrete 无收缩混凝土
nonskid ['nɔn'skid] *a.* 不滑的,防滑的
  ~ brick 防滑砖
  ~ carpet 防滑地毯
  ~ chain 防滑链
  ~ concrete surface 混凝土(抗滑)糙面
  ~ finish 防滑面层
  ~ ground surface 防滑地面
  ~ mat 防滑垫
  ~ paint 防滑涂料
  ~ property 防滑性
  ~ quality 防滑特性
  ~ rib(bed) tile 防滑螺纹瓦(砖)
  ~ surface 不滑路面,不滑面
  ~ treatment 防滑处理
non-slip ['nɔn'slip] *a.* 防滑的
  ~ batten 防滑条
  ~ cleat 防滑木
  ~ floor 防滑楼板

~ granolithic 防滑混凝土铺面
~ lath 防滑条
~ nosing 防滑(突缘)条
~ tread 防滑踏板,防滑(轮胎)花纹
non-specification *n.* 非规格化
non-sprayable *a.* 不可喷涂的
non-staining *a.* 不着色的
~ cement 白色水泥
non-standard *a.* 不标准的
~ material 非标准材料
non-stationary *a.* 不稳定的
non-stickness *n.* 无黏性
non-stripping *a.* 防剥落的
nonstructural *a.* 非结构[构造建筑]的
~ (bonding) adhesive 非结构胶粘剂
~ top screed (混凝土)铺面,表面覆盖层
nontoxic *a.* 无毒的
non-transparency *n.* 不透明度
non-uniform ['nɔn'juːnifɔːm] *a.* 不均匀的,不一致的
~ beam 变截面梁
~ bearing structure 非均匀承重结构
~ sand 非均质砂
~ settlement 不均匀沉陷[降]
~ shrinkage 不均匀收缩
non-ventilated *a.* 不通风的
~ flat roof 不通风平顶
nonviscous *a.* 非黏性的
nonvitreous ceramic tile 非玻璃质陶瓷瓦,非透明陶瓷瓦
nonvoided concrete beam 实心混凝土梁
nonvolatile ['nɔn'vɔlətail] *a.* 不挥发的,不挥发性的,永久的
non-walking-way gutter 屋檐滴水槽
non-watertiht *a.* 透水的,漏水的
~ material 透水材料
non-weight-carrying *a.* 非承重的
~ floor block 非承重楼板
~ floor brick 非承重楼面砖

~ panel 非承重间壁,非承重护墙板
~ partition (wall) 非承重隔墙
nonwelding *a.* 不焊合的
non-woven fabric 粘着纤维布,无纺布
non-yellowing *a.* 不发黄的
non-yielding *a.* 不屈服的
~ retaining wall 刚性挡土墙
noodle ['nuːdl] *n.* 面条
~ style 面条式(建筑)
nook [nuk] *n.* 角,隅,转角处
noon [nuːn] *n.* 正午,中午;顶点,全盛期 *a.* 正午的,子午的
nopal ['nəupəl] *n.* 仙人掌
Norfolk latch (=**thumb latch**) 诺福克插销,门窗插销
norm [nɔːm] *n.* 规范,标准;模方,范数
technical ~s 技术标准
Normagal *n.* 铝镁耐火材料
normal ['nɔːməl] *n.* 法线,垂直线 *a.* 正常的;中性的;标准的;垂直,正交的
~ aggregate 普通集料
~ block 普通砖块
~ bond 普通(砖石)砌合
~ breaking 正常破裂
~ brightness 正射(法向)亮度
~ cast 普通(混凝土)预制件
~ cast block 普通预制块
~ cast concrete 普通预制混凝土
~ cast tile 普通预制(空心)砖瓦
~ cement 普通水泥
~ concrete build element 普通混凝土建筑构件
~ concrete cast member 普通混凝土预制件
~ concrete floor 普通混凝土楼板
~ concrete hollow block 普通混凝土空心块
~ concrete lintel 普通混凝土过梁
~ condition 常规条件,正常状况
~ consistency 标准[正常]稠度
~ fold 正褶皱
~ format 标准格式
~ heating degree days 标准采暖

度日
~ heavy concrete 普通重混凝土,普通混凝土
~ hollow block 普通空心块
~ hydrated lime 正常水化石灰
~ illumination 垂直照度
~ intensity of light 法向发光强度
~ length 标准长度
~ (masonry) bond 标准(砖石)砌合
~ moisture 标准湿度
~ mortar 普通砂浆
~ opening 标准开度[宽度]
~ pitch 标准(螺,轨)间距,常用屋面坡度
~ portland cement 普通波特兰水泥,普通硅酸盐水泥
~ position 正常[静止]位置
~ sand 标准砂
~ section 法向截面
~ slump 标准坍落度
~ spoilage 正常破损
~ width 标准宽度

normalised = normalized【美】a. 正规的,标准的
~ form 标准型

normalising n. = normalizing 正规化;正常化

normalizable a. 可规范化的

Norman [ˈnɔːmən] n. 诺曼人 a. 诺曼的
~ architecture 诺曼式建筑
~ brick 诺第砖
~ crypt 诺曼式地穴
~ Gothic style 诺曼式哥特风格
~ roofing tile 诺曼式屋顶瓦
~ style 诺曼风格
~ vault 诺曼连拱式圆屋顶
~ window 诺曼式窗户

normative [ˈnɔːnətiv] a. 标准的,规范的,正常的

north [nɔːθ] n. ;a. 北,北方(的)
~ aisle 北侧通道
~ arrow 指北针
~ levation 北立面图
~ light cylindrical shell 北面采光圆筒形薄板
~ light roof 北面采光屋盖
~ porch 北侧门廊

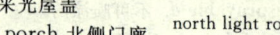
north light roof

northeast [ˌnɔːθˈiːst] n. ;a. 东北方(的)

northeastward [nɔːθˈiːstwəd] n. 东北方向 a. ;ad. 向东北

northern [ˈnɔːðən] a. 北方的,北部的

north-facing a. 朝北的,面北的
~ facade 建筑物向北的正面
~ wall 北墙,朝北的墙
~ window 北窗,朝北的窗户

northwest [ˌnɔːθˈwest] n. ;a. 西北方(的)

nose [nəuz] n. 鼻端,凸头,墩端,前缘,艇头,楼梯踏步突缘
~ in 倾斜
~ key 暗楔
~ section 前缘剖面
~ wedge 暗楔
non-pointed ~ 钝头部
parabolic ~ 抛物线型头部
rounded ~ 圆头
thin ~ pliers 扁嘴钳

nosing [ˈnəuziŋ] n. 突缘饰;踏步前缘
~ bead 突缘圆饰
~ strip 凸缘条
half-round ~ 半圆形突边

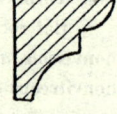
nosing strip

no-slump concrete 无坍落度混凝土,干硬性混凝土

notation [nəuˈteiʃən] n. 记号,标号;符号标记法
drawing ~ 图形标志

notch [nɔtʃ] n. 凹,凹口,缺口;豁[切,槽]口;峡谷;标记; v. 刻凹槽;放入凹槽
~ board 楼梯搁板,凹槽板
~ connection 凹槽连接

~ groove 凹槽
~ joist 凹形桁条
~ pin 缺口销
adjustment ~ 安装[调整]标记
tuyere ~ 风口孔
notchboard *n.* 凹板,(楼梯的)搁板
notched [nɔtʃt] *a.* 切口的,凹槽的;有切口的,凹槽的
~ beam 开槽梁
~ brick 开槽砖
~ chisel 带槽凿子
~ column 有槽柱,阶形柱
~ flange 有槽突缘
~ joint 插榫接合
~ molding 凹口线脚,齿形线脚
~ sill 齿槛,开槽槛
~ talcon 齿榫
notcher *n.* 开槽机,刻痕器
notching *n.* 切口,槽口
~ curve 下凹曲线
~ joint 槽口缝
oblique ~ 开斜槽
single ~ 单面刻槽

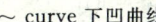

notching

note [nəut] *n.* 笔记,摘录,注解;票据,货币 *v.* 加记号;注意
~s on drawings 图注
notice [ˈnəutis] *n.* 注意,标记,告示,预告 *v.* 注意,通知;评价
~ board 布告板
noumenal [ˈnɔuminəl] *a.* 实体的,本体的
nouveau [ˈnuːvəu] 【法】新艺术派形式
novaculite *n.* 均密石英岩
Novalite *n.* 一种铜铝合金
novel [ˈnɔvəl] *a.* 新奇的,珍奇的 *n.* 小说
~ design 新颖设计
novelty [ˈnɔvəti] *n.* 新奇,新颖,新事物
~ siding 企口壁板
novice [ˈnɔvis] *n.* 新手,生手
novolac = novolak [ˈnəuvəlæk] *n.* 酚醛清漆
~ expoxy 酚醛环氧树脂
noxious [ˈnɔkʃəs] *a.* 有害的,有毒的
nuance [njuːˈɑːns] *n.* 色调、色彩等微细的差别,色调的变化
nude [njuːd] *a.* 裸体的,肉色的,光秃的
~ wire 裸线
nudity [ˈnjuːdity] *n.* 裸露,露出墙体画
nugatory [ˈnjuːgətəri] *a.* 无价值的,无效的,没有的
Nu-gild *n.* 装饰用黄铜
number [ˈnʌmbə] *n.* 数(目);数字;数值;号码,第……号 *v.* 计算,数,编号
~ of stories 楼层数
drawing ~ 图号
model ~ 型号
numberless *a.* 无数的,不可胜数的;没号码的,无号数的
numeral [ˈnjuːmərəl] *n.* 数字
~ dial lock 号码锁
numeroscope [ˈnjuːmərəskəup] *n.* 示数器
numerous [ˈnjuːmərəs] *a.* 为数公多的,大批的,多次的,许多的,无数的
nunatak [ˈnʌnətæk] *n.* 受冰河包围的丘陵或石山
nun buoy *n.* 纺锤形浮标
nunnery [ˈnʌnəri] *n.* (*pl.* -neries) 修道院,尼庵
~ church 修道院
nurag(h)e *n.* 撒丁岛史前石建筑
Nural *n.* 努拉尔铝合金
Nuremberg gold 铜铝金装饰用合金(纽伦堡合金)
nursing [ˈnəːsiŋ] *a.* 养育的
~ facilities 保育设施
nut [nʌt] *n.* 螺母,螺帽,螺套;坚果 *v.* 上紧螺钉
~ anchorage 螺帽锚固
~ bolt 带帽螺栓
~ cap 螺帽盖

nurag(h)e

nutschfilter

~ collar 螺帽垫圈
~ -lock washer 止松[锁紧]垫圈
~ oil 胡桃油
~ runner 螺母紧松器
~ wrench 螺母扳手
acorn ~ 螺母,螺帽
adjusting ~ 调整螺母
anchor (bolt) ~ 系紧螺母
anchor pin lock ~ 锁销防松螺母
arresting ~ 止动螺母
attaching ~ 配合螺母
back ~ 支承[限动]螺母
binding ~ 夹紧螺母
blocking ~ 防松螺母
cap ~ 螺母
capstan ~ 带孔螺母
captive ~ 外加螺母
castellated ~ 槽顶螺母
castle ~ 槽形螺母
check ~ 防松螺母
circular ~ 圆螺母
clamp ~ 花螺母
collar ~ 环形螺母,凸缘螺母
connecting ~ 连接螺母
countersunk ~ 埋头螺母
coupling ~ 连接螺母
cover ~ 螺帽
dome ~ 圆盖螺母
double ~ 双螺母
eared ~ 元宝螺母
elastic stop ~ 弹性防松螺母
eye ~ 有眼螺母
feed screw ~ 丝杆螺母
finger ~ 指形螺母
fixing ~ 固定螺母
flange ~ 凸缘螺母
flush ~ 平顶螺母

fly ~ 蝶形[元宝]螺帽
four pin driven ~ 四锥孔螺母
full ~ 全高螺母
grip ~ 夹紧螺母
guiding ~ 导螺母
hex ~ 六角螺母
holding ~ 支承螺母
knurled ~ 滚花螺母
mother ~ 螺母
pull rod ~ 拉杆螺母
quick adjusting ~ 自动锁紧弹簧螺母
retaining ~ 扣紧螺母
right-and-left ~ 连接螺母
self-locking ~ 自锁螺母
slotted ~ 有槽螺母
stamped ~ 模压螺母
terminal ~ 接线柱螺母
thumb ~ 元宝螺帽
tommy bar ~ 带孔螺母
unit bolt ~ 螺栓螺母组合
wing ~ 元宝螺母

nutschfilter [ˈnʌtʃfiltə] *n.* 吸滤器
Nykrom *n.* 高强度低镍铬合金钢
Nylatron *n.* 石墨填充酰胺纤维
nylon [ˈnailən] *n.* 尼龙,耐纶
~ cloth 尼龙布
~ fabric 尼龙结构,尼龙编织物
~ filter antiscour system 尼龙滤层防冲系统
~ grille 尼龙格栅
~ rope 尼龙绳(索)
~ tube 尼龙管
~ wire 尼龙丝
~ woven fabric 尼龙织编物
~ yarn 尼龙纱
Nytron *n.* 碳氢化合物,硫酸钠清洁剂

O

oak [əuk] *n.* 栎,橡;栎木家具 *a.* 栎木的,栎木制的
~ block 橡木(垫)块
~ floor cover(ing) 橡木地板
~ panels 橡木镶板
~ parquet(ry) 橡木拼花地板
~ shingle 橡木屋顶板,橡木瓦
~ sill 橡木窗台板
~ slat fence 橡木条,栅栏[篱笆]
~ threshold 橡木槛
red ~ 红栎木

oaken ['əukən] *a.* 栎木制的
oakum ['əukəm] *n.* 麻絮[丝];填絮
obelisk ['ɔbilisk] *n.* 方尖石塔;方尖形柱;方尖碑形物
object ['ɔbdʒikt] *n.* 物件,物体;目标,目的,客观 *v.* 反对,抗议
~ cost 目标成本
cylindrical ~ 圆柱体

objectify [ɔb'dʒektifai] *vt.* 具体化,体现
objection [əb'dʒekʃən] *n.* 障碍;缺点;反对
objectionable [əb'dʒekʃənəbl] *a.* 不满意的,有异议的,不能采用的,不合适的
objective [əb'dʒektiv] *n.* 物镜;目标,目的 *a.* 客观的;目标的
oblate ['ɔbleit] *a.* 长圆的,扁圆的
oblateness *n.* 扁圆形;扁率
obligate ['ɔbligeit] *vt.* 使负责,使负义务;强迫 *a.* 受约束的;主要的,必要的
oblique [ə'bli:k] *vi.* 倾斜,歪曲 *a.* 斜的,斜交的
~ angle 斜角
~ -angled 斜角的;斜向的
~ arch 斜拱
~ bond 斜纹砌块

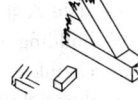
oblique butt joint

~ butt joint 斜对接
~ cocking 斜向翘起的
~ compression joint 斜接头
~ crossing 斜交,斜形交叉
~ cylinder 斜圆筒
~ dovetail 斜鸠尾
~ fillet weld 斜交角焊缝
~ grain 斜纹理
~ halving 斜嵌
~ joint 斜接;斜节理,斜接头
~ line 斜线
~ masonry bond 斜纹砌块
~ notching 开斜槽
~ offset 斜支距
~ profile 斜剖面
~ projection 斜投影
~ reflected lighting 斜反射照明法
~ scarf joint 斜嵌接,斜接榫
oblique scarf joint
~ scarf with wedge 楔形斜嵌
~ section 斜断面,斜截面
~ tabled scarf 斜叠嵌接
~ tenon 倾榫舌,斜榫舌
~ T-joint 斜接 T 形接头
~ view 斜视图

obliquity [ə'blikwiti] *n.* (*pl.* -ties)斜度;倾斜
obliteration [əb,litə'reiʃən] *n.* 磨损,涂去,清除
obliterator *n.* 涂抹器
oblong ['ɔblɔŋ] *n.* ; *a.* 长方形(的);长椭圆形(的)
~ dome 长椭圆形穹顶
~ ovate 长椭圆形
~ -shaped 长方形的
~ slot 长圆孔

obpyramidal [ˌɔbpiˈræmidəl] *a.* 倒金字塔形的
obpyriform [ɔˈpiərifɔ:m] *a.* 倒梨形的
obscure [əbˈskjuə] *a.* 含糊的;难解的;黑暗的 *vt.* 使暗,使不分明 *vi.* 隐藏
~d glass 毛玻璃,不透明玻璃
obsequent [ˈɔbsikwənt] *a.* 逆向的
observation [ˌɔbzəˈveiʃən] *n.* 观察;观测;观察报告
~ hole 观察[检查]孔
~ line 观测线
~ procedure 观测程序
~ room 观测室
~ storey (= ~ story (美)) 观测楼层,观测层
~ window 观察窗
obsidian [əbˈsidiən] *n.* 黑曜岩
obsolescence [ˌɔbsəˈlesns] *n.* 报废,过时,陈货,废弃
obsolescent [ˌɔbsəˈlesnt] *a.* 需废弃的
obsolete [ˈɔbsəli:t] *a.* 已废弃的;过时的
~ equipment 陈旧设备
obstacle [ˈɔbstəkl] *n.* 障碍,阻碍,妨害
obstreperous [əbˈstrepərəs] *a.* 无秩序的
obstruct [əbˈstrʌkt] *vt.* 阻碍,妨碍;塞,遮拦
obstruction [əbˈstrʌkʃən] *n.* 阻碍;闭塞;障碍物
~ -guard 排障器;护栏
obtrude [əbˈtru:d] *vt.* 强加;挤出;伸出,突出 *vi.* 闯入
obturate [ˈɔbtjuəreit] *vt.* 闭塞,紧塞……口
obtuse [əbˈtju:s] *a.* 钝的,钝角的,圆头的
~ angle 钝角
~ angle arch 钝角拱
~ arch 钝拱
~ quoin (砖墙)钝形对角,钝形楔块

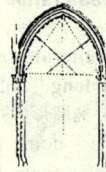
obtuse angle arch

obverse [ˈɔbvə:s] *n.* 正面,前面 *a.* 正向的;相对的;底部窄于顶部的
obversion [əbˈvə:ʃən] *n.* 转向;变换,折算
obviate [ˈɔbvieit] *vt.* 排除,避免,消除
obvious [ˈɔbviəs] *a.* 明显的,显著的
ocala *n.*【美】欧卡拉石灰石
occasional [əˈkeiʒənl] *a.* 临时的;偶然的
occlude [əˈklu:d] *vt.* 吸留,闭塞,堵塞,使光透不过
occlusion [əˈklu:ʒən] *n.* 闭塞,堵塞
occult [ɔˈkʌlt] *a.* 隐蔽的;秘密的;神秘的
~ing light 明暗相间灯,断续蔽光灯
occultation [ˌɔkʌlˈteiʃən] *n.* 掩蔽,隐藏
occupancy [ˈɔkjupənsi] *n.* (*pl.* occupancies)占用,占有,被占用的建筑物
~ per person in dwelling 居住密度
~ rate 居住率
~ standard 居住标准
occupant [ˈɔkjupənt] *n.* 占有人,居住人
occupation [ˌɔkjuˈpeiʃən] *n.* 职业,工作,占有,占有地
occupy [ˈɔkjupai] *vt.* 占有,充满,从事
occurrence [əˈkʌrəns] *n.* 发生,出现,存在,事件
random ~ 随机事件
ochre [ˈəukə] =ocher【美】*n.* 赭石;赭色,黄褐色
red ~ 赭石红
yellow ~ 赭石黄
ocpan *n.* 锡基白合金
octagon [ˈɔktəgən] *n.* 八角形,八边形,八角形建筑物 *a.* 八角形的,八边形的
~ column 八面柱
octagonal [ɔkˈtægənl] *a.* 八角形的,有八边的
~ aisle 八角耳堂
~ building 八角楼,八角形建筑物
~ chamber 八边形房间
~ dome 八角穹顶
~ keep 八角楼

~ lantern 八角形天窗
~ mosaic tile 八边形陶瓷锦砖
~ tower 八角塔
~ window 八角窗
octastyle ['ɔktəstail] *n.* 八柱式建筑 *a.* 八柱式的
~ temple 八柱式庙宇
octastylos [ˌɔktə'stailəs] *n.* 八柱式建筑物
octet [ɔk'tet] *n.* 八角体,八隅体
octobolite [ɔk'təubəlait] *n.* 辉石
octofoil *n.* 八瓣形,八叶饰
octopartite vault 八节头筒拱
oculus ['ɔkjuləs] *n.* (*pl.* -oculi)眼洞窗
odd [ɔd] *a.* 奇[单]数的
~ -pitch roof 不规则坡度屋顶
oddside *n.* 假箱,副箱;假型
~ board 模板
~ pattern 向外凸的半个型板

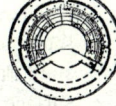

odeum

ode(i)on [əu'di:ɔn] *n.* 希腊和罗马式戏院,音乐厅
odeum [əu'di:əm] *n.* (*pl.* odeums, odea [əu'di:ə])(古希腊、罗马)奏乐堂;音乐厅
odontoid [ə'dɔntɔid] *a.* 齿形的
odorant ['əudərənt] *n.* 有气味物质 *a.* 有气味的
odorous ['əudərəs] *a.* 有气味的
odo(u)r ['əudə] *n.* 气味,香味,臭味
~ free 除臭
penetration ~ 刺激性气味
oecus *n.* 古希腊式房屋,古罗马式房屋中的主要起居室
oeil-de-boeuf [ˌə:jdə'bæf]【法】*n.* 圆窗,牛眼窗
off [ɔ:f] *a.* 断的,停的;休闲的,不工作的,不佳的 *ad.* 脱下,掉下,离开;停止 *prep.* 脱落,离开;少于,低于,在……之外
~ -colo(u)r 颜色不合格的

~ -cut 截片;边角料;下脚料;截片不合规格的
~ -design 偏离设计条件的,超出设计规定的
~ -gauge 不按规格的,不标准的
~ gauge plate 不合格板
~ grade 等外品,不合格品,次品
~ normal 不正常的,离位的
~ -rating 非标准条件,不正常条件
~ -shuttering concrete 混凝土表层处理
~ -size 不合尺寸的
~ -sorts 等外品
~ -specification 不合格
~ -the-shelf 流行的,现用的
~ -white 米黄色,灰白色
laying ~ 放样
offal ['ɔfəl] *n.* 碎屑,垃圾;废物;次料,下脚料
~ wood 次木料
offer ['ɔfə] *n.*;*v.* 提出,提供;奉献,贡献;报价,出售
office ['ɔfis] *n.* 办公室[处],事务所
~ area 办公区
~ block 办公楼,办事处集中的街区
~ building 办公楼,写字楼
~ entrance 办公楼入口处
~ floor 办公楼层
~ hut 临时木板办公房
~ illumination 办公室照明
~ landscape 景观式办公室布置
~ layout 办公室室内设计
~ lighting 办公室照明
~ partition (wall) 办公室隔墙
~ portion 办公楼侧壁
~ room 办公室
~ skyscraper 摩天办公大楼
~ storey (= ~ story (美))办公楼层
~ tower 办公大厦
~ wing 办公室侧厅
field ~ 工地办公室,现场办公室
front ~ 旅馆前厅办公室

information ~ 问讯处
offset ['ɔfset] *n.* 支距,垂距;壁上凸出部分,水平断错;支脉;阴阳榫接缝;迂回管 *v.* 抵消,补偿,偏置;形成分支
~ angle 偏斜角
~ course 阶宽墙基,突层
~ distance 支距,偏距
~ drilling 边界钻孔,补气钻孔
~ footing 宽阶底脚,大方基脚
~ hinge 长翼铰链
~ joints (砖墙) 错缝
~ -mounted 错开装置的
~ screw 弯头螺钉旋具
~ screwdriver 弯头螺钉旋具
offshore [ˈɔːfˈʃɔː] *n.* 海滨;*a.* 离开海岸的,向海面的 *ad.* 向海面
ogee ['əudʒiː] *n.* 双弯曲线,S形曲线;葱形饰

ogee

~ arch 葱形拱,尖拱
~ curve 双弯曲线,S形曲线
~ doorway 葱形门道
~ moulding 双弯线条,波状花边
~ plane 凹槽刨,浅脚刨
~ roof S形屋顶
~ roof gutter S形屋顶雨水沟
~ washer S形边垫片
~ window 葱头形窗
ogival ['əuˈdʒaivəl] *a.* 尖顶式的
~ arch 葱形拱,尖拱
~ dome 尖顶穹窿
ogive ['əudʒaiv] *n.* 尖形穹窿;葱形饰;卵形线;尖顶部;分布曲线,累积频率曲线
tangent ~ 正切卵形线
oil [ɔil] *n.* 油,石油 *v.* 浇油,涂油,加油;润滑
~ alkyd 油改性醇酸树脂
~ alkyd paint 油醇酸树脂漆
~ base paint 油基漆
~ -bound distemper 油料色粉涂饰
~ cloth 油布

~ -colo(u)r 油画颜料,油画
~ enamel 油性瓷漆[搪瓷]
~ -fast 防油的
~ fastness 抗[耐]油性
~ gloss 油光泽
~ gloss paint 油光调合漆
~ gun 喷漆枪
~ -impregnated wood 油浸过的木材
~ -in-water emulsion 水包油型乳剂
~ -modified alkyd resin 油改性树脂
~ paint 油画颜料,油漆
~ painting 油画,油画术
~ paper 油纸
~ primer 油质底漆,油质底层涂料
~ -proof rendering 防油底涂
~ putty 油灰,油质腻子
~ resistance 耐[抗]油性
~ scumble glaze 油基薄涂釉面
~ stain 油斑,透明色料
~ surface 油基面涂料
~ varnish 油基清漆
~ white 混合白颜料
Chinese (wood) ~ 桐油
concrete form ~ 混凝土模板用油
core ~ 泥心油,漆匠用油
dust laying ~ 防尘油
linoleum ~ 油地毡用油
rustproof ~ 防锈油,抗腐蚀油
rust resisting ~ 防锈油
tung ~ 桐油
turpentine ~ 松节油
wood ~ 桐油
oil-base(d) *a.* 油基的
~ mastic 油基胶粘剂,油基胶粘水泥
~ paint 油基漆
oiled [ɔiled] *a.* 涂油的,浇油的,浇沥青的
~ linen 油布
oktastylos [ˌɔktəˈstailɔs] *n.* (古希腊神庙的) 八柱式
old [əuld] *a.* 年老的;年久[古老]的;陈旧[过时]的
~ civilization 古代文明

~ dwelling [unit] 老年人住宅
~ -hand 熟练工人,老手
~ people's community centre 老年人活动中心
~ people's home 老年人之家
~ silver 旧银器色
~ stock 陈货
oleate [ˈəulieit] *n.* 油酸盐,油酸脂
oleo-alkyd resin 含油醇酸树脂
oleoresin [ˌəuliəuˈrezin] *n.* 含油树脂
oleoresinous varnish 油树脂清漆
olive [ˈɔliv] *n.* 橄榄；橄榄色,淡绿 *a.* 黄绿色的
~ acanthus 橄榄叶形装饰
~ black 黑橄榄色
~ brown 黄褐色的
~ building 橄榄形大楼
~ drab 草绿色,草黄色
~ green 橄榄绿
~ green clear 淡橄榄绿
~ green deep 深橄榄绿
~ knuckle butts 松轴颌,橄榄形肘节铰链
~ knuckle hinge 橄榄形肘节铰链

olive knuckle hinge

olivine [ˌɔliˈviːn] *n.* 橄榄石
~ brick 橄榄石砖
~ rock 纯橄榄石
ollite [ˈɔlait] *n.* 滑石
omnifarious [ˌɔmniˈfɛəriəs] *a.* 各种各样的,五花八门的
omnigraph *n.* 缩图器
omnium [ˈɔmniəm] *n.* 总额,全部
~ -gatherum 混合物(剂,气)
omphalos [ˈɔmfələs] *n.* (*pl.* omphali) 中心点,中枢,核,脐；(古希腊)盾中央的浮雕饰
on [ɔn] *ad.* 接通,开动 *prep.* 在……上,关于
~ -and off switch 通-断开关
~ end laying 竖砌,立砌陡砌
~ ga(u)ge 标准的,合格的
~ -off control 开关控制,离合控制
~ request 承索
~ shade 近似颜色
~ test 合格
once-through flowsheet 一次循环工艺图
oncoming [ˈɔnˌkʌmiŋ] *n.* ;*a.* 接近,来临；新兴起(的)
one [wʌn] *n.* ;*a.* 单,一
~ -and-half brick wall 一砖半墙
~ balcony type auditorium 单眺台式观众厅
~ -bolt fastening 单螺栓固定
~ brick wall 单砖墙
~ -centred arch 单心拱,圆弧拱
~ coat work 单层粉刷
~ -component primer 单一成分底漆
~ DK 由一居室和餐室、厨房组成的住户
~ -family dwelling 私人住宅,单家独户
~ -hinged arch 单铰拱
~ layer 单层
~ -leaf shutterdoor 单扇卷帘门
~ -leaf sliding shutter door 单扇推拉式卷帘门
~ -off design 特种[专门]建筑物设计
~ panel 单门板
~ -piece panel 单镶板
~ -pin(ned) arch(ed) girder 单铰拱梁
~ point perspective 一点透视
~ -room system 每户以一大室作灵活分间的住宅体系
~ -row brick-on-edge arch 单排砖缘拱
~ -sach(ed) window 单框格窗
~ sole course 单层
~ storey house 单层房屋,平房
~ -tier wall 单层墙
~ -turn stairs 单转向楼梯

one-bay
~ -way slab 单向板
~ -wing(ed) block 单翼楼
one-bay *a.* 单跨的
~ beam 单跨梁
~ slab 单跨板
one-bed *a.* 单人的
~ (guest) room 单人(客)房
~ dwelling unit 单人居住单元
~ flat 单人公寓
one-coat *a.* 单层的,表层的
~ plaster(ed) ceiling 表层涂有灰泥的顶棚
~ work 单层粉刷
one-floor *a.* 一层的,平房的
~ house 平房
one-pack *a.* 单一成分的,纯的
~ (bonding) adhesive 单成分粘合剂
~ cement 纯水泥
~ coating 单涂层
one-pipe *a.* 单管的
~ plumbing 单管排气
one-quarter *n.* 四分之一
~ brick bond 1/4 砖砌合
~ clay brick 四分之一砖
~ bend 90°弯头
one-shot ['wʌnʃɔt] *n.*;*a.* 一次使用(的),只有一次(的)
~ operation 单步操作
~ type 单层式表面处理
one-side *a.* 单侧的,一面的
~ butt welding 单面对焊
~ formwork 单面模板
~ polishing 单面抛光
one-stor(e)y 一层(的)
~ house 平房
one-way ['wʌn'wei] *n.* 单向的;单方面的
~ arch 单向拱
~ slab 单向板
onion dome 圆顶,洋葱形圆屋顶
onlay ['ɔnlei] *n.* 盖板;接合板;贴胶;修整,修饰 *vt.* [ɔn'lei] 覆盖

onsite ['ɔn'sait] *a.* 现场的,就地的
~ construction 就地建造
~ mixing 现场拌合
~ prefabrication 工地预装配
ontariolite *n.* 中柱石
onyx ['ɔniks] *n.* 石华,缟[条纹]玛瑙 *a.* 黑的,乌黑的
~ marble 黑大理石,条纹大理石
oolite ['əuəlait] *n.* 【美】鲕状岩,鱼卵石
oolitic [ˌəuə'litik] *a.* 鲕状岩的,鱼卵状的
~ texture 卵状结构
opacifier [əu'pæsifaiə] *n.* 不透明剂,遮光剂
opacity [əu'pæsiti] *n.* (*pl.* -opacities) 不透光;不透明度;不透明体;浑浊度;阻光度;不反光
opal ['əupəl] *n.* 蛋白石 *a.* 乳白的
~ bulb 乳白灯泡
~ glass 乳白玻璃
opalescence [ˌəupə'lesns] *n.* 乳光;乳白色
opalescent [ˌəupə'lesnt] *a.* 乳色的,乳光的
~ glass 乳白色玻璃
opalesque [ˌəupə'lesk] *a.* 似蛋白石的;发乳白光的
opaline ['əupəli:n] *n.* 乳白玻璃 *a.* 乳色的,乳光的
~ glass 乳光玻璃
opalite *n.* 蛋壳釉面[材料]
opalize ['əupəlaiz] *vt.* 使成乳色
opaque [əu'peik] *a.* 不透明的;不透光的;暗的
~ coat 不透明层
~ colo(u)r 不透明色,暗色
~ finish 无光饰面
~ glass 不透明玻璃
~ glaze 不透明釉面
~ glazed coat(ing) 不透明釉层
~ paint 不透明油漆
open ['əupən] *n.* 空地;户外 *v.* 开;断;订约 *a.* 开放的,开的;无遮盖的,开阔

的;多孔的,有孔隙的
~ arc lamp 敞式弧光灯
~ arch 明拱
~ beam construction 露梁结构
~ boarding 开缝铺板法
~ butt joint 明对接,开口对接
~ cornice 敞檐(外露桁架人字木端部的)
~ -deck garage 无侧墙式车库
~ dovetail 明鸠尾榫
~ dovetailing 明鸠尾榫接合

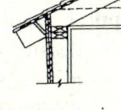
open cornice

~ fish ladder 开敞式鱼梯
~ floor 明露搁栅楼板,敞肋楼板
~ -flume 明槽
~ fold 敞开褶皱
~ frame 空腹架构
~ frame girder 空腹梁
~ grain 木材软纹
~ ground 天然地面,无保护层地面
~ (-)joint 开缝接头,明缝
~ joist 明露搁栅
~ layout 自由式设计
~ light 可开可关的窗户
~ line 开路线;明缝
~ newel stairs 露井楼梯
~ outside court type 外院式平面布置
~ peristyle court 周围列柱的院场
~ -plan system 开敞式平面布置方法
~ rafter 露明椽
~ riser stairs 露明楼梯
~ roof 无顶棚屋顶
~ sand 松砂,多孔隙砂
~ shed 敞开式棚架
~ sided building 侧敞开式大楼
~ slating 疏铺石板
~ spandrel arch 空腹拱,敞肩拱
~ stack 开架式书库
~ stair 露天梯
~ steel flooring 露钢梁楼板
~ T joint 开口T形接头

~ timber roof 露梁屋顶
~ -top mixer 敞顶拌和机
~ trellis bond 格构式砌合
~ trellis wall 格构式砌墙
~ truss steel joist 露桁楼板搁栅
~ valley 屋面天沟
~ (-)web truss 空腹桁架
~ (-)web girder 空腹大梁
~ window 活动窗,可开可关的窗户
day lighting ~ 采光口
proscenium ~ 舞台台口
openable [ˈəupənəbl] *a.* 能开的
open-air [ˈəupənˈɛə] *a.* 户外的,露天的,野外的
~ basilica 露天式大会堂
~ cinema 露天电影院
~ corridors 露天回廊
~ museum 露天博物馆
open-end(ed) [ˈəupənˈend(id)] *a.* 无限制的,开口的,无底的
~ block H 形空心砌块
~ bucket 活底斗,活底吊桶
~ hangar 敞口木库
~ shed 敞口木棚
~ wrench 开口扳手
opening [ˈəupniŋ] *n.* 开放,开始;孔,口,穴,隙 *a.* 开放的,开始的
~ area 开口面积,空洞面积
~ for lighting 采光口
~ in (门、窗)内开
~ of bids 开标
~ of tender 开标
~ out (门、窗)对开
adit ~ 坑道口,出入口,平洞洞口
admission ~ 进口
air ~ 通风洞
air induction ~ 进气口
emergency ~ 安全出口
exit ~ 太平门,屋顶窗
manhole ~ 人孔口
proscenium ~ 舞台门
return ~ 回风洞

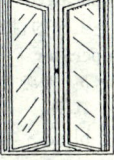
opening out

openwork [ˈəupnwəːk] *n.* 透雕细工；露天开采 *a.* 有网状小孔的
~ gable 漏空山墙
~ gablet 漏空小山墙
~ rosette 漏空圆花窗，透雕圆花饰
~ tracery 漏空窗花格
~ed balustrade 格子栏杆
opera [ˈɔpərə] *n.* 歌剧，歌剧院
operant [ˈɔpərənt] *n.* 操作人员，工作人员 *a.* 工作的，发生作用的
operate [ˈɔpəreit] *v.* 操作，工作，经营
operation [ˌɔpəˈreiʃən] *n.* 工序；操作；运行；营业
~ capacity 工作量
~ cycle 操作周期
~ requirements 操作要求
~ rule 操作规程
~ schedule 作业计划表
~ specification 操作规程
combination ~ 组合操作
cycle ~ 循环作业
finishing ~ 精整理[修饰]
safe ~ 安全操作
site ~ 现场施工
working ~ 工序
operational [ˌɔpəˈreiʃənəl] *a.* 可使用的；操作上的；业务上的；运算上的
~ checkout 作业检查
~ instructions 操作规程
operative [ˈɔpərətiv] *n.* 职工，技工 *a.* 工作的，操作的；运转的；有效力的
operator [ˈɔpəreitə] *n.* 作业员，操作员，驾驶员，报务员；运算符，运算数，算子
blind ~ 百叶窗片控制器
operatory *n.* 工作室
opisthodomos [ˌɔpisˈθɔdəməs] *n.* 后厅（古典寺庙后进的小室）
opportunity [ˌɔpəˈtjuːniti] *n.* 机会，时机
oppose [əˈpəuz] *v.* 对抗，反对；相对，相反，使……对向
~d-blade damper 反向板条缓冲器

opposite [ˈɔpəzit] *n.*；*a.* 相对，相反，对向 *prep.* 在……对面
~ direction 对向，相反方向
~ hand 对面，另方面，反面
~ joint 对缝
~ side 对边
opposition [ˌɔpəˈziʃən] *n.* 对抗，对立，对向；障碍物
optical [ˈɔptikəl] *a.* 光(学)的，视觉的
~ density 不透明度
~ glass 光学玻璃
~ illusion 视错觉
optimal [ˈɔptiməl] *a.* 最适宜的，最优的，最佳的
~ choice 最优选择
optimality *n.* 最优性
optimization [ˌɔptimaiˈzeiʃən] *n.* 适宜，合宜；优选法，最佳化
~ calculation 优化计算
structural ~ 结构优化（设计）
optimize [ˈɔptimaiz] *vt.* 最佳化，最优化
~ing decision 最佳决策
optimum [ˈɔptiməm] *n.*；*a.* 最佳，最适宜，最适度
~ condition 最佳条件，最适宜情况
~ design 最优设计
~ illumination 最佳照度
~ mix design 最佳配合比设计
~ moisture content 最佳含水量
~ seeking method 优选法
~ temperature 最佳温度
option [ˈɔpʃən] *n.* 选择，任意，随意
optional [ˈɔpʃnəl] *a.* 任[随]意的
opus [ˈəupəs]【拉】*n.* 著作，作品
~ alexandrinum 大理石块铺面
~ antiquum 有混凝土填心的块石工，粗毛石砌体
~ emplectum 空斗石墙
~ incertum 有混凝土填心的块石工
~ isodomum of granite blocks 大理石块端砌
~ latericium 嵌砖混凝土墙

~ -listatum 砖石砌墙
~ mixtum 砖面圬工墙
~ musivum 墙面镶嵌装饰
~ quadratum 方石筑墙
~ reticulatum 带有水平缝的方格混凝土墙
~ rusticum 粗面石工
~ sectile 陶瓷锦砖
~ signinum 有几何图形的罗马式陶瓷锦砖
~ spicatum 人字形嵌砖细工
~ tesselatum 嵌云石块细工
~ testaceum 混凝土墙的三角形石片面饰

opus spicatum

opus testaceum

orange [ˈɔrindʒ] *n.*; *a.* 橙(的);橙色(的)
~ -brown 橙褐色
~ -chestnut 橙色
~ clear 浅橙黄色
~ lac 虫漆片
~ light 淡橙黄色
~ pale 淡白橙色
~ -peel effect 橘皮纹
~ red 橙红色的
~ yellow 橙黄色的
china ~ 橙色
oranium bronze 铝青铜
oratory *n.* (*pl.* -ries)(宗教)祷告所,小礼拜堂
orb [ɔ:b] *n.* 球,球体;环;圆;球形饰;轨道
orbicular [ɔ:ˈbikjulə] *a.* 球形构造的;正圆形的
order [ˈɔ:də] *n.* 次序;阶;定货;命令;柱型[式]
~ goods 订货
architectural ~ 建筑柱式
attic ~ 顶层角柱式
columnar ~ 柱形
composite ~ 复合柱式
Greek Doric ~ 古希腊多立克式柱式
Roman Corinthian ~ 罗马科林斯柱式
Roman ~s 罗马柱式
Roman Doric ~ 罗马多立克式柱式
Roman Ionic ~ 罗马爱奥尼柱式
Tuscan ~ (古罗马)塔司干柱式
ordinal [ˈɔ:dinl] *n.* 序数 *a.* 顺序的,次序的
ordinary [ˈɔ:dnri] *a.* 通常的,普通的,正常的;编制内的
~ form (work) 固定模板,普通模板
~ lime mortar 普通灰浆
~ rubble masonry 粗石墙,毛石圬工
~ structural concrete 普通结构混凝土
~ structural steel 结构用普通钢
Oregon pine 美国黄松,花旗松
organic [ˈɔ:gænik] *n.* 有机物 *a.* 有机体的,器官的;生来的,固有的
~ anticorrosion agent 有机系防腐剂
~ architecture 有机建筑
~ binding agent 有机粘合剂
~ corrosion inhibitor 有机防腐剂
~ glass 有机玻璃
~ pigment 有机颜料
~ space 有机空间
organism [ˈɔ:gənizəm] *n.* 生物体,有机体;组织,机关
organization [ˌɔ:gənaiˈzeiʃən] *n.* 组织,构造;机构,结构
organogel *n.* 有机凝胶
organolite *n.* 离子交换树脂
oricycle *n.* 极限圆
oriel [ˈɔ:riəl] *n.* 凸肚窗
~ window 凸肚窗
orient [ˈɔ:riənt] *v.* 定向,定方位;使……向东

oriental [ˌɔːriˈentl] *a.* 东方的;优质的,珍贵的
~ alabaster 条带状大理岩
~ architecture 东方式建筑
~ dome 东方式圆顶
~ metal 沥青罩面防水钢板

Orientalism [ˌɔːriˈentəlizəm] *n.* 东方民族特征,东方风格

orientation [ˌɔːrienˈteiʃən] *n.* 定向,定位;方位,向东方
~ diagram 方位[定向]图
~ of building 房屋方位,房屋定向
disordered ~ 不规则取向;无规则排列
preferred ~ 择优[最佳]取向

orifice [ˈɔrifis] *n.* 孔口;管口;孔板;隔板
circular ~ 圆隔板
rectangular ~ 矩形孔口
round ~ 圆孔
slit ~ 裂缝,槽口
square ~ 方形口

origin [ˈɔridʒin] *n.* 起源,来源;成因;原点,起点
~ cohesion 初始凝聚[粘结]力
noise ~ 噪声源

original [əˈridʒənl] *n.* 原型,原物 *a.* 原始[状,生]的
~ asphalt 天然沥青,原生沥青
~ bed 原河床
~ cleavage 原生劈理
~ cohesion 天然凝聚力
~ design 初步设计,原始设计
~ drawing 底图
~ form 原型
~ form survey 现场测量
~ interstice 原始缝隙,原生间隙
~ map 原图
~ position 初始位置

originality [əˌridʒiˈnæliti] *n.* 创造力,独创性;珍品

originate [əˈridʒineit] *v.* 开创,开始,发端

origination [əˌridʒiˈneiʃən] *n.* 起点;起因;创办

O-ring *n.* O形圈,橡皮圈
~ packing O型环密封件,橡皮圈

orle *n.* 平橡

orlon [ˈɔːlɔn] *n.* 腈纶,奥纶

ormolu [ˈɔːməluː] *n.* 锌锡铜合金;装饰用的仿金铜箔;镀金物

ornament [ˈɔːnəmənt] *n.* 装饰物,装饰品 [ˈɔːnəment] *vt.* 装饰
~ed architecture 盛饰建筑
~ed rafter end 装饰椽头
Assyrian ~ 亚西利亚装饰
capital ~ 柱头饰
flax ~ 麻叶形纹饰
garden ~ 庭园小品点缀
Honeysuckle ~ 忍冬卷草纹,忍冬饰
radiator ~ 散热器饰件
relief ~ 浮雕装饰
ridge ~ 脊饰
side ~ 侧面装饰品
stone ~ 点缀石
window ~s 窗饰

ornamental [ˌɔːnəˈmentl] *n.* 装饰品 *a.* 装饰的,观赏的
~ acoustical gypsum waffle slab 装饰性音响石膏镶板
~ aluminium 装饰铝
~ arch 装饰拱,拱形饰
~ architecture 华丽建筑
~ archivolt 拱门饰,拱门墙侧装饰线条
~ area 装饰区
~ art 装潢艺术
~ artificial stone 人工点缀石
~ band 花边嵌条,扁带饰
~ barrel vault 穹形饰
~ board 装饰板
~ bond 装潢砌筑
~ bracket 装饰墙基
~ brick 装饰砖,釉面砖
~ brickwork 图案砌筑
~ bronze 装饰用铜
~ cast block 装饰用预制块
~ ceiling 顶棚装饰

~ coat(ing) 装饰面层
~ column 装饰柱
~ concrete 装饰混凝土
~ design 装饰图案设计
~ door 装饰门
~ finish 花样修饰
~ fittings 金属装饰品
~ floor(ing) 装潢地板
~ foil 装饰箔
~ form 装潢式样
~ gable 装饰山墙
~ gilding 装饰性镀金
~ glass 装饰玻璃
~ grille 铁花格,格栅饰
~ hardware 五金装饰品
~ iron 装饰用铁件
~ joint 装饰缝
~ lake 装饰染料
~ light fitting (英)装饰灯具
~ margin 装饰边缘,镶边
~ motif 装饰花纹,花纹饰
~ moulding 艺术造型
~ painting 装饰性绘画
~ partition 装饰隔断
~ plant 观赏植物
~ plastic slab 装饰用塑料板
~ relief (混凝土表面)装饰花纹
~ sheet 装饰板
~ steel bar (围墙用)饰花钢条
~ surface 饰面
~ wall 装饰墙
~ work 装饰工作
Assyrian ~ 亚西利亚装饰
beak-head ~ 鸟嘴头饰
bird's beak ~ 外窗台装饰,披水板装饰
dog tooth ~ 犬齿饰
drop ~ 吊悬装饰
geometrical ~ 几何形装饰
hatched ~ 绫纹装饰
national ~ 民族图案,民族装饰
ornamentalism *n.* 装饰派

ornamentalist *n.* 装饰家
ornamentation [ˌɔːnəmenˈteiʃən] *n.* 装饰(术);装饰品
ornate [ɔːˈneit] *a.* (装饰)华丽的,雕琢过的
oroide [ˈəurəuaid] *n.* 铜锌锡合金
orthochromatic plate 正色片
orthoclase [ˈɔːθəkleis] *n.* 正长石
 ~ porphyry 正长斑岩
orthodox [ˈɔːθədɔks] *a.* 正统的,传统的
 ~ construction method 传统建筑法
 ~ material 正常材料
orthodrome *n.* 大圆弧,大圆圈线
orthogon [ˈɔːθəgən] *n.* 矩形
orthogonal [ɔːˈθəgənl] *a.* 正交的,直交的,矩形的
 ~ anisotropic plate 正交各向异性板
 ~ joint 横节理,正交接合
orthograph [ˈɔːθəgrɑːf] *n.* 正投影图
orthographic(al) [ˌɔːθəˈgræfik(əl)] *a.* 正交的,直交的;用直线投射的
 ~ projection 正射投影,平行投影
orthometric drawing 正视画法
orthonormal [ˌɔːθəˈnɔːməl] *a.* 正规化的,标准化的;正交的
orthophyre [ˈɔːθəfaiə] *n.* 正长斑岩
orthostat 丁边砌法
orthostyle *n.* 直线形列柱式建筑,柱廊式建筑
oscillate [ˈɔsileit] *v.* 摆动,振动,振荡
oscillating *a.* 摆动的,振动的
 ~ agitator 摇摆式拌和机
 ~ insulation 隔振,防振
oscillation [ˌɔsiˈleiʃən] *n.* 摆动,振动振荡
osier [ˈəuʒə] *n.* 柳树,柳条,柳枝
osmium [ˈɔsmiəm] *n.* 锇(Os)
 ~ lamp 锇丝灯
Osmosalts [ˈɔzməsɔːlts] *n.* (=Osmosar)渗透盐剂(一种木材防腐剂)
osmosis = osmose [ɔzˈməusis] *n.* 渗透(作用)
osmotic [ɔzˈmɔtik] *a.* 渗透的
 ~ membrane 渗透隔膜

osram lamp 钨丝灯
ossein ['ɔsiin] *n.* 骨胶原；生胶质
oster chaser *n.* 管子板牙，管子丝口扳钳
otter board 网板
Ottoman architecture 土耳其式建筑
O-type dowel 环型暗销[榫]，圆形齿环
out-and-in bond 凹凸砌合
outboard ['autbɔːd] *n.* 外侧 *a.* 外侧的
outbond ['autbɔnd] *a.* 外砌的，横叠式的，顺砖砌合的
outbound ['autbaund] *a.* 往外地的；对墙面纵砌的
outbreak ['autbreik] *n.*；[aut'breik] *v.* 破裂，断裂；流行；爆发
outbuilding ['aut͵bildiŋ] *n.* 外围建筑，附属建筑，外屋
outcrop ['autkrɔp] *n.*；[aut'krɔp] *v.* 露头，露出
outcut *n.* 切口
outdated [aut'deitid] *a.* 过时的，老式的，不流行的
　~ specification 旧规范
outdiffusion *n.* 外扩散
outdoor ['autdɔː] *a.* 户外的，野外的
　~ architecture 室外建筑
　~ corrider 室外阳台
　~ lighting 室外照明
　~ opening 采光口
　~ paint 室外油漆
　~ rail 外门围栏，门轨
　~ room 户外房间(露台，帐篷等)
　~ theater 露天剧场
　~ varnish 室外用清漆
outer ['autə] *a.* 外(面)的，外部的
　~ aisle 外侧廊
　~ belt rail 窗台外镶条
　~ coating 表面层，面层，外部涂层
　~ column 外柱，后柱
　~ deck sash 外顶栅窗
　~ diameter 外径
　~ door 外门
　~ garden 外苑，外花园，前园
　~ inspection 外观检查
　~ race 外座圈，外环
　~ reveal (窗)外侧壁
　~ shoe 外支块，外托板
　~ string 楼梯外侧斜梁
outermost ['autəməust] *a.* 最外方的，最远的
outfire *n.* 灭火
outfit ['autfit] *n.*；*v.* 装配，设备，装备，配备
　repair ~ 修理工具
　tool ~ 成套工具
outgas [aut'gæs] *vt.* 除气
outgate ['autgeit] *n.* 出口
outgoing ['autgəuiŋ] *n.*；*a.* 外出，流出，排出；高程；(*pl.*)开支
　~ air 排气
outhouse ['authaus] *n.* 外屋，附属小屋，(户外)厕所
outlay ['autlei] *n.* 经费，费用，支出
outlet ['autlelt] *n.* 出口，排水口；出口管
　~ box 接线盒
　~ conduit 出水管道
　~ portal 出口门
　~ structure 泄水建筑物
　louvered air ~ 透气百叶窗
　roof ~ 屋面排水口
outline ['autlain] *n.* 轮廓，略图；外形线 *vt.* 画轮廓，描略图
　~ drawing 略图
　~ map 略图
　~ sketch 略图
　front ~ 正[前]视图
outlook ['autluk] *n.* 景色，形势；望楼
outlying ['aut͵laiiŋ] *a.* 在外的
out-of-pocket (expenses) *a.* 实际花费，现款支付的
out-of-position *a.* 不在适当位置
out-of-repair *a.* 失修
out-of-shape *a.* 变形的，走样的
out-of-square *a.* 倾斜的，歪的
out-opening-door 向外回旋门
outporch *n.* 外门廊，外走廊
outrigger ['aut͵rigə] *n.* 悬臂梁，斜撑；

平衡器;外伸支架
~ base 平衡板,悬臂板
~ beam 挑出梁,悬臂梁
~ scaffolds 挑出脚手架
outshot ['aut-ʃɔt] = outshut *n.* 侧屋,房屋之一翼;凸出部分
outside ['aut'said] *n.;a.;ad.* 在外,外面,外界
~ architrave 外部框缘
~ base 外墙底板
~ caliper 外卡钳
~ casing 窗外框
~ cellar steps 户外地下室阶梯
~ coating 外涂层,外部涂刷
~ column 外柱
~ corner molding 外角线脚
~ diameter 外径
~ dimension 外包尺寸
~ drawing 外观图,外形图
~ face 外部立面
~ finish 屋外整修,外饰面
~ frame 外门框
~ lighting 室外照明
~ lining 外衬砌
~ measurement 外包[外侧]尺寸
~ micrometer 外径千分尺
~ plank 曲面板,(椅子的)背板
~ radius 外半径
~ shutter 外开百叶窗
~ stair 室外楼梯
~ stile 门窗边梃
~ string 楼梯外栅栏
~ surface 木材离心面,边材面
~ view 外观,外形图
~ widening 外侧加宽
~ work 室外工作

outside corner molding

outsize ['aut-saiz] *n.;a.* 特大;缺尺寸(的),非标准尺寸(的)
outskirts ['aut-skəːts] *n.* 郊外,外围,外缘

outspread ['aut'spred] *n.;a.* 扩张,展开,散布
outstanding [aut'stændiŋ] *a.* 突出的;未解决的;未付款的
~ feature 特色,突出的特点
~ flange 突缘,伸出肢
outstep [aut'step] *vt.* 超过,越过
~ boring 边缘镗孔
outstretch [aut'stretʃ] *vt.* 扩张,伸开,伸长
out-swinging *a.* 外旋转的
~ casement 外开式窗扇
~ side hung casement 边轴外旋转门式窗
~ window 外推窗
top-hinged ~ 上悬外撑式(窗)
outtake ['autteik] *n.* 通气管;通风道;烟道
outward ['autwəd] *n.;a.* 向外(的),外部(的)
~ bound 向外开出
~ opening (窗、门)外开式
oval ['əuvəl] *n.;a.* 卵形,椭圆形
~ arch 椭圆形拱
~ court 椭圆形庭院
~ cross-section 蛋形[椭圆]截面
~ head rivet 椭圆头铆钉
~ scale 椭圆尺,卵形板
~ tubing 卵形管
ovalisation *n.* 成椭圆形
oven-dried *a.* 烘干的
~ aggregate 烘干骨料
~ wood 烘干木材
ovenproof glass 防火玻璃
ovenware ['ʌvnwɛə] *n.* 抗高温玻璃
ovenstone *n.* 耐火石
overall ['əuvərɔːl] *n.* 外衣,罩衣 *a.* 全部的,总的;(*pl.*)工作裤
~ arrangement 总体布置
~ design 总体设计
~ dimension 总尺寸,外形(轮廓)尺寸
~ floorage 总建筑面积

~ height 总[全]高
~ length 总[全]长
overarch [ˈəuvərˈɑ:tʃ] v. 覆以拱,成拱形,似拱形地弯曲,架设拱圈
over arm 横杆,悬梁
over arm support 撑杆,支架
overbend n.;v. 下垂,下陷,下弯
overbending a. 过度弯曲的
over brace 横杆,支架
overbreak [ˈəuvəˈbreik] v. 过度断裂;裂面过大;塌方;超挖
overburnt brick 过火砖
overcoat [ˈəuvəkəut] vt. 多涂一层油漆于,涂饰 n. 大衣
overcoating [ˈəuvəˈkəutiŋ] n. 保护涂层,外敷层
overcolo(u)r [əurəˈkʌlə] vt. 着色过浓,夸张
overcover [ˈəuvəˈkʌvə] n. 遮盖物
overcrowd [ˌəuvəˈkraud] v. 使太拥挤
~ing sleeping 过密就寝(每人 3 平方米以下)
over-decorated a. 过分装饰
overdesign [ˈəuvəˈdizain] n.;v. 保险设计,留有余地的设计
overdimensioning [ˈəuvədiˈmenʃəniŋ] n.;a. 尺寸过大(的);选择参数的裕度
overdoor [ˈəuvəˈdɔ:] n. 门头饰 a. 置于门上的
over-dwelling n. (住房)过密拥挤,居住区过密
overdye [ˈəuvəˈdai] vt. 重染色;过度染色
overestimate [ˈəuvərˈestimeit] vt. 估计过高;过度估价
overfelt n. 压制毛毡
overfinishing [ˈəuvəˈfiniʃiŋ] n. 过度修整
overfold [ˈəuvəˈfəuld] n. 倒转褶皱
overfulfil(l) [ˌəuvəfulˈfil] vt. 超额完成

overdoor

overgauge [ˈəuvəgeidʒ] a. 等外尺寸的
overgild [ˈəuvəˈgild] vt. 镀金于……
overglass n. 玻璃灯罩
overglaze [ˈəuvəgleiz] n. 面釉,覆釉 vt. [ˌəuvəˈgleiz] 覆釉
~ color 重釉色彩
overgrainer 木纹漆刷
overgrind v. 研磨过度,过度粉碎
overground [ˈəuvəgraund] a. 地上的
overhand [əuvəˈhænd] a. 支撑的 ad. 用于内脚手架从内墙砌砖
overhang [ˈəuvəhæŋ] n. 突出物,悬重物;突出;灯具悬伸距 v. [ˌəuvəˈhæŋ] 伸出,突出 a. 倒台阶的
~ door 吊门
~ sash 悬挑窗扇,上悬外撑窗扇
~ sash window 上悬外撑窗
overhanging a. 伸出的,突出的
~ bank 陡岸,悬崖
~ beam 悬臂梁,伸出梁
~ eave(s) 飞檐
~ lenght 悬挑长度
~ roof 外伸屋顶,挑檐
~ stair 悬挑式楼梯
overhardening n.;a. 过硬(的),硬化
overhaul [ˈəuvəhɔ:l] n. 超运,大修,检修 [ˌəuvəˈhɔ:l] vt. 修检
annual ~ 年修
general ~ 大修
major ~ 大修
master ~ 大修
minor ~ 小修
top ~ 大修
overhead [ˈəuvəhed] a. 头上的,上面的;架空的
~ concealed closer 架空暗闭锁器
~ door 升降门,上升卷门
~ illumination 顶部照明
~ structure 上部结构

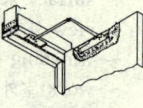

overhead concealed closer

~ suspension 架空悬置
~ view 俯视图
~ welding 仰焊
overheads [ˈəuvəhedz] *n*. 管理费；杂支；附加费
overheat [ˌəuvəˈhiːt] *vt*. 过热，过分加热
~ allowance （涂膜的）加热容许性
overhung [ˈəuvəˈhʌŋ] *a*. 悬挂的，外伸的
~ door 吊门
roof ~ 挑檐
overladen [ˈəuvəˈleidn] *a*. 过载的，装饰过度的
overlap [ˈəuvəlæp] *n*.；[ˌəuvəˈlæp] *v*. 重叠，互搭，搭接
~ joint 搭接
~ length 搭接长度
~ ping astragal 互搭圈线
~ ping tile 搭接瓦
tapered ~ 两斜面接头
overlay [ˈəuvəlei] *n*. 外罩，覆盖层，装饰层；套图塑料膜；[ˌəuvəˈlei] *vt*. 铺，盖；镀
~ flooring 镶木地板，拼花地板
~ life 墁面使用年限
~ paper （混凝土）覆盖纸
overlaid plywood 贴面胶合板
overlaid seam 搭缝
over-limed cement 石灰过多的水泥
overlength *n*. 过长，剩余长度
overlook [ˌəuvəˈluk] *n*. 疏忽，忽略；可俯视之处 *vt*. 俯视，忽视；监视
overlying [ˌəuvəˈlaiiŋ] *a*. 上覆的
~ bed 覆盖层
overmantel [ˈəuvəˈmæntl] *n*. 壁炉架额饰
overmeasure [ˈəuvəˈmeʒə] *vt*. 高估，计量过高
overmixing [ˈəuvəˈmiksiŋ] *a*. 拌和过度的
overnight [ˈəuvəˈnait] *n*.；*a*. 昨晚；晚上（的）；终夜
~ pond 小水池

over-ornamented *a*. 过分装饰的
overpaint [ˈəuvəˈpeint] *vt*. 全面涂漆于
overpickling *n*. （板材等的）过酸洗
overpulverization *n*. 过度粉碎
overramming *a*. 过度捣实的
overrate [ˈəuvəˈreit] *vt*. 过高估计
overrefine [ˈəuvəriˈfain] *vt*. 过度精制
overrich *a*. 过富的，过浓的
~ mixture 多水泥拌合物
overripe [ˈəuvəˈraip] *a*. 过熟的，颓废的
~ wood 过老木材，过熟木材
overroof [ˈəuvəˈruːf] *vt*. 覆以屋顶，顶盖
overs *n*. 筛渣，筛余物
oversailing 突出层，连续突腰层
~ course 突出砖层，腰线
oversand [ˈəuvəˈsænd] *v*. 多砂
~ed mix (ure) 多砂混合料
~ing concrete 多砂混凝土
oversaturated [ˈəuvəˈsætʃəreitid] *a*. 过饱和的
overseer [ˈəuvəsiə] *n*. 监工，监督，工头，管理人
oversew [ˈəuvəˈsəu] *vt*. 对缝，缝合
overshadow [ˌəuvəˈʃædəu] *vt*. 保护，遮蔽
oversize [ˈəuvəsaiz] *a*. 过大的，加大尺寸的
~ brick 大型砖
~ material 不合格材料
~ stone 过大石
overstory [ˈəuvəˌstɔːri] *n*. 上层，顶层
overtamping [ˈəuvəˈtæmpiŋ] *n*. 捣固过度
overwet [ˈəuvəˈwet] *a*. 过湿的
oviform [ˈəuviˈfɔːm] *a*. 卵形的
ovoid [ˈəuvɔid] *n*.；*a*. 卵形体（的）
ovolo [ˈəuvələu] *n*. （建筑物）凸出四分之一圆（饰），馒形饰；圆凸线脚
ovum [ˈəuvəm] *n*. (*pl*. **ova**) 卵形装饰，卵饰

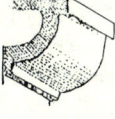
ovolo

owner ['əunə] *n.* 所有者,物主
 ~'s engineer 业主工程师,甲方代表
 ~'s inspector 业主监工员,甲方代理人
ownership ['əunəʃip] *n.* 所有权
oxalate ['ɔksəleit] *n.* 草酸盐
oxalic acid 草酸,乙二酸
oxbow ['ɔksbəu] *n.* 马蹄形弯道 *a.* 呈马蹄形弯的
Oxford blue 暗蓝色
oxidant ['ɔksidənt] *n.* 氧化剂
oxidation [ɔksi'deiʃən] *n.* 氧化作用
 ~ agent 氧化剂
 ~ resistance 抗氧化能力
 ~ -resistant steel 抗氧化钢,不锈钢
oxide ['ɔksaid] *n.* 氧化物
 ~ ceramic 氧化陶瓷
 barium ~ 氧化钡
 ferric ~ 红铁粉;铁丹
oxidize ['ɔksidaiz] *v.* 氧化
 ~d asphalt 氧化沥青
oxidizing agent 氧化剂
oxidizing anticorrosion agents 氧化防腐剂
oxidizing flux 氧化焊剂
oxyacetylene ['ɔksiə'setili:n] *a.* 氧炔的
oxychloride cement 氯氧化水泥
oxygen ['ɔksidʒən] *n.* 氧
 ~ cutting 氧气切割
 ~ lance 氧气吹管
oxygenated asphalt 氧化沥青
oylet *n.* 孔眼,视孔
oyster ['ɔistə] *n.* 蛎
 ~ gray 浅灰色
 ~ (shell) lime 贝[蛎]壳石灰
ozocerite *n.* (ozokerite),地蜡,石蜡
ozone ['əuzəun] *n.* 臭氧,新鲜空气
ozonide ['əuzənaid] *n.* 臭氧化物

P

pace [peis] *n.* 步(子),步速,步距,一步长(=0.75米);梯步,楼梯平台,梯台 *v.* 步行量距;配合
 ~ out 步测出(一段距离)
pachometer [pə'kɔmitə] *n.* 测厚计
pachymeter [pə'kimitə] 测厚计
pacific [pəsifik] *a.* 和平的;太平洋的 *n.* 太平洋
pack [pæk] *n.* 包,捆,包裹;组合(件),部件;容器;接头;毛石砌体 *v.* 包[组]装,装箱,组合;装填,填塞;密封,压紧,夯实,堆积[垛];安排
 ~ of bricks 砖砌体
 ~ed bed 填充层
 ~ed joint 填塞接缝,填实缝
package ['pækidʒ] *n.* 包,外壳;密封装置;机组,部件,单元,成套设备;程序包,数据包 *vt.* 打包,装箱;封[组,集]装,装配 *a.* 整批的
 ~ of veneer 镶面板装饰
 ~ type construction 预制装配构造,轻便式构造
packer ['pækə] *n.* 包装机,包装商[者];装填机,压土机,密垫,栓[灌浆]塞
packing ['pækiŋ] *n.* 包装,装配;包装材料;填充[密],密封,压缩,堆砌,灌注[筑],夯实,填料,衬垫,密封件,垫木
 ~ block 垫块,填塞块
 ~ concrete 捣实混凝土
 ~ course 填层,堆砌层
 ~ felt 毡垫,毡衬
 ~ filler 填充料
 ~ of pipe joint 管缝填塞
 ~ piece 衬片
 ~ plate 垫板
 ~ space 填密空间
 ~ washer 密封垫圈
 joint ~ 接合填密

leather ~ 皮垫
metallic ~ 金属垫料
rubber ~ 橡皮垫
packwork *n.* 填塞作业
pact [pækt] *n.* 合同,协定,契约,公约
pad [pæd] *n.* 衬垫,垫板,法兰盘,基[底]座;踏板;(工具的)垫座;发射台,缓冲器,衰减器,焊接区[点];底漆 *vt.* 装填,铺垫
　~ -eye 垫板孔眼
　~ foundation 独立基础
　~ -lock 挂锁,扣锁
　~ -stone 垫石,(圆柱顶部的)顶板
　~ ded door 装有衬垫的门
　concrete ~ 混凝土座
　felt ~ 毡垫
　rubber ~ 橡胶垫
padding ['pædiŋ] *n.* 大石块,填料;装填,填塞,填满
　~ data 装填数据
paddle ['pædl] *n.* 桨(状物),桨叶,轮叶;闸门[板],开关;踏板;(搅拌用的)桨形棒,搅棒 *v.* 划(桨);涉[戏]水
　~ agitator 叶片搅拌器
　~ mixer 转臂式混砂机,桨叶式拌和机
padstone *n.* 承梁垫石,座石
pagan ['peigən] *n.* (非基督教的,罗马与希腊均为)异教徒 *a.* 异教的
　~ basilica (罗马)教堂
pagoda [pə'gəudə] *n.* 宝塔,浮屠,塔状建筑物
　many-storied ~ (多层)塔
pagodite ['dəgəudait] *n.* 宝塔石,寿山石,冻石
paid [peid] *a.* 已付的;有工资的
　~ cash book 现金支出账
pailou ['pai'ləu] *n.* 牌楼
paint [peint] *n.* 颜料,油漆,涂料;绘画作品 *v.* 喷[涂]漆,着色;描绘,画
　~ base 油漆打底,底漆
　~ brush 漆刷
　~ bucket 油漆提桶

~ coat 涂层
~ coat cushion 涂油垫层
~ consistency 油漆稠度
~ display 油漆涂绘手法
~ drier 油漆干燥剂
~ filler 油漆填料
~ film 漆膜
~ from nature 写生(画)
~ grinder 涂料研磨机
~ gun 油漆喷枪
~ ingredient 油漆成分,涂料成分
~ in oil (画)油画
~ mist 漆雾
~ mixer 涂料混合器
~ naphtha 调漆油
~ primer 髹漆底涂,油漆底层
~ remover 清洗油漆剂
~ rock 铁赭石
~ roller 油漆辊
~ scheme 油漆配色
~ scraper 油漆刮刀
~ scrubber 洗漆剂
~ shop 油漆间
~ skin 涂层,漆膜
~ spray(er) 喷漆器
~ spraying gun 喷漆枪
~ spraying system 喷漆法
~ thinner 涂料稀释剂
~ed decoration 油漆装饰,油漆饰面
~ed enamel 着色瓷釉,搪瓷
~ed enrichment 油漆饰面
~ed film 漆膜
~ed glass 涂色玻璃
~ed ornamental finish 油漆饰面
aluminium ~ 铝粉涂料,银色漆
antifouling ~ 防腐油漆
antiglare ~ 无光漆
antirusting ~ 防锈漆
carburization-preventing ~ (金属)防渗漆涂料
celluloid ~ 透明油漆
luminous ~ 发[夜]光涂料

mural ~ 壁画
red ~ 红漆[涂料,铅油],铅丹漆
solder ~ 焊药膏
thermal ~ 示温漆
zinc ~ 锌粉涂料
painter ['peintə] ***n.*** 油漆工具,油漆工,画家
painting ***n.*** 油漆,颜料;着色,绘画;图画
~ brush 画笔
~ in free-sketch style 写意画
~ in oil 油画
~ in watercolour 水彩画
~ -room 画室
~ workshop 喷漆车间
abstract ~ 抽象画
academic ~ 学院派绘画
cave ~ 洞窟壁画
Chinese ink and wash ~ 中国水墨画
electrostatic ~ 静电涂装
gallery of ~ 画廊
ink ~ without colour 水墨画
metaphysical ~ 无形抽象画
mural ~ 壁画
Post-conceptual ~ 后期概念绘画
traditional Chinese ~ of mountains and water 中国传统山水画
paintwork ***n.*** 油漆(工程);油画
pair [pɛə] ***n.*** (***pl.*** pairs or pair) 一对[双,副];(对,配)偶;成[配]对;配合;(电缆)对绞
~ of columns 双柱
~ of gateways 门口成对建筑物
~ of glasses 双层玻璃
~ of minarets (回教寺院建筑的)成对邦克楼[光塔]
~ of nippers 钳子,剪丝钳
~ of pliers 一把钳子
~ of pylons (埃及)庙门道(楼)的成对塔形建筑
~ of steps 双折梯子,绳梯
~ of turrets 成对角楼

~ ed columns 成对柱,并置柱
~ ed pilasters 成对壁柱
paktong ['pæktɔŋ] ***n.*** 白铜
palace ['pælis] ***n.*** 宫,宫殿,大型公共建筑;地下仓库;派力斯织物
~ lantern 宫灯
~ of culture 文化宫
movie ~ 电影院
palacheite ***n.*** 赤铁矾
palaestra [pə'lestrə] ***n.*** = palestra(古希腊的)练习角力及各种竞技的公共场所;角力场,运动场;角力学校;健身房
palafitte ['pæləfit] ***n.*** 瑞士与意大利史前高架在湖上的住屋
palagonite ***n.*** 橙玄玻璃
palatial [pə'leiʃəl] ***a.*** 宫殿(似)的;富丽堂皇的
~ architecture 宫殿式建筑
~ furnishings 宫室内陈列品
~ hall 富丽堂皇的厅堂
~ house 皇宫
~ style 宫殿风格
palatium [pə'leiʃiəm] ***n.*** (***pl.*** -tia)宫殿(尤指古罗马之皇宫)
palazzo [pɑ:lɑ:tsəu] ***n.*** (***pl.*** -zi)宫殿,庄严的大府邸;城中大厦
pale [peil] ***a.*** 淡色的,暗淡的,弱光的 ***n.*** 栅.(板),围篱,尖板条;界限,范围;圆柱 ***v.*** 使变淡(暗);用栅围住
~ brick 未烧透砖,红砖
~ fencing 栅栏,围墙
~ layer 灰白层
blue ~ 淡蓝
cerulean ~ 淡青
green ~ 淡绿
grey ~ 淡灰
lemon yellow ~ 浅柠檬黄
lilac ~ 淡紫藤色
mauve ~ 淡红紫色
rose ~ 淡玫瑰色
turquoise green ~ 淡碧绿色
paleo-clay ***n.*** 老黏土

paleo-loess n. 老黄土

palette ['pælit] n. 调色盘[板]（调色盘上的）一套颜色
  ~ knife 调色刀

paling ['peiliŋ] n. （造）栅，（筑）垣，围篱，打桩

palisade [,pæli'seid] n. =palisado 栅，栅栏,围篱,桩,(pl.)断崖 v. 用栅围绕

pali(s)sander [pæli'sændə] n. 红木 a. 用红木作的
  ~ wood 红木

palium n. 铝基轴承合金(铜 4.5%,铅 4%,锡 2.6%,镁 0.6%,锰 0.3%,锌 0.3%,其余铝)

Palladian [pə'leidjən] a. (16 世纪意大利建筑家)帕拉第奥(Andrea Palladio)的;帕拉第奥建筑型式的
  ~ architecture 帕拉第奥式建筑
  ~ motif 帕拉第奥式建筑处理手法[特色]
  ~ window 帕拉第奥式窗

Palladianism n. 帕拉第奥建筑主义

Palladium [pə'leidjəm] n. 智慧女神雅典娜(Pallas Athena)的神像

pallet ['pælit] n. 制模板,托板[架,盘];泥刀,抹子;抹灰盘,镘板;调色板;榫;棘爪;锤垫;集装箱[架]
  ~ board 承砖坯板
  ~ -molding 砂模制砖法
  flat ~ 平台板
  frame ~ 架[框]式板台

palliative ['pæliətiv] n. 减轻[剂];减尘剂;防腐剂

palm [pɑ:m] n. 手掌,掌尺(宽约 4 英寸,长约 8 英寸);掌状物;棕榈树 v. 混用;用手抚弄
  ~ capital 棕叶饰柱头
  ~ grip hand knob 星形手钮
  ~ leaf ornament 棕叶装饰
  ~ stay 掌形撑条
  ~ vaulting 棕榈叶形拱顶,扇形穹顶
  ~ wax 棕榈蜡,椰子蜡

palm capital

palmetle n. 棕叶状花饰

pampre n. 葡萄饰

palmetle

pan [pæn] n. 盘(状物),(不同形状的金属制的)厨房用具;池,槽,容器;底壳,底土,硬土层,母岩;浇灌混凝土用的金属或木质模板[壳];墙隔块;全,总,泛 v. 扫视,拍摄全景
  ~ closet 盘式便器,便盆
  ~ construction 肋板结构
  ~ head bolt 平头螺栓
  ~ head screw 大柱头螺钉
  ~ soil 硬土,坚土
  ~ tile 波形屋面瓦
  clay ~ 黏土薄层
  shaker ~ 平板振动筛

panalarm n. (有灯光和振铃的)报警系统

pancake ['pænkeik] n. 盘形混凝土块,(浇灌沥青不均匀而形成的)油饼 a. 扁平的 vt. 使扁平

panchromatic [,pænkrə'mætik] a. 全[泛]色的

pandal ['pændl] n. 公共集会用的临时栅舍

pane [pein] n. 门窗上的方格玻璃;顶棚上的方格;嵌板;螺帽的侧面;锤顶[头,尖] v. 嵌玻璃
  ~ of glass 玻璃板
  cross- ~ hammer 横头锤

panel ['pænl] n. 镶[嵌,壁,底,护墙]板;仪表[配电,控制]板;操纵台[盘,屏];油画板;门窗上的方格,节间;(飞机的)翼片,叶片;小组,组,批;人员名单;对群众抽样访查 vt. 给……镶板,嵌板于
  ~ bolt 面板螺栓
  ~ brick 护墙砖板
  ~ ceiling 镶板[格子式]顶棚
  ~ ed door 镶板[格板]门
  ~ ed ceiling 嵌板平顶
  ~ girder 格子[花格]梁

~ joint 节点
~ lighting 区格式[板式]照明
~ moulding 镶板线条饰
~ of vibrated brickwork 振动砖砌护墙板
~ planer 刨板机
~ point 桁架节点
~ radiator 嵌入[辐射]式散热器
~ shuttering 模板
~ strip 嵌条,盖[压]缝条
~ system 大板结构体系
~ -type construction 预制墙板式结构
~ -type house 预制墙板房屋
~ wall 板式墙
~ work 构架[镶格]工程
braced ~ 斜撑节段
door ~ 门心板
drapery ~ 门心装饰板
glass-reinforced ~ 增强玻璃板
jali ~ 雕刻镶板
lay(ing) ~ 横纹镶板
leaded ~ (铅条镶嵌的)彩色玻璃板,花窗
load-bearing wall ~s 承重墙板
mounting ~ 安装板,接线板
multi-core ~ 多层预制板
prefabricated ~s 预制板
reinforced foam concrete ~ 钢筋泡沫混凝土镶板
sound-proof ~ 隔声板
spandrel ~ 隔声板
subdivided ~ 复分镶板,再分节间
sunk ~ 埋入式镶板,藻井
T-beam floor ~ T字梁楼板节间
vaulted ~ 弓形板
veneered ~ 胶合镶板
wall ~ 护墙板
paneling ['pænəliŋ] *n.* 镶板,门心板;嵌板细工;分段法
~ door 镶板门

paneling

wood ~ 木镶板
panellization *n.* (建筑)大板化
panhead ['pænhed] *n.* 截锥头,盘形头
~ bolt 盘头铆钉(螺栓)
~ rivet 盘头铆钉(螺栓)
~ screw 盘头螺钉
panic [pænik] *n.* 惊慌 *a.* 恐慌的;可从里面推开的(门闩等) *v.* 使恐慌
~ bolt 太平门栓
~ button 紧急保险开关,紧急按钮
panlite *n.* 聚碳酸酯树脂
~ G 玻璃纤维加强聚碳酸酯树脂
panne [pæn] *n.* 一种天鹅绒织物,平绒
~ velvet 平绒
pannier ['pæniə] *n.* 驮篮,背筐,篾筐
panorama [ˌpænə'nɑːmə] *n.* 全景(图),概观,风景的全貌,(舞台的)活动画景,全景装置
panoramic [ˌpænə'ræmik] *a.* 全景的
~ perspective 全景透视
~ view 全景
pantechnicon [pæn'teknikən] *n.* 大型仓库、货栈,家具仓库;家具搬运车
Pantheon [pæn'θiːən] *n.* 罗马万神庙(建于120～124年);伟人祠;大教堂
pantile ['pæntail] *n.* 波形瓦
~d roof 瓦屋顶
pantiling *n.* 以波形瓦盖屋顶
pantograph ['pæntəɡrɑːf] *n.* = pantagraph 缩放仪,放大尺,比例绘图器;(电车顶上的)导电杆架
~ ratio 缩放比
pantology [pæn'tɔlədʒi] *n.* 百科全书
pantry ['pæntri] *n.* 食品(餐具)室,备餐间,冷菜间
~ sink 备餐间污水池
~ window 备餐间送货窗口
paper [peipə] *n.* 纸,报纸;论文,文件;证券,纸币,票据 *a.* 纸做的;书面的;理论上的 *v.* 用纸包[盖];用砂纸擦[磨光];裱糊
~ -ceiling 纸糊顶棚
~ clay 薄层黏土

~ cure（混凝土的)纸板养护
~ gasket 纸衬垫
~ hanger 裱糊壁纸的工人
~ hangings 壁纸
~ method 纸上[室内]作业法
~ mould 纸模
~ roller 滚纸筒
~ sculpture 纸制装饰
~ shale 薄层页岩
~ sheathing board 厚纸质衬板，纸制盖板
~ sheeting （建筑用)纸板
~ sliding-door 糊纸推拉门，纸隔扇
~ sliding-screen 纸隔扇
~ window 纸糊窗
antirust ~ 防锈纸
asbestos ~ 石棉纸
atlas ~ 印画纸，绘图纸
blue-(out) ~ 蓝晒纸
building ~ 防潮纸，建筑纸
carborundum ~ （金刚)砂纸
cellophane ~ 玻璃纸，赛璐玢透明纸
coordinate ~ 坐标纸[方格]纸
curing ~ 混凝土养护纸
deadening felt ~ 隔声毡纸
design ~ 设计图纸
ferroprussiate ~ 蓝晒图纸
fibre ~ 纤维板纸
filter ~ 滤纸
fireproofed ~ 防火纸
glass ~ 玻璃砂纸
glass fibre ~ 玻璃纤维纸
glassine ~ 玻璃纸
granite ~ 花岗石纹纸
grease-proof ~ 防油纸
heliographic ~ 晒图纸
incombustible ~ 耐火纸
marble ~ 大理石纸
roofing ~ 屋面油纸，铺顶油毡纸
sand ~ 砂纸
saturating ~ 油纸
sheathing ~ 柏油（绝热)纸，衬纸

sized ~ 涂胶纸
tar ~ 焦油纸，沥青油纸
torchon ~ 粗面水彩画纸
tracing ~ 描图纸
waterproof ~ 防水[潮]纸
wall- ~ 壁纸

papercore plywood 纸心胶合板

papier colle [pæpjei'kɔlei]【法】图片剪贴艺术，贴纸图案

papier-mache [pæpjeimɑ:ʃei]【法】n. 纸壳子，混凝纸（用于制造盒、盆、盘的纸质可塑材料）；印刷纸型 a. 制型纸做的，人造的，假的
~ facade 虚饰的门面

papreg ['peipreg] n. 层压纸板

papyrus [pə'paiərəs] n. (pl. papyri)纸(莎)草，纸草制成的纸；古写本
~ capital （古埃及的)莎草花柱头
~ column 莎草花圆柱

papyrus capital

parabasalt [ˌpærə'bæsɔ:lt] n. 普通玄武岩

parabola [pə'ræbələ] n. (pl. -las) 抛物线[面]；反射器
~ cupola 抛物线穹形屋顶
~ dome 抛物线穹形屋顶
~ vault 抛物线拱顶
~ (roof) truss 抛物线屋面桁架

parabolic(al) [ˌpærə'bɔlik(əl)] a. 抛物线[面]的
~ arch 抛物线拱
~ chord truss 抛物线形折弦桁架
~ cylinder 抛物线形柱面
~ frame 抛物线构架
~ girder 抛物线型大梁
~ mirror 抛物面镜
~ rib 抛物线拱肋

paraboloid [pə'ræbəlɔid] n. 抛物面，抛物体，抛物面天线，抛物面反射器
~ roof 抛物面屋顶[面]
~ umbrella roof 抛物面伞形屋顶

paraboloidal [pəræbə'lɔidl] *a.* 抛物面的,抛物线体的
　～ surface 抛物柱[面,体]曲面
parachute ['pærəʃu:t] *n.* 降落伞,(竖井筒内的)防坠器 *v.* 跳伞,空投
　～ vault 伞形圆形屋顶
paraclase ['pærəkleis] *n.* 断层
paracril *n.* 丁腈橡胶
paradise ['pærədais] *n.* 天堂,乐园,极美的地
　～ of children 儿童乐园
paraffin(e) ['pærəfi(:)n] *n.* 石蜡(油),煤油,链烷(属)烃 *vt.* 涂以石蜡,用石蜡处理
　～ oil 石蜡油
　～ paper 石蜡纸
　～ radiation shielding wall 石蜡防射线屏蔽墙
　～ wax (硬)石蜡
paragutta *n.* 合成树胶,假橡胶
parallel ['pærəlel] *a.* 平[并]行的,同一方向[目的]的;同类的;并联的; *n.* 平行线;并联(线路);纬度线 *vt.* 平行于,类似于;与……相等;对比;(使)同步
　― boom truss 平行弦杆桁架
　― chord truss 平行弦杆桁架,梯形桁架
　― coping 平行盖顶
　― curve rib arch 平行曲线肋拱
　― lattice dome 平行构格穹顶

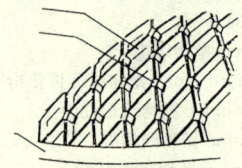

parallel lattice dome

　～ lattice girder 平行构格大梁
　～ perspective 平行透视
　～ strips 平行板条
　～ thread 平行螺纹
　～ to grain 顺纹

parapet ['pærəpit] *n.* 女儿墙,胸墙;栏杆,矮墙;防浪墙
　～ brick 砌砖墙
　～ facing 护墙饰面
　～ grille 格栅
　～ gutter 箱形檐槽,压檐墙天沟
　～ masonry wall 砖砌檐墙
　～ skirting 女儿墙脚泛水
　～ stone 拦墙石
　～ wall 女儿[压檐]墙
　～ed terrace 有栏杆的阳台
　brick-cap ～ 砖顶压檐[女儿]墙
　castellated ～ 城垛墙,城垛
paratactic(al) [pærə'tæktik(əl)] *a.* 罗列的,并列的
paraton(e) *n.* 巴拉东(一种黏度添加剂)
paravent *n.* 挡风屏障[帘]
parbuckle ['pɑ:bʌkl] *n.* (拉上或放下大桶等用的)套拉绳,上下索,重物起吊具 *vt.* 以套拉绳拉上或放下
parchment ['pɑ:tʃmənt] *n.* 羊皮纸,仿羊皮纸;垫衬沥青纸毡
　～ paper 假羊皮纸,硫酸纸
　animal ～ 动物羊皮纸
　glazed ～ 蜡光仿羊皮纸
　imitation ～ 仿羊皮纸
　transparent ～ 透明羊皮纸
　vegetable ～ 仿羊皮纸
　virgin ～ 上等羊皮纸
parclose ['pɑ:klauz] *n.* = perclose(作间隔用的)屏障,隔幕
pare [pɛə] *vt.* 削去,逐渐减少,修剥,刮
　～ down expenses 节省开支
parenchyma [pə'reŋkimə] *n.* 软细胞组织;主质,实质
　～ of wood 木材的薄壁组织
parergon [pæ'rə:gɔn] *n.* 附属装饰,附录,附属物;副业(产品)
parget ['pɑ:dʒit] *n.* = parge 石膏,灰泥,墙粉 *vt.* 粉刷,涂灰泥
　～ coat 粗涂灰泥
pargetry *n.* 装饰抹灰,拉毛凸粉刷

parging [ˈpɑːdʒiŋ] *n.* 墙面的薄灰泥
Parian [ˈpɛəriən] *n.*; *a.* (以产大理石著名的)巴罗斯(Paros)岛(的);似白色大理石的,白色瓷器
- ~ cement 仿云石水泥(含硼砂和石膏的水泥)
- ~ marble 巴罗斯大理石
- ~ plaster 巴罗斯石膏
- ~ ware 白色瓷器

paring [ˈpɛəriŋ] *n.* 刨花,切片(屑)
- ~ chisel 削凿刀,扁铲,扁凿
- ~ gouge 削面凹凿

Paris [ˈpæris] *n.*; *a.* 巴黎(的),巴黎式的
- ~ blue 巴黎蓝,绀青
- ~ green 巴黎绿,翠绿
- ~ plaster 熟石膏
- ~ white 巴黎白

parish [ˈpæriʃ] *n.* 教区;教区的全体居民
- ~ hall 教堂
- ~ house 小[地方性]教堂
- ~ room 教堂

park [pɑːk] *n.* 公园,停车场,停车坪;材料库 *vt.* 停车于,置于,将……围起 *vi.* 停车
- ~ architecture 园林建筑(艺术)
- ~ block 多层停车库
- ~ -land 公共绿地

Parker's cement 帕克水泥
Parker truss 帕克尔桁架(上弦呈多边形)
PAR lamp 平行光源
parlo(u)r [ˈpɑːlə] *n.* 客厅,接待[会客]室,起居室,营业室 *a.* 适用于客厅的
- beauty ~ 美容院
- beer ~ 啤酒店
- hairdresser's ~ 女子理发店,美容院
- ice cream ~ 冷饮室
- sun ~ 日光室

parodos *n.* 古希腊剧场舞台和观众席之间的侧廊,过道
paronite *n.* 石棉橡胶板
parpend *n.* 穿墙石块
parquet [ˈpɑːkei] *n.* 镶木[席纹,拼花]地板;正厅座 *vt.* 铺镶木地板 *a.* 镶木细工的
- ~ block 镶嵌木块[料]
- ~ floor 拼花(席纹)地板
- ~ polishing machine 镶木地板打光机
- ~ sealing 镶嵌层
- ~ strip 拼花小板条
- ~ timber 镶嵌木料
- ~ varnish 镶木地板清漆
- ~ work 镶嵌木工,镶木细工

parquetry [ˈpɑːkitri] *n.* 镶木细工[地板]
- boarded ~ 木板镶嵌
- herringbone ~ 席纹地板,人字拼木地板

part [pɑːt] *n.* 部分,片段;成分,要素;零[部,元,工,配]件;角色,作用;(几)分之一,建筑中柱下部半径的1/30 *v.* 分开,切断,剖截 *a.* 部分的 *ad.* 部分地
- ~ of a building 建筑配件
- ~ sectioned view 局部剖面图
- embedded ~ 嵌入[埋置]部分

parterre [pɑːˈtɛə] *n.* (庭院中的)花坛;剧场池座,剧院楼下正厅后座
- ~ boxes 剧院楼下正厅后排包厢

Parthenon [ˈpɑːθinən] *n.* 希腊雅典女神之神殿,帕提农神庙(公元前438年建于希腊雅典)
Parthian architecture 安息建筑(现伊朗东北部之古建筑)
parti [pɑːˈtiː] 【法】*n.* 建筑设计的总方案;图解;计划
partial [ˈpɑːʃəl] *a.* 部分的,局部的,单独的
- ~ hip 半斜屋脊
- ~ splice 局部拼接
- ~ view 局部图
- ~ vitrification 半透明

partially [ˈpɑːʃəli] *ad.* 部分地,局部地,不完全地
- ~ exposed basement 半地下室
- ~ fixed joint 半嵌固连接
- ~ transparent 半透明的

particle [ˈpɑːtikl] *n.* 微粒,颗粒,质点,粒

子,土粒
~ -board 碎屑胶合板
particoloured [ˈpɑːtikʌləd] *a.* 杂色的,斑驳的,各色各样的
paticular [pəˈtikjulə] *a.* 特殊的,特有的,卓著的;详细的 *n.* 项目,特色,摘要,细节;(*pl.*)详细资料 *ad.* 特别地,细致地
particularist Gothic 哥特式建筑风格
parting [ˈpɑːtiŋ] *n.* 分离,剖截,切断,劈开;道岔,错车道;夹层 *a.* 分离的
~ agent 脱模剂
~ bead 隔条
~ clay 黏土夹层
~ face 分离面
~ lath 薄隔板
~ rail 中横档
~ slip 隔片
~ stop (双悬窗)分隔木条
~ wall 分隔墙

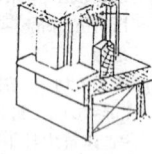
parting bead

partition [pɑːˈtiʃən] *n.*, *vt.* 划分,区分,分配,分割开,隔开,瓜分,分块;隔板墙,间壁
~ block 隔墙块
~ board 隔板
~ brick 隔墙砖
~ component 隔墙构件
~ panel 隔墙板
~ plate 隔板
~ studding board 骨架式隔墙板
~ tile 隔墙用空心砖
~ wall 隔墙,间壁
~ wall beam 隔墙托梁
~ window 间隔窗
~ing glazing 玻璃隔断
~ing system 隔墙体系
accordion ~ 折摺隔屏
air ~ 隔风墙
air-tight ~ 气密隔墙
battened ~ 薄板隔墙
bearing ~ 承重(隔)墙
brick nogged ~ 砖填木架隔墙
dwarf ~ 矮隔墙

folding ~ 屏风隔断
framed ~ 构架间壁
glazed ~ 玻璃间壁
isolating ~ 隔离间隔,绝缘隔板
non-bearing ~ 非承重隔墙
rolling ~ 卷帘隔断,移动式隔断
screen ~ 屏风隔断
slag-alabaster ~ 矿渣石膏隔墙
solid ~ 实心隔墙
structural clay ~ 建筑黏土隔墙
stud ~ 立楞隔断
par-tox *n.* 窗框防腐剂
party [ˈpɑːti] *n.* (*pl.* -ties)用户;班,组,队;宴会,集会;*a.* 参与的
~ arch 共用拱门
~ coloured 杂色的
~ corbel (悬挑出的)防火隔墙
~ fence 共用隔篱[墙]
~ wall 共用界墙,户界墙
parvis [ˈpɑːvis] *n.* 建筑物前庭,天井,院子;寺院前庭;(教堂前面的)柱廊
~ turret 教堂前庭的塔楼
pass [pɑːs] *n.* 狭路,通[航]道,隘[垭]口,通行证,护照,免票;合格;焊道 *v.* 通过,传递;合格;流通;(程序)扫描
~ -through 厨房与餐厅间递菜用的长方形口
passage [ˈpæsidʒ] *n.* 出入口,通路,走廊
~ aisle 通道,过道
~ for freight handling 货运通道
~ space 建筑物内部的通道(走廊,楼梯等)
~ way 通道,走廊
passementerie [pɑːsˈmentəri] *n.* 金银饰带,饰球
passing [ˈpɑːsiŋ] *a.* 穿过的,越过的;通过的,合格的;短暂的 *n.* 经过,超越
passive [ˈpæsiv] *a.* 被动的,消极的,钝态的;无源的 *n.* 被动语态
paste [peist] *n.* (浆)糊,膏,胶,柔软的塑性混合物;含铅量高的晶亮玻璃 *vt.* 粘贴,裱糊,涂胶

~ -board 纸板,硬纸板
~ brush 浆糊刷
~ drier 灰浆干燥器
~ for wall paper hanging 墙纸粘贴剂
~ mold 衬碳模
~ paint 厚漆
~ polish 虫胶清漆
~ -pot 浆糊桶,烧沥青桶
~ resin 膏状树脂
aluminium ~ 银灰漆,铝涂料
cement ~ 水泥浆
lime ~ 石灰膏
neat-cement ~ 净水泥浆
violet ~ 蓝(铅)油

pasteboard ['peistbɔːd] *n.* 胶纸板,硬纸卡 *a.* 纸板的;人造的

pastel [pæs'tel] *n.* 彩色粉笔(画),蜡笔(画);大青(染料);中间色,轻淡色彩 *a.* 色彩轻淡的,蜡笔画的
~ colour 中间色
~ drawing 蜡笔画
~ shade 清淡优美的色调

paster ['peistə] *n.* 涂胶纸,贴笺纸;贴片;贴纸工,花砖工

pasties *n.* 乳饰

pasting ['peistiŋ] *n.* 涂,裱糊;贴面薄纸

pasty ['peisti] *a.* 浆糊状的,黏性的;苍白的,(面色)发青的

pat [pæt] *v.* 轻拍; *n.* 扁块,小块试样,(水泥安定性试验的)试饼,馒头形水泥试块 *a.* 恰当的 *ad.* 合适地

patand *n.* 地槛,底木,地梁,压条

patch [pætʃ] *n.* 补钉,缀片;盖[搭,连接]板,斑点、一小块土地 *v.* 修补,补充,临时接线
~ bolt 连接件螺栓
~ compound 修补剂
~ concrete 修补混凝土
~ mortar 修补砂浆
~ plaster 修补灰浆[石膏]
~ plate 修补板料

patching *n.* 修补,(路面)补坑;(临时性)接线
~ hole 修补坑洞,补孔
~ material 修补材料
~ operation 修补工作
~ technique 修补技术
surface ~ 补面

patent ['peitənt] *n.* 专集,专利权[证],执照 *a.* 专利的,特许的,上等的
~ (bush) hammer 薄刃石锤
~ claw 薄爪凿
~ cloth 蜡布
~ glass 平板玻璃
~ glazing 无油灰窗玻璃
~ hammer 面石修饰锤
~ leather 漆皮,黑漆皮
~ roof glazing 无油灰屋顶玻璃
~ stone floor covering 铸石地面饰面
~ stone tile 铸石贴面砖

patent hammer

patera *n.* 古典建筑上一种小的扁平的圆形或椭圆形班瑰花瓣装,圆盘饰;插座;接线盒

patera

paterage *n.* 圆花饰

pateriform *a.* 浅碟形的

pate-sur-pate 泥浆堆花浮雕

patina ['pætinə] *n.* (*pl.* patinae)铜绿[锈],(金属或矿物的)氧化表层

patination [ˌpæti'neiʃən] *n.* 生锈,布满铜绿

patinous ['pætinəs] *a.* 生锈的

patio ['pɑːtiəu] *n.* (*pl.* patioes)天井,庭院

patten ['pætən] *n.* 柱基[脚];平板,狭条板

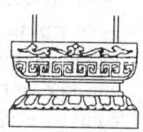
patten

patter ['pætə] *n.* (泥

瓦工用的)镘板,木抹子

pattern ['pætən] *n.* 模型,木模;标本,样品;型式,图形,图案,花样,花纹,建筑设计风格 *v.* 模仿,仿造;以图案装饰,加花样,构图
~ bond (砖墙或圬工)砌筑型式
~ brick masonry work (有装饰性的)砖的砌式
~ cracking 网状裂缝
~ die 型模
~ floor cover tile 花式地面砖
~ glass 装饰花纹玻璃
~ plate 模板
~ veneer 装饰镶板
~ wire glass 嵌丝玻璃
basket weave ~ brick 席纹铺砖
geometric ~ 几何图样
ground ~ 底样,有色背景图案
spatial ~ 空间图形
streamline ~ 流线型

patterning *n.* 图案装饰,图案结构,图形

paulin ['pɔːlin] *n.* 防水[焦油]帆布,油布,篷布
roof ~ 篷布

pave [peiv] *vt.* 铺砌[设],密布;铺路面
~(ed) apron 铺面停机坪;铺面墙裙板
~ed ditch 铺砌沟
~ed invert 有衬砌的倒拱底板
~ed light 从穹形玻璃屋顶进入的光线
~ed with tiles 瓷砖铺面

pavement ['peivmənt] *n.* 铺砌层,路[护,铺]面,铺地材料;铺道,人行道,(机场跑道)道面
~ light 沿人行道的房屋地下室采光窗,路面灯
~ of cobble stone 大卵石路面
~ of riprap 乱石路面
blockwood ~ 木块铺面
checker-work ~ 方格式铺面
coursed ~ 成层铺面
earthenware tile ~ 陶砖铺面
herring bone ~ 人字式铺地
mosaic ~ 拼花铺面,锦砖铺面
terrazzo ~ 水磨石铺面

paver ['peivə] *n.* = pavier 铺路工,铺路机;铺筑材料

pavilion [pə'viljən] *n.* (尖顶)大帐篷,更衣室,(运动场)休息室,亭,阁,馆;天幕 *v.* 搭帐篷
~ roof 棱锥形(方攒尖)屋顶
~ type 分隔式(平面布置)

paving *n.* 铺路(的),铺砌(的);铺面材料
~ brick 铺地砖
~ course 铺砌层
~ expansion joint 铺面伸胀缝
~ finisher 铺面修整机
~ in echelon 梯形铺砌
~ material 铺面材料
~ tile 铺面砖
~ with pebbles 卵石铺面
~ wood 铺面木块
brick-on-edge ~ 侧砖铺面
diagonal ~ 斜铺砌
dry ~ 无灰浆铺砌,干砌
encaustic ~ 瓷砖面层

pavio(u)r ['peivjə] *n.* 铺路工,砌砖工,铺路机;一种坚硬的铺面砖

pawn [pɔːn] *n.* 长廊,有顶的通道;典当,抵押物

Paxboard 或 **Paxfelt** *n.* 一种绝缘材料

Paxolin *n.* 一种酚醛层压塑料

pay [pei] *n.* 报酬,工资;含油的沙层或地带 *a.* 收费的;富矿的;有效的;自动收款的 *v.* 支付,偿还,付出[款];合算,值得

payasat *n.* 缅甸塔

payawut *n.* 缅甸佛堂

payment ['peimənt] *n.* 付款,报酬;惩罚

pea [piː] *n.* 豌豆;12~18mm 无烟煤块
~ gravel 豆砾石,绿豆砂
~ green 豆青色,淡绿色
~ grit 砾石,绿豆砂

~ stone 砾石
peace [pi:s] n. 和平,安静;和约
peach [pi:tʃ] n. 桃子[树],略带黄色的粉红色,桃色 a. 粉红色的
~ blossom 桃红色
~ tan 桃棕色
peachblow ['pi:tʃbəu] n. 中国瓷器上的桃红色釉;紫红色
peacock ['pi:kɔk] n. 孔雀;孔雀蓝色 v. 炫耀
~ blue 孔雀蓝
~ stone 孔雀石
peak [pi:k] n. 山顶[巅],峰端;峰值,最高负荷,峰负荷值;突出部波峰 v. 耸起,以……为顶点,峰化 a. 巅峰的
~ arch 尖顶拱
~ed roof 尖形屋顶
peanut ['pi:nʌt] n. 花生
~ gallery 剧场的上层楼厅
pear [pɛə] n. 梨(树,木,形,物)
~ -push 悬吊式按钮
~ switch 悬吊开关,梨形拉线开关
pearl [pə:l] n. 珍珠(状物),微粒;珍品;珍珠(白)色 a. 珍珠(状,制)蝗,淡蓝灰色的 vi.(使)呈珍珠状 vt. 结成珍珠
~ ash 粗碳酸钙,珍珠灰
~ beading 串珠饰
~ braid 波状花线编带
~ glue 颗粒胶
~ molding 串珠线脚

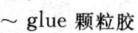 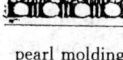
pearl molding

~ spar 白云石(珍珠白云石)
~ white 锌钡白
pearled [pə:ld] a. 用珍珠装饰的;有珍珠色彩的
pearlite ['pə:lait] n. 珠光体,珠层钱,珠层[粒]体,珍珠岩,层片形组织
~ absorbent ceiling 珍珠岩吸声平顶
~ asphalt concrete 珍珠岩沥青混凝土
~ (insulating) concrete 珍珠岩(保温、隔热、隔声)混凝土
~ concrete roof slab 珍珠岩混凝土屋面板
~ loose insulation fill 珍珠岩疏松绝缘填充料
~ plaster 珍珠岩灰浆
peavey ['pi:vi] n. 长撬棍,(翻木头用的)钩棍,尖头搬钩
pebble ['pebl] n. 卵石,砾石,透明水晶;(皮革,纸张面上的)粗纹 vt. 铺卵[砾]石;研碎;用球磨机粉碎
~ -dash 灰泥卵石涂层,(干贴)卵石饰面
~ filling 卵石填层
~ pavement 卵石(铺砌)路面
~ roundstone 圆卵石
~ stone 卵石
~ ware 斑纹陶器
facetted ~ 棱石
pebbly ['pebli] a. 从石子[卵石]的,粗纹[面]的
pecker ['pekə] n. 鹤嘴锄,十字镐,穿孔器,尖头器具;接续器,簧片
pecky ['peki] a. (木材)有灰斑的
pectin ['pektin] n. 果胶,黏质胶
pede window 十字架底的窗
pedestal ['pedistl] n. 基座,基础,柱脚,台,垫,支架,支座,塑像座,轴承座[架];焊接凸点 vt. 加台脚支持,置于台座上
~ footing 柱脚,座墩基脚
~ frieze (建筑上的)基座腰线
~ lamp 座灯,台灯
~ lavatory 台座式[柱脚式]洗脸盆
~ moulding 底座线脚
~ rock 基岩,支柱石
~ -type drinking fountain 台座式饮水喷口
~ urinal 台座式[柱脚式]小便池
pediment ['pedimənt] n. (建筑物之)山形墙,人字墙,三角楣饰

pediment

~ apex 人字脊

~ arch 三角形或弧形拱饰
peel [pi:l] *n*. (16世纪英格兰,苏格兰边境的)堡塔,碉楼;*v*. 剥(皮),剥落,凿净,粗加工
peen [pi:n] *n*. 锤头[尖、顶] *vt*. 用锤头锤击,冷锻;喷砂(表面强化)
~ hardening 弹射[锤击,喷射]硬化
peep [pi:p] *n*. 偷看;窥视孔;吉普车 = jeep(美) *v*. 偷看,往里瞧
~ hole 窥视孔,观察孔
~ window 窥视窗
peg [peg] *n*. 木栓[钉、桩],销,标桩,轴柱;标记,标高,方位物;测标 *v*. 钉[栓,系]牢;以标桩划界;限定
~ stake 木桩,木钉
~ stay (窗的)套栓式风撑
~ top paving 小方石铺面
pegamoid ['pegəmɔid] *n*. 人造革,防水布
pegging *n*. 销子连接
pegmatolite [peg'mætəlait] *n*. 正长石
pein *n*. (= peen)锤头
pek *n*. 油漆,涂料
pelhamite ['peləmait] *n*. 蛭石
pellet ['pelit] *n*. 丸,小球,颗粒,粉末,片,片状器件,圆形木楔,装饰品上圆形浮雕 *vt*. 压丸,做成丸[片]状

pellet moulding

~ moulding 链珠形线脚
~ ornament 圆形浮雕装饰品
pellmell ['pel'mel] *n*. 极度混乱,杂乱 *a*. 杂乱的 *v*. 使混杂 *ad*. 杂乱地
~ construction 杂块砌的建筑
pellouxite ['peluksait] *n*. 生石灰
pellucidity [,pelju:'siditi] *n*. 透明度[性],明晰性,透明状态
pelmet ['pelmit] *n*. 门(窗)帘盒
~ board 门窗帘板

pelmet board

pemosors = multilayer copolymer 多层共聚物

pen [pen] *n*. (钢)笔,笔尖;小围栏;小水闸[坝];畜舍,畜牧场;修理潜艇的船坞[台],隐蔽码头;丘陵,小山;栏,圈,棚;种植园;填充带;临时风巷;风眼 *v*. 写作;栏,围
a play ~ for babies 幼儿游戏场
drawing ~ 绘图笔,鸭嘴笔
pencatite *n*. 水滑大理石
pencil ['pensl] *n*. 铅笔,色笔,写[记录]头,光(线)锥,光束 *vt*. 用铅笔写[画,标出];成射束状
~ diamand 玻璃刀
~ of rays 光束
~ rod 细棒材
~ rounded (木工)细磨圆角
~ sketch 铅笔画,草图
~ stone 滑石,石笔石
~ -work 铅笔图画
coloured ~ 彩色铅笔
glass ~ (在玻璃上写字用)蜡笔
grease ~ 纸皮油彩画笔,石印笔
light ~ 光束
pend [pend] *vi*. 悬垂
pendant ['pendənt] *n*. 悬耳[挂],悬垂物,悬垂形花饰;挂钩,悬[吊]架,吊灯;附录,附属物 *a*. 悬吊的,下垂的;未定的
~ chain 吊链
~ fixture 悬吊装置,吊灯
~ light fitting 吊灯
~ point 悬挂点
~ post 墙架立柱
~ push 悬吊按钮[开关]
~ signal 吊灯信号
cord ~ 电灯吊线
cord ~ lamp 吊灯
lamp ~ 灯垂饰
pipe ~ lamp 管吊灯
raise and fall ~ (吊灯的)升降悬线
pendantswitch *n*. 悬挂开关
pendeloque *n*. 梨形宝石;枝形吊灯的梨形玻璃垂饰
pendent ['pendənt] *a*. = pendant 悬垂

的,悬而未决的
~ lamp 吊灯
pendentive
['pen'dentiv] *n.* 穹隅(圆屋顶过渡到支柱之间渐变曲面),方墙四角穹窿,斗拱
~ dome 三角穹(上)圆(屋)顶

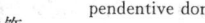
pendentive dome

pendicle *n.* 垂饰
pendulation *n.* 摆动,振动
pendulum ['pendjuləm] *n.* 摆,振动体 *a.* 摆动的
penetrability [penitrə'biliti] *n.* 穿透性,贯空力,贯入性,渗透性,透明性
penetrable ['penitrəbl] *a.* 可贯入的,可穿过的,可渗透的
penetralia [ˌpeni'treiljə] *n.* (复)最深处(尤指房屋的内院,高宇的内殿);密室
penetrance [ˌpenitrəns] *n.* 贯穿性,透射
penetrant ['penitrənt] *n.* 贯入料,渗透 *a.* 渗透的,透彻的
~ method (超声波探伤的)透过法
penetrate ['penitreit] *v.* 贯穿;灌入,贯入;洞视;浸染,充满
~d mortise and tenon 长榫
~d mortise hole 穿横板的榫眼
penetrating *n.* ;*a.* 渗透(的);贯穿(的)
~ aid 防水剂
~ asphalt 渗入沥青
~ finish 浸漆处理
~ luminous flux 透射光通量
~ oil 贯入用沥青
~ sealer 渗透保护层,防水层
~ type sealant 混凝土防冻用的透入式封面料
~ varnish 渗透清漆
penetration [peni'treiʃən] *n.* 渗透,贯入,灌入,穿透率,渗透性,贯[针]入度,(焊接)熔深
~ course 灌浇层

~ grade bitumen 膏体(石油)沥青
~ of dampness 透湿度
~ sleeve 穿通套筒[管]
bituminous ~ 沥青灌注
capillary ~ 毛细渗透
penning ['peniŋ] *n.* 石块铺砌,护坡
penny ['peni] *n.* 便士(英货币单位=1/100镑),分(美国,加拿大之铜币)
~ arcade 花费最少的游乐场所
~ gaff (英国)低级戏院或音乐厅
pennystone *n.* 扁平(小块)石
pension ['penʃən] *v.* ;*n.* (发给)养老[抚恤]退休金 *a.* 养老金的
pension ['pɑːŋsiɔŋ]【法】*n.* 供膳的宿舍,公寓,寄宿学校
pent roof 单坡屋顶
pentacle ['pentəkl] *n.* 五角星(形),五角星饰
pentadecagon *n.* 十五边形
pentagon ['pentəgən] *n.* 五角形
pentagonal [pen'tægənl] *a.* 五边[角]形的,五边形数
pentagram ['pentəgræm] *n.* 五角星(形)
pentahedron [pentə'hiːdrən] *n.* (*pl.* -drons,-dra)五面体
pentangular [pen'tæŋgjulə] *a.* 五角的
pentastyle ['pentəstail] *n.* =pentastylos(古希腊神庙)五柱式建筑物 *a.* 五柱式的
penthouse ['penthaus] *n.* =pentice 屋顶房间,搁楼,电梯机房,坡屋,遮檐,雨篷,遮阳篷
~ apartment 屋顶公寓
~ roof 单坡屋顶
pentile *n.* =pantile 波形瓦
penton *n.* 片通(一种氯化聚醚塑料)
penton-rubber *n.* 氯化聚醚塑料橡胶
pentroof *n.* 单坡屋顶,檐下披屋
pentrough *n.* 水槽
pepper ['pepə] *n.* 胡[辣,花]椒;*vt.* 密布
~ -and-salt 椒盐色的,有黑白色相间的

peptizate *n.* 胶溶体
peptizator [peptaizeitə] *n.* 胶溶[化]剂
peptizer *n.* 塑解剂
pepyrus column 有纸莎草茎图案的圆柱
per [pə:] *prep.* 每,按,由,以
  ~ annum 每年
  ~ diem 每天,按日
perbunan ['pə:bju:nən] *n.* 丁腈橡胶;别布橡胶
percentage [pə'sentidʒ] *n.* 百分率,比率;(按百分率计算的)利息,佣金;折扣
  ~ elongation 伸长百分率,延伸率
  ~ increase 增长率
  ~ humidity (湿)饱和度
  ~ of completion 完工百分率
  ~ of bright sunshine 日照率
  ~ of moisture 含水率
  ~ of saturated water content 饱和含水率
  ~ of void 空隙率
  ~ of wear and tear 折旧率
perception [pə'sepʃən] *n.* 感觉,理解,体会
  ~ of depth 景深感觉
  ~ of relief 立体感
  ~ of solidity 实体感
  ~ threshold 感觉阈[限]
  stereo ~ 立体感觉
perceptivity [pə'sep'tiviti] *n.* 知觉,理解力
perceptron [pə'septrən] *n.* 视感控制器,感知器
perceptual [pə'septʃuəl] *a.* 知觉的,感性的
perch ['pə:tʃ] *n.* 棒,杆,主轴;(英国丈量单位)杆
percolate ['pə:kəleit] *v.* 渗透[滤],渗流,砂滤 *n.* 渗[滤]出液
  ~ing water 渗透[漏]水
percolation [pə:kə'leiʃən] *n.* 渗透作

pepyrus column

用,深层渗透,地面渗入
  ~ loss 渗漏损失
  ~ pressure 渗透压力
  ~ ratio 渗透率
  adsorptive ~ 吸附渗透
  capillary ~ 毛细渗透
percolator ['pə:kəleitə] *n.* 渗滤[流]器,过滤池
percussion [pə:'kʌʃən] *n.* 碰撞,冲击,敲击,击发,激动,振动
  ~ riveting 冲击式铆接
  ~ welding 冲击焊
perennial [pə'renjəl] *a.* 常年的,多年的 *n.* 多年生植物
perfect ['pə:fikt] *a.* 完全的,理想的,准确的,熟练的 *n.* (文法)完成式 *vt.* [pə'fekt] 使完善,使熟练
  ~ transmitting body 完全透明体
perfectly ['pə:fiktli] *ad.* 完全
  ~ water-proofing plywood 全防水胶合板
perfectness *n.* 完全[整]性
perflation *n.* 通风,换气
perfoot *n.* 极点,顶点
perforate ['pə:f:əeit] *v.* 穿[钻]孔,打眼,贯穿 *a.* 有孔的,凿穿的
perforated ['pə:fəreitid] *a.* 钻孔的,穿孔的,贯通的,打眼的
  ~ asbestos cement 多孔石棉胶泥
  ~ backing 多孔垫层[板]
  ~ bottom 穿孔底板
  ~ brick 多孔砖,空心砖
  ~ calcium silicate brick 多孔灰砂硅酸盐砖
  ~ cap 穿孔盖板
  ~ clay pipe 多孔瓦管
  ~ concrete block 多孔混凝土砌块
  ~ concrete tube 多孔混凝土管
  ~ glass 泡沫玻璃,多孔玻璃
  ~ grids 格孔栅
  ~ gypsum board 多孔石膏板
  ~ metal screen 大孔筛
  ~ panel air diffuser 穿孔板送风口

~ panel air outlet 多孔板散流器
~ pipe 多[穿]孔管
~ plate sieve 板孔筛
~ roller 多孔滚筒
~ stone 透水石
~ strainer 多孔滤管
~ tracery 漏花窗格
~ wall 穿孔墙,有漏窗的墙

perforating tool 打眼工具,穿孔
perforator ['pə:fəreitə] *n.* 钻孔器,穿孔机
performance [pə'fɔ:məns] *n.* 运行,行为;性能;实行,表演;操作,作业;生产力
~ figure 质量指标
performeter *n.* 工作监视器
pergament ['pə:gəmənt] *n.* （假）羊皮纸
pergamyn paper 耐油纸
pergola ['pə:gələ] *n.* 凉亭,阳台廊,藤架,藤荫小径
peribolos *n.* (*pl.* -loi)圣地[殿]之围墙,庙宇庭院
periclase *n.* =periclasite 方镁石
~ brick 方镁石砖
periclinal [peri'klainəl] *a.* 穹状的
peridot ['peridɔt] *n.* 橄榄石
peridrome *n.* （古希腊建筑的）围廊
perimeter [pə'rimitə] *n.* 周围,周界,周长;（光学）视野计
~ area ratio 周ექ面积比
~ heating （建筑物）周边采暖
~ of a circle 圆的周长
~ zone 周边区
period ['piəriəd] *n.* 周期,循环;时间,时段,时间间隔;时代,（地质年代）纪 *a.* 具有某一时代之特征的
~ architecture 当代建筑
~ of curing 养护期
~ of design 设计年限
~ style 时代风格
decorated ~ （建筑）盛饰时期
periodic [,piəri'ɔdik] *a.* 周期（性）的,定时的,间歇的,断续的

~ maintenance 定期养护
periodical [piəri'ɔdikl] *a.* 周期的 *n.* 期刊,杂志
~ room 期刊阅览室
peripheral [pə'rifərəl] *a.* 边的,外围[部]的,周的;非本质的;外部[辅助]设备
~ beam 圈梁
~ blocks 周边砌块
~ equipment 外部[辅助]设备
~ loading 周边荷载
~ tie beam 周边联系梁,圈梁
periphery [pə'rifəri] *n.* (*pl.* -eries)周围,边缘,圆周,圆柱体表面,外围
~ beam 外墙地脚梁
~ wall 外围墙
peripteral [pə'riptərəl] *a.* 列柱式的 *n.* 列柱式建筑
peripteros [pə'riptərəs] *n.* (*pl.* -teroi [-tərəi])列柱式建筑
periptery [pə'riptəri] *n.* 列柱式建筑
perisphere ['perisfiə] *n.* 球形建筑物
peristerite *n.* 钠长石
peristyle ['peristail] *n.* 周柱式,柱列,列柱廊,列柱中庭
peristylium [peri'stailiəm]【拉】*n.* =peristyle(古希腊,古罗马住宅中)有柱廊的庭院,列柱中庭
perk [pə:k] *vt.* 竖起;详细调查;修饰;快速动作 *a.* 趾高气扬的
perlite ['pə:lait] *n.* 珠光体,珍珠岩
~ aggregate 珍珠岩骨料
~ concrete 珍珠岩混凝土
permaclad *n.* 覆盖不锈钢板的合成钢板
permanence ['pə:mənəns] *n.* 永[耐]久,耐久性,稳定性,安定度
permanent ['pə:mənənt] *a.* 永[持]久的,恒定的,不变的,常设的
~ assets 固定资产
~ building 永久[性]建筑
~ construction 永久性建筑
~ contraction 永久性收缩
~ deformation 永久变形,剩余变形

~ forms 固定模板
~ humidity 恒定湿度
~ sash 固定窗扇[框],亮子
~ structure 永久性结构,永久性建筑物

permanently ['pə:mənəntli] *ad.* 永久地

permeability [pə;miə'biliti] *n.* 渗透性[率];透气性;导磁性[率];穿透率
~ cell 透气管
~ reducing agent 减渗剂
~ test 渗透试验
acoustic ~ 传声性
air ~ 透气性

permeable ['pə:miəbl] *a.* 可渗透的,渗透性的
~ bed 透水层
~ layer 透水层
~ material 渗透性材料
~ plastics 可透塑料

permeate ['pəmieit] *v.* 弥漫,扩散,充满,渗透

permeation [pə;mi'eiʃən] *n.* 渗透作用,扩散,弥漫

permendur(e) *n.* 一种铁钴磁性合金(钴50%,钒 1.8% ~ 2.1%,其余铁)

permenorm *n.* 一种铁镍合金(镍 50%,铁 50%)

permissible [pə'misəbl] *a.* 容[准]许的,许可的
~ deviation 容许偏差
~ error 容许误差
~ tolerance 容许公差
~ value 容许值

permissive [pə'misiv] *a.* 容许的,随意

permit [pə'mit] *v.* 容许,准许 *n.* ['pə:mit] 许可,许可证,执照

permoglaze *n.* 一种透明的防水墙壁涂料

perpend ['pə:pend] *n.* 穿墙石;贯石;(复)砖石砌体的垂直缝 *v.* [pə'pend] 仔细考虑,注意
~ stone 穿墙石

~ wall 单石(薄)墙

perpendicular [pə;pən'dikjulə] *a.* 正交的,与……垂直的 *n.* 垂直,竖直,垂线,垂直面
~ architecture 哥特式建筑
~ cathedral 哥特式大教堂
~ cut 垂直切割
Perpendicular Style 哥特式建筑风格,垂直式建筑风格
~ tracery 垂直窗格

perpendicular tracery

perpendicularity [pə;pəndikju'læriti] *n.* 垂直(性),直立,正交

perpetual [pə'petjuəl] *a.* 永久的,不间断的
~ shadow 全年阴影

perpetuate [pə;'petʃueit] *vt.* 维持,保全,使不朽 *a.* 已成为永恒的

perpetuity [pə;pə'tju(:)ti] *n.* 永恒,永存物

perron [perən] *n.* (建筑物入口)露天梯级;作讲台用的大石块,石台座

Persian ['pə:ʃən] *a.*;*n.* 波斯(的),伊朗的,波斯人(的);波斯语(的);(复)男像柱,人形柱
~ architecture 波斯建筑艺术
~ blinds 百叶窗
~ carpet [rug] 波斯地毯
~ column 波斯(人形)柱
~ corbeled arch 波斯突拱
~ red 波斯红,格红
~ style 波斯式建筑

Persian

persiennes [pə;sienz] *n.* (复)百叶窗

persist [pə'sist] *v.* 持久,耐久;持续;主张

persistence [pə(:)'sistəns] 或 **persistency** [pə(:)'sistənsi] *n.* 持久(性)稳定(性),保留时间,(时间)常数

persistent [pə'sistənt] *a.* 持久的,不变的,稳固的

person [pə:sn] *n.* 人(员),个体
personal ['pə:sənl] *a.* 个人的,本人的,私[自]的,专用的
~ error 人为(操作)误差
~ factor 人为因素
personality [pə:sə'næliti] *n.* (*pl.* -ties) 品格,个性
personification [pə(:)sənifi'keiʃən] *n.* 人格化;典型[范];体现
personnel [pə:sə'nel] *n.* 全体人员[职员],人事部门 *a.* 人事的
~ block 办公大楼
~ -carrying hoist 乘人电梯
~ changing room 职工更衣室
~ house 职工住宅
engineering technical ~ 工程技术人员
persorption [pə(:)'sə:pʃən] *n.* 多孔性吸附
perspective [pə'spektiv] *n.* 透视(图,画法),投影,配景;*a.* 透视的,配景的
~ center 透视中心
~ drawing 透视图
~ geometry 投影几何
~ grid 透视格网
~ model 透视模型
~ picture 透视画
~ projection 立体[透视]投影
~ view 透视图
angular ~ 斜透视
architectural ~ 建筑透视图
bird's eye ~ 鸟瞰图
isometric ~ 等角透视
one point ~ 一点透视
panoramic ~ 全景透视
two point ~ 二点透视
perspectivity [pə:spek'tiviti] *n.* 透视,明晰度
perspectograph [pə'spektəgrɑ:f] *n.* 透视绘画器
perspex ['pə:speks] *n.* 有机玻璃,透明塑料,塑胶玻璃
perspire [pəs'paiə] *v.* 排出,蒸发

pertain [pə:tein] *vi.* 附属于;与……有关系;适于
pertaining *a.* 附属的 *n.* 附属物
perthite ['pə:θait] *n.* 条纹长石
pertinax *n.* 电木,胶纸板,酚醛塑料
pertinent ['pə:tinənt] *a.* 相应的,与……有关的 *n.* (*pl.*)附属物
perturb [pə'tə:b] *vt.* 干扰,扰动,使紊乱
pervial *a.* 可透水的,能透过的
pervious ['pə:vjəs] *a.* 能透过的,可透水的,有孔的
~ concrete 透水混凝土
~ backill material 透水性回填材料
~ bed 透水层,渗透层
~ course 透水层
~ material 透水材料
~ overburden 透水覆盖层
~ sand gravel 透水砂砾层
~ to air [moisture, water] 透气[湿,水]
perviousness ['pə:vjəsnis] *n.* 透水[过]性
petaloid ['petəloid] *a.* 花瓣状的
petcock ['petkɔk] *n.* 小型旋塞,小龙头,手压[减压,排泄]开关
petralol *n.* 液体石油膏
petrean [pi'tri:ən] *a.* 石质的,岩[化]石的,硬化的
petrify ['petrifai] *v.* 转化为石质,硬化
~ing liquid (喷墙)防潮液
petrobenzene *n.* 石油苯
petroglyph ['petrəuglif] *n.* 岩石雕刻(尤指史前的),摩崖石刻
petrol ['petrəl] *n.* 石油,汽油 *vt.* 加汽油
~ asphalt 石油沥青
petroleum [pi'trəuljəm] *n.* 石油
~ asphalt 石油沥青
~ bitumen 石油沥青
~ coal 固体石油
~ grease 石油润滑剂
~ jelly 矿脂,凡士林,石油膏

~ lubricant 石油润滑剂
~ oil 石油
~ paraffin 石油石蜡
~ resin 石油树脂
lacquer ~ 漆用汽油

petty ['peti] *a.* 细小的,次要的,小规模的

pew [pju:] *n.* 教堂中的一排座位(靠背固定的长凳),教堂内某一家族专用之厢席

pewter ['pju:tə] *n.* 锡蜡(锡,铅合金);锡蜡器皿 *a.* 锡蜡制的

phaeton ['feitn] *n.* 敞篷汽车,游览车,一种轻快的四轮马车

phantasm ['fæntæzəm] *n.* 幻影[觉、想],假象

phantom ['fæntəm] *n.* 幻象[影,觉]仿真;(部分)剖视图 *a.* 空幻的,虚的,假想的;部分剖视的
~ drawing 幻想画
~ view 部分剖视图,幻想景色

pharos ['fɛərɔs] *n.* 灯塔,航标灯

pharosage *n.* 辐射[照明]照度

phenol ['fi:nɔl] *n.* (苯)酚,石炭酸
~ formaldehyde resin (苯)酚(甲)醛树脂
~ oil 苯酚润滑油
~ resin bonding 酚醛树脂胶粘剂

phenolic *a.* (苯)酚的,酚醛的
~ aniline resin 酚醛苯胺树脂
~ adhesive 酚醛胶粘剂
~ compound 酚型化合物
~ (-formaldehyde) resin 酚醛树脂
~ laminated board 酚醛层压板
~ plastics 酚醛塑料
~ varnish 酚醛清漆
~ vehicle 酚醛溶剂
~ synthetic resin cement 酚醛合成树脂胶
~ putty 酚醛合成树脂腻子

phenolplast *n.* 酚醛塑料

phenomenon [fi'nɔminən] *n.*(*pl.* phenomena [fi'nɔminə])现象;不平常的事物、奇迹,珍品

phenoplast *n.* (苯)酚醛塑料

phenoweld *n.* 改性酚醛树脂胶粘剂

phenoxy resin 苯氧基树脂

phiale *n.* (教堂前的)喷水池

philharmonic [filɑ:'mɔnik] *n.* 交响乐团,(交响乐团主持的)管弦乐队;音乐会 *a.* 交响乐的,音乐团体的
~ hall (演奏交响乐的)音乐厅

Philip ['filip] *n.* 菲利普
~s driver 十字螺钉旋具(螺丝刀)

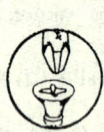   

Philips driver   phoenix

~s screw 带十字槽头的螺钉

phoenix ['fi:niks] *n.* 凤凰,凤凰饰
~ column 弓形槽钢组合圆柱

phon [fɔn] *n.* (声音的响度单位,=1分贝)

phone [fəun] *n.* 电话,送受话器,耳机 *v.* 打电话
~ booth 公用电话亭,电话间
~ installation 电话设施,电话台
~ line 电话线
~ point 电话插座
~ socket 电话接线匣

phosphate ['fɔsfeit] *n.* 磷酸盐[酯]
~ chrome 磷铬黄
~ crown glass 磷(酸盐)晃玻璃

phosphor ['fɔsfə] *n.* 黄磷,磷光体,荧光物质
~ bronze 磷青铜
~ copper 磷铜
~ screen 荧光屏,荧光板

photetch *n.* 光蚀剂,光刻技术

photic ['fəutik] *a.* 光的,发光的,感光的,透(日)光的
~ zone 透光层

photoelectric(al) [ˌfəutəuiˈlektrik(əl)] *a.* 光电的
~ material 光电材料

photo-fabrication [ˈfəutəufæbrikeiʃən] *n.* 光加工, 光镂, 照相化学腐蚀制造法

photogalvanic effect 光电效应

photoluminescence [ˈfəutəuluːmiˈnesns] *n.* 光致发光, 荧光

photomural [ˌfəutəˈmjuərəl] *n.* (装饰性)大幅照片, 壁画式照片

photosensitive [ˈfəutəˈsensitiv] *a.* 光敏的, 能感光的
~ glass 光敏玻璃

photosensitizer [ˌfəutəˈsensitaizə] *n.* 光敏剂, 感光剂

photostudio *n.* 照相馆, 摄影棚

phthalic acid 酞酸, 苯二酸

phthalocyanine [ˌθæləuˈsaiəniːn] *n.* 酞菁染料
~ blue 酞菁蓝
~ dyestuff 酞菁染料
~ green 酞菁绿

physical [ˈfizikəl] *a.* 物质的, 物理(学的); 自然界的; 身体的, 体力的 *n.* 体格检查
~ design (机械的)结构设计; 城市(物质环境)设计
~ demension 实际尺寸
~ environment 自然环境
~ facilities 物质设施
~ factor 自然因素
~ life-time (of building) (房屋)使用期限
~ wear and tear 实际磨损

piauzite [piˈɔːzait] *n.* 板沥青

piazza [piˈætsə] *n.* 广场, 市场; 有(拱形)顶的长廊; 游廊, 外廊

pick [pik] *n.* 鹤嘴锄, 丁字镐 *v.* 掘, 挖, 凿; 琢; 挑选, 抓取, (绘画的)修改
~ axe 丁字斧, 鹤嘴斧
~ed dressing 粗琢(石面)凿平修饰

picket [ˈpikit] *n.* 尖木桩 *vt.* 用木桩[栅]围住
~ fence 尖桩篱栅

picking [ˈpikiŋ] *n.* 掘, 刨, 挑选
~ -up coat 拉毛粉刷层

pickle [ˈpikl] *n.* 酸洗液, 浸渍盐水 *vt.* 浸, 泡, 酸洗
~ing oil 防锈油

picture [ˈpiktʃə] *n.* 图画, 照片, 电影, 摄影镜头, 画面 *vt.* 描绘[述], 画
~ drome 电影院
~ frame 画框
~ gallery 画廊, 美术陈列馆
~ hall 电影院
~ house 电影院
~ moulding 墙上挂镜线
~ palace 电影院
~ rail (rod) 挂镜线, 挂画条
~ show 画展, 电影(院)
~ theatre 电影院
~ window 大型景窗, 借景窗
~ wire 挂图像用金属线
watercolour block printing ~ 水印画

piece [piːs] *n.* 片, 块, 件; 零[部]件, 构件, 段 *vt.* 拼成, 接合, 修理
adapting ~ 连接件
apron ~ 遮檐[挨水]板
camber ~ 起拱构件
centre ~ (顶棚的)中心装饰
corbel ~ 支[托]架, 支臂, 悬臂; 挑出块
discharging ~ (门窗)过梁
drop end ~ (檐沟的)端部下水口
head ~ 木隔断的压顶木
head ~ of column 柱头
knee ~ 曲橡
nogging ~ 压墙槛
plancier ~ 支承屋顶檐口衬板的水平梁
pole ~ 立柱, 支柱
protecting ~ 护板

sprocket ～ 屋架斜梁，短人字木，小椽
standard test ～ 标准试件
stop end ～ 檐沟的端盖
strutting ～ 斜撑，拉条，支柱
T ～ T 形板，T 形接头
turning ～ 单拱架
valley ～ 无沟椽
wall ～ 搁墙撑头木

piedroit *n.* 与墙结合的平柱

pier [piə] *n.* (桥，墙，闸)墩，窗间墙，户间壁，扶壁柱，角柱，支柱
～ arch 支柱拱
～ bath 单间浴室
～ capping beam 墩盖梁
～ column 墩柱，支柱
～ fender 码头护木，桥墩护栏
～ glass 大穿衣镜，窗间镜
～ stud 墩柱
brick ～ 砖柱[墩]
compound ～ 组合柱
engaged ～ 墙垛
gate ～ 墙门墩柱
supporting ～ 撑墙垛，支墩
wall ～ 墙墩

pierce [piəs] *v.* 贯穿，渗透，钻入；
～ -ed brick 穿孔砖
～ -ed buttress 穿孔式护壁

piercing ['piəsiŋ] *a.* 锐利的，融穿的 *n.* 贯穿，钻孔
～ mill 穿孔机
～ saw 弓锯，钢丝锯

pietra dura *n.* 镶嵌用的硬细石

pigment ['pigmənt] *n.* 颜料，色素 *v.* 加颜色，变色
～ed finish 着色装饰[涂料]
～ed glass 有色玻璃
body ～ 浓缩颜料，底质颜料
loading ～ 颜料填充剂
mars ～ 合成氧化铁颜料
traditional Chinese painting ～s 中国画颜料

pigmentation [pigmən'teiʃən] *n.* 颜料淀积(作用)，色素形成(作用)
～ of cement 水泥着色

pilaster [pi'læstə] *n.* 扶壁柱，半露方柱
～ capital 壁柱帽
～ strip 无帽壁柱
corner ～ 转角壁柱
paired ～s 对半柱

pilastrade [piləs'treid] *n.* 一排壁柱

pile [pail] *n.* 桩[柱]一堆，一排，层积 *v.* 堆积，打桩，用桩支撑
～ crown 桩头

pillar ['pilə] *n.* 柱，柱装物，墩 *v.* 用柱支持
～ bearing 墩[柱]支座
～ -bolt 柱形螺栓
～ stone 奠基石
～ support 柱支座，柱基
～ switch 柱式开关
～ type basin tap 立柱式洗脸盆水龙头
cloud ～ 华表
ornamental ～ 华表
window ～ 窗棂

pillar

pillaret ['pilərit] *n.* 小柱

pillbox [pilbɔks] *n.* 小屋，碉堡，掩体

pillow ['piləu] *n.* 轴枕，垫座[板] *v.* 枕，靠，搁
～ joint 球形接合

piloti [pi'lɔt] *n.* (*pl.* pilotis) 鸡腿式建筑，桩基，高层建筑架空底层的柱

pimple ['pimpl] *n.* 小突起，小高处

pimpling *n.* 粗糙度

pin [pin] *n.* 枢轴，销，钉，螺栓，铰 *a.* 销的 *vt.* 钉[按]住；牵制
～ bolt 销钉，带销螺栓
～ -connected construction 铰接结构
～ connection 铰结合
～ filler 栓钉垫
～ fin 钉头
～ hinge 销铰铰链
～ joint 铰接
～ plate 栓接板

~ rail 衣帽架,带栓横梁
~ seal 销钉连接
~ splice 销钉拼接板
~ spot-light 细光束聚光灯
glazing ~ (窗上固定玻璃的)玻璃卡钉
pinacotheca [pinəkəu'θi:kɔ] *n.* =pinacothek(*pl.* -cas)美术馆,艺术陈列馆
pinboard *n.* 插销[接线]板
pincers ['pinsəz] *n.* 钳子,镊子
pinchbeck ['pintʃbek] *n.*; *a.* 金色黄铜,铜锌合金;伪造品
pine [pain] *n.* 松树[木],凤梨(菠萝)
~ tar 松木焦油,松木焦油沥青
~ tar pitch 松木焦油脂[硬沥青]
~ -tree resin 松[树]脂,松香
pink [piŋk] *n.* 粉红色;石竹 *a.* 粉红色的,石竹色的 *vt.* 刺,戳,穿小孔;装饰
pinnacle ['pinəkl] *n.* 小尖塔,尖顶[柱]尖峰,尖,极点 *vt.* 置于尖顶上;筑小尖塔[顶、阁]
~ terminating in conical form 圆锥体尖顶
~ terminating in pyramidal form 金字塔形尖顶
~ed canopy 尖拱棚顶
pinner ['pinə] *n.* (支撑大块石材的)小块石
pinoleum blind *n.* 木条卷帘
pin-up *n.* 钉在墙上的东西
~ lamp 壁灯
pipe [paip] *n.* 管子[道],导管;筒 *v.* 给……装管子,用管子输送
~ arrangement 管道布置
~ drain 排水管,管式排水
~ for conducting air 通风管道
~ line 管线,管路
~ pendant lamp 管吊灯
~ radiator 管式[盘管]散热器
~ storm drain 雨水管
blow ~ 排气管,风管
drain ~ 排水管
heating ~ 暖气管
intake ~ 进水管
mud ~ 排污管
offtake ~ 排水管
pressure water ~ 自来水管
radiating ~ 散热管
vent ~ 通风管
warm-air 暖气管
water service ~ 供水管
pipeline ['paiplain] *n.* 管线[路],供给系统,输油管,商品供应线 *v.* 装管道,用管道输送
piping ['paipiŋ] *n.* 管道系统,配管,管系总长 *a.* 尖声的,高音的
~ diagram 管道布置图
pisolite ['paisəlait] *n.* 豆石
pisspot ['pispɔt] *n.* 室内便器
pissasphalt ['pisæsfælt] *n.* 软沥青
pistol ['pistl] *n.* 手枪,手持喷枪,枪式焊接器 *vt.* 以手枪射击,喷射
spray ~ 喷雾枪
pit [pit] *n.* 坑,槽,洞,纹孔;(煤)矿井,地窖;地下温室;(剧场正厅)后排 *vt.* 型凹,挖坑
~ foyer (剧院)正厅后排休息室
pitch [pitʃ] *n.* 节[齿、螺]距,铆钉距;高跨比,矢高,斜度,斜坡;(硬)沥青,树脂,塔脂 *v.* 涂沥青,铺砌
~ face 斜凿面
~ faced masonry 凿面石建筑
~ -faced stone 凿面石
~ of arch 拱高,拱的高跨比
~ of centers 顶尖[轴心]高度
~ of roof 屋顶高跨比[斜度]
~ of stairs 楼梯斜度
~ point 节点
~ stone 琢石
~ed dressing 琢边石工
~ed felt 油毡
~ed paper 油纸
~ed roof 斜屋顶
full ~ 高跨相等的屋面坡度
Gothic ~ 哥特式屋面坡度(60°坡)

pitchy

Grecian ~ 希腊式屋面坡度(15°)
ordinary ~ 普通(屋面)坡度(37.5°)
quarter ~ 26.5°的屋顶坡度
Roman ~ 罗马式屋面坡度(25°)
roof ~ 屋面坡度
square ~ 45°的屋顶坡度
third ~ 33.5°的屋顶坡度

pitchy ['pitʃi] *a.* 沥青的,涂有沥青的；漆黑的；多树脂的
~ lumber 多树脂的木材

pivot ['pivət] *n.* 枢(轴),支点,转轴,中枢,中心点,基准点 *a.* 在枢轴上转动的,枢轴的
~ gate 旋转门
~ hinge 轴铰链
~ light 摇窗
~ pin 中心承枢
~ed casement 旋转窗
~ed end column 铰支柱
~ed sash 旋转窗扇
~ed window 摇窗
~ing bearing 中心支承

place [pleis] *n.* 地方[区,段],区域；广场；空间；位置 *vt.* 放置,配置；灌筑；安排
~ of abode 居住区
~ of amusement 娱乐场所
~ of historical interest 古迹
~ of interest 名胜
~ of public resort 公共休养所
~ed rockfill 干砌块石
working ~ 工作空间[区],操作位置

placeability [pleisə'biliti] *n.* 混凝土的可灌筑性,和易性,工作度
~ of concrete 混凝土和易性
~ of mix 混凝土拌合料的和易性

placement ['pleismənt] *n.* 方位,位置,布局,填筑
~ policy 布局规则

plafond [plafɔ] *n.* 【法】(*pl.* -fonds)有装饰的顶棚,顶棚上的彩画[雕刻]

plagioclase ['pleidʒiəukleis] *n.* 斜长石

plaid [plæd] *n.* ; *a.* 方格花纹(的)

plain [plein] *a.* 简单的；单调[色]的,(朴)素的,普通的；平滑的 *n.* 平原[地] *ad.* 平(易);清楚
~ arch 单拱,光面拱
~ batten door 平板门
~ boltel 简单凸圆饰
~ butt joint 平接缝
~ cement 纯水泥
~ concrete 素[纯,无筋]混凝土
~ dressing 光面修整
~ flat strip moulding 平腰线,压缝条
~ mitre joint 平斜接
~ pendant 素垂饰
~ round moulding 纯圆饰
~ tile 平瓦
~ work 平缝,凿平的石面

plan [plæn] *n.* 计[规]划,设计,方案,平面[设计,规划布置]圆 *v.* 计划,设计,绘制……图
~ layout 平面布置
~ of site 总平面布置图
~ view 平面图
block ~ 区划图,位置图
construction ~ 施工平面[布置]图
first floor ~ 第二层平面图(英),第一层平面图(美)
floor ~ 楼面布置图
foundation ~ 基础平面图
general ~ 总图,总计划
general arrangement ~ 总体[总平面]布置图
ground floor ~ 底层平面图
key ~ 建筑位置图,索引图
location ~ 位置图
longitudinal ~ 纵剖面图
master ~ 总平面图,总体规划图
outline ~ 初步计划,提纲
perspective ~ 透视图
rough ~ 设计草图,初步计划
sheer ~ 侧面图

wiring ~ 线路敷设图
working ~ 工程程序图,施工图
planar ['pleinə] *a.* 平面的,在平面内的,平的;二维的
plane [plein] *n.* 平面;投影;刨;镘;法国梧桐树;飞机 *a.* 平的,平面的 *vi.* 弄平,整平,刨平;乘机
~ figure 平面图
~ framework 平面框[构]架
~ of illumination 受照面
~ of incidence 入射面
~ of projection 投影面
~ of reference 参考面,基准面
~ of structure 结构平面
~ section 剖面
~ tree 法国梧桐
cut ~ 剖面
datum ~ 基准面,水准基面
profile ~ 侧立面
wood ~ 木工刨
planform ['plænfɔ:m] *n.* 平面图
planish ['plæniʃ] *vt.* 弄[刨,磨]平,锤[抛]光
plank [plæŋk] *n.* 厚木板(厚5~15cm,宽大于23cm),垫板,支撑物 *vt.* 铺厚木板
~ base 木基层
~ floor 木地板
~ frame (门窗)木框
~ nail 木板钉
~ platform 木板平台
~ roof 木屋顶
~ sheathing 木护壁
adjacent ~ 踢脚板,门楣,门框边梃
factory ~s 门窗板条
rough ~ 毛板
planking ['plæŋkiŋ] *n.* 铺板工作;板材,地板
planner ['plænə] *n.* 设计人员,规划工作者
planning ['plæniŋ] *n.* 计划,规划;设计;分布
~ commission 计划[规划]委员会

~ grid (平面草图)设计网格
~ of civic center 市中心规划
~ of park and green 公园绿地规划
city ~ 城市规划
complex town ~ 城市整体规划,城市综合布置
integrated ~ 整体规划
urban ~ 都市规划
plant [pla:nt] *n.* 工厂,(电)站,车间,(机械)设备,装置;植物,苗木 *vt.* 设置,种植
~ed slope 绿化的边坡
concrete ~ 混凝土厂
construction ~ 施工工地
plantagenet style *n.* 英国金雀花王朝式(建筑)
planting *n.* 种植,绿化;基础底层
plaque [pla:k] *n.* (金属、陶瓷等制成的)装饰板、片;送风口散流板
~ rail (墙上的)装饰品
plaster ['pla:stə] *n.* 灰泥,灰浆,墙粉,涂层,熟石膏 *v.* 粉刷,抹灰,粘贴,涂抹
~ base 抹灰底层
~ board nail 灰泥板钉
~ cast 石膏模型
~ ceiling 抹灰顶棚
~ figure (mould) 石膏像
~ finish 抹灰饰面
~ kiln 石膏窑
~ material 塑性材料,塑料
~ model 石膏模型
~ of paris 熟石膏($CaSO_4 \cdot 1/2H_2O$)
~ on metal lath 钢丝网抹灰
~ stone 生石膏
~ tablet 石膏板
~ wainscot cap 护墙板抹灰压顶
calcium ~ 石膏灰浆
hard wall ~ 水泥粉刷
lath and ~ 板条抹灰
wall ~ 墙面粉刷
plasterboard nail 糊墙纸板钉
plastering ['pla:stəriŋ] *n.* 抹灰工作,泥

水工作,粉刷,石膏制品
plastic ['plæstik] *a.* 可塑的,塑性的,塑料的,塑造的,造型的 *n.* (常用 *pl.*)塑料,合成树脂,电木;塑料制品
~ article 塑料制品
~ arts 造型艺术(指雕塑,陶瓷品制造等)
~ colour 造型配色
~ effect 立体感,浮雕效应
~ floor 塑料地面
~ foam 泡沫乳胶
~ glues 合成树脂胶
~ insulated conductor 塑料绝缘导线
~ language 造型语言
~ paint 塑料漆
~ pipe 塑料管
~ refractory 耐火材料
~ sign 造型语言
~ thinking 造型构思
~ veneer 塑料饰面
~ window lights 塑料玻璃窗
cellular ~ 泡沫塑料
optical ~ 光学塑料,有机玻璃
plasticine ['plæstisi:n] *n.* 造型材料,蜡泥塑料,型砂
plasticizer ['plæstisaizə] *n.* 增塑剂,塑化剂,柔韧剂
plasticon *n.* 聚苯乙烯薄膜
plastigel *n.* 塑性[增塑]凝胶
plastilock *n.* (用合成橡胶改性的)酚醛树脂胶粘剂
plastisol ['plæstisɔl] *n.* 塑料溶胶
vinyl ~ 乙烯塑料溶胶
plasto-concrete *n.* 塑性混凝土
plastoelastic *a.* 弹塑性的
plat [plæt] *n.* 地段,地区;地段图 *vt.* 绘地图;编织
platan(e) ['plætən] *n.* 悬铃木,法国梧桐
platband ['plætbænd] *n.* 平边,长条地;(建筑立面上的)横线条
plate [pleit] *n.* 板,盘,平板,板材,钢板,金属板,玻璃板,电极板,感光板;整版的插图 *vt.* 镀,涂覆,铺板,打成薄板
~ and angle column 钢板角钢组合柱
~ and channel column 钢板槽钢组合柱
~ arch (平)板拱
~ closet 壁橱
~ connector 组合插座
~ girder 板梁
~ glass 平板玻璃
~ rack 餐具橱
~ scullery 洗碟室
~ shelf 墙上装饰架支架
~ tracery 板制雕窗花格
~ -type dome 板型圆顶
~ -type absorber 隔板式吸声装置
back ~ 背面板,后板
ceiling ~ 顶棚
crown ~ 梁垫顶板
division ~ 隔板
face ~ 面板
floor ~ 铺板
flush ~ 门心板,平贴盖板
ground-glass ~ 毛玻璃板
metal-on-glass ~ 敷金属玻璃板
mild steel ~ 低碳钢板
muffler ~ 减声板
name ~ 铭牌
partition ~ 隔墙板
running ~ 地板
steel ~ 钢板

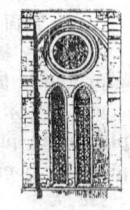

plate tracery

plated *a.* 电镀的,镀……的
~ beam 叠板梁
platfond *n.* 顶棚
platform ['plætfɔ:m] *n.* 平[站,工作]台,楼梯休息平台,台地 *a.* 平台式的 *v.* 把……放在台上
platina ['plætinə] *n.* 铂,白金(Pt.)
platine *n.* (装饰用)锌铜合金(锌57%,其余铜)

plating ['pleitiŋ] *n.* 电镀,喷镀,镀金属,外覆金属板,熨平
~ paper 印纹纸

platoom ['plætinəm] *n.* 铂(Pt)白金 *a.* 白金色的

platy ['pleiti] *a.* 板状的,扁平状的
~ joint 板状节理
~ structure 板状结构[构造]

play [plei] *v.* 游戏,演奏;起作用 *n.* 游戏,戏剧,活动,缝隙,间距,冲程
~ ground (学校的),操场,运动场,儿童游戏场
~ house 剧场,儿童游戏室
~ room 儿童游戏室,文娱活动室

playing ['pleiiŋ] *n.* 游戏,竞技,比赛,演奏
~ field 运动场,竞技场,(儿童)游戏场

plaz(z)a ['plɑ:zə]【西】*n.* 广场,集市[会]场所

pleasure ['pleʒə] *n.*;*v.* (使)高兴,(使)愉快
~ dome 富丽堂皇的大厦
~ garden 公园,游乐场
~ ground 娱乐场
~ -house 娱乐场所

plenish ['pleniʃ] *vt.* (给房屋)安装设备

plexiglass [pleksiglɑ:s] *n.* 耐热有机玻璃

pliers ['plaiəz] *n.* 钳子,夹钳,扁嘴钳,老虎钳

plinth [plinθ] *n.* 底座,勒座,柱础
~ block 基底石块
~ course 底座层
~ stone 底石
~ wall 柱脚,台脚

plot [plɔt] *n.* 地块,基址,(小)区;曲线(图),图表;绘图板;计划 *vt.* 绘图,测定(点、线的位置);区划,划分,策划
~ line 地基界线,建筑红线
~ plan 地区[位置,坝址,平面]图
~ planning 小区规划,总平面布置

plotter ['plɔtə] *n.* 绘图机[器],绘图员
~ field 绘图板

plotting ['plɔtiŋ] *n.* 测绘,绘图,划曲线
~ board 曲线板
~ office 绘图室
~ paper 方格绘图纸,比例纸
~ scale 比例尺

plough [plau] *n.* =plow 犁,刨煤机,扫雪机,开沟器,平地机,(木工)槽刨,刮板 *v.* (木工)开槽
~ bolt 防松螺栓
~ed and tongued joint 嵌榫拼接
~ed bead moulding 沟槽圆饰

plug [plʌg] *n.* 塞,栓,插入物,电插头[销],消防龙头 *v.* 堵塞,插入 *a.* 柱形的,插入式的
~ board 插线板
~ of felt 油毛毡填充料
~ receptacle [socket] 插座
~ tap 转塞式龙头
~ tenon 塞榫,尖榫
fire ~ 消火栓
lamp ~ 电灯插头
telephone ~ 电话插塞

plugging *n.* 堵塞;隔声材料
~ agent 堵塞[止水]剂
~ chisel 嵌缝刀
~ -up line 用户(闭塞)线

plum [plʌm] *n.* (混凝土用)毛石块料

plumb [plʌm] *n.* 铅锤,线铊;垂线,垂直 *a.* 垂直的;公正的 *v.* 使垂,用铅锤检查垂直度,铅锤测水深;安装铅管
~ bob 铅[测]锤,垂球
~ level (有铅垂线的)水准仪
~ line 铅垂线,准绳

plumbago [plʌm'beigəu] *n.* 石墨(粉),炭精
~ refractory 石墨耐火材料

plumbeous ['plʌmbiəs] *a.* 含[似]铅的,铅色的,重的

plumbing ['plʌmiŋ] *n.* 卫生管道[设备]工程;燃料管道;液压系统管路;铅锤测量

~ equipment 给排水设备,卫生设备
~ fittings 卫生管道配件
~ fixtures 卫生设备[器具],给排水设备
~ installation 卫生工程管道施工法
~ system 室内管道系统
~ trap (卫生)管道存水弯
~ work 卫生设备工程
house ~ 住宅卫生设备

plummet ['plʌmit] *n.* 测[铅]锤,垂线 *vt.* 垂直落下,骤然跌落
~ level 定垂线尺

plums *n.* 填料,圬工石

plunger ['plʌndʒə] *n.* 柱塞,插棒,压实器,撞针;跳水者

plural ['pluərəl] *n.*;*a.* 复数(的),多于一个的
~ gel 复合凝胶

pluramelt *n.* 包(不锈钢)层钢板

plus [plʌs] *n.* 正[加]号,正数,附加额[物] *a.* 正的,略大[高]的;阳性的,附加的 *prep.* 外加
~ driver 十字螺钉旋具
~ material 面料
~ screw 十字槽头螺钉
~ sight (测量)后视

plus-minus *n.* 正负,加减,调整
~ screw 调整螺钉

ply [plai] *n.* 层,股,叠加 *v.* 使胶[绞]合,折,弯
~ bamboo 多层竹板
~ cushion 垫层
~ extra 附加层
~ form 胶合板模板
~ glass 纤维夹层玻璃
~ max 镶铅装饰胶合板
~ metals 包层金属板
~ wood 胶合板,层压木板
three- ~ wood 三合板

plyer *n.* 拔管台,拔管小车;(*pl.*)钳子,手[老虎]钳

plywood ['plaiwud] *n.* 胶合板 层压板

~ adhesive 胶合板用胶粘剂
~ formwork 胶合板模板
~ metal sheet 金属面胶合板
~ sheathing 胶合板衬板
~ square 胶合板镶块

pneumatic [nju:'mætik] *a.* 空气的,气压的;气动的;风动的;充气的;气体(力)学的
~ architecture 充气建筑
~ ejector 气压喷射器
~ gun 喷射枪,气锤,风动铆钉枪
~ mortar 喷射砂浆,压力喷浆
~ paint brush 气压喷漆器
~ placing 压气浇筑(混凝土)
~ sewage system 气压排水系统
~ structure 充气结构
~ water supply system 压气供水系统

pneumatically *ad.* 靠压缩空气
~ applied mortar 喷浆
~ placed concrete 喷混凝土

pocket ['pɔkit] *n.* 袋,囊,匣;窝,坑,矿穴;小块地区,死胡同 *a.* 袖珍的,小型的,压缩的 *v.* 装入袋内,封入
~ butt 凹槽平缝铰链
~ calculator 袖珍计算器
~ of air 气穴,气坑
~ piece 吊窗锤箱(修理)口
~ tape 钢卷尺
~ rule 折尺

pocket-handkerchief *n.* 手帕,小型物
a ~ lawn 小块草地

podium ['pəudiəm] *n.* 垫块,列柱,墩座,墩座墙(古代圆形竞技场地与观众席隔开的矮墙);交通警指挥台
~ block 墩座石块

point [pɔint] *n.* 点,小数点,地点,测点,交点;插座;道岔;论[观]点 *v.* 面[朝]向,瞄准,琢石,嵌填,勾缝
~ building 塔式建筑
~ design 符合规定要求的设计
~ house 点式住宅
~ light source 点光源

~ of fixity 固定点
~ sound source 点声源
~ source 点光源,点污染源
~ vanishing (透视)消灭点
conjugate ~ 共轭点
critical ~ 临界点
light ~ 照明插座
pointed *a*. 尖的,头锐的;有所指的 *v*. 琢石,勾缝
~ arch 尖拱,内二心桃尖拱
~ architecture 尖拱式建筑,哥特式建筑
~ dome 尖圆屋顶
~ dressing 尖凿修整
~ drill 尖凿
~ element 尖头构件
~ finish 点凿面
~ in white cement 白水泥勾缝
~ joint 勾缝
~ joint of random rubble (成层砌筑的)乱石砌法
~ opening 尖拱式门洞
~ pediment 三角形檐饰
~ pile 尖头桩
~ saracenic arch 撒拉逊尖券
~ shell dome 葱头形薄壳穹顶

pointed arch

pointing *n*.;*a*. 指瞄准,削尖砌砖勾缝,嵌填
~ chisel 点凿
~ joint 勾缝,嵌缝
~ masonry 勾缝[嵌灰]圬工
~ rule 勾缝尺
~ stuff 勾缝料
~ trowel 勾缝镘刀
halfround ~ 半圆勾缝
square recessed ~ 方形凹缝

pointing trowel

pointless ['pɔintlis] *a*. 钝的,无尖头的;无目标的
pointolite ['pɔintəlait] *n*. 点光源
poke [pəuk] *n*.;*v*. 戳,捅,插,搅拌

~ hole 搅拌孔
~ welding 手压点焊
poker ['pəukə] *n*. 搅拌杆,火钳,烙画用具
~ -picture 烙画,焦笔画
pole [pəul] *n*. 极,极点;杆,棒,桩,柱 *v*. 支撑,架线杆
~ bracket 悬臂,撑架
~ frame construction 木构架结构
~ plate 挑檐檩,挑檐枋
~ scaffolds 撑杆脚手架
gum ~ 圆木
roof ~ 屋顶支柱
straddle ~ 撑架梁
polish ['pɔliʃ] *n*. 磨光,擦亮,精加工,擦光漆,虫胶清漆,泡立水 *v*. 磨光,擦亮,研磨
~ face 磨光面
~ finish (of stone) (石料)磨光面
~ed brass 抛光黄铜
~ed dressing 磨光,研磨
~ed finish of stone 打光石面
~ed glass 磨光玻璃
~ed surface 抛光面,精加工表面
~ing cloth 抛光布
~ing varnish 抛光漆
polka dot 圆点花纹
pollutant [pəu'lju:tənt] *n*. 污染物,污染物散布者 *a*. 污染的
~ substance 污染物质
air ~ 大气污染物
pollution [pə'lju:ʃən] *n*. 污染,浑浊
~ control 污染控制
atmospheric ~ 大气污染
environmental ~ 环境污染
poly ['pɔli] *n*. 多,聚
polyamide [pɔli'æmaid] *n*. 聚酰胺,尼龙
~ plastics 聚酰胺塑料
~ resin 聚酰胺树脂
polychloroprene [ˌpɔli'klɔ(:)rəpri:n] *n*. 聚氯丁烯,氯丁橡胶
~ rubber 氯丁橡胶

polychrome ['pɔlikrəum] *a.* = polychromatic 多色的,多色彩饰的 *n.* 多色画;彩色花瓶;色彩配合 *v.* 以多色绘画;彩画
~ finish 彩色装修
polychromy ['pɔliˌkrəumi] *n.* 彩色装饰,多色画(法)
polydirectional [ˌpɔlidi'rekʃənl] *a.* 多方向性的
polyene ['pɔliiːn] *n.* 聚烯
polyester [pɔli'estə] *n.* 聚酯
~ bath 聚酯塑料浴盆
~ fibres 聚酯纤维
~ film 聚酯软片
~ -glassfibre sheet 聚酯胶结玻璃纤维板
~ paint 聚酯漆
~ plasticizers 聚酯增塑料剂
~ resin 聚酯树脂
polyethylene [ˌpɔlie'θiliːn] *n.* 聚乙烯
~ film 聚乙烯薄膜
~ pipe 聚乙烯管
~ plastic(s) 聚乙烯塑料
~ sheeting 聚乙烯薄膜
polyfoam ['pɔlifəum] *n.* 泡沫塑料
polyfunctional [pɔli'fʌnkʃnəl] *a.* 多功能的,多重性的
polygon ['pɔligən] *n.* 多边[角]形
polygonal [pɔ'ligənl] *a.* 多边[角]形的
~ column 多边[角]柱
~ ground 龟裂地面
~ masonry 多角石(砌)圬工

polygonal masonry

~ rubble 多角毛石砌体
polyhedral ['pɔli'hedrəl] *a.* 多面体的
~ shell 多面体薄壳
polyhedron ['pɔli'hedrən] *n.* (*pl.* -dra, drons) 多面体;可剖分空间

polylaminate *a.* 多层的
polymer ['pɔlimə] *n.* 聚合物[体]
~ coating 聚合物涂层
~ sealer 聚合物密封剂
~ stabilization 聚合物加固
polypropylene [pɔli'prəupiliːn] *n.* 聚丙烯
~ corrugated pipe 聚丙烯波纹管
polystyle ['pɔlistail] *n.* 多柱式建筑(物) *a.* 多柱的
polystyrene [pɔli'staiəriːn] *n.* 聚苯乙烯
~ -plywood laminate 聚苯乙烯胶合板
~ resin 聚苯乙烯树脂
~ sheet 聚苯乙烯片[板]
polytechnic [pɔli'teknik] *a.* 多种工艺的,多种科技的 *n.* 综合性工艺学校,工业学校[大学]
polythene ['pɔliθiːn] *n.* = polyethylene 聚乙烯
polytropic [ˌpɔli'trɔpik] *a.* 多元[向]性的,多变性的
~ change 多元[因素]变化
~ compression 多向压缩
polyurethane *n.* 聚氨基甲酸酯
polyvinyl [ˌpɔli'vainil] *n.* ;*a.* 聚乙烯化合物(的)
~ acetate bonding agent 聚醋酸乙烯酯结合料
~ alcohol 聚乙烯醇
~ butyral 聚乙烯缩丁醛
~ chloride 聚氯乙烯(P.V.C)
~ chloride ceiling 聚氯乙烯塑料顶棚
~ fluoride 聚氟乙烯
~ resin 聚乙烯树脂
pommel ['pʌml] *n.* ;*v.* 圆头;球形端饰 *v.* 打击
pompous ['pɔmpəs] *a.* 豪华的,浮夸的,铺张的
ponceau ['pɔnsəu] *n.* 深[朱]红,丽春红,鲜红色染料

pony ['pəuni] *a.* 小型的 *n.* 小马驹 *v.* 付钱
~ girder 矮大梁
~ -size 小尺寸的
~ truss 矮桁架

pool [puːl] *n.* 水池[塘,潭];石油层 *v.* 形成小池;合办,联营,共享
swimming ~ 游泳池

poor [puə] *a.* 贫穷[瘠,乏]的;粗劣的,稀薄的,弱的
~ aggregate 劣质集料
~ concrete 贫混凝土
~ graded 级配不良的
~ graded sand 不良级配砂
~ lime 劣质石灰

pop [pɔp] *v.* 爆裂,发出爆裂声;突然行动 *n.* 爆裂声,汽水 *ad.* 突然 *a.* 通俗的,流行的,大众的
~ art 流行[通俗]艺术
~ safety valve 紧急安全阀

poplar ['pɔplə] *n.* 白杨树[木]
white ~ 白杨

poppy ['pɔpi] *n.* (*pl.* -pies) 深红色;罂粟花
~ head 顶花饰,罂粟花饰

popular ['pɔpjulə] *a.* 大众的,通俗的,能[流]行的

populate ['pɔpjuleit] *v.* 居住;增殖;提供
~ ed area 居住区

populous ['pɔpjuləs] *a.* 人口稠密的

porcelain ['pɔːslin] *n.* 瓷,瓷器 *a.* 瓷制的,精美的,易碎的
~ brick 瓷砖
~ carving 瓷雕
~ clay 瓷土
~ enamel 搪瓷
~ insulator 瓷绝缘子,瓷瓶
~ pipe 瓷管
~ radiator 陶瓷散热器
~ tube 瓷管
Art of pottery and ~ 陶瓷艺术

poppy head

biscuit ~ 素瓷
decorative ~ 彩瓷

porch [pɔːtʃ] *n.* 门[走,游]廊;入口处;大门内停车处
~ column 门廊柱
painted ~ 彩画游[门]廊
screen ~ 纱窗阳台
sun ~ 日光室
wind ~ 风廊

pore [pɔː] *n.* 孔隙,毛[细]孔,气孔
~ space 孔隙

porosint *n.* 多孔材料

porosity [pɔːˈrɔsiti] *n.* (*pl.* -ties) 孔隙率,多孔性,孔隙度
~ ratio 孔隙比

porous ['pɔːrəs] *a.* 多孔的,疏松的
~ brick 多孔砖
~ concrete 多孔混凝土
~ material 多孔材料
~ plastics 泡沫塑料
~ plate 多孔板
~ stone 透水石
~ structure 多孔结构
~ wall 多孔壁

porphyry ['pɔːfiri] *n.* (*pl.* -ries) 斑岩

port [pɔːt] *n.* 港口,港口城市,入口 *v.* 入港
~ structure 港口建筑物

portable ['pɔːtəbl] *a.* 轻便的,可携带的,手提式的
~ bath 移动式浴盆
~ dust suctor 轻便吸尘器
~ electric saw 手提式电锯,轻便电锯
~ extinguisher 手提式灭火器,轻便灭火器
~ fence 活动栅栏
~ house 活动房屋
~ ladder 轻便梯

portal ['pɔːtl] *n.* 大门,入口,桥[洞,隧]道门;门架
~ building 入口建筑
~ column 正门柱

portative ['pɔːtətiv] *a.* 轻便的,可搬运

的,可拆卸的
porter ['pɔ:tə] n. 看门人,清洁工,搬运工,列车员
～ house 酒店
～'s lodge 传达室
～'s room 传达室
porthole ['pɔ:thəul] n. 采光[通风]口,观察孔
portico ['pɔ:tikəu] n. (pl. -coes, cos) 门[柱,回]廊

portico

portion ['pɔ:ʃən] n. 部分,区划[段] v. 分配
Portland ['pɔ:tlənd] n. 波特兰(美国一港口地名)
～ blast-furnace cement 波特兰(硅酸盐)矿渣水泥
～ cement concrete 普通水泥混凝土
～ cement paint 硅酸盐水泥涂料
～ clinker 硅酸盐水泥熟料
～ -pozzolan cement [硅酸盐]火山灰水泥
～ -slag cement 硅酸盐矿渣水泥
～ stone 波特兰石(黄白色石灰石)
position [pə'ziʃən] n. 位置,地点,布局;座席,台 v. 布置,定位
positive ['pɔzətiv] a. 正的,阳(性)的;刚性的 n. 正面,正数,正压,照相正片
～ draft 人工通风
～ moment reinforcement 正弯矩钢筋
possess [pə'zes] v. 具有,拥有;支配
possibility [pɔsə'biliti] n. 可能性
possible ['pɔsəbl] a. 可能的
post [pəust] n. 柱,杆,桩,支撑;职位;邮政 ad. 在后 v. 公布,揭示;贴(布告等)
～-and-beam framing 柱梁构架

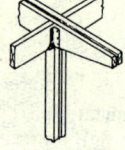
post-and-beam framing

～ -and-block fence 柱板围墙
～ and girder 梁柱结构
～ -and-lintel 连梁柱,过梁柱
～ -and-pan 砖木房屋
～ -and panel structure 立柱镶板式结构
～ and railing 立柱栏杆
～ fence 柱式护栏
～ light 柱灯
Post Modern Architecture 后现代建筑
Post Modern Formalism 后现代形式主义
base of ～ 柱基,柱基石
bearing ～ 支承柱
gate ～ 门柱
newel ～ 旋梯中柱
storey ～ 楼梯栏杆木柱
wall ～ 壁柱
poster ['pəustə] n. 广告[宣传]画
～ color 广告画颜料
posterior [pɔs'tiəriə] n. 后部 a. 后面的,后验的
postern ['pəustə:n] n. 便门,后门,暗道
postimpressionism ['poustim'preʃənizəm] n. 后期印象画派
post-installation a. 安装后的
～ review 安装后检测
pot [pɔt] n. 壶,罐,瓶,槽,坩埚 v. 装入壶[罐]中
～ clay 陶土
～ floor 空心砖楼板
～ -ted landscape 盆景
～ metal 低级黄铜
～ metal glass 有色玻璃
～ plant 盆栽植物
potential [pə'tenʃəl] a. 潜在的;位差的,势差的;电位的 n. 潜力[能],(动力)资源;势[位]能;电势
potstone n. 块滑石,粗皂石
potter ['pɔtə] n. 陶工,制陶工人
pottery ['pɔtəri] n. (pl. -teries) 陶器,

制陶术,制陶厂
~ clay 陶土
~ kiln 陶器窑
ancient painted ~ 石代彩陶
Black Pottery Culture 黑陶文化
colour-painted ~ 彩陶
glazed coloured ~ 彩釉陶
Stamped Pottery Culture 印纹陶文化
Tri-coloured glazed ~ of Tang Dynasty (中国瓷器)唐三彩
white ~ 白陶

pour [pɔː] *v.*; *n.* 浇注,浇筑,倾倒
~ concrete 浇筑混凝土
~ -in-place concrete 现场浇筑混凝土
~ joint filler 灌注式缝料
~ing pot (混凝土)浇筑罐
~ing tower (混凝土)浇筑塔

powder ['paudə] *n.* 粉末,香粉粉剂;炸药 *v.* 研磨成粉,粉化,施粉于
~ room 化妆室,盥洗室
~(ed) lime 粉状石灰
~ed asphalt 粉状地沥青
~ed pumice 浮石粉
red lead ~ 红铅粉

power ['pauə] *n.* 动力,电力;功率,效率;乘方,幂;发动机 *v.* 发动;装发动机
~ -off protection 断电保护
~ point 电力插座
~ skirting 电力插座护板
~ switch 电源开关
~ to sign 代表签字权
~ -transfer relay 故障继电器
main ~ 电源

power-consuming *a.* 消耗动力的,耗电的

power-operated *a.* 自动的,机动的

pozzolan [ˌpɔtsəˈlɑːn] *n.* = pozzuolana 火山灰
~ -cement 火山灰水泥

pozzolanic [ˌpɔtsəˈlɑːnik] *a.* = pozzuolanic 火山灰(质)的
~ admixture 火山灰质掺合料
~ cement 火山灰水泥
~ lime 火山灰质石灰,水硬石灰
~ material 火山灰质材料

practical ['præktikəl] *a.* 实际[践]的,实用的,实事求是的,经验丰富的
~ experience 实际经验
~ life 实际寿命
~ unit 实用单位

practice ['præktis] *n.* 实践,实施;实习,演习;习惯(作法),惯例;熟练;开业 *v.* = practise 实行,实习,开业
accepted ~ 常例,习惯做法
regular ~ 习惯[常规]做法
routine ~ 操作规程
safe ~ 安全技术[规章]
standard ~ 标准操作规程

practise ['præktis] *v.* 实施,养成……习惯;训练,实习;开业

practitioner [prækˈtiʃnə] *n.* 专业人员,开业医生,老手

pragmatic(al) [prægˈmætik(əl)] *a.* 实用主义的,实际的,实效的

prank [præŋk] *n.*; *v.* 装饰,点缀

prase [preiz] *n.* 绿石英

prearrange [ˈpriːəˈreindʒ] *v.* 预先安排,预定

preassemble [ˈpriːəˈsembl] *v.* 预装(配),预先组装
~d section 预装构件

preassigned [ˌpriːəˈsaind] *a.* 预先指定的,预先分配的

prebatched *a.* 预先配料的

prebend [priˈbend] *v.* 预弯(曲)

preblend [ˈpriːˈblend] *n.*; *v.* 预拌,预先混合,预拌混合料

precalculated *a.* 预先计算好的

precast [priˈkɑːst] *a.* 预浇(的),预制(的),装配式的 *v.* 将(建筑材料)预浇制成方块形
~ bathroom and toilet 预制卫生间
~ block 预制砌块

～ concrete 预制混凝土
～ concrete floor 预制混凝土楼板
～ curb 预制路缘石
～ joint filler 预制填缝料[嵌缝板]
～ light concrete slab 预制轻混凝土板
～ panel 预制墙板
～ pile 预制混凝土桩
～ prestressed concrete 预制预应力混凝土
～ products 预制件
～ reinforced concrete construction 预制钢筋混凝土(结构)
～ ribbed slab 预制肋形板
～ unit 预制构件

precaution [pri'kɔ:ʃən] *n.*; *v.* 预防(措施),保护(措施),警戒
～ against fire 防火措施
safety～s 安全保障措施

precautionary [pri'kɔ:ʃənəri] *a.* 预防的,警戒的
～ measure 预防措施

precedence [pri'si:dəns] *n.* 领先,优先,在前,先例

precedent *n.* 先例,惯例,条件

precept ['pri:sept] *n.* (技术)规则,方案

precheck [pri:'tʃək] *v.*; *n.* 预先检查

precious ['preʃəs] *a.* 珍贵的,过分讲究的,非常的
～ alloy 精密合金
～ metals 贵金属
～ stone 宝石

precise [pri'sais] *a.* 精密的,准确的,严谨的

precision [pri'siʒən] *n.* 精度,精细,正确 *a.* 精密的,精密的
～ instrument 精密仪器
～ measuring instrument 精密量器
～ square 检验角尺

precoat ['pri:'kəut] *v.*; *n.* 预涂,预浇
～ed base 预涂基层

precoating *n.* 预浇面层,底漆,预涂层

preconcert ['pri:kən'sə:t] *v.* 预定,事先同意

precondition ['pri:kən'diʃən] *n.* 前提,先决条件 *v.* 预处理,预先安排好

preconsolidate *v.* 预先[前期]固结

preconstruction stage *n.* 施工前阶段[准备阶段]

precontamination *n.* 初期污染

predella [pri'delə] *n.* 祭坛台阶,祭坛油画,祭坛雕刻

predesign *v.*; *n.* 初步[草图]设计

predetermine ['pri:di'tə:min] *v.* 预定,注定
～d cost 预计成本

predominant [pri'dɔminənt] *a.* 主要的,显著的,流行的,占优势的
～ current 主流,优势流

predominate [pri'dɔmineit] *v.* 占优势,统治,支配

pre-drawing *n.* 预拉伸

prefab ['pri:fæb] *a.* 预制的 *n.* 预制品,活动房屋
～ form 预制模板

prefabricate ['pri:'fæbrikeit] *v.* 工厂预制,装配 *n.* 预制品
～d building 预制装配式建筑
～d formwork 预制模板
～d house unit 预制装配式住宅构件
～d parts 预制构件
～d structure 预制装配结构
～d unit 预制构件

preferred [pri'fə:d] *a.* 优先选用的,择优的
～ dimension 选用尺寸
～ plan 最佳方案
～ sizes 选用尺寸

prefill *v.* 预装填

prefiltering *v.* 预先过滤

prefinishing *n.* 预先修饰,预饰面

preform ['pri:'fɔ:m] *v.* 预制,初步加工,预先决定 ['pri:fɔ:m] *n.* 初加工成品,预型件
～ joint 预制缝
～ joint filler 预制填缝料[板]

preframe *v.* 预装配
pregwood *n.* 浸胶压缩木材
pre-hardening *n.* 初凝,预硬化
preliminary [pri'liminəri] *n.* (*pl.* naries) 准备工作,初步行动 *a.* 初步的,预先的,开端的
  ~ acceptance 初步验收
  ~ computation 概算
  ~ design 初步设计,设计方案
  ~ design review 初步设计审查
  ~ dimension 初步尺寸,预定尺寸
  ~ drawing 初步[方案]设计图
  ~ investigation 初步调研
  ~ location 初步定线
  ~ measures 初步措施
  ~ plan 初步计划
  ~ report 初步报告
  ~ scheme 初步规划
  ~ sketch 初步设计,草图
  ~ treatment 初步处理
  ~ works 预备工程,前期工作
preload ['pri:ləud] *v.*; *n.* 预加荷载
premature [premə'tjuə] *a.* 过早的,早熟的
  ~ drying-out 过早干燥
  ~ hardening 早期硬化,假凝
premises ['premisis] *n.* 房产,围墙内场地,境内
  business ~ 办公室,事务所
premix ['pri:miks] *v.* 预拌,预拌混合料
  ~ed concrete 预拌混凝土
premo(u)ld ['pri:məuld] *v.* 预塑,预制 *n.* 塑料片
  ~ed asphalt plank 预塑地沥青板
  ~ed asphalt sealing strip 预塑沥青嵌缝板
  ~ed joint 预塑缝
  ~ed joint filler 预塑嵌缝板,预制填缝料
prepack ['pri:'pæk] *n.*; *v.* 预先装填
  ~ed aggregate concrete 预填集料灌浆混凝土
preparation [prepə'reiʃən] *n.* 预加工,配制,(*pl.*) 准备工作
  ~ of land 土地平整
  ~ of programs 程序设计
  ~ of specimen 试样制备
preparatory [pri'pærətəri] *a.* 预[筹]备的,初步的
prepare [pri'pɛə] *v.* 预备,准备,训练,制备;精制
  ~ed paint 调合涂料,调合漆
  ~ing specification 制订规范
preplace [pri:'pleis] *v.* 预置
  ~d-aggregate concrete 预填骨料混凝土
preplan [pri:plæn] *v.* 规划,预先计划
pre-process *v.* 预加工,预处理
Pre-Romanesque style 罗马风格以前的形式
presbyterium *n.* (教堂的)内堂;十字架坛,(社寺建筑的)中央大殿
prescription [pris'kripʃən] *n.* 规定,法则,方案,惯例,传统
presence ['prezns] *n.* 出席,到场;风度
  ~ chamber 接见厅
present ['prezənt] *a.*; *n.* 出席(的),在场(的),当今(的);礼物;[pri'zent] *v.* 提出,呈现,引起,介绍
presentation [prezen'teiʃən] *n.* 提出,展示,显象,描绘,建筑渲染,报告书,文献,赠品
preservation [ˌprezə'veiʃən] *n.* 保存,维护,防腐
preservative [pri'zə:vətiv] *a.* 保存的,防腐的,预防的 *n.* 防腐剂,预防法
  ~ agent 防腐剂
  ~ fluid 防腐液
  ~ liquid 防腐液
  ~ process 防腐法
  ~ substance 防腐剂
  ~ treatment 防腐处理
  wood ~ 木材防腐剂
preservatize [pri'zə:vətaiz] *v.* 防腐;加防腐剂
preservatory [pri'zə:vətəri] *a.* 保存的

n. 储藏所

preserve [pri'zə:v] *n.* ; *v.* 保存,防腐
~d plywood 防腐胶合板
~d timber 防腐木材

preset [pri:'set] *v.* 预置,预先调整,安装程序,初凝

preshaping ['pri:ʃeipiŋ] *n.* 预先成形,预形成

press [pres] *n.* , *v.* 压缩,冲压,承压;制模机,压床,印刷(品,机,厂);出版社;柜橱;夹具
~ -board 压制板,纸板
~ button 按钮
~ -in connector [dowel] 压入榫,结合暗销
~ed brick 压制砖
~ed cement tile roofing 厚石棉水泥瓦屋面
~ed concrete 压制混凝土
~ed finish 压光,滚光
~ed glass 压制玻璃
~ed machine brick 机制砖
glue ~ (木工的)胶夹

presspahn *n.* 纸板,压板,木浆压制板

pressuretightness *n.* 气密性,不渗透性

prestressed ['pri:strest] *a.* 预(加)应力的
~ cable 预应力钢索
~ component 预应力构件
~ concrete bar 预应力混凝土钢筋
~ concrete roof 预应力混凝土屋面
~ concrete structure 预应力混凝土结构
~ reinforcement 预应力钢筋
~ shell 预应力薄壳
~ unit 预应力构件
~ wire 预应力钢丝

prestressing ['pri:stresiŋ] *a.* ; *n.* 预(加)应力(的)
~ cable 预应力钢索
~ strand 预应力钢缆线
~ tendon 预应力钢丝束

prestretching *n.* 预先拉伸

pretension [pri'tenʃən] *n.* 预张拉,预应力 *v.* ['pri:'tenʃən] 预拉伸,先张

pretest ['pri:test] *n.* 预先试[检]验
~ treatment 试验前准备

pretreat [pri:'tri:t] *v.* 粗加工,预清理;pretreatment, *n.*
~ed water 预处理水

pretreatment ['pri:'tri:tmənt] *n.* 粗加工,预处理,预清理

prettify ['pritifai] *v.* 装饰,美化

prevail [pri'veil] *v.* 占优势,胜过,盛地,克服;流行,盛行

preventer [pri'ventə] *n.* 防护设备,预防法
leak ~ 防漏剂
rust ~ 防锈剂

prevention [pri'venʃən] *n.* 预防,防护,阻止,防碍
~ of accidents 安全技术,故障[事故]预防
corrosion ~ 耐腐蚀,防锈
fire ~ 防火
rust ~ 防锈

preventive [pri'ventiv] *a.* 预防的 *n.* 预防剂,预防措施
~ maintenance 预防性维修

prevulcanized latex 预硫化胶乳

price [prais] *n.* 价格,价值 *v.* 标价,定价
~ adjustment 价格调整
~ index 物价指数
cost ~ 成本
current ~ 现行价格
factory ~ 出厂价格
floor ~ 最低价,底价

prick [prik] *v.* 刺,戳,穿;标出,缝合 *n.* 刺痕
~ up 打底子,漆底子
~ing up 抹灰底层,括糙层

prill [pril] *n.* 金属颗粒 *a.* 散装的 *v.* 使固体变颗状

primal ['praiməl] *a.* 最初的,原始的,主要的

primary ['praiməri] *a.* 原始的,初等的,基本的,主要的 *n.* 主要事物,底色,原色
~ beam 主梁
~ coat 底涂层,结合层
~ colours 原色,基色
~ crack 主要裂缝
~ light source 原始光源
~ lining 底层衬砌
~ materials 原材料
~ member 主要构件

prime [paim] *a.* 主要的,原始的,上等的 *n.* 初期;精华;字码右上角的撇号 *v.* 使准备好;涂底色,打底子;灌注,装填
~ attribute 主属性
~ coat 底涂层,结合层,头道抹灰
~ cost 成本,原价
~ lacquer 上底漆
~ material 底层材料,底漆
~ paint 底漆,底层涂料

primer ['praimə] *n.* 第一层,首[底]涂层,底漆;雷管,导火线
~ coat 底漆
alkali-resistant ~ 抗碱底漆
asphalt ~ 沥青底漆
etching ~ 腐蚀性涂料
oil ~ 油质底漆
red lead ~ 红丹底漆
synthetic resin ~ 合成树脂底漆
wash ~ process 涂料蚀洗处理
zinc chlorinate ~ 氯化锌底漆

primeval [prai'mi:vl] *a.* 远古的,原始的

priming ['praimiŋ] *n.* 涂底漆,打底子,涂油,装雷管,起爆
~ application 第一遍浇油
~ coat 底涂层,结合层
~ colour （有色的）底层油漆,底涂颜料
~ material 底漆
~ operation 浇沥青透层,涂底层

~ paint 底层漆
~ tube 雷管
~ varnish 底漆

primitive ['primitiv] *a.* 原始的,基本的,简陋的,朴素的,不发达的 *n.* 原始（人,事物）,原色
~ colour 原色

primrose ['primrəuz] *n.* 樱草 *a.* 浅黄色的

principal ['prinsəpəl] *a.* 主要的,第一的;资本的 *n.* 负责人;(主)屋架,主构[材,梁];资本,基本财产
~ base 主要基面
~ beam 主梁
~ building 主要建筑物
~ component 主要成分
~ cornice 屋檐,顶部挑檐
~ floor 主楼层[板]
~ member 主要构[杆]件
~ molding 主要线脚
~ post 主柱
~ rafter 屋架上弦杆,人字木
~ rafter beam 人字木之间的系梁
~ reinforcement 主钢筋
~ section 主截面
~ system （结构的）基本体系

principium [prin'sipjəm] *n.* (*pl.* principia)原理,原则,基础,初步

principle ['prinsəpl] *n.* 原则,原理,定律,法则,因素,本质
~ layout 总布置图
~ of design 设计原则

print [print] *v.* 印刷,晒图 *n.* 印刷（品）,印刷业;照片,插图,版画
~ing frame 晒图架
~ing machine 晒图机
~ing room 复印间
blue ~ 蓝图
paper ~ 纸质蓝图
plastic ~ 塑料蓝图
sepia ~ 深棕色蓝图

printer ['printə] *n.* 印刷机,印相机,晒

图机

prior ['praiə] *a.* 居先的,优先的,先验的

priority [prai'ɔriti] *n.* (*pl.* -ties) 先前,优先;重点
～ construction 首期建设,优先建筑

prism ['prizm] *n.* 棱镜,三棱形[体],棱柱,(*pl.*)光谱之七色

prismatic [priz'mætik] *a.* 棱柱的,角柱的,棱镜形的,等截面的,五光十色的
～ barrel vault 角柱筒形穹顶
～ bilet molding 棱形长方带线脚
～ colours 光谱之七色
～ joint 柱状节理
～ surface 棱柱曲面

pristine ['pristain] *a.* 原始的,早期的,质朴的

privacy ['praivəsi] *n.* (*pl.* -cies) 隐密,私用室,不受干扰的环境;保密性

private ['paivit] *a.* 私人的,非公用的,专用的,秘密的
～ door 便门
～ room 个人用房间
～ space 个人占用空间[面积]
～ volume 专用卷宗

privy ['privi] *a.* 私人的,隐蔽的 *n.* 厕所

prize ['praiz] *n.* 奖赏,奖品,奖金 *a.* 得奖的 *v.* 珍视,评价
～ competition design 有奖设计竞赛

probabilistic [prəbəbi'listik] *a.* 概率的,随机的

probability [prɔbə'biliti] *n.* 概率,可能性,可能发生的事情

probable ['prɔbəbl] *a.* 概率的,大概的,可能的,假定的
～ error 概率误差

problem ['prɔbləm] *n.* 问题,课题,难题 *a.* 成问题的,难对付的

procedure [prə'si:dʒə] *n.* 程序,工艺规程,生产过程;措施
～ charts 程序图
safety ～ 安全规程

set-up ～ 装配程序

proceed [prə'si:d] *v.* 进行,着手,发生 *n.* (*pl.*)收入,结果

proceeding [prə'si:diŋ] *n.* 程序处置,方法

process ['prəuses] *n.* 过程,流程,工序,步骤,工艺规程;操作 *v.* 加工,处理 *a.* 处理过的
～ chart 工艺流程图
～ of consolidation 固结过程
～ of setting 凝结过程
～ of weathering 风化过程

processing ['prəusesiŋ] *n.* 处理,加工,配制,操作;工艺过程

processor ['prəusesə] *n.* 加工机械,加工程序,信息处理机

proclaim [prə'kleim] *v.* 宣告,声明,公布

procure [prə'kjuə] *v.* 获得,采购;实现,达成

produce [prə'dju:s] *v.* 生产,制造,引起,延长(线段) *n.* ['prɔdju:s] 产品,产量,成果

producer [prə'dju:sə] 发生器,煤气发生炉;发电机;生产井,生产厂
～ gas 发生炉煤气
～ -gas (coal) tar 发生炉煤沥青

product ['prɔdəkt] *n.* 产品,作品;生产,制造;乘积
～ description 产品说明
～ summary 产品目录
art ～ 艺术作品
timber ～ 木材制品

production [prə'dʌkʃən] *n.* 生产,制造;产品,产量;生产力
～ bay (工厂)生产区
～ chart 施工进度表
current ～ 流水作业

productive [prə'dʌktiv] *a.* 生产的,生产性的,有成果的

productivity [prəu'dʌktiviti] *n.* 生产率[力,量],多产性

profane [prə'fein] *a.* 凡俗的 *v.* 玷污;

误用
~ architecture 非宗教性建筑,世俗建筑
profession [prəˈfeʃən] *n.* 职业,专业
professional [prəˈfeʃənəl] *a.* 专业的,业务的 *n.* 专业人员,专家
~ architect 职业建筑师
~ engineer 专业工程师
~ etiquette 行规
~ skill 专门技术
professionalize [prəˈfeʃənəlaiz] *v.* 职业化,专门化
professor [prəˈfesə] *n.* 教授
proficient [prəˈfiʃənt] *a.* 熟练的,精通的 *n.* 专家,能手
profile [ˈprəufail] *n.* 纵断[剖]面,纵断面图,侧[立]面(图) *v.* 画纵断面[侧面]图
~ board 模板,侧板
~ gage 样板,曲线板,轮廓量规
~ line 纵断面线
~ map 纵断面图
~ paper 纵断面图纸
~ steel 型钢
~ed bar 异形钢材
~ed sheet iron 成型薄钢板
cross-sectional ~ 横断面图
transversal ~ 横剖面
profit [ˈprɔfit] *n.* 利益,(*pl.*)赢利,利润(率) *v.* 有益于,获利
program [ˈprəugræm] *n.* 程序,次序表;纲领,提纲;方案,说明书,节目单 *v.* 拟定程序[计划],制定大纲
~ control 程序控制
running ~ 运转程序
program(m)ing [ˈprəugræmiŋ] *n.*;*a.* 计划(的),大纲(的),编程序(的)
progress [ˈprəugres] *n.*;[prəˈgres] *v.* 进展,进程,发展,进步
~ chart 进度表
~ map 进度图
~ schedule 进度时间表
progressive [prəˈgresiv] *a.* 前进的,发展的,递增的
prohibit [prəˈhibit] *v.* 禁止,阻止
~ed area 禁区
project [ˈprɔdʒekt] *n.* 设计,规划,方案;工程,建筑;草图;突状物;[prəˈdʒekt] *v.* 投射,投影,凸出;设计,计划
~ appraisal 工程评估,项目审查
~ area 规划区,工程面积
~ cost 工程费用
~ designer 项目设计师
~ director 工程负责人
~ formulation 制订项目
~ management 工程[项目]管理
~ manual 工程手册
~ planning 工程规划
~ scheduling 工程进度计划
~ supervision 工程质量管理
~ team 设计方案小组
~ed area 投影面积
~ed concrete 喷射混凝土
~ed costs 预定造价,计划成本
~ed (sash) window 滑开窗
turn-key ~ 包建工程
projecting [prəˈdʒektiŋ] *n.* 设计,计划,投影 *a.* 突(凸)出的,投影的
~ bricks 外突式砖
~ quoin 突出隅石
~ scaffold 悬挑式脚手架
projection
[prəˈdʒekʃən] *n.* 投射,突出部分,投影,计划,设计,估计

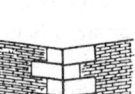
projecting quoin

~ booth 放映室
~ course 挑出层
~ drawing 投影图,投影法
~ lamp 投射灯
~ lantern 幻灯,映画器
~ of cornice 檐口突出部
~ receiver 投影式电视接收器
~ screen 银幕,投影屏
~ surface 投影面

roof ~ 挑檐
projector [prə'dʒektə] *n.* 放映机,投影机,幻灯机,探照灯;投影[射]线;设计者
projecture [prə'dʒektʃə] *n.* 突出物
prolate ['prəuleit] *a.* 伸长,扁长的,扩展的
~ ellipsoid 长椭球,长椭圆面
prolong [prə'lɔŋ] *v.* 延长,拖延 *n.* 冷凝管
prolongation [ˌprəu'lɔŋgeiʃən] *n.* 延长,拓展,拉长
promenade [ˌprɔmi'nɑːd] *n.* 散步;散步场所(如剧场走廊,大街,海滨大道等);宽廊 *v.* 散步,开车
~ tile 铺面缸砖
prominence ['prɔminəns] *n.* 起伏度;凸出物;突起
promise ['prɔmis] *n.* 约定,诺言 *v.* 允诺,订约
promnard ['prɔmnɑːd] *n.* 散步小路
promote [prə'məut] *v.* 促进,激励,提倡,推销,提升
promoter [prə'məutə] *n.* 发起人,建筑物业主;催化剂,加速器
promotional [prə'məuʃənəl] *a.* 促进,提倡,推销
prone [prəun] *a.* 倾斜的,陡的
proof [pruːf] *n.* 证明,证据;试验;不穿透性 *a.* 耐……的,防……的 *v.* 试验;校对;使……穿透
~ sample 试样,样品
~ test 验收试验
damp ~ 防潮
moisture ~ 防湿
oil ~ 耐油
shock ~ 防震
sound ~ 隔声
steam ~ 隔蒸汽
vapour ~ 防水汽
vermin ~ 防虫蛀
water ~ 防水
proofing [pruːfiŋ] *n.* 证明,试验,使不透……防护剂
damp ~ 防潮
moisture ~ 防湿[潮]
ray ~ 防辐射
water ~ 防水处理
prop [prɔp] *n.* 支撑[柱,架] *v.* 支撑,用支柱加固
~ and sill 支柱与地梁
~ stay 支柱
~ wall 密集支柱
~ped beam 加撑梁
~ped cantilever beam 有支撑悬臂梁
breaking ~ 复式支架
propagation [ˌprɔpəgeiʃən] *n.* 传播,扩散
propane ['prəupein] *n.* 丙烷
~ asphalt 丙烷地沥青
~ burner 丙烷加热器
propanone ['prəupənəun] *n.* 丙酮
propene ['prəupiːn] *n.* 丙烯
proper ['prɔpə] *a.* 特征的,固有的,常态的
~ length 真长度
~ operation 正确操作
architecture ~ 狭义建筑学
property ['prɔpəti] *n.* 性质,特征,财产,器材,(戏剧)道具(=prop)
~ corner 地界标石
~ line 地界线,用地线,建筑红线
~ line post 地界标
~ line wall 界墙,围墙
proportion [prə'pɔːʃən] *n.* 比例[率],均衡,配合(*pl.*)大小(长、宽、厚),容积,面积 *v.* 使成比例
~ of ingredients 成分比例
~ of mixture 混合比例
~ of resin present 树脂含量
~ of size 尺寸比例
proportional [prə'pɔːʃənl] *a.* 成比例的,相称的,平衡的 *n.* 比例量
~ illumination 均匀照明
~ scale 比例尺
proportioning [prə'pɔːʃəniŋ] *n.* 使成比

例,按比例定量调合
~ by volume 按体积配合
~ by water-cement ratio 水灰比配合(混凝土)
~ by weight 按重量配合
~ of concrete 混凝土配合比
~ of mortar 灰浆配合比
proposal [prə'pəuzəl] *n.* 提议,申请,投标
sealed ~s 投标
proprietary [prə'praiətəri] *a.* 专利(权)的,业主的。*n.* (*pl.* -taries)业主,所有权
~ material 专利材料
propylaea [ˌprɔpi'li(ː)ə] *n.* (神殿等的)入口

propylaea

propylon [prɔpilən] *n.* (*pl.* -lons, -ls) 古埃及寺庙进口处及前方(与寺庙不相连)之大门
proscenium [prəu'siːnjəm] *n.* (*pl.* -nia)舞台前部装置,舞台台口,显著位置
~ arch 舞台口
~ box 舞台前侧包厢
~ bridge 舞台口上部灯光通廊
~ opening 舞台台口
~ stage type 台口式剧场布置
~ wall 舞台台口侧墙
proskenion *n.* 舞台台口
prospect ['prɔspekt] *n.* 景色,风景,视野,境界;前景,前途,勘探;林荫路
prospectus [prə'spektəs] *n.* (*pl.* -tuses)计划任务书,说明书,大纲
prostyle ['prəustail] *a.*;*n* 柱廊式建筑,柱廊(式的)

prostyle

prostylos *n.* 古希腊神庙的前柱式
protean [prəu'tiːən] *a.* 变幻莫测的,多方面的
~ stone 石膏制人造石
protect [prə'tekt] *v.* 保护,防止,保存
~ switch 盒装开关
~ed apron 防护墙
~ed cover 保护罩
~ed device 防护设备
~ing wall 防护墙,挡土墙,胸墙
protectant [prə'tektənt] *n.* 防护剂
protection [prə'tekʃən] *n.* 保护,防御,保护措施[装置]
~ course 保护层
~ device 保护装置
~ fence 护栏
~ from the wind 防风设施
~ of face 护面
corner ~ strip 护角条
corrosion ~ 防腐
environmental ~ 环境保护
fire ~ 防火
flood ~ 防洪
labor ~ 劳动保护
lightening ~ 避雷
screen ~ 屏蔽保护
water proofing ~ 防水层
protective [prə'tektiv] *a.* 防护的,保安的,屏蔽的
~ agent 防护剂,抗氧化剂
~ covering 保护层
~ equipment 保护设施[用品]
~ guard (混凝土墙或柱的)护角
~ layer 保护层
~ measure 保护措施
protector [prə'tektə] *n.* 保护装置,保护层,防腐剂
lightening ~ 避雷装置
proto-Doric 古希腊建筑之形式
proto-Doric column 多立克柱
proto-Doric order 多立克柱式的原型
protohistory [ˌprəutəu'histəri] *n.* 史前

时期
Proto-Ionic capital 爱奥尼柱头
prototype [ˈprəutətaip] *n.* 原型,样品,足尺模型;范例,标准 *a.* 实验性的
~ structure 原型结构

Proto-Ionic capital

protract [prəˈtrækt] *v.* 延长,伸出,(用量角器或比例尺)绘图
protractor [prəˈtræktə] *n.* 量角器,分角器,分度规,半圆规
angle ~ 量角规
bevel ~ 活动量角器
protrude [prəˈtruːd] *v.* 伸出,凸出,耸立
protrusion [prəˈtruːʒən] *n.* 突出,隆起物(部)
protuberance [prəˈtjuːbərəns] *n.* 突出(物),节疤,疙瘩
provide [prəˈvaid] *v.* 预备,供应,准备
provision [prəˈviʒən] *n.* 设备,装置
provisional [prəˈviʒənəl] *a.* 临时的,假定的
~ agreement 临时协议
~ contract 临时契约[合同]
~ estimate 概算
~ sum 不可预见费用
proviso [prəˈvaizəu] *n.* 附带条件,限制性条款
provisory [prəˈvaizəri] *a.* 附有条件的,临时性的
~ clause 附文,附带条款
proximal [ˈprɔksiməl] *a.* 近端的,基部的
proximate [ˈprɔksimit] *a.* 紧接的,近似的
proximity [prɔkˈsimiti] *n.* 邻近,近似
proxy [ˈprɔksi] *n.* (*pl.* proxies) 代理人,代表权,委托书
prune [pruːn] *n.* 深紫红色
Prussian blue *n.* 普鲁士蓝,深蓝色

pseudo [ˈ(p)sjuːdəu] *a.* 假的,伪的,冒充的
~ -dipteros (古希腊神庙的)仿双重周柱式
~ -peripteros (古希腊神庙的)仿周柱式,假列柱围廊式

pseudo-peripteros

pseudoclassic [ˈ(p)sjuːdəklæsik] *a.* 伪[仿]古典的
~ architecture 仿古建筑
~ style 仿古式
pseudoclassicism 伪[仿]古典主义
psychological [ˌsaikəˈlɔdʒikəl] *a.* 心理学的,精神的
psychology [saiˈkɔlədʒi] *n.* (*pl.* -gies) 心理学,心理状态
~ of design 设计心理(学)
psychrometer [saiˈkrɔmitə] *n.* (干湿球)湿度计,干湿计
public [ˈpʌblik] *a.* 公共的,公用的,全国的,各国的 *n.* 公众,社会
~ -address system 有线广播系统
~ area 公共场所
~ assembly hall 公共会堂
~ bath 公共浴室
~ bidding 公开投标
~ building 公共房屋[建筑]
~ canteen 公共食堂
~ comfort station 公共厕所
~ corridor 公共走廊
~ drinking fountain 公共饮水处
~ hall 大会堂
~ house 小酒馆,饮食店,居住小区集会处
~ latrine (lavatory) 公共厕所
~ library 公共图书馆
~ lighting 街道照明
~ lodging house 简易宿舍[住宿处]
~ office 行政机关,办事处
~ operated house 公营住宅

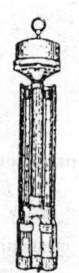
psychrometer

~ parking garage 公共车库
~ space 公共场所,公用面积
~ works 公共工程,公共建筑,市政工程
puce [pju:s] *a.* 紫褐色(的)
pucker ['pʌkə] *v.* 折叠,皱起 *n.* 皱折[纹]
pudding ['pudiŋ] *n.* 布丁(状物)
~ granite 球粒花岗岩
pugging ['pʌgiŋ] *n.* 隔声层;隔声材料
pull [pul] *v.;n.* 拉伸,牵引,抽曳;拉力,张力,吸引力;拉手柄
~ bar 拉杆
~ box 引线盒
~ -rope (窗用)拉绳,牵引索
~ switch 拉线开关
pulley ['puli] *n.* (*pl.*-leys) 滑轮,皮带轮,滚筒 *v.* 用滑车推动
guide ~ 导向滑轮
winding ~ 绕线滑轮

guide pulley

pulling ['puliŋ] *n.* 拉,拔,牵引,拉力,张力
~ device 牵引装置
pulp [pʌlp] *n.* 砂[泥,纸]浆 *v.* 制浆,把……捣成浆状
~ board 纸浆板
~ -wood 制浆木材
pulpit ['pulpit] *n.* 控制[操纵]室[台],讲坛
pulverize ['pʌlvəraiz] *v.* 磨碎,粉碎;喷雾;疏松
~d asbestos 石棉粉
~d lime 石灰粉
~ing mixer 粉碎拌和机
pulverizer ['pʌlvəraiz] *n.* 粉碎机,喷雾器
pulvimix ['pʌlvimiks] *n.;v.* 松土拌和;经粉碎拌和的混合料
pulvin *n.* 柱头垫石
pulvinus *n.* 柱头垫石

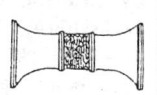

pulvinus

pumeconcrete *n.* 浮石混凝土
pumicate ['pju:mikeit] *n.* 浮[轻]石 *v.* 用浮石磨光
pumice ['pʌmis] *n.* 浮[轻]石,浮[泡沫]岩 *v.* 用浮石磨光
~ aggregate concrete 浮石骨料混凝土
~ concrete 浮石混凝土
~ sand 浮石砂
~ -slag brick 浮石渣砖
~ stone 浮石,泡沫岩
pumicite ['pʌmisait] *n.* 浮石,轻石,泡沫岩
pummel ['pʌml] *n.* 球端,圆头
pump [pʌmp] *n.* 泵,抽(水,气)机,打气筒;*v.* 用泵抽吸,用泵增压
~ concrete 泵送混凝土
pumpcrete ['pʌmpkrit] *n.* 泵送混凝土
~ machine 混凝土泵
punch [pʌntʃ] *n.* 穿孔器,冲压机,冲床,剪票夹 *v.* 穿孔
puncheon ['pʌntʃən] *n.* 短柱,短木料,半圆木料,一面凿光的石板,石凿,钻工
puppet ['pʌpit] *n.* 木偶,傀儡
~ theatre 木偶剧院
pure [pjuə] *a.* 纯净的,完美的,纯理论的,抽象的 *v.* 纯化,提纯
~ colour 纯色
purfle ['pə:fl] *v.;n.* 镶边,边饰,装饰,美化
~d work 镶边细工
purge [pə:dʒ] *v.;n.* 清净,提纯,精炼
~ unit 清洗装置
purification [pjuərifikeiʃən] *n.* 纯净(作用),提纯,精制
~ device 净化设备
purifier ['pjuərifaiə] *n.* 清洗装置,纯化装置
purify ['pjuərifai] *v.* 净化,提纯,使清洁
~ing agent 净化剂
purity ['pjuəriti] *n.* 纯净,纯度,品位
purlin(e) ['pə:lin] *n.* 檩条,桁条
~ brace 檩条撑

~ cleat 檩条支托楔块
~ post 檩条下柱
~ roof 檩条屋架
angle ~ 角钢檩条
eaves ~ 檐檩
intermediate ~ 中间檩
ridge ~ 脊檩

purple ['pə:pl] *n.* 紫色,紫色颜料 *v.* 染成紫色
~ black 紫黑色
~ brown 紫褐色
royal ~ 蓝紫色

purse [pə:s] *n.* 款项,国库,财力 *v.* 起皱,缩拢
~ strings 财力,金钱
public ~ 国库

push [puʃ] *v.* ; *n.* 推,按,推进,推压力
~ and pull brace (门上的)推拉把手,推拉杆
~ button 按钮
~ contact 按钮开关
~ pull tape 卷尺
~ -through geyser 通过式热水器
bell ~ 电铃按钮

puteal *n.* (古罗马)井栏

putlock ['putlək] *n.* 脚手架跳板横木
~ holes 墙上脚手架孔

putlog ['putlɔg] *n.* 脚手架跳板横木
~ holes 墙上脚手架横木支承孔

putrefaction [pju:tri'fækʃən] *n.* 腐烂[作用]

putrid ['pju:trid] *a.* 腐烂的

putto ['putəu] *n.* (*pl.* -ti)裸体小儿雕像饰,肥胖年轻男天使或神童的形象

putty ['pʌti] *n.* (*pl.* -ties)油灰,腻子 *v.* 嵌油灰
~ chaser 油灰刮刀
~ glazing 油灰装玻璃
~ joint 油灰缝
~ knife 油灰刀
~ oil 油灰油
~ less glazing 无灰装玻璃

back ~ 面灰
front ~ 底灰
glass ~ 窗用油灰
glaziers' ~ 镶玻璃用油灰
plasterers' ~ 粉刷用外层细料
tinted ~ 有色油灰

puzzolane *n.* = puzzolana 白榴火山灰
~ cement 火山灰水泥

pyknostylos *n.* (古希腊,古罗马神庙的)密柱式(柱间距为柱径之1.5倍)

pyller *n.* 塔门,标塔

pylon ['pailən] *n.* 塔门,机场标塔,高压电缆塔,塔状物,柱台,支架,标杆,定向起重机,牌楼门
~ antenna 铁塔天线

pyramid ['pirəmid] *n.* 金字塔,棱锥,四面体 *v.* 使成角锥形
~ construction 金字塔式建筑
~ -shaped roof 尖攒屋顶

pyramidal [pi'ræmidl] *a.* = pyramidical 金字塔形的;锥形的
~ roof 锥形屋顶

Pyrex ['paiəreks] *n.* 硼硅酸玻璃(一种耐热玻璃)

pyrifrom ['pirifɔ:m] *a.* 梨形的
~ profile 拱梨形轮廓

pyrifrom

pyroceram [pairə'serəm] *n.* 耐高温玻璃;耐高温陶瓷粘结剂

pyroligneous [pairəu'ligniəs] *a.* 干馏木材而得的
~ acid 焦木酸
~ alcohol 焦木酒精

pyroplasticity *n.* 热塑性

pyroxene [pairɔksi:n] *n.* 辉石

pyroxenite *n.* 辉岩

pyroxylin(e) [pai'rɔksilin] *n.* 低氮硝化纤维,火棉

Pyruma *n.* 一种耐火黏土水泥

Q

Q alloy 镍铬合金
Q factor = quality factor 品质[质量]因数
Q-tempering 淬火回火
quad [kwɔd] *n.* ; *a.* 四边形；象限；四个一套；方庭，四合院，内院；四心线缆
  ~ cable 四芯(扭绞)电缆
  ~ pair cable 八芯电缆
quadra ['kwɔdrə] *n.* (*pl.* -rae) 勒脚；方形框架
quadrangle ['kwɔ'dræŋgl] *n.* 四边形；方庭，方院，四周围着建筑物的院子；四合院，内围
  ~ roof 四边形屋顶
  complete ~ 完全四边形
quadrangular [kwɔ'dræŋgjulə] *a.* 四边形的，方形的 *n.* 四棱柱
  ~ truss 四角形桁架
quadrant ['kwɔdrənt] *n.* 象限，象限仪；四分之一圆；鱼鳞板；扇形体[板]
  ~ stay 扇形窗撑杆
  ~ step 扇形踏步
quadrantal [kwɔ'dræntl] *a.* 象限的；扇形的；鱼鳞板的
quadrate ['kwɔdrit] *a.* ; *n.* 长方形(的)，正方形(的)；方块物；平方，二次 *vt.* 使成正方形；使一致；平方
  ~ lobe 方叶
quadratic [kwɔ'drætik] *a.* 二次的，平方的
  ~ prism 方棱柱
quadratrix [kwɔ'dreitriks] *n.* 割圆曲线
quadrature ['kwɔdrətʃə] *n.* 平方面积；错视装饰画
quadrel ['kwɔdrit] *n.* 方块石，方砖，方瓦；方形草皮

quadratic prism

quadriad ['kwɔdriæd] *n.* 包括四个的一组
quadrifid ['kwɔdrifid] *a.* 分成四部分的，四裂的[叶或花瓣]
  ~ petal 四裂花瓣
quadrifoliate [,kwɔdri'fəulit] *a.* 有四叶的，叶成四片一组
quadrifrontal [,kwɔdri'frʌntəl] *a.* 有四个正面的
quadriga [kwɔ'driːgə] *n.* (*pl.* -gae) (古希腊罗马建筑中的)四马拖车雕饰
quadrilateral [,kwɔdri'lætərəl] *a.* ; *n.* 四边形(的)，四方面(的)，四角(的)
  ~ peristyle 四边列柱廊式
  ~ pyramid 四棱锥
quadrille [kwɔ'dril] *n.* 正方形或长方形模式 *a.* 方眼[格]的
  ~ pattern 方格图样
quadripartite [,kwɔdri'paːtait] *a.* 四分的，由四部分组成的，一分为四的
  ~ vault 四分穹顶

quadripartite vault

quadriporticus *n.* 方厅，方庭，内院
quadruped ['kwɔdruped] *n.* 四足兽 *a.* 有四足的
qualification [,kwɔlifi'keiʃən] *n.* 技能，熟练程度；资格，合格证，执照；技术指标；限制条件
  ~ test 质量鉴定[合格]试验
  ~ test specification 质量鉴定试验规范
qualify ['kwɔlifai] *v.* (使)合格，限制，适合，修饰
quality ['kwɔliti] *n.* (*pl.* -ties) 品质，特性，质量，性能；纯度，精度；品位，品

级 *a.* 优质的,高级的
~ assurance 质量保证
~ audit 质量检查(评估)
~ certificate 质量证明书
~ check 质量检查
~ concrete 优质混凝土
~ control 质量控制
~ control standard 质量控制标准
~ criteria 质量标准
~ inspection 质量检验
~ requirements 质量要求
~ specification 质量标准,技术规格
artistic ~ 艺术性
enduring ~ 持久性能
lasting ~ 耐久性
quantity ['kwɔntiti] *n.* (*pl.* -ties) 数量,定量,值,参数,定额
~ estimate sheet (建筑成本)数量估算表
~ sheet 工程数量表
feedback ~ 反馈值
sample ~ 样本数量
quaquaversal [ˌkweikwə'və:səl] *a.* (地质中)由中心向四方扩散的, *n.* 穹状(隆起)物
~ structure 穹状结构
quarl ['kwɔ:l] *n.* 异形耐火砖
quarrel ['kwɔrəl] *n.* 方形瓦,菱形花砖,菱板,菱形装饰,小四边形构件 *v.* 争吵,争论
~ pane 菱形玻璃块
quarry ['kwɔri] *n.* (*pl.* -ries) 采石场,石矿,露天矿;小方面砖,菱形瓦[片]源泉猎物 *v.* 采石,开采;钻掘
~ -faced (建筑石块)表面未琢磨的
~ -faced stone 粗石,原开石
~ glass 方形玻璃
~ light (铝条嵌)菱形玻璃窗
~ pitched stone 粗凿[琢]石
~ rock 毛石,粗石
~ stone 粗石,毛石
~ stone masonry 粗[毛]石污工[砌体]

~ tile 方砖,缸砖
quarter ['kwɔ:tə] *n.* 四分之一;一刻钟;一季度;(东南西北的)一个方面象限;寓所,宿舍,营房;路肩(四分之一的路宽);船尾部,中部船舷 *v.* 分为四等分
~ bat 四分砖,二五砖
~ beam 四开木梁
~ closer 四开填塞砖
~ -cut 四开木材
~ -foil 四叶饰,四瓣形饰

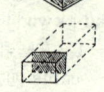
quarter closer

~ grain 四开木材纹理,板面木纹
~ landing 直角转弯楼梯平台
~ light 边窗,侧窗
~ newelled staircase 直角转弯楼梯
~ pace 直角转弯的楼梯平台
~ partition 方木隔断,立楞隔墙
~ round (moulding) 1/4 圆装饰线条
~ sawn grain 木材径面纹理
~ screw 调节螺钉
~ space landing (楼梯的)直角转弯平台
~ turn stair 直角转弯楼梯
close ~s 狭窄的居住空间
industrial ~ 工业区
residential ~ 住宅区
staff ~ 职员宿舍
three ~s 六分头,3/4 砖
quartersawn *n.* 1/4 径向锯木
alternate ~ 1/4 交错锯木
common ~ 普通 1/4 锯木
radial ~ 1/4 辐射锯木
triple-cut ~ 三重叠切割锯木
quartz [kwɔ:ts] *n.* 石英,水晶
~ cloth 玻璃布
~ crystal 石英晶体
~ fibre 石英纤维
~ glass 石英玻璃
~ mercury lamp 石英汞灯
~ plate 水晶片

~ sand 石英砂
~ slate 石英板
~ syenite 石英正长岩
~ trachyte 石英粗面岩
~ tube 石英管
~ wool 石英纤维(棉)

quartzite ['kwɔːtsait] *n.* 石英岩,硅岩
~ brick 硅砖
~ fireproofing materials 硅耐火材料
~ sand stone 石英质砂岩

quartzlite glass 透紫外线玻璃

quarzal *n.* 铝基轴承合金

quasi ['kwɑːzi(ː)] *a.* 类似的,就是的;[调头]准,半,拟,伪,类
~ -adiabatic 准绝热的
~ satellite city 准卫星城市

quaterfoil ['kætrəfɔil] *n.* 四叶式,四叶饰

quattrocento [ˌkwætrəuˈtʃentəu] *a.;n.* 文艺复兴初期(15世纪)的(文学艺术)

quay ['kiː] *n.* 顺岸码头;岩壁;基体

quebracho *n.* 破斧木,铁木,极坚硬的木材
white ~ 白坚木

queen [kwiːn] *n.* 女王,王后;(*pl.*)大石板
Queen Anne Style 英国18世纪安娜女王时代的建筑、家具式样
Queen Anne arch 安娜女王拱
~ bolt 拉杆
~ closer 小接砖,大开条砖,对开顺砖
~ 's metal 锡基合金
~ post 双柱式
~ post roof 双柱式屋顶
~ post truss 双柱(上撑式)桁架,双竖杆桁架
~ 's-ware 上乳白色釉的英国瓷器
~ trussed beam 双柱桁架撑梁

quench [kwentʃ] *v.;n.* 淬火,淬冷;灭火,冷却
~ alloy steel 淬硬合金钢
~ed and tempered steel 调质钢

que-ti(**column bracket**)雀替(我国古典木建筑中的梁柱交接处的托座)

quick [kwik] *a.;ad.* 快速的,短时间的,敏捷的;流动的
~ bead 齐平圆线脚
~ -break switch 速断开关
~ cement 快凝水泥
~ curing 快凝的
~ -detachable 易拆卸的
~ drying enamel 磁釉
~ drying paint 快干漆[涂料]
~ drying varnish 快干漆
~ -erection method 快速安装法
~ -fence 树篱
~ hardening cement 快硬水泥
~ -lime pile 石灰桩
~ lunch stand 快餐食堂,快餐座
~ setting cement 快凝水泥

quiescent [kwaiˈesnt] *a.* 静止的,静态的
~ load 静荷载

quiet ['kwaiət] *a.;n.* 寂静(的),静止的,静态,平静(的),单调的

quilt [kwilt] *n.* 棉被,填褥;隔层填料
v. 缝被子,用垫料填塞
~ed figure (木板的)褶裥花纹
~ed plastic panel 塑料芯板

quinary ['kwainəri] *a.;n.* 五(个,倍,元)的;五进制的
~ alloy 五元合金

quincuncial [kwinˈkʌnʃəl] *a.* =quincunxial 五点形的,梅花式的
~ piles 梅花桩

quincunx ['kwinkʌŋks] *n.* 梅花形(式)

quinquangular [kwinˈkwæŋgjulə] *a.* 五角形的,五边形的

quint ['kwint] *n.* 五件一套
~ -point arch 五点拱

quintefoil ['kwintfɔil] (=cinquefoil) *n.* 五叶形装饰

quire [kwaiə] *n.* (=choir)一刀纸,未装订成册的;教堂中唱诗班的席位
~ aisle 教堂侧厢走廊

~ arcade 教堂拱廊
~ arch 教堂拱券
~ architecture 教堂建筑艺术
~ loft 教堂楼座
~ screen 围隔唱诗班的栏杆
quirk [kwə:k] *n.* 深槽,小沟,凹部;斜角镶条,菱形窗玻璃;边花花纹,花体美术字

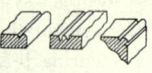
quirk beads

~ bead 刻槽串珠线脚
~ board 串珠线脚板
~ mo(u)lding 鸟喙形线脚,方沟线脚
quisqueite *n.* 硫沥青
quitch [kwitʃ] *n.* 茅草
quoin [kɔin] *n.* 楔形石,楔子;墙角,砌角石块,隅石 *v.* 用楔子固定

~ bonding 墙角砌合
~ brick 墙角砖
~ header 丁砖(屋角)石
~ stone 屋隅石,墙角石
quonset ['kwɔnsit] *n.* 活动房屋(屋顶为半圆形瓦楞铁皮构件)
~ hut 金属结构掩藏室,半圆形活动房屋
quota ['kwəutə] *n.* 份额,定额,定量
~ quantities 配额数量
~ management system 定额管理制度
~ sample 限额样本
production ~ 生产指标
quote [kwəut] *v.;n.* 引用,列出;报价
~ favourable terms 报优惠价
quotient ['kwəuʃənt] *n.* 系数,份额

R

rabbet ['ræbit] *n.* (木板等的)榫头;槽口[舌];半槽边,企口缝 *v.* 开槽口;嵌接;槽舌[口]接合
~ joint 槽舌接合,嵌接;企口接合
~ machine 开槽机
~ plane 凸边刨;槽刨
~ed lock 槽口门锁
door ~ 门边半槽
glass ~ 玻璃槽口
keel ~ 龙骨镶口
window ~ 窗边半槽
rabbit ['ræbit] *n.* 样品容器;兔
~ ear faucet 兔耳式龙头
rabbler ['ræblə] *n.* 铲子;刮刀;搅拌器
Rabitz *n.* 拉比兹(人名)
~ finish 拉比兹灰浆饰面
~ type board 拉比兹式平板
~ type ceiling paster 拉比兹式顶棚抹灰
~ type gypsum 拉比兹式石膏
~ type vault 拉比兹式圆顶

rack [ræk] *n.* 齿条(板,杆,轨);导[滑]轨;支架;拦污栅;栅格 *vt.* 装架;推压;转[移,振]动;变形;倾斜;(圬工)阶梯形砌接
~ bar screen 格栅
~ course 斜砌(砖)层
~ed joint 深凹接缝
~ed timbering 斜角接缝
book ~ 书架
stake ~ 栅栏
supporting ~ 支持[承]架
racking ['rækiŋ] *n.* 推压动作,挤压(运动);(墙的)阶梯形砌接
~ course 斜砌砖层
raddle ['rædl] *n.* 编枝,编条;(树枝编的)篱笆;泥壁内的枝条;赭土
radial ['reidjəl] *a.* 径向的,辐向[射]的 *n.* 径向,辐向;放射部;垂直于圆弧部分的杆[臂]
~ -arch-roof 辐射拱形屋顶
~ brick 扇形砖

~ chimney brick 扇形烟囱砖
~ crack 辐射形裂缝
~ deformation 径向变形
~ displacement 径向位移
~ drill machine 旋臂钻床
~ gap 径向缝隙
~ -rib cupola 径向肋圆(屋)顶
~ roof fan 径向吊扇
~ sawing 长向锯木
~ shrinkage （木材）径向干缩[量]

radial-arch-roof

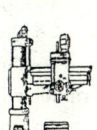
radial drill machine

radiance ['reidjəns] n. = radiancy ['reidjənsi] 发光,辐射;面辐射强度;深粉红色
radiant ['reidjənt] a. 辐[放,发]射的;放热的;发光的;灿烂的,光辉的 n. 光[热,辐射]源,发[热]的物体,光点[体]
~ illumination 辐(射)照度
~ panel heating 辐射板供热
radiate ['reidieit] v. 发射(光线,电磁波);放射(热量);射出;(从一点)扩散,传[广]播,播送; a. 辐射的
radiating brick 辐射型砖
radiating joint 径向缝
radiation [ˌreidi'eiʃən] n. 发射[光,热,散],放[辐]射;照射[作用];放射线[物],辐射线[能,热];散热器;辐射状排列,放射形
~ heat transfer 辐射传热
~ intensity (照)辐射强度
~ loss (热)辐射损失
~ protection 辐射防护
~ shield 防辐射(屏)
~ shield design 防辐射设计
~ shield door 防辐射门
radiator ['reidieitə] n. 辐射体[器,源];放[发]射器;散热器[片];暖气片[管,装置];冷却器,(汽车等)水箱;发射天线

~ block 散热器组
~ box 散热器罩
~ cover 散热器罩
~ fin 散热片
~ guard 散热器护罩
~ of sound 声源,声辐射器,扬声器
~ protection 防辐射保护
~ section 暖气片
~ shutter 散热器百叶窗
radio ['reidiəu] n. 无线电;无线电收发机;收音机;无线电电台,广播电台;无线电报 v. 拍发无线电;广播
~ phone 无线电话
radius ['reidjəs] n. (pl. radii) 半径(距离,范围);(活动)范围,界限;辐(条);辐[径向]射线,辐射光线;辐射状部分,放射状,倒圆 v. 使(切)成圆角
~ brick (砌烟囱等的)扇形砖
~ of extrados 外拱圈半径
~ of intrados 内拱圈半径
~ of soffit 内拱圈半径
radwood n. 辐射处理木料
raft [ra:ft] n. 筏,木排[筏],筏形基础,排基 a. 筏式的
~ foundation 筏(式)基(础)
concrete ~ 混凝土排基
rafter [ra:ftə] n. 椽子;木材筏运工 vt. 装椽子
~ bearer notch 承椽槽口
~ connection 椽木连接
~ roof 椽木屋顶
~ set 人字木屋架
angle ~ 角椽
common ~ 普通[共用]椽木
compass ~ 轮椽(山墙装饰)
curb ~ 侧椽
fly ~ 飞椽板
hip ~ 脊椽,(四坡屋顶)戗椽
hip jack ~ 脊椽,(四坡屋顶的)端坡椽
intermediate ~ 中间椽

jack ~ 小椽,(四坡屋顶的)面坡椽
laminated ~ 层叠椽
main ~ 主椽
nailed trussed ~ 钉合屋架
open ~ 明椽
principal ~ 主椽
trimming (trimmed) ~ 整饰过的椽
valley ~ (四坡屋顶的)沟椽

rafting *n.* 合金,熔金物

rag [ræg] *n.* 毛刺,飞边;条石,石板瓦,硬质(石灰)岩;破[碎,擦]布;无价值的东西 *v.* 除去毛刺;压[滚]花;(轧辊)刻纹[槽];划伤
~ felt (以破布为原料的)油毡,(建筑用)粗制毡
~ paving 石板铺面
~ -rolled finish 滚花抹面
~ rubble 粗毛石
~ rubble wall 粗面毛石墙
~ work 石板砌筑

ragged ['rægid] *a.* 凹凸不平的,参差不齐的;粗糙的,锯齿形的;不规则的

raggle ['rægəl] *n.* (固定屋面用的)墙上槽口,(石)水槽

raglan ['ræglən] *n.* 平顶[楼板]搁栅

raglet *n.* 墙上凹槽,披水槽

raglin *n.* 楼板搁栅

Rahmen【德】*n.* 框架结构

rail [reil] *n.* 轨(道,条);钢[导]轨;铁路;(*pl.*)铁路网;栏杆,围栏;横木,门窗横档[冒头] *v.* 铺铁轨;由铁路运输;装栏杆
~ post 栏杆柱
~ stanchion 栏杆柱
chair ~ 靠椅扶手;护墙板
curtain ~ 窗帘轨
lock ~ 门锁横档
middle ~ (门的)中档,腰冒头
shutter ~ 卷帘轨
top ~ (门的)上横梃

railing ['reiliŋ] *n.* 栏杆
~ material 栏杆建筑材料

railway ['reilwei] *n.* 铁道(部门),铁路(设施,系统)
~ roof 站台篷,月台篷

rain [rein] *n.* 雨(水,天),下雨;雨季;电子流 *v.* 降(下)雨;(雨水般)淌下
~ chamber 喷水吸[除]尘室
~ gutter 雨水沟,檐沟,天沟
~ leader 雨水[水落]管
~ -repellent 防雨的
~ screen 防雨屏
~ spout 水落管,排水口

rainbow ['reinbəu] *n.* 虹(霓),彩虹
~ arch 同心圆拱
~ roof 弓形屋顶

raincloth *n.* 防雨布

rainfall ['reinfɔ:l] *n.* 降水[雨];雨量

rainless *a.* 无雨的

rain-out *n.* 冲洗,清除(指雨从大气中清除尘埃等);(因雨)阻碍[中断]

rainspout *n.* 水落管,排水口

rainwater *n.* 雨[软]水
~ conductor 雨水管
~ inlet 雨水口
~ pipe 雨水管

raise [reiz] *v.* 升起,举起;抬高;竖立;建立,兴建;引起,扬起(灰砂等);提出(问题),筹集(资金) *n.* 升起,举起
~d arch 突起拱
~d chord truss 起拱桁架
~d cottage 架空住房
~d face flange 突面法兰
~d floor 架空地板
~d grain 浮起纹理
~d head 凸头
~d moulding 隆起式(装饰)线条
~d panel 鼓起的镶板
~d piece 槛,浮雕制品
~d skylight 隆起天窗

rake [reik] *n.* 耙,长柄耙;倾斜,倾角 *v.* 耙,刮;收集;倾斜
~ bond 对角[斜纹]接合
~ cornice 斜挑檐

~d joint 刮缝,清缝
raking *a.* 斜的;刮;耙除
~ arch 高低脚拱,斜拱
~ balustrade 斜面[楼梯]栏杆
~ beam 斜梁
~ bond 对角砌合,斜纹接合
~ coping 斜面压顶石,山墙压顶石
~ cornice 斜挑檐,斜檐口线
~ course 斜砌砖层
~ element 斜撑
~ flashing 斜遮雨板
~ molding 斜面线脚
~ -out (of joints) (砌体勾缝前)刮灰缝,清缝
~ out joint 勾突缝
~ pile 斜桩
~ shore 斜撑
~ strecher bond 斜纹顺砖砌合
raker ['reikə] *n.* 撑脚(杆),支柱,斜撑;耙路机;(清理灰缝等用的)刮刀
ram [ræm] *n.*;*v.* 夯(具),夯实,桩锤;(液压机等的)柱塞;冲头;(牛头刨床)滑枕;水锤泵,水轮泵 *v.* 夯;冲,冲压
~ med clay 夯实黏土
~ med earth 夯实土
~ med foundation 夯实地基
~ med in layer 分层夯实
~ med soil 夯实土
~ med wall 夯筑墙
ramie ['ræmi] *n.* 麻
rammability *n.* 可压[夯]实性
rammer ['ræmə] *n.* 夯具,夯锤,桩锤;冲头,压力 *v.* 冲压,捣紧
~ compactor 夯实[土]机
ramming *n.* 夯实,夯筑,锤击
ramp [ræmp] *n.* 斜坡道,斜弯道;(立体交叉的)匝道,盘旋引道;斜面;楼梯扶手的弯曲部分;斜面升降机,缆车;(上下船,飞机等的)梯架;(有坡度的)跳板,引桥;(地质的)对冲断层 *v.* 倾斜;做成斜坡;(草木)丛生
~ road 斜坡道
slope ~ 斜坡道

rampant ['ræmpənt]
a. 蔓延的;猛烈的;具有一个比一个高的拱座[桥台]的
~ arch 跛拱,斜拱
~ vault 斜拱顶,跛拱顶

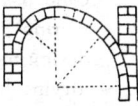
rampant vault

rance [ræns] *n.* 支柱 *v.* 支撑,闩住
ranch [ræntʃ] *n.* =rancho 牧场,农场
~ house 农场房屋,平房建筑
rand [rænd] *n.* 边,缘;卡圈;凸缘
randanite *n.* 硅藻土
random ['rændəm] *a.* 随机的,任意的;抛弃的,抛填的;不整齐的,不规则的 *n.* 乱,随便;随机
~ ashlar facing 乱砌琢面石
~ bond 乱砌合
~ colour 任意色
~ course work 乱砌层
~ coursed rubble 不规则成层毛石(砌体)
~ crack(ing) 不规则裂缝
~ grain 不规则纹理
~ masonry 乱砌砌体
~ material 未筛选的材料
~ range masonry 乱砌圬工
~ rubble 乱石,乱砌毛石
~ rubble masonry 乱砌毛石圬工,乱石圬工
~ stone work 不规则毛石砌体,乱石工
~ work 不规则石工
range [reindʒ] *v.* 排列,整理,把……排成行;(距离)调整;定向;(把……)分[归]类,编[列]入,分等,评定;分布 *n.* 幅度,范围,区域;波幅,潮差,水位差;极[界]限,限度;(时间)间隔,(生存)期间;行(系,排)列;炉灶
~ masonry 层砌琢石
~ work 层砌石工
rapid ['ræpid] *a.* 急的,快的;(坡度)陡的 *n.*(*pl.*) 激流,急流,险滩
~ cement 快硬水泥

~ condensation 快速凝结
~ -curing 快凝,速干
~ design method 快速设计方法
~ -drying 快干的
~ -setting cement 快凝水泥
Rap-rig *n.* 一种能快速架设的轻便脚手架
rare [rɛə] *a.* 稀少的,稀有的;稀薄的;珍贵的
rascle ['reiskl] *n.* (地质的)灰岩参差蚀面
rasher ['ræʃə] *n.* 薄片
rasp [ra:sp] *n.* 粗[木]锉,木工锉; *v.* (用精锉刀)锉,粗刮,摩擦
~ -cut file 木锉
raster ['ræstə] *n.* 光栅,网板,屏面
ratch [rætʃ] *n.* 棘轮,齿杆
ratchel ['ra:tʃəl] *n.* 砾石,毛石
rate [reit] *n.* 率,比率;速率;价格,利率,税率;等级;评价 *v.* 评价,评定,定额;判断
~ of decay 风化率;分解率
~ of deformation 变形速率
~ of spoiled products 废品率;破损率
~ of wear 磨损率
accuracy ~ 精确度
occupation ~ 居住密度,居住率(每间居室的人数)
rateable ['reitəbl] *a.* 可估[评]价的;按比例的;应纳税的
~ value 征税价格
rated ['reitid] *a.* 额定的,鉴定的;适用于……的
man- ~ 适于人用的
ratification [ˌrætifi'keiʃən] *n.* 批准,认可
rating ['reitiŋ] *n.* 额定(技术性能),额定值;定额;率定,评价,鉴定;测流;分等;评级;级别,等级
~ schedule 检定程序表
merit ~ 质量评定
ratio ['reiʃiəu] *n.* 比,比率

~ error 比例误差
~ of mixture 配合比
~ of rise to span 高跨比
~ of similitude (柱的)细长比
cement-water ~ 灰水比
correlation ~ 相关比
daylight ~ 天然采光率
density ~ 相对密度
depth-to-span ~ 高跨比
depth-to-width ~ 高宽比
mixture ~ 混合比;(燃料)成分比
mortar-voids ~ 灰浆空隙比
slenderness ~ 长细比;柔性系数
spacing ~ 空间比(两灯间的距离对其至工作面距离之比)
voids ~ 孔隙比
water ~ 含水率
water cement ~ 水灰比
wet /dry tenacity ~ 湿/干韧度比
rational ['ræʃənl] *a.* 合[有]理的;理论的;推理的
~ construction 合理施工
~ design 合理设计
rationality [ˌræʃə'næliti] *n.* 合理性;理由
ratofkite *n.* 萤石
rattan [rə'tæn] *n.* =ratan 藤,藤条
~ chair 藤椅
~ rope 藤索
~ work 藤制品
rattler ['rætlə] *n.* (转筒式)磨耗试验机;磨砖机;货运列车
~ loss (磨耗试验的)磨耗率
rattrap bond 咬合(鼠夹)砌砖
rauchwacke *n.* 糙面白云石
raunchy ['rɔ:ntʃi] *a.* 不够标准的;破旧的
ravage ['rævidʒ] *n.*;*v.* 破坏,荒废;毁坏,劫掠
ravelin [rævlin] *n.* 半月形城堡
raw [rɔ:] *a.* 生的,粗制的,未经加工的,粗糙的
~ aggregate 生骨料,原状骨料

~ brick 砖坯
~ condition 粗糙状态,原状,未加工状态
~ copper 粗铜,泡铜
~ glass 粗玻璃
~ grinding 初[粗]磨
~ lacquer 生漆
~ material 原材料
~ stuff 原材料
ray [rei] n. 光线;射线;(木材)射髓
~ flower 边花
~ of light 光线
~ -proof 防射线[辐射]的
infrared ~s 红外线
ultraviolet ~s 紫外线
rayless ['reilis] a. 无光线的,黑暗的
rayon ['reiən] n. 人造丝;粘胶
~ fabric 人造丝织物
rayonnant ['reiənənt] a. 辐射状线条的
Rayonnant style 辐射式 (指14世纪法国哥特式以辐射式窗格为特色的建筑)

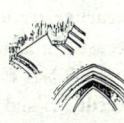

Rayonnant style

raze [reiz] v. 铲平,夷为平地,拆毁;刮[削,抹]去;消除(印象)
react [ri:'ækt] v. 反应,起反作用;反抗;再做,重演
~ing steel 再结晶钢
reaction [ri'ækʃən] n. 反应,反作用;反力
~ cement 活性胶结剂
reactionless a. 无反应的,惯[惰]性的
reactivate [ri:'æktiveit] v. 再生,再激活
reactive [ri'æktiv] a. 反应的,反作用的;活性的;反动的;无功的;电抗性的
~ aggregate 活性骨料,(与水泥起)反应的骨料
~ pigment 活性颜料
reading ['ri:diŋ] n. 读数;读;读物
~ lamp 台灯
~ room 阅览室

readjust ['ri:ə'dʒʌst] v. 调整,整顿
readjustment ['ri:ə'dʒʌstmənt] n. 调整,整顿;(仪表等)重校,重调
ready ['redi] a. 有准备的;现成的;轻便的;简易的
~ coating 简易浇面
~ cut house 预制木结构房屋
~ -made units 预制部构件
~ -mix(ed) concrete 预拌混凝土
~ -mix(ed) paint 调合漆
~ roofing 组成屋面料
real [riəl] a. 实的,真实[正]的,现实的,实际的
~ crystal 全晶[含铅晶]玻璃
~ price 实际价格
realism [riəlizəm] n. 现实主义;真实[性,感]
realistic [riə'listik] a. 现实(主义)的;逼真的;实际的
~ rendering 逼真的渲染(建筑绘画)
~ testing specimen 仿真试样
realization ['riəlai'zeiʃən] n. 实现;认识,体会
realm [relm] n. 范围,领域;(生物的)类
reams n. 非均质材料叠层
reappraisal ['ri:ə'preizəl] n. 重新估价

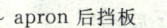

rear vault

rear [riə] n.; a. 后部[面],尾部;背面
~ apron 后挡板
~ arch 背拱
~ elevation 后视图,背面立视图
~ vault 背拱顶
~ view 后视
rearrangement ['riə'reindʒmənt] n. 重新整理[编排],调整;整顿
rearward ['riəwəd] a.; ad. 在后背,向后面 n. 在后部
reasonable ['ri:znəbl] a. 合理的,适当的
~ size 适当尺寸

rebate ['ri:beit] *n.* 半槽(边),凹凸榫;回[折]扣 *v.* 减少;打折扣
~ bead 槽口圆线
~ of glazing 镶嵌玻璃的槽口
~d and filleted joint 槽口圆贴角接缝
~d cement slab 半槽口水泥板
~d grooved and tongued joint 槽口接合,企口接合
~d joint 半槽接合,凹凸榫接合
~d lock 槽口门锁
~d plane 企口刨
~d siding 凹凸榫盖板

rebated joint

reblade *v.* (用平地机)重复整型,重新作成横断面
reblending ['ri:blendiŋ] *n.* 再拌和[混合]
rebound [ri'baund] *v.* 回弹,跳开
rebuild ['ri:'bild] *v.* 重建;修复;改造;改建
recalking *n.* =recoulking 重凿缝
recapping ['ri:kæpiŋ] *n.* 重修面层,翻修路面
recarburizer *n.* 增碳剂,渗碳剂
recast [ri:'ka:st] *n.*; *vt.* 重铸,重浇;改造;重新计算;(计划等)改订,重改
recede [ri'si:d] *v.* 退回,退却,退潮,(水位等)回落;撤回;向后倾斜,缩入
receding colour 后退色
receding line (建筑物)退缩线
receive [ri'si:v] *v.* 接[收]到;容纳;接[会]见,接待;遭受,受到
receiving antenna 接收天线
receiving basin 进水池
recent ['ri:snt] *a.* 新的,新[最]近的;(地质的)现代
receptacle [ri'septəkl] *n.* 容[贮]器;插孔,插座;贮藏所
~ plug 插头
female ~ 插孔板

plug ~ 插座
reception [ri'sepʃən] *n.* 接收,接受[纳,见]欢迎,招待会
~ chamber 会客室
~ desk 接待处
~ room 会客室
recess [ri'ses] *n.*; *v.* 退潮,退水,凹处,凹座;开槽;(隧道内的)避车洞;(船闸等的)闸门槽,闸门库;深处;体会;收纳
~ basin 三面嵌入式浴缸
~ engraving 凹版雕刻
~ lighting fitting 埋入式照明器具
~ tub 凹入墙内的浴盆
~ed arch 凹进的层叠拱
~ed head screw 埋头螺钉
~ed heater 暗装取暖器
~ed joint 凹[方槽]缝
~ed ledge 凹进(的)壁架
~ed luminaire 嵌装式照明设备
~ed pointing 凹槽嵌填
~ed portal 凹形门,凹形入口
~ed soap dish 嵌入式肥皂盆
~ed toilet paper holder 嵌入式手纸架
~ed V-joint 尖槽缝
circular ~ 圆形槽
window ~ 窗壁凹
recheck [ri:'tʃek] *v.* 再核对;复核;复查
rechipper [ri:'tʃipə] *n.* 复切(木片)机;精削机
Recidal *n.* 一种易切削高强度铝合金
recipe ['resipi] *n.* 处方,制法;方法,诀窍
reciprocal [ri'siprəkəl] *a.* 往复的;相互的,相反的;倒数的(条约等)互惠的 *n.* 倒数;互相起作用的事物
~ agreement 互惠协定
~ contract 互惠合同
~ friction test 磨损试验
reciprocating *n.*; *a.* 往复(的);摆动(的)
recirculate [ri:'sə:kjuleit] *v.* 再循环,再周转

~d air intake 循环空气进口
~d water 循环水
~ing cooling system 循环冷却系统
~ing heating system 循环供暖系统
reclad [ˌriːˈklæd] v. 再包上一层金属；在砖、石上再做上一层贴面
reclaim [riˈkleim] vt. 围垦，垦拓，填筑；取料；回收；要求归还
~ed asphaltic mixture 复拌沥青混合料
reclaimer [riˈkleimə] n. 取料机；挖掘机；(旧废料)回收设备
recloser [riːˈkləuzə] n. 自动开关[装置]
reclothe [ˈriːˈkləuð] vt. 使再穿[包，覆盖]上
~d stone (墙)包石
recoat [ˈriːˈkəut] vt. (用油漆等)再[重新]涂

reclothed stone

recognizance [riˈkɔŋizəns] n. 保证书[金]；具结；抵押金
recolour [ˈriːˈkʌlə] v. 重新着色
recombination [ˈriːkɔmbiˈneiʃən] n. 复合；恢复；再化[结]合
surface ~ 表面复合
recombine [ˌriːkəmˈbain] v. 重新结[组，联]合，复合
recommend [ˌrekəˈmend] vt. 推荐，介绍；委托；劝告
~ed illumination 推荐照度
~ed standard 推荐标准
recommendation [ˌrekəmenˈdeiʃən] n. 推荐，介绍(信)；建筑及维护规则
recompose [ˈriːkɔmˈpəuz] vt. 重新组合[安排]；改组[作]
recompress [ˌriːkəmˈpres] v. 再(次)压(缩)；空气压力增加
recondition [ˌriːkənˈdiʃən] v. 修理，修复；纠正
~ing of timber 木材的蒸汽处理
reconfiguration [ˈriːkənˌfigjuˈreiʃən] n. 改变外形；结构变形

reconnaissance [riˈkɔnisəns] n. 踏勘，搜索，草测
~ map 踏勘地图，草测图
~ survey 踏勘；草测
~ trip 现场踏勘
reconnoiter [ˌrekəˈnɔitə] v. = reconnoitre 踏[查]勘；搜索
reconsolidation n. 再固结
reconstruct [ˈriːkənsˈtrʌkt] v. 重[再]建，恢复；(按原状)修复；改建[造]
~ed stone 人造石
reconstruction [ˈriːkənsˈtrʌkʃən] n. 重[再]建；改建[造]
reconversion [ˈriːkənˈvəːʃən] n. 恢复；恢复平时生产，恢复原状，复旧
reconvert [ˈriːkənˈvəːt] v. 恢复；恢复平时生产，复旧
recover [riˈkʌvə] v. 恢复；回收；再生；分离；补偿；弥补；重新发现 [ˈriːˈkʌvə] vt. 重新盖；改装封面
recoverable [riˈkʌvərəbl] a. 可恢复的；可回收的
recovery [riˈkʌvəri] n. (pl. -eries) 复，复原；收回，回收；再生；开采
~ of elasticity 弹性恢复
recreate [ˈriːkriˈeit] vt. 再[改]造；重做；再创造 [ˈrekrieit] v. 得到休养；娱乐
recreation [ˌrekriˈeiʃən] n. 改造；重做；再[重新]创造；休养；娱乐；游览
~ area 游览区，游憩地
~ center 旅游中心
~ complex 综合旅游区
~ desk 休憩平台，屋顶康乐场
~ facility 娱乐设施
~ ground 娱乐场
~ room 娱乐室
rectangle [ˈrektæŋgl] n. (长)方形，矩形；直角
rectangular [rekˈtæŋgjulə] a. 矩形的，长方形的；直交的；成直角的
~ flat plate 矩形平板
~ prism 矩形棱柱体，直角棱镜

~ skylight 矩形天窗
~ timber 方木(材),锯材
rectification [ˌrektifi'keiʃən] *n.* 调整;校正;整治;整流;精馏;(曲线)求长度
rectify ['rektifai] *vt.* 调整;校正;整治;整流;精馏;(曲线)求长度
rectilineal [ˌrekti'liniəl] *a.* = rectilinear 直线的;直线运动的 *n.* 环箍筋
~ motion 直线运动
~ polygon 直线多边形
~ style 直线条式建筑
rectitude ['rektitju:d] *n.* 正直;正确;笔直
recumbent [ri'kʌmbənt] *a.* 躺着的
~ fold 伏褶皱
recurrence [ri'kʌrəns] *n.* 重现,复发;循环
recurvation *n.* 反(向)弯(曲)
recurvature [ri'kə:vətʃə] *n.* 反[后]弯;(风)转向
recurve [ri:'kə:v] *v.* 反弯;折回
red [red] *a.* 红(色)的;(磁石)指北(极)的 *n.* 红(色);赤字;磁铁北极
~ beds 红色岩层
~ birch 赤杨,桦木
~ brass 锡锌合金;红(色黄)铜(低锌)
~ brick 红砖
~ carpet 红地毯
~ clay 红黏土(中国)
~ cypress 红柏木,红丝柏
~ earth 红壤[土]
~ fir 红杉
~ lead (oxide) 红丹,红铅粉
~ lead primer 红丹底漆
~ lead putty (封水管接头用的)红丹油灰
~ loan (soil) 红壤(土)
~ metal (含铜大于80%的)红色黄铜
~ minium 红铅(丹)
~ mud 红泥
~ -mud sulphated cement 赤泥硫酸盐水泥
~ oak 红橡木
~ ochre 红赭石
~ ooze 红软泥
~ oxide 铁丹
~ phosphorus 红[赤]磷
~ pine 红松
~ pottery 红陶
~ prussiate 赤血盐,铁氰化物
~ sandal wood 紫檀
~ wood 红木,红杉
dark ~ 暗红
deep ~ 深红
iron ~ 铁红,红色氧化铁
redden ['redn] *v.* 使[变]红
reddish ['rediʃ] *a.* 带红色的
reddle ['redl] *n.* 红土,代赭石 *v.* 用代赭石涂
redecorate ['ri:'dekəreit] *v.* 重新装饰[油漆]
redeem [ri'di:m] *vt.* 挽[收,赎]回;恢复;补救;偿还;履行;兑现
redesign [ˌri:di'zain] *v.*;*n.* 再[重新]设计;重算[建]
redevelopment ['ri:di'veləpmənt] *n.* 再发展;改建;复兴
~ scheme 重建计划
redintegration [reˌdinti'greiʃən] *n.* 更新,重建
redix *n.* 环氧类树脂
Redo *n.* 雷度(一种乙烯树脂涂胶织物)
redress [ri'dres] *v.*;*n.* 矫正,修正;调整;重新修整;使再平衡
~ing of setts 重新修整小方石铺砌
reduce [ri'dju:s] *v.* 减少;简化;约分;还原,折合;换算,处理;破碎;分解;归纳
~d density 折算密度
~d scale 缩尺
~d stone 碎石
~d tee 缩径三通管
reducer [ri'dju:sə] *n.* 减速器;异径管节,渐缩管;减压阀;还原剂;还原器;稀释剂
self-flow ~ 自流式节流活门
shock ~ 减振器

union ~ 渐缩接头管
reducing [ri'dju:siŋ] n.;a. 减少(的);缩小(的);下降;压延;减轻;还原;简化;消退;脱轻质油
~ bend 变径弯头
~ coupling 变径管节
~ cross 变径十字头(四通)
~ flange 异径[缩口]法兰
~ joint 缩径接头
~ lateral 缩径支管
~ pipe joint 变径管节
~ tee 变径三通
reductant [ri'dʌktənt] n. (燃料的)成分;试剂;还原剂
reduction [ri'dʌkʃən] n. 缩减,减少,简化;约分;还原,整理;换算;归纳[并];破碎;分解
~ printer 缩小仪
~ range 压缩范围
percent ~ 减缩率
size ~ 粉碎;磨细
reductive [ri'dʌktiv] a. 减少[小]的;缩小[减]的;还原的,恢复的 n. 还原剂;脱氧剂
~ agent 还原剂
Redux n. 一种树脂粘结剂
redwood ['redwud] n. 红杉;红木
~ furniture 红木家具
redye [ri:'dai] v. 再[重]染
reed [ri:d] n. 芦苇;簧片;扣齿;(簧式)扒钉;卡子;小凸嵌线脚,芦苇形线脚
~ and bead 芦苇饰
~ mat 芦席
~ mould 小凸嵌线脚
~ roof 芦苇屋顶,麦秸屋顶
~ slab wall 芦苇加筋土墙
~ thatch 芦苇屋面
reeded ['ri:did] a. 有沟的,有凹槽的
~ glass 槽纹玻璃
~ tile (for stairs) (楼梯的)防滑踏步砖
re-edify ['ri:'edifai] vt. 重建;恢复

reed mould

reefer ['ri:fə] n. 冰箱;冷藏室[车];冷藏集装箱
reek [ri:k] n. 热气;水蒸气;烟雾 v. 冒气;冒烟
reel [ri:l] n. 卷;盘;卷筒;卷轴 v. 卷绕;缠
~ tape 卷尺
asbestos ~ 石棉卷筒
re-engineering n. 重[改]建,再设计
reentrant [ri:'entrənt] a.;n. 再[重新]进入,凹入
~ angle 凹角
~ corner 阴角,凹角,内隅角
~ part 凹角,内角
re-equip ['ri:'kwip] vt. 重新装备
reestablish ['ri:is'tæbliʃ] vt. 重建,恢复
reface ['ri:'feis] vt. 整修墙面;光[修]面
refacer n. 光面器;整修工具;整修工
refashion ['ri:'fæʃən] vt. 再[重]作;改变[造];给以新形式
refectorium n. [拉] (教堂,神学院的)食堂,饭厅
refectory [ri'fektəri] n. (修道院内的)食堂,饭厅
reference ['refrəns] n. 参考[照];依据;参考资料;提到;关系;基[标]准
~ area 基准(面);参考面(积)
~ datum 基准面;基准零点
~ dimension 参考尺寸
~ drawing 参考图
~ ga(u)ge 基准水位标尺;(校正用)基准标尺;基准规尺
~ illumination 标准照度
~ library 资料室,图书参考室
~ line 基准[参照]线
~ mark 基准标点;参照符号
~ room 参考书室
~ standard 参考标准
~ surface 基准面
refigure [ri'figə] v. 重新描绘[塑造];恢复形状

refine [ri'fain] *v.* 精制[炼];提炼;改进;改善
~d gold 纯金
~d metal 精炼金属
~d net 加细网格

refinement [ri'fainmənt] *n.* 精制[炼];提炼;洗炼;修订

refinish [ˌriː'finiʃ] *vt.* 返工修光;整修……表面

refit ['riː'fit] *v.*;*n.* 修理;改装,修缮,重新装配

reflect [ri'flekt] *v.* 反射[映];考虑;回顾
~ed light 反射光
~ing coating 反光涂面
~ing layer 反射层

reflectance [ri'flektəns] *n.* 反射(比,率,能力,系数)

reflection [ri'flekʃn] *n.* = reflexion 反射[映];反射波;回顾;深思;考虑
~ of sound 回声
~ pool 倒影池

reflective [ri'flektiv] *a.* 反射的;反映的;回顾的
~ coating 反射涂膜

reflector [ri'flektə] *n.* 反光镜;反射镜[物;器];反光罩
~ lamp 反光灯
~ marker 反光[射]路标
~ material 反光材料
~ sign 反光[射]标志

reflectorized paint 反光漆

reflex ['riːfleks] *n.* 反射[映] *a.* 反射的 [ri'fleks] *v.* 把……折转[回]
~ lamp trough 反射灯槽

reflexive [ri'fleksiv] *a.* 反射(性)的;折转[回]的;自反的

reflexless *a.* 无反射[映,照]的

refloor *v.* 重新铺面;重铺楼板

refract [ri'frækt] *vt.* 折射;测定……的折射度
~ing prisms 折射棱镜

refraction [ri'frækʃn] *n.* 折射

refractory [ri'fræktəri] *a.* 耐[熔]的;难控制的 *n.* 耐火材料;耐腐蚀材料
~ aggregate 耐火骨料
~ alumina cement 耐火铝[矾土]水泥
~ brick 耐火砖
~ ceramics 耐火陶瓷
~ clay 耐火黏土,耐火泥
~ coating 耐火涂料
~ fireclay block 黏土耐火砌块
~ glass 耐火玻璃
~ (insulating) concrete 耐火(隔热)混凝土
~ insulation 耐火材料绝缘
~ lime 耐火石灰
~ lining 耐火衬砌
~ mortar 耐火泥[砂浆]
~ quotient 耐火率
~ wall 隔热墙,耐火墙
acid ~ 酸性耐火材料
aluminous ~ 高铝耐火材料
basic ~ 碱性耐火材料

refrax *n.* 碳化硅耐火材料,金刚砂砖

refreshment [ri'freʃmənt] *n.* 恢复;更新;休息;点心;饮料
~ room 小吃部,茶点室
~ kiosk 小食亭

refrigerant [ri'fridʒərənt] *n.* 制冻剂;冷冻 *a.* 制冷的,冷冻的

refrigerator [ri'fridʒəreitə] *n.* 冷冻机;冰箱;冷却车;(蒸馏)冷却装置

refurbish ['riːfəːbiʃ] *vt.* 重新磨光;再刷新;整修

refurnish ['riː'fəːniʃ] *vt.* 再供给;重新装备

refusal [ri'fjuːzəl] *n.* 拒绝;取舍权;(桩的)埋入度;(打桩)阻力

refuse [ri'fjuːz] *v.* 拒绝;再熔化 *n.* 废物;垃圾 *a.* 无用的,废弃的
~ chute 垃圾井筒,垃圾槽

regal [ri'gel] *a.* 国王的;豪华的

regency ['iːdʒənsi] *n.* (*pl.* -cies)摄政政治

Regency style 摄政王式(建筑)

regenerative [ri'dʒenəreitiv] *a.* 再生的；反馈的；回热的
～ hear exchanger 再生式换热器

regime [rei'ʒi:m] *n.* （正常）状态；河况，水情；(河道)冲淤平衡；系统；统治
～ analysis 状况分析

regional ['ri:dʒənl] *a.* 区域性的；地方性的；局部的
～ center 区域中心
～ metamorphism 局部变质作用
～ planning 区域规划
～ survey 区域勘测

register ['redʒistə] *n.* 登记，注册，登记证；自动记录器；节气门；通风装置；寄存器 *v.* 登记，注册；对齐，定位
～ed architect 注册的建筑师
～ed trademark 注册商标

reglet *n.* (建筑中的)平嵌线

regnant ['regnənt] *a.* 占优势的，流行的
～ fashion 流行样式

regolith ['regəliθ] *n.* 表[浮]土，风化层，土被

regrade [ri'gra:d] *v.* 再分类；改变坡度；修整(道路等的)表面
regrading skin 石面琢新，修整表面

rgeula(e) *n.* (建筑中的)方嵌条，三槽板下短条线脚，扁带饰

regular ['regjulə] *a.* 有规律的；正规的，经济的；普通的；顺直的 *n.* 老顾客；长工 *ad.* 规则地；十分
～ bond 普通砌合
～ cement 普通水泥
～ coursed rubble 成层砌筑毛石
～ dome 正规球形穹顶
～ reflection 单向反射
～ reinforcement 普通钢筋
～ size 正规尺寸
～ stone 整齐石(料)
～ thumb screw 对称翼形螺钉

regulate ['regjuleit] *vt.* 调整[节]；整治；校准；管理

～ expenditure 控制用费
～d power supply 稳压电源
～d-set cement 控凝水泥
regulating apparatus 调整装置
regulating plate 调节板

regulation [,regju'leiʃən] *n.* 调整；调节；校准；管理；规章 *a.* 规定的；正式的；普通的
～ for technical operation 技术操作规程

regulator ['regjuleitə] *n.* 调节器；校正器；调整者；整理者；标准钟
～ tube 稳压管，稳流管
current ～ 电流调节器

Reich's bronze 一种铝青铜

reign [rein] *n.*, *vi.* 领域；盛行
～ing winds direction 盛行风向

rein [rein] *v.* 控制 *n.* 手柄；(建筑中的)拱底石

reinforce [,ri:in'fɔ:s] *vt.* 加强，增强，支援；(混凝土)加筋，配筋
～ provision 增加供应
～d aerated concrete 加气钢筋混凝土
～d bar (螺纹)钢筋
～d bituminous coating 加筋的沥青涂层
～d brick construction 配筋砖结构
～d brick masonry 加筋砖砌体
～d brickwork 加筋砖砌
～d butt weld 补强的对接焊缝
～d concrete 钢筋混凝土
～d concrete arch 钢筋混凝土拱
～d concrete beam floor 钢筋混凝土梁楼板
～d concrete columns 钢筋混凝土柱
～d concrete constuction 钢筋混凝土构造
～d concrete facing 钢筋混凝土护面层
～d concrete foundation 钢筋混凝土基础
～d concrete frame 钢筋混凝土构架

~d concrete lined tunnel 钢筋混凝土衬砌
~d concrete mat 钢筋混凝土垫层
~d concrete pipe 钢筋混凝土管
~d concrete retaining wall 钢筋混凝土挡土墙
~d concrete sandwich slab 钢筋混凝土夹层板
~d concrete skeleton construction 钢筋混凝土骨架结构
~d concrete work 钢筋混凝土工程
~d glass 钢化玻璃
~d grillage 钢筋网
~d grouted brick masonry 加筋灌浆砌体
~d gunite layer 加筋喷浆层
~d joint 加筋(接)缝
~d material 加固[筋]材料
~d plastic mortar 强化塑料砂浆
~d plastics 增强塑料
~d stock 加固橡胶
~d tile 加固瓦
reinforcing mesh 钢筋网
reintegrate [ˌriːˈintigreit] *vt.* 恢复；重建
rejoining *n.* （砌合）勾缝
rejoint [ˌriːˈdʒɔint] *v.* 再填缝
relative [ˈrelətiv] *a.* 相对的；相[有]关的；成比例的
~ brightness 相对亮度
~ concentration 相对浓度
~ consistency 相对稠度，稠度指数
~ density 相对密度
~ humidity 相对湿度
release [riˈliːs] *v.* 释放；放松；脱钩；免除；降压
reliability [riˌlaiəˈbiliti] *n.* 可靠性
~ design 可靠性设计
~ design analysis report 可靠性设计分析报告
~ index 可靠指标
~ of material 材料可靠性
~ trial 可靠性试验
relics [ˈreliks] *n.* 纪念物；遗迹；出土文物

ancient ~ 古代遗物
cultural ~ 文物
unearthed ~ 出土文物
relief [riˈliːf] *n.* 地势凹凸不平；安全释放；卸荷；援救；浮雕
~ decoration 浮雕装饰
~ drawing 地形原图
~ feature 地形要素
~ model 地形模型
~ needle nozzle 安全针形喷嘴
~ ornament 浮雕装饰
relieve [riˈliːv] *vt.* 减轻，解除；卸载；救援；替换；衬托
relieving arch 辅助拱，肋拱
relieving wall 辅助墙
relievo [riˈliːvəu] *n.* 浮雕，浮雕品
reline [ˈriːˈlain] *vt.* 重砌内衬；重新划线，更换衬套
remain [riˈmein] *vi.* 剩余；保持
~ing arch 残余拱
remap *v.* 重测图
remedial [riˈmiːdjəl] *a.* 补救的；矫正的；治疗的
~ treatment 补强处理
~ works 补救工程
remedy [ˈremidi] *n.* (*pl.* -dies) *v.* 补强；加固；修理
remix [ˌriːˈmiks] *vt.* 再混合
remodel [ˈriːˈmɔdl] *vt.* 改装[型]
remote [riˈməut] *a.* 遥远的；遥控的；非主要的；间接的
~ control 遥控
~ effects 间接影响
~ system 遥控系统
remould [ˈriːˈməuld] *vt.* 重塑
removable [riˈmuːvəbl] *a.* 可拆卸的；活动的
~ casement 可装卸窗扉
~ rail 活动栏杆
removal [riˈmuːvəl] *n.* 移去；清除
~ of form 拆模
~ of mould 拆模

~ of shuttering 拆模
remover [ri'mu:və] *n.* 拆卸工具；洗漆剂；搬运工人
　paint ~ 去漆剂
　rust ~ 除锈剂
renaissance [rə'neisəns] *n.* 文艺复兴；文艺复兴时期的风格
　~ architecture 文艺复兴时期的建筑，文艺复兴式建筑
　~ church 文艺复兴式教堂
　~ palace 文艺复兴式宫殿
render ['rendə] *v.* 抹灰，粉刷；打底；提炼；提供
　~ and set 两层抹灰
　~ float and set 三层抹灰
　~ing coat 抹灰底涂层
　~ing lath 抹灰板条
　~ing sand 抹灰用砂
　~ing work 抹灰工程
rendezvous ['rɔndivu:] *n.* (*pl.* -vous [-vu:z]) 集合；约会；公共集会场所
rendizina *n.* 黑色石灰土，腐殖质石灰土
rendoll *n.* 黑色石灰软土
rendu ['rəndu] *n.* (渲染了的)建筑(学)设计图，已渲染的设计
renew [ri'nju:] *v.* 重新，重新开始；修复；重订(合同等)
renewable [ri'nju:əbl] *a.* 可更新[换]的
　~ contract 可继续的契约
renewal [ri'nju:əl] *n.* 更新；更换
　~ of air 换气
renovation [ˌrenəu'veiʃən] *n.* 革新，刷新；改造；修整
　~ cost 修缮费
repaint [ri:'peint] *vt.* 重新涂(漆)，重画 ['ri:peint] *n.* 重新涂漆(的东西)，重画(的部分)
repair [ri'pɛə] *vt.* ; *n.* 修理，维修；纠正；补救
　~ equipment 修理设备
　~ing expense of house 房屋维修费
　capital ~ 大修
　current ~ 小修
　first-aid ~ 抢修
　operating ~ 日常维护检修
　running ~ 现场修理
reparation [repə'reiʃən] *n.* 赔偿；补救；(*pl.*) 赔款
repellent [ri'pelənt] *a.* 排斥的；防水的； *n.* 防虫剂；防水物
repeller [ri'pelə] *n.* 反射极；弹回装置；栏板
repetition [repi'tiʃən] *n.* 重复，再现；拷贝，复制品
replacement [ri'pleismənt] *n.* 置换，接替；补充；代替物
　~ material 代用材料
　~ of foundation 基础替换
　~ period 更新期限，可用年限
repoint ['ri:'pɔint] *vt.* 重嵌灰缝，再勾缝
　~ing masonry 勾缝砌体
repolish [ri:'pɔliʃ] *vt.* 再磨[抛]光
repository [ri'pɔzitəri] *n.* (*pl.* -tories) 仓库，陈列室；资源丰富地区
repousse [rə'pu:sei] *n.* 敲花艺术，敲花细工，凸纹面
repress [ri'pres] *vt.* 加压，抑制；镇压
　~ brick 加压砖
reprocess [ri:'prəuses] *v.* 重新处理，再加工；精制，改造
reproduction [ri:prə'dʌkʃən] *n.* 再生产；复制(品)；繁殖，仿制
requirement [ri'kwaiəmənt] *n.* 需要；要求
　detail ~s 详细规格，详细技术要求
requisition [ˌrekwi'ziʃən] *n.* 需要，征用；申请 *vt.* 要求；征用；索取
rere-arch ['riəa:tʃ] *n.* (建筑的)背拱
reredos ['riədɔ:s] *n.* 屏风；壁炉背面；供坛背壁
reregister *v.* 再对准；再定位

reredos

research ['ri'sə:tʃ] *n.*; *v.* 研究；周密调查；勘察
～ center 研究中心
～ institute 研究[院]
～ laboratory 研究试验室

reserve [ri'zə:v] *v.*; *n.* 保留；储备；预订 *a.* 保留的；备用的
～ capacity 储备[备用]能力；保存量
～ depot 备用仓库
～ parts 备件
～ storage 备用仓库

reshape ['ri:ʃeip] *v.* 整形；修改

residence ['rezidəns] *n.* 住宅；居住

resident ['rezidənt] *a.* 居住的；驻扎的 *n.* 居民；驻外代表；住院医生
～ architect 工地建筑师

residential [,rezi'denʃəl] *a.* 住宅的
～ area 住宅区
～ belt 居住区
～ building 居住建筑[房屋]
～ construction 住宅建设
～ density 居住密度
～ district 住宅区
～ quarter 居住区
～ standard 居住标准
～ street 居住区街道
～ suburb 郊外住宅区

residual [ri'zidjuəl] *n.* 残[剩]余；差额；误差 *a.* 剩[残]余的
～ material 剩余材料
～ shrinkage 残余收缩

resilience [ri'ziliəns] *n.* 回弹；弹性；回弹能；弹力

resilient [ri'ziliənt] *a.* 弹性的
～ connector 弹性接头
～ floor 弹性地板
～ hanger 弹性吊钩
～ joint material 弹性接缝料

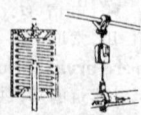
resilient hanger

resin ['rezin] *n.* 树脂 *v.* 用树脂处理
～ bonded plywood 树脂胶合板
～ chip board 树脂胶结木屑板
～ concrete 树脂混凝土
～ emulsion cement 乳化树脂粘结料
～ emulsion paint 树脂涂料，树脂漆
～ glass 树脂玻璃
～ glue 树脂胶
～ oil 松香油
～ treated wood 经树脂处理的木材
acrylic ～ 丙烯酸树脂
alkyd ～ 醇酸树脂
epozy ～ （EP）环氧树脂
phenolic ～ 酚醛树脂
polyester ～ 聚酯树脂
resole ～ 甲阶[可溶]酚醛树脂
synthetic ～ 合成树脂
thermoplastic ～ 热塑性树脂
thermosetting ～ 热固性树脂

Resistal *n.* 一种铝青铜；耐蚀硅砖

resistance [ri'zistəns] *n.* 抗[阻]力；电阻；抵抗
～ bond 抗粘性，粘着阻力
～ corrosion 耐腐蚀性
～ to fouling 抗侵蚀性
～ to heat 耐热性
fatigue ～ 抗疲劳性[强度]
flame ～ 耐火性
weather ～ 耐老化；耐风雨侵蚀能力

resistant [ri'zistənt] *a.* 有抵抗力的；耐……的 *n.* 反对者；防腐
～ metal 耐蚀金属

resit *n.* ＝resite 丙阶[不溶]酚醛树脂

resitol *n.* ＝resolite 乙阶[半溶]酚醛树脂

resiweld *n.* 环氧树脂类粘合剂

resol *n.* 可溶酚醛树脂（甲阶）

resonance [,reznəns] *n.* 共[谐]振；共鸣

resonant ['reznənt] *a.* 共[谐]振的；共鸣的

resorcin(e) ＝resorcinol *n.* 间苯[苯间]二酚
～ resin 间苯二酚树脂
～ resin glue 间苯二酚树脂胶

resort [ri'zɔ:t] *v.*; *n.* 依靠;求助于;诉诸;常去(的地方)
～ health 休养胜地
resource [ri'sɔ:s] *n.* 手段;方法(*pl*) 资源
natural ～s 自然资源
respace *v.* 重新隔开,重间隔
respond [ris'pɔnd] *v.* 答复;响应;起反应 *n.* 壁联,用作拱支座的墩式壁柱
response [ris'pɔns] *n.* 反[响]应;灵敏度;特性曲线
responsibility [risˌpɔnsə'biliti] *n.* 责任,职责;任务;可靠性
～ range 职责范围
ressaut *n.* 墙面凸出部分
rest [rest] *v.*;*n.* 停[静]止;休息;放置;基于;*n.* 托架,支柱
～ area 休息场所
～ centre 疗养中心
～ -house (途中)旅馆,客栈
～ pier 支墩
～ quarter 生活间
～ room 休息室
arm ～ 扶手
elbow ～ 扶手
foot ～ 脚架,搁脚,脚踏板
tripod ～ 三角架
restaurant ['restərɔ:ŋ] *n.* 餐馆
restitution [ˌresti'tju:ʃən] *n.* 恢复,归还;重建
restoration [ˌrestə'reiʃən] *n.* 恢复;修复;归还
restore [ris'tɔ:] *vt.* 恢复;修复;归还
～d building 重新翻修的建筑
～ing force 恢复力
restrain [ris'trein] *vt.* 抑制;约束;克制
restraint [ris'treint] *n.* 抑制;限制;约束;约束力;克制
resurface ['ri:'sə:fis] *vt.* 重做面层,重铺路面
resurfacing by addition 加料翻修(路面)
resurfacer *n.* 表面修整器
retail ['ri:teil] *n.*;*v.* 零售 *a.* 零售的 *ad.* 以零售方式
～ salesroom 零售处
～ store 零售店
retard [ri'ta:d] *vt.*;*n.* 减速;推迟;妨碍;阻止;延缓;停滞
～ed cement 缓凝水泥
～ing agent (混凝土)缓凝剂
retarder [ri'ta:də] *n.* (混凝土)缓凝剂;减速器
retemper *v.* (砂浆等)加水重拌[塑]
reticulate [ri'tikjulit] *v.* 分成网格;分配 *a.* 网状的
～ molding 网状线脚

reticulate molding

～d ashlar 网状纹石面
～d bond 网状砌合
～d dressing 网状纹修琢(石面)
～d masonry 网状圬工
～d shell 网状薄壳
～d structure 网状结构
～d tracery 网状窗格
retiform ['ri:tifəm] *a.* 网状的,有交叉线的
retimber *v.* 重新支撑,修理木支架
retractable [ri'træktəbl] *a.* 可收缩[回]的
retrench [ri:'trentʃ] *v.* 减少;紧缩;删除
return [ri'tə:n] *n.*;*v.* 返回;归还;回程;报酬;回答;报告 *a.* 返回的;回程的;报答的
～ -air course 回风巷道
～ air duct 回气管道
～ -air grill 回气花格栅
～ bead 转角处的联珠线脚
～ corner block 转角砌块
～ nosings 踏步转角延伸的一个凸沿

~ed molding 回转线脚
revamp [ri'væmp] vt. 翻新,修复;改订;重修,重建
reveal [ri'vi:l] vt. 揭露;揭示;泄露;展现 n. 侧墙;窗沿
　inner ~ （窗的）内侧壁
　outer ~ （窗的）外侧壁
reverse [ri'və:s] a. 颠倒的,反[换]向的 n.;v. 颠倒,倒[反]转,换向,倒退

reversed zigzag molding

~d arch 倒拱
~d zigzag molding 反锯齿形线脚
reversible [ri'və:səbl] a. 可逆的
　~ key 两向键
　~ lock 双向锁
　~ mortise lock 可反向插锁
revert [ri'və:t] vi. 回复;复原;恢复
revetment [ri'vetmənt] n. 护岸;砌面;护坡[堤],铺[砌]面
　~ wall 护岸,护墙
revolution [ˌrevə'lu:ʃən] n. 回,[旋]转;革命
　~ door 转门
　~ window 转窗
revolve [ri'vɔlv] v. 回[旋]转
　revolving chair 转椅
　revolving door 转门
　revolving ladder 旋梯
　revolving light 旋转灯
　revolving shutter 卷筒百叶窗
　revolving stage 旋转式舞台
　revolving window 旋转窗
Rex steel 一种耐热耐蚀高合金钢
Rhenish brick 莱茵式砖（轻质硅酸盐砖）
Rhenish style 莱茵式（建筑）
rhodonite n. 蔷薇辉石
rhomb [rɔm] n. 斜方六面体;菱形
rhombohedron [ˌrɔmbə'hi:drən] n. 菱形(六面)体,菱面体

rhomboid ['rɔmbɔid] n. 长斜方形,扁菱形
rhombus ['rɔmbəs] n. 菱形,斜方形
rhone n. 雨水槽,排水槽
rhus lacquer 漆树漆
rhus verniciflua 漆树
rhyolite ['raiəlait] n. = rhyolith 流纹岩
rib [rib] n. 肋,肋材;拱肋 v. 加肋
　~ and panel vault 肋板支承的穹顶
　~ arch 肋拱
　~ flange 肋凸缘
　~ floor slab 肋形楼板
　~ of dome 圆穹顶肋
　~ bed bar 竹节钢筋
　~ bed floor 肋构楼板,密肋楼板
　~ bed glass 起肋玻璃,木条固定的玻璃
　~ bed slab 肋形板,密肋楼板
　~ bed vault 扇形肋穹顶
　angle ~ 角肋
　bearing ~ 凸缘
　rib
riband ['ribənd] n. 缎丝带（装饰用）
　~ stone 条纹砂岩
ribband ['ribbænd] n. 板条;木桁 v. 用木桁固定
ribbon ['ribən] n. 饰[色]带,带状物;板条
　~ board 条板
　~ building 带式房屋,连片式房屋
　~ course 带状瓦层
　~ strip 条板
　~ window （横向）统长窗,通窗
rich [ritʃ] a. 富的,丰富的;贵重的;有意义的
　~ colour 浓艳色彩
　~ concrete 富混凝土
　~ -glittering 金碧辉煌的
　~ mortar 多灰砂浆,浓砂浆
ricker n. 脚手架立柱,堆垛工
riddle ['ridl] v. 筛;穿过;解（难题）;

（使）迷惑 *n*. 粗筛
～ drum 圆筒筛
ridge [ridʒ] *n*. 岭；山脊；屋脊；埂
～ beam 屋脊梁
～ board 脊板
～ cap （屋）脊瓦
～ corner tile 脊角瓦
～ course 屋脊瓦层
～ crest 脊顶饰
～ piece 栋梁，脊木
～ -pole 栋梁，屋脊梁
～ purlin 脊檩
～ rib 脊肋
～ rod 圆脊檩
～ roll 脊瓦卷形饰
～ roof 人字[有脊]屋顶
～ spike 脊尖饰
～ tile 脊瓦
rift [rift] *n*. 裂缝，裂口；正常断层；长狭谷 *v*. 劈开；分裂，渗入
～ crack 木材干裂
～ grain 顺纹
～ sawn timber 四开木材
～ trough 正断层谷
rigesity *n*. 糙度
riggot *n*. 地面排水沟
right [rait] *v*. 扶直；恢复平稳 *n*. 正确；法权 *a*. 直的；右面的；对的；适[正]当的
～ ahead 正前方
～ -angled bend 直角弯管
～ arch 正拱
～ elevation 右视图
～ -hand screw 右旋螺钉
～ hand stair 右扶手楼梯
～ lane 右侧车道
rigid [ridʒid] *a*. 刚性的；硬质的；坚固的；固定的；严格的
～ arch 刚性拱
～ expanded plastics 硬泡沫塑料
～ material 刚性材料
～ plastic foam 硬泡沫塑料

rigorous ['rigərəs] *a*. 严格的，严密的
rilievo [ˌrili'eivon] *n*. (*pl*. -vi) 浮雕
rim [rim] *n*. 边缘，框；轮箍[圈]
～ latch 弹簧锁
～ lock 弹簧锁
～ saw 圆锯
rinceau *n*. 树枝状的装饰
ring [riŋ] *n*. 环，圈；集团
～ arch 环拱
～ beam （房屋的）圈梁，环梁
～ bolt 带环螺栓
～ connection 环接合
～ course 拱圈层
～ flange 法兰盘
～ girder 环形主梁
～ heating system 环形供暖系统
～ joint 环接
～ valve 环形阀
～ wall 围墙
elastic ～ 弹性垫圈
rubber ～ 橡胶垫圈
seal ～ 密封环
ringstone *n*. 砌拱用的楔形砖或块材
rink [riŋk] *n*. (室内)溜冰场
rinser *n*. 冲洗器，清洗装置
rip [rip] *v*. 劈[裂]形 *n*. 裂口；激流；清管器；刮刀
ripe [raip] *a*. 成熟的；准备好的；熟练的
～ wood 成熟木材
ripping ['ripiŋ] *n*. 撕裂；折除；割挖；凿开
～ chisel 细长凿
～ saw 粗齿锯
ripple ['ripl] *n*. 微波；波纹
～ finish 皱纹面饰
～ glass 波纹玻璃
～ surface 波纹面
～ varnish 波纹清漆
rise [raiz] *v*. 升起，上涨；发生 *n*. 上升；增高；增长量；矢高；(楼梯)级高
～ and run ratio 踏步级高与踏步宽之比
～ of arch 拱高，拱矢

~ of span 起拱(度)
~ of vault 穹顶矢高
~ -span ratio 高跨比
riser ['raizə] n. 竖[立]管；楼梯踏步竖板；升降器；垂直井
~ board (梯级)竖板
~ height 竖板高度
~ vent 透气孔
~ seat ~ 座位升降器
~ stair ~ 楼梯起步板
rising ['raiziŋ] n. 上升,增长；高地 a. 上升的
~ arch 跂拱
risk [risk] n. 危险,冒险 vt. 冒险
~ analysis 风险分析
rive [raiv] n. 裂缝,碎片 v. 劈[裂,撕]开,拆断
riving hammer 劈锤
rivelling 皱纹,条纹
river ['rivə] n. 河,江,水道
~ sand 河砂
rivet ['rivit] n. 铆钉 vt. 铆接
~ -back plate 铆钉垫圈
~ head 铆钉头
~ hole 铆钉孔
~ joint 铆接
~ pitch 铆钉间距
~ spacing 铆接间距
~ truss 铆合桁架
~ed bond 铆接
~ed construction 铆接结构
~ed girder 铆接大梁
~ed joint 铆接
~ing gun 铆钉枪
~ing hammer 铆钉锤
rivet(t)er n. 铆工；铆钉枪[机]
rivulet ['rivjulit] n. 小河,溪流
road [rəud] n. 道路,公路,航路,铁路；途径；方法
roadside ['rəudsaid] n. 路边,路旁地带

rivet

~ delineator 路边(夜间)反光标志
~ embellishment 路旁绿化
roadway ['rəudwei] n. 车行道,道路
~ light 路灯
roak n. 发裂,表面缺陷
roast [rəust] v.；n. 烘烤,煅烧
robe [rəub] n. 外袍；衣橱
~ ward ~ 壁橱
robing room 化妆室
rock [rɔk] n. 岩石,石块,礁石
~ block 大块石
~ bolt 地脚螺栓
~ faced dressing 粗琢石
~ garden 岩石花园,假山庭园
~ sand 岩砂
~ wool 石棉,玻璃纤维
~ work 粗面石工,石块砌筑
rockery ['rɔkəri] n. 假山,石园
rocking ['rɔkiŋ] n.；a. 摆动,处于不稳定状态
~ chair 摇椅
rococo architecture (欧洲 17 ~ 18 世纪的)洛克克式建筑

rococo architecture

rod [rɔd] n. 杆,棒；标尺；避雷针；竿
~ crack 纵裂纹
~ spacing 钢筋间距
~ hanging ~ 挂衣棍
rodent ['rəudənt] a. 侵蚀性的
~ proof 防蚀的,防鼠咬的
rodman ['rɔdmən] n. 钢筋工

roll-and-fillet molding

roll [rəul] n.；v. 滚压,转动碾压；卷起
~ -and-fillet molding 外圆线脚
~ bar 碾杆
~ billet molding 滚筒错交线脚架
~ molding 漩涡形线脚,卷线脚
~ -out 压平,拉长,延伸
~ rim 卷边

~ roofing 屋面卷材
~ top desk 带折叠盖板的写字台
~ -up door 卷升门
glass wool ~ 玻璃棉毡
toilet ~ holder 手纸架
rolled [rəuld] *a*. 轧制的,辊压的
~ angle 轧制角钢
~ bar 条钢
~ glass 滚轧玻璃
~ shape 型钢,型材
~ sheet steel 轧制钢板
roller [ˈrəulə] *n*. 碾压机;滚筒;辊
~ shutter 卷帘
~ track 滚轴支座,滚轮轨
grinding ~ 砂轮
paint ~ 油漆辊筒
rollerarm *n*. 翻窗撑档
rolling [ˈrəuliŋ] *n*.;*a*. 滚动;碾压;轧制(的)
~ blind 卷帘
~ door 滑动门
~ partition 滚动式隔断
~ shutter door 卷升百叶门
Roman [ˈrəumən] *n*. 罗马人;罗马字 *a*. 罗马(人)的
~ arch 半圆拱
~ mosaic 嵌镶玻璃,罗马式锦砖镶嵌
~ order 罗马柱式
~ ornament 罗马浮雕装饰
~ Renaissance 罗马文艺复兴时期
~ style 罗马式
~ tile 罗马式瓦
~ wagon vault 罗马筒形穹顶
Romanesque [ˌrəuməˈnesk] *a*.;*n*. 罗马式的(建筑)
romanticism [rəˈmæntisizm] *n*. 浪漫主义
rondure [ˈrɔndʒə] *n*. 圆形(物);优美的弧度
rood [ru:d] *n*. 十字架
~ beam 十字架支承梁
~ loft 教堂中十字架坛
roof [ru:f] *n*. 屋顶,屋面;顶盖 *vt*. 盖屋顶;保护
~ aggregate 屋顶铺面集料
~ arch 屋顶拱
~ batten 挂瓦条
~ beam 屋顶梁
~ boarding 屋面板
~ covering 屋面覆盖层
~ deck 屋顶平台,屋面板
~ dormer 屋顶[老虎]窗
~ drain 雨水斗,屋顶下水管
~ drainage 屋面排水
~ framing 屋顶框架
~ garden 屋顶花园
~ glazing bar 天窗芯子
~ gutter 屋顶落水沟;天沟
~ insulation 屋顶隔热
~ joist 屋顶托梁
~ light 天窗
~ light flashing 天窗泛水
~ outlet 屋面排水口
~ overhang 挑檐
~ panel 屋面板
~ paper (铺屋面的)油毡
~ pitch 屋顶坡度
~ plate 屋面盖板
~ projection 挑檐
~ purline 屋面桁条
~ rafter 屋面椽条,大瓦条
~ ridge 屋脊
~ scuttle 屋顶天窗
~ sheathing 屋顶盖板
~ sheets 屋面板材
~ slab 屋顶板
~ slate 屋面石板瓦
~ slope 屋面坡度
~ span 屋架跨度
~ structure 屋顶结构
~ tank 屋顶水箱
~ terminal 屋顶盖板
~ tile 屋面瓦
~ timber 屋顶用木料
~ tree 脊檩,栋梁

~ truss 屋架
~ valley 屋顶排水沟,天沟
~ ventilator 屋顶通风机
~ woodwork 木屋架
barrel ~ 筒形屋顶
broach ~ 尖塔层顶
built-up flat ~ 卷材平屋面
cable suspension ~ 悬索屋顶
cantilever ~ 悬臂屋顶
collar-and-tie ~ 系梁三角屋架
collar beam ~ 系梁屋顶
compass ~ 半圆形屋顶
composite ~ (钢木)组合屋面
concave ~ 凹面屋顶
conical ~ 圆锥体屋顶
conoid ~ 圆锥体屋顶
couple ~ 双坡屋顶
curb ~ 复折形屋顶
deck ~ 平台式屋顶
deck glass ~ 玻璃砖平台屋顶
dome-shaped ~ 圆盖屋顶,穹顶屋顶
double bent shell ~ 双曲薄壳屋顶
double-gable ~ 双人字屋顶
double-pitch ~ 双坡屋顶
fence ~ 棚顶
flat ~ 平屋顶
framed ~ 构架屋顶
Gothic ~ 哥特式屋顶
groined ~ 穹棱屋顶
half-pitch ~ 45°斜坡屋顶
hip ~ 四坡屋顶
jerkinhead ~ 半山头屋顶(山墙尖呈斜坡的两坡式屋顶)
lamella ~ 叠层薄板屋面
monitor ~ 采光(通风)屋顶
mushroom ~ 伞形屋顶
northern-light ~ 锯齿形屋顶
Ogee ~ 渥奇式屋顶,S形屋顶
open ~ 开敞式屋顶
pent ~ 单坡屋顶
pitched ~ 坡屋顶
polygonal ~ 多边形屋顶

purlin ~ 檩支屋面
pyramidal ~ 棱锥屋顶
quadrangle ~ 方形屋顶
rafter ~ 橼子承重屋顶,人字木屋顶
ribbed ~ 加肋穹顶
ridge ~ 有脊屋顶
rounded ~ 圆形屋顶
saddle ~ 鞍形屋顶
sawtooth ~ 锯齿形屋顶
shed ~ 单坡屋顶
shell structure ~ 壳体屋顶
single pitch ~ 单坡屋顶
span ~ 等斜[起脊]屋顶(双坡屋顶)
sprung ~ 砌拱屋顶
tent ~ 锥形屋顶
tiled ~ 瓦屋顶
trussed ~ 桁架屋顶
valley ~ 带天沟屋顶
vaulted ~ 拱状屋顶
roofage [ˈruːfidʒ] *n.* 盖屋顶的材料
roofing [ˈruːfiŋ] *n.* 盖瓦,屋面,卷材
~ asphalt 屋面沥青
~ cement 屋面水泥
~ felt (铺屋面的)油毡
~ iron 屋面薄铁皮
~ malthoid 屋面油毡
~ material 屋面材料
~ membrane 屋面薄膜
~ paper 屋面油纸
~ slate 屋面石板,石板瓦
~ tile 瓦
rookery [ˈrukəri] *n.* 贫民窟
room [rum] *n.* 房间;场所;空间,地位;机会
~ air conditioner 室内空调器
~ antenna 室内天线
~ layout 室内平面布置
~ neck 进入洞室的巷道
~ separation wall 房间隔墙
~ thermostat 室内恒温调节器
bath ~ 浴室
cloak ~ 衣帽间
conference ~ 会议室

control ~ 控制室
dining ~ 餐室
dressing ~ 化妆室
living ~ 起居室
printing ~ 复印室
store ~ 贮藏室
strong ~ （防火防盗）保险库
tea ~ 饮茶室
telex ~ 电话交换台间
utensil ~ 工具间
utility ~ 杂务间
roomette [ru:m'et] *n.* 小房间
roomy ['ru:mi] *a.* 宽敞的
rope [rəup] *n.* 绳，索，缆；一串 *vi.* 扎；缚；捆
rosace ['rəuzeis] *n.* 蔷薇花饰；圆花窗
rose [rəuz] *n.* 玫瑰；圆花窗；玫瑰图，频率图；滤网
~ coloured 玫瑰红色的
~ molding 玫瑰花线脚

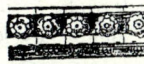
rose molding

~ window 圆花窗
shower ~ 淋浴喷头
wind ~ 风（向力）图，风玫瑰图
rosette [rəu'zet] *n.* 圆花饰；天花板电线匣；插座；玫瑰图
~ fracture 星[花]状断口
rosin ['rɔzin] *n.* 松香，树脂
rostral column 有船头装饰的纪念柱
rostrum ['rɔstrəm] *n.* 讲坛，城楼
rotary ['rəutəri] *n.* 环形交叉（道路）*a.* 旋转的；环形的；轮换的
rotate [rəu'teit] *v.* 旋转；转动；循环；更替
rotor ['rəutə] *n.* 转子；转轮
rot-proofness *n.* 耐腐性
rotten ['rɔtn] *a.* 腐烂的；腐败的；坏的
~ stone 磨石
~ wood 朽木，腐木
rotund [rəu'tʌnd] *a.* 圆形的；华丽的
rotunda [rəu'tʌndə] *n.* 大圆厅；圆顶建筑物

rouge [ru:ʒ] *n.* 红（铁）粉，三氧化二铁
rough [rʌf] *a.* 粗糙的；不平的；粗制的；近似的
~ ashlar 毛石块
~ board 毛板
~ brick 粗面砖
~ carriage 楼梯支承斜梁
~ cast 粗糙粉刷
~ cast glass 毛玻璃
~ coat （粗糙的）底涂层
~ coating 粗灰泥打底子
~ draft 草图
~ drawing 草图
~ dressing 粗琢石
~ edged 毛边
~ finish 粗加工
~ floor 粗糙地面
~ grinding stone 粗磨石
~ guess 粗略估计
~ hardware 粗五金
~ log 未加工圆木
~ lumber 粗原木
~ -machining 粗加工
~ plan 初步计划
~ plank 毛板
~ rock 粗石，毛石
~ rubble 毛石，乱石
~ sill 未加工的窗门槛
~ sketch 草图
~ stucco 拉毛粉刷
~ surface 粗糙表面，毛石
~ wall 乱[毛]石墙
roughage ['rʌfidʒ] *n.* 粗材料
roughcast ['rʌfkɑ:st] *n.* 拉毛；粗糙表面，打底用灰；初步方案 *v.* 拉毛；粗刷，粗涂，打底子
roughen ['rʌfn] *v.* 使粗糙，凿毛
roughly ['rʌfli] *ad.* 粗糙地；粗略地
~ -squared stone 粗方石
roughness ['rʌfnis] *n.* 粗率，粗糙度；不平整度；崎岖
~ coefficient 粗糙系数
~ of surface 表面粗糙度

~ ratio 糙率比
round [raund] *n.* 一圈,一周;(一个)循环;回合 *v.* 弄圆;环绕;旋转;取整 *a.* 圆的;取整的
~ arch (半)圆拱
~ bed 圆形花坛
~ church 圆形教堂
~ molding 圆线脚
~ pedimend 半圆三楣饰
~ timber 圆木材
~ trifoliated arch 三叶形拱
~ window 小圆窗

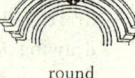
round pedimend

rounded ['raudid] *a.* 圆形的,圆拱的,全面的
~ angle 圆角
~ material 圆形材料
~ roof 圆屋顶
roundel ['raundl] *n.* 圆物;圆玻璃;圆窗
roundlet ['raundlit] *n.* 小圆,小圆形物
routine [ruː'tiːn] *n.* 程序;常规;例行工作 *a.* 常规的
~ inspection 日常[例行]检查
~ maintenance 例行保养,日常维修
~ test 常规试验
~ work 日常工作
row [rəu] *n.* 排;横列;行列
~ -house 成排房屋中的一幢
~ -lock 竖砌砖
~ -lock arch 砖砌圆拱
~ -lock cavity wall 竖砌砖空心墙
~ lock course 竖砌砖层

row-lock

rub [rʌb] *n.* ;*v.* 摩擦;擦,擦亮
~ bed finish 磨光面
~ bed joint 胶结缝
~ bed surface 磨光面
~ bing board 镘子,抹子

~ bing varnish 耐磨清漆
rubber ['rʌbə] *n.* 橡皮;擦具;摩擦物;碰垫 *a.* 橡胶[皮]的
~ bond 橡胶粘结料
~ -bitumen sealing compound 橡胶沥青填缝料
~ buffer 橡皮缓冲垫
~ cement 橡胶胶合剂
~ cloth 橡胶布
~ covered 包橡皮的
~ emulsion 橡胶乳液,胶乳
~ hose 橡胶管
~ packing 橡皮衬垫[填料]
~ paving 橡胶铺面
~ seal 橡胶止水[密封]
~ strip 橡皮防水条
~ varnish 橡胶清漆
~ washer 橡胶垫圈
flexible ~ 可伸缩橡皮管
gum ~ 天然橡胶
synthetic ~ 人造[合成]橡胶
rubberize ['rʌbəraiz] *vt.* 涂(橡)胶
~d asphalt sealing strip 橡胶地沥青嵌缝板
~d waterproof seal 橡胶防水层
rubble ['rʌbl] *n.* 毛石,块石
~ arch 粗石拱
~ ashlar 毛方石
~ backing 毛石底层
~ concrete 毛石混凝土
~ concrete wall 毛石混凝土墙
~ filling 抛石,堆石,毛石填筑
~ masonry 毛石砌体
~ -mound 抛石棱体,抛石基床
~ -mound foundation 抛石基床
~ paving 块石铺面
~ pitching 块石护坡
~ retaining wall 毛石挡土墙
~ vault 毛石穹顶
~ wall 毛石墙
~ with binding material 浆砌块石
~ work 毛石砌体

regular coursed ~ 整层砌的毛石
square ~ 方块毛石
rubbly ['rʌbli] *a.* 毛石的;块石的
ruby ['ru:bi] *n.* (*pl.* -bies)红宝石 *a.* 鲜红的
~ glass 红宝石玻璃
selenium ~ glass 硒(宝石)红玻璃
ruckle [rʌkl] *v.* 变皱 *n.* 砾石堆
rude [ru:d] *a.* 粗糙的;未加工的,天然的
~ drawing 草图
rudiment ['ru:dimənt] *n.* 初步,入门,基本原理
rufous ['ru:fəs] *a.* 带红色的,橘红色的,赤褐色的
rug [rʌg] *n.* 毛毯,地毯
rugged ['rʌgid] *a.* 崎岖的;结实的;粗糙的,(天气)恶劣的
~ face 粗糙面
ruggedize ['rʌgidəiz] *vt.* 加强,加固
rugosity [ru:'gɔsiti] *n.* 粗糙度,皱纹
rule [ru:l] *n.* 尺;规章,惯例;统治 *v.* 统治;决定;(用尺)划
~ for implementation 实施细则
~ joint 肘节形接头
~ of thumb 经验方法
carpenter's ~ 木工尺
convex ~ 凸面卷尺
copy ~ 仿形[放大]尺
drawing ~ 绘图尺
flexible ~ 卷尺
folding pocket ~ 折尺
ruler ['ru:lə] *n.* 尺;划线板;直规
run [rʌn] *n.* 跑;行程;楼梯踏步;航程;运行;试验 *v.* 跑;流动;航[运]行;经营
~ a moulding 拉线脚
~ board 楼梯踏脚板
rung [rʌŋ] *n.* 梯级;轮辐;舵轮的把手 *a.* 加箍的
~ ladder 直爬梯
running ['rʌniŋ] *a.* 运行的;流动的;连续的;例行的 *n.* 运行[转]
~ bond 顺砖压缝砌合
~ maintenance 日常检修

~ ornament 流水形[装]饰件

running ornament

~ repair 日常检修
~ rule (抹灰用)准尺,靠尺
~ screed 定墙灰厚度用的准条
~ surface 波状表面
~ trap U形存水弯

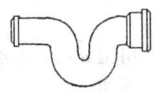
running trap

rupture ['rʌptʃə] *v.* ; *n.* 破裂,破坏;断裂
rural ['ruərəl] *a.* 乡村的;农业的
~ architecture 乡村建筑
rust [rʌst] *n.* 锈 *v.* 生锈
~ inhibitor 防锈剂
~ preventer 防锈剂
~ -preventive paint 防锈漆
~ -proof 不[防]锈的
~ proof steel 不锈钢
~ removal 除锈器
~ -resisting 防锈的
rustic ['rʌstik] *a.* 粗面的(建筑);乡村的;朴素的
~ brick 粗面砖
~ column 无装饰柱,粗面柱
~ dressing 粗琢面
~ finish 水刷石饰面
~ home 农村房舍
~ joint (石砌体的)凹入粗缝
~ masonry 粗面砌体
~ siding 粗面壁板
~ slate 粗面石板
~ stone 毛石,粗石
~ terrazzo 粗面水磨石
~ work 粗面砌筑

rustic work

rusticate ['rʌstikeit] *v.* 粗琢
~d dressing 粗琢面
~d joint 粗琢[明显]缝
rustics *n.* 粗面砖
rustless ['rʌstlis] *a.* 不锈的

~ steel 不锈钢
rusty ['rʌsti] a. ruster, rustiest 生锈的
~ spot 锈斑
rut [rʌt] n. 压痕,凹槽;常规,惯例

S

sabicu n. (古巴出产的)一种质地坚硬的贵重木材
sable ['seibl'sebl] n. 黑貂(皮) a. 黑的,深褐色的
sabot ['sæbəu] n. 桩靴,垫板
sack [sæk] n. 麻袋,粗布袋,牛皮纸袋,邮袋,包,罩,vt. 装袋
  ~ cloth 麻袋布
  ~ rubbed finish 布纹粉刷
  ~ed cement 袋装水泥
sacrarium [sə'krɛəriəm] n. 圣堂
sacred ['seikrid] a. 神圣的;献给……的;专供……用的
sacrificial [ˌsækri'fiʃəl] a. 牺牲的,损失的
sacristy ['sækristi] n. (pl. -ties) 圣器收藏室;教堂祭器室
saddle ['sædl] n. 鞍(架,座,部,状物);座(板,架),凹座;台盘,(门口的)踏板;管托,支管架,托架,圆枕木;(峰脊线中的)凹壳[点],鞍点 v. 装鞍(于);强加(于);加(重物等)于
  ~ arch 鞍形拱
  ~ back 鞍背,鞍形山脊
  ~ back coping 鞍形盖顶
  ~ backed girder 鞍背形大梁
  ~ back roof 鞍形屋顶
  ~ bar 撑棍,嵌玻璃的铅条
  ~ board 鞍形板,屋脊板
  ~ cheek (英)高背靠椅
  ~ roof 鞍形屋顶,两山头房顶
  ~ stairs 鞍形楼梯
saddle-tree ['sædltri:] n. 鞍架,(美)百合木
safe [seif] a. 安全的,无危险的,可靠的,稳妥的,保险的,有把握的 n. 保险柜[箱],冷藏柜
  ~ allowable load 安全容许荷载
  ~ clearance 安全净空
  ~ -deposit vault 安全储存室,保险库
  ~ gap 安全空隙
  ~ guard 安全措施,防护设施
  ~ guarding 安全防护
  ~ light 安全灯
  ~ parapet wall 矮围墙
  ~ service life 安全使用年限
  ~ sign colour 安全标志色
safety ['seifti] n. 安全(性,措施),保险,稳定,可靠性;保险装置,安全设备,防护器材 v. 保护,防护,使保险 a. 保障安全的,保险的
  ~ arch 安全拱,分载拱
  ~ barrier 安全栅栏
  ~ bolt 安全销,安全螺栓
  ~ catch 安全挡[销]
  ~ clearance 安全净空,安全间隙
  ~ coefficient 安全系数
  ~ cover 防护罩
  ~ cut-off 保安开关
  ~ design 安全设计
  ~ dog 安全挡块
  ~ glass 安全[不碎]玻璃
  ~ goggles 护目镜
  ~ guard 护栏,保险板
  ~ helmet 安全帽
  ~ installation 安全装置[设备]
  ~ ladder 安全梯
  ~ lamp 安全灯
  ~ latch 安全插销
  ~ lock 安全锁

~ nosing 楼梯踏步上的防滑条
~ nut 安全螺帽
~ pin 安全销钉
~ screen 安全隔层,安全网
~ switch 保险开关

safflower ['sæflauə] *n.* 草红花,江花;红花染料
~ oil 红花子油

saffron ['sæfrən] *n.* 番红花;(取自番红花的)一种橘黄色染料 *a.* 橘黄色 *vt.* 以番红花染料着色
~ yellow 橘黄色

sag [sæg] *v.* 下垂,弯下来,凹下,沉降,下沉;下跌[降],萧条 *n.* 垂[挠]度,经济萧条,跌价
~ bolt 防毛螺栓
~ ratio 垂跨比
~ s and crests 凹凸不平

saggar ['sægə] *n.* = sagger(烧瓷用的)烧箱[盆],耐火泥土 *v.* 用烧箱烘
~ clay 泥[烧]箱土,火泥,烧钵土

sail [seil] *n.* 帆,篷,帆船,帆状物;航行,航程 *v.* 扬帆行驶;张帆,开船,驾船,滑翔
~ -cloth 帆布
~ varnish 帆布清漆

sala【意】*n.* 大厅,大会厅

salamander ['sæləmændə] *n.* 耐火保险箱,能耐高热的东西;烤火钢盆,一种石棉制的防火板
~'s wool 石棉

salesroom ['seilzru:m,'selzrum] *n.* 售货场[处]

salience ['seiljəns] *n.* 凸出,突起

salient ['seiljənt] *a.* 凸[突]出的;显著的 *n.* 凸角,突出部
~ angle 房屋的外角,阳角
~ corner 凸出角
~ features 特征[点,色]

sally ['sæli] *n.* 凸出部,纯角;橡头 *v.* 冲出
~ port 暗门,太平门,保垒,地道

salmon ['sæmən] *n.* 赭色 *a.* 橙红色(的)

~ brick 未烧透的红砖

saloon [sə'lu:n] *n.*【法】= salon 大[客]厅,沙龙,酒吧间,交谊室,画廊
~ bar 酒吧间
~ cabin 头等舱
~ car 头等列车,客厅式车厢,轿车
billiard ~ 弹子房,撞球场
dancing ~ 跳舞厅
dining ~ 餐厅
refreshment ~ 饮食店

salt [sɔ:lt;sɒlt] *n.* 盐 *a.* 含盐的,咸的,盐渍的 *v.* 撒[加]盐,盐渍
~ -box type (美)两坡不对称的硬山顶住宅
~ screen 荧光(增感)屏

saltation [sæl'teiʃən] *n.* 跳跃,突变,跃移,脉动;河底滚

slat-glazed *a.* (上)盐釉的
~ brick 瓷砖
~ structural facing unit 盐釉建筑贴面块
~ ware 陶土制品

salvage ['sælvidʒ] *n.* 打捞;(工程)抢修;(废物)利用;废品(处理) *vt.* 打捞沉船
~ department 废料(利用)间

samel *n.* 半烧砖
~ birck 半烧砖,粗制砖

sample ['sa:mpl, 'sæmpl] *n.* 样品(本),子样,试件;标本,模型,实例 *v.* 取[抽,采]样;试验样品性能,抽查,试用
~ room 样品(陈列)室
~ window 样品陈列窗

San Sophia Church 圣索菲亚教堂

San Peter Cathedral church 圣彼德主教堂

sanatorium [ˌsænə'tɔ:riəm] = sanitarium, sanatariam 疗养院,休养地

sancturary ['sæŋktjuəri] *n.*(宗教的)圣[教]堂,圣殿,寺院内殿,庇护所

sanctum ['sæŋktəm] *n.*(*pl.* -tums;-ta)密室,内殿,书房

sand [sænd] *n.* 砂,矿砂,矿潭;沙地[漠,滩] *v.* 铺[填,喷,掺]砂,用砂(纸)擦[磨]
~ asphalt 沥青砂,砂质沥青
~ bag wall 砂袋护墙
~ blast finish 喷砂饰面
~ blasting 喷砂(冲毛),砂冲
~ -cloth (金刚)砂布
~ coated wall 砂面墙
~ -dry surface 砂干表面(油漆面干到不粘砂)
~ -faced brick 砂面砖
~ finish 砂饰面
~ finished brick 砂面砖
~ grout (水泥)砂浆
~ -haydite concrete (砂)陶粒混凝土
~ -lime brick 灰砂砖
~ moulded brick 砂模砖
~ paper (金刚)砂纸
~ plaster 细砂粉刷,掺砂墙泥
fine ~ 细砂
fire ~ 防火砂
normal ~ 标准砂

sandal ['sændl] *n.* 檀香木
~ -wood 檀香木
red ~ 紫檀

sandstone ['sændstəun] *n.* 砂石[岩]
filter ~ 透水砂岩[石]

sandwich ['sænwidʒ] *n.* 夹层,分层,[叠层,蜂窝夹层]结构;复合[夹层]板;夹心面包,三明治 *a.* 层状[多层]的 *v.* 夹在当中,夹上,插入
~ arrangement 交错重叠布置
~ beam 多层叠合梁,夹层梁
~ board 夹层板
~ damp course 复合防潮层
~ frame 双构架
~ girder 夹合梁,夹层大梁
~ glass 夹层玻璃
~ material 夹层材料
~ panel 多[夹]层板
~ plate 多[夹]层板

~ -plate method 多层板法
~ structure 多层叠合结构
~ wall 夹层墙
~ winding 叠层[交错多层]

sandy ['sændi] *a.* (砂)质的,含[多]沙的,流沙似的,不稳固的
~ aggregate 砂质集料
~ gravel 砂砾石
~ marl 砂质泥灰
~ -size (像)砂粒大小(的尺寸)
~ -skeletal 粗骨砂

sangar ['sæŋgə] *n.* =sanga(原始)木桥,胸墙

sanidine ['sænidi:n] *n.* 透长石,玻璃长石
~ trachyte 透长石粗面岩

sanitarium [ˌsæni'tɛəriəm] *n.* 疗养院,休养地

sanitary ['sænitəri] *a.* (环境)卫生的,保健的,清洁的 *n.* (厕所等)卫生设备
~ appliance 卫生设备
~ corner (为保持清洁而做成的)光[圆]面墙角
~ door 光面门
~ drinking fountain 卫生饮水喷泉
~ earthenware 卫生陶器
~ facilities 卫生设备[装置]
~ fittings [installation, provision] 卫生设备[装置]
~ fixture 卫生设备[装置]
~ installation 卫生设备
~ ware 卫生器具

sanitation [ˌsæni'teiʃən] *n.* (环境)卫生,卫生(设备)下水道设备
~ equipment 卫生设备(尤指排水设备)
~ of buildings 建筑卫生

santal *n.* 檀香(木)

sap [sæp] *n.* 树液[浆,汁];(树皮下的)白木质;(渗碳钢)软心;风化岩石;坑[地]道 *vt.* (逐渐)削弱[浸蚀,损坏]下陷;耗竭,挖掘(坑道)
~ green 暗绿色

~ stain 木材变色
~ -wood 液材,边材,白木质
sapling ['sæpliŋ] *n.* 树苗,小树
sapparite ['sæpərait] *n.* 蓝晶石
sapphire ['sæfaiə] *n.* 青玉,蓝宝石 *a.* 青玉色;蔚蓝色的,蓝宝石色的
~ filament 蓝宝石纤维(增强剂)
~ quartz 深蓝色石英
~ substrate 蓝宝石衬底
sappy ['sæpi] *a.* 多汁液的,多[似]白木质的
saprobic [sə'prəubik] *a.* 腐生的,污水生的
sapstain *n.* 木材变色
sapwood ['sæpwu:d] *n.* (树皮下较软的)白木质,边材(心材外增生的木质部),液材
~ rot 边材腐朽
Saracenic [ˌsærə'senik] *a.* 回教徒的,穆斯林的
~ arch 萨拉森式拱(马蹄形拱)
~ architecture 阿拉伯回教建筑,穆斯林建筑
saran ['sə'ræn] *n.* 莎纶(聚偏氯乙烯纤维)萨冉树脂,耐火塑胶
sarcophagus [sa:'kɔfəgəs] *n.* (古希腊,罗马,埃及之)雕刻精美的石棺
sarking ['sa:kiŋ] *n.* 衬垫材料,衬垫板,镶板
~ boards 屋面衬板,望板
~ felt 油毛毡衬层
sash [sæʃ] *n.* 框(格),窗框[扇],推拉窗扇 *v.* 装上窗格[窗框]
~ angle 窗扇角钢
~ balance 吊窗平衡吊锤
~ bar 窗框条,钢窗料
~ bead 中旋窗窗压条
~ bolts 窗插销
~ center 窗扇转动轴
~ chain 吊窗链
~ door 玻璃门
~ fastener 窗风扣
~ frame 窗框

~ handle 窗拉手
~ holder 窗提手,窗风钩
~ latch 窗锁闩
~ lift 窗框提手
~ pulley 吊窗滑轮
~ rail 窗框横冒头,横框
~ roller 窗框滑轮
~ weight 吊窗锤
~ window (上下)直拉窗,框格窗

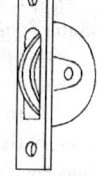

sash pulley

sassafras ['sæsəfræs] *n.* 黄樟,擦树[木]
~ oil 黄樟油
sassanian architecture 萨珊式建筑
satellite ['sætəlait] *n.* (人造)卫星;伴生矿物;追随者,附属物 *a.* 附属的,辅助的,伴(随)的,卫星的

sassanian architecture

~ city 卫星城市
~ equipment 卫星装置[设备]
~ terminal 卫星车站
~ town 卫星城镇
satin ['sætin] *n.* 缎子 *a.* 缎子做[似]的 *vt.* 轧光,加光泽
~ finishing 擦亮,抛[研]光
~ -glazed tile 有光泽的釉面砖[瓦]
~ paper 蜡光纸
~ white 缎光白,白色颜料
~ walnut 核桃木
~ wood 椴木
saturate ['sætʃəreit] *vt.* 使饱和[浸透] *a.* 饱和的,浸透的 *n.* 饱和物
~ colour 鲜纯色
~ felt 浸油毡
~ surface dry 饱和表面干燥(状态)
~ -surface-dry condition (of aggregate) (骨料)饱和面干状态
saturation [ˌsætʃə'reiʃən] *n.* 饱和(度,状态),浸透[润]
~ capacity 饱和容量,饱和含水量

~ degree 饱和度
~ moisture content 饱和含水量
saucer ['sɔːsə] *n.* 茶碟,小碟,碟状物; 大型漂浮物,底盘
~ shapped dome 碟[盘]形穹顶
~ spring 盘状弹簧
sauna ['saunə] *n.* 一种芬兰蒸汽浴(以水冲击加热石使化为蒸汽),桑那浴,蒸汽浴室
~ bath 蒸汽浴,桑那浴
~ bath installation 蒸汽浴设施
sausage ['sɔsidʒ] *n.* 香[腊]肠
~ construction 铅丝网石笼(护岸)建筑
save [seiv] *v.* 节省,存贮,保存,储蓄,救助 *n.* 救助,保存
~ all 节省器,节约装置;防溅器,挡雾罩,承(烛)油碟,安全网
saving ['seiviŋ] *a.* 节约的,保存[留]的;储蓄的,救助的 *n.* 节省,保存,储蓄,救助
~ arch 辅助拱,安全拱
saw [sɔː] *n.* 锯(子,床),锯齿状部 *v.* 锯(开,成),用[拉]锯
~ block 锯木架,马凳
~ dust [powder] 锯屑[末]
~ -dust brick 锯屑砖
~ -dust concrete 木[锯]屑混凝土
~ lumber 锯(成的)材
~ -tooth roof 锯齿形屋顶
~ -tooth skylight 锯齿式天窗
~ -tooth truss 锯齿形桁架
band ~ 带锯
head [log] ~ 原木[圆材]锯
strap ~ 带锯
string ~ 线锯
taper back hand ~ 斜背手锯
veneer cutting ~ 单板锯,镶板锯
web ~ 排锯,框锯
whip ~ 狭边[钩齿]粗木锯,框锯,弓锯
sawn [sɔːn] *v.* 锯
~ shingle 锯齿形屋顶

~ timber 锯成木材,锯材
sax [sæks] *n.* 修琢石板的尖锤,瓦刀,板斧
Saxon ['sæksn] *n.* 撒克逊人;一种烟火 *a.* 撒克逊人[式]的
~ architecture 撒克逊式建筑
~ facade 撒克逊式(建筑)饰面
~ masonry work 撒克逊式砖石工程
~ style (英国的)撒克逊式建筑
scab [skæb] *n.* 疤,疵痂,斑点,眼孔,铸件表面粘砂;拼接板;凸块 *v.* 结疤,拼接,凿平石料
scabbed [skæbid] *a.* 有疤的,拼接的
scabble ['skæbl] *vt.* (将石)粗琢,粗加整修
~d dressing 粗琢面
~d rubble work 粗琢石工作
scabbling *n.* 粗琢(石)工作,小石片,石屑
~ hammer 粗琢锤
scaffold ['skæfəld] *n.* 脚手架,鹰架;吊盘 *vt.* 搭脚手架
~ board 脚手板
~ bridge 脚手架跳板
~ cradle 悬挂脚手架
~ lashing 绑扎脚手架
~ standard 脚手架立杆
bracket ~ 悬臂[挑出]脚手架
cradle ~ 挑出[吊挂]式脚手架
flying ~ 挑出[悬挂]式脚手架
framing ~ 脚手架
gabbart ~ 方木脚手架
hanging ~ 吊盘,悬空脚手架
needle ~ 横架支托的脚手架
raise ~ 升降式脚手架
straddle ~ 跨立式脚手架
suspended ~ 悬吊式脚手架
swinging ~ 回转式脚手架
trestle ~ 排架式脚手架
tubular ~ 钢管脚手架
scagliola [skæˈljəulə] *n.* 人造大理石,仿云石,仿石粉刷

scalariform [skəˈlærifɔːm; skəˈlærəˌfɔːm] *a.* 梯子状的,阶状的

scale [skeil] *n.* 刻度,标度;比例[标,缩]尺;尺度,规模,程度;秤,天平;鳞(片,皮,屑);结垢,水垢 *v.* 按比例缩小,用缩尺制图;约略估计,换算,度量;去锈
～ board 极薄的木板,镶面薄木片,玻璃框镜背板
～ gutter tile 鳞状沟瓦
～ -like 鳞状的
～ map 比例图
～ model 按比例缩小的模型
～ wax 粗石蜡,片[鳞]状蜡
～ wood 木片板,层板
drafting ～ 制图尺
full ～ 足尺(比例),全尺寸,满标
full ～ construction 全面施工
full ～ model 实尺模型

scaled [skeild] *a.* 成比例的,有刻度的,鳞片状的

scalene [ˈskeiliːn] *a.* 不规则[不等边]的
～ triangle 不规则[不等边]三角形

scalenohedron *n.* 偏三角面体

scaling [ˈskeiliŋ] *n.* 定标;定比例;计数;起鳞,剥落,剥除(表层);生成水锈,结垢;除鳞[铁锈,水锈];(按)比例描绘,换算
～ brush 除锈刷
～ -down 按比例缩小
～ of concrete 混凝土剥落
～ system 计算[设计]图
～ up 比例加大

scaliola *n.* 仿云石

scall [skɔːl] *n.* 松软岩石,碎岩

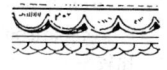
scallops

scallop [ˈskɔləp] *n.* 扇贝,贝壳;饰扇形皱褶;壳形图案 *vt.* 做(成)扇形皱褶,弄[切]成扇形
～ed arch 扇形皱褶拱,贝壳形拱
～ed capital 贝壳饰柱头
～ed surface 贝壳面

scalper [ˈskælpə] *n.* 筛机,初级碎石机;雕刻刀

scaly [ˈskeili] *a.* 鳞状的,多鳞的
～ coating 鳞片状涂层
～ pattern 鳞形,三角形

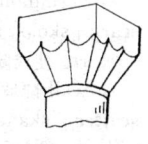
scalloped capital

scamillus *n.* 多立克柱式颈部小槽,爱奥尼亚柱式柱础下面的底座

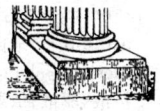
scamillus

scan [skæn] *v.* 浏览;扫描;搜索,观[检]测 *n.* 全景摄影;记录

Scandinavian [ˌskændiˈneiviən] *a.* ; *n.* 斯堪的纳维亚的[人],北欧的[人]
～ architecture 斯堪的纳维亚建筑,北欧建筑
～ plaster 斯堪的纳维亚灰浆[油灰],薄墙灰浆,瑞典砂灰浆

scantling [ˈskæntliŋ] *n.* 小方木;建筑材料尺寸;草图;样品
～ gauge 样板
～ lumber 小方木

scape [skeip] *n.* 柱身,柱底放大部分

scapolite *n.* 方柱石

scapus *n.* 柱身

scarcement [ˈskɛəsmənt] *n.* 壁阶;梯架

scarf [skɑːf] *n.* (*pl.* scarfs or scarves) 斜嵌槽;嵌接,斜切口,榫面接,榫接;围巾,领带,披肩 *v.* 嵌接
～ connection 斜嵌连接
～ joint 嵌接,斜接
～ splice 镶嵌拼接
～ together 嵌接合
flat ～ 平接
half (lap) ～ 半叠接
lap ～ 互搭楔接
oblique ～ with wedge 楔式斜嵌
oblique tabled ～ 斜叠嵌接

scarlet [ˈskɑːlit] *n.* ; *a.* 深[鲜,猩,绯]红色
～ red 深红,猩红

~ vermilion 深朱红
scarp [skɑ:p] *n.* 成陡坡,悬崖,沟内斜坡 *vt.* 使成陡坡
~ wall 陡坡墙
scatter ['skætə] *v.* ;*n.* 散布[开],散射;分[扩,耗,疏]散
~ed beam 散射束
~ed light 散射光
~ed trap 散射格栅
ground ~ 地面散射
scattering ['skætəriŋ] *n.* ;*a.* 散射[布],扩散,驱散;散射物;散射的,扩散的
acoustic ~ 声波散射
air ~ 空气散射
scavenge ['skævindʒ] *v.* 换[扫]气,回油;从(废物)中提取有用物质;吹[扫]除,净化,除垢
~ port 换气口,排泄口
scavenger ['skævindʒə] *n.* 清除机(具),换气管;洗池;清洁工
~ fan 换气[风]扇
~ pipe 排出管
scena *n.* 古代剧场舞台
scene [si:n] *n.* 事件,史实;出事地点,现场;布景,景色,场面
~ dock 道器布景储存室
~ storage 布景储存室
scenery ['si:nəri] *n.* (*pl.* -eries)景,风景,景色,风光,全景
scenic ['si:nik] *a.* 天然景色的,布[风,背]景的;舞台的,戏剧性的
~ easement 点(缀)景(色)建筑物
~ light 舞台照明
~ overlooks 观景
~ route 游览路线
~ spot 风景区,名胜地
~ vintage point 景色眺望点
scenograph *n.* 透视图
scenography [si:'nɔgrəfi] *n.* 透视图法
scent [sent] *n.* 线索,嗅觉;香气,气味,嗅[出],闻出
schamotte ['ʃɑ:mɔt] *n.* 耐火黏土
schedule ['ʃedju:l, 'skedju:l] *n.* 程序,工艺流程;目录,图表,一览表,清单 *v.* 排定,预定,计划,制表
~ control 进度控制
~ drawing 工程[序]图
~ engineering time 安排的工程时间
~ of construction 施工进度表,建筑一览表
~ overhaul 定期大修
~ time 预定时间
~d completion date 计划完工日期
design ~ 设计计划[进度]表
inspection ~ 检查图表
master ~ 主要作业,标准工艺过程
operating ~ 工作时间表
scheduler ['ʃedju:lə] *n.* 程序机(专用于生产上的一种计算机);生产计划员,调度程序
~ program 调度程序
scheduling ['ʃedju:liŋ] *n.* 编目录,(编)制(时间,进度)表;工序,程序
~ system 调度[程序]系统
schema ['ski:mə] *n.* (*pl.* schemata)大纲,概要,图解
schematic [ski(:)'mætik] *a.* 图解的,示意的,概略的,计划性的 *n.* 简[略]图
~ design 草图设计,初步设计
~ diagram 示意图,流程图
~ drawing 简图,示意图,原理图
~ plan 平面方案,平面示意图
~ section 示意剖面
scheme [ski:m] *n.* 方案,计划;系统;简图;接线图;电[线]路;分类表,图表[解,纸] *v.* 计划,设计
~ arch 平弧拱,弓形拱
~ comparison 方案比较
~ drawing 计划[方案]图
~ of color 配色法,色调
~ of wiring (电气)安装图,接[布]线图
block ~ 方块[框]图

scheme arch

colo(u)r ~ 配色法,色彩设计
flow ~ 工艺流程图
original ~ 原始方案
preliminary ~ 初步规划
wiring ~ (电气)安装图,接线图
working ~ 工作计划

schist [ʃist] *n.* 片岩,页[板]岩,晶片岩

school [sku:l] *n.* (学,流)派;学校,学院,研究所;学业,功课 *v.* 锻炼,训练,教授;*a.* 学校的
~ architecture 学校建筑
~ building 校舍
~ caravan 临时活动教室
~ table 课桌

schorl [ʃɔ:l] *n.* 黑电气石
~ blanc 白榴石

sciagraphy [si'ægrəfi] *n.* 投影(几何)法;房屋纵断面图

science ['saiəns] *n.* 科学,学科;科学研究,理论知识
~ building 科学馆
~ museum 科学博物馆

scientific [ˌsaiən'tifik] *a.* 科学(上)的,学术(上)的,应用科学的,系统的
~ administration 科学管理
~ management 科学管理

scintigram ['sintiˌgræm] *n.* 闪烁曲线,闪烁图

scission ['siʒən;'siʃən] *n.* 切[割,剪]断,裂[切]开;裂变,分离

scissors ['sizəs] *n.* (复)剪刀(起落架的)剪形装置
~ stairs 剪式楼梯
~ truss 剪式桁架
~ type 剪刀式布置

scobs [skɔbz] *n.* 锯屑(末),刨花,锉屑

sconcheon ['skɔntʃən] *n.* (木梁的)嵌槽口,(门窗洞口安装门窗框的)槽沟
~ arch 嵌槽拱

scone [skɔn] *n.* 锭剂; 扁平的圆饼
~ brick 平砖

scontion *n.* 门窗框内屋角石

scoop [sku:p] *n.* 杓(子),勺,戽斗,铲(斗),收集器,(砂型用)汤匙形墁刀;洞,穴,凹处;特快新闻,独家新闻 *v.* 挖,掘
~ light 杓状聚光灯
ventilator ~ 通风斗

scope [skəup] *n.* 范围,眼界,视野,目标;工作域,场所;示波器,观测设备,显微镜
~ of work 工作范围

scorch [skɔ:tʃ] *v.*;*n.* 烧[烤]焦;焦化,(橡胶)过早硫化;枯萎;高速行驶,飞跑
~ed pencil 炭笔,烧画笔

score [skɔ:] *n.* 刻痕,裂缝,划线器,两脚规 *v.* 斫刻,划痕;划线,打记号;计算
~ cutter 刻痕面砖
~d surface 刻痕面
~d tile 有槽空心砖

scoring ['skɔ:riŋ] *n.* 划痕,划线;作记号;擦伤,凹槽,沟纹
~ knife 刻痕刀
~ of tiles 砖面沟纹
~ stage 录音室

scotch [skɔtʃ] *n.* 刻痕,擦伤,切口;制动棒,车轮的止转棒 *vt.* 加刻痕于,轻切,浅刻;压碎
~ glue 透明(反射)胶
~ light 反射光线
~ tape (粘贴用)透明胶带

scotchlite ['skɔtʃlait] *n.* 一种反光玻璃材料(用于道路反射标志)

scotia ['skəuʃə] *n.* 柱基的凹弧边饰,凹弧饰线脚
~ scaper 凹圆形柱身

scotopic [skəu'tɔpik] *a.* 微光的

Scott cement 透明石膏水泥(生石灰加 5% 石膏)

scour ['skauə] *v.*;*n.* 疏

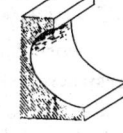

scotia

浚;沟蚀;擦(亮,净,掉);洗刷[净,涤],冲刷,打磨,消除
~ prevention 防冲(设备)

scourer ['skauərə] *n.* 洗刷器,擦洗物,打光机
household ~ 家庭用的洗擦物

scouring ['skauəriŋ] *n.* 冲刷(作用),侵蚀(作用);洗净
~ action 冲刷作用
~ equipment 冲刷设备
~ pad 擦洗用的钢丝围

scrap [skræp] *n.* 碎片,铁屑,残渣石边角料 *v.* 废弃,报废,作碎 *a.* 碎片的,废的
~ a contract 撕毁合同
~ build 设备改装,改新

scraper ['skreipə] *n.* 铲运机,刮土机;刮泥器;刮具,电耙,扒矿机;消字器,橡皮擦
~ blade 铲运机铲刀,刮刀
~ plane 刮刨
hook ~ 钩形刮刀
mark ~ 划线器
sand ~ 混砂刮板,刮砂板

scraping ['skreipiŋ] *n.* 刮(研),刮削(加工);耙运,电耙运输;刮泡
~ cutter 刮刀
~ straightedge 刮尺

scrappage ['skræpeidʒ] *n.* 废物,废材,报废(率)

scrappy ['skræpi] *a.* 碎料的,零碎的,剩余的

scratch [skrætʃ] *n.*;*v.* 刻[划,刮]痕,刮伤;乱涂
~ board 刮板
~ brush 钢丝刷
~ coat 划痕打底,涂灰打底
~ tool 刻线工具,刮刀
~ work 乱涂

scratcher ['skrætʃə] *n.* 抹灰用的划痕器

screed [skri:d] *n.* (压实混凝土的)整平板,样板;定墙上灰泥厚薄的准条,匀泥尺;找平层,抹灰的冲筋 *v.* 取样板压实刮平
~ board 样板,刮板
~ -coat 找平层
~ finish (混凝土面)找平饰面
~ rail 抹灰样板
~ strip 抹灰靠尺[冲筋]
cement ~ 水泥砂浆面层
granolithic ~ 细石混凝土面层

screen [skri:n] *n.* 屏,幕,帘;荧光屏,投影屏,屏幕;隔板,屏障,烟幕;筛子,筛网 *v.* 筛(选,分);审查,甄[鉴]别
~ door 纱门,屏蔽门
~ door latch 纱门门闩
~ molding 纱窗线脚,纱窗压条
~ partition 屏风隔断
~ porch 纱窗阳台,纱门门廊
~ sash 纱窗窗扇
~ table 折叠桌
~ wall 花格墙,挡墙
~ window 纱窗

screened ['skri:nd] *a.* 过筛的,筛出的;屏蔽的
~ material 筛分材料

screening ['skri:niŋ] *n.* 筛选[分];遮蔽,屏蔽(物);筛屑

screw [skru:] *n.* 螺旋[钉,丝,杆,孔],螺旋桨 *v.* 拧(紧)
~ anchor 锚定螺栓
~ base 螺口插座
~ bolt 螺栓
~ bulb 螺口灯泡
~ dowel 定缝螺钉
~ driver 螺钉旋具
~ed pipe 螺纹管
~ed plug 螺旋塞
~ flange coupling 螺旋法兰连接
~ joint 螺旋接头[合]
~ key 螺丝扳钳
~ nail (木)螺钉
~ nipple 螺纹连接管
~ nut 螺母
~ socket 螺口插座

~ stairs 盘旋梯,螺旋式楼梯
~ thread 螺纹
attachment ~ 装配[紧固]螺钉
calibration ~ 校准螺钉
coach ~ 方头木螺钉
dome capped ~ 圆头螺钉
flat head ~ 平头螺钉
foot ~ 地脚螺栓
log ~ 方头木螺钉
plain ~ 平螺钉,普通螺钉
wood ~ 木螺钉
scribe [skaib] v. (木工)雕合,合缝;划痕,划割,划线
~ joint 合缝接头
scribing block 划针盘,画线架
scriber ['skraibə] n. 划针,划线器(具)
scrim [skrim] n. (窗帘用)麻棉布
scroll [skrəul] n. 卷轴;涡管;旋涡形,旋涡花样;涡形道 v. 卷成卷轴形;用旋涡花样装饰
~ moulding 漩涡形线脚
~ saw 锯曲线用的带形锯
~ steps 盘梯级
~ work 漩涡形饰
scroll
scrub [skrʌb] n. 擦(洗);灌木,丛林 v. 擦(洗,净),涤气,气体洗涤
~ brush 擦洗刷
~ plane 粗刨
~ bing brush 硬毛刷,板刷
scrubber ['skrʌbə] n. 擦洗[洗涤,除尘]器;拖板刷,擦布,刮刀;擦洗者
dust ~ 除尘器
scrutator [skru:'teitə] n. 观察者,检查者
scudding ['skʌdiŋ] n. 刮面;顺风行驶
~ knife 切纸刀
scullery ['skʌləri] n. 餐具洗涤室,洗碗槽
sculptor ['skʌlptə] n. 雕刻[塑]家
sculptress ['skʌlptris] n. 女雕刻[塑]家
sculpture ['skʌlptʃə] n. 雕刻(术,品),雕塑(品);刻蚀,浸蚀[风化]的痕迹 v. 雕(刻,塑);刻蚀,侵蚀,风化
~ and architecture 雕塑与建筑
~ art 雕塑艺术
~ gallery 雕塑馆
~ garden 雕塑公园
scum [skʌm] n. 浮渣,泡沫,渣滓,水垢 v. 去浮渣[泡沫];形成泡沫
~ board 浮渣隔板
~ rubber 泡沫橡胶
scumble ['skʌmbl] v. 涂暗色,涂不透明色 a. 涂有暗色的
scutcheon ['skʌtʃən] n. 钥匙孔盖;盾饰
scuttle ['skʌtl] n. 煤斗,筐;小舱口,雕窗,天[气,舷]窗
sea [si:] n. 海(洋,水,面);(波)浪,涛海(上,岸,滨)的,航海的,近海的
~-board 沿海地区
~ green 海绿色
~ wax 海蜡
~ weed 海藻,海草
seal [si:l] n. 密封(垫,剂,装置),垫圈;封层,图章,印;海豹 v. (密)封,密闭,给……封口;盖印于
~ coat 封(闭)层
~ course 密封层
~ edge 封边
~ gum 密封胶
~ up 密封,封固
~ed concrete 密封混凝土,填缝混凝土
~ing compounds 封口胶,油灰
~ing device 止水设备
~ing layer 防水层
~ing material 封填材料
~ing ring 密封环
~ing rope 密封条
~ing washer 密封垫圈
~ing wax 封口蜡,火漆
sliding ~ 滑动式密封
wedge-type ~ 楔形密封

sealant ['si:lənt] *n.* 密封胶[剂];密封层
tape ~ 密封带

sealer ['si:lə] *n.* 保护层,封闭器
lacquer ~ 漆封剂
sanding ~ 嵌贝腻子

seam [si:m] *n.* (焊)缝,接合缝;咬口,卷边;发裂,皱纹;*v.* 缝合,接合,裂开口
~ -faced 细纹凿面的
boundary layer ~ 附面层表面
double ~ (金属薄板的)双咬口
lap ~ 搭接缝
transverse ~ 圆周接缝

seamless ['si:mlis] *a.* 无缝的,轧制的
~ door 无缝门
~ floor 无缝地面
~ pipe 无缝管
~ steel pipe 无缝钢管
~ tube 无缝管

seamy ['si:mi] *a.* 有(接,裂)缝的,有伤痕的

search [sə:tʃ] *n.*;*v.* 搜索,探索,搜[调,检]查,探测[查],研究,寻找,寻优(法);进入,侵入

searching ['sə:tʃiŋ] *a.*;*n.* 搜索[查](的),探查,调查,透彻的,严密[格]的
~ for site (工程)地址调查

seashore ['si:ʃɔ:] *n.* 海滨,海岸
~ industrial reservations 临海工业地带

seaside ['si:said] *n.*;*a.* 海滨(的),海边(的)
~ hotel 海滨旅馆
~ promenade 海滨游戏场

season ['si:zn] *n.* 季,季节,时节,旺季,流行期;(木材)风干;月[季]票;时化,时效 *v.* 使适用[适应];风干(木材),晾干,老化;贮放
~ timber 风[晾]干木材

seasonal ['si:zənl] *a.* 季节(性)的,随季节而变化的

seasoning ['si:zəniŋ] *n.* 篜燥(法,处理),风干,晾干

artificial ~ 人工干燥
natural ~ 自然干燥(法)
steam ~ 蒸气干燥(法)

seat [si:t] *n.* 座,支座[面];位置,座位,场所,中心,所在地;座板 *v.* 使固定,安置;修理[安装],底座
~ angle 座角钢
~ clay 耐火黏土,火泥
~ connection (梁柱的)托座连接
~ cover 座罩,座套
~ cushion 座垫
~ cut 支柱嵌切
~ pad 座垫
~ stand 座台,座椅看台
back ~ 后座
soft ~ 软底板

seat cut

seating ['si:tiŋ] *n.* (底,支,插)座,支架,基(础),座位;装置,设备 *a.* 座位的
~ shoe 柱脚,垫座

secern [si'sə:n] *v.* 区分,鉴别;分开[离]

secession *n.* 直线式,直线派

secessionist *n.* 直线派建筑家

seclude [si'klu:d] *vt.* 隔绝,隔[分]离

second ['sekənd] *n.* (角度,时间)秒,片刻;第二 *a.* 第二的,其他的;次的,副的,后属的,辅助的,另外的
~ coat of plaster 二道抹灰
~ floor (美)二楼,(英)三楼
~ floor plan (美)二层平面图;(英)三层平面图
~ growth timber 次生木材
~ -hand brick 旧砖
~ house 别墅,周末休息的郊外住宅
~ interface (层状体系的)第二层间面
~ storey 第二层楼的

secondary ['sekəndəri] *a.* 第二(性,级,阶段)的,再生的;次要[辅助,补充]的,中等的 *n.* 副手,代理人,中间产品
~ beam 次梁

~ coat 第二道粉刷[涂层]
~ color 次色调和
~ compensation 二次补偿
~ compression 次固结
~ girder 次梁
~ reinforcement 辅助钢筋,次要钢筋
~ tint 次色调
~ treatment 二次处理
~ truss 次要桁架
~ ventilation 辅助通风

secret ['si:krit] *a.* 秘[机,神]密的 *n.* 机[秘]密
~ door 隐蔽门
~ dovetail joint 暗鸠尾榫接头
~ miter joint 暗榫斜接
~ nailing 暗钉,埋头钉
~ passage 暗道
~ screwing 暗螺钉拼接
~ valley 屋顶排水暗沟

secret dovetail joint

secretary ['sekrətri] *n.* (*pl.* -taries) 秘书,书记,部长,大臣

sectile ['sektail] *a.* 可切的,可分的,可剖开的
~ opus 可分土墙

section ['sekʃən] *n.* 断[截,剖]面;断面图;路段,工段;部分,区域,部[零]件;舱;章节 *v.* 拆,截,切断,区分;作截[剖]面图
~ iron 型铁[钢]
~ plane 剖面平视图
~ steel 型钢
~ view 剖视图
bar ~ 型材
box ~ 箱形截面
bulb (angle) ~ 圆形角材
business ~ 商业区
center ~ 中心剖面
die ~ 拼合模块
free ~ 可拆部分
heavy ~ 大型材

hollow ~ 空心截面
longitudinal ~ 纵剖面(图)
oblique ~ 斜剖面
principal ~ 主截面
solid ~ 实心截面
transverse ~ 横断[截]面
uniform cross ~ 等截面
vertical ~ 垂直截面

sectional ['sekʃənl] *a.* 分区[段]的,部分的,地方性的,局部的;截[断]面的
~ drawing 断[剖]面图
~ elevation 立剖面,截面立视图
~ ladder 组合式楼梯
~ plan 部分[分段]平面图
~ plane 剖视面
~ view 剖视图

sector ['sektə] *n.* 扇形面,齿弧;区段,方面;两脚规
~ -type 扇形

secular ['sekjulə] *a.* 长期的,现世的;非宗教的
~ architecture 非宗教性建筑

secure [si'kjuə] *a.* 可靠的,安全的,稳定的 *v.* 使安全[可靠],保险;紧闭[固],固定

security [si'kjuəriti] *n.* 安全,可靠,稳固;保护[障],保险;保证(物),担保(物)
~ device 安全装置
~ glass 保险玻璃,防弹玻璃
~ window 保险窗

sedile [si'daili] *n.* (*pl.* -dila) (内殿圣坛近侧的)司祭席

sediment ['sedimənt] *n.* 沉积(物),沉淀(物);渣滓,残渣,水垢,淀积

sedile

sedimentation [,sedimen'teiʃən] *n.* 沉积物,沉淀,沉淀作用
~ plant 沉淀设备
~ process 沉淀过程,沉淀处理

seed [si:d] *n.* 种子,根源 *v.* 播种

~ed strip 绿化[植草]带

seepage ['si:pidʒ] *n.* 渗漏[流], 过滤, 渗出

seepy ['si:pi] *a.* 漏[透]水的, 排水不良的

~ material 透水材料

segment ['segmənt] *n.* 段节; (计算机的)程序段, 数据段; 弓形, 扇形体 *v.* 分割[裂]

~ block 弧形砌块

segmental [seg'məntl] *a.* 部分的, 扇形的, 圆缺的; 分割的, 片断的

~ arch 扇形拱

~ arch culvert 弓形拱洞

~ barrel vault 弧形筒穹顶

~ billet 弓形棒嵌饰

~ construction 预制(节)段拼装施工法; 分段构造

~ die 组合[可拆]模

~ girder 弓形大梁

~ truss 弓形桁架

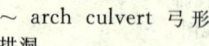

segmental arch

segregate ['segrigeit] *v.* 分离[开], 隔离[开]; 分凝, 凝离

~d concrete 离析的混凝土

segregation [segri'geiʃən] *n.* 分[隔]开; 离析

seismic ['saizmik] *a.* 地震的

~ design (结构)抗震设计

seize [si:z] *v.* 捕捉, 夺取, 抓住, 卡紧; 擦伤, 磨损

select [si'lekt] *v.* 选择, 选出, 挑选, 精选 *n.* 精选品

~ed lump lime 精选块状石灰

~ed quick lime 精选生石灰

selection [si'lekʃən] *n.* 选择, 挑选; 精选物, 淘汰; 提取, 分离

~ of design 设计评选

selective [si'lektiv] *a.* 有选择性的, 选择的, 挑选的, 淘汰的

~ absorption of sound 选择吸声

~ corrosion 选择性腐蚀

~ assembly 选择装配

selectron [si'lektrɔn] *n.* 不饱和聚酯树脂(商品名)

selenite ['seli:nait] *n.* 透明石膏

~ cement 石膏水泥

~ lime 含5%熟石膏的石灰

self [self] *n.* 自己, 本身, 自身

~-acting 自动的

~-acting thermostat 自动恒温器

~-adjusting 自动调节的

~-closing faucet 自关龙头

~-closing fire door 自关闭防火门

~-colored 原色的, 天然色的

~-contained flat 设备齐全的居住单位

~-cooking place 自炊处

~-cost 成本

~-faced 未修整的(石板)表面

~-healing (裂缝的)自动闭合

~-help housing 自助(公建)住宅

~-locking nut 防松螺母, 自锁螺帽

~-luminescent 自发光的

~-solidifying 自[动]凝[固]的

~-sticking coefficient 固有粘附系数

~-supporting scaffold 自撑式脚手架

~-supporting wall 自承重墙

~-ventilation 自行通风[换气]

semiannual ['semi'ænjuəl] *a.* 每半年的, 一年两次的

semi-arid ['semi'ærid] *a.* 半干旱[燥]的

semiasphaltic *a.* 半沥青的

semiautomatic ['semiɔ:tə'mætik] *a.* 半自动化

semi-bungalow ['semi'bʌŋgələu] *n.* 附有阁楼的平房

semi-cantilever *n.* 半悬臂

semicircle

semicircular arch

['semi'səːkl] *n.* 半圆
semicircular ['semi'səːkjulə] *a.* 半圆形的
～ arch 半圆拱
～ column 半圆柱
～ vault 半圆形穹顶
semi-closed ['semi'kləuzd] *a.* 半封闭的，半闭合的

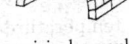
semicirular vault

semi-colloid ['semi'kɔlɔid] *n.* 半胶体
semicolumn ['semi'kɔləm] *n.* 半圆柱
semiconic ['semi'kɔnik] *a.* 半圆锥体的
semicylinder ['semi'silində] *n.* 半圆柱体
semidetached ['semidi'tætʃt] *a.* 半分离的
～ house 有公隔墙的两毗连的房屋
semidirect ['semidi'rekt] *a.* 半直接的
～ lighting 半直接照明
semidome ['semidəum] *n.* 半穹窿式屋顶
semi-double strength 2.8～3.0mm 厚度玻璃，中厚窗玻璃
semidry [semi'drai] *a.* 半干的
～ press process 半干压制(砖)法
～ state 半干状态
semi-elastic ['semii'læstik] *a.* 半弹性的
～ deformation 半弹性变形
～ subgrade 半弹性地基
semielliptical ['semii'liptikl] *a.* 半椭圆(形)的
～ arch 半椭圆拱
～ sewer 半椭圆形断面下水道
semifinished ['semifiniʃt] *a.* 半完成的,半成品的,半加工的
～ product 半成品
semi-fireproof ['semi'faiəpruːf] *a.* 半耐火的
～ construction 半防火建筑[结构]
semiflexible ['semi'fleksibl] *a.* 半柔性的

～ joint 半柔性接头
～ pavement 半柔性路面
semifluid ['semi'fluːid] *a.* 半流体的
semi-girder ['semi'gəːdə] *n.* 悬臂梁
semi-globe ['semi'gləub] *n.* 半球形
semi-Gothic ['semigɔθik] *a.* 半哥特式的
～ arch 半哥特式拱
semi-gravel *a.* 半砾石的
semi-gravity type 半重力式的
～ abutment 半重力式桥台
semi-grouting ['semi'grautiŋ] *a.* 半泥灌浆的
semi-hand ['semi'hænd] *a.* 半手工的
semihyaline ['semihaiəlin] *a.* 半透明的
semi-impervious *a.* 半透水的
semi-infinite ['semi'infinit] *a.* 半无限的
～ bar 半无限长杆件
～ beam 半无限长梁
～ cylinder 半无限圆柱
～ plate 半无限平板
semi-insulator ['semi'insjuleitə] *n.* 半绝缘体
semi-intrados 半拱的凹面
semiliquid ['semi'likwid] *a.* 半流体的,半液体的
semilunar ['semiluːnə] *a.* 新月形的,月牙形的
～ deformed bar 月牙纹钢筋
semi-manufactured ['semimænju'fæktʃəd] *a.* 半制成的
～ material 半制成品
semi-monolithic ['semimɔnə'liθik] *a.* 半整体的
～ construction 半整体式构造
seminatural ['semi'nætʃrəl] *a.* 半自然的
semiopaque ['semiəu'peik] *a.* 半透明的
semipaste paint 半厚油漆
semipermanent ['semi'pəːmənənt] *a.*

半永久的
~ structure 半永久性结构

semipermeable ['semi'pəːmiəbl] *a.* 半渗透性的
~ material 半透水材料
~ membrane 半透膜

semiproduct ['semi'prɔdəkt] *n.* 半制成品

semiround ['semi'raund] *a.* 半圆的; *n.* 半圆

semisolid ['semi'sɔlid] *a.* 半固体的; *n.* 半固体
~ asphalt 半固体沥青
~ flush door 半实心夹板门

semispheric ['semi'sferik] *a.* 半球形的

semi-stable ['semi'steibl] *a.* 半稳定的

semi-transparent ['semitræns'pɛərənt] *a.* 半透明的

semivault ['semivɔːlt] *n.* 半穹窿顶

semiwall ['semi'wɔːl] *n.* 半墙

semivault

senior ['siːnjə] *a.* 年长的,前辈的;高年级的; *n.* 年长者,前辈
~ engineer 高级工程师

sense [sens] *v.* 感觉,知觉;了解; *n.* 意义,意识
~ of harmony 和谐感
clockwise[cw] ~ 顺时针方向

sensibility [sensi'biliti] *n.* 感觉(性),知觉(性);灵敏度,敏感性,感光度

sensibilization ['sensibəlai'zeiʃən] *n.* 敏化(作用)

sensible ['sensəbl] *a.* 敏感的,可感觉的;明显的,通情达理的
~ plan 切合实际的计划

sensistor ['sensistə] *n.* 硅[正温度系数热敏]电阻

sensitive ['sensitiv] *a.* 灵敏的,敏感的;高度机密的; *n.* 对……敏感的物质

[材料]
~ color 敏感色
~ paper 感光纸
~ to heat 热敏(材料)
light ~ 光敏的,感光的
pressure ~ 压敏材料
super ~ 超灵敏的
temperature ~ 对温度灵敏的,热敏的

sensitiveness ['sensitivnis] *n.* 灵敏度[性];感应性
~ to light 感光性

sensitivity [sensi'tiviti] *n.* 灵敏度,敏感性;感光性
~ level 灵敏度,敏感度,响应级
~ to light 感光灵敏度

sensor ['sensə] *n.* 敏感元件,传感器,探测设备,敏感装置
light ~ 光传感器
thermal ~ 热传感器

separable ['sepərəbl] *a.* 可分开的,可拆开的,能区分的
~ attachment plug 连接插头

separate ['sepəreit] *v.* , -rated,-rating;分离,插入; ['sepərit] *a.* 分离的,分开的,不相连的,独立的
~ air conditioner system 分隔空调方式
~ board 分隔板
~ contract 单项承包,分项承包
~ curb 分式路缘
~ed aggregate 分级骨料
~ elements tread 分格纹
~ flat 独立单元住房
~ house 独立式房屋
~ type room air conditioner 分体式房间空调器
~ vent pipe 单独透气管
separating wall 挡板,隔墙

separation [sepə'reiʃən] *n.* 分离[开,裂];分类,区分;离析
~ partition 分隔墙
~ wall 分隔墙

bearer frame ~ (indoor) 搁架分隔（室内）
screen ~ 筛分,屏风分隔

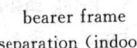
bearer frame separation (indoor)

separative ['sepərətiv] *a.* 分离的,分离性的

separator ['sepəreitə] *n.* 分离[选,隔]器,离析器;分隔符,(道路上)分隔带,分车设备;隔离物,隔板
dust ~ 除尘器

sepia ['si:pjə] *n.* 深褐色,棕黑色
~ print 深棕色蓝图

sepiolite ['sipiə/lait] *n.* 海泡石

septal ['septl] *a.* 中隔的,隔膜的

septarium [sep'tɛəriəm] *n.* 龟背石

septate ['septeit] *a.* 有隔膜的

septum ['septəm] *n.* 隔膜[板],中隔,隔墙[壁]

sepulchral [si'pʌlkrəl] *a.* 坟墓的,阴沉的
~ architecture 坟墓建筑
~ chapel 坟堂

sequential [si'kwenʃəl] *a.* 连续的,顺序的;序列[贯]的;
~ construction 序列施工
~ operation 时序操作

serial ['siəriəl] *a.* 连续的,串联的;顺序的;按期分版的;
~ construction 连续施工
~ production 批量[成批]生产

series ['siəri:z] *n.* 连贯,连续;系列,组序;级数;串联物;*a.* 成批的,串接的,串行的
~ connection 串联

serpentine ['sə:pəntain] *n.* 蛇纹石;蛇形线;*a.* 蛇形的,螺旋形的,盘旋的
~ marble 蛇纹大理石
~ wall 蛇纹形(砖)墙

serpentuator [sə:'pentʃueitə] *n.* 蛇形管

serrate ['serit] *a.* 齿形的,细齿的

service ['sə:vis] *n.*;*v.*,-iced,-icing,*n.* 服务,业务,工作;设备;*vt.* 供给,使用,操作,看管;维修,保养,使……适于使用;*a.*,服务的,有用的;修理服务的
~ area 服务区,供应区
~ box 进线箱
~ branch 入户支管
~ flat 服务层,共用厨房的住宅
~ layout 屋内管线布置
~ slide 供餐用的推拉窗
~ stairs 旁门楼梯
~ table 配餐台

serviceability [ˌsə:visə'biliti] *n.* 使用能力,服务能力;使用的可靠性;功能,耐用性;维护保养方便性

serviceable ['sə:visəbl] *a.* 合用的,耐用的

servo ['sə:vəu] *n.*,伺服机构,随动系统,伺服传动装置;*vt.* 补偿,修正;*a.* 伺服的,随动的,补偿的

set [set] *v.*;*n.*, set, setting; *v.* 安置[装],树立,固定;凝固[结];安排,着手;下沉,嵌,镶; *n.* 套,组,副;装置;沉陷,残留变形;凝固,硬化;定位; *a.* 固[预、规]定的
~ -back building (由红线)后退进去的房屋
~ bolt 定位螺栓
~ chisel 尖凿
~ collar 棚顶梁
~ in bookcase 附在墙上的书橱
~ in broken stones 用碎石嵌实(铺砌)
~ of furniture 一套家具
~ -off 墙的凸出物,装饰品
~ out 测定,定线,放样
~ piece 定位块,立体布景
~ scene 立体布影
~ screw 定位[固定]螺钉,制动螺丝
~ -square 三角板
~ -work 板条上抹灰
arch ~ 拱形棚子

facing ~ 平面刮刀
setscrew ['setsktru:] *n.* 定位[止动]螺钉
sett [set] *n.* 石块,铺石,小方石
settee [se'ti:] *n.* 长靠背椅,长沙发
setter ['setə] *n.* 装定器,调节器;安置者;定位器
~ -out 放样细木工,出版者
setting ['setiŋ] *n.* 安置,安装,凝固[结];下沉;开动,起动;背景,环境
~ block (玻璃)镶嵌垫块,整锯器
~ coat 抹灰罩面层,第三道油漆
~ glass 装置玻璃
~ -out work 放样定线
~ plan 装配平面图
~ plate 垫板
~ stake 定线桩
~ -up 硬化,建立,组装
brick ~ 砌砖
settle ['setl] *v.*;*n.*, -tled,-tling; *v.* 安排,整理;沉陷[降,积,淀];解决,决定;结算,付清; *n.* 高背长靠椅
settlement ['setlmənt] *n.* 沉陷[降];沉积物;决[确]定,解决,居留地,住所;付清,清算
~ crack (由)沉陷(引起的)裂缝
~ joint 沉降缝
settling ['setliŋ] *n.* 沉淀[积]物;稳定,回复,还原
sever ['sevə] *v.* 切[割]断,分开,隔离;断绝,终止
several ['sevrəl] *a.* 若干,几个,一些;个别的,各自的,不同的
severe [si'viə] *a.*, -verer,-verest;严格[厉]的,激烈的,艰难的,繁重的;紧凑的;纯朴的
severy ['sevəri] *n.* 筒拱顶隔块,穹顶的分隔间
~ bay (哥特式建筑)穹顶的分隔间
sewage ['sju(:)idʒ] *n.* 污水,下水道,阴沟水
~ disposal 污水处理
~ purification 污水净化
~ treatment 污水处理
sewer ['sjuə] *n.* 污水管,下水道,阴沟; *v.* 修暗沟,用下水道排污水
~ arch 下水道拱顶
~ brick 阴沟砖
main ~ 污水干管,总排水管
storm ~ 雨水管,雨水沟
sewerage ['sjuəridʒ] *n.* 污水[排水,下水]工程,排水[沟渠]系统;下水道,沟渠;污水;阴沟水
sewing ['səuiŋ] *n.* (塑料)熔合;缝合,缝合物
sexangle ['seks,æŋgl] *n.* 六角形
sexfoil ['seksfɔil] *n.* 六叶形(装饰图案)
~ window 六叶形窗
sexpartite [seks'pa:tait] *a.* 分成六部分的
~ vault 六肋拱穹顶
sextant ['sekstənt] *n.* 六分仪;圆的六分之一

sexfoil window

sextic *a.*;*n.* 六次(的),六次(曲)线
sextile *n.* 六十度角距
sextuple ['sekstjupl] *a.*;*v.*, -pled, -pling; *a.* 六倍的,六重的,六度的,六部分的; *n.* 六倍的量; *v.* (使)成六倍
~ space 六维(度)空间
sgabello *n.* 矮木椅
sgraffito(-iato) *n.* 釉雕,用釉雕装饰的陶瓷
shabby ['ʃæbi] *a.*, -bier,-biest;破旧的,破烂的;失修的
shack ['ʃæk] *n.* 棚屋,窝棚,棚房
shackle ['ʃækl] *n.* 钩环,钩链;绝缘器;带销 U 形环,铁扣;枷形装饰
~ bolt 钩环螺栓
shade [ʃeid] *n.*;*v.*, shaded, shading; *n.* 色调,色泽;遮阳板[篷],护板;阴处,阴影;罩帘; *vt.* 遮蔽
~ color 暗色
~ holder 灯罩座
~ shed 凉棚

~ tree 行道树,成荫树木
full ~ 饱和色
heavy ~ 饱和色
lamp ~ 灯罩
pale ~ 色
rich ~ 强色
shadow ['ʃædəu] *n.* 阴影,影子;微量,一点点;*vt.* 遮;伪装,保护;预示
　~ area 投影面积
　~ box 玻璃盖匣
　~ graph 投影画
　~ suface 阴影面
shadowgraph ['ʃædəuˌgra:f] *n.* X 光照片;逆光摄影;阴影法,阴影照相;-shadowgraphic, *a.*;-shadowgraphist, *n.*
shadowy ['ʃædəui] *a.* 有影的,阴影的;模糊的,朦胧的
shaft [ʃa:ft] *n.* (传动)轴;矿井,竖井;支柱,炉身,塔尖,烟囱身
　~ of column 柱身
　~ ring 柱环饰(绕柱环状线脚)
annulate ~ 环柱
column ~ 柱身
shallow ['ʃæləu] *a.* 浅的,薄的,薄层的;*n.* 浅水,浅滩;*v.* 使变浅
　~ arch 浅拱,坦拱
　~ beam 浅梁,矮梁
　~ cellular deck 浅空心板
　~ foundation 浅基础
　~ layer 薄层
sham [ʃæm] *n.*;*a.*;*v.*, shammed, shamming;*n.* 赝品,伪装物;*a.* 假的,仿制的,劣等的,假装;*v.* 假装
　~ beam 假梁
　~ door 假门
shank [ʃæŋk] *n.* 柄,末梢,后部;螺钉的无螺纹部分,轴耳;支柱,开沟器
shanty ['ʃænti] *n.* 简陋的小屋
shape [ʃeip] *n.*;*v.*, shaped, shaping; *n.* 形状[形态],轮廓,模型,样子;种类;钢材,模型塑胶;成[定]形
　~d bar (小尺寸)异形钢材
　~d brick 型砖,异形砖
　~d steel 型钢
oblong ~ 长方形,长圆形
streamline ~ 流线形
shapen ['ʃeipən] *a.* 作成[绘以]一定形状的
shaping ['ʃeipiŋ] *n.* 整形,成形,整平;做出横断面(形状);修刨
　~ plate 样板
sharp [ʃa:p] *a.* 尖锐的,锐利的,成尖角的;敏锐的,剧烈的
　~ aggregate 多角骨料
　~ bit 木工凿
　~ edged timber 方边木
　~ -toothed 尖齿的
sharpen ['ʃa:pən] *v.* 磨快,变锐利,变尖;削
shatter ['ʃætə] *n.* 碎裂片,破片;破损,岩屑,废石;粉碎,震裂;*v.* 打碎,破碎,击碎
　~ crack 破碎,裂缝
　~ -index 粉碎系数,震裂系数
　~ -proof glass 防震玻璃,防碎玻璃
　~ing effect (爆破时的)破碎效应
shaving ['ʃeiviŋ] *n.* 削下薄片;刨花;刮平,修整,整平,刮面
shear [ʃiə] *n.*;*v.*, sheared, sheared 或 shorn [ʃɔ:n], shearing;剪,切,割
sheath [ʃi:θ] *n.* (预应力混凝土的)钢筋鞘管;外皮,膜,套管;挡板
sheathing ['ʃi:ðiŋ] *n.* 盖板,望板,夹衬板;壳层,屋面底层,覆盖层
　~ nail 壁板钉
　~ paper 柏油纸,绝热纸
gypsum board ~ 石膏衬板
roof ~ 屋面板
shed [ʃed] *n.*;*v.*, shed, shedding; *n.* 车库;棚,小屋,坡屋;*v.* 流出,泻,脱落
　~ cover 棚顶
　~ dormer 棚顶窗
　~ roof 单坡屋顶
car ~ 车棚
maintenance ~ 保状态养性车库

sheen [ʃi:n] *n.* 光泽,光辉,光彩; sheenful, *a.*; sheenless, *a.*; sheenly, *ad.*
sheep [ʃi:p] *n. pl.* sheep; 羊,羊皮
sheer [ʃiə] *a.* 绝粹的,绝对的,真正的; 极薄的,透明的;无斜坡的,陡峭的
sheet [ʃi:t] *n.* 片,张,层,薄板;图表; *vt.* 扩张,铺设
~ asbestos 石棉片[板]
~ bar 薄板坯,板料
~ brass 黄铜片
~ copper 红铜片
~ glass 平板玻璃
~ iron 铁皮
~ iron encasing 铁皮镶板
~ lead 铅皮
~ line 图纸中线
~ metal 金属薄板
~ metal roofing 金属板屋面
~ -metal screws 自攻丝螺钉
~ paper 硬纸板
~ pile 板桩
~ tin 马口铁皮,白铁皮
~ zinc 薄锌板
~ backing ~ 底板
~ bitumen ~ 沥青毯
~ bracing ~ 加强板
~ cork ~ 软木片
~ corrugated ~ 瓦楞铁皮,波纹板
~ corrugated asbestos ~ 波形石棉瓦
~ fibre ~ 纤维板
~ fluted ~ 槽形铁皮
~ galvanized ~ 镀锌铁皮
~ operation ~ 施工说明书,工艺规程
~ packing ~ 垫片
~ polystyrene ~ 聚苯乙烯片[板]
~ PVC ~ 聚氯乙烯片[板]
~ veneer ~ 层压板,胶合板
sheeting [ˈʃi:tiŋ] *n.* 薄片,薄膜;帐篷;挡板,板栅,护墙板
~ cofferdam 板桩围堰
shelf [ʃelf] *n.*; *pl.* shelves; 大陆架,陆棚;搁架,搁板,格
~ location 架上安装
~ nog 支搁板的木砖
~ rest 搁板座
adjustable ~ 活动架
book ~ 搁书板
cable ~ 电缆支架
plate glass ~ 玻璃搁板
shell [ʃel] *n.* 壳,壳体,薄壳,外壳,地壳;外框,骨架,轮廓;贝壳饰
~ arch 薄壳拱
~ construction 壳体结构[建筑]
~ dome 薄壳穹顶
~ of hyperbolic parabloids 双曲抛物面壳,马鞍壳
~ of tile 空心砖外壁
~ pavement 贝壳路面
~ -proof structure 避弹结构
~ roof 薄壳屋顶
~ sand 贝壳砂
~ sculpture 贝雕
~ -slab 薄壳板
~ structure 薄壳结构
~ with ribs 带肋薄壳

shell arch

shell with ribs

shellac [ʃəˈlæk] *n.* 虫胶,虫胶漆,紫胶
~ bond 虫胶粘合剂
~ varnish 虫胶清漆
shelter [ˈʃeltə] *n.* 工棚,风雨棚,掩护物;保护,隐蔽; *v.* 保护,隐蔽
~ed refuge 有棚站台
shelve [ʃelv] *vt.*, shelved, shelving; 将……置于架上,搁置;辞退,解雇
shield [ʃi:ld] *n.* 盾,挡泥板,保护物,罩,屏,防卸; *v.* 保护
~ room 隔离室
~ -shaped 盾形的
dust ~ 防尘板
flame ~ 耐火墙
joint ~ 防渗接缝
mud ~ 挡泥板

protecting ～ 防护板
test ～ 测试护套
shift [ʃift] *n.* 移动,变动,换班,轮班,变位,移程,平移; *v.* 移动,替换
～ bar 换挡杆,开关槽
～ -boss 当班工长
～ing bearing 活动支座
shim [ʃim] *n.*; *v.*, shimmed, shimming; *n.* (楔形)填隙片,垫片; *vt.* 用垫隙片垫,垫补
～ ming plate 填隙板
laminated ～ 叠层垫片
spring ～ 弹簧垫片
shimmer [ˈʃimə] *vi.*; *n.* 闪[发]光,闪烁,微光
shimmy [ˈʃimi] *n.*; *pl.* -mies; *v.* mied,-mying; *n.* 摇摆,跳动,摆动; *v.* 摆动,横向滑动
shingle [ˈʃiŋgl] *n.*; *v.*, -gled,-gling; *n.* 木瓦,盖板,屋顶板;砾石; *v.* 搭迭木瓦
～ constructiion 木瓦结构
～ lap 木瓦搭接
～ nail 墙面板钉;木瓦钉
～ roof covering 木板铺屋顶
～ roofing 木板屋面
shiplap [ˈʃiplæp] *n.* 搭叠,鱼鳞板
～ flooring 错缝铺接地板
～ joint 搭接
～ lagging 搭叠壁板
～ sheet piling 搭叠板桩
shive [ʃiv] *n.* 碎片;下脚麻 *v.* 切,戳
shoddy [ˈʃɔdi] *a.*, -dier,-diest. 劣质的,冒充的
～ products 次品,劣品
shoe [ʃu:] *n.* 鞋,桩靴,柱架;瓦形物,闸瓦;导向板;底板
～ plate 底座板
～ store 鞋店
cross head ～ 十字滑块
seating ～ 底座
slipper ～ 滑瓦
supporting ～ 支撑块

shook [ʃuk] *n.* 可拼成(桶,箱等的)成套板料
shop [ʃɔp] *n.*; *v.*, shopped, shopping, *n.* 车间,工场;商店; *vt.* 购物
～ coat 出厂前油漆涂层
～ drawing 加工图
～ fabricated 工厂预制的
～ front decorating 店面装饰
～ -made 定做的
～ window 橱窗
shopping [ˈʃɔpiŋ] *n.* 购物
～ arcade 购物廊
～ center 购物中心
～ street 商业街道
shore [ʃɔ:] *n.*; *v.*, shored, shoring *n.* 岸,滨,涨潮与落潮地带;支柱,顶柱,柱支撑
～ arm 撑臂,支撑
adjustable ～ 可调顶柱
raking ～ 斜支撑
side ～ 侧支撑
shoring [ˈʃɔ:riŋ] *n.* 支撑,临时撑,加固撑,支柱
short [ʃɔ:t] *a.* 短的,不足的;低矮的;浅陋的,易碎的; *v.* 命名短路,缩减
～ column 短柱
～ leaf pine 短叶松
～ lived materials 非耐用物料

shoring

shortage [ˈʃɔ:tidʒ] *n.* 不足,缺点,缺乏,缺额
shortcoming [ʃɔ:tˈkʌmiŋ] *n.* 不足,缺点,缺乏,忽略
shortcutting [ˈʃɔ:tkʌtiŋ] *n.* 简化
shorten [ˈʃɔ:tn] *v.* 变短,缩短,减少
shot [ʃɔt] *n.*; *pl.* shots or shot. 小球,细粒;冲击;爆炸;应付之款
～ glass 小酒杯
～ -plastering 喷涂抹灰
long ～ 远景,远摄
short ～ 中景

shotpin ['ʃɔtpin] *n.* 止销,制动销
shoulder ['ʃəuldə] *n.* 肩,路肩,肩状物;
v. 挑起,肩起
~ed arch 并肩形拱
~ ed tenon 肩榫,突榫

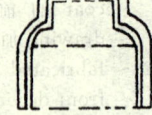
shouldered arch

show [ʃəu] *n.*;*v.*, showed, shown, showing; *n.* 展览会,展览场,陈列,表现;*v.* 展览,陈列,显示,表明
~ bill 招贴,广告
~ -case 陈列柜窗
~ piece 展品
~ rafter 外露椽
~ room 陈列室
~ stand 展销台
~ window 橱窗
showcase ['ʃəukeis] *n.* 陈列橱,陈列窗
shower ['ʃauə] *n.* 阵雨,暴雨;淋浴;指示器,展出者,*v.* 下阵雨,使湿透;showerless, *a.* showerlike, *a.*
~ (bath) room 淋浴室
~ compartment 淋浴间
~ curtain 淋浴帐帘
~ head 莲蓬头
~ pan 淋浴盆
~ partition 淋浴隔断
~ rose 淋浴莲蓬头
~ stall 淋浴隔间

shower rose

shred [ʃred] *n.*;*v.*, shredded or shred, shredding; *n.* 碎片,细条,微量,少量;*v.* 扯碎,切碎
~ed wood fiber board 碎木胶合板
shrill [ʃril] *a.* (声音)尖锐的,刺耳的;*v.* 发出尖锐的声音
shrine [ʃrain] *n.* 神龛,神殿,庙,祠堂,古迹
shrivel varnish 皱纹清漆
shroud [ʃraud] *n.* 覆板,护罩,套罩;套管,屏,幕

~ ring 箍环,包箍
canvas ~ 帆布罩
cooled ~ 冷却套
dirt ~ 防尘罩
shut [ʃʌt] *v.*, shut, shutting; 关闭,关拢;封闭; *n.* 停止,闭塞
~ ting post 关门模木
~ ting shoe 门栓座
~ ting stile 装锁门梃
shutter ['ʃʌtə] *n.* 挡空气板,百叶窗;风门片;卷帘式铁门;模板;(照相机)快门;*v.* 装上百叶窗,装上快门
~ bar 百叶闩
~ blade 百叶窗板
~ blind 百叶帘
~ -board 模板
~ door 百叶门
~ rail 卷帘轨
absorbent ~ 吸水模板
roller ~ 卷帘
sliding ~ 推拉百叶窗
spring-steel ~ 弹簧钢卷帘门

shutter bar

shuttering ['ʃʌtəriŋ] *n.* 模壳,模板;装百叶窗
~ boards 模板的木板
~ form 模板
siccative [si'ketiv] *a.* 干燥的;*n.* 催干剂
~ varnish 快干凡立水,快干清漆
sickle ['sikl] *n.* 镰刀,切割器,小镰刀
~ -shaped arch 镰刀形拱,新月形拱
side [said] *n.*;*a.*;*v.*, sided, siding; *n.* 边,旁边,侧面;方面;*a.* 旁边的,侧面的;副的,次(要)的,后属的;*v.* 站在……一边;刨平侧面,装上侧面
~ -aisle 侧道,侧廊
~ angle 侧角
~ arch 边拱
~ arm (气窗两侧的)推拉杆
~ attached 侧悬挂(式)的
~ band 边带
~ board 餐具架,侧板
~ box 剧院边厢

~ corridor 外廊
~ door 边门,便门
~ elevation 侧面[视]图
~ entrance 侧门
~ face 侧面
~ faced brick 侧边面砖
~ -hinged window 侧铰链窗
~ hung window 平开窗
~ jamb 门侧框
~ light 侧窗,边灯
~ lighting 侧向照明
~ plate 侧板
~ shoring 边撑
~ surface 侧面
~ table 条桌,靠墙小桌
~ tenon 边榫
~ timber 侧支柱
~ view 侧视图,侧面形状
~ wall 边墙,八字墙
gable ~ 侧山墙
hanging ~ 顶板,上盘

side jamb

sidelong ['saidlɔŋ] *a.* 横的,斜的;侧面的,间接的
sidewise ['saidwaiz] *ad.* 傍,横斜着,后旁边;*a.* 向侧的,沿边的,斜的
siding ['saidiŋ] *n.* (铁路)侧线,(房屋)板壁,披叠板,滑轨
~ board 护墙板
~ machine 边缘修整机
~ shingle 护墙盖板
~ steel 护壁钢板
wood ~ 木墙板
sienna [si'enə] *n.* (富铁)黄土;赭色
sift [sift] *v.* 筛分,挑选,淘汰;详细调查
~ed sand 过筛砂
~ing screen 细分筛
sifter ['ʃiftə] *n.* 筛子,细筛;筛分机
sight [sait] *n.* 见解,意见,看法;视力,视觉;观察孔;风景;*v.* 瞄准,照准;观察, sightable, *a.*
~ hole 检查孔,窥视孔
~ rod 测杆,花杆
sighting ['saitiŋ] *n.* 看见;瞄准,调准;

视界,视线;可见度
sigmoid ['sigmɔid] *a.* S形的,反曲的,*n.* S形,反曲形
~ curves S形曲线
sign [sain] *n.* 标志,记号,符号;信号;招牌,广告;*v.* 用标志表示,签字,订(契约)
~ post 标杆,指示牌
identification ~ 识别标志
shop ~ 招牌
signal ['signəl] *n.* 信号,征象,记号,目标,符号,指令;*v.* 打信号;*a.* 信号的,显著的
identification ~ 识别信号
steering ~ 控制信号,控制指令
significance [sig'nifikəns] *n.* 意义;重要性,显著性;有效数字,有效数
significant [signifikənt] *a.* 有意义的,重要的;有影响的, significantly, *ad.*
silastic 硅橡胶[密封物]
silastomer *n.* 硅塑料
silence ['sailəns] *n. v.* 消声,无声;抑制,静化
~ cabinet 隔声室
silencing of noise 消声(音)
silencer ['sailənsə] *n.* 消声器
exhaust ~ 排气消声器
noise ~ 噪声消声器
silent ['sailənt] *a.* 无声的,寂静的
~ block 隔声装置
~ falling ball tap 消声形龙头
~ fan 无噪声风扇
silex ['saileks] *n.* 石英
~ glass 石英玻璃
silhouette [ˌsilu(:)'et] *n.*;*v.*, -etted, -etting; *n.* 轮廓,侧面影像;*v.* 映出影子,映出轮廓
~ effect 背景效果,轮廓线效果
silica ['silikə] *n.* 硅石;硅土,二氧化硅
~ brick 硅砖,硅酸盐砖
~ cement (火山灰)硅石水泥
~ gel 硅胶
~ glass 石英玻璃

silicate

~ refractories 硅工耐火制件
~ sand 石英砂,硅质砂
~ soil 硅土

silicate ['silikit] *n.* 硅酸盐
~ brick 硅酸盐砖
~ cotton 矿棉
~ crown 硅酸盐冕玻璃
~ flux 硅酸盐焊剂
~ wool 矿棉
~ of alumina 硅酸铝
~ of lime 硅酸钙
~ of soda 硅酸钠(水玻璃)
~ paint 硅胶漆
~d macadam 硅化碎石

sailication [,sili'keiʃən] *n.* 硅化,硅化作用

siliceous [si'liʃəs] *a.* 含硅的,硅质的
~ aggregate 硅质骨料
~ brick 硅质砖
~ lime 含硅[硅酸]石灰
~ sand 硅质砂
~ sandstone 硅质砂岩,玻璃砂

silicon ['silikən] *n.* 硅(Si)
~ bronze 坚铜,硅青铜
~ carbide 金刚砂,碳化硅
~ carbide brick 碳化硅砖
~ carbide refractories 碳化硅耐火材料
~ earth 硅(质)土
~ liner 耐火衬垫
~ plastic 硅塑料
~ resin 硅树脂
~ varnish 硅树脂清漆

silicone ['silikəun] *n.* 有机硅树脂,聚硅氧
~ rubber 硅氧橡胶

silk [silk] *n.* 丝,丝织品
art[artificial fibre] ~ 人造丝
~ paper 薄纸

sill [sil] *n.* 基石;底木;门槛,窗台,底槛;床,潜坝
~ anchor 地脚螺栓
~ beam 槛梁

~ course 窗台层
~ elevation 基石标高
~ joint 垫板接缝
~ lip 槛档,坝槛
~ plate 窗台栏
~ timber 底木,槛木

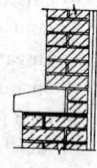
sill course

bank ~ 岸梁
cap ~ 顶梁
external G. S. Z ~ 镀锌铁皮外窗台
floor ~ 底基
pointing ~ 活闸门槛

floor sill

square-set ~ 方框支架底梁
stone ~ 石槛
stope ~ 底柱
straining ~ 二重桁架

silt [silt] *n.*; *a.* 粉土,粉砂(的),淤泥(的)
~ loam 粉土壤
~-pelite 粉砂泥岩
~ slurry 淤泥浆

siltstone ['siltstəun] *n.* 粉砂石,泥砂岩,粉土岩

silumin [si'lju:min] *n.* 铝硅合金,高硅铝合金

silundum *n.* 硅碳刚石

silver ['silvə] *n.* 银(Ag);银制物;银色; *a.* 银制的,似银的,银白的; *v.* 镀银,使变为银色
~ birch 银桦
~ bridge 乐池周围的通道
~-clad 包银
~-faced 镀银的
~ fir 银枞松,银枞杉
~ foil 银箔
~ leaf 银箔
~ grey 银灰色
~ plate 银器,镀银器皿
~ plating 镀银
~ plated glass 镀银玻璃
~ screen 银幕
~-ware 银器

~ -work 银饰品

similar ['similə] *a.*, 相似的,类似的,同样的

similitude [si'militju:d] *n.* 相似性,相似,类似物,对应物,副本,复制品

simple ['simpl] *a.* 简单[易]的;单纯的,率直的
~ beam 简支梁
~ frame construction 简单框架结构
~ glass 普通玻璃
~ roof 单坡屋顶
~ square crossing 简单十字形交叉

simplex ['simpleks] *n.*;*a.* 单形[纯],单体[缸];单一[纯]化的,简化的

simulate ['simjuleit] *v.*,-lated,-lating;模拟,模仿;假装,伪装

simulation [ˌsimju'leiʃən] *n.* 模拟,模仿,仿真;假装,伪装

simultaneous [ˌsiməl'teinjəs] *a.* 同时的,同时发生的,同时做的,联立(方程)的

sinaite ['sainəait] *n.* 正长岩

single ['siŋgl] *a.* 单的,独的;单纯的,单层的;一次的
~ action door 单向门
~ -aisle building 单过道房屋
~ -armed lever 单臂杠杆
~ beam 单跨梁
~ beam stairs 单梁楼梯
~ bedroom 单人卧室
~ bridging 简单斜撑
~ bull nose 单圆角
~ -casement window 单扇窗
~ coat 单层
~ course 单层
~ cyclone 单旋风除尘器
~ door 单扇门
~ double-swing door 单扇双开弹簧门
~ -family house 独户住宅

single beam stairs

~ flemish bond 单面荷兰式砌合,每层楔顺砖交替砌合
~ -flight stairs 单段楼梯,单跑楼梯
~ floor 单层地板
~ floor type auditorium 无楼座观众厅
~ handed saw 单手锯
~ -headed eagle 单头鹰徽饰
~ hinge 单铰链
~ hinged arch 单铰拱
~ -lap joint 单面搭接
~ lap tile 平搭瓦
~ latticed 单格子的
~ -layer 单层的
~ layout 单面布置
~ -leaf door 单扇门
~ -measure door 双面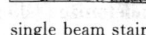
包夹板门   single lap tile
~ pedestal desk 一头带抽屉的桌子
~ -pitch roof 单坡屋顶
~ -pole knife switch 单极开关
~ pole scaffold 单排立杆脚手架
~ -purpose building 专用房屋
~ -rabbet frame 单槽口框
~ rivet 单行铆钉
~ rivet joint 单行铆接
~ room apartment 单室户公寓
~ -row rivet joint 单行铆接
~ -sashed window 单扇窗
~ -skew notch 单斜凹槽接合
~ -story building 平房
~ -strength glass 普通(强度)玻璃,1.8 ~ 2mm 厚玻璃
~ -tenon 榫舌
~ tier grillage 单层格床
~ use room 专用室
~ wall 单层墙
~ wall construction 单层墙体结构
~ window 单扇窗

singular ['siŋgjulə] *a.* 奇异的,异常的;单一的,单独的, singularity, *n.*

sink [siŋk] *n.*;*v.*, sank, sunk; *n.* 渗坑,落水洞;水槽,洗涤盆,水斗;*v.* 沉

下,埋入,插进
~ bib （厨房）水池龙头
~ drain board 污洗池排水挡板
~ faucet 洗涤盆的横向龙头
~ -float process 浮沉冲洗法
~ grating 污洗池篦子
~ mat 污洗池底板

sink faucet

sinkage ['siŋkidʒ] *n.* 沉,沉陷
sinker ['siŋkə] *n.* 冲钻,钻孔器
~ drill 冲钻,钻孔器
sinter ['sintə] *n. v.* 熔渣,烧结,sinterer, *n.*
~ed aggregate 陶粒集料
~ed carbide 烧结碳化物,硬质合金
~ed clay 烧结黏土
~ed glass 烧结玻璃
~ed metal 金属陶瓷
sinuous ['sinjuəs] *a.* 弯曲的,蜿蜒的
sinus ['sainəs] *n.* 海湾,穴凹地
siphon ['saifən] *n.* 虹吸管,虹吸;存水湾; *v.* 虹吸,用虹吸管抽上
~ closet 虹吸气便桶
site [sait] *n.; v.*, sited, siting; *n.* 地点,位置,工地,现场,部分,地区; *v.* 设置
~ cleaning 现场清理
~ exploration 现场勘探
~ investigation 现场调查,工地勘察
~ -mixed concrete 现场拌和的混凝土,工地拌制混凝土
~ plan 总平面图
~ welded 现场焊接
building ~ 建筑工地
sitting ['saitiŋ] *n.* 建筑地点定位
~ room 起居室
situ ['saitju] 同 (in situ) *n.* 就地,原地,现场
cast-in ~ 就地浇筑 (混凝土)
situation [sitju'eiʃən] *n.* 位置,地点,形势,情况,环境
sitz [sits] *n.* 【德】坐,坐浴,坐浴室
~ bath 坐浴盆
sizable ['saizəbl] *a.* 大的,广大的,大小相当的
size [saiz] *n.; v.*, sized, sizing; *n.* 大小,尺寸,尺码,体积;胶水; *v.* 量尺寸,测大小,涂胶水
~ colour 水粉颜色
~ distemper 水粉色,胶水粉涂饰
~d lumber 规格木材
~d slate 规格石板,条石
~ water 胶水
actual ~ 实际尺寸
aperture ~ 孔径尺寸
basic ~ 基本尺寸
effective ~ 有效尺寸
just ~ 正确尺寸
sample ~ 样本大小
standard ~ 标准尺寸
under ~ 尺寸不足
skeleton ['skelitn] *n.* 骨架,构架;轮廓,梗概
~ construction 骨架构造,框架结构
~ core （空心门的）内架
~ diagram 概略[轮廓]图
~ drawing 原理图,结构图
~ frame 骨架,框架
~ of building 房屋骨架
~ sketch 轮廓草图,构架图
~ structure 框[骨]架结构
skeletonize ['skelitənaiz] *v.*, -ized, izing;成为骨架,记成概要;绘草图
skene【希】= scene *n.* 舞台,布景
sketch [sketʃ] *n.* 简图,草图;草稿,素描,概要; *v.* 作草图,记概要 sketchability
~ drawing 草图,简图
~ map （地形）草图,示意图
~ plan 草图,初步计划
~ plate 异形板,非标准板
~ing technique 技术草图

diagrammatic ~ 示意图
eye ~ 目测图
free hand ~ 徒手画
outline ~ 轮廓图
preleminary ~ 方案草图
skew [skju:] ***n.*** 斜交,斜砌石；***a.*** 斜交的,歪的,非对称的
~ angle 斜交角
~ arch 斜拱
~ -back 拱座(斜块)

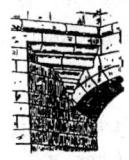
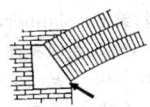

skew-arch　　skew-back

~ butt 压顶石,升降铰链
~ corbel 斜座石
~ curve 不对称曲线
~ scarf joint 斜嵌接
~ slab 斜板
~ table 斜座石,山墙盖板
ski [ski:] ***n.*** 滑雪,雪橇；***v.*** 滑雪
~ lodge 滑雪场小屋
skid [skid] ***n.*** ；***v.*** ,skidded,skidding；***v.*** 滑动,打滑；***n.*** 溜滑,滑行器
~ prevention 防滑
skidding ['skidiŋ] ***n.*** 滑行,滑溜；集材装置
skill [skil] ***n.*** 技巧,技能；(特殊)技术
skim [skim] ***n.*** ；***v.*** , skimmed, skimming；***v.*** 撇去,撇取；铲削；去沫；***n.*** 扒渣,去沫,渣沫
~ -grading 刮整表面
skin [skin] ***n.*** ；***v.*** , skinned, skinning；***n.*** 表皮,表面,外壳薄膜；***v.*** 剥落,脱皮
~ area 表皮面积
~ coat 表层
~ covering 表层,罩面,盖层
~ drying 表面干燥
~ friction 表面摩擦
~ glue 皮胶
~ of paint 油漆层
~ patching 表面修补
~ plate 罩面板
~ treatment 表面处理
external ~ 砖墙的外皮(外层墙)
skirt [skə:t] ***n.*** 侧缘,边；踢脚板；裙板；***v.*** 环绕,和……接界,位于……的边缘；-skirtless, ***a.***
~ beam 边梁,圈梁
~ board 边缘裙板,踢脚板,壁脚板
~ -roof 缠腰屋顶
skirting ['skə:tiŋ] ***n.*** 踢脚板,壁脚板
~ board 踢脚板
~ level 踢[裙]脚板标高
~ tile 踢脚砖
coved ~ 凹圆踢脚
hollow rounded ~ 凹圆踢脚
power ~ 电器插座护板
timber ~ 木踢脚板
skive [skaiv] ***v.*** , skived, skiving；切成薄片,刮,削,磨
sky [skai] ***n.*** 天空
~ -blue 天蓝色
~ brightness 天空亮度
~ deck 高楼顶的瞭望层
~ factor 天空投射系数,日光数
skylight ['skailait] ***n.*** 天窗,天棚照明,天空(扩散)光
~ glass 天窗玻璃
~ net 天窗网格
~ purlin 天窗檩
~ window 天窗

skylight window

slab [slæb] ***n.*** ；***v.*** , slabbed, slabbing；***n.*** 平板,石板,厚片；板岩；***vt.*** 铺石板；slabber, ***n.***
~ and girder floor 板梁式楼板
~ door 厚板门,平板玻璃门
~ floor 石板地面
~ footing 平板底脚
~ foundation 板式基础
~ iron 铁板

~ paving 石板铺砌
~ soffit 平板底面
~ type apartment house 板式住宅大楼
apron ~ 护坦底板
base ~ 基础底板
canlitever ~ 悬臂板
cover ~ 盖板
crest ~ 顶板
fibrous ~ 纤维板
fixed-edge ~ 嵌固板
flat ~ 无梁楼板
floating ~ 浮置板
floor ~ 楼板
foundation ~ 基础底板
groined ~ 井字形楼板
gunite ~ 喷浆混凝土板
landing ~ 平台板
mushroom ~ 伞形柱顶,无梁楼板
one way ~ 单向板
precast ~ 预制板
ribbed ~ 肋板
roof ~ 屋面板
suspended ~ 简支板
terrazzo ~ 水磨面板
two-way ~ 双向板
waist ~ 楼梯梯段板
wall ~ 墙板
wood-wool ~ 木丝板

ribbed slab

slabby ['slæbi] *a*. 板状的,片状的,黏稠的; slabbily, *ad*.
slabstone [slæbstəun] *n*. 石板,片岩
slag [slæg] *n*. 矿渣,熔渣,炉渣
~ -alabaster partition 矿渣石膏隔墙
~ concrete 矿渣混凝土
~ concrete block 矿渣混凝土砌块
~ felt 矿渣毡
~ portland-cement 矿渣普通水泥,矿渣硅酸盐水泥
~ -sulfate cement 硫酸盐水泥,石膏矿渣水泥
~ wool 矿渣棉[绒]
slagmac ['slægmæk] *n*. 黑色矿渣碎石(一种冷铺沥青拌矿渣混合料)
slamming stile 碰锁门梃
slamming strip 碰锁门梃木条
slant [slɑːnt] *a*. 倾斜的,歪的; *n*. 倾斜,(岩层)斜向; *v*. 使倾斜

slamming stile

~ bar 弯起钢筋
~ drilling 斜钻
~ed strut 斜撑
~ing roof 倾斜屋顶
slap [slæp] *v*., slapped, slapping;涂刷;敲击, slapper, *n*.
~ dash 粗涂抹面
slash [slæʃ] *n*. 湿地,多沼泽的;砍切; *v*. 削减,大大减少;slasher, *n*.
~ grain 平锯木纹
slat [slæt] *n*. 板条,木板,石片,平板条
~ crimper 百叶卷褶机

slat crimper

slate [sleit] *n*.;*v*., slated, slating; *n*. 板岩,石板,石板瓦; *v*. 铺石板,用石板瓦盖屋顶
~ cement 板岩水泥
~ hanging 挂石板瓦
~ lime 页岩石灰
~ roofing 铺石板瓦屋面
~ slab 厚石板
slating gage 铺石板准尺
slating nail 屋面钉
empress ~ 大石板瓦
slater ['sleitə] *a*. 石板瓦工
slaty ['sleiti] *a*. 石板状的,石质的,板岩质的
sleek [sliːk] *a*. 光滑的,有光泽 *v*. 弄滑,修光; *n*. 修型,镘刀;sleekly, *ad*.
sleeper ['sliːpə] *n*. 小搁栅,枕木,轨枕;卧车
~ clip 小搁栅固定夹
~ joist 小栏栅
~ slab (接缝下的)垫板
~ support 搁栅托梁下的横木

~ wall 地龙墙
sleeping ['sli:piŋ] *n.* 睡眠,休息
～ porch 凉台
～ room 卧室
sleeve [sli:v] *n.* 套管,套筒;插塞套,塞孔套;袖子
～ barrel 套筒
adaptive ～ 紧定套,接头套
clamping ～ 夹紧连接轴套
wall ～ 墙套管
sleight [slait] *n.* 技巧,手法
slender ['slendə] *a.* 细长的,薄弱的,狭窄的,微小的
～ column 细长柱
slew [slu:] *n.*;*v.* 回转,旋转
slice [slais] *n.*;*v.*, sliced, slicing; *n.* 片,薄片;泥刀,长柄火铲;一部分; *v.* 切成片,切去(一部分)
slick [slik] *a.* 光滑的,平滑的; *v.* 使光滑,整齐; *n.* 平滑面
～ surface 光滑的表面
slickenside ['slikənsaid] *n.* 擦痕面;滑面
slide [slaid] *n.*;*v.*, slided, sliding; *n.* 滑坍,滑坡,坍坡,坍方,滑板,幻灯片; *v.* 滑坍,滑动,溜掉
～ block 滑块
～ fastener 拉链
～ gate 滑动门,拉门
～ fire escape 火警太平滑道
～ glass 玻璃片
cross ～ 横向滑板,横刀架
socket ～ 环,圈,套
sliding ['slaidiŋ] *n.* 滑动,滑过; *a.* 滑动[走]的
～ damper 活挡
～ door 滑门,拉门
～ door rail 拉门滑轨
～ folding door 滑动折叠门
～ gate 推拉式门
～ hinge 滑动铰
～ -plate bearing 滑板支座
～ sash window 推拉窗

～ shutter 推拉百叶窗
～ stage 活动舞台
～ stay 活动窗撑
～ window 推拉窗
slight [slait] *a.* 轻微的,纤细的,少量的;细长的
slim [slim] *a.* 细长的,细小的;微弱的,不充足的; slimly, *a.*; slimness, *n.*
slime [slaim] *n.* 黏质物,软泥,黏液;沥青
slimy ['slaiml] *a.* 黏土的,黏液的,泥浆状的,泥泞的,糊状的
sling [sliŋ] *n.*;*v.*, slung, slinging; *n.* 吊环,吊具,链钩,吊重装置;抛掷装置; *v.* 吊,悬,投,掷
～ chair 帆布躺椅
～ -dog 吊钩
slip [slip] *n.*;*v.*, slipped, slipping; *n.* 滑动[滑移]纸片,板条; *v.* 滑动[移]
～ bar 滑杆
～ bolt 伸缩螺栓
～ butt hinge 活销铰链
～ cover 沙发套
～ hook 活钩,滑环
～ mortise 滑动榫眼
～ rail 活动栏栅
～ sill 滑槛,门窗滑轨
～ -sheet 薄衬纸
slit [slit] *n.*;*v.*, slit, slitting; *n.* 缝,缝隙,裂缝,长条切口; *v.* 切开,扯裂
～ deals 薄松板
slitting saw 开槽锯
sliver ['slivə] *n.* 长条,裂片,碎料; *v.* 切成长条,切开
slop [slɔp] *n.*; *v.*, slopped, slopping; *n.* 水坑,污水,泥浆,弄湿的方;廉价现成服装; *v.* 泼出,溢出
～ basin 脏水盆
～ built 建筑粗劣的
～ moulded brick 湿

sliding shutter

slop sink

模制砖
~ sink 污水盆
~ wax 粗蜡,原料[未经过滤的]石蜡
slope [sləup] *n.*; *v.*, sloped, sloping; *n.* 坡度,斜度,边坡,斜率; *a.* 倾斜的; *v.* 使成斜坡,使倾
~ of roof 屋顶坡度
~ of sheathing 望板坡度,屋面板坡度
~ of stair 楼梯坡度
~ paving 斜面铺砌,砌坡
sloping ['sləupiŋ] *n.* 倾斜,成斜坡; *a.* 倾斜的,倾面的,-slopingly *ad.*
~ beam 斜梁
~ ground 倾斜地面
~ roof 斜坡屋顶
sloppy ['slɔpi] *a.* 湿透的,泥泞的,草率的,散漫的, sloppily, *ad.*
slot [slɔt] *v.*; *v.*, slotted, slotting; *n.* 狭槽,缝,长孔,狭窄通道,轨迹; *vt.* 切槽,开缝
~ joint 槽沟接合
~ mortise 狭槽榫接
~ pitch 槽距,槽缝尺寸
~ wall 开缝墙
~ welding 槽焊
bevelled ~ 斜槽
closed ~ 闭口槽
dovetail ~ 燕[鸽]尾槽
drift ~ 斜形槽,出销槽
key ~ 键槽
open ~ 开口槽
skewed ~ 斜槽
ventilation ~ 通风孔
~ ted angle 开槽角钢
~ ted head screw 槽头螺钉
~ ted hole 长圆孔
~ ted plate 有长槽的板
slow [sləu] *a.*; *ad.* 慢的,迟钝的,低速的; *v.* 放慢,减速,滞后
~ -burning construction 耐火建筑
~ cement 慢凝水泥
~ closing faucet 慢闭水龙头

~ curing cutback asphalt 慢凝稀释地沥青
~ -setting cement 慢凝水泥
~ -taking cement 慢凝水泥
sludge [slʌdʒ] *n.* 污泥,淤泥淤渣
~ asphalt 酸渣沥青
~ blanked 泥渣层,污泥层
sluggish ['slʌgiʃ] *a.* 停滞的,缓慢的;黏滞的,流动性低的
sluiceway ['slu:swei] *n.* 排水道,泄水道
slum [slʌm] *n.* 陋巷,破房,贫民区,贫民窟
~ clearance 拆除违章建筑
~ district 陋巷区域,贫民区
~ -land 贫民区
slump [slʌmp] *n.* 滑移,滑动,沉陷;坍落,(混凝土)坍落度
slurry ['slə:ri] *n.* 泥浆,淤泥,稀砂浆,沥青砂浆
~ concrete 稀混凝土
~ concrete wall 混凝土防渗墙
~ wall 泥浆墙
slush [slʌʃ] *n.* 软泥,泥浆;污水;油灰; *v.* 灌泥浆,填油灰; slusher, *n.*
~ ed -up joint 砂浆填缝
slype [slaip] *n.* (教堂中的)有覆盖的走廊
small [smɔ:l] *a.* 小的,细小的,少的,窄的; *n.* 狭小部分,腰部; smallness, *n.*
~ ashlar 小块琢石,墙面小块石板
~ pole 小杆材树
~ -scale 小型的,小规模的
~ tenon 小榫
smalls [smɔ:lz] *n.* 细料,粉末,细末
smalt [smɔ:lt] *n.* 大青,蓝玻璃
smalto 做[镶嵌细工用的]有色玻璃
smaragdite *n.* 绿闪石
smash [smæʃ] *v.* 打碎,破碎;猛撞; smashable, *a.*
smear [smiə] *v.* 涂,抹,弄脏; *n.* 油渍,污斑; smearer, *n.*; smearless, *a.*
smectic ['smektik] *a.* 使清洁的,净化

的,纯净的

smirch [smə:tʃ] *v.* 沾污,弄脏;*n.* 污点,瑕斑; smircher, *n.*, smirchless, *a.*

smithereens ['smiðˈriːnz] *n.* (=smithers)碎片,细屑,碎屑

smog [smɔg] *n.* 烟雾
~ -control 烟雾控制

smoke [smauk] *n.*;*v.*, smoked, smoking;*n.* 烟,烟气,冒烟;*v.* 发烟,冒烟;弥漫
~ baffle plate 隔烟板
~ bell 烟罩
~ damper 烟道调节板
~ detector 火灾烟火探测器
~ pipe 烟囱筒
~ pipe bonding 烟囱的砌筑

smokemeter ['smaukˌmiːtə] *n.* 测烟仪,测烟表;烟雾指示器

smokescope ['smaukskəup] *n.* 烟尘浓度测定器

smoky ['smauki] *a.* 发烟的,多烟的,烟雾弥漫的;如烟的,烟色的; smokily, *ad.*; smokiness, *n.*

smooth [smuːð] *a.* 平滑的,光滑的;平稳的,顺当的;*v.* 弄平滑,使容易,消除; smoothable, *a.* smoothness, *n.*
~ ashlar 光面琢石
~ contour 光滑的外形
~ curve 平滑曲线
~ -face clay tile 光面黏土砖[瓦]
~ file 细锉
~ finish 光面修整
~ millboard 光面纸板
~ plastic pipe 平滑塑料管
~ -surfaced roofing 光面屋顶

smoother ['smuːðə] *n.* 路面整平机,刮路机;整平工具

smoothing ['smuːðiŋ] *n.* 粉光,镘光
~ belt (混凝土路面)镘光带
~ board 粉光板,镘光板
~ board finish 用镘板粉光
~ cement 腻子,油灰
~ hammer 平锤
~ plane 木工细刨
~ trowel 粉光用抹子,粉光镘

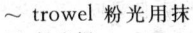
smoothing trowel

smoothness ['smuːðnis] *n.* 平滑度,平整度

snake [sneik] *n.*;*v.*, snaked, snaking;*n.* 蛇;清除管道污垢的铁丝 *a.* 蛇形的 *v.* 拖拉,曲折前进
~ bend 蛇形弯头[弯管]
~ fence 蛇形栅栏
~ -like 蛇形的
~ mark 蛇形斑点
~ -stone 蛇纹石

snap [snæp] *v.*, snapped, snapping;迅速闭合;突然折断;(猛地)咬住,抓住;排出 *n.* 揿钮,弹性凸膜片;速断 *a.* 急速的,突然的
~ bolt 自动门闩
~ head bolt 圆头螺栓
~ head river 圆头铆钉
~ hook 弹簧扣
~ lock 弹簧锁
~ ping line 弹线
~ -lock 弹簧锁
~ switch 速动开关
~ terminal 弹簧夹

snapper ['snæpə] *n.* 揿钮,瞬动咬合器;拍快照者

snarl [snɑːl] *v.* 弄乱,缠乱;(在金属薄片上)打出浮雕花纹

snecked [snekt] *a.* 用乱石砌筑的
~ rubble 杂乱毛石

snib [snib] *n.*;*v.*, snibbed, snibbing; *n.* 闩,插销,门窗钩 *v.* 插上插销

snip [snip] *n.*;*v.*, snipped, snipping; *v.* 剪去,剪断 *n.* 剪切;小片;剪刀,铁丝剪
bench~s 台剪
circular~s 圆剪
straight~s 平剪,直剪

snoot [snuːt] *n.* 鼻,喷嘴,喷口,小孔

snow [snəu] *n.* 雪,积雪,雪状物 *v.* 下雪,被雪封住
- ~ barrier 防雪栅,防雪板
- ~ board 挡[防]雪板
- ~ cornice 防雪挑檐
- ~ protection hedge 挡雪栅栏
- ~ screen 雪栅
- ~ shed 防雪篷
- ~ shield 防雪栅,防雪板
- ~ wall 挡雪墙
- ~ white 雪白,锌白

snubber ['snʌbə] *n.* 缓冲器,减震器,减声器

sock [sɔk] *n.* 软的保护套
- plastic ~ 塑料保护软管

socket ['sɔkit] *n.* 套节,插座,承口; -socketless, *a.*
- ~ bend 管节弯头
- ~ connector 套接接头
- ~ pipe 套节管,套管
- ~ power 插座电源
- ~ screw 凹头螺钉
- ~ spanner 套筒扳手
- ~ wrench 套筒扳手
- bayonet ~ 卡口插座
- cable ~ 电缆端头
- cushion ~ 弹簧插座,减震插座
- ebonite ~ 橡皮座
- external ~ 外电源插座
- plug ~ 插座,插口

socle ['sɔkl] *n.* 管脚,管座;支架,柱脚,建筑上外壁的基部,承放美术品的平台
- column ~ 柱基座

sod [sɔd] *n.*,*v.*, sodded, sodding; *n.* 草皮,草泥 *vt.* 铺草皮于,覆以草泥

soda ['səudə] *n.* 纯碱,碳酸钠,碱
- ~ alum 钠明矾
- ~ -baryta glass 钠钡玻璃
- ~ glass 钠玻璃
- ~ -granite 钠花岗岩
- ~ -lime glass 钠钙玻璃
- ~ -peal ash glass 钠钾玻璃
- ~ -potash glass 钠钾玻璃
- baking ~ 碳酸氢钠
- caustic ~ 苛性钠,烧碱

sodden ['sɔdn] *a.* 水浸的,浸润的 *v.* 浸透,弄湿; soddenly, *ad.*; soddenness, *n.*

sodium ['səudjəm] *n.* 钠(Na)
- ~ silicate 硅酸钠,水玻璃
- ~ silicate cement 硅酸钠胶结料
- ~ -vapour lamp 钠光灯,钠汽灯

sofa ['səufə] *n.* 长沙发
- ~ bed 坐卧两用沙发,沙发床

soffit ['sɔfit] *n.* 拱腹,梁底面,桃檐底面
- ~ block 拱腹形块
- ~ board 背面板,橡檐底板
- ~ cusp 拱顶石
- ~ finish 底面抹灰
- ~ formwork 拱腹模板,底面模板
- ~ level 拱腹标高,板底标高
- ~ lining 底衬
- ~ of girder 梁腹
- ~ scaffolding 拱腹架,砌拱支架
- eaves ~ 檐腹板,檐口托板
- haunched ~ 加腋拱腹
- stair ~ 楼梯段背面

soffit scaffolding

soft [sɔft] *a.* 柔软[和]的; *ad.* 柔软[和]地; softish, *a.*; softly, *ad.*
- ~ asphalt 软地沥青
- ~ -board 轻质(纤维)板
- ~ brick 未烧透砖,欠烧砖
- ~ clay 软黏土
- ~ finishes (建筑)软质面层
- ~ goods 纺织品
- ~ lighting 柔光照明
- ~ -mud process 软泥制坯法
- ~ pitch 软焦油沥青,软柏油脂
- ~ rib 柔性肋
- ~ sand 细砂,软砂
- ~ soil 软土,松土
- ~ steel 低碳钢,软钢

~ stone 软石
~ wood 软质木材,针叶树
softening ['sɔfniŋ] *n.* 弄软,变软,软化
~ cracking（混凝土的）开裂软化
softener ['sɔfnə] *n.* 软化剂,软化器,垫木,垫衬,硬水软化器
softness ['sɔftnis] *n.* 柔软度,柔软性;柔和,柔弱
soggy ['sɔgi] *a.* 潮湿的,湿润的,湿透的; soggily, *ad.*
~ aggregate 集料土,碎石土,土团料
solar ['səulə] *a.* 太阳的,日光的
~ control film（贴在玻璃上的）遮光膜
~ control glass 遮光玻璃
~ orientation（房屋的）光线方位
~ screen 遮阳篷,遮光板
~ water heater 太阳能热水器
solarium [soə'lɛəriəm] *n.* 日光浴室
solder ['sɔldə] *n.* 焊料,焊剂,焊锡, *v.* 焊接
~ed joint 焊接,焊缝
soldering ['sɔldəriŋ] *n.* 低温焊,软焊,锡焊
soldier ['səudʒə] *n.* 立砌砖;竖桩,模板支撑
~ arch 门窗顶上的立砌砖拱,立砌过梁
~ course 立砌砖层
~ piles 竖桩,立桩
sole [səul] *n.* 底基,底脚 *a.* 单一的,单独的,本身的 soleness, *n.*
~ timber 垫木
solid ['sɔlid] *n.* 固体,固态;立体,实体 *a.* 固体的,实体的,坚实的; solidly, *ad.*; solidness, *n.*
~ arch 实体拱
~ -barrel arch 实心板拱
~ bituminous materials 固体沥青材料
~ brick 坚实砖
~ cast door 实体浇铸门
~ column 实体柱

~ -core door 实芯门
~ drawn tube 无缝管
~ floor 实心楼板
~ molding 雕木线脚
~ newel 螺旋梯的实心中柱
~ newel staircase 实心中柱式螺旋楼梯

solid-core door

~ panel 厚镶板
~ partition 实心隔墙
~ retaining wall 实体挡土墙
~ rib arch 实肋拱
~ rock 坚石,基岩
~ roll 实心轧辊
~ section 实心部分
~ spandrel arch 实腹拱
~ steel door 实心钢门
~ strutting 坚固的支撑
~ -walled structure 承重墙结构
~ wood floor 实心木楼板
solidification [sə‚lidifi'keiʃən] *n.* 凝固,固化
solidify [sə'lidifai] *v.* -fied,-fying 固化,固结,凝固,变硬
solubility [‚sɔlju'biliti] *n.*; *pl.* -ties 溶解性[度];可溶性
soluble ['sɔljubl] *a.* 可溶的,可解释的; solubleness, *n.*; solubly, *ad.*
~ drier 溶性干燥剂
~ glass 可溶玻璃
~ oil 溶性油,调水油
~ tar 轻木焦油
solvent ['sɔlvənt] *a.* 有溶解能力的;有偿付能力的 *n.* 解决办法;溶剂; solvently, *ad.*
~ action 溶解作用
~ naphtha 轻汽油溶剂
~ paint 溶媒漆
sommer ['sʌmə] *n.* 大梁,地[过]梁,基石,柱顶石

sommer

sonigage ['sɔnigeidʒ] 超声波测厚仪

soot [sut] *n.* 烟灰,积炭 *v.* 熏黑,覆以煤烟; sootless, *a.*
sophisticate [sə'fistikeit] *v.* 改进,采用先进技术,使完善;成熟;掺杂,掺和;伪造; sophisticator, *n.*
~d equipment 先进设备
sopping ['sɔpiŋ] *a.*;*ad.* 湿的,浸透的,彻底地
soppy ['sɔpi] *a.* 浸湿的,湿透的,潮湿的,多雨的; soppiness, *n.*
sorb [sɔ:b] *v.* 吸着,吸收,吸附
~ed film 吸附膜
~ing layer 吸附层
~ing material 吸附材料
sorbate ['sɔ:beit] *n.* 吸附物,吸着物
~ layer 吸附层
Sorel's cement 镁石水泥
sorption ['sɔ:pʃən] *n.* 吸着作用,吸收
~ agent 吸附剂
~ capacity 吸附能力
sorrel ['sɔrəl] *a.* 红褐色的,粟色的
sort [sɔ:t] *n.* 种类,类别;品质 *v.* 分类,整理,区分
sound [saund] *n.* 声音,声学;探针;海峡,海湾 *a.* 健全的,完善的;可靠的,坚固定的 *v.* 发声音;探测,测量水深; soundable, *a.*
~ absorber 吸声体[器]
~ absorbing material 吸声材料
~ absorbing paint 吸声油漆
~ absorption 吸声力,声吸收
~ amplifying system 扩音系统
~ arrester 隔声装置
~ attenuator 消声器
~ baffle 阻声屏
~ barrier 声障
~ board 共鸣板
~ boarding (楼板下的)隔声板
~ box 吸声箱,共鸣箱
~ cement 安定性水泥
~ chamber 声室
~ control material 声控材料
~ damper 消声器
~ -deadening material 吸声材料
~ deading paint 隔声涂料
~ elimination 消声
~ filter 滤声
~ hole 传声窗
~ insulation 隔声
~ insulator 隔声材料
~ lock 播音室的隔声前室
~ reflecting board 反射声板
~ resonance 共鸣,反响声
~ retardance 减声
~ reverberation 回声,声干扰,混声
~ screen 声幕
~ wood 良木,坚硬木
sounding ['saundiŋ] *n.* 测探,钻探;测深,水深测量; *a.* 作声的
~ board 共鸣板,共振板
soundness ['saundnis] *n.* 安定性,固定性;健全
~ of cement 水泥安定性
sound proof ['saund'pru:f] *a.* 防声的,隔声的
~ chamber 隔声室
~ construction 隔声构造
~ cover 隔声罩
~ door 隔声门
~ floor 隔声地面
~ wall 隔声墙
source [sɔ:s] *n.* 水源;本源,来源,起源;出处,原始资料
~ nipple 螺纹接口,短管接头
arc ~ 弧光源
point ~ 点能源,点光源
sourdine [suə'di:n] *n.* 消声器,噪声抑制器
south [sauθ] *n.* 南,南方[部] *a.*;*ad.* 南方的,向南的,南来的 *v.* 转向南方
~ aisle 朝南走廊
southeast ['sauθ'i:st] *n.* 东南部 *a.*;*ad.* 向东南的,在东南的
southern ['sʌðən] *a.* 在南方的,南部

的,向南的

southernmost ['sʌðənməust] *a.* 最南端的,极南的

southward ['sauθwəd] *a.*; *ad.* 南方(的),向南(的) *n.* 南方(地区)

southwest ['sauθ'west] *n.* 西南(部) *a.*; *ad.* 从西南地,向西南地

southwester ['sauθ'westə] *n.* 西南大风;防水帽,油布长雨衣

southwestern ['sauθ'westən] *a.* 在西南的,向西南的,从西南的

space [speis] *n.*; *v.*, spaced, spacing. *n.* 空间,场所,间隔,距离;时段;宇宙; *vt.* 隔开,分隔
- ~ band 间隔嵌条
- ~ bar 空间杆
- ~ between beams 梁间距
- ~ character 间隔符号
- ~ column 空腹柱格架柱
- ~ configuration 空间形态[状]
- ~ construction 空间构造
- ~ coordinate 空间坐标
- ~ diagonal 空间对角线
- ~ diagram 空间图,立体图
- ~ dome 空间圆穹
- ~ for balance 推拉窗之平衡锤箱
- ~ frame 空间构[框]架
- ~ grid 空间网格
- ~ layout (indoor) 空间设计,(室内)空间布置
- ~ orientation 空间定位
- ~ panelling pattern 空间嵌板图案
- ~ perspective 空间透视
- ~ structure 空间结构
- ~ truss 空间桁架,网架
- ~d column 格构柱,空腹柱
- floor ~ 楼面面积
- quarter ~ 楼梯转变处之梯台

spacemanship ['speismənʃip] *n.* 建筑中充分利用空间的方法

spacer ['speisə] *n.* 间隔物;定位件;垫片
- ~ block (抹灰用的)定准砌块,垫块

spacial ['speiʃəl] *a.* =spatial 空间的

spacing ['speisiŋ] *n.* 间隔[距],距离,跨度;净空,空格
- ~ board 定位板,间隔板
- ~ strip 间隔条,分格嵌条
- ~ timber 定位木,隔条

spacious ['speiʃəs] *a.* 宽广的,宽裕的

spackling ['spækliŋ] *n.* 抹泥修墙

spalling ['spɔ:liŋ] *n.* 剥落,破碎,散裂
- roof ~ 顶板剥落

span [spæn] *n.*; *v.*, spanned, spanning. *n.* 跨度[距]孔[墩]距;净空[跨];全长; *v.* 跨越;跨度为;(数学中的)生成
- ~ by span construction 逐跨施工法
- ~ length 跨长
- ~ of arch 拱跨
- ~ piece 拉杆,击梁,横木,檩条
- ~ roof 等斜坡屋顶
- ~ saw 框锯

spandrel ['spændrəl] = spandril ['spændril] *n.* 拱肩[墙,侧],上下层窗空间;三角壁

spandrel

Spanish ['spæniʃ] *a.* 西班牙的
- ~ blind 西班牙式遮帘
- ~ style 西班牙式建筑
- ~ tile 半筒瓦

spanner ['spænə] *n.* 扳手[钳];横拉条,交叉支撑
- adjustable ~ 活动扳手
- bent ~ 弯头扳手
- box ~ 套筒扳手
- shifting ~ 活动扳手

spar [spɑ:] *n.* 圆木,小梁;椽子;桁条;桅杆;晶石
- ~ finish 光亮面,反光面
- ~ varnish 清光漆
- auxiliary ~ 副[辅助]梁
- rear ~ 后翼梁

spare [spɛə] *a.* 备用的;多余的; *n.* 备件
- ~ room 备用房,客房

sparge [spɑːdʒ] *v.* sparged, sparging 洒；撒；溅 *n.* 喷水口
sparse [spɑːs] *a.* sparser, sparsest. 稀疏的，稀少的
spatial ['speiʃəl] *a.* 三向的，空间的，立体的；-spatiality, *n.* -spatially, *ad.*
～ art 空间艺术
～ distribution 空间分布
～ grid 空间网格
～ grid structure 空间网架结构
～ model 立体模型；三维模式
～ structure 空间[立体]结构
spatter ['spætə] *n.*；*v.* 溅；洒
～ dash 抹面
spatula ['spætjulə] *n.* 刮刀
spec [spek] *n.* 说明书，加工单，材料表
special ['speʃəl] *a.* 特别[种、制]的；专用的；*n.* 专用部件，专车；特刊
～ cement 特种水泥
～ concrete 特种混凝土
～ coupling 特种管接头
～ emergency staircase 专用太平梯
～ engineering structure 特种工程结构
～ joint 典型接缝[头]
～ kitchen 特餐厨房
～ mortar 特种沙浆
～ -purpose equipment 专用设备
～ refractory 特种耐火材料
～ rigid frame 不规则刚架
～ rolled-steel bar 异形钢
～ room 专用房间
～ splice plate 异型接缝板
～ varnish 特种清漆
specialist ['speʃəlist] *n.* 专家；专门研究者；专业(技术)人员
species ['spiːʃiːz] *n.*；*pl.* -cies. 种，类；形式，物质
wood ～ 木材品种
specific [spi'sifik] *a.* 比率的；专门的；特殊的
～ absorption 吸水率

～ indent 特定订单
～ weight 比重
specification [ˌspesifi'keiʃən] *n.* 规范[程]；细则；说明书；规格；明细表；技术要求
～ cement 标准水泥
～ for design 设计规范
～ for materials 材料规格
～ for workmanship 操作规程
design ～ 设计要求[任务书]
operation ～ 操作规程
production ～ s 生产技术条件
quality ～ 质量技术条件
specified ['spesifaid] *a.* 指定的；规定的
～ amount 规定量
～ by building specification 建筑规范规定的
～ compressive strength (混凝土的)规定抗压强度
～ grading 指定级配
～ head 规定水头，设计水头
～ load 规定荷载
～ mix 指定配合比
～ project 按规范进行设计
specimen ['spesimən] *n.* 试件；样品，样本
speck [spek] *n.* 斑[污]点，瑕疵
speckle [spekl] *n.*；*v.*, -led, -ling. *n.* 斑点[纹]；*vt.* 加以斑点
～ pattern 斑纹图样
speckstone ['spekstəun] *n.* 滑石
spectacle ['spektəkl] *n.* 景象；展品；(*pl.*)眼镜
～ plate 双孔板
spectator [spek'teitə] *n.* 观众
～ stand 观众席，看台
specular ['spekjulə] *a.* 镜面的；反射的
～ finish 镜面加工，表面平滑加工
speculum ['spekjuləm] *n.* 金属反射镜，镜用合金
speed [spiːd] *n.*；*v.*, sped [sped] or speeded, speeding. *n.* 速度[率]；*v.*

加[调]速;急行
~ cement 速凝水泥
spelter ['speltə] *n.* 锌;锌焊料
~ coating 锌涂层,锌涂料
speos ['spi:ɔs] *n.* (古埃及)石窟墓室,石窟神庙
sphenoid ['sfi:nɔid] *a.* 楔形的
sphere [sfiə] *n.* 球体[面];球状体,地球仪;天体;范围,领域
spheric(al) ['sferik(əl)] *a.* 球[面、状]的
~ cupola 半球壳
~ dome 球形屋顶
~ illumination 球面照度
~ shell 球面壳体
~ vault 圆形穹[拱]顶
spherule ['sferju:l] *n.* 小球体,球粒
sphinx [sfiŋks] *n.*; *pl.* sphinxes ['sfiŋksiz] 斯芬克斯狮身人面像
spiciform ['spaisifɔ:m] *a.* 穗状(排列)的
spider ['spaidə] *n.* 多脚撑;十字叉;三脚架;星形轮;蜘蛛
spigot ['spigət] *n.* 插口[头];阀门,龙头;套管
spike [spaik] *n.* 道[大,长]钉;峰值
~ and ferrule installation 钉和箍
~ rod 道钉型钢
spile [spail] *n.*; *v.*, spiled, spiling. -*n.* 小塞子,插管;木桩 *v.* 加塞子,装导管
spiling *n.* 木桩
spill [spil] *v.* spilled or spilt [spilt] spilling. *n.* 使(水等)流出或落下;溢出,溢流,倒出,漏失
~ shield louver lighting 隔栅照明,格片照明
spillage ['spilidʒ] *n.* 溢出[洒落,倒出,溅出](的物质)泄漏(量),溢出量
~ of material 材料损耗
spin [spin] *v.*, spun [spʌn], spinning, *n.* 旋[绕]转;疾驰;拔丝,旋压成形

spindle ['spindl] *n.* 轴;(栏杆的)纺锤(状物),纺锤形立柱,螺旋梯的中柱
~ -cyclide 纺锤形圆纹曲面
~ sleeve 轴套
spine [spain] *n.* 脊柱[骨]
~ wall 内墙,隔墙
spinning ['spiniŋ] *a.* 旋转的,离心的
spiracle ['spaiərəkl] *n.* 通气孔,通风口
spiral ['spaiərəl] *a.*; *n.*; *v.*, -ral(l)ed, ral(l)ing. *a.* 螺旋(形)的;螺线[纹]的;盘旋(上升)的; *n.* 螺旋形物; *v.* (使)成螺旋形;盘旋
~ car park 以螺旋坡道连接的室内多层停车场
~ column 螺旋式柱
~ curve 螺旋曲线
~ -fluted column 饰以螺旋状凹纹的立柱
~ minaret 螺壳状的回教寺院的尖塔
~ reinforcement 螺旋状钢筋
~ rope 绞绳
~ spring 发条,盘簧
~ staircase 盘梯,螺旋梯
~ stairs 盘梯
~ stirrup column 螺箍筋柱
~ stringer 螺旋形楼梯边梁

spiral stair

spire [spaiə] *a.* 塔尖,尖顶
~ roof 尖塔形屋顶
~ tower 尖顶塔
spirelet ['spaiəlit] *n.* 小尖塔
spirit ['spirit] *n.* 酒精,醇;燃料;精神,潮流
~ circular level 圆水平仪
~ level 气泡水准仪
~ of wine 酒精,乙醇
~ stain 酒精着色剂,酒精污斑

spirit level

~ varnish 凡立水,挥发漆
white ~ (用于油漆的)石油溶剂,白

节油

splash [splæʃ] *n.*;*v.* 飞溅,喷溅
~ apron 挡板,防溅板,挡泥板
~ -back 防溅挡板(特指浴缸或水池四周涂衬不透水材料的墙壁)
~ baffle 挡板,防溅板,挡泥板
~ block 防溅挡板
~ erosion 雨淋冲刷
~ guard 挡板,防溅板,挡泥板
~ -proof 防溅水的

splat [splæt] *n.* 接口盖板,板间盖条;椅背中部纵板

splay [splei] *n.* 斜角,倾斜面,八字面 *v.* 弄斜,展宽,使成喇叭形
~ bonder 斜面砖一端削成45°的特种砖
~ brick 斜面砖
~ ceiling 八字形平顶
~ knot (木材)斜节
~ piece 鸡尾形键
~ skirting 上缘抹角墙脚板
~ wall 翼墙,八字墙
~ed abutment 八字形桥台
~ed arch 八字形拱
~ed ceiling 八字顶棚,槽形顶棚
~ed coping 墙压顶斜面
~ed door 八字[式]门
~ed footing 八字形基脚
~ed ground 斜木托条
~ed heading joint 八字形接头
~ed jambs 八字侧墙,八字形(门窗)竖框
~ed joint 斜角[面]连接,楔形接缝
~ed niche 八字形壁龛
~ed scarf 八字楔榫

splice [splais] *v.*,spliced, splicing *vt.* 叠接(两块木材);*n.* 拼[镶]接
~ angle 拼接角钢
~ bar 连接筋,连接板,鱼尾板
~ block 拼接块
~ bolt 拼接螺栓
~ joint 拼接接头,鱼尾板接头
~ pad 拼接衬垫

~ piece 拼[镶]接板
~ plate joint 二侧拼接板接合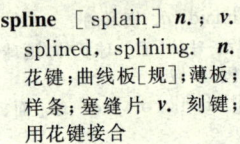
~d pile 拼接桩
~d pole 叠接(电)杆
~d strut 附加短柱

splice plate joint

spline [splain] *n.*;*v.* splined, splining. *n.* 花键;曲线板[规];薄板样条;塞缝片 *v.* 刻键;用花键接合
~ batten 活动曲线规
~ joint 填实缝,花键连接
~ shaft 多槽[花键]轴

spline

splint [splint] *n.* 固定木楔;固定销,键;夹板,薄片

splinter ['splintə] *n.* 碎片;薄片;*v.* 裂开,劈开
~ -proof 防弹的,防破的

split [split] *v.* split, splitting *v.* 割裂;分裂,裂开;*n.* 分裂;(*pl.*) 较标准砖稍薄的砖
~ across 对裂开
~ bolt 开尾螺栓
~ bond 半厚砖砌筑
~ brick 半厚砖
~ cotter 开尾销
~ course 削砖砌层,半混砖砌层
~ die[məuld] 组合[可拆]模
~ face 分块饰面,分块面,条纹面
~ holding ring 对卡圈
~ leg tripod 伸缩三脚架
~ -level 房屋夹层,搁楼层
~ -level house 错层式住宅
~ pin 开口[尾]销
~ plate 裂口垫板
~ ring connector 开口环连接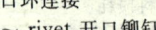
~ rivet 开口铆钉
~ saw 粗齿锯
~ -segment die 组合[可拆]模

split-ring connector

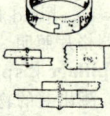

splitter ['splitə] *n.* 分裂器[机];气流分隔片
 ~ plate type sound absorber 隔板式[片式]消声器
spoil [spɔil] *n.* 弃土;挖出的土石
 ~ disposal 废料处理
spoke [spəuk] *n.* 扶梯踏蹬,扶梯棍;辐条
 ~ shave 小圆刨,辐刨
sponge [spʌndʒ] *n.* 海绵,泡沫[多孔]材料
 ~ glass 毛玻璃
 ~ plastic 泡沫[多孔]塑料
 ~ rubber 海绵橡皮
 ~ structure 多孔性结构
spongy ['spʌndʒi] *a.*, -gier, -giest. 海绵状的,多孔的
 ~ rubber 海绵状橡皮
 ~ soil 橡皮土,松软土
 ~ structure 海绵状结构
spoon [spu:n] *n.* 挖土机,挖泥铲,吊斗,匙[勺]状物
 ~ and square 小墁刀
sporadic [spə'rædik] *a.* 分散的,散见的,零星的
 ~ building 无规则的建筑
sport [spɔ:t] *n.* 运动,游戏
 ~s center 体育中心,综合性运动场,大型体育馆
 ~s club 体育俱乐部
 ~s facility 体育设施
spot [spɔt] *n.*;*v.* spotted, spotting *n.* 斑点;污点;地点;场所;*v.* 沾污,加以斑点
 ~ board 灰泥板
 ~ footing 柱式基础
 ~ lamp 聚光灯
 ~ level 水准点
 ~ light 聚光灯;点光源
 ~ punch 点式手动穿孔,点式穿孔器
 ~ welding 点焊
 focal ~ 焦点
spout [spaut] *n.* 喷口[嘴],喷水管,水落管,出水口,斜槽;*vi.* 喷;涌
sprag [spræg] *n.* 斜撑,拉条,制轮木
 bottom ~ 底部支撑
 face ~ 工作面支柱
spray [sprei] *v.* 喷射[洒,涂];*n.* 喷雾器;水沫,浪花,枝状饰
 ~ aerator 喷洒通气器
 ~ application 喷涂
 ~ can 喷雾器
 ~ cement coating 水泥喷射涂层
 ~ coating 喷涂
 ~ curing 洒[喷]水养护
 ~ fountain 喷水池
 ~ gun 水泥喷枪,喷漆枪
 ~ jet 喷雾器,喷雾口
 ~ lacquering 喷漆
 ~ lance 喷枪,喷雾器
 ~ metal coating 金属喷涂法[层]
 ~ nozzle 喷嘴
 ~ painting 喷漆
 ~ pistol 喷[漆]枪
 ~ plate 隔沫板
 ~ed acoustical ceiling 喷涂隔声顶棚
 ~ed concrete 喷射混凝土
 ~ed mineral wool 喷涂石棉
 ~ed mortar 喷砂浆
 power ~ 粉末喷镀[涂]
 shower ~ 淋浴喷头
spread [spred] *v.*, spread, spreading. 伸展,扩展,展开;*n.* 伸[扩]展;摊铺
 ~ mortar 喷布砂浆
 ~ing rate 涂面率
sprig [sprig] *n.*;*v.*, sprigged, sprigging *n.* 扁头钉,无头钉;嫩枝;(*pl.*) 用铁皮做成镶嵌玻璃用的三角条 *v.* 用小钉固定,用小枝装饰
 ~ bit 打眼钻
spring [spriŋ] *v.*, sprang [spræŋ] *n.*, or sprung [sprʌŋ], sprung, springing, *v.* 跳跃,涌出,发生;*n.* 弹簧,簧片;起拱点[面];潮水,泉水;春天
 ~ action cock 弹簧旋塞

~ an arch 砌成拱形,砌拱圈
~ balancing sach 弹簧平衡窗
~ beam 弹性梁,系梁
~ block 拱座
~ bolt 弹簧螺栓
~ butt (hinge) 弹簧合页[铰链]
~ catch 弹簧销
~ clamp 弹簧夹
~ coil 弹簧圈
~ cotter 开口销,弹簧制销
~ hinges 弹簧铰链
~ line 起拱线
~ lock 弹簧锁
~ pad 弹簧垫
~ ring dowel 弹性环销
~ rod 弹簧杆
~ shackle 弹簧挂钩
~ snib 弹簧插销[闩]
~ stop 弹簧门闩
~ -style 草图设计
~ support 弹簧支承
~ washer 弹簧垫圈
air ~ 气垫
extension ~ 拉簧
floor ~ 地弹簧

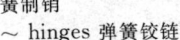
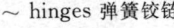
spring hinges

springer ['spriŋə] *n.* 拱底石,拱脚石,起拱石

springing ['spriŋiŋ] *n.* 起拱;弹性装置
~ block 拱底石,拱座
~ course 起拱层
~ height 起拱高度
~ line 起拱线
~ needle (托换基础用的)支柱
~ of curve 起拱点,曲线起点

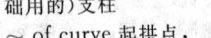
springing line

sprinkle ['spriŋkl] *v.* -kled,-kling 洒[撒];喷撒
sprinkling basin 喷水池

sprocket ['sprɔkit] *n.* 接椽,檐椽接长木,檐椽支撑木;链轮;扣链齿
~ gear 扣齿链轮

~ hole 定位孔
~ piece 小椽,短人字木

spruce [spru:s] *n.* 云杉,针枞
~ pine 枞松

sprung [sprʌŋ] *v.* spring 的过去分词, *a.* 支在弹簧上的
~ arch 起拱点

spud [spʌd] *n.* 定位柱,定位桩,销钉,钻头;草铲

spun [spʌn] *adj.* 旋制的,离心制造的
~ column 离心法制成的立柱
~ glass 玻璃纤维,玻璃丝

spur [spə:] *n.* 凸壁,有垛墙;角撑,齿轮;截水墙;挑水坝,丁坝;迹;支线
~ pile 斜桩
~ post 斜柱
~ stone 护角石
nonslipping ~ 防滑板

sputter ['spʌtə] *n.*,*v.* 真空喷镀

squamose ['skweiməus] = **squamous** ['skweiəs] *a.* 鳞状的,有鳞的
~ structure 鳞片状结构

square [skwɛə] *n.*;*a.* squarer, squarest, *v.*, squared, squaring, *n.* 正方形;方形物;广场;平方;丁字尺;直角尺;街区; *a.* 正方形的;平方的; *vt.* 使成方形,使成直角
~ and flat panel 方块平镶板
~ bar 方钢,方形杆
~ basket (镶木地板的)方格形
~ bend [bent] 直角弯头
~ cogging 齿榫接合
~ column 方柱
~ dressed pavement 方石块路面
~ -edged board 方边板,刨方板
~ (edge) timber 方木材
~ file 方锉
~ grid 方网架
~ head bolt 方头螺栓
~ head screw 方头螺钉
~ headed window 方头窗
~ jamb 方角侧墙,(门窗)方竖框

~ joint (平头)对接
~ man 石匠,木匠
~ nail 方钉
~ neck bolt 方颈螺栓
~ nut 方螺母
~ pier 方形墩
~ pitch 45°房面坡度
~ quoin 方隅石,(房间)方角落
~ roll 方形卷饰
~ roof 两边45°坡的屋顶
~ rubble 方块毛石
~ rule 直角尺
~ section 方形截面
~ splice 方块搭接
~ staff 方角缘
~ steel 方钢
~ stone (细)方石,琢石
~ thread 方螺纹
~ tile 方形瓷砖
~ timber 方木
~ trench 方形沟槽
~ washer 方形垫圈[板]
~ window 方窗
~d log 方木,方材
~d rubble 方块毛石
beam ~ 角尺
bevel ~ 斜角规
set ~ 三角板
T ~ 丁字尺

squat [skwɔt] *a.* 短而粗的,蹲踞的
~ pier 短粗型墩
~ ting water closet 蹲式洗大便池

squatter ['skwɔtə] *n.* 无权而定居公地者
~ house 违章建筑

squinch [skwintʃ] *n.* 内[突]角拱,支承上层结构的墙角拱

squint [skwint] *n.* 斜孔小窗,窥视窗
~ brick 边斜砖
~ pier 转角处墩柱
~ quoin of wall 墙边斜隅石
~ window 窥视窗

squirrel ['skwirəl] *n.* 松鼠
~ tail 鼠尾形
~ tail pipe jointer 鼠尾形管子连接器

stab [stæb] *v.*, stabbed [stæbd], stabbing 把砖墙凿粗糙(以涂灰泥)

stabile ['steibil] *n.* (抽象主义的)静态雕刻

stability [stə'biliti] *n. pl.* -ties 稳定性,耐久性
corrosion ~ 耐[抗]蚀性
thermal ~ 耐热性

stabilization [steibilai'zeiʃən] *n.* 稳定(作用);稳定化处理

stable ['steibl] *a.* 稳定的;*n.* 厩,畜舍
~ door (上下可以分别开关的)两截门;带小门的门

stack [stæk] *n.* 烟囱,烟突;通风管;竖管;(化工的)塔;货[书]架;书库;堆栈;叠式存储器;*vt.* 堆积(起)
~ bond 通风栅
~ moulding (机械铸造的)叠箱造型
~ of bricks 砖垛
~ pipe 立管,竖管
~ed plate 堆积式承载板
soil ~ 生活污水立管
vent ~ 通风立管

stad(d)le ['stædəl] *n.* 支柱;拉条;支架;基础
~ stone 蘑菇状的矮立柱

stadion ['steidiən] *n.* (古希腊的)赛跑运动场;有看台的露天大型运动场

stadium ['steidiəm] *n. pl.* stadiums or stadia 周围有多层看台的大型露天体育[运动]场
~ construction 体育场建筑[施工]
~ facilities 体育场设施
~ type auditorium 阶梯式观众厅

staff [stɑ:f] *n. pl.* staffs or staves [stɑ:vz] 纤维灰浆,纸筋灰,麻刀灰,标尺;测尺;杆;(*pl.*)职员
~ angle 纤维灰浆抹灰的墙角
~ bead 纤维灰浆抹灰的圆(球)形

线条
~ building 职工宿舍楼
~ plate 水平尺垫
~ quarter 职工宿舍
~ rod 标尺
leveling ~ 水准标尺
stage [steidʒ] *n.* 坛,台,舞台;剧场;鹰架,脚手架;阶段;步骤
~ box 舞台前部的包厢
~ construction 分期施工;多层面施工
~ door (演员进入舞台的)后台门
~ lighting 舞台照明
~ of construction 施工期
~ of work 工程进度
~ -scene 舞台布景
~ strip light 舞台带形灯
~ wagon 装舞台布景用的车台
~ wall 舞台的台口墙
~ with trap 有升降台板的舞台
object ~ 载物台架
specimen ~ 样品台
stagger ['stægə] *n.*;*v.* 错列,间隔
~ angle 交错角
~ arrangement 交错排列
staggered ['stægəd] *a.* 交错的
~ air heater 拐折空气加热器
~ course 交错瓦层
~ piling 打梅花桩,错列打桩
~ rivet joint 错列铆接
~ seating 前后排座位错开的布置形式
~ stud partition 错列间柱隔板
staging ['steidʒiŋ] *n.* 脚手架,构架,工作平台
~ of pipeline 管架
stahlton slab ['stɑ:ltən slæb] *n.* (瑞士创制的)砖铺底浇灌砂浆的预应力槽形板
stain ['stein] *n.* 污点;颜料;染料 *v.* 着色,染色,沾污,生锈
~ etch 染色腐蚀
~ paper 彩色墙纸

~ -proof paint 防污涂料,防污漆
~ resistance 耐腐蚀性、耐锈蚀性
~ed glass 彩色玻璃
~ed glass window 彩色玻璃窗
~ed wood 着色木材
~ing power (对油漆等的)着色能力
stainer ['steinə] *n.* 着色剂,色料;着色工,染工
stainless ['steinlis] *a.* 不锈的
~ clad steel (plate) 包不锈钢板,镀防锈层的钢板
~ paint 防污涂料,防污漆
~ steel 不锈钢
stair [stɛə] *n.* 阶梯之一级[段](*pl.*)楼梯;浮码头,趸船
~ carpet 铺在楼梯上的地毯
~ clip 固定楼梯地毯压条
~ exit 楼梯安全出口
~ flight 单段梯
~ hall 楼梯间[厅]
~ handrail 楼梯扶手
~ head 楼梯顶口
~ horse 楼梯架
~ landing 楼梯平台
~ light 楼梯灯
~ nosing 梯级凸沿
~ platform 楼梯(中间)平台
~ post 楼梯柱
~ railing 楼梯栏杆
~ reeding 梯级上的防滑条
~ rise 楼梯踏步高度
~ riser 楼梯踢脚板;梯级起步
~ rod 楼梯地毯压条
~ seat 阶梯形坐席
~ soffit 楼梯背面
~s resting on arches 拱支楼梯
~ step 楼梯踏步
~ stringer 楼梯(斜)梁
~ tread 楼梯踏步平板
~ trimmer 楼梯段端部托梁
~ wall 楼梯承重墙
~ well 楼梯井

~ winder 楼梯的斜踏步
winding ~s 盘旋楼梯
staircase ['stɛəkeis] *n.* 楼梯,楼梯间
~ of bifurcated type 双分叉式楼梯
~ of dog-legged type 无梯井式楼梯,狗腿式楼梯
~ of half-turn type 180°转角楼梯
~ of helical type with newel 中柱螺旋式楼梯,中柱支承悬臂楼梯
~ of open-well type 带梯[梯]井式楼梯
~ of quarter-turn type 90°转角楼梯
~ of straight flight type 单跑式[单梯段式]楼梯
~ of three-quarter turn type 270°转角楼梯,三跑楼梯
~ window 楼梯窗
~ with several flights 多梯段楼梯
grand ~ 剧场等公共场所之正面大楼梯
spiral ~ 螺旋式楼梯
suspended ~ 悬挑楼梯
stairway ['stɛəwei] *n.* 楼梯,楼梯间
~ enclosure 有围护的楼梯
travelling ~ 自动扶梯,移动梯
stake [steik] *n.* ; *v.*, staked, staking *n.* (测量用)标桩,柱杆 *v.* 立桩,加桩
-staker, *n.*
~ rack 栅栏
rack ~ 栏栅杆
staking-out 放样,定线,立桩
stalactite ['stæləktait] *n.* 钟乳石(状物)
~ ornament 钟乳石饰
~ vault 钟乳石形饰穹顶
stalagmite ['stæləgmait] *n.* 石笋[状物]
stale [steil] *a.* staler, stalest. 陈腐[旧]的,变坏的
~ lime 已潮的石灰,陈石灰
~ -proof 不腐的
stalk [stɔːk] *n.* (梁)腹板;高烟囱;植物的茎、杆;茎梗饰

~ plate 腹板
~ stiffener 腹板加劲肋
stall [stɔːl] *n.* 厩,畜舍;货摊;礼拜堂的教士或唱诗班人的坐位;图书馆中没有坐位的小角落;小隔间;停车场中的停车位;*v.* 失速,停车
~ board 铺面栏板
~ urinal 立式小便器
shower ~ 淋浴间
stalloy ['stɔːlɔi] *n.* 硅钢片,薄钢片
stalpeth cable 钢、铝、聚乙烯组合铠装电缆
stambha ['stɑːmbhə] *n.* (印度的)石雕纪念像柱
stamp [stæmp] *n.* 捣碎(机),冲模;图章,标记;邮票 *vt.* 印上,捣碎,标明
~ sand 压碎砂,人造砂
~ed nail 压制钉
stancheon ['stɑːnʃən] = stanchion *n.* (型)钢柱,支柱,标柱,撑杆
~ sign (可移动的)柱座标志
stand [stænd] *v.*, stood [stud], standing, *n.*; *v.* 竖起,使直立;位于;*n.* 讲台;看台;台架;摊棚,地点
~ bar 站立进食的酒吧
~ -by instrument 备用仪表
~ by plant 备用设备
~ -down 停工
~ -pipe 立管,水鹤
~ -post hydrant 柱式消火栓
~ sheet 不开窗,封死窗
~ wear and tear 耐磨损
clamp ~ 固定支架
service ~ 工作梯(架)
standard ['stændəd] *n.* 标准,准则,判据;支架,构架;样品;本位
~ bolt 标准螺栓
~ brick 标准砖
~ cement 标准水泥
~ consistency 标准稠度
~ cylinder 标准圆柱体
~ design 标准设计

~ detail 标准细部图
~ dimension 标准尺寸
~ drawing 标准图
~ error 标准误差
~ floor (房屋设计中的)标准层
~ illumination 标准照度
~ ladder 梯架立柱,脚手架立柱
~ lamp 落地灯
~ metal window 标准金属窗
~ model 标准样品
~ of potable water 饮用水标准
~ of scaffolds 脚手架立柱
~ orifice 标准孔口
~ part 标准零件
~ pile 支承板桩的柱,标准桩
~ pipe 标准管
~ portland cement 标准波特兰水泥
~ sample 标准样品
~ sand 标准砂
~ scale 标准尺
industrial ~ 工业标准
reference ~ 参考标准
tentative ~ 暂行标准

standardization [ˌstændədaiˈzeiʃən] *n.* 标准化,规格化

standardize [ˈstændədaiz] *v.* 使合标准,使标准化
~d house 标准[定型]住宅
~d-shape member 标准(形)构件

standing [ˈstændiŋ] *a.* 直立的,站的;静止的,停滞的;固定的,持续的
~ bolt 双帽螺栓
~ geyser 间歇喷泉
~ gutter 立式水槽
~ ladder 直立梯
~ leaf (折叠门的)固定门页
~ operating procedure 标准操作规程
~ panel 立式门板
~ room (戏院)站席
~ urinal 立式小便器

stannum [ˈstænəm] [拉] *n.* 锡(sn)
stanzaite *n.* 红柱石

staple [ˈsteipl] *n.*;*v.*, -pled,-pling. *n.* U形钉,钉书钉,扒钉;纤维;商业中心, *vt.* 用(U形)钉钉住
~ glass fibre 人造[标准]玻璃纤维
stapling tacker 勾钉
hasp and ~ (桌、门上用的)搭扣

star [stɑː] *n.* 星;星形;星号
~ house 星形平面的塔式住宅
~ knob 星形捏手
~ moulding 星形线脚饰
~ pattern 星形
~ section 十字截面
~ shake (木材的)星形裂纹
~ vault 星状肋的拱顶
five-pointed ~ 五角星

starch [stɑːtʃ] *n.* 淀粉,浆糊
~ glue 淀粉粘合剂,浆糊

starling [ˈstɑːliŋ] *n.* 桥墩尖端分水桩;桥墩四周的防护桩
~ coping 凸出顶层;墙帽

start [stɑːt] *n.*;*v.* 开始,出发,启动
starter [ˈstɑːtə] *n.* 起动装置,起动器
starting [ˈstɑːtiŋ] *n.* 起动,发动
~ material 原材料

starve [stɑːv] *v.*, starved, starving 缺乏;饥饿
~d joint (粘结不良的)失效接缝,缺胶接头[节点]

stash [stæʃ] *v.* 隐藏,停止 *n.* 隐藏[贮存]处[物]

state [steit] *n.* 情形,状态;国家,州
~ apartment 大礼堂,大厅
~ room 大厅,(车、船上的)特等舱,包厢

statement [ˈsteitmənt] *n.* 报告书,财务报表;语句;声明

static(al) [ˈstætik(əl)] *a.* 静止[力态]的
~ grounding rod 接地杆

statically [ˈstætikəli] *a.* 静力地,静态地
~ determinate beam 静定梁
~ determinate structure 静定结构

~ determinate truss 静定桁架
station ['steiʃən] *n.* 车站,电站,电厂,测站,电台;位置,场所;测桩间标准距离(100 或 600 尺)
~ hall （车站)候车大厅
~ restaurant 车站餐厅
public convenient ~ 公共厕所
public ~ 公用电话亭
stationary ['steiʃnəri] *a.* 静止的,不变的;固定的,不动的;稳定的,稳态的
~ window 固定窗
stationery ['steiʃneri] *n.* 文具
~ room 文具室
statistic(al) [stə'tistik(əl)] *a.* 统计的
statuary ['stætjuəri] *n.* 雕像,塑像;雕塑术;雕刻家
~ art 雕塑艺术
~ column 雕像柱
~ marble 雕像大理石

statuary column

statue [stætju:] *n.* 雕像,铸像,塑像; *vt.* 用雕[塑]像装饰
~ niche 雕像龛
statuette [ˌstætju'et] *n.* 小雕像,小塑像
status ['steitəs] *n.* 情形,状态
~ report （工程)情况报告
statute ['stætju:t] *n.* 法规;章程,条例
staunch [stɔ:ntʃ] *v.* = stanch. 使不漏水,密封;止血; *a.* 密封的,不漏水的,不透气的
~ing bead 止水片[缝]
stave [steiv] *n.* 梯级横木,狭板条
~ church 板条教堂
~ jointing 拼板接合
~ pipe 木质板拼成管
stay [stei] *n.* 撑条,拉条,支柱
~ bar 拉条钢筋,拉杆;撑杆
~ cable 拉索
~ collar 拉环,牵环
~ guy 固定拉索
~ hook 撑钩

~ log 支撑木
~ piece 屋架系杆
~ pile 锚固桩,支承桩
~ pipe 撑管
~ plate 坐板,垫板
~ post 撑柱
~ tightener 拉线张紧器
~ tube 支撑管,牵拉管
~ed pole 牵拉杆
~ing wire 拉线,拉索
steadier ['stediə] *n.* 支座[架],底座
steady ['stedi] *a.* steadier, steadiest;稳定的,稳态的,不变的,持续的
~ rest 固定中心架
steam [sti:m] *n.* 蒸汽; *v.* 蒸;用蒸汽动力开动;steamlike, *a.*
~ atomizing 蒸汽喷雾
~ bending （木材)蒸汽弯曲
~ coil heated 蒸汽盘管供暖的
~ heating system with dry (wet) return 干[湿]式回水蒸汽供暖系统
~ hose 蒸汽软管
~ pressure type cascade heating system 蒸汽加压(串列)采暖系统
~ -proof 不[防]透汽的
~ purifier 蒸汽净化器
~ room 蒸汽浴室
~ -tight 不漏气的,汽密性的
~ed concrete 蒸汽养护的混凝土
~ing of wood 木材蒸干
stearin(e) ['stiərin] *n.* 硬脂,甘油硬脂酸酯
steatite ['stiətait] *n.* 块滑石,皂石
~ ceramics 块滑石陶瓷
steel [sti:l] *n.* 钢;钢制品;steellike, *a.*
~ alloy 钢合金
~ and reinforced concrete 钢结构和钢筋混凝土,混合结构
~ angle 角钢
~ arch support 钢拱支架
~ -band tape 钢卷尺
~ bar 钢筋,条钢
~ beam 钢梁

~ beam enclosed in concrete 外包混凝土钢梁
~ beam grillage 钢梁格栅
~ bearing 钢支座
~ bearing skeleton 承重钢骨架
~ bracket 钢托架,钢牛腿
~ brush 钢丝刷
~ building 钢建筑物
~ cable roof 钢缆屋顶
~ casement 钢窗框
~ centering 钢拱架
~ channel 槽钢
~ -clad wire rope 包钢钢丝绳
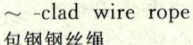

steel column
~ column 钢柱
~ conduit 布线钢管
~ construction 钢结构
~ core concrete column 钢管混凝土柱
~ cupola 钢穹顶
~ deck plate 钢(桥)面板
~ design 钢(结构)设计
~ diaphragm 钢隔板,钢隔墙
~ dog 蚂蝗钉,扒钉
~ dome 钢(圆)拱顶
~ door frame 钢门框
~ engraving 钢板雕刻,钢板印刷
~ fabric reinforcement 钢丝网片,网状钢筋
~ fabric sheet 钢丝网片
~ facing 钢板护面
~ fastener 钢固定件,钢扣,钢钩
~ fiber concrete 钢纤维混凝土
~ file 钢锉
~ flat roof truss 钢平顶桁架
~ floor 钢板楼面
~ floor plate 楼面用钢板
~ floor tile 楼面用钢板
~ folded plate roof 钢摺板屋顶
~ form 钢模板
~ form panel 钢制定型模板
~ form work 钢模板(工程)
~ frame 钢[框]架

~ framed house 钢框架房屋
~ framed reinforced concrete column 钢框钢筋混凝土柱
~ framed window 钢框窗
~ -frame structure 钢[框]架结构
~ girder 钢大梁
~ -glass system 钢-玻璃结构体系
~ grey 钢青色
~ grid 钢筋网格
~ grid floor 钢格楼板
~ grid footing 钢格基础
~ grillage 钢筋网格
~ grit 钢砂,硬砂粒
~ grit concrete 钢砂混凝土
~ hanger 钢吊架
~ H-column H 形断面钢柱
~ hinge 钢铰链,钢铰
~ hipped-plate roof 钢折板屋顶
~ hook 钢钩,钢弯钩
~ hoop 钢箍,钢环
~ jacket 钢套
~ joist (小)钢梁,钢托梁
~ ladder 钢梯
~ lagging 钢套,钢挡板
~ lamella dome 层状钢拱
~ lathing (抹灰用)钢丝网
~ lattice truss 钢格构桁架
~ latticed column 格构式钢柱
~ latticed girder 格构钢梁
~ light structure 轻型钢结构
~ -lined concrete pipe 钢板衬砌混凝土管
~ lining 钢板衬砌
~ lintel 钢过梁
~ loop 钢环
~ measuring tape 钢卷尺
~ member 钢构件
~ mesh 钢丝网
~ -mesh lagging 钢丝网衬板
~ mesh reinforcement 钢丝网,网状钢筋
~ mould 钢模

~ movable form 钢滑动模板
~ nickel chrome 镍铬钢
~ panel 钢护板
~ pantile 钢波形瓦
~ partition 钢隔板[墙]
~ penstock 压力钢管
~ pile 钢桩
~ pipe 钢管
~ pipe butter 钢管端部固定杆
~ pipe column 钢管柱
~ pipe pile 钢管桩
~ pipe reinforced concrete structure 钢管混凝土结构
~ pipe scaffold 钢管脚手架
~ pipe structure 钢管结构
~ pipe support 钢管支柱
~ plain fish-plate 平面钢鱼尾板
~ plain splice 钢板平面拼接
~ plate 钢板
~ plate floor 钢板楼[桥]面
~ plate girder 钢板大梁
~ plate structure 钢板结构
~ plate washer 钢垫片
~ -ply form 折叠式钢模板
~ products 钢制品,钢材
~ purlin 钢檩条
~ radiation shielding wall 钢制辐射防护墙
~ raking tool (清缝用)钢刮刀
~ reinforced concrete column 钢筋混凝土柱
~ reinforced concrete structure 钢筋混凝土结构
~ reinforcement 钢筋
~ requirements 钢筋[钢材]需要量
~ -ribbed cupola 钢肋穹顶
~ -ribbed dome 钢肋穹顶
~ rivet 钢铆钉
~ rod 钢筋,钢杆
~ rolling shutter 钢制卷帘
~ roof covering 钢制屋面层
~ roof truss 钢屋架
~ rope 钢丝绳,钢缆

~ rope roof 悬索屋顶
~ rule 钢尺
~ sash 钢推拉窗扇,钢窗

steel rope roof

~ saw 钢锯
~ scale 钢尺,钢锈斑
~ screw 钢螺钉,钢螺杆
~ screw pile 钢螺旋桩
~ section 型钢截面[形状],型钢
~ segment 钢管片段
~ shape 型钢截面[形状],型钢
~ sheathing 钢面板,钢模板
~ sheet 薄钢板
~ sheet facing 钢板衬砌
~ shell 钢壳体
~ shielding wall 钢制防护墙
~ shore 钢撑,钢柱
~ shore strut 钢支撑,钢顶撑
~ shoring column 钢柱
~ shot concrete 钢砂混凝土
~ shutter 钢模板,钢百叶窗
~ shutter door 钢门
~ shuttering 钢模板
~ skeleton 钢骨架
~ skeleton building 钢骨架房屋
~ skeleton (multistorey) car park (多层)钢骨架停车场
~ skeleton structure 钢骨架结构
~ skin 钢板衬砌
~ skyscraper 钢制摩天大楼
~ slab 钢板
~ slab prop 钢板撑
~ sleeve 钢套管
~ spigot ring 钢插口环
~ stack 钢烟囱,钢通风管
~ stanchion 钢柱
~ strand 钢绞线
~ strap 钢条,钢片
~ string(er) 钢板楼梯斜梁
~ strip 带钢,簧片
~ structural engineering 钢结构工程

~ structure 钢结构
~ support structrue 钢支撑结构
~ tape 钢带,钢卷尺

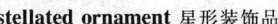
steel tape

~ trowel 钢泥刀,钢抹子,钢馒刀
~ truss 钢桁架
~ tube 钢管
~ wheel glass cutter 钢砂轮玻璃刀
~ wire 钢丝
~ wire lashing 钢丝绑扎
~ wire rope 钢丝绳
~ -yard 提秤,杆秤
alloy ~ 合金钢
angle ~ 角钢
flat ~ 扁钢
glass-hard ~ 特硬钢
H ~ 宽缘工字钢,H形钢
hard ~ 硬钢
high carbon ~ 高碳钢
low carbon ~ 低碳钢
medium carbon ~ 中碳钢
stainless ~ 不锈钢
structural mild ~ 结构用低碳钢
tool ~ 工具钢
steel-cored *a.* 钢心的
steel-framed *a.* 钢[框]架的
steelwork ['sti:lwə:k] *n.* 钢铁工程,钢结构,钢制品
~ details 钢结构详图
steen [sti:n] *n.* (=stean)古陶罐
steep [sti:p] *a.*, steeper, steepest; *a.* 陡峭的,急剧(升降)的; *n.* 陡坡,悬崖,绝壁; *v.* 浸泡,浸湿 steeply, *ad.*
~ arch 陡拱
~ gable 陡(双)坡屋顶
~ roof 陡坡屋顶
~ing brick 浸湿砖
steeple ['sti:pl] *n.* (礼拜堂的)尖阁,尖塔; steepleless; steeplelike, *a.*
~ -crowned 尖塔形的,尖顶的
~ -head rivet 尖头铆钉
steepletop ['sti:pltɔp] *n.* 尖塔状顶部

steepness ['sti:pnis] *n.* 陡度,斜[坡]度
steeve [sti:v] *n.*; *v.*, steeved, steeving; *n.* 吊杆,起重桅; *v.* 用起重桅装(货)
stele [sti:li:] *n.*; *pl.* -lae, les; 石碑;建筑物或岩石上留作刻字用的平面
stellar ['stelə] *a.* 星的,星状的
~ plan 星形布置图
~ pattern 星形
~ vault 星肋穹顶,带肋穹顶

stele

stellated ornament 星形装饰品
stellite ['stelait] *n.* 钨铬钴硬质合金
stem [stem] *n.* 杆;梁腹;管;推拉用小把手;茎;船首
~ section (T形梁等的)梁腹截面,心柱断面
stemple ['stempl] *n.* (矿井)梯级横木,巷道横梁,嵌入梁
stench [stentʃ] *n.* 臭气,恶臭; stenchful, *a.*
~ trap (连接便池等的)防臭瓣,存水弯
stencil ['stensl] *n.*; *v.*, -cil(l)ed,-cil(l)-ing; *n.* (镂花)模板,型板,蜡纸; *vt.* 以镂花模版印刷
~ paint 镂空板印花用涂料
~ paper 油印用蜡纸
~ plate (镂花)型版,蜡版
~ sheet 蜡纸
~ led decoration (用镂花模板的)印花装饰
step [step] *n.* 台阶,踏步,梯级,步长; (*pl.*)一段楼梯,一段梯级
~ block 台阶形垫铁
~ board 踏步板,踏面
~ bolt 半圆头方颈螺栓,上杆螺钉
~ brass 轴瓦,轴铜衬
~ -by-step design 逐步设计
~ component (楼梯)梯级
~ footing 阶状基础
~ iron 铁爬梯,铁踏步

~ ladder 阶梯,梯凳,活梯
~ member （楼梯的）梯级
~ profile 阶形断面
~ pyramid 阶梯形金字塔
~ stone 阶梯石级,阶沿石
~ unit （楼梯的）梯级
~ ping iron 金属踏步
~ ping pedestal 台架
~ ping stone 踏步石,楼梯石级
~ ping wood 木楼梯的厚踏步板

step footing

stepped [stept] *a.* 阶梯形的,分级的
~ arch 阶梯形拱
~ footing 阶形底脚
~ foundation 阶形基础
~ hillside house 坡地住宅
~ roof 阶形屋顶
~ stalls type auditorium 阶梯式观众厅
~ stringer 阶形楼梯斜梁
~ tenon 阶梯式榫
~ voussoirs 阶形拱石

stere [stiə] *n.* 立方米(m^3)

stereo ['stiəriəu] *n.* 立体,立体声,立体照片

stereognosis [ˌsteriəg'nəusis] *n.* 形体[实体]感觉

stereogram ['steriəgræm] *n.* 立体图,三维图

stereographic [ˌsteriəg'ræfic] *a.* 立体平画的,立体照相的

stereomodel [ˌsteriə'mɔdel] *n.* 立体模型

stereophonic [ˌsteriə'fɔnik] *a.* 立体声的,立体音响的;
~ effect 立体音响效果
~ television 立体声电视

stereoplanigraph [ˌsteriəplænigrɑ:f] *n.* 立体伸缩绘图仪,精密立体测图仪

stereoplotter [ˌsteriə'plɔtə] *n.* 立体绘图仪

stereoscope ['steriəskəup] *n.* 立体镜,立体显微镜

stereoscopic [ˌsteriə'skɔpik] *a.* 立体的,体视的
~ image 立体像
~ model 立体模型

stereotomy [ˌsteri'ɔtəmi] *n.* 切石法;实体切割术;石雕刻 stereotomist, *n.*; stereotomic, stereotomical, *a.*

sterile ['sterail] *a.* 贫瘠的;消过毒的;无效果的
~ chamber 无菌室
~ working 无菌操作

sterling ['stə:liŋ] *n.* 英国货币;纯银 *a.* 纯粹的
~ white 纯白颜料

stiacciato [stjɑ:t'tʃɑ:təu] *n.* 浅浮雕

stick [stik] *n.*, *v.*, stuck, sticking; *n.* 棒,杆,炸药棒,棒状炸药;(建筑物的)一部分;粘结物 *v.* 粘住,胶粘,伸出 -stickable, *a.*
~ out 突出部分
~ing plaster 橡皮膏
abrasive ~ 油石,磨条

stickability [stikə'biliti] *n.* 粘着性,粘着能力

sticker ['stikə] *n.* 粘着剂;背面有粘胶的标签;尖刀,尖物

sticky ['stiki] *a.*, stickier, stickiest; 粘的,胶粘的; stickily, *ad.*; stickiness, *n.*
~ cement 胶粘水泥
~ clay 胶粘土
~ point 粘点,胶粘点

stiff [stif] *a.* 不易弯曲的,劲性的;刚硬的;密实的,稠的
~ blue clay 硬青黏土
~ brush 硬刷
~ clay 硬黏土
~ concrete 硬稠[干硬(性)]混凝土
~ connection 刚性接头
~ (consistency) concrete 干性混凝土

stiffen

~ (consistency) mortar 干硬性砂浆
~ fissured clay 硬裂纹黏土
~ gel 硬化凝胶
~ girder 加劲梁
~ joint 刚性接头
~ -jointed flat frame 刚接平面框架
~ leaf 密叶饰
~ mix 干硬性混合料
~ mud brick 硬泥砖
~ mud process 砖泥制砖法
~ paste paint 干稠性油漆
~ structure 刚性结构

stiffen ['stifn] *v.* 加劲[强],硬化,稠化
stiffened ['stifnd] *a.* 加劲的,强化的,稠化的
~ arched girder 加劲拱形梁
~ bracket 加劲托座
~ junction 加劲接头
~ shell structure 加劲壳结构
~ trussed arch 加劲桁架拱
~ wire lath 加劲钢丝网

stiffener ['stifnə] *n.* 加劲杆,支肋,刚性构件,增稠剂
stiffening ['stifniŋ] *n.* 加劲,劲化
~ agent 硬化剂,增稠剂
~ angle 加劲角钢
~ beam 加劲梁
~ diaphragm 加劲隔板
~ frame 加劲框架
~ grillage 加劲格栅
~ truss 加劲桁架

stiffness ['stifnis] *n.* 劲度,刚度,硬度,稠度,稳定性
stile [stail] *n.* 门梃,窗梃;竖框,主框条;十字形旋门;横路栅栏;梯磴
~ of elevator 电梯门框
door ~ 门框条
meeting~s (双扇门的) 碰头边梃
still [stil] *n.* 静止,寂静;蒸馏所,酿酒场,蒸馏器; *a.* 静止的,寂静的,静物的
Stillson wrench ['stilsnr'entʃ] *n.* 可调管子钳,活动扳手

stilt [stilt] *n.* 支撑,支材;(架于水上等建筑物之)支柱;耐火垫片;(装窑用的)承坯架
~ed arch 高矢拱,上心拱
~ed vault 高矢穹顶,上心穹顶
sting [stiŋ] *n.* 架杆,探臂式支杆
stink [stiŋk] *n.* 臭味,臭气
~ cupboard 通气橱
~ trap 防臭存水弯,防臭瓣
stipple [stipl] *n.* 点彩饰面,凹凸纹饰面
~ paint 点彩涂料,凹凸纹涂料
stipulation [ˌstipju'leiʃən] *n.* 契约,合同,规定,条款
stir [stə:] *v.*, stirred, stirring, *n.* 搅拌;拌动,搅和;吸取,抽送; stirrable *a.*
~ machine 搅拌机[器]
stirrer ['stə:rə] *n.* 搅拌器
stirring ['stə:riŋ] *n.* 搅拌,搅动,泵送
stirrup ['stirəp] *n.* 钢箍,箍筋,U 形卡,镫形铁件

stirrup

~ bolt 镫形夹螺栓,管夹,扒钉
~ frame 框式机架
stitch [stitʃ] *n.* 缝合,针脚; stitcher, *n.*, stitchlike, *a.*
~ bonding 连续点焊焊接,自动点焊,跳焊
~ brake lining 制动闸边皮
~ rivet 缀合铆钉
~ welding 缝合点焊
stithy ['stiði] *n.*; *pl.* stithies,锻工场,打铁铺;铁砧
stoa ['stəuə] *n. pl.* stoae, stoai, stoas;(希腊建筑中的) 柱廊或拱廊
stochastic [stə'kæstik] *a.* 随机的,不确定的
stock [stɔk] *n.* 原料,备料,存货;股本,股票;台,架,柄,把;饰面砖 *a.* 库存的,普通的,现有的,标准的
~ board 砖模底板
~ brick 普通砖

~ -brick work 普通砖工程
~ core 长条泥芯
~ door 机制门
~ heap 贮料堆
~ house 料仓,库房
~ lock 门外锁
~ lumber 锯成木材
~ mold 砖模
~ room 贮藏室,商品展览室
~ stand 材料台,器材架
~ steel window 标准钢窗
coil ~ 卷材,卷料
feed ~ 原材料
loaded ~ 填料
silent ~ tube 塑料管
stockpile ['stɔkpail] *n.* 储存,存货;资源
stoma ['stəumə] *n. pl.* -mata,-mas;小孔,气孔
stone [stəun] *n. pl.* stones ;石,石料,碑; stonelike, *a.*
~ aggregate 粗骨料
~ arcade 石拱廊
~ ashlar 琢石,方石
~ axe 凿石斧
~ balustrade 石栏杆,石扶手
~ beam 石梁
~ block 石块
~ blue 灰蓝色
~ bolt 底脚螺栓,棘螺栓
~ bond 圬工,砌石
~ building 石砌建筑物,石屋
~ carving 石雕,石刻
~ cladding work 石料镶面工程
~ column 石柱;碎石桩
~ concrete 块石混凝土
~ construction 石结构,石建筑
~ corbel 石支托,石牛腿
~ counterfort 石扶壁,石拥壁
~ cutting 琢石,切石
~ dressing 石料砌面[饰面],琢石面
~ -faced arch 石饰面拱

~ -faced masonry 石镶面圬工
~ facing 石料砌面[饰面,镶面]
~ flag 石板,扁石
~ footing 毛石基础
~ foundation 石基础
~ fragments 碎石,石屑,石片
~ garden 叠石庭园
~ house 石屋
~ lantern 石灯(笼)
~ lifting tongs 石块吊升夹具
~ lime 高钙石,石灰石
~ lined 块石衬砌的
~ lining 块石衬砌
~ lintel 石过梁
~ lintol 石过梁
~ marker 标石
~ mason 石工,泥瓦工
~ mesh construction 铁丝石笼结构
~ monument 石碑,石标
~ mullion window 石直棂窗
~ ornament 石头点缀
~ packing 块石基层,填石
~ paving 石块铺面
~ pier 石墩
~ pillar 石柱,石墩
~ -pine 五针松,石松
~ pitched facing 砌石护面
~ railing 石栏杆,石扶手
~ riprap 乱石堆,抛石
~ saw 石锯
~ sett 小方石(块)
~ setter 砌石工
~ shingle 石板瓦
~ sill 石窗台
~ sill of door 门槛石
~ slab 石板
~ slab cladding 石板衬砌
~ slab facing 石板衬砌
~ slab flooring 石板楼面
~ slab roofing 石板屋面
~ staircase 石阶梯
~ step 石阶

~ surfacing 石料砌面
~ threshold 石门槛
~ tile 石瓦
~ transom （门窗）石横档
~ vault 石穹
~ -vaulted ceiling 石穹屋顶
~ veneer 石料砌面[饰面]
~ wall 石墙，石壁
man-made ~ 人造石
stoneware ['stəunwɛə] *n.* （常用作化工容器或管道的）粗陶器
~ article 粗陶制品
~ ceiling tile 粗陶瓦
~ clay 粗陶土
~ (discharge) pipe 粗陶（排水）管
~ facing 粗陶护面
~ for sewer pipes 粗陶污水管
~ tile 缸瓷砖，缸砖
stonework ['stəunwək] *n.* 石方[砌石]工程；石板
~ decoration 石板装饰
stoney gate ['stəuni geit] *n.* 提升式平板闸门，模造大理石门
stony ['stəuni] *a.* 多石的
~ clay 含石黏土
~ soil 石质土，含石黏土
stool [stu:l] *n.* 凳；内窗台；垫板；托架；便器
~ plate 垫板，窗台找平板
insulation ~ 绝缘座
stoop [stu:p] *n.* 门廊，门阶，门口的平台
stop [stɔp] *v.*, stopped, stopping, *n.* 障碍物；门闩，门止块；制动器；停车站；停止；*v.* 抹平；停止
~ bead 窗止条，钉在门窗框上的压条
~ bracket 挡铁，止动块
~ butt hinge 半开铰链
~ button 停止[制动]按钮
~ cock 止水栓，小龙头
~ dog 挡块，碰停块
~ end piece 端盖
~ gear 停止装置

stop cock

~ joint 嵌固接头
~ level 自动停止水位
~ lever 制动操作杆
~ log 选梁
~ nut 防松螺母
~ pin 止动销
~ plate 止动片
~ ring 止动环
~ signal 停止标志
~ stone 止门石
~ped dado 半开榫槽
door ~ 门制，止门器
stoppage ['stɔpidʒ] *n.* 阻塞；停止；停机
~ period 停工（时）期
stopper ['stɔpə] *n.* 制止器，塞，阀，门挡
~ screw 止动螺钉，紧固螺钉
stopping ['stɔpiŋ] *n.* 停止，制动；填塞（物）；腻子
~ knife 油灰刀，抹灰刀
storage ['stɔridʒ] *n.* 仓库；贮藏所；存储；库容；蓄水
~ basin 蓄水池
~ bin 贮料仓，贮料罐
~ cell 存储单元，蓄电池
~ furniture 贮物用家具
~ room 储藏室
~ wall 壁橱
~ water heater 储水热水器
store [stɔ:] *n.* 商店；仓库，栈房；存储器；储藏
~ closet 贮藏室
~ for samples 样品贮藏间
~ front （商店的）铺面
~ front finish 店面装饰
~ room 贮藏室
~ space 贮藏处，堆料场；存贮空间
storey ['stɔ:ri] *n.* (= story), *pl.* -reys, 楼层；-storeyed, *a.*
~ band 台口线
~ frame 楼层框架
~ height 层高

~ post 楼层柱
single- ~ house 平房
the first ~ (英)第二层,二楼;(美)底层,一楼
storied ['stɔːrid] *a.* 有……层楼的
~ house 楼房
two- ~ house 两层楼房子
storm [stɔːm] *n.* 暴风雨,风暴,(十级)狂风;-stormlike, *a.*
~ door 防风暴门
~ inlet 雨水进口,雨水井
~ louver 防雨百叶窗
~ outfall sewer 暴雨排水管
~ pavement 防暴雨护面
~ -proof building 防风暴建筑物
~ sash 防暴风雨的外重窗
~ sewer 雨水沟,雨水管
~ water flume 雨水槽
~ window 防风窗(双层窗的外层)
story ['stɔːri] *n. pl.* -ries; storm sewer 层楼(= storey);故事,传记,情况
~ height 楼层高度
stout [staut] *a.* 粗大的,结实的,坚[牢]固的
stove [stəuv] *n.*; *v.*, stoved, stoving; *n.* 温室;火炉,电炉; *v.* 烘,烤
~ finish 烘干的油漆
~ heating 火炉供暖
~ plant 温室植物
~ plate 厨房灶台板
~ tile 瓷砖,面砖
gas ~ 煤气炉
stow [stəu] *vt.* 装载;整理仓库;stowable *a.*; stower, *n.*
~ wood 楔木,垫木
stowage ['stəuidʒ] *n.* 贮藏处;装载(物)
~ space 装载空间,有效空间
straddle ['strædl] *n.*; *v.* -dled,-dling; *n.*; *v.*, 双腿分开立或坐; straddler,

n.;straddlingly, *ad.*
~ cutter 双面铣刀
~ scaffold 跨立式脚手架
straggling ['stræɡliŋ] *n.* 分散,离散,误差
straight [streit] *a.* 直的,水平的,连续的,纯净的,光面的;
~ angle 平角
~ arch 平拱
~ bar 直杆
~ beam 直梁
~ bituminous filler 纯沥青填塞料
~ brick 普通砖,标准砖
~ brick lintel 直砖过梁
~ burr 直纹
~ cement 纯水泥(不加掺和料)
~ cement concrete (不加掺和料的)纯水泥混凝土
~ cement mortar 纯水泥沙浆
~ chair 直靠背椅子
~ dovetail 直鸠尾榫
~ down gutter 直通雨水管
~ edge 直尺,直规
~ grain board 直纹木板
~ grained (wood) 直纹理的,直纹(木材)
~ -halved joint 对半接合
~ joint 直缝对接,直线接合
~ joint floor 直拼地板
~ joint tile 平接瓦,对接瓦
~ line block 板式建筑物,直线型建筑物
~ line distribution 直线分布
~ -line graticule 直角坐标网
~ pin 圆柱销
~ pipe 直管
~ scarf joint 直嵌接头
~ sided column 直边柱,棱柱体柱
~ sliding (shutter) door 直线型推拉(百叶)门
~ stair 直跑楼梯,单跑楼梯
~ staircase 直跑楼梯

~ tenon 直榫
~ timber 直木材
~ tongue 直雄榫,直榫舌
~ vault 平穹顶
~ welded joint 对缝焊接
straighten ['streitn] v. 弄平,矫直;整理
~ing device 整直装置
~ing roll 矫直辊
straightener ['streitnə] n. 矫直机
coil ~ 卷材矫直机
straightforward [streit'fɔ:wəd] a. 顺向的,流水作业的
strain ['strein] n. 应变,变形,形变,延伸率,单位伸长;拉紧;滤波; v. 拉紧;过滤
~ clamp 耐拉线夹
strainer ['streinə] n. 应变器,拉杆,拉紧装置,松紧螺扣;滤网,粗滤器
straining ['streiniŋ] n. 应变,张拉,过滤
~ arch 扶拱
~ beam 跨腰梁,拉[系]梁
~ frame 加载架,应变架
~ piece 系杆,拉杆
strait ['streit] n. 海峡
~ flange 窄凸缘
strand [strænd] n. 钢绞线,多芯绞线,软钢绳;(绳索的)股,束;多股电缆线;岸,滨 v. 绞合;搁浅
~ cable 绞合电缆,钢绞线
~ grip 钢绞线的固定夹具
~ rope 绞股绳
~ vise 钢索钳
~ wire 多股绞线
~ing connection 绞接,绞合
stranded ['strændid] a. 多股的,绞合的;搁浅的
~ rope 绞合绳索
~ steel wire 钢绞线[索]
S-trap S形存水弯
~ tube S形存水管

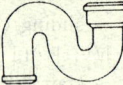
s-trap tube

strap ['stræp] n. 带,条;盖板
~ beam 带形梁
~ bolt 带形螺栓,扁尾螺栓
~ clamp 带夹,卡板
~ cover 条状盖板
~ hanger 檐槽吊钩
~ hinge 带式铰链
~ iron 扁铁,条钢
~ joint 盖板接头
~ lap joint 夹板接合
backing ~ 条状垫板
butt ~ 对接搭板
strapwork ['stræpwək] n. 带箍线条饰,交织凸起带装饰
strass [stræs] n. 有光彩的铅质玻璃,假金刚石
straticulate [strə'tikjulit] a. (成)薄层的,分层的
stratified ['strætifaid] a. 成层的,有层次的
~ material 层状材料
~ mixture 分层混合料
~ plastics 层压塑料
~ soil 层状土,成层土
stratiform ['strætifɔ:m] a. 层状的
stratigraphic a. 地层分离的
~ analysis 地层(分离法)分析
straw [strɔ:] n. 禾秆,稻草,麦秸
~ color 稻草色,淡黄色
~ mat 草席
~ mattress 草席,草垫
~ roofing 稻草屋顶
~ rope 草绳
~ thatch 稻草屋顶
streak [stri:k] n. 条纹,纹理,矿脉;(油漆)起条
~ flaw 条状裂痕
~ line 条纹线,流线
streamlined ['stri:mlaind] a. 流线型的,流水般的;合理的
street [stri:t] n. 街道
~ architecture 街道建筑

~ building 沿街建筑,市区建筑
~ construction 街道建筑
~ crown 街道路拱
~ curb 街道路缘石
~ design 街道设计
~ door 临街大门
~ illumination 街道照明
~ lighting column 路灯柱
~ lighting luminaire 街道照明设备
strength [streŋθ] *n.* 强度;浓度
~ of cement 水泥强度等级
~ rivet 强力铆钉
strengthen ['streŋθən] *v.* 加强,强化
~ing beam 加固梁
~ing corner 加固角
~ing soil 加固土
stressed [strest] *a.* 预加应力的
~ connection 预应力接合
~ menbrane 预应力薄膜
~ -skin construction 加载式结构,薄壳建筑,夹层建筑
stressing ['stresin] *a.* 预加应力的,先张拉的
~ bar 预应力钢筋
~ cable 预应力钢索
~ element 预应力构件
~ reinforcement 预应力钢筋
~ rod 预应力钢筋
~ strand 预应力钢绞线
stretch [stretʃ] *n.;v.* 拉伸,展开
~ forming 张拉成形
~ed connection 预应力接头
stretchability [stretʃə'biliti] *n.* 可伸性
stretcher ['stretʃə] *n.* 顺砌(砖),露侧砖;拉伸器;(薄板)矫直机
~ bond 顺砖砌合
~ brick 顺砌砖
~ face (砖的)侧面
~ forming 拉伸成形法
~ paving 顺砖铺地
stretching [stretʃiŋ] *n.* 张拉,延伸
~ bond 顺砖砌合

~ course 顺砖砌层
~ reinforcement 预应力钢筋
~ screw 张拉螺丝
~ tendon 预应力钢丝束
~ wire 预应力钢丝

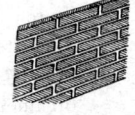
stretcher bond

stretchy ['stretʃi] *a.*, stretchier, stretchiest; 能伸长的,有弹性的; stretchiness, *n.*
~ nylon 弹性尼龙
strew [struː] *vt.*, strewed,strewed or strewn,strewing;撒,铺 strewer, *n.*
~ sand 撒砂,铺砂
stria ['straiə] *n.;pl.* striae;柱沟;柱身凹槽或突筋;条纹,裂纹,擦痕
striate ['straieit] *v.*, -ated,-ating;在……上加线纹
~d column 细沟装饰柱
striation [strai'eiʃən] *n.* 条纹,细沟,擦痕
strickle ['strikl] *n.* 刮平,磨光;刮平器;磨石,油石
~ board 造型刮板
strict [strikt] *a.* 精确的,严格的;-strictness, *n.*
striga ['straigə] *n.* 柱槽
strike [straik] *n.;v.*, struck, struck or stricken,striking; *n.; v.* 拆模勾缝,弹粉线;(地层)走向;打击,敲
~ board 刮板
~ line 走向线
~ -off 整平,刮平
~ -off beam (混凝土摊铺机的)整平梁
~ -off blade 整平板
~ -off board 刮板
~ -off method 刮平法
~ plate 锁舌片,门鼻子
striker ['straikə] *n.* 大铁锤;撞针
striking ['straikiŋ] *n.* 拆模;刮平;勾缝;放电
~ board 样板,刮平板

strikle

~ knife 三棱刮刀
~ of arch 拆除拱架
~ of shuttering 拆模
~ -off 勾缝,刮平
~ -off lines 勾划线
~ plate 拆除支架板;(门窗锁的)锁舌片
~ times 拆模时间

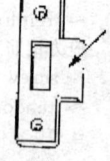
strike plate

strikle ['strikl] *n.* =strickle

string [striŋ] *n.* 楼梯斜梁;细绳,弦,索

~ board 楼梯斜梁侧板,楼基盖板
~ course (外墙的)带状线脚,腰线;层拱
~ level 悬挂式水准管
~ line 放样麻线
~ lining 麻线放样
~ measurement 绳测

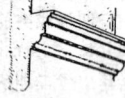
string course

~ of casing 套管柱
~ packing 线形填缝
~ pendulum 线摆
~ wall (支承楼梯踏步的)梯墙
~ wire 钢丝,钢弦
~ -wire concrete 钢弦[丝]混凝土

stringer [striŋə] *n.* 纵梁,桁条;楼梯斜梁,楼梯基;纵向加强肋

~ bracing 纵梁斜撑
~ bracket 纵梁托架
~ wall (支承楼梯踏步的)梯墙
spiral ~ 螺旋形楼梯边梁

strip [strip] *n.*; *v.*, stripped or stript, stripping; *n.* 板条,带状物,细长片; *v.* 剥离

~ bar 窄扁钢
~ ceiling 板条顶棚
~ film 可剥膜
~ floor 狭条木地板
~ flooring 条木地板
~ footing 条形基脚
~ forms 条形模板
~ foundation 条形基础
~ iron 带钢
~ lamp 管状灯,灯管

~ line 传送带,带状线
~ pilaster 狭条扁柱
~ steel 带钢,钢带
apron ~ 盖缝条
corner ~ 压缝条
edging ~ 盖檐条
felt ~ 油毡条
flush beam ~ 板梁
sealing ~ 密闭条,止水条
terminal ~ 接线条,端子板
washleather ~ 软皮带
wear ~ 防磨条,防磨损板

stripe [straip] *n.*; *v.*, striped, striping; *n.* 车道;条纹,镶条; *v.* 划线

~ figure (木材的)条纹

striplight ['striplait] *n.* 带形照明器,长条状灯

strippable ['stripəbl] *a.* 可剥离的

~ coating 可剥性涂层
~ paint 可剥性油漆

stripper ['stripə] *n.* 脱模机;涂层消除剂

~ plate 脱模板,挤压板
fixed ~ 固定模板
paint ~ 洗漆剂

stripping ['stripiŋ] *n.* 脱模;涂料剥除;表土剥离,清基

~ knife 剥离刀
~ line 清基线
~ machine 脱模机
~ strength 脱模强度
~ test 剥落试验

strix [striks] *n. pl.* striges, strixes;柱上凹槽间的隆起线,柱沟

stroboscopic [strəubə'skɔpik] *a.* 频闪的

~ illumination 频闪照明

strong [strɔŋ] *a.* 坚固的;有力的;强烈的;

~ clay 强黏土
~ concrete 高强度混凝土
~ mortar 高强度砂浆

struck [strʌk] *a.* 刮平的
~ joint 刮平缝
structural ['strʌktʃərəl] *a.* 结构的;构造的;建筑的;-stucturally, *ad.*
~ adhesive 结构胶粘剂
~ architecture 房屋建筑
~ attachment 结构附件
~ clay tile 空心砖
~ clay tile partition 自承重的空心砖隔墙
~ concrete panel 结构混凝土铺板
~ concrete topping 结构混凝土面层
~ construction 承重建筑
~ design of building 建筑结构设计
~ design drawing 结构设计图
~ detail 结构细部
~ diagram 结构简图
~ division 结构隔板
~ drawing 结构图,构造图
~ features 结构特征,构造细部
~ field concrete 现浇结构混凝土
~ glass 大块建筑用玻璃
~ glass panel 结构玻璃板
~ glue 结构粘合剂
~ in situ cast concrete 现浇结构混凝土
~ joint 结构结点,结构接缝
~ lumber 建筑用木材
~ material 结构[建筑]材料
~ member 结构构件[杆件]
~ partition wall 结构隔墙
~ profile 结构型体
~ quality 结构性质
~ space 结构空间
~ steel 结构钢
~ stone 建筑用石
~ timber 建筑用木材
~ tube 结构管材
~ wall 结构墙,承重墙
~ -wall tile 承重墙用空心砖
~ wood 建筑用木材

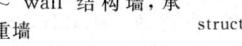
structural wall

structurals ['strʌktʃərəlz] *n.* 建筑用的重型钢架、钢梁等
structure ['strʌktʃə] *n.* 构造,结构,建筑物
~ bore (地质)构造钻孔
~ nose 建筑物突出部分
~ of wood 木材结构
~ plane 构造面
~ preservation coat 建筑物防护层
antiseismic ~ 抗震结构
brick ~ 砖结构
composite material ~ 复合材料结构
concrete ~ 混凝土结构
flexible ~ 柔性结构
frame ~ 框架结构
laminated ~ 层状结构
load bearing ~ 承重结构
load carrying ~ 承重结构
prefabricated ~ 预制装配结构
prestressed concrete ~ 预应力混凝土结构
reinforced concrete ~ 钢筋混凝土结构
roof ~ 屋顶结构
sandwich ~ 夹层结构
shell ~ 壳体结构,壳式建筑
space ~ 空间结构
space frame ~ 空间框架结构
space truss ~ 空间桁架结构
spatial grid ~ 空间网架结构
spheroidal ~ 球形结构
steel ~ 钢结构
steel and concrete composite ~ 钢筋[管]混凝土组合结构
suspension ~ 悬挂结构
thin shell ~ 薄壳结构
underground ~ 地下结构
wood ~ 木结构
strum [strʌm] *n.* (吸入)滤网
strut [strʌt] *n.* 支柱,支撑,压杆
~ beam 支梁
~ bracing 支撑,压杆

~ of roof truss 屋架斜撑
~ rail 横撑
~ tenon 撑榫
angle ~ 角铁支撑
bracing ~ 支柱,斜撑
diagonal ~ 斜撑
shock ~ 减振柱
strutted ['strʌtid] *a.* 支撑式,撑杆式
~ roof 撑杆式屋顶
strutting ['strʌtiŋ] *n.* 支撑物,加固件
~ board 支撑板
~ piece 支撑件
stub [stʌb] *n.* 粗短支柱,柱墩;树桩
~ -bar 露头钢筋,预埋连接钢筋
~ cut-off 防渗矮墙
~ line 短截线
~ mortise(m) 短粗木榫眼
~ nail 短粗钉
~ pipe 短管
~ pole 短木柱
~ post 短粗柱
~ supported 短线[轴,管]支撑
~ teeth 短齿
~ tenon 粗榫
~ wall 短墙
sleeve ~ 套管短柱
stuc [stʌk] *n.* 砌石灰浆
~ mixture 砌石灰浆
~ stuff 砌石灰浆
stucco ['stʌkəu] *n. pl.* -coes,-cos;粉墙灰泥,灰墁,拉毛粉刷;stuccoer, *n.*
~ ceiling 抹灰顶棚
~ cornice 拉毛粉饰檐
~ embossed surface 外墙面上浮雕装饰
~ -encrusted wall 灰泥抹面墙
~ finished building 抹灰饰面建筑,混水墙建筑
~ pattern 拉毛刷型(板)
~ work(er) 拉毛粉刷工(人)
stuccoist ['stʌkəuist] *n.* 拉毛粉刷工人
stuccowork ['stʌkəuwəːk] *n.* 拉毛粉刷工作
stud [stʌd] *n.* 支柱,壁骨,板墙筋;门窗梃;房间净高;双头螺栓
~ bolt 双头螺栓,地脚螺栓,柱头螺栓
~ dowel 合缝钉,暗榫
~ gun 射钉枪
~ -link chain 日字环节链
~ nut 柱螺栓螺母
~ partition 木柱隔墙,立筋隔断
~ pin 大头钉
~ stave 标桩
~ wall 木柱隔墙
~ work 安装板墙筋
binder ~ 接合柱螺栓
contact ~ 接触钉
locating ~ 定位销
shock absorbing ~ 减振支柱
studding ['stʌdiŋ] *n.* 间柱,支承;灰板墙筋(材料)
studio ['stjuːdiəu] *n. pl.* -dios;画室;工作室;播音室;只有单房及厨房和浴室的公寓住宅
~ apartment 小型公寓房间
~ broadcast 室内广播
~ couch 可以当床用的沙发
studwork ['stʌdwək] *n.* 有间柱小桁的建筑物
study ['stʌdi] *n. pl.* studies, *v.*, studied, studying; *n.*; *v.* 学习,研究; studiable, *a.*; studier, *n.*
~ room 书房
stuff [stʌf] *n.* 建筑材料,原料,填料,资料;混合涂料;织物
~ goods 呢绒,毛织品
inch ~ 一英寸厚的木板
thick ~ 厚(四英寸以上)木板
stuffing ['stʌfiŋ] *n.* 填塞料
~ box 填料盒[函]
~ gland 填料盒压盖,密封套
stuffy ['stʌfi] *a.*, stuffier, stuffiest; 不通气的,闷热的; stuffily, *ad.*; stuffiness.
~ room 不通风的房间,闷热房间

stull [stʌl] *n.* 支柱;横梁,横撑,顶梁
~ -divided 横撑分隔的
stump [stʌmp] *n.* 短柱,柱墩;树桩;
-stumplike, *a.*
~ mortise 浅榫眼
~ tenon 短粗榫
stunt [stʌnt] *vt.* 阻碍
~ -head (浇灌混凝土时用的)堵头板
stupa ['stu:pə] *n.* 塔,佛塔,浮图,窣堵波
~ base 塔基
~ shrine 塔庙
~ stylobate 塔柱座
stylar ['stailə] *a.* 尖笔状的,尖的,针状的
style [stail] *n.* 风格,格式,类型
~ -book 样本,式样书
~ character 风格特征
~ characteristic 风格特征
~ device (建筑上的)特点
Style Empire 帝国式
~ feature (建筑上的)特点
~ of ornamentation 装饰风格
national ~ 民族风格
the Gothic ~ of architecture 哥特式的建筑
styliform ['stailifɔ:m] *a.* 尖的,针状的
~ ornament 针状装饰
stylistic [stai'listik] *a.* 风格上的;stylistically, *ad.*
~ conception 风格观念
~ development 风格形成[演变]
~ feature 风格特征
stylization [ˌstailai'zeiʃən] *n.* (风格的)仿效,因袭
stylobate ['stailəbeit] *n.* 柱基,柱座
stylolite ['stailəlait] *n.* 缝合岩面;(石灰石等中的)小石柱
stylolitic [stailə'litik] *a.* 柱状的
~ structure 柱状构造

stylobate

styrene ['stairi:n] *n.* =styrol 苯乙烯
~ foam 泡沫聚苯乙烯
~ mat 苯乙烯垫层
~ plastic 苯乙烯塑料
~ polymer 苯乙烯聚合物
sub [sʌb] *n.* 地道;代用品;下属,局部,副,微
~ floor 底层地板,板面底层
~ joint 辅助接头,副接头
subacid ['sʌb'æsid] *a.* 微酸性的;subacidly, *ad.*; subacidness, *n.*
subaerial ['sʌb'ɛəriəl] *a.* 地面的,陆上的,露天的;subaerially, *ad.*
subalkaline [sʌb'ælkəlain] *a.* 略带碱性的
subangular [sʌb'æŋgjulə] *a.* 略带棱角的,半多角形的(砂)
subaquatic ['sʌbə'kwætik] *a.* 水下的,水底的,适于水下的
~ concreting 水下浇筑混凝土
subarch [sʌb'ɑ:tʃ] *n.* 副拱,子拱
subarcuation [sʌb'ɑ:ku:ʃən] *n.* 副拱结构,该结构中的副拱
subassembly ['sʌbə'sembli] *n.* *pl.* -blies;分部装配;组件
subbase ['sʌbˌbeis] *n.* 基础下卧层,底基层;(数学)子基;subbasal, *a.*
~ course 底基层,基础下卧层
~ course drainage 底基层排水
~ course material 底基层材料
subbasement ['sʌbˌbeismənt] *n.* 底层下的地下室,下层地下室
subbeam ['sʌb'bi:m] *n.* 副梁,次梁
subbox ['sʌbɔx] *n.* 小格子
subcellar ['sʌbˌselə] *n.* 下层地下室
subchannel *n.* 分流道,支[辅助]通道
subcontract ['sʌb'kɔntrækt] *n.* 分包[转包]契约[合同]
subcylindrical ['sʌbsi'lindrikə] *a.* 接近[近似于]圆柱形的
subdivide [sʌbdi'vaid] *v.*, -vided, -viding;再分,细分;subdividable, *a.*
~d panel 再分节间,分格镶板

subdividing partition 分块隔板
subdivision [sʌbdi'viʒən] *n*. 再分，细分；分格；subdivisional, *a*.
window ~ 窗分格
subdue *vt*. 使缓和，使柔和
~ light 柔光
suber [su'ber] *n*. 软木
subface ['sʌbfeis] *n*. 底面
subfloor ['sʌbflɔː] *n*. 底层地板，桥面底层，毛地板
subflooring ['sʌbflɔːriŋ] *n*. 底层地板（用的材料）

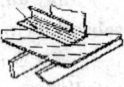
subflooring

subforeman ['sʌb'fɔːmən] *n*. 副领班，副工长
subfoundation [sʌbfaun'deiʃən] *n*. 基础底层
subframe [sʌbfreim] *n*. 下支架，辅助构架
subintradosal block [sʌbintrə'dɔsəl blɔk] *n*. 内拱砌块
subjacent [sʌb'dʒeisənt] *a*. 在下面的，下层的；subjacency, *n*.
~ bed 下卧层
~ support 下面的支承
subject ['sʌbdʒikt] *n*. 学科，主题，对象；*a*. 服从的，易受……的；*v*. 使服从，使经受
~ construction cost 主体工程费
subjective [sʌb'dʒektiv] *a*. 主观的；subjectively, *ad*.; subjectiveness, *n*.
sublayer ['sʌb'leiə] *n*. 下层，底层
~ laminar 次层流
subletting *vt*. 转租，转包
submain ['sʌbmein] *n*. 次干管，辅助干线；地下总管
~ line 副干线
submarine ['sʌbməriːn] *a*. 海底的，水下的；*n*. 潜水艇
~ cable 海底电缆，水下电缆
~ conduit 海底管道

~ light 水下照明
~ -line 海底管线[线路]
~ pipe column foundation 水下管柱基础
~ pipe line 海底管道
~ work 潜水工作，水底工程
submember [sʌb'membə] *n*. 副构件，辅助构件
submerged [səb'məːdʒd] *a*. 水中的，浸没的
~ condition 浸水[浸没]状态
~ displacement 排水量
~ earth 漫水土层
~ meter 淹没式水表
~ orifice 淹没孔口
~ outlet 埋没式出水口
~ pipeline 海底管道，水下管线
~ structure 水下建筑物
~ type construction joint 暗式施工缝
~ type contraction joint 暗式收缩缝
submergible [səb'məːdʒibl] *a*. 可浸入水中的，淹没的
submersible [səb'məːsəbl] *a*. 淹没的，可浸入水中的
~ roller 过水滚筒
submersion [səb'məːʃən] *n*. 淹没，沉没，潜在水中
~ -proof 可潜水的，防水的
subminiaturization ['sʌbˌminjətʃərai'zeiʃən] *n*. 超小型化
suboffice [sʌb'ɔːfis] *n*. 分办事处，分局，支局
subordinate [sə'bɔːdnit] *a*. 次要的，辅助的，附属的
~ building 附属建筑
~ entrance 次要出入口
subpanel ['sʌbpænl] *n*. 副面板，辅助板，底板
subparagraph [sʌb'pærəgrɑːf] *n*. 小段，小节，附属条款

subplan ['sʌbplæn] *n.* 辅助方案
subplate ['sʌbpleit] *n.* 底板,连接板
subpost ['sʌbpəust] *n.* 副柱,小柱
subprincipal *a.* 贴近构架主材的系梁
subpurlin *n.* 副檩条,副椽
subquadrate [sʌb'kwɔdrit] *a.* 近正方形的,正方而带圆角的
subquality products 不合格产品,次级品
subsample ['sʌbsɑ:mpl] *n.*;*v.*,-pled,-pling;*n.* 子样本;*v.* 两次抽样
subscriber [səb'skraibə] *n.* 订用,用户
subscription [səb'skripʃən] *n.* 签名;预订
subseal [sʌb'si:l] *v.* 基础处理;封底
～ing treatment 封底处理
subsidence [səb'saidəns] *n.* 沉降,沉陷,下沉,沉淀;消退
～ of a building 建筑物的沉陷
～ skylight (window) 下沉式天窗
subsidiary [səb'sidjəri]

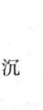
subsidence skylight

a.;*n.*;*pl.* -aries;*a.* 辅助的,次要的,补充的;*n.* 辅助物;子公司,附属机构
～ beam 辅助梁
～ building 辅助[附属]建筑
～ business center 辅助商业中心
～ company 子[附属]公司
～ condition 附加[辅助]条件
～ conveyer 辅助输送机
～ criteria 附则
～ drain 排水支沟,集水暗管
subsiding [səb'saidiŋ] *n.* 下沉
～ ground 沉降[下陷]地面
subsidize ['sʌbsidaiz] *vt.*,-dized,-dizing;资助,津贴;subsidizable, *a.*
～d apartment 公助的住房
subsill ['sʌbsil] *n.* 窗框托,副槛
subsoil ['sʌbsɔil] *n.* 下层土,地基底土; 底土;*v.* 掘松或翻起底土
～ compacting 下层土压实
～ drain 地下排水沟
～ drainage 地下排水
～ puddling 基土捣固
subspan ['sʌbspæn] *n.* 子[部分]跨度
substance ['sʌbstəns] *n.* 物质,材料;实质,本质,要义
anisotropic ～ 各向异性物质
antiknock ～ 抗爆剂
foreign ～ 杂质
isotropic ～ 各向同性物质
magnetic ～ 磁性材料
solid ～ 固态物质
substandard [sʌb'stændəd] *a.* 低于标准的,次等的;*n.* 低标准
～ cement 水泥次品,低标号水泥
substantial [səb'stænʃəl] *a.* 本质的,真实的,重要的,牢固的,实体的
～ pillar 实体柱
substantiate [səb'stænʃieit] *vt.* 证实,证明,使具体化
substation [sʌb'steiʃən] *n.* 变电所[站];分局[所,站]
～ of district heat supply network 集中供热系统分站
outdoor ～ 室外变电站
substitute ['sʌbstitju:t] *n.* 代替,替换,代用品
～ goods 代用品
～ material 代用材料
substitution [sʌbsti'tju:ʃən] *n.* 代替,代换
substope ['sʌbstəup] *n.* 分段工作面
substore [sʌb'stɔ:] *n.* 辅助仓库
substrata ['sʌb'streit] *n. pl.* substratum;下层,底层
～ formation 下层[底层]构造
substrate ['sʌbstreit] *n.* 基片,衬底,感光胶层,基质;地层,基层
substratum ['sʌb'strɑ:təm] *n.* 下层地层,下卧层,基层;基底
～ for waterproofing 防水底层

substruction [sʌb'strʌkʃən] *n.* 下部结构,基础

substructure ['sʌbstrʌktʃə] *n.* 下部结构,底层结构,基础;子结构
~ work 下部结构工程

substrut ['sʌbstrʌt] *n.* 副撑

subsurface [sʌb'sə:fis] *a.* 地面下的,表面下的; *n.* 下层面
~ boring 地下钻探
~ coverage 地下覆盖段
~ drainage 地下排水
~ drainage check 暗沟排水节制闸
~ earth exploration 地下土质查勘
~ engineering 地下工程
~ material 下层材料
~ structure 下层[地下]结构

subterrane ['sʌbtərein] *n.* 下层,地下室,洞穴

subterranean [ˌsʌbtə'reinjən] *a.* 地下的,隐藏的,秘密的
~ chamber 地下室
~ heat 地热
~ line 地下线路[管路]
~ network 地下管线网,地下道路网
~ outcrop 掩盖露头
~ room 地下室
~ work 地下工程

subterraneous [ˌsʌbrə'reinjəs] = subterranean;地下的,隐藏的,秘密的

subterrestrial [ˌsʌbtə'restriəl] *a.* 地下的

subtle ['sʌtl] *a.* 精细的,微妙的,稀薄的;

subtractive [sʌb'træktiv] *a.* 减法的,负号的
~ colour mixture 减色混色法

subtranslucent [ˌsʌbtrænz'lu:snt] *a.* 微透明的

subtransparent [sʌbtræns'pεərənt] *a.* 半透明的

subtriangular [sʌbtrai'æŋgjulə] *a.* 近似三角形的

subtriple ['sʌbtriple] *a.* 三分之一的

sub-truss ['sʌbtrʌs] *n.* 支撑桁架

subtype ['sʌbtaip] *n.* 子[分]型,辅助型

subulate ['sju:bjulit] *a.* 锥形的,钻状的

suburb ['sʌbə:b] *n.* 市郊,郊区
~ populated area 市郊居住区
~ residential area 市郊居住区

subvertical [sʌb'və:tikəl] *n.* 副竖杆,辅助竖杆
~ member 副竖杆

subwalk ['sʌbwɔ:k] *n.* 人行隧道

subwater ['sʌbwɔtə] *a.* 水下的

subway ['sʌbwei] *n.* 地下铁道;地下管道,地道
~ crossing 地下交叉口;地下人行过道
~ station 地铁车站

subzone ['sʌbzəun] *n.* 分区,小区

succeeding [sək'si:diŋ] *a.* 随后的,接连的
~ lift 后继浇筑层

successive [sək'sesiv] *a.* 连续的,顺序的,逐次的
~ courses 邻接层
~ cuts 连续铲挖,连续切割
~ layers 相继各层,逐层

suck [sʌk] *v.* ;*n.* 吸入,抽吸

sucker ['sʌkə] *n.* 吸管,进油管,吸泥管

sucking ['sʌkiŋ] *a.* 吸的,抽的
~ tube 吸筒

suction [sʌkʃən] *n.* 吸收,抽空;真空度,吸力
~ and force pump 吸压泵
~ apparatus 吸入装置
~ attachment 吸附件,吸管接头
~ bend 吸水管弯头
~ blower 吸风机,吸入式通风机
~ check valve 吸入单向阀
~ cleaner 吸尘器
~ cock 抽气旋塞
~ conduit 吸管
~ cutter dredge 旋浆吸泥机
~ fan 吸[抽]风机

~ filter 吸滤器,真空滤池
~ flue 吸烟道
~ grid 吸水网
~ head 虹吸水头,吸升水头
~ hose 吸水软管,真空皮管
~ overfall 虹吸溢水道
~ pipe 抽[吸]入[水]管
~ plant 吸尘装置,抽吸泵站
~ pump 抽吸泵,抽水机
~ rate (砖的)吸水率
~ sweeper 街道吸尘器,真空除尘器
~ tube 吸管,吸筒
~ valve 吸入[水]阀
~ ventilation 吸入通风
~ water column 吸力[负压]水柱

suctorial [sʌk'tɔːriəl] *a.* 吸附的
sudatorium [sjuːdə'tɔːriəm] *n.* 蒸汽浴[室]
sudden ['sʌdn] *a.* 突然的,骤加的
~ contraction 突然收缩
~ drawdown 骤降
~ settlement 突然沉陷

suffuse [sə'fjuːz] *vt.* 充满,弥漫
suggestion [sə'dʒestʃən] *n.* 暗示,提示,启发,建议
suitability [sjuːtə'biliti] *n.* 适合,相配,适用
suitable ['sjuːtəbl] *a.* 适合的,合宜的
suitcase ['sjuːtkeis] *n.* 手提箱
suite [swiːt] *n.* 套房;随员;一套(家具)
~ of apartments 一套房间
~ of furniture 一套家具
~ of room 整套房间
~ room 套间房
WC ~ 厕所设备

sulcate ['sʌlkeit] *a.* 有平行深槽的,有槽的,有裂缝的
sulcus ['sʌlkəs] *n.* 沟,槽,裂缝
sulfate ['sʌlfeit] *n.* 硫酸盐,硫酸酯
~ of copper 硫酸铜,胆矾
~ of iron 硫酸铁,绿矾
~ of lime 钙类硫酸盐
~ -resistant cement 抗硫酸盐水泥

~ sludge 硫酸污泥
sulfoaluminate [sʌlfəuə'ljuːmineit] *n.* 硫铝酸盐
~ cement 硫铝酸盐水泥(即膨胀水泥)

sulfur ['sʌlfə] *n.* 硫(S),硫磺
~ -aggregate 含硫集料
~ band (钢材)硫带,硫偏摺带
~ cement 硫磺胶合剂
~ concrete 硫磺混凝土
~ crack 硫裂(一种焊接缺陷)
~ dioxide 二氧化硫
~ impregnated concrete 融硫混凝土
~ trioxide 三氧化硫

sulfuric [sʌl'fjuːrik] *a.* 硫的
~ cement 硫磺胶泥

sum [sʌm] *n.* 总和,总数
~ total 总额[计,数]

summary ['sʌməri] *n.* 摘要,概略,一览;*a.* 简明的,概括的;总计的
~ of reinforcement 钢筋明细表
~ punch 计总穿孔
~ sheet 观察记录表
~ statistics 概括统计量,主要统计指标

summation [sʌ'meiʃən] *n.* 求和;总计;
summer ['sʌmə] *n.* 大梁,檩条;柱顶石,楣石;加法器;夏季

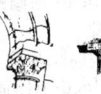
summer

~ beam 大梁
~ house (公园中的)凉亭
~ stone 柱顶石
~ wood 夏材,大木材
brace ~ 支撑梁

summing ['sʌmiŋ] *n.* 合计,总计,摘要
summit ['sʌmit] *n.* 顶点,极点;峰值;凸处
~ curve 顶[凸]曲线

sump [sʌmp] *n.* 集水坑,集水井;贮油槽

～ basin 集水池
～ tank 贮水槽
sun [sʌn] n. 日,太阳
～ blind 百叶窗,遮帘
～ crack 干裂,晒裂
～ dried drick 晒干砖坯
～ -glass 有色玻璃,遮光玻璃
～ lamp 日光灯
～ light 日光
～ louver 遮阳板
～ parlour 向阳间,日光浴室
～ porch 阳台
～ -proof 耐晒的,不透日光的
～ -proof paint 耐晒漆
～ screen 日光屏,遮阳百叶板,遮阳栅
～ shade 遮阳罩,遮阴棚
～ shield 遮阳板
～ snake S 形装饰花纹
～ steel 压花钢板
～ visor 遮光板
sunbaked ['sʌnbeikt] a. 晒干的
～ soil 晒干土
sunbeam ['sʌnbi:m] n. 日光,日光束
sunbreaker ['sʌnbreikə] n. 百叶窗,遮阳板
sunk [sʌŋk] a. 沉没的,水底的,凹下去的,埋头的
～ bead 凹圆形线脚;凹焊缝
～ draft 房角埋石
～ face 下凹面
～ fillet 凹线条
～ fillet mould 平条埋嵌线饰
～ foundation 沉埋式基础
～ moulding 凹沉线条[饰条]
～ panel 嵌进镶板,藻井
～ relief 凹凸浮雕
～ rivet 埋头铆钉
～ screw 埋头螺丝
sunken ['sʌŋkən] a. 沉没的,水底的,埋头的
～ tube 沉管

sunk fillet

sunroom ['sʌnrum] n. 日光浴室
sunshine ['sʌnʃain] n. 太阳辐射,日照
superactivity [sju:pərək'tiviti] n. 超活性
superadiabatic [sju:pərədiə'bætik] a. 超绝热的
superalloy [sju:pər'æləi] n. 高级合金;超耐热不锈钢
supercapillary [sju:pəkə'piləri] a. 超毛细的
～ interstice 超毛细空隙
～ percolation 超毛细渗透
super-capital [sju:pə'kæpitəl] n. 副柱头,拱基
supercarbon [sju:pə'ka:bən] a. 超碳的
～ steel 超碳钢
supercement [sju:pəsi'ment] n. 超级水泥
supercolumnar [sju:pəkə'lʌmnə] a. 重列柱式
supercolumniation [sju:pəkəlʌmni'eiʃən] n. 重列柱
super-compactor [sju:pəkəm'pæktə] n. 重型压实机
superconductivity [sju:pəkəndʌk'tiviti] n. 超导性
super-conductor [sju:pəkən'dʌktə] n. 超导体
supercooled [sju:pə'ku:ld] a. 过冷却的
～ soil 过冷土
superduty [sju:pə'dju:ti] a. 超级的,超重型的
Super Dylan 高密度聚乙烯
superelasticity [sju:pərelæs'tisiti] n. 超弹性
superelevation [sju:pəreli'veiʃən] n. 超高
superette ['sju:pə'ret] n. 小型自动售货商店
superface ['sju:pəfeis] n. 顶面
super-fast-setting ['sju:pə'fa:st'setiŋ] a. 超级快凝的
～ cement 超级快凝水泥

superficial [sjuːpəˈfiʃəl] *a.* 表面的,地面的,外部的
~ coat 表层
~ compaction 表面压实
~ fold 表层摺曲
~ layer 表层
~ treatment of timber 木材表面处理

superficies [ˌsjuːpəˈfiʃiːz] *n.* 表面,表面积;外观;地上权;地上建筑物

superfine [ˈsjuːpəˈfain] *a.* 极细的;特级的
~ aggregate 极细骨料
~ cement 超细水泥
~ filler 极细填充物
~ powder 极细粉末
~ sand 超细砂

superfines [ˈsjuːpəˈfainz] *n.* 超细粉末 (<10υ)

superfinishing [ˈsjuːpəˈfiniʃiŋ] *n.* 超精加工

superfoundation [ˈsjuːpəfaunˈdeiʃən] *a.* 基础以上的
~ structure 基础以上的建筑物

superhard [ˈsjuːpəhɑːd] *a.* 超硬的

superheavy [ˈsjuːpəhevi] *a.* 超重的

superimpose [ˌsjuːpəimˈpəuz] *vt.*, -posed,-posiong;叠加;加在上面;

superimposed [ˌsjuːpəimˈpəuzd] *a.* 叠加的,加于其上的
~ arches 层叠拱
~ layer 叠加层

superinsulator [ˌsjuːpəˈinsjuːleitə] *n.* 超绝缘体

superintendent [ˌsjuːpəinˈtendənt] *n.* 管理员;指挥人;总段长;所长;监造师

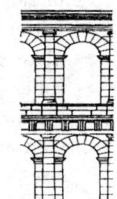
superimposed layer

superior [sjuːˈpiəriə] *a.* 高级的,优等的;在上的
~ limit 上限

supermatic [ˌsjuːpəˈmætik] *a.* 完全自动化的,高度自动化的

superminiature [ˌsjuːpəˈminjətʃə] *a.* 超小型的

supernatant [ˌsjuːpəˈneitənt] *a.* 浮在表面的,浮于上层的
~ layer 清液层

superperformance [ˌsjuːpəpəˈfɔːməns] *n.* 超级性能

superplasticizer [ˌsjuːpəˈplæstisaizə] *n.* 高效塑化剂

superpolymer [ˌsjuːpəˈpɔlimə] *n.* 高聚合物

super-rapid hardening [ˈsjuːpəˈræpid ˈhɑːdniŋ] *a.* 超快硬化的
~ cement 超快硬水泥

superscript [ˈsjuːpəskript] *n.* 上标,上角标

supersede [ˌsjuːpəˈsiːd] *vt.*, -seded, -seding;替代,更换,废弃

superstratum [ˌsjuːpəˈstreitəm] *n. pl.* -ta, tums;上层,覆盖层

super-strength [ˌsjuːpəˈstreŋθ] *n.; a.* 超强度(的)

superstruction [ˌsjuːpəstˈrʌkʃən] *n.* 上部结构,上层建筑

superstructure [ˈsjuːpəˌstrʌktʃə] *n.* 上部结构,上层建筑
~ work 上部结构工程

supersulphated [ˌsjuːpəsʌlˈfeitid] *a.* 高硫酸盐的
~ cement 高硫酸盐水泥
~ slag cement 石膏矿渣水泥

superterranean [ˌsjuːpətəˈreiniən] *a.* 地上的,地表的,架空的,天上的

superthermal [ˌsjuːpəˈθəːməl] *a.* 超热的

supervarnish [ˌsjuːpəˈvɑːniʃ] *n.* 超级清漆,桐油清漆

supervise [ˈsjuːpəvaiz] *v.* 监督,管理,控制,操纵
~ing architect 监理建筑师
~ing engineer 监理工程师

supervision [ˌsjuːpəˈviʒən] *n.* 监督,管理,领导,指导
～ of works 施工管理
～ program 监控程序
construction ～ 工程[施工]监督
project ～ 工程质量管理
site ～ 现场质量管理

supervisor [ˈsjuːpəvaizə] *n.* 管理人员,导师;管理程序
～ mode 管理方式
site ～ 工地负责人

supplement [ˈsʌplimənt] *n.* 增补,添加,补充物,附录;补角
～ of an angle 补角

supplemental [ˌsʌpliˈmentəl] = supplementary
～ area 公寓辅助面积(包括大厅、会议室、卸货台、游泳池等)

supplementary [ˌsʌpliˈmentəri] *a.* 补充的,附加的,补角的
～ direction sign 辅助指路标志
～ means 辅助手段[工具]
～ valve 辅助阀门

supplier [səˈplaiə] *n.* 供给者,供应厂商

supply [səˈplai] *n. pl.* -plies, *v.*, -plied, -plying; *n.; v.* 供应,供给
～ air duct 供气管道
～ air outlet 送风口
～ and disposal services 上下水道设备
～ cistern 给水池
～ conduit 进[给]水管道
～ equipment 供水设备
～ failure 断电
～ fan 进气风扇
～ flume 供水槽
～ grille 进气格栅,进气花格板
～ hose 供水软管
～ line 供[给]水管线,供应线
～ main 给水总管,供应总管,输电干线
～ network 供应[供电]网络

supply grille

～ of materials 材料供应
～ pipe 供水管
～ riser 供水立管
～ transformer 电源变压器
power ～ 动力供应,电[能]源
water ～ system 供水系统

support [səˈpɔːt] *n.; v.* 支持,支撑,支座
～ bar 支杆
～ base 柱基
～ bracket 支承牛腿,支托
～ cap 柱帽
～ connection 柱连接
～ design 支柱设计
～ dimension 支柱尺寸
～ facing 支柱饰面
～ footing 支柱基础
～ forms 支柱模板
～ foundation 支柱基础
～ frame 支承框架
～ guide 支架导杆
～ hardware 支持硬件
～ head 柱帽
～ height 支柱高
～ ing floor joists 支承地板搁栅
～ leg 支柱,支脚
～ of ridge purlin 脊檩支承
～ pattern 支承方式
～ plate 支承板
～ roller 支承滚柱,支承轮
～ section 支柱截面
～ shuttering 支柱模板
～ size 支柱尺寸
～ spacing 支柱间距
～ stiffness 支柱刚度
～ strength 支柱强度
～ structure 支承[下部]结构
～ -to-floor connection 柱板连接
～ -to-footing connection 柱基连接
～ -to-support connection 柱柱连接
～ unit 支承元件,单体支柱

support footing

~ width 柱宽
fixed ~ 固定支座
free ~ 自由支座
handrail ~ 扶手托座
pipe ~ 管座
supportability [səˌpɔːtəˈbiliti] *n.* 承载能力
supportable [səˈpɔːtəbl] *a.* 能支承的
supported [səˈpɔːtid] *a.* 被支承的
~ along four sides 四边支承的
~ at both ends 两端支承的
~ at circumference 周边支承的
~ at edges 周边支承的
~ boom 支承臂
~ diaphragm 支承隔板
supporting [səˈpɔːtiŋ] *a.* 支承的
~ angle iron 支承角钢
~ area 支承面积
~ bead 支承垫圈
~ beam 支承梁
~ body 承载体
~ brick 承重砖
~ capacity 承载能力
~ construction 支承结构
~ course 承重层,持力层
~ frame 支承框架
~ layer 承托层
~ load 承重层,承载能力
~ masonry wall 承重圬工墙
~ out wall 承重外墙
~ part 支承部分
~ partition 承重隔墙
~ pier 支墩
~ pillar 支柱,承重柱
~ plate 支承板,支座板
~ point 支点
~ post 支柱
~ shoe （加工工件）托架
~ skeleton 承重骨架
~ space structure 承重空间结构
~ structure 支承[下部]结构
~ system 承重体系
~ value 承重量,承载能力

~ wall 承重墙
~ wire 支撑线,吊线
supremum [sjuːˈpriːməm] *n.* 上确界
surbase [ˈsəːbeis] *n.* 腰板,柱脚花线,柱基饰
surbased [ˈsəːbeist] *a.* 扁拱形的,扁的;有柱脚花线的;-surbasement, *n.*
~ arch 扁拱
~ dome 扁圆屋顶
~ spherical vault 有座穹顶

surbase

surcharge [ˈsəːtʃɑːdʒ] *n.* 超载,过载;附加荷载;充电过度;洪水超高;附加费用;-surcharger, *n.*
~ depth 超高水深
~ load 超载荷重
~ storage 超额容量
surcharged [səːˈtʃɑːdʒəd] *a.* 过载的,超额的
~ earth 超载地
~ wall 超填挡土墙
surface [ˈsəːfis] *n.*; *v.*, -faced,-facing;面;表面;地面;水面; *v.* 镶面,表面加工,饰面,铺面
~ ablation 表面消融[烧蚀]
~ abrasion 表面磨耗
~ absorption 表面吸收
~ -active material 表面活性材料
~ activity 表面活性
~ alignment 地面定线测量
~ analysis 故障面[故障范围]分析
~ angularity 表面凹凸性
~ appearance 外貌,表面形状
~ application 表面处理,敷面
~ arcade 墙面假连拱
~ area 表面积
~ bearing 表面支承
~ blemish 表面缺陷
~ bolt 明装插销
~ bond 表面粘结[结合]
~ capacity 表面承载力
~ carrier 地面交通工具

~ cavity 表面凹坑
~ channel 明沟
~ checking 表面起网状裂纹,表面龟裂
~ coat 面层
~ -coated wall paper 墙纸
~ coating 表面涂层
~ coefficient of friction 表面摩擦系数
~ compaction 表土夯实
~ configuration 表面形状
~ construction 路面建筑
~ contamination 表面污染
~ content 表面积
~ contraction 表面收缩
~ course 面层,表层
~ crack 表面裂缝
~ creep 表层蠕动[塌滑]
~ decoration 表面装修
~ defect 表面缺陷
~ deformation 表面变形
~ density 表面密度
~ deposit 表面沉积物
~ dislocation 表面位移
~ ditch 明沟,阳沟
~ drain 明沟
~ drainage 表面排水
~ dressing 表面整修,敷面料
~ drying 表面干燥
~ duct 地面管道
~ dump 地面堆放物
~ elevation 地面高程
~ enrichment 表面装修
~ feature 地形,地貌
~ filtration 表面过滤
~ finish 表面修琢[饰]
~ flaw 表面发裂[缺陷]
~ hardener 表面硬化剂
~ indicator 平面规,表面找平器
~ load 表面荷载
~ lubricant 表面润滑剂,脱模剂
~ maintenance 表面养护

~ material 面层材料
~ metal raceway 面装金属电线管
~ -mounted astragal 面装圈带
~ mounted luminaire 吸顶灯具

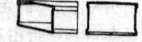
surface metal raceway

~ movement 表层移动
~ of adherence 粘附面
~ of contact 接触面
~ ornament 表面装修
~ paint 面层涂料
~ penstock 露天压力管道
~ pipe 地表管
~ pittings 表面凹孔,(混凝土)表面点蚀
~ plane 地面平面图
~ plate 平台[板],划线台
~ preparation 表面处理
~ protection 表面防护
~ repelling agent 表面防水剂
~ retardant 缓凝表层
~ retarder (混凝土)表面缓凝剂
~ retention 表面滞留
~ reverberation 表面交混回响,表面反射
~ roughness 表面糙度
~ shrinkage 表面收缩
~ smoothness 表面平整度
~ spalling 表面剥离
~ state 表面状态
~ stratum 表层
~ temperature 表[地]面温度
~ treatment 表面处理
~ trench 明沟,敞露排水沟
~ voids (混凝土)麻面,表面坑穴
~ washing 表面冲洗[刷]
~ water drain 地面排水沟
~ water proofing 表面防水
~ waterproof agent 表面防水剂
~ welding 表面焊接
~ wiring switch 拉线开关

surfacing [ˈsəːfisiŋ] *n.* 铺面,表面修整,表面加工,面层,路面;堆焊
  ～ board 铺面板
  ～ concrete slab 铺面混凝土板
  ～ course 面层
  ～ electrode 堆焊焊条
  ～ material 铺面材料
  ～ of lumber 木料的表面处理
  ～ sheet 铺面板
  ～ welding rod 堆焊焊丝
  arc ～ 电弧堆焊

surfactant [səˈfæktənt] *n.* 表面活化剂
  ～ effect 表面活化作用

surficial [səːˈfiʃəl] *a.* 表面的,地面的,地表的

surmount [səːˈmaunt] *n.* 饰顶;罩面;克服
  ～ed arch 超半圆拱
  ～ed dome 超半圆穹顶

surplus [ˈsəːpləs] *n.* 过剩,剩余,超过额; *a.* 过剩的,剩余的
  ～ earth 剩余土方
  ～ material 剩余材料

surround [səˈraund] *n.* 外包层,围绕物; *v.* 围绕,包围;surrounder, *n.*

surrounding [səˈraundiŋ] *a.* 周围的,环绕的; *n.* (*pl.*)环境,外界
  ～ masonry wall 围绕的圬工墙,围墙
  ～ rock 围岩

survey [səːvei] *n. pl.* -veys ;调查,测量,观察,查勘,综述,评价;-surveyable, *a.*
  ～ area 测区面积
  ～ beacon 测量标志
  ～ crew 测量[勘测]人员
  ～ data 测量[勘测]资料
  ～ drawing 测量图
  ～ field notes 野外测量手簿
  ～ group 测量队,查勘队
  ～ instrument 测量仪器
  ～ map 测量图

  ～ marker 测量标志,测标
  ～ meter 探测器,普查仪器
  ～ monument 测量标石
  ～ network 测图网
  ～ of ground water 地下水调查
  ～ of site 工地测量
  ～ party 测量队
  ～ pegs 测桩
  ～ point 测点
  ～ report 测量[查勘]报告
  ～ sheet 测量原图
  ～ stake 标桩
  ～ station 测站
  ～ targets 测量觇标
  ～ team 测量队
  ～ traverse 测量导线
  ～ed drawing 实测图
  aerial ～ 航空测量
  expert's ～ 专家评价[意见]
  field ～ 野外测量,勘测
  levelling ～ 水准测量
  rough ～ 草测;初步勘测
  sample ～ 样品鉴定

surveying [səˈveiiŋ] *n.* 测量,调查,观测
  ～ aneroid 空盒气压计
  ～ calculation 测量计算
  ～ computation 测量计算
  ～ gang 测量队
  ～ instrument 测量仪器
  ～ map board 测图板
  ～ marker 测标
  ～ office 测量局
  ～ panel 测量平板仪
  ～ plane table 测量平板仪
  ～ range 测量范围
  ～ rod 测杆,测量标尺
  ～ sheet 测量图,地形图
  ～ stake 测量标桩
  ～ specification 测量工作规程
  ～ work 测量工作,勘测

surveyor [səˈveiə] *n.* 测量员,调查员,鉴定人
  ～'s chain 测链(长 66 英尺)

~'s compass 测量罗盘仪
~'s dial 测量罗盘仪
~'s level 水平测量仪
~'s pole 测量标杆
~'s report 鉴定证明书
~'s rod 水准尺
~'s table 测量平板仪
~'s transit 经纬仪
quantity ~ 预算(工程)师

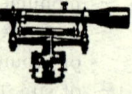
surveyor's level

susceptibility [səseptəˈbiliti] *n. pl.* -ties;敏感性,灵敏度,磁化率
~ to heat 热敏性

susceptible [səˈseptəbl] *a.* 敏感的,灵敏的,易感的; susceptibleness, *n.*; susceptibly, *ad.*
~ to moisture 对湿度敏感的

suspend [səsˈpend] *v.* 悬,吊,挂;悬浮;暂停;延缓

suspended [səsˈpendid] *a.* 悬挂的,悬浮的,暂停的
~ absorber 空间吸声体
~ acoustical plaster ceiling 隔声泥灰吊顶
~ acoustical tile ceiling 隔声砖吊顶
~ beam 简支梁
~ building 悬挂式建筑
~ cable structure 悬索结构
~ catwalk 悬挂脚手架
~ ceiling 吊顶,悬式顶棚
~ concrete floor 悬式混凝土桥[楼]面
~ counterweight 悬式平衡重
~ deck 悬桥面
~ fender 悬挂式防撞物
~ floor 悬吊桥[楼]面
~ forms 悬吊式模板
~ -frame weir 活动式悬架堰
~ ground floor 悬吊式地面板
~ joint 悬式接头,浮接
~ load 悬荷,吊载,悬移质[泥沙]
~ luminaire 悬吊式照明器具
~ metal ceiling 悬式金属顶棚
~ metal lath 悬式金属条板
~ plaster ceiling 灰泥吊顶
~ railroad 高架铁路,悬架铁道

suspended metal lath

~ reinforced roof 悬吊式钢筋混凝土屋顶
~ roof 悬吊式屋顶
~ sash 吊窗
~ scaffold 悬式脚手架
~ sediment 悬浮泥沙
~ shuttering 悬式模板
~ slab 悬式板,简支板
~ span 悬跨,悬吊跨度
~ sprayed acoustical ceiling 喷涂隔声吊顶
~ staircase 悬挑楼梯
~ state 暂停状态
~ stiffening truss 悬索加劲桁架
~ structure 悬挑结构
~ truss 悬式桁架
~ wall 悬墙
~ water 悬着水,上层滞水,曝气(范围)的水

suspender [səsˈpendə] *n.* 吊杆,悬索,吊索,挂钩
~ cable 悬索,吊索

suspending [səsˈpendiŋ] *a.* 悬吊的,悬浮的
~ agent 悬浮剂
~ velocity 悬浮速度

suspension [səsˈpenʃən] *n.* 悬吊,悬浮;暂停,中止
~ anchor cable 悬吊锚定索
~ bridge 悬索桥,吊桥

suspension bridge

~ cable 悬索,吊索
~ cable roof construction 悬索屋顶施工
~ cable structure 悬索结构

~ center 吊点,悬点
~ colloid 悬浮胶体
~ construction 悬吊构造
~ dredger 悬挂式挖泥机
~ effect 悬浮作用
~ floor 悬桥[楼]面
~ girder 悬梁,吊梁
~ grout 悬液浆
~ joint 悬式接头
~ light 吊灯
~ link 悬杆
~ member 吊杆,吊索
~ of work 工程停顿
~ post 吊杆
~ rod 悬杆,吊杆
~ roof 悬式屋顶,吊顶
~ rope 悬索
~ structure 悬吊结构
~ time 中止时间
~ -truss bridge 悬式桁架桥
~ type 悬挂式

suspensoid [səs'pensɔid] *n.* 悬胶体

sussex garden-bond 三顺一顶砌合

sustain [səs'tein] *v.* 支持,支承;持续,维持,忍受,经受,sustainable, *a.* sustainer, *n.*

sustained [ses'teind] *a.* 被支持的;持续不断的;一样的;sustainedly, *ad.*
~ fault 持续故障
~ grade 持续坡度
~ growth 持续增长
~ load 持续荷载
~ modulus of elasticity 持久弹性模量
~ oscillation 等幅振荡,持续摆动
~ overload 持续超载
~ velocity 持续速度
~ vibration 持续振动
~ yield 持续生产量

sustainer [səs'teinə] *n.* 支撑,支座,支点
~ engine 主发动机

sustaining [səs'teiniŋ] *a.* 支持的,持续的; sustainingly, *ad.*
~ power 支持能力
~ slope 连续坡度
~ wall 扶壁

suttle ['sʌtl] *a.* ;*n.* 净重(的)
~ weight 净重

suture ['sjuːtʃə] *n.* ;*v.* , -tured,-turing; *n.* ;*v.* 接缝,缝合,连接

swab ['swɔb] *v.* , swabbed, swabbing, *n.* ;*v.* 擦洗,打扫; *n.* 拖把,墩布,扫帚

swag [swæg] *n.* 垂花饰,装饰用的帷幕
~ leaf 垂花饰叶

swage [sweidʒ] *n.* 型钢,锻模;swager, *n.*
~ block 型模块
~ die 型模

swale [sweil] *n.* 低洼地,低湿草地

swallet ['swɔlit] *n.* 落水洞,灰岩渗水坑,地下水

swallow ['swɔləu] *n.* 燕子
~ tail 燕尾榫
~ -tailed 燕尾形的

swamp [swɔmp] *n.* 沼泽,湿地
~ area 沼泽区
~ drainage 沼泽地排水
~ land 沼泽地
~ soil 沼泽土

swan [swɔn] *n.* 天鹅
~ base 插口式灯头,卡口接头
~ neck bend 鹅颈弯头
~ neck downpipe 双弯头水落管
~ neck hand-rail 鹅颈形扶手
~ socket 卡口灯座

sward [swɔːd] *n.* 草地; *v.* 铺草皮

swastika ['swæstikə] *n.* 乐(万字)纹连续图案;swastikaed, *a.*

S-wave = secondary wave 次[s]波,地震横波

sway [swei] *n.* ;*v.* 侧移;摇晃,倾斜; swayable, *a.* ; swayer, *n.* ;

swayingly, ad.
~ bar 摆杆,稳定杆
~ bracing 斜支撑,防侧移支撑
~ rod 斜撑杆,防倾杆
~ strut 斜支撑,阻摇支撑
sweat [swet] *n.* 湿气,汗
sweating ['swetiŋ] *n.*（未凝混凝土）析水
~ dross 热析浮渣
~ rate 析水率
Swedish ['swi:diʃ] *a.* 瑞典的,瑞典人的
~ method 瑞典法(滑坡稳定分析)
~ slip-circle method 瑞典滑弧法
~ window 瑞典式窗（中间装有活动百叶的双层玻璃窗）
sweep [swi:p] *n.*; *v.*, swept, sweeping; *n.*; *v.* 扫描,冲刷,疏浚,扫除,弯曲
~ elbow 巨肘弯头
~ lock 弯头锁
~ of duct 管道弯头
~ record 扫描纪录
~ saw 弧线锯
~ tee 弯曲三通管

sweep lock

sweeper ['swi:pə] *n.* 清扫机
road ~ 扫路机
snow ~ 扫雪机
sweet ['swi:t] *a.* 甜的,甘的
~ camber 平坦路拱
swell [swel] *n.*; *v.*; sewlled, swelled or swollen, swelling; *n.*; *v.* 膨胀,湿胀,隆起,涨水,涌浪
~ factor 膨胀系数
~ index 膨胀指数
~ measurement 隆起测定
~ of a mould 胀砂
~ -shrink characteristics （土的）胀缩性
~ -shrinking soil 胀缩土
~ test 膨胀试验
~ value 膨胀值
~ wall 膨胀曲壁
swelling ['sweliŋ] *n.* 膨胀,湿胀,隆起

~ agent 泡胀剂
~ capacity 膨胀[湿胀]量
~ ground 膨胀地基
~ index 膨胀指数,回弹指数
~ isotherm 等温膨胀性
~ limit 膨胀极限
~ load 递增荷载
~ material 膨胀性材料
~ of soil 土壤膨胀
~ of the parttern 模型膨胀
~ potential 膨胀潜能,膨胀势
~ pressure 膨胀压力
~ property 膨[湿]胀性
~ resistance 抗胀性
~ rock 膨胀性岩石
~ soil 膨胀土
~ strain 膨胀变形
~ stress 膨胀应力
~ test 膨胀试验
~ value 膨胀值
swift [swift] *n.* 急流,湍流
~ change 突然变化
~ running water 急流水,激流
swimming ['swimiŋ] *n.* 游泳
~ bath 游泳池(室内)
~ beach 海滨游泳池
~ place 天然游泳场
~ pool 游泳池
swing [swiŋ] *n.*; *v.*, swung, swinging; *n.*; *v.* 摆动,摇晃,侧移; swingable, *a.*; swingably, *ad.*; swinger, *n.*
~ angle 旋变角度
~ area 回转水面
~ bar 摇杆,吊杆
~ bearing 摇座
~ bolt 铰链螺栓
~ bridge 旋开桥,平旋桥

swing check valve

~ check valve 旋转回止阀
~ cock 旋转龙头[阀门]
~ crane 旋臂[回转式]起重机
~ diffuser 旋转喷嘴

~ door 双开式弹簧门,双动自止门
~ excavator 全回转式挖土机
~ faucet 旋转龙头[阀门]
~ gate 旋转门
~ hammer crusher 旋锤式破碎机
~ hose rack 旋转式软管架
~ jib crane 动臂式起重机
~ joint 旋转式连接
~ link 摆杆
~ offset 旋转位移
~ sash 翻窗,摇窗,双动自止窗
~ saw 摆锯
~ span （旋开桥）平旋跨
~ wire 摆动缆

swinging ['swiŋiŋ] *n.* 摆动,晃动,回转
~ boom 起重机旋转臂
~ circle 回转底盘
~ conveyer 回转式运输机
~ hopper 悬吊式料斗,吊斗
~ -in casement 内开窗扇
~ mooring 单点系泊
~ motion 摇摆,摆动
~ post （门的）悬吊柱
~ radius （起重机臂）工作半径
~ scaffold 摇摆脚手架
~ screen 摇筛,振动筛
~ sieve 摇筛
~ staging 悬吊式脚手架
~ valve 平旋阀

swirler ['swəːlə] *n.* 旋流器,涡旋式喷嘴

switch [switʃ] *n.* 电键,电开关,分支开关
~ board 配电盘,开关柜,交换台
~ box 开关箱[盒]
~ cabinet 开关箱
~ cock 开关旋塞
~ fuse 开关保险丝
~ house 开关室,配电室
~ identifier 开关标识符
~ off 切断
~ on 接通
~ order 开关指令
~ plate 开关板
~ room 开关室,配电室
~ station 开关站
~ yard 露天开关站;调车场,编组站
compound ~ 组合开关
dimmer ~ 调光开关
intermediate ~ 中间开关
one way ~ 单头开关
pull ~ 拉线开关
two way ~ 双头开关

swivel ['swivl] *n.* 旋开桥,旋转,旋轴
~ arm 旋臂
~ bridge 平旋桥,旋开桥
~ chair 转椅
~ chute 回转溜槽
~ cowl 旋转式通风帽
~ damper 转动式烟道
~ frame 中旋式窗框
~ hook 转环钩
~ joint 转环接合
~ pen 曲线笔
~ pulley 动滑轮
~ rod 旋转杆
~ sash （摇）转窗

swiveling ['swivliŋ] *n.* 旋转
~ nozzle 旋转喷嘴
~ pile driver 旋转打桩机

swollen ['swəulən] *a.* 膨胀的,冻胀的
~ soil 膨胀土

syenite ['saiənait] *n.* 正长岩,黑花岗石; syenitic, *a.*

syllabus ['siləbəs] *n. pl.* -buses,-bi; 摘要,大纲,提纲

symbol ['simbəl] *n.* 符号,记号,象征
~ word marking （道路的）文字标牌

symmetric(al) [si'metrik(əl)] *a.* 对称的,平衡的,调和的
~ arch 对称式拱
~ axis 对称轴
~ balance 对称平衡
~ body 对称体

symmetry

~ center 对称中心
~ deformation 对称变形
~ distribution 对称分布
~ frame 对称框架
~ fret 对称回文饰
~ load 对称荷载
~ matrix 对称矩阵
~ one-legged multistoried frame 对称单柱多层框架
~ relation 对称关系
~ section 对称断面
~ state 对称状态
~ stress 对称应力
~ tensor 对称张量
~ transformation 对称变换
~ treatment 对称处理
~ two-legged multistoried frame 对称型双柱多层框架

symmetry ['simitri] *n.* 对称性
~ plane 对称平面
~ transformation 对称变换
axis of ~ 对称轴

symmetric(al) two-legged multistoried frame

symmetroid ['simitrɔid] *n.* 对称曲面
sympathetic(al) [ˌsimpə'θetik(əl)] *a.* 共鸣的,共振的,感应的; sympatheticness, *n.*
~ cracking 感应开裂
~ earthquakes 和应地震
~ resonance 共鸣,共振
~ vibration 共振,共鸣

sympathy ['simpəθi] *n. pl.* -thies; 共振,共鸣;同情,同感
synchro ['siŋkrəu] *n.* 同步; *a.* 同步的
~ control 同步控制
~ coupling 同步耦合
~ drive 同步传动
~ generator 同步发电机
~ -switch 同步开关
~ -system 同步系统

synchromesh ['siŋkrəu'meʃ] *n.* 同步配合,同步装置
synchronization [ˌsiŋkrənai'zeiʃən] *n.* 同步,整步
~ control 同步控制
~ operation 同步操作
~ regulator 同步调速器

synchronous ['siŋkrəuəs] *a.* 同步的,同时的
~ gate 同步门
~ generator 同步发电机
~ motor 同步电动机
~ oscillations 同步振荡
~ speed 同步速度
~ supervision mechanism 同步监控装置
~ system 同步系统
~ vibration 同步振动

synclastic [sink'læstik] *a.* 同向的,同向弯曲的; *n.* 顺裂碎面
~ curvature 同向曲率
~ surface 同向曲面

syncline ['siŋklain] *n.* 向斜(层)
syneresis [sini'ərəsis] *n.* (胶体)脱水收缩作用
~ of grease 润滑脂的脱水收缩
syngenesis [sin'dʒenisis] *n.* 同生,共生; syngenetic, *a.*
synopsis [si'nɔpsis] *n. pl.* -ses;摘要,大纲;说明书
syntactic [sin'tæktik] *a.* 合成的
~ foam 复合泡沫塑料
synthal ['sinθəl] *n.* 合成橡胶
synthesis ['sinθisis] *n. pl.* -ses;合成,综合
~ method 合成法
synthesize ['sinθisaiz] *vt.*, -sized, sizing;合成,综合; synthesization, *n.*
synthesizer ['sinθisaizə] *n.* 合成装置
electronic ~ 电子合成装置
synthetic [sin'θetik] *a.* 合成的,综合的,人造的

~ adhesive 合成胶粘剂
~ aggregate light concrete 合成骨料轻质混凝土
~ building product 合成建材
~ city 综合性城市
~ concrete curing agent 合成混凝土养护剂
~ construction product 合成建材
~ copal 合成酚醛树脂
~ corundum 人造刚玉
~ data 综合数据[资料]
~ detergent 合成洗涤剂
~ enamel 合成磁漆
~ fibre 合成纤维
~ fibre fabric 合成纤维织物
~ fibre material 合成纤维材料
~ glue 合成胶结剂
~ leather 人造革,合成皮革
~ material 合成材料
~ natural gas 合成天然煤气
~ paint 合成涂料,合成漆
~ plastic coating 合成塑料涂层
~ polymer coagulant 合成高分子混凝剂
~ resin 合成[人造]树脂
~ -resin adhesives 合成树脂胶
~ -resin cement 合成树脂胶合剂
~ -resin coating 合成树脂(防水)涂料
~ -resin external coat 合成树脂外墙涂料
~ -resin floor covering 合成树脂楼面涂层
~ resin primer 合成树脂底漆
~ rubber 合成[人造]橡胶
~ rubber contact adhesive 合成橡胶接触胶粘剂
~ sludge 合成污泥
~ stone 人造石
~ structural product 合成建筑用品
~ sunshade 综合式遮阳
~ system 综合系统
~ varnish 合成[人造]清漆

synthon [ˈsinθɔn] *n.* 合成纤维
syphon [ˈsaifən] *n.* 虹吸,虹吸管
~ hood 虹吸罩
~ mouth 虹吸管进口
~ spillway 虹吸式溢洪道
~ tube 虹吸管
syphonage [ˈsaifənidʒ] *n.* 虹吸作用
syringe [ˈsirindʒ] *v.*, -ringed,-ringing; 注水,洗涤
system [ˈsistim] *n.* 系统,体系,体制,方式,次序,分类法
~ analysis 系统分析
~ -builtconcrete structure 预制混凝土结构
~ construction method 预制施工法
~ control center 系统控制中心
~ design 系统设计
~ diagram 系统图,流程图
~ engineering 系统工程
~ ensemble 综合系统
~ management 系统管理
~ model 系统模式
~ of fit 配合[装配]方式
~ of sewerage 下水道系统
~ of sewers 下水道系统
~ of units 单位制
~ of water supply 给水系统
~ optimization 系统优化
dome ~ 拱形体系
double stack ~ 双立管系统
hot water circulation ~ 热水循环系统
single stack ~ 单立管系统
systematic [sistiˈmætik] *a.* 系统的,有规则的,有次序的
systematize [ˈsistimətaiz] *v.* 系统化,体系化
systyle [ˈsistail] *a.* 两径间排柱式(即两柱间距等于柱直径的两倍); *n.* 此种柱式的建筑物
systylos [sisˈtailəs] *n.* 窄柱式(间距为柱径的二倍)

T

T [ti:] ***n.*** T[丁]字形(物),三通(管)接头,T形接头;[T]绝对温度的符号;塔的伞形(镀金)顶饰;***a.*** T[丁]字形的

tab [tæb] ***n.***;***v.***, tabbed, tabbing; ***n.*** 接头(片),薄片,窄带,供悬挂[手拉]用的小突出部;(卡片)索查突舌,牌子;附[记]录,账目; ***vt.*** 加上小突出部,装以薄片;选出,指定;把……列表
~ assembly 翼片安装[组合]
current connection ~ 导电接头
end ~ 引弧[引出]板
locating ~ 定位梢
spring ~ 弹簧调整片
trimming ~ 配平调整片

tabac [ˌtaˈba] 【法】***n.*** 烟草;黄褐色

tabby [ˈtæbi] ***a.***;***n.***;***v.***, -bied,-bying; ***n.*** 黏土,砂和碎石混合料(即土质混凝土),贝壳(石)灰砂; ***vt.*** 使……起波纹; ***a.*** 平纹的,起波纹的,有斑条的

tabbyite ***n.*** 韧沥青

tabergite ***n.*** 叶绿泥石,钠黑蛭石

tabernacle [ˈtæbənækl] ***n.*** 临时房屋,幕屋,帐篷,壁龛,教堂
~ work 华盖装饰,以华盖为特点的建筑设计

tabernacle

tabia ***n.*** 三合土

tabid [ˈtæbid] ***a.*** 溶解的,腐烂的

tablature [ˈtæblətʃə] ***n.*** 壁画,刻有铭文的碑

table [ˈteibl] ***n.***;***v.***, -bled,-bling; ***n.*** (工作)台,桌,架;平[石]板,平盘[面](薄)片,牌子;扣捭,飞檐,檐板;表(格),图表,目录;高原,陆台,地块;(地下水)面; ***v.*** (木工)嵌接[合],榫接;(把……)放在桌上;把……制成表格,列表,造册;把……列入议事日程,提出(报告)

~ area 工作台面积
~ balance 托盘天平[秤]

table balance
~ band saw 小带锯
~ base 工作台,底座
~ chair 一种桌、椅两用家具
~ cloth 台布
~ control device 控制台装置
~ fan 台扇
~ feeder 平板给料机
~ flap (折叠式桌面的)折板
~ -hinges 台铰
~ -inking 调墨板
~ joint 嵌接
~ lamp 台灯
~ land 高原,台地
~ look(-)up 一览表,查表
~ -lookup instruction 查表指令
~ of contents 目录
~ planing machine 龙门刨床
~ runner (装饰用)狭长台布
~ saddle 工作台滑座
~ settle 高背长靠椅
~ (-)slate 屋面(平)石板
~ standard 台上柱形灯
~ stop 工作台定位器
~ tap 台用插头
~ type jack 台座式千斤顶
~ vibrator 台式振动器,振动台
~ vice[vise] 台虎钳
~ viewer 台式投影仪
~ ware 餐具
~d fish plate 带扣榫的接合板,嵌接鱼尾板
~d joint 嵌接,榫接
~d scarf 叠嵌接

adjustable drawing ~ 活动制图桌
automatic spacing ~ 自动定距钻孔台
conveyor ~ 输送机,转运台
crest ~ 墙帽
drafting ~ 制图台
dressing ~ 化妆台,整修台
drilling machine ~ 钻台
feed ~ 送料盘
filing ~ 锉工台,钳工工作台
fiter operating ~ 滤池操作台
finishing ~ 精加工台,研磨台;精选摇床
flow ~ 流水槽
jarring ~ 振动台
knee ~ (工具机上的)三角桌
laying-out ~ 划线台
layout ~ 设计图表
marking-out ~ 划线台
mast ~ 桅杆柱脚
night ~ 床头柜
plane ~ 平板(仪),标图板
plot ~ 标图板
plotting ~ 图表
side ~ 茶几
slewing ~ 回转台,转盘
surveyors ~ 平板仪
traverse ~ (测量用)小平板
turn ~ 转台[盘]
underground water ~ 地下水面[位],潜水位
universal ~ 万能工作台
water(-) ~ 地下水位,潜水位
wire tension ~ 紧线台,绞车
yardage ~ 土方表(英制)
tableau ['tæbləu] *n. pl.* -leaus,-leaux;表(格);动人的布局[场面、画]
~ curtain (由中间向上方两侧提起的)大幕
tablet ['tæblit] *n.* 签石,顶层;碑,匾;小片(状物),小块;便笺[报告纸]簿; *vt.*

把……压制成片[块]
~ (-arm) chair (课堂用)扶手上有写字板的椅
memorial ~ 纪念碑
tabling ['teibliŋ] *n.* 盖顶,墙帽;(木工)嵌合[接];制表,造册
tablinum *n.* (古罗马)家谱室
tablite ['tæblait] *n.* (钠)板石
tabo(u) ret ['tæbərit] *n.* 小儿,矮桌[凳];绣框[架]
tabula ['tæbjulə] *n. pl.* -lae [-liː];写字板
~ rasa 【拉】光板,石板;洁白状
tabular ['tæbjulə] *a.* 平板状的,扁平的,薄层的;表格式的,列成表的; *n.* 表(格),表(列)值
~ book 表式账簿
~ data 表列数据[资料]
~ figure 表列数据
~ ledger 表式分类账
~ presentation 图表
~ structure 板状构造
~ surface 平(坦)面
tabulate ['tæbjuleit] *v.*, -lated,-lating;(把……)制成表,列(入)表(内),用表格表示;使成平面[平板状];精简,概括; *a.* 平面[板状]的;薄片构成的
tabulation [ˌtæbjuˈleiʃən] *n.* 列表,造册;表列结果
~ of mixture 混合物配料表
tabulator ['tæbjuleitə] *n.* 制表机[仪],列[制]表人,打字机的制表键
tacciometer *n.* 表面粘力计
tach = **tachometer** *n.* 转速表,指示转速的仪器
tache [tætʃ] *n.* 斑点,黑点;钩,环,扣
tacheometric(al) *a.* 视距的
~ rule 视距计算尺
~ survey(ing) 视距测量
tachometer [tæˈkɔmitə] *n.* 转速[转数,流速]计,转速表,测速仪
~ generator 测速发电机;转数表传感器

chronometric ～ (计)时序式转速计
drag-cup ～ 托杯式感应测速仪
resonant-reed ～ 谐振舌片式测速仪
viscosity ～ 黏性转数计
tachylite ['tækilait] (= tachylyte) *n.*
玄武玻璃
～ basalt 玻璃玄武岩
tachymeter [tæ'kimitə] *n.* (快速测定
距离,方位等用的)视距仪,测速仪;速
度计
tachymetry [tæ'kimitri] *n.* 视距测量
(学),快速测距术
tack [tæk] *n.* 平头钉,图钉;胶粘性;航
向,行动步骤;方法;Z字型移动；*v.* 钉
住,拼拢,定位搭焊,缝合
～ board 布告板
～ bolt 定位螺栓
～ claw 钉爪,平头钉拔除器
～ coat (of priming) 粘结层,(沥青)
粘层
～ course 粘结层
～ dry 粘干状态
～ driver 平头钉敲打机
～ eliminator 防粘剂
～ (-) free (涂层)固着干燥
～ free time (涂层)指触干燥时间
～ (-) hammer 平头钉锤
～ line 粘结线
～ point 粘着点,粘着温度
～ puller 拔图钉的拔钉机
～ range 粘性期
～ rivet 定位[平头]铆钉
～ screw 装配螺栓
～ system 流水作业
～ (-) weld (ing) 定位焊接,点固焊
(接),平头焊接
canvas ～ 帆布输送带钉卡箍
dry ～ 半干粘性
thumb ～ 图钉
tacker *n.* 定位搭焊工
tackifier *n.* 增粘[胶]剂
tackiness ['tækinis] *n.* 胶粘性,粘附力

～ agent 胶粘剂
～ paper tape 浸胶纸带,胶带纸
tackle ['tækl] *n.;v.,* -led,-ling； *n.*
(复)滑车,辘轳;用[索]具; *v.* 用滑车
固定[拉上来];装滑车;抓住;处理,对
付,解决
～ (-) block 滑车[轮]组
～ burton 复滑车,辘轳
～ -fall 复滑车的通索
～ -house 装有起重滑
车设备的房屋
differential ～ 差动滑车
differential chain ～ 差
动链滑车
fleeting ～ 辅助滑车,
水平滑车
hoisting ～ 起重滑车,
辘轳
tackle(-)block
pulley ～ 滑车组,复式滑车,起重滑车
single-whip ～ 定[导向]滑车
suspension ～ 吊钩,悬架[挂]
tacky ['tæki] *a.* 发粘的,胶粘的,(胶,
漆)未干的
～ -dry 干(后)粘性
～ resin 胶粘用树脂
～ surface 胶粘面,粘面层
tact [tækt] *n.* 间歇(式),(自动加工线
的)生产节拍,拍子;触觉
～ system 流水作业(线)
～ timing 生产节拍时间计算
tactile ['tæktail] *a.* (有)触觉的,能触
知的
～ impression 触感
taenia [ti:niə] *n. pl.*
-niae;束带饰,带形
花边,Doric 式建筑
的带状饰
taffeta ['tæfitə] *n.*
taffety ['tæfiti] *n.*
taenia
波纹绸、塔夫绸;黄色防油纤维膜
～ weave 平纹组织
tag [tæg] *n.;v.,* tagged, tagging； *n.*
(金属)箍,销钉,标签[牌];标记,标识
(符);特征;辅助信息;电缆终端接头

vt. 装金属箍;加标签于;添加;紧[尾]随;结合,连接
~ card 特征卡片
~ end 末尾,末端
~ mark(ing) 特征记号
~ ging method 示踪法,标记法
connectin ~ 连接销
contact ~ 接触金属箍,触针,接片
dog ~ 识别标志
garment ~ 外表特征
parts ~ 零件标签
shipping ~ 货运标签

tagger ['tægə] *n.* 装箍的人,加标签的人;附加物;(*pl.*)薄铁片
~ plate 极薄(0.18毫米以下)镀锡钢板

tagman *n.* 起重操作工

tail [teil] *n.* 尾状[垂下]物,尾[后]部,末端,结尾,下游;(*pl.*)瓦当;尾矿[渣](砖、石的)嵌入部分;*vt.*(把木材、砖石的一端)嵌进[砌入],(使)架住[搭牢],接[添]上;位于……后部;*a.* 尾[后]部的,后面(来)的
~ beam 梁尾端(墙内挑出部分),尾梁
~ bearing 尾轴承,梁尾端支承
~ block 末端滑轮
~ board scraper 尾板式铲运机
~ core print 榫尾芯头
~ cup 尾罩[帽]
~ end 末端[尾],结尾部,结束时期
~ -hood 尾盖
~ hook 尾钩
~ -in work 收尾工作
~ joist 挑出的搁栅尾端
~ loader 后挂装载机
~ rope 尾索,拖索
~ section 尾部
~ stock 尾架,尾座
~ swing 尾摆度,(旋转式起重机等)后方所需的净空
~ tower 尾塔,缆式起重机的移动塔
~ trimmer 千斤搁栅,尾[短]托梁
pig ~ 输[引]出端

rat ~ 天线升高部分和引下线的连接线束
rivet ~ 铆钉镦头
union ~ 连接尾管

tailing ['teiliŋ] *n.* 砌置墙上挑出块,(半嵌在墙内砖、石的)嵌入部分;尾部操纵,跟踪;(*pl.*)尾砂[渣、矿],筛余物,石屑[碴],渣滓
~ brick 尾矿砖,[矿渣]砖
~ iron 悬臂构件固定铁
screen~s 筛余物,筛屑

tailoff *n.* 关机,发动机关闭,尾推力中止

tailor ['teilə] *n.* 缝纫工;*vt.* 改装[编],修整[琢]制造[作]缝制,裁剪
~ -made 定做的,专用的
~ -made oil 精炼油

tailover *n.* 筛渣,筛除物

tailpiece ['teilpi:s] *n.* 一端嵌入墙内的短梁,端件[块];附属物;尾管;管接头,套筒[连轴节]连接,接线头

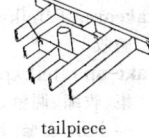
tailpiece

tailstock ['teilstɔk] *n.* 尾座[架];托柄的尾部;滑轮活轴

taint [teint] *n.* 污染,污痕;*v.* 弄脏,沾染

Taiwan cypress 台湾杉

take [teik] *vt.*, took, taken, taking;取,拿,接受,采用;测[读]出,记[摘]录,(花)费,订(购);假定,推断,处理,研究;(河道)封冻,凝固,吸收;传递;以……为例
~ (-)apart 拆卸,分[拆]开
~ back 收回,撤回
~ bearings 定方位
~ charge of 负责,担任
~ effect 生效,(被)实施
~ -in device 夹具,接线夹
~ levels 抄平,水平测量
~ load up 装载
~ means 采取手段
~ measure of 测定

~ -off pipe 放水管
~ off thd slag 出渣
~ samples 取[抽]样
~ sights 水平测量,找平
~ spell 换班
~ stock 盘存,清理存货
~ the altitude 确定标高
~ time 费时,需时
~ up water 吸(收)水(分)

takedown ['teik‚daun] *n.* 拆卸,取[记]下,移去,扫尾; *a.* 可拆卸的(部件)
~ time 拆卸时间

takeoff ['teikɔːf] *n.* 取[引]出,卸掉,牵引,移去[送],分出,放水(沟),出发(点),起飞,估计[量];摹仿
~ pipe 放水管

takeout *n.* 取[拿]出的东西[数量];自动取出装置

takeover ['teik‚əuvə] *n.* 接收[管],验收

take-up ['teikʌp] *n.* 拉[张]紧,缠绕,聚集,收缩,调整,补偿;提升装置
~ bolt 夹紧[扣紧]螺栓
~ reel 接收[卷带,卷线]盘,引出轴
~ set (钢筋)预加应力装置;夹具
~ spool 卷片[带]轴
spring ~ 弹簧张紧装置

taking ['teikiŋ] *n.* 取样;取得[出],开采;摄影;(*pl.*)收入,利息 *a.* 动人的,吸引人的
~ down (work) 拆卸[除,毁](作业)

Talbot process 塔尔博特方法(用于铸铁等的一种防腐蚀方法)

talc [tælk] *n.*,*v.*, tale(k)ed, tale(k)ing; *n.* 滑石(粉);(矽光用的)云母 *v.* 用滑石处理
~ flour 滑石粉
~ plastics 云母雕刻
~ powder 滑石粉,爽身粉
~ -schist 滑石片岩

talcite ['tælsait] *n.* 滑块石,变白云母

talcky ['tælki] (=talcose ['tælkəus]) *a.* (含)滑石的

talcum ['tælkəm] *n.* 滑石
~ powder 滑石粉

talik *n.* 层间不冻层(在永久冻土之间),(多年冻土上的)融区

talk [tɔːk] *v.*,*n.* 讨论,交谈,(用讯号等)通讯[话];(*pl.*)会谈;滑石
~ -back circuit 工作联络通话电路
~ box 讲话箱(一种防止噪声的设备)
~ channel 通话电路
~ key 通话电键
~ -listen button 通话按钮
~ing machine 唱机

talker ['tɔːkə] *n.* 扬声器;电话[扩音器],传令人员,谈话者
~ key 通话电键

tall [tɔːl] *a.* 高(大)的,巨大的
~ block 高大楼房,高层建筑(物),摩天[塔]楼,大厦
~ block building
facade 高层建筑[塔楼]正[立]面

tall block

~ -case clock 立地钟
~ flats (英)高层[塔楼]公寓
~ price 高价
~ guyed tower 拉索加固的高塔
~ industrialized block (英)工业化建筑[预制]高层[塔楼]公寓
~ slender column 高细长柱
~ structure 高层建筑(物)

tallboy ['tɔːlbɔi] *n.* (烟囱顶部)罩帽;高脚橱柜

tallow ['tæləu] *n.* (动物)脂油,牛[油]脂
~ compound 调配牛油(固体润滑剂)
~ dip 动物油脂做的蜡烛
piney ~ 松木硬脂

tally ['tæli] *n.*, *pl.* -lies, *v.*, -lied, -lying; *n.* 标签,筹码,牌[标]号;点[计]数,运[计,结]算;计数符,单位数,计算单位; *vt.* 计[结]算,点数,清点,记录;加标签于;使符合

~ card （材料）计算卡
~ clerk 理货员
~ cliker 计数器
~ counter 计数器(手执)
~ keeper 推销员,理货员
~ plan （英）分期付款
~ register 计数器
~ sheet 计数单[纸],理货单
~ shop （英）分期付款方式售货的商店
~ slates 论块出售石板
~ system 赊卖
~ the cargo 点货
~ the trade 赊销
buy on[upon] ~ 赊购
hand ~ （手摇）计数器
tallyman ['tælimən] *n.* 推销[点筹]员,理货[记账]员
talon ['tælən] *n.* 爪饰,爬形条纹;爪(状物),手(指)
talus ['teiləs] *n.* 竖斜面(如城墙的斜面);坝脚抛石;废料;山麓碎石堆,岩屑堆
~ wall 斜面墙,单侧坡面墙,堆石护岸,陡壁
tamarack ['tæməræk] *n.* 北美落叶松
tamarisk ['tæmərisk] *n.* 柽柳
tambour ['tæmbuə] *n.* 圆形屋顶的柱间墙[鼓形座],支持圆顶的下层结构,拱顶座
tamis ['tæmis] *n. pl.* -ises；滤布,（布格）筛
tammy ['tæmi] *n.* 滤布,格筛
tamo *n.* 日本产的一种棕色木材
tamp [tæmp] *v.* 夯实,捣固[实],用炮泥填塞[实](装有炸药的炮孔);*n.* 夯(具),捣棒
~ed backfill 夯实回填土
~ed concrete 捣实混凝土
~ed finish 夯实整修面
~ed joint 浇筑混凝土的施工缝
~ed well （用炮泥）填塞炮孔
tamper ['tæmpə] *n.* 夯(具),碾,捣锤,打夯[捣固]机,羊足[夯击式]辗;填塞工具,捣(实)棒,炮棍;填[捣]实者,装炮工;*v.* 夯实,捣固[塞]
~ finisher 夯板平整机
~ roller 羊足辗
ballast ~ 砸道机,夯道机,石碴捣固工具
compressed air ~ 压缩气夯
hammer-blow ~ 锤击夯
mechanical ~ 机械砸道器
power ~ 动力夯
sheep foot ~ 羊足辗[压路机]
tamping ['tæmpiŋ] *n.* 夯实,捣固;填塞(物),炮泥;填(炸药)孔
~ bag 填塞袋
~ bar 炮棍;捣棒,夯棍;(铁路)砸道棍
~ beam finisher 夯板平整机
~ -crane rammer 落锤式夯实机,机械夯板
~ devices (混凝土)捣实工具
~ drum 辗压滚筒
~ foot (辗压滚筒上突出的)羊足
~ hammer 捣固锤
~ in layers 分层夯实
~ -levelling finisher (混凝土)整平捣固机
~ machine 打夯[捣固,砸道]机
~ material 填[堵]塞材料,炮泥
~ pick 捣镐,砸道镐
~ plate 夯板
~ rod 炮眼封泥棒;捣棒,夯棍;(铁路)砸道棍
~ stick 炮棍,装药棒
~ -type roller 夯击式压路机,压路辗
~ weight 夯击重量
tampion ['tæmpiən] *n.* 塞子
tampon ['tæmpən] *n.* 塞子,(涂漆等用)软布团
tan [tæn] *n.; a.* 鞣料(树皮),鞣酸皮渣(可用于筑路);黄褐[棕黄]色(的),担(中国重量单位)=50kg
~ ned jute 精制黄麻

~ yard 制革厂
spent ~ 废鞣料,鞣酸皮渣
Tanalith ['tænəliθ] *n.* 一种木材防腐剂(可防止白蚁危害)
tandem ['tændəm] *n.* 串联压路机,载重拖车;双轴; *a.*, *ad.*; 串联的,串[纵]列的,一前一后排列的
~ axle(s) (前后)串列轮轴,双轴
~ axle(s) load 串列轴[双轴]荷载
~ -bowl scraper 联斗铲运机
~ drive 串联式驱动,双轴驱动
~ elevator (美)双层电梯
~ knife-switch 串联闸刀开关
~ lift (英)双层电梯
~ mixer 联列(混凝土)拌和机,复式拌和机
~ roller 串联式[双轮]压路[辗压]机
tandour ['tænduə] *n.* (置于方桌下的取暖)火炉
tang [tæŋ] *n.* (工具的)柄脚[根],刀根
tang
Tang G (T&G)(舌槽,企口)匹配,企口[舌槽](接合)
~ ceiling 企口顶棚
~ connection 企口接合
~ floor 企口楼板
~ joint 企口接缝,舌槽接合
~ subfloor 企口地板
tangency ['tændʒənsi] *n.* 相切,(在一点上)接触,邻接
tangent ['tændʒənt] *n.* 正切,切线;(铁路,道路的)直线区间; *a.* 正切[相切]的,切线的
~ method 切线法
~ plane 切面
~ ray 切向光线
~ screw 微调[微动]螺钉
~ wedge 切向楔
tangential [tæn'dʒenʃəl] *a.* 切向的,相[正]切的,切线[面]的
~ bearing 弧形板支座
~ deformation 切向变形

~ force 切向力
~ grain type (木材层积组合方式中的)环弧型
~ motion 切向运动
~ path (曲线间的)直[共切]线段
~ sawn 弦锯的(木材)
~ section 弦切面
tangerine [ˌtændʒə'ri:n] *n.*; *a.* 橘红色(的)
tank [tæŋk] *n.* (液体,气体)容器,桶,箱,槽,柜,池;坦克 *v.* 把……储在箱[槽,池,容器]内,把……放在箱[槽,池,容器]内处理
~ block 箱[罐]座;池窑壁砖[耐火砖],槽炉用砌块
~ cock 水箱龙头[旋塞]
~ for flushing 冲水箱
~ furnace 池窑,浴炉
~ kiln 池[槽]窑
~ shell 罐板[壳]
~ sprayer (沥青)喷洒机
~ treatment (废水)在水池中(沉淀)处理
cable ~ 电缆槽
conservancy ~ 储存池
contact ~ (污水处理)接触池
detention ~ (污水)停留池
detritus ~ (下水道)存砂池
feed ~ 流量箱,进料槽
flush ~ 冲洗池
gas ~ 气柜[箱]
gas-holder ~ 储气罐
liming ~ (加)石灰槽
sedimentation ~ 沉淀[沉沙]池
settling ~ 沉淀[沉沙]池,澄清槽[箱]
storage ~ 储(油)罐,蓄水池,容器
lupflow ~ 上流式沉淀池
wash-water ~ 冲洗水箱
tankage ['tæŋkidʒ] *n.* 油箱[贮槽]容量;容器设备;用槽贮藏(法);贮藏费

tank for flushing

tankdozer *n.* 坦克推土机

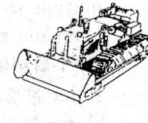
tankdozer

tanker ['tæŋkə] *n.* 油[水]槽车,加[储]油车,供水车,油船,(沥青)喷洒机[车]

tanking *n.* 池[槽]阻水层[止水],地下(室)防水[潮]层

tannic ['tænik] *a.* 鞣质的,由富于鞣酸的树皮取出的
~ acid 单宁酸,鞣酸

tannin ['tænin] *n.* 鞣酸(处理混凝土模板用),单宁酸
~ extract 烤胶

tantiron *n.* 高硅耐热耐酸铸铁

Taoist ['tɑːəuist] *n.* 道家,道教徒; *a.* 道教[家]的
~ temple 道教庙宇

tap [tæp] *n.; v.*, tapped, tapping; *n.* (旋)塞,开关,龙头;管嘴;分支[流],分接头,(中间)抽头;丝锥,螺丝攻; *vt.* 开孔,去塞,放出(水等);安装,堵塞;攻螺纹
~ bolt (有帽)螺栓
~ bore 螺纹底孔
~ borer 螺孔钻,开塞子锥

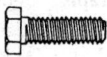
tap bolt

~ -changing switch 抽头切换开关,分接头切换开关
~ chuck 丝锥夹头
~ density (粉状物料的)振实密度
~ die holder 丝锥扳牙两用夹头
~ drill 螺孔钻
~ handle 丝锥扳手
~ hole 塞孔,放液[出渣]孔
~ house 酒店,客栈
~ post 水栓柱
~ room 酒吧间
~ switch 分接开关,抽头切换开关
~ water 自来[饮用]水
~ wrench 丝锥[螺纹]扳手,绞杠
cold ~ 冷水龙头
grease ~ 润滑孔[嘴]
hot ~ 热水龙头
pressure ~ 测压点[孔],放压孔
regulator ~ 控制旋塞,可控开关
screw ~ 螺钉攻,丝锥;螺钉口龙头;内螺纹模型
wall ~ 靠墙水龙头

tape [teip] *n., v.,* taped, taping; *n.* 卷[皮]尺;(胶,绝缘,磁,皮,纸)带;传送带,箍带;带[扁]钢 *vt.* 用卷尺量;用带系[绑,扎]
~ correction 卷尺修正
~ -drive mechanism 皮带传动机构

tape measure

~ joint 带条粘接缝
~ library 录音带馆,录音资料馆
~ -line 卷[皮,带]尺
~ measure 卷[皮,带]尺
~ perforator 纸带穿孔机
~ printer 带式打印机
~ punch 纸带穿孔机
~ -rod 卷动标尺
~ standardization 卷尺检定
~ verifier 纸带校对机
~ yarn 扁丝
adhesive ~ 绝缘[胶粘]带
band ~ 卷尺
black friction ~ (黑)绝缘胶(布)带
chain ~ 测量链
end ~ 端点尺
guidance ~ 导板[带]
gummed ~ 胶纸带
leveling ~ 水准带尺
proofed ~ 涂上橡胶的带
Scotch ~ 透明胶带
steel ~ 钢带
teletype ~ 电传打字带
transaction ~ (数据)变动带,处理带

tapeman ['teipmən] *n.* 持尺员

taper ['teipə] *n.* 锥形[体],圆锥;锥[斜]坡度;(宽度)渐变(路段);尖头;尖削; *a.* 锥形的,渐尖的斜削的; *v.* 递[渐]减;锥[楔,斜]削

~ bit 锥形铰刀[钻头]
~ bolt 锥形螺栓
~ cone 圆锥
~ file 圆锥锉,斜面锉
~ fit 锥度配合
~ -foot roller 羊足辗[压路机],羊蹄滚筒
~ gage 锥度量规
~ key 斜[楔]键
~ liner 斜垫,楔形塞垫
~ male threads 锥形阳螺纹
~ pile 变截面桩,斜桩
~ pin 锥形针
~ pipe 楔形管,大小头管
~ reducer 锥形异径管
~ sectioning 斜剖面[法]
~ sheet pile 尖头板桩
~ sleeve 锥套管
~ -stick 烛台
~ tap 锥形丝锥
~ thread 锥形螺纹
~ trowel 刮平镘刀

tapered ['teipəd] *a.* 锥形[度]的,楔形的,渐缩[减]的,斜[尖]削的
~ beam 加腋梁,变截面梁
~ bend 锥形弯头[管]
~ bolt 锥形螺栓
~ column 锥形柱,渐细柱
~ coupling 锥形管接头
~ core bit 锥形取芯钻
~ die 锚锥,锥形模(冷拔钢丝用)
~ dowel 锥形销钉
~ draft tube 锥形尾水管
~ -end joint 锥头形榫接
~ enlargement (管道)锥形扩大
~ file 尖锉,斜面锉
~ -flange beam 楔形翼缘梁
~ -foot roller 羊足辗[压路机],羊蹄滚筒
~ inlet 喇叭形进口
~ joint 锥形接头
~ pier 锥形[斜面]墩
~ pin 锥销

~ pintype jack 锥锚式千斤顶
~ plain tile 楔形平瓦
~ rivet 斜梢铆钉
~ roller 锥形滚筒
~ shims 楔形填隙片
~ slot 斜沟
~ thimble 锥形套管
~ washer 锥形垫圈,斜垫圈
~ wire rope 锥头索,(变截面)钢丝绳

tapestry ['tæpistri] *n.* 挂[花]毡[毯]，绒绣;织锦 *v.* 用挂毡装饰
~ brick 饰面砖
~ -brick wall 缀锦砖墙
~ carpet 织绒地毯
~ portrait 绣像
artistic ~ 艺术挂毯
Gobelin ~ 壁挂,壁饰花毡
Mial ~ 苗锦
silk ~ 织锦
Song dynasty ~ 宋锦
storied ~ 历史画挂毯
Zhuang ~ 壮锦

tapis ['tæpi:]【法】*n.* 桌[地,挂]毯
tapper ['tæpə] *n.* 螺母攻丝机;攻螺纹工人,修整工
~ tap 螺母丝锥
tappet ['tæpit] *n.* 挺[捣,推]杆,凸轮从动件
~ rod 挺杆
tapping ['tæpiŋ] *n.* 攻(螺)丝;开[冲,钻]孔;放液;抽头,分接头,缠绝缘带
~ attachment 攻丝装置[夹头]
~ clamp 钻孔夹具,丝锥夹头
~ drill 螺孔钻头
~ machine 开[钻]孔机,攻(螺)丝机
~ screw 自攻螺钉
~ sleeve 接水套管

tapping screw

tar [tɑː] *n.*; *v.* tarred, tarring; *n.* 焦油(沥青),柏油,*vt.* 浇[涂]柏油[焦油](于); *a.* (涂有)焦[柏]油的
~ asphalt 焦油沥青

~ barrel 柏油桶
~ base 柏油打底,柏油浇底[基]层
~ binder 柏油结合料
~ boiler 柏油熔化炉
~ bound surface 柏油结合(碎石路)面
~ brush 柏油帚[刷]
~ carpet 柏油路面磨耗层
~ -coated pipe 涂柏油管
~ -coated road 柏油路(面)
~ concrete 柏油混凝土,焦油(沥青)混凝土
~ dipping ladle 柏油舀勺[挖斗]
~ distillate 焦油溜出物
~ dolomite brick 焦油(沥青)白云石砖
~ -dressing (machine) 浇柏油机
~ emulsion 柏油乳液,乳化柏油
~ felt 柏油[油毛]毡
~ filler 柏油填(缝)料
~ finisher 柏油铺面机
~ flux 柏油稀释剂
~ -gravel roofing 柏油砂石屋顶
~ (-grouted) macadam 灌柏油碎石路
~ -grouted surfacing 柏油浇灌碎石路面
~ heater 柏油加热炉[器]
~ heavy oil 厚[重质]柏油
~ -impregnated paper 柏油纸
~ -impregnated roofing paper 盖屋顶用柏油纸
~ kettle 柏油烧锅
~ light oil 轻质柏油
~ mixer 柏油调合机
~ oil 煤焦油
~ -oil preservative 柏油防腐剂
~ paint 柏油涂料
~ paper 焦油纸
~ paper cover 柏油纸覆面
~ -paved (路面)铺柏油的
~ pitch 柏油,焦油沥青
~ pot 柏油烧锅
~ prepared roofing 柏油成品屋面材料,柏油尾面毡[纸]
~ -sand 沥青砂,柏油砂混合料
~ -saturated 浸柏油的,柏油浸制的
~ sheet 柏油帆布,(防水)油布,盖舱布
~ (iron) slag chip (ping)s 柏油矿渣
~ spray can 浇柏油箱
~ sprayer 柏油喷洒机
~ -spraying 喷洒柏油,柏油表面处治
~ spraying machine 柏油喷洒机
~ stabilization 柏油加固处理
~ -treated 柏油处治的
~ -water 焦油冷冷剂
~ red board 柏油纸(板)
~ red felt 油毛[柏]毡
~ red joint runner 柏油填缝条
~ red macadam 柏油碎石(路)
~ red marline 柏油麻绳
~ red paper 柏[焦]油纸
~ red road 柏油路
~ red roofing felt 屋面油毛毡,屋面柏油纸板
~ red rope sealing 柏油麻绳止水[填料]
acid ~ 酸柏油[焦油]
coal ~ 煤焦油
coke ~ 焦油
heavy ~ 厚[重质]柏油,重焦油沥青
mineral ~ 矿质焦油
rock ~ 原油

tarare *n.* 麻帆布
taraspite *n.* 白云石
tare [tɛə] *n.* 皮重,包装,[容器]重,车身自重;包装箱 *v.* 称[标出]……的皮重,除皮重;修正,校准
target ['tɑːgit] *n.* (测量用)觇标[板],标志;目标,指标
~ cost 目标[标准]成本
~ date 预定(开始,结束)日期
~ price 标准价格

~ strength (混凝土)试配强度,期望(的)强度
diamond ~ 菱形舰板

tariff ['tærif] *n.* 税,关税;税率[则],费[运价]率,收费[价目,工资]表; *v.* 定收费标准[税率],征收关税
preferential ~ 特惠(关)税率,优惠税则

tarmac ['tɑːmæk] (= tarmacadam [ˌtɑːməˈkædəm] *n.* 柏油碎石(路),柏油路面材料,铺地用沥青

tarnish ['tɑːniʃ] *n.* 表面变色;生锈,污点;管子接头材料; *v.* 变色;生锈

tarp [tɑːp] *n.* 柏油[焦油]帆布,(防水)油布,盖舱布;(油布)防水衣[帽,罩]
~ paper 防潮纸

tarpaulin [tɑːˈpɔːlin] *n.* = tarp

tarry ['tɑːri] *a.*, -rier, -riest;柏油[焦油]状[质]的;涂柏油的; *n.* 煤焦物质; *vi.* 滞留,停留
~ cut 焦油馏分

tarsia ['tɑːsiə] *n.* 嵌[镶,拼花]木[制品]

tartan ['tɑːtn] *n.* (运动场铺地面用)合成橡胶材料,格子呢

tartar ['tɑːtə] *n.* 酒石(酸氢钾)
~ emetic 吐酒石,酒石酸氧锑钾

tartaric [tɑːˈtærik] *a.* (含)酒石(酸)的
~ acid 酒石酸

tarungar *n.* 装碎砖石的编框

tarus *n.* 复折屋顶筒形脊

tarvia ['tɑːviə] *n.* 一种筑路用柏油
~ macadam 柏油碎石(路)

Tarvialithic [ˌtɑːviəˈliθik] *n.* 一种冷铺柏油混凝土混合料

taseometer [ˌtæsiˈɔmitə] *n.* 应力计

task [tɑːsk] *n.* 作业;工作,任务;职务; *v.* 派给……工作,使辛劳
~ equipment 专用设备
~ management 任务管理,定量作业管理
~ master 工头,监工
~ time 工时定额

taskwork ['tɑːskwəːk] *n.* 计件[包干]工作,件工;重活

taspinite *n.* 杂块花岗石

tassel ['tæsl] *n.* 缨,流苏饰,垂花饰
~ bush 流苏形灌木
~ flower 流苏花

taste [teist] *n.*; *v.*, tasted, tasting *n.* 味;风格[韵],鉴别[审美]力,经验; *v.* 品尝
~ and odor control 除味去臭

tatami(tatami-type met) 稻草垫席,日本式铺地席垫,榻榻米

tatty ['tæti] *n. pl.* -ties 湿帘(保持潮湿的门帘或窗帘,用以降低室内气温) *a.* 简陋的

tau [tɔː 或 tau] *n.* T 字形(物)
~ cross T 字形十字架

taula *n.* 毛石平台

taupe [təup]【法】*n.* 暗灰色
~ brown 灰棕色
~ gray 褐灰色

taut [tɔːt] *a.* 拉紧的,弹性的;整齐的;严格的
~ -line cableway 紧索道,架空索道

tautness *n.* 拉紧,紧固度
~ meter 伸长[拉力]计

tavern ['tævən] *n.* 酒馆;客栈,旅店

tawny ['tɔːni] *n.*; *a.* 茶色(的),黄褐色(的)

tax [tæks] *n.* 税(款);重负 *v.* 对……征税;使负担重;要[计,开]价
~ abatement 减税
~ delivery 上交税金
~ exempt 免税的
~ on value added 增值税
~ rate 税率
additional ~ 附加税
income ~ 所得税

taxi ['tæksi] *n. pl.* taxie 出租汽车[飞机,艇] *v.* (飞机在地面或水上)滑

行,使……滑行;乘出租汽车
～ strip 滑行道
～ track 滑行道
taxilane *n.* 滑行道
taxiway ['tæksiwei] *n.* (飞机)滑行道
～ light 滑行道照明灯
tazza *n.* 喷水池的台座
T-bar T型钢,丁字型钢,丁[T]型梁
T-beam T型[丁字型]梁,板梁
～ bridge T型梁桥
～ floor T型梁楼板
～ slab 肋形板

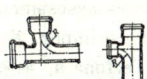
T-beam floor
T-branch pipe

T-bend T形管,三通管
T-branch pipe 三通管
T-connection T形连接
T-crossing 丁字街,丁字形截面
T-cross section T形截面
T-dike 丁字堤
T-distribution T形分布
tea [ti:] *n.* 茶树[叶],茶,茶点
～ bar 茶馆[室]
～ -ceremony house (日本)单幢的茶室,茶室风格的建筑物
～ china 瓷杯,茶杯
～ garden 设有茶点部的公园;茶(树种植)园
～ -house 茶馆[室]
～ kitchen 茶房[站],茶点厨房
～ oil 茶油
～ plant 茶树
～ room 茶室,饮茶间;公共餐室,小饭馆
～ service 茶具
～ set 一套茶具
～ station 茶房[站]
～ table 茶桌[几]
～ -things 茶具
～ tray 茶盘
～ urn 水壶
～ wagon (有脚轮的)茶具柜
teaberry *n.* 冬青树
teacherage ['ti:tʃəridʒ] *n.* 教师宿舍,教师之家
teaching ['ti:tʃiŋ] *n.* 教学,讲授,训练;(*pl.*)学说,教导;*a.* 教学的,教导的
～ aid 教学用具
～ block 教学大楼
～ building 教学大楼
～ lab(oratory) 教学实验室
～ pool (室内)教学游泳池
teagle ['ti:gl] *n.* 滑车,滑轮组;卷扬机
tea-green marl 茶绿(色)泥灰岩
teak [ti:k] *n.* 柚木(树)
～ tread 柚木踏梯
～ trimming 柚木装修
～ wood 柚木
team [ti:m] *n.* 队,组,全体作业人员;*v.* 组成一队;协作,包给承包人;*a.* 队[组]的
～ design 成套设计
～ work 协力,配合,集体工作
～ yard 货车装卸场
teapoy ['ti:pɔi] *n.* 茶几,三脚几;三条腿装饰台
tear [tɛə] *v.* 撕[扯](裂,破),刺[戳,划]破;磨[破]损
～ line (包装袋)撕裂线
～ proof 抗撕裂的,耐磨的
～ resistance 抗撕[耐磨]力,抗撕强度,抗撕性
～ -shaped 梨形的
～ strength 抗撕强度
～ test 撕裂试验
tearing ['tɛəriŋ] *a.* 撕[扯]裂的;*n.* 撕裂(作用),断[破]裂,撕开;磨损
～ rupture 撕裂
～ strain 撕裂应变
～ strength 扯裂强度
tearout *n.* 撕断力;*v.* 撕下
tearproof ['tɛəpru:f] *a.* 抗撕裂的;耐磨的
tease [ti:z] *v.* 梳理,使表面起毛

teat [ti:] n. 突出部,凸缘;轴颈,枢轴
tebi n. 埃及泥砖
tecassir ['tekəsiə] n. 伊斯兰教寺院中为妇女设的席位,清真寺中之妇女楼座
technic ['teknik] n. 技巧[能,艺]
technical ['teknikəl] a. 技术[能]的,工艺[程]的;专门(性)的;学术上的; n.(pl.)技术术语[细则]
~ center 技术中心
~ characteristic 技术特征
~ classifcation 技术分类
~ code 技术规范
~ conditions 技术条件
~ cooperation 技术合作
~ design 技术设计
~ division of labo(u)r 技术分工
~ hangar 维修棚[车间]
~ institute 工艺学院,技术研究院,工业专科学校
~ mechanics 工程力学
~ process 工艺过程
~ regulation 技术规范
~ requirement 技术要求
~ skill 专门[技术]技能
~ specification 操作规范,技术[工艺]规范
~ term[word] 术语,技术名词
technicality [ˌtekniˈkæliti] n. 技术[专门]性;技术细节;术语
technician [tekˈniʃən] n. 技术(人)员,技师[工],专家
technicist ['teknisist] n. 技术人员,技师[工],专家
technicolor ['tekniˌkʌlə] n. 彩色(印片法,电影) a. 彩色的,天然色的
technics ['tekniks] n. 工艺(学),工程,专门技术;术语
technique [tekˈni:k] n. (工艺或艺术)技术,技能[巧];方法;工程[艺];技术装备
technocracy [tekˈnɔkrəsi] n. 专家管理,技术统制[治]

technologic(al) [ˌteknəˈlɔdʒik(əl)] a. 工艺(学)的,技术的
~ approach 技术表现手法
~ breakthrough 技术性突破
~ capability 技术能力
~ process 工艺过程
~ property 工艺特性
~ transformation 技术改造
technologist [tekˈnɔlədʒist] n. 工艺学家,技术专家
technology [tekˈnɔlədʒi] n. 工艺(学)的,(生产)技术;术语(学),专门语
~ assessment 技术评价[评定,估价]
~ import 技术引进
Tectona n. 柚木属
~ grandis 柚木
tectonics [tekˈtɔniks] n. 筑造学,工学;构造地质学;大地构造学
tectonism ['tektənizəm] n. 构造作用
tectorial [tekˈtɔ:riəl] a. 构成覆盖物的;房顶形的
tedge n. 直立物,竖管
tee [ti:] = T
~ (building) component T 形预制(建筑)构件
~ floor slab T 形楼板[桥面板]
~ hinge T 形铰链
~ joint T 形接头
~ junction 三通,T 形接头[连接,交叉]
~ (-)square 丁字尺
~ steel T 形钢,丁字钢
~ weld T 形焊缝
brace ~ 丁字撑杆
bulb ~ 圆头丁字铁
double sweep ~ 双弯三通
reducing ~ 缩径 T 形[三通]管

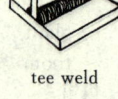
tee weld

teflon ['teflɔn] n. 聚四氟乙烯(塑料,绝缘材料),特氟隆
~ -coating (混凝土钢模接触面)特氟隆涂层
~ -neoprene bearing 氟板橡胶支座

~ plate bearing 氟板支座
tego *n.* 铅基轴承合金;酚醛树脂
　~ film 酚醛树脂胶膜[片](可用来做薄的胶合板)
tegula *n.* 瓦
tegular ['tegjulə] *a.* (似)瓦的,如瓦片一样排列的
teil [tiːl] *n.* 菩提树
telamon ['teləmən] *n. pl.* telamones (建筑)男像柱
telecamera [ˌteli'kæmərə] *n.* 电视摄像机
telechirics *n.* 遥控系统
telecontrol ['telikən'trəul] *n.; v.* 遥控,远距离操纵[控制]
telegraphone [ti'legrəfəun] *n.* 录音电话机
telemeasurement *n.* 遥测,远距离测量
telemeter ['telimiːtə] *n.* 遥测计,测距仪; *v.* 遥测
　~ equipment 遥测设备
　~ system 遥测系统,远距离控制方式
telemetering ['telimiːtəriŋ] *n.* 遥测(技术),沿无线电遥测线路传递信息
　~ depth meter 遥测深度仪
telemonitor *v.* 遥控,远距离监测
telemotion [teli'məuʃən] *n.* 无线电操纵
telephone ['telifəun] *n.* 电话(机) *v.* 打电话
　~ booth (公用)电话间[亭]
　~ box (公用)电话间[亭]
　~ cabinet 电话橱
　~ cable 电话电缆
　~ exchange room 电话总机室[交换台]
　~ installation 电话装置
　~ kiosk (公用)电话间
　~ line 电话线路
　~ plant 电话设备
　~ point 电话连接点,电话接线插口
　~ shelf 电话搁板
　extension ~ 电话分机

　inter (communicating) ~ 对讲[内部]电话
telescopic(al) [ˌteli'skɔpik(əl)] *a.* 望远镜(式)的,套筒(式)的,套管[叠]的,可伸缩的可抽[拉]出的
　~ gate 叠合滑动门
　~ leg (可)伸缩柱
　~ pipe 伸缩(套)管
　~ screw 套筒螺丝
　~ tripod 套管式三脚架
telescopiform [telis'kɔpifɔːm] *a.* 套叠式的,可伸缩的
telescoping *a.* 可伸缩的,伸缩式
　~ shuttering 套筒式模板,滑升模板
　~ tube 伸缩套管
　~ -type tremie 套筒式水下灌注混凝土导管
teleset ['teliset] *n.* 电视(接收)机,电话机
teleswitch ['teliswitʃ] *n.* 遥控开关
telesynd *n.* 遥测设备,远程同步遥控装置
television ['teliviʒən] *n.* 电视(学,术);电视(接收)机
　~ aerial 电视天线
　~ antenna 电视天线
　~ building 电视(台)大厦
　~ set 电视(接收)机
　~ studio 电视广播室
　~ surveillance 电视监控
　close(d) circuit ~ 闭路电视
televox ['telivɔks] *n.* 声控机器人,声控装置
telex ['tiliks] *n.* 用户[自动电传打字]电报;国际交换电报
　~ room 电报房,电传打字机室
telford ['telfəd] *n.; a.* 用几层压紧的石子铺成的地[路]面(的),大石块基道路
　~ stone 基层大块石
tellite *n.* 指示灯
telltale ['tel-teil] *n.* (液面,风向,航位等的)指示器;参照点,水准控制点;信号装置;(测应变杆 *a.* 起警告[监督]

作用的
~ board 控制[操作]信号盘
~ cock 警号龙头
Tellurium [te'ljuəriəm] *n.* 碲(Te)
~ lead 碲铅合金
temper ['tempə] *v.* (使)回[退]火,使软化,调合,捏(黏土); *n.* (钢的)韧[硬]度,(灰,泥等)硬性;钢中含碳量;回火色[度];调合物,掺和剂
~ screw 调节(用定位)螺杆
temperature ['tempəritʃə] *n.* 温度
~ sensitive paint 热敏油漆
curing ~ 养护温度
freezing ~ 凝固[冻结]温度
tempered ['tempəd] *a.* 经过回火的;调合的
~ hardboard 热处理硬质纤维板
~ (safety)glass 钢化玻璃,回火玻璃
~ steel 回火[淬硬]钢
temperer ['tempərə] *n.* 搅拌石灰、水泥等的机器;调合黏土的工人
tempering ['tempəriŋ] *n.* 回火(处理);人工老化;混合,调和
~ glass 钢化玻璃
Temperite *n.* (混凝土)氯化钙防冻剂
Tempietto *n.* 1502年意大利罗马圣彼得教堂院内的围廊式圆形建筑物
template ['templit] *n.* (= templet)承框,梁托,梁端垫块,承梁木[枕木];穿台[盘]板,样板,样[卡]规,刮尺;透明绘图纸; *v.* 放样
~ beam 垫梁
~ frame 放样架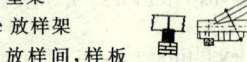
~ shop 放样间,样板车间,放大样地
circle ~ 画图样板
ellipse ~ 椭圆样板
rail gauge ~ 道尺,轨距规
stone ~ 石垫
temple ['templ] *n.* 神庙[宇,殿],神庙建筑,庙(宇),寺(院),圣堂,大教堂

~ architecture 神庙[寺院]建筑
Temples of Egypt 埃及神庙
cave ~ 石窟
temporal ['tempərəl] *a.* 瞬时[间]的,暂时的
temporality [ˌtempə'ræliti] *n.* 暂时性,短暂性
temporarily ['tempərərili] *ad.* 暂[临]时
~ erected 临时装设的
temporary ['tempərəri] *n.* (= tempo)临时建筑物,临时工; *a.* 暂[临]时的
~ access road 临时通道[便道]
~ assembly 临时装配
~ block 临时房屋,工棚
~ building 临时建筑物,工棚
~ bulk head 封端墙
~ cofferdam 临时围堰
~ construction 临时建筑(物),暂设工程
~ dwelling 暂设[应急]住宅,临时简易住宅
~ enclosure 临时性围墙
~ erection 临时架设[安装,竖立],试组装
~ expedients 临时维修措施
~ flush board 临时挡水板
~ ground[site] 临时用地
~ joint 临时接头[合,缝]
~ maintenance expedients 临时维修措施
~ paving 临时铺装,临时性路面
~ scaffold (简易)脚手架
~ seal for joint 临时封缝
~ shed 简易[临时性]工棚
~ stage (简易)脚手架,临时舞台
~ structure 临时结构
~ support 临时支柱[撑]
~ worker 临时工
~ works 临时工程,临时准备设施
tenacious [ti'neiʃəs] *a.* 粘(滞)的,粘着(力强)的;坚韧的,可延的;-tenaciously, *ad.*

~ clay 黏土
~ wood 坚韧木料
tenacity [ti'næsiti] *n.* 黏性,韧性[度],极限抗拉强度
tenaille *n.* (筑城)钳堡,凹角堡
tenantless ['tenəntlis] *a.* 无人居住的,空的
~ room (未使用的)空闲房间,(未租出的)空房
tenantry ['tenəntri] *n.* 租赁,承租人
tender ['tendə] *n.* 投标,报价(单),标件;看管者;煤水车;供应船[站],拖驳,交通艇;货币;脆化; *v.* 投标;报价;申请;偿付;脆化; *a.* 脆弱的,稳性小的
~ design 投标计划[设计]
~ documents 交单
~ of specified contractors 指名投票
~ for the construction of 投标承建
~ contractors 指名投标
~ sum 发包[投标]价格
~ system 投标制度
call for~s 招标
open sealed ~ 开标
tenderer ['tendərə] *n.* 投标者,申请人,提供者
tenement ['tenimənt] *n.* 住房,寓所,(公寓中的)一套房间,租用房屋,租地,地产
~ house[building] 经济公寓,长排式建造的住宅
~ style 长排建造式(数户住宅连接建造的方式)。联立式
tener *n.* 大型聚光灯
ten-footer *n.* 10 英尺见方的小屋侧房
tenia ['ti:niə] *n.* (多立克式建筑的)挑檐处带形花边饰
tenon ['tenən] *n.* 榫(舌),凸榫,销 *v.* 制榫,用榫接合
~ and mortise 雌雄榫
~ -bar splice 榫条拼接
~ cutter 榫凿
~ -cutting machine 制[开]榫机
~ hole 榫孔

~ joint(ing) 榫接(合)
~ saw (截)榫锯,手锯
~ through 穿[斗]
haunched ~ 加腋榫
oblique ~ 斜榫
tapered ~ 楔削榫,斜榫
tusks ~ 多牙榫

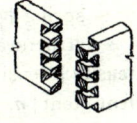
tusks tenon

tenoner *n.* 开榫机;接榫者
tenpenny nail (三英寸)大钉
tense [tens] *a.* 拉紧的,有应力的; *v.* 拉紧
tensibility [ˌtensi'biliti] *n.* 可拉[伸长]性
tenside *n.* 表面活性剂
tensile ['tensail] *a.* 拉[张]力的,受[张]拉的,抗拉[张]的,可延伸的
~ -socket 张拉索
~ strength 抗拉[拉伸]强度
~ strength across the grain (木材)横纹抗拉强度
~ strength paralleled to grain (木材)顺纹抗拉强度
tensility [ten'siliti] *n.* 可拉伸性,延性
tension ['tenʃən] *n.* 拉[张,压,应,弹]力;电压;拉[绷]紧;拉伸; *vt.* 拉[绷]紧,拉伸;tensional, *a.*
~ bar 拉杆
~ brace 拉撑,拉条
~ cable 受拉缆索
~ chord 受拉弦杆
~ crack 张拉裂缝,拉伸裂缝
~ diagonal 拉(力)斜撑,斜拉杆
~ element 挠性件(钢丝绳、链条等),受拉构件
~ fissure 张拉裂缝
~ fold 张褶皱
~ joint 受拉接头[合];张节理
~ member 受拉杆件[构件],拉杆
~ piece 拉杆,缀条
~ reinforcement 受[抗]拉钢筋,拉力钢筋

~ rod 拉杆
~ sleeve 拉紧套管,带丝扣的U形环(钢丝绳用)
~ spring 拉簧
~ structure 张拉[受拉]结构
tensity ['tensiti] *n.* 紧张(度)
tent [tent] *n.* 帷[帐]幕;帐篷(状物);住处,寓所 *v.* 搭[住]帐篷,用帐篷遮盖
~ -bed 行军床,(能调节温湿度的)帐篷式卧床
~ fly 帐篷盖
~ -guy 帐篷索
~ peg 帐篷桩
~ roof 锥形屋顶,帐篷顶
~ structure 张拉[受拉]结构
~ed arch 四心外心桃尖拱
tentative ['tentətiv] *a.* 试验(性)的,试行的,暂行[定]的;假定的,推测的;临时的;初步的; *n.* 试验;推测; (*pl.*)试用标准,临时[暂行]规定
~ assembling 工地装配,现场组装
~ specification 暂行(技术)规程[规范]
~ standard 暂[试]行标准
tenure ['tenjuə] *n.* (不动产的)占有(权,期,条件),(土地的)使用(权,期);(职务的)任期
tephrite *n.* 碱玄岩
tepidarium [,tepi'dɛəriəm] *n. pl.* daria;温水浴室
teram *n.* 楼梯踏步端头旋涡形装饰
Terco *d.* 碳化硅耐火材料
terebene ['terəbi:n] *n.* 松节油与松油精之混合物(用作防腐剂,涂料等),芸香烯
term [tə:m] *n.* 术语,专门名词;项;期限[间];边界,极限,界石[标];(谈判,合同)条款;费用,价钱;胸像柱
~ day 付款日期
~ of life 使用期限
~ of service 使用期限
~ of validity 有效期限
~ of works 工期
terminal ['tə:minl] *n.* 端饰,胸像柱;末端,端点,总站,终点站,转运基地,码头,港口,机场,车站;接线柱,引线,电极; *a.* 终点(站)的,末端的,电极的
~ area (机场)终点区;焊盘,接点
~ block 端接块;接线板;电缆管块
~ bolt 端螺栓
~ box 接线盒
~ building 街道末端的建筑物,终点站[车站,港口,码头]建筑物
~ department store 终点站[大站]百货商店
~ facilities 港口[码头]设施,终点站[车站]设备
~ feature 街道尽端布置,尽头街景
~ hotel 机场,车站的旅馆
~ pedestal 胸像台(座)
~ pin 尾销
~ point of base 基线端点
air ~ 机场,候机楼,避雷针
connecting ~ 夹具,接头

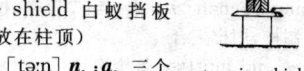
terminal pedestal

terminate ['tə:mineit] *v.* 终[停]止,收尾;满期;限定; *a.* 有限的,有尽的
termination [,tə:mi'neiʃən] *n.* 终点(站),终端(装置,设备);终止,满期;地界,界限;端接(法);辞职,解雇
~ of agreement 协定结束,契约满期
~ report 总结报告
terminus ['tə:minəs] *n. pl.* -ni,-nuses;终点(站);界限[标];极限
termite ['tə:mait] *n.* 白蚁
~ -proof 防白蚁(的)
~ resisting 防白蚁的
~ shield 白蚁挡板(放在柱顶)

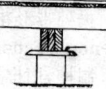
termite shield

tern [tə:n] *n.; a.* 三个一套(的);三重(的)
ternary ['tə:nəri] *n.; a.* 三个(一套,构成)的;三进制[变数,元,重]的
terne [tə:n] *n.* 镀铅锡合金薄铁[钢]板;

vt. 镀锡[铅]

terneplate ['tə:npleit] *n.* 镀铅锡合金薄钢板

terra ['terə][拉] *n.* 土,(土)地;地球
~ alba 石膏粉
~ cariosa (板状)硅藻土
~ house 街坊房屋
~ nera 黑土
~ nova wall plaster 墙面有色抹灰粉饰
~ ponderosa 重土,重晶石
~ rossa 红土
~ sigillata 黄土,古代红或黑色陶器
~ verde[verte] 绿土

terrace ['terəs] *n.* ; *a.* ; *v.* , -raced,-racing; *n.* 阳[平,露]台,平屋顶晒台,地坪;阶[台]地;斜坡上的排屋;里弄;街坊;公园式林阴路;路堤; *vt.* 做成台阶,筑坛; *a.* (成)平顶的;(成)梯形的
~ garden housing 台阶式花园住宅
~ house (各户有专用院子和分界墙的)联列式房屋,(建在坡地的)阶梯式房屋,高台基房屋,街坊房屋
~ -house development 联列式住宅区
~ roof 平台屋顶
~ slope 上下坡踏步台阶
~ steps 平台踏步
~ walk 梯级式步道
~ wall 梯级式挡土墙
~d aisle (剧院等的)台阶式通道
~d garden 露台花园
~d riprap 阶状抛石体

terra(-) cotta ['terə'kɔtə] *n.* 玻璃砖[瓦],陶瓷砖[瓦],空心[饰面]砖;赤陶[土];赤土陶器,陶塑制品;赤褐色; *a.* (陶器)用赤土制成的;赤褐色的
~ block finish 陶砖饰面
~ clay 陶塑用黏土
~ pipe 陶瓦管
~ sewage pipe 陶瓦下水管
~ tile 陶砖[瓦]

~ vase 赤陶花瓶
~ work 陶塑[瓷](制品)工程

terrain ['terein] *n.* 地形[势]地面[区]段;地层[体];岩层;领域; *a.* 地[形]的
~ intelligence data 地形资料
~ interpretation 土质[地层]鉴别

terrarium [tə'rεəriəm] *n. pl.* -ia;小动物饲养园

terra-rossa ['terə'rɑ:sə] *n.* 石灰石风化而成的红土

terratolite *n.* 密高岭土

terrazzo [te'rɑ:tsəu]【意】*n.* 水磨石,磨石子(地),预制[现浇]的陶瓷锦砖和灰浆制的面层
~ block 水磨石板
~ finish 水磨石饰面
~ floor 水磨石地面[坪]
~ layer 水磨石铺嵌工
~ -mix 水磨石混合料
~ tile (预制)水磨石板[砖]
~ work 水磨石工程

terrene [te'ri:n] *n.* 地区[段];地表[形];陆地,地球; *a.* 土(质)的;陆地的,地球的; terrenely, *ad.* terreneness, *n.*

terrestrial [ti'restriəl] *a.* 地面上的,陆(上,地,生)的;地球的; *n.* 地球;地球上的人[生物]

terre-verte ['terivə:t] *n.* 绿土;灰绿色

territorial [ˌteri'tɔ:riəl] *a.* 土地的;领土的

territory ['teritəri] *n. pl.* -ries;地区,范围;领土,领域;土地

terro-metallic clinker 黑黏土缸砖

terron *n.* 草源坯块

Tertiarium *n.* 特蒂锡铅焊料

tessella ['tesələ] *n. pl.* -lae;小块大理石,玻璃镶嵌物

tessellate ['tesileit] *v.* , -lated,-lating; *a.* ; *vt.* 把(地板等)镶嵌成(棋盘)花纹; *a.* 镶嵌成花纹的,镶嵌细工的; tessellation, *n.* ; tessellated, *a.*

~d pavement 嵌花(棋盘形)铺面,嵌装图案的人行道
tessellation [tesi'leiʃən] *n.* 嵌石装饰,棋盘形铺嵌
tessera ['tesərə] *n. pl.* tesserae;(小块大理石;玻璃、砖瓦等作成的)镶嵌物;嵌石铺面,镶嵌地块
tesseral ['tesərəl] *a.* 镶嵌物(似)的;等轴(晶系)的

tesserae

test [test] *n.*; *v.* 试[检,化]验,测验[试],检查;研究;准则,检验标准;试验品;试验方法;试验的事例或结果; testability, *n.*; testable, *a.*
~ beam 试验梁
~ bed 试验台座[面]
~ chart 草图
~ cube (混凝土或石材)立方体试块
~ cylinder (混凝土)圆柱体试块
~ floor 试验台座[面]
~ for soundness (水泥)安定性试验
~ in place 现场[工地]试验
~ on site 现场试验,工地试验
~ piece 试件[样,块]
~ -pit digging 试挖
~ plot 初步方案图
~ specimen 试件[块,样]
attrition ~ 磨损试验
cone penetration ~ 圆锥贯入度试验
testimonial [,testi'məunjə] *n.* 鉴定(证明)书;奖状,纪念品; *a.* 证明的,纪念的
testing ['testiŋ] *n.* 试[检,化]验,测验[验,量],检查,试车,研究,试验过程; *a.* 试验的;作试验用的; testingly, *a.*
testudo [tes'tju:dəu] *n. pl.* -dines;(罗马建筑)弓形屋顶
tete de pont 【法】*n.* 桥头堡
tether ['teðə] *n.* 系绳[链];界限,范围; *vt.* (用绳、链等)系,拴
tetrad ['tetræd] *n.* 四个(一组,一套);四位二进制

tetragon ['tetrəgən] *n.* 四边[角]形
tetrahedron [,tetrə'hi:drən] *n. pl.* -drons,-dra;四面体[形]
tetrapod ['tetrəpɔd] *n.* 四脚[角]锥体(防波堤用);四角支座
~ wall 四脚锥体堆墙
tetrapylon [,tetrə'pailən] *n.* 古建筑十字交叉通道的拱门
tetrastyle ['tetrəstail]
a. 建筑物正面有四根柱子的; *n.* 前面有四根柱子的建筑(如庙宇等)
tetrastylos
[,tetrə'stailəs] *n.* 前面有四根柱子的建筑(如庙宇等)

tetrastyle

tewel *n.* 烟囱,烟道
texrope *n.* 三角皮带
textile ['tekstail] *n.* (纺)织品[物];织物原料; *a.* 纺织的,织物的
~ belt 布皮带
~ block style 织物纹样砌块风格
~ carpet 织物[花纹]地毯
~ coating 织物涂布
~ fibres 纺织纤维
~ fabrics 织物,纺织品
~ finishing 纤维板贴面[饰面],织物整饰
~ glass 玻璃纤维
~ glass fabric 玻璃丝布
~ glass mat 玻璃丝毯
~ glass staple fibre 玻璃短纤维
~ tyre 编织轮胎
textolite ['tekstəlait] *n.* (层压)胶布板,夹布胶木
texture ['tekstʃə] *n.* 组织,结构,构造,纹理,刻花;织物;(材料的)质感,材质,特微; *vt.* 使具有某种结构[组织]
~ brick 粗面砖
~ paint 起纹漆,纹理漆
~d finish 织纹状饰面,粗饰面
~d pattern 纹理图案

~d paving 起纹道面
~d pile 花式绒头(指地毯)
texturization n. 纹饰,刻花
T-fillet welding T形接头贴角焊

T-fillet welding

T-fitting 三通,T形支管
T-fixture 丁字形夹具
T-gauge T形划线规
T-girder T形(大)梁
T-grade separation T形立体交叉
Thai style 泰国式(建筑)
thalamus ['θæləməs] n. pl. -ami;(古希腊)内室,闺房;花托
thallite ['θælait] n. 绿帘石
T-handle 丁字把手,丁字形手柄
tharandite n. 铁白云石
thatch [θætʃ] n. (盖屋顶用的)茅草;茅(草)屋顶,稻草屋顶;vt. 用茅草[稻草]盖屋顶
~ roofing (盖)茅草屋顶
~ed cottage 茅屋
~ed roof 茅草屋顶
grass ~ 茅草屋顶
thatchboard n. 草秸板,茅[稻]草板;胶合茅草板
thatchy ['θætʃi] a., thatchier, thatchiest;(像)茅草屋顶的
thaumasite n. 硅灰石膏,托马钙石
T-head 顶端有横档的梁,丁字头
~ bolt T形螺栓
~ post 丁字形柱
The American Institute of Planners (AIP)
美国规划工作者协会
The Architects Collaboratives(TAC)
(1964年格罗庇乌斯组织的)建筑师协作组织
theater, theatre ['θiətə] n. 剧院[场],电影院;(阶梯)教室,会场;舞台
~ dimmer 剧场调光器
~ -in-the-round 表演场设在观众坐席中央的剧院

~ restaurant 有演出节目的餐馆
drama ~ 戏剧院
lecture ~ 阶梯教室
open-air ~ 露天剧场
The Institution of civil Engineers Proceeding 土木工程师协会会刊
therblig(s) ['θə:blig(z)] n. (工艺操作中的)基本[分解]动作,(动作研究中的)基本元素
thermae ['θə:mi:] n. pl. 温泉(浴场);古罗马之公众浴场

thermae

thermal ['θə:məl] a. 热的,热量[力]的;温泉的;温度的;n. 上升(暖)气流;thermally, ad.
~ foil 保温箔片
~ glass 耐热玻璃
~ insulating course[layer] 隔热层
~ insulating material 隔热[保温,绝热]材料
~ insulation 隔热(材料),热绝缘
~ material 保温材料
~ switch 热控开关
~ wallboard 保温墙板
thermalloy n. 铁镍耐热耐蚀合金
thermic ['θə:mik] a. 热的,由于热(造成)的;-thermically, ad.
~ boring[drilling] (混凝土)热法钻孔
thermit ['θə:mit] = **thermite** ['θə:mait] n. 铝热剂;铝粉焊接剂
~ joint 铝热剂,焊接缝
~ welding 铝热剂焊
thermocolo(u)r ['θə:məukʌlə] n. 热敏油漆示温涂料,色温标示
thermocompression ['θə:məukəm'preʃən] n. 热压
~ bonding 热压焊接[粘合]
thermocutout ['θə:məu'kʌtaut] n. 热保险装置,热断流器
thermoduric [,θə:məu'djuərik] a. 耐热的

thermo-hygrograph [ˈθəːməuˈhaigrəgrɑːf] *n.* 温湿计,温度湿度记录器

thermoindicator paint 示温漆,变色漆

thermolabile [ˌθəːməuˈleibil] *a.* 不耐热的

thermolith *n.* 耐火水泥

thermopaint [ˈθəːməupeint] *n.* 示温涂料,测温漆,彩色温度标示漆

thermopermalloy *n.* 铁镍合金

thermophilic [θəːˈmɔfilik] *a.* (细菌等的)嗜热性的,耐热(性)的

thermoplastic [ˌθəːməuˈplæstik] *a.* 热塑(性)的,加热软化的;*n.* 热塑塑料[材料]
~ adhesive 热塑性粘合剂
~ film 热塑性塑料薄膜
~ laminate 热塑性层压,[复合]材料
~ material 热塑性材料,热塑料
~ putty 热塑性油灰
~ resin 热塑性树脂
~ tile 热塑性屋面料
~ vinyl tile 热塑乙烯基饰面砖

thermosensitive [ˌθəːməuðˈsensitiv] *a.* 热敏的
~ paint 热敏油漆

thermoset [ˈθəːməset] *a.* 热固[凝]的; *n.* 热固(性),热凝,热固(性)塑料
~ plastics 热固塑料

thermosetting [ˈθəːməuˌsetiŋ] *a.* 加热成形后即硬化的;*n.* 热固(性),热固化
~ adhesives 热固性胶,热固性胶粘剂
~ cement 热固性粘合胶
~ material 热固性材料,热固塑料
~ plastics 热固(性)塑料
~ resin 热固性树脂

thermostable [θəːməuˈsteibəl] *a.* 耐热的,热稳定的

thermostat [ˈθəːməstæt] *n.* 恒温器[箱],热动开关
~ varnish 耐热漆,热稳定漆

thermostatic(al) [θəːməuˈstætik(əl)] *a.* 恒温(器)的,热静力学的
~ -chamber 恒温室

thermoswitch [ˈθəːməuswitʃ] *n.* 热敏[控]开关

thermotolerance *n.* 耐热性

thermoweld *n.* 热焊接,熔焊

thesaurus [θi(ː)ˈsɔːrəs] *n. pl.* -ri, -ruses; 仓库,宝库;词典,百科全书

Theseion [θiˈsaiɔn] *n.* 古希腊神庙

thick [θik] *a.* 厚[粗,深]的;浓[稠,密集]的;不透明的,多雾的; *n.* 厚度;最厚[浓,密,强烈]部分; *ad.* 厚[浓]地,不清晰地;-thickly, *ad.*
~ china 贴面瓷板
~ cylinder 厚壁圆筒
~ darkness 漆黑
~ -film lubrication 厚油膜润滑
~ glass 厚玻璃
~ lift paving 厚层铺筑法
~ line 粗线
~ oil 稠油
~ plank 厚板
~ wall 厚壁

thick china

thicken [ˈθikən] *v.* (使)变厚[粗,密,浓,浊];复杂化
~ ed-edge slab 厚边板

thickener [ˈθikənə] *n.* 浓缩[增稠]器[剂],(污泥)浓缩池,沉淀浓缩装置

thickening [ˈθikəniŋ] *n.* 加厚,浓缩,稠化;(按比例重)分级;增稠剂,被加厚的东西
~ agent 增稠剂

thicket [ˈθikit] *n.* 灌木丛[林],障密集的东西

thickness [ˈθiknis] *n.* 厚(度),粗(度),稠密;浓(度),黏稠[性];混浊,多雾[烟];最厚[粗,密,浓]处;(一)层
~ moulding 厚线条饰
~ of coating 抹灰[粉侧,油漆]厚度
~ of lining 衬砌厚度
~ shear 厚度切变

thicknesser *n.* 刨板机,单面压刨;测榫器;划线盘;平面规
double ~ 双面压刨
thill [θil] *n.* (车)杠,辕(杆)
thimble ['θimbl] *n.* (活动)套管[筒,环],铁[嵌]环,梨形圈;盲管道;thimbled, *a.*
~ joint 套筒连接
steel ~ 钢套管
thin [θin] *n.;v.;ad.;a.*, thinner, thinnest; *a.;ad.* 薄[细,稀]的;(颜色)淡的; *n.* 薄处,细小[稀薄]部分; *v.* (使)变薄[细,稀,淡],削[磨]去
~ board 薄板
~ concrete overlay 薄层混凝土罩面
~ film 薄膜
~ flattish chips 薄片石屑,扁石屑
~ iron sheet 薄铁板
~ -lift 薄层
~ membrane box structrue 薄壁箱形结构
~ pipe 薄壁圆筒[管]
~ plaster 稀墙泥,涂墙稀灰泥
~ -plate element 薄板单元[构件]
~ plate-glass 薄(平)板玻璃
~ plate structure 薄壁[板]结构
~ (steel) sheet 薄钢板,钢皮
~ sheet metal 薄金属片[板]
~ shell 薄壳(结构)
~ -shell barrel roof 筒形薄壳屋顶
~ shell constructure 薄壳结构
~ shell surface 薄壳面
~ slab 薄板
~ slab constructure 薄板结构
~ -walled constructure 薄壁结构
~ -walled structure 薄壁结构
~ -webbed girder 薄腹梁

thin-webbed girder

T-hinge T形铰链,T形折页
thinner ['θinə] *n.* 稀释[冲淡]剂,(油漆)稀料,溶剂

paint ~ 涂料稀释剂
third [θə:d] *n.;a.* 第三档[速率];第三(的);三分之一(的); thirdly, *ad.*
~ floor[story] (英)四楼,(美)三楼
~ programme 第三期工程
thirsty ['θə:sti] *a.*, thirstier, thirstiest;(干燥)的,有高度吸水性的;渴(望)的; thirstily, *ad.*; thirstiness, *n.*
~ soil 干燥土
thistle ['θisl] *n.* 蓟(一种草)
~ board 轻质板,(纸面)石膏板
thixotrope *n.* 触变胶
thixotropic [θiksə'trɔpik] *a.* 触变的,摇溶的
~ agent 触变[摇溶]剂
~ paint 触变漆
thole [θəul] *n.* 壁龛
tholobate ['θɔləbeit] *n.* 圆屋顶座,支持圆顶之下层结构,穹窿顶底座
tholos ['θɔuləs] *n. pl.* tholoi;(古建筑)陵园形建筑物,蜂窝形圆形陵墓
tholus ['θəuləs] *n. pl.* tholi;(古建筑)圆形建筑物,蜂窝形圆形陵墓
thorough ['θʌrə] *n.* 通过墙身的顶砖; *a.* 彻底的,完全的,详尽的,周到的,精确的,细节的;
~ bond 穿通砌石
~ consolidation 完全凝固
thread [θred] *n.* 螺纹[线,齿];线,细丝,纤维;流线,(河流的)中泓线; *v.* 刻螺纹,穿绳[线];通过
~ cutter 丝板,螺纹刀具
~ cutting 切削螺纹
~ driver 螺钉旋具
~ fraise 螺纹铣刀
~ gauge 螺纹规
~ glass 线纹饰玻璃
~ stay 螺旋撑条
~ stripping 螺纹滑扣
~ed bolt 螺栓[杆]
~ed connection 螺纹[丝扣]接合
~ed outlet (井点总管上的)螺丝接头

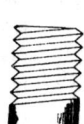
thread

~ed pipe 螺纹管
~ed rope 螺纹杆
~ed sleeve 螺纹套筒
~ed stud 螺杆,双头螺栓
glass ~ 玻璃丝
threaten ['θretn] *v.* 有……的危险,威胁; threatener, *n.*; threatening, *a.*; threateningly, *ad.*
three [θri:] *n.*;*a.* 三(个),三个一组,一组[系列]中的第三个
~ -aisled basilica (古罗马)三条通道的教堂
~ attributes of colour 色彩的三项属性(指色相、明度、彩度)
~ -centered arch 三(中)心拱
~ -dimensional design 立体造型设计
~ -dimensional frame 空间构架
~ -element arch 三心[分]拱
~ -hinged arch 三铰拱
~ -layer particle board 三层碎料板
~ -level structure (立体交叉的)三层式[三平面]构筑物
~ -light window 三分格窗,三扇窗
~ -pinned arch 三铰拱
~ -pin plug 三脚插头
~ piping system 三管道(热水管、冷水管和回水管)配置式
~ -ply 三夹板;三层(厚);三重[股]的
~ -ply board 三夹板
~ -ply butt 三折合页
~ -pole switch 三级开关
~ primary colours 三原色
~ -prong plug 三脚插头
~ -quarter brick 3/4 砖
~ -quarter header 3/4 丁砖
~ -quarter column 露出墙面 3/4 柱径的壁柱
~ quarters bat[header]3/4 砖
~ -space 立体的,空间的
~ -square 三角[棱]的,截面成等边三角形的
~ -square file 三角锉

~ -start screw 三头螺纹
~ -storied house 三层楼房屋
~ -way cock 三向龙头,三通旋塞
~ -way flat slab 三向无梁楼板
~ -way switch 三向开关
~ -way valve 三通阀
throat [θrəut] *n.* 滴水槽;(水道等的)咽喉段,缩窄段;入口,狭道,缩口管;喷管临界截面;焊缝(厚度);锯沟;喉部[管]
~ of hinge 铰颈
~s of crossing 交叉口入口,广场入口
~ width 入口宽度
throating ['θrəutiŋ] *n.* 滴水槽[沟]
~ plate 拨水板
throttle ['θrɔtl] *n.*,*v.*, -tled,-tling, *n.* 节流;风[油]门,节流[气]阀,节[扼]流圈;气管;光圈[闸,阑],快门; *vt.* 节流[气],调节,阻塞; throttler, *n.*
~ governor 节流调速器
~ lever 油门杆,节气阀杆
~ nozzle 节流喷嘴
~ steam 用节气阀调节蒸气
~ valve 节流阀,风门,调节阀,减压阀
through [θru:] *prep.*;*ad.* 通[穿,经,透]过,贯穿,直通[达],自始至终,经由;完全[毕],彻底,全部地; *a.* 下承的;直达[联运,连续,通过,贯穿]的
~ arch 下承拱
~ beam 下承[连续]梁
~ bolt 贯穿[双头]螺栓
~ bond 横向砌合
~ -carved work 镂花雕刻
~ hole 通孔,透眼
~ lintel 穿透过梁,整墙过梁
~ metal 金属支架
~ -pass 过道
~ shake (顺木纹方向的)贯透裂,劈裂
~ stone 系石,穿墙石
~ tenon 贯穿雄榫,穿透榫舌

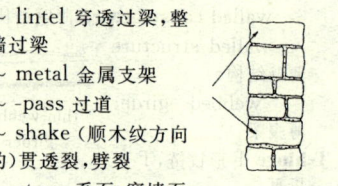

through stones

~ -voided 全空心的
~ -wall flashing 穿墙挡水板

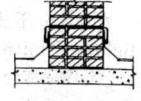
through-wall flashing

thrower [ˈθrəuə] *n.* (散货)平舱机,抛料机;投掷[喷射]器

thrum [θrʌm] *n.*;*v.*, thrummed, thrumming; *n.* 纱[线]头,碎线绳屑;*v.* 把绳屑嵌入(防擦,堵漏)

thrust [θrʌst] *n.*;*v.*, thrust, thrusting; *n.* 推[拉,牵引]力,侧向压[拉]力,轴向压力,反推力;推进,冲,插;*v.* 推[冲插]入,延伸; thruster, *n.*
~ ring 圆屋顶支承环,止推环
~ washer 止推垫圈

thumb [θʌm] *n.* 馒形饰;拇指;*vt.* 翻阅;(用拇指)压,揿;thumblike, *a.*
~ latch [lock] 门窗插销
~ moulding 拇指饰,扁平装饰线条
~ nut 元宝[翼形,蝶形]螺帽
~ -piece 指动碰簧销
~ pin [tack] 图钉,揿钉
~ rope 手捻草绳
~ -turn (室内)指旋销,指旋器
~ wheel 旋钮,拨盘

thumber [ˈθʌmə] *n.* 制动器

thumbnail [ˈθʌmˈneil] *n.* 略图,短文;拇指甲;*a.* 简略的,很小的;*vt.* 作简略说明
~ sketch 略[草]图

thumbscrew [ˈθʌmskruː] *n.* 指旋[翼形,蝶形]螺钉

thumbtack [ˈθʌmtæk] *n.* 揿[图]钉;*vt.* 用揿钉钉上

thwach [θwæk] *n.*;*v.* 夯,捣实,重击,猛烈

thymele *n.* (古希腊)剧院舞台前的讲坛

thyroid [ˈθairoid] *n.*;*a.* 盾形(的)

Tibetan [tiˈbetən] *a.* 西藏的
~ style (中国)西藏式(建筑)

ticker [ˈtikə] *n.* 断续装置,振动器;收报机,钟摆;作滴答响的东西

ticket [ˈtikit] *n.* 票,券;单据;标签;执照,许可证,证明书;计[规]划;*vt.* 加标签于;为……购票
~ counter 售票台
~ lobby 售票厅

tie [tai] *n.*;*v.*, tied, tying; *n.* 拉杆[条],系杆;角撑架;锚碇;绳,箍;结,扣,系结法;联系;(美)枕木,轨枕;*v.* 系,拴,扎,结,扣,加箍筋;拉紧,束缚,联系;把(轨)固定在轨枕上,给(铁路)铺枕木
~ -back 拉杆[条],锚碇
~ -back wall 锚定挡墙
~ bar 拉[系]杆,连接[转向]杆
~ bolt 连接[系紧]螺栓,地脚螺栓
~ brick 束砖
~ cable 连接电缆
~ coat 粘结层
~ column 系杆加劲柱
~ element 连系杆件
~ line 直接连线,拉线
~ member 系件,拉杆
~ pile 拉桩
~ plate 系[垫,固定,格构]板
~ pluy 塞孔栓,木塞;键,销
~ rod 拉杆[条],系杆,(轨距)联杆
~ -strut 系[撑]杆
~ wall 拉结墙,横墙
~ wire 绑扎用铁丝,火烧丝(绑钢筋用)
~d arch 系杆拱(有拉杆的拱),弦系拱
~d arch frame 有拉杆的拱式构架
~d retaining wall 锚定式挡土墙
~d rib arch 有拉杆的肋拱
~d structure 有系[拉]杆的结构
diagonal ~ 斜拉杆
dragon ~ 拉杆,角铁联系;支承脊橡梁
frame cross ~ 底架横撑
land ~ 着地拉杆
lateral ~s of column 钢筋柱箍,钢架柱的侧面联条

tie

say ~ 吊杆
wall ~ 系墙铁
tieback ['taibæk] *n.* 牵索,拉条[杆],横[系]梁
~ anchor 拉条[牵索]锚杆,牵索锚具
~ method 锚[后]拉挡土墙施工法
tie-bed *n.* (铺铁路的)碎石层
tie-down ['taidaun] *n.* 系紧,栓系;捆索,系固装置
~ fitting 系固装置
tiehole ['taihəul] *n.* 系孔
tie-in ['taiin] *n.* 连[绑]接;捆成束;打结;关连;相配(物)
tieline ['tai'lain] *n.* 系线,连线,对角线,联络[转接]线,直达[专用]通信线路
tiepiece *n.* 系梁
tier [tiə] *n.* 层,排,(钢丝绳)盘,单层砖墙,剧场中的渐升座位行列;[taiə] 包扎工[装置],包扎工; *v.* 堆积成层,分层布置,堆垛[叠]
~ building 多层房屋
~ pole 排杆
~ table 宝塔形台
tierceron ['tiəsrən] *n.* 居间(拱)肋,放射肋,枝肋(哥特式建筑拱顶交叉肋条间的辅助肋条)

tierceron

tige *n.* 柱身[干],顶杆
tiger ['taigə] *n. pl.* -gers,-ger; tiger-like, *a.*
~ board 泰格板(一种石膏板)
~ grain (木材)虎斑纹理
~ wood 非洲核桃木,虎斑木
tight [tait] *n.* 水密钻孔;坚固岩层; *a.* 紧(固,密,封)的,不透[漏](水,油,气)的;拉[绷]紧的;坚固的,密集的
~ -fit 紧配合
~ joint 紧密结合
~ knot (木材的)紧节
~ riveting 紧密铆合
~ rope 绷紧绳索

light ~ 不漏光
wind ~ 不透风
tighten ['taitn] *v.* (使)变紧,固定;拉[扎]紧;密合[封],(使)不漏
~ing pulley 张紧轮
~ing screw 固紧螺钉
tightener ['taitnə] *n.* 张紧工具[装置]
tightly ['taitli] *ad.* 紧(密实),紧紧地
tightness ['taitnis] *n.* 紧密度[性],张紧度,不透水[气]性
water ~ 水密性
tightwire ['taitwaiə] *n.* (钢制的)绷索,钢索
tigna *n.* 木屋顶的系梁
til [til] *n.* 浪线(即 ~),波形号
tile [tail] *n.*; *v.*, tiled, tiling; *n.* 瓦(管,筒,沟);瓷[面,花,空心,铺地]砖,(软木、橡胶等制成用于铺地面的)弹性砖片; *vt.* 铺瓦,装[埋]瓦管,贴(瓷)砖;tilelike, *a.*
~ -and-half tile 宽瓦(较普通瓦宽一半)
~ batten 挂瓦条
~ clip 固瓦弹簧夹
~ drainage 瓦管排水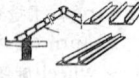
tile batten
~d rainage 瓦管排水
~d valley 瓦管屋面天沟
~ facing 贴瓷砖,铺瓷砖面
~ fillet 瓦口
~ floor 瓷[花]砖地面
~ hanging 缠结瓦片,挂瓦,盖墙瓦
~ hook 挂瓦钩
~ -hung wall 贴瓦砖墙
~ laying 埋瓦管,铺瓦
~ lining (瓷)砖(内)衬
~ lintel floor 空心砖密肋楼板
~ nail 瓦钉,固瓦木销
~ of wood (企口)木瓦,屋面板
~ paving 瓷[花]砖铺面,铺瓦[砖]
~ pin 瓦钉
~ roof 瓦屋顶[面]

~ roofing 铺瓦,瓦屋面[顶]
~ stone 石瓦,石板
~ walk 花砖人行道
acoustic(al) ~ 隔声板
acoustolith ~ 吸声贴砖
asbestic half-round ~ 石棉(水泥)脊瓦,半圆形石棉瓦
asbestos[asbestic] ~ 石棉瓦,石棉水泥板
backup ~ 空心垫砖
bent ~ 曲瓦,槽瓦
bond ~ 搭盖瓦
book ~ 屋脊瓦,(带凹凸边的)书形空心砖
building ~ 空心砖,(板)瓦
ceramic ~ 瓷砖
channel ~ 槽瓦
clay facing ~ (美)特制陶瓦
corrugated ~ 波纹瓦
distribution ~ 配水瓦管
double gutter ~ 双企口瓦
dry mo(u)lded ~ 干塑瓦管
Dutch ~ 饰瓦;彩砖,荷兰砖
eaves course ~ 屋檐瓦
fibreglass ~ 玻璃钢瓦
flap ~ 凹瓦
floor ~ 地面砖
furring ~ 墙面瓷砖
glass ~ 玻璃瓦板
glazed (facing) ~s 釉面[琉璃]瓦
grey ~ 青瓦[砖]
gutter ~ 沟瓦
hip ~ 肩脊瓦
hollow ~ 空心砖
interlocking ~ 咬接瓦
mission ~ 半圆形截面瓦,拱形瓦,阴阳瓦
mosaic ~ 彩色镶嵌瓷砖
overlapping ~ 搭接瓦
pan ~ 波形[槽形]屋面瓦
partition ~ 内隔墙用的空心砖
plain ~ 平[无楞]瓦

promenade ~ 铺面缸砖
quarry ~ 缸砖
quilt ~ 瓷砖
reeded ~ for stairs 防滑踏步砖
ridge ~ 脊瓦
Roman ~ 罗马式瓦(仰瓦为槽形,覆瓦为筒形)
roof ~ 屋面瓦
single-lap ~ 平瓦
skirting ~ 壁脚砖,踢脚(板)砖
slate ~ 面砖,石板
Spanish ~ 筒瓦,西班牙式屋面瓦
stove ~ 搪炉砖,瓷[面]砖
terrazzo ~ 水磨石面砖
valley ~ 天沟瓦
ventilating ~ 通风瓦,换气瓦
vinyl ~ 乙烯基(楼面)板
vitrified ~ 陶(土)瓦(管),上釉[琉缸]瓦管
wall ~ 墙面砖
water-rib ~ 挑檐瓦
weather ~ 墙顶盖瓦
Y- ~ 三通瓦管

tileboard *n*. 花砖式饰板,花砖式饰面拼板
tilecoping on gable 封檐瓦垄[压顶]
tilecreasing *n*. 墙头挑出瓦
tiler ['tailə] *n*. 砖瓦工,制砖瓦者,铺瓦[贴砖]工
~ work 贴面工程
roof ~ 铺瓦工
tilery ['tailəri] *n. pl.* -eries;制瓦厂,瓦窑;装饰性砖瓦铺贴术
tilestone ['tailstəun] *n*. 石瓦[板]
tiling ['tailiŋ] *n*. 盖瓦,贴[铺]砖;瓦[花]砖;瓦[砖]面,砖瓦结构
~ batten 挂瓦条
~ work 贴砖工程
tilt [tilt] *n*. 倾斜[倒];斜坡[面,率];仰[倾]角;车[帐,帆布]篷,覆布;木屋;落锤; *v*. (使)倾斜[卸],斜置;用篷盖;摇[摆,上下]动;锤击; tiltable, *a*.; tilter, *n*.

~ and turn window 倾斜转动式窗
~ angle 倾(斜)角
~ hammer 落锤,跳动锤
~ roof 帐篷式屋顶
~ -up construction method 预制构件装配施工法,预制大型墙板装配施工法
~ -up door 上翻门
~ing board 椽上封檐板条,檐垫板
~ing door[gate] 卧倒门,倾倒式闸门
~ing fillet 披水条,檐板条
~ing seat (可调整靠背斜度的)转动式座椅

timber ['timbə] *n.* 木材[料],栋木,肋材;原木;树木,森林; *vt.* 用木材造,用木料支撑; timbery, *a.*
~ -and-tile roof 木托瓦屋顶
~ apron 木底板[护坦]
~ assortment 木材材种
~ beetle 蛀木筒蠹
~ bent 木排架
~ blocking 木垫块,枕木垛
~ board 木板
~ boarding 木地板
~ brick 木砖,木块
~ building 木房屋,木结构建筑
~ cladding 木板贴面
~ cleat 木楔,木屋架上承托檩条的三角木
~ component 木构件
~ connector 木结构结合件,合缝销,暗销,结合环

timber connector

~ construction 木结构
~ dog (木)扒钉,蚂蝗钉
~ dryer [drier] 木材干燥设备,木材烘干炉
~ drying kiln 木材干燥窑
~ finishes 木面层,木地面
~ floor 木地板[铺面]
~ form 木模板
~ frame 木构架

~ -framed brick construction 木构架包砖结构
~ grain 木纹
~ -grillage 木格床
~ header 木丁头
~ house 木结构的住宅
~ joints 木连接
~ knot 木节
~ lagging 挡土板
~ meeting piece 止水木条
~ partition 木隔墙[板]
~ penelling 木镶板
~ pile 木桩
~ plain splice 木材平面拼接
~ planking 木(面)板
~ plate girder 木板梁
~ preservation 木材防腐
~ preservative agent 木材防腐剂
~ products 木材制品
~ saddle 木支座
~ saw 木锯
~ scaffold 木脚手架
~ seal 木质止水[水封],止水木条

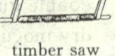
timber saw

~ separator 木隔板
~ sheeting 木板(桩);木挡板
~ shuttering 木模板
~ sill (beam) 木槛,门槛木
~ skirting 木踢脚板
~ spacer 定位木块
~ splicing 木材拼接(法)
~ structure 木结构
~ strut 木支柱[撑],撑木
~ed stope 有木支撑的工作面
creosoted ~ 油浸木材,用杂酚油防腐处理过的木材
dressed ~ 刨光木材
glued ~ 胶合木材
jack ~ 小椽
laminated ~ 叠层[胶合]木材
paint on ~ 木材面油漆
roof structures of ~ 各式木屋顶结构

sawn ~ 锯材
seasoned ~ 风干木材
sole ~ 垫木
squared ~ 方木
structural ~s 建筑用木料
unwrought ~ 未加工材

timberer ['timbərə] ***n.*** 木材工人,支架工

timbering ['timbəriŋ] ***n.*** 木[结构]材；木结构,用木制造；木模；加固,支撑[架],拱鹰架
~ gallery 木框支撑式廊道
~ machine 支架安装机
~ materials 支撑材料
~ of a cut 挖方支撑
~ of trenches 沟槽木支撑
~ set (隧道)木支撑

timberman ['timbəmən] ***n. pl.*** -men；(搭脚手架的)架子工,支架工；木材商人

timberwork ['timbəwəːk] ***n.*** 木结构(尤指屋架,栋梁等),结构材；木工作业,木材加工厂

timbo ***n.*** 树藤；用作家具和内部细工的红木板材

time [taim] ***n.*** ；***a.*** ；***v.*** , timed, timing；***n.*** 时(间,刻,期),时代；工作[占用,所需]时间；次,回；节奏；计时工资率；(***pl.***)倍,乘；***v.*** 安排[测定,记录]……的时间,选择时间[机]；***a.*** 定时[期]的,(记录)时间的；分期(付款)的
~ and materials 工时和材料
~ of final setting (混凝土的)终凝时间
~ of initial setting (混凝土)初凝时间
~ of setting 凝固时间

timely ['taimli] ***ad.*** ；***a.*** , -lier,-liest；及[适]时的,正好；timeliness, ***n.***
~ completion 按期(或提前)竣工

time-out ['taim'aut] ***n.*** 窝工[停工]时间；临时停工；时间已过

tin [tin]***n.*** ；***a.*** ；***v.*** , tinned, tinning；***n.*** 锡；白铁(皮)；马口铁；厅,罐,桶；***a.*** (含)锡的,白铁皮[马口铁,锡]制的；***vt.*** 镀锡,包白铁皮,装成罐头；tinlike, ***a.***
~ bar 锡块[条],白铁皮原板
~ coated steel plate 镀锡钢板
~ foil [leaf] 锡箔[纸]
~ -lined pipe 衬锡管
~ pipe 锡管
~ plate [sheet] 镀锡铁皮,白铁皮,马口铁
~ roofing 白铁皮屋面
~ snips 铁皮剪刀
~ solder 锡焊料,焊锡
~ ned plate 镀锡铁皮,白铁皮,马口铁
~ ned sheet iron 镀锡铁皮,白铁皮,马口铁

tincal ['tiŋkɑːl] ***n.*** (原,粗)硼砂

tincture ['tiŋktʃə] ***n.*** ；***v.*** , -tured,-turling；***n.*** 酊剂；色泽；气味；染料；***vt.*** (浸)染

tinge [tindʒ] ***n.*** ；***v.*** , tinged, tingeing or tinging；***n.*** (较淡的)色调[彩]；气味；***vt.*** (较淡地)着色于,染；使带气息；tinger, ***n.*** ；tingible, ***a.***

tingle ['tiŋgl] ***n.*** (砌砖用)线垫；固定夹片；-tingler, ***n.***

tinker ['tiŋkə] ***n.*** 白铁工,修补(工)；***v.*** 做白铁工,修补,调整；tinkerer, ***n.***

tinman ['tinmən] ***n. pl.*** -men；白铁工

tinner ['tinə] ***n.*** 白铁工

tinny ['tini] ***a.*** , -nier,-niest；(含,似)锡的；不耐久的；tinnily, ***a.*** ；tinniness, ***n.***
~ glaze 含锡釉

tinol ***n.*** 锡焊膏

tinplate ['tinpleit] ***n.*** ；***v.*** , -plated,-plating；***n.*** 马口铁,镀锡铁[铜]皮；***vt.*** 镀锡,包马口铁；tinplater, ***n.***

tinsel ['tinsəl] ***n.*** ；***a.*** ；***v.*** , -sel(l)ed, sel-(l)ing；***n.*** (金属)箔,金属丝[片]；

华而不实的(东西);*a.* 以金属箔[丝]做成[装饰]的;华而不实的;*vt.* 用金箔装饰;tinsellike, *a.*
~ cord 箔线

tinsmith ['tinsmiθ] *n.* 白铁工,锡工
~'s solder 锡焊条

tint [tint] *n.* 色调[彩,度],浓淡,浅[淡]色;(雕刻中的)线晕;*vt.* 着[涂,染]色于,微染;tinter, *n.*
~ colour 淡[浅]色
~ed glass 有色玻璃
~ed granolithic 有色细石混凝土
~ed paint 有色油漆
~ing strength 着色能力,色度

T-intersection T 形(平面)交叉,丁字形交叉

tinty ['tinti] *a.*, tintier, tintiest;色彩不调和的

tiny ['taini] *a.*, -nier, niest;微[极]小的,很少的;tinily, *ad.*; tininess, *n.*

tip [tip] *n.*; *a.*; *v.*, tipped, tipping; *n.* 倾斜[覆,卸];翻车机,卸货机桥;垃圾[弃土]场;尖(端),末梢,触点,刃尖;刀片;接[管]头,预测; *a.* 用倾斜角度卸货的; *v.* (使)倾斜[卸],翻倒[转],倒出;装尖头[龙头];暗示;tippable, *a.*
~ diameter (圆木等的)小头直径
~ end (原木的)小头,原木梢,梢径
~ jack 塞[插]孔,尖端[单孔]插座
~ ping bucket 倾斜(料)斗,倾倒[翻转]式吊桶,翻斗
~ping gear 倾倒[倾卸]装置
~ping hopper 卸料车,倾倒漏斗
~ping truck 翻斗卡车
~ping tube 溜管
~ping unit 倾翻[翻斗]装置
~ping wagon 翻斗料车

tipcart ['tipkɑ:t] *n.* 倾卸车,翻斗车

tip-off ['tipɔ:f] *n.*; *v.* 脱[拆]焊,拆封;翻倒

tip-over *v.* 翻倒,倾覆

tipper ['tipə] *n.* 倾卸装置,倾翻机构;自(动倾)卸车,翻斗车

~ -hopper 翻斗
automatic ~ 自动倾卸槽
end ~ 后倾自卸车
wagon ~ 翻车机

tippler *n.* 翻车机,翻笼;翻车工;自卸卡车
~ lift 自动倾卸升降机

tiradaet *n.* 粘结[合]剂

T-iron T 形钢[铁],丁字形钢[铁]
double ~ 工字钢[铁],双 T 形钢[铁],σ 形钢铁

tissue ['tisju:] *n.* 织物,薄绢;薄[棉]纸;

titanium [tai'teiniəm] *n.* 钛(Ti); titanitic, *a.*
~ dioxide 二氧化钛;钛白(一种白色颜料)
~ oxide porcelain (二)氧化钛陶瓷
~ paint 钛涂层
~ white 钛白(即二氧化钛)

title ['taitl] *n.* 题目,图标;职别,学术,所有权;锦标
~ block 工程图明细表
~ insurance 产权保证

T-joint T 形管,三通,T 形接头,T 形焊接

T-junction T 形交叉,丁字形交叉
~ box T 形套筒

toadback handrail 哈蟆背形(楼梯)扶手

toadstone ['təudstəun] *n.* 玄武斑岩

to-and-fro ['tu(:)ən'frəu] *v.*; *ad.*; *v.*, *n. pl.* -fros;往复(的,地),来回(的,地)
~ (aerial) ropeway 往复(架空)索道
~ oscillations 来回摆动

toaster ['təustə] *n.* 烘炉,烤面包器

toat *n.* 台刨把手

tocsin ['tɔksin] *n.* 警钟[报],警戒信号

toe [təu] *n.*; *v.*, toed, toeing; *n.* 坡[柱,堤]脚,(基,坡)趾,脚趾(状物);尖头,柄[柱]尖;(炮膛,钻孔)底;斜钉;焊边;轴踵; *v.* 斜敲(钉子);(轮子)斜向;toelike, *a.*
~ board 踏[搁]脚板;底部围(护)板

~ dog 小撑杆
~ drain 坡脚排水
~ of beam 工字梁梁边
~ of wall 墙脚
~ protection 坡脚保护
~ slab （挡土墙）趾板
~(d) nail 斜钉
~d voussoir 带榫的拱石
toenail ['təuneil] *n.* 斜钉; *vt.* 斜钉
钉子,用斜钉钉牢
~ing wood-strip flooring 地板斜钉
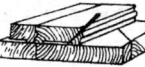
toenailing wood-strip flooring
toft [tɔft] *n.* 屋[宅]基; 小丘
toggle ['tɔgl] *n. pl.* -gles, *v.*, -gled, -gling; *n.* 肘(节,杆,板),肘环套接,套环; 套索桩[栓,钉]; (系在绳索末端的)横把,把手; 曲柄杠杆机构; *vt.* 用……系紧,栓牢; 供以套环; 备用肘节; toggler, *n.*
~ bolt 系墙螺栓
~ joint 肘节,肘接接合
~ lever [link] 肘节杆
~ mechanism 肘节机构
~ switch 肘式[板扭]开关

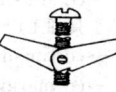
toggle bolt

toggle switch

toggler *n.* 栓紧装置
toilet ['tɔilit] *n.* 盥洗室,卫生间,厕所; 梳妆(室,台); 装饰
~ articles 梳妆用具
~ bowl 抽水马桶
~ cubicle 厕所间
~ paper holder 便纸架
~ roll holder （转动式）手纸架
~ room 盥洗室,化妆室
~ seat 马桶座圈
~ set 盥洗设备
~ soap 香皂
~ table 梳妆台
~ unit 卫生间单元
~ -ware 化妆物品
~ water 化妆用香水,花露水
men's ~ 男厕
waterless ~ 干厕(化学消毒的厕所)
women's ~ 女厕
token ['təukən] *n.* 标志[记,识],记号,象征; 特征
tolerance ['tɔlərəns] *n.* （配合）公差,容（许极）限,允许误差[公差,间隙]; 超长[余量]尺寸,毛长; 忍耐力[度]
~ deviation 容许偏差
~ limits 公差极限
~ of [on] fit 配合公差
~ of salinity 耐盐度
fit ~ 配合公差
manufacturing ~ 制造公差
toluene ['tɔljui:n] = toluol ['tɔljuəl] *n.* 甲苯
toluidine ['tə'lu:idi:n] *n.* 甲苯胺
~ red 甲苯胺红
tomb [tu:m] *n.* 墓(碑),坟; tomblike, *a.*; tombal *a.*
tombstone ['tu:mstəun] *n.* 墓碑(石)
tommy ['tɔmi] *n. pl.* -mies; 螺钉旋杆,圆螺帽扳手; 定位销钉; 实物工资制
~ bar （套筒扳手、铰盘等）插杆,旋转棒; 螺丝钻
~ hole 扳手孔
~ screw 贯头螺钉,虎钳丝杠
hooked ~ 圆螺帽钩头扳手
ton [tʌn] *n.* （重量单位）吨: 公吨(metric ~ =1000公斤),英[长]吨(long ~ 或 gross ~ =2240磅),美[短]吨(short ~ =2000磅); 货物容积单位(如木材等一般为40立方尺,石料为16立方尺)
tonal ['təunl] *a.* 音调的,声音的; 色调的; -tonally, *ad.*
tonality [təu'næliti] *n. pl.* -ties; 调子; 色调,调色; tonalitive, *a.*
tondino [təun'di:nəu] *n.* 圆盘形图[浮雕]

tondo [ˈtɔndə]【意】*n. pl.* -dos 或-di; 圆形浮雕[瓷盘]; 圆画

tone [təun] *n.*; *v.*, toned, toning; *n.* 音调, 色调, 光度, 明暗, 风格; (市场)供销[价格]情况; *v.* 调色, 上色
~ colour 音色[品]; 色调
~ quality 音质, 音色
cold ~ 冷色调
essential ~ 基本色调
tender ~ 柔和色调
warm ~ 暖色调

tongs [tɔŋz] *n. pl.* 夹钳[具]

tongue [tʌŋ] *n.*; *v.*, tongued, tonguing; *n.* 舌(状物), 雄榫, (企口板)榫舌; 舌形饰, (铁路的)尖轨; *v.* 做榫, 榫[嵌, 销]接; 呈舌形突出; tonguelike, *a.*
~ -and-groove (connection [joint]) 企口[舌槽]接合
~ -and-groove floor(ing) 企口地板

tongue-and-groove joint

~ -and-groove form 企口模板
~ -and-groove joint 企口接缝, 舌槽接合
~ -and-groove sheet pile 企口[锁口]板桩
~ -and-groove siding 企口(侧)墙板
~ -and-lip joint 饰边企口板接缝
~ bar (墙壁加固用)尖铜条, 尖铜棍
~ joint 舌榫接合, 嵌接
~ miter [mitre] 企口斜角缝, 舌榫斜拼合
~ piece 舌片
~ scraper 舌板式铲运机, 整平器
~ -shaped sand flood 舌状积砂
tonguing (and grooving) plane 榫舌刨, 槽刨
tonguing chisel 企口凿

joint ~ 榫舌

tonnage [ˈtʌnidʒ] *n.* 吨位, 吨数; (船、货的)吨税; 货运量
~ capacity 通过[运输]能力; (船的容积)吨位, 货车标记载重

tonne [tʌn] *n.* = metric ton(公)吨(= 1000公斤)

tool [tu:l] *n.* 工[器, 量, 刀]具; 仪器; 设备; 方法, 手段; *v.* 用工具加工, 用凿刀修整; 配备工具设备
~ -faced masonry 琢石砌体
~ kit 工具匣, 成套工具
~ outfit 工具装备(箱), 工具箱
~ rack 工具架
~ set 成套工具, 工具箱
~ sharpener 磨工具的设备, 砂轮
~ s register 工具登记簿, 工具底账
~ steel 工具钢
~ed finish (石料)用工具修琢面层, 工具加工面
~ ed-finish of stone (平行)琢石面
~ ed margin of stone 斧琢石边
bricklayer's ~ 瓦工的工具
formwork ~ 立模工具
grafting ~ 锹, 铲
graving ~ 雕刻刀
jointing ~ 接榫工具, 接缝器
knurling ~ 压花滚轮
pitching ~ 斧凿
soldering ~ 焊接工具

toolbox [ˈtu:lbɔks] *n.* 工具箱

tooler [ˈtu:lə] *n.* (石工用)阔凿

toolholder [ˈtu:l¦həuldə] *n.* 工具柄, 刀把[杆, 夹]

tooth [tu:θ] *v.*; *n. pl.* teeth; *n.* 出头砖(石); (牙, 轮)齿, 齿状物; *v.* 使成锯齿状, 使表面粗糙; 待(齿)接; 装[加]齿; 啮[咬]合; -toothlike, *a.*
~ chisel finish of stone 齿凿石面
~ ornament 齿形装饰
~ed ring dowel 齿形

tooth ornament

环榫[接]
~ed sleeve 有齿套筒
dog ~ 十字花饰,犬齿饰
toother ['tu:θə] *n.* 齿接砖,(直)叉砖
top [tɔp] *n.;a.;v.*, topped, topping; *n.* 顶(部,点,端)上部,表面,盖,篷;梢,尖端;最前面,最高级,首位;(石油的)轻馏分;*v.* 盖,给……加顶,到……顶上;给……涂保护层;去梢,截顶;使平坦;超过,完成,用分馏法提炼;*a.* 上[顶]部的;最重要的
~ aggregate size 最大骨料尺寸
~ -and-butt (木板)大小头拼合的
~ beam 顶梁
~ board 盖板,顶(端)板;柜台;桌子面板
~ bracing 顶撑,顶[上]拉条
~ coat 面[顶]层,外涂[保护]层,面层涂料
~ column 悬臂顶柱
~ course 顶层铺砌,面层
~ drawing 顶视图
~ end (原木的)小头,原木梢,顶端
~ flange plate 上翼缘板
~ flight 最高一层(房屋);高级的,优秀的
~ floor (楼的)顶层
~ -hat structure 顶部加劲结构
~ heading (隧道的)顶部导洞,上导洞;顶撑
~ -hinged outswinging (窗扇的)上悬外撑
~ hung window 吊窗,上悬窗
~ lateral bracing 上弦横向水平支撑
~ layer 顶层,上部钢筋,表层
~ level 顶面标高
~ light 顶部采光;(平)天窗;桅杆[电视塔]顶灯
~ longitudinal bracing 上弦纵向(水平)支撑
~ nailing 顶部钉固,上部加钉,顶面施钉
~ panel (plate)顶板

~ rail 压顶木,栏杆扶手;(门窗的)上档,上冒头
~ reinforcement 上部钢筋,架立钢筋
~ sand 粗砂
~ section 顶截面;拱顶段
~ side lighting 屋顶侧面光,顶部侧光
~ sod 面层草皮
~ spit (粗)废料
~ story (楼的)顶层
~ -timber 顶肋材
~ transom (栏杆)上档
~ view 俯[顶]视图
chimney ~ 烟囱帽
mast ~ 柱[杆]顶
smoke stack ~ 烟囱帽
straight ~ 平顶
wagon ~ 筒形顶,车顶
topaz ['təupæz] *n.* 黄玉,黄晶
tope [təup] *n.* (印度)塔,林园
topek ['təupek] *n.* 爱斯基摩房屋
tophaceous *a.* 砂[石]质的,凝灰岩的,粗糙的
topiary ['təupiəri] *a.;n. pl.* -aries;修整树形[态](的),剪修装饰
~ work (树木的)修剪整形,定型修剪
topless ['tɔplis] *a.* 无顶[盖,篷]的
toplighting *n.* 顶部照明
topmost ['tɔpməust] *a.* 最高处,最高[上]的
~ stor(e)y 楼顶层
top-notch [tɔp'nɔtʃ] *a.* 最高质量的,第一流的;*n.* 顶点
topographic(al) [,tɔpə'græfik(əl)] *a.* 地形(学)的,地形测量的;-topographically, *ad.*
~ details 地形细部[详图]
~ drawing 地形图
~ map 地形图
~ mapping 地形测绘,绘制地形图
top-out ['tɔpaut] *n.* 上部[层]建筑
topple ['tɔpl] *v.*, -pled,-pling;向前倾

[倒],使倾覆[倒塌]
toran ['təurən] *n.* (印度)高耸的庙门
torch [tɔ:tʃ] *n.* 焊[气]炬,焊枪[灯],喷灯,吹管;火炬[焰];*v.* 用焊枪烧焊;用喷灯烧去旧漆;用泥灰填塞;torch-like, *a.*
 painter's ~ 漆工火[气]炬
torchen *n.* 为石板屋顶塞泥灰的工人
tore [tɔ:] *n.* 环状半圆(凸)线脚装饰,柱脚圆盘线脚;(圆,椭圆)环
toreutics [tɔu'ru:tiks] *n.* 金属浮雕工艺
toric ['tɔ:rik] *a.* 复曲面的
 ~ surface 复曲面
toripherical *a.* 准球形的
tormentor = tormenter: [tɔ:'mentə] *n.* (舞台的)第一道侧幕,舞台后面的固定幕帷,回声防止幕
 ~ tower (舞台的)第一道侧幕背面塔架
toroid ['tɔ:rɔid] *n.* (圆,螺)环,环形线[室],复曲面,球或椭圆球形体,螺旋管; toroidal, *a.*; toroidally, *ad.*
toroidal [tɔu'rɔidl] *a.* 环形(线)的,喇叭口形的,螺旋管形的; *n.* 圆环,复曲面
 ~ shell 圆或椭圆形壳
torpedo [tɔ:'pi:dəu] *n. pl.* -does, *v.*, -doed, -doing; *n.* (铁路用)告警信号雷管,鱼雷(形汽车或装置); *v.* 破坏,废弃; torpedolike, *a.*
 ~ gravel 尖砾石
 ~ sand 粗砂
torse [tɔ:s] *n.* 可展曲面,扭曲面;扭花环装饰
torsel ['tɔ:sl] *n.* 墙中梁端垫块,梁托[垫],承梁木游沿木,(沿墙的垫梁木);漩涡花形装饰
torso ['tɔ:səu] *n. pl.* -sos, si; (无头和四肢的)躯干雕像
tortile ['tɔ:til] *a.* 扭曲的,盘卷的
tortuosity [tɔ:tu'ɔsiti] *n. pl.* -ties; 弯曲(度),曲折(性),扭曲
tortuous ['tɔ:tjuəs] *a.* 扭[弯]曲的,曲

折的,盘旋的;不在一个平面内的;
torture ['tɔ:tʃə] *v.*, -tured, -turing; 使翘[弯]曲
torus ['tɔurəs] *n. pl.* -ri, 座盘饰,环状半圆(凸)线脚装饰,柱脚圆盘线脚;(圆,椭圆)环
 ~ element 安全环套(具有吸能特性的防碰撞装置)
 ~ shell 环壳

torus

total ['təutl] *n., a., v.*, -tal(l)ed, -tal(l)ing; *n.* 总数[计],全体; *vt.* 计算……的总数,加起来,总计,求积; *a.* 总(计,体)的,全部的,完全的,绝对的
 ~ absorption 室内总吸声量
 ~ coverage 总建筑(占地)面积系数(总建筑占地面积和建筑用地总面积之比)
 ~ floor area 总建筑面积
totara *n.* 新西兰罗汉松
totem ['təutəm] *n.* 图腾(像),物像,标志
 ~ pole [post] 图腾[像]柱
touch [tʌtʃ] *n.; v.* (草图的)笔触,笔势;(使)接触,邻接;修整[饰];试[测]验;相切
 ~ down point 接地点
 ~ up 修整[饰],局部补修,补抹
 ~ing button 触钮
 finishing ~ 最后修整
tough [tʌf] *a.* 坚韧[固]的,不易磨损的;(泥灰等)粘的,稠的
 ~ iron 韧铁
 ~ wood 韧木
toughen ['tʌfn] *v.* (使)变坚韧,(使)变粘稠;-toughener, *n.*
 ~ed glass 韧化[钢化]玻璃
 ~ed 韧化聚苯乙烯
toughness ['tʌfnis] *n.* 韧性[度],延性,刚性[度],粘(稠)性[度]
tour [tuə] *n.; v.* 游览,旅行;参观,考察;转动,旋转;周转,流通;车[航]次

~ garden 环游[游赏]式庭园
touristry ['tuəristri] *n.* 旅游(者)
tow [təu] *n.* 拖,拉,牵[曳]引;拖索[缆],牵绳,拖车[船];顶堆船队,方头推轮;麻屑,短纤维;*vt.* 拖,拉,牵[曳]引;towable, *a.*
~ sack 麻袋[包]
~ing bridle 带钩拖索,钢缆套索
~ing device 拖曳装置
towel ['tauəl] *n.* 毛巾,抹布,擦手[脸]纸;*vt.* 用毛巾擦
tower ['tauə] *n.* 塔(楼,架),塔式建筑,高楼,阁,(高大的塔形)支墩,柱;视标;堡垒,城堡;*vi.* 高耸,越出;tower-like, *a.*
~ bolt 管插销
~ building 塔式建筑,摩天楼
~ clock 塔钟,屋顶钟
~ column 桥塔立柱
~ dwelling 塔式住宅,塔楼
~ structure 塔式结构,塔式建筑物
gate ~ 门楼
king ~ 承重柱
town [taun] *n.* (市,城)镇,(都,城)市,小城市,市区;街道,商业中心
~ drainage 城市(地面)排水
townlet ['taunlit] *n.* 小城镇
townsite ['taunsait] *n.* 城[镇]址,城镇计划用地
toxicity [tɔk'sisiti] *n. pl.* -ties;毒性[度]
toy [tɔi] *n.* 玩具;游戏;小型的事物(比类似事物小)
~ theater 木偶戏剧场
T-piece T形管,三通,T形接头,丁字管节
T-plate T形板
T-pipe T形管,三通
T-post T形支柱
T-profile T形钢[截面]
trabeate(d) ['treibieit(id)] *a.* 横梁的,过梁的,楣式(结构)的
~ construction 横檐梁式结构,楣式构造
~ style 横(檐)梁式,过梁式,楣式
trabeation [ˌtreibi'eiʃən] *n.* 横檐梁,柱顶上部(包括飞檐、雕带及横梁三部分)
trabs *n.* 墙板
trace [treis] *n.*;*v.*, traced, tracing; *n.* 迹线,轨[痕]迹;圆形[样],曲线图;交点[线];扫描,微量;*v.* 描绘[图];自动记录;跟[示]踪;探测[索]
~ of rake (饰面抹灰)扫尋痕纹
tracery ['treisəri] *n. pl.* -eries;窗花格,(哥特式建筑的)花饰窗格,窗饰
curvilineal ~ 曲线窗饰
geometrical ~ 几何图形窗饰
wall ~ 墙花格

geometrical tracery

trachelium *n.* (多立克柱)柱头饰
trachybasalt [ˌtræki'bæsɔːlt] *n.* 粗玄岩
trachyte ['treikait] *n.* 粗面岩
~ porphyry 粗面斑岩
tracing ['treisiŋ] *n.* 描绘[图];线雕;自动记录;示[跟]踪,显迹,线路寻迹,故障探测
~ instrument 描图仪器
~ paper [sheet] 描图纸,透明纸
tractable ['træktəbl] *a.* (材料等)易处理[加工]的;tractableness, *n.*;tractably, *ad.*
trade [treid] *n.*,*v.*, traded, trading; *n.* 贸易,商业;行[职]业,手工艺;同行[业];(the trades)信风,贸易风;*v.* 经营,交易[换];购物
~ association 专业组织,同业工会
~ hall 工会会所
~ mark 商标,标志
~ mark registration 商标注册
~ relation 贸易关系
building ~ 建筑工种
tradesman ['treidzmən] *n. pl.* -men;手工工人,技工

traditional [trəˈdiʃənl] *a.* 传统的,惯例的; traditionality, *n.*; traditionally, *ad.*
~ architecture 传统建筑

traditionary [trəˈdiʃənəri] *a.* 传统的,惯例的

transbeam [ˈtrænzbiːm] *n.* 横梁

transection [trænˈsekʃən] *n.* 横切;横断面

transenna 庙堂前栅栏围墙

transept [ˈtrænsept] *n.* 建筑翼部,(教堂的)交叉甬道,十字形教堂的两翼 transeptal, *a.*; transeptally, *ad.*

transenna

transfiguration [ˌtrænsfigjuˈreiʃən] *n.* 变形

transient [ˈtrænziənt] *n.* 瞬变现象[过程],瞬态,过渡现象[过程,状态];瞬变[暂态]值; *a.* 瞬变[时,态]的,暂时[态]的,过渡[路]的; transiently, *ad.*; transientness, *n.*
~ deformation 瞬时变形
~ service 临时检修,小修

transite *n.* 石棉水泥(防火)板,石棉水泥管

translucence [trænzˈljuːsns] = **translucency** [trænzˈljuːsnsi] *n.* 半透明(性,度)

translucent [trænzˈljuːsnt] *a.* (半)透明的,(水)清澈的; translucently, *ad.*
~ body (半)透明体
~ colo(u)r 半透明颜料
~ concrete construction 半透明混凝土结构,玻璃混凝土组合结构
~ glass 半透明玻璃
~ mosaic 半透明镶嵌

translucent glass

transmissibility [trænsˌmisiˈbiliti] *n.* 含水层输水能力[率],传输率,可透[传]性

transmission [trænzˈmiʃən] *n.* 传递[导,播];传动(装置),变速(箱,器);输电送[送];透射,渗透
~ bar 传动杆
~ of humidity 透湿,渗潮

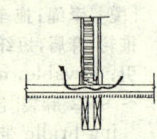
transmission of humidity

transmissivity [ˌtrænzmiˈsiviti] *n.* 透射系数[比,率],导水系数,透明[光]度

transmit [trænzˈmit] *v.*, -mitted,-mitting;传送[递,动,轮,导]透射[光];发射[报]
~ted light 透射光
~ted lighting 透射照明

transom [ˈtrænsəm] *n.* 楣[门,顶,气,摇头]窗,亮子;(门窗的)横档;(车辆的)横梁;船尾肋材
~ bar 门、窗的横档
~ catch 气窗[翻窗]插销
~ cleat 拉绳挂铁,系绳铁角
~ lifter 气窗梃销,捣棒
~ light 气窗
~ window 气窗;有横楣分隔的窗
top ~ (栏杆)上档
bottom ~ (栏杆)下档

transparence [trænsˈpɛərəns] *n.* 透明(性),透明[光]度

transparency [trænsˈpɛərənsi] *n. pl.* -cies;透明(性),透明[能见]度;透明体,幻灯片,透明画[花样]
colour ~ 彩色幻灯片

transparent [trænsˈpɛərənt] *a.* (半)透明的;某种辐射线可透过的;transparently, *ad.*
~ colour 透明色(透光介质的色彩),透明颜料
~ finish clear coating 透明涂面[饰]
~ frit 透明玻璃料,纯玻璃料
~ lacquer 透明亮漆

~ medium 透明介质
~ pigment 透明颜料
~ plate glass 透明厚板玻璃
~ sheet 透明纸
~ window 透明窗(采用各种透明材料的窗)

transport [v. træns'pɔːt, n. 'trænspɔːt] vt.；n. 运输[送]，输送，搬[转]运，传递；输[推]移；运输工具[船，飞机]

transportable [træns'pɔːtəbl] a. 可运输[搬运，移动]的，轻便的
~ black-top plant 轻便沥青拌和设备

transportation [ˌtrænspɔː'teiʃən] n. 运输[送]，搬运[移]，输[推]移；客[货]运；运费；运输工具；transportational, a.

transversal [trænz'vəːsəl] a. 横向的，横断[切]的；n. 截断线，贯线；transversally, ad.

transverse ['trænzvəːs] a. 横(向，放)的，横断[切]的；n. 横向物，横梁[墙，轴]，(椭圆)长轴；transversely, ad.；transverseness, n.
~ arch 横向拱
~ architrave 横线脚
~ beam 横(向)梁
~ bent 横向排架
~ brace [bracing] 横(向支)撑，横拉条
~ bracket 横向托座[牛腿]
~ bulkhead 横隔墙
~ cable (预应力配筋中的)横向钢丝绳
~ crack 横向断裂，横向裂纹[缝]，贯穿裂缝
~ distribution 横向分布
~ fissure 横裂(缝)
~ girder 横[向大]梁
~ grade [slope] 横(向)坡(度)
~ profile [section] 横断面(圆)
~ rib 横肋
~ section 横断面，横截面

transverse arch

~ strut 横杆[撑]
~ wall 山墙，横向外墙，(沉箱中的)横隔墙，(壳体结构)侧端的加固墙体

trap [træp] n.；v., trapped, trapping; n. 存水湾(井)，(泥沙等)截留装置；活板[底开]门，(舞台面上的)活动升降台板；舞台升降装置；凝汽[防臭，滤水]阀，吸尘罩，回收[分离]器；暗色岩；(pl.)家具，轻便梯架；v. 截留，拦住，陷入；使分离；安装存水湾

transverse rib

~ door (舞台等的)地板门，活门；(房顶的)活动天窗；调节风门
~ seal 存水弯水封，(存水弯的)水封弯管
~ valve 除污阀，滤阀
~ window 活板窗，通气窗
~ ping structure 栏砂建筑物
bath ~ 浴盆存水弯
dust ~ 除[集]尘器
grease ~ 隔油器，油脂分离器
sink ~ 洗碗池存水弯
stench ~ 防臭瓣，防臭存水弯

trapeze [trə'piːz] n. (灯具)吊架；梯形，不规则四边形

trapezium [trə'piːzjəm] n. pl. -ziums, -zia ;(英)梯形；(美)不规则四边形；-trapezial, a.

trapezoid ['træpəzɔid] n.；a. (英)不规则四边形(的)；(美)梯形(的)
~ pedestal 裙形[钟形]垫底

trapezoidal [ˌtræpə'zɔidəl] a. (英)不规则四边形的；(美)梯形的
~ brick 梯形砖
~ masonry 多角石圬工
~ notch 梯形槽口
~ rigid frame 梯形刚架
~ section 梯形断面[截面]

traprock ['træprɔk] n. 暗色岩(一种火成岩)

trash [træʃ] n. 废料[物]，垃圾，残[碎]

屑;vt. 除去废料
~ can 金属制垃圾箱,垃圾筒
~ disposal 污物处理
~ pump 排污泵
~ rack 拦污栅;护板,挡泥板
~ -removal device 除污设备
~ screen 拦污栅
trashery ['træʃəri] n. 废物,垃圾
trass [træs] n. (浮石)火山灰,粗面凝灰岩
~ cement 火山灰水泥
trave [treiv] n. 横木,小横梁;天花板格(此词现较少用)
travelable ['trævləbl] a. 可移动的
travel(l)er ['trævlə] n. 移动式[桥式]起重机,活动起重[运物,脚手]架;横拉幕
~ curtain (可拉入舞台两旁的)横拉幕
~ gantry 移动式龙门架
travel(l)ing ['trævliŋ] a. 移动[传播]的,移动式,旅行(用)的; n. 旅行
~ centering 移动式拱鹰架[脚手架]
~ framework 移动式脚手架
~ roller 移动滚筒
~ shutter(ing) 移动式模板
~ staircase [stairs] 自动楼梯

travel(l)ing staircase [stairs]

traverse ['trævə:s] n.;a.;v., -ersed,-ersing; n. (测量)导线;横断(物),横梁[栏,墙],窗帘,隔室;(与对面建筑物相通的)通廊;横动;横截线;横向分布; v. 横断[切,刨,截],(横向)往返移动;详细考察[讨论]; a. 横(断,放,动)的;曲线的
~ rod (挂窗帘等的)水平滑杆
travertin(e) ['trævətin] n. 石灰华,钙华,凝灰石
traviated ['treivi:eitid] a. (天花板)横向分隔的

tray [trei] n. 盆,槽,盘,浅抽屉,低浅容器;底板
~ ceiling 盘形顶棚
~ table 盘桌
trayle n. (英国古代用的)葡萄串形装饰
treacliness n. 粘(滞)性[度]
tread [tred] n.;v., trod, trodden or trod, treading; n. (楼梯等的)踏板,梯级,级宽,(车轮,履带)着地面,轮胎面(花纹),轨顶,轮迹;(左右轮)轮距, v. 踩(碎),踏(成);treader, n.
~ return 踏板延侧面
trun ~ 扇形踏步,斜(踏)步

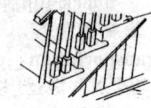

tread return

treadboard ['tredbɔ:d] n. (楼梯)踏步板,踏阶
treadle ['tredl] n.;v., -dled,-dling; n. 踏板,轨道接触器; vi. 踩踏板; -treadler, n.
treasure ['treʒə] n.;v., -ured,-uring; n. 财富[产]珍宝[品]; vt. 珍藏[重]
~ -house 宝库
art~s 文物,珍贵艺术品
treasury ['treʒəri] n. pl. -uries;金[宝]库,库房[存]经费,基[资]金
Eight treasuries 八宝(佛教图案)
treat [tri:t] v. (防腐)处理,浸渍,为……涂保护层;加工;净化;论述,探讨;对待;交涉; treatable, a.; treater, n.
~ed pile 浸渍木桩,防腐处理木桩
~ed timber 防腐(处理过的)木材,已加工木料
treater n. 处理[净化]器,处理设备[装置]
cottrell ~ 电收尘器
treatment ['tri:tmənt] n. 处理(方法),浸渍,加工,作业;分析,论述
~ before hardening of concrete 混凝土硬化前凿毛(处理)
~ of elevation 立面处理

architectural ~ 建筑(术的)处理
brush ~ (木材)涂刷防腐剂
open tank ~ (木材防腐)敞罐浸油处理法

treaty ['tri:ti] *n. pl.* -ties；条约,协议[商],谈判
~ articles 协定条款
~ condition 协定条件
~ wording 协定文本
provisional ~ 临时条约

tree [tri:] *n.* 树,树[乔]木；木料[材]；木制构件,轴,支柱,纵梁
~ nail 木栓[钉],定缝销钉
~ of Buddha 菩提树
~ peony 牡丹
~ wart 树瘤
fir ~ 冷杉,枞
globe-shaped ~ 球形树
mantel ~ 壁炉过梁
pine ~ 松树
roof ~ 栋梁
willow ~ 柳树

treenail ['tri:neil] *n.* 木栓[钉,键],木销钉[堵头]

trefoil ['trefɔil] *n.* 三叶形(饰),三叶花样
~ arch 三心花瓣拱

treillage ['treilidʒ] *n.* 格构,花木架格栅,格子墙[篱]

trellis ['trelis] *n.* 格构,格子结构,花格墙[篱],(花格)棚,架,格构凉亭[拱道],格子遮板,棚架式拱道；*vt.* 装格子等,用棚架支撑,使交织成格状；*a.* 格构[网状]的;-trellislike, *a.*
~ drainage 格形排水系统
~ girder 格构(大)梁
~ post 格构柱
~ window 格构窗,花格窗
~ed veranda(h) 花格走廊

tremolite ['treməlait] *n.* 透闪石；-tremolitic, *a.*

trenail ['tri:neil] *n.* 木销钉[堵头],木钉[栓,键]

iron ~ 大铁钉

T-rest T 形托板

triangle ['traiæŋgl] *n.* 三角(形,板,架)
~ belt 三角皮带
~ file 三角锉
~ mesh 三角形网眼
~ mesh wire fabric 三角形网眼钢丝网
~ tile 三角面砖
equilateral ~ 等边三角形
isosceles ~ 等腰三角形

triangular [trai'æŋgjulə] *a.* 三角(形)的,三脚的,三者间的
~ arch 三角拱
~ compasses 三脚规
~ point (测量导线网的)三角点

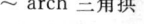
triangular arch

~ prism 三角形棱柱
~ Pratt roof 三角形普拉特式屋顶
~ Pratt truss 三角形普拉特屋架
~ prism 三棱柱
~ pyramid 三棱锥
~ rabbet joint 三角槽舌接合,三角凹凸榫接头
~ scale 比例尺,三棱尺

triangulate [*v.* trai'æŋgjuleit; *a.* trai'æŋgjulit] *a.*；*v.*,-lated,-lating; *vt.* 进行三角测量；分成[组成]三角形；*a.* (由)三角形(组成)的,有三角形花样的; triangulately, *ad.*; triangulator *n.*

triaxial [trai'æksiəl] *a.* 三轴[向,元]的,空间的;-triaxiality, *n.*
~ diagram 三轴(坐标)图

tribar ['traibɑ:] *n.* (混凝土预制)三柱块体,山字块体

triblet ['triblit] *n.* 心轴

tribrach ['tribræk] *n.* 三脚台；tribrachic, *a.*; tribrachial, *a.*

tribune ['tribju:n] *n.* 讲(演)台；观礼台

trice [trais] *n.*；*v.*, triced, tricing; *n.* 吊索；瞬息[间]；*vt.* 捆,绑；(缚住并)拉起,吊起,卷起

trichlorethylene ['traiˌkləu'reθəli:n] n. 三氯乙烯

trichromatic [traikrə'mætik] a. 三(原)色的, 天然色的

trick [trik] n. 班次; 窍门, 策略, 手段; v. 换班
~ lock 对字锁, 暗锁
~ paint 图案颜料, 纹理油漆
~ scene 旋转舞台(布景)

tricorn ['traikɔ:n] a. 三角的
~ bit 三角钻头

trident ['traidənt] n. 三叉曲线; a. 三叉的

tridimensional [ˌtraidi'menʃənl] a. 三维[向]的, 立体[空间]的

triethanolamine ['traiˌeθənə'læmi:n] n. 三乙醇胺

triforium [trai'fɔ:riəm] n. pl. -ria; 教堂拱门上面的拱廊

trig [trig] n.; a.; v., trigged, trigging; n. (砌砖时用以控制拉线的)垫砖[块]; 刹车(垫块), 楔子; v. 用垫块挡车, 用楔块垫紧; 支撑, 撑住; 修饰; a. 坚牢的

trigger ['trigə] n. 起动器[装置, 电路], 触发器; 扳[闸]柄; 雷管, 引爆器; 制动[轮, 滑]器, 锁定装置, 垫块; v. 触发, 起动
door ~ 门开关

triglyph ['traiglif] n. (多立克建筑上的)三槽[陇]板, 三联线槽饰; -triglyphed, a.

trigonum ['trigənəm] n. pl. -na; 三角形镶嵌物

trihedral [trai'hi:drəl] n.; a. 三面形(的), 三面体的

trihedron n. 三面体

tri-hinges [trai'hindʒiz] n. 三联铰

tri-level a. 三层的

trilinear [trai'liniə] a. 三线的, 由三条线围成的

trilit(e) n. 三硝基甲苯

trilith ['triliθ] n. (=trilithon)两石柱上架石梁的纪念碑

trim [trim] n.; a.; ad. v. trimmed, trimming; n. 整理, 修饰, 装饰(物); 书框, (带装饰的)框缘, (门窗)贴脸(板), 门[窗]头线, 镶边; 剪[修]枝, 修剪; (船)纵倾度, 吃水差; v. 整理, 修[装]饰; 修剪; 刨平, 细凿; 调整(纵倾度); a.; ad. 整齐的; trimly, ad.; trimness, n.
~ in 嵌[镶]入
~ joist 托[承接]梁
~ stone 镶边石
~ med rafter 装饰过的椽子
~ med surface 修琢面
~ ming joist 加劲搁栅, 过[托]梁, 承接梁
~ ming plate 切去边缘的板
interior ~ 内部装修
window ~ 窗缘饰

trimetric [trai'metrik] a. 斜方(晶)的
~ drawing 正三轴测图

trimmer ['trimə] n. 托梁, 承接梁; 剪切具, 修剪[整]机; 修整[剪]者, 装饰者; 整边炮机; 异形配件; 调整器; 平舱机, 装货者
~ arch 壁炉前拱
~ beam (支承几根小梁端的)托梁
~ joint 镶榫接头, 托梁(与小梁)接头
~ joist 托梁, 承接梁
brick ~ 砖托梁

T ring T形断面活塞环; 径向法面的断面为T字形的环

tringle ['tringl] n. 狭直条饰; 挂帘子的横杆, 支撑杆

trinol ['trainɔl] n. 三硝基甲苯

trip [trip] n.; v., tripped, tripping; n. (往返)行程, 行驶; 断开[跳闸, 脱扣](装置); (自动)停止[分离]机构; 倾翻器; v. 断开, 松开, 跳闸, 脱扣, 关闭; 倾翻[斜], (输送机)卸料

~ bolt 紧固螺栓
~ holder 夹[压]紧模座
~ pin 松放销
~ relay 切断[脱扣]继电器
tripping-bar 脱钩[跳闸]杆
door ~ 门开关
safety ~ 安全断路
time-lag ~ 延迟断开
tripartite ['trai'pɑ:tait] *a.* 分成三部分的,三个一组的,一式三份的,(涉及)三方面的; tripartitely, *ad.*
~ vault 三个成一组的穹顶
triple ['tripl] *a.*; *n.*; *v.*, -pled, -pling; *a.* 三倍[重,层]的,(由)三部分(组成的); *n.* 三倍,三个一组; *v.* 三倍于,(使)增至三倍
~ articulation arch 三连[铰]拱
~ -casement window 三连[扇]窗
~ course (屋檐)三层瓦
~ hinged arch 三铰拱
~ laminate 三层夹板
~ -lap pile 三叠接桩
~ -pole switch 三极开关
~ -rivet 三行式铆接
~ screw 三纹螺钉
~ valve 三通阀
~ -window 三层的窗扇[框]
triplet ['triplit] *n.* 三个一组,三件一套;三通(管),T形接头
triplex ['tripleks] *n.* 由三部分组成的房屋(如三层一套的公寓或有三套住房的房屋); *a.* 三倍[重,联,层]的,三部分的
~ apartment 三层楼公寓
~ building 有三套住房的房屋
~ glass 夹层玻璃
triply ['tripli] *ad.* 三重[倍]地
~ wood 三夹板
tripod ['traipɔd] *n.* 三脚架[台,凳,支撑物];三角桩群;三脚块体;鼎; *a.* 有三脚的,以三脚架支撑的
~ leg 三脚架
~ rest 三脚架

~ socket 三脚架底座
tripoli(te) *n.* 硅藻土,硅藻板
tripper ['tripə] *n.* (自动)倾卸车;卸料装置,(带式运输机的)卸料小车;钩杆,脱钩[解锁]装置,自动分离机构
tripping-bar *n.* 脱钩[跳闸,跳动]杆
triptych ['triptik] *n.* 三(幅相)联(的图)画,三个相联的雕刻
triquetra *n.* 三角形饰
triquetrous [trai'kwi:trəs] *a.* 三角形的,有三角形横断面的,三面形的,三棱的
trisect [trai'sekt] *vt.* 三等分;分成三份,截成三段
trishores *n.* 三脚支撑
triskelion [tris'keliən] *n. pl.* triskelia; 三腿[枝]形饰
trislab *n.* 三肋板
trisquare *n.* 曲尺
tri-tee beam 三T形梁
tri-tee slab 三T形板
Triton ['traitn] *n.* [希腊神话]人头人身鱼尾的海神,美人鱼雕像
triturate ['tritjureit] *n.*; *v.*, -rated, -rating; *n.* 磨碎物; *vt.* 磨[捣]碎,研成粉末
~d clay (研成)粉状黏土
triumphal [trai'ʌmfl] *a.* 凯旋(式)的;(庆祝)胜利的
~ arch 凯旋门
~ column 凯旋柱

Triton

triuphal arch

trivet ['trivit] *n.* 三脚架[台],矮脚金属架
~ table 三脚桌
trochilus ['trɔkiləs] *n. pl.* -li; 凹环形线脚,凹线饰
troffer ['trɔfə] *n.* 暗灯槽,平顶暗装管形照明器;槽形支架
trolite ['trɔlit] *n.* 苯乙烯绝缘材料

trolitul ['trɔlitəl] *n.* 苯乙烯塑料材料
trolley ['trɔli] *n. pl.* -leys, *v.*, -leyed, -leying; *n.* 手推车;矿车,倾卸式货车皮;手摇车;缆车,空中吊运车,(桥吊等的)起重小车;(英)无轨电车,(美)有轨电车;(电车上的)触轮; *v.* 用手推车[矿车]载运,乘电车[手摇车]
～ batcher 触轮式分批称料斗[配料器]
～ bucket 吊车戽斗[料罐]
～ track 滚轮滑轨
Trollflower ['trɔːlflauə] *n.* 金莲花
trommel ['trɔml] *n.* 滚筒(筛),旋转式圆筒筛; *v.* 旋转滚筒筛分
～ screen 滚筒筛
trona ['trəunə] *n.* 天然碱,碳酸钠石,天然苏打
tropical ['trɔpikl] *a.* 热带(地区)的,酷热的;tropically, *ad.*
～ finish 热带漆
～ timber [wood] 热带木材
troubleproof ['trʌblpruːf] *a.* 不出故障的,安全的,不停顿的
～ -saving 预防故障[事故]的
trough [trɔf] *n.* 槽,沟,盆,池,凹处;雨水[檐]槽,水落[导水]管;道[电缆]沟;槽钢;(曲线)凹点;(波)谷;地槽;低压槽; *v.* 开槽[沟]; *a.* 槽形的;trough-like, *a.*
～ beam 槽形[双山形]梁
～ gutter 槽[匣]形天沟,檐沟,雨槽
～ iron 槽钢[铁],U形钢
～ plate 槽形[钢]板
～ roof M 形屋顶
～ -shaped 槽式断面
～ urinal 小便槽
～ vault 长方形反水槽式穹顶
discharge ～ 排[污]水沟
eaves ～ 檐槽
trowel ['trauəl] *n.*; *v.*, trowel(l)ed, trowel(l)ing; *n.* 抹子,镘刀,泥[灰]刀; *vt.* 以抹子[泥刀]抹平[拌和],勾缝; troweler, *n.*

～ adhesive 高粘度粘合剂
～ coating 抹涂层
～ face 抹光面,镘平面
～ finish 抹(光)面,抹子压光
～(l)ed surface 抹光面
～(l)ing machine 抹[镘]平机
brick layer's ～ 砖工镘
buttering ～ (砌砖)涂灰镘
circle ～ 圆泥刀
hand float ～ 镘刀,抹子
laying ～ 大灰抹子
plaster ～ 抹灰刀
pointing ～ 勾缝刀
smoothing ～ 粉光镘

brick layer's trowel

trowelman ['trauəlmən] *n. pl.* -men;抹灰工,镘平工
truck [trʌk] *n.* 卡车,运货[载重](汽)车;(铁路)敞车;手推车;转向架;实物工资; *v.* 以货车运送,驾驶货车
true [truː] *n.*, *v.*, trued, tru(e)ing, *a.*, truer, truest, *ad.*; *n.* (安装,调整等)精确; *vt.* 装[配,校]榫,配齐,(使)正确; *a.* 真的,实际的;榫[正]确的;纯粹的;平衡的; *ad.* 真实[准确]地;
～ arch 纯拱
～ cost 真正成本
～ earning 真正收益
～ fresco 一种壁画技术
～ height 标高,海拔;实际高度
～ length 实际长度
trug [trʌg] *n.* (灰浆,水)槽
trumeau [truːˈməu] 【法】 *n. pl.* trumeaux;门窗口的中央柱,窗间墙,(二窗或二门之间的)间壁或间柱
trump [trʌmp] *n.* 王牌;最后手段;喇叭
～ stone 主景石(在功能、美观及手法上起重要作用的假山石)
～ tree 主景树(在功能、美观及手法上起重要作用的树木)
trumpet ['trʌmpit] *n.* 喇叭管,漏斗状筒

~ joint 融叭形管接头,扩口管接头
~ vault 喇叭形穹顶
speaking ~ 扩音器,喇叭筒
truncated [ˈtrʌŋkeitid] *a.* 截(短,头,平)的,斜截的;削蚀的;不完全的
~ angular arch 截顶的角形拱
~ roof 平塔顶
trunk [trʌŋk] *n.* (主,树)干,主要部分;柱身;干线,总[导,象鼻]管,筒;线槽;中继线,信息通路; *a.* 主要的,干线的;箱形的,有筒管的
elephant ~ (浇灌混凝土用的)象鼻(形溜)管,混凝土溜[输送]管;巨大柱身
truss [trʌs] *n.* 桁[构]架;捆,束; *vt.* 用桁[构]架支持;捆,扎,箍
~ arch 拱形桁架
~ bolt 桁架螺栓;地脚螺栓,锚栓
~ camber 桁架起拱
~ depth 桁架高度
~ frame 桁架

truss girder

~ girder 桁架梁,平行弦桁架
~ in space 空间桁架
~ joint (桁架的)节点
~ member 桁架杆件
~ post 桁架式柱
~ principal 人字木,桁架主杆件
~ rod 桁架(对角系)杆
~ed beam 桁架(式)梁
~ed frame 桁架式构架
~ed girder 桁架梁,平行弦桁架
~ed purlin (拉杆)加劲檩条
~ed roof 桁架屋顶
~ed structure 桁架式结构
~ing piece 捆扎件
arch ~ 拱(形桁)架
bowstring roof ~ 弓弦屋架,带上曲弦和拉杆的屋架
built-up ~ 组合桁架
cantilever ~ 悬臂桁架
cantilever braced ~ 悬臂支承桁架
complex ~ 复式桁架

double bowstring ~ 鱼形桁架,双弓弦桁架
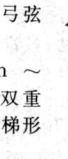
double Warren ~ 复式华伦桁架(双重三角形腹杆系的梯形桁架)
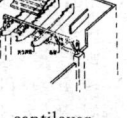
cantilever braced truss

English ~ 英国式桁架(三角桁架)
girder ~ 桁架梁,梁构桁架
hog-back ~ 弓形桁架
hog-chain ~ 双柱下撑式桁架,链式桁架
inverted king post ~ 倒单柱桁架
inverted queen post ~ 倒双柱桁架
inverted triangular ~ 倒(置)三角形桁架
joggle ~ 拼接桁架
king post ~ 单柱(倒三角)桁架
lattice ~ 格构桁架
lenticular ~ 鱼形式桁架
Linville ~ 双 N 形腹杆系桁架
mansard roof ~ 折线形桁架
north light ~ 单斜式屋架
pitched ~ 倾斜桁架,坡顶屋架
plane ~ 平面桁架
queen(post) ~ 双柱桁架
roof ~ 屋架
shed roof ~ 单坡屋顶桁架
space ~ 空间桁架
suspended stiffening ~ 悬索加劲桁架
Town ~ 方格(子)梁,方格式桁架
vierendeel ~ 空腹桁架
Warren ~ 华伦桁架,斜腹杆桁架
Whipple ~ 惠伯桁架,双斜腹杆桁架
Whipple-Murphy ~ 惠伯-莫菲式桁架,多腹杆桁架
wind ~ 抗风桁架
trussframed *a.* 桁架的,构架的
trussing *n.* 桁架系统,捆扎
truth [tru:θ] *n., pl.* truths;真理;真值;事实,实际情况;精确性[度]

try [trai] *n. pl.* tries，*v.*，tried，trying；*n.* 试验；试图；*v.* 试(验，用)；试图，力求，设法；校准；为……最后加工刨平
　～(ing) plane 细刨，平[大，长]刨
　～(ing) square 矩[曲，直角]尺，验方(角)尺

try(ing) square

tryout ['trai'aut] *n.* (示范性)试验，试用，检验
T-section T形截面，T形钢
　～ cramp T形截面窗钳夹器
　～ girder T形大梁
T-shaped T形的，丁字形的
T-slot T形槽
T-socket T形套筒，接通头
T-square 丁字尺
T-steel T形钢
tub [tʌb] *n.*；*v.*，tubbed，tubbing；*n.* 桶，木盆，一桶之量；沐浴；*v.* 在盆中洗物[澡]，把……装进桶(盆)中
　～ cock 浴盆龙头
tubal ['tju:bl] *a.* 管的
tubate ['tju:beit] *a.* 有管的，形成管的
tube [tju:b] *n.* 管，管材，隧道，地下铁道，筒；*v.* 敷设管道；用管子输送，把……装管；把……做成管状
　～ adapter 管接头
　～ and coupler scaffolding 铜管脚手架，单管脚手架
　～ caisson foundation 管柱基础
　～ construction 管形结构
　～ -in-tube structure 套筒结构
　～ orifice 管孔，管内小孔
　～ radiator 管式散热器
　～ structure 管筒式结构
　～ing brick 空心砖
　fluorescent ～ 荧光灯管
　luminous ～ 灯管
　seamless steel ～ 无缝钢管
tubeaxial fan 轴流式风扇
tubular ['tju:bjulə] *a.* 管状的，由管构成的
　～ brick 管砖，空心砖
　～ construction 筒形结构，管状建筑
　～ furniture 钢管家具
　～ girder 管式大梁，管腹工字梁
　～ rivet 管形铆钉
　～ scaffolding 管式[子]脚手架
　～ space structure 管式空间结构
　～ splice 套管连接
　～ vault 筒形穹顶
　～ welded frame scaffold 管状焊接框架式脚手架
tubulate ['tju:bjuleit] *a.*；*v.*，-lated，-lating；*a.* 管状的，成管的，筒状的；*vt.* 制成管，装管；
tuck [tʌk] *n.* 褶缝；*v.* 打褶；塞进；卷起
　～ and pat pointing 凸扁嵌缝
　～ pointing (砖砌体)嵌凸缝，勾凸缝
Tudor ['tju:də] *n.* 都铎建筑式；*a.* 都铎建筑式的
　～ apple (英国)都铎式球形饰
　～ arch 尖顶拱门
　～ architecture 都铎式建筑
　～ flower 三叶花饰
tufa ['tju:fə] *n.* 凝灰岩，石灰质
　～ cement 凝灰岩水泥
tuff [tʌf] *n.* (火山)凝灰岩
　～ loam 红土，凝灰岩壤土
　～ sandstone 凝灰质砂岩
tuffcrete ['tʌfkri:t] *n.* 凝灰岩水泥混凝土
tuft [tʌft] *n.* 一簇[束，圈]；*v.* 形成一簇
　～ed carpet 立毛地毯
tulipwood ['tju:lipwud] *n.* 郁金香木，带颜色条纹之木材，鹅掌揪木
tulle [tju:l] *n.* (丝质或尼龙)薄纱
tumblast 转筒喷砂
tumble ['tʌmbl] *n.*；*v.*，-bled，-bling；*n.* 跌落，翻滚；*v.* 跌倒，倒坍，翻腾，滚下
　～ polish 滚转抛光
　～ing course 嵌砌砖层

~ing mill 翻转机,滚筒式磨机
tumbled-in course 嵌砌砖层,对角砌层
tumbler ['tʌmblə] *n.* 肘式开关,搬把开关;齿轮换向器,转向轮;平底大玻璃杯
 ~ bearing 铰式[摆动]支座
 ~ switch 拨动式开关,倒扳开关
tumbling course 嵌砌砖层
tumbling -in 向内倾斜,嵌进(木材)
tumbling mixer 转筒拌和机
tumulus ['tjuːmjuləs] *n. pl.* -luses,-li; *n.* 古墓,古坟
tungate *n.* 桐油制成的催干剂
tung oil 桐油
tungsten ['tʌŋstən] *n.* 钨(W)
 ~ arc bulb 钨丝弧光灯泡
 ~ arc lamp 钨丝弧光灯
 ~ bronze 钨青铜
 ~ carbide 碳化钨(一种硬质合金)
 ~ chrome steel 钨铬钢
 ~ filament 钨丝
 ~ iron 钨
 ~ lighting 钨丝灯照明
 ~ steel 钨钢
T-union 三通接头
tunnel ['tʌnl] *n.* ;*v.* ,-nel(l)ed,nel(l)ing;*n.* 隧道,隧洞,烟囱;*v.* 凿隧道,掘地道
 ~ invert 隧洞底拱
 ~ lining 隧道衬砌
 ~ roof 隧洞顶板
 ~ vault 筒形拱[穹]顶

turf [təːf] *v.* ;*n. pl.* turfs 或 turves; *n.* (剪齐的)草皮,草坪,跑马场,泥炭;*v.* 铺草皮

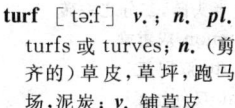

tunnel vaults

 ~ wall 草皮土墙
Turkey umber 土耳其赭土(褐色矿物颜料)
Turkish bath 土耳其浴[浴室],蒸汽浴[浴室]
Turkish mortar 用1/3砖粉和2/3石灰粉拌合的砂浆
turn [təːn] *n.* 旋转,弯曲,转弯,转变;*v.* 转弯,旋转,扭曲,倾倒,翻转
 ~ stair 回转式楼梯
 ~ tread 斜[踏]步,扇形踏步
 ~ed bolt 精[加工]螺栓
 ~ed wooden articles 旋床加工的木配件(如栏杆柱等)
 ~ed work 旋木工件,旋床加工的木配件
 ~ing-bar 壁炉拱的支承铁杆
turnbuckle *n.* 松紧螺钉扣[螺套],螺钉接头,花篮螺钉
 ~ screw 旋扣螺钉,花篮螺钉
turnbutton *n.* 旋转式按钮
turnkey ['təːnkiː] *n. pl.* -keys ;监狱的看守
 ~ contract 全部承包合同,包括规划、设计和管理的施工合同
 ~ delivery 承包(建筑安装工程的)安装及启用
 ~ job 承包(使建筑安装工程达到投产或使用要求)
 ~ project 包建工程
turn-pin 硬式旋钉(管工用)
turn-screw ['təːnskruː] *n.* 螺钉旋具
turnstile ['təːnstail] *n.* 十字回转门,(入口)旋转式栅门,回转栏
turntable ['təːnteibl] *n.* 舞台转盘(旋转舞台的圆形地板),转盘,转车台
turpentine ['təːpəntain] *n.* ;*v.* ,-tined,-tining; *n.* 松节油,松脂,松油;*v.* 涂松节油;-turpentiner, *n.*
 ~ oil 松节油
 ~ varnish 松节油清漆
turquois(e) ['təːkwɑːz] *n.* 绿蓝[天蓝]色,绿松石; *a.* 绿蓝[天蓝]色的,绿松石的
 ~ blue 翠蓝
turret ['tʌrit] *n.* 塔楼,角楼,角塔,(建筑物之)小塔
 ~ step 塔楼梯级,螺旋梯梯级
 ~ed roof 塔状屋顶,有塔楼[角楼]的屋顶
turriculate(d) [təˈrikjuleit(id)] *a.* 有小

塔的,小塔状的,形似小塔的

Tuscan order (古罗马)塔司干柱式

Tuscan style (古罗马)塔司干式建筑

tusk [tʌsk] *n.* 榫眼,凹榫,齿状物; *v.* 以长牙挖掘或刺戳

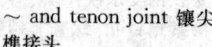
Tuscan order

~ and tenon joint 镶尖榫接头

~ tenon 多牙尖榫

twin [twin] *n.*; *a.*; *v.*, twinned, twinning; *n.* 豪华[高级]的套间,双生子之一; *a.* 成对的,孪生的; *v.* 使成对,给……提供配对物

~ brick 方砖

~ columns 并柱,双柱

~ elbow 双弯管

~ flood lighting 两向泛光照明

~ house 成对房屋,拼连的两所房屋

~ pier 双桥墩,成对的窗间墙

~ room (有两张单人床的)双人房间

~ tenon 双榫舌,双雄榫

~ twisted bars 双股扭合钢筋,双股螺纹钢筋

twine [twain] *n.*; *v.*, twined, twining; *n.* 细绳,合股线,盘绕,编织,捻,搓; *v.* 捻,搓,编织,盘绕

twist [twist] *n.* 线,索,绳,扭转而成之物,扭曲,扭歪; *v.* 扭转,弯曲,缠绕,捻,编

~ warp 扭转翘曲

~ed bar 扭转钢筋,螺纹钢(筋)

~ed column 麻花[螺旋]形柱

~ed grain (木材的)扭转[螺旋]纹理

~ed growth 木材节疤,扭转纹

~ed rope 搓捻的绳索,纹绳

twisted columns

~ed steel 扭转钢筋,螺纹钢[筋]

two [tuː, tu] *a.* 二,两个,双; *n.* 二,两,双

~ aisle building 带两披间的房屋,双走廊的房屋

~ -and-a-half brick wall 两砖半墙

~ -bed (guest)room 双人(客)房

~ brick wall 两砖墙

~ centered arch 双心拱

~ coat lime plaster 石灰砂浆抹面二遍做法

~ columm radiator 双柱散热器,双柱暖气片

~ course concrete pavement 双层混凝土铺面

~ cusped arch 内拱弧梅花雕饰的小双尖拱

~ DK(dining kitchen) 两居室并带厨房兼餐室的住宅

~ fold door 双摺门

~ fold window 双摺窗扇

~ hinged arch 双铰拱

~ hinged structure 双铰结构

~ leafed door 双摺门

~ leafed window 双摺窗

~ leg lavatory 双柱洗脸盆

~ light frame 两扇窗框

~ panelled door 两格式门

~ pin plug 两脚插头

~ pinned arch 双铰拱

~ plywood 薄层镶面胶合板,双层胶合板

~ point suspension scaffold 两点悬吊[摇摆]式脚手架

~ quarters 半头砖,二分之一砖

~ stepped tenon 双重榫

~ storied gate 双层门

~ tone paint 双色油漆

~ way beams 双向梁

~ way cock 双向龙头,两通旋塞

~ way joists 双向格式搁栅

~ way switch 双向[路]开关

tympan ['timpən] *n.* 门楣中心,拱与楣间的部分

tympanum ['timpənəm] *n.* 门楣中心,拱与楣间的部分

type [taip] *n.*; *v.*, typed, typing; *n.* 形式,式样,类型,(印刷)铅字,在讲坛上装置的声音响的天篷; *v.* 打字,拍发电报,成为……的典型
～ house 定型房屋
～ section 定型剖面
typewriter ['taipraitə] *n.* 打字机[员]
～ stand 打字台[桌]

tympanum

typical ['tipikəl] *a.* 典型的,通常的,标准的
～ design 标准[典型]设计
～ detail drawing 标准详图
～ floor 标准层
～ floor plan (高层建筑中)标准层平面图
～ -house 定型房屋
～ model 定型
～ -sample 典型样式
typification [tipifi'keiʃən] *n.* 典型化

U

U-abutment *n.* U形桥台[拱座]
U-bar *n.* 槽钢
U-beam *n.* 槽钢,U形梁
U-bend *n.* 马蹄〔U〕形弯头
U-bolt ['ju:bəult] *n.* U形螺栓,马蹄螺栓,锚环
U-channel *n.* U型槽钢
U-clamp [ju:klæmp] *n.* U形夹,U形压板
U-hanger *n.* U形吊钩
uinta(h)ite [ju'intəait] *n.* 天然硬沥青
U-iron *n.* U形铁,水落管卡,槽铁[钢]
U-line lamp U形(荧光)灯
U-link *n.* U形夹[钩],马蹄钩,U形连接环
ultimate ['ʌltimit] *a.* 最终的,极限的,最大的; *n.* 极限,终极,顶点
ultra ['ʌltrə] *a.* 极端的,过度的,超的
ultra-basic rock 超基性岩
ultra-clay *n.* 超黏土粒
ultramarine [ˌʌltrəmə'ri:n] *n.* 群青(一种颜料),深蓝色; *a.* 深蓝色的,海外的
ultramodern ['ʌltrə'mɔdən] *a.* 极其现代化的,最新(式)的,尖端的; *n.* 极端

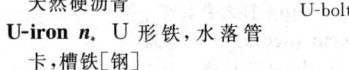
U-bolt

现代主义者;
ultraportable *a.* 极轻便的
ultrasonic ['ʌltrə'sɔnik] *a.* 超声的,超音速的;
～ static painting 超声波静电涂漆
ultraviolet ['ʌltrə'vaiəlit] *a.* 紫外线的; *n.* 紫外线(辐射)
～ absorbing glass 吸收紫外线玻璃
～ intercepting glass 防紫外线玻璃
～ lamp 紫外线灯
ultraviscoson *n.* 超声黏度计
umber ['ʌmbə] *n.* 棕土,赭土,赭色颜料; *a.* 赭色的,红棕色的,棕土的; *v.* 把……着红棕色
umbrage ['ʌmbridʒ] *n.*; *v.*, -braged, -braging; *n.* 树荫,阴影; *vt.* 使入阴影中
umbrella [ʌm'brelə] *n.* 伞,伞形物,烟囱顶罩; *a.* 伞状的,似伞的; *v.* 用伞遮盖,用伞保护
～ arch 隧道护拱,顶拱
～ shape 伞形(罩)
～ -shaped roof 伞形屋顶
～ shell 伞形屋面壳
～ type 伞形(罩)

umbrella-shaped roofs

umbriferous [ʌmˈbrifərəs] *a.* 投影的，有阴影的，成荫的

unadorned [ˈʌnəˈdɔːnd] *a.* 没有[未加]装饰的，自然的，原来的

unaired [ˈʌnˈɛəd] *a.* 不通风的
～ room 不通风房子

unalloyed [ˈʌnəˈlɔid] *a.* 纯金属的，非合金的，无杂物的，完全的，真实的

unartificial [ˈʌnɑːtiˈfiʃəl] *a.* 非人工的，自然的

unartistic [ˈʌnɑːˈtistik] *a.* 非艺[美]术的

unassembled [ˈʌnəˈsembld] *a.* 未装配[组装]的

unbacked [ˈʌnˈbækt] *a.* 无靠背的，无衬的，无支持的

unbalance [ˈʌnˈbæləns] *n.*, *v.*, -anced, -ancing; *n.* 不平衡(性)，不对称(性); *v.* 使不平衡，使不均衡，使失去平衡

unbecoming [ˈʌnbiˈkʌmiŋ] *a.* 不相称的，不合适的，不相配的

unbend [ˈʌnˈbend] *v.* -bent 或 bended, -bending; 弄直，放松，解开，卸下
～ing column 坚硬的直柱，笔直的柱

unbias(s)ed [ˈʌnˈbaiəst] *a.* 无偏见的，不偏的，无[未加]偏压的，公正的

unbind [ˈʌnˈbaind] *vt.*, -bound, -binding; 解开，松开，拆散，释放

unbleached [ˈʌnˈbliːtʃt] *a.* 未经漂白的，原色的

unblemished [ʌnˈblemiʃt] *a.* 无(瑕)疵的，没有污[缺]点的

unblended [ˈʌnˈblendid] *a.* 未掺合[混合]的
～ asphalt 未掺配(过的)地沥青

unbolt [ˈʌnˈbəult] *v.* 开启，旋开螺栓，拔开门闩

unbolted [ˈʌnˈbəultid] *a.* 未筛过的，粗糙的，卸掉螺栓的，未上螺栓的

unbonded [ˈʌnˈbɔndid] *a.* 未粘[砌]合的，无束缚的，自由的
～ concrete overlay 隔离式混凝土加厚层

～ prestressed bar（后张法中）未粘结的预应力钢筋

unbound [ˈʌnˈbaund] *a.* 非结合的，未连接的，无约束的，自由的，未装订的(书)
～ aggregate 未粘结骨料，无结合料骨料

unbounded [ˈʌnˈbaundid] *a.* 无界的，无限(制)的，非固定的，自由的; unboundedly, *ad.*; unboundedness, *n.*

unbrace [ˈʌnˈbreis] *vt.*, -braced, -bracing; 放松，松弛,不加支撑

unbreakable [ˈʌnˈbreikəbl] *a.* 打不破的，不易破碎的
～ glass 不碎玻璃

unbroke(n) [ˈʌnˈbrəuk(ən)] *a.* 完整的，未破坏的，继续不断的;
～ curve 连续曲线
～ wall 完整墙

unbuffered [ˈʌnˈbʌfəd] *a.* 无缓冲(装置)的

unbuilt [ˈʌnˈbilt] *a.* 未建造[建筑]的，无建筑物的，尚未建筑好的

unburnt brick 欠火砖，未烧透砖

uncased [ˈʌnˈkeisd] *a.* 无套管[外壳]的，未装箱的，露出的
～ drill hole 无套管钻孔
～ pile 无壳套现浇桩，无壳套桩
～ piling 打无壳套桩

uncemented *a.* 未胶结的
～ tile 外孔瓦管

uncertainty [ʌnˈsəːtnti] *n. pl.* -ties; 不确定(性)，不可靠，不固定

uncharted [ˈʌnˈtʃɑːtid] *a.* 图上没有标明的，未经探查和绘图的，未知的，不详的

unchecked [ˈʌnˈtʃekt] *a.* 未校核的，未经检查的

unchipped surfacing 不撒石屑面层

unchock *v.* 除去楔子[塞块]

unciform [ˈʌnsifɔːm] *a.* 钩形的，钩状的

unclamp [ˈʌnˈklæmp] *vt.* 松开(夹子等)

unclasp [ˌʌnˈklɑːsp] v. 打开，解开，放开

unclean surface 不清洁表面，有缺陷表面

unclench [ˌʌnˈklentʃ] v. 弄开，撬开，松开，放开

unclog [ˌʌnˈklɔɡ] vt., -clogged, -clogging; 清除油污[障碍物，阻碍]

unclosed [ˌʌnˈkləuzd] a. 不闭合的，未封闭的，打开的，未终结[完成]的

uncoated [ˌʌnˈkəutid] a. 无覆盖的，未涂面层的
～ cast-iron pipe（未涂保护层的）光面铸铁管

uncoil [ˌʌnˈkɔil] v. 解开(盘绕之物)，开捆，松开(弹簧)，开卷

uncoined [ˌʌnˈkɔind] a. 非铸造的，天然的

uncolo(u)red [ˌʌnˈkʌləd] a. 未着[染]色的，本色的，没有修饰的，原样的

uncompacted [ˌʌnkəmˈpæktid] a. 未压实的，不密实的
～ filling 不密实填土[料]
～ rockfill 未压实堆石[体]

uncompleted construction 未竣工工程，未完工程

uncomplicated [ˌʌnˈkɔmplikeitid] a. 简单的，不复杂的，未被复杂化的

unconnected [ˌʌnkəˈnektid] a. 不连贯的，分离的，支离破碎的，不相关的；unconnectedly, ad.；unconnectedness, n.

unconservative [ˌʌnkənˈsəːvətiv] a. 不防腐的，不稳健的

unconsolidated [ˌʌnkənˈsɔlideitid] a. 松散的，未固结的
～ material 未固结土料，松散材料
～ surface layer 未固结表层

uncontaminated [ˌʌnkənˈtæmineitid] a. 未污染的，不脏的，未沾染的

uncontinuous a. 不连续的
～ construction 不连续施工(法)

uncontrollable [ˌʌnkənˈtrəulbl] a. 难控制的，不可调节的，无法管束的；

uncontrolled [ˌʌnkənˈtrəuld] a. 未受管束的，自由的，未受抑制的，未经检查的；uncontrolledly, ad.
～ mosaic 无控制点镶嵌图

unconverted [ˌʌnkənˈvəːtid] a. 不变的，无变化的，未改变的

unco-ordinated [ˌʌnkəuˈɔːdineitid] a. 未经协调的，未调整的

uncorruptible [ˌʌnkəˈrʌptəbl] a. 不易腐蚀的，难损坏的

uncouple [ˌʌnˈkʌpl] v., -coupled, -coupling; 拆[解，断]开，解除……间的连接，拆散；
～d lever chain 起钩杆链

uncoursed [ˌʌnˈkɔːst] a. 乱砌的，不分层的
～ random rubble 不分层乱石圬工
～ rubble 不分层毛石，乱砌毛石
～ rubble masonry 不分层毛石圬工
～ square rubble 不分层方毛石圬工

uncoursed rubble masonry

uncover [ʌnˈkʌvə] v. 除去……的覆盖物，暴露，揭开……的盖子，剥离，(使)露出
～ed reservoir 露天水池

uncracked [ʌnˈkrækt] a. 未裂开的，无裂缝的

unction [ˈʌŋkʃən] n. 涂油，油脂，软膏

unctuous [ˈʌŋktjuəs] a. 油性的，含油脂的，塑性的
～ clay 油性黏土
～ paint 油质涂料

uncurl [ˌʌnˈkəːl] v. 弄直，(使)伸长[直]，(把卷着的东西)展开

uncurtained [ˌʌnˈkəːtənd] a. 无窗帘的，无幕的，帘[幕]被拉起的

uncut [ˌʌnˈkʌt] a. 未切[割]的，未雕刻的，未加工的，毛边的

undamaged [ˌʌnˈdæmidʒd] a. 未受损坏的，没有破损的

undark [ʌnˈdɑːk] n. 夜明涂料

undecagon [ʌn'dekəgɔn] *n.* 十一边形

undeformed [ˌʌndi'fɔːmd] *a.* 未变形的

under ['ʌndə] *prep.* 在……下面[底下],在……之中,少于,未满; *ad.* 在下,以下,少于; *a.* 下面的,次位的,标准以下的
~ beam girder 梁下纵桁
~ bed 底面,下端
~ coating 内涂层,上底漆,涂底漆
~ coating paint 底层涂料
~ construction 在施工[建造]中,正在施工
~ cover 遮盖,保护起来
~ crossing 下穿式(立体)交叉
~ cutting 刨削 T 形槽,凹割
~ load 欠载,轻[不满]负载;承受荷载
~ size 尺寸不足,过小,减小尺寸

under-ageing [ˌʌndə'reidʒiŋ] *n.* 硬化不足,凝固不足

underbaked *n.* 欠烘的,未烘透的

underbed ['ʌndəbed] *n.* 底架[座]

underboarding *n.* 垫板,衬板

under-bracing [ˌʌndə'breisiŋ] *n.* 下支撑,帮桩,(电杆)杆根横木

underbrush ['ʌndəbrʌʃ] *n.* 小树丛,矮丛林

under-burned brick 欠火砖,未烧透砖

undercarriage ['ʌndəˌkærid ʒ] *n.* 底盘,底架,起落架

undercoat ['ʌndəkəut] *vt.* 涂底漆,涂一层防锈物质于……上; *n.* 内涂层,衬里,底衬,底漆
~ing varnish 打[上]底漆

undercourse 底层,垫层

undercroft ['ʌndekrɔft] *n.* (教堂中的)地下室,地窖

undercure ['ʌndekjuə] *n.* 欠处理,(胶合剂)硬化不足,养护不够

undercut [*v.* ˈʌndəˈkʌt; *n.*,*a.* ˈʌndəkʌt] *v.*,-cut,-cuting,*n.*;*a.*;*v.* 从下部切开,暗掘,基础淘刷,(焊接的)咬边,雕出,浮雕,削低(价格); *n.* 从下部削砍; *a.* 下边部分被削掉的,挖底的;

-undercutter, *n.*
~ side wall 带凹圆线脚的边墙
~ swing saw 落地式摆锯
~ tenon 欠削榫舌

underdeveloped [ˌʌndədi'veləpt] *a.* 不发达的,落后的,未充分发展[开发]的

underdrainage ['ʌndəˌdreinidʒ] *n.* 暗沟[地下]排水

underdrained setting basin 渗透池,渗水塘

underearth [ˌʌndərə:θ] *n.* 地面下层土,地面下之物,地下; *a.* 地(面)下的
~ -eaves course 檐下瓦层

underfelt *n.* 底层油毡

underfilling ['ʌndəˌfiliŋ] *n.* 底层填料,未充满

underfloor ['ʌndəflɔː] *a.* 地板下面的
~ heating 地板下供暖
~ venthole 地板下通风口
~ wiring work 地板下布线工程

underframe ['ʌndəfreim] *n.* 底盘架[框],底构架
box section ~ 箱形底架

undergird [ˌʌndəˈgəːd] *v.* 从底层加固,支持,加强,巩固

underglaze ['ʌndəgleiz] *a.* 釉底的,上釉前着色的; *n.* 上釉之前所着之颜色
~ colour 釉底颜料

undergrade ['ʌndəgreid] *a.* 等外(材)的,劣等的
~ lumber 劣等木料

underground [*a.*;*n.* ˈʌndəgraund; *ad.* ˌʌndəˈgraund] *a.* 地下的,秘密的; *n.* 地下,地下铁道; *ad.* 在地下,秘密地
~ cellar 地下室,地窖
~ chamber 地下室
~ circuit 地下线路
~ diaphragm wall 地下连续墙
~ duct 地下管道
~ hydrant 地下消火栓
~ market 地下商场
~ openings 地下孔洞
~ structure 地下建筑

~ work 地下工程
underguard ['ʌndəgɑːd] *n.* 下部护板,下部保护物
underhung ['ʌndə'hʌŋ] *a.* 自下支承的
~ door （靠轮子在轨道上滚动的）拉门
underlay [*v.* ˌʌndə'lei; *n.* 'ʌndəlei]
v.,-laid,-laying, *n.*; *v.* 将（某物）置于他物之下,铺底层,做垫层; *n.* 置于下面之物,衬底纸,衬底,底基层
~ foil 垫底薄铁皮
underlayer ['ʌndəˌleiə] *n.* 垫层,底基层
underlayment [ʌndə'leimənt] *n.* 衬垫之物
underlining felt 层面衬毡
underlustred *a.* 光泽不够的
underlying [ˌʌndə'laiiŋ] *a.* 在下面的,基础的,下伏的,潜在的
undermixing ['ʌndə'miksiŋ] *n.* 拌和不足,混合料的不均匀性
underpart ['ʌndəpɑːt] *n.* 下部（结构）,非重要构件
underpavement [ˌʌndə'peivmənt] *n.* 下层铺面［路面］
underpitch vault 隧道拱顶
underplate ['ʌndəpleit] *n.* 底板,底座,垫板
underproof ['ʌndə'pruːf] *a.* 不合格的,低于标准的
underprop ['ʌndə'prɔp] *n.*; *v.*,-propped,-propping; *n.* 顶撑,支柱,支撑物; *vt.* 支撑,支持,用立柱加固; underpropper, *n.*
underreinforced [ˌʌndəriːin'fɔːst] *a.* 配筋不足的,低配筋的
~ concrete 配筋不足的混凝土,少（钢）筋混凝土
under-ridge tile 脊下瓦
undersanded mix 少砂拌合料
undersanding ['ʌndə'sændiŋ] *n.* （混凝土）含砂不足
undersea ['ʌndəsiː] *a.* 海底的,在海面下（进行）的; *ad.* 在海底,在海面下
~ cable 海底电缆
~ device 水下［海底］设备
underseal work 底封,基础处理
underset ['ʌndə'set] *vt.*,-set,-setting; 支撑,支持,放在……下面
undersize ['ʌndə'saiz] *n.* 尺寸过小,尺寸不足,逊径,筛底料,筛下; *a.* 尺寸不足的,不够大的,筛下的
understructure ['ʌndəˌstrʌktʃə] *n.* 基础,下层结构
undersurface ['ʌndə'səːfis] *n.* 底面,下表面; *a.* 表面下的,水面下的
underwater ['ʌndəwɔːtə] *a.* 水下的,潜水的,海中的; *ad.* 在水下; *n.* 水下,水底
~ construction 水下施工
~ lighting 水下照明
~ operation 水下作业
underway ['ʌndə'wei] *n.* 下穿道,水底通道,未完成的阶段; *a.* 行进［驶］中的; *ad.* 进行中
undisturbed [ˌʌndis'təːbd] *a.* 未扰动的,原状的;
~ sand 原状砂
undivided ['ʌndi'vaidid] *a.* 未分开的,不可分割的,连绵的,完整的;
undrained [ʌn'dreind] *a.* 不排水的,无排水设置的
undulate [*v.* 'ʌndjuleit; *a.* 'ʌndjulit]
v.,lated,-lating, *a.*; *v.* （使）波动,起伏,成波浪形; *a.* 波浪形的,波状的,起伏的
~d sheet iron 波纹铁皮,瓦垄薄钢板
unenclosed [ˌʌnin'kləuzd] *a.* 未（用墙）围起来的,公共的
unequal [ˌʌn'iːkwəl] *a.* 不等的,不均匀的,不对称的; *n.* 不等同的事物
~ angle iron 不等边角钢
~ gable roof 不等长双坡屋顶
uneven [ˌʌn'iːvən] *a.* 不平（坦）的,不均匀的,不规则的,不直的,奇数的

~ gauge 不均厚度
~ grain （木材的）不均匀纹理
~ ground 不平整地（面）
~ surface 粗糙表面
unevenness [ˌʌnˈiːvənnis] *n.* 不平整（度），凹凸（度），粗糙（度），不均匀（度）
unfenced *a.* 没有防御的，没有篱笆的
unfilled [ʌnˈfild] *a.* 未充填的，空的，未满的
~ section 欠缺断面
unfinished [ʌnˈfiniʃt] *a.* 未完（成）的，未琢磨[修整]的，粗加工的，粗糙的
~ bolt 粗制[毛面]螺栓
~ building 未完工建筑物
~ construction 未完工程
~ work 未完工作，未成品
unflanged [ʌnˈflændʒd] *a.* 无突缘的
unfluted [ʌnˈfluːtid] *a.* 无(凹)槽的
~ shaft of column 无槽立柱
unformed [ʌnˈfɔːmd] *a.* 无模板的，不用模板的，未成形的，无定形的，未做成的
unfreezable tap 防冻龙头[水嘴]
unglazed [ˌʌnˈgleizd] *a.* 未上釉的，素烧的，没有装玻璃的，(纸)无光的
~ earthenware 素烧陶器
~ pipe 未上釉(瓦)管
~ porcelain 素瓷，素烧陶瓷制片
~ roofing tile 素烧瓦，无釉瓦
~ roof-tile 无釉屋面瓦[砖]
~ tile 无釉瓦[砖]
ungraded aggregate 无级配骨料
unhardened concrete 未硬化混凝土
unhydrous plaster 干灰膏，未水化灰泥
unicell block 单孔[单槽]砌块
uni-dimensional consolidation 单向固结
unidirectional solidification 单向凝固
unified [ˈjuːnifaid] *a.* 统一的，统一标准的，联合的
~ budget 统一预算
~ precast elements 通用预制构件
~ thread 统一标准螺纹

uniform [ˈjuːnifɔːm] *a.* 均匀的，相等的，一致的，统一标准的；*v.* 使一致
~ beam 等截面梁
~ brightness 等亮度
~ coefficient of sand 砂的均匀系数
~ cost control 统一成本控制
~ distribution 均匀分布
~ gage 均匀厚度
~ illumination 均匀照明[度]
~ light 均匀光线[照明]
~ thickness 均匀厚度
uniformly [ˈjuːnifɔːmli] *ad.* 均匀地，无变化地
unilateral [ˌjuːniˈlætərəl] *a.* 单向的，一方的，片面的；*n.* 单向作用
~ load 单侧荷载
uninhabitable [ˌʌninˈhæbitəbl] *a.* 不适于居住的，不能居住的
uninterrupted [ˈʌnintəˈrʌptid] *a.* 不间断的，连续的，不停的
~ concreting 混凝土连续浇筑
union [ˈjuːnjən] *n.* 管接头，连接，联合，协会，工会，同盟
~ cock 接头旋塞
~ elbow 弯管接头
~ joint 管接头，连接管
~ nut 连接螺母
flanged ~ 凸缘管接
male ~ 外螺纹连管节
nipple ~ 外螺纹连管节
swivel ~ 旋转联结，旋转连管节
uniplanar [ˌjuːniˈpleinə] *a.* (发生在)同平面的，共平面的
~ bending 平面弯曲
unique [juːˈniːk] *a.* 独特的，惟一的，极好的，珍奇的，只有一种可能的；*n.* 独一无二之物
uniserial [ˌjuːniˈsiəriəl] *a.* 同系列的，单系列的，排成一列的
unit [ˈjuːnit] *n.* 单位，单元，组合，极组，元[部,组]件；*a.* 单位的，单元的，一套的
~ area 单位面积

～ body 建筑群中的个体建筑,单体,个体,组合体的基本单位
～ bolt nut 螺栓螺母组合
～ bond 直缝的砌砖(法)
～ -cast 整体浇注的
～ cost 单位成本,单价
～ design 单元设计
～ furniture 成套家具   unit bond
～ heater 单位式散热器,供热机组,单位供暖机
～ kitchen 单元式厨房,定型厨房的基本单位
～ of construction 结构单元,构件
～ of illumination 照度单位
～ of structure 构件,结构单元
～ plan 单位平面(建筑物平面构成的基本单位)
～ switch 组合开关,单位开关
building ～ 建筑构件
index ～ 提示装置
precast roof ～ 预制屋顶构件
ribbed heating ～ 肋式供暖器
universal [ˌjuːniˈvəːsəl] *a.* 普遍的,万能的,万[通]用的,多方面的,全世界的; *n.* 普遍原则;
～ cock 万向龙头,通用旋塞
～ plane 万能刨
～ saw bench 通用圆锯台
～ screw wrench 万能螺旋扳手
～ switch 通用开关
unjammable *a.* 防[抗]干扰的
unjoint [ˈʌnˈdʒɔint] *vt.* 脱开,松开接头,拆开……之连接处; unjointed, *a.*
unkempt [ˈʌnˈkempt] *a.* 未雕琢[修饰]的,不整洁的; unkemptly, *ad.*; unkemptness, *n.*
unlable(l)ed [ˌʌnˈleibld] *a.* 无标号的,未标记的,未分类的
unlace [ˈʌnˈleis] *vt.*, -laced,-lacing; 解开(带子),松开
unlade [ˈʌnˈleid] *v.*, -laded,-laing; 卸料,卸载,卸下……之货
unlapped [ʌnˈlæpt] *a.* 未覆盖[包住]的,非重叠的
unlined [ʌnˈlaind] *a.* 未镶面的,未衬砌的,无衬里的
unlit [ʌnˈlit] *a.* 不发光的,未点燃[亮]的
unload [ˈʌnˈləud] *v.* 卸(货,料,载),清除,取下,去载,转存(信息); unloader, *n.*
unlock [ˈʌnˈlɔk] *v.* 拔去销钉,拆开,开锁,打开
unmake [ˈʌnˈmeik] *vt.*, -made,-making; 破坏,毁坏,废除,撤消; unmaker, *n.*
unmarked [ʌnˈmɑːkt] *a.* 无标记的,未做记号的,未被注意到的
unmendable *a.* 不可修理的
unmixed [ˈʌnˈmikst] *a.* 未掺杂[混合]的,纯粹的;-unmixedly, *ad.*
unmodified [ˈʌnˈmɔdifaid] *a.* 未改建的,未改[变]的
～ resin 净[原,未改性]树脂
unmounted [ˈʌnˈmauntid] *a.* 未安装[镶嵌]的,未装框[架]的
unnail [ʌnˈneil] *v.* 拆除……上的钉子,拆除钉子以松开
unnotched [ʌnˈnɔtʃt] *a.* 无凹[缺,槽]口的,无刻痕的
～ bar 光面钢筋
unobstructive *a.* 小巧精微的
unoccupied [ˈʌnˈɔkjupaid] *a.* 未被占据的,空的,没有人居住的,没有人使用的
unopened [ˈʌnˈəupənd] *a.* 未开口的,封着的,没有拆开的,未开放的
unoriented [ʌnˈɔːrientid] *a.* 无定向的,无一定位置[方向,目的]的
U-notch U 形缺口[刻槽]
unpackaged [ʌnˈpækidʒd] *a.* 未包装的,散装的
～ sand 散装砂
unpaved [ʌnˈpeivd] *a.* 未铺砌的,未铺

路(面)的

unpeeled ['ʌn'pi:ld] *a.* 未剥皮[剥落]的

~ log dump 未剥皮的原木堆

unpeg ['ʌn'peg] *vt.*, -pegged, -pegging; 拔去……的钉子,除去木栓[钉子]以松开,使解冻

unperfect [ʌn'pə:fikt] *a.* 有缺陷的,不完整[全,美]的

unpermeability *n.* 不透水性

unpolished ['ʌn'pɔliʃt] *a.* 未磨光的,无光泽的,未抛光的,粗糙的,没有擦亮的

unprotected ['ʌnprə'tektid] *a.* 无保护(层)的,无防护设备的,未加保护的,无关税保护的(工业等);

unpunched ['ʌn'pʌntʃt] *a.* 未穿孔的,不打孔的,无孔的

unqualified ['ʌn'kwɔlifaid] *a.* 不合格的,无资格的

~ material 经检查不合格材料

unreactive aggregate 惰性骨料,(与水泥)无反应骨料

unreinforced ['ʌnri:in'fɔ:st] *a.* 无(钢)筋的,不加固的

unrendered *a.* 未抹灰[粉刷]的,未着色的

unrivet [ʌn'rivit] *v.* 拆除铆钉

unroofed [ʌn'ru:ft] *a.* 无屋顶的,露天的

unsafe ['ʌn'seif] *a.* 不安全的,不可靠的,危险的; unsafely, *ad.*; unsafeness, *n.*

~ foundation depth 不安全的基础埋置深度

unsanded gypsum plaster 无砂石膏灰浆

unsaturated [ʌn'sætʃəreitid] *a.* 非[未]饱和的; unstauration, *n.*

~ polyester coating 非饱和聚酯涂料

~ polyester resin 不饱和聚酯树脂

unsawn timber 原木,粗材

unscreened [ʌn'skri:nd] *a.* 未过筛[分级,过滤]的,未经检查的,无保护的

~ gravel 未过筛的砾石,混砂砾石

unscrew ['ʌn'skru:] *v.* 拧松螺钉,旋下螺丝

unsealed joint 未封缝,无封缝料的接缝

unseasoned ['ʌn'si:znd] *a.* 未风干的,未干透[燥]的,未成熟的,无经验的

~ timber 未干燥木材,未处理过的木材

~ wood 新伐木材,未干燥木材

unseptate *a.* 无隔壁的,无分隔物的

unserviceable ['ʌn'sə:visəbl] *a.* 不适用的,无用的,不耐用的,使用不可靠的

unset ['ʌn'set] *a.* 未装配的,未镶嵌的,未凝固的

unshaded [ʌn'ʃeidid] *a.* (窗等)无遮蔽的

unshaped ['ʌn'ʃeipt] *a.* 不成型状的,未定型的,尚未完工的,粗制的; unshapedness, *n.*

unshatterable *a.* 不碎的

unshored *a.* 无支撑的

~ composite beam 无支撑叠合梁

unshrinkable ['ʌn'ʃriŋkəbl] *a.* 防缩的,不收缩的

unshrinking [ʌn'ʃriŋkiŋ] *a.* 不收缩的,坚定的,不退缩的

unsized [ʌn'saizd] *a.* 未筛分的,未分大小的,未筛过的

unskilled ['ʌn'skild] *a.* 不熟练的,没有经验的,不灵巧的,没有特殊技能的

~ labour 粗(活)工作,粗工[活],生手工人

~ man 普工

unslaked ['ʌn'sleikt] *a.* 未消毒或沸化的,生的(石灰),未满足的

~ lime 生石灰,未消石灰

unsmooth *a.* 不光滑的,粗糙的

unsoaked *a.* 未浸透的

unsoddy *a.* 未铺草皮的

unsoil [ʌn'sɔil] *v.* 剥离表土,除去表土

unsoiled *a.* 未弄脏的,洁净的

unsolder ['ʌn'sɔldə] *vt.* 焊[烫]开,使分离[开],拆开

unsorted ['ʌn'sɔːtid] *a.* 未经分类的,未经分[挑]选的,未加整理的

unsound ['ʌn'saund] *a.* 不健全的,有缺点的,腐烂的,不可靠的,不坚实的; unsoundly, *ad.* ;unsoundness, *n.*
~ cement 变质水泥,不安定水泥
~ knot 腐朽木节,朽节

unsplinterable *a.* 不碎的

unsplit [ʌn'split] *a.* 无裂口的,整体的,不可拆卸的

unstable ['ʌn'steibl] *a.* 不稳定的,不坚固的,易变的; unstableness, *n.*; unstably, *ad.*
~ condition 不稳定条件
~ constitution 不稳定构造[状态]
~ equilibrium 不稳定平衡
~ ground 不稳(定)地面
~ state 不稳定状态
~ structure 不稳定结构

unstained [ʌn'steind] *a.* 未染色的,纯净的,无污点的

unstayed *a.* 未固定的,未加支撑的

unsteady ['ʌn'stedi] *a.* 不稳定的,非恒定的,易变的; *vt.* 使变得不稳定或不安定; unsteadily, *ad.*; unsteadiness, *n.*

unsteamed concrete 非蒸汽养护混凝土

unsteel [ʌn'stiːl] *vt.* 使失去钢性,使软化

unstick ['ʌn'stik] *vt.*, -stuck,-sticking;扯开,松开,使不再粘合[附着]

unstiffened [ʌn'stifənd] *a.* 未变硬的,未加劲的

unstop ['ʌn'stɔp] *vt.*, -stopped,-stopping;除去……的障碍,拔去……的塞子,打开……的口,开放

unstrained ['ʌn'streind] *a.* 未拉紧的,无应变的,不紧张的
~ pile 自由桩头

unstratified [ʌn'strætifaid] *a.* 不成层的,非层状的,不是按层排列的

unstressed [ʌn'strest] *a.* 不受力的,无应力[张力]的

~ member (构架的)无应力杆件

unstuck [ʌn'stʌk] *a.* 未粘牢的,未系住的,不固着的

unsubstantial ['ʌnsəb'stænʃəl] *a.* 不坚固[结实]的,空心的

unsupported ['ʌnsə'pɔːtid] *a.* 自由的,无支承[持]的
~ distance 自由长度,无支撑长度,净跨
~ height 无支撑挖掘深度,自由高度
~ length 自由长度
~ width 自由宽度

unsurveyed *a.* 未测量的

unswerving [ʌn'swəːviŋ] *a.* 不歪的,直的,不偏离的;unswervingly, *ad.*

unsymmetrical ['ʌn-si'metrikl] *a.* 不对称,不匀称的,不平衡的
~ beam 不对称梁
~ bending 不对称弯曲

unsymmetry ['ʌn'simitri] *n.* 不对称(性),不匀称,不调和

untangle ['ʌn'tæŋgl] *vt.*, -gled,-gling;整理,清理,解决,排解(纠纷)

untapered pile 无斜度的桩,平头桩

untenanted ['ʌn'tenəntid] *a.* 无人租住的

untight *a.* 未密封的,不紧密的

untrammel(l)ed [ʌn'træmld] *a.* 无阻碍的,不受束缚的,自由的,未受限制的;

untreated [ʌn'triːtid] *a.* 未经处理的,不处理的,未浸渍过的
~ aggregate 未处治骨料,无结合料骨料
~ timber pile 未经(防腐)处理的木桩

untuck ['ʌn'tʌk] *vt.* 拆散,展开

untwine [ʌn'twain] *v.*, -twined,-twining;解开(缠绕物),散开,分开

untwisting *n.* 解开(缆索)

unused ['ʌn'juːzd] *a.* 不(使)用的,从未用过的,空着的,未消耗的

unusual [ʌn'juːʒuəl] *a.* 特殊的,例外

的,异常的,稀罕的,非常的;unusually, *ad.*;unusualness, *n.*

U-nut U 型螺母

unvariable expense 不变费用

unvarnished [ˈʌnˈvɑːniʃt] *a.* 未油漆的,未修饰的

unwater [ˈʌnˈwɔːtə] *v.* 排水,疏干,去湿,使干燥
~ing pipe 排水管
~ing system 排水系统

unwieldy [ʌnˈwiːldi] *a.* 难使用的,不便利的,笨重的;unwieldily, *ad.*;unwieldiness, *n.*

unwonted material 不常用的材料

unworkable concrete 不易浇筑的混凝土,和易性差的混凝土

unwrought [ˈʌnˈrɔːt] *a.* 未最后成形的,未制造[加工]的,粗糙的,未开发的
~ timber 未经加工的木料,原料,坯料

unyielding [ʌnˈjiːldiŋ] *a.* 不能弯曲的,无沉陷的,不可压缩的,顽强的,坚固的,不屈服的;unyieldingly, *ad.*;unyieldingness, *n.*
~ support 不沉陷的支座

up and down rod 上下联杆

up and over door 上翻门

update [ʌpˈdeit] *v.*, -dated, -dating;使……适合新的要求,不断改进;*n.* 现代化

upgrade [ʌpˈgreid] *n.*;*v.*, -graded, -grading, *a.*;*ad.*;*n.*;*v.*, 上坡,加固,加浓,提高(等级,质量);*a.*;*ad.*, 上坡的[地]

upholster [ʌpˈhəulstə] *vt.* 装饰,装潢,布置,摆设,为……装垫子[套子,弹簧];upholsterer, *n.*
~ed chair 软垫[弹簧]坐椅
back~ing 靠背

upholstered *a.* 有软垫的;装饰过的

upholstery [ʌpˈhəulstəri] *n. pl.* -steries;室内装饰品[业],家具覆盖饰物

U-pipe U 形管

upkeep [ˈʌpkiːp] *n.*(房屋,土地,汽车等之)保养,(生活之)维持,保养[费],生活[费],维修[费]
~ and mending 房屋维修[修缮]

upper [ˈʌpə] *a.* 上(面,部,游)的,较高的
~ beam 顶梁
~ bed 上端,顶部,顶端
~ brick paving 立式砖块铺砌
~ cat bar 上推插销
~ class 上部
~ coat 上涂层
~ croft 教堂拱门上的走廊
~ fascia 柱顶横梁的顶条,柱顶盘座面
~ flange of girder 大梁上翼,板梁上翼(缘)
~ floor 上层,上层楼面,楼上
~ gallery 上层楼座
~ limit (构配件的)最大容许尺寸,上限,上限尺寸
~ part 上部,顶部,顶端
~ plate 顶板,上盘
~ side 上端,顶部
~ surface 上端,顶面
~ wing wall 上游翼墙

upper-frame *n.* 顶架[框]

upraise [ʌpˈreiz] *vt.*, -raised, -raising;举起,抬起,升高;upraiser, *n.*

upright [ˈʌprait] *n.* 立柱,立杆;*a.* 直立的,竖立的,正直的;*v.* 立起,竖立;uprightly, *ad.*;uprightness, *n.*
~ board 竖立式(绘画)板
~ course 竖立砌层
~ of frame 构架立柱,框腿
~ panel 竖钉对接护墙板;竖向铺钉的木墙板
~ post 竖杆,直立杆,脚手架立杆,立柱
~ stanchion 立杆,支柱
~ ventilating slit door 竖式百叶门,

棂条门
~ wall 直立墙,直立外堤

uprise [*n.* 'ʌpraiz; *v.* ʌp'raiz] *n.*; *v.*, -rose,-risen,-rising; *n.* 直立管,涌高,升起; *v.* 起立,升起,起浪,涌高
~ pipe 上升立管

upset [*v.* ʌp'set; *n.* ʌp'set, 'ʌpset] *v.*, -set,-setting, *n.*; *a.*; *v.*; *n.* 缩锻,加压,翻转,木材纤维板断裂,倾覆,干扰, *a.* 倾覆的; upsetting, *a.*
~ bolt 膨径螺栓
~ frame (砂)箱框
~ rivet 膨径铆钉
clamp ~ 弯压铁

upstand beam (穿透楼板的)直立梁,上翻梁

upswept frame 弓形框架,上弯构架

uptown ['ʌp'taun] *n.* 市近郊,住宅区; *a.* 市郊的,住宅区的; *ad.* 在近郊,在住宅区

upturned [ʌp'tə:nd] *a.* 向上翻的,雕刻的,翻转的
~ eaves 飞檐

upward ['ʌpwəd] *a.* 向上的,朝上的,上升的; *ad.* 向上,在上面; upwardly, *ad.*; upwardness, *n.*
~ view 仰视图

upward-acting door 上开门

upwarp *n.*; *v.* 向上翘曲,翘起

urban ['ə:bən] *a.* 市的,都市的,市区的
~ architecture 城市建筑
~ art 城市艺术
~ decoration 城市装饰
~ design 城市设计
~ environment 城市环境
~ green 城市绿地
~ lighting 城市照明
~ park 城市公园
~ place 城市空间
~ space 城市空间

urea ['juəriə] *n.* 尿素,脲
~ resin 尿素树脂
~ resin adhesive 尿素树脂胶粘剂
~ resin varnish 尿素树脂清漆

urea-formaldehyde 脲醛
~ foam 脲醛泡沫(绝热保温材料)
~ glue 脲醛胶
~ resin 脲醛树脂

urethane ['juərəθein, ju'reθein] *n.* 脲酯,氨基甲酸(乙)酯
~ adhesive 氨基甲酸(乙)酯胶粘剂
~ foam 氨基甲酸(乙)脂泡沫
~ foam filler 脲烷泡沫填缝料
~ resin 氨基甲酸(乙)酯树脂,脲烷树脂

urinal ['juərinl] *n.* 小便池[槽,斗,处],漏斗形小便器
~ channel (小便器的)排尿槽
~ gutter 小便槽
~ range 一排小便池
~ spreader 洒水器,小便器
~ stall 小便器
~ strainer 小便器滤网
~ trap 小便器存水弯
~ water closet 大小便两用便器
~ with trap 带存水弯的小便器[池]
bowl ~ 小便器
stall ~ 立式小便器

urinal

urn [ə:n] *n.* 缸,瓮,(茶水)壶

urushiol [u:'ru:fiɔl] *n.* 漆酚
~ resin coating 漆酚树脂涂料

usable ['ju(:)zəbl] *a.* 可用的,有效的,适用的; usableness, *n.*
~ floor area 可用地板面积,建筑中的使用面积
~ value 使用价值

use zoning 功能[用途]分区

used [ju:st] *a.* 用过的,旧的

useful ['ju:sful] *a.* 有效的,有用的,实用的; usefully, *ad.*; usefulness, *n.*
~ floor area 可用地板面积
~ height 有效高度
~ life 使用[有效]期

usefulness of a project 工程效益
user ['ju:zə] n. 用户,使用者
U-shaped beam 槽形梁
U-shaped bolt U形螺栓
U-shaped section U形断面
ushers room 服务员室,传达室
U-steel U形钢,槽钢
U-strap (木结构)U接网条
utensil ['ju(:)'tensl] n. 用具,器皿
　lubricating ～ 润滑设备
　painting ～s 绘画用具
utilidors 保温管道
utility [ju(:)'tiliti] n. pl. -ties ,a.; n. 公用事业设备,有益,实用; a. 有多种用途的,通用的,公用事业的
　～ control console 使用控制台
～ distribution 管道间
～ grade 可用材
～ program 实用程序
～ room 杂用间,杂品存储室
utilization [,ju:tilai'zeiʃən] n. 利用,应用
～ of debris 废物[品,料]利用
U-trap U形存水弯
U-tube U形管
U-turn U形[180°]转弯,急转弯
U-washer U形垫圈,开口垫圈
uviol glass 透紫外线的玻璃
uvioresistant a. 抗紫的,不透紫外线的
～ glass 不透紫外线玻璃
U-washer n. 开口[U形]垫圈

V

vacancy ['veikənsi] n. pl. -cies ;空,空位[隙,处,地]
vacant ['veikənt] a. 空的,没有被占用的,空着的,无人住的房子; vacantly, ad.; vacantness, n.
　～ house 空房(指无人住的房屋)
　～ land 空地
　～ line 虚线
　～ lot 空地
　～ room (未使用的)空闲房间,(未租出的)空房
vacate [və'keit] v. -cated,-cating; 弄空,搬出,腾出,作废,休假,撤去,取消
vac-sorb 真空吸附
vacu-forming 真空造型
vacuous ['vækjuəs] a. 空的,真的
vacuum ['vækjuəm] n. pl. vacuums 或 vacua ; v.; a.; n. 真空,真空吸尘器,空处; v. 吸尘; a. 真空的,产生[利用]真空的
　～ cleaner 真空吸尘器
　～ collector 真空集尘器,真空罩
～ deposition 真空喷涂法,真空沉积
～ fan 真空排气扇[抽气机]
～ form 真空模板
～ forming 真空成型
～ heater 真空加热器
～ heating 真空供暖
～ heating system 真空供暖系统
～ line 真空管线
～ mat 真空模板
～ moulding 真空压塑[成型]
～ pad 真空垫
～ panel 真空模板
～ process 真空处理,真空作业法
～ shutter 真空模板
～ spraying 真空喷涂
～ steam heating 真空式蒸汽供暖
～ steel 真空钢
～ ventilation system 真空式通风系统
valance ['væləns] n. 窗帘上部的框架,帷幔,布帘

valance

valet room 随从室
valid ['vælid] *a.* 有效的,有根据的,正确的;validly, *ad.*; validness, *n.*
  ～ contract 有效合同
validate ['vælideit] *vt.*, -dated,-dating;使生效,批准,确认;
validity [və'liditi] *n. pl.* -ties;有效,确实性,合法性
  ～ check 确实[有效]性检查
  duration of ～ 有效期间
  term of ～ 有效期
valley ['væli] *n. pl.* -leys;天沟,屋顶排水沟,屋谷,谷,沟
  ～ board 天沟底板,斜沟底[槽]板
  ～ flashing 斜沟槽,天沟[斜沟]雨水管,天沟泛水;天沟防雨板
  ～ gutter 天沟[斜沟]雨水管,斜沟槽
  ～ jack 斜沟(处抬高的)椽(子)
  ～ of corrugation 瓦楞槽,波谷
  ～ rafter 天沟[斜沟]椽,(四坡屋顶的)沟椽
  ～ roof 有天沟的屋顶
  ～ tile 天沟[斜沟]瓦

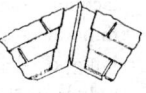
valley flashing

valley-side batten 天沟[斜沟]两侧挂瓦条
valuable ['væljuəbl] *a.* 有价值的,重要的,有用的;-valuableness, *n.*;-valuably, *ad.*
valuation [ˌvælju'eiʃən] *n.* 评价,估价,赋值;-valuational, *a.*
value ['vælju:] *n.*;*v.*, -ued,-uing; *n.*(色彩的)明度,价值,数值,颜色的浓淡程度; *v.* 估价,评价
  ～ contrast 明度对比
  ～ harmony 明度谐调[调和]
  design ～ 设计参数值
  optimal ～ 最佳值
  tonal ～ 色调,影象的明晰度
valve [vælv] *n.*; *v.*, valved, valving; *n.* 阀,阀[活,气,截]门; *v.* 给……装阀门;-valveless, *a.*
  ～ control 阀门控制
  ～ washer 阀门垫圈
  ～ wrench 阀门扳手
  air vent ～ 通气阀
  feed water ～ 给水阀
  fire ～ 消防阀
  relief ～ 安全阀
vamure *n.* 墙顶路,屋顶女儿墙后的通道
vanadium [və'neidjəm] *n.* 钒(V)
  ～ iron 钒铁合金
  ～ steel 钒钢
vandex ['vændeks] *n.* 一种混凝土防水剂
Vandyke [væn'daik] *n.* 锯齿形装饰边
  ～ brown 深褐色,铁棕色
  ～ pieces 锯齿边料
vang [væŋ] *n.* 支索,张索
vanish ['væniʃ] *v.* 消失,消散
  ～ing door 隐门(隐入墙内的拉门)
  ～ing point 没影点,灭点(透视画中平行线条的会聚点)
vapo(u)r ['veipə] *n.* 蒸汽,水蒸气; *v.* (使)蒸发,(使)汽化;vaporlike, *a.*
  ～ barrier 防潮层,隔汽层[具]
  ～ heating system (低压)蒸汽供暖系统
  ～ permeability 透湿性
  ～ permeance 透湿系数,(材料的)蒸汽渗透系数
  ～ proof 防水汽
  ～ proof barrier 防水汽层
  ～ seal 汽封
  ～ system 蒸汽式,蒸汽系统
  ～ -tight 汽密的,不漏汽的
vapo(u)rarium [ˌveipə'reəriəm] *n. pl.* -rariums 或 -raria;蒸汽浴(室)
vapo(u)rblast 蒸汽喷砂
vaporific inhibitor 挥发性防腐蚀剂,汽化性防锈剂
vapo(u)r-proof layer 防潮层,隔汽层
vapo(u)r-proofing material 防湿[潮]

材料
variable ['vɛəriəbl] *n*. 变量,变数; *a*. 可变的,不定的
~ -radius arch 变半径拱
~ -thickness arch 变厚度拱
design ~s 设计参数
variator ['vɛərieitə] *n*. 伸缩缝,胀缝
varicolo(u)red ['vɛəriˌkʌləd] *a*. 杂色的,五颜六色的,各色各样的

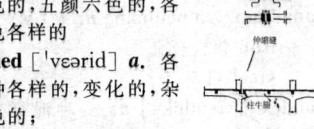
variator

varied ['vɛərid] *a*. 各种各样的,变化的,杂色的;
variegate ['vɛərigeit] *vt*. -gated,-gating; 染色, 弄成杂色, 加彩色; 使有差异; variegation, *n*.; variegator, *n*.
variety [və'raiəti] *n. pl*. -ties; 多样性,变化,变形,种类
variety hall 杂技场,曲艺场
various ['vɛəriəs] *a*. 各种各样的,多方面的,杂色的; variousness, *n*.
variplotter ['vɛəriplɔtə] *n*. 自动作图仪,自动曲线绘制器
varisized ['vɛərisaizd] *a*. 各种大小的,不同尺寸的
varnish ['vɑːniʃ] *n*. 清漆,凡立水,釉子,光泽面,粉饰; *vt*. 给……上漆,给……上釉,使有光泽,装[修,粉]饰; varnisher, *n*.; varnishy, *a*.
~ brush 清漆毛刷
~ cover 清漆涂层
~ cure 清漆固化[熟化]
~ for deadening sound 隔声清漆
~ for foil 金属薄片用清漆
~ formation 漆膜形成
~ paint 清漆
~ paper 涂漆绝缘纸
~ resin 清漆树脂
~ silk 浸漆绸
~ silk tape 浸漆[绝缘]绸带
~ spray gun 喷漆枪
~ stain 清漆着色剂
~ tube 浸漆纤维管
~ work 涂刷清漆
acid-proof ~ 耐酸漆
adhesive ~ 胶性漆
Chinese ~ 中国漆,火漆
air-drying ~ 风干漆
alcoholic ~ 凡立水
anticorrosive ~ 防锈漆
baking ~ 烤漆
clear ~ 透明清漆
enamel ~ 瓷漆
epoxy resin ~ 环氧(树脂)清漆
flat ~ 无光漆
glossy ~ 亮漆
lac ~ 光漆
lacquer ~ 亮漆,凡立水
mineral ~ 石漆
Ningpo ~ 金漆
polish ~ 打光清漆
rubbing ~ 磨光漆
spirit ~ 挥发[清]漆
stoving ~ 烤漆,烘漆
thermosetting ~ 热固性油漆
undercoating ~ 打底漆
varve clay 成层黏土,纹泥(薄层相间的各种泥土)
vary ['vɛəri] *v*., varied, varying;(使)变化,改变,修改
vase [vɑːz] *n*. 瓶饰,(花)瓶,哥林辛柱头的钟形饰; vaselike, *a*.
carved ivory ~ 牙雕素瓶
china ~ 瓷花瓶
porcelain ~ 瓷瓶
vaseline ['væsiliːn] *n*. 凡士林,矿脂
vat [væt] *n*.; *v*., vatted, vatting; *n*. (大)桶,(大)缸,容器; *vt*. 把……装入大桶,在大桶里处理
mixing ~ 混料桶
vault [vɔːlt] *n*. 拱顶,拱形圆顶,筒拱,拱穹,地下室; *v*. 做成拱形,覆以拱顶; vaultlike, *a*.

~ ceiling 拱顶式[穹窿]顶棚
~ construction 拱的施工
~ head 圆拱顶盖
~ light 地下室照明
~ roof 穹窿屋顶
~ rib 穹窿拱肋
~ed dam 双曲拱坝,穹窿坝
~ed roof 拱状屋顶
~ed shell 双曲壳体,穹壳
~ing capital 穹窿柱顶
~ing course 穹窿的拱脚石层
~ing horse 拱柱或拱墩
~ing masonry 筒拱圬工
~ing shaft (支承拱顶肋条的)承肋柱,(拱墩上的)装饰性支柱
acoustic ~ 吸声穹窿
annular ~ 椭圆拱顶
barrel ~ 筒形拱顶
brick ~ 砖拱
cavetto ~ 凹形拱顶
cloister ~ 回廊穹窿
concrete ~ 混凝土地下室(掩蔽所)
double ~ 双重穹窿
elliptic ~ 椭圆穹窿
fan ~ 扇状拱顶
groined ~ 交叉筒拱穹窿拱顶

vaulting shaft

helical ~ 螺旋拱顶
lierne ~ 扇形拱
masonry ~ 砌筑穹窿
net ~ 网状拱顶
quadripartite ~ 四区穹窿
rib ~ 穹棱拱顶
screw ~ 斜拱顶
simple ~ 无缝穹腹
stellar ~ 墨状拱顶
storage ~ 贮存[储藏]室
surbased spherical ~ 有座穹顶
surmounted ~ 超半圆拱
thin ~ 薄壳拱顶
trough ~ 凹槽拱顶
trumpet ~ 喇叭状穹窿

wagon ~ 筒形拱顶
V-block V 形块,三角槽块
V-clamp V 形夹
V-cut V 形开挖,V 形切割[法],楔形掏槽
V-drag 拖式 V 形刮铲
V-drain V 形排水沟
vee [vi:] *n.* V 形槽饰; *a.* V 形的,三角形的

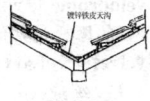
V-drain

~ -grooved (带)V 形槽的
~ gutter V 形槽
~ notch V 形缺口
~ roof V 形屋面
~ -shaped V 形的,三角形的
~ support V 形支座
~ -thread 三角螺纹
~ -trough V 形槽
vegetable ['vedʒitəbl] *n.* ; *a.* 植物(的),蔬菜(的)
~ fibre board 植物纤维板,(木)纤维板
~ glue 植物胶,树胶
~ parchment 假羊皮纸,硫酸纸
~ silk 丝状植物纤维,植物丝
~ wax 树蜡,植物蜡
vegetation [ˌvedʒi'teiʃən] *n.* 植物生长,植被,草木
vegetative lining 植草衬护
veil [veil] *n.* 幕,幔,遮蔽用物,面罩; *vt.* 遮盖,掩饰
~ ing reflection 光膜[光帏,眩光]反射
~ ing glare 光帷[光幕]眩光
vein [vein] *n.* 纹理,木纹,裂缝,缝隙; *v.* 象脉络般分布于;veinal, *a.*; veined, *a.*; veinless, *a.*; veinlike, *a.*
~ ed ebony 纹理黑檀木
~ ed marble 有纹理的大理石
~ ed wood 纹理木
velarium [vi'lɛəriəm] *n. pl.* velaria;天幕,露天剧场的篷帐

vellum ['veləm] *n.*; *a.* (精制)犊皮纸的,上等皮纸的,仿羊皮纸的
~ paper 厚牛皮纸,仿羔皮纸
velocity [vi'lɔsiti] *n. pl.* -ties;速度,流速;*a.* 速度的
velodrome ['vi:lədrəum] *n.* 室内赛车场,摩托车竞赛场
velvet ['velvit] *n.* 天鹅绒,丝绒; *a.* 天鹅绒(似,制)的,柔软的;-velvetlike, *a.*

velvet carpet

~ carpet 天鹅绒地毯
~ tapestry carpet 天鹅绒(挂)毯
velveteen ['velvi'ti:n] *n.* 绒布,棉绒;*a.* 棉制天鹅绒做的
veneer [və'niə] *n.* 饰面板,胶合板,薄木板,表层;*vt.* 镶面,镶饰,砌面,胶合 veneerer, *n.*
~ board 胶合[夹,镶]板
~ brick 镶面砖,饰面砖
~ construction 镶面结构
~ core plywood 薄芯板的胶合板
~ door 贴面板[夹板,胶合板]门
~ knife 胶合[镶面]板刨刀
~ lathe 旋板机
~ of mortar 灰浆胶层
~ of wall 镶[贴]墙面,墙表层
~ on 罩面
~ panel 镶板
~ planer 胶合板刨床
~ plywood 胶合板,镶木
~ saw 胶木板锯
~ shaver 胶合[镶面]板刨刀
~ slicing 镶板切制
~ wood 镶木,胶合板
~ed brick 陶[釉]面砖
~ed door 镶胶合板门
~ed panel 镶板
~ed plastics board 镶塑料板
~ed wall 镶面墙
~ing press 粘合用压力机
birch ~ 桦夹板

brick ~ 砖镶面
commercial ~ 表面薄板
cross banding ~ 内层薄板
face ~ 表面薄板,出面夹板
fancy ~ 装饰薄板
sliced ~ 平切单板
stone ~ 石镶衬,石衬砌
venenous ['veninəs] *a.* 有毒的
venerable ['venərəbl] *a.* 历史悠久的,古老的;venerability, *n.*; venerableness, *n.*; venerably, *ad.*
~ building 古建筑物
Venestra 一种胶合板
Venetian [vi'ni:ʃən] *a.* 威尼斯(式)的; *n.* 软百叶帘
~ arch 威尼斯拱
~ awning 威尼斯式遮篷,上悬外撑百叶遮阳板
~ blind 软百叶窗,威尼斯百叶窗
~ dentil 威尼斯排齿饰
~ door 两侧供采光的门
~ mast (装饰街市的)彩色饰柱
~ mosaic 威尼斯嵌镶砖
~ pointed arch 威尼斯尖拱
~ Renaissance 威尼斯文艺复兴式(15及16世纪)
~ school 威尼斯画派(文艺复兴时代)
~ shutters 固定百叶窗,遮阳窗
~ window 三尊窗,附有两侧窗的窗子
vent [vent] *n.* 出口,通风,排气,通风孔,排气道,漏洞,孔隙,裂口;*v.* 排出,泄放,钻孔,开孔;ventless, *a.*
~ branch 通[透]气支管
~ cap 通气孔盖
~ fan 排气扇,通风扇
~ gutter 通风道
~ hole 排气孔,通风孔
~ pipe 透[通]气管,排气管,通风管
~ stack 通风立管,通[透]气竖管
air ~ 气孔,通风口

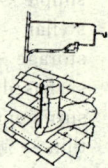
vent pipe

ventage ['ventidʒ] *n.* 小孔,出口,通风孔[管],孔隙
ventiduct ['ventidʌkt] *n.* 通气管,通风管[道]
ventilate ['ventileit] *vt.*, -lated,-lating;通风,排气,换气,装以通风设备;ventilable, *a.*
　～d insulation 通风绝热
　～ing brick 通风空心砖
　～ing device 通风设备,换气装置
　～ing duct 通风管道
　～ing fan 通风扇[机],排气风扇
　～ing grille 通风格子窗
　～ing pipe 通风管道
　～ing system 通风系统
　～ing trunk 通风道
　～ing window 通风[气]窗
ventilation [,venti'leiʃən] *n.* 通风,换气
　～ casement 气窗,通风窗
　～ facilities 通风设备
　～ hole 通风孔
　～ installation 通风装置
　～ louver 通风百叶窗
　～ slot 通风口
ventilator ['ventileitə] *n.* 风机[器],通风装置,通风口[管]
　～ hood 通风罩,通风管[器]罩
Venus ['vi:nəs] *n.* 维纳斯(女神,雕像)
veranda(h) [və'rændə] *n.* 阳台,游廊,走廊;
　long ～ 长廊
verdancy ['və:dənsi] *n.* 翠绿,生疏,不老练
verd(e)antique (古)铜绿,(古)绿石,杂蛇纹石
verdigris ['və:digris] *n.* 铜绿,铜锈
verditer ['və:ditə] *n.* 铜盐颜料
verdure ['və:dʒə] *n.* 翠绿,新鲜
verge [və:dʒ] *n.*;*v.*, verged, verging;*n.* 滴水瓦,檐口瓦,山墙突瓦,道路路边分隔带,边缘;*vi.* 接近,濒于,趋向;*vt.* 予以界限,设范围,作……疆界

　～ board 封檐板,山头封檐板
　～ fillet 山墙挂瓦条
　～ rafter 檐口椽
　～ tile 檐瓦
verge-perforated *a.* 边缘穿孔的
verge-punched *a.* 边缘穿孔的
vermeil ['və:meil] *n.* 朱红色,朱砂;*a.* 朱红色的,鲜红的;镀金铜[银]
vermiculate [və'mikjuleit;və'mikjulit] *a.*;*v.*, -lated,-lating;*a.* 有弯曲线条的,蠕虫状的,早蛀形的,早蛀的;*vt.* 使成虫蚀状[虫爬迹形]之装饰
　～d dressing 虫蚀状錾琢
　～d rustic work 虫迹状粗琢石面工作,虫迹状粗木房子
　～d work 虫蚀状雕塑,虫迹状装饰
vermiculation [və:,mikju'leiʃən] *n.* 虫迹饰(蠕虫爬迹形装饰)

vermiculated work

vermiculite [və:'mikjulait] *n.* 蛭石(绝热材料)
　～ concrete 蛭石混凝土
　～ mortar 蛭石砂浆
　～ plaster 蛭石灰浆
　～ plaster finish 蛭石灰浆抹面
vermil(l)ion [və'miljən] *n.* 朱砂,朱红色;*a.* 朱红的;*vt.* 涂[染]成朱红色
　Chinese ～ 中国朱砂
vermilionette *n.* 朱砂代用品
vermin-proof 防虫的
vernacular arts 地方工艺
vernier ['və:njə] *n.* 游标(尺),(游标)卡尺;*a.* 微调[动]的
　～ cal(l)iper 游标卡尺[钳]
　～ micrometer 游标千分尺,游标测微器
versatile ['və:sətail] *a.* 通用的,多用途的,万能的;versatilely, *ad.*;versatileness, *n.*
　～ digital computer 通用数字计算机
　～ pile frame 多用途桩架

versatility [ˌvəːsəˈtiliti] *n.* 适应性,通用性,多功能性

versicolo(u)r(ed) [ˈvəːsiˌkʌlə(d)] *a.* 杂色的,(受光照射)颜色多变化的

version [ˈvəːʃən] *n.* 型(式),种(类),方案;-versional, *a.*
　modern ~ 新型,现代化的(型式)
　simplified ~ 简化的型式,示意图,草图

vertex [ˈvəːteks] *n. pl.* -taxes. -tices；顶点,角顶,最高点
　~ angle 顶角

vertical [ˈvəːtikəl] *n.* 垂线,竖杆,竖向; *a.* 垂直的,铅直的,竖向的;verticalness, verticalism, *n.*
　~ abutment joint 竖向接缝
　~ accent 突出竖线条的建筑(哥特式)
　~ alignment 纵断面设计线
　~ angle 竖角,垂直角
　~ bar of truss 桁架竖杆
　~ bar window 直棂窗,棂子窗
　~ batter stud 竖向支撑,竖撑条
　~ blind 竖向百叶窗
　~ cat bar 竖插销
　~ column 垂直柱
　~ curb 直立式路缘石
　~ deflection 竖向挠度
　~ damp proof course 垂直防潮层
　~ drainage 垂直排水
　~ extent 深度,竖向延伸
　~ face 垂直面
　~ -fibre lug brick 竖凸沿砖,竖凸沿缸砖,竖筋砖
　~ glue joint 垂直粘合[胶结]
　~ grain (四开木材的)直纹理
　~ grain timber 直纹理木材
　~ hair 竖丝
　~ hinge revolving (门轴)竖гиб旋转
　~ illumination 垂直面照度
　~ interval 垂直间隔
　~ laddle 垂直梯,竖梯,直立梯
　~ lamination (木材)垂直叠合(法)
　~ letter 立体字
　~ -lift door 直升门
　~ line 铅垂(垂直)线,垂线
　~ louver 竖向百叶窗[遮阳板],垂直(固定)百叶窗
　~ masonry joint 砖[石墙]竖缝,(圬工)竖缝,直缝
　~ mulching 竖向覆盖层[防护层]
　~ pipe 立管
　~ pivoted sash 竖轴窗扇
　~ plane 垂直(平)面
　~ reinforcement 竖向[纵向]钢筋
　~ sash 竖式窗框
　~ saw 竖[立]锯,制板用锯
　~ shingling 竖向护墙板
　~ siding 竖向墙板
　~ sliding door 竖拉门,吊门
　~ sliding sash 竖拉窗扇
　~ spring-pivot hinge 竖向弹簧支枢铰链
　~ sunshade 垂直遮阳
　~ thread 竖丝,纵丝
　~ tiling 竖向贴砖
　~ view 俯视图,立面视图
　~ wainscot 竖向护墙板
　~ wire 竖直丝

vertical sunshade

verticality [ˌvəːtiːˈkæliti] *n.* 垂直度,铅直,竖向(性)

vertically [ˈvəːtikəli] *ad.* 竖直地,直立地
　~ pivoted window 立转窗
　~ sliding (窗开启方式的)垂直推拉
　~ sliding door 直升拉门
　~ suspended door 竖[纵]吊门

vesica-piscis 古建筑哥特式双尖圆形光轮

vesicular [vəˈsikjulə] *a.* 有气泡的,多孔的,蜂窝状的;vesicularly, *ad.*
　~ lava 多孔状熔岩

vesica-piscis

~ structure 多孔[多泡]构造
vesiculation [viˌsikjuˈleiʃən] *n.* 形成气泡
vestibular [vesˈtibjulə] *a.* 前[门]厅的,前庭的
vestibule [ˈvestibjuːl] *n.*; *v.*, -buled, -buling; *n.* 门[前]厅,连[通]廊,挡风门斗; *vt.* 用通廊连接,为……设置门廊
~ door 避风门,门厅门
vestopal binder 聚脂树胶粘合料
V-grooved plywood V形槽胶合板
V-grooving and tonguing V形企口,V形雌雄榫
V-gutter V形截面雨水槽,V形槽,V形(砌)沟
viability [ˌvaiəˈbiliti] *n.* 耐久性,寿命,服务期限,生存性
vibracast concrete 浇捣混凝土
vibrate [vaiˈbreit] *v.*, -brated, -bating; 振[颤]动,振捣; vibratingly, *a.*
~d brick panel 振动(过的)砖墙板
~d concrete 振捣(过的)混凝土
~ing board 平板振捣器,振动板
~ing compaction 振动捣实
~ing concrete float 混凝土振捣整平器
~ing float 振动抹灰板
~ing joint cutter 振动式混凝土切缝机
vibration [vaiˈbreiʃən] *n.* 振动(作用),颤动,摇动
vibration-absorbing base 吸振[减振]基础
vibration-absorbing layer 减振层
vibration-proof *a.* 防振[抗振]的
~ foundation 防振基础
~ material 防振材料
~ rubber 防振橡胶
vibrocast concrete 振捣混凝土
vibro-concrete pile 振实式混凝土桩
vibrofinisher 振动整平[平整]机
vibroshock [ˈvaibrəʃɔk] *n.* 减振[震]

器,缓冲器,阻尼器
vibrosieve *n.* 振动筛
vibrosmoothin trowel 振动抹平镘板
vice [vais] *n.* 台虎钳,台钳,英国古建筑中螺旋楼梯
hand ~ 手虎钳
parallel-jaw ~ 平口台钳
pipe ~ 台管钳
screw ~ 螺旋虎钳
table ~ 台虎钳
tube ~ 管子虎钳
wood working ~ 木工虎钳
vicinage [ˈvisinidʒ] *n.* 附近(地区),近邻,邻居
Victorian style 英女皇维多利亚时代建筑形式(1837～1901)
Victoric joint 维克托利克型接头,抗震[防漏]接头
video sign 荧光灯广告牌
Vierendeel girder 空腹大梁
Vierendeel truss 空腹桁架,连框桁架
view [vjuː] *n.* 视图,观察,景色,视野; *v.* 看,望,观察
~ing screen 银幕,电视屏
~ing window 观察窗
aerial ~ 鸟瞰图,俯瞰
auxiliary ~ 辅助视图
back ~ 后视图
bird ~ 鸟瞰图
bird's-eye ~ 鸟瞰图
bottom ~ 下[底]视图
close-up ~ 近视图,全貌图
cross-sectional ~ 横断面图
cut-away ~ 剖视图
cut-open 剖视图
diagrammatic(al) ~ 简图,图示
dorsal ~ 背视图
elevation 立[正]视图
end ~ 端视图
expanded ~ 透视图,展开图
exploded ~ 分解图,立体影象
external ~ 外视图

field of ~ 视野,视界
front ~ 前[主]视图
full ~ 全视图
general ~ 全（视）图
isometric ~ 轴测图
lateral ~ 侧视图
left ~ 左视图
part sectioned ~ 局部剖面图
perspective ~ 透视图,鸟瞰图,远景图
plan ~ 平面图,俯视图,顶视图
rear ~ 后[背]视图
right ~ 右视图
sectional ~ 剖视图,断面图
side ~ 侧视（图）
stretch-out ~ 展开图
top ~ 上[顶]视图
upward ~ 仰视图
vertical ~ 俯视图
worm's-eye ~ 仰视图
viewer ['vju:ə] *n.* 观察器,观测仪器
over-all ~ 全貌窥视窗
vignette [vi'njet] *n.*; *v.*, -gnetted, -gnetting; *n.* 都德式葡萄饰,蔓叶花饰; *vt.* 使（图画,照片等）渐次暗淡; vignettist, *n.*
vihara [vi'hɑ:rə] *n.* 佛教僧院[寺庙],佛教僧侣集会所,精舍（印度僧院）
villa ['vilə] *n.* 别墅,城市小屋;-villa-like, *a.*
~ marina 海滨房舍[别墅]
~ of countryside 乡间别墅
~ rustica 乡间别墅

villa of countryside

villadom ['vilədəm] *n.* 别墅和居住别墅的人们
village ['vilidʒ] *n.* 村庄,农村; *a.* 乡村的,村庄的
~ exchange 农村（自动）电话局
~ green 村庄广场,村内活动绿地
~ hospital 疗养院,农村医院

~ shops 乡间的店铺
villanette *n.* 小别墅
vimana [vi'mɑ:nə] *n.* 婆罗门寺院上的塔,印度方[尖]庙塔
Vinal Haven granite 浅灰细纹花岗石
vine black 葡萄色
vinery ['vainəri] *n. pl.* -eries 葡萄温室,葡萄园
vinette 叶形及卷须形饰
vineyard ['vinjəd] *n.* 葡萄园

vinette

vinoleum *n.* 乙烯地毯,乙烯布
vinsol resin 松香皂热塑性树脂
vinyl ['vainil] *n.* 乙烯基,乙烯树脂
~ acetate 醋酸乙烯酯
~ acetate adhesive 醋酸乙烯酯粘合剂
~ acetate finish 醋酸乙烯酯涂面料
~ butyral resin 乙烯丁醛树脂,聚乙烯醇缩丁醛树脂
~ capping 聚氯乙烯扶手
~ chloride 氯乙烯,乙烯基氯
~ chloride pipe 氯乙烯管
~ chloride plywood 氯乙烯胶合板
~ chloride resin 氯乙烯树脂
~ chloride rubber （聚）氯乙烯[合成]橡胶
~ chloride wire 氯乙烯绝缘线
~ disc 乙烯塑料盘
~ duct 乙烯基（塑料通风）管道
~ flooring 乙烯基[塑料]铺地料
~ fluoride resin 氟乙烯树脂
~ leather 乙烯基皮革
~ paint 乙烯基树脂涂料[油漆]
~ pipe 乙烯塑料管
~ plastic 乙烯基塑料
~ resin 乙烯基树脂
~ steel plate 乙烯基饰面钢板
~ tape 聚氯乙烯绝缘带
~ tile 聚氯乙烯板,乙烯基（楼面）板
~ wall cladding 乙烯塑料壁纸

vinyl-asbestos tile 乙烯基石棉板,石棉聚氯乙烯板
vinyl-coated sheet 乙烯基塑料覆面薄板
vinyl-coated steel plate 涂氯乙烯钢板
vinyloid sheet 乙烯薄板
vinylon ['vainilən] *n.* 维尼纶,聚乙烯醇缩甲醛纤维
vinyon ['vinjɔn] *n.* 聚乙烯塑料
violaceous [ˌvaiə'leiʃəs] *a.* 紫罗兰色的; violaceously, *ad.*
violet ['vaiəlit] *n.* 紫(罗兰)色; *a.* 紫罗兰色的,紫色的; violetlike, violety, *a.*
~ wood 紫(色)硬木
virgin forest 原始森林
virgin land 处女地,未开垦地
virgin rock 原生岩石
viridian *n.* 翠绿色(颜料)
virtual ['və:tjuəl] *a.* 虚的,实质上的,有效的; virtuality, *n.*
virulence ['viruləns] *n.* 毒性,毒度,有毒
viscid ['visid] *a.* 黏滞的,胶粘的,半流体的,浓厚的;
viscoelastic [ˌviskəui'læstik] *a.* 黏弹性的; viscoelasticity, *n.*
~ plate 黏弹性板
viscoelasticity ['viskəuˌelæs'tisiti] *n.* 粘弹性(力学)
viscogel *n.* 黏性凝胶
viscose ['viskəus] *n.* 粘胶(丝); *a.* 黏滞[性胶]的
~ glue 粘胶,胶水
~ paper 粘胶纸
~ staple fibre 粘胶短纤维
viscosity [vis'kɔsiti] *n. pl.* -ties; 黏滞性,黏度,内摩擦
viscous ['viskəs] *a.* 黏滞的
~ refractory 黏性耐火材料
vise [vais] *n.* (老)虎钳,台钳; *v.* (用钳)钳住,夹[压]紧
visible ['vizəbl] *a.* 可见的,看得见的,有形的; -visibleness, *n.*

~ crack 可见裂缝
~ line 轮廓线,外形线
~ marker 可见标记
~ ray 可见光线
vision ['viʒən] *n.* 视线[力,觉],景象; *vt.* 幻见,想像,显示
~ cable 电视电缆
~ distance 明视距离

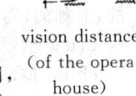
vision distance (of the opera house)

visit ['vizit] *n.*;*v.* 访问,参观,巡视,调查,游览,视察
visor ['vaizə] *n.* 护目镜,风挡,遮阳板,观察孔; *v.* (用护目镜等)遮护; visored, *a.*; visorless, *a.*
sun ~ 遮光[阳]板
vista ['vistə] *n.* 视景线,透视线,远景,街景,路旁风景线,深景,视景,树列远景; vistaed, *a.*; vistaless, *a.*
visual ['vizjuəl] *a.* 视觉[力]的,直观的,形象化的; visually, *ad.*
~ analysis (街景设计的)视觉分析
~ brightness 视觉亮度
~ center 视中心,主点
~ colorimeter 目视色度[比色]计
~ colourimetry 视感[觉]测色
~ degree 视度
~ design 视觉对象设计(指图表、阵列、标志、包装等的设计),图案[造型]艺术设计
~ environment 视觉环境
~ examination 外部[外表]检查,表观检验
~ inspection 外观[目力]检查
~ observation 目测,外部观察
~ photometry 视觉[目视]测光,目视光度测定
visualize ['vizjuəlaiz] *v.,*-ized,-izing 目测,(用肉眼)检验,观察,使可见; visualization, *n.*; visualizable, *a.*; visualizer, *n.*
visualized *a.* 直观的,具体的,形象化的
vitaglass ['vaitəgla:s] *n.* 维他玻璃,

(透)紫外线玻璃
vitalight lamp 紫外线灯,太阳灯
vitiate ['viʃieit] v.,-ated,-ating;使污浊,污染,弄脏,损坏
~d air 污浊的空气
vitrail n. 门窗彩花玻璃
vitreous ['vitriəs] a. 玻璃状的,透明的,上釉的,陶化的
~ brick 釉瓷砖,琉璃砖
~ china 玻璃陶瓷,陶器
~ coatings 涂釉层
~ enamel 搪瓷,釉瓷,珐琅
~ enameling sheet 玻璃搪瓷用薄板
~ luster 玻璃光泽
~ material 玻璃质材料
~ silica 透明石英
~ tiles 釉瓷瓦,玻璃瓦
~ ware 瓷器
vitrics ['vitriks] n. 玻璃器,玻璃状物质[制品]
vitrification [ˌvitrifi'keiʃən] n. 瓷化,玻璃化,透明化
vitrified a. 玻璃化的,陶瓷的,上釉的
~ block 陶瓷块
~ bond 陶瓷结合剂
~ brick 陶[缸,瓷]砖
~ clay 缸化[上釉]黏土
~ enamel 搪瓷
~ paving brick 铺路缸砖
~ pipe 陶管
~ sewer pipe 缸瓦下水管
~ tile 陶(土)瓦,釉面瓦管
vitrolite n. 瓷板[砖]
vitrophyre n. 玻(基)斑岩
vitrophyric glass (玻基)斑状玻璃地
vitruvian scroll or nament 波状涡纹装饰
vivid green 鲜[嫩]绿色
V-joint V形缝,V形焊接,V形接合
V-jointed panel V形镶板
V-junction V形交叉
V-molding V形线脚
V-notch V形缺口

void [void] n. 空隙(率),孔隙,孔率,内腔;a. 空的,无人占用的,无效的;v. 排泄,使无效,把……作废
~ cement ratio 孔隙与水泥容积比
~ filling capacity 填隙能力
~ method of proportioning 混凝土孔隙配料法
~ plywood 中空胶合板
~ slab 空心(楼)板
~ test 空隙试验
~ed slab 空心板

void slab

void-cement ratio 水泥孔隙比,隙灰比
void-free a. 密质的,无孔的,无空隙的
volatile ['vɔlətail] a. 挥发性的,易变的; n. 挥发性物质; volatileness, n.
~ covering 挥发性覆盖层
~ flux 挥发性助熔剂
~ matter 挥发物
~ rust-proofing agent 挥发性防锈剂
~ solid 挥发(性)物质,挥发性固体
~ spacing agent 挥发增孔剂
~ thinner 挥发性冲淡剂[稀释剂]
~ varnish 挥发性清漆,醇溶性清漆
volatility [ˌvɔlə'tiliti] n. pl. -ties; 挥发性[度]
volatilizable ['vɔlətilaizəbl] a. 可[易]挥发的
volatilize [vɔ:'lætilaiz] v.,-ized,-izing;(使)挥发,蒸发; volatilizer, n.
volcanic [vɔl'kænik] n. 火山岩; a. 火山的,火成的; volcanically, ad.
~ glass 黑耀石,火山玻璃
volleyball court 排球场
volt [vəult] n. 伏(特)(电压单位)
voltage ['vəultidʒ] n. 电压,电位差
~ divider 分压器
rated ~ 额定电压
voltmeter ['vɔltˌmi:tə] n. 电压表,伏特计
volume ['vɔljum] n. 体积,容积,音量,响度; a. 大量的; v. 把……装订成册
~ batching 按体积比配料

～ damper 风量调节阀[板]
～ density 体积密度
volumetric [ˌvɔljuˈmetrik] *a.* 体[容]积的；volumetrically, *ad.*; volumetry, *n.*
～ mixing 体积配合[混合]
～ weight 表观密度，密度
volute [vəˈljuːt] *n.* 螺旋饰品，卷涡纹； *a.* 螺旋形的； voluted, *a.*
～ casing 蜗壳
～ spring 螺旋形弹簧

volute

vomitorium [ˌvɔmiˈtɔːriəm] *n. pl.* -toria ；剧场或运动场看台入口
vomitory [ˈvɔmitəri] *n. pl.* -ries ；剧场或运动场看台入口
vortex [ˈvɔːteks] *n. pl.* -texes 或 -tices；旋涡，涡流，旋风
～ agitator 涡动搅动器
～ cone 旋锥体，涡流锥
～ line 涡线
vortical [ˈvɔːtikəl] *a.* 旋涡的，旋风的，旋转的； vortically, *ad.*
～ activator of cement 旋涡式水泥活化器
voucher [ˈvautʃə] *n.* 凭单，收据[条]； *v.* 为……准备凭据，证实……的可靠性
voussoir [vuːˈswɑː] *n.* 拱楔石[块]，拱石
～ arch 楔块拱
～ brick 拱楔砖
～ joint 拱块砌缝

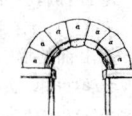
voussoir arch

～ stone 楔形石
V-rest V 形支架
V-roof V 形屋顶
V-shaped V[锥，楔，三角，漏斗]形的
～ joint （砖砌体的）V 形缝，（木板的）人字形缝拼接
～ pier V 形墩
～ reinforcement V 形钢筋
V-slot V 形槽
V-strap V 形带钢
V-sweep V 形铲
V-thread V 形[三角，牙形，管]螺纹
V-tool V 形工具，三角凿
V-tooled joint V 形勾缝
V-type notch V 形缺口[凹]
vuggy rock *a.* 多孔岩
vulcalose *n.* 同硬橡皮一样的绝缘材料
vulcanite [ˈvʌlkənait] *n.* 硬橡胶皮，胶木，硫化橡胶； *a.* 用硬橡胶皮做成的
vulcanize [ˈvʌlkənaiz] *v.,* -ized, -izing；（高温加硫使橡胶）硫化，硬化
～d fiber 硬化纸板
～d rubber 硫化橡胶
～d rubber cable 硫化橡胶包皮电缆
vulnerability [ˌvʌlnərəˈbiliti] *n.* 脆弱性，易损性
vultite *n.* 一种同沥青乳液、水泥、砂和水组成的混合防滑罩面材料
V-weld V 形焊缝
vycor glass 维克玻璃，高硼硅酸耐热玻璃，石英玻璃
Vynitop *n.* 涂聚氯乙烯钢板

W

wabble [ˈwɔbl] *v.,* -bled, -bling, *n.*；摆动，摇摆，震颤；摆动角；（声音）变量； wabblingly, *ad.*
wad [wɔd] *n.; v.,* wadded, wadding；*n.* 填块，填料，锰土，石墨，土地测量标布； *v.* 填塞，把……弄成小块，把……

压成一叠
bas ～ 底垫，基垫
wadding [ˈwɔdiŋ] *n.* 填塞物；填料，衬料
wafer [ˈweifə] *n.* 晶[薄，板]片；封缄纸；刮板； *v.* 压片，切成薄片

waffle

~ switch 晶片开关
waffle ['wɔfl] *n.* 华夫饼干
  ~ floor （华夫饼干式的）格纹楼面［桥面］,井字梁楼板
  ~ slab 格子板,双向密肋板
  ~ type 格纹式,华夫饼干式（表面有小格子）
wafter ['wɑːftə] *n.* 转盘风扇
wag [wæg] *v.*, wagged, wagging *n.*; 摆动,上下移动;变迁,推移; wagge, *n.*
wage [weidʒ] *n.*; *v.*, waged, waging; （常用 *pl.*）工资,报酬
wagon ['wægən] *n.* （铁路）货车;（四轮）拖车;旅行车,小型客车; *v.* 用货车运输;
  ~ headed 斜顶形的
  ~ headed ceiling 斜顶形顶棚
  ~ headed dormer 斜顶形老虎窗
  ~ -headed vault 斜顶拱顶
  ~ roof 斜顶形屋顶,筒形屋顶
  ~ top 斜顶,筒形顶
  ~ vault 斜穹窿屋顶,筒形拱顶

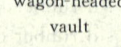
wagon-headed vault

wainscot ['weinskət] *n.*; *v.*, -scot(t)ed,-scot(t)ing; *n.* 护墙板,壁板,腰板; *v.* 装（上）板壁,用护壁（镶）板装饰
  ~ cap 墙板压顶
  ~ oak 护壁橡木
  ~ rail 下墙板镶条
wainscot(t)ing ['weinskətiŋ] *n.* 护墙板材料,壁板
waist [weist] *n.* 腰（部）,中间细部,收敛部分; *v.* 收敛,减少直径; waistless, *a.*
wait [weit] *n.* 等（候,待）;期待,拖延,暂缓;服侍
  ~ bay 短时停车湾;避车道
  ~ing on cement to set 候凝期,等待水泥凝固
  ~ room 候车［机］室,候诊室
wale [weil] *n.*; *v.*, waled, waling, *n.*

横撑［挡］,腰（护）板,腰梁,（凸起的）条纹,精华; *v.* 撑［箍］住;挑选
  ~ girder 横挡梁
  ~ pieces 横撑板

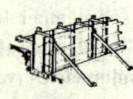
wale

waling ['weiliŋ] *n.* 横撑,横夹木,水平木,支腰梁,支横挡,围令
  ~ strip 围令,横夹木
  ~ timber 横撑木
walk [wɔːk] *n.*; *v.* 人行道,走道;散步;移动,游动
  ~ path 人行道
  ~ -up apartment house 无电梯公寓楼
  ~ -up building 无电梯楼房
  ~ -up domestic block 无电梯的住宅楼
  ~ -up story 无电梯楼层
nonslip safety ~ 无电梯楼层
walking ['wɔːkiŋ] *n.*; *a.* 步行（式的）,移动式的,可移动的
  ~ beam 摇梁,摆动梁
walkway ['wɔːkwei] *n.* 人行道,通道
wall [wɔːl] *n.* （围,城）墙,（墙）壁;间层壁,内侧,分界物,屏障; *a.* 墙（上）的,靠墙的; *v.* 用墙转住,堵住(塞);挂镜线上部的墙
  ~ arcade 实心连拱廊
  ~ arch 墙拱
  ~ base 墙座
  ~ beam 墙梁
  ~ bearing construction 承重墙结构
  ~ bed 墙基
  ~ board 墙板,壁板
  ~ box 墙上（安装）梁（的凹）穴;暗线盒
  ~ bracket 墙上托架
  ~ brander 沿墙栏栅
  ~ brick 墙砖
  ~ bushing 穿墙套管
  ~ carpet 墙挂毯,壁毯
  ~ cast in situ 现场浇灌墙壁

~ chase 墙槽
~ cladding 墙面覆盖层
~ clearance 孔壁间隙
~ clock 挂钟
~ coating 墙面涂料
~ coefficient 墙面吸收(声)系数
~ column 墙墩,壁柱
~ corner 墙角
~ cornice 墙挑檐
~ covering 墙面涂料
~ crest 墙顶(拱)
~ crown 墙顶(拱)
~ dowel 墙木栓,墙内预埋木砖
~ drinking fountain 墙上饮水器
~ entrance insulator 穿墙进线绝缘导管
~ facing 墙衬,墙表面修整
~ fan 墙上排气(风)扇
~ fence 护墙
~ fiberboard sheathing 墙衬纤维板
~ finish 墙面涂料[饰面]
~ fire warning device 墙上火警装置
~ fitting 壁灯
~ flashing 墙面泛水
~ footing 墙(基)脚
~ -forming 墙的造型

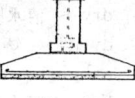

wall foundation

~ foundation 墙(基)脚
~ frame 壁式框架,承托墙的框架
~ friction 墙面摩擦
~ furnace 壁炉
~ furring 墙上板条
~ garden 墙顶花池
~ girder 墙梁
~ hanger 墙上梁托[承梁件]
~ hanging 墙帷
~ heating 火墙
~ heating panel 火墙护板
~ hole 墙孔
~ hollow 壁龛,墙的门槽
~ hook 墙钩,壁钩

~ -hung boiler 墙挂式锅炉
~ -hung closet 墙挂壁橱
~ -hung cupboard 墙挂碗橱
~ -hung fixtures 墙挂器具
~ -hung lavatory 墙挂式洗脸盆
~ -hung toilet 墙挂壁橱
~ -hung urinal 墙挂式小便器
~ -hung water closet 墙挂冲水箱
~ illumination 墙上照明设备
~ joint 墙[上灰]缝,砖缝
~ light fitting 固定在墙上的照明设备
~ lining 墙衬
~ moulding 墙上线脚
~ mounted heater 壁装式加热器
~ -mounted lighting 嵌墙灯,墙上挂灯
~ -mounted oven 挂在墙上的烘箱
~ niche 壁龛
~ of a basin 水池墙
~ of partition 隔墙,隔断(壁)
~ opening 墙洞
~ oriel 墙上凸肚窗
~ ornament 墙上装饰品
~ outlet 壁装电源插座
~ paint 墙上涂料
~ painter 墙壁油漆工
~ painting 壁画
~ panel 护墙板,镶板,墙面格间
~ paper 墙纸,壁纸
~ partition 隔墙,间壁
~ passage 墙式通道
~ pier 扶垛,堵墙;支墩,壁柱;附墙方柱(部分砌在墙内的柱身);窗间墙,扶壁
~ pillar 墙支墩,墙的扶垛扶壁
~ pipe 穿墙管
~ plank 墙板
~ plants 长在墙上的植物
~ plaster 墙面粉刷
~ plastering 墙面抹灰
~ plate 砌入墙内的托梁垫板

~ plug 墙上插头
~ pocket 墙上安装梁的凹穴
~ pointing machine 墙勾缝机
~ post 壁柱
~ primer 墙壁[打]底漆
~ profile 墙面轮廓
~ protector 室内护墙拦杆,缓冲挡
~ quartz clock 石英壁钟
~ radiant tubes 墙壁辐射管;壁暖气管
~ radiator 墙挂式散热器
~ recess 壁龛,墙的门槽
~ reflection 墙壁反射
~ roughness 墙粗糙度
~ rubber 墙橡胶
~ rosace 圆浮雕墙
~ scraper 墙刮刀
~ screw 墙螺钉,棘螺栓,墙锚栓
~ settle 嵌墙长椅
~ shaft 墙面支承弯肋的小柱
~ shelve 架墙,(安在墙上的)吊架
~ shingle 墙面板
~ shower 墙挂式淋浴器
~ sill 窗台
~ slab 壁板
~ sleeve 穿墙套管
~ socket 墙壁插座,墙上灯座
~ space (门窗之间的)墙面
~ sprinkler 墙上喷水器
~ standing by itself 独立墙
~ static pressure tap 壁式静压龙头
~ stopping 隔墙
~ string(er) 贴墙楼梯小梁
~ stud 管线支架,壁柱
~ supports 墙座
~ surface finish 墙面修饰
~ surface line 墙面位置线,建筑基线
~ tie closer 镶墙边的砖石
~ tile 墙面贴砖,面砖
~ tracery 墙花格

wall string(er)

~ trimming 壁面修整
~ wash basin 墙上洗手盆
~ washer 撑墙支架板
adiabatic ~ 绝热壁
arcade type ~ 拱廊式墙
arch ~ 拱墙
ashlar ~ 琢石墙
baffle ~ 砥墙,隔墙
barrier ~ 围墙,屏障
base ~ 底层墙
base of ~ 墙基
batten ~ 板条墙
bearing ~ 承重墙
bench ~ 承拱墙,台阶墙
blank ~ 平壁,无门窗的墙
boulder ~ (卵)石墙,萤石墙
brick veneer ~ 砖贴面墙
column-and-panel ~ 柱夹镶板墙
containing ~ 防护外壁
cross ~ 横墙,隔墙
curtain ~ 幕墙,悬墙
decorative ~ 装饰墙
divide ~ 隔墙
double face ~ 双面墙
dry ~ 清水[干砌]墙
dwarf ~ 矮隔屏
exhaust partition ~ 排气隔断
fire ~ 隔火墙
flare wing ~ 斜翼墙,八字塔
flat ~ 光面墙
frame ~ 构架墙
furred ~ (有粉面的或嵌钉薄板的)泥水墙
gable ~ 人字墙,山墙
glass ~ 玻璃幕墙[壁]
half brick ~ 半砖墙
half tiled ~ 半砖墙
half-timber ~ 露明木[骨]架墙
head ~ 端面墙,山墙
hollow brick ~ 空心砖墙
hollow (core) ~

hollow brick wall

空心墙
insulating ~ 绝热[保温]墙
internal ~ 内墙
jacketed ~ 双层墙
jamb ~ 门窗边框侧墙
load ~ 承重墙
loam ~ 板筑土墙,夯土墙
masonry ~ 圬工墙
mortarless ~ 干砌墙
non-bearing ~ 非承重墙
one-brick ~ 单砖墙
panel ~ 节间墙,大墙板
parapet ~ 女儿墙,压檐墙
partition ~ 隔墙,隔断
party ~ 共用隔墙通[隔]墙
perforated ~ 穿孔墙,有漏窗的墙
perimeter ~ 周边墙
perpend ~ 单石[薄]墙
piling ~ 桩墙
pinion ~ 扶垛墙
pipe ~ 管壁
plinth ~ 勒脚,台脚
porous ~ 多孔壁
screen ~ 花隔墙
self-supporting ~ 自承重墙
separation ~ 分隔墙
serpentine ~ 蛇纹形砖[石]墙
shear ~ 剪力墙
sheet-pile ~ 板桩墙
skew ~ 斜歪面墙(播音室用,避免连续声反射),拱座斜墙
spandrel ~ 拱肩墙;上下层窗间墙,窗肚墙
splay ~ 八字形墙
stair ~ 楼梯承墙
suspended ~ 吊壁
sustaining ~ 扶壁,扶墙
tail ~ 丁字堤基凸部
talus ~ 单侧坡面墙
tapestry-brick ~ 缀锦砖墙
tile-hung ~ 贴面砖墙
veneered ~ 镶面墙
wallboard *n.* 壁[墙]板,遮着墙壁或顶棚用的人造板
~ corner beads 墙板角护条,墙板圆线条
~ masonry 砖砌墙板
~ nails 墙板钉
gypsum ~ 石膏墙板
walling ['wɔːliŋ] *n.*；*a.* 筑墙(的),墙(砌体)
~ board 坑壁横撑板
~ cast in situ 现场浇筑墙
~ component 墙体建筑单元
~ hammer 筑墙
~ lining 墙衬
~ piece 护壁板板件
~ slab [预制]墙板,壁板,搁墙撑头木
~ thread 墙线
~ unit [预制]墙体砌块
walnut ['wɔːlnʌt] *n.* 核桃木,胡桃木[树]
walt *a.* 空心的;不坚固的
wan [wɔn] *a.*, wanner, wannest, *v.*, wanned, wanning. *a.* 苍白的;(光)淡弱的,暗淡的; *v.* (使)变苍白
wand [wɔnd] *n.* 棍,棒,杆
wane [wein] *v.*, waned, waning. *n.* 减少,圆棱;缺角方木

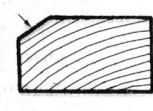
wane

waney ['weini] *a.* wanier, waniest, = wany 不等径的;缺棱的,缺角损的,宽窄不齐的
~ edge 缺损边缘
~ log 不等径圆木
ward [wɔːd] *n.*; *v.* 挡住,防止;市政区;病房,监护室
wardrobe ['wɔːdrəub] *n.* 藏衣室,衣橱
ware [wɛə] *n.* 制品,器具,商品; *v.* 当[留,小]心,注意
~ crack (木材)干裂
brown ~ (褐色)陶器
china ~s 瓷器
coarse sandy ~s 粗砂陶
enamel ~ 搪瓷器皿

fine sandy clay ~s 细砂陶器
glass ~ 玻璃器皿
hard ~ 小五金,金属器皿
ivory ~s (of imitation antique) (仿古)牙雕
jade-inlaid blackwood~s 红木嵌玉石制品
jade~s 玉器
lacquer ~ 漆器
refractory ~ 耐火器材
white ~ 白色陶器

warehouse ['wɛəhauzə] *n. pl.* -houses, *v.*, -housed, -housing; *n.* 仓库,货栈,储存室; *v.* 入库

wareroom ['wɛərum] *n.* 货物陈列室,货物贮藏室

warm [wɔ:m] *n.* 变暖,烤火,保暖的东西; *v.* 使变热,取暖; *a.* (温,保)暖的; warmly, *ad.* warmness, *n.* warmer, *n.*
~ air duct 管送暖气
~-air feed 暖气供给,暖风供给
~-air furnace 热风炉,空气加热器
~ air heating system 热风供暖系统
~-air pipe 热空气管,热风管道
~-air stove 暖气火炉
~ bituminous mixes 温铺沥青混合料
~ colo(u)r 暖色
~ concrete 热混凝土
~ hardening 加热硬化,人工硬化
~ heating 供暖,加热
~ heating installation 供暖设备
~ glue 热胶
~ hardening 加温人工硬化
~ house 温室,暖房
~-laid 温铺的(沥青混合料等)
~ pool 暖水池
~ register 热通风装置
~ rising duct 加热管道
~ roof 保温屋面
~ spraying 热喷雾[洒]

~ spring 温泉
~ water curing 热水养护

warning ['wɔ:niŋ] *n.*; *a.* 警告(的),预报,探测; warningly, *ad.*
~ board 警告牌
~ color 警告(颜)色
~ device 报警装置
~ light 警告灯
~ pipe 溢水管
~ signal 警告标志,警号
~ system 报警系统

warp [wɔ:p] *v.*, *n.* 曲折,弯翘,屈折,放淤,淤填[灌],沉积物,冲积土; warp-age, *n.*
~ clay 淤积黏土,淤泥(尤指潮水河的)
~ satin 经缎纹织物
~ed surface 扭[挠]曲面
~ing joint 铰接缝

warrant ['wɔrənt] *v.* 保证,担保,保险; *n.* 保证;保险(期);执照,委托书,栈单;耐火[煤层下]黏土
~ period 保用[保证]期

Warren truss *n.* 华伦桁架,斜腹杆桁架

wash [wɔʃ] *n.*; *v.* 洗涤,冲洗,洗浆;洗涤剂;冲积物,旧河床
~ basin 洗脸盆
~ board 洗衣板;壁脚板
~ bowel 洗脸盆
~ clay 黏土纯化,黏土漂洗;黏土浆
~ down closet 抽水马桶
~ drawing 水彩画
~ house 洗衣房
~ pipe 冲洗管
~-prime coat 防冲[涂]层
~ room 盥洗室
~ sink 洗涤盆
~ stand 脸盆架
~ed sand 洗净砂,净砂
lime ~ 刷[涂]白

washable ['wɔʃəbl] *a.* 可[耐]洗的,洗得掉的

~ wallpaper 可洗的墙纸
~ water paint 可用水洗掉的水溶液性颜料
washboard ['wɔʃbɔ:d] *n.* 洗衣板；(道路)搓板(现象)；踢脚板；防浪板
~ effect (路面)搓板现象
washer ['wɔʃə] *n.* 洗净器，洗涤设备，洗槽，衬垫，垫圈；洗衣者；洗浆池；洗涤器，洗衣机，洗矿机，洗煤机；垫片，衬垫
washing ['wɔʃiŋ] *n.*；*a.* 洗涤(用的)，冲洗，冲刷；洗涤剂；涂料，薄涂层
~ chamber 洗涤[盥洗]室
~ mark 冲刷痕
washleather glazing 软皮镶玻璃
washout ['wɔʃaut] *n.* 冲洗[刷，去]，洗净；冲溃[蚀]
~ urinal stall 冲洗型立式小便器
~ thread 不完整螺纹
washroom ['wɔʃrum] *n.* 厕所，洗涤间，盥洗室
wash-wear *a.* 耐洗的
wastage ['weistidʒ] *n.* 损耗(量)，磨损(量)，渗漏，(冰雪)消融；废品，(木材等的)干缩
waste [weist] *v.*, wasted, wasting, *n.*；*a.* 废弃的，荒(芜)的；无用的，剩余的，排泄的；*n.*；*v.* 消耗(量)，耗[烧]损，浪费；废物，残渣，垃圾，污水
~ collecting chamber 废物储藏室
~ material 废料
watch [wɔtʃ] *n.*；*v.* (挂，手)表；注意，看守，监视，警戒，值班，等待；看人，值班时间，班[更]次
~ box 守望亭，岗亭
~ crystal 表面皿，表(面)玻璃
~ house 哨所，守望所，岗房
~ loft 屋顶瞭望所
~ room 警卫室
~ -tower 岗楼，瞭望塔
~ turret (小)瞭望台，墙外吊楼
watchdog ['wɔtʃdɔg] *n.*；*v.*, -dogged, -dogging. *n.* 监视器，监控设备；*v.* 看管；监视

water ['wɔ:tə] *n.* 水；水深，水位，潮位，水面；(常用 *pl.*)水体，水域，海域；*v.* 灌溉，洒水，注水，掺水，冲淡
~ -absorbed 吸水的
~ absorbent 吸水剂
~ absorption 吸水性[量]
~ after-treatment 湿治，湿养护
~ bar (窗台塞缝的)挡水条，止水条
~ based paint 水溶性涂料
~ bath 水浴(器)，水[恒温]槽
~ -binder ratio 水-结合料之比，水灰比
~ bosh 水封
~ cellar 水窖，地下水池
~ cement 水硬[水凝]水泥
~ -cement ratio (混凝土)水灰比
~ chamber 水箱，水套
~ check 阻水活栓，逆流截门；(平屋顶)挡水立缘
~ cistern 水槽，水池
~ clarifier 净水器；水的澄清剂
~ closet (W.C.) 厕所，冲水厕所
~ closet flush tank 厕所冲水箱
~ closet pan 抽水马桶
~ cock 水旋塞，水龙头
~ colo(u)r 水彩画；水彩画颜料
~ conditioner 净水器

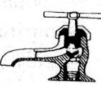
water clost pan

~ conditioning 水净化，水质处理
~ -cured concrete 水养护的混凝土
~ curtain 水幕
~ discharge pipe 排水管，泄流管
~ drainage pipe 泄水[排水]管
~ drip 滴水槽；披水屋檐
~ emulsion paint 乳液涂料，浮胶漆
~ escape 泄[弃]水道
~ exhaust 排水沟
~ faucet 水嘴，水龙头
~ feed(er) 给水[器管]
~ filled radiator 水暖气片，水散热器

water faucet

~ filter 滤水器
~ finish 水纹面饰
~ fountain 喷水,喷泉
~ -free material 无水材料
~ -front structure 沿岸建筑物
~ gas 水煤气
~ gas pipe line 水煤气管线
~ gauge cock 水表旋塞
~ glass 水玻璃,硅酸钠
~ glass coat 水玻璃胶
~ glass mastic 水玻璃胶
~ glass paint 水玻璃颜料
~ glass putty 水玻璃油灰
~ heater 热水炉,水加热器
~ -holding material 蓄[含]水材料
~ hose 水龙带;软管,皮带管
~ inlet 进水口,取水口
~ inlet pipe 进水管
~ insoluble 不溶于水的
~ joint 防水接头,水封接头
~ leaf 幌菊饰
~ leak 漏水
water leaf
~ lime 水硬石灰
~ -line paint 水线漆
~ lock 存水湾,水闸
~ meter 水表,流量计
~ nozzle 喷水嘴
~ of mixing 拌合用水
~ oozing 渗水,漏水
~ outlet 出水口
~ paint 水溶性颜料
~ percolation 渗水
~ permeability 透水性
~ pipe 水管
~ piping 敷设水管;水管线路
~ plug 消防栓;[抽水马桶]抽水装置
~ -potash glass 水钾玻璃
~ purifier 净水器
~ purifying tank 净水池
~ -rejecting 防水性,憎[疏]水性
~ -repellent 憎[抗,防]水的,防水剂
~ -repellent admixture 憎水[外加]剂
~ -repellent cement 憎水水泥
~ repellent concrete 憎水性混凝土
~ -repellent motar 憎水砂浆
~ -repellent preservative 憎水防腐剂
~ repellent treatment 抗水[防水]处理
~ -resistant 抗[防]水的
~ -resistant glue 防水粘合剂
~ resistant gypsum board 防水石膏板
~ -resistant paint 防水油漆
~ -resisting admixture 防水(添加)剂
~ resisting layer 隔水层
~ -resisting enamel 防水瓷漆
~ -resisting paint 防水油漆
~ -resisting varnish 防水清漆
~ -retaining capacity 持水量,保水能力
~ retention 保水性
~ -retentive portland cement 保水性硅酸盐水泥
~ rheostat 水电阻器
~ riser pipe 给水立管
~ roll 水辊
~ screen 水帘
~ seal 水封,止水
~ -sealed joint 止水缝
~ -seasoning (木材)浸水两周后再风干(法)
~ section (厨房,浴室等的)用水间
~ sediment complex 水砂混合体,含泥砂的水
~ seepage 渗水
~ service 供水,供水设备,给水工程
~ service installation 给水装置
~ sheet 水帘
~ shield 水屏,水防护屏
~ shoot (屋檐)排水槽,滴水石
~ -side structure 滨水建筑物
~ slaked lime 水解石灰

~ -soaked 水浸透的,饱水的
~ -soap solution 肥皂水溶液
~ soda ash glass 水玻璃
~ softener 软水剂,软水器
~ softening 水的软化
~ -soluble 水溶(性)的
~ -sorted material 水选材料
~ -spray nozzle 喷水嘴
~ stable aggregate 水稳性骨料
~ stain 水污斑;水溶性颜料
~ sterilization 消毒水
~ structure 不透水结构
~ suction hose 吸水软管
~ supply facility 给水设施
~ -supply line 给水管线,上水道
~ supply pipe 给水管
~ -supporting deck 挡水面板
~ tank 水池[箱,柜,塔],水槽
~ tap 水龙头
~ temperature regulator 水温调节器
~ -tight 不漏水的,防水的;水密的
~ tight basement 防水地下室
~ -tight concrete 不透水[水密性]混凝土
~ -tight diaphragm 防水薄层,防水膜,阻水层,防渗斜墙
~ -tight door 防水门,水密门
~ -tight facing 防渗[阻水]面层
~ -tight joint 水密接合,不透水缝
~ -tight layer 不透水层,隔水层
~ -tight seal 止水,水封
~ trap 存水弯
~ tube 水管
~ valve 水阀,水门
~ vapor permeability 透湿性
~ wall 防水墙,不透水墙
~ -worn gravel 水磨砾石
calcareous ~ 石灰水
city ~ 自来水
distilled ~ 蒸馏水
lime ~ 石灰水
ordinary ~ 普通水
service ~ 家[生活]用水

waterless ['wɔːtəlis] *a.* 无水的,干的
~ toilet (用化学方法消毒)干厕

waterlog ['wɔːtəlɔg] *v.*, -logged, -logging. *v.* 使漏水;使泡水

waterlogged ['wɔːtəlɔgd] *a.* 吸饱水的, 浸透了的,水涝的,泥泞的

waterproof ['wɔːtəpruːf] *a.* 防水的; *n.* 防水布,雨衣; *v.* 使不透水
~ adhesive 防水粘胶剂
~ agent 防水剂
~ cement 防水水泥
~ cloth 防水布
~ coating 防水涂层
~ concrete 防水混凝土
~ course 防水层
~ emulsion 防水乳剂
~ felt 防水(油)毛毡
~ finish 防水涂层装修
~ing admixture 防水添加剂
~ing agent 防水剂
~ing compound 防水化合物
~ing floor 防水地板
~ing powder 防水粉

waterproofing floor

~ing wall 防水墙
~ layer 防水层
~ material 防水材料
~ -membrane 防水(隔)膜
~ mortar 防水砂[灰,泥]浆
~ paint 防水漆
~ paper 防潮纸,油毡
~ plaster 防水灰泥
~ wax 防水蜡

waterproofer *n.* 防[隔]水层,防水材料
water -shoot *n.* 屋檐排水槽,滴水石
waterside ['wɔːtəsaid] *a.*; *n.* 水边;水滨,海滨
watersoaked *a.* 水浸透的,浸水的

waterspout ['wɔːtəspaut] *n.* 水龙卷；水落管，排水口，水口，水柱

waterstone *n.* 水磨石

waterstop *n.* 止水

watertight ['wɔːtətait] *a.* 不透水的；watertightness, *n.*
~ material 不透水材料

watertightness ['wɔːtətaitnis] *n.* 不透水性；水密性；闭水性

waterworn ['wɔːtəwɔːn] *a.* 水蚀的，被水磨平了的

watery ['wɔːtəri] *a.* 水的；多水分的；水一般的；溃的；潮湿的；waterily, *ad.* waterliness, *n.*

Watt [wɔt] *n.* 瓦(特)(电功率单位)
~ -hour meter 电(度)表
~ -meter 瓦特计，电(功率)表
~ per candle 瓦/烛光

wattle ['wɔtl] *n.; v.*, -tled, -tling, *a.*
n. 篱笆；枝条；柴排；*v.* 编篱笆；wattled, *a.*
~ and daub 抹灰篱笆墙
~ fence 篱笆围墙
~ gum 红色阿拉伯胶
~ house 篱笆房屋

wattling *n.* 柴捆，柴排

waughammer *n.* 架柱式凿岩机

waughoist *n.* 柱式绞车，带架绞车

wave [weiv] *n.; v.*, waved, waving.
n. 波(浪)；波动，起伏；*v.* 波动，起伏；加波纹
~ arch 波形拱
~ curve 波形曲线
~ moulding 波形线条，波形饰
~ ornament 波形装饰品，波形装饰
~ overtopping （防波堤）波涛越顶
~ pattern 波型，波谱；波型花纹
~ scroll 波形装饰品，波形涡卷
~ d roof 波状屋顶
~ d shell roof 波形薄壳屋顶
~ d tube 波形管
shape ~ 薄钢板的浪形缺陷

waver ['weivə] *v.* 摇摆；摇晃；*n.* 波段开关

waving groin 波状穹顶肋

wavy ['weivi] *a.*, wavier, waviest. 波状的，起伏的；多浪的；摇摆的；动摇的
~ cord 波形绳索
~ dressing 波形装饰
~ edge （薄钢带的）波形边
~ fibred 波纹硬纸板
~ -fibred growth 波状乱纹（木材）
~ -grained wood 波状纹理木
~ grown timber 波纹木材，扭曲木材
~ line 波形线
~ trace 蜿蜒[起伏]路线

wax [wæks] *n.* 蜡；蜡状物；*v.* 打蜡，封蜡；*a.* 蜡制的
~ cloth 涂蜡漆布(铺地板用)
~ (concrete) curing compound (混凝土)养护蜡涂料
~ figure 蜡像
~ -free 无蜡的
~ -like 似蜡
~ painting 蜡画
~ paper 蜡纸
~ polish 上光蜡，打蜡
~ -sealed 用蜡密封的
~ shale 油页岩
~ stain 蜡染剂，蜡溶颜料
~ -tree 蜡树(女贞；白蜡树)
~ wall 黏土密封墙
~ wire 打蜡线
~ yellow 淡黄色
ceresin(e) ~ (纯)地蜡
mineral ~ 石蜡
ozokerite [earth] ~ 地蜡
sealing ~ (密)封蜡，火漆
synthetic ~ 人造蜡，合成石蜡

waxen ['wæksən] *a.* 蜡制的，上[涂]蜡的

waxwork ['wækswəːk] *n.* 蜡像；蜡制品

waxy ['wæksi] *a.* waxier, waxiest 蜡状的；涂蜡的；蜡制的；可塑的

~ bitumen 蜡质沥青
~ luster 蜡光泽
way [wei] ***n.*** 道路;途径,路程,路线;方法,样子,状态
~ beam (桥梁)纵梁
~ -board (两厚层中的)薄隔板
~ mark 路标
~ out 出口,(戏院)太平门;标新立异的
aerial rope ~ 架空索道
cable ~ 索道
flange ~ 凸缘沟
guide ~ 导轨
key ~ 键槽
rope ~ 索道
slide ~ 滑道
three ~ 三通管接头
two- ~ valve 二[双]通阀,双行程活门
wire ~ 钢丝提升绳道,电线槽
wayleave ['weili:v] ***n.*** 道路使用权,通行权
~ -lease 通行权契约
wayside ['weisaid] ***a.;n.*** 路旁;路边的
weak [wi:k] ***a.*** 弱的,软的;稀薄的
~ concrete 低强度[贫]混凝土
~ formation 软弱构造,软弱层
~ ground 软弱地基
~ links 薄弱环节
~ point 弱点
~ sand 瘦(型)砂
~ soil 软土
~ solution 稀溶液
weaken ['wi:kən] ***v.*** 削弱,弄薄,减轻,降低; weakener, ***n.***
~ plane joint (路面)弱面缝,假缝,半缝
wealth [welθ] ***n.*** 财产[富],资源;丰富,大量
wear [wɛə] ***v.***, wore, worn, wearing, ***n.*** 磨耗,磨损;耐用,耐磨
~ and tear 磨耗及损伤
~ inhibitor 防磨损剂
~ plate 防磨耗板

~ -resistance 抗磨力,耐磨性
~ ring 耐磨环
~ing capacity 耐磨性;磨损量
~ing carpet 磨耗[损]层
~ing floor 楼面磨耗层
~ing layer 磨耗[损]层
~ing plate 衬板,衬砌板
~ing strength 抗磨强度
~ing surface 磨耗[损]面
wearproof ['wɛəpru:f] ***a.*** 不磨损的,耐磨的,抗磨的
weather ['weðə] ***n.*** 天气,气候[象];***v.*** 风化;经受风雨;曝于大气中;晾干,通风;***a.*** 挡风的,顶风的
~ bar 挡水条(如窗台或门槛塞缝的)
~ -board 挡风板,封檐板,护壁板
~ check 滴水槽,屋檐
~ checked tie 风化辐裂的枕木
~ cloth 防雨布
~ door 外重门,防暴风雨门
~ fillet 挡风雨贴角压条;防泛水条
~ groove 泻水槽
~ -resistant 抗风化的
~ shake (木材等)干冻裂隙,风干裂纹
~ shingling 防风的薄板,贴面薄板
~ slating (防天气侵蚀的)墙顶石盖板,铺石板瓦
~ stained 因风雨褪色的
~ strip 挡风雨条(门窗口沿边的贴角压条)
~ -struck joint 防雨刮缝
~ -tight 不透风雨的
~ tile 盖墙瓦
weathered ['weðəd] ***a.*** 风化的;作坡泻水的,晾干的
~ clay 风化黏土
~ effect 风化效应;天然(造成的)效果
~ pointing 泻(水)勾缝
~ slope (窗台等)泻水坡度
~ back (烟囱或桥墩的)泻水坡

weatherproof ['weðəpru:f] *a.* 抗风化的,防风雨的; *v.* 使防风雨[日晒] *n.* 耐风雨的材料
~ glue 抗老化胶

weave [vi:v] *v.* ; *n.* 织,编(成,排),构成;编织型式;摆动,摇晃;迂回,曲折

weaving *n.* 编织,(横向)摆动
~ mattress 柴排
~ pole 木排
~ wire 编织用钢丝

weavy grain 卷纹,织纹

web [web] *n.* 腹板,梁腹,T形材的立股;蛛网;织品; *v.* 织成蛛网
~ bar 腹筋
~ cleat (钢梁的)腹板锚固夹板
~ core sandwich panel 腹板夹芯板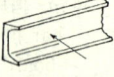
~ covers 梁腹连接盖板,翼缘板
~ -crack (混凝土梁的)腹剪裂缝
~ of tile 空心砖腹隔
~ plate 腹板
~ plate connection 腹板连接
~ plate joint 腹板接缝
~ reinforcement 抗剪钢筋,腹筋
~ -type spar 腹板式翼梁
twist ~ 麻花钻心

wedge [wedʒ] *n.* ; *v.* , wedged, wedging, *n.* 楔,楔形(物);尖劈; *v.* 楔入,楔牢;劈开;挤进
~ analysis 土楔分析法
~ anchorage 楔形锚具
~ and shims 插楔开石工具
~ -and-sleeve bolt 插楔和套筒门闩[插销]
~ block 楔块
~ boarding 楔接板
~ bonding 楔形接合,楔形焊点
~ brick 楔形砖
~ clamp 楔形压板
~ clamp joint 楔嵌接
~ clip 楔形扣板
~ coping 楔形盖顶石

~ coupler 楔形联结装置
~ curb 楔形壁座
~ cut 楔式开挖,楔式切割
~ friction wheel 楔形磨擦轮
~ gate valve 楔形闸门
~ gauge 楔形槽
~ groove 楔形规
~ hammer 楔锤
~ joint 楔(形)接(缝)
~ key 楔形键
~ of earth 土压楔体
~ pile 楔形桩
~ sewer brick 楔形阴沟砖
~ -shape tread 楔形楼梯踏步板
~ -shaped curb 楔形路缘石
~ -shaped dam 楔形防水墙
~ -shaped step 楔形楼梯踏步
~ sheet pile 楔形板桩
~ stone 拱楔石
~ structure 楔形结构
~ tile 楔形砖
~d slot 楔形狭槽,楔形长孔
~d tenon 楔榫
~ing action 楔合作用
~ing block 楔合块
adjuster ~ 调整楔
double ~ 双面楔
fox ~ 紧榫楔
gate ~ 斜铁,门楔

weed [wi:d] *n.* 杂草;废物; *v.* 扫清;淘汰;除草
~ cleaner 杂草清除器
~ cutter 割草机

weedy ['wi:di] *a.* 杂草似的;多杂草的;无价值的

week [wi:k] *n.* 一星期,一周,工作周
~ -day chapel 平日小教堂
~ end 周末

weekday ['wi:kdei] *n.* 星期日以外的任何一天,平日,工作日

weep [wi:p] *v.* , wept, weeping, *n.* ; *v.* 分泌,泌水,漏水,滴落; *n.* (水或液体)

渗出
～ drain 排[分]水管,排水盲沟
～ -hole (挡土墙或女儿墙上的)排水孔
～ pipe 滴水管,排水管(混凝土砌体背后排水用)
weeper *n.* 泄水孔,渗水孔
weeping ['wi:piŋ] *n.* 分泌,泌[渗,漏]水,滴落; *a.* 垂下的;滴水的
～ pipe (挡土墙的)渗水管
～ willow 垂柳
weft [weft] *n.* 纬,织物;信号旗,(求救)信号
weigh [wei] *v.*; *n.* 称量,对比,权衡,估计;重(若干); weighter, *n.*
～ batcher 称重配料器,工地用配料秤
～ batching (混凝土拌合中)按重量配料
～ -batching hopper (按重量)配料斗
～ batching plant 按重量配料厂,分配称料设备
～ -beam 天平杆,秤杆
～ box 称重器[斗]
～ bridge 台秤,地磅,桥秤
～ bucket (混凝土拌合)称料斗
～ by volume 按体积称重
～ feeder 称量加料器
～ gear 称重设备
～ house 过磅处
～ installation 称重设备
～ plant 冲器厂
～ tank 称量桶
weighing ['weiiŋ] *n.* 称量;重压;加权,权重;衡量,评价
～ apparatus 称量装置
～ appliance 称量装置
～ batch box 按重量配料箱,分批称料箱
～ device 称量设备;称
～ equipment 称量设备
～ hopper 称料漏斗
～ lever 称杆
～ platform 称台

～ room 称料间
～ tank 称量桶
weigher ['weiə] *n.* 司磅员,验称员,衡器,称
weight [weit] *n.* 重量;重力;权;法码,称锤;衡; *v.* 加重,装载,加负荷,加权
～ bearing capability 承载能力
～ bearing power 荷载能力
～ box 窗的[平衡]锤箱
～ coefficient 加权系数,重量系数
～ distribution factor 重量分配系数
～ empty 皮重,空载
～ factor 权重因数
～ floor block 承载地面(黏土)砖
～ frame 承载刚架
～ lever regulator 称杆调节器
～ limit sign 重量限制标志
～ -loaded regulator 重锤式调节阀
～ partition wall 承重隔墙
～ wall 承重墙
bare ～ 空重,皮重
bulk ～ 空重,散重
center ～ 中心锤
net ～ 净重
rough ～ 毛重
safety ～ 安全负荷
sash ～ 吊窗锤
soaking ～ 湿重
weightometer *n.* 重量计,自动秤,自动秤重仪
weights *n.* 砝码
welcome mat 欢迎垫(门前擦鞋棕垫)
weld [weld] *n.* 焊(接),焊接点;焊缝; *v.* 焊(接);焊合,熔接
～ assembly 焊接部件
～ bead 焊道(熔敷一次所形成的单道焊缝)
～ -bonding 粘接点焊
～ crack 焊接裂纹
～ crosswise 交叉焊接
～ decay 焊接侵蚀
～ defects 焊接缺陷

~ edgewise 沿边焊接
~ length 焊缝长度
~ metal 焊接金属
~ ripples 焊缝的波纹
~ rotation 焊缝转角
~ seam 焊缝
~ size 焊接尺寸
~ metal zone 焊缝区
~ed all round 沿周焊接的
~ed connection 焊接连接
~ed construetion 焊接结构
~ed flange 焊接翼缘
~ed frame 焊接骨架
~ed girder 焊接大梁
~ed joint 焊接缝,焊接接头

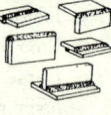
welded joints

~ed (metal)mesh 焊接(金属)网
~ed pipe 焊接管
~ed pipe construction 焊管结构
~ed plate girder 焊接板梁
~ed reinforcement 焊接钢筋
~ed rigid frame 焊接刚架
~ed seam 焊缝
~ed steel box-girder 焊接箱形钢梁
~ed steel fabric 焊接钢筋网
~ed steel structure 焊接钢结构
~ed truss 焊接桁架
~ed tube 焊接管
~ed wire fabric 焊接钢丝网
alternating current arc ~ 交流电弧焊
atomic hydrogen ~ 原子氢焊
autogenous ~ 乙炔焊
automatic electric ~ 自动电焊
butt ~ 对焊,对接焊缝
weldability [weld'biliti] *n.* 可焊性
welder ['weldə] *n.* 焊工,焊机
~'s gloves 焊工手套
~'s goggles 焊工护目镜
~'s gauntlets 焊工的长手套
~'s hand shield 焊工手持护目罩
~'s helmet 焊工帽
~'s spats 焊工护脚
arc ~ 电弧焊机
laser ~ 激光焊机
electric-driven ~ 电动焊机
gas ~ 气焊工
spot ~ 点焊机
welding ['weldiŋ] *n.* 熔焊,熔接;*a.* 焊接的,熔接的
~ alloy 焊接合金
~ and cutting torch 焊割两用气焊枪
~ bench 焊接工作台
~ booth 焊接工作间
~ cut bit 焊割工具
~ deformation 焊接变形
~ electrode 电焊条
~ electrode hold 焊条夹
~ equipment 焊接设备,电焊机
~ flame 焊接火焰
~ flux 焊剂
~ generator 电焊机
~ glass 黑玻璃,滤色玻璃
~ goggles 焊工护目镜
~ gun 焊枪
~ head 焊头,烙铁头
~ helmet 焊工帽
~ installation 焊接设备
~ machine 焊机
~ manipulator 焊接机械手
~ metal 焊接金属
~ paste 焊接涂料
~ plant 焊接设备,焊接车间
~ powder 焊熔剂,焊粉
~ robot 焊接机器人
~ rod 焊条
~ shop 焊接车间
~ wire 焊丝
~ worker 电焊工
weldless ['weldlis] *a.* 无焊缝的
~ steel tube 无缝钢管
weldment ['weldmənt] *n.* 焊(接)件,

焊接装配

weldwood *n.* 特制胶合板

welfare ['welfɛə] *n.* 福利
~ building 生活间
~ facilities 福利设施

well [wel] *n.*; *v.*; *ad.*, better, best, *a.*, better, best. *n.* 竖坑,楼梯井;源泉; *v.* 涌出,流出; *a.* 好的,适当的,充分的;井的;像井的;wellness, *n.*
~ -appointed 装备好了的,设备齐全的
~ -balanced 匀称的,平衡的
~ beam floor 井梁楼板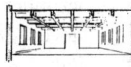
well beam floor
~ -behaved 性能良好的
~ -bonded 充分粘结
~ burned brick 烧透砖,优质砖
~ column 钻井支柱
~ crib pattern 井字形花纹
~ -defined 轮廓分明的
~ -designed 设计完美的,精心设计的
~ -graded aggregate 级配良好的骨料
~ graded sand (良好)级配砂
~ -hole 井孔,升降机井道,楼梯井
~ -known 著名的
~ lining 井衬砌
~ -preserved 保养很好的
~ -proportioned 相当匀的;良好配合
~ -rounded 很圆熟的;流线型的;圆角的
~ screen 井管防砂罩,井管滤网
~ -shaped 很好修整过的,外形精美的
~ timbered 用木材加固的
~ winch 矿井绞车
loading ~ 供料(井)室

Welsh arch *n.* 威尔士拱,有楔形拱心石的平拱

Welsh groin *n.* 威尔士穹顶肋;两高低穹顶直交形成的交叉拱

Welsh vault *n.* 威尔士穹顶

welt [welt] *n.* (装在物件上的)贴边;盖缝条,衬板; *v.* 装上贴边
~ ed edge 搭接缝[边],摺边,贴边

west [west] *n.* 西;西方,西部; *a.* 西方的,向西的
~ African Copal 西非树脂
~ Asiatic architecture 西亚建筑
~ (-facing) wall 西墙
~ (-facing) window 西窗
~ Garden 苏州西园
~ Indies architecture 西印度群岛建筑
~ longitude 西径

wester ['westə] *n.* 西风; *v.* 转向西面

western ['westən] *a.* 西方的,西部的
~ architecture 西方建筑
~ Byzantine 西方拜占庭(建筑)样式
~ facade 西方式(房屋)正面
~ gallery 西方式外廊
~ method 西部砌筑法,抹灰砌砖法
~ pediment 西方式柱廊上三角形或弧形檐饰
~ porch 西方式门廊
~ red cedar 西部红杉

westphalt *n.* 用沥青粉拌和的冷铺沥青混合料

westward ['westwəd] *a.* 西方的; *n.* 西方

wet [wet] *a.*, wetter, wettest, *v.*, wet or wetted, wetting; *n.* 潮湿;湿度;雨天;水分; *v.* 弄湿,湿润; *a.* 湿的;湿式的
~ abrasive blasting 湿喷砂
~ adhesive bonding 湿粘合
~ -aggregate process 湿砂搅拌
~ area 潮湿地区
~ ashing 湿法灰比
~ batch 湿配料
~ batch method (混凝土)湿料分批法,湿料计盘法,湿料运送法
~ batch rating 湿法搅拌批量,[混凝

土搅拌机]按湿料数量计算的生产率
~ blacking 黑色涂料,碳素涂料
~ blast 湿喷砂
~ blast(ing) 水砂抛光处理;湿喷(含有磨蚀剂的水)
~ blasting process 湿喷砂法
~ cleaner 湿式除尘[滤清]器,湿清刷机
~ combustion 湿法氧化燃烧
~ comminution 湿磨
~ concrete (未结硬的)塑性混凝土,流态混凝土
~ condense pipe 湿式冷凝管
~ connection 湿连接
~ consistency (of concrete) (混凝土)塑性稠度
~ construction 湿法施工,浆砌施工
~ cooling 湿冷却
~ -crushed 湿法破碎的
~ crushing mill 湿碎磨
~ curing 湿养护;湿烘焙
~ cut 湿开挖;水下开挖
~ cyclone 湿式旋风除尘器
~ density 湿容重[密度]
~ dust collector 湿式集尘器
~ excavation 湿开挖;水下开挖
~ feed method 湿送料法
~ -fed cement kiln 湿式给料水泥回转窑
~ filter 湿式过滤器
~ -fuel 液体燃料
~ grinding 湿磨,湿研
~ grout 湿状灰浆
~ lime 湿石灰
~ milling 湿磨
~ mix 湿拌
~ mix aggregate 湿拌(混凝土)骨料
~ mix shotcrete 湿拌喷射混凝土
~ mixer 湿拌机
~ mud brick 湿泥砖,湿法压制砖
~ paint 油漆未干(警告用语)
~ pan 湿式轮碾机;拌泥器的转筒
~ -pit pump 排水泵

~ process 湿法生产(水泥)
~ -proof 防潮的
~ return line 回水管
~ rock cuttings 湿凿[碎]石
~ rot (木材)湿腐[朽]
~ sand-binder constuction 湿砂粘结施工
~ sand blasting 湿砂喷射
~ sand cure (of concrete) (混凝土)湿砂养护
~ screened 湿筛的
~ scrubber 喷水除尘装置
~ sieve 湿格筛
~ soil 湿土
~ stamping 湿碎法
~ steam 湿蒸汽
~ stock 湿材料;黏状浆
~ strength 湿强度
~ suction fan 湿式吸尘器
~ trade 湿工序,灰浆工序
~ treatment 湿处理,湿选
~ vent 排湿气孔

wetted $a.$ 湿润的
~ area 受潮面积
~ cross section 湿润断面
~ surface 湿润[浸水]面

wetting ['wetiŋ] $n.$ 浸湿,润湿
~ agent 湿润剂
~ power 湿润能力
~ properties 湿润性,受潮性
~ system (道路)洒水设备,洒水系统

wettability [ˌwetə'biliti] $n.$ 受湿性,湿润性,湿润度

whalebone brush $n.$ 鲸须刷(清沟渠用)

Whatman paper $n.$ 瓦特曼纸(高级图画纸)

what-not ['wɔtnɔt] $n.$ 陈列书籍[古董]的架子

wheel [wi:l] $n.$ 轮;车轮; $v.$ 滚[转,推]动;旋转; wheeled, $a.$
~ and axle 差动滑轮,轮轴
~ -barrow 手推车,独轮车
~ brace drill 手摇轮钻

~ cutter 轮式切削器,轮式铰刀
~ step (楼梯的)斜踏步
~ tracery 轮形花格窗
~ type hanging roof 圆形悬索屋顶
~ window 轮形窗

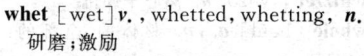
wheel barrow

whet [wet] v., whetted, whetting, n. 研磨;激励
~ slate 磨石

whetstone ['wetstəun] n. 砂轮,磨刀石,砥[油]石

whim [wim] n. 绞盘,辘轳,卷扬机

whin [win] n. 暗色岩,粗玄岩

whinstone ['winstəun] n. 粗玄岩

whip [wip] n.;v., whipped or whipt, whipping. 搅起;抖动,急(骤)动(作); -whiplike, a. 绕,捆,绞;滑轮吊车,鞭状天线;垂曲
~ crane 动臂起重机
~ hoist 动臂起重机

whippletree ['wipltri:] n. 横杠

whipsaw ['wipsɔ:] n. 狭边(钩齿)粗木锯,双人横切锯

whirl [wə:l] n.;v. 旋转;急转;卷成旋涡;涡流[动],旋涡
~ coating 旋转涂覆法
~ crane 旋转式起重机
~ mix 摆轮式混砂机
~ spreader (松散材料)旋涡式摊铺机

whispering ['wispəriŋ] n. 耳语,沙沙声
~ cupola 回音廊(反射低音的圆廊)
~ dome 耳语穹顶,混响穹顶
~ gallery 回音廊

whistler ['wislə] n. 通气孔,排气道

white ['wait] n.;a., whiter, whitest, v.;ad. n. 白色;白色颜料;蛋白; a. 白(色)的;无色透明的
~ alkali soil 白碱土
~ alloy 白色合金,假银
~ alum 明矾

~ architecture 白色建筑风格
~ asbestos 白石棉
~ asbestos-cement board 白水泥石棉板
~ bark pine 白皮松
~ base 白色基层(沥青路面下的混凝土或碎石基层)
~ basswood 白椴
~ body 白黏土坯体,白石头
~ bole (=kaolin) 高岭土
~ brass 白铜
~ brich 欧洲白桦
~ brick 白砖
~ bronze 白青铜
~ cast iron 白口(铸)铁
~ cedar 白杉,金钟柏
~ cement 白水泥
~ (chalk) lime 高钙石灰,白石灰
~ city 游艺场,娱乐场所
~ coat 白灰涂面
~ collar class 白领阶层
~ concrete 白色混凝土
~ copper 白铜
~ copperas 皓钒;硫酸锌
~ deal 挪威杉木
~ deposit (混凝土表面的)霜斑,渗斑
~ dune 白砂丘
~ (finish) coat 白灰罩面
~ fir 白枞(树),白冷杉
~ frost 白霜,霜
~ glazed finish 白釉饰面
~ glue 白胶
~ gold 白金
~ goods 漂白的床单桌布;大型家用电器
~ heat 白热
~ iron 白铁皮,镀锡薄钢皮
~ joint mortar 白色接缝砂浆
~ lac 白漆片,白虫(胶)漆
~ lead 白铅(粉),铅白
~ lead paint 白铅漆[涂料]
~ -lead putty 白铅油灰

~ light 白光
~ lime 熟石灰
~ lime mortar 白石灰砂浆
~ lump lime [纯]块石灰
~ marble 白大理石
~ mastic 白色玛碲脂,白色胶合铺料
~ metal 白色金属,白合金,巴氏合金
~ mica 白云母
~ mineral oil 石蜡油
~ mulberry 桑树
~ noise 白噪声,混合噪声
~ oak 白栎(木),白橡(木)
~ peat 白泥炭
~ pig iron 白生铁,白口铁
~ pigment 白色颜料
~ pointing mortar 白勾缝砂浆
~ poplar 银白杨
~ portland cement(WPC) 白色硅酸盐水泥
~ quicklime 浓[纯]生石灰
~ radiation 白色光
~ room 绝尘室,极清洁的房间(如精密仪器室,手术室等)
~ rot 白斑腐(木材缺陷)
~ sand 白砂
~ spirit (用于油漆的)石油溶剂,白节油
~ vitriol 皓矾;硫酸锌
~ wax 白蜡
~ way 白路,白道(灯火灿烂的街道或大路)
~ zinc 白锌粉,锌白,氧化锌
~ zinc paint 白锌漆
whiten ['waitn] v. 弄白,刷白,漂白;变白; whitener, n.
~ed surface 刷白表面
~ing 刷白,漂白镀锡
whiteware ['waitwɛə] n. 白色陶瓷,卫生陶瓷
whitewash ['waitwɔʃ] v. 白灰粉饰; n. 石灰水
~ brush 白灰刷帚

whitewood ['waitwud] n. 白色木材;白杨
whiting ['waitiŋ] n. 白垩[铅]粉
~ putty 铅白油灰
Whithey key n. 半圆键,月牙键
Whitworth die n. 惠氏螺纹钢板
whitworth screw thread n. 惠氏螺纹
Whitworth wire gauge n. 惠氏线规
whizzer ['wizə] n. 离心干燥机
whole [həul] a.;n. 整体的,全部的;总数
~ brick 整砖
~ brick wall 整砖墙,砖厚墙
~ colored 纯色的
~ deals 整块松板($1^{1}/_{4}$ 英寸厚)
~ hip 四坡屋顶
~ section member 整体式构件,非拼接构件
~ timber 整材,非拼接木材
~ tine grab 带齿抓斗
wholesale ['həulseil] n.;a.;ad.;v.;-saled,-saling. 批发(的);大批(的);全部的
~ cut-off 连续[全面]截弯
wholesome ['həulsəm] a. 卫生的,有益的;安全的
~ water 卫生饮用水
Wichart truss 茨铰桁架
wick [wik] n. 灯心,烛心,吸油绳;庄,村,乡;导火线
~ lubricator 油绳润滑器
~ screw 灯心螺丝
wicker ['wikə] n. 柴束;柳条;柳条制品; a. 柴束的;枝条编织的
~ mat 柳条垫
~ work 柴束工作;编枝工;柳枝编物
wicket ['wikit] n. (装在大门上的)小门,便门;放水门;(只有开下半扇的)半门,(附有拉门的)小窗口,售票处[窗]
~ door 便门,边门
~ gate 便门;旋闸,导叶
~ -type gate 旋转式闸门
wickiup ['wikiʌp] n. 美国印第安人的

小屋子;窝棚
wide [waid] *a*., wider, widest, *ad*. 宽的,阔的;(离得)远的;广泛的
~ -angle 大角度的,广角的
~ -angle multiplex projector 广角多倍投影测圆仪
~ -angle scanning 广角扫描
~ -bottom flange rail 宽底钢轨
~ -bucket excavator 宽斗挖土机
~ corridor 宽檐廊
~ crack 宽裂缝
~ -flange beam 宽翼缘梁
~ -flange girder 宽翼缘梁
~ flanged section 宽翼缘工字钢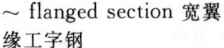
wide-flange beam
~ -ledge tub 宽边浴盆
~ -meshed 宽网孔的,大筛孔的
~ open 全开的
~ planting 疏植
~ return bend 宽径回转的弯头
~ -ringed timber 疏年轮木材
~ screen 宽银幕
~ -spaced 疏柱式建筑物(柱距约等于柱径的 4～5 倍)
~ span 大跨度
~ -span precast beam 大跨度预制梁
~ -span roof 大跨度屋顶
~ -spread rain 大面积降雨
~ strip 宽带材[钢]
widen ['waidn] *v*. 加宽,变阔
widespread ['waidspred] *a*. 广泛的,普遍的
widow's walk *n*. 面对大海的屋顶阳台
width [widθ] *n*. 宽度,广度
~ between centers 中心距
~ of step 阶宽
~ -thickness ratio (集料的)宽厚度
~ -to-depth ratio (截面)宽高比
wigwam ['wigwæm] *n*. 印第安人树皮覆盖的锥形小屋,拱形顶草棚
Williot (deflection) diagram *n*. 威洛(变位)图,桁架变位图
Williot graphical method *n*. 威洛图解法(用于分析桁架变位)
willow ['wiləu] *n*. 柳树,柳木制品
~ brush 柳梢
~ pattern (中国瓷器的)白底蓝色柳树图案
wimble ['wimbl] *n*.; *v*., -bled, -bling. *n*. 锥,钻;手摇钻,螺旋钻; *v*. 以钻或锥子钻
winch [wintʃ] *n*. 绞车,绞盘,卷扬机;曲柄; *v*. 用绞车举起
~ attachment 绞车附件
~ barrel 绞车卷筒
~ cable 绞车缆索
~ drum 绞车卷筒
~ hoist 绞车起重[卷扬]机
~ line 绞车绳[索]
~ rope 绞车绳[索]
boring ~ 钻机绞车
builder's ~ 建筑绞车
crab ~ 绞车
crane ~ 起重绞车
double-drum ~ 双滚筒绞车
double purchase ~ 双卷筒绞车
electric ~ 电力绞车
hand ~ 手绞车
haulage ~ 运输绞车
hauling ~ 升降绞车,升降卷扬机
hydraulially powered ~ 液力传动绞车
linkage ~ 车悬挂式绞车
motor ~ 机动绞车
steam ~ 蒸汽绞车
trailer-mounted ~ 装在拖车上的绞盘
wind [wind] *n*.; *v*., winded, winding. *n*. 风,气流; *v*. 吹,通风; [waind] *v*., wound, winding, *n*. 缠绕,绞起,绞车;转动,吊起
~ beam 圈梁
~ braced boom 有抗风支撑的吊杆
~ cowl 烟囱风帽

~ stop 挡风条
~ tie 抗风联杆,抗风索系杆
~ truss 抗风桁架
winder ['waində] *n.* 卷线器;提升机;
(楼梯)斜踏步;盘梯
automatic ~ 自动化提升机
coil ~ 线材拉拔机
counter-weight ~ 平衡锤提升机
ground-type twin-rope friction ~
地面式双绳摩擦提升机
mobile ~ 移动式提升机
multi-rope ~ 多绳提升机
reel ~ 绞轮提升机
winding ['waindiŋ] *n.* 缠绕;卷;线圈;
a. 缠绕的;蜿蜒的
~ barrel 绞车卷筒
~ drum 提升绞筒
~ engine 卷扬机
~ gear 卷扬机齿轮,升降装置
~ rope 提升绳索,起重索
~ stairs 盘旋式楼梯
~ staircase 盘旋式楼梯
windlass ['windləs] *n.* 卷扬机,绞盘,
小绞车;*v.* 用卷扬机绞起
windmill ['windmil] *n.* 风车(般旋转);
风力发动机
~ -palm 棕榈
window ['windəu] *n.* 窗,窗口,窗洞;
v. 开窗
~ accessories 窗的附件
~ air conditioner 窗上空气调节器
~ aperture 窗口,窗格,窗孔
~ apron 窗台板
~ arch 窗的拱弧
~ awning 遮阳篷
~ back 窗腰
~ band 带形窗
~ bar 窗闩,窗梃,窗栅
~ -blind 窗帘,遮阳帘
~ -blind holland(window holland)
窗帘布
~ board 窗台板

~ bolt 窗插销
~ bottom rail 窗扇下冒头
~ box 吊窗锤箱;窗台外的种花箱子
~ breast rail 窗腰栏
~ brightness 窗亮度
~ butt 窗铰链
~ casement (平开的)窗扇
~ casing 窗线脚
~ catch 窗钩
~ cellar 酒窖
~ check 窗边槽口
~ cleaner 玻璃刮水器
~ column 窗间柱
~ cornice 窗口檐板
~ cross 窗横撑
~ -curtain 窗帘
~ decoration fitting 窗装饰配件
~ dial 窗梯
~ display 橱窗陈设
~ drape paper 窗帘纸
~ dressing 橱窗布置[装饰]
~ efficiency rato 窗面采光效率比,
采光系数
~ fan 窗户扇
~ fastener 窗闩销
~ fittings 窗配件
~ flanning 窗框两侧斜边
~ flaring 八字形窗洞
~ form(window shape) 窗形
~ frame 窗框
~ furniture 窗配件
~ gearing (气)窗开关联动装置
~ glass 窗玻璃
~ glass run channel 窗玻璃滑道
~ glazing 窗配玻璃
~ glazing bar 窗芯横木
~ grating 窗栅,防护铁栅
~ grille 窗栅拦,窗花
~ groove 窗槽口
~ guard 窗铁栅
~ guide rail 窗玻璃导轨
~ hanging 窗帘,窗帷

~ hardware 窗(小五金)配件
~ hatch 镶玻璃的舱口
~ head 窗框上槛
~ head moulding 窗头线
~ hinge 窗铰链
~ hood 窗罩
~ in work stone 料石窗
~ jamb 窗樘边梃；窗口侧墙
~ lead 镶玻璃铅条
~ lead-in 穿窗入线
~ ledge 窗台,窗槛
~ lift 吊窗提手
~ lining 窗框镶板
~ lintel 窗过梁
~ lock 窗插销
~ lowering crank 开窗摇柄
~ manufacture 窗门生产,窗门制造业
~ measurement 窗口测量
~ module 门窗模数
~ mullion 窗间直梃,窗的竖框,窗间小柱
~ niche 床框或窗帘凹槽
~ niche arch 窗壁龛拱
~ of the slope 斜率窗口
~ of tube 荧光屏
~ opener 窗开关,开窗装置
~ opening 窗口,窗孔[洞]
~ openwork gablet 花格窗小山墙
~ pan [下落式]窗玻璃腔
~ -pane 窗玻璃
~ panel 窗镶板
~ parapet 窗心墙,窗下墙
~ part 窗配件
~ paster 糊窗纸
~ pier 窗墩
~ pillar 窗口中柱
~ post 窗间柱
~ profile 窗(纵)剖面图
~ pull [开启]窗玻璃拉带
~ rabbet 窗边槽口
~ rail 窗横挡
~ recess 窗框或窗帘凹槽

~ recess with slipping walls 有外层墙的窗框
~ regulator 车窗开闭调节器
~ reveal 窗帮,窗口侧墙
~ sandwich 夹层玻璃窗
~ sash 上下拉的窗扇
~ sash stripping 窗框挡风条
~ screen 纱窗
~ sealing gasket 窗框衬垫
~ seat （窗槛下的）窗座

window seats

~ section 窗形式,窗纵剖面
~ -shade 遮阳篷,遮阳窗帘
~ shade blind 遮光窗帘
~ -shop 逛商店,浏览橱窗
~ shutter 窗挡板,百叶窗
~ sill 窗槛,窗台
~ sill brick 砌窗台砖
~ sill unit 窗台板单元
~ spandrel 上下层窗之间的空间
~ splay 八字形(斜槽)窗洞
~ spring bolt 窗门弹簧螺栓
~ stay 窗支撑
~ stile 窗边框或中框
~ stool 窗台[板]
~ strap 窗子开关带
~ strip 窗玻璃密封条
~ stud 窗间柱,窗板墙筋
~ style 窗(边)框
~ sunblind 窗遮阳
~ tax 窗税(一种房屋税,按窗计税)
~ template 窗框垫块
~ tier 分层窗
~ top rail 窗扇上冒头
~ tracery 花格窗,窗饰
~ transom 窗框中槛[横挡]
~ trim 窗饰,饰窗花格,修理窗扇
~ type air conditioner 窗装型温度调节器,窗装式空调器
~ ventilator 通气窗,车窗通风器
~ visor 遮阳板

~ -washing equipment 窗清洗设备
~ weight 窗的平衡锤
~ with sashes opening outwards 外向窗框口窗
~ with wings opening inwards 内向侧翼窗
air vent ~ 通气窗
arched ~ 拱窗
awning type ~ 篷天窗
bay ~ 凸出墙外的窗
blank ~ 假窗
blind ~ 百叶窗
casement ~ 竖铰链窗,双扇窗
catherine wheel ~ 轮形圆窗,玫瑰形窗
center-hung swivel ~ 中旋窗
center pivoted ~ 中枢支承旋转窗
compass ~ 圆肚窗
continuous sash ~ 连续框格窗
continuous sliding ~ 连续拉窗
counterbalanced ~ 均衡窗
dormer ~ 老虎窗,屋顶窗
double-hung counterbalanced ~ 双侧平衡铊式吊窗
drop slide ~ 上下滑动吊窗
dummy ~ 假窗
entrance ~ 入射窗
false ~ 盲窗,假窗
fire ~ 防火窗
folding ~ 折叠窗
gabled ~ 山墙窗
gemel ~s 对窗
guillotine ~ 吊升窗,闸刀式窗
hanging ~ 吊窗
hinged ~ 铰链窗
horizontally pivot hung ~ 中悬窗
jut ~ 突出窗
inward opening ~ 内开窗
lancet ~ 尖头窗
lattice ~ 花格窗
louver ~ 气窗,固定百叶窗

louvered panel ~ 百叶窗
lunette ~ 弧形窗,弦月窗
mansard dormer ~ 折线形屋顶窗
observation ~ 观察窗
out ward opening ~ 外推窗
partition ~ 间隔窗
refrigerated ~ 冷藏橱窗
rose ~ 圆花窗
round-headed ~ 圆头形窗
sample ~ 陈列窗
sash ~ 框格窗
screen ~ 纱窗
sexfoil ~ 六(瓣)叶窗
shielding ~ 防护窗
shop ~ 铺面橱窗
show ~ 橱窗,展览窗
side hang ~ 平开窗
sidelight ~ 边窗
single ~ 单窗
skylight ~ 天窗
sliding ~ 扯窗,滑窗
stain glass ~ 彩色玻璃镶嵌窗
tilt and turn ~ 斜转窗
top hung ~ 上悬窗
transom ~ 气窗
trap ~ 活板窗,通气窗
two-fold ~ 双折窗
ventilating ~ 通风窗
vertically sliding ~ 垂直推拉窗

windowless ['windəulis] *a.* 无窗的; -windowlessness, *n.*
~ building 无窗房屋

windscreen ['windskri:n] *n.* 风挡,挡风板,汽车挡风玻璃

windshield ['windʃi:ld] *n.* 风挡,挡风板;(汽车的)挡风玻璃

windward ['windwəd] *n.* 上风,向风;向风的一方; *a.* 上风的,迎风的;-windwardness, *n.*
~ side 向风面,上风面,向风一侧
~ truss 迎风面桁架

louver window

wing [wiŋ] *n.* 翼,翼板;侧厅;厢房,耳房;建筑物在边侧突出之部分
~ chair 高背椅
~ door 侧门
~ fence 翼形栅栏
~ light 侧窗
~ groin 翼状穹顶肋
~ of door 门侧扇
~ panel 翼板
~ wall 翼墙,八字墙

winged [wiŋd] *a.* 有翼的;迅速的; **wingedly,** *ad.* **wingedness,** *n.*
~ disc 古建筑中带翅膀形的托盘装饰
~ mouldboard 装翼模板

wingless ['wiŋdlis] *a.* 无翼的;winglessness, *n.*

winglet ['wiŋlit] *n.* 小翼

winter ['wintə] *n.* 冬季; *v.* 过冬,越冬; **winterer,** *n.*
~ building construction 建筑物冬季施工
~ concreting 冬季浇筑混凝土
~ air-conditioning installation 冬季空调设备
~ garden 冬景花园(玻璃暖房)
~ hardiness 耐冬性,耐寒性
~ -proofing 防寒,防冻
~ resistance 防冻性,防寒性
~ window 双层窗

winterization [wintərai'zeʃən] *n.* 准备过冬;安装防寒设置

wintery ['wintri] *a.* -trier,-triest. 冬天的;寒冷的;**wintrily,** *ad.* **wintriness,** *n.*

wipe [waip] *v.*, wiped, wiping, *n.* 拭,揩,擦
~ out 拭去,洗掉
~ed joint 拭接,热[焊]接;焊接点

wiper ['waipə] *n.* 揩抹器;(汽车风挡的)刮水器;揩布

wire ['waiə] *n.*; *a.*; *v.*, wired, wiring. *n.* 金属线,钢丝; *v.* 布[架,配]线,敷设导线; **wirer,** *n.*

~ back-tie 钢丝拉条(模板的),反拉钢丝
~ bars 线材坯,线锭
~ basket 铁丝笼[篮]
~ binders 绑扎铁丝,绑钢筋用的退火钢丝
~ bolster 填石铁丝笼
~ box 铅丝笼,铁丝笼
~ brad 铁丝制的无头钉
~ broadcasting 有线广播
~ broom 钢丝路刷
~ brush 钢丝刷
~ cable 钢丝绳,钢缆,钢丝束
~ cable grease 钢缆润滑油
~ -cable lubricant 缆索润滑剂
~ cable spacer 钢丝束定位板
~ casing 钢丝套管,线槽
~ clip 钢丝剪[夹]
~ cloth 钢[铜]丝布
~ comb 钢丝梳
~ -cut brick 丝切砖
~ cutter 钢丝钳[剪]
~ fabric 钢丝网,铁纱
~ fence 铁丝围拦
~ flue brush 铁丝烟筒刷子
~ galvanizing 镀锌钢丝
~ ga(u)ge 钢丝号码;线规
~ gauze 金属丝网,铁丝网
~ glass 嵌丝玻璃,铁丝网玻璃
~ laminated glass 金属丝叠层玻璃
~ lath 金属丝网,钢丝网
~ lath and paster suspended ceiling 钢丝网灰泥吊顶棚
~ lath and paster wall 钢丝网灰泥墙
~ mesh 钢丝网,金属网
~ -mesh fence 钢丝网围拦
~ -mesh reinforcement 钢丝网配筋
~ -mesh screen 金属丝网筛
~ nail 圆头钉
~ netting 做窗纱的金属丝网,金属栅栏
~ nut 金属线帽,钢丝帽

~ pattern glass 嵌丝图案玻璃
~ -puller 预留在线管中拉电线的铁丝
~ reinforced 钢丝配筋的

wire nut

~ rope 钢丝缆索,钢丝绳
~ scratcher 钢丝刷
~ section 钢丝截面
~ solder 焊条
~ staple U 形钉,马蹄钉
~ strand 钢绞线
~ stretcher 拉丝机,钢丝拉伸机
~ tack 钢丝平头钉
~ terminal 钢丝接头
~ -wrapping connection 绞接
common ~ 公共导线;中性线
compound ~ 复合导线
concealed ~ 暗线
core ~ 芯线
cotton-insulated ~ 纱包线
earth ~ 接地线
enamel-covered ~ 漆包线
fuse ~ 保险丝
ground ~ 接地线
lead ~ 铅线;引线
open ~ 明线
pilot ~ 引示线
twin ~ 双(芯)线
twisted ~ 绞合线
varnished ~ 漆包线

twin wire

Wollaston ~ 极细的导线,渥拉斯顿线

wired ['waiəd] *a.* 装有电线的,有铁丝网的;用金属丝缚的
~ glass caisson 嵌丝玻璃藻井
~ glass rooflight 嵌丝玻璃天窗
~ -opaque white glass 嵌丝乳白玻璃
~ ornamental glass 嵌丝装饰玻璃

wiring ['waiəriŋ] *n.* 接线,布线;加网状钢筋;配线; *a.* 架线的,线缝的;用于架线的,用于线缝的
~ board 插接线板
~ case 接线箱[盒]
~ clip 钢丝剪;线夹
~ conduit 电线管道
~ diagram 线路(装配)图,配[布,接]线图
~ grommet 电线索眼,电线环状接头
~ layout 线路设计[布置]
~ plug 线路插头
~ regulation 布线规程,接线规程
~ point 接线点
~ switch 线路开关

wirecore *n.* 线芯

wireman ['waiəmən] *n. pl.* -men 装线工

wireway ['waiəwei] *n.* 布线槽;电线导管

withe [wið] *n.* ; *v.* , (= wythe) withed, withing, *n.* 烟道隔板;空斗墙的立砌隔砖;枝条; *v.* 用柱条捆束

withdrawable *a.* 可更换的,可拆[卸,装]的,活[络,动]的
~ bit 活络钻头

withstand [wið'stænd] *v.* , -stood,-standing. 抵抗,抗拒;经受;withstander, *n.*
~ fire 耐火

wobble ['wɔbl] *v.* ,-bled,-bling, *n.* 摆动;摇动;震动;-wabbler, *n.*
~ pump 手摇泵
~ saw 摇摆锯

wood [wud] *n.* 木材;树林,林地; *a.* 木制的; *v.* 供木材给……;植树,造林
~ and brick-clad home 砖木结构住宅,木结构架和砖填充墙住宅
~ alcohol 甲醇,木精
~ arch 木拱
~ base broad 木踢脚板
~ -base materials 木基材料(指纤维板和碎料板等人造板)
~ -base (fibre and particle) panel material 木质[纤维和碎料]人造板
~ batten 板条,压缝条,木挂瓦条

~ block 木砖,木块
~ block filler 木垫块
~ block floor 镶木楼板[地板]
~ block pavement 木块路[铺]面

wood block floor

~ board 木板壁;木墙板
~ borer 木工钻床,蛀木虫
~ brick 木砖,木落砖
~ buffer 木缓冲器
~ carving 木刻,木雕
~ ceiling blocks 顶棚木块
~ cell 木垛,木笼
~ cement 木材水泥
~ cement board 木丝水泥板
~ cement roof 木材水泥屋顶
~ centering 木拱架
~ char 木炭
~ chip 木片,木屑,刨花
~ chip absorbent ceilling 碎木片吸声顶棚
~ chipboard 碎木板
~ chisel 木凿
~ clamp 木夹
~ clapper [木]踏脚板
~ coal 木炭
~ column 木柱
~ concrete composite beam 木混凝土合成梁
~ concrete roof slab 木丝混凝土屋顶板
~ connection 木结构联结
~ construction 木结构,木建筑
~ consumption 木材消耗量
~ cotton tree 木棉树
~ culvert 木质涵洞
~ dado 木墙裙
~ door 木门

wood door

~ drilling machine 钻木机
~ drying 木材干燥

~ dust 木屑
~ dye 植物染料
~ engraving 木刻[木板]画
~ excelsior 木丝,木刨花
~ excelsior absorbent sheet 木丝吸声板
~ excelsior concrete 木丝[木屑]混凝土
~ excelsior concrete slab 木屑混凝土板
(~) excelsior hollow filler 木丝空心垫衬
(~) excelsior insulation 木丝绝缘
(~) excelsior partition 木丝板隔墙
~ fender pile 木护栏[舷]桩(码头)
~ fiber =fibre 木纤维,木丝
~ fibre concrete 锯末混凝土
-fiber plaster 木纤维灰浆粉刷
-fibered plaster 木纤维泥
-fiber plaster baseboard 抹灰木纤维护壁板
~ fiberboard acoustic tiled ceiling 木纤维吸声顶棚
~ fiberboard (sound) absorptive ceiling 木纤维吸声顶棚
~ fibre concrete 木丝混凝土
~ file 木锉
~ filler block 木填块
~ filling 木填料,油灰
~ float finish (混凝土表面的)木镘修整
~ floated 木抹子粉平的
~ floding rule 木折尺
~ floor 木地板
~ flour 木屑
~ flume 木渡槽
~ for furniture 家具用材
~ former 木模
-frame construction 木框架结构
~ frame house 木框架房屋
~ framing 木框架
~ fret 木回纹花饰;木格子细工
~ fungus 腐木菌

~ furniture 木质隔空材料
~ furring 钉木板条
~ girder 木大梁
~ grabber 抓木工具
~ grain 木纹
~ grating 木踏板,木栅板(浴室地面用)
~ gum 木胶,树脂
~ gutter 木檐[槽]
~ house 木屋
~ joint 木接合；木榫头
~ joist 木搁栅
~ key 木键
~ lagging 木隔板
~ -land 林地
~ linedpipe 木衬管
~ lining 木镶面
~ lath 木灰条,灰板条
~ lintel 木过梁
~ mallet 小木槌
~ meal 木粉填料
~ moulding 木线脚饰
~ moulding machine 制木线脚机
~ mosaic 镶木马赛克,镶木细工
~ nog 木榫,木栓,木钉
~ of commerce 商品材
~ of coniferous trees 针叶树木材
~ of deciduous trees 落叶树木材
~ of ripe age 成熟材,可用材
~ oil 桐油
~ oil varnish 桐油清漆
~ panelled partition 木格板隔墙
~ panelling 镶木板
~ paper 木制纸
~ particle material 碎木材料
~ pattern 木模
~ pavement 木块(铺砌)路面；木块铺面
~ paving 木块铺砌；木块路面；木材铺面
~ pavior 木铺砌工人

wood gutter

~ -peeling machine 薄板锯床
~ pile 木桩
~ pile cluster 木群桩,木桩群
~ plane 木刨
~ pipe 木制管
~ planing machine 刨木机
~ plank 木板[模]
~ planking 木板
~ planer 机刨,刨木机
~ plastic product 木塑制品
~ plastic window 木塑窗
~ plug 木栓,木钉
~ preservative 木材防腐剂
~ preserving oil 木材防腐油
~ pulp 木纸浆
~ ray 木心髓线
~ roll 木滚筒；木制漩涡形柱头装饰
~ roof 木屋顶
~ resion 木材树脂
~ rosin 木[蒸]松香
~ rubbing strip (护舷木桩前的)木擦条
~ rule 木尺
~ sash putty 木窗框油灰
~ saw 木锯
~ sawer 锯木工人
~ screw 木螺钉
~ sculpture 木雕,木雕塑物
~ separater 木隔板
~ shake 木材环裂
~ shaper 木牛头刨
~ shaving 刨花
~ sheathing 木屋面板；木护墙板
~ shed 柴薪棚
~ shingle 木墙面板,木板屋面
~ siding 木壁[墙]板,木披叠板
~ sill 木门槛,木窗台
~ -slat 木板条
~ sleeper 枕木
~ slip 衬板
~ spirit 甲醇,木精
~ springing-type fender 木和弹簧

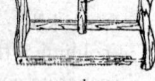
wood saw

式舷木
~ stain 木染料
~ stair 木楼梯
~ stave 木制排气管
~ -stave flume 木板[渡]槽
~ -stave pipe 木条水管
~ stilt 木垫板
~ storage 贮木场
~ strip flooring 木条地板
~ structure 木结构
~ struct 木撑
~ stud 木墙筋
~ -stud partition 木立筋隔墙
~ studding 木间柱
~ studs construction 木立筋构造
~ tar 木焦油,木柏油
~ -tar pitch 木焦油沥青
~ tile 木瓦
~ tool grinder 木工磨床
~ trim 木门窗镶边
~ turning lathe 木车床
~ turpentine 松节油
~ used for bridges (building) 桥梁材
~ used for in foundation's buildings 基础工程用材
~ varnish 木材清漆
~ vault 木拱形圆顶
~ veneer 木镶板,木饰面板
~ waling 木围楦,木横挡
~ waste 废木
~ window 木窗
~ wool 木丝;(木)刨花
~ woolslab 木丝板
~ -yard 贮木场
~ -worm 蛀木虫
absolutely dry ~ 全干材
acid ~ 干馏材,酸性木材
air-dried ~ 风干木材
bamboo ~ 竹林
bituminous ~ 沥青木(有木质外形的褐煤)

clip ~ 薄木片
compregnated ~ 渗(胶)压(缩)木材
coniferous ~ 针叶树材
dense ~ 密纹木材
desiccated ~ 干木材
drift ~ 浮木
fat ~ 轻木,松木
fire resisting ~ 耐火木材
glued ~ 胶合木[板]
green ~ 新伐木材
hard ~ 硬木
heart ~ 芯材
heat-stabilized compressed ~ 加热压缩材
imitation ~ 人造材
impregnated ~ 浸渍防腐木材
inlaid ~ 镶嵌木
knaggy ~ 多节木料
knotty ~ 多节木材
laminated ~ 胶合层压板
lath ~ 板条
light ~ 轻木
multi-ply ~ 多层板
ply ~ 胶合板
prop ~ 抗木
round ~ 圆木;圆材
sap ~ 白木质;边树
sound ~ 坚硬木
standing ~ (未伐倒的)立木
three-ply ~ 三合板
timber ~ 用材木(直径3英寸以上者)
unbarked ~ 原木,未剥皮木
unseasoned ~ 未干燥木材
veined ~ 纹理木
veneer ~ 镶木,胶合板
wooded ['wudid] *a.* 树林繁多的,多树木的
~ area 产木地区;森林面积
~ stand 木台,木架
wooden ['wudn] *a.* 木(质)的,木制的;
woodenly, *ad.*;woodenness, *n.*

~ astragal 木制玻璃格条,木制窗芯子(镶嵌玻璃)
~ baluster 木栏杆(小)柱
~ banister 栏杆小木柱
~ barrack (临时的)木板房
~ baseplate 木踢脚板,(用以刮平混凝土面的)木制刮板
~ bathtub 木浴盆
~ batten 木板条,木挂瓦条
~ beam 木梁,木枋
~ beam bridge 木梁桥
~ beam floor 木梁楼板
~ binder 木联结梁
~ binding beam 木联结梁
~ block 木块;铺木;木滑车
~ board covering 木板盖面
~ board lining 木板衬里
~ -box drain 木箱式暗沟
~ brake block 木闸瓦
~ brake shoe 木闸瓦
~ bracket 木托座
~ broach 木尖塔
~ building 木材建筑物
~ casement window 木制竖铰链窗
~ chipboard 碎木胶合板
~ chock 木楔
~ cistern 木水槽
~ coaming 木围板
~ concrete form 木制混凝土模板
~ crib 木屋,牛栏;木笼叠木笼,填石木笼
~ cushion pads 木垫板
~ dam 木坝,木堰
~ dark slide 木制暗盒
~ dome 木制圆屋顶
~ door frame 木门框架
~ (door) threshold 木制门槛
~ dowel 木梢钉,木暗榫
~ eave(s) gutter 木檐沟[槽]
~ engraving 木刻术;木刻画,木板画
~ facing 木板饰面
~ fastener 木扣件
~ fence 木围[栅]栏
~ fence post 木护栏桩
~ fender 木围墙,木栅;护舷木;木碰垫
~ filler 木填缝料
~ fillet 木嵌条;木楞条
~ flat roof 木平屋顶
~ floating floor 架空木地板
~ folded plate roof 木折板屋顶
~ folding door 木折叠门
~ form (木)模板
~ formwork 木模板
~ frame 木框架
~ framework building 木框架建筑

wooden formwork

~ furring 钉木板条
~ gate 木大[闸]门
~ glazing bar 木窗芯,玻璃窗木嵌条
~ grid 木窗格,木格栅
~ grillage 木格排
~ grillage footing 木格床基础
~ ground 木地撑
~ ground floor 木地板
~ gutter 屋顶木雨水槽,屋顶木檐槽
~ hammer 木锤
~ handle 木把手
~ handrail 木扶手
~ header 木顶梁,露头木砖
~ hipped-plate roof 四坡木板屋顶
~ hoist tower 木吊扣塔,木起重塔
~ house 木屋
~ hut 木棚屋,木制临时营房
~ joist 木搁栅
~ joist upper floor 木搁栅上层盖板
~ key 木键,木楔
~ ladder 木梯子
~ lamella cupola 薄板条叠层圆屋顶
~ lamella dome 薄板条叠层穹顶
~ lath 抹灰板条
~ latticed beam 引格构式梁
~ latticed cupola 木格构圆顶
~ lining 木隔板,木衬板

~ mallet 木槌
~ mast 木桅,木柱
~ mat 木柴排,木垫块
~ maul 大木锤
~ mitring gate 木制人字形闸门
~ mold 木模型
~ molding 木制线脚饰
~ mopboard 木踢脚板
~ mosaic 镶嵌细木工
~ mo(u)ld 木线脚
~ moulding 木制线脚装饰
~ nailing plug (墙上)受钉木塞
~ packing 木衬垫
~ palisade fence 木围篱
~ pattern 木模
~ pavement 木材铺面;木路面
~ peg 木钉,木检
~ pile 木桩
~ pile joint 木桩接头
~ pile staging 木桩,工作平台
~ pillar 木支柱[墩]
~ pin 木钉
~ plain (web) girder 木制实腹梁,木板梁
~ plate-girder 木板梁
~ prefab(ricated) construction 预制装配木结构
~ principal post 木主柱
~ prismatic shell roof 木棱柱壳屋顶,木折板屋顶
~ pulpit 木制讲坛
~ punner 木夯,木捣棒
~ purlin(e) 木檩(条)
~ quoin 木楔
~ rafer 木椽
~ railroad tie 木轨枕
~ rails 木横挡
~ rainwater gutter 木天[檐]沟
~ rammer 木夯,木捣棒
~ revolving door 木旋转门
~ roller shutter 木制卷帘百叶窗
~ roof cladding 木屋面盖板

~ roof sheathing 木屋面盖板
~ roof spire 木尖塔屋顶
~ roof truss 木屋顶架
~ roofed 木屋顶
~ runner 木滑道;木滑条
~ sash bar (镶嵌玻璃)木窗芯子
~ screed (抹灰)木刮板
~ scrub board 木擦洗板
~ separator 木隔板
~ sheet pile 木板桩
~ shingles 木板瓦
~ sleeper 枕木
~ spar 木质翼梁
~ square 木制直角尺
~ staircase 木楼梯
~ stage 木吊盘
~ stairs 木楼梯
~ stanchion 木支柱,窗间小木柱
~ stand 木架
~ steeple 木尖塔
~ step 木阶梯
~ string(er) 木楼梯斜梁
~ strut 木支柱,木顶撑
~ stud 木立筋
~ sub-floor 底层木地板
~ support 木支架
~ swing door 木制双开式弹簧门
~ tamper 木夯,木捣棒
~ tie 枕木
~ timbering 木支架
~ tilted-slab roof 斜木板屋顶
~ tower 木塔,木井架
~ transom 木摇头窗,木气窗,(门窗的)木横挡
~ tread 木楼梯踏步板
~ trestle 木栈桥
~ tripod 木制三角架
~ truss 木桁架
~ truss bridge 木桁架桥
~ truss button 木窗搭闩
~ underframe 木底盘架
~ valley gutter 木斜沟槽

~ Venetian blind 木制(威尼斯)软百叶窗
~ wall plug 墙上木质插头,可受钉木塞
~ -ware 木器
woodcraft [ˈwudkrɑːft] *n.* 木材加工,木工技术
woodcut [ˈwudkʌt] *n.* 木刻,版画
woodcutter [ˈwudˌkʌtə] *n.* 伐木工人,木刻[版画]家
woodwork [ˈwudwəːk] *n.* 细木工,细木作;木制品
woodworking [ˈwudˌwəːkiŋ] *a.* 木工的,制造木制品的; *n.* 木材加工
~ band sawing 木工带锯
~ industry 木制品加工业
~ instruments 木工工具
~ rip saw 解木直锯
woodwork's vise 木工台钳
woody [ˈwudi] *a.*, woodier, woodiest 木(质)的;木制的,树木茂盛的
~ fracture 木纹状断口

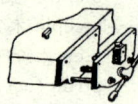
woodwork's vise

~ peat 木质泥炭
wool [wul] *n.* 羊毛(状物),绒[毛]线;纤维,渣棉
~ blanket 毛毯
~ carpet 羊毛地毯
~ felt (建筑用)粗制毯
~ pack cloud 卷毛云
work [wəːk] *n.*; *a.*; *v.*, worked or wrought, working *n.* 工作,操作,职业;作功;制品; *v.* 工作,运行,操纵,开动;加工,处理,制造
~ abstract 工程摘要表
~ bench 工作台[架]
~ box 工具箱
~ breakdown structure(WBS) 作业[工作]分解结构(把最终目标按统筹方法分解为小分支)
~ deck 工作台

~ equipment 工具;施工设备
~ inmid-air 高空作业
~ joint 施工缝
~ load 资用荷载,工作荷载,工作量
~ map 工作图
~ method 操作法
~ noise 工作噪声
~ of art 工艺品
~ of defence 防护建筑物,防护设施

work of art

~ of erosion 侵蚀作用
~ of protection 防护建筑物,防护设施
~ organization plan 施工部署[组织]计划
~ permit 工作许可证
~ piece 工件;分部工程
~ program 工作计划
~ -rest 工作台
~ -yard 施工[工作]场地
~ed lumber 加工过的木材
~ed material 加工材料
black and white ~ 木石结构(木构架中填石灰泥)
boxing ~ 模板工程
carpentry ~ 木工工程
cold ~ 冷作
concrete ~ 混凝土工程
construction ~ 施工工程
development ~ 研制工作
face ~ 镶石工作,抹面工作
facebrick ~ 饰面砖工程
false ~ 脚手架,膺架
field ~ 现场工作,野外作业
finishing ~ 结尾[修整]工作
fitting ~ 设备安装工作
floor laying ~ 地面铺设工程
glazing ~ 玻璃安装工作
hand ~ 手工加工
head ~ 拱顶石饰
inlaid ~ 镶嵌细工
installation ~ 安装工程

insulating ～ 隔热保温工程
iron-mongery ～ 五金安装工程
joggle ～ 细木工程,拼接工作
lattice ～ 构架工作,格子细工
line ～ 线路工作
long and short ～ (砖)长短砌合
maintenance ～ 日常维修工作
metal ～ 金工
painting ～ 油漆工作
panel ～ 镶格工作
parallel ～ 平行作业
parquet ～ 镶木工作,镶木细工
pavior ～ 铺砌工程
pilot ～ 试运行
pipe ～ 管道工程
plaster ～ 抹灰工程
plumbing ～ 管道工程
rag ～ 砌石板
rendering ～ 抹灰工程
retaining ～s 蓄水工程,挡土[水]建筑物
routine ～ 常规作业
rubble ～ 毛石圬工,砌石工程
shuttering ～ 模板工程
struck-joint ～ 随砌随手勾缝
surface ～ 地面工作
tiling ～ 贴砖工程
timber ～ 木作工程
water-proofing ～ 防水工程

workability [ˌwəːkə'biliti] *n.* 可加工[使用]性,施工性能,可塑性,和易性
～ agent (混凝土)增塑剂,塑化剂

water-proofing work

workaday ['wəːkədei] *a.* 工作日的,平日的,日常的
workbag ['wəːkbæg] *n.* 工具袋
workbin *n.* 零件盒,料箱[斗]
workblank *n.* 毛坯
workbook ['wəːkbuk] *n.* 工作手册,规划手册,笔记本

workbox ['wəːkbɔks] *n.* 工具箱
workday ['wəːkdei] *n.* 工作日
worker ['wəːkə] *n.* 工作者,工人,职工
～ occupancy rate 工人工作面积率
～'s barrack 临时性(木板)工房
～'s dwelling unit 工人居住处
～'s housing 工人住房
concrete ～ 混凝土工
sheetmetal ～ 白铁工
workhand ['wəːkhænd] *n.* 人手
working ['wəːkiŋ] *n.* 活动,劳动;(*pl.*)巷道; *a.* 工作的;施工用的;运转的
～ agreement 合作协议
～ aloft 高空作业
～ chamber 工作室
～ condition 生产[工作]条件
～ control 实际控制
～ costs 工作成本;经营费,加工费
～ current 工作电流
～ datum 施工基准面
～ design 施工设计
～ details 施工详图
～ diagram 作业图,施工图
～ district 作业区
～ drawing 施工(详)图,加工图,工程图
～ edge (木材的)加工边缘
～ element 施工构件
～ expenses 工作费用;经营费
～ expenditure 工作费用;经营费
～ face 规划面;工作面;作用面
～ gallery 施工通道
～ gangway 常用通道
～ hours 工作时间,工时
～ instruction 操作规程
～ joint 工作缝
～ life 使用期,使用寿命
～ lights 作业照明灯
～ load 室用荷载,工作荷载
～ load design 使用荷载设计(法),正常荷载设计
～ operation 工序

~ order 工作正常运转状态
~ -out 规划,制订计算
~ paper 工作底稿[文件]
~ place 工作地点,工作区,工地,现场
~ plan 工作计划;施工图
~ plane (照明设备的)工作面
~ platform 工作平台
~ point 作用点,受力点
~ position 工作位置,工作状态
~ pressure 工作压力
~ progress 施工进度
~ scale 操作规模
~ scafford 作业脚手架
~ schedule 施工计划
~ scheme 工作计划
~ season 施工[工作]季节
~ shaft 工作(竖)井,施工竖井
~ specification 操作规程
~ speed 工作速度
~ standard 通用[现行,工作]标准
~ stress design 容许应力设计(法),工作应力设计
~ substance 可用物质
~ system 作业法
~ -table 工作台
~ zone 工作区段

workman ['wə:kmən] *n. pl.* -men. 工人,劳动者,职工,专业工人
~'s compensation 工人伤残赔偿

workmanlike ['wə:kmənlaik] *a.* 熟练的,有技能的

workmanship ['wəkmənʃip] *n.* 工作质量;手艺,工艺;作品,工艺品

works *n.* 工厂,工程,工事;著作
~ canteen 工地食堂,工地小卖部
~ drawing 施工图
~ fire brigade 工地消防队

workshop ['wə:kʃɔp] *n.* 工场,工厂
~ assembly 工场[车间]装配
~ building 厂房
~ truss 厂房桁架,锯齿形桁架

worksite ['wə:ksait] *n.* 工地

workspace ['wə:kspeis] *n.* 工作空间

worm [wə:m] *n.* 蚯蚓;蜗杆;螺旋杆,蛇管;*v.* 爬行,蠕动
~ adjustment 蜗杆调节
~ auger 螺旋钻
~ channel 蛀眼,蛀孔
~ conveyer 螺旋式输送机
~ -drive 蜗杆传动
~ -eaten 虫蛀的,多蛀孔的
~ elevator 螺旋提升机
~ feeder 螺旋加料器
~ fence 弯弯曲曲的栅栏

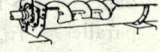
worm feeder

~ geared hoist 蜗轮卷扬机
~ geared winch 蜗轮绞车
~ hole 蛀孔,蛀洞
~ lift 螺旋提升机
~ -operated bar-bender 螺旋式弯(钢)筋机
~ pipe 蜗轮管
~ wheel 蜗轮
~ work dressing 虫蚀状錾琢

worn [wɔ:n] *a.* 磨损的,用旧的;耗尽的
~ flat 磨平
~ -out surface 磨损表面
~ pavement 磨损路[地]面

wornout ['wɔ:n'aut] *a.* 磨损的;用旧的,用旧的
~ soil 贫瘠土壤

worst [wə:st] *a.* 最坏的,最不利的
~ -case condition 最坏条件
~ -case design 按最坏情况设计
~ -case noise 最大的噪声

worth [wə:θ] *n.* 价值;性能,效用;*a.* 值得的,有价值的

worthless ['wə:θlis] *a.* 无价值的

wove paper *n.* 布纹纸

woven ['wəuvən] *a.* 织[编]成的,纺织的
~ fabric 纺织品
~ mattress 编梢沉排,编织衬垫
~ steel fabric 钢丝网

～ tape 布卷尺
～ willow mat 柳条编的柴排
～ wire 钢丝网
～ wire fence 铁丝网围栏
～-wire reinforcement 钢丝网配筋
wrack [ræk] **n.** 劣等材
wrap [ræp] **v.**, wrapped or wrapt, wrapping, **n.** 卷[卷]，包；缠绕；隐蔽
～ angle 包角
～-around hanger 半合成油管挂
～ attachment 吊线装置
～ bending test 卷弯试验
wrappage ['ræpidʒ] **n.** 封套，包装（材料），包皮
～ed cable 绕扎电缆
～ed joint 缠绕接线头
wrapper ['ræpə] **n.** 封套，外皮；包裹料；包装板
wrapping ['ræpiŋ] **n.** 包装；(常用 **pl.**) 包装材料
～ machine 包装机
～ paper 包装纸
～ plane 包装板
wreath [ri:θ] **n. pl.** wreaths 花圈[环]，圈[环]状物，螺旋形物；(楼梯扶手的)涡卷
wreathe [ri:ð] **v.**, wreathed, wreathed or (古) wreathen, wreathing, 编成环，绕成圈
～d column 扭形柱，螺旋柱
～d hand-rail 扭曲形扶手
～d stringer 盘旋楼梯斜梁
wreck [rek] **n.**；**v.** 失事；破坏；沉船；破损物，残骸
～ crane 救险起重机
wrecker ['rekə] **n.** 急救车，急救机，（路面）捣碎设备
wrecking ['rekiŋ] **n.**；**a.** 拆除房屋；遭难，（船只）失事；救险的，打捞的
～ and breaking hammer 拆毁破碎锤
～ bar 起钉杆，撬棍
～ block 特大滑车（用来架设临时吊

重货专用吊杆的大型滑车
～ cable 特大缆绳（由大绳搓成的缆绳,周径14 ～ 16英寸）
～ car 救险车，吊拖车
～ company 打捞（沉船）公司，(旧建筑物)拆除公司
～ contract 拆除合同
～ crane 救险起重机，拆卸(建筑物)起重机
～ crew 拆除房屋工程队，打捞[营救]队
～ gang 拆除房屋工作队，救险队，打捞[营救]队
～ permit 允许拆除
～ site 拆除现场
～ tool 拆毁工具；救险车[船]；打捞车
～ truck 救险车
wrench [rentʃ] **n.** 扳手；扳钳，扳头；**v.** 拧，扭转
～ jaw 钳子叉头[牙口]
～ socket 扳手套筒
adjustable ～ 活动扳手，活扳手
alligator ～ 管钳，管扳手
all ～ 通用扳手
box ～ 套筒扳手
coach ～ 活动扳手
construction ～ 大型安装扳手
fork ～ 叉形扳手
pipe ～ 管扳手，管钳(子)
set-screw ～ 固定螺钉扳手
socket ～ 套筒扳手
speed ～ 快速扳手

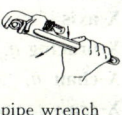
pipe wrench

wring [riŋ] **v.**, wrung, wringing, **n.** 拧，绞，扭；粘合；歪曲,曲解
wringer ['riŋə] **n.** 绞拧器,榨干机
wringing n. 粘合；绞接
wrinkle ['riŋkl] **n.**；**v.**, -kled, -kling. **n.** 皱纹，褶皱；**v.** 使[起,折]皱，折叠
～ finish 皱褶饰面
～ resistance 抗皱性能
～ varnish 皱纹清漆

wrinkling *n.* 褶皱,(薄壁构件的)局部扭曲
wrist [rist] *n.* 肘节,肘杆
　～ pin 活塞销,曲柄销
　～ plate 肘板
writing ['raitiŋ] *n.* 书写,记录,登记,书法
　～ room 写作室
　～ table 写字台,书桌
wrought [rɔ:t] *v.* work 的过去式和过去分词 *a.* 可锻的,精炼的,精制的
　～ alloy 锻制合金
　～ board 装修用板,精制的板
　～ (concret) formwork 加工混凝土模板
　～ grounds 钉在墙上的刨光木砖
　～ iron 熟铁,锻铁
　～ iron pipe 熟(锻)铁管
　～ iron sectional boiler 熟铁片式锅炉
　～ iron window grille 熟铁窗栅
　～ nail 锻制钉
　～ steel 熟钢,锻钢
　～ timber 精制木材
W-truss W 形桁梁
wurtzilite *n.* 韧沥青
wychelm ['witʃ'elm] *n.* 榆木
wye [wai] *n. pl.* wyes, Y 形接头, Y 形支架, 三通
　～ branch Y 形支管
　～ connection(＝Y-connection) Y 形接法,星形接法
　～ level Y(形)水准仪,回转水准仪
　～ pipe(＝Y-pipe) Y 形管,三通管
　～ track 三岔轨道

wye level

wythe *n.* 枝条,柳条;空斗墙的立砌隔砖;半砖墙

X

X alloy *n.* 铜铝值金
xanthic ['zænθik] *a.* 黄(色)的,黄原酸的
xat [za:t] *n.* (印第安人的)雕刻纪念柱,物像柱
X-axis ['eks,æksis] *n.* X 轴,横坐标轴
X-bracing *n.* 剪刀撑,交叉支撑
X-chair *n.* 交叉折椅
X-cross member X 形横梁
xenidium 胶合板
xenodocheum [,zenədə'ki:əm] *n.* 古希腊的客栈
xenodochium [,zenədə'kaiəm] *n. pl.* -chia. (中世纪的)收容院
xenon [,zenɔn] *n.* 氙(Xe)
　～ lamp 氙气灯,氙(气)管
　～ filled quartz discharge tube 充氙(气)石英放电管
xeric ['zirik] *a.* 干旱的,旱生的
xerochore *n.* 无水沙漠区
xeroll *n.* 干热软土
xerogel *n.* 干凝胶
xerophyte ['ziərəfait] *n.* 旱生植物
xerosol *n.* 干旱土
X-form cross member X 形构件, X 形横梁
X-frame *n.* 叉形架, X 形构架
X-frame brace 框架对角撑木
xing *n.* 交叉(点),横过
　～ pedestrian 人行横道
X-line *n.* 横轴线
X-member *n.* 叉形杆件
xoanon ['zɔuənɔn] *n.* 希腊原始木雕神像
X-plate *n.* X〔水平偏转〕板
X-quadripole *n.* 斜格形四端网络[架]
X-ray ['eks'rei] *n.* X 射线,伦琴射线
　～ defectoscopy X 射线探伤法

~ flaw detector X 射线探伤器
~ inspection X 射线检查
~ protection X 射线防护设备
~ protective concrete 防辐射混凝土
~ shielding 防辐射屏
~ roads 交叉道路
XSECT = **cross section** 横截面
X-stretcher *n.* 家具的十字花形横挡
X-type *a.* X 形的,交叉形的
~ frame 交叉形框架
~ groove (熔缝的)X 形坡口
~ member X 形梁
xylan ['zailæn] *n.* 木聚糖,木糖胶
xylanthrax [zai'lænθræks] *n.* 木炭
xylary ['zailəri] *a.* 木的,木质的
xylem ['zailem] *n.* 木质部
xylene ['zaili:n] *n.* 二甲苯
xylogen ['zailədʒən] *n.* 木质,木纤维

xyloglyphy 艺术木雕
xylograph ['zailəgra:f] *n.* 木刻;木板画;木纹图案
xylolite *n.* 菱苦土木屑板,(水泥和锯屑制成的)木屑板
xylometer [zai'lɔmitə] *n.* 木材比重计;测木仪
xylon *n.* 木质,木纤维
xylonite = **celluloid** ['zailənait] *n.* 赛洛璐,假象牙
xylophagan *n.* 蚀木虫,蛀木虫
xylophagous *a.* 蚀木虫,蛀木的
xylophyta *n.* 木本植物
X-Y plotter X-Y 绘图仪
X-Y recorder X-Y 记录器
xyst = **xystus** [zist] *n.* (古希腊的)室内运动场;沿花园周围柱廊,庭院内散步道

Y

yacca *n.* 浅黄色细纹木材(做家具用);罗汉松
yacht [jɔt] *n.* 轻舟,游艇
~ chair 折叠帆布椅
~ club 游艇俱乐部
yard [ja:d] *n.* 码(长度单位,=3 英尺 = 91.44 厘米);庭院,工(作)场,调车场 *v.* 把(木材)暂时集中堆放
~ catch basin 宅院渗水井,庭院集水井
~ crane 移动吊车,场内(移动)起重机
~ development cost 堆场建设费,场地建设费
~ drain 场地排水
~ drainage system 场地排水系统
~ dried lumber 场干木材,(工地)风干木材
~ gulley 有格栅的下水道进口
~ lumber 场堆木材,堆放木材
~ masonry wall 宅院圬工墙

~ measure 码尺
~ pavement 宅院铺面
~ slip 下横桁吊索
~ space 庭院空地
~ trap 进水口防臭设备
back ~ 后院
building ~ 建筑场地;造船厂
fabricating ~ 施工现场;制作场
formation ~ 装配场,装配车间
goods ~ 货物堆场
hold ~ 停车场;停留场
repairing ~ 修理厂
saw ~ 锯木场
service ~ 杂作坊
stack ~ 堆料场,成品仓库
yardage ['ja:didʒ] *n.* 方码数,土方
~ distribution 土方分配
~ meter [英制]土方计
~ table 土方表
yardarm ['ja:da:m] *n.* 桁端,横杆端,帆

横端
~ iron 横端铁箍
yarder *n.* 集材绞盘机；木场曳引机
yardstick ['jɑːdstik] *n.* 码尺，尺度，标准
yarn [jɑːn] *n.* 沙[毛]线；细股(绳)
cable ~ 钢索股绳
glass ~ 玻璃丝
yarrah *n.* 澳大利亚一种适于水工用的硬红木
yate tree *n.* 澳大利亚一种富有弹韧性的木材
Y-axis ['waiæksis] *n.* Y 轴，纵坐标轴
Y-bend *n.* 分叉弯头，Y 形叉管
Y-branch *n.* 分叉管，Y 形支管
Y-clean-outs *n.* 分叉清除；Y 形扫清设备
Y-connection *n.* Y 形接头；(电路) Y 形接线法，星形接法
Y-divider *n.* 分水岔
year [jəː] *n.* 年，年度
~ of completion 竣工年份
yearly ['jəːli] *a.* 年度的，每年的
~ budget 年度预算
~ installments 按年摊付
~ maintenace 年度保养[维修]
yellow [jeləu] *a.* ; *n.* 黄的，黄色(的) *v.* (使)变黄，染黄，发黄
~ brass 黄铜
~ deal (黄)松板
~ earth 赫黄土
~ gum (澳洲)桉树
~ metal 黄铜
~ oak 栎树
~ ochre 赫黄(土)
~ organic pigment 黄色有机颜料
~ paint 黄色油漆
~ pigment 黄色颜料
~ pine 黄松
~ sand 黄砂
~ sandal 黄檀木
Yellow sea *n.* 黄海
yeso *n.* (南美洲)刷墙用石膏

yew [juː] *n.* 水松，紫杉
Y-grade separation Y 形立体交叉
yield [jiːld] *n.* 产量，产额，供水量，屈服极限，塑流 *v.* 生产，出产；屈服子
~ of lime 石灰产浆量
yieldable arch 可缩性拱形支架
~ ring 可缩性环形支架
~ steel set 可缩性钢支架
~ support 可缩性支架
yielding ['jiːdiŋ] *a.* ; *n.* 屈服的，易变形的，流动性的
~ ground 易沉陷的软土
~ of foundation *n.* 基础沉陷
~ of supports 支座下沉，基础沉陷
~ seat 下沉支座
~ soil 流动土，软土
Y-intersection *n.* (道路) Y 形交叉(口)
Y-joint 分叉管接，Y 形接头
Y-level Y 形水准仪，活镜水准仪
yoke [jəuk] *n.* ; *v.*, yoked, yoking. *n.* 轭状物，轭架；北窗框的上槛，窗头板；护轨夹 *v.* 加轭；结[配]合
~ lever 叉形杠杆
~ trunnion 叉形十字头
~ venting 轭管通气
axle ~ 轴轭
truss ~ 桁箍
Yorkshire ['jɔːkʃə] *n.* (英国)约克郡
~ bond 跳丁砖砌合，每层二顺一丁砌合
~ light 横扯窗
young [jʌŋ] *a.* 年幼的，年轻的 *n.* 年轻者
youth [juːθ] *n. pl.* youths, youth. 青年，初期
~ centre 青年中心
~ hotel 青年旅店
Y-pipe *n.* 斜叉三通
Y-shaped *a.* Y 形的
~ bend Y 形支管
~ building Y 形(砌)块，星形建筑
~ fitting Y 形接头

~ valve Y形阀
Y-stay *n.* Y形拉线
Y-strut *n.* Y形支柱
Y-system *n.* （三相系统）星形接法
Y-tile *n.* 三通陶管
Y-trach *n.* 分叉线，Y形轨道
Y-tube *n.* Y形管

ytong *n.* 轻质混凝土，多孔混凝土
ytterbium [i'tə:bjəm] *n.* 镱（Yb）
yttrium ['ifriəm] *n.* 钇（Y, Yt）
Y-tube *n.* Y形管
Yueyang Tower 岳阳楼
yurt [juət] *n.* 蒙古包，圆顶帐篷

Z

zaccab *n.* 白泥石灰浆
Zapata scorpion *n.* 扎帕塔式蝎尾型结构（钻井台架）
Zapata tripod structure *n.* 扎帕塔式三脚结构
zapon lacquer *n.* 硝化纤维清漆，硝基清漆
zare(e)ba [zə'ri:bə] *n.* 木栅；篱笆；（非洲苏丹地区之）防御栅
zax [zæks] *n.* 石工用的斧子
Z-bar Z形钢
  ~ column Z形钢柱
  ~ column with covers 覆有翼缘板的Z形钢柱
Z-beam *n.* Z型梁
Z-connection *n.* Z型接法，曲折接线
Z-crank *n.* Z型曲柄
Z-direction *n.* Z轴方向
zebra ['zi:brə] *n.* 斑马 *a.* 有斑马纹的
  ~ crossing 黑白相间的人行横道线，斑马线
zed [zed] *n.* （英字母）Z；Z形铁[钢]
zee [zi:] *n.* （英字母）Z；Z形钢；Z形截面
  ~ -bar Z形钢
  ~ -bar pass Z字钢孔型
  ~ -type pile Z型板桩
zenith ['zeniθ] *n.* 天顶；顶点，最高点
  ~ angle 天顶角
  ~ distance 天顶距

zeolite ['zi:əlait] *n.* 沸石
  ~ exchanger 沸石（滤沙）软水交换器
  ~ filter 沸石软水池
  ~ process 沸石软水法
  ~ softener 沸石软水剂
  ~ tuff 沸石[火山]凝灰岩
zerk *n.* 加油嘴
zero ['ziərəu] *n. pl.* -ros-roes, *a*; *v.*; *n.* 零；零点，坐标起点，坐标零点 *v.* 调整
  ~ line 零位线；基准线
  ~ slope 平坡
  ~ -slump concrete (= no slump concrete) 干硬性混凝土，无坍落度混凝土
  ~ thrust pitch 无推力螺距
ziggurat ['ziguræt] (=zikkurat) *n.* 古代亚述及巴比伦之宝塔和建筑，锥形塔，叠级方尖塔
zigzag ['zigzæg] *a.; ad.; v.* -zagged, -zagging, *n.* 之字线，盘旋线；Z形装饰 *a.* Z字形的，曲折的，锯齿形的 *v.* 使锯齿形，使曲折
  ~ anchor bar 交错锚杆[筋]
  ~ arch 曲折拱，锯齿拱
  ~ bond (herringbone wirk or bond) 人字砌合，锯齿形砌合
  ~ connection 交错[曲折]连接
  ~ cracks 不规则裂缝
  ~ development 之字[盘旋]展线法
  ~ fluting 曲折柱槽

~ fault 曲折断层
~ fold 曲折褶皱
~ form 锯齿形;"之"字形
~ frieze 曲折雕带
~ girder 三角形孔梁(即 Warrent truss 华伦桁架)
~ kiln 曲折窑
~ line 曲折线,Z 形线
~ moulding 曲折的回纹形线条,锯齿形花饰
~ ornament 人字形饰线条,锯齿形花饰
~ pattern Z 形图案
~ riveting 交错铆接
~ rule 曲尺
~ sill 曲折槛,锯齿槛
~ watershed 锯齿状分水岭

zinc [zink] *n.* 锌 *v.* 镀锌
~ alloy 锌合金
~ barium-lead glass 锌钡铅玻璃
~ batten roof 锌板条屋顶
~ (box) rainwater gutter (匣形)锌皮雨水槽沟
~ chloride 氯化锌
~ chromate primer 铬化锌底漆
~ coated steel tile 镀锌钢瓦
~ coated wire 镀锌金属线
~ covered flat roof 锌皮平屋顶
~ covering 锌皮披水,锌皮屋面
~ crown 锌冕玻璃,上等锌玻璃
~ die-casting 压铸锌
~ detonator 锌起爆雷管
~ dust pigment 锌粉颜料
~ finish 镀锌
~ gage 锌板厚度
~ galvanizing 镀锌
~ gray 锌灰漆(一种由锌粉配制的油漆)
~ ore 锌矿
~ oxide 氧化锌
~ paint 锌漆
~ plating 镀锌
~ powder 锌粉

~ primer 富锌底漆
~ -rich paint 富锌涂料
~ roof 锌皮屋顶,马口铁屋顶
~ rule 锌质嵌线
~ sheet 薄锌片,锌皮
~ shingle 锌盖板;锌屋顶板
~ -silicate glass 硅酸锌玻璃,锌冕玻璃
~ spraying 喷涂锌
~ strip 锌板条
~ sulphate 硫酸锌
~ white 锌白,氧化锌,锌华
~ yellow 锌黄(粉)

zip-fastener ['zip,faːsnə] *n.* 拉链[锁]
zipper ['zipə] *n.* 拉链[锁]; *v.* 以拉练扣紧
zircon *n.* 锆(英)石,锆土
~ cement 锆-镁耐火水泥
~ sand 锆砂
zirconium [zəˈkəunjəm] *n.* 锆(Zr)
Z-iron Z 形铁
zoccola *n.* 座石,柱脚
zonal ['zəunl] *a.* 地带的,区域的
zone [zəun] *n.* ; *v.* , zoned, zoning. *n.* 区域,地带,范围 *v.* (将……)分成区,划分地区[带]
~ grouting 分区灌浆
~ of aeration 通气带,通气层,包气带,含气带
~ of cementation 胶结带
~ of eluviation 淋滤(土)层
~ of fracture 破碎带,断裂带
~ of heating 加热区,加热层
~ of protection 保护区
~ of residential 住宅区
~ of swelling 膨胀区
~ of transition 过渡区[带]
~ sampler 分层取样器
~ system 分区式(供暖)系统,分区[平行]下水道系统

zoning ['zəuniŋ] *n.* 分区;分区规划

zoccola

～ a drawing 将图纸分区
～ plan 分区规划,区划平面图
zonolite *n.* 烧蛭石(可做建筑材料)
zoo [zuː] *n.* 动物园
zoological garden *n.* 动物园
zoomorphic [ˌzəuəˈmɔːfik] *a.* 动物形象的,表现动物形象的
zoomorphic column *n.* 兽形柱
zoomorphism *n.* 动物图案
zoophorus *n.* 古建筑挑檐腰线上的人或兽形装饰
zotheca *n.* 套间,起居凹室
Z-piling bar Z形钢板桩
Z-section Z形断面
Z-steel Z形钢
Z-tie Z形系墙铁
Z-truss Z形桁架
Z-type piling bar Z字板桩
zwinger *n.* 保卫城市的要塞